T0180505

Lecture Notes in Computer Science 12348

More information about this subseries at http://www.springer.com/series/7412

Andrea Vedaldi · Horst Bischof ·
Thomas Brox · Jan-Michael Frahm (Eds.)

Computer Vision – ECCV 2020

16th European Conference
Glasgow, UK, August 23–28, 2020
Proceedings, Part III

Editors
Andrea Vedaldi
University of Oxford
Oxford, UK

Horst Bischof
Graz University of Technology
Graz, Austria

Thomas Brox
University of Freiburg
Freiburg im Breisgau, Germany

Jan-Michael Frahm
University of North Carolina at Chapel Hill
Chapel Hill, NC, USA

ISSN 0302-9743 ISSN 1611-3349 (electronic)
Lecture Notes in Computer Science
ISBN 978-3-030-58579-2 ISBN 978-3-030-58580-8 (eBook)
https://doi.org/10.1007/978-3-030-58580-8

LNCS Sublibrary: SL6 – Image Processing, Computer Vision, Pattern Recognition, and Graphics

This Springer imprint is published by the registered company Springer Nature Switzerland AG
The registered company address is: Gewerbestrasse 11, 6330 Cham, Switzerland

Foreword

Hosting the European Conference on Computer Vision (ECCV 2020) was certainly an exciting journey. From the 2016 plan to hold it at the Edinburgh International Conference Centre (hosting 1,800 delegates) to the 2018 plan to hold it at Glasgow's Scottish Exhibition Centre (up to 6,000 delegates), we finally ended with moving online because of the COVID-19 outbreak. While possibly having fewer delegates than expected because of the online format, ECCV 2020 still had over 3,100 registered participants.

Although online, the conference delivered most of the activities expected at a face-to-face conference: peer-reviewed papers, industrial exhibitors, demonstrations, and messaging between delegates. In addition to the main technical sessions, the conference included a strong program of satellite events with 16 tutorials and 44 workshops.

Furthermore, the online conference format enabled new conference features. Every paper had an associated teaser video and a longer full presentation video. Along with the papers and slides from the videos, all these materials were available the week before the conference. This allowed delegates to become familiar with the paper content and be ready for the live interaction with the authors during the conference week. The live event consisted of brief presentations by the oral and spotlight authors and industrial sponsors. Question and answer sessions for all papers were timed to occur twice so delegates from around the world had convenient access to the authors.

As with ECCV 2018, authors' draft versions of the papers appeared online with open access, now on both the Computer Vision Foundation (CVF) and the European Computer Vision Association (ECVA) websites. An archival publication arrangement was put in place with the cooperation of Springer. SpringerLink hosts the final version of the papers with further improvements, such as activating reference links and supplementary materials. These two approaches benefit all potential readers: a version available freely for all researchers, and an authoritative and citable version with additional benefits for SpringerLink subscribers. We thank Alfred Hofmann and Aliaksandr Birukou from Springer for helping to negotiate this agreement, which we expect will continue for future versions of ECCV.

August 2020

Vittorio Ferrari
Bob Fisher
Cordelia Schmid
Emanuele Trucco

Preface

Welcome to the proceedings of the European Conference on Computer Vision (ECCV 2020). This is a unique edition of ECCV in many ways. Due to the COVID-19 pandemic, this is the first time the conference was held online, in a virtual format. This was also the first time the conference relied exclusively on the Open Review platform to manage the review process. Despite these challenges ECCV is thriving. The conference received 5,150 valid paper submissions, of which 1,360 were acccpted for publication (27%) and, of those, 160 were presented as spotlights (3%) and 104 as orals (2%). This amounts to more than twice the number of submissions to ECCV 2018 (2,439). Furthermore, CVPR, the largest conference on computer vision, received 5,850 submissions this year, meaning that ECCV is now 87% the size of CVPR in terms of submissions. By comparison, in 2018 the size of ECCV was only 73% of CVPR.

The review model was similar to previous editions of ECCV; in particular, it was double blind in the sense that the authors did not know the name of the reviewers and vice versa. Furthermore, each conference submission was held confidentially, and was only publicly revealed if and once accepted for publication. Each paper received at least three reviews, totalling more than 15,000 reviews. Handling the review process at this scale was a significant challenge. In order to ensure that each submission received as fair and high-quality reviews as possible, we recruited 2,830 reviewers (a 130% increase with reference to 2018) and 207 area chairs (a 60% increase). The area chairs were selected based on their technical expertise and reputation, largely among people that served as area chair in previous top computer vision and machine learning conferences (ECCV, ICCV, CVPR, NeurIPS, etc.). Reviewers were similarly invited from previous conferences. We also encouraged experienced area chairs to suggest additional chairs and reviewers in the initial phase of recruiting.

Despite doubling the number of submissions, the reviewer load was slightly reduced from 2018, from a maximum of 8 papers down to 7 (with some reviewers offering to handle 6 papers plus an emergency review). The area chair load increased slightly, from 18 papers on average to 22 papers on average.

Conflicts of interest between authors, area chairs, and reviewers were handled largely automatically by the Open Review platform via their curated list of user profiles. Many authors submitting to ECCV already had a profile in Open Review. We set a paper registration deadline one week before the paper submission deadline in order to encourage all missing authors to register and create their Open Review profiles well on time (in practice, we allowed authors to create/change papers arbitrarily until the submission deadline). Except for minor issues with users creating duplicate profiles, this allowed us to easily and quickly identify institutional conflicts, and avoid them, while matching papers to area chairs and reviewers.

Papers were matched to area chairs based on: an affinity score computed by the Open Review platform, which is based on paper titles and abstracts, and an affinity score computed by the Toronto Paper Matching System (TPMS), which is based on the paper's full text, the area chair bids for individual papers, load balancing, and conflict avoidance. Open Review provides the program chairs a convenient web interface to experiment with different configurations of the matching algorithm. The chosen configuration resulted in about 50% of the assigned papers to be highly ranked by the area chair bids, and 50% to be ranked in the middle, with very few low bids assigned.

Assignments to reviewers were similar, with two differences. First, there was a maximum of 7 papers assigned to each reviewer. Second, area chairs recommended up to seven reviewers per paper, providing another highly-weighed term to the affinity scores used for matching.

The assignment of papers to area chairs was smooth. However, it was more difficult to find suitable reviewers for all papers. Having a ratio of 5.6 papers per reviewer with a maximum load of 7 (due to emergency reviewer commitment), which did not allow for much wiggle room in order to also satisfy conflict and expertise constraints. We received some complaints from reviewers who did not feel qualified to review specific papers and we reassigned them wherever possible. However, the large scale of the conference, the many constraints, and the fact that a large fraction of such complaints arrived very late in the review process made this process very difficult and not all complaints could be addressed.

Reviewers had six weeks to complete their assignments. Possibly due to COVID-19 or the fact that the NeurIPS deadline was moved closer to the review deadline, a record 30% of the reviews were still missing after the deadline. By comparison, ECCV 2018 experienced only 10% missing reviews at this stage of the process. In the subsequent week, area chairs chased the missing reviews intensely, found replacement reviewers in their own team, and managed to reach 10% missing reviews. Eventually, we could provide almost all reviews (more than 99.9%) with a delay of only a couple of days on the initial schedule by a significant use of emergency reviews. If this trend is confirmed, it might be a major challenge to run a smooth review process in future editions of ECCV. The community must reconsider prioritization of the time spent on paper writing (the number of submissions increased a lot despite COVID-19) and time spent on paper reviewing (the number of reviews delivered in time decreased a lot presumably due to COVID-19 or NeurIPS deadline). With this imbalance the peer-review system that ensures the quality of our top conferences may break soon.

Reviewers submitted their reviews independently. In the reviews, they had the opportunity to ask questions to the authors to be addressed in the rebuttal. However, reviewers were told not to request any significant new experiment. Using the Open Review interface, authors could provide an answer to each individual review, but were also allowed to cross-reference reviews and responses in their answers. Rather than PDF files, we allowed the use of formatted text for the rebuttal. The rebuttal and initial reviews were then made visible to all reviewers and the primary area chair for a given paper. The area chair encouraged and moderated the reviewer discussion. During the discussions, reviewers were invited to reach a consensus and possibly adjust their ratings as a result of the discussion and of the evidence in the rebuttal.

After the discussion period ended, most reviewers entered a final rating and recommendation, although in many cases this did not differ from their initial recommendation. Based on the updated reviews and discussion, the primary area chair then made a preliminary decision to accept or reject the paper and wrote a justification for it (meta-review). Except for cases where the outcome of this process was absolutely clear (as indicated by the three reviewers and primary area chairs all recommending clear rejection), the decision was then examined and potentially challenged by a secondary area chair. This led to further discussion and overturning a small number of preliminary decisions. Needless to say, there was no in-person area chair meeting, which would have been impossible due to COVID-19.

Area chairs were invited to observe the consensus of the reviewers whenever possible and use extreme caution in overturning a clear consensus to accept or reject a paper. If an area chair still decided to do so, she/he was asked to clearly justify it in the meta-review and to explicitly obtain the agreement of the secondary area chair. In practice, very few papers were rejected after being confidently accepted by the reviewers.

This was the first time Open Review was used as the main platform to run ECCV. In 2018, the program chairs used CMT3 for the user-facing interface and Open Review internally, for matching and conflict resolution. Since it is clearly preferable to only use a single platform, this year we switched to using Open Review in full. The experience was largely positive. The platform is highly-configurable, scalable, and open source. Being written in Python, it is easy to write scripts to extract data programmatically. The paper matching and conflict resolution algorithms and interfaces are top-notch, also due to the excellent author profiles in the platform. Naturally, there were a few kinks along the way due to the fact that the ECCV Open Review configuration was created from scratch for this event and it differs in substantial ways from many other Open Review conferences. However, the Open Review development and support team did a fantastic job in helping us to get the configuration right and to address issues in a timely manner as they unavoidably occurred. We cannot thank them enough for the tremendous effort they put into this project.

Finally, we would like to thank everyone involved in making ECCV 2020 possible in these very strange and difficult times. This starts with our authors, followed by the area chairs and reviewers, who ran the review process at an unprecedented scale. The whole Open Review team (and in particular Melisa Bok, Mohit Unyal, Carlos Mondragon Chapa, and Celeste Martinez Gomez) worked incredibly hard for the entire duration of the process. We would also like to thank René Vidal for contributing to the adoption of Open Review. Our thanks also go to Laurent Charling for TPMS and to the program chairs of ICML, ICLR, and NeurIPS for cross checking double submissions. We thank the website chair, Giovanni Farinella, and the CPI team (in particular Ashley Cook, Miriam Verdon, Nicola McGrane, and Sharon Kerr) for promptly adding material to the website as needed in the various phases of the process. Finally, we thank the publication chairs, Albert Ali Salah, Hamdi Dibeklioglu, Metehan Doyran, Henry Howard-Jenkins, Victor Prisacariu, Siyu Tang, and Gul Varol, who managed to compile these substantial proceedings in an exceedingly compressed schedule. We express our thanks to the ECVA team, in particular Kristina Scherbaum for allowing open access of the proceedings. We thank Alfred Hofmann from Springer who again serve as the publisher. Finally, we thank the other chairs of ECCV 2020, including in particular the general chairs for very useful feedback with the handling of the program.

August 2020

Andrea Vedaldi
Horst Bischof
Thomas Brox
Jan-Michael Frahm

Organization

General Chairs

Vittorio Ferrari	Google Research, Switzerland
Bob Fisher	University of Edinburgh, UK
Cordelia Schmid	Google and Inria, France
Emanuele Trucco	University of Dundee, UK

Program Chairs

Andrea Vedaldi	University of Oxford, UK
Horst Bischof	Graz University of Technology, Austria
Thomas Brox	University of Freiburg, Germany
Jan-Michael Frahm	University of North Carolina, USA

Industrial Liaison Chairs

Jim Ashe	University of Edinburgh, UK
Helmut Grabner	Zurich University of Applied Sciences, Switzerland
Diane Larlus	NAVER LABS Europe, France
Cristian Novotny	University of Edinburgh, UK

Local Arrangement Chairs

Yvan Petillot	Heriot-Watt University, UK
Paul Siebert	University of Glasgow, UK

Academic Demonstration Chair

Thomas Mensink	Google Research and University of Amsterdam, The Netherlands

Poster Chair

Stephen Mckenna	University of Dundee, UK

Technology Chair

Gerardo Aragon Camarasa	University of Glasgow, UK

Tutorial Chairs

Carlo Colombo	University of Florence, Italy
Sotirios Tsaftaris	University of Edinburgh, UK

Publication Chairs

Albert Ali Salah	Utrecht University, The Netherlands
Hamdi Dibeklioglu	Bilkent University, Turkey
Metehan Doyran	Utrecht University, The Netherlands
Henry Howard-Jenkins	University of Oxford, UK
Victor Adrian Prisacariu	University of Oxford, UK
Siyu Tang	ETH Zurich, Switzerland
Gul Varol	University of Oxford, UK

Website Chair

Giovanni Maria Farinella	University of Catania, Italy

Workshops Chairs

Adrien Bartoli	University of Clermont Auvergne, France
Andrea Fusiello	University of Udine, Italy

Area Chairs

Lourdes Agapito	University College London, UK
Zeynep Akata	University of Tübingen, Germany
Karteek Alahari	Inria, France
Antonis Argyros	University of Crete, Greece
Hossein Azizpour	KTH Royal Institute of Technology, Sweden
Joao P. Barreto	Universidade de Coimbra, Portugal
Alexander C. Berg	University of North Carolina at Chapel Hill, USA
Matthew B. Blaschko	KU Leuven, Belgium
Lubomir D. Bourdev	WaveOne, Inc., USA
Edmond Boyer	Inria, France
Yuri Boykov	University of Waterloo, Canada
Gabriel Brostow	University College London, UK
Michael S. Brown	National University of Singapore, Singapore
Jianfei Cai	Monash University, Australia
Barbara Caputo	Politecnico di Torino, Italy
Ayan Chakrabarti	Washington University, St. Louis, USA
Tat-Jen Cham	Nanyang Technological University, Singapore
Manmohan Chandraker	University of California, San Diego, USA
Rama Chellappa	Johns Hopkins University, USA
Liang-Chieh Chen	Google, USA

Yung-Yu Chuang	National Taiwan University, Taiwan
Ondrej Chum	Czech Technical University in Prague, Czech Republic
Brian Clipp	Kitware, USA
John Collomosse	University of Surrey and Adobe Research, UK
Jason J. Corso	University of Michigan, USA
David J. Crandall	Indiana University, USA
Daniel Cremers	University of California, Los Angeles, USA
Fabio Cuzzolin	Oxford Brookes University, UK
Jifeng Dai	SenseTime, SAR China
Kostas Daniilidis	University of Pennsylvania, USA
Andrew Davison	Imperial College London, UK
Alessio Del Bue	Fondazione Istituto Italiano di Tecnologia, Italy
Jia Deng	Princeton University, USA
Alexey Dosovitskiy	Google, Germany
Matthijs Douze	Facebook, France
Enrique Dunn	Stevens Institute of Technology, USA
Irfan Essa	Georgia Institute of Technology and Google, USA
Giovanni Maria Farinella	University of Catania, Italy
Ryan Farrell	Brigham Young University, USA
Paolo Favaro	University of Bern, Switzerland
Rogerio Feris	International Business Machines, USA
Cornelia Fermuller	University of Maryland, College Park, USA
David J. Fleet	Vector Institute, Canada
Friedrich Fraundorfer	DLR, Austria
Mario Fritz	CISPA Helmholtz Center for Information Security, Germany
Pascal Fua	EPFL (Swiss Federal Institute of Technology Lausanne), Switzerland
Yasutaka Furukawa	Simon Fraser University, Canada
Li Fuxin	Oregon State University, USA
Efstratios Gavves	University of Amsterdam, The Netherlands
Peter Vincent Gehler	Amazon, USA
Theo Gevers	University of Amsterdam, The Netherlands
Ross Girshick	Facebook AI Research, USA
Boqing Gong	Google, USA
Stephen Gould	Australian National University, Australia
Jinwei Gu	SenseTime Research, USA
Abhinav Gupta	Facebook, USA
Bohyung Han	Seoul National University, South Korea
Bharath Hariharan	Cornell University, USA
Tal Hassner	Facebook AI Research, USA
Xuming He	Australian National University, Australia
Joao F. Henriques	University of Oxford, UK
Adrian Hilton	University of Surrey, UK
Minh Hoai	Stony Brooks, State University of New York, USA
Derek Hoiem	University of Illinois Urbana-Champaign, USA

Timothy Hospedales	University of Edinburgh and Samsung, UK
Gang Hua	Wormpex AI Research, USA
Slobodan Ilic	Siemens AG, Germany
Hiroshi Ishikawa	Waseda University, Japan
Jiaya Jia	The Chinese University of Hong Kong, SAR China
Hailin Jin	Adobe Research, USA
Justin Johnson	University of Michigan, USA
Frederic Jurie	University of Caen Normandie, France
Fredrik Kahl	Chalmers University, Sweden
Sing Bing Kang	Zillow, USA
Gunhee Kim	Seoul National University, South Korea
Junmo Kim	Korea Advanced Institute of Science and Technology, South Korea
Tae-Kyun Kim	Imperial College London, UK
Ron Kimmel	Technion-Israel Institute of Technology, Israel
Alexander Kirillov	Facebook AI Research, USA
Kris Kitani	Carnegie Mellon University, USA
Iasonas Kokkinos	Ariel AI, UK
Vladlen Koltun	Intel Labs, USA
Nikos Komodakis	Ecole des Ponts ParisTech, France
Piotr Koniusz	Australian National University, Australia
M. Pawan Kumar	University of Oxford, UK
Kyros Kutulakos	University of Toronto, Canada
Christoph Lampert	IST Austria, Austria
Ivan Laptev	Inria, France
Diane Larlus	NAVER LABS Europe, France
Laura Leal-Taixe	Technical University Munich, Germany
Honglak Lee	Google and University of Michigan, USA
Joon-Young Lee	Adobe Research, USA
Kyoung Mu Lee	Seoul National University, South Korea
Seungyong Lee	POSTECH, South Korea
Yong Jae Lee	University of California, Davis, USA
Bastian Leibe	RWTH Aachen University, Germany
Victor Lempitsky	Samsung, Russia
Ales Leonardis	University of Birmingham, UK
Marius Leordeanu	Institute of Mathematics of the Romanian Academy, Romania
Vincent Lepetit	ENPC ParisTech, France
Hongdong Li	The Australian National University, Australia
Xi Li	Zhejiang University, China
Yin Li	University of Wisconsin-Madison, USA
Zicheng Liao	Zhejiang University, China
Jongwoo Lim	Hanyang University, South Korea
Stephen Lin	Microsoft Research Asia, China
Yen-Yu Lin	National Chiao Tung University, Taiwan, China
Zhe Lin	Adobe Research, USA

Haibin Ling	Stony Brooks, State University of New York, USA
Jiaying Liu	Peking University, China
Ming-Yu Liu	NVIDIA, USA
Si Liu	Beihang University, China
Xiaoming Liu	Michigan State University, USA
Huchuan Lu	Dalian University of Technology, China
Simon Lucey	Carnegie Mellon University, USA
Jiebo Luo	University of Rochester, USA
Julien Mairal	Inria, France
Michael Maire	University of Chicago, USA
Subhransu Maji	University of Massachusetts, Amherst, USA
Yasushi Makihara	Osaka University, Japan
Jiri Matas	Czech Technical University in Prague, Czech Republic
Yasuyuki Matsushita	Osaka University, Japan
Philippos Mordohai	Stevens Institute of Technology, USA
Vittorio Murino	University of Verona, Italy
Naila Murray	NAVER LABS Europe, France
Hajime Nagahara	Osaka University, Japan
P. J. Narayanan	International Institute of Information Technology (IIIT), Hyderabad, India
Nassir Navab	Technical University of Munich, Germany
Natalia Neverova	Facebook AI Research, France
Matthias Niessner	Technical University of Munich, Germany
Jean-Marc Odobez	Idiap Research Institute and Swiss Federal Institute of Technology Lausanne, Switzerland
Francesca Odone	Università di Genova, Italy
Takeshi Oishi	The University of Tokyo, Tokyo Institute of Technology, Japan
Vicente Ordonez	University of Virginia, USA
Manohar Paluri	Facebook AI Research, USA
Maja Pantic	Imperial College London, UK
In Kyu Park	Inha University, South Korea
Ioannis Patras	Queen Mary University of London, UK
Patrick Perez	Valeo, France
Bryan A. Plummer	Boston University, USA
Thomas Pock	Graz University of Technology, Austria
Marc Pollefeys	ETH Zurich and Microsoft MR & AI Zurich Lab, Switzerland
Jean Ponce	Inria, France
Gerard Pons-Moll	MPII, Saarland Informatics Campus, Germany
Jordi Pont-Tuset	Google, Switzerland
James Matthew Rehg	Georgia Institute of Technology, USA
Ian Reid	University of Adelaide, Australia
Olaf Ronneberger	DeepMind London, UK
Stefan Roth	TU Darmstadt, Germany
Bryan Russell	Adobe Research, USA

Mathieu Salzmann	EPFL, Switzerland
Dimitris Samaras	Stony Brook University, USA
Imari Sato	National Institute of Informatics (NII), Japan
Yoichi Sato	The University of Tokyo, Japan
Torsten Sattler	Czech Technical University in Prague, Czech Republic
Daniel Scharstein	Middlebury College, USA
Bernt Schiele	MPII, Saarland Informatics Campus, Germany
Julia A. Schnabel	King's College London, UK
Nicu Sebe	University of Trento, Italy
Greg Shakhnarovich	Toyota Technological Institute at Chicago, USA
Humphrey Shi	University of Oregon, USA
Jianbo Shi	University of Pennsylvania, USA
Jianping Shi	SenseTime, China
Leonid Sigal	University of British Columbia, Canada
Cees Snoek	University of Amsterdam, The Netherlands
Richard Souvenir	Temple University, USA
Hao Su	University of California, San Diego, USA
Akihiro Sugimoto	National Institute of Informatics (NII), Japan
Jian Sun	Megvii Technology, China
Jian Sun	Xi'an Jiaotong University, China
Chris Sweeney	Facebook Reality Labs, USA
Yu-wing Tai	Kuaishou Technology, China
Chi-Keung Tang	The Hong Kong University of Science and Technology, SAR China
Radu Timofte	ETH Zurich, Switzerland
Sinisa Todorovic	Oregon State University, USA
Giorgos Tolias	Czech Technical University in Prague, Czech Republic
Carlo Tomasi	Duke University, USA
Tatiana Tommasi	Politecnico di Torino, Italy
Lorenzo Torresani	Facebook AI Research and Dartmouth College, USA
Alexander Toshev	Google, USA
Zhuowen Tu	University of California, San Diego, USA
Tinne Tuytelaars	KU Leuven, Belgium
Jasper Uijlings	Google, Switzerland
Nuno Vasconcelos	University of California, San Diego, USA
Olga Veksler	University of Waterloo, Canada
Rene Vidal	Johns Hopkins University, USA
Gang Wang	Alibaba Group, China
Jingdong Wang	Microsoft Research Asia, China
Yizhou Wang	Peking University, China
Lior Wolf	Facebook AI Research and Tel Aviv University, Israel
Jianxin Wu	Nanjing University, China
Tao Xiang	University of Surrey, UK
Saining Xie	Facebook AI Research, USA
Ming-Hsuan Yang	University of California at Merced and Google, USA
Ruigang Yang	University of Kentucky, USA

Technical Program Committee

Mahsa Baktashmotlagh
Kavita Bala
Yogesh Balaji
Guha Balakrishnan
V. N. Balasubramanian
Federico Baldassarre
Vassileios Balntas
Shurjo Banerjee
Aayush Bansal
Ankan Bansal
Jianmin Bao
Linchao Bao
Wenbo Bao
Yingze Bao
Akash Bapat
Md Jawadul Hasan Bappy
Fabien Baradel
Lorenzo Baraldi
Daniel Barath
Adrian Barbu
Kobus Barnard
Nick Barnes
Francisco Barranco
Jonathan T. Barron
Arslan Basharat
Chaim Baskin
Anil S. Baslamisli
Jorge Batista
Kayhan Batmanghelich
Konstantinos Batsos
David Bau
Luis Baumela
Christoph Baur
Eduardo Bayro-Corrochano
Paul Beardsley
Jan Bednavr'ik
Oscar Beijbom
Philippe Bekaert
Esube Bekele
Vasileios Belagiannis
Ohad Ben-Shahar
Abhijit Bendale
Róger Bermúdez-Chacón
Maxim Berman
Jesus Bermudez-cameo
Florian Bernard
Stefano Berretti
Marcelo Bertalmio
Gedas Bertasius
Cigdem Beyan
Lucas Beyer
Vijayakumar Bhagavatula
Arjun Nitin Bhagoji
Apratim Bhattacharyya
Binod Bhattarai
Sai Bi
Jia-Wang Bian
Simone Bianco
Adel Bibi
Tolga Birdal
Tom Bishop
Soma Biswas
Mårten Björkman
Volker Blanz
Vishnu Boddeti
Navaneeth Bodla
Simion-Vlad Bogolin
Xavier Boix
Piotr Bojanowski
Timo Bolkart
Guido Borghi
Larbi Boubchir
Guillaume Bourmaud
Adrien Bousseau
Thierry Bouwmans
Richard Bowden
Hakan Boyraz
Mathieu Brédif
Samarth Brahmbhatt
Steve Branson
Nikolas Brasch
Biagio Brattoli
Ernesto Brau
Toby P. Breckon
Francois Bremond
Jesus Briales
Sofia Broomé
Marcus A. Brubaker
Luc Brun
Silvia Bucci
Shyamal Buch
Pradeep Buddharaju
Uta Buechler
Mai Bui
Tu Bui
Adrian Bulat
Giedrius T. Burachas
Elena Burceanu
Xavier P. Burgos-Artizzu
Kaylee Burns
Andrei Bursuc
Benjamin Busam
Wonmin Byeon
Zoya Bylinskii
Sergi Caelles
Jianrui Cai
Minjie Cai
Yujun Cai
Zhaowei Cai
Zhipeng Cai
Juan C. Caicedo
Simone Calderara
Necati Cihan Camgoz
Dylan Campbell
Octavia Camps
Jiale Cao
Kaidi Cao
Liangliang Cao
Xiangyong Cao
Xiaochun Cao
Yang Cao
Yu Cao
Yue Cao
Zhangjie Cao
Luca Carlone
Mathilde Caron
Dan Casas
Thomas J. Cashman
Umberto Castellani
Lluis Castrejon
Jacopo Cavazza
Fabio Cermelli
Hakan Cevikalp
Menglei Chai
Ishani Chakraborty
Rudrasis Chakraborty
Antoni B. Chan

Kwok-Ping Chan
Siddhartha Chandra
Sharat Chandran
Arjun Chandrasekaran
Angel X. Chang
Che-Han Chang
Hong Chang
Hyun Sung Chang
Hyung Jin Chang
Jianlong Chang
Ju Yong Chang
Ming-Ching Chang
Simyung Chang
Xiaojun Chang
Yu-Wei Chao
Devendra S. Chaplot
Arslan Chaudhry
Rizwan A. Chaudhry
Can Chen
Chang Chen
Chao Chen
Chen Chen
Chu-Song Chen
Dapeng Chen
Dong Chen
Dongdong Chen
Guanying Chen
Hongge Chen
Hsin-yi Chen
Huaijin Chen
Hwann-Tzong Chen
Jianbo Chen
Jianhui Chen
Jiansheng Chen
Jiaxin Chen
Jie Chen
Jun-Cheng Chen
Kan Chen
Kevin Chen
Lin Chen
Long Chen
Min-Hung Chen
Qifeng Chen
Shi Chen
Shixing Chen
Tianshui Chen
Weifeng Chen
Weikai Chen
Xi Chen
Xiaohan Chen
Xiaozhi Chen
Xilin Chen
Xingyu Chen
Xinlei Chen
Xinyun Chen
Yi-Ting Chen
Yilun Chen
Ying-Cong Chen
Yinpeng Chen
Yiran Chen
Yu Chen
Yu-Sheng Chen
Yuhua Chen
Yun-Chun Chen
Yunpeng Chen
Yuntao Chen
Zhuoyuan Chen
Zitian Chen
Anchieh Cheng
Bowen Cheng
Erkang Cheng
Gong Cheng
Guangliang Cheng
Jingchun Cheng
Jun Cheng
Li Cheng
Ming-Ming Cheng
Yu Cheng
Ziang Cheng
Anoop Cherian
Dmitry Chetverikov
Ngai-man Cheung
William Cheung
Ajad Chhatkuli
Naoki Chiba
Benjamin Chidester
Han-pang Chiu
Mang Tik Chiu
Wei-Chen Chiu
Donghyeon Cho
Hojin Cho
Minsu Cho
Nam Ik Cho
Tim Cho
Tae Eun Choe
Chiho Choi
Edward Choi
Inchang Choi
Jinsoo Choi
Jonghyun Choi
Jongwon Choi
Yukyung Choi
Hisham Cholakkal
Eunji Chong
Jaegul Choo
Christopher Choy
Hang Chu
Peng Chu
Wen-Sheng Chu
Albert Chung
Joon Son Chung
Hai Ci
Safa Cicek
Ramazan G. Cinbis
Arridhana Ciptadi
Javier Civera
James J. Clark
Ronald Clark
Felipe Codevilla
Michael Cogswell
Andrea Cohen
Maxwell D. Collins
Carlo Colombo
Yang Cong
Adria R. Continente
Marcella Cornia
John Richard Corring
Darren Cosker
Dragos Costea
Garrison W. Cottrell
Florent Couzinie-Devy
Marco Cristani
Ioana Croitoru
James L. Crowley
Jiequan Cui
Zhaopeng Cui
Ross Cutler
Antonio D'Innocente

Rozenn Dahyot
Bo Dai
Dengxin Dai
Hang Dai
Longquan Dai
Shuyang Dai
Xiyang Dai
Yuchao Dai
Adrian V. Dalca
Dima Damen
Bharath B. Damodaran
Kristin Dana
Martin Danelljan
Zheng Dang
Zachary Alan Daniels
Donald G. Dansereau
Abhishek Das
Samyak Datta
Achal Dave
Titas De
Rodrigo de Bem
Teo de Campos
Raoul de Charette
Shalini De Mello
Joseph DeGol
Herve Delingette
Haowen Deng
Jiankang Deng
Weijian Deng
Zhiwei Deng
Joachim Denzler
Konstantinos G. Derpanis
Aditya Deshpande
Frederic Devernay
Somdip Dey
Arturo Deza
Abhinav Dhall
Helisa Dhamo
Vikas Dhiman
Fillipe Dias Moreira de Souza
Ali Diba
Ferran Diego
Guiguang Ding
Henghui Ding
Jian Ding
Mingyu Ding
Xinghao Ding
Zhengming Ding
Robert DiPietro
Cosimo Distante
Ajay Divakaran
Mandar Dixit
Abdelaziz Djelouah
Thanh-Toan Do
Jose Dolz
Bo Dong
Chao Dong
Jiangxin Dong
Weiming Dong
Weisheng Dong
Xingping Dong
Xuanyi Dong
Yinpeng Dong
Gianfranco Doretto
Hazel Doughty
Hassen Drira
Bertram Drost
Dawei Du
Ye Duan
Yueqi Duan
Abhimanyu Dubey
Anastasia Dubrovina
Stefan Duffner
Chi Nhan Duong
Thibaut Durand
Zoran Duric
Iulia Duta
Debidatta Dwibedi
Benjamin Eckart
Marc Eder
Marzieh Edraki
Alexei A. Efros
Kiana Ehsani
Hazm Kemal Ekenel
James H. Elder
Mohamed Elgharib
Shireen Elhabian
Ehsan Elhamifar
Mohamed Elhoseiny
Ian Endres
N. Benjamin Erichson
Jan Ernst
Sergio Escalera
Francisco Escolano
Victor Escorcia
Carlos Esteves
Francisco J. Estrada
Bin Fan
Chenyou Fan
Deng-Ping Fan
Haoqi Fan
Hehe Fan
Heng Fan
Kai Fan
Lijie Fan
Linxi Fan
Quanfu Fan
Shaojing Fan
Xiaochuan Fan
Xin Fan
Yuchen Fan
Sean Fanello
Hao-Shu Fang
Haoyang Fang
Kuan Fang
Yi Fang
Yuming Fang
Azade Farshad
Alireza Fathi
Raanan Fattal
Joao Fayad
Xiaohan Fei
Christoph Feichtenhofer
Michael Felsberg
Chen Feng
Jiashi Feng
Junyi Feng
Mengyang Feng
Qianli Feng
Zhenhua Feng
Michele Fenzi
Andras Ferencz
Martin Fergie
Basura Fernando
Ethan Fetaya
Michael Firman
John W. Fisher

Matthew Fisher
Boris Flach
Corneliu Florea
Wolfgang Foerstner
David Fofi
Gian Luca Foresti
Per-Erik Forssen
David Fouhey
Katerina Fragkiadaki
Victor Fragoso
Jean-Sébastien Franco
Ohad Fried
Iuri Frosio
Cheng-Yang Fu
Huazhu Fu
Jianlong Fu
Jingjing Fu
Xueyang Fu
Yanwei Fu
Ying Fu
Yun Fu
Olac Fuentes
Kent Fujiwara
Takuya Funatomi
Christopher Funk
Thomas Funkhouser
Antonino Furnari
Ryo Furukawa
Erik Gärtner
Raghudeep Gadde
Matheus Gadelha
Vandit Gajjar
Trevor Gale
Juergen Gall
Mathias Gallardo
Guillermo Gallego
Orazio Gallo
Chuang Gan
Zhe Gan
Madan Ravi Ganesh
Aditya Ganeshan
Siddha Ganju
Bin-Bin Gao
Changxin Gao
Feng Gao
Hongchang Gao
Jin Gao
Jiyang Gao
Junbin Gao
Katelyn Gao
Lin Gao
Mingfei Gao
Ruiqi Gao
Ruohan Gao
Shenghua Gao
Yuan Gao
Yue Gao
Noa Garcia
Alberto Garcia-Garcia
Guillermo Garcia-Hernando
Jacob R. Gardner
Animesh Garg
Kshitiz Garg
Rahul Garg
Ravi Garg
Philip N. Garner
Kirill Gavrilyuk
Paul Gay
Shiming Ge
Weifeng Ge
Baris Gecer
Xin Geng
Kyle Genova
Stamatios Georgoulis
Bernard Ghanem
Michael Gharbi
Kamran Ghasedi
Golnaz Ghiasi
Arnab Ghosh
Partha Ghosh
Silvio Giancola
Andrew Gilbert
Rohit Girdhar
Xavier Giro-i-Nieto
Thomas Gittings
Ioannis Gkioulekas
Clement Godard
Vaibhava Goel
Bastian Goldluecke
Lluis Gomez
Nuno Gonçalves
Dong Gong
Ke Gong
Mingming Gong
Abel Gonzalez-Garcia
Ariel Gordon
Daniel Gordon
Paulo Gotardo
Venu Madhav Govindu
Ankit Goyal
Priya Goyal
Raghav Goyal
Benjamin Graham
Douglas Gray
Brent A. Griffin
Etienne Grossmann
David Gu
Jiayuan Gu
Jiuxiang Gu
Lin Gu
Qiao Gu
Shuhang Gu
Jose J. Guerrero
Paul Guerrero
Jie Gui
Jean-Yves Guillemaut
Riza Alp Guler
Erhan Gundogdu
Fatma Guney
Guodong Guo
Kaiwen Guo
Qi Guo
Sheng Guo
Shi Guo
Tiantong Guo
Xiaojie Guo
Yijie Guo
Yiluan Guo
Yuanfang Guo
Yulan Guo
Agrim Gupta
Ankush Gupta
Mohit Gupta
Saurabh Gupta
Tanmay Gupta
Danna Gurari
Abner Guzman-Rivera

JunYoung Gwak
Michael Gygli
Jung-Woo Ha
Simon Hadfield
Isma Hadji
Bjoern Haefner
Taeyoung Hahn
Levente Hajder
Peter Hall
Emanuela Haller
Stefan Haller
Bumsub Ham
Abdullah Hamdi
Dongyoon Han
Hu Han
Jungong Han
Junwei Han
Kai Han
Tian Han
Xiaoguang Han
Xintong Han
Yahong Han
Ankur Handa
Zekun Hao
Albert Haque
Tatsuya Harada
Mehrtash Harandi
Adam W. Harley
Mahmudul Hasan
Atsushi Hashimoto
Ali Hatamizadeh
Munawar Hayat
Dongliang He
Jingrui He
Junfeng He
Kaiming He
Kun He
Lei He
Pan He
Ran He
Shengfeng He
Tong He
Weipeng He
Xuming He
Yang He
Yihui He
Zhihai He
Chinmay Hegde
Janne Heikkila
Mattias P. Heinrich
Stéphane Herbin
Alexander Hermans
Luis Herranz
John R. Hershey
Aaron Hertzmann
Roei Herzig
Anders Heyden
Steven Hickson
Otmar Hilliges
Tomas Hodan
Judy Hoffman
Michael Hofmann
Yannick Hold-Geoffroy
Namdar Homayounfar
Sina Honari
Richang Hong
Seunghoon Hong
Xiaopeng Hong
Yi Hong
Hidekata Hontani
Anthony Hoogs
Yedid Hoshen
Mir Rayat Imtiaz Hossain
Junhui Hou
Le Hou
Lu Hou
Tingbo Hou
Wei-Lin Hsiao
Cheng-Chun Hsu
Gee-Sern Jison Hsu
Kuang-jui Hsu
Changbo Hu
Di Hu
Guosheng Hu
Han Hu
Hao Hu
Hexiang Hu
Hou-Ning Hu
Jie Hu
Junlin Hu
Nan Hu
Ping Hu
Ronghang Hu
Xiaowei Hu
Yinlin Hu
Yuan-Ting Hu
Zhe Hu
Binh-Son Hua
Yang Hua
Bingyao Huang
Di Huang
Dong Huang
Fay Huang
Haibin Huang
Haozhi Huang
Heng Huang
Huaibo Huang
Jia-Bin Huang
Jing Huang
Jingwei Huang
Kaizhu Huang
Lei Huang
Qiangui Huang
Qiaoying Huang
Qingqiu Huang
Qixing Huang
Shaoli Huang
Sheng Huang
Siyuan Huang
Weilin Huang
Wenbing Huang
Xiangru Huang
Xun Huang
Yan Huang
Yifei Huang
Yue Huang
Zhiwu Huang
Zilong Huang
Minyoung Huh
Zhuo Hui
Matthias B. Hullin
Martin Humenberger
Wei-Chih Hung
Zhouyuan Huo
Junhwa Hur
Noureldien Hussein
Jyh-Jing Hwang
Seong Jae Hwang

Sung Ju Hwang
Ichiro Ide
Ivo Ihrke
Daiki Ikami
Satoshi Ikehata
Nazli Ikizler-Cinbis
Sunghoon Im
Yani Ioannou
Radu Tudor Ionescu
Umar Iqbal
Go Irie
Ahmet Iscen
Md Amirul Islam
Vamsi Ithapu
Nathan Jacobs
Arpit Jain
Himalaya Jain
Suyog Jain
Stuart James
Won-Dong Jang
Yunseok Jang
Ronnachai Jaroensri
Dinesh Jayaraman
Sadeep Jayasumana
Suren Jayasuriya
Herve Jegou
Simon Jenni
Hae-Gon Jeon
Yunho Jeon
Koteswar R. Jerripothula
Hueihan Jhuang
I-hong Jhuo
Dinghuang Ji
Hui Ji
Jingwei Ji
Pan Ji
Yanli Ji
Baoxiong Jia
Kui Jia
Xu Jia
Chiyu Max Jiang
Haiyong Jiang
Hao Jiang
Huaizu Jiang
Huajie Jiang
Ke Jiang
Lai Jiang
Li Jiang
Lu Jiang
Ming Jiang
Peng Jiang
Shuqiang Jiang
Wei Jiang
Xudong Jiang
Zhuolin Jiang
Jianbo Jiao
Zequn Jie
Dakai Jin
Kyong Hwan Jin
Lianwen Jin
SouYoung Jin
Xiaojie Jin
Xin Jin
Nebojsa Jojic
Alexis Joly
Michael Jeffrey Jones
Hanbyul Joo
Jungseock Joo
Kyungdon Joo
Ajjen Joshi
Shantanu H. Joshi
Da-Cheng Juan
Marco Körner
Kevin Köser
Asim Kadav
Christine Kaeser-Chen
Kushal Kafle
Dagmar Kainmueller
Ioannis A. Kakadiaris
Zdenek Kalal
Nima Kalantari
Yannis Kalantidis
Mahdi M. Kalayeh
Anmol Kalia
Sinan Kalkan
Vicky Kalogeiton
Ashwin Kalyan
Joni-kristian Kamarainen
Gerda Kamberova
Chandra Kambhamettu
Martin Kampel
Meina Kan
Christopher Kanan
Kenichi Kanatani
Angjoo Kanazawa
Atsushi Kanehira
Takuhiro Kaneko
Asako Kanezaki
Bingyi Kang
Di Kang
Sunghun Kang
Zhao Kang
Vadim Kantorov
Abhishek Kar
Amlan Kar
Theofanis Karaletsos
Leonid Karlinsky
Kevin Karsch
Angelos Katharopoulos
Isinsu Katircioglu
Hiroharu Kato
Zoltan Kato
Dotan Kaufman
Jan Kautz
Rei Kawakami
Qiuhong Ke
Wadim Kehl
Petr Kellnhofer
Aniruddha Kembhavi
Cem Keskin
Margret Keuper
Daniel Keysers
Ashkan Khakzar
Fahad Khan
Naeemullah Khan
Salman Khan
Siddhesh Khandelwal
Rawal Khirodkar
Anna Khoreva
Tejas Khot
Parmeshwar Khurd
Hadi Kiapour
Joe Kileel
Chanho Kim
Dahun Kim
Edward Kim
Eunwoo Kim
Han-ul Kim

Hansung Kim
Heewon Kim
Hyo Jin Kim
Hyunwoo J. Kim
Jinkyu Kim
Jiwon Kim
Jongmin Kim
Junsik Kim
Junyeong Kim
Min H. Kim
Namil Kim
Pyojin Kim
Seon Joo Kim
Seong Tae Kim
Seungryong Kim
Sungwoong Kim
Tae Hyun Kim
Vladimir Kim
Won Hwa Kim
Yonghyun Kim
Benjamin Kimia
Akisato Kimura
Pieter-Jan Kindermans
Zsolt Kira
Itaru Kitahara
Hedvig Kjellstrom
Jan Knopp
Takumi Kobayashi
Erich Kobler
Parker Koch
Reinhard Koch
Elyor Kodirov
Amir Kolaman
Nicholas Kolkin
Dimitrios Kollias
Stefanos Kollias
Soheil Kolouri
Adams Wai-Kin Kong
Naejin Kong
Shu Kong
Tao Kong
Yu Kong
Yoshinori Konishi
Daniil Kononenko
Theodora Kontogianni
Simon Korman
Adam Kortylewski
Jana Kosecka
Jean Kossaifi
Satwik Kottur
Rigas Kouskouridas
Adriana Kovashka
Rama Kovvuri
Adarsh Kowdle
Jedrzej Kozerawski
Mateusz Kozinski
Philipp Kraehenbuehl
Gregory Kramida
Josip Krapac
Dmitry Kravchenko
Ranjay Krishna
Pavel Krsek
Alexander Krull
Jakob Kruse
Hiroyuki Kubo
Hilde Kuehne
Jason Kuen
Andreas Kuhn
Arjan Kuijper
Zuzana Kukelova
Ajay Kumar
Amit Kumar
Avinash Kumar
Suryansh Kumar
Vijay Kumar
Kaustav Kundu
Weicheng Kuo
Nojun Kwak
Suha Kwak
Junseok Kwon
Nikolaos Kyriazis
Zorah Lähner
Ankit Laddha
Florent Lafarge
Jean Lahoud
Kevin Lai
Shang-Hong Lai
Wei-Sheng Lai
Yu-Kun Lai
Iro Laina
Antony Lam
John Wheatley Lambert
Xiangyuan lan
Xu Lan
Charis Lanaras
Georg Langs
Oswald Lanz
Dong Lao
Yizhen Lao
Agata Lapedriza
Gustav Larsson
Viktor Larsson
Katrin Lasinger
Christoph Lassner
Longin Jan Latecki
Stéphane Lathuilière
Rynson Lau
Hei Law
Justin Lazarow
Svetlana Lazebnik
Hieu Le
Huu Le
Ngan Hoang Le
Trung-Nghia Le
Vuong Le
Colin Lea
Erik Learned-Miller
Chen-Yu Lee
Gim Hee Lee
Hsin-Ying Lee
Hyungtae Lee
Jae-Han Lee
Jimmy Addison Lee
Joonseok Lee
Kibok Lee
Kuang-Huei Lee
Kwonjoon Lee
Minsik Lee
Sang-chul Lee
Seungkyu Lee
Soochan Lee
Stefan Lee
Taehee Lee
Andreas Lehrmann
Jie Lei
Peng Lei
Matthew Joseph Leotta
Wee Kheng Leow

Gil Levi
Evgeny Levinkov
Aviad Levis
Jose Lezama
Ang Li
Bin Li
Bing Li
Boyi Li
Changsheng Li
Chao Li
Chen Li
Cheng Li
Chenglong Li
Chi Li
Chun-Guang Li
Chun-Liang Li
Chunyuan Li
Dong Li
Guanbin Li
Hao Li
Haoxiang Li
Hongsheng Li
Hongyang Li
Houqiang Li
Huibin Li
Jia Li
Jianan Li
Jianguo Li
Junnan Li
Junxuan Li
Kai Li
Ke Li
Kejie Li
Kunpeng Li
Lerenhan Li
Li Erran Li
Mengtian Li
Mu Li
Peihua Li
Peiyi Li
Ping Li
Qi Li
Qing Li
Ruiyu Li
Ruoteng Li
Shaozi Li
Sheng Li
Shiwei Li
Shuang Li
Siyang Li
Stan Z. Li
Tianye Li
Wei Li
Weixin Li
Wen Li
Wenbo Li
Xiaomeng Li
Xin Li
Xiu Li
Xuelong Li
Xueting Li
Yan Li
Yandong Li
Yanghao Li
Yehao Li
Yi Li
Yijun Li
Yikang LI
Yining Li
Yongjie Li
Yu Li
Yu-Jhe Li
Yunpeng Li
Yunsheng Li
Yunzhu Li
Zhe Li
Zhen Li
Zhengqi Li
Zhenyang Li
Zhuwen Li
Dongze Lian
Xiaochen Lian
Zhouhui Lian
Chen Liang
Jie Liang
Ming Liang
Paul Pu Liang
Pengpeng Liang
Shu Liang
Wei Liang
Jing Liao
Minghui Liao
Renjie Liao
Shengcai Liao
Shuai Liao
Yiyi Liao
Ser-Nam Lim
Chen-Hsuan Lin
Chung-Ching Lin
Dahua Lin
Ji Lin
Kevin Lin
Tianwei Lin
Tsung-Yi Lin
Tsung-Yu Lin
Wei-An Lin
Weiyao Lin
Yen-Chen Lin
Yuewei Lin
David B. Lindell
Drew Linsley
Krzysztof Lis
Roee Litman
Jim Little
An-An Liu
Bo Liu
Buyu Liu
Chao Liu
Chen Liu
Cheng-lin Liu
Chenxi Liu
Dong Liu
Feng Liu
Guilin Liu
Haomiao Liu
Heshan Liu
Hong Liu
Ji Liu
Jingen Liu
Jun Liu
Lanlan Liu
Li Liu
Liu Liu
Mengyuan Liu
Miaomiao Liu
Nian Liu
Ping Liu
Risheng Liu

Helmut Mayer
Amir Mazaheri
David McAllester
Steven McDonagh
Stephen J. Mckenna
Roey Mechrez
Prakhar Mehrotra
Christopher Mei
Xue Mei
Paulo R. S. Mendonca
Lili Meng
Zibo Meng
Thomas Mensink
Bjoern Menze
Michele Merler
Kourosh Meshgi
Pascal Mettes
Christopher Metzler
Liang Mi
Qiguang Miao
Xin Miao
Tomer Michaeli
Frank Michel
Antoine Miech
Krystian Mikolajczyk
Peyman Milanfar
Ben Mildenhall
Gregor Miller
Fausto Milletari
Dongbo Min
Kyle Min
Pedro Miraldo
Dmytro Mishkin
Anand Mishra
Ashish Mishra
Ishan Misra
Niluthpol C. Mithun
Kaushik Mitra
Niloy Mitra
Anton Mitrokhin
Ikuhisa Mitsugami
Anurag Mittal
Kaichun Mo
Zhipeng Mo
Davide Modolo
Michael Moeller
Pritish Mohapatra
Pavlo Molchanov
Davide Moltisanti
Pascal Monasse
Mathew Monfort
Aron Monszpart
Sean Moran
Vlad I. Morariu
Francesc Moreno-Noguer
Pietro Morerio
Stylianos Moschoglou
Yael Moses
Roozbeh Mottaghi
Pierre Moulon
Arsalan Mousavian
Yadong Mu
Yasuhiro Mukaigawa
Lopamudra Mukherjee
Yusuke Mukuta
Ravi Teja Mullapudi
Mario Enrique Munich
Zachary Murez
Ana C. Murillo
J. Krishna Murthy
Damien Muselet
Armin Mustafa
Siva Karthik Mustikovela
Carlo Dal Mutto
Moin Nabi
Varun K. Nagaraja
Tushar Nagarajan
Arsha Nagrani
Seungjun Nah
Nikhil Naik
Yoshikatsu Nakajima
Yuta Nakashima
Atsushi Nakazawa
Seonghyeon Nam
Vinay P. Namboodiri
Medhini Narasimhan
Srinivasa Narasimhan
Sanath Narayan
Erickson Rangel Nascimento
Jacinto Nascimento
Tayyab Naseer
Lakshmanan Nataraj
Neda Nategh
Nelson Isao Nauata
Fernando Navarro
Shah Nawaz
Lukas Neumann
Ram Nevatia
Alejandro Newell
Shawn Newsam
Joe Yue-Hei Ng
Trung Thanh Ngo
Duc Thanh Nguyen
Lam M. Nguyen
Phuc Xuan Nguyen
Thuong Nguyen Canh
Mihalis Nicolaou
Andrei Liviu Nicolicioiu
Xuecheng Nie
Michael Niemeyer
Simon Niklaus
Christophoros Nikou
David Nilsson
Jifeng Ning
Yuval Nirkin
Li Niu
Yuzhen Niu
Zhenxing Niu
Shohei Nobuhara
Nicoletta Noceti
Hyeonwoo Noh
Junhyug Noh
Mehdi Noroozi
Sotiris Nousias
Valsamis Ntouskos
Matthew O'Toole
Peter Ochs
Ferda Ofli
Seong Joon Oh
Seoung Wug Oh
Iason Oikonomidis
Utkarsh Ojha
Takahiro Okabe
Takayuki Okatani
Fumio Okura
Aude Oliva
Kyle Olszewski

Björn Ommer
Mohamed Omran
Elisabeta Oneata
Michael Opitz
Jose Oramas
Tribhuvanesh Orekondy
Shaul Oron
Sergio Orts-Escolano
Ivan Oseledets
Aljosa Osep
Magnus Oskarsson
Anton Osokin
Martin R. Oswald
Wanli Ouyang
Andrew Owens
Mete Ozay
Mustafa Ozuysal
Eduardo Pérez-Pellitero
Gautam Pai
Dipan Kumar Pal
P. H. Pamplona Savarese
Jinshan Pan
Junting Pan
Xingang Pan
Yingwei Pan
Yannis Panagakis
Rameswar Panda
Guan Pang
Jiahao Pang
Jiangmiao Pang
Tianyu Pang
Sharath Pankanti
Nicolas Papadakis
Dim Papadopoulos
George Papandreou
Toufiq Parag
Shaifali Parashar
Sarah Parisot
Eunhyeok Park
Hyun Soo Park
Jaesik Park
Min-Gyu Park
Taesung Park
Alvaro Parra
C. Alejandro Parraga
Despoina Paschalidou
Nikolaos Passalis
Vishal Patel
Viorica Patraucean
Badri Narayana Patro
Danda Pani Paudel
Sujoy Paul
Georgios Pavlakos
Ioannis Pavlidis
Vladimir Pavlovic
Nick Pears
Kim Steenstrup Pedersen
Selen Pehlivan
Shmuel Peleg
Chao Peng
Houwen Peng
Wen-Hsiao Peng
Xi Peng
Xiaojiang Peng
Xingchao Peng
Yuxin Peng
Federico Perazzi
Juan Camilo Perez
Vishwanath Peri
Federico Pernici
Luca Del Pero
Florent Perronnin
Stavros Petridis
Henning Petzka
Patrick Peursum
Michael Pfeiffer
Hanspeter Pfister
Roman Pflugfelder
Minh Tri Pham
Yongri Piao
David Picard
Tomasz Pieciak
A. J. Piergiovanni
Andrea Pilzer
Pedro O. Pinheiro
Silvia Laura Pintea
Lerrel Pinto
Axel Pinz
Robinson Piramuthu
Fiora Pirri
Leonid Pishchulin
Francesco Pittaluga
Daniel Pizarro
Tobias Plötz
Mirco Planamente
Matteo Poggi
Moacir A. Ponti
Parita Pooj
Fatih Porikli
Horst Possegger
Omid Poursaeed
Ameya Prabhu
Viraj Uday Prabhu
Dilip Prasad
Brian L. Price
True Price
Maria Priisalu
Veronique Prinet
Victor Adrian Prisacariu
Jan Prokaj
Sergey Prokudin
Nicolas Pugeault
Xavier Puig
Albert Pumarola
Pulak Purkait
Senthil Purushwalkam
Charles R. Qi
Hang Qi
Haozhi Qi
Lu Qi
Mengshi Qi
Siyuan Qi
Xiaojuan Qi
Yuankai Qi
Shengju Qian
Xuelin Qian
Siyuan Qiao
Yu Qiao
Jie Qin
Qiang Qiu
Weichao Qiu
Zhaofan Qiu
Kha Gia Quach
Yuhui Quan
Yvain Queau
Julian Quiroga
Faisal Qureshi
Mahdi Rad

Filip Radenovic
Petia Radeva
Venkatesh
B. Radhakrishnan
Ilija Radosavovic
Noha Radwan
Rahul Raguram
Tanzila Rahman
Amit Raj
Ajit Rajwade
Kandan Ramakrishnan
Santhosh
K. Ramakrishnan
Srikumar Ramalingam
Ravi Ramamoorthi
Vasili Ramanishka
Ramprasaath R. Selvaraju
Francois Rameau
Visvanathan Ramesh
Santu Rana
Rene Ranftl
Anand Rangarajan
Anurag Ranjan
Viresh Ranjan
Yongming Rao
Carolina Raposo
Vivek Rathod
Sathya N. Ravi
Avinash Ravichandran
Tammy Riklin Raviv
Daniel Rebain
Sylvestre-Alvise Rebuffi
N. Dinesh Reddy
Timo Rehfeld
Paolo Remagnino
Konstantinos Rematas
Edoardo Remelli
Dongwei Ren
Haibing Ren
Jian Ren
Jimmy Ren
Mengye Ren
Weihong Ren
Wenqi Ren
Zhile Ren
Zhongzheng Ren
Zhou Ren
Vijay Rengarajan
Md A. Reza
Farzaneh Rezaeianaran
Hamed R. Tavakoli
Nicholas Rhinehart
Helge Rhodin
Elisa Ricci
Alexander Richard
Eitan Richardson
Elad Richardson
Christian Richardt
Stephan Richter
Gernot Riegler
Daniel Ritchie
Tobias Ritschel
Samuel Rivera
Yong Man Ro
Richard Roberts
Joseph Robinson
Ignacio Rocco
Mrigank Rochan
Emanuele Rodolà
Mikel D. Rodriguez
Giorgio Roffo
Grégory Rogez
Gemma Roig
Javier Romero
Xuejian Rong
Yu Rong
Amir Rosenfeld
Bodo Rosenhahn
Guy Rosman
Arun Ross
Paolo Rota
Peter M. Roth
Anastasios Roussos
Anirban Roy
Sebastien Roy
Aruni RoyChowdhury
Artem Rozantsev
Ognjen Rudovic
Daniel Rueckert
Adria Ruiz
Javier Ruiz-del-solar
Christian Rupprecht
Chris Russell
Dan Ruta
Jongbin Ryu
Ömer Sümer
Alexandre Sablayrolles
Faraz Saeedan
Ryusuke Sagawa
Christos Sagonas
Tonmoy Saikia
Hideo Saito
Kuniaki Saito
Shunsuke Saito
Shunta Saito
Ken Sakurada
Joaquin Salas
Fatemeh Sadat Saleh
Mahdi Saleh
Pouya Samangouei
Leo Sampaio
Ferraz Ribeiro
Artsiom Olegovich
Sanakoyeu
Enrique Sanchez
Patsorn Sangkloy
Anush Sankaran
Aswin Sankaranarayanan
Swami Sankaranarayanan
Rodrigo Santa Cruz
Amartya Sanyal
Archana Sapkota
Nikolaos Sarafianos
Jun Sato
Shin'ichi Satoh
Hosnieh Sattar
Arman Savran
Manolis Savva
Alexander Sax
Hanno Scharr
Simone Schaub-Meyer
Konrad Schindler
Dmitrij Schlesinger
Uwe Schmidt
Dirk Schnieders
Björn Schuller
Samuel Schulter
Idan Schwartz

William Robson Schwartz
Alex Schwing
Sinisa Segvic
Lorenzo Seidenari
Pradeep Sen
Ozan Sener
Soumyadip Sengupta
Arda Senocak
Mojtaba Seyedhosseini
Shishir Shah
Shital Shah
Sohil Atul Shah
Tamar Rott Shaham
Huasong Shan
Qi Shan
Shiguang Shan
Jing Shao
Roman Shapovalov
Gaurav Sharma
Vivek Sharma
Viktoriia Sharmanska
Dongyu She
Sumit Shekhar
Evan Shelhamer
Chengyao Shen
Chunhua Shen
Falong Shen
Jie Shen
Li Shen
Liyue Shen
Shuhan Shen
Tianwei Shen
Wei Shen
William B. Shen
Yantao Shen
Ying Shen
Yiru Shen
Yujun Shen
Yuming Shen
Zhiqiang Shen
Ziyi Shen
Lu Sheng
Yu Sheng
Rakshith Shetty
Baoguang Shi
Guangming Shi
Hailin Shi
Miaojing Shi
Yemin Shi
Zhenmei Shi
Zhiyuan Shi
Kevin Jonathan Shih
Shiliang Shiliang
Hyunjung Shim
Atsushi Shimada
Nobutaka Shimada
Daeyun Shin
Young Min Shin
Koichi Shinoda
Konstantin Shmelkov
Michael Zheng Shou
Abhinav Shrivastava
Tianmin Shu
Zhixin Shu
Hong-Han Shuai
Pushkar Shukla
Christian Siagian
Mennatullah M. Siam
Kaleem Siddiqi
Karan Sikka
Jae-Young Sim
Christian Simon
Martin Simonovsky
Dheeraj Singaraju
Bharat Singh
Gurkirt Singh
Krishna Kumar Singh
Maneesh Kumar Singh
Richa Singh
Saurabh Singh
Suriya Singh
Vikas Singh
Sudipta N. Sinha
Vincent Sitzmann
Josef Sivic
Gregory Slabaugh
Miroslava Slavcheva
Ron Slossberg
Brandon Smith
Kevin Smith
Vladimir Smutny
Noah Snavely
Roger D. Soberanis-Mukul
Kihyuk Sohn
Francesco Solera
Eric Sommerlade
Sanghyun Son
Byung Cheol Song
Chunfeng Song
Dongjin Song
Jiaming Song
Jie Song
Jifei Song
Jingkuan Song
Mingli Song
Shiyu Song
Shuran Song
Xiao Song
Yafei Song
Yale Song
Yang Song
Yi-Zhe Song
Yibing Song
Humberto Sossa
Cesar de Souza
Adrian Spurr
Srinath Sridhar
Suraj Srinivas
Pratul P. Srinivasan
Anuj Srivastava
Tania Stathaki
Christopher Stauffer
Simon Stent
Rainer Stiefelhagen
Pierre Stock
Julian Straub
Jonathan C. Stroud
Joerg Stueckler
Jan Stuehmer
David Stutz
Chi Su
Hang Su
Jong-Chyi Su
Shuochen Su
Yu-Chuan Su
Ramanathan Subramanian
Yusuke Sugano

Masanori Suganuma
Yumin Suh
Mohammed Suhail
Yao Sui
Heung-Il Suk
Josephine Sullivan
Baochen Sun
Chen Sun
Chong Sun
Deqing Sun
Jin Sun
Liang Sun
Lin Sun
Qianru Sun
Shao-Hua Sun
Shuyang Sun
Weiwei Sun
Wenxiu Sun
Xiaoshuai Sun
Xiaoxiao Sun
Xingyuan Sun
Yifan Sun
Zhun Sun
Sabine Susstrunk
David Suter
Supasorn Suwajanakorn
Tomas Svoboda
Eran Swears
Paul Swoboda
Attila Szabo
Richard Szeliski
Duy-Nguyen Ta
Andrea Tagliasacchi
Yuichi Taguchi
Ying Tai
Keita Takahashi
Kouske Takahashi
Jun Takamatsu
Hugues Talbot
Toru Tamaki
Chaowei Tan
Fuwen Tan
Mingkui Tan
Mingxing Tan
Qingyang Tan
Robby T. Tan
Xiaoyang Tan
Kenichiro Tanaka
Masayuki Tanaka
Chang Tang
Chengzhou Tang
Danhang Tang
Ming Tang
Peng Tang
Qingming Tang
Wei Tang
Xu Tang
Yansong Tang
Youbao Tang
Yuxing Tang
Zhiqiang Tang
Tatsunori Taniai
Junli Tao
Xin Tao
Makarand Tapaswi
Jean-Philippe Tarel
Lyne Tchapmi
Zachary Teed
Bugra Tekin
Damien Teney
Ayush Tewari
Christian Theobalt
Christopher Thomas
Diego Thomas
Jim Thomas
Rajat Mani Thomas
Xinmei Tian
Yapeng Tian
Yingli Tian
Yonglong Tian
Zhi Tian
Zhuotao Tian
Kinh Tieu
Joseph Tighe
Massimo Tistarelli
Matthew Toews
Carl Toft
Pavel Tokmakov
Federico Tombari
Chetan Tonde
Yan Tong
Alessio Tonioni
Andrea Torsello
Fabio Tosi
Du Tran
Luan Tran
Ngoc-Trung Tran
Quan Hung Tran
Truyen Tran
Rudolph Triebel
Martin Trimmel
Shashank Tripathi
Subarna Tripathi
Leonardo Trujillo
Eduard Trulls
Tomasz Trzcinski
Sam Tsai
Yi-Hsuan Tsai
Hung-Yu Tseng
Stavros Tsogkas
Aggeliki Tsoli
Devis Tuia
Shubham Tulsiani
Sergey Tulyakov
Frederick Tung
Tony Tung
Daniyar Turmukhambetov
Ambrish Tyagi
Radim Tylecek
Christos Tzelepis
Georgios Tzimiropoulos
Dimitrios Tzionas
Seiichi Uchida
Norimichi Ukita
Dmitry Ulyanov
Martin Urschler
Yoshitaka Ushiku
Ben Usman
Alexander Vakhitov
Julien P. C. Valentin
Jack Valmadre
Ernest Valveny
Joost van de Weijer
Jan van Gemert
Koen Van Leemput
Gul Varol
Sebastiano Vascon
M. Alex O. Vasilescu

Subeesh Vasu
Mayank Vatsa
David Vazquez
Javier Vazquez-Corral
Ashok Veeraraghavan
Erik Velasco-Salido
Raviteja Vemulapalli
Jonathan Ventura
Manisha Verma
Roberto Vezzani
Ruben Villegas
Minh Vo
MinhDuc Vo
Nam Vo
Michele Volpi
Riccardo Volpi
Carl Vondrick
Konstantinos Vougioukas
Tuan-Hung Vu
Sven Wachsmuth
Neal Wadhwa
Catherine Wah
Jacob C. Walker
Thomas S. A. Wallis
Chengde Wan
Jun Wan
Liang Wan
Renjie Wan
Baoyuan Wang
Boyu Wang
Cheng Wang
Chu Wang
Chuan Wang
Chunyu Wang
Dequan Wang
Di Wang
Dilin Wang
Dong Wang
Fang Wang
Guanzhi Wang
Guoyin Wang
Hanzi Wang
Hao Wang
He Wang
Heng Wang
Hongcheng Wang
Hongxing Wang
Hua Wang
Jian Wang
Jingbo Wang
Jinglu Wang
Jingya Wang
Jinjun Wang
Jinqiao Wang
Jue Wang
Ke Wang
Keze Wang
Le Wang
Lei Wang
Lezi Wang
Li Wang
Liang Wang
Lijun Wang
Limin Wang
Linwei Wang
Lizhi Wang
Mengjiao Wang
Mingzhe Wang
Minsi Wang
Naiyan Wang
Nannan Wang
Ning Wang
Oliver Wang
Pei Wang
Peng Wang
Pichao Wang
Qi Wang
Qian Wang
Qiaosong Wang
Qifei Wang
Qilong Wang
Qing Wang
Qingzhong Wang
Quan Wang
Rui Wang
Ruiping Wang
Ruixing Wang
Shangfei Wang
Shenlong Wang
Shiyao Wang
Shuhui Wang
Song Wang
Tao Wang
Tianlu Wang
Tiantian Wang
Ting-chun Wang
Tingwu Wang
Wei Wang
Weiyue Wang
Wenguan Wang
Wenlin Wang
Wenqi Wang
Xiang Wang
Xiaobo Wang
Xiaofang Wang
Xiaoling Wang
Xiaolong Wang
Xiaosong Wang
Xiaoyu Wang
Xin Eric Wang
Xinchao Wang
Xinggang Wang
Xintao Wang
Yali Wang
Yan Wang
Yang Wang
Yangang Wang
Yaxing Wang
Yi Wang
Yida Wang
Yilin Wang
Yiming Wang
Yisen Wang
Yongtao Wang
Yu-Xiong Wang
Yue Wang
Yujiang Wang
Yunbo Wang
Yunhe Wang
Zengmao Wang
Zhangyang Wang
Zhaowen Wang
Zhe Wang
Zhecan Wang
Zheng Wang
Zhixiang Wang
Zilei Wang
Jianqiao Wangni

Anne S. Wannenwetsch
Jan Dirk Wegner
Scott Wehrwein
Donglai Wei
Kaixuan Wei
Longhui Wei
Pengxu Wei
Ping Wei
Qi Wei
Shih-En Wei
Xing Wei
Yunchao Wei
Zijun Wei
Jerod Weinman
Michael Weinmann
Philippe Weinzaepfel
Yair Weiss
Bihan Wen
Longyin Wen
Wei Wen
Junwu Weng
Tsui-Wei Weng
Xinshuo Weng
Eric Wengrowski
Tomas Werner
Gordon Wetzstein
Tobias Weyand
Patrick Wieschollek
Maggie Wigness
Erik Wijmans
Richard Wildes
Olivia Wiles
Chris Williams
Williem Williem
Kyle Wilson
Calden Wloka
Nicolai Wojke
Christian Wolf
Yongkang Wong
Sanghyun Woo
Scott Workman
Baoyuan Wu
Bichen Wu
Chao-Yuan Wu
Huikai Wu
Jiajun Wu
Jialin Wu
Jiaxiang Wu
Jiqing Wu
Jonathan Wu
Lifang Wu
Qi Wu
Qiang Wu
Ruizheng Wu
Shangzhe Wu
Shun-Cheng Wu
Tianfu Wu
Wayne Wu
Wenxuan Wu
Xiao Wu
Xiaohe Wu
Xinxiao Wu
Yang Wu
Yi Wu
Yiming Wu
Ying Nian Wu
Yue Wu
Zheng Wu
Zhenyu Wu
Zhirong Wu
Zuxuan Wu
Stefanie Wuhrer
Jonas Wulff
Changqun Xia
Fangting Xia
Fei Xia
Gui-Song Xia
Lu Xia
Xide Xia
Yin Xia
Yingce Xia
Yongqin Xian
Lei Xiang
Shiming Xiang
Bin Xiao
Fanyi Xiao
Guobao Xiao
Huaxin Xiao
Taihong Xiao
Tete Xiao
Tong Xiao
Wang Xiao
Yang Xiao
Cihang Xie
Guosen Xie
Jianwen Xie
Lingxi Xie
Sirui Xie
Weidi Xie
Wenxuan Xie
Xiaohua Xie
Fuyong Xing
Jun Xing
Junliang Xing
Bo Xiong
Peixi Xiong
Yu Xiong
Yuanjun Xiong
Zhiwei Xiong
Chang Xu
Chenliang Xu
Dan Xu
Danfei Xu
Hang Xu
Hongteng Xu
Huijuan Xu
Jingwei Xu
Jun Xu
Kai Xu
Mengmeng Xu
Mingze Xu
Qianqian Xu
Ran Xu
Weijian Xu
Xiangyu Xu
Xiaogang Xu
Xing Xu
Xun Xu
Yanyu Xu
Yichao Xu
Yong Xu
Yongchao Xu
Yuanlu Xu
Zenglin Xu
Zheng Xu
Chuhui Xue
Jia Xue
Nan Xue

Tianfan Xue
Xiangyang Xue
Abhay Yadav
Yasushi Yagi
I. Zeki Yalniz
Kota Yamaguchi
Toshihiko Yamasaki
Takayoshi Yamashita
Junchi Yan
Ke Yan
Qingan Yan
Sijie Yan
Xinchen Yan
Yan Yan
Yichao Yan
Zhicheng Yan
Keiji Yanai
Bin Yang
Ceyuan Yang
Dawei Yang
Dong Yang
Fan Yang
Guandao Yang
Guorun Yang
Haichuan Yang
Hao Yang
Jianwei Yang
Jiaolong Yang
Jie Yang
Jing Yang
Kaiyu Yang
Linjie Yang
Meng Yang
Michael Ying Yang
Nan Yang
Shuai Yang
Shuo Yang
Tianyu Yang
Tien-Ju Yang
Tsun-Yi Yang
Wei Yang
Wenhan Yang
Xiao Yang
Xiaodong Yang
Xin Yang
Yan Yang
Yanchao Yang
Yee Hong Yang
Yezhou Yang
Zhenheng Yang
Anbang Yao
Angela Yao
Cong Yao
Jian Yao
Li Yao
Ting Yao
Yao Yao
Zhewei Yao
Chengxi Ye
Jianbo Ye
Keren Ye
Linwei Ye
Mang Ye
Mao Ye
Qi Ye
Qixiang Ye
Mei-Chen Yeh
Raymond Yeh
Yu-Ying Yeh
Sai-Kit Yeung
Serena Yeung
Kwang Moo Yi
Li Yi
Renjiao Yi
Alper Yilmaz
Junho Yim
Lijun Yin
Weidong Yin
Xi Yin
Zhichao Yin
Tatsuya Yokota
Ryo Yonetani
Donggeun Yoo
Jae Shin Yoon
Ju Hong Yoon
Sung-eui Yoon
Laurent Younes
Changqian Yu
Fisher Yu
Gang Yu
Jiahui Yu
Kaicheng Yu
Ke Yu
Lequan Yu
Ning Yu
Qian Yu
Ronald Yu
Ruichi Yu
Shoou-I Yu
Tao Yu
Tianshu Yu
Xiang Yu
Xin Yu
Xiyu Yu
Youngjae Yu
Yu Yu
Zhiding Yu
Chunfeng Yuan
Ganzhao Yuan
Jinwei Yuan
Lu Yuan
Quan Yuan
Shanxin Yuan
Tongtong Yuan
Wenjia Yuan
Ye Yuan
Yuan Yuan
Yuhui Yuan
Huanjing Yue
Xiangyu Yue
Ersin Yumer
Sergey Zagoruyko
Egor Zakharov
Amir Zamir
Andrei Zanfir
Mihai Zanfir
Pablo Zegers
Bernhard Zeisl
John S. Zelek
Niclas Zeller
Huayi Zeng
Jiabei Zeng
Wenjun Zeng
Yu Zeng
Xiaohua Zhai
Fangneng Zhan
Huangying Zhan
Kun Zhan

Xiaohang Zhan
Baochang Zhang
Bowen Zhang
Cecilia Zhang
Changqing Zhang
Chao Zhang
Chengquan Zhang
Chi Zhang
Chongyang Zhang
Dingwen Zhang
Dong Zhang
Feihu Zhang
Hang Zhang
Hanwang Zhang
Hao Zhang
He Zhang
Hongguang Zhang
Hua Zhang
Ji Zhang
Jianguo Zhang
Jianming Zhang
Jiawei Zhang
Jie Zhang
Jing Zhang
Juyong Zhang
Kai Zhang
Kaipeng Zhang
Ke Zhang
Le Zhang
Lei Zhang
Li Zhang
Lihe Zhang
Linguang Zhang
Lu Zhang
Mi Zhang
Mingda Zhang
Peng Zhang
Pingping Zhang
Qian Zhang
Qilin Zhang
Quanshi Zhang
Richard Zhang
Rui Zhang
Runze Zhang
Shengping Zhang
Shifeng Zhang
Shuai Zhang
Songyang Zhang
Tao Zhang
Ting Zhang
Tong Zhang
Wayne Zhang
Wei Zhang
Weizhong Zhang
Wenwei Zhang
Xiangyu Zhang
Xiaolin Zhang
Xiaopeng Zhang
Xiaoqin Zhang
Xiuming Zhang
Ya Zhang
Yang Zhang
Yimin Zhang
Yinda Zhang
Ying Zhang
Yongfei Zhang
Yu Zhang
Yulun Zhang
Yunhua Zhang
Yuting Zhang
Zhanpeng Zhang
Zhao Zhang
Zhaoxiang Zhang
Zhen Zhang
Zheng Zhang
Zhifei Zhang
Zhijin Zhang
Zhishuai Zhang
Ziming Zhang
Bo Zhao
Chen Zhao
Fang Zhao
Haiyu Zhao
Han Zhao
Hang Zhao
Hengshuang Zhao
Jian Zhao
Kai Zhao
Liang Zhao
Long Zhao
Qian Zhao
Qibin Zhao
Qijun Zhao
Rui Zhao
Shenglin Zhao
Sicheng Zhao
Tianyi Zhao
Wenda Zhao
Xiangyun Zhao
Xin Zhao
Yang Zhao
Yue Zhao
Zhichen Zhao
Zijing Zhao
Xiantong Zhen
Chuanxia Zheng
Feng Zheng
Haiyong Zheng
Jia Zheng
Kang Zheng
Shuai Kyle Zheng
Wei-Shi Zheng
Yinqiang Zheng
Zerong Zheng
Zhedong Zheng
Zilong Zheng
Bineng Zhong
Fangwei Zhong
Guangyu Zhong
Yiran Zhong
Yujie Zhong
Zhun Zhong
Chunluan Zhou
Huiyu Zhou
Jiahuan Zhou
Jun Zhou
Lei Zhou
Luowei Zhou
Luping Zhou
Mo Zhou
Ning Zhou
Pan Zhou
Peng Zhou
Qianyi Zhou
S. Kevin Zhou
Sanping Zhou
Wengang Zhou
Xingyi Zhou

Yanzhao Zhou
Yi Zhou
Yin Zhou
Yipin Zhou
Yuyin Zhou
Zihan Zhou
Alex Zihao Zhu
Chenchen Zhu
Feng Zhu
Guangming Zhu
Ji Zhu
Jun-Yan Zhu
Lei Zhu
Linchao Zhu
Rui Zhu
Shizhan Zhu
Tyler Lixuan Zhu
Wei Zhu
Xiangyu Zhu
Xinge Zhu
Xizhou Zhu
Yanjun Zhu
Yi Zhu
Yixin Zhu
Yizhe Zhu
Yousong Zhu
Zhe Zhu
Zhen Zhu
Zheng Zhu
Zhenyao Zhu
Zhihui Zhu
Zhuotun Zhu
Bingbing Zhuang
Wei Zhuo
Christian Zimmermann
Karel Zimmermann
Larry Zitnick
Mohammadreza Zolfaghari
Maria Zontak
Daniel Zoran
Changqing Zou
Chuhang Zou
Danping Zou
Qi Zou
Yang Zou
Yuliang Zou
Georgios Zoumpourlis
Wangmeng Zuo
Xinxin Zuo

Additional Reviewers

Victoria Fernandez Abrevaya
Maya Aghaei
Allam Allam
Christine Allen-Blanchette
Nicolas Aziere
Assia Benbihi
Neha Bhargava
Bharat Lal Bhatnagar
Joanna Bitton
Judy Borowski
Amine Bourki
Romain Brégier
Tali Brayer
Sebastian Bujwid
Andrea Burns
Yun-Hao Cao
Yuning Chai
Xiaojun Chang
Bo Chen
Shuo Chen
Zhixiang Chen
Junsuk Choe
Hung-Kuo Chu
Jonathan P. Crall
Kenan Dai
Lucas Deecke
Karan Desai
Prithviraj Dhar
Jing Dong
Wei Dong
Turan Kaan Elgin
Francis Engelmann
Erik Englesson
Fartash Faghri
Zicong Fan
Yang Fu
Risheek Garrepalli
Yifan Ge
Marco Godi
Helmut Grabner
Shuxuan Guo
Jianfeng He
Zhezhi He
Samitha Herath
Chih-Hui Ho
Yicong Hong
Vincent Tao Hu
Julio Hurtado
Jaedong Hwang
Andrey Ignatov
Muhammad Abdullah Jamal
Saumya Jetley
Meiguang Jin
Jeff Johnson
Minsoo Kang
Saeed Khorram
Mohammad Rami Koujan
Nilesh Kulkarni
Sudhakar Kumawat
Abdelhak Lemkhenter
Alexander Levine
Jiachen Li
Jing Li
Jun Li
Yi Li
Liang Liao
Ruochen Liao
Tzu-Heng Lin
Phillip Lippe
Bao-di Liu
Bo Liu
Fangchen Liu

Hanxiao Liu
Hongyu Liu
Huidong Liu
Miao Liu
Xinxin Liu
Yongfei Liu
Yu-Lun Liu
Amir Livne
Tiange Luo
Wei Ma
Xiaoxuan Ma
Ioannis Marras
Georg Martius
Effrosyni Mavroudi
Tim Meinhardt
Givi Meishvili
Meng Meng
Zihang Meng
Zhongqi Miao
Gyeongsik Moon
Khoi Nguyen
Yung-Kyun Noh
Antonio Norelli
Jaeyoo Park
Alexander Pashevich
Mandela Patrick
Mary Phuong
Bingqiao Qian
Yu Qiao
Zhen Qiao
Sai Saketh Rambhatla
Aniket Roy
Amelie Royer
Parikshit Vishwas Sakurikar
Mark Sandler
Mert Bülent Sarıyıldız
Tanner Schmidt
Anshul B. Shah
Ketul Shah
Rajvi Shah
Hengcan Shi
Xiangxi Shi
Yujiao Shi
William A. P. Smith
Guoxian Song
Robin Strudel
Abby Stylianou
Xinwei Sun
Reuben Tan
Qingyi Tao
Kedar S. Tatwawadi
Anh Tuan Tran
Son Dinh Tran
Eleni Triantafillou
Aristeidis Tsitiridis
Md Zasim Uddin
Andrea Vedaldi
Evangelos Ververas
Vidit Vidit
Paul Voigtlaender
Bo Wan
Huanyu Wang
Huiyu Wang
Junqiu Wang
Pengxiao Wang
Tai Wang
Xinyao Wang
Tomoki Watanabe
Mark Weber
Xi Wei
Botong Wu
James Wu
Jiamin Wu
Rujie Wu
Yu Wu
Rongchang Xie
Wei Xiong
Yunyang Xiong
An Xu
Chi Xu
Yinghao Xu
Fei Xue
Tingyun Yan
Zike Yan
Chao Yang
Heran Yang
Ren Yang
Wenfei Yang
Xu Yang
Rajeev Yasarla
Shaokai Ye
Yufei Ye
Kun Yi
Haichao Yu
Hanchao Yu
Ruixuan Yu
Liangzhe Yuan
Chen-Lin Zhang
Fandong Zhang
Tianyi Zhang
Yang Zhang
Yiyi Zhang
Yongshun Zhang
Yu Zhang
Zhiwei Zhang
Jiaojiao Zhao
Yipu Zhao
Xingjian Zhen
Haizhong Zheng
Tiancheng Zhi
Chengju Zhou
Hao Zhou
Hao Zhu
Alexander Zimin

Contents – Part III

SIZER: A Dataset and Model for Parsing 3D Clothing and Learning Size Sensitive 3D Clothing

Garvita Tiwari[1(✉)], Bharat Lal Bhatnagar[1], Tony Tung[2], and Gerard Pons-Moll[1]

[1] MPI for Informatics, Saarland Informatics Campus, Saarbrücken, Germany
{gtiwari,bbhatnag,gpons}@mpi-inf.mpg.de
[2] Facebook Reality Labs, Sausalito, USA
tony.tung@fb.com

Fig. 1. *SIZER* dataset of people with clothing size variation. (*Left*): 3D scans of people captured in different clothing styles and *sizes.* (*Right*): T-shirt and short pants for sizes small and large, which are registered to a common template.

Abstract. While models of 3D clothing learned from real data exist, no method can predict clothing deformation as a function of garment size. In this paper, we introduce SizerNet to predict 3D clothing conditioned on human body shape and garment size parameters, and ParserNet to infer garment meshes and shape under clothing with personal details in a single pass from an input mesh. SizerNet allows to estimate and visualize the dressing effect of a garment in various sizes, and ParserNet allows to edit clothing of an input mesh directly, removing the need for scan segmentation, which is a challenging problem in itself. To learn these models, we introduce the *SIZER* dataset of clothing size variation which includes 100 different subjects wearing casual clothing items in various sizes, totaling to approximately 2000 scans. This dataset includes the scans, registrations to the SMPL model, scans segmented in clothing parts, garment category and size labels. Our experiments show better

Electronic supplementary material The online version of this chapter (https://doi.org/10.1007/978-3-030-58580-8_1) contains supplementary material, which is available to authorized users.

A. Vedaldi et al. (Eds.): ECCV 2020, LNCS 12348, pp. 1–18, 2020.
https://doi.org/10.1007/978-3-030-58580-8_1

parsing accuracy and size prediction than baseline methods trained on *SIZER*. The code, model and dataset will be released for research purposes at: https://virtualhumans.mpi-inf.mpg.de/sizer/.

1 Introduction

Modeling how 3D clothing fits on the human body as a function of size has numerous applications in 3D content generation (e.g., AR/VR, movie, video games, sport), clothing size recommendation (e.g., e-commerce), computer vision for fashion, and virtual try-on. It is estimated that retailers lose up to $600 billion each year due to sales returns as it is currently difficult to purchase clothing online without knowing how it will fit [2,3].

Predicting how clothing fits as a function of body shape and garment size is an extremely challenging task. Clothing interacts with the body in complex ways, and fit is a non-linear function of size and body shape. Furthermore, *clothing fit differences with size are subtle*, but they can make a difference when purchasing clothing online. Physics based simulation is still the most commonly used technique because it generalizes well, but unfortunately, it is difficult to adjust its parameters to achieve a realistic result, and it can be computationally expensive.

While there exist several works that learn how clothing deforms as a function of pose [30], or pose and shape [22,30,34,37,43], there are few works modeling how garments drape as a function of size. Recent works learn a space of styles [37, 50] from physics simulations, but their aim is plausibility, and therefore they can not predict how a *real garment* will deform on a real body.

What is lacking is (1) a 3D dataset of people wearing the same garments in different sizes and (2) a data-driven model *learned from real scans* which varies with sizing and body shape. In this paper, we introduce the *SIZER* dataset, the first dataset of scans of people in different garment sizes featuring approximately 2000 scans, 100 subjects and 10 garments worn by subjects in four different sizes. Using the *SIZER* dataset we learned a Neural Network model, which we refer to as *SizerNet*, which given a body shape and a garment, can predict how the garment drapes on the body as a function of size. Learning *SizerNet* requires to map scans to a registered *multi-layer meshes* – separate meshes for body shape, and top and bottom garments. This requires segmenting the 3D scans, and estimating their body shape under clothing, and registering the garments across the dataset, which we obtain using the method explained in [14,38]. From the multi-layer meshes, we learn an encoder to map the input mesh to a latent code, and a decoder which additionally takes the body shape parameters of SMPL [33], the size label (S, M, L, XL) of the input garment, and the desired size of the output, to predict the output garment as a displacement field to a template.

Although visualizing how an existing garment fits on a body as a function of size is already useful for virtual try-on applications, we would also like to change the size of garments in existing 3D scans. Scans however, are just pointclouds, and parsing them into a multi-layer representation at test time using [14,38] requires segmentation, which sometimes requires manual intervention. Therefore, we propose *ParserNet*, which automatically maps a single mesh registration (SMPL deformed to the scan) to multi-layer meshes with a single feed-forward pass. *ParserNet*, not only segments the single mesh registration, but it reparameterizes the surface so that it is coherent with common garment templates. The output multi-layer representation of *ParserNet* is powerful as it allows simulation and editing meshes separately. Additionally, the tandem of *SizerNet* and *ParserNet* allows us to edit the size of clothing directly on the mesh, allowing shape manipulation applications never explored before.

In summary, our contributions are:

- *SIZER* dataset: A dataset of clothing size variation of approximately 2000 scans including 100 subjects wearing 10 garment classes in different sizes, where we make available, scans, clothing segmentation, SMPL+G registrations, body shape under clothing, garment class and size labels.
- SizerNet: The first model learned from real scans to predict how clothing drapes on the body as a function of size.
- ParserNet: A data-driven model to map a single mesh registration into a multi-layered representation of clothing without the need for segmentation or non-linear optimization.

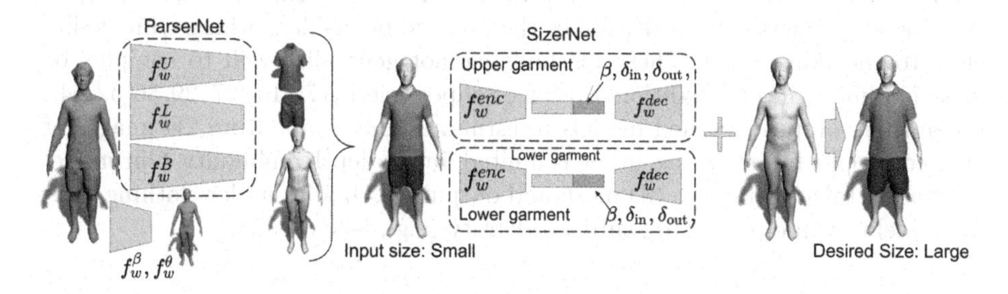

Fig. 2. We propose a model to estimate and visualize the dressing effect of a garment conditioned on body shape and garment size parameters. For this we introduce *ParserNet* (f_w^U, f_w^L, f_w^B), which takes a SMPL registered mesh $M(\boldsymbol{\theta}, \boldsymbol{\beta}, \mathbf{D})$ as input and predicts the SMPL parameters ($\boldsymbol{\theta}, \boldsymbol{\beta}$), parsed 3D garments using predefined templates $T^g(\boldsymbol{\beta}, \boldsymbol{\theta}, \mathbf{0})$ and predicts body shape under clothing while preserving the personal details of the subject. We also propose *SizerNet*, an encoder-decoder ($f_w^{\text{enc}}, f_w^{\text{dec}}$) based network, that resizes the garment given as input with the desired size label ($\delta_{\text{in}}, \delta_{\text{out}}$) and drapes it on the body shape under clothing.

2 Related Work

Clothing Modeling. Accurate reconstruction of 3D cloth with fine structures (e.g., wrinkles) is essential for realism while being notoriously challenging. Methods based on multi-view stereo can recover global shape robustly but struggle with high frequency details in non-textured regions [6,16,32,44,47,51]. The pioneering work of [8,9] demonstrated for the first time detailed body and clothing reconstruction from monocular video using a displacement from SMPL, which spearheaded recent developments [7,10,23–25,42]. These approaches do not separate body from clothing. In [14,26,30,38], the authors propose to reconstruct clothing as a layer separated from the body. These models are trained on 3D scans of real clothed people data and produce realistic models. On the other hand, physics based simulation methods have also been used to model clothing [21,22,35,37,43,45,46,48,49]. Despite the potential gap with real-world data, they are a great alternative to obtain clean data, free of acquisition noise and holes. However, they still require manual parameter tuning (e.g., time step for better convergence, sheer and stretch for better deformation effects, etc.), and can be slow or unstable. In [21,22,43] a pose and shape dependent clothing model is introduced, and [37,50] also model garment style dependent clothing using a lower-dimensional representation for style and size like PCA and garment sewing parameters, however there is no direct control on the size of clothing generated for given body shape. In [53], authors model the garment fit on different body shapes from images. Our model *SizerNet* automatically outputs realistic 3D cloth models conditioned on desired features (e.g., shape, size).

Shape Under Clothing. In [11,57,60], the authors propose to estimate body shape under clothing by fitting a 3D body model to 3D reconstructions of people. An objective function typically forces the body to be inside clothing while being close to the skin region. These methods cannot generalize well to complex or loose clothing without additional prior or supervision [17]. In [27–29,36,52,54], the authors propose learned models to estimate body shape from 2D images of clothed people, but shape accuracy is limited due to depth ambiguity. Our model *ParserNet* takes as input a 3D mesh and outputs 3D bodies under clothing with high fidelity while preserving subject identity (e.g., face details).

Cloth Parsing. The literature has proposed several methods for clothed human understanding. In particular, efficient cloth parsing in 2D has been achieved using supervised learning and generative networks [18–20,55,56,58]. 3D clothing parsing of 3D scans has also been investigated [14,38]. The authors propose techniques based on MRF-GrabCut [41] to segment 3D clothing from 3D scans and transfer them to different subjects. However the approach requires several steps, which is not optimal for scalability. We extend previous work with *SIZER*, a fully automatic data-driven pipeline. In [13], the authors jointly predict clothing and inner body surface, with semantic correspondences to SMPL. However, it does not have semantic clothing information.

3D Datasets. To date, only a few datasets consist of 3D models of subjects with segmented clothes. 3DPeople [40], Cloth3D [12] consists of a large dataset of synthetic 3D humans with clothing. None of the synthetic datasets contains realistic cloth deformations like the SIZER dataset. THUman [61] consists of sequences of clothed 3D humans in motion, captured with a consumer RGBD sensor (Kinectv2), and are reconstructed using volumetric SDF fusion [59]. However, 3D models are rather smooth compared to our 3D scans and no ground truth segmentation of clothing is provided. Dyna and D-FAUST [15,39] consist of high-res 3D scans of 10 humans in motion with different shape but the subjects are only wearing minimal clothing. BUFF [60] contains high-quality 3D scans of 6 subjects with and without clothing. The dataset is primarily designed to train models to estimate body shape under clothing and doesn't contain garments segmentation. In [14], the authors create a digital wardrobe with 3D templates of garments to dress 3D bodies. In [26], authors propose a mixture of synthetic and real data, which contains garment, body shape and pose variations. However, the fraction of real dataset (~300 scans) is fairly small. DeepFahsion3D [62] is a dataset of real scans of clothing containing various garment styles. None of these datasets contain garment sizing variation. Unlike our proposed *SIZER* dataset, no dataset contains a large amount of pre-segmented clothing from 3D scans at different sizes, with corresponding body shapes under clothing.

3 Dataset

In this paper, we address a very challenging problem of modeling garment fitting as a function of body shape and garment size. As explained in Sect. 2, one of the key bottlenecks that hinder progress in this direction is the lack of real-world datasets that contain calibrated and well-annotated garments in different sizes draped on real humans. To this end, we present *SIZER* dataset, a dataset of over 2000 scans containing people in diverse body shapes in various garments styles and sizes. We describe our dataset in Sects. 3.1 and 3.2.

3.1 SIZER Dataset: Scans

We introduce the *SIZER* dataset that contains 100 subjects, wearing the same garment in 2 or 3 garment sizes (S, M, L, XL). We include 10 garment classes, namely shirt, dress-shirt, jeans, hoodie, polo t-shirt, t-shirt, shorts, vest, skirt, and coat, which amounts to roughly 200 scans per garment class. We capture the subjects in a relaxed A-pose to avoid stretching or tension due to pose in the garments. Figure 1 shows some examples of people wearing a fixed set of garments in different sizes. We use a Treedy's static scanner [5] which has 130+ cameras, and reconstruct the scans using Agisoft's Metashape software [1]. Our scans are high resolution and are represented by meshes, which have different underlying graph connectivity across the dataset, and hence it is challenging to use this dataset directly in any learning framework. We preprocess our dataset, by registering them to SMPL [33]. We explain the structure of processed data in the following section.

3.2 SIZER Dataset: SMPL and Garment Registrations

To improve general usability of the *SIZER* dataset, we provide SMPL+G registrations [14,31] registrations. Registering our scans to SMPL, brings all our scans to correspondence, and provides more control over the data via pose and shape parameters from the underlying SMPL. We briefly describe the SMPL and SMPL+G formulations below.

SMPL represents the human body as a parametric function $M(\cdot)$, of pose ($\boldsymbol{\theta}$) and shape ($\boldsymbol{\beta}$). We add per-vertex displacements ($\mathbf{D}$) on top of SMPL to model deformations corresponding to hair, garments, etc. thus resulting in the SMPL model. SMPL applies standard skinning $W(\cdot)$ to a base template $\mathbf{T}$ in T-pose. Here, $\mathbf{W}$ denotes the blend weights and $B_p(\cdot)$ and $B_s(\cdot)$ models pose and shape dependent deformations respectively.

$$M(\boldsymbol{\beta}, \boldsymbol{\theta}, \mathbf{D}) = W(T(\boldsymbol{\beta}, \boldsymbol{\theta}, \mathbf{D}), J(\boldsymbol{\beta}), \boldsymbol{\theta}, \mathbf{W}) \tag{1}$$

$$T(\boldsymbol{\beta}, \boldsymbol{\theta}, \mathbf{D}) = \mathbf{T} + B_s(\boldsymbol{\beta}) + B_p(\boldsymbol{\theta}) + \mathbf{D} \tag{2}$$

SMPL+G is a parametric formulation to represent the human body and garments as separate meshes. To register the garments we first segment scans into garments and skin parts [14]. We refine the scan segmentation step used in [14] by fine-tuning the Human Parsing network [20] with a multi-view consistency loss. We then use the multi-mesh registration approach from [14] to register garments to the SMPL+G model. For each garment class, we obtain a template mesh which is defined as a subset of the SMPL template, given by $T^g(\boldsymbol{\beta}, \boldsymbol{\theta}, \mathbf{0}) = \mathbf{I}^g T(\boldsymbol{\beta}, \boldsymbol{\theta}, \mathbf{0})$, where $\mathbf{I}^g \in \mathbb{Z}_2^{m_g \times n}$ is an indicator matrix, with $\mathbf{I}^g_{i,j} = 1$ if garment g vertex $i \in \{1 \ldots m_g\}$ is associated with body shape vertex $j \in \{1 \ldots n\}$. m_g and n denote the number of vertices in the garment template and the SMPL mesh respectively. Similarly, we define a garment function $G(\boldsymbol{\beta}, \boldsymbol{\theta}, \mathbf{D}^g)$ using Eq. (3), where $\mathbf{D}^g$ are the per-vertex offsets from the template

$$G(\boldsymbol{\beta}, \boldsymbol{\theta}, \mathbf{D}^g) = W(T^g(\boldsymbol{\beta}, \boldsymbol{\theta}, \mathbf{D}^g), J(\boldsymbol{\beta}), \boldsymbol{\theta}, \mathbf{W}). \tag{3}$$

For every scan in the *SIZER* dataset, we will release the scan, segmented scan, and SMPL+G registrations, garment category and garment size label.

This dataset can be used in several applications like virtual try-on, character animation, learning generative models, data-driven body shape under clothing, size and(or) shape sensitive clothing model, etc. To stimulate further research in this direction, we will release the dataset, code and baseline models, which can be used as a benchmark in 3D clothing parsing and 3D garment resizing. We use this dataset to build a model for the task of garment extraction from single mesh (*ParserNet*) and garment resizing (*SizerNet*), which we describe in the next section.

4 Method

We introduce *ParserNet* (Sect. 4.2), the first method for extracting garments directly from SMPL registered meshes. For parsing garments, we first predict the underlying body SMPL parameters using a pose and shape prediction network (Sect. 4.1) and use *ParserNet* to extract garment layers and personal features like hair, facial features to create body shape under clothing. Next, we present *SizerNet* (Sect. 4.3), an encoder-decoder based deep network for garment resizing. An overview of the method is shown in Fig. 2.

4.1 Pose and Shape Prediction Network

To estimate body shape under clothing, we first create the undressed SMPL body for a given clothed input single layer mesh $M(\boldsymbol{\beta}, \boldsymbol{\theta}, \mathbf{D})$, by predicting $\boldsymbol{\theta}, \boldsymbol{\beta}$ using $f_w^{\boldsymbol{\theta}}$ and $f_w^{\boldsymbol{\beta}}$ respectively. We train $f_w^{\boldsymbol{\theta}}$ and $f_w^{\boldsymbol{\beta}}$ with L_2 loss over parameters and per-vertex loss between predicted SMPL body and clothed input mesh, as shown in Eqs. (4) and (5). Since the reference body under clothing parameters $\boldsymbol{\theta}, \boldsymbol{\beta}$ obtained via instance specific optimization (Sect. 3.2) can be inaccurate, we add an additional per-vertex loss between our predicted SMPL body vertices $M(\hat{\boldsymbol{\theta}}, \hat{\boldsymbol{\beta}}, \mathbf{0})$ and the input clothed mesh $M(\boldsymbol{\beta}, \boldsymbol{\theta}, \mathbf{D})$. This brings the predicted undressed body closer to the input clothed mesh. We observe more stable results training $f_w^{\boldsymbol{\theta}}$ and $f_w^{\boldsymbol{\beta}}$ separately initially, using the reference $\boldsymbol{\beta}$ and $\boldsymbol{\theta}$ respectively. Since the $\boldsymbol{\beta}$ components in SMPL are normalized to have $\sigma = 1$, we un-normalize them by scaling by their respective standard deviations $[\sigma_1, \sigma_2, \ldots, \sigma_{10}]$ as given in Eq. (5).

$$\mathcal{L}_\theta = w_{\text{pose}}||\hat{\boldsymbol{\theta}} - \boldsymbol{\theta}||_2^2 + w_v||M(\boldsymbol{\beta}, \hat{\boldsymbol{\theta}}, \mathbf{0}) - M(\boldsymbol{\beta}, \boldsymbol{\theta}, \mathbf{D})|| \tag{4}$$

$$\mathcal{L}_\beta = w_{\text{shape}} \sum_{i=1}^{10} \sigma_i(\hat{\boldsymbol{\beta}}_i - \boldsymbol{\beta}_i)^2 + w_v||M(\hat{\boldsymbol{\beta}}, \boldsymbol{\theta}, \mathbf{0}) - M(\boldsymbol{\beta}, \boldsymbol{\theta}, \mathbf{D})|| \tag{5}$$

Here, w_{pose}, w_{shape} and w_v are weights for the loss on pose, shape and predicted SMPL surface. $(\hat{\boldsymbol{\theta}}, \hat{\boldsymbol{\beta}})$ denote predicted parameters. The output is a *smooth* (SMPL model) body shape under clothing.

4.2 ParserNet

Parsing Garments. Parsing garments from a single mesh ($\mathbf{M}$) can be done by segmenting it into separate garments for each class ($\mathbf{G}_{\text{seg}}^{g,k}$), which leads to different underlying graph connectivity ($\mathcal{G}_{\text{seg}}^{g,k} = (\mathbf{G}_{\text{seg}}^{g,k}, \mathbf{E}_{\text{seg}}^{g,k})$) across all the instances (k) of a garment class g, shown in Fig. 3 (right). Hence, we propose to parse garments by deforming vertices of a template $T^g(\boldsymbol{\beta}, \boldsymbol{\theta}, \mathbf{0})$ with fixed connectivity $\mathbf{E}^g$, obtaining vertices $\mathbf{G}^{g,k} \in \mathcal{G}^{g,k}$, where $\mathcal{G}^{g,k} = (\mathbf{G}^{g,k}, \mathbf{E}^g)$, shown in Fig. 3 (middle).

Our key idea is to predict the deformed vertices $\mathbf{G}^g$ directly as a convex combination of vertices of the input mesh $\mathbf{M} = M(\boldsymbol{\beta}, \boldsymbol{\theta}, \mathbf{D})$ with a learned sparse regressor matrix $\mathbf{W}^g$, such that $\mathbf{G}^g = \mathbf{W}^g\mathbf{M}$. Specifically, *ParserNet* predicts the sparse matrix ($\mathbf{W}^g$) as a function of input mesh features (vertices and normals) and a predefined per-vertex neighborhood ($\mathcal{N}_i$) for every vertex i of garment class g. We will henceforth drop $(.)^{g,k}$ unless required. In this way,

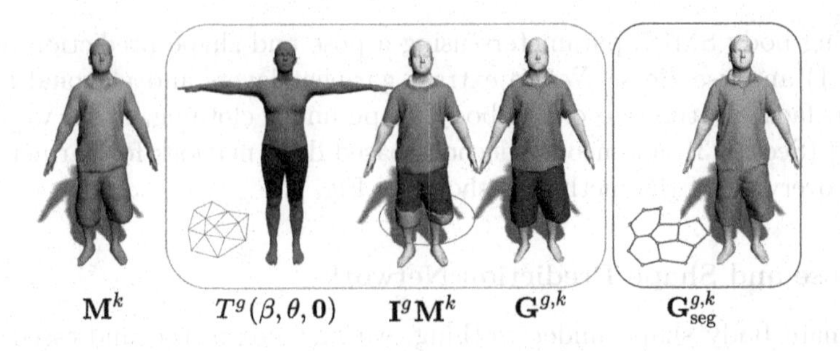

Fig. 3. Left to right: Input single mesh ($\mathbf{M}^k$), garment template ($T^g(\boldsymbol{\beta},\boldsymbol{\theta},\mathbf{0}) = \mathbf{I}^g T(\boldsymbol{\beta},\boldsymbol{\theta},\mathbf{0})$), garment mesh extracted using $\mathbf{G}^{g,k} = \mathbf{I}^g\mathbf{M}^k$, multi-layer meshes ($\mathbf{G}^{g,k}$) registered to SMPL+G, all with garment class specific edge connectivity $\mathbf{E}^g$, and segmented scan $\mathbf{G}^{g,k}_{\text{seg}}$ with instance specific edge connectivity $\mathbf{E}^{g,k}_{\text{seg}}$.

the output vertices $\mathbf{G}_i \in \mathbb{R}^3$, where $i \in \{1,\ldots,m_g\}$, are obtained as a convex combination of input mesh vertices $\mathbf{M}_j \in \mathbb{R}^3$ in a predefined neighborhood ($\mathcal{N}_i$).

$$\mathbf{G}_i = \sum_{j\in\mathcal{N}_i} \mathbf{W}_{ij}\mathbf{M}_j. \tag{6}$$

Parsing Detailed Body Shape Under Clothing. For generating detailed body shape under clothing, we first create a *smooth body mesh*, using SMPL parameters $\boldsymbol{\theta}$ and $\boldsymbol{\beta}$ predicted from $f_w^{\boldsymbol{\theta}}, f_w^{\boldsymbol{\beta}}$ (Sect. 4.1). Using the same aforementioned convex combination formulation, *Body ParserNet* transfers the visible skin vertices from the input mesh to the smooth body mesh, obtaining hair and facial features. We parse the input mesh into upper, lower garments and detailed shape under clothing using 3 sub-networks (f_w^U, f_w^L, f_w^B) of *ParserNet*, as shown in Fig. 2.

4.3 SizerNet

We aim to edit the garment mesh based on garment size labels such as S, M, L, etc., to see the dressing effect of the garment for a new size. For this task, we propose an encoder-decoder based network, which is shown in Fig. 2 (right). The network f_w^{enc}, encodes the garment mesh $\mathbf{G}_{in}$ to a lower-dimensional latent code $\boldsymbol{x}_{\text{gar}} \in \mathbb{R}^d$, shown in Eq. (7). We append ($\boldsymbol{\beta}, \delta_{\text{in}}, \delta_{\text{out}}$) to the latent space, where $\delta_{\text{in}}, \delta_{\text{out}}$ are one-hot encodings of input and desired output sizing and $\boldsymbol{\beta}$ is the SMPL $\boldsymbol{\beta}$ parameter for underlying body shape.

$$\boldsymbol{x}_{\text{gar}} = f_w^{\text{enc}}(\mathbf{G}_{in}), \quad f_w^{\text{enc}}(.) : \mathbb{R}^{m_g\times 3} \rightarrow \mathbb{R}^d \tag{7}$$

The decoder network, $f_w^{\text{dec}}(.) : \mathbb{R}^{|\boldsymbol{\beta}|} \times \mathbb{R}^d \times \mathbb{R}^{2|\delta|} \longrightarrow \mathbb{R}^{m_g\times 3}$ predicts the displacement field $\mathbf{D}^g = f_w^{\text{dec}}(\boldsymbol{\beta}, \boldsymbol{x}_{\text{gar}}, \boldsymbol{\delta}_{\text{in}}, \boldsymbol{\delta}_{\text{out}})$ on top on template. We obtain the output garment $\mathbf{G}_{out}$ in the new desired size δ_{out} using Eq. (3).

4.4 Loss Functions

We train the networks, *ParserNet* and *SizerNet* with training losses given by Eqs. (8) and (9) respectively, where $w_{3\text{D}}$, w_{norm}, w_{lap}, w_{interp} and w_{w} are weights for the loss on vertices, normal, Laplacian, interpenetration and weight regularizer term respectively. We explain each of the loss terms in this section.

$$\mathcal{L}_{\text{parser}} = w_{3\text{D}}\mathcal{L}_{3\text{D}} + w_{\text{norm}}\mathcal{L}_{\text{norm}} + w_{\text{lap}}\mathcal{L}_{\text{lap}} + w_{\text{interp}}\mathcal{L}_{\text{interp}} + w_{\text{w}}\mathcal{L}_{\text{w}} \tag{8}$$

$$\mathcal{L}_{\text{sizer}} = w_{3\text{D}}\mathcal{L}_{3\text{D}} + w_{\text{norm}}\mathcal{L}_{\text{norm}} + w_{\text{lap}}\mathcal{L}_{\text{lap}} + w_{\text{interp}}\mathcal{L}_{\text{interp}} \tag{9}$$

- **3D vertex loss for garments.** We define $\mathcal{L}_{3\text{D}}$ as L_1 loss between predicted and ground truth vertices

$$\mathcal{L}_{3\text{D}} = ||\mathbf{G}_{\text{P}} - \mathbf{G}_{\text{GT}}||_1. \tag{10}$$

- **3D vertex loss for shape under clothing.** For training f_w^B (ParserNet for the body), we use the input mesh skin as supervision for predicting personal details of subject. We define a garment class specific geodesic distance weighted loss term, as shown in Eq. (11), where $\mathbf{I}^s$ is the indicator matrix for skin region and $\boldsymbol{w}_{\text{geo}}$ is a vector containing the *sigmoid* of the geodesic distances from vertices to the boundary between skin and non-skin regions. The loss term is high when the prediction is far from the input mesh $\mathbf{M}$ for the visible skin region, and lower for the cloth region, with a smooth transition regulated by the geodesic term. Let $\text{abs}_{ij}(\cdot)$ denote an element-wise absolute value operator. Then the loss is computed as

$$\mathcal{L}_{3\text{D}}^{\text{body}} = ||\boldsymbol{w}_{\text{geo}}^T \cdot \text{abs}_{ij}(\mathbf{G}_{\text{P}}^s - \mathbf{I}^s\mathbf{M})||_1. \tag{11}$$

- **Normal Loss.** We define $\mathcal{L}_{\text{norm}}$ as the difference in angle between ground truth face normal ($\mathbf{N}_{\text{GT}}^i$) and predicted face normal ($\mathbf{N}_P^i$).

$$\mathcal{L}_{\text{norm}} = \frac{1}{N_{\text{faces}}} \sum_{i}^{N_{\text{faces}}} (1 - (\mathbf{N}_{\text{GT},i})^T \mathbf{N}_{P,i}). \tag{12}$$

- **Laplacian smoothness term.** This enforces the Laplacian of predicted garment mesh to be close to the Laplacian of ground truth mesh. Let $\mathbf{L}^g \in \mathbb{R}^{m_g \times m_g}$ be the graph Laplacian of the garment mesh $\mathbf{G}_{\text{GT}}$, and $\boldsymbol{\Delta}_{\text{init}} = \mathbf{L}^g\mathbf{G}_{\text{GT}} \in \mathbb{R}^{m_g \times 3}$ be the differential coordinates of the $\mathbf{G}_{\text{GT}}$, then we compute the Laplacian smoothness term for a predicted mesh $\mathbf{G}_{\text{P}}$ as

$$\mathcal{L}_{\text{lap}} = ||\boldsymbol{\Delta}_{\text{init}} - \mathbf{L}^g\mathbf{G}_{\text{P}}||_2. \tag{13}$$

- **Interpenetration loss.** Since minimizing per-vertex loss does not guarantee that the predicted garment lies outside the body surface, we use the interpenetration loss term in Eq. (14) proposed in GarNet [22]. For every vertex $\mathbf{G}_{\text{P},j}$, we find the nearest vertex in the predicted body shape under clothing ($\mathbf{B}_i$) and define the body-garment correspondences as $\mathcal{C}(\mathbf{B}, \mathbf{G}_{\text{P}})$. Let $\mathbf{N}_i$ be the normal of the ith body vertex $\mathbf{B}_i$. If the predicted garment vertex $\mathbf{G}_{\text{P},j}$ penetrates the body, it is penalized with the following loss

$$\mathcal{L}_{\text{interp}} = \sum_{(i,j)\in\mathcal{C}(\mathbf{B},\mathbf{G}_{\text{P}})} \mathbb{1}_{d(\mathbf{G}_{\text{P},j},G_{\text{GT},j})<d_{tol}} ReLU(-\mathbf{N}_i(\mathbf{G}_{\text{P},j} - \mathbf{B}_i))/m_g, \quad (14)$$

where notice that $\mathbb{1}_{d(\mathbf{G}_{\text{P},j},G_{\text{GT},j})<d_{tol}}$ activates the loss when the distance between predicted garment mesh vertices and ground truth mesh vertices is small *i.e.* $< d_{tol}$.

- **Weight regularizer.** To preserve the fine details when parsing the input mesh, we want the weights predicted by the network to be sparse and confined in a local neighborhood. Hence, we add a regularizer which penalizes large values for $\mathbf{W}_{ij}$ if the distance between of $\mathbf{M}_j$ and the vertex $\mathbf{M}_k$ with largest weight $k = \arg\max_j \mathbf{W}_{ij}$ is large. Let $d(\cdot,\cdot)$ denote Euclidean distance between vertices, then the regularizer equals

$$\mathcal{L}_w = \sum_{i=1}^{m_g} \sum_{j\in\mathcal{N}_i} \mathbf{W}_{ij} d(\mathbf{M}_k, \mathbf{M}_j), \;\; k = \arg\max_j \mathbf{W}_{ij}. \quad (15)$$

4.5 Implementation Details

We implement f_w^{θ} and f_w^{β} networks with 2 fully connected and a linear output layer. We implement *ParserNet* f_w^U, f_w^L, f_w^B with 3 fully connected layers. We use neighborhood ($\mathcal{N}_i$) size of $|\mathcal{N}_i| = 50$, for our experiments. We first train the network for garment classes which share the same garment template and then fine-tune separately for each garment class g. To speed up training for *ParserNet*, we train the network to predict $\mathbf{W}^g = \mathbf{I}^g$, where $\mathbf{I}^g$ is the indicator matrix for garment class g, explained in Sect. 3.2. This initializes the network to parse the garment by cutting out a part of the input mesh based on the constant per-garment indicator matrix, shown in Fig. 3.

For *SizerNet* we use $d = 30$ and we implement f_w^{enc}, f_w^{dec} with fully connected layers and skip connections between encoder and decoder network. We held out 40 scans for testing in each garment class, which includes some cases with unseen subjects and some with unseen garment size for seen subjects. For pose-shape prediction network, *ParserNet* and *SizerNet* we use batch-size of 8 and learning rate of 0.0001.

5 Experiments and Results

5.1 Results of 3D Garment Parsing and Shape Under Clothing

To validate the choice of parsing the garments using a sparse regressor matrix ($\mathbf{W}$), we compare the results of *ParserNet* with two baseline approaches: (1) A linearized version of *ParserNet* implemented with LASSO, and (2) A naive FC network, which has the same architecture as *ParserNet*. However, instead of predicting the weight matrix ($\mathbf{W}$), the FC network directly predicts the deformation ($\mathbf{D}^g$) from the garment template ($T^g(\boldsymbol{\beta}, \boldsymbol{\theta}, \mathbf{0})$) for a given input ($\mathbf{M}$).

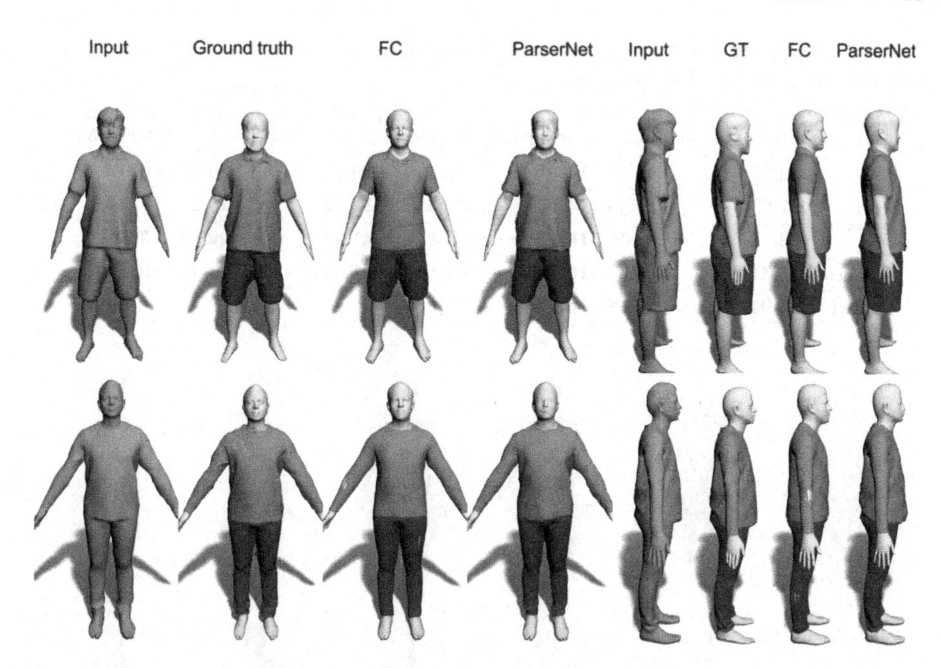

Fig. 4. Comparison of *ParserNet* with a FC network from front and lateral view.

We compare the per-vertex error of *ParserNet* with the aforementioned baselines in Table 1. Figure 4 shows that *ParserNet* can produce details, fine wrinkles, and large garment deformations, which is not possible with a naive FC network. This is attainable because *ParserNet* reconstructs the output garment mesh as a localized sparse weighted sum of input vertex locations, and hence preserves the geometry details present in the input mesh. However, in the case of naive FC network, the predicted displacement field ($\mathbf{D}^g$) is smooth and does not explain large deformations. Hence, naive FC network is not able to predict loose garments and does not preserve fine details. We show results of *ParserNet* for more garment classes in Fig. 5 and add more results in the supplementary material.

5.2 Results of Garment Resizing

Editing garment meshes based on garment size label is an unexplored problem and, hence there are no well defined quantitative metrics. We introduce two quantitative metrics, namely change in mesh *surface area* (A_{err}) and *per-vertex error* (V_{err}) for evaluating the resizing task. *Surface area* accounts for the scale of a garment, which only changes with the garment size, and *per-vertex error* accounts for details and folds created due to the underlying body shape and looseness/tightness of the garment. Moreover, subtle changes in garment shape with respect to size are difficult to evaluate. Hence, we use heat map visualizations for qualitative analysis of the results.

Table 1. Average per-vertex error V_{err} of proposed method for parsing garment meshes for different garment class (in mm).

Garment	Linear model	FC	*ParserNet*	Garment	Linear model	FC	*ParserNet*
Polo	32.21	17.25	**14.33**	Shorts	29.78	20.12	**16.07**
Shirt	27.63	19.35	**14.56**	Pants	34.82	18.2	**17.24**
Vest	28.17	18.56	**15.89**	Coat	41.27	22.19	**15.34**
Hoodies	37.34	23.69	**15.76**	Shorts2	31.38	23.45	**16.23**
T-Shirt	26.94	15.98	**13.77**				

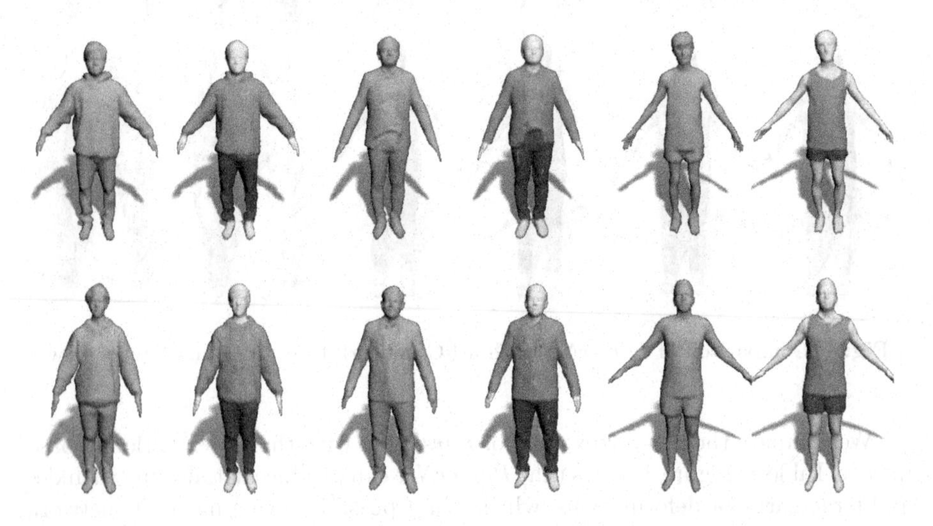

Fig. 5. Input single mesh and *ParserNet* results for more garments.

Since there is no other existing work for garment resizing task to compare with, we evaluate our method against the following three baselines.

1. *Error margin* in data: We define error margin as the change in *per-vertex location* (V_{err}) and *surface area* (A_{err}) between garments of two consecutive size for a subject in the dataset. Our model should ideally produce a smaller error than this margin.
2. *Average prediction*: For every subject in the dataset, we create the average garment (G_{avg}), by averaging over all the available sizes for a subject.
3. *Linear scaling + Alignment*: We linearly scale the garment mesh, according to desired size label, and then align the garment to the underlying body.

Table 2 shows the errors for each experiment. *SizerNet* results in lower errors, as compared to the linear scaling method, which reflects the need for modelling the non-linear relationship between garment shape, underlying body shape and garment size. We also see that network predictions yield lower error as compared to average garment prediction, which suggests that the model is learning the size variation, even though the differences in the ground truth itself are subtle. We present the results of *SizerNet* for common garment classes in Table 2, Fig. 6, 7 and add more results in the supplementary material.

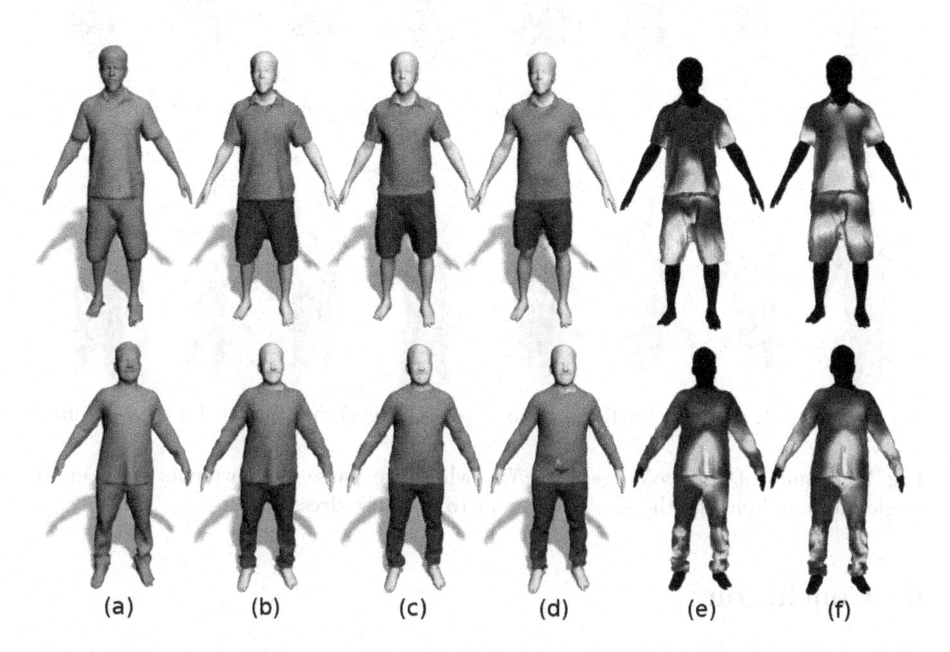

Fig. 6. (a) Input single mesh. (b) Parsed multi-layer mesh from ParserNet. (c), (d) Resized garment in two subsequent smaller sizes. (e), (f) Heatmap of change in per vertex error on original parsed garment for two new sizes.

Table 2. Average per vertex error (V_{err} in mm) and surface area error (A_{err} in %) of predicted of proposed method for garment resizing.

Garment	Error-margin		Average-pred		Linear scaling		Ours	
	V_{err}	A_{err}	V_{err}	A_{err}	V_{err}	A_{err}	V_{err}	A_{err}
Polo t-shirt	33.25	24.56	23.86	3.63	35.05	8.45	**16.42**	**1.79**
Shirt	36.52	19.57	21.95	2.76	34.53	7.01	**15.54**	**1.41**
Shorts	43.21	27.21	24.79	5.41	35.77	4.99	**16.71**	**2.38**
Pants	30.83	15.15	21.54	4.73	38.16	7.13	**19.26**	**2.43**

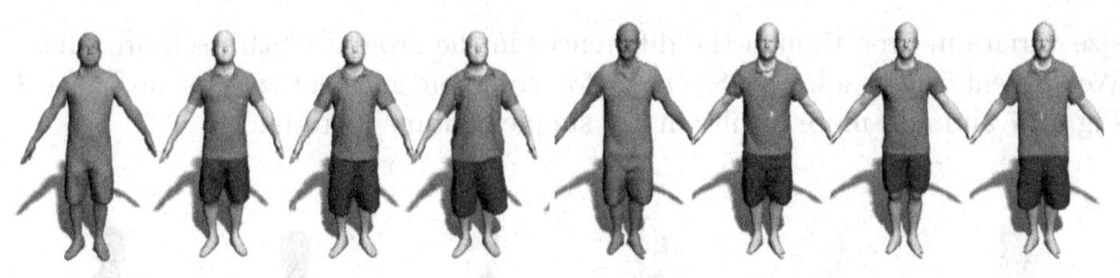

(a) Small(input, parsed), Medium, Large (b) Medium(input, parsed), Small, Large

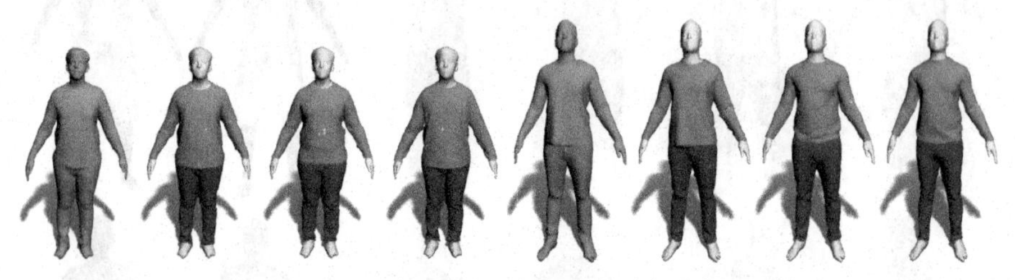

(c) Large(input, parsed), Medium, XLarge (d) XLarge(input,parse), Large, Medium

Fig. 7. Results of *ParserNet* + *SizerNet*, where we parse the garments from input single mesh and change the size of garment to visualise dressing effect.

6 Conclusion

We introduce *SIZER*, a clothing size variation dataset and model, which is the first real dataset to capture clothing size variation on different subjects. We also introduce *ParserNet*: a 3D garment parsing network and *SizerNet*: a size sensitive clothing model. With this method, one can change the single mesh registration to multi-layer meshes of garments and body shape under clothing, without the need for scan segmentation and can use the result for animation, virtual try-on, etc. *SizerNet* can drape a person with garments in different sizes.

Since our dataset only consists of roughly aligned A-poses, we are limited to A-pose. We only exploit geometry information (vertices and normals) for 3D clothing parsing. In future work, we plan to use the color information in *ParserNet* via texture augmentation, to improve the accuracy and generalization of the proposed method. We will release the model, dataset, and code to stimulate research in the direction of 3D garment parsing, segmentation, resizing and predicting body shape under clothing.

Acknowledgements. This work is funded by the Deutsche Forschungsgemeinschaft (DFG, German Research Foundation) - 409792180 (Emmy Noether Programme, project: Real Virtual Humans) and a Facebook research award. We thank Tarun, Navami, and Yash for helping us with the data capture and RVH team members [4], for their meticulous feedback on this manuscript.

References

1. Agisoft metashape. https://www.agisoft.com/
2. The high cost of retail returns. https://www.thebalancesmb.com/the-high-cost-of-retail-returns-2890350
3. IHL Group. https://www.ihlservices.com/
4. Real virtual humans, Max Planck Institute for Informatics. https://virtualhumans.mpi-inf.mpg.de/people.html
5. Treedy's scanner. https://www.treedys.com
6. de Aguiar, E., Stoll, C., Theobalt, C., Ahmed, N., Seidel, H., Thrun, S.: Performance capture from sparse multi-view video. ACM Trans. Graph. **27**(3), 98:1–98:10 (2008)
7. Alldieck, T., Magnor, M., Bhatnagar, B.L., Theobalt, C., Pons-Moll, G.: Learning to reconstruct people in clothing from a single RGB camera. In: IEEE Conference on Computer Vision and Pattern Recognition (CVPR) (2019)
8. Alldieck, T., Magnor, M., Xu, W., Theobalt, C., Pons-Moll, G.: Detailed human avatars from monocular video. In: International Conference on 3D Vision (3DV) (2018)
9. Alldieck, T., Magnor, M., Xu, W., Theobalt, C., Pons-Moll, G.: Video based reconstruction of 3D people models. In: IEEE Conference on Computer Vision and Pattern Recognition (CVPR) (2018)
10. Alldieck, T., Pons-Moll, G., Theobalt, C., Magnor, M.: Tex2shape: detailed full human body geometry from a single image. In: IEEE International Conference on Computer Vision (ICCV). IEEE (2019)
11. Bălan, A.O., Black, M.J.: The naked truth: estimating body shape under clothing. In: Forsyth, D., Torr, P., Zisserman, A. (eds.) ECCV 2008. LNCS, vol. 5303, pp. 15–29. Springer, Heidelberg (2008). https://doi.org/10.1007/978-3-540-88688-4_2
12. Bertiche, H., Madadi, M., Escalera, S.: CLOTH3D: clothed 3D humans. vol. abs/1912.02792 (2019)
13. Bhatnagar, B.L., Sminchisescu, C., Theobalt, C., Pons-Moll, G.: Combining implicit function learning and parametric models for 3D human reconstruction. In: Vedaldi, A., Bischof, H., Brox, T., Frahm, J.M. (eds.) European Conference on Computer Vision (ECCV), vol. 12347, pp. 311–329. Springer, Cham (2020). https://doi.org/10.1007/978-3-030-58536-5_19
14. Bhatnagar, B.L., Tiwari, G., Theobalt, C., Pons-Moll, G.: Multi-garment net: learning to dress 3D people from images. In: IEEE International Conference on Computer Vision (ICCV). IEEE (2019)
15. Bogo, F., Romero, J., Pons-Moll, G., Black, M.J.: Dynamic FAUST: registering human bodies in motion. In: IEEE Conference on Computer Vision and Pattern Recognition (2017)
16. Bradley, D., Popa, T., Sheffer, A., Heidrich, W., Boubekeur, T.: Markerless garment capture. ACM Trans. Graph. **27**, 99 (2008)
17. Chen, X., et al.: Towards 3D human shape recovery under clothing. CoRR abs/1904.02601 (2019)
18. Dong, H., Liang, X., Wang, B., Lai, H., Zhu, J., Yin, J.: Towards multi-pose guided virtual try-on network. In: International Conference on Computer Vision (ICCV) (2019)
19. Dong, H., et al.: Fashion editing with adversarial parsing learning. In: Conference on Computer Vision and Pattern Recognition (CVPR) (2020)

20. Gong, K., Liang, X., Li, Y., Chen, Y., Yang, M., Lin, L.: Instance-level human parsing via part grouping network. In: Ferrari, V., Hebert, M., Sminchisescu, C., Weiss, Y. (eds.) ECCV 2018. LNCS, vol. 11208, pp. 805–822. Springer, Cham (2018). https://doi.org/10.1007/978-3-030-01225-0_47
21. Guan, P., Reiss, L., Hirshberg, D., Weiss, A., Black, M.J.: DRAPE: DRessing any PErson. ACM Trans. Graph. (Proc. SIGGRAPH) **31**(4), 35:1–35:10 (2012)
22. Gundogdu, E., Constantin, V., Seifoddini, A., Dang, M., Salzmann, M., Fua, P.: GarNet: a two-stream network for fast and accurate 3D cloth draping. In: IEEE International Conference on Computer Vision (ICCV). IEEE (2019)
23. Habermann, M., Xu, W., Zollhoefer, M., Pons-Moll, G., Theobalt, C.: Livecap: real-time human performance capture from monocular video (2019)
24. Habermann, M., Xu, W., Zollhoefer, M., Pons-Moll, G., Theobalt, C.: DeepCap: monocular human performance capture using weak supervision. In: IEEE Conference on Computer Vision and Pattern Recognition (CVPR). IEEE (2020)
25. Huang, Z., Xu, Y., Lassner, C., Li, H., Tung, T.: ARCH: animatable reconstruction of clothed humans. In: Proceedings of the IEEE/CVF Conference on Computer Vision and Pattern Recognition, pp. 3093–3102 (2020)
26. Jiang, B., Zhang, J., Hong, Y., Luo, J., Liu, L., Bao, H.: BCNet: learning body and cloth shape from a single image. arXiv preprint arXiv:2004.00214 (2020)
27. Kanazawa, A., Black, M.J., Jacobs, D.W., Malik, J.: End-to-end recovery of human shape and pose. In: Computer Vision and Pattern Regognition (CVPR) (2018)
28. Kolotouros, N., Pavlakos, G., Black, M.J., Daniilidis, K.: Learning to reconstruct 3D human pose and shape via model-fitting in the loop. In: International Conference on Computer Vision (2019)
29. Kolotouros, N., Pavlakos, G., Daniilidis, K.: Convolutional mesh regression for single-image human shape reconstruction. In: CVPR (2019)
30. Lähner, Z., Cremers, D., Tung, T.: DeepWrinkles: accurate and realistic clothing modeling. In: Ferrari, V., Hebert, M., Sminchisescu, C., Weiss, Y. (eds.) ECCV 2018. LNCS, vol. 11208, pp. 698–715. Springer, Cham (2018). https://doi.org/10.1007/978-3-030-01225-0_41
31. Lazova, V., Insafutdinov, E., Pons-Moll, G.: 360-degree textures of people in clothing from a single image. In: International Conference on 3D Vision (3DV) (2019)
32. Leroy, V., Franco, J., Boyer, E.: Multi-view dynamic shape refinement using local temporal integration. In: IEEE International Conference on Computer Vision, ICCV, Venice, Italy, pp. 3113–3122 (2017)
33. Loper, M., Mahmood, N., Romero, J., Pons-Moll, G., Black, M.J.: SMPL: a skinned multi-person linear model. ACM Trans. Graph. (Proc. SIGGRAPH Asia) **34**(6), 248:1–248:16 (2015)
34. Ma, Q., et al.: Learning to dress 3D people in generative clothing. In: IEEE Conference on Computer Vision and Pattern Recognition (CVPR). IEEE (2020)
35. Miguel, E., et al.: Data-driven estimation of cloth simulation models. Comput. Graph. Forum **31**(2), 519–528 (2012)
36. Omran, M., Lassner, C., Pons-Moll, G., Gehler, P., Schiele, B.: Neural body fitting: unifying deep learning and model based human pose and shape estimation. In: International Conference on 3D Vision (2018)
37. Patel, C., Liao, Z., Pons-Moll, G.: The virtual tailor: predicting clothing in 3D as a function of human pose, shape and garment style. In: IEEE Conference on Computer Vision and Pattern Recognition (CVPR). IEEE (2020)
38. Pons-Moll, G., Pujades, S., Hu, S., Black, M.: ClothCap: seamless 4D clothing capture and retargeting. ACM Trans. Graph. **36**(4), 1–15 (2017)

39. Pons-Moll, G., Romero, J., Mahmood, N., Black, M.J.: Dyna: a model of dynamic human shape in motion. ACM Trans. Graph. **34**, 120 (2015)
40. Pumarola, A., Sanchez, J., Choi, G., Sanfeliu, A., Moreno-Noguer, F.: 3DPeople: modeling the geometry of dressed humans. In: International Conference in Computer Vision (ICCV) (2019)
41. Rother, C., Kolmogorov, V., Blake, A.: GrabCut: Interactive foreground extraction using iterated graph cuts. ACM Trans. Graph. (TOG) **23**, 309–314 (2004)
42. Saito, S., Huang, Z., Natsume, R., Morishima, S., Kanazawa, A., Li, H.: PIFu: pixel-aligned implicit function for high-resolution clothed human digitization. In: Proceedings of the IEEE International Conference on Computer Vision, pp. 2304–2314 (2019)
43. Santesteban, I., Otaduy, M.A., Casas, D.: Learning-based animation of clothing for virtual try-on. Comput. Graph. Forum (Proc. Eurograph.) **38**, 355–366 (2019)
44. Starck, J., Hilton, A.: Surface capture for performance-based animation. IEEE Comput. Graph. Appl. **27**(3), 21–31 (2007)
45. Stuyck, T.: Cloth Simulation for Computer Graphics. Synthesis Lectures on Visual Computing. Morgan & Claypool Publishers, San Rafael (2018)
46. Tao, Y., et al.: SimulCap: single-view human performance capture with cloth simulation. In: IEEE Conference on Computer Vision and Pattern Recognition (CVPR) (2019)
47. Tung, T., Nobuhara, S., Matsuyama, T.: Complete multi-view reconstruction of dynamic scenes from probabilistic fusion of narrow and wide baseline stereo. In: IEEE 12th International Conference on Computer Vision, ICCV, Kyoto, Japan, pp.1709–1716 (2009)
48. Wang, H., Hecht, F., Ramamoorthi, R., O'Brien, J.F.: Example-based wrinkle synthesis for clothing animation. ACM Trans. Graph. (Proc. SIGGRAPH) **29**(4), 107:1–107:8 (2010)
49. Wang, H., Ramamoorthi, R., O'Brien, J.F.: Data-driven elastic models for cloth: modeling and measurement. ACM Trans. Graph. (Proc. SIGGRAPH) **30**(4), 71:1–71:11 (2011)
50. Wang, T.Y., Ceylan, D., Popovic, J., Mitra, N.J.: Learning a shared shape space for multimodal garment design. ACM Trans. Graph. **37**(6), 1:1–1:14 (2018)
51. White, R., Crane, K., Forsyth, D.A.: Capturing and animating occluded cloth. ACM Trans. Graph. **26**(3), 34 (2007)
52. Xiang, D., Joo, H., Sheikh, Y.: Monocular total capture: posing face, body, and hands in the wild. In: Proceedings of the IEEE Conference on Computer Vision and Pattern Recognition, pp. 10965–10974 (2019)
53. Xu, H., Li, J., Lu, G., Zhang, D., Long, J.: Predicting ready-made garment dressing fit for individuals based on highly reliable examples. Comput. Graph. **90**, 135–144 (2020)
54. Xu, Y., Zhu, S.C., Tung, T.: DenseRaC: joint 3D pose and shape estimation by dense render and compare. In: International Conference on Computer Vision (2019)
55. Yamaguchi, K.: Parsing clothing in fashion photographs. In: Proceedings of the 2012 IEEE Conference on Computer Vision and Pattern Recognition (CVPR). CVPR 2012, pp. 3570–3577. IEEE Computer Society, USA (2012)
56. Yamaguchi, K., Kiapour, M.H., Berg, T.L.: Paper doll parsing: retrieving similar styles to parse clothing items. In: IEEE International Conference on Computer Vision, ICCV 2013, Sydney, Australia, 1–8 December 2013, pp. 3519–3526. IEEE Computer Society (2013)

57. Yang, J., Franco, J.-S., Hétroy-Wheeler, F., Wuhrer, S.: Analyzing clothing layer deformation statistics of 3d human motions. In: Ferrari, V., Hebert, M., Sminchisescu, C., Weiss, Y. (eds.) ECCV 2018. LNCS, vol. 11211, pp. 245–261. Springer, Cham (2018). https://doi.org/10.1007/978-3-030-01234-2_15
58. Yang, W., Luo, P., Lin, L.: Clothing co-parsing by joint image segmentation and labeling (2014)
59. Yu, T., et al.: Doublefusion: real-time capture of human performances with inner body shapes from a single depth sensor. In: The IEEE International Conference on Computer Vision and Pattern Recognition(CVPR). IEEE (2018)
60. Zhang, C., Pujades, S., Black, M., Pons-Moll, G.: Detailed, accurate, human shape estimation from clothed 3D scan sequences. In: IEEE CVPR (2017)
61. Zheng, Z., Yu, T., Wei, Y., Dai, Q., Liu, Y.: DeepHuman: 3D human reconstruction from a single image. In: The IEEE International Conference on Computer Vision (ICCV) (2019)
62. Zhu, H., et al.: Deep fashion3D: a dataset and benchmark for 3D garment reconstruction from single images. arXiv preprint arXiv:2003.12753 (2020)

LIMP: Learning Latent Shape Representations with Metric Preservation Priors

Luca Cosmo[1,2](✉), Antonio Norelli[1], Oshri Halimi[3], Ron Kimmel[3], and Emanuele Rodolà[1]

[1] Sapienza University of Rome, Rome, Italy
cosmo@di.uniroma1.it
[2] University of Lugano, Lugano, Switzerland
[3] Technion - Israel Institute of Technology, Haifa, Israel

Abstract. In this paper, we advocate the adoption of metric preservation as a powerful prior for learning latent representations of deformable 3D shapes. Key to our construction is the introduction of a geometric distortion criterion, defined directly on the decoded shapes, translating the preservation of the metric on the decoding to the formation of linear paths in the underlying latent space. Our rationale lies in the observation that training samples alone are often insufficient to endow generative models with high fidelity, motivating the need for large training datasets. In contrast, metric preservation provides a rigorous way to control the amount of geometric distortion incurring in the construction of the latent space, leading in turn to synthetic samples of higher quality. We further demonstrate, for the first time, the adoption of differentiable intrinsic distances in the backpropagation of a geodesic loss. Our geometric priors are particularly relevant in the presence of scarce training data, where learning any meaningful latent structure can be especially challenging. The effectiveness and potential of our generative model is showcased in applications of style transfer, content generation, and shape completion.

Keywords: Learning shapes · Generative model · Metric distortion

1 Introduction

Constructing high-fidelity generative models for 3D shapes is a challenging problem that has met with increasing interest in recent years. Generative models are applicable in many practical domains, ranging from content creation to shape exploration, as well as in 3D reconstruction. As a new generation of methods, they come to face a number of difficulties.

Electronic supplementary material The online version of this chapter (https://doi.org/10.1007/978-3-030-58580-8_2) contains supplementary material, which is available to authorized users.

A. Vedaldi et al. (Eds.): ECCV 2020, LNCS 12348, pp. 19–35, 2020.
https://doi.org/10.1007/978-3-030-58580-8_2

Fig. 1. Disentangled interpolation of FAUST shapes, obtained with our generative model trained under metric preservation priors. The yellow shapes at the two corners are given as input; the remaining shapes are generated by bilinearly interpolating the latent codes of the input, and decoding the resulting codes. Our model allows to disentangle pose from identity, illustrated here as different dimensions. (Color figure online)

Most existing approaches address the case of static or *rigid* geometry, for example, man-made objects like chairs and airplanes, with potentially high intra-class variability; see the ShapeNet [7] repository for such examples. In this setting, the main focus has been on the abstraction capabilities of the encoder and the generator, describing complex 3D models in terms of their core geometric features via parsimonious part-based representations. Shapes generated with these techniques are usually designed to have valid part semantics that are easy to parse. Concurrently, several recent efforts have concentrated on the definition of convenient representations for the 3D *output*; these methods find broader application in multiple tasks, where they enable more efficient and high-quality synthesis, and can be often plugged into existing generative models.

To date, relatively fewer approaches have targeted the *deformable* setting, where the generated shapes are related by continuous, non-rigid deformations. These model a range of natural phenomena, such as changes in pose and facial expressions of human subjects, articulations, garment folding, and molecular flexibility to name but a few. The extra difficulties brought by such non-rigid deformations can be tackled, in some cases, by designing mathematical or parametric models for the deformation at hand; however, these models are often violated in practice, and can be very hard to devise for general deformations – hence the need for learning from examples.

The framework we propose is motivated by the observation that existing data-driven approaches for learning deformable 3D shapes, and autoencoders (AE) in particular, do not make use of any *geometric prior* to drive the construction of the latent space, whereas they rely almost completely on the expressivity of the training dataset. This imposes a heavy burden on the learning process, and further requires large annotated datasets that can be costly or even impossible to acquire. In the absence of additional regularization, limited training data leads to limited generalization capability, which is manifested in the generated 3D shapes exhibiting unnatural distortions. Variational autoencoders (VAE) provide a partial remedy by modeling a distributional prior on the data via a parametrized density on the latent space. This induces additional regularization, but is still insufficient to guarantee the preservation of geometric properties in the output 3D models.

In this paper, we introduce *Latent Interpolation with Metric Priors* (LIMP). We propose to explicitly model the local *metric* properties of the latent space by enforcing metric constraints on the decoded output. We do this by phrasing a metric distortion penalty that has the effect to promote naturally looking deformations, and in turn to significantly reduce the need for large datasets at training time. In particular, we show that by coupling the Euclidean distances among latent codes (hence, along linear paths in the latent space) to the metric distortion among decoded shapes, we obtain a strong regularizing effect in the construction of the latent space. Another novel ingredient of the proposed approach is the backpropagation of intrinsic (namely, geodesic) distances during training, which is made possible by a recent geodesic computation technique. Using geodesics makes our approach more flexible, and enables the successful application of our generative model to style and pose transfer applications. See Fig. 1 for an example of novel samples synthesized with our generative model.

2 Related Work

Our method falls within the class of AE-based generative models for 3D shapes. In this Section we cover methods from this family that are more closely related to ours, and refer to the recent survey [8] for a broader coverage.

In the 3D computer vision and graphics realms, generative models for part-compositional 3D objects play the lion's share. Such approaches directly exploit the hierarchical, structural nature of 3D man-made objects to drive the construction of encoder and generator [22,29,31,32]. These methods leverage on the insight that objects can be understood through their components [26], making an interpretable representation close to human parsing possible. In this setting, a continuous exploration of the generated latent spaces is not always meaningful; the mechanism underlying typical operations like sampling and interpolation happen instead in *discrete* steps in order to generate plausible intermediate shape configurations (e.g., for transitioning from a 4-legged chair to a 3-legged stool). For this reason, with rigid geometry one usually deals with "structural blending" rather than continuous deformations. Structural blending has been realized, for instance, by learning abstractions of symmetry hierarchies via spatial arrangements of oriented bounding boxes [22], or by explicitly modeling part-to-part relationships [29]; generative-adversarial modeling has been applied on volumetric object representations [45]; structural hierarchies have been applied for the generation of composite 3D scenes [23] and building typologies [28] as well. Contributing to their success, is the fact that all these methods train on ShapeNet-scale annotated datasets with $>50K$ unique 3D models, and the recent publication of dedicated benchmarks like PartNet [30] testify to the increasing interest of data-driven models for structure-aware geometry processing. In this paper, we address a different setting; we do not assume part-compositionality of the 3D models since we deal with deformable shapes, where continuous deformations are well-defined, and where annotated datasets are not as prominent.

A second thread of research revolves around the definition of a meaningful representation for the generated 3D output. While many approaches mostly use polygonal meshes with predefined topology or directly synthesize point clouds [1,41], the focus has been recently shifting towards more effective representations in terms of overall quality, fidelity, and flexibility. These include approaches that predict implicit shape representations at the output, requiring an ex-post isosurface extraction step to generate a mesh at the desired resolution [16,27,33]; isosurfacing has been replaced by binary space partitioning in [9]; while in [18], shapes are represented by a set of parametric surface elements. In this work, we focus on learning a better *latent* representation for deformable shapes, rather than on constructing a better representation for the output.

More closely related to ours are some recent methods from the area of geometric deep learning. A graph-convolutional VAE with dynamic filtering convolutional layers [44] was introduced in [24] for the task of deformable shape completion of human shapes. The method is trained on $\sim$7000 shapes from the DFAUST dataset of real human scans [4]; due to the lack of any geometric prior, the learned generator introduces large distortions around points in the latent space that are not well represented in the training set. Geometric regularization was injected in [17] in the form of a template that parametrizes the surface. The method shows excellent performance in shape matching, however, it crucially relies on a large and representative dataset of 230, 000 shapes, and performance drops significantly with smaller training sets or bad initialization. More recently, a geometric disentanglement model for deformable point clouds was introduced in [2]. The proposed method uses Laplacian eigenvalues as a weak geometric prior to promote the separation of intrinsic and extrinsic shape information, together with several other de-correlation penalties, and a training set of $>40K$ shapes. In the absence of enough training examples, the approach tends to produce a "morphing" effect between point clouds that does not correspond to a natural motion; a similar phenomenon was observed in [1]. Finally, in [43], a time-dependent physical prior was used to regularize interpolations in the latent space with the goal of obtaining a convincing simulation of moving tissues.

In particular, our approach bears some analogies with the theory of shape spaces [20], in that we seek to synthesize geometry that minimizes a deformation energy. For example, in [14] it was shown how to *axiomatically* modify a noisy shape such that its intrinsic measures would fit a given prior in a different pose. Differentiating the geodesic distances was done by fixing the update order in the fast marching scheme [21]. Our energy is not minimized over a fixed shape space, but rather, it drives the construction of a novel shape space in a data-driven fashion.

In this paper, we leverage classical ideas from shape analysis and metric geometry to ensure that shapes on the learned latent space correspond to plausible (i.e., low-distortion) deformations of the shapes seen at training time, even when only few training samples are available. We do this by modeling a geometric prior that promotes deformations with bounded distortion, and show that this model provides a powerful regularization for shapes *within* as well as *across* different classes, e.g., when transitioning between different human subjects.

3 Learning with Metric Priors

Our goal is to learn a latent representation for deformable 3D shapes. We do this by training a VAE on a training set $\mathcal{S} = \{\mathbf{X}_i\}$ of $|\mathcal{S}|$ shapes, under a purely geometric loss:

$$\ell(\mathcal{S}) = \ell_{\text{recon}}(\mathcal{S}) + \ell_{\text{interp}}(\mathcal{S}) + \ell_{\text{disent}}(\mathcal{S}). \tag{1}$$

The loss is composed of three terms. The first is a geometric reconstruction loss on the individual training shapes, as in classical AE's; the second one is a pairwise interpolation term for points in the latent space; the third one is a disentanglement term to separate intrinsic from extrinsic information.

The main novelty lies in (1) the interpolation loss, and (2) the disentanglement loss *not* relying upon corresponding poses in the training set. The interpolation term provides control over the encoding of each shape *in relation to the others*. This induces a notion of proximity between latent codes that is explicitly linked, in the definition of the loss, to a notion of metric distortion between the decoded shapes. As we show in the following, this induces a strong regularization on the latent space and rules out highly distorted reconstructions.

The disentanglement loss promotes the factorization of the latent space into two orthogonal components: One that spans the space of isometries (e.g., change in pose), and another that spans the space of non-isometric deformations (e.g., change in identity). As in the interpolation loss, for the disentanglement we also exploit the metric properties of the decoded shapes.

3.1 Losses

We define $\mathbf{z} := \text{enc}(\mathbf{X})$ to be the latent code for shape $\mathbf{X}$, and $\mathbf{X}' := \text{dec}(\mathbf{z})$ to be the corresponding decoding. During training, the decoder (dec) and encoder (enc) are updated so as to minimize the overall loss of Eq. (1); see Sect. 3.3 for the implementation details.

Geometric Reconstruction. The reconstruction loss is defined as follows:

$$\ell_{\text{recon}}(\mathcal{S}) = \sum_{i=1}^{|\mathcal{S}|} \|\mathbf{D}_{\mathbb{R}^3}(\mathbf{X}'_i) - \mathbf{D}_{\mathbb{R}^3}(\mathbf{X}_i)\|_F^2, \tag{2}$$

where $\mathbf{D}_{\mathbb{R}^3}(\mathbf{X})$ is the matrix of pairwise Euclidean distances between all points in $\mathbf{X}$, and $\|\cdot\|_F$ denotes the Frobenius norm. Equation (2) measures the cumulative reconstruction error (up to a global rotation) over the training shapes.

Metric Interpolation. This loss is defined over all possible pairs of shapes $(\mathbf{X}_i, \mathbf{X}_j)$:

$$\ell_{\text{interp}}(\mathcal{S}) = \sum_{i \neq j}^{|\mathcal{S}|} \|\mathbf{D}(\text{dec}(\underbrace{(1-\alpha)\mathbf{z}_i + \alpha\mathbf{z}_j}_{\substack{\text{interpolation of} \\ \text{latent codes}}})) - \underbrace{((1-\alpha)\mathbf{D}(\mathbf{X}'_i) + \alpha\mathbf{D}(\mathbf{X}'_j))}_{\substack{\text{interpolation of} \\ \text{geodesic or local distances}}}\|_F^2, \quad (3)$$

where $\alpha \sim \mathcal{U}(0,1)$ is a uniformly sampled scalar in $(0,1)$, different for each pair of shapes. In the equation above, the matrix $\mathbf{D}(\mathbf{X})$ encodes the pairwise distances between points in $\mathbf{X}$. We use two different definitions of distance, giving rise to two different losses which we sum up together. In one loss, $\mathbf{D}$ contains *geodesic* distances between *all* pairs of points. In the second loss, we consider *local Euclidean* distances from each point to points within a small neighborhood (set to 10% of the shape diameter); the rationale is that local Euclidean distances capture local detail and tend to be resilient to non-rigid deformations, as observed for instance in [39]. All distances are computed on the fly, on the decoded shapes, at each forward step.

Since the error criterion in Eq. (3) encodes the discrepancy between pairwise distance matrices, we refer to it as a *metric preservation prior*. We refer to Sect. 3.2 for a more in-depth discussion from a continuous perspective.

Disentanglement. We split the latent codes into an intrinsic and an extrinsic part, $\mathbf{z} := (\mathbf{z}^{\text{int}}|\mathbf{z}^{\text{ext}})$. The former is used to encode "style", i.e., the space of non-isometric deformations; the latter is responsible for changes in pose, and is therefore constrained to model the space of possible isometries.

The loss is composed of two terms:

$$\ell_{\text{disent}}(\mathcal{S}) = \ell_{\text{int}}(\mathcal{S}) + \ell_{\text{ext}}(\mathcal{S}), \quad \text{with} \quad (4)$$

$$\ell_{\text{int}}(\mathcal{S}) = \sum_{\substack{i \neq j \\ \text{iso}}}^{|\mathcal{S}|} \|\mathbf{D}_{\mathbb{R}^3}(\text{dec}(\underbrace{(1-\alpha)\mathbf{z}_i^{\text{int}} + \alpha\mathbf{z}_j^{\text{int}}}_{\substack{\text{interpolation of} \\ \text{style}}}|\mathbf{z}_i^{\text{ext}})) - \mathbf{D}_{\mathbb{R}^3}(\mathbf{X}_i)\|_F^2 \quad (5)$$

$$\ell_{\text{ext}}(\mathcal{S}) = \sum_{\substack{i \neq j \\ \text{non-iso}}}^{|\mathcal{S}|} \|\mathbf{D}_g(\text{dec}(\mathbf{z}_i^{\text{int}}|\underbrace{(1-\alpha)\mathbf{z}_i^{\text{ext}} + \alpha\mathbf{z}_j^{\text{ext}}}_{\substack{\text{interpolation of} \\ \text{pose}}})) - \mathbf{D}_g(\mathbf{X}_i)\|_F^2 \quad (6)$$

The ℓ_{int} term is evaluated only on isometric pairs (i.e., just a change in pose), for which we expect $\mathbf{z}_i^{\text{int}} = \mathbf{z}_j^{\text{int}}$. For a pair $(\mathbf{X}_i, \mathbf{X}_j)$, it requires that $\mathbf{X}_i$ can be reconstructed exactly even when its intrinsic part $\mathbf{z}_i^{\text{int}}$ is interpolated with that of $\mathbf{X}_j$. This enforces $\mathbf{z}_i^{\text{int}} = \mathbf{z}_j^{\text{int}}$, thus all the pose-related information is forced to move to $\mathbf{z}^{\text{ext}}$.

The ℓ_{ext} term is instead evaluated on *non*-isometric pairs. Here we require that the *geodesic* distances of $\mathbf{X}_i$ are left untouched when we interpolate its pose with that of $\mathbf{X}_j$. This way, we force all the style-related information to be moved to $\mathbf{z}^{\text{int}}$. We see that by having direct access to the metric on the decoded shapes, we can phrase the disentanglement easily in terms of distances.

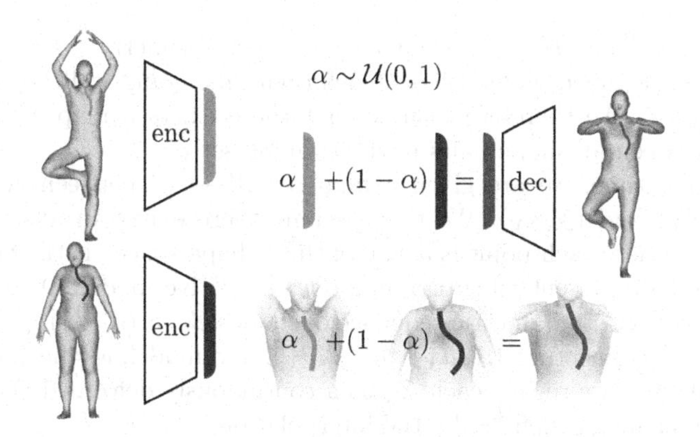

Fig. 2. Our architecture is a standard VAE, with PointNet as the encoder and a fully connected decoder. Our loss asks that the geodesic distances on the decoded convex combination of latent codes (middle row) are equal to the convex combination of the input distances.

The assumption that the metric is nearly preserved under pose changes is widely used in many shape analysis applications such as shape retrieval [37], matching [11,12,19,38] and reconstruction [5,10].

Relative Error. In practice, we always measure the error on the Euclidean distances (appearing in Eqs. (2), (3), (5)) in a *relative* sense. Let $\mathbf{A}$ be the "ground truth" Euclidean distance matrix computed on the input shape, and let $\mathbf{B}$ be its predicted reconstruction. Instead of taking $\|\mathbf{A}-\mathbf{B}\|_F^2 = \sum_{ij}(\mathbf{A}_{ij}-\mathbf{B}_{ij})^2$, we compute the relative error $\sum_{ij} \frac{(\mathbf{A}_{ij}-\mathbf{B}_{ij})^2}{\mathbf{A}_{ij}^2}$. In our experiments, this resulted in better reconstruction of local details than by using the simple Frobenius norm.

3.2 Continuous Interpretation

In the continuous setting, we regard shapes as metric spaces $(\mathcal{X}, d_\mathcal{X})$, each equipped with a distance function $d_\mathcal{X} : \mathcal{X} \times \mathcal{X} \to \mathbb{R}_+$. Given two shapes $(\mathcal{X}, d_\mathcal{X})$ and $(\mathcal{Y}, d_\mathcal{Y})$, a map $\phi : \mathcal{X} \to \mathcal{Y}$ is an *isometry* if it is surjective and preserves distances, $d_\mathcal{X}(x,x') = d_\mathcal{Y}(\phi(x),\phi(x'))$ for all $x, x' \in \mathcal{X}$. Isometries play a fundamental role in 3D shape analysis, since they provide a mathematical model for natural deformations like changes in pose. In practice, however, isometry is rarely satisfied exactly.

Why Interpolation? Our approach is based on the insight that *non*-isometric shapes are related by sequences of near-isometric deformations, which, in turn, have a well defined mathematical model. In our setting, we do not require the training shapes to be near-isometric. Instead, we allow for maps ϕ with *bounded metric distortion*, i.e., for which there exists a constant $K > 0$ such that:

$$|d_\mathcal{X}(x,x') - d_\mathcal{Y}(\phi(x),\phi(x'))| \leq K \tag{7}$$

for all $x, x' \in \mathcal{X}$. For $K \to 0$ the map ϕ is a near-isometry, while for general $K > 0$ we get a much wider class of deformations, going well beyond simple changes in pose. We therefore assume that there exists a map with bounded distortion between all shape pairs in the training set.

At training time, we are given a map $\phi : \mathcal{X} \to \mathcal{Y}$ between two training shapes $(\mathcal{X}, d_{\mathcal{X}})$ and $(\mathcal{Y}, d_{\mathcal{Y}})$. We then assume there exists an abstract metric space $(\mathcal{L}, d_{\mathcal{L}})$ where each point is a shape; this "shape space" is the latent space that we seek to represent when training our generative model. Over the latent space we construct a parametric sequence of shapes $\mathcal{Z}_\alpha = (\mathcal{X}, d_\alpha)$, parametrized by $\alpha \in (0, 1)$, connecting $(\mathcal{X}, d_{\mathcal{X}})$ to $(\mathcal{Y}, d_{\mathcal{Y}})$. By modeling the intermediate shapes as $(\mathcal{X}, d_\alpha)$, we regard each $\mathcal{Z}_\alpha$ as a continuously deformed version of $\mathcal{X}$, with a different metric defined by the interpolation:

$$d_\alpha(x, x') = (1 - \alpha) d_{\mathcal{X}}(x, x') + \alpha d_{\mathcal{Y}}(\phi(x), \phi(x')), \tag{8}$$

for all $x, x' \in \mathcal{X}$. Each $\mathcal{Z}_\alpha$ in the sequence has the same points as $\mathcal{X}$, but the shape is different since distances are measured differently.

It is easy to see that if the training shapes $\mathcal{X}$ and $\mathcal{Y}$ are isometric, then $d_\alpha(x, x') = d_{\mathcal{X}}(x, x')$ for all $x, x' \in \mathcal{X}$ and the entire sequence is isometric, i.e., we are modeling a change in pose. However, if $\phi : \mathcal{X} \to \mathcal{Y}$ has bounded distortion without being an isometry, each intermediate shape $(\mathcal{X}, d_\alpha)$ also has bounded distortion with respect to $(\mathcal{X}, d_{\mathcal{X}})$, with $K_\alpha < K$ in Eq. (7); in particular, for $\alpha \to 0$ one gets $K_\alpha \to 0$ and therefore a near-isometry. In other words, by using the metric interpolation loss of Eq. (3), as α grows from 0 to 1 we are modeling a general non-isometric deformation as a sequence of approximate isometries.

Flattening of the Latent Space. Taking a linear convex combination of latent vectors as in Eq. (3) implies that distances between codes should be measured using the Euclidean metric $\|\cdot\|_2$. This enables algebraic manipulation of the codes and the formation of "shape analogies", as shown in the inset (real example based on our trained model). By the connection of Euclidean distances in the latent space with intrinsic distances on the decoder's output, our learning model performs a "flattening" operation, in the sense that it requires the latent space to be as Euclidean as possible, while absorbing any embedding error in the decoder. A similar line of thought was followed, in a different context, in the purely axiomatic model of [40].

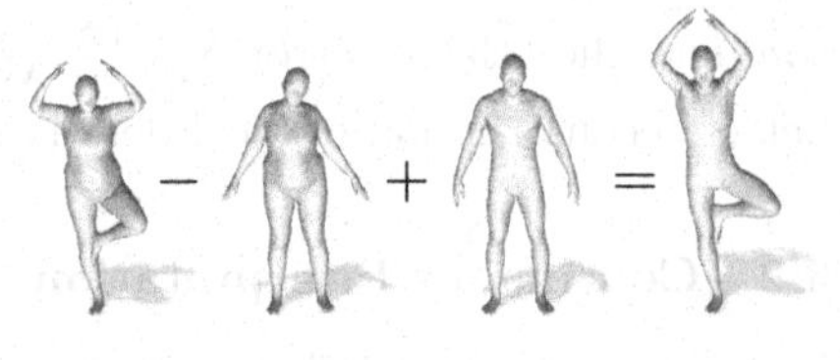

3.3 Implementation

We design our deep generative model as a VAE (Fig. 2). The input data is a set of triangle meshes; each mesh is encoded as a matrix of vertex positions $\mathbf{X} \in \mathbb{R}^{n \times 3}$, together with connectivity encoded as a $n \times n$ adjacency matrix. We anticipate here that mesh connectivity is never accessed directly by the network.

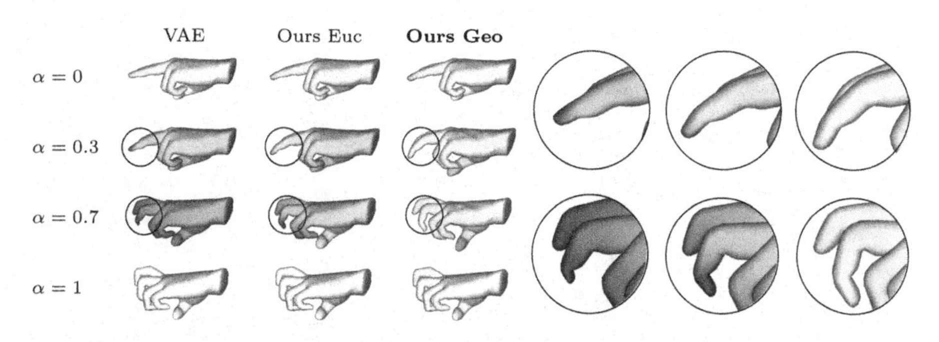

Fig. 3. Interpolation example on a small training set of just 5 shapes, where the deformation evolves from top ($\alpha = 0$) to bottom ($\alpha = 1$). Color encodes the per-point metric distortion, growing from white to red; changes in pose as in this example should have distortion close to zero. We show the results obtained by three different networks: baseline VAE; ours with Euclidean metric regularization only; ours with Euclidean *and* geodesic regularization (i.e., the complete loss). (Color figure online)

Architecture. The encoder takes vertex positions $\mathbf{X}$ as input, and outputs a d-dimensional code $\mathbf{z} = \text{enc}(\mathbf{X})$. Similarly, the decoder outputs vertex positions $\mathbf{Y} = \text{dec}(\mathbf{z}) \in \mathbb{R}^{n \times 3}$. In order to clarify the role of our priors versus the sophisticacy of the architecture, we keep the latter as simple as possible. In particular, we adopt a similar architecture as in [2]; we use PointNet [34] with spatial transform as the encoder, and a simple MLP as the decoder. We reserve 25% of the latent code for the extrinsic part and the remaining 75% for the intrinsic representation, while the latent space and layer dimensions vary depending on the dataset size. A detailed description of the network is deferred to the Supplementary Material. We implemented our model in PyTorch using Adam as optimizer with learning rate of 1e–4. To avoid local minima and numerical errors in gradient computation, we start the training by optimizing just the reconstruction loss for 10^4 iterations, and add the remaining terms for the remaining epochs.

Geodesic Distance Computation. A crucial ingredient to our model is the computation of geodesic distances $\mathbf{D}_g(\text{dec}(\mathbf{z}))$ during training, see Eq. (3). We use the heat method of [13] to compute these distances, based on the realization that its pipeline is fully differentiable. It consists, in particular, of two linear solves and one normalization step, and all the quantities involved in the three steps depend smoothly on the vertex positions given by the decoder (we refer to the Supplementary Material for additional details).

To our knowledge, this is the first time that on-the-fly computation of geodesic distances appears in a deep learning pipeline. Previous approaches using geodesic distances, such as [19], do so by taking them as pre-computed input data, and leave them untouched for the entire training procedure.

Supervision. We train on a collection of shapes with known pointwise correspondences; these are needed in Eq. (3), where we assume that the distance matrices

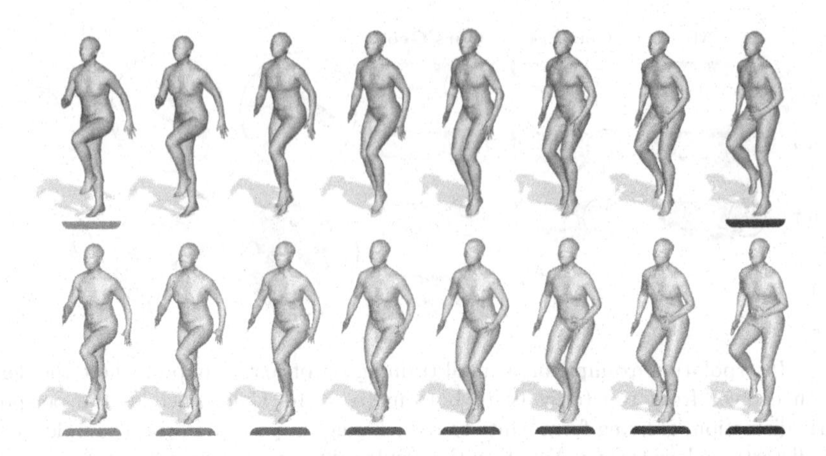

Fig. 4. *Top row*: A 4D sequence from the real-world dataset DFAUST. We train our generative model on the left- and right-most keyframes (indicated by the orange and blue bar respectively), together with keyframes extracted from other sequences and different individuals. *Bottom row*: The 3D shapes generated by our trained model. Visually, both the generated and the real-world sequences look plausible, indicating that geometric priors are well-suited for regularizing toward realistic deformations. (Color figure online)

have compatible rows and columns. From a continuous perspective, we need maps for the interpolated metric of Eq. (8) to be well defined. Known correspondences are also needed by other approaches dealing with deformable data [17,24,25]. In practice, we only need few such examples (we use <100 training shapes), since we rely for the most part on the regularization power of our geometric priors.

Differently from [17,24] we do *not* assume the training shapes to have the same mesh, since the latter is only used as an auxiliary structure for computing geodesics in the loss; the network only ever accesses vertex positions. Further, we do not require training shapes with similar poses across different subjects.

4 Results

4.1 Data

To validate our method, we performed experiments using 5 different datasets (3 are obtained from real-world scans, 2 are fully synthetic). **FAUST** [3] is composed of 10 different human subjects, each captured in 10 different poses. We train our network on 8 subjects (thus, 80 meshes in total) and leave out the other 2 subjects for testing. **DFAUST** [4] is a 4D dataset capturing the motion of 10 human subjects performing 14 different activities, spanning *hundreds* of frames each. As training data **we only use 4 representative frames from each subject/sequence pair. COMA** [36] is another 4D dataset of human faces; it is composed of 13 subjects, each performing 13 different facial expressions

Table 1. Ablation study in terms of interpolation and disentanglement error on 4 datasets. Our full pipeline (denoted by 'Ours Geo') achieves the minimum error in all cases, and is **more than one order of magnitude** better than the baseline VAE on the interpolation. We do not report the disentanglement error for HANDS, since the dataset only contains one hand style.

	Interpolation error			Disentanglement error		
	VAE	Ours Euc	Ours Geo	VAE	Ours Euc	Ours Geo
FAUST	3.89e−2	5.08e−3	**3.82e−3**	7.16	4.04	**3.48**
DFAUST	9.82e-2	3.43e-3	**2.89e−4**	6.15	4.90	**4.11**
COMA	1.32e−3	1.03e−3	**7.51e−4**	1.55	1.30	**1.22**
HANDS	6.01e−3	8.12e−4	**4.62e−4**	–	–	–

Fig. 5. Interpolation example on the *cat* shapes of TOSCA dataset [6]. On the left, we show an interpolation sequence between two shapes of the training set (yellow shapes on the right). On the right, we manually selected the most similar shapes present in the training set, composed in total by just 11 shapes. You can appreciate how shapes in the middle of the interpolated sequence significantly differ from the training shapes.

represented as a sequence of 3D meshes. As opposed to the test split proposed in [36], where 90% of the data is used for training, we only select 14 frames for each subject (one representative for each of the 13 expressions, plus one in a neutral pose), thus **training with less than 1% of the dataset. TOSCA** [6] is a synthetic dataset containing both animals and human bodies. In our experiments we use only the *cat* class, containing 11 shapes in different poses. The last dataset, which we refer to as **HANDS**, is also completely synthetic and consists of 5 meshes depicting one hand in 5 different poses. For all the datasets, we subsample the meshes to 2500 vertices by iterative edge collapse [15].

4.2 Interpolation

We first perform a classical interpolation experiment. Given two shapes $\mathbf{X}$ and $\mathbf{Y}$, we visualize the decoded interpolation of their latent codes, given by $\mathrm{dec}((1-\alpha)\mathrm{enc}(\mathbf{X}) + \alpha\mathrm{enc}(\mathbf{Y}))$ for a few choices of $\alpha \in (0,1)$. We measure the interpolation quality via the *interpolation error*, defined as the average (over all surface points) geodesic distortion of the interpolated shapes.

Two examples of interpolation are shown in Figs. 3 and 5. In these examples, the training sets consist of just **5** and **11 shapes** respectively, meaning that

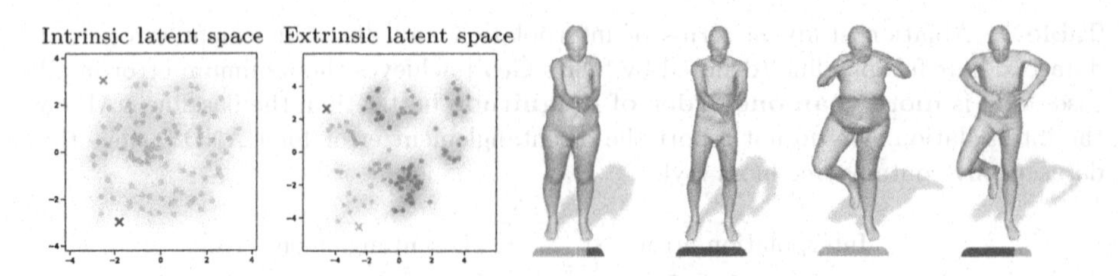

Fig. 6. *Plots on the left*: Planar embedding of the intrinsic and extrinsic parts of the latent codes from FAUST. Colors identify gender (left) and pose (ten different poses; right). We observe cohesive clusters in either case, suggesting that the encoder has generalized the projection onto each factor. The four small crosses are random samples. *Right*: Decoded shapes from the four combinations of the random samples; the specific combinations are illustrated by compatible colors between the crosses and the bars below each shape.

the intermediate poses have never been seen before. In this few-shot setting, proper regularization is crucial to get meaningful results. In the experiment in Fig. 3, we also conduct an ablation study. We disable all the interpolation terms from our complete loss, resulting in a baseline VAE; then we disable the geodesic regularization only; finally we keep the entire loss intact, showing best results. Quantitative results on 4 different datasets are reported in Table 1 (first 3 columns), showing that best results are obtained when our full loss is used.

As an additional qualitative experiment, in Fig. 4 we show the decoded shapes in-between two keyframes of a 4D sequence from DFAUST. We remark that none of the intermediate shapes were seen at training time, nor was any similar-looking shape present in the training set. We then compare our reconstructed sequence with the original sequence of real-world scans. The purpose of this experiment is to show that our geometric priors are essential for the generation of realistic motion; apart from a perceptual evaluation, any quantitative comparison here would not be meaningful – there is not a unique "true" way to transition between two given poses.

4.3 Disentanglement

Our second set of experiments is aimed at demonstrating the effectiveness of our geometric priors for the disentanglement of intrinsic from extrinsic information. We illustrate this in different ways.

In Fig. 6, we show disentanglement for a generator trained on the FAUST dataset. For visualization purposes, for each vector $\mathbf{z} := (\mathbf{z}^{\text{int}}|\mathbf{z}^{\text{ext}})$ in the latent space (here comprising both training and test shapes), we embed the $\mathbf{z}^{\text{int}}$ and $\mathbf{z}^{\text{ext}}$ parts *separately* onto the plane (via multidimensional scaling), and attribute different colors to different gender and poses. We then randomly sample two new $\mathbf{z}^{\text{int}}$ and two new $\mathbf{z}^{\text{ext}}$, and compose them into four latent codes by taking all the combinations. The figure illustrates the four decoded shapes.

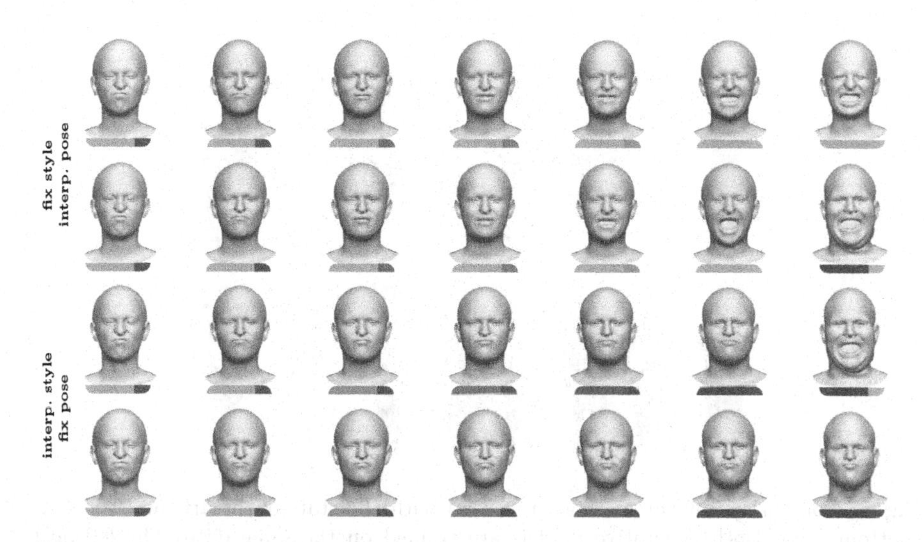

Fig. 7. Disentanglement + interpolation examples on the COMA dataset; the source shape is always the same. Each row presents a different scenario, with interpolation happening left-to-right. Please refer to the color code below each shape as a visual aid; for example, for the first column we have (style|pose).

In Fig. 7 we show the simultaneous action of disentanglement and interpolation. Given a source and a target shape, we show the interpolation of pose while fixing the style, and the interpolation of style while fixing the pose. We do so with different combinations of source and target. In all cases, our generative model is able to synthesize realistic shapes with the correct semantics, suggesting high potential in style and pose transfer applications.

As we did with the case of interpolation, we also provide a notion of *disentanglement error*, defined as follows. Given shapes $\mathbf{X}_i$ and $\mathbf{X}_j$ with latent codes $(\mathbf{z}_i^{\text{int}}|\mathbf{z}_i^{\text{ext}})$ and $(\mathbf{z}_j^{\text{int}}|\mathbf{z}_j^{\text{ext}})$, we swap $\mathbf{z}_i^{\text{ext}}$ with $\mathbf{z}_j^{\text{ext}}$ and then measure the average point-to-point distance between $\text{dec}(\mathbf{z}_i^{\text{int}}|\mathbf{z}_j^{\text{ext}})$ and the corresponding ground-truth shape from the dataset. In Table 1 (last 3 columns) we report the disentanglement error on all 4 datasets, together with the ablation study.

Finally, in Fig. 8 we show a qualitative comparison with the recent state-of-the-art method [2] (using public code provided by the authors), which uses Laplacian eigenvalues as a prior to drive the disentanglement, together with multiple other de-correlation terms. Similarly to other approaches like [24,42], the quality of the interpolation of [2] mostly depends on the smoothness properties of the VAE, on the complexity of the deep net, or on the availability of vast training data. For this comparison, both generative models were trained on the same 80 FAUST shapes.

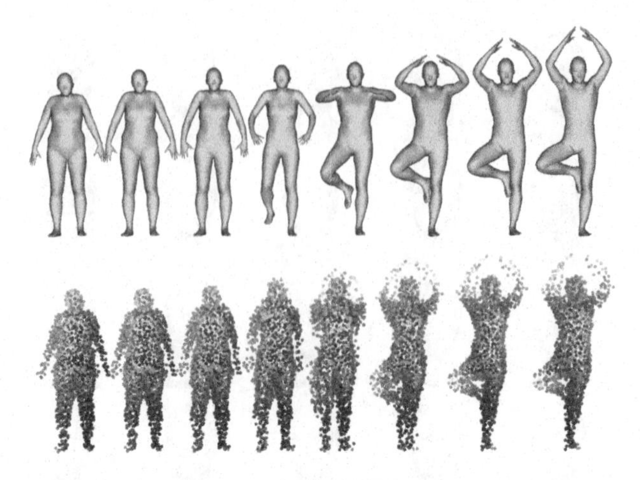

Fig. 8. Comparison of our method (top row) with the state-of-the-art method of [2] (bottom row). Both generative models are trained on the same data. The leftmost and rightmost shapes are from the training set, while the intermediate shapes are decodings of a linear sequence in the latent space. Observe that source and target are *not* isometric; according to our continuous interpretation of Sect. 3.2, our trained model decomposes the non-isometric deformation into a sequence of approximate isometries.

5 Conclusions

We introduced a new deep generative model for deformable 3D shapes. Our model is based on the intuition that by directly connecting the Euclidean distortion of latent codes to the metric distortion of the decoded shapes, one gets a powerful regularizer that induces a well-behaved structure on the latent space. Our idea finds a theoretical interpretation in modeling deformations with bounded metric distortion as sequences of approximate isometries. Under the manifold hypothesis, our metric preservation priors explicitly promote a flattening of the true data manifold onto a lower-dimensional Euclidean representation. We demonstrated how having access to the metric of the decoded shapes during training enables high-quality synthesis of novel samples, with practical implications in tasks of content creation and style transfer.

Perhaps the main **limitation** of our method, which we share with other geometric deep learning approaches, lies in the requirement of labeled pointwise correspondences between the training shapes. These can be hard to obtain in certain settings, for example, when dealing with shapes from the same semantic class but with high intra-class variability. Few interesting directions of future work may consist in a self-supervised variant of our model, where dense correspondences are not needed for the training, but are estimated during the learning process or in the exploitation of spectral properties of the reconstructed shape, that has been shown [10, 35] to contain important information of the embedding geometry.

Finally, while in this paper we showed that even a simple prior such as metric distortion can have a significant effect, we foresee that bringing techniques from the areas of shape optimization and analysis closer to deep generative models will enable a fruitful line of stimulating research.

Acknowledgments. LC, AN and ER are supported by the ERC Starting Grant No. 802554 (SPECGEO) and the MIUR under grant "Dipartimenti di eccellenza 2018–2022" of the Department of Computer Science of Sapienza University. OH and RK are supported by the Israel Ministry of Science and Technology grant number 3-14719, the Technion Hiroshi Fujiwara Cyber Security Research Center and the Israel Cyber Directorate.

References

1. Achlioptas, P., Diamanti, O., Mitliagkas, I., Guibas, L.: Learning representations and generative models for 3D point clouds. In: International Conference on Machine Learning, pp. 40–49 (2018)
2. Aumentado-Armstrong, T., Tsogkas, S., Jepson, A., Dickinson, S.: Geometric disentanglement for generative latent shape models. In: Proceedings of the IEEE International Conference on Computer Vision, pp. 8181–8190 (2019)
3. Bogo, F., Romero, J., Loper, M., Black, M.J.: FAUST: dataset and evaluation for 3D mesh registration. In: Proceedings IEEE Conference on Computer Vision and Pattern Recognition (CVPR). IEEE, Piscataway, NJ, USA (2014)
4. Bogo, F., Romero, J., Pons-Moll, G., Black, M.J.: Dynamic FAUST: registering human bodies in motion. In: IEEE Conference on Computer Vision and Pattern Recognition (CVPR) (2017)
5. Boscaini, D., Eynard, D., Kourounis, D., Bronstein, M.M.: Shape-from-operator: recovering shapes from intrinsic operators. Comput. Graph. Forum **34**, 265–274 (2015)
6. Bronstein, A.M., Bronstein, M.M., Kimmel, R.: Numerical Geometry of Non-rigid Shapes. Springer, New York (2008)
7. Chang, A.X., et al.: ShapeNet: an information-rich 3D model repository. Technical report. arXiv:1512.03012 [cs.GR]. Stanford University – Princeton University – Toyota Technological Institute at Chicago (2015)
8. Chaudhuri, S., Ritchie, D., Xu, K., Zhang, H.R.: Learning generative models of 3D structures. In: Jakob, W., Puppo, E. (eds.) Eurographics 2019 - Tutorials. The Eurographics Association (2019). https://doi.org/10.2312/egt.20191038
9. Chen, Z., Tagliasacchi, A., Zhang, H.: BSP-Net: generating compact meshes via binary space partitioning. arXiv preprint arXiv:1911.06971 (2019)
10. Cosmo, L., Panine, M., Rampini, A., Ovsjanikov, M., Bronstein, M.M., Rodolà, E.: Isospectralization, or how to hear shape, style, and correspondence. In: Proceedings of the IEEE Conference on Computer Vision and Pattern Recognition, pp. 7529–7538 (2019)
11. Cosmo, L., Rodola, E., Albarelli, A., Mémoli, F., Cremers, D.: Consistent partial matching of shape collections via sparse modeling. Comput. Graph. Forum **36**, 209–221 (2017)
12. Cosmo, L., Rodola, E., Masci, J., Torsello, A., Bronstein, M.M.: Matching deformable objects in clutter. In: 2016 Fourth International Conference on 3D Vision (3DV), pp. 1–10. IEEE (2016)

13. Crane, K., Weischedel, C., Wardetzky, M.: Geodesics in heat: a new approach to computing distance based on heat flow. ACM Trans. Graph. **32**(5), 1–11 (2013)
14. Devir, Y.S., Rosman, G., Bronstein, A.M., Bronstein, M.M., Kimmel, R.: On reconstruction of non-rigid shapes with intrinsic regularization. In: 2009 IEEE 12th International Conference on Computer Vision Workshops, ICCV Workshops, pp. 272–279. IEEE (2009)
15. Garland, M., Heckbert, P.S.: Surface simplification using quadric error metrics. In: Proceedings of the 24th Annual Conference on Computer Graphics and Interactive Techniques, pp. 209–216 (1997)
16. Gropp, A., Yariv, L., Haim, N., Atzmon, M., Lipman, Y.: Implicit geometric regularization for learning shapes. arXiv preprint arXiv:2002.10099 (2020)
17. Groueix, T., Fisher, M., Kim, V.G., Russell, B.C., Aubry, M.: 3D-coded: 3D correspondences by deep deformation. In: Ferrari, V., Hebert, M., Sminchisescu, C., Weiss, Y. (eds.) Proceedings of the European Conference on Computer Vision (ECCV), pp. 230–246. Springer, Cham (2018). https://doi.org/10.1007/978-3-030-01216-8_15
18. Groueix, T., Fisher, M., Kim, V.G., Russell, B.C., Aubry, M.: A Papier-mâché approach to learning 3D surface generation. In: Proceedings of the IEEE Conference on Computer Vision and Pattern Recognition, pp. 216–224 (2018)
19. Halimi, O., Litany, O., Rodolà, E., Bronstein, A.M., Kimmel, R.: Unsupervised learning of dense shape correspondence. In: Proceedings of the IEEE Conference on Computer Vision and Pattern Recognition, pp. 4370–4379 (2019)
20. Heeren, B., Rumpf, M., Schröder, P., Wardetzky, M., Wirth, B.: Exploring the geometry of the space of shells. Comput. Graph. Forum **33**(5), 247–256 (2014)
21. Kimmel, R., Sethian, J.A.: Computing geodesic paths on manifolds. Proc. Natl. Acad. Sci. **95**(15), 8431–8435 (1998)
22. Li, J., Xu, K., Chaudhuri, S., Yumer, E., Zhang, H., Guibas, L.: Grass: generative recursive autoencoders for shape structures. ACM Trans. Graph. (TOG) **36**(4), 1–14 (2017)
23. Li, M., et al.: Grains: generative recursive autoencoders for indoor scenes. ACM Trans. Graph. (TOG) **38**(2), 1–16 (2019)
24. Litany, O., Bronstein, A., Bronstein, M., Makadia, A.: Deformable shape completion with graph convolutional autoencoders. In: Proceedings of the IEEE Conference on Computer Vision and Pattern Recognition, pp. 1886–1895 (2018)
25. Litany, O., Remez, T., Rodolà, E., Bronstein, A., Bronstein, M.: Deep functional maps: structured prediction for dense shape correspondence. In: Proceedings of the IEEE International Conference on Computer Vision, pp. 5659–5667 (2017)
26. Liu, Z., Freeman, W.T., Tenenbaum, J.B., Wu, J.: Physical primitive decomposition. In: Ferrari, V., Hebert, M., Sminchisescu, C., Weiss, Y. (eds.) ECCV 2018. LNCS, vol. 11216, pp. 3–20. Springer, Cham (2018). https://doi.org/10.1007/978-3-030-01258-8_1
27. Mescheder, L., Oechsle, M., Niemeyer, M., Nowozin, S., Geiger, A.: Occupancy networks: learning 3D reconstruction in function space. In: Proceedings CVPR, pp. 4460–4470 (2019)
28. de Miguel, J., Villafane, M.E., Piskorec, L., Sancho-Caparrini, F.: Deep form finding - using variational autoencoders for deep form finding of structural typologies. In: Architecture in the Age of the 4th Industrial Revolution - Proceedings of the 37th eCAADe and 23rd SIGraDi Conference, pp. 71–80 (2019)
29. Mo, K., et al.: StructureNet: hierarchical graph networks for 3D shape generation. ACM Trans. Graph. **38**(6), 1–19 (2019)

30. Mo, K., et al.: PartNet: a large-scale benchmark for fine-grained and hierarchical part-level 3D object understanding. In: Proceedings of the IEEE Conference on Computer Vision and Pattern Recognition, pp. 909–918 (2019)
31. Nash, C., Ganin, Y., Eslami, S., Battaglia, P.W.: Polygen: an autoregressive generative model of 3D meshes. arXiv preprint arXiv:2002.10880 (2020)
32. Nash, C., Williams, C.K.: The shape variational autoencoder: a deep generative model of part-segmented 3D objects. Comput. Graph. Forum **36**(5), 1–12 (2017)
33. Park, J.J., Florence, P., Straub, J., Newcombe, R., Lovegrove, S.: DeepSDF: learning continuous signed distance functions for shape representation. In: Proceedings of the IEEE Conference on Computer Vision and Pattern Recognition, pp. 165–174 (2019)
34. Qi, C.R., Su, H., Mo, K., Guibas, L.J.: PointNet: deep learning on point sets for 3D classification and segmentation. In: Proceedings of the IEEE Conference on Computer Vision and Pattern Recognition, pp. 652–660 (2017)
35. Rampini, A., Tallini, I., Ovsjanikov, M., Bronstein, A.M., Rodolà, E.: Correspondence-free region localization for partial shape similarity via Hamiltonian spectrum alignment. In: 2019 International Conference on 3D Vision (3DV), pp. 37–46. IEEE (2019)
36. Ranjan, A., Bolkart, T., Sanyal, S., Black, M.J.: Generating 3D faces using convolutional mesh autoencoders. In: Ferrari, V., Hebert, M., Sminchisescu, C., Weiss, Y. (eds.) ECCV 2018. LNCS, vol. 11207, pp. 725–741. Springer, Cham (2018). https://doi.org/10.1007/978-3-030-01219-9_43
37. Reuter, M., Wolter, F.E., Peinecke, N.: Laplace-Beltrami spectra as 'Shape-DNA' of surfaces and solids. Comput. Aided Des. **38**(4), 342–366 (2006)
38. Rodola, E., Bronstein, A.M., Albarelli, A., Bergamasco, F., Torsello, A.: A game-theoretic approach to deformable shape matching. In: 2012 IEEE Conference on Computer Vision and Pattern Recognition, pp. 182–189. IEEE (2012)
39. Rodolà, E., Cosmo, L., Bronstein, M.M., Torsello, A., Cremers, D.: Partial functional correspondence. Comput. Graph. Forum **36**(1), 222–236 (2017)
40. Shamai, G., Kimmel, R.: Geodesic distance descriptors. In: Proceedings of the IEEE Conference on Computer Vision and Pattern Recognition, pp. 6410–6418 (2017)
41. Shu, D.W., Park, S.W., Kwon, J.: 3D point cloud generative adversarial network based on tree structured graph convolutions. In: Proceedings of the IEEE International Conference on Computer Vision, pp. 3859–3868 (2019)
42. Tan, Q., Gao, L., Lai, Y.K., Xia, S.: Variational autoencoders for deforming 3D mesh models. In: Proceedings of the IEEE Conference on Computer Vision and Pattern Recognition, pp. 5841–5850 (2018)
43. Tan, Q., Pan, Z., Gao, L., Manocha, D.: Realtime simulation of thin-shell deformable materials using CNN-based mesh embedding. IEEE Robot. Autom. Lett. **5**(2), 2325–2332 (2020)
44. Verma, N., Boyer, E., Verbeek, J.: FeastNet: feature-steered graph convolutions for 3D shape analysis. In: Proceedings of the IEEE Conference on Computer Vision and Pattern Recognition, pp. 2598–2606 (2018)
45. Wu, J., Zhang, C., Xue, T., Freeman, B., Tenenbaum, J.: Learning a probabilistic latent space of object shapes via 3D generative-adversarial modeling. In: Advances in Neural Information Processing Systems, pp. 82–90 (2016)

Unsupervised Sketch to Photo Synthesis

Runtao Liu[1], Qian Yu[1,2(✉)], and Stella X. Yu[1]

[1] UC Berkeley/ICSI, Berkeley, USA
qianyu@buaa.edu.cn
[2] Beihang University,
Xueyuan Rd. No. 37, Haidian District, Beijing, China

Abstract. Humans can envision a realistic photo given a free-hand sketch that is not only spatially imprecise and geometrically distorted but also without colors and visual details. We study unsupervised sketch to photo synthesis for the first time, learning from *unpaired* sketch and photo data where the target photo for a sketch is unknown during training. Existing works only deal with either style difference or spatial deformation alone, synthesizing photos from edge-aligned line drawings or transforming shapes within the same modality, e.g., color images.

Our insight is to decompose the unsupervised sketch to photo synthesis task into two stages of translation: First shape translation from sketches to grayscale photos and then content enrichment from grayscale to color photos. We also incorporate a self-supervised denoising objective and an attention module to handle abstraction and style variations that are specific to sketches. Our synthesis is sketch-faithful and photo-realistic, enabling sketch-based image retrieval and automatic sketch generation that captures human visual perception beyond the edge map of a photo.

1 Introduction

Sketches, i.e., rapidly executed freehand drawings, make an intuitive and powerful visual expression (Fig. 1). There is much research on sketch recognition [7,35], sketch parsing [26,27], and sketch-based image or video retrieval [21,28,36]. We study how to imagine a realistic photo given a sketch that is spatially imprecise and missing colorful details, by learning from *unpaired* sketches and photos.

Sketch to photo synthesis is challenging for three reasons.

1) Sketches of objects often do not match their shapes in photos, since sketches commonly drawn by amateurs have large spatial and geometrical distortion. Translating a sketch to a photo thus requires shape rectification. However, it is not trivial to rectify shape distortion in a sketch, as line strokes are only

R. Liu and Q. Yu—equal contribution. http://sketch.icsi.berkeley.edu.

Electronic supplementary material The online version of this chapter (https://doi.org/10.1007/978-3-030-58580-8_3) contains supplementary material, which is available to authorized users.

A. Vedaldi et al. (Eds.): ECCV 2020, LNCS 12348, pp. 36–52, 2020.
https://doi.org/10.1007/978-3-030-58580-8_3

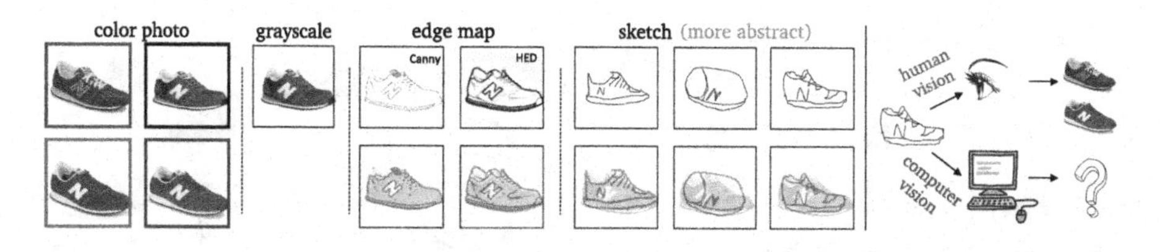

Fig. 1. Comparisons of image types and challenges of sketch to photo synthesis. **Left:** A single object shape could have multiple distinctive colorings yet a common or similar grayscale. Edges extracted by Canny and HED detectors lose colorful details but align well with boundaries in the color photo, whereas sketches are more abstract lines drawn with deformations and style variations. Row 2 shows their lines overlaid on the grayscale photo. **Right:** Human vision can imagine a realistic photo given a free-hand sketch. Our goal is to equip computer vision with the same imagination capability.

suggestive of the actual shapes and locations, and the extent of shape fidelity varies widely between individuals. In Fig. 1, the three sketches for the same shoe are very different both overall proportions and local stroke styles.

2) Sketches are color-less and lacking details. Drawn in black strokes on white paper, sketches outline mostly object boundaries and characteristic interior markings. To synthesize a photo, shading and colorful textures must be filled in properly. However, it is not trivial to fill in details either. Since a sketch could depict multiple photos, any synthesizer must have the capability to produce not only realistic but also diverse photos for a single sketch.

3) Sketches may not have corresponding photos. Free-hand sketches can be created from observation, memory, or pure imagination; they are not so widely available as photos, and those with corresponding photos are even rarer. A few sketch datasets exist in computer vision. TU-Berlin [6] and QuickDraw [11] contain sketches only, with 20,000 and 50 million instances over 250 and 345 categories respectively. Contour Drawing [19] and Scenesketchy [39] have sketch-photo image pairs at the scene level; their sketches are either contour tracings or cartoon-style line drawings, neither representative of real-world free-hand sketches. Sketchy [28] has only 500 sketches paired with 100 photos in each of 125 categories. ShoeV2 and ChairV2 [36] contain 6,648/2,000 and 1,297/400 sketches/photos in a single semantic category of shoes and chairs respectively. To enable data-driven learning of sketch to photo synthesis, we must handle limited sketch data and *unpaired* sketches and photos.

Existing works focus on either shape or color translation alone (Fig. 2). **1)** Most image synthesis that deals with shape transfiguration tends to stay in the same visual domain, e.g. changing the picture of a dog to that of a cat [15,22], where visual details are comparable in the color image. **2)** Sketches are a special case of line drawings, and the most studied case of line drawings in computer vision is the edge map extracted automatically from a photo. Such an edge map based drawing to photo synthesis task does not have the spatial deformation problem between sketches and photos, and realistic photos can be synthesized *with* [16,31] or *without* [38] paired training data between drawings and photos.

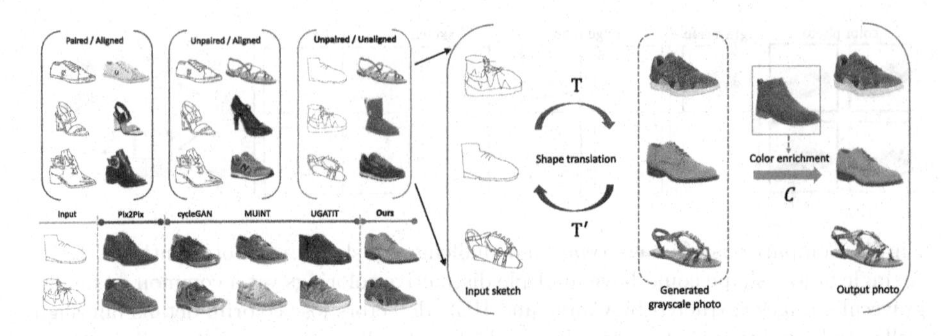

Fig. 2. Comparison of sketch to photo synthesis settings and results. **Left:** Three training scenarios on whether line drawings and photos are provided as paired training instances and whether line drawings are spatially aligned with the photos. Edges extracted from photos are aligned, whereas sketches are not. The bottom panel compares synthesis results from representative approaches in each setting, indicated by the same line/bracket color. Ours are superior to unsupervised edge map to photo methods (cycleGAN [38], MUINT [15], UGATIT [18]) and even supervised methods (Pix2Pix [16]) trained on paired data. **Right:** Our unsupervised sketch-to-photo synthesis model has two separate stages handling spatial deformation and color enrichment respectively: Shape translation learns to synthesize a grayscale photo given a sketch, from unpaired sketch set and photo set, whereas color enrichment learns to fill the grayscale with colorful details given an optional reference photo.

We will show that existing methods fail in sketch to photo synthesis when both shape and color translations are needed simultaneously.

We consider learning sketch to photo synthesis from sketches and photos of the same object category such as *shoes*. There is no pairing information between individual sketches and photos; these two sets can be independently collected.

Our insight for unsupervised sketch to photo synthesis is to decompose the task into two separate translations (Fig. 2). Our two-stage model performs first shape translation in grayscale and then content fill-in in color. **Stage 1)** Shape translation learns to synthesize a grayscale photo given a sketch, from unpaired sketch set and photo set. Geometrical distortions are eliminated at this step. To handle abstraction and drawing style variations, we apply a self-supervised learning objective to noise sketch compositions, and also introduce an attention module for the model to ignore distractions. **Stage 2)** Content enrichment learns to fill the grayscale with details, including colors, shading, and textures, given an *optional* reference image. It is designed to work with or without reference images. This capability is enabled by a mixed training strategy. Our model can thus produce diverse outputs on demand.

Our model links sketches to photos and can be used directly in sketch-based photo retrieval. Another exciting corollary result from our model is that we can also synthesize a sketch given a photo, even from unseen semantic categories. Strokes in a sketch capture information beyond edge maps defined primarily on intensity contrast and object exterior boundaries. Automatic photo to sketch generation could lead to more advanced computer vision capabilities and serve as a powerful human-user interaction device.

Our work makes the following contributions. **1)** We propose the first two-stage unsupervised model that can generate diverse, sketch-faithful, and photo-realistic images from a single free-hand sketch. **2)** We introduce a self-supervised learning objective and an attention module to handle abstraction and style variations in sketches. **3)** Our work not only enables sketch-based image retrieval but also delivers an automatic sketcher that captures human visual perception beyond the edge map of a photo. See http://sketch.icsi.berkeley.edu.

2 Related Works

Sketch-Based Image Synthesis. While much progress has been made on sketch recognition [6,35,37] and sketch-based image retrieval [9,13,20,21,28,36], sketch-based image synthesis remains under-explored.

Prior to deep learning (DL), Sketch2Photo [4] and PhotoSketcher [8] compose a new photo from photos retrieved for a sketch. Sketch2Photo [4] first retrieves photos based on the class label, then uses the given sketch to filter them and compose a target photo. PhotoSketcher [8] has a similar pipeline but retrieves photos based on a rather restrictive sketch and hand-crafted features.

The first DL-based free-hand sketch-to-photo synthesis is SketchyGAN [5], which trains an encoder-decoder model conditioned on the class label for sketch and photo pairs. Contextual GAN [23] treats sketch to photo synthesis as an image completion problem, using the sketch as a weak contextual constraint. Interactive Sketch [10] focuses on multi-class photo synthesis based on incomplete edges or sketches. All of these works rely on paired sketch and photo data and do not address the shape deformation problem.

Sketches are often used in photo editing [1,25,34], e.g., line strokes are drawn on a photo to change the shape of a roof. Unlike our sketch to photo synthesis, these works mainly address a constrained image inpainting problem.

Synthesis from the opposite direction, photo to sketch, has also been studied [19,29]: The former proposes a hybrid model to synthesize a sketch stroke by stroke given a photo, whereas the latter aims to generate boundary-like drawings that capture the outline of the visual scene. Both models require paired data for training. While photo to sketch is not our focus, our model trained only on *shoes* can generate realistic sketches from photos in other semantic categories.

Generative Adversarial Networks (GAN). GAN has a generator (G) and a discriminator (D): G tries to fake instances that fool D and D tries to detect fakes from reals. GAN is widely used for realistic image generation [17,24] and translation across image domains [15,16].

Pix2Pix [16] is a conditional GAN that maps source images to target images; it requires paired (source,target) data during training. CycleGAN [38] uses a pair of GANs to map an image from the source domain to the target domain and then back to the source domain. Imposing a consistency loss over such a cycle of mappings, it allows both models to be trained together on unpaired

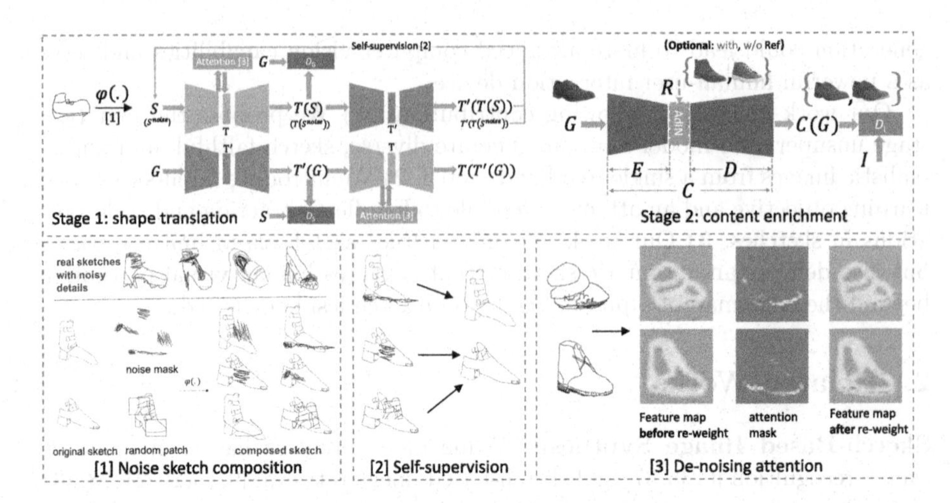

Fig. 3. Our two-stage model architecture (top) and three major technical components (bottom) that tackle abstract and style-varying strokes: noise sketch composition for training data augmentation, a self-supervised de-noising objective, and an attention module to suppress distracting dense strokes.

source and target images in two different domains. UNIT [22] and MUNIT [15] are variations of CycleGAN, both achieving impressive performance.

None of these methods work well when the source and target images are spatially poorly aligned (Fig. 1) and across different appearance domains.

3 Unsupervised Two-Stage Sketch-to-Photo Synthesis

In our unsupervised learning setting, we are given two sets of data in the same semantic category such as shoes, and no instance pairing is known or available. Formally, all we have are n sketches $\{S_1, \ldots, S_n\}$ and m color photos $\{I_1, \ldots, I_m\}$ along with their grayscale versions $\{G_1, \ldots, G_m\}$.

Compared to photos, sketches are spatially imprecise and colorless. To synthesize a photo from a sketch, we deal with these two aspects at separate stages: We first translate a sketch into a grayscale photo and then translate the grayscale into a color photo filled with missing details on texture and shading (Fig. 3).

3.1 Shape Translation: Sketch $S \rightarrow$ Grayscale G

Overview. We first learn to translate sketch S into grayscale photo G. The goal is to rectify shape deformation in sketches. We consider *unpaired* sketch and photo images, not only because paired data are scarce and hard to collect, but also because heavy reliance on paired data could restrict the model from recognizing the inherent misalignment between sketches and photos.

A pair of mappings, $T : S \rightarrow G$ and $T' : G \rightarrow S$, each implemented with an encoder-decoder architecture, are learned with cycle-consistency objectives: $S \approx$

$T'(T(S))$ and $G \approx T(T'(G))$. Similar to [38], we train two domain discriminators D_G and D_S: D_G tries to tease apart G and $T(S)$, while D_S teases apart S and $T'(G)$ (Fig. 3). The predicted grayscale $T(S)$ goes to content enrichment next.

The input sketch may exhibit various levels of abstraction and different drawing styles. In particular, sketches containing dense strokes or noisy details (Fig. 3) cannot be handled well by a basic CycleGAN model.

To deal with these variations, we introduce two strategies for the model to extract style-invariant information only: **1)** We compose additional noise sketches to enrich the dataset and introduce a self-supervised objective; **2)** We introduce an attention module to help detect distracting regions.

Noise Sketch Composition. In a rapidly drawn sketch, strokes could be deliberately complex, or simply careless and distractive (Fig. 3). We augment limited sketch data with more noise. Let $S^{\text{noise}} = \varphi(S)$, where $\varphi(.)$ represents composition. We detect dense strokes and construct a pool of noise masks. We randomly sample from these masks and artificially generate *complex* sketches by inserting these dense stroke patterns into original sketches. We generate *distractive* sketches by adding a random patch from a different sketch on an existing sketch. The noise strokes and random patches are used to simulate irrelevant details in a sketch. We compose such noise sketches on the fly and feed them into the network with a fixed occurrence ratio.

Self-supervised Objective. We introduce a self-supervised objective to work with the synthesized noise sketches. For a composed noise sketch, the reconstruction goal of our model is to reproduce the *original clean* sketch:

$$L_{ss}(T, T') = \left\| S - T'\left(T(S^{\text{noise}})\right) \right\|_1 \tag{1}$$

This objective is different from the cycle-consistency loss used on untouched original sketches. It makes the model ignore irrelevant strokes and put more efforts on style-invariant strokes in the sketch.

Ignore Distractions with Active Attention. To identify distracting strokes, we also introduce an attention module. Since most areas of a sketch are blank, the activation of dense stroke regions is stronger than others. We can thus locate distracting areas and *suppress* the activation there accordingly. That is, the attention module generates an attention map A to be used for re-weighting the feature representation of sketch S (Eq. 2):

$$f_{\text{final}}(S) = (1 - A) \odot f(S) \tag{2}$$

where $f(.)$ refers to the feature map and $\odot$ denotes element-wise multiplication. Our attention is used for area suppression instead of the usual area highlight.

Our total objective for training a shape translation model is:

$$\min_{T,T'} \max_{D_G,D_S} \lambda_1 (L_{adv}(T, D_G; S, G) + L_{adv}(T', D_S; G, S))$$
$$+ \lambda_2 L_{cycle}(T, T'; S, G) + \lambda_3 L_{identity}(T, T'; S, G) + L_{ss}(T, T'; S^{noise}).$$

We follow [38] to add an identity loss $L_{identity}$, which slightly improves the performance. See the details of each loss in the Supplementary.

3.2 Content Enrichment: Grayscale $G \rightarrow$ Color I

Now that we have a predicted grayscale photo G, we learn a mapping C that turns it into color photo I. The goal at this stage is to enrich the generated grayscale photo G with missing appearance details.

Since a color-less sketch could have many colorful realizations, many fill-in's are possible. We thus model the task as a style transfer task and use an *optional* reference color image to guide the selection of a particular style.

We implement C as an encoder (E) and decoder (D) network (Fig. 3). Given a grayscale photo G as the input, the model outputs a color photo I. The input G and the grayscale of the output I, specifically the L-channel in CIE Lab color space of the output should be the same. Therefore we use a self-supervised intensity loss (Eq. 3) to train the model:

$$L_{it}(C) = \|G - \text{grayscale}\,(C\,(G))\|_1 \tag{3}$$

We train discriminator D_I to ensure that I is also as photo-realistic as $I_1, \ldots, I_m$.

To achieve the output diversity, we introduce a conditional module that takes an optional reference image for guidance. We follow AdaIN [14] to inject style information by adjusting the feature map statistics. Specifically, the encoder E takes the input grayscale image G and generates a feature map $\mathbf{x} = E(G)$, then the mean and variance of $\mathbf{x}$ are adjusted by the reference's feature map $\mathbf{x}^{ref} = E(R)$. The new feature map is $\mathbf{x}^{new} = AdaIN(\mathbf{x}, \mathbf{x}^{ref})$ (Eq. 4), which is subsequently sent to the decoder D for rendering the final output image I:

$$AdaIN(\mathbf{x}, \mathbf{x}^{\text{ref}}) = \sigma(\mathbf{x}^{\text{ref}})(\frac{\mathbf{x} - \mu(\mathbf{x})}{\sigma(\mathbf{x})}) + \mu(\mathbf{x}^{\text{ref}}) \tag{4}$$

Our model can work with or without reference images, in a *single* network, enabled by a mixed training strategy. When there is no reference image, only intensity loss and adversarial loss are used while $\sigma(\mathbf{x}^{ref})$ and $\mu(\mathbf{x}^{ref})$ are set to 1 and 0 respectively; otherwise, a content loss and style loss are computed additionally. The content loss (Eq. 5) is used to guarantee that the input and output images are consistent perceptually, whereas the style loss (Eq. 6) is to ensure the style of the output is aligned with that of the reference image.

$$L_{cont}(C; G, R) = \|E(D(t)) - t\|_1 \tag{5}$$

$$L_{style}(C; G, R) = \sum_{i=1}^{K} \|\mu\,(\phi_i(D(t))) - \mu\,(\phi_i(R))\|_2 + \sum_{i=1}^{K} \|\sigma\,(\phi_i(D(t))) - \sigma\,(\phi_i(R))\|_2 \tag{6}$$

$$\text{where} \quad t = AdaIN(E(G), E(R)) \tag{7}$$

$\phi_i(.)$ denotes a layer of a pre-trained VGG-19 model. In our implementation, we use *relu*1_1, *relu*2_1, *relu*3_1, *relu*4_1 layers with equal weights to compute the style loss. Equation 8 shows the total loss for training the content enrichment model. Network architectures and further details are provided in the Supplementary.

$$\min_{C}\max_{D_I} \lambda_4 L_{adv}(C, D_I; G, I) + \lambda_5 L_{it}(C) + \lambda_6 L_{style}(C; G, R) + \lambda_7 L_{cont}(C; G, R) \quad (8)$$

4 Experiments and Applications

4.1 Experimental Setup and Evaluation Metrics

Datasets. We train our model on two single-category sketch datasets, ShoeV2 and ChairV2 [36], with 6,648/2,000 and 1,297/400 sketches/photos respectively. Each photo has at least 3 corresponding sketches drawn by different individuals. Note that we do not use pairing information at training. Compared to QuickDraw [11], Sketchy [28], and TU-Berlin [6], sketches in ShoeV2/ChairV2 have more fine-grained details. They demand like-kind details in synthesized photos and are thus more challenging as a testbed for sketch to photo synthesis.

Baselines for Image Translation. 1) Pix2Pix [16] is our supervised learning baseline which requires paired training data. **2) CycleGAN** [38] is an unsupervised bidirectional image translation model. It is the first to apply cycle-consistency with GANs and allows unpaired training data. **3) MUNIT**[15] is also an unsupervised model that could generate multiple outputs given an input. It assumes that the representation of an image can be decomposed into a content code and a style code. **4) UGATIT** [18] is an attention-based image translation model, with the attention to help the model focus on the domain-discriminative regions and thereby improve the synthesis quality.

Training Details. We train our shape translation network for 500 (400) epochs on shoes (chairs), and train our content enrichment network for 200 epochs. The initial learning rate is 0.0002, and the input image size is 128×128. We use Adam optimizer with batch size 1. Following the practice by CycleGAN, we train the first 100 epochs at the same learning rate and then linearly decrease the rate to zero until the maximum epoch. We randomly compose *complex* and *distractive* sketches with the possibility of 0.2 and 0.3 respectively. The random patch size is 50×50. When training the content enrichment network, we feed reference images into the network with possibility 0.2.

Evaluation Metrics. 1) Fréchet Inception Distance (FID). It evaluates image quality and diversity according to the distance between synthesized and real samples according to the statistics of activations in layer pool3 of a pre-trained Inception-v3. A lower FID value indicates higher fidelity. **2) User study** (Quality). It evaluates subjective impressions in terms of similarity and realism. As in [30], we ask the subject to compare **two** generated photos and select the one better fitting their imagination for a given sketch. We sample 50 pairs for each comparison (more details in Supplementary). **3) Learned perceptual image patch similarity** (LPIPS). It measures the distance between two images. As in [15,38], we use it to evaluate the *diversity* of synthesized photos.

Fig. 4. Our model can produce high-fidelity and diverse photos from a sketch. **Top:** Result comparisons. Most baselines cannot handle this task well. While UGATIT can generate realistic photos, our results are more faithful to the input sketch, e.g., the three chair examples. **Bottom:** Results without (Column 2) or with (Column 3) the reference image. Our single content enrichment model can work under both settings, with or without a reference photo (shown in the top right corner).

Table 1. Benchmarks on ShoeV2/ChairV2. '*' indicates paired data for training.

Model	ShoeV2			ChairV2		
	FID ↓	Quality ↑	LPIPS ↑	FID ↓	Quality ↑	LPIPS ↑
Pix2Pix*	65.09	27.0	0.071	177.79	13.0	0.096
CycleGAN	79.35	12.0	0.0	124.96	20.0	0.0
MUNIT	92.21	14.5	**0.248**	168.81	6.5	**0.264**
UGATIT	76.89	21.5	0.0	107.24	19.5	0.0
Ours	**48.73**	**50.0**	0.146	**100.51**	**50.0**	0.156

Fig. 5. Left: With different references, our model can produce diverse outputs. **Middle:** Given sketches of similar shoes drawn by different users, our model can capture their commonality as well as subtle distinctions. Each row shows input sketch, synthesized grayscale image, synthesized RGB photo. **Right:** Our model even works for sketches at different completion stages, delivering realistic closely looking shoes. (Color figure online)

4.2 Sketch-Based Photo Synthesis Results

Table 1 shows that: **1)** Our model outperforms all the baselines in terms of FID and user studies. Note that all the baselines adopt one-stage architectures. **2)** All the models perform poorly on ChairV2, probably due to more shape variations but far fewer training data for chairs than for shoes (1:5). **3)** Ours outperforms MUNIT by a large margin, indicating that our task-level decomposition strategy, i.e., two-stage architecture, is more effective than feature-level decomposition for this task. **4)** UGATIT ranks the second on each dataset. It is also an attention-based model, showing the effectiveness of attention in image translation tasks.

Comparisons in Fig. 4 and Varieties in Fig. 5 (Left). Our results are more realistic and faithful to the input sketch (e.g., buckle and logo); our synthesis with different reference images produces varieties.

Robustness and Sensitivity in Fig. 5 (Middle & Right). We test our ShoeV2 model under two settings: 1) sketches corresponding to the same photo, 2) sketches at different completion stages. Given sketches of similar shoes drawn by different users, our model can capture their commonality as well as subtle distinctions and translate them into photos. Our model also works for sketches

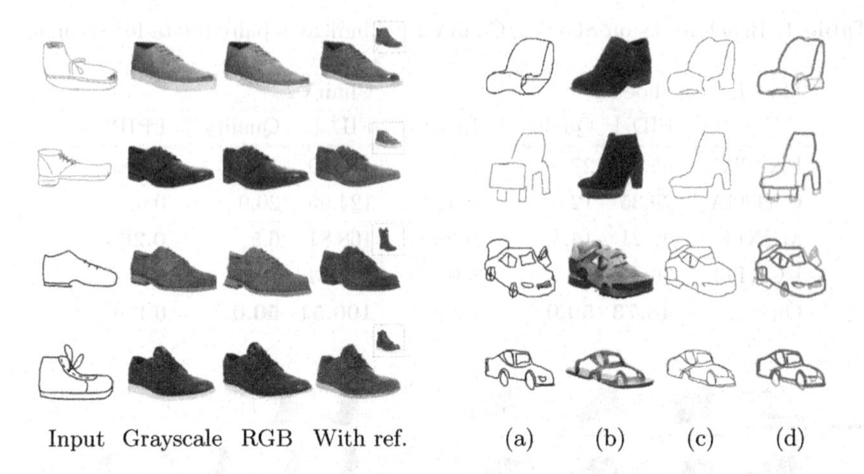

Fig. 6. Left: Generalization across domains. Column 1 are sketches from two unseen datasets, Sketchy and TU-Berlin. Columns 2–4 are results from our model trained on ShoeV2. **Right:** Our shoe model can be used as a shoe detector and generator. It can generate a shoe photo based on a non-shoe sketch. It can further turn the non-shoe sketch into a more shoe-like sketch. (a) Input sketch; (b) synthesized grayscale photo; (c) re-synthesized sketch; (d) Green *(a)* overlaid over gray *(c)*. (Color figure online)

Table 2. Comparison of different architecture designs.

FID ↓	CycleGAN (1-stage)	CycleGAN (2-stage)	Edge Map	Grayscale (Ours)
ShoeV2	79.35	51.80	96.58	**48.73**
ChairV2	124.96	109.46	236.38	**100.51**

at different completion stages (obtained by removing strokes according to their orderings), synthesizing realistic closely-looking shoes for partial sketches.

Generalization Across Domains in Fig. 6 (Left). When sketches are randomly sampled from different datasets such as TU-Berlin [6] and Sketchy [28], which have greater shape deformation than ShoeV2, our model trained on ShoeV2 can still produce good results (see more examples in the Supplementary).

Sketches from Novel Categories in Fig. 6 (Right). While we focus on a single category training, we nonetheless feed our model sketches from other categories. When the model is trained on shoes, the shape translation network has learned to synthesize a grayscale shoe photo based on a *shoe* sketch. For a non-shoe sketch, our model translates it into a shoe-like photo. Some fine details in the sketch become a common component of a shoe. For example, a car becomes a trainer while the front window becomes part of a shoelace. The superimposition of the input sketch and the re-*shoe*-synthesized sketch reveals which lines are chosen by our model and how it modifies the lines for re-synthesis.

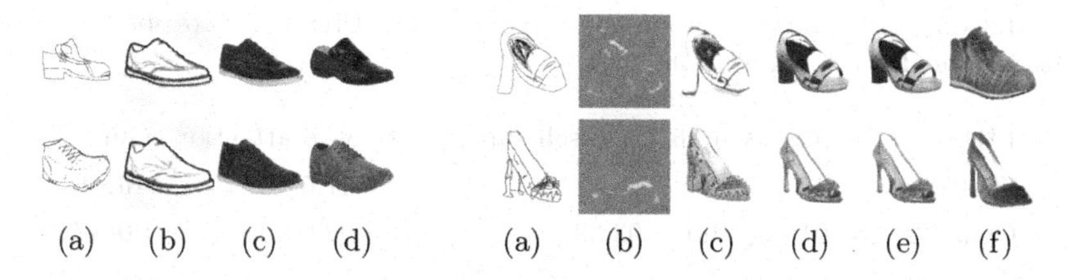

Fig. 7. Left: Synthesized results when the edge map is used as the intermediate goal instead of the grayscale photo. (a) Input sketch; (b) Synthesized edge map, (c) Synthesized RGB photo using the edge map; (d) Synthesized RGB photo using grayscale (Ours). **Right:** Our model can successfully deal with noise sketches, which are not well handled by another attention-based model, UGATIT. For an input sketch (a), our model produce an attention mask (b); (c) and (d) are grayscale images produced by vanilla and our model. (e) and (f) compare ours with the result of UGATIT. (Color figure online)

Fig. 8. Comparisons of paired and unpaired training for shape translation. There are four examples. For each example, the 1st one is the input sketch, the 2nd and the 3rd are grayscale images synthesized by Pix2Pix and our model respectively. Note that for each example, although the input sketches are different visually, Pix2Pix produces a similar-looking grayscale image. Our results are more faithful to the sketch.

4.3 Ablation Study

Two-Stage Architecture. Two-stage architecture is the key to the success of our model. This strategy can be easily adapted by other models such as cycleGAN. Table 2 compares the performance of the original cycleGAN and its two-stage version (i.e., cycleGAN is used only for shape translation while the content enrichment network is the same as ours). The two-stage version outperforms the original cycleGAN by 27.55 (on ShoeV2) and 68.33 (on ChairV2), indicating the significant benefits brought by this architectural design.

Edge Map vs. Grayscale as the Intermediate Goal. We choose *grayscale* as our intermediate goal of translation. As shown in Fig. 1, *edge maps* could be an alternative since it does not have shape deformation either. We can first translate sketch to an edge map, and then fill the edge map with colorful details.

Table 2 and Fig. 7 show that using the edge map is worse than using the grayscale. Our explanations are: **1)** Grayscale images contain more visual details thus can provide more learning signals for training shape translation network; **2)** Content enrichment is easier for grayscale as they are closer to color photos than edge maps. The grayscale is also easier to obtain in practice.

Deal with Abstraction and Style Variations. We have discussed the problem encountered during shape translation in Sect. 3.1, and further introduced 1) a self-supervised objective along with noise sketch composition strategies and

Table 3. Contribution of each proposed component. The FID scores are obtained based on the results of *shape translation stage*.

FID ↓	Pix2Pix	Vanilla	w/o self-supervision	w/o attention	Ours
ShoeV2	75.84	48.30	46.88	47.0	**46.46**
ChairV2	164.01	104.0	93.33	92.03	**90.87**

Table 4. Exclude the effect of paired data. Although the paired information is not used during training, they indeed exist in ShoeV2. We compose a new dataset where pairing does not exist to train the model again. Results obtained on the same test set.

Dataset	Paired exist?	Use pair info.	FID ↓
ShoeV2	Yes	No	**48.7**
UT Zappos50K	No	No	**48.6**

2) an attention module to handle the problem. Table 3 compares FID achieved at the first stage by different variants. Our full model can tackle the problem better than the vanilla model, and each component contributes to the improved performance. Figure 7 shows two examples and compares the results of UGATIT.

Paired vs. Unpaired Training. We train a Pix2Pix model for shape translation to see if paired information helps. As shown in Table 3 (*Pix2Pix*) and Fig. 8, It turns out the performance of Pix2Pix is much worse than ours (FID: 75.84 vs. 46.46 on ShoeV2 and 164.01 vs. 90.87 on ChairV2). It is most likely caused by the shape misalignment between sketches and grayscale images.

Exclude the Effect of Paired Information. Although pairing information is not used during training, they do exist in ShoeV2. To eliminate any potential pairing facilitation, we train another model on a composed dataset, created by merging all the sketches of ShoeV2 and 9,995 photos of UT Zappos50K [33]. These photos are collected from a different source than ShoeV2. We train this model in the same setting. In Table 4, we can see this model achieves similar performance with the one trained on ShoeV2, indicating the effectiveness of our approach for learning the task from entirely **unpaired** data.

4.4 Photo-to-Sketch Synthesis Results

Synthesize a Sketch Given a Photo. As the shape translation network is bidirectional (i.e., T and T'), our model can also translate a photo into a sketch. This task is not trivial, as users can easily detect a fake sketch based on its stroke continuity and consistency. Figure 9 (Top) shows that our generated sketches mimic manual line-drawings and emphasize contours that are perceptually significant.

Sketch-Like Edge Extraction. Sketch-to-photo and photo-to-sketch synthesis are opposite processes. We suspect that our model can create sketches from photos in broader categories as it may require less class priors.

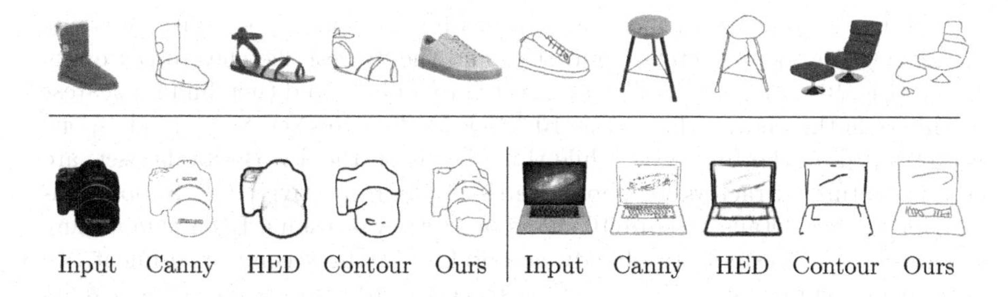

Fig. 9. Our results on photo-based sketch synthesis. **Top**: each sketch-photo pair: left: input photo, right: synthesized sketch. Results obtained on ShoeV2 and ChairV2. **Bottom**: Results obtained on ShapeNet [3]. The column 1 is the input photo, Column 2–5 are lines generated by Canny, HED, Photo-Sketching [19] (*Contour* for short), and our model. Our model can generate line strokes with a hand-drawn effect, while HED and Canny detectors produce edge maps faithful to the original photos. Ours emphasize perceptually significant contours, not intensity-contrast significant as in edge maps.

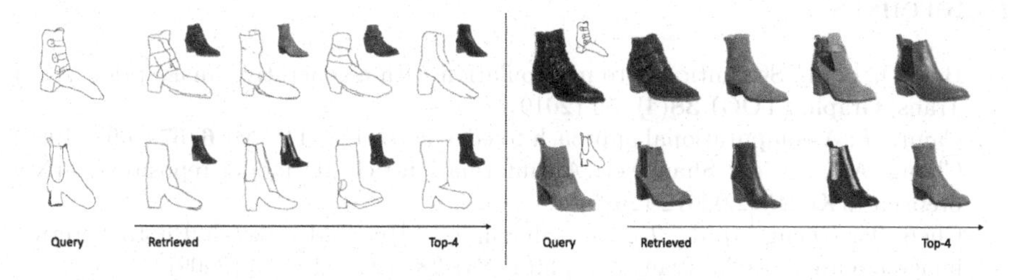

Fig. 10. Sample retrieval results. Our synthesis model can map photo to sketch domain and vice versa. Cross-domain retrieval task can thus be converted to intra-domain retrieval. **Left:** All candidate photos are mapped to sketches, thus both query and candidates are in the sketch domain. **Right:** The query sketch is translated to a photo, so the matching is in the photo domain. Top right shows the original photo or sketch.

We test our shoe model directly on photos in ShapeNet [3]. Figure 9 (Bottom) lists our results along with those from HED [32] and Canny edge detector [2]. We also compare with Photo-Sketching [19], a method specifically designed for generating boundary-like drawing from photos. 1) Unlike HED and Canny producing an edge map faithful to the photo, ours presents a hand-drawn style. 2) Our model can dub as an edge+ extractor on unseen classes. This is an exciting corollary product: A promising automatic sketch generator that captures human visual perception beyond the edge map of a photo (more results in Supp.).

4.5 Application: Unsupervised Sketch-Based Image Retrieval

Sketch-based image retrieval is an important application of sketch. One of its main challenges is the large domain gap. Existing methods either map sketches and photos into a common space or use edge maps as the intermediate representation. However, our model enables direct mapping between these two domains.

We thus conduct experiments in two possible mapping directions: **1)** Translate gallery photos to sketches, and then find the nearest sketches to the query sketch (Fig. 10 (Left)); **2)** Translate a sketch to a photo and then find its nearest neighbors in the photo gallery (Fig. 10 (Right)). Two ResNet18 [12] models, one is pretrained on the ImageNet while the other is on the TU-Berlin dataset, are used as feature extractors for photos and sketches respectively (see Supplementary for further details). Figure 10 shows our retrieval results. Even *without* any supervision, the results are already acceptable. In the second experiment, we achieve an accuracy of 37.2% (65.2%) at top5 (top20) respectively. These results are higher than the results from *sketch to edge map*, which are 34.5% (57.7%).

Summary. We propose the first unsupervised two-stage sketch-to-photo synthesis model that can produce photos of high fidelity, realism, and diversity. It enables sketch-based image retrieval and automatic sketch generation that captures human visual perception beyond the edge map of a photo.

References

1. Bau, D., et al.: Semantic photo manipulation with a generative image prior. ACM Trans. Graph. (TOG) **38**(4), 59 (2019)
2. Canny, J.: A computational approach to edge detection. TPAMI **6**, 679–698 (1986)
3. Chang, A.X., et al.: ShapeNet: An information-rich 3D model repository. arXiv preprint arXiv:1512.03012 (2015)
4. Chen, T., Cheng, M.M., Tan, P., Shamir, A., Hu, S.M.: Sketch2Photo: internet image montage. ACM Trans. Graph.(TOG) **28**, 124:1–124:10 (2009)
5. Chen, W., Hays, J.: SketchyGAN: towards diverse and realistic sketch to image synthesis. In: CVPR (2018)
6. Eitz, M., Hays, J., Alexa, M.: How do humans sketch objects? ACM Trans. Graph. (TOG) **31**, 44:1–44:10 (2012)
7. Eitz, M., Hildebrand, K., Boubekeur, T., Alexa, M.: An evaluation of descriptors for large-scale image retrieval from sketched feature lines. Comput. Graph. **34**(5), 482–498 (2010)
8. Eitz, M., Richter, R., Hildebrand, K., Boubekeur, T., Alexa, M.: Photosketcher: interactive sketch-based image synthesis. IEEE Comput. Graph. Appl. **31**, 56–66 (2011)
9. Eitz, M., Hildebrand, K., Boubekeur, T., Alexa, M.: Sketch-based image retrieval: benchmark and bag-of-features descriptors. TVCG **17**(11), 1624–1636 (2011)
10. Ghosh, A., et al.: Interactive sketch & fill: multiclass sketch-to-image translation. In: CVPR (2019)
11. Ha, D., Eck, D.: A neural representation of sketch drawings. arXiv preprint arXiv:1704.03477 (2017)
12. He, K., Zhang, X., Ren, S., Sun, J.: Deep residual learning for image recognition. In: Proceedings of the IEEE Conference on Computer Vision and Pattern Recognition, pp. 770–778 (2016)
13. Hu, R., Barnard, M., Collomosse, J.: Gradient field descriptor for sketch based retrieval and localization. In: ICIP (2010)
14. Huang, X., Belongie, S.: Arbitrary style transfer in real-time with adaptive instance normalization. In: Proceedings of the IEEE International Conference on Computer Vision, pp. 1501–1510 (2017)

15. Huang, X., Liu, M.Y., Belongie, S., Kautz, J.: Multimodal unsupervised image-to-image translation. In: Ferrari, V., Hebert, M., Sminchisescu, C., Weiss, Y. (eds.) ECCV 2018. LNCS, vol. 11207. Springer, Cham (2018). https://doi.org/10.1007/978-3-030-01219-9_11
16. Isola, P., Zhu, J.Y., Zhou, T., Efros, A.A.: Image-to-image translation with conditional adversarial networks. In: Proceedings of the IEEE Conference on Computer Vision and Pattern Recognition, pp. 1125–1134 (2017)
17. Karras, T., Aila, T., Laine, S., Lehtinen, J.: Progressive growing of GANs for improved quality, stability, and variation. arXiv preprint arXiv:1710.10196 (2017)
18. Kim, J., Kim, M., Kang, H., Lee, K.: U-GAT-IT: unsupervised generative attentional networks with adaptive layer-instance normalization for image-to-image translation. CoRR abs/1907.10830 (2019)
19. Li, M., Lin, Z., Mech, R., Yumer, E., Ramanan, D.: Photo-sketching: inferring contour drawings from images. In: 2019 IEEE Winter Conference on Applications of Computer Vision (WACV) (2019)
20. Li, Y., Hospedales, T., Song, Y.Z., Gong, S.: Fine-grained sketch-based image retrieval by matching deformable part models. In: BMVC (2014)
21. Liu, L., Shen, F., Shen, Y., Liu, X., Shao, L.: Deep sketch hashing: Fast free-hand sketch-based image retrieval. arXiv preprint arXiv:1703.05605 (2017)
22. Liu, M.Y., Breuel, T., Kautz, J.: Unsupervised image-to-image translation networks. In: Advances in Neural Information Processing Systems, pp. 700–708 (2017)
23. Lu, Y., Wu, S., Tai, Y.-W., Tang, C.-K.: Image generation from sketch constraint using contextual GAN. In: Ferrari, V., Hebert, M., Sminchisescu, C., Weiss, Y. (eds.) ECCV 2018. LNCS, vol. 11220, pp. 213–228. Springer, Cham (2018). https://doi.org/10.1007/978-3-030-01270-0_13
24. Mirza, M., Osindero, S.: Conditional generative adversarial nets. arXiv preprint arXiv:1411.1784 (2014)
25. Portenier, T., Hu, Q., Szabo, A., Bigdeli, S.A., Favaro, P., Zwicker, M.: FaceShop: deep sketch-based face image editing. ACM Trans. Graph. (TOG) **37**(4), 99 (2018)
26. Qi, Y., Guo, J., Li, Y., Zhang, H., Xiang, T., Song, Y.: Sketching by perceptual grouping. In: ICIP, pp. 270–274 (2013)
27. Qi, Y., et al.: Making better use of edges via perceptual grouping. In: CVPR (2015)
28. Sangkloy, P., Burnell, N., Ham, C., Hays, J.: The sketchy database: learning to retrieve badly drawn bunnies. In: SIGGRAPH (2016)
29. Song, J., Pang, K., Song, Y.Z., Xiang, T., Hospedales, T.M.: Learning to sketch with shortcut cycle consistency. In: Proceedings of the IEEE Conference on Computer Vision and Pattern Recognition, pp. 801–810 (2018)
30. Wang, T.C., Liu, M.Y., Zhu, J.Y., Tao, A., Kautz, J., Catanzaro, B.: High-resolution image synthesis and semantic manipulation with conditional GANs. In: Proceedings of the IEEE Conference on Computer Vision and Pattern Recognition, pp. 8798–8807 (2018)
31. Xian, W., et al.: TextureGAN: controlling deep image synthesis with texture patches. In: CVPR (2018)
32. Xie, S., Tu, Z.: Holistically-nested edge detection. In: ICCV (2015)
33. Yu, A., Grauman, K.: Fine-grained visual comparisons with local learning. In: Proceedings of the IEEE Conference on Computer Vision and Pattern Recognition, pp. 192–199 (2014)
34. Yu, J., Lin, Z., Yang, J., Shen, X., Lu, X., Huang, T.S.: Free-form image inpainting with gated convolution. arXiv preprint arXiv:1806.03589 (2018)
35. Yu, Q., Yang, Y., Song, Y., Xiang, T., Hospedales, T.: Sketch-a-net that beats humans. In: BMVC (2015)

36. Yu, Q., Liu, F., Song, Y.Z., Xiang, T., Hospedales, T.M., Loy, C.C.: Sketch me that shoe. In: CVPR (2016)
37. Yu, Q., Yang, Y., Liu, F., Song, Y.Z., Xiang, T., Hospedales, T.M.: Sketch-a-net: a deep neural network that beats humans. JICV **122**(3), 411–425 (2017)
38. Zhu, J.Y., Park, T., Isola, P., Efros, A.A.: Unpaired image-to-image translation using cycle-consistent adversarial networks. In: Proceedings of the IEEE International Conference on Computer Vision, pp. 2223–2232 (2017)
39. Zou, C., et al.: SketchyScene: richly-annotated scene sketches. In: Ferrari, V., Hebert, M., Sminchisescu, C., Weiss, Y. (eds.) ECCV 2018. LNCS, vol. 11219, pp. 438–454. Springer, Cham (2018). https://doi.org/10.1007/978-3-030-01267-0_26

A Simple Way to Make Neural Networks Robust Against Diverse Image Corruptions

Evgenia Rusak[1,2(✉)], Lukas Schott[1,2], Roland S. Zimmermann[1,2], Julian Bitterwolf[2], Oliver Bringmann[1], Matthias Bethge[1,2], and Wieland Brendel[1,2]

[1] University of Tübingen, Tübingen, Germany
{evgenia.rusak,lukas.schott,roland.zimmermann, oliver.bringmann,matthias.bethge,wieland.brendel}@uni-tuebingen.de
[2] International Max Planck Research School for Intelligent Systems, Tübingen, Germany
julian.bitterwolf@uni-tuebingen.de

Abstract. The human visual system is remarkably robust against a wide range of naturally occurring variations and corruptions like rain or snow. In contrast, the performance of modern image recognition models strongly degrades when evaluated on previously unseen corruptions. Here, we demonstrate that a simple but properly tuned training with additive Gaussian and Speckle noise generalizes surprisingly well to unseen corruptions, easily reaching the state of the art on the corruption benchmark ImageNet-C (with ResNet50) and on MNIST-C. We build on top of these strong baseline results and show that an adversarial training of the recognition model against locally correlated worst-case noise distributions leads to an additional increase in performance. This regularization can be combined with previously proposed defense methods for further improvement.

Keywords: Image corruptions · Robustness · Generalization · Adversarial training

1 Introduction

While Deep Neural Networks (DNNs) have surpassed the functional performance of humans in a range of complex cognitive tasks [2,12,30,38,44], they still lag

E. Rusak, L. Schott and R. S. Zimmermann—Joint first authors.

O. Bringmann, M. Bethge and W. Brendel—Joint senior authors.

Electronic supplementary material The online version of this chapter (https://doi.org/10.1007/978-3-030-58580-8_4) contains supplementary material, which is available to authorized users.

A. Vedaldi et al. (Eds.): ECCV 2020, LNCS 12348, pp. 53–69, 2020.
https://doi.org/10.1007/978-3-030-58580-8_4

behind humans in numerous other aspects. One fundamental shortcoming of machines is their lack of robustness against input perturbations. Even minimal perturbations that are hardly noticeable for humans can derail the predictions of high-performance neural networks.

For the purpose of this paper, we distinguish between two types of input perturbations. One type are minimal image-dependent perturbations specifically designed to fool a neural network with the smallest possible change to the input. These so-called *adversarial perturbations* have been the subject of hundreds of papers in the past five years, see e.g. [11,21,35,39]. Another, much less studied type are *common corruptions.* These perturbations occur naturally in many applications and include simple Gaussian or Salt and Pepper noise; natural variations like rain, snow or fog; and compression artifacts such as those caused by JPEG encoding. All of these corruptions do not change the semantic content of the input, and thus, machine learning models should not change their decision-making behavior in their presence. Nonetheless, high-performance neural networks like ResNet50 [12] are easily confused by small deformations [1]. The juxtaposition of adversarial examples and common corruptions was explored in [8] where the authors discuss the relationship between both and encourage researchers working in the field of adversarial robustness to cross-evaluate the robustness of their models towards common corruptions.

We argue that in many practical applications, robustness to common corruptions is often more relevant than robustness to artificially designed adversarial perturbations. Autonomous cars should not change their behavior in the face of unusual weather conditions such as hail or sand storms or small pixel defects in their sensors. Not-Safe-For-Work filters should not fail on images with unusual compression artifacts. Likewise, speech recognition algorithms should perform well regardless of the background music or sounds.

Besides its practical relevance, robustness to common corruptions is also an excellent target in its own right for researchers in the field of adversarial robustness and domain adaptation. Common corruptions can be seen as distributional shifts or as a weak form of adversarial examples that live in a smaller, constrained subspace.

Despite their importance, common corruptions have received relatively little attention so far. Only recently, a modification of the ImageNet dataset [34] to benchmark model robustness against common corruptions and perturbations has been published [13] and is referred to as ImageNet-C. Now, this scheme has also been applied to other common datasets resulting in Pascal-C, Coco-C and Cityscapes-C [25] and MNIST-C [29].

Our contributions are as follows:

- We demonstrate that data augmentation with Gaussian or Speckle noise serves as a simple yet very strong baseline that is sufficient to surpass almost all previously proposed defenses against common corruptions on ImageNet-C for ResNet50. We further show that the magnitude of the additive noise is a crucial hyper-parameter to reach optimal robustness.
- Motivated by our strong results with baseline noise augmentations, we introduce a neural network-based *adversarial noise generator* that can learn

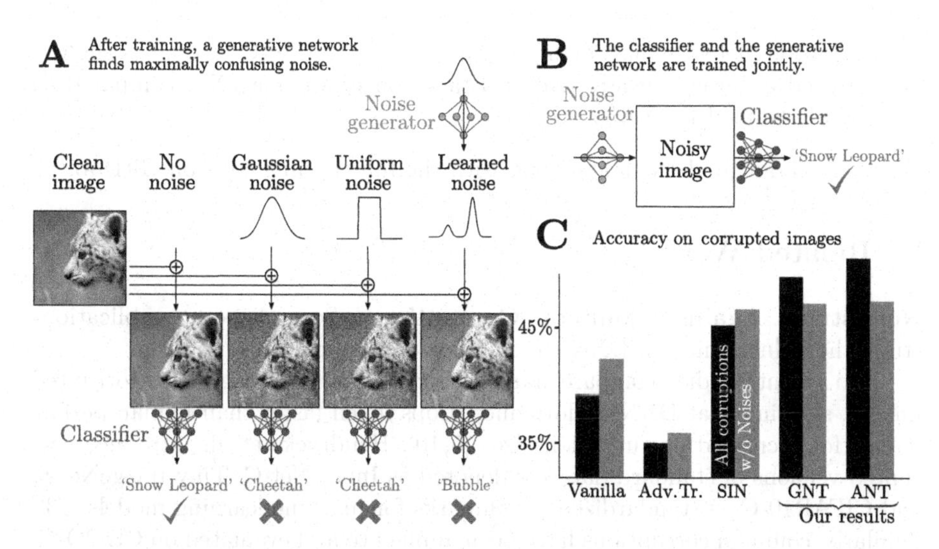

Fig. 1. Outline of our approach. A: First, we train a generative network against a vanilla trained classifier to find the adversarial noise. B: To achieve robustness against adversarial noise, we train the classifier and the noise generator jointly. C: We measure the robustness against common corruptions for a vanilla, adversarially trained (Adv. Tr.), trained on Stylized ImageNet (SIN), trained via Gaussian data augmentation (GNT) and trained with the means of Adversarial Noise Training (ANT). With our methods, we achieve the highest accuracy on common corruptions, both on all and non-noise categories.

arbitrary uncorrelated noise distributions that maximally fool a given recognition network when added to their inputs. We denote the resulting noise patterns as *adversarial noise.*

- We design and validate a constrained Adversarial Noise Training (ANT) scheme through which the recognition network learns to become robust against adversarial i.i.d. noise. We demonstrate that our ANT reaches state-of-the-art robustness on the corruption benchmark ImageNet-C for the commonly used ResNet50 architecture and on MNIST-C, even surpassing the already strong baseline noise augmentations. This result is not due to overfitting on the noise categories of the respective benchmarks since we find equivalent results on the non-noise corruptions as well.
- We extend the adversarial noise generator towards locally correlated noise thereby enabling it to learn more diverse noise distributions. Performing ANT with the modified noise generator, we observe an increase in robustness for the 'snow' corruption which is visually similar to our learned noise.
- We demonstrate a further increase in robustness when combining ANT with previous defense methods.
- We substantiate the claim that increased robustness against regular or universal adversarial perturbations does not imply increased robustness against common corruptions. This is not necessarily true vice-versa: Our noise trained

recognition network has high accuracy on ImageNet-C and also slightly improved accuracy on adversarial attacks on clean ImageNet compared to a vanilla trained ResNet50.

We released our model weights along with the full training code on GitHub.[1]

2 Related Work

Robustness Against Common Corruptions. Several recent publications study the vulnerability of DNNs to common corruptions.

Two recent studies compare humans and DNNs on recognizing corrupted images, showing that DNN performance drops much faster than human performance for increased perturbation sizes [5,10]. Hendrycks et al. introduce corrupted versions of standard datasets denoted as ImageNet-C, Tiny ImageNet-C and CIFAR10-C as standardized benchmarks for machine learning models [13]. Similarly, common corruptions have been applied to and evaluated on COCO-C, Pascal-C, Cityscapes-C [25] and MNIST-C [29].

There have been attempts to increase robustness against common corruptions. Zhang et al. integrate an anti-aliasing module from the signal processing domain in the ResNet50 architecture to restore the shift-equivariance which can get lost in deep CNNs and report an increased accuracy on clean data and better generalization to corrupted image samples [45]. Concurrent work to ours demonstrates that having more training data [22,43] or using stronger backbones [18,25,43] can significantly improve model performance on common corruptions.

A popular method to decrease overfitting and help the network generalize better to unseen data is to augment the training dataset by applying a set of (randomized) manipulations to the images [26]. Furthermore, augmentation methods have also been applied to make the models more robust against image corruptions [9]. Augmentation with Gaussian [8,19] or uniform noise [10] has been tried to increase model robustness. Conceptually, Ford et al. is the closest study to our work, since they also apply Gaussian noise to images to increase corruption robustness [8]. They use a different architecture (InceptionV3 versus our ResNet50). Also, they train a new model from scratch solely on images perturbed by Gaussian noise whereas we fine-tune a pretrained model on a mixture of clean and noisy images. They observe a low relative improvement in accuracy on corrupted images whereas we were able to outperform all previous baselines on the commonly used ResNet50 architecture.[2] Lopes et al. restrict the Gaussian noise to small image patches, which improves accuracy but does not yield state-of-the-art performance on the ResNet50 architecture [19]. Geirhos et al. train ImageNet classifiers against a fixed set of corruptions but find no generalized robustness against unseen corruptions [10]. However, they considered vastly higher noise levels than us. Considering the efficacy of Gaussian or uniform data

[1] github.com/bethgelab/game-of-noise.

[2] To compare with Ford et al., we evaluate our approach for an InceptionV3 architecture, see our results in Appendix H.

augmentation to increase model robustness, the main difference to our work is that other works have used either munch larger [10] or smaller [8,19] values for the standard deviation σ. A too large σ leads to an overfitting to the used noise distribution whereas a too small σ leads to noise levels that are not different enough from the clean images. We show that taking σ from the intermediate regime works best for generalization both to other noise types and non-noise corruptions.

Link Between Adversarial Robustness and Common Corruptions. There is currently no agreement on whether adversarial training increases robustness against common corruptions in the literature. Hendrycks et al. report a robustness increase on common corruptions due to adversarial logit pairing on Tiny ImageNet-C [13]. Ford et al. suggest a link between adversarial robustness and robustness against common corruptions, claim that increasing one robustness type should simultaneously increase the other, but report mixed results on MNIST and CIFAR10-C [8]. Additionally, they also observe large drops in accuracy for adversarially trained networks and networks trained with Gaussian data augmentation compared to a vanilla classifier on certain corruptions. On the other hand, Engstrom et al. report that increasing robustness against adversarial ℓ_∞ attacks does not increase robustness against translations and rotations, but they do not present results on noise [7]. Kang et al. study robustness transfer between models trained against ℓ_1, ℓ_2, ℓ_∞ adversaries/elastic deformations and JPEG artifacts [17]. They observe that adversarial training increases robustness against elastic and JPEG corruptions on a 100-class subset of ImageNet. This result contradicts our findings on full ImageNet as we see a slight decline in accuracy on those two classes for the adversarially trained model from [42] and severe drops in accuracy on other corruptions. Jordan et al. show that adversarial robustness does not transfer easily between attack classes [16]. Tramèr et al. [40] also argue in favor of a trade-off between different robustness types. For a simple and natural classification task, they prove that adversarial robustness towards l_∞ perturbations does neither transfer to l_1 nor to input rotations and translations, and vice versa and support their formal analysis with experiments on MNIST and CIFAR10.

3 Methods

3.1 Training with Gaussian Noise

As discussed in Sect. 2, several researchers have tried using Gaussian noise as a method to increase robustness towards common corruptions with mixed results. In this work, we revisit the approach of Gaussian data augmentation and increase its efficacy. We treat the standard deviation σ of the distribution as a hyper-parameter of the training and measure its influence on robustness.

To formally introduce the objective, let $\mathcal{D}$ be the data distribution over input pairs $(\boldsymbol{x}, y)$ with $\boldsymbol{x} \in \mathbb{R}^N$ and $y \in \{1, \ldots, k\}$. We train a differentiable classifier $f_\theta(\boldsymbol{x})$ by minimizing the risk on a dataset with additive Gaussian noise

$$\mathop{\mathbb{E}}_{\boldsymbol{x},y\sim\mathcal{D}} \mathop{\mathbb{E}}_{\boldsymbol{\delta}\sim\mathcal{N}(\mathbf{0},\sigma^2\mathbb{1})} \left[\mathcal{L}_{\mathrm{CE}}\left(f_\theta(\mathrm{clip}(\boldsymbol{x}+\boldsymbol{\delta})), y\right)\right], \tag{1}$$

where σ is the standard deviation of the Gaussian noise and $\boldsymbol{x}+\boldsymbol{\delta}$ is clipped to the input range $[0,1]^N$. The standard deviation is either kept fixed or is chosen uniformly from a fixed set of standard deviations. In both cases, the possible standard deviations are chosen from a small set of nine values inspired by the noise variance in the ImageNet-C dataset (cf. Sect. 3.3). To maintain high accuracy on clean data, we only perturb 50% of the training data with Gaussian noise within each batch.

3.2 Adversarial Noise

Learning Adversarial Noise. Our goal is to find a noise distribution $p_\phi(\boldsymbol{\delta})$, $\boldsymbol{\delta} \in \mathbb{R}^N$ such that noise samples added to $\boldsymbol{x}$ maximally confuse the classifier f_θ. More concisely, we optimize

$$\max_{\phi} \mathop{\mathbb{E}}_{\boldsymbol{x},y\sim\mathcal{D}} \mathop{\mathbb{E}}_{\boldsymbol{\delta}\sim p_\phi(\boldsymbol{\delta})} \left[\mathcal{L}_{\mathrm{CE}}\left(f_\theta(\mathrm{clip}(\boldsymbol{x}+\boldsymbol{\delta})), y\right)\right], \tag{2}$$

where clip is an operator that clips all values to the valid interval (i.e. $\mathrm{clip}(\boldsymbol{x}+\boldsymbol{\delta}) \in [0,1]^N$) and restricts their norm $||\boldsymbol{\delta}||_2 = \epsilon$.[3]

We follow the literature of implicit generative models [4,28] as we do not have to explicitly model the probability density function $p_\phi(\boldsymbol{\delta})$ since optimizing Eq. (2) only involves samples drawn from $p_\phi(\boldsymbol{\delta})$. We model the samples from $p_\phi(\boldsymbol{\delta})$ as the output of a neural network $g_\phi : \mathbb{R}^N \rightarrow \mathbb{R}^N$ which gets its input from a normal distribution $\boldsymbol{\delta} = g_\phi(\boldsymbol{z})$ where $\boldsymbol{z} \sim \mathcal{N}(\mathbf{0}, \mathbb{1})$. We enforce the independence property of $p_\phi(\boldsymbol{\delta}) = \prod_n p_\phi(\delta_n)$ by constraining the network architecture of the noise generator g_ϕ to only consist of convolutions with 1×1 kernels. Lastly, the projection onto a sphere $||\boldsymbol{\delta}||_2 = \epsilon$ is achieved by scaling the generator output with a scalar while clipping $\boldsymbol{x}+\boldsymbol{\delta}$ to the valid range $[0,1]^N$. This fixed size projection (hyper-parameter) is motivated by the fact that Gaussian noise training with a single, fixed σ achieved the highest accuracy.[4]

The noise generator g_ϕ has four 1×1 convolutional layers with ReLU activations and one residual connection from input to output. The weights of the layers are initialized to small numbers; for this initialization, the input is passed through the residual connection to the output. Since we use Gaussian noise as input, the noise generator outputs Gaussian noise at initialization. During training, the weights change and the generator learns to produce more diverse distributions.

[3] We apply the method derived in [32] and rescale the perturbation by a factor γ to obtain the desired ℓ_2 norm; despite the clipping, the squared ℓ_2 norm is a piece-wise linear function of γ^2 that can be inverted to find the correct scaling factor γ.

[4] We also experimented with an adaptive sphere radius ϵ which grows with the classifier's accuracy. However, we did not see any improvements and followed Occam's razor.

Adversarial Noise Training. To increase robustness, we now train the classifier f_θ to minimize the risk under adversarial noise distributions jointly with the noise generator

$$\min_\theta \max_\phi \mathop{\mathbb{E}}_{\boldsymbol{x},y\sim\mathcal{D}} \mathop{\mathbb{E}}_{\boldsymbol{\delta}\sim p_\phi(\boldsymbol{\delta})} \left[\mathcal{L}_{\mathrm{CE}}\left(f_\theta(\mathrm{clip}(\boldsymbol{x}+\boldsymbol{\delta})), y\right)\right], \tag{3}$$

where again $\boldsymbol{x}+\boldsymbol{\delta} \in [0,1]^N$ and $||\boldsymbol{\delta}||_2 = \epsilon$. For a joint adversarial training, we alternate between an outer loop of classifier update steps and an inner loop of generator update steps. Note that in regular adversarial training, e.g. [21], $\boldsymbol{\delta}$ is optimized directly whereas we optimize a constrained distribution over $\boldsymbol{\delta}$.

To maintain high classification accuracy on clean samples, we sample every mini-batch so that they contain 50% clean data and perturb the rest. The current state of the noise generator is used to perturb 30% of this data and the remaining 20% are augmented with samples chosen randomly from previous distributions. For this, the noise generator states are saved at regular intervals. The latter method is inspired by experience replay from reinforcement learning [27] and is used to keep the classifier from forgetting previous adversarial noise patterns. To prevent the noise generator from being stuck in a local minimum, we halt the Adversarial Noise Training (ANT) at regular intervals and train a new noise generator from scratch. This noise generator is trained against the current state of the classifier to find a current optimum. The new noise generator replaces the former noise generator in the ANT. This technique has been crucial to train a robust classifier.

Learning Locally Correlated Adversarial Noise. We modify the architecture of the noise generator defined in Eq. 2 to allow for local spatial correlations and thereby enable the generator to learn more diverse distributions. Since we seek to increase model robustness towards image corruptions such as rain or snow that produce locally correlated patterns, it is natural to include local patterns in the manifold of learnable distributions. We replace the 1×1 kernels in one network layer with 3×3 kernels limiting the maximum correlation length of the output noise sample to 3×3 pixels. We indicate the correlation length of noise generator used for the constrained adversarial noise training as $\mathrm{ANT}^{1\mathrm{x}1}$ or $\mathrm{ANT}^{3\mathrm{x}3}$.

Combining Adversarial Noise Training with Stylization. As demonstrated by [9], using random stylization as data augmentation increases the accuracy on ImageNet-C due to a higher shape bias of the model. We combine our ANT and the stylization approach to achieve robustness gains from both in the following way: we split the samples in each batch into clean data (25%), stylized data (30%) and clean data perturbed by the noise generator (45%).

3.3 Evaluation on Corrupted Images

Evaluation of Noise Robustness. We evaluate the robustness of a model by sampling a Gaussian noise vector $\boldsymbol{\delta}$ (covariance $\mathbb{1}$). We then do a line search along the direction $\boldsymbol{\delta}$ starting from the original image $\boldsymbol{x}$ until it is misclassified. We denote the resulting minimal perturbation as $\boldsymbol{\delta}_{\min}$. The robustness of a model is then denoted by the median[5] over the test set

$$\epsilon^* = \underset{\boldsymbol{x},y\sim\mathcal{D}}{\text{median}}\, ||\boldsymbol{\delta}_{\min}||_2, \tag{4}$$

with $f_\theta(\boldsymbol{x} + \boldsymbol{\delta}_{\min}) \neq y$ and $\boldsymbol{x} + \boldsymbol{\delta}_{\min} \in [0,1]^N$. Note that a higher ϵ^* denotes a more robust classifier. To test the robustness against adversarial noise, we train a new noise generator at the end of the Adversarial Noise Training until convergence and evaluate it according to Eq. (4).

ImageNet-C. The ImageNet-C benchmark[6] [13] is a conglomerate of 15 diverse corruption types that were applied to the validation set of ImageNet. The corruptions are organized into four main categories: noise, blur, weather, and digital. The MNIST-C benchmark is created similarly to ImageNet-C with a slightly different set of corruptions [29]. We report the Top-1 and Top-5 accuracies as well as the 'mean Corruption Error' (mCE) on both benchmarks. We evaluate all proposed methods for ImageNet-C on the ResNet50 architecture for better comparability to previous methods, e.g. [9,19,45]. The clean ImageNet accuracy of the used architecture highly influences the results and could be seen as an upper bound for the accuracy on ImageNet-C. Note that our approach is independent of the used architecture and could be applied to any differentiable network.

4 Results

For our experiments on ImageNet, we use a classifier that was pretrained on ImageNet. For the experiments on MNIST, we use the architecture from [21] for comparability. All technical details, hyper-parameters and the architectures of the noise generators can be found in Appendix A–B. We use various open source software packages for our experiments, most notably Docker [24], scipy and numpy [41], PyTorch [31] and torchvision [23].

[5] Samples for which no ℓ_2-distance allows us to manipulate the classifier's decision contribute a value of ∞ to the median.

[6] For the evaluation, we use the JPEG compressed images from github.com/hendrycks/robustness as is advised by the authors to ensure reproducibility. We note that Ford et al. report a decrease in performance when the compressed JPEG files are used as opposed to applying the corruptions directly in memory without compression artifacts [8].

(In-)Effectiveness of Regular Adversarial Training to Increase Robustness Towards Common Corruptions. In our first experiment, we evaluate whether robustness against regular adversarial examples generalizes to robustness against common corruptions. We display the Top-1 accuracy of vanilla and adversarially trained models in Table 1; detailed results on individual corruptions can be found in Appendix C. For all tested models, we find that regular ℓ_∞ adversarial training can strongly decrease the robustness towards common corruptions, especially for the corruption types Fog and Contrast. Universal adversarial training [37], on the other hand, leads to severe drops on some corruptions but the overall accuracy on ImageNet-C is slightly increased relative to the vanilla baseline model (AlexNet). Nonetheless, the absolute ImageNet-C accuracy of 22.2% is still very low. These results disagree with two previous studies which reported that (1) adversarial logit pairing[7] (ALP) increases robustness against common corruptions on Tiny ImageNet-C [13], and that (2) adversarial training can increase robustness on CIFAR10-C [8].

Table 1. Top-1 accuracy on ImageNet-C and ImageNet-C without the noise category (higher is better). Regular adversarial training decreases robustness towards common corruptions; universal adversarial training seems to slightly increase it.

Model	IN-C	IN-C w/o noises
Vanilla RN50	39.2%	42.3%
Adv. training [36]	29.1%	32.0%
Vanilla RN152	45.0%	47.9%
Adv. training [42]	35.0%	35.9%
Vanilla AlexNet	21.1%	23.9%
Universal adv. training [37]	22.2%	23.1%

We evaluate adversarially trained models on MNIST-C and present the results and their discussion in Appendix E. The results on MNIST-C show the same tendency as on ImageNet-C: adversarially trained models have lower accuracy on MNIST-C and thus indicate that adversarial robustness does not transfer to robustness against common corruptions. This corroborates the results of Ford et al. [8] on MNIST who also found that an adversarially robust model had decreased robustness towards a set of common corruptions.

Effectiveness of Gaussian Data Augmentation to Increase Robustness Towards Common Corruptions. We fine-tune ResNet50 classifier pretrained on ImageNet with Gaussian data augmentation from the distribution $\mathcal{N}(\mathbf{0}, \sigma^2 \mathbb{1})$ and vary σ. We try two different settings: in one, we choose a single noise level σ

[7] Note that ALP was later found to not increase adversarial robustness [6].

while in the second, we sample σ uniformly from a set of multiple possible values. The Top-1 accuracy of the fine-tuned models on ImageNet-C in comparison to a vanilla trained model is shown in Fig. 2. Each black point shows the performance of one model fine-tuned with one specific σ; the vanilla trained model is marked by the point at $\sigma = 0$. The horizontal lines indicate that the model is fine-tuned with Gaussian noise where σ is sampled from a set for each image. For example, for the dark green line, as indicated by the stars, we sample σ from the set $\{0.08, 0.12, 0.18, 0.26, 0.38\}$ which corresponds to the Gaussian corruption of ImageNet-C. Since Gaussian noise is part of the test set, we show both the results on the full ImageNet-C evaluation set and the results on ImageNet-C without noises (namely blur, weather and digital). To show how the different σ-levels manifest themselves in an image, we include example images in Appendix G.

There are three important results evident from Fig. 2:

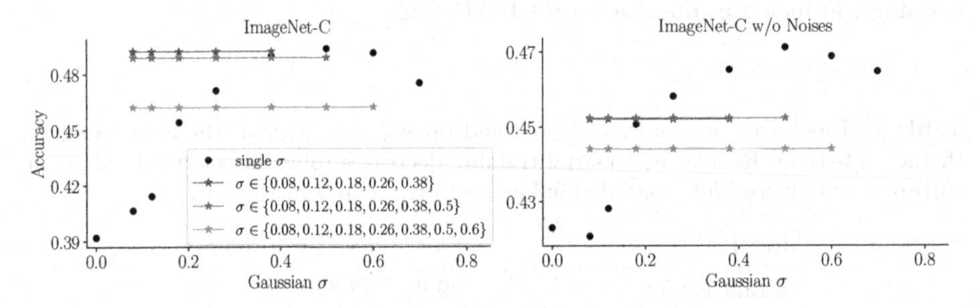

Fig. 2. Top-1 accuracy on ImageNet-C (left) and ImageNet-C without the noise corruptions (right) of a ResNet50 architecture fine-tuned with Gaussian data augmentation of varying σ. Each dot or green line represents one model. We train on Gaussian noise sampled from a distribution with a single σ (black dots) and on distributions where σ is sampled from different sets (green lines with stars). We also compare to a vanilla trained model at $\sigma = 0$. (Color figure online)

1. Gaussian noise generalizes well to the non-noise corruptions of the ImageNet-C dataset and is a powerful baseline. This is surprising as it was shown in several recent works that training on Gaussian or uniform noise does not generalize to other corruption types [10,19] or that the effect is weak [8].
2. The standard deviation σ is a crucial hyper-parameter and has an optimal value of about $\sigma = 0.5$ for ResNet50.
3. If σ is chosen well, using a single σ is enough and sampling from a set of σ values is detrimental for robustness against non-noise corruptions.

In the following Results sections, we will compare Gaussian data augmentation to our Adversarial Noise Training approach and baselines from the literature. For this, we will use the models with the overall best-performance: The model $GN_{0.5}$ that was trained with Gaussian data augmentation with a single $\sigma = 0.5$ and the model GN_{mult} where σ was sampled from the set $\{0.08, 0.12, 0.18, 0.26, 0.38\}$.

Evaluation of the Severity of Adversarial Noise as an Attack. In this section, we focus on the question: Can we learn the most severe uncorrelated additive noise distribution for a classifier? Following the success of simple uncorrelated Gaussian noise data augmentation (Sect. 4) and the ineffectiveness of regular adversarial training (Sect. 4) which allows for highly correlated patterns, we restrict our learned noise distribution to be sampled independently for each pixel.

To measure the effectiveness of our adversarial noise, we report the median perturbation size ϵ^* that is necessary for a misclassification for each image in the test set as defined in Sect. 3.3. We find $\epsilon^*_{\text{GN}} = 39.0$ for Gaussian noise, $\epsilon^*_{\text{UN}} = 39.1$ for uniform noise and $\epsilon^*_{\text{AN}} = 15.7$ for adversarial noise (see Fig. 1 for samples of each noise type). Thus, we see that our AN is much more effective at fooling the classifier compared to Gaussian and uniform noise.

Table 2. Accuracy on clean data and robustness of differently trained models as measured by the median perturbation size ϵ^*. A higher ϵ^* indicates a more robust model. We compute standard deviations for ϵ^*_{AN} for differently initialized generator networks. To provide an intuition for the perturbation sizes indicated by ϵ^*, we show example images for Gaussian noise below and a larger Figure for different noise types in Appendix I.

Model	Clean acc.	ϵ^*_{GN}	ϵ^*_{UN}	$\epsilon^*_{\text{AN1x1}}$
Vanilla RN50	**76.1%**	39.0	39.1	15.7 ± 0.6
GNT$\sigma_{0.5}$	75.9%	74.8	74.9	31.8 ± 3.9
GNT$_{\text{mult}}$	**76.1%**	130.1	130.7	24.0 ± 2.2
ANT$^{1\text{x}1}$	76.0%	**136.7**	**137.0**	**95.4 ± 5.7**

Evaluation of Adversarial Noise Training as a Defense. In the previous section, we established a method for learning the most adversarial noise distribution for a classifier. Now, we utilize it for a joint Adversarial Noise Training (ANT$^{1\text{x}1}$) where we simultaneously train the noise generator and classifier (see Sect. 3.2). This leads to substantially increased robustness against Gaussian, uniform and adversarial noise, see Table 2. The robustness of models that were trained via Gaussian data augmentation also increases, but on average much less compared to the model trained with ANT$^{1\text{x}1}$. To evaluate the robustness against adversarial noise, we train four noise generators with different random seeds and measure $\epsilon^*_{\text{AN1x1}}$. We report the mean value and the standard deviation over the four runs. To visualize this effect, we visualize the temporal evolution of the probability density function $p_\phi(\delta_n)$ of uncorrelated noise during $\text{ANT}^{1\text{x}1}$ in Fig. 3A. This shows that the generator converges to different distributions and therefore, the classifier has been trained against a rich variety of distributions.

Comparison of Different Methods to Increase Robustness Towards Common Corruptions. We now revisit common corruptions on ImageNet-C and compare the robustness of differently trained models. Since Gaussian noise is part of ImageNet-C, we train another baseline model with data augmentation using the Speckle noise corruption from the ImageNet-C holdout set. We later denote the cases where the corruptions present during training are part of the test set by putting corresponding accuracy values in brackets. Additionally, we compare our results with several baseline models from the literature:

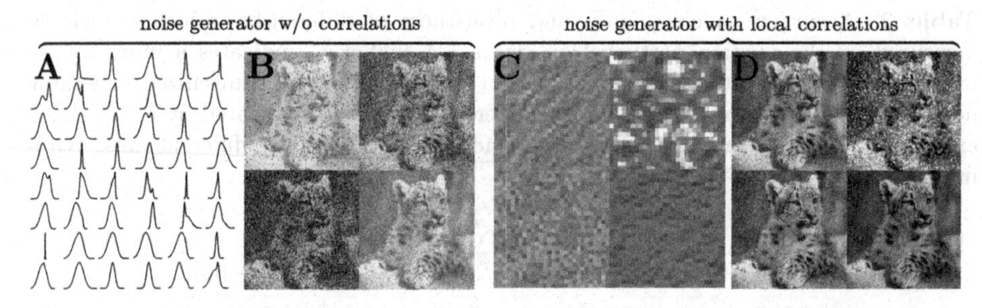

Fig. 3. A: Examples of learned probability densities over the grayscale version of the noise $\boldsymbol{\delta}_n$ during $\text{ANT}^{1\text{x}1}$ where each density corresponds to one local minimum; B: Example images with sampled uncorrelated adversarial noise; C: Example patches of locally correlated noise with a size of 28×28 pixels learned during $\text{ANT}^{3\text{x}3}$; D: Example images with sampled correlated adversarial noise.

1. Shift Inv: The model is modified to enhance shift-equivariance using anti-aliasing [45].[8]
2. Patch GN: The model was trained on Gaussian patches [19].[9]
3. SIN+IN: The model was trained on a stylized version of ImageNet [9].[10]
4. AugMix: [14] trained their model using diverse augmentations.[11] They use image augmentations from AutoAugment [3] and exclude contrast, color, brightness, sharpness, and Cutout operations to make sure that the test set of ImageNet-C is disjoint from the training set. We would like to highlight the difficulty in clearly distinguishing between the augmentations used during training and testing as there might be a certain overlap. This can be seen by the visual similarity between the Posterize operation and the JPEG corruption (see Appendix J).

[8] Weights were taken from github.com/adobe/antialiased-cnns.
[9] Since no model weights are released, we include the values reported in their paper.
[10] Weights were taken from github.com/rgeirhos/texture-vs-shape.
[11] Weights were taken from github.com/google-research/augmix.

The Top-1 accuracies on the full ImageNet-C dataset and ImageNet-C without the noise corruptions are displayed in Table 3; detailed results on individual corruptions in terms of accuracy and mCE are shown in Tables 3 and 4, Appendix D. We also calculate the accuracy on corruptions without the noise category since we observe that the generated noise can sometimes be close to the i.i.d. corruptions of ImageNet-C raising concerns about overfitting. Additionally, the expressiveness of the generated i.i.d. noise is quite limited compared to natural corruptions like 'snow'. We hence extend the $\text{ANT}^{1\text{x}1}$ procedure to include spatially correlated noise over 3×3 pixels. Samples are shown in Fig. 3C and Fig. 3D.

The results on full ImageNet-C are striking (see Table 3): a very simple baseline, namely a model trained with Speckle noise data augmentation, beats almost all previous baselines reaching an accuracy of 46.4% which is larger than the accuracy of SIN+IN (45.2%) and close to AugMix (48.3%). The $\text{GN}\sigma_{0.5}$ surpasses SIN+IN not only on the noise category but also on almost all other corruptions, see a more detailed breakdown in Table 3, Appendix D.

Table 3. Average accuracy on clean data, average Top-1 and Top-5 accuracies on ImageNet-C and ImageNet-C without the noise category (higher is better); all values in percent. We compare the results obtained by the means of Gaussian (GNT) and Speckle noise data augmentation and with Adversarial Noise Training (ANT) to several baselines. Gray numbers in brackets indicate scenarios where a corruption from the test set was used during training.

Model	IN	IN-C		IN-C w/o noises	
	Clean acc.	Top-1	Top-5	Top-1	Top-5
Vanilla RN50	76.1	39.2	59.3	42.3	63.2
Shift Inv [45]	77.0	41.4	61.8	44.2	65.1
Patch GN [19]	76.0	(43.6)	(n.a.)	43.7	n.a.
SIN+IN [9]	74.6	45.2	66.6	46.6	68.2
AugMix [14]	**77.5**	48.3	69.2	50.4	71.8
Speckle	75.8	46.4	67.6	44.5	65.5
GNT_{mult}	76.1	(49.2)	(70.2)	45.2	66.2
$\text{GNT}\sigma_{0.5}$	75.9	(49.4)	(70.6)	47.1	68.3
$\text{ANT}^{1\text{x}1}$	76.0	(51.1)	(72.2)	47.7	68.8
$\text{ANT}^{1\text{x}1}$+SIN	74.9	(52.2)	(73.6)	49.2	70.6
$\text{ANT}^{1\text{x}1}$ w/o EP	75.7	(48.9)	(70.2)	46.5	67.7
$\text{ANT}^{3\text{x}3}$	76.1	50.4	71.5	47.0	68.1
$\text{ANT}^{3\text{x}3}$+SIN	74.1	**52.6**	**74.4**	**50.6**	**72.5**

The ANT^{3x3}+SIN model produces the best results on ImageNet-C both with and without noises. Thus, it is slightly superior to Gaussian data augmentation and pure ANT^{3x3}. Comparing ANT^{1x1} and ANT^{3x3}, we observe that ANT^{3x3} performs better than ANT^{1x1} on the 'snow' corruption. We attribute this to the successful modeling capabilities of locally correlated patterns resembling snow of the 3×3 noise generator. We perform an ablation study to investigate the necessity of experience replay and note that we lose roughly 2% without it (ANT^{1x1} w/o EP vs ANT^{1x1}). We also test how the classifier's performance changes if it is trained against adversarial noise sampled randomly from $p_\phi(\delta_n)$. The accuracy on ImageNet-C decreases slightly compared to regular ANT^{1x1}: 51.1%/71.9% (Top-1/Top-5) on full ImageNet-C and 47.3%/68.3% (Top-1/Top-5) on ImageNet-C without the noise category. We include additional results for ANT^{1x1} with a DenseNet121 architecture [15] and for varying parameter counts of the noise generator in Appendix K.

For MNIST, we train a model with Gaussian data augmentation and via ANT^{1x1}. We achieve similar results with both approaches and report a new state-of-the-art accuracy on MNIST-C: 92.4%, see Appendix E for details.

Table 4. Adversarial robustness on ℓ_2 ($\epsilon = 0.12$) and ℓ_∞ ($\epsilon = 0.001$) compared to a Vanilla ResNet50 on ImageNet.

Model	Clean acc. [%]	ℓ_2 acc. [%]	ℓ_∞ acc.[%]
Vanilla RN50	**76.1**	41.1	18.1
GNT$\sigma_{0.5}$	75.9	49.0	28.1
ANT^{1x1}	76.0	50.1	28.6
Adv. training [36]	60.5	**58.1**	**58.5**

Robustness Towards Adversarial Perturbations. As regular adversarial training can decrease the accuracy on common corruptions, it is also interesting to check what happens vice-versa: How does a model which is robust on common corruptions behave under adversarial attacks?

Both our ANT^{1x1} and GNT models have slightly increased ℓ_2 and ℓ_∞ robustness scores compared to a vanilla trained model, see Table 4. We tested this using the white-box attacks PGD [20] and DDN [33]. Expectedly, an adversarially trained model has higher adversarial robustness compared to ANT^{1x1} or GNT. In this experiment, we only verify that we do not unintentionally reduce adversarial robustness compared to a vanilla ResNet50. For details, see Appendix E for MNIST and Appendix F for ImageNet.

5 Conclusions

So far, attempts to use simple noise augmentations for general robustness against common corruptions have produced mixed results, ranging from no generalization from one noise to other noise types [10] to only marginal robustness increases [8,19]. In this work, we demonstrate that carefully tuned additive noise patterns in conjunction with training on clean samples can surpass almost all current state-of-the-art defense methods against common corruptions. By drawing inspiration from adversarial training and experience replay, we additionally show that training against simple uncorrelated or locally correlated worst-case noise patterns outperforms our already strong baseline defense, with additional gains to be made in combination with previous defense methods like stylization [9].

There are still a few corruption types (e.g. Motion or Zoom blurs) on which our method is not state of the art, suggesting that additional gains are possible. Future extensions of this work may combine noise generators with varying correlation lengths, add additional interactions between noise and image (e.g. multiplicative interactions or local deformations) or take into account local image information in the noise generation process to further boost robustness across many types of image corruptions.

References

1. Azulay, A., Weiss, Y.: Why do deep convolutional networks generalize so poorly to small image transformations? (2018)
2. Campbell, M., Hoane Jr., A.J., Hsu, F.: Deep blue. Artif. Intell. **134**(1–2), 57–83 (2002). https://doi.org/10.1016/S0004-3702(01)00129-1
3. Cubuk, E.D., Zoph, B., Mane, D., Vasudevan, V., Le, Q.V.: AutoAugment: Learning augmentation policies from data. arXiv preprint arXiv:1805.09501 (2018)
4. Diggle, P.J., Gratton, R.J.: Monte Carlo methods of inference for implicit statistical models. J. Roy. Stat. Soc.: Ser. B (Methodol.) **46**(2), 193–212 (1984)
5. Dodge, S.F., Karam, L.J.: A study and comparison of human and deep learning recognition performance under visual distortions. CoRR abs/1705.02498 (2017). http://arxiv.org/abs/1705.02498
6. Engstrom, L., Ilyas, A., Athalye, A.: Evaluating and understanding the robustness of adversarial logit pairing. CoRR abs/1807.10272 (2018). https://arxiv.org/abs/1807.10272
7. Engstrom, L., Tsipras, D., Schmidt, L., Madry, A.: A rotation and a translation suffice: fooling CNNs with simple transformations. In: ICML (2019)
8. Ford, N., Gilmer, J., Carlini, N., Cubuk, D.: Adversarial examples are a natural consequence of test error in noise. In: ICML (2019)
9. Geirhos, R., Rubisch, P., Michaelis, C., Bethge, M., Wichmann, F.A., Brendel, W.: ImageNet-trained CNNs are biased towards texture; increasing shape bias improves accuracy and robustness. In: International Conference on Learning Representations (2019). https://openreview.net/forum?id=Bygh9j09KX
10. Geirhos, R., Temme, C.R.M., Rauber, J., Schütt, H.H., Bethge, M., Wichmann, F.A.: Generalisation in humans and deep neural networks. In: Bengio, S., Wallach, H., Larochelle, H., Grauman, K., Cesa-Bianchi, N., Garnett, R. (eds.) Advances in Neural Information Processing Systems, vol. 31, pp. 7538–7550. Curran Associates, Inc. (2018). http://papers.nips.cc/paper/7982-generalisation-in-humans-and-deep-neural-networks.pdf

11. Gilmer, J., et al.: Adversarial spheres. CoRR abs/1801.02774 (2018). http://arxiv.org/abs/1801.02774
12. He, K., Zhang, X., Ren, S., Sun, J.: Deep residual learning for image recognition. In: Proceedings of the IEEE Conference on Computer Vision and Pattern Recognition, pp. 770–778 (2016)
13. Hendrycks, D., Dietterich, T.: Benchmarking neural network robustness to common corruptions and perturbations. In: International Conference on Learning Representations (2019). https://openreview.net/forum?id=HJz6tiCqYm
14. Hendrycks, D., Mu, N., Cubuk, E.D., Zoph, B., Gilmer, J., Lakshminarayanan, B.: AugMix: a simple data processing method to improve robustness and uncertainty. In: International Conference on Learning Representations (2020). https://openreview.net/forum?id=S1gmrxHFvB
15. Huang, G., Liu, Z., Weinberger, K.Q.: Densely connected convolutional networks. In: CVPR (2017)
16. Jordan, M., Manoj, N., Goel, S., Dimakis, A.G.: Quantifying perceptual distortion of adversarial examples. arXiv preprint arXiv:1902.08265 (2019)
17. Kang, D., Sun, Y., Brown, T., Hendrycks, D., Steinhardt, J.: Transfer of adversarial robustness between perturbation types. CoRR abs/1905.01034 (2019). http://arxiv.org/abs/1905.01034
18. Lee, J., Won, T., Hong, K.: Compounding the performance improvements of assembled techniques in a convolutional neural network. arXiv preprint arXiv:2001.06268 (2020)
19. Lopes, R.G., Yin, D., Poole, B., Gilmer, J., Cubuk, E.D.: Improving robustness without sacrificing accuracy with patch Gaussian augmentation. CoRR abs/1906.02611 (2019). http://arxiv.org/abs/1906.02611
20. Madry, A., Makelov, A., Schmidt, L., Tsipras, D., Vladu, A.: Towards deep learning models resistant to adversarial attacks. arXiv preprint arXiv:1706.06083 (2017)
21. Madry, A., Makelov, A., Schmidt, L., Tsipras, D., Vladu, A.: Towards deep learning models resistant to adversarial attacks. In: International Conference on Learning Representations (2018). https://openreview.net/forum?id=rJzIBfZAb
22. Mahajan, D., et al.: Exploring the limits of weakly supervised pretraining. In: Ferrari, V., Hebert, M., Sminchisescu, C., Weiss, Y. (eds.) ECCV 2018. LNCS, vol. 11206, pp. 185–201. Springer, Cham (2018). https://doi.org/10.1007/978-3-030-01216-8_12
23. Marcel, S., Rodriguez, Y.: Torchvision the machine-vision package of torch. In: ACM International Conference on Multimedia (2010)
24. Merkel, D.: Docker: lightweight Linux containers for consistent development and deployment. Linux J. **2014**(239), 2 (2014)
25. Michaelis, C., et al.: Benchmarking robustness in object detection: Autonomous driving when winter is coming. arXiv preprint arXiv:1907.07484 (2019)
26. Mikołajczyk, A., Grochowski, M.: Data augmentation for improving deep learning in image classification problem. In: 2018 International Interdisciplinary PhD Workshop (IIPhDW), pp. 117–122 (2018)
27. Mnih, V., et al.: Human-level control through deep reinforcement learning. Nature **518**(7540), 529 (2015)
28. Mohamed, S., Lakshminarayanan, B.: Learning in implicit generative models. arXiv preprint arXiv:1610.03483 (2016)
29. Mu, N., Gilmer, J.: MNIST-C: A robustness benchmark for computer vision. arXiv preprint arXiv:1906.02337 (2019)
30. OpenAI: OpenAI Five. https://blog.openai.com/openai-five/ (2018)

31. Paszke, A., et al.: Automatic differentiation in PyTorch. In: NIPS Autodiff Workshop (2017)
32. Rauber, J., Bethge, M.: Fast differentiable clipping-aware normalization and rescaling. arXiv preprint arXiv:2007.07677 (2020). https://github.com/jonasrauber/clipping-aware-rescaling
33. Rony, J., Hafemann, L.G., Oliveira, L.S., Ayed, I.B., Sabourin, R., Granger, E.: Decoupling direction and norm for efficient gradient-based L2 adversarial attacks and defenses. In: Proceedings of the IEEE Conference on Computer Vision and Pattern Recognition, pp. 4322–4330 (2019)
34. Russakovsky, O., et al.: ImageNet large scale visual recognition challenge. CoRR abs/1409.0575 (2014). http://arxiv.org/abs/1409.0575
35. Schott, L., Rauber, J., Bethge, M., Brendel, W.: Towards the first adversarially robust neural network model on MNIST. In: International Conference on Learning Representations (2019). https://openreview.net/forum?id=S1EHOsC9tX
36. Shafahi, A., et al.: Adversarial training for free! arXiv preprint arXiv:1904.12843 (2019)
37. Shafahi, A., Najibi, M., Xu, Z., Dickerson, J.P., Davis, L.S., Goldstein, T.: Universal adversarial training. CoRR abs/1811.11304 (2018). http://arxiv.org/abs/1811.11304
38. Silver, D., et al.: Mastering the game of go without human knowledge. Nature **550**, 354–359 (2017)
39. Szegedy, C., et al.: Intriguing properties of neural networks. arXiv preprint arXiv:1312.6199 (2013)
40. Tramèr, F., Boneh, D.: Adversarial training and robustness for multiple perturbations. In: NeurIPS (2019). http://arxiv.org/abs/1904.13000
41. Virtanen, P., et al.: SciPy 1.0: fundamental algorithms for scientific computing in Python. Nat. Meth. **17**, 261–272 (2020). https://doi.org/10.1038/s41592-019-0686-2
42. Xie, C., Wu, Y., van der Maaten, L., Yuille, A.L., He, K.: Feature denoising for improving adversarial robustness. In: CVPR (2019)
43. Xie, Q., Hovy, E., Luong, M.T., Le, Q.V.: Self-training with noisy student improves ImageNet classification. arXiv preprint arXiv:1911.04252 (2019)
44. Xiong, W., et al.: Achieving human parity in conversational speech recognition. In: IEEE/ACM Transactions on Audio, Speech, and Language Processing (2016)
45. Zhang, R.: Making convolutional networks shift-invariant again. In: ICML (2019)

SoftPoolNet: Shape Descriptor for Point Cloud Completion and Classification

Yida Wang[1(✉)], David Joseph Tan[2], Nassir Navab[1], and Federico Tombari[1,2]

[1] Technische Universität München, München, Germany
yida.wang@tum.de
[2] Google Inc., Menlo Park, USA

Abstract. Point clouds are often the default choice for many applications as they exhibit more flexibility and efficiency than volumetric data. Nevertheless, their unorganized nature – points are stored in an unordered way – makes them less suited to be processed by deep learning pipelines. In this paper, we propose a method for 3D object completion and classification based on point clouds. We introduce a new way of organizing the extracted features based on their activations, which we name soft pooling. For the decoder stage, we propose regional convolutions, a novel operator aimed at maximizing the global activation entropy. Furthermore, inspired by the local refining procedure in Point Completion Network (PCN), we also propose a patch-deforming operation to simulate deconvolutional operations for point clouds. This paper proves that our regional activation can be incorporated in many point cloud architectures like AtlasNet and PCN, leading to better performance for geometric completion. We evaluate our approach on different 3D tasks such as object completion and classification, achieving state-of-the-art accuracy.

1 Introduction

Point clouds are unorganized sparse representations of a 3D point set. Compared to other common representations for 3D data such as 3D meshes and voxel maps, they are simple and flexible, while being able to store fine details of a surface. For this reason, they are frequently employed for many applications within 3D perception and 3D computer vision such as robotic manipulation and navigation, scene understanding, and augmented/virtual reality. Recently, deep learning approaches have been proposed to learn from point clouds for 3D perception tasks such as point cloud classification [4,14,18,19] or point cloud segmentation [11,13,14,17,23]. Among them, one of the key breakthroughs in handling unorganized point clouds was proposed by PointNet [18], introducing the idea of a max pooling in the feature space to yield permutation invariance.

Electronic supplementary material The online version of this chapter (https://doi.org/10.1007/978-3-030-58580-8_5) contains supplementary material, which is available to authorized users.

A. Vedaldi et al. (Eds.): ECCV 2020, LNCS 12348, pp. 70–85, 2020.
https://doi.org/10.1007/978-3-030-58580-8_5

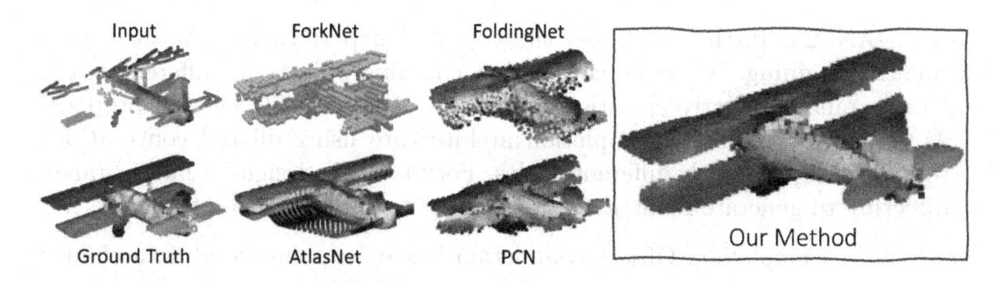

Fig. 1. This paper proposes a method that reconstructs 3D point cloud models with more fine details.

An interesting emerging research trend focusing on 3D data is the so-called 3D completion, where the geometry of a partial scene or object acquired from a single viewpoint, *e.g.* through a depth map, is completed of the missing part due to (self-)occlusion as visualized in Fig. 1. This can be of great use to aid standard 3D perception tasks such as object modeling, scene registration, part-based segmentation and object pose estimation. Most approaches targeting 3D completion have been proposed for volumetric approaches, since 3D convolutions are naturally suited to this 3D representation. Nevertheless, such approaches bring in the limitations of this representation, including loss of fine details due to discretization and limitations in scaling with the 3D size. Recently, a few approaches have explored the possibility of learning to complete a point cloud [9,25,26].

This paper proposes an encoder-decoder architecture called SoftPoolNet, which can be employed for any task that processes a point cloud as input in order to regress another point cloud as output. One of the tasks and a main focus for this work is 3D object completion from a partial point cloud.

The theoretical contribution of SoftPoolNet is twofold. We first introduce soft pooling, a new module that replaces the max-pooling operator in PointNet by taking into account multiple high-scoring features rather than just one. The intuition is that, by keeping multiple features with high activations rather than just the highest, we can retain more information while keeping the required permutation invariance. A second contribution is the definition of a regional convolution operator that is used within the proposed decoder architecture. This operator is designed specifically for point cloud completion and relies on convolving local features to improve the completion task with fine details.

In addition to evaluating SoftPoolNet for point cloud completion, we also evaluate on the point cloud classification to demonstrate its applicability to general point cloud processing tasks. In both evaluations, SoftPoolNet obtains state of the art results on the standard benchmarks.

2 Related Work

Volumetric Completion. Object [7] and scene completion [20,22] are typically carried out by placing all observed elements into a 3D grid with fixed resolution. 3D-EPN [7] completes a single object using 3D convolutions while

3D-RecGAN [24] further improves the completion performance by using discriminative training. As scene completion contains objects in different scales and more random relative position among all of them, SSCNet [20] proposes a 3D volumetric semantic completion architecture using dilated convolutions to recognize objects with different scales. ForkNet [22] designs a multi-branch architecture to generate realistic training data to supplement the training.

Point Cloud Completion. Object completion based on point cloud data change partial geometries without using a 3D fixed grid. They represent completed shapes as a set of points with 3D coordinates. For instance, FoldingNet [25] deforms a 2D grid from a global feature such as PointNet [18] feature to an output with a desirable shape. AtlasNet [9] generates an object with a set of local patches to simulate mesh data. But overlaps between different local patches makes the reconstruction noisy. MAP-VAE [10] predicts the completed shape by joining the observed part with the estimated counterpart.

CNNs for Point Clouds. Existing works like PointConv [23] and PointCNN [13] index each point with k-nearest neighbour search to find local patches, where they then apply the convolution kernels on those local patches. Regarding point cloud deconvolutional operations, FoldingNet [25] uses a 2D grid to help generate a 3D point cloud from a single feature. PCN [26] further uses local FoldingNet to obtain a fine-grained output from a coarse point cloud with low resolution which could be regarded as an alternative to point cloud deconvolution.

3 Soft Pooling for Point Features

Given the partial scan of an object, the input to our network is a point cloud with N_{in} points written in the matrix form as $\mathbf{P}_{\text{in}} = [\mathbf{x}_i]_{i=1}^{N_{\text{in}}}$ where each point is represented as the 3D coordinates $\mathbf{x}_i = [x_i, y_i, z_i]$. We then convert each point into a feature vector $\mathbf{f}_i$ with N_f elements by projecting every point with a point-wise multi-layer perceptron [18] (MLP) $\mathbf{W}_{\text{point}}$ with three layers. Thus, similar to $\mathbf{P}_{\text{in}}$, we define the $N_{\text{in}} \times N_f$ feature matrix as $\mathbf{F} = [\mathbf{f}_i]_{i=1}^{N_{\text{in}}}$. Note that we applied a softmax function to the output neuron of MLP so that the elements in $\mathbf{f}_i$ ranges between 0 and 1.

The main challenge when processing a point cloud is its unstructured arrangement. This implies that changing the order of the points in $\mathbf{P}_{\text{in}}$ describes the same point cloud, but generates a different feature matrix that flows into our architecture with convolutional operators. To solve this problem, we propose to organize the feature vectors in $\mathbf{F}$ so that their k-th element are sorted in a descending order, which is denoted as $\mathbf{F}'_k$. Note that k should not be larger than N_f. A *toy example* of this process is depicted in Fig. 2(a) where we assume that there are only five points in the point cloud and arrange the five feature vectors from $\mathbf{F} = [\mathbf{f}_i]_{i=1}^{5}$ to $\mathbf{F}'_k = [\mathbf{f}_i]_{i=\{3,5,1,2,4\}}$ by comparing the k-th element of each vector. Repeating this process for all the N_f elements in $\mathbf{f}_i$, all $\mathbf{F}'_k$ together result to a 3D tensor $\mathbf{F}' = [\mathbf{F}'_1, \mathbf{F}'_2, \dots \mathbf{F}'_{N_f}]$ with the dimension of $N_{\text{in}} \times N_f \times N_f$. As a result, any permutation of the points in $\mathbf{P}_{\text{in}}$ generate the same $\mathbf{F}'$.

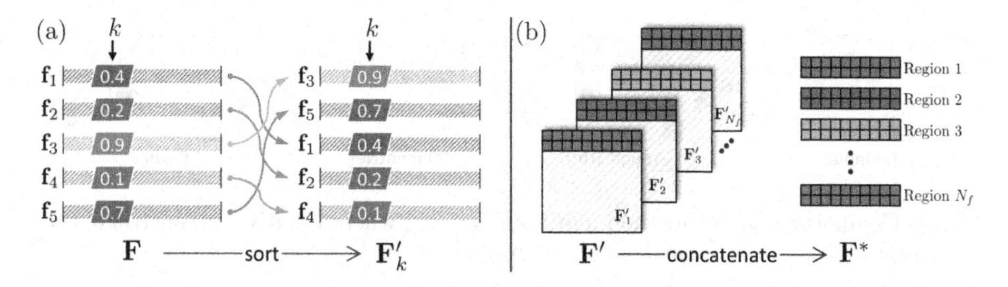

Fig. 2. Toy examples of (a) sorting the the k-th element of the vectors in the feature matrix $\mathbf{F}$ to build $\mathbf{F}'_k$ and consequently $\mathbf{F}'$ and (b) concatenation of the first N_r rows of $\mathbf{F}'_k$ to construct the 2D matrix $\mathbf{F}^*$ which corresponds to the regions with high activations.

Sorting the feature vectors in a descending order highlights the ones with the highest activation values. Thus, by selecting the first N_r feature vectors from all the $\mathbf{F}'_k$ as shown in Fig. 2(b), we assemble $\mathbf{F}^*$ that accumulates the features with the highest activations. Altogether, the output of soft pooling is the $N_f \cdot N_r$ point features. Since each feature vector corresponds to a point in $\mathbf{P}_{\text{in}}$, we can interpret the first N_r feature vectors as a region in the point cloud. The effects of the activations on the 3D reconstruction are illustrated in Fig. 3, where the point cloud is divided into N_f regions. Later in Sect. 6, we discuss on how to learn $\mathbf{W}_{\text{point}}$ by incorporating these regions. That section introduces several loss functions which optimize towards entropy, Chamfer distance and earth-moving distance such that each point is optimized to fall into only one region and to be selected for $\mathbf{F}^*$ by maximizing the k-th element of the feature vector associated to the same region.

Similar to PointNet [18], we also rely on MLP to build the feature matrix $\mathbf{F}$. However, PointNet directly applies max-pooling on $\mathbf{F}$ to produce a vector while we try to generalize this approach and sort the feature vectors in order to assemble a matrix $\mathbf{F}^*$ as illustrated in Fig. 2. Considering the distinction between the two approaches, we refer our approach as *soft pooling*. Fundamentally, in addition to the increased amount of information from our feature vectors, the advantage of our method is the ability to apply regional convolutional operations to $\mathbf{F}^*$, as discussed in Sect. 4. The differences are evident in Fig. 4, where the

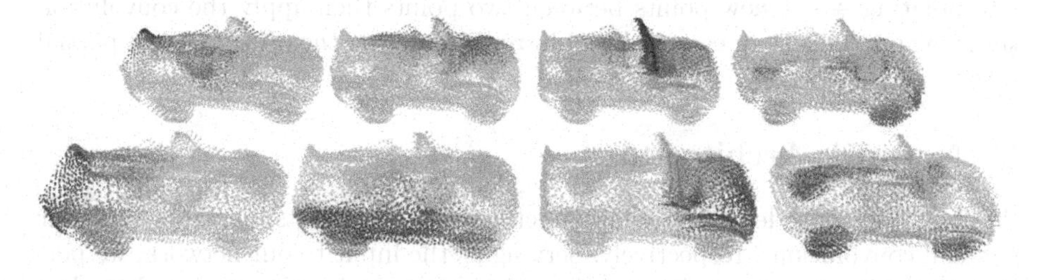

Fig. 3. Deconstructing the learned regions (unsupervised) that correspond to different parts of the car.

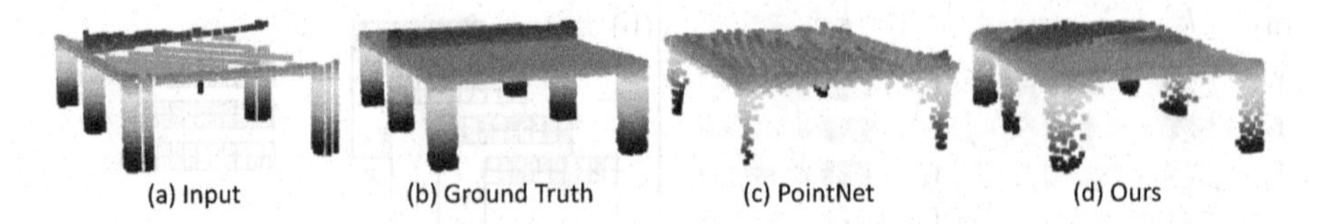

Fig. 4. Comparison of our method and PointNet [18] where PointNet reconstructs the more typical four-leg table instead of six in (c).

proposed method achieves detailed results on reconstructing all the six legs while PointNet follows the more generic structure of the table with four. This proves that *soft pooling* makes our decoder able to take all observable geometries into account to complete the shape, while the max-pooled PointNet feature cannot reveal the rarely seen geometry.

4 Regional Convolution

Operating on $\mathbf{F}^*$, we introduce the convolutional kernel $\mathbf{W}_{\text{conv}}$ that transforms $\mathbf{F}^*$ to a new set of points $\mathbf{P}_{\text{conv}}$ by taking several point features into consideration. We structure $\mathbf{W}_{\text{conv}}$ with a dimension of $N_p \times N_f \times 3$ where N_p represent the number of points which are taken into consideration such that

$$\mathbf{P}_{\text{conv}}(i,j) = \sum_{l=1}^{N_f} \sum_{k=1}^{N_p} \mathbf{F}^*(i+k,l)\mathbf{W}_{\text{conv}}(j,k,l) \ . \tag{1}$$

Here, the kernel slides only inside each region of features without taking features from two different regions in one convolutional operation. As the kernel size allows it to cover N_p features, we pad each region with $N_p - 1$ duplicated samples at the end of each region in order to keep the output resolution the same as N_{in}. Experimentally, we tried different numbers of N_p ranging from 4 to 64 and evaluated that 32 generates the best results. Learning the values in $\mathbf{W}_{\text{conv}}$ is discussed in Sect. 6.

In addition, we use a convolution stride which is set as a value smaller than N_p to change the output resolution in terms of the number of point features. With a stride of S, we then take samples every S point feature in $\mathbf{F}^*$. Notice that, by using a stride which is smaller than 1, we can also upsample $\mathbf{F}^*$ by interpolating $\frac{1}{S} - 1$ new points between two points then apply the convolution kernel again. This is an essential tool in reconstructing the object from a partial scan.

5 Network Architecture

We build an encoder-decoder architecture which consists of MLP and our regional convolutions, respectively. Serving as the input to our network, we permutate the input scans and resample 1,024 points. If the partial scans have less than 1,024 points, we then duplicate the missing samples.

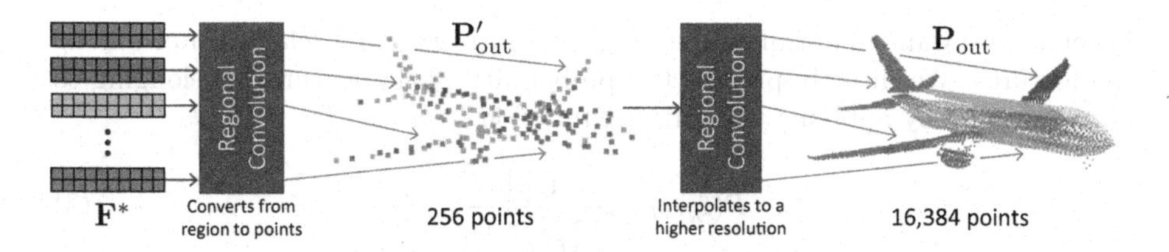

Fig. 5. Decoder architecture of SoftPoolNet with two regional convolution that converts the features from the regions to point clouds and interpolates from the coarse 256 points to a higher resolution with 16,384 points.

Our encoder is a point-wise MLP that generates the output neuron with a dimension of $[512, 512, 8]$. We then perform soft pooling as described in Sect. 3 that produces $\mathbf{F}^*$ with the size of $[256, 8]$ by setting N_r to 32 and N_f to 8, resulting an output of $N_f \cdot N_r = 256$ features.

Finally, for the decoder, we propose a two-stage point cloud completion architecture which is trained end-to-end. The output of the first is used as the input of the second point cloud completion network. Both of them produces the completed point cloud but with different resolutions. Illustrated in Fig. 5, we construct the decoder with two regional convolutions from Sect. 4. The first output $\mathbf{P}'_{\text{out}}$ is fixed at 256 while the second $\mathbf{P}_{\text{out}}$ produces a maximum resolution of 16,384.

6 Loss Functions

During learning, we evaluate whether the predicted point feature $\mathbf{P}_{\text{out}}$ matches the given ground truth $\mathbf{P}_{\text{gt}}$ through the Chamfer distance. Similar to [9,25,26], we use the regression loss function for the shape completion from a point cloud

$$\mathcal{L}_{\text{complete}}(\mathbf{W}_{\text{point}}, \mathbf{W}_{\text{conv}}) = \text{Chamfer}(\mathbf{P}_{\text{out}}, \mathbf{P}_{\text{gt}}) \ . \tag{2}$$

We observed that there are two major drawbacks in using this loss function alone – the reconstructed surface tends to be either curved on the sharp edges such FoldingNet [25] or having noisy points appear on flat surfaces such as AtlasNet [9] and PCN [26]. In this work, we tackle these problems by finding local regions first, then by optimizing the inter- and intra-regional relationships.

Moreover, while FoldingNet [25] sacrifices local details to present the entire model with a single mesh having smooth surface, AtlasNet [9] and PCN [26] use local regions (or patches) to increase the details in the 3D model. However, both of them [9,26] have severe overlapping effects between adjacent regions which makes the generated object noisy and the regions discontinuous. To solve this problem, we aim at reducing the overlaps between two adjacent regions.

6.1 Learning Activations Through Regional Entropy

Considering that the dimension of a single feature is N_f, we can directly define N_f regions for all features. Given the probabilities of regions to which the feature

$\mathbf{f}_i$ belong, we want to optimize the inter- and intra-regional relationships among the features. We directly present the probability of the feature $\mathbf{f}_i$ belonging to all N_f regions by applying the softmax function to $\mathbf{f}_i$ as

$$P(\mathbf{f}_k, i) = \frac{\mathbf{f}_k[i]}{\sum_{j=1}^{N_f} \mathbf{f}_k[j]} . \tag{3}$$

Since the information entropy evaluates both the distribution and the confidence of the probabilities of a set of data, we define the feature entropy and the regional entropy based on the regional probability of the feature.

The goal of the inter-regional loss function is to similarly distribute the number of points throughout the regions. We define the regional entropy as

$$\mathcal{E}_r = -\frac{1}{B}\sum_{j=1}^{B}\sum_{i=1}^{R}\left[\left(\frac{1}{N}\sum_{k=1}^{N} P(\mathbf{f}_k, i)\right) \cdot \log\left(\frac{1}{N}\sum_{k=1}^{N} P(\mathbf{f}_k, i)\right)\right] \tag{4}$$

where B is the batch-size. Here, we want to maximize $\mathcal{E}_r$. Considering that the upper-bound of $\mathcal{E}_r$ is $-\log\frac{1}{R} = \log(R)$, we can then define the inter-regional loss function as

$$\mathcal{L}_{\text{inter}}(\mathbf{W}_{\text{point}}) = \log(R) - \mathcal{E}_r \tag{5}$$

in order to acquire a positive loss function. Once $\mathcal{E}_r$ is close to $\log(R)$, each region would contain similar amount of point features. Interestingly, we can select the number of regions by evaluating how much the regional entropy $\mathcal{E}_r$ differs from its upper-bound. The best number of regions should be the one with a small $\mathcal{L}_{\text{inter}}$. This is evaluated later in Table 6.

On the other hand, the goal of the intra-regional loss function is to boost the confidence of each feature to be in a single region. The intra-regional loss function then minimize the feature entropy

$$\mathcal{L}_{\text{intra}}(\mathbf{W}_{\text{point}}) = -\frac{1}{N}\frac{1}{B}\sum_{k=1}^{N}\sum_{j=1}^{B}\sum_{i=1}^{N_f} P(\mathbf{f}_k, i)\log P(\mathbf{f}_k, i) . \tag{6}$$

The optimum case of the feature entropy is for each feature to be a one-hot code, *i.e.* when only one element is 1 while the others are zero.

6.2 Reducing the Overlapping Regions

Although $\mathcal{L}_{\text{intra}}$ tries to make each point feature confident about the region to which it belongs, instances exist where many adjacent points would fall under different regions. For example, we observe in Fig. 6 that patches from different regions are stacked on top of each other, producing noisy reconstructions. Notably, this introduces unexpected results when fitting a mesh to the point cloud. Thus, we want to minimize region overlap by optimizing the network to restrict the connection between adjacent regions to their boundaries.

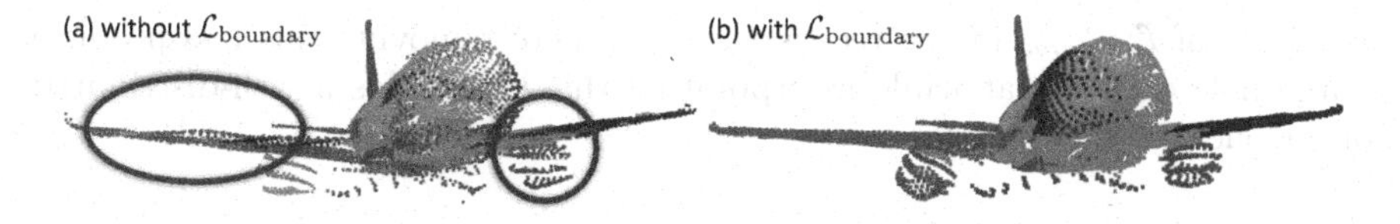

Fig. 6. Effects of without and with $\mathcal{L}_{\text{boundary}}$ where the wings are not planar and the engines are less visible in (a). Note that the colors represent different regions.

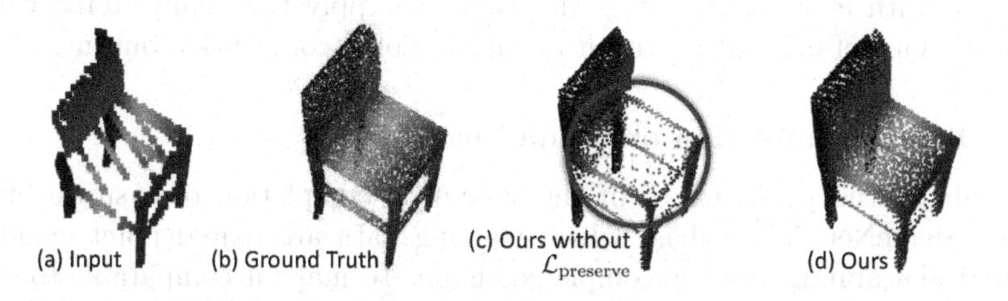

Fig. 7. Effects of without and with $\mathcal{L}_{\text{preserve}}$ where the seat is missing in (c).

First, each point is assigned to a region with the highest activation. All points that belong to region i but has activation for region j larger than a threshold τ are included in the set $\mathcal{B}_i^j$. Inversely, the points that belong to region j but have activation for region i larger than τ are added in the set $\mathcal{B}_j^i$. Note that, if both sets $\mathcal{B}_i^j$ and $\mathcal{B}_j^i$ are not empty, the regions i and j are then adjacent. Thus, by minimizing the Chamfer distance between $\mathcal{B}_i^j$ and $\mathcal{B}_j^i$, we can make the overlapping sets of points smaller such that the optimal result is a line. We then define the loss function for the boundary as

$$\mathcal{L}_{\text{boundary}}(\mathbf{W}_{\text{point}}, \mathbf{W}_{\text{conv}}) = \sum_{i=1}^{N_f} \sum_{j=i}^{N_f} \text{Chamfer}(\mathcal{B}_i^j, \mathcal{B}_j^i) \tag{7}$$

where both $\mathbf{W}_{\text{point}}$ and $\mathbf{W}_{\text{conv}}$ are optimized. After experimenting on different values of τ from 0.1 to 0.9, we set τ to be 0.3.

6.3 Preserving the Features from MLP

After sorting and filtering the features to produce $\mathbf{F}^*$, some feature vectors in $\mathbf{F}^*$ are duplicated while some vectors from $\mathbf{F}$ are missing in $\mathbf{F}^*$. To avoid these, we introduce the loss function

$$\mathcal{L}_{\text{preserve}}(\mathbf{W}_{\text{point}}) = \text{Earth-moving}(\mathbf{F}^*, \mathbf{F}) \ . \tag{8}$$

Since the earth moving distance [12] is not efficient when the size of the samples is large, we then randomly select 256 vectors from $\mathbf{F}$ and $\mathbf{F}^*$. Considering that the feature dimensions in $\mathbf{F}$ and $\mathbf{F}^*$ are both N_f, the earth moving distance then takes features with N_f dimension as input. In practice, Fig. 7 visualizes the effects of $\mathcal{L}_{\text{preserve}}$ in the reconstruction, where removing this loss produce a large hole on the seat while incorporating this loss builds a well-distributed point cloud.

7 Experiments

For all evaluations, we train our model with an NVIDIA Titan V and parameterize it with a batch size of 8. Moreover, we apply the Leaky ReLU with a negative slope of 0.2 on the output of each regional convolution output.

7.1 Object Completion on ShapeNet

We evaluate the performance of the geometric completion of a single object on the ShapeNet [5] database where training data are paired point clouds of the partial scanning and the completed shape. To make it comparable to other approaches, we adopt the standard 8 category evaluation [26] for a single object completion. As rotation errors are common in the partial scans, we further evaluate our approach against other works on the ShapeNet database with rotations. We also evaluate the performance on both high and low resolutions which contain 16,384 and 2,048 points, respectively.

We compare against other point cloud completion approaches such as PCN [26], FoldingNet [25], AtlasNet [9] and PointNet++ [19]. To show the advantages over volumetric completion, we also compare against 3D-EPN [24] and ForkNet [22] with an output resolution of $64 \times 64 \times 64$. Notably, we achieve the best results on most objects and in all types of evaluations as presented in Table 1, Table 2 and Table 3.

An interesting hypothesis is the capacity of $\mathcal{L}_{\text{boundary}}$ to be integrated in other existing approaches. Thus, Table 1 and Table 2 also evaluate this hypothesis and prove that this activation helps FoldingNet [25], PCN [26] and AtlasNet [9] perform better. Nevertheless, even with such improvements, the complete version of the proposed method still outperforms them.

Table 1. Completion evaluated by means of the Chamfer distance (multiplied by 10^3) with the output resolution of 16,384.

Output resolution = 16,384

Method	Plane	Cabinet	Car	Chair	Lamp	Sofa	Table	Vessel	*Avg.*
3D-EPN [7]	13.16	21.80	20.31	18.81	25.75	21.09	21.72	18.54	20.15
ForkNet [22]	9.08	14.22	11.65	12.18	17.24	14.22	11.51	12.66	12.85
PointNet++ [19]	10.30	14.74	12.19	15.78	17.62	16.18	11.68	13.52	14.00
FoldingNet [25]	5.97	10.80	9.27	11.25	12.17	11.63	9.45	10.03	10.07
FoldingNet + $\mathcal{L}_{\text{boundary}}$	5.79	10.61	8.62	10.33	11.56	11.05	9.41	9.79	9.65
PCN [26]	5.50	10.63	8.70	11.00	11.34	11.68	8.59	9.67	9.64
PCN + $\mathcal{L}_{\text{boundary}}$	5.13	9.12	7.58	9.35	9.40	9.31	7.30	8.91	8.26
Our method	**4.01**	**6.23**	**5.94**	**6.81**	**7.03**	**6.99**	**4.84**	**5.70**	**5.94**

Table 2. Completion evaluated using the Chamfer distance (multiplied by 10^3) with the output resolution of 2,048.

Output resolution = 2,048

Method	Plane	Cabinet	Car	Chair	Lamp	Sofa	Table	Vessel	*Avg.*
FoldingNet [25]	11.18	20.15	13.25	21.48	18.19	19.09	17.80	10.69	16.48
FoldingNet + $\mathcal{L}_{\text{boundary}}$	11.09	19.95	13.11	21.27	18.22	19.06	17.62	10.10	16.30
AtlasNet [9]	10.37	23.40	13.41	24.16	20.24	20.82	17.52	11.62	17.69
AtlasNet + $\mathcal{L}_{\text{boundary}}$	9.25	22.51	12.12	22.64	18.82	19.11	16.50	11.53	16.56
PCN [26]	8.09	18.32	10.53	19.33	18.52	16.44	16.34	10.21	14.72
PCN + $\mathcal{L}_{\text{boundary}}$	6.39	16.32	9.30	18.61	16.72	16.28	15.29	9.00	13.49
TopNet [21]	5.50	12.02	8.90	12.56	9.54	12.20	9.57	7.51	9.72
Our method	**4.76**	**10.29**	**7.63**	**11.23**	**8.97**	**10.08**	**7.13**	**6.38**	**8.31**
– *without* $\mathcal{L}_{\text{inter}}$	10.82	20.45	15.21	20.19	18.05	18.58	15.65	8.81	15.97
– *without* $\mathcal{L}_{\text{intra}}$	5.23	16.10	12.49	14.62	13.90	12.37	12.96	5.72	11.67
– *without* $\mathcal{L}_{\text{inter}}$, $\mathcal{L}_{\text{intra}}$	10.91	20.54	15.27	20.28	18.16	18.66	15.75	8.91	16.06
– *without* $\mathcal{L}_{\text{boundary}}$	5.46	10.98	8.27	11.95	9.51	10.92	7.78	7.40	9.03
– *without* $\mathcal{L}_{\text{preserve}}$	10.29	19.75	14.13	19.35	17.88	18.21	15.23	8.11	15.37

Table 3. Completion results using the Earth-Moving distance (multiplied by 10^2) with the output resolution of 1,024. We report the values of DeepSDF [16] from their original paper by rescaling according to the difference of point density.

Output resolution = 1,024

Method	Plane	Cabinet	Car	Chair	Lamp	Sofa	Table	Vessel	*Avg.*
3D-EPN [7]	6.20	7.76	8.70	7.68	10.73	8.08	8.10	8.17	8.18
PointNet++ [19]	5.96	11.62	6.69	11.06	18.58	10.26	8.61	8.38	10.14
FoldingNet [25]	15.64	22.13	17.46	29.74	32.00	24.57	18.99	21.88	22.80
PCN [26]	3.88	7.07	5.50	6.81	8.46	7.24	6.01	6.27	6.40
DeepSDF [16]	3.88	–	–	5.63	–	**4.68**	–	–	–
LGAN [3]	3.32	–	–	5.59	–	–	–	–	–
MAP-VAE [10]	3.23	–	–	5.57	–	–	–	–	–
Our method	**2.52**	**5.49**	**4.08**	**5.20**	**6.17**	5.25	**4.61**	**5.80**	**4.89**

7.2 Car Completion on KITTI

The KITTI [8] dataset present partial scans of real-world cars using Velodyne 3D laser scanner. We adopt the same training and validating procedure for car completion as proposed by PCN [26]. We train a car completion model based on the training data generated from ShapeNet [5] and test our completion method on sparse point clouds generated from the real-world LiDAR scans. For each sample, the points within the bounding boxes are extracted with 2,483 partial point clouds. Each point cloud is then transformed to the box's coordinates to be completed by our model then transformed back to the world frame. PCN [26] proposed three metrics to evaluate the performance of our model: (1) *Fidelity*, *i.e.* the average distance from each point in the input to its nearest neighbour in the output (*i.e.* measures how well the input is preserved); (2) *Minimal Matching*

Table 4. Car completion on LiDAR scans from KITTI.

Method	Fidelity	MMD	Consistency
FoldingNet [25]	0.03155	0.02080	0.01326
AtlasNet [9]	0.03461	0.02205	0.01646
PCN [26]	0.02800	0.01850	0.01163
Our method	**0.02171**	**0.01465**	**0.00922**
PCN [26] (rotate)	0.03352	0.02370	0.01639
Our method (rotate)	**0.02392**	**0.01732**	**0.01175**

Distance (MMD), *i.e.* the Chamfer distance between the output and the car's point cloud nearest neighbor from ShapeNet (*i.e.* measures how much the output resembles a typical car); and, (3) *Consistency*, the average Chamfer distance between the completion outputs of the same instance in consecutive frames (*i.e.* measures how consistent the network's outputs are against variations in the inputs).

Table 4 shows that we achieve state of the art on the metrics compared to FoldingNet [25], AtlasNet [9] and PCN [26]. When we introduce random rotations on the bounding box in order to simulate errors in the initial stages, we still acquire the lowest errors.

7.3 Classification on ModelNet and PartNet

We evaluate the performance of the features in term of classification on ModelNet10 [27], ModelNet40 [27] and PartNet [15] datasets. ModelNet40 contains 12,311 CAD models in 40 categories. Here, the training data contains 9,843 samples and the testing data contains 2,468 samples. Following RS-DGCNN [2], a linear Support Vector Machine [6] (SVM) is trained on the representations learned in an unsupervised manner on the ShapeNet dataset. RS-DGCNN [2] divides the point cloud of the objects into several regions by positioning the object in a pre-defined voxel grid, then use the regional information to help train latent feature. In Table 5, the proposed method outperforms RS-DGCNN [2] by 1.64% accuracy on ModelNet40 dataset, which shows that our feature contains better categorical information. Notably, similar results are also acquired from ModelNet10 [27] and PartNet [15] with their respective evaluation strategy.

7.4 Ablation Study

Loss Functions. In the reconstruction and classification experiments, Table 2 and Table 5 also include the ablation study that investigates the effects of the loss functions from Sect. 6. For both experiments, we notice all loss functions are critical to achieve good results since each of them focuses on different aspects.

Table 5. Object classification on ModelNet40 [27], ModelNet40 [27] and PartNet [15] datasets in terms of accuracy.

Method	ModelNet40 [27]	ModelNet10 [27]	PartNet [15]
VConv-DAE	75.50%	80.50%	–
3D-GAN	83.30%	91.00%	74.23%
Latent-GAN	85.70%	95.30%	–
FoldingNet	88.40%	94.40%	–
VIP-GAN	90.19%	92.18%	–
RS-PointNet [2]	87.31%	91.61%	76.95%
RS-DGCNN [2]	90.64%	94.52%	–
KCNet [1]	91.0%	94.4%	–
Our method	**92.28%**	**96.14%**	**84.32%**
– *without* $\mathcal{L}_{\text{inter}}$	89.40%	95.75%	81.13%
– *without* $\mathcal{L}_{\text{intra}}$	83.70%	90.21%	79.28%
– *without* $\mathcal{L}_{\text{inter}}$, $\mathcal{L}_{\text{intra}}$	82.97%	90.02%	78.41%
– *without* $\mathcal{L}_{\text{boundary}}$	88.26%	95.01%	80.86%
– *without* $\mathcal{L}_{\text{preserve}}$	86.09%	92.27%	79.05%

Activations. Since the number of regions is one of the hyper-parameters in our approach, we evaluate on the performance with different number of regions quantitatively in Table 6. These results demonstrate that the accuracy for the shape completion is increasing as the number of regions increases from 2 to 8, then the performance gradually drops as the number of regions continues to increase from 8 to 32. By observing $\mathcal{L}_{\text{inter}}$ at the same time, we find that it achieves the minimum value of 0.20 when there are 8 regions as well. This proves that $\mathcal{L}_{\text{inter}}$ can be used as an indicator for whether the expected number of regions could be used or not.

Moreover, Fig. 8 shows the regional activations when we shuffle the sequence of points in the partial scan. We can see that both the reconstructed geometry relative sub-regions are identical. So, it illustrates that, by using the proposed regional activations, our model is permutation invariant, which indicates that the reordered point cloud is suitable to perform convolutions.

Table 6. Influence of N_f and N_r on the Chamfer distance (multiplied by 10^3) and $\mathcal{L}_{\text{inter}}$.

(N_f, N_r)	$(2, 128)$	$(4, 64)$	$(8, 32)$	$(16, 16)$	$(32, 8)$
Chamfer distance	7.80	6.31	**5.94**	6.27	6.75
$\mathcal{L}_{\text{inter}}$	0.41	0.67	**0.20**	0.49	1.33

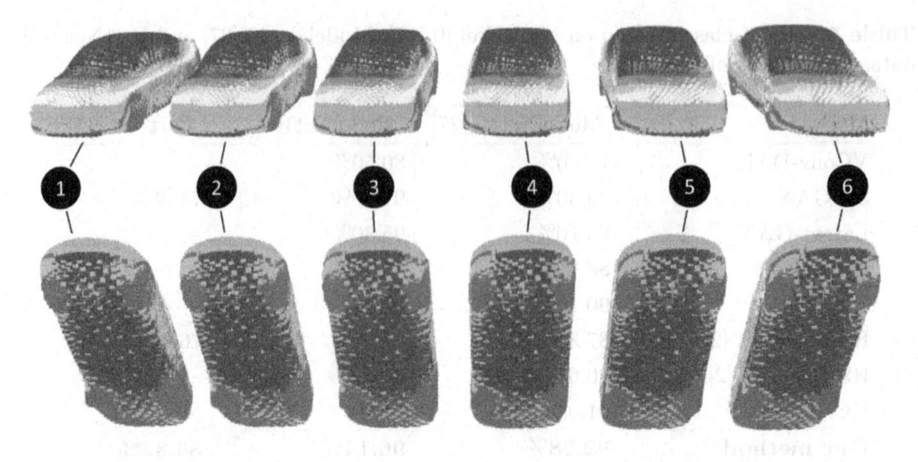

Fig. 8. With identical results, this evaluation shows the robustness of the reconstruction when we randomly shuffle the input point cloud.

Point Cloud Versus Volumetric Data. In addition to achieving worse numerical results in Sect. 7.1, volumetric approaches have smaller resolutions than the point cloud approaches due to the memory constraints. The difference becomes more evident in Fig. 9, where ForkNet [22] is limited by a $64 \times 64 \times 64$ grid. Nevertheless, both the volumetric and point cloud approaches have difficulty in reconstructing thin structures. For instance, the volumetric approach tends to ignore the joints between the wheels and car chassis in Fig. 9 while FoldingNet [25] and AtlasNet [9] only use large surface to cover the area of wheels. In contrast, our approach is capable of reconstructing the thin structures quite well. Moreover, in Table 7, we also achieve the lowest inference time compared to all point cloud and volumetric approaches.

Table 7. Overview of the object completion methods. The inference time is the amount of time to conduct inference on a single sample.

Method	Size (MB)	Inference time (s)	Closed surface	Type of data
3D-EPN [7]	420	–	Yes	Volumetric
ForkNet [22]	362	–	Yes	Volumetric
FoldingNet [25]	19.2	0.05	Yes	Points
AtlasNet [9]	2	0.32	No	Points
PCN [26]	54.8	0.11	No	Points
DeepSDF [16]	7.4	9.72	Yes	SDF
Our method	37.2	0.04	Yes	Points

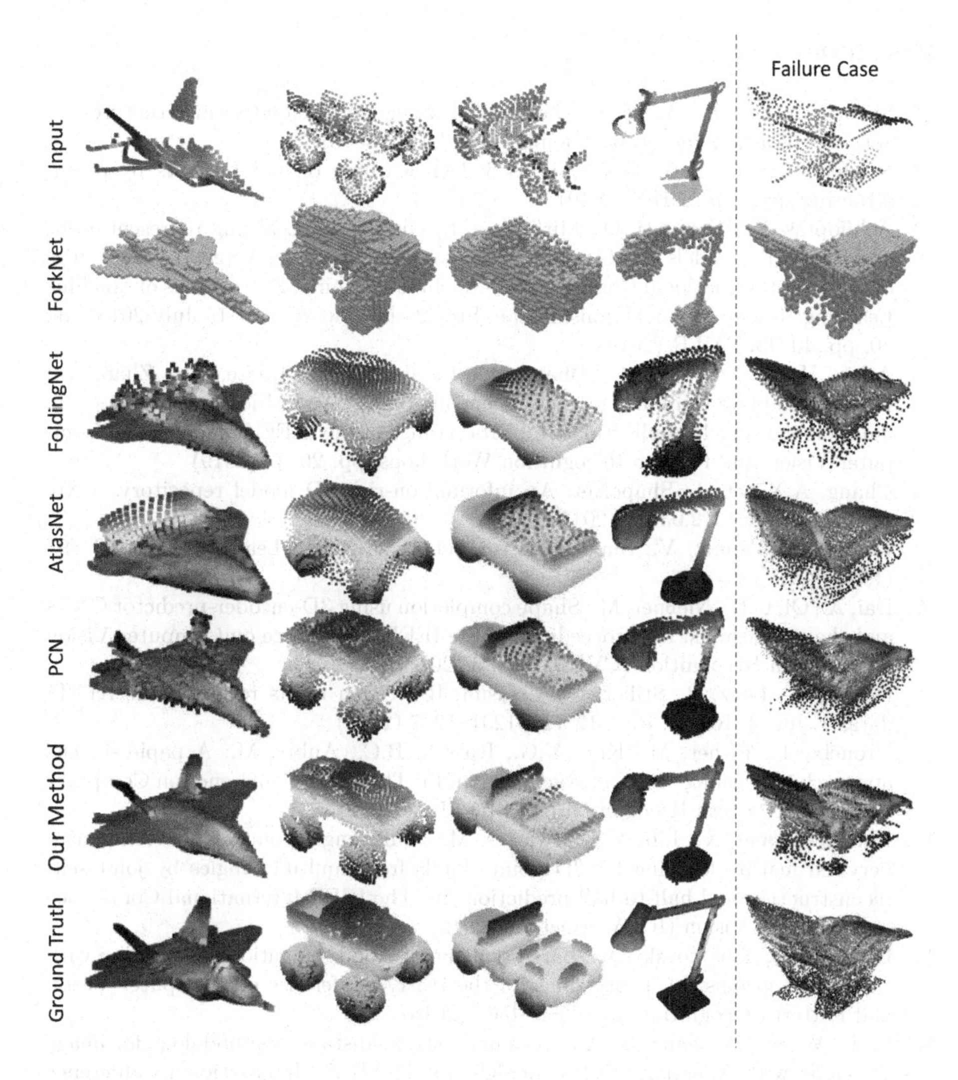

Fig. 9. Evaluated on ShapeNet [5], comparison of shape completion based on ForkNet [22], FoldingNet [25], AtlasNet [9] and PCN [26] against our method.

8 Conclusion

This paper introduced the SoftPool idea as a novel and general way to extract rich deep features from unordered point sets such as 3D point clouds. Also, it proposed a state-of-the-art point cloud completion approach by designing a regional convolution network for the decoding stage. Our numerical evaluation reflects that our approach achieves the best results on different 3D tasks, while our quantitative results illustrate the reconstruction and completion ability of our method with respect to ground truth.

References

1. Shen, Y., Feng, C., Yang, Y., Tian, D.: Mining point cloud local structures by kernel correlation and graph pooling. In: CVPR (2018)
2. Sauder, J., Sievers, B.: Self-supervised deep learning on point clouds by reconstructing space. In: NIPS (2019)
3. Achlioptas, P., Diamanti, O., Mitliagkas, I., Guibas, L.: Learning representations and generative models for 3D point clouds. In: Dy, J., Krause, A. (eds.) Proceedings of the 35th International Conference on Machine Learning. Proceedings of Machine Learning Research, Stockholmsmässan, Stockholm, Sweden, 10–15 July 2018, vol. 80, pp. 40–49. PMLR (2018)
4. Arief, H.A., Arief, M.M., Bhat, M., Indahl, U.G., Tveite, H., Zhao, D.: Density-adaptive sampling for heterogeneous point cloud object segmentation in autonomous vehicle applications. In: Proceedings of the IEEE Conference on Computer Vision and Pattern Recognition Workshops, pp. 26–33 (2019)
5. Chang, A.X., et al.: ShapeNet: An information-rich 3D model repository. arXiv preprint arXiv:1512.03012 (2015)
6. Cortes, C., Vapnik, V.: Support-vector networks. Mach. Learn. **20**(3), 273–297 (1995)
7. Dai, A., Qi, C.R., Nießner, M.: Shape completion using 3D-encoder-predictor CNNs and shape synthesis. In: Proceedings of the IEEE Conference on Computer Vision and Pattern Recognition (CVPR), vol. 3 (2017)
8. Geiger, A., Lenz, P., Stiller, C., Urtasun, R.: Vision meets robotics: the KITTI dataset. Int. J. Robot. Res. **32**(11), 1231–1237 (2013)
9. Groueix, T., Fisher, M., Kim, V.G., Russell, B.C., Aubry, M.: A papier-mâché approach to learning 3D surface generation. In: The IEEE Conference on Computer Vision and Pattern Recognition (CVPR) (June 2018)
10. Han, Z., Wang, X., Liu, Y.S., Zwicker, M.: Multi-angle point cloud-VAE: unsupervised feature learning for 3D point clouds from multiple angles by joint self-reconstruction and half-to-half prediction. In: The IEEE International Conference on Computer Vision (ICCV) (October 2019)
11. Landrieu, L., Simonovsky, M.: Large-scale point cloud semantic segmentation with superpoint graphs. In: Proceedings of the IEEE Conference on Computer Vision and Pattern Recognition, pp. 4558–4567 (2018)
12. Li, P., Wang, Q., Zhang, L.: A novel earth mover's distance methodology for image matching with Gaussian mixture models. In: The IEEE International Conference on Computer Vision (ICCV) (December 2013)
13. Li, Y., Bu, R., Sun, M., Wu, W., Di, X., Chen, B.: PointCNN: convolution on x-transformed points. In: Advances in Neural Information Processing Systems, pp. 820–830 (2018)
14. Liu, Y., Fan, B., Xiang, S., Pan, C.: Relation-shape convolutional neural network for point cloud analysis. In: Proceedings of the IEEE Conference on Computer Vision and Pattern Recognition, pp. 8895–8904 (2019)
15. Mo, K., et al.: PartNet: a large-scale benchmark for fine-grained and hierarchical part-level 3D object understanding. In: The IEEE Conference on Computer Vision and Pattern Recognition (CVPR) (June 2019)
16. Park, J.J., Florence, P., Straub, J., Newcombe, R., Lovegrove, S.: DeepSDF: learning continuous signed distance functions for shape representation. In: Proceedings of the IEEE Conference on Computer Vision and Pattern Recognition, pp. 165–174 (2019)

17. Pham, Q.H., Nguyen, T., Hua, B.S., Roig, G., Yeung, S.K.: JSIS3D: joint semantic-instance segmentation of 3D point clouds with multi-task pointwise networks and multi-value conditional random fields. In: Proceedings of the IEEE Conference on Computer Vision and Pattern Recognition, pp. 8827–8836 (2019)
18. Qi, C.R., Su, H., Mo, K., Guibas, L.J.: PointNet: deep learning on point sets for 3D classification and segmentation. In: Proceedings of the IEEE Conference on Computer Vision and Pattern Recognition, pp. 652–660 (2017)
19. Qi, C.R., Yi, L., Su, H., Guibas, L.J.: PointNet++: deep hierarchical feature learning on point sets in a metric space. In: Advances in Neural Information Processing Systems (NIPS) (2017)
20. Song, S., Yu, F., Zeng, A., Chang, A.X., Savva, M., Funkhouser, T.: Semantic scene completion from a single depth image. In: Proceedings of the IEEE Conference on Computer Vision and Pattern Recognition (CVPR). IEEE (2017)
21. Tchapmi, L.P., Kosaraju, V., Rezatofighi, H., Reid, I., Savarese, S.: TopNet: structural point cloud decoder. In: Proceedings of the IEEE Conference on Computer Vision and Pattern Recognition, pp. 383–392 (2019)
22. Wang, Y., Tan, D.J., Navab, N., Tombari, F.: ForkNet: multi-branch volumetric semantic completion from a single depth image. In: Proceedings of the IEEE International Conference on Computer Vision, pp. 8608–8617 (2019)
23. Wu, W., Qi, Z., Fuxin, L.: PointConv: deep convolutional networks on 3D point clouds. In: Proceedings of the IEEE Conference on Computer Vision and Pattern Recognition, pp. 9621–9630 (2019)
24. Yang, B., Rosa, S., Markham, A., Trigoni, N., Wen, H.: Dense 3D object reconstruction from a single depth view. IEEE Trans. Pattern Anal. Mach. Intell. **41**, 2820–2834 (2019)
25. Yang, Y., Feng, C., Shen, Y., Tian, D.: FoldingNet: point cloud auto-encoder via deep grid deformation. In: Proceedings of the IEEE Conference on Computer Vision and Pattern Recognition, pp. 206–215 (2018)
26. Yuan, W., Khot, T., Held, D., Mertz, C., Hebert, M.: PCN: point completion network. In: 2018 International Conference on 3D Vision (3DV), pp. 728–737. IEEE (2018)
27. Wu, Z., et al.: 3D ShapeNets: a deep representation for volumetric shapes. In: 2015 IEEE Conference on Computer Vision and Pattern Recognition (CVPR), pp. 1912–1920 (2015)

Hierarchical Face Aging Through Disentangled Latent Characteristics

Peipei Li[1,3], Huaibo Huang[1,3,4], Yibo Hu[1], Xiang Wu[1], Ran He[1,2,3(✉)], and Zhenan Sun[1,2,3]

[1] Center for Research on Intelligent Perception and Computing, NLPR, CASIA, Beijing, China
{peipei.li,huaibo.huang}@cripac.ia.ac.cn, huyibo871079699@gmail.com, alfredxiangwu@gmail.com, {rhe,znsun}@nlpr.ia.ac.cn

[2] Center for Excellence in Brain Science and Intelligence Technology, CAS, Beijing, China

[3] School of Artificial Intelligence, University of Chinese Academy of Sciences, Beijing, China

[4] Artificial Intelligence Research, CAS, Jiaozhou, Qingdao, China

Abstract. Current age datasets lie in a long-tailed distribution, which brings difficulties to describe the aging mechanism for the imbalance ages. To alleviate it, we design a novel facial age prior to guide the aging mechanism modeling. To explore the age effects on facial images, we propose a Disentangled Adversarial Autoencoder (DAAE) to disentangle the facial images into three independent factors: age, identity and extraneous information. To avoid the "wash away" of age and identity information in face aging process, we propose a hierarchical conditional generator by passing the disentangled identity and age embeddings to the high-level and low-level layers with class-conditional BatchNorm. Finally, a disentangled adversarial learning mechanism is introduced to boost the image quality for face aging. In this way, when manipulating the age distribution, DAAE can achieve face aging with arbitrary ages. Further, given an input face image, the mean value of the learned age posterior distribution can be treated as an age estimator. These indicate that DAAE can efficiently and accurately estimate the age distribution in a disentangling manner. DAAE is the first attempt to achieve facial age analysis tasks, including face aging with arbitrary ages, exemplar-based face aging and age estimation, in a universal framework. The qualitative and quantitative experiments demonstrate the superiority of DAAE on five popular datasets, including CACD2000, Morph, UTKFace, FG-NET and AgeDB.

Keywords: Facial age analysis · Variational autoencoder

P. Li, H. Huang and R. He—Equal contribution.

Electronic supplementary material The online version of this chapter (https://doi.org/10.1007/978-3-030-58580-8_6) contains supplementary material, which is available to authorized users.

A. Vedaldi et al. (Eds.): ECCV 2020, LNCS 12348, pp. 86–101, 2020.
https://doi.org/10.1007/978-3-030-58580-8_6

1 Introduction

Facial age analysis, including face aging, exemplar-based face aging and age estimation, is one of the crucial components in modern face analysis for entertainment and forensics. Face aging aims to aesthetically render the facial appearance based on the given age, while exemplar-based face aging aims to render the facial appearance according to the age of the exemplar face. In recent years, with the developments of the generative adversarial network (GAN) [8], impressive progress [17,18,20,32,35,40] has been made on face aging. Since all the existing age datasets, such as CACD2000, Morph and UTKFace, perform a long-tailed age distribution, it is difficult for current methods to describe the aging mechanism for the imbalanced age distribution. Researchers often employ the time span of 10 years as the age clusters for face aging. This age cluster strategy potentially limits the diversity of aging patterns, especially for the younger with the large inter-class appearance variance.

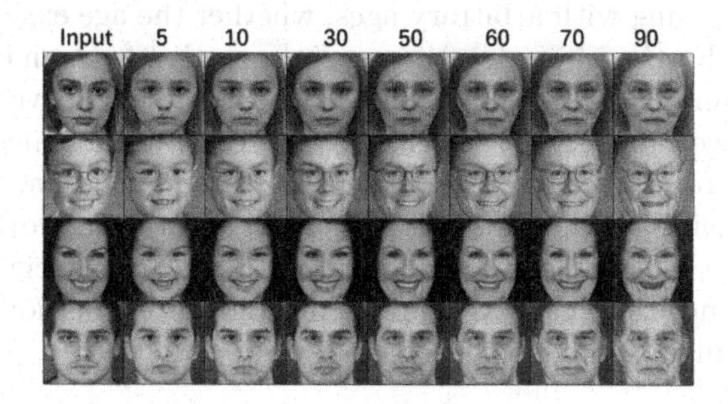

Fig. 1. Continuous face aging results on UTKFace. The first column are the inputs and the rest columns are the synthesized faces from 5 to 90 years old.

Recently, Variational Auto-Encoder (VAE) [15] shows the promising ability in discovering the underlying data distribution in the latent space [2,12,34]. Conditional Adversarial Autoencoder (CAAE) [40] proposes to learn a face manifold in the latent space. With the given age label and the learned face manifold, CAAE achieves continuous face aging. However, there are still four limitations with CAAE: 1) It only focuses on the learning of the image representation and ignores the age representation. The possibility of utilizing it to handle more age-related analysis, such as age estimation and exemplar-based face aging, is limited. 2) CAAE conducts face aging on the cropped faces and reckons without the extraneous information, such as hair. 3) The identity and age embeddings are only injected into the input layer of the generator, which leads to the "wash away" of them during generation. 4) It still adopts a group-based training strategy.

To address the mentioned issues in CAAE, we propose a Disentangled Adversarial Autoencoder (DAAE). We first design a facial age distribution as the facial age prior to directly learn the age representation from the image. A well-trained identity classifier is also employed to supervise the identity feature learning in a knowledge distilling way [10]. Besides, we discover it is also important to disentangle the extraneous information, including hairstyle and pose, from the given image. Following most VAE based methods, we introduce the Gaussian distribution as the prior for extraneous information. Supervised by the variational evidence lower bound (ELBO) [15,25] and identity knowledge distillation, the facial images are disentangled into three independent factors: age, identity and extraneous information. This disentangling manner makes DAAE more flexible and controllable for facial age analysis. To avoid the "wash away" of identity and age information during generation, we propose a hierarchical aging architecture by passing the disentangled identity and age vectors to the high-level and low-level layers with class-conditional BatchNorm. To synthesize photo-realistic facial images, a disentangled adversarial learning mechanism is introduced to optimize the inference network and the generator jointly and adversarially. Inspired by IntroVAE [12], DAAE is expected to have the capability to self-estimate the age accuracy, identity preserving and generation quality of the synthesized images.

Finally, by manipulating the mean value of age distribution, DAAE can easily realize facial face aging with arbitrary ages, whether the age exists or not in the training dataset. Further, by extracting age information from an exemplar facial image, DAAE can achieve exemplar-based face aging. Moreover, given a face image as input, we can easily obtain its age representation, which indicates the ability of DAAE to achieve age estimation. As stated above, we can implement three different facial age analysis tasks in a universal framework. To the best of our knowledge, DAAE is the first attempt to achieve facial age analysis, including face aging, exemplar-based face aging and age estimation, in a universal framework. The main contributions of DAAE are as follows:

- We propose a novel Disentangled Adversarial Autoencoder (DAAE) for facial age analysis tasks, including face aging, exemplar-based face aging and age estimation. We design two different priors as well as identity knowledge distillation to assist disentangling the facial images into three independent factors: age, identity and extraneous information.
- To fully utilize the disentangled low-level age, high-level identity and mixed-level extraneous information, we propose a hierarchical conditional generator with class-conditional BatchNorm.
- We propose a disentangled adversarial learning mechanism by training the encoder and generator with age preserving regularization and identity knowledge distillation in an introspective adversarial manner.
- Extensive qualitative and quantitative experiments demonstrate that DAAE successfully formulates the facial age prior in a disentangling manner, obtaining state-of-the-art results on the five popular datasets.

2 Related Work

2.1 Face Aging

Recently, deep conditional generative models have shown considerable ability in face aging [17,18,32,40]. Zhang et al. [40] propose a Conditional Adversarial

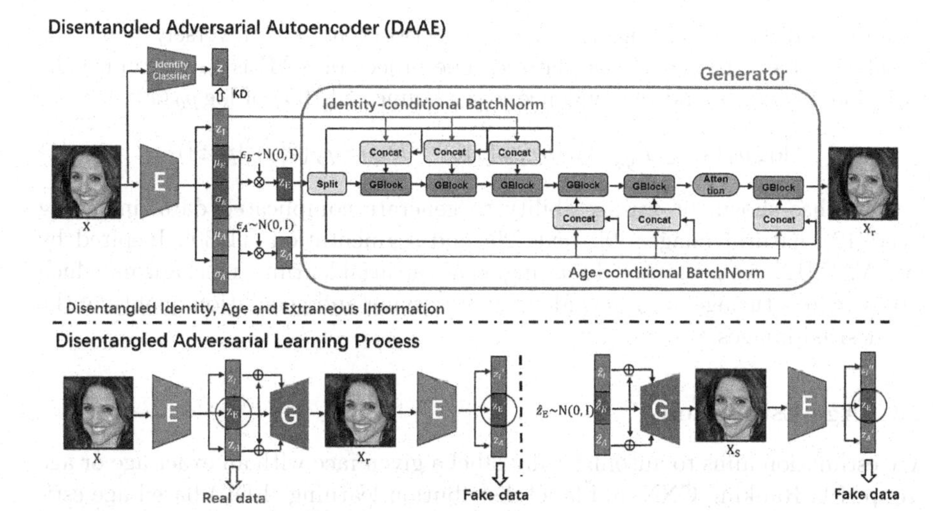

Fig. 2. Overview of the architecture and training flow of our approach. Our model contains two components, the inference network E and the hierarchical generative network G. X, X_r and X_s denote the real sample, the reconstruction sample and the new sample, respectively. Please refer to Sect. 3 for more details.

Autoencoder (CAAE) to transform an input facial image to the target age. Yang et al. [35] propose a pyramid GAN to simulate the aging effects in a finer manner. To capture the rich textures in the local facial parts, Li et al. [18] propose a Global and Local Consistent Age Generative Adversarial Network (GLCA-GAN). Meanwhile, Identity-Preserved Conditional Generative Adversarial Networks (IPCGAN) [32] introduces an identity-preserved term and an age classification term into face aging. Liu et al. [20] imposes attribute information to guide the aging process and proposes a wavelet-based discriminator to encourage generation quality. Although these methods have achieved promising visual results, they have limitations in discovering the disentangled factors in face aging. Besides, the image and age label are only injected into the input layer of the generator, leading to the "wash away" of image and age information during generation. For non-deep learning methods, [28] demonstrates that disentangling the common and individual components in facial images is crucial for face aging. [22] proposes Multi-Attribute Robust Component Analysis (MA-RCA) that incorporates knowledge from age and identity for age progression. In this paper, benefiting from the proposed VAE-based method, we focus on disentangling facial images into three independent factors: age, identity and extraneous information.

2.2 Variational Autoencoder

Variational Autoencoder (VAE) [15,25] consists of two networks: an inference network $q_\phi(z|x)$ maps the data x to the latent variable z, which is assumed as a gaussian distribution, and a generative network $p_\theta(x|z)$ reversely maps the latent variable z to the visible data x. The object of VAE is to maximize the variational lower bound (or evidence lower bound, ELBO) of $\log p_\theta(x)$:

$$\log p_\theta(x) \geq E_{q_\phi(z|x)} \log p_\theta(x|z) - D_{KL}(q_\phi(z|x) || p(z)) \tag{1}$$

VAE has shown promising ability to generate complicated data, including faces [12], natural images [9], text [29] and segmentation [11,30]. Inspired by IntroVAE [12], we propose a disentangled adversarial learning mechanism, which self-estimates the age accuracy, identity preserving and generation quality of the synthesized images.

2.3 Age Estimation

Age estimation aims to automatically label a given face with an exact age or age group [19]. Ranking-CNN and label distribution learning (LDL) based age estimation achieve state-of-the-art performance. Ranking-CNN consists of a series of CNNs, each of which learns a binary classification. Then these binary values are aggregated for the final result [4]. To model the correlations among different ages, label distribution learning [6] utilizes a specific distribution to formulate the aging mechanism. Inspired by it, we design a new age distribution as the facial age prior to directly learn the age information from the image. In this way, the mean value of the learned age posterior distribution can be treated as an age estimator.

3 Approach

In this paper, we propose a Disentangled Adversarial Autoencoder (DAAE) for face aging, examplar-based aging and age estimation. The key idea is to disentangle the facial image into three independent factors, i.e., age, identity and extraneous information. A hierarchical aging generator is introduced to produce photo-realistic images by transferring the identity and age information sequentially. As depicted in Fig. 2, two different priors as well as identity distillation are assigned to regularize the inferred representations. The inference network E and the generator network G are trained in an introspective disentangling manner.

3.1 Disentangled Variational Representations

In the original VAE [15], a probabilistic latent variable model is learned by maximizing the variational lower bound to the marginal likelihood of the observable variables. However, the latent variable z is difficult to interpret and control, since each element of z is treated equally in training. To alleviate this, we manually split z into three parts, i.e., z_A representing the age information, z_I representing the identity information, and z_E representing the extraneous information.

Assume that z_A, z_I and z_E are independent on each other, then the posterior distribution can be written as: $q_\phi(z|x) = q_\phi(z_A|x) q_\phi(z_I|x) q_\phi(z_E|x)$. The prior distribution $p(z) = p_A(z_A) p_I(z_I) p_E(z_E)$, where $p_A(z_A)$, $p_I(z_I)$ and $p_E(z_E)$ are the prior distributions for z_A, z_I and z_E, respectively. According to Eq. (1), the optimization objective for the modified VAE is to maximize the lower bound of $\log p_\theta(x)$:

$$\begin{aligned}\log p_\theta(x) \geq & E_{q_\phi(z_A, z_I, z_E|x)} \log p_\theta(x|z_A, z_I, z_E) \\ & - D_{KL}(q_\phi(z_A|x) \| p_A(z_A)) \\ & - D_{KL}(q_\phi(z_I|x) \| p_I(z_I)) \\ & - D_{KL}(q_\phi(z_E|x) \| p_E(z_E)),\end{aligned} \tag{2}$$

where the first item regularizes the reconstruction accuracy, and the last three regularize the latents, i.e., z_A, z_I and z_E, to learn different types of facial information.

To capture facial aging characteristics, the prior $p_A(z_A)$ for z_A is designed as a facial aging-specific distribution. It is set to be a centered isotropic multivariate Gaussian, i.e., $p_A(z_A) = \mathbb{N}(\mathbf{y}, \mathbf{I})$, where $\mathbf{y}$ is a vector filled by the age label y of x. We assume the posterior $q_\phi(z_A|x)$ also follows a centered isotropic multivariate Gaussian, i.e., $q_\phi(z_A|x) = \mathbb{N}(z_A; \mu_A, \sigma_A^2)$. As depicted in Fig. 2, μ_A and σ_A are the output vectors of the inference network E. The input z_A for the generator G is sampled from $\mathbb{N}(z_A; \mu_A, \sigma_A^2)$ using a reparameterization trick, i.e., $z_A = \mu_A + \epsilon_A \odot \sigma_A$, where $\epsilon_A \sim \mathbb{N}(\mathbf{0}, \mathbf{I})$. The negative version of the second term in Eq. (2) can be formed as

$$L_{kl}^{(age)} = \frac{1}{2} \sum_{i=1}^{C_A} ((\mu_A^i - y)^2 + (\sigma_A^i)^2 - \log((\sigma_A^i)^2) - 1), \tag{3}$$

where y is the age label of the input x and C_A denotes the dimension of z_A. Noted that $(\mu_A^i - y)^2$ in Eq. (3) can be viewed as an L2 constraint between the predicted μ_A and the age label y, which leads to the capability of the proposed method to estimate facial age.

For the difficulty in modeling the identity space with a simple distribution, the prior $p_I(z_I)$ is obtained through a well-pretrained identity classifier C. Assume that for each $z \in p_I(z_I)$, there exists a facial image x satisfying that $z = F(x)$, where $F(x)$ is the extracted feature before the softmax layer in C. We employ a paired L1 loss function between the predicted z_I and the extracted feature $F(x)$ to describe the relations between the posterior $q_\phi(z_I|x)$ and the prior $p_A(z_A)$. The paired L1 loss distills the identity knowledge from the identity classifier C to the inference network E, where C and E can be regarded as the teacher net and student net in knowledge distilling [10], respectively. This loss function is formed as

$$L_{kd}^{(id)} = \sum_{i=1}^{C_I} |z_I^i - F(x)|, \tag{4}$$

where z_I and $F(x)$ are extracted from the same image x by the inference network E and the identity classifier C, and C_I denotes the dimension of z_I.

Following the original VAE [15], we set the prior $p_E(z_E) = \mathbb{N}(\mathbf{0}, \mathbf{I})$ and the posterior $q_\phi(z_E|x) = \mathbb{N}(z_E; \mu_E, \sigma_E^2)$. Similar to Eq. (3), the negative version of the last term in Eq. (2) can be reformed as

$$L_{kl}^{(ext)} = \frac{1}{2}\sum_{i=1}^{C_E}((\mu_E^i)^2 + (\sigma_E^i)^2 - \log((\sigma_E^i)^2) - 1), \tag{5}$$

where μ_E and σ_E are the output vectors of E, C_E denotes the dimension of z_E. The reconstruction term in Eq. (2) can be optimized by the following form

$$L_{rec} = \frac{1}{2}\|x - x_r\|_F^2, \tag{6}$$

where x and x_r are the input and output images, respectively.

In summary, the optimization object in Eq. (2) can be rewritten in the negative version:

$$L_{vae} = L_{rec} + L_{kl}^{(age)} + L_{kd}^{(id)} + L_{kl}^{(ext)}. \tag{7}$$

3.2 Hierarchical Conditional Generator

In order to effectively utilize the disentangled low-level age, high-level identity and mixed-level extraneous information (e.g., pose, skin color), we propose a hierarchical conditional generator, borrowing from Conditional Batch Normalization (CBN) [5] literature. We regard all of the age, identity and most extraneous information as the conditions for face aging.

We first split the extraneous information into several parts and the first part is regarded as the input of the generator. Since extraneous information contains both high-level (e.g., pose) and low-level (e.g., skin color) information, the rest parts are concatenated with identity or age information and used as the condition information at each residual block. As shown in Fig. 2, identity with extraneous information is passed into the first few layers for high-level identity generation, while age with another extraneous information is passed into the last few layers for low-level texture generation. With the proposed hierarchical conditional generator, the proposed DAAE enables higher intuition and interpretability.

3.3 Disentangled Adversarial Learning

To further disentangle the inferred representations, i.e., z_A, z_I and z_E, and improve the quality of generation, a disentangled adversarial learning mechanism is proposed to optimize the inference network E and the generator network G jointly and adversarially. Inspired by IntroVAE [12], the model is expected to have the capability to self-estimate the age accuracy, identity preserving and image quality of the produced images.

As illustrated in Fig. 2, there exist two types of generated images, i.e., the reconstructed image $x_r = G(z_A, z_I, z_E)$ and the sampled image $x_s = G(\hat{z}_A, \hat{z}_I, \hat{z}_E)$. z_A, z_I and z_E are the inferred representations of the input x, while $\hat{z}_A$, $\hat{z}_I$ and $\hat{z}_E$ are sampled from three marginal product distribution, i.e., $p_A(\hat{z}_A) = p_A(z_A)q_\phi(z_A|x)$, $p_I(\hat{z}_I) = p_I(z_I)q_\phi(z_I|x)$, $p_E(\hat{z}_E) = p_E(z_E)q_\phi(z_E|x)$, respectively. This sampling strategy makes a random combination of age, identity and extraneous information from different sources. Along with the introduced constraints in the following, the learned representations z_A, z_I and z_E can be well disentangled.

To preserve the aging and identity characteristics accurately, two regularization terms are introduced for the generator G. They are computed as

$$L_{reg}^{(age)} = \frac{1}{C_A}\sum_{i=1}^{C_A}\|z_A^{'i} - z_A\| + \frac{1}{C_A}\sum_{i=1}^{C_A}\|z_A^{''i} - \hat{z}_A\| \tag{8}$$

$$L_{reg}^{(id)} = \frac{1}{C_I}\sum_{i=1}^{C_I}\|z_I^{'i} - z_I\| + \frac{1}{C_I}\sum_{i=1}^{C_I}\|z_I^{''i} - \hat{z}_I\| \tag{9}$$

where $z_A^{'}$ and $z_I^{'}$ are the inferred representations from the generated images x_r, while $z_A^{''}$ and $z_I^{''}$ are inferred from the generated images x_s.

To alleviate the problem of generating blurry samples in VAEs, the KL distance in Eq. (5) is employed as the adversarial signal to train the inference network E and the generator G adversarially [12]. When training E, the model minimizes the KL-distance of the posterior $q_\phi(z_E|x)$ from its prior $p_E(z_E)$ for the real data and maximize it for the generated samples. When training G, the model minimizes this KL-distance for the generated samples. The adversarial training objects for E and G are defined as below:

$$\begin{aligned} L_E^{(adv)} = L_{kl}^{(ext)}(\mu_E, \sigma_E) + \alpha\{\left[m - L_{kl}^{(ext)}(\mu_E', \sigma_E')\right]^+ \\ + \left[m - L_{kl}^{(ext)}(\mu_E'', \sigma_E'')\right]^+\}, \end{aligned} \tag{10}$$

$$L_G^{(adv)} = L_{kl}^{(ext)}(\mu_E', \sigma_E') + L_{kl}^{(ext)}(\mu_E'', \sigma_E''), \tag{11}$$

where m is a positive margin, α is a weighting coefficient, (μ_E, σ_E), (μ_E', σ_E') and (μ_E'', σ_E'') are computed from the real data x, the reconstruction sample x_r and the new samples x_s, respectively. $[]^+ = max(0, .)$, which has the same meaning in hinge loss.

The total objective function is a weighted sum of the above losses, defined as

$$L_E = L_{rec} + \lambda_1 L_{kl}^{(age)} + \lambda_2 L_{kd}^{(id)} + \lambda_3 L_E^{(adv)}, \tag{12}$$

$$L_G = L_{rec} + \lambda_4 L_{reg}^{(age)} + \lambda_5 L_{reg}^{(id)} + \lambda_6 L_G^{(adv)}, \tag{13}$$

where $\lambda_{1\sim6}$ are the weighted parameters to balance the importance of each loss.

3.4 Inference and Sampling

By regularizing the disentangled representations with the age prior $p_A(z_A) = \mathbb{N}(\mathbf{y}, \mathbf{I})$, identity knowledge prior $p_R(z_I)$ and extraneous prior $p_E(z_E) = \mathbb{N}(\mathbf{0}, \mathbf{I})$, DAAE is thus a universal framework for face aging, exemplar-based face aging and age estimation.

Face Aging. We concatenate the identity variable z_I, the extraneous variable z_E and a target age variable $\hat{z}_A$ as the input of the generator G, where z_I and z_E are the inferred identity and extraneous information from the input x, while $\hat{z}_A$ is sampled from a distribution $p_A(z_A) = \mathbb{N}(\mathbf{y}, \mathbf{I})$. The face aging result $\hat{x}$ is written as:

$$\hat{x} = G(\hat{z}_A, z_I, z_E) \tag{14}$$

Exemplar-Based Face Aging. We remain the identity and extraneous information unchanged, and transfer the age information from the given exemplar x_e. Specifically, we concatenate the identity variable z_I, the extraneous variable z_E and the age variable z_{A_e} as the input of G, where z_A, z_I and z_{A_e} are from the posterior distribution $q_\phi(z_A|x)$, $q_\phi(z_I|x)$ and $q_\phi(z_{A\_e}|x_e)$, respectively. The exemplar-based face aging result is formulated as:

$$\hat{x} = G(z_{A\_e}, z_I, z_E) \tag{15}$$

Age Estimation. We calculate the mean value of C-dimension vector μ_A as the age estimation result, defined as:

$$\hat{y} = \frac{1}{C} \sum_{i=1}^{C} \mu_A^i \tag{16}$$

where μ_A is one of the output vectors of the inference network E.

4 Experiments

4.1 Datasets and Settings

Datasets. We conduct experiments on five popular datasets. **CACD2000** [3] consists of 163,446 color facial images of 2,000 celebrities, where the ages range from 14 to 62 years old. However, there are many dirty data in it, which leads to a challenging model training. **Morph.** [26] is the largest publicly available dataset collected in the constrained environment. It contains 55,349 color facial images of 13,672 subjects with ages ranging from 16 to 77 years old. **UTKFace** [38] is a large-scale facial age dataset with a long age span, which ranges from 0 to 116 years old. It contains over 20,000 facial images in the wild. We employ classical 80-20 split on CACD2000, Morph and UTKFace. **FG-NET.** [16] contains 1,002 facial images of 82 subjects. We employ it as the testing set to evaluate the generalization of DAAE. **AgeDB.** [21] is a manually collected database, which consists of 16,488 images of 568 subjects from 0 to 101 years old.

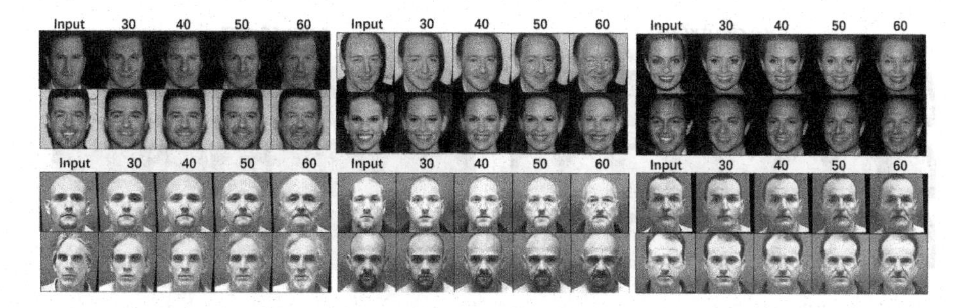

Fig. 3. Face aging results on CACD2000 (the first two rows) and Morph (the last two rows). For each subject, the first column is the input and the rest four columns are the synthesized results in 30, 40, 50 and 60 years old, respectively.

Experimental Settings. Following [17], we employ the multi-task cascaded CNN [37] to detect the faces. All the facial images are cropped and aligned into 224×224. Our model is implemented with Pytorch. During training, we choose Adam optimizer [14] with β_1 of 0.9, β_2 of 0.99, a fixed learning rate of 2×10^{-4} and batch size of 16. The trade-off parameters $\lambda_{1\sim6}$ are all set to 1, 100, 1, 100, 100, 1, respectively. Besides, m is set to 200 and α is set to 0.5. More details of the network architectures and training processes are provided in the supplementary materials.

4.2 Qualitative Evaluation of DAAE

Face Aging. By manipulating the mean value μ_A and sampling from age distribution, the proposed DAAE can generate facial images with arbitrary ages based on the input. Figure 3 presents the face aging results on CACD2000 and Morph, respectively. We observe that the synthesized faces are getting older and older with ages growing. Specifically, the face contours become longer, the beards turn white and the nasolabial folds are deepened. Since both CACD2000 and Morph lack of images of children, we conduct face aging on UTKFace. Figure 4 (a) describes the aging results on UTKFace from 0 to 110 years old. Obviously, from birth to adulthood, the aging effect is mainly shown on craniofacial growth, while the aging effect from adulthood to elder is reflected on the skin aging, which is consistent with human physiology. To evaluate the model generalization, we train our DAAE on UTKFace and test it on FG-NET. The aging results are shown in Fig. 4 (b). The left image of each subject is the input and the rest seven are the generated results from 5 to 100 years old.

The comparison results with previous works, including IAAP [13], RFA [31], RJIVE [28], MA-RCA [22], IPCGANs [32], CAAE [38], Yang et al. [35], GLCA-GAN [18], waveletGLCA-GAN [17] and Liu et al. [20] are depicted in Fig. 5. We can see that our DAAE generates more obvious or comparable age effects on the input. Besides, previous face aging methods roughly divide the data into four or nine age groups to four or nine times increase the training data for specific age groups, while our DAAE is trained with original age labels.

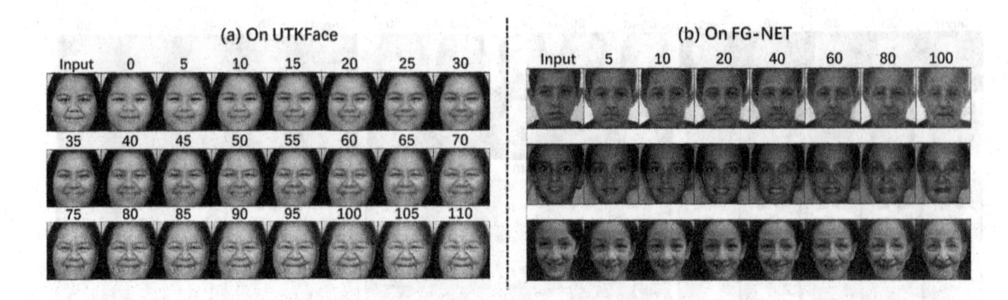

Fig. 4. Face aging results on UTKFace and FG-NET. (a) shows the aging results on UTKFace from 0 to 110 years old. The first image (top left) is the input, the rest are the synthesized results. (b) shows cross-dataset face aging results on FG-NET.

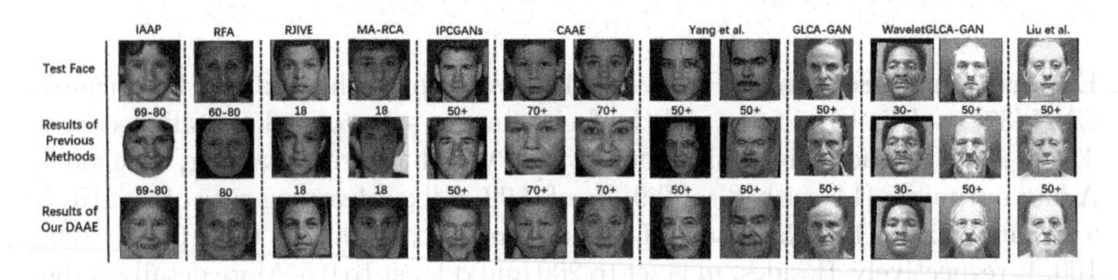

Fig. 5. Comparison with the previous works. The first row is the input face. The second row are the synthesized results of previous methods. The last row are the synthesized results by our DAAE.

4.3 Quantitative Evaluation of DAAE

Aging accuracy is an essential quantitative metric for face aging. Following [18, 35], we utilize the online face analysis tool of Face++ [1] to evaluate the ages of the synthesized results on Morph and CACD2000. We divide the testing data of the two datasets into four age groups: 30-(AG0), 31-40(AG1), 41-50(AG2), 51+(AG3). We choose AG0 as the input and synthesize images in AG1, AG2 and AG3. Then we estimate the ages of the synthesized images and calculate the average ages for each group. As shown in Table 1, we compare the DAAE with previous works on Morph and CACD2000. We observe that the generated ages by DAAE are closer to the real data than by CAAE [40] as well as GLCA-GAN [18], and comparable to [20,35]. Note that [20,35] need to train a specific model for each age group, while DAAE trains a unified model for arbitrary age synthesis, as well as other tasks.

Identity Preserving. Identity preserving is another important quantitative metric for face aging. We evaluate this performance of DAAE by face verification. We also choose AG0 as the input and synthesize images in AG1, AG2 and AG3. For each testing face in AG0, we evaluate the verification rates between it and its corresponding aging results: [testing face → Age1], [testing face → Age2] and [testing face → Age3]. We adopt Light-CNN [33] as the identity extractor.

Table 1. Comparisons of the aging accuracy on Morph and CACD2000.

Method	(a) on Morph				(b) on CACD2000			
	Input	AG1	AG2	AG3	Input	AG1	AG2	AG3
CAAE [40]	–	28.13	32.50	36.83	–	31.32	34.94	36.91
Yang et.al [35]	–	42.84	50.78	59.91	–	44.29	48.34	52.02
GLCA-GAN [18]	–	43.00	49.03	54.60	–	37.09	44.92	48.03
Liu et al. [20]	–	38.47	47.55	56.57	–	38.88	47.42	54.05
Ours	–	37.46	49.40	59.67	–	39.21	46.38	51.66
Real Data	28.19	38.89	48.10	58.22	30.73	39.08	47.06	53.68

Table 2. Comparisons of the face verification results (%) on Morph and CACD2000.

Method	(a) on Morph				(b) on CACD2000			
	Input	AG1	AG2	AG3	Input	AG1	AG2	AG3
CAAE [40]	–	15.07	12.02	8.22	–	4.66	3.41	2.40
Yang et al. [35]	–	100.00	98.91	93.09	–	99.99	99.81	98.28
GLCA-GAN [18]	–	97.66	96.67	91.85	–	97.72	94.18	92.29
Liu et al. [20]	–	100.00	100.00	98.26	–	99.76	98.74	98.44
Ours	–	99.48	99.36	99.36	–	99.24	99.19	99.19

Following [20,35], we adopt thresholds = 76.5 and FAR = 1e−5 in our face verification experiments. The comparison results on Morph and CACD2000 are reported in Table 2. It is worth noting that it is unfair to directly compare DAAE with [20]. Because [20] utilizes extra attribute labels, including gender and race, to improve aging performance. Besides, [20,35] need to train a specific model for each age group.

Age-Invariant Face Verification. Following the testing protocol in [28], we evaluate our method on AgeDB. As shown in Table 3, DAAE achieves promising performance of face verification on AgeDB. The qualitative results are reported in the supplementary materials.

4.4 More Facial Age Analysis by DAAE

The previous face aging methods [17,18,32,40] directly concatenate an age label to control the aging process, which are limited in handling various age analysis. Benefiting from the disentangling and modeling of age, identity and extraneous representations in the latent space, the proposed DAAE is able to realize more age-related tasks, such as exemplar-based face aging and age estimation.

Table 3. Comparisons of mean AUC and accuracy on AgeDB.

		5 years	10 years	20 years	30 years
RJIVE [28]	AUC	0.686	0.654	0.633	0.584
	Accuracy	0.637	0.621	0.598	0.552
Ours	AUC	0.989	0.988	0.986	0.981
	Accuracy	0.969	0.969	0.956	0.953

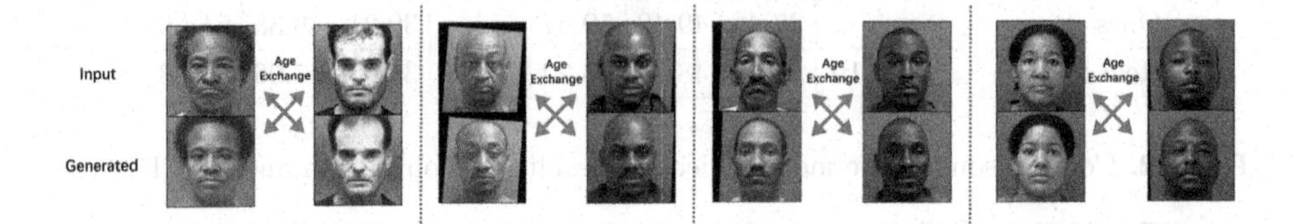

Fig. 6. Exemplar-based face aging results on Morph. For each image group, the first row are the input and the second row are the aging results with age information z_A exchanged in the group.

Exemplar-Based Face Aging. Given an exemplar image x_e, the DAAE first extracts its age information z_{A_e} and then transfers it to the input x. Figure 6 presents some results under this situation on Morph. We observe that the identity and extraneous information are preserved across rows, and the age information, such as wrinkles and beards, is changed according to the given exemplar. This demonstrates that our DAAE effectively disentangles age and age-irrelevant representations in the latent space.

Age Estimation. To further demonstrate the disentangling ability of DAAE, we conduct age estimation on Morph. We detail the evaluation metrics in the supplementary materials. Following [24], we report the mean absolute error (MAE). As shown in Table 4, the age estimation result of DAAE on Morph is nearly as

Table 4. Comparisons with state-of-the-art methods on Morph. Lower MAE is better.

Methods	Pre-trained	Morph
OR-CNN [23]	–	3.34
DEX [27]	IMDB-WIKI	2.68
Ranking [4]	Audience	2.96
Posterior [39]	–	2.87
SSR-Net [36]	IMDB-WIKI	2.52
M-V Loss [24]	–	2.51
ThinAgeNet [7]	MS-Celeb-1M	2.35
Ours	–	2.23

good as the state-of-the-arts, which demonstrates that the age representation is well learned from the given image.

4.5 Ablation Study

Table 5. Face verification results (%) of the ablation study on Morph.

Testing face	AG1	AG2	AG3
w/oL_{adv}	95.74	95.61	95.60
$w/oL_{reg}^{(age)}$	97.84	97.89	97.89
$w/oL_{reg}^{(id)}$	96.21	93.01	93.09
Setting I	89.04	87.35	87.35
Setting II	76.10	72.20	72.15
Ours	99.48	99.36	99.36

We report face verification results of DAAE and its five variants for a comprehensive comparison as the ablation study. Table 5 presents the comparison results. For the setting I, age with extraneous information is passed into the first few layers of the generator, while identity with another extraneous information is passed into the last few layers. For the setting II, we concatenate the age, identity and extraneous vectors and send it to the input layer of the generator. We observe that the face verification accuracy will decrease when one of the three losses is removed or the generator's architecture is changed. These phenomena indicate that each component in our method is essential for face aging.

5 Conclusion

This paper proposes a Disentangled Adversarial Autoencoder (DAAE) for facial age analysis. Specifically, we assign two different priors as well as identity distillation to assist disentangling the facial images into three independent factors: age, identity and extraneous information. A hierarchical conditional generator is introduced to produce photo-realistic images by transferring the identity and age information layer-by-layer. Finally, we propose a disentangled adversarial learning mechanism by training encoder and generator with age preserving regularization and identity knowledge distillation in an introspective adversarial manner. To the best of our knowledge, DAAE is the first attempt to achieve facial age analysis, including face aging, exemplar-based face aging and age estimation in a universal framework. This indicates that DAAE can efficiently formulate the facial age prior, which contributes to interpretable facial age manipulation. The qualitative and quantitative experiments demonstrate the superiority of the proposed DAAE on five popular datasets.

Acknowledgement. This work is partially funded by National Natural Science Foundation of China (Grant No. U1836217) and Shandong Provincial Key Research and Development Program (Major Scientific and Technological Innovation Project) (NO.2019JZZY010119).

References

1. Face++ research toolkit. megvii inc. http://www.faceplusplus.com ([Online])
2. Burgess, C.P., et al.: Understanding disentangling in beta-VAE. arXiv preprint arXiv:1804.03599 (2018)
3. Chen, B.C., Chen, C.S., Hsu, W.H.: Face recognition and retrieval using cross-age reference coding with cross-age celebrity dataset. IEEE Trans. Multimed. **17**(6), 804–815 (2015)
4. Chen, S., Zhang, C., Dong, M., Le, J., Rao, M.: Using ranking-CNN for age estimation. In: CVPR (2017)
5. De Vries, H., Strub, F., Mary, J., Larochelle, H., Pietquin, O., Courville, A.C.: Modulating early visual processing by language. In: NeurIPS (2017)
6. Gao, B.B., Zhou, H.Y., Wu, J., Geng, X.: Age estimation using expectation of label distribution learning. In: IJCAI (2018)
7. Gao, B.B., Zhou, H.Y., Wu, J., Geng, X.: Age estimation using expectation of label distribution learning. In: IJCAI (2018)
8. Goodfellow, I., et al.: Generative adversarial nets. In: NeurIPS (2014)
9. Gulrajani, I., et al.: Pixelvae: a latent variable model for natural images. arXiv preprint arXiv:1611.05013 (2016)
10. Hinton, G., Vinyals, O., Dean, J.: Distilling the knowledge in a neural network. arXiv preprint arXiv:1503.02531 (2015)
11. Hou, X., Shen, L., Sun, K., Qiu, G.: Deep feature consistent variational autoencoder. In: WACV (2017)
12. Huang, H., He, R., Sun, Z., Tan, T., et al.: Introvae: introspective variational autoencoders for photographic image synthesis. In: NeurIPS (2018)
13. Kemelmacher-Shlizerman, I., Suwajanakorn, S., Seitz, S.M.: Illumination-aware age progression. In: CVPR (2014)
14. Kingma, D.P., Ba, J.: Adam: a method for stochastic optimization. arXiv preprint arXiv:1412.6980 (2014)
15. Kingma, D.P., Welling, M.: Auto-encoding variational Bayes. In: ICLR (2014)
16. Lanitis, A., Taylor, C.J., Cootes, T.F.: Toward automatic simulation of aging effects on face images. IEEE Trans. Pattern Anal. Mach. Intell. **24**(4), 442–455 (2002)
17. Li, P., Hu, Y., He, R., Sun, Z.: Global and local consistent wavelet-domain age synthesis. IEEE Trans. Inf. Forensics Secur. **14**, 2943–2957 (2018)
18. Li, P., Hu, Y., Li, Q., He, R., Sun, Z.: Global and local consistent age generative adversarial networks. In: ICPR (2018)
19. Li, P., Hu, Y., Wu, X., He, R., Sun, Z.: Deep label refinement for age estimation. Pattern Recogn. **100**, 107–178 (2020)
20. Liu, Y., Li, Q., Sun, Z.: Attribute-aware face aging with wavelet-based generative adversarial networks. In: CVPR (2019)
21. Moschoglou, S., Papaioannou, A., Sagonas, C., Deng, J., Kotsia, I., Zafeiriou, S.: AgeDB: the first manually collected, in-the-wild age database. In: CVPRW (2017)
22. Moschoglou, S., Ververas, E., Panagakis, Y., Nicolaou, M.A., Zafeiriou, S.: Multi-attribute robust component analysis for facial uv maps. IEEE J. Sel. Top. Signal Process. **12**(6), 1324–1337 (2018)

23. Niu, Z., Zhou, M., Wang, L., Gao, X., Hua, G.: Ordinal regression with multiple output CNN for age estimation. In: CVPR (2016)
24. Pan, H., Han, H., Shan, S., Chen, X.: Mean-variance loss for deep age estimation from a face. In: CVPR (2018)
25. Rezende, D.J., Mohamed, S., Wierstra, D.: Stochastic backpropagation and approximate inference in deep generative models. In: ICML (2014)
26. Ricanek, K., Tesafaye, T.: Morph: a longitudinal image database of normal adult age-progression. In: FGR (2006)
27. Rothe, R., Timofte, R., Van Gool, L.: Deep expectation of real and apparent age from a single image without facial landmarks. Int. J. Comput. Vis. **126**(2–4), 144–157 (2018)
28. Sagonas, C., Ververas, E., Panagakis, Y., Zafeiriou, S.: Recovering joint and individual components in facial data. IEEE Trans. Pattern Anal. Mach. Intell. **40**(11), 2668–2681 (2017)
29. Semeniuta, S., Severyn, A., Barth, E.: A hybrid convolutional variational autoencoder for text generation. arXiv preprint arXiv:1702.02390 (2017)
30. Sohn, K., Lee, H., Yan, X.: Learning structured output representation using deep conditional generative models. In: NeurIPS (2015)
31. Wang, W., et al.: Recurrent face aging. In: CVPR (2016)
32. Wang, Z., Tang, X., Luo, W., Gao, S.: Face aging with identity-preserved conditional generative adversarial networks. In: CVPR (2018)
33. Wu, X., He, R., Sun, Z., Tan, T.: A light CNN for deep face representation with noisy labels. IEEE Trans. Inf. Forensics Secur. **13**(11), 2884–2896 (2018)
34. Wu, X., Huang, H., Patel, V.M., He, R., Sun, Z.: Disentangled variational representation for heterogeneous face recognition. In: AAAI (2019)
35. Yang, H., Huang, D., Wang, Y., Jain, A.K.: Learning face age progression: a pyramid architecture of GANs. In: CVPR (2018)
36. Yang, T.Y., Huang, Y.H., Lin, Y.Y., Hsiu, P.C., Chuang, Y.Y.: SSR-NET: a compact soft stagewise regression network for age estimation. In: IJCAI (2018)
37. Zhang, K., Zhang, Z., Li, Z., Qiao, Y.: Joint face detection and alignment using multitask cascaded convolutional networks. IEEE Signal Process. Lett. **23**(10), 1499–1503 (2016)
38. Zhang, Z., Song, Y., Qi, H.: Age progression/regression by conditional adversarial autoencoder. In: CVPR (2017)
39. Zhang, Y., Liu, L., Li, C., et al.: Quantifying facial age by posterior of age comparisons. arXiv preprint arXiv:1708.09687 (2017)
40. Zhang, Z., Song, Y., Qi, H.: Age progression/regression by conditional adversarial autoencoder. In: CVPR (2017)

Hybrid Models for Open Set Recognition

Hongjie Zhang[1], Ang Li[2], Jie Guo[1], and Yanwen Guo[1(✉)]

[1] State Key Laboratory for Novel Software Technology, Nanjing University, Nanjing 210023, China
hjzhang@smail.nju.edu.cn, {guojie,ywguo}@nju.edu.cn
[2] DeepMind, Mountain View, CA, USA
anglili@google.com

Abstract. Open set recognition requires a classifier to detect samples not belonging to any of the classes in its training set. Existing methods fit a probability distribution to the training samples on their embedding space and detect outliers according to this distribution. The embedding space is often obtained from a discriminative classifier. However, such discriminative representation focuses only on known classes, which may not be critical for distinguishing the unknown classes. We argue that the representation space should be jointly learned from the inlier classifier and the density estimator (served as an outlier detector). We propose the OpenHybrid framework, which is composed of an encoder to encode the input data into a joint embedding space, a classifier to classify samples to inlier classes, and a flow-based density estimator to detect whether a sample belongs to the unknown category. A typical problem of existing flow-based models is that they may assign a higher likelihood to outliers. However, we empirically observe that such an issue does not occur in our experiments when learning a joint representation for discriminative and generative components. Experiments on standard open set benchmarks also reveal that an end-to-end trained OpenHybrid model significantly outperforms state-of-the-art methods and flow-based baselines.

Keywords: Flow-based model · Density estimation · Image classification

1 Introduction

Image classification is a core problem in computer vision. However, most of the existing research is based on the closed-set assumption, *i.e.*, training set is assumed to cover all classes that appear in the test set. This is an unrealistic assumption. Even with a large-scale image dataset, such as ImageNet [15], it is impossible to cover all scenarios in the real world. When a closed-set model encounters an out-of-distribution sample, it is forced to identify it as a known class, which can cause issues in many real-world applications. We instead study the "open-set" problem where the test set is assumed to contain both known and unknown classes. So the model has to classify samples into either known (inlier)

A. Vedaldi et al. (Eds.): ECCV 2020, LNCS 12348, pp. 102–117, 2020.
https://doi.org/10.1007/978-3-030-58580-8_7

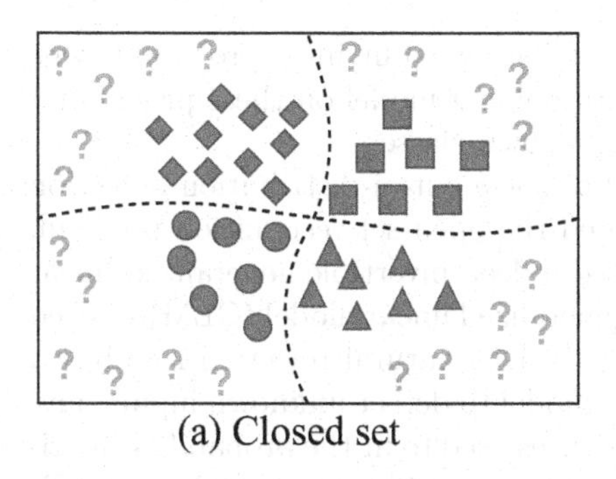

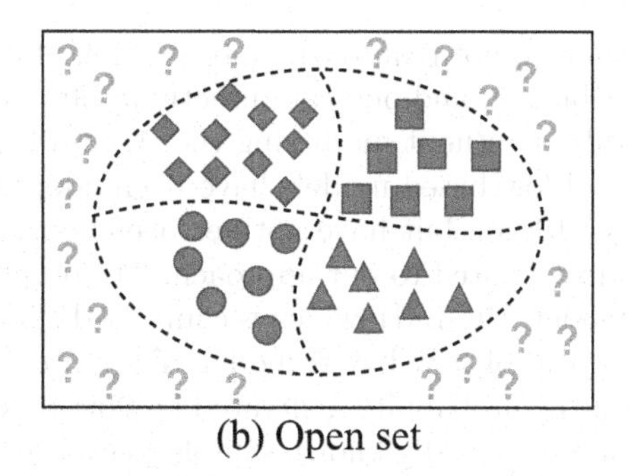

Fig. 1. Decision boundaries of a closed set classifier (a) and an open set classifier (b). Green symbols indicate known samples (different shapes represent different classes), and orange question marks indicate unknown samples. Dashed lines indicate the decision boundaries. (a) Closed set leads to unbounded decision boundaries of a typical 4-class classifier. Unknown samples are forced to be classified into one of known classes. (b) open set results in bounded decision boundaries for a 5-class classifier, which can classify both known and unknown samples.

classes or the unknown (outlier) category. Figure 1 illustrates the difference of classification decision boundaries under open set and closed set assumptions.

Identifying unknown samples is naturally challenging because they are not observed during training. Existing approaches fit a probability distribution of the training samples at their embedding space, and detect unknown samples according to such distribution. Since the feature representation of unknown classes is unknown, most of the methods operate on a discriminative feature space obtained from a supervised classifier trained on known classes. A thresholding on this probability distribution is then used to detect samples from unknown classes. A common approach along this direction is to threshold on SoftMax responses, but [2] has conducted experiments to show that it reaches only sub-optimal solutions to open set recognition. Some variants have been proposed to better utilize the SoftMax scores [7,22,33]. These methods modify the SoftMax scores to perform both unknown detection while maintaining its classification accuracy. It is extremely challenging to find a single score measure on the SoftMax layer, that can perform well on both the generative and discriminative tasks. We believe the discriminative feature space learned by classification of inlier classes may not be sufficiently effective for identifying outlier classes. So we propose to employ a flow-based generative model for outlier detection, and learn a joint feature space in an end-to-end manner from both the classifier and the density estimator.

Flow-based models have recently emerged [1,4–6,13], allowing a neural network to be invertible. They can fit a probability distribution to training samples in an unsupervised manner via maximum likelihood estimation. The flow models can predict the probability density of each example. When the probability density of an input sample is large, it is likely to be part of the training distribution (known classes). And the outlier samples (unknown class) usually have a small probability density value. The advantage of flow-based models is that they do not require the intervention of a classifier when fitting a probability distribution, and one can directly apply a thresholding model on these probability values without modifying the scores of any known classes.

Flow-based models have been adopted to solve out-of-distribution detection [10,19,20], but have not yet been considered the open set recognition problem. Most related to our approach, [20] proposed a deep invertible generalized linear model (DIGLM), which is comprised of a generalized linear model (GLM) stacked on top of flow-based model. They use the model's natural rejection rule based on the probability generated by flow-based model to detect unknown inputs, and directly classify known samples with the features used to fit the probability distribution. Our work differs in that instead of adding a classifier on top of flow model's embedding, we propose to learn a joint embedding for both the flow model and the classifier. Our insight is that the embedding space learned from only flow-based model may not have sufficient discriminative expressiveness.

We empirically observe in our experiments that learning a joint embedding space resolves a common issue in flow-based model that the flow-based model may assign higher likelihood to OOD inputs (mentioned in [10,19,26]). This issue was considered in [12], the underlying factor of which is believed to be to the inconsistency between a uni-modal prior distribution and a multi-modal data distribution. In our framework, the deep network can well represent the multi-modal distribution of the input data, which is probably the reason for the improved performance of flow models.

We perform extensive experiments on various benchmarks including MNIST, SVHN, CIFAR10 and TinyImageNet. The proposed OpenHybrid model outperforms both state-of-the-art methods [2,7,22,24,37] and hybrid model baselines [10,20] in these benchmarks. We further compare our method with an additional baseline which uses a pre-trained encoder and the result suggests the importance of jointly training both the classifier and the flow-based model.

Contribution. The contribution of this paper can be summarized as follows:

1. To the best of our knowledge, we are the first to incorporate a generative flow-based model with a discriminative classifier to address open set recognition, while most of the existing open set approaches focus on either using the softmax logits or adversarial training.
2. We propose the OpenHybrid model that learns a joint representation between the classifier and flow density estimator. Our approach ensures that the inlier classification is unaffected by outlier detection. We find joint training an important contributing factor, according to the ablation study.
3. A known issue of flow-based models is that they may assign higher likelihood to unknown inputs. However, we do not observe such phenomenon in OpenHybrid, possibly because our encoder fits the multi-modal input distribution to a latent space suitable to the unimodal assumption of flow models.
4. We conducted extensive experiments on various open set image classification datasets and compared our approach against state-of-the-art open set methods and flow-based baseline models. Our approach achieves significant improvement over these baseline methods.

2 Related Work

2.1 Open Set Recognition

Open set recognition has been surprisingly overlooked, though it has more practical value than the common closed set setting. Existing methods on this topic can be broadly classified into two categories: discriminative and generative models.

Discriminative Methods. Before the deep learning era, most of the approaches [11,30,31,38] are based on traditional classification models such as Support Vector Machines (SVMs), Nearest Neighbors, Sparse Representation, etc. These methods usually do not scale well without careful feature engineering. Recently, deep learning based models have shown more appealing results. The first among them is probably [2], which introduced Weibull-based calibration to augment the SoftMax layer of a deep network, called OpenMax. Since then, the OpenMax is further developed in [7,27]. [37] presented the classification-reconstruction learning algorithm for open set recognition (CROSR), which utilizes latent representations for reconstruction and enables robust unknown detection without harming the known classification accuracy. [24] proposed the C2AE model for open set recognition, using class conditioned auto-encoders with novel training and testing methodology. Open set recognition principles have been applied to text classification [33,35], and semantic segmentation [3].

Generative Methods. Unlike discriminative models, generative approaches generate unknown samples based on Generative Adversarial Network (GAN) [9] to help the classifier learn decision boundary between known and unknown samples. [7] proposed the Generative OpenMax (G-OpenMax) algorithm, which uses a conditional GAN to synthesize mixtures of known classes and finetune the closed-set classification model. G-OpenMax improves the performance of both SoftMax and OpenMax based deep network. Although G-OpenMax effectively detects unknowns in monochrome digit datasets, it fails to produce significant performance improvement on natural images. Different from G-OpenMax, [22] introduced a novel dataset augmentation technique, called counterfactual image generation (OSRCI). OSRCI adopts an encoder-decoder GAN architecture to generate the synthetic open set examples which are close to knowns. They further reformulated the open set problem as classification with one additional class containing those newly generated samples. GAN-based methods also have been used to solve open set domain adaptation problem recently [29,39].

Out-of-Distribution Detection. The open set recognition is naturally related to some other problem settings such as out-of-distribution detection [18,32,36], outlier detection [28], and novelty detection [25], etc. They can be incorporated in the concept of open set classification as an unknown detector. However, they do not require open-set classifiers because those models does not have discriminative power within known classes. We focus in this paper on the broader open set recognition problem.

2.2 Flow-Based Methods

Flow-Based (also called invertible) models have shown promises in density estimation. The original representative models are NICE [5], RealNVP [6] and Glow [13]. The design ideas of these flow-based models are similar. Through the ingenious design, the inverse transformation of each layer of the model is relatively simple, and the Jacobian matrix is a triangular matrix, so the Jacobian determinant is easy to be calculated. Such models are elegant in theory, but there exists an issue in practice, *i.e.*, the nonlinear transformation ability of each layer becomes weak. Apart from these flow-based models, [1] proposed an Invertible Residual Network (I-ResNet), which adds some constraints to the ordinary ResNet structure to make the model invertible. The I-ResNet model still retains the basic structure of a ResNet and most of its original fitting ability. So previous experience in ResNet design can basically be re-used. Unfortunately, the density evaluation requires computing an infinite series. The choice of a fixed truncation estimator used by [1] leads to substantial bias which is tightly coupled with the expressiveness of the network. It cannot be used to perform maximum likelihood because the bias is introduced in the objective and gradients. [4] improved I-ResNet, and introduced the Residual Flows, a flow-based generative model that produces an unbiased estimate of the log density. Residual Flows allows memory-efficient backpropagation through the log density computation. This allows model to use expressive architectures and train via maximum likelihood in many tasks, such as classification, density estimation and generation, etc. Our work differs from existing flow-based models in that we explicitly address a broader open-set problem, where the flow model is a sub-component.

2.3 Flow-Based Methods for Out-of-Distribution Detection

Flow based models have been applied to out-of-distribution (OOD) detection, which is relevant to open set problem. Nalisnick et al. [20] presented a neural hybrid model created by combining deep invertible features and GLMs to filter out-of-distribution (OOD) inputs, using the model's natural "reject" rule based on the density estimation of the flow-based component. However, this rejection rule is not guaranteed to work in all settings. The main reason is that deep generative models can assign higher likelihood to OOD inputs. Nalisnick et al. [19] find that the density learned by flow-based models cannot distinguish images of common objects such as dogs, trucks, and horses (i.e. CIFAR-10) from those of house numbers (i.e. SVHN), assigning a higher likelihood to the latter when the model is trained on the former. [26] also observed that likelihood learned from deep generative models can be confounded by background statistics (e.g. OOD input with the same background but different semantic component). [10] proposed a simple technique that uses out-of-distribution samples to teach a network heuristics to detect out-of-distribution examples, namely Outlier Exposure (OE). But this improvement is limited and sensitive to the selection of OE dataset. [12] showed that a factor underlying this phenomenon is a mismatch between the nature of the prior distribution and that of the data distribution.

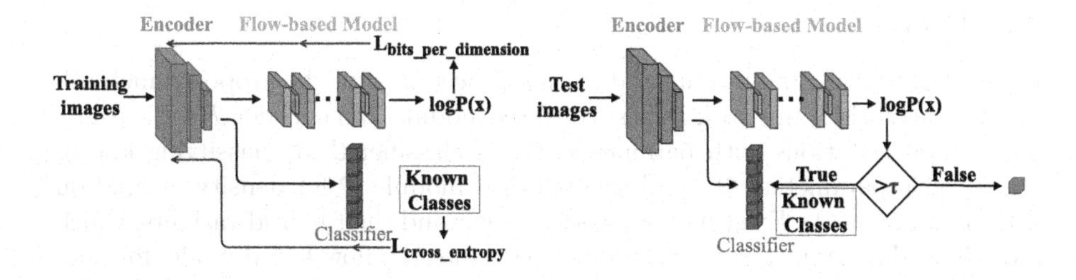

Fig. 2. Proposed architecture for open set recognition. During the training phase (left), images are mapped into a latent feature space by the encoder, then the encoded features are fed into two branches for learning: One is typical classification learning with a classifier via cross entropy loss, and the other is density estimation with a flow-based model via its log likelihood. The whole architecture is trained in an end-to-end manner. In testing phase (right), the $\log p(x)$ of each image is computed and then compared with the lowest $\log p(x)$ taken over the training set. If it is greater than the threshold τ, it is sent to the classifier to identify its specific known class, otherwise it is rejected as an unknown sample.

They proposed the use of a mixture distribution as a prior to make likelihoods assigned by deep generative models sensitive to out-of-distribution inputs. [21] explained the phenomenon through typicality and proposed a typicality test based on batches of inputs which solves many of the failure modes. While we also follow the same hybrid modeling direction, our work differs from [20] in that we choose to share a common visual representation for both the classifier and the flow model and [20] uses the output of the flow model as the input to the classifier. It is observed from our experiments that the proposed representation sharing approach is effective in our setup.

3 Our Approach

We start this section by defining the open set problem and introducing the notations. Following this is an overview of our proposed approach which we call "OpenHybrid". After an explanation to details of each module, we introduce how to achieve open set recognition using OpenHybrid.

3.1 Problem Statement and Notation

For open set recognition, given a labeled training set of instances $\mathbf{X} \in \mathbb{R}^{m \times n}$ and their corresponding labels $\mathbf{y} \in \{1, \ldots, k\}^n$ where k is the number of known classes, n is the total number of instances and m is the dimension of each instance, we learn a model $f : \mathbf{X} \rightarrow \{1, \ldots, k+1\}^n$ such that the model accurately classify an unseen instance (in test set, not in $\mathbf{X}$) to one of the k classes or an unknown class (or the "none of the above" class) indexed using $k+1$.

3.2 Overview

Figure 2 overviews the training and testing procedures for the proposed method. The OpenHybrid framework consists of three modules: an encoder $\mathcal{F}$ for learning latent representations with parameters Θ_f, a classifier $\mathcal{C}$ for classifying known classes with parameters Θ_c, and a flow-based module $\mathcal{D}$ for density estimation with parameters Θ_d. Existing flow-based models and their hybrid variants, which directly feed as input the original image data into the flow-based model for density estimation. Different from these works, our OpenHybrid framework directly uses the latent representation (the output of encoder $\mathcal{F}$) as the input to the flow model $\mathcal{D}$. The reason for this is that density estimation directly on the original image is susceptible to the population level background statistics (*e.g.*, in MNIST, the background pixels that account for most of the image are similar), which makes it hard to detect unknown samples via exact marginal likelihood. Even in some settings with different backgrounds, unknown samples are assigned higher likelihoods than known samples, and this behavior still exists and has not been explained so far. We propose to estimate the density of latent representations instead of the original input. We find our method to be effective in all of our experimental benchmarks and we do not observe the "higher outlier likelihood" issue using such framework.

For classification, the classifier $\mathcal{C}$ is directly connected to the output of the encoder $\mathcal{F}$ instead of the output of the invertible transformation $\mathcal{D}$. We choose to remove the dependency of the classifier on the flow model because we believe the output of the invertible transform loses the discriminative power. We find this approach allows both the detection of unknown classes and the classification of known classes are effective.

3.3 Training

We define the training loss function in this section.

Classification Loss. Given images in a batch $\{X_1, X_2, \ldots, X_N\}$ and their corresponding labels $\{y_1, y_2, \ldots, y_N\}$. Here N is the batch size and $\forall y_i \in \{1, 2, \ldots, k\}$. Encoder $\mathcal{F}$ and classifier $\mathcal{C}$ are trained using the following cross entropy loss.

$$\mathfrak{L}_C(\{\Theta_f, \Theta_c\}) = -\frac{1}{N}\sum_{i=1}^{N}\sum_{j=1}^{k}\mathbb{I}_{y_i}(j)\log p(y_j|x_i;\Theta_f,\Theta_c) \tag{1}$$

where $\mathbb{I}_{y_i}$ is an indicator function for label y_i, and $p(y_j|x_i;\Theta_f,\Theta_c)$ is the probability of the jth class from the probability score vector predicted by $\mathcal{C}(\mathcal{F}(x_i))$.

Density Estimation Loss. For unknown detection, unlike general open set methods, flow-based models directly fit the distribution of the training set, and compute the probability $p(x_i;\Theta_d)$ of each training sample from the training distribution (also can be treated as the distribution of known classes) through the maximum likelihood estimation. Then, they use the model's natural reject rule based on $p(x_i;\Theta_d)$ to filter unknown inputs. Although this is intuitively feasible, there are still problems as mentioned above. We suspect the problems come from the difficulty of flow models representing the original input space. So we instead estimate the density of learned latent representations $\mathcal{F}(x_i)$.

Flow-based model are the first key building block in our approach. These are simply high-capacity, bijective transformations with a tractable Jacobian matrix and inverse. The bijective nature of these transforms is crucial as it allows us to employ the change-of-variables formula for exact density evaluation:

$$\begin{aligned} \log p(x_i;\Theta_f,\Theta_d) &= \log p(\mathcal{F}(x_i;\Theta_f);\Theta_d) \\ &= \log p(\mathcal{D}(\mathcal{F}(x_i;\Theta_f);\Theta_d)) + \log \left|\det \frac{\partial \mathcal{D}(\mathcal{F}(x_i;\Theta_f);\Theta_d)}{\partial \mathcal{F}(x_i;\Theta_f)}\right|. \end{aligned} \tag{2}$$

Please note here we slightly abuse the notation for simplicity since the output of the flow model is not exactly the density of input x but instead the density of its latent embedding $\mathcal{F}(x;\Theta_f)$. A simple base distribution such as a standard normal distribution is often used for $p(\mathcal{D}(\mathcal{F}(x_i;\Theta_f);\Theta_d))$. Tractable evaluation of Eq. 2 allows flow-based models to be trained using the maximum likelihood with the loss function:

$$\mathfrak{L}_D(\{\Theta_f,\Theta_d\}) = -\frac{1}{N}\sum_{i=1}^{N} \log p(x_i;\Theta_f,\Theta_d). \tag{3}$$

In training, we map the loss $\mathfrak{L}_D(\{\Theta_f,\Theta_d\})$ to bits per dimension results by normalizing the loss by the dimensionality of the flow input. In our OpenHybrid framework, there are multiple choices for the flow-based module. Considering the stability of the density estimation, we use a tractable unbiased estimate of the log density, called residual flow [4].

Full Loss. The complete loss function of our method is:

$$\mathfrak{L}(\{\Theta_f,\Theta_c,\Theta_d\}) = \mathfrak{L}_C(\{\Theta_f,\Theta_c\}) + \lambda\mathfrak{L}_D(\{\Theta_f,\Theta_d\}) \tag{4}$$

where λ is a scaling factor on the contribution of $p(x)$. In all of our experiments in this paper, we empirically set it to 1.

3.4 Inference

Outlier Threshold. At test time, we use the probability density estimated by flow-based module to detect unknown samples from probability distributions. This value corresponds to the probability of a sample being generated from the distribution of the training classes (known classes). Theoretically, the minimum boundary of this probability distribution in the training set is the maximum value of the outlier threshold. We assume that the known samples of the training set and the test set are from the same domain, then the outlier threshold is calculated as $\tau = \min_{x_i \in \mathbf{X}} \log p(x_i;\Theta_f,\Theta_d) + s$, where s is a free parameter providing slack in the margin. We estimate the outlier threshold using training samples without data augmentation.

Open Set Recognition. Open set recognition is a classification over $k+1$ class labels, where the first k labels are from the known classes the classifier $\mathcal{C}$ is trained on, and the $k+1$-st label represents the unknown class that signifies that an instance does not belong to any of the known classes. This is performed using the outlier threshold τ and the score estimated in Eq. 2. The outlier threshold is first calculated on training data. If the estimated probability is smaller than outlier threshold, the test instance is classified as $k+1$, which in our case corresponds to the unknown class, otherwise the appropriate class label is assigned to the instance from among the known classes. More formally, the prediction of a sample x is define as

$$pred(x) = \begin{cases} k+1, & \mathcal{D}(\mathcal{F}(x_i;\Theta_f);\Theta_d) < \tau, \\ \arg\max_{j\in\{1,\ldots,k\}} \; p(y_j|x;\Theta_f,\Theta_c), & \text{otherwise.} \end{cases} \tag{5}$$

4 Experiments

We evaluate our OpenHybrid framework and compare it with the state-of-the-art non-flow-based and flow-based open set methods. We follow other methods' protocols for fair comparisons. That is, we compare with non-flow-based open set methods without considering operating threshold while we set an unified threshold value during the comparison with flow-based methods.

4.1 Experiment Setups

Implementation. In our experiments, the encoder, decoder, and classifier architectures are similar to those used in [22]. The last layer of encoder in [22] maps 512d to 100d. We moved this layer in our model to the classifier since we do not want the input dimension of flow model to be too small. So the output of our encoder is 512d instead. For flow-based model, we use the standard setup of passing the data through a logit transform [6], followed by 10 residual blocks. We use activation normalization [13] before and after every residual block. Each residual connection consists of 6 layers (*i.e.*, LipSwish [4] → InducedNormLinear → LipSwish → InducedNormLinear → LipSwish → InducedNormLinear) with hidden dimensions of 256 (the first 6 blocks) and 128 (the next 4 blocks) [20]. We use the Adam optimizer with a learning rate 0.0001 for the encoder and flow-based module to learn log probability distribution. For training classification, we use the Stochastic Gradient Descent (SGD) with momentum 0.9 and learning rate 0.01 for TinyImageNet data, 0.1 for other data. Gradients are updated alternatively between the flow model and the classifier. The parameter s is empirically set to 80. Another important factor affecting open-set performance is openness of the problem. we define the openness based on the ratio of the numbers of unique classes in training and test sets, *i.e.*, $openness = 1 - \sqrt{k_{\text{train}}/k_{\text{test}}}$ where k_{train} and k_{test} are the number of classes in the training set and the test set, respectively. In following experiments, we will evaluate performance over multiple openness values depending on different dataset settings.

Datasets. We evaluate open set classification using multiple common benchmarks, such as MNIST [17], SVHN [23], CIFAR10 [14], CIFAR+10, CIFAR+50 and TinyImageNet [16] datasets. We reuse the data splits provided by [22].

- *MNIST, SVHN, CIFAR10*: All three datasets contain 10 categories. MNIST are monochrome images with hand-written digits, and it has 60k 28×28 gray images for training and 10k for testing. SVHN are street view house numbers, consisting of ten digit classes each with between 9981 and 11379 32×32 color images. To validate our method on non-digital images, we apply the CIFAR10 dataset, which has 50k 32×32 natural color images for training and 10k for testing. Each dataset is partitioned at random into 6 known and 4 unknown classes. In these settings, the openness score is fixed to 22.54%.
- *CIFAR+10, CIFAR+50*: To test the method in a range of greater openness scores, we perform CIFAR+U experiments using CIFAR10 and CIFAR100 [14]. 4 known classes are sampled from CIFAR10 and U unknown classes are drawn randomly from the more diverse CIFAR100 dataset. Openness scores of CIFAR+10 and CIFAR+50 are 46.54% and 72.78%, respectively.
- *TinyImageNet*: For the larger TinyImagenet dataset, which is a 200-class subset of ImageNet, we randomly sampled 20 classes as known and the remaining classes as unknown. In this setting, the openness score is 68.37%.

The out-of-distribution (OOD) detection community often evaluates methods on cross-dataset setups [10,20,21,26], such as training on CIFAR10 and testing on CIFAR100. So we perform extra experiments on two such settings between CIFAR10 and CIFAR100 and report results comparable to OOD literature.

Metrics. Open set classification performance can be characterized by F-score or AUROC (Area Under ROC Curve) [8]. AUROC is commonly reported by both open set recognition and out-of-distribution detection literature. So we mainly use AUROC to compare with existing methods. We adopt F-score in some of our experiments as it also measures the in-distribution classification performance. For both metrics, higher values are better.

4.2 Results

Comparison with Non-flow-based Methods. We compare OpenHybrid against the following non-flow-based baselines:

1. *SoftMax*: A standard confidence-based method for open-set recognition by using SoftMax score of a predicted class.
2. *OpenMax* [2]: This approach augments the baseline classifier with a new OpenMax layer replacing the SoftMax at the final layer of the network.
3. *G-OpenMax* [7]: A direct extension of OpenMax method, which trains networks with synthesized unknown data by using a Conditional GAN.
4. *OSRCI* [22]: An improved version of G-OpenMax work, which uses a specific data augmentation technique called counterfactual image generation to train the classifier for the $k+1$-st class.

Table 1. AUROC for comparisons of our method with recent open set methods. Results averaged over 5 random class partitions. The best results are highlighted in **bold**.

Method	MNIST	SVHN	CIFAR10	CIFAR+10	CIFAR+50	TinyImageNet
SoftMax	0.978	0.886	0.677	0.816	0.805	0.577
OpenMax [2]	0.981	0.894	0.695	0.817	0.796	0.576
G-OpenMax [7]	0.984	0.896	0.675	0.827	0.819	0.580
OSRCI [22]	0.988	0.910	0.699	0.838	0.827	0.586
C2AE [24]	0.989	0.922	0.895	0.955	0.937	0.748
CROSR [37]	0.991	0.899	0.883	0.912	0.905	0.589
OpenHybrid (ours)	**0.995**	**0.947**	**0.950**	**0.962**	**0.955**	**0.793**

5. *C2AE* [24]: This approach uses class conditioned auto-encoders with novel training and testing methodologies for open set recognition.
6. *CROSR* [37]: A deep open set classifier augmented by latent representation learning which jointly classifies and reconstructs the input data.

Table 1 presents the open set recognition performance of our method and non-flow-based baselines on six datasets. Our approach OpenHybrid outperforms all of the baseline methods, which demonstrates the effectiveness of our approach. It is interesting to note that our method on MNIST dataset produces a minor improvement compared to the other methods. The main reason is that the MNIST is relatively simple, and the results of all methods on it are almost saturated. But for other relatively complex databases, our method performs significantly better than the baseline methods, especially for natural images, such as CIFAR (6% better than the second best) and TinyImageNet (5% better than the second best).

Comparison with Flow-Based Methods. We compare our approach against our implementations of the following flow-based approaches:

1. *DIGLM* [20]: A neural hybrid model consisting of a linear model defined on a set of features computed by a deep invertible transformation. It uses the model' natural reject rule based on the generative component $p(x)$ to detect unknown inputs. The threshold is setted as $\min_{\mathbf{x}\in\mathbf{X}} p(x;\theta) - c$, where the minimum is taken over the training set and c is a free parameter providing slack in the margin.
2. *OE* [10]: A training method leveraging an auxiliary dataset of unknown samples to improve unknown detection. The framework is the same as DIGLM, except that during training, a margin ranking loss on the log probabilities of training and outlier exposure samples is used to update the flow-based model. In this experiment, we use counterfactual images generated by [22] from training samples as its outlier exposure dataset.

Table 2 shows the AUROC of our method and the flow-based baselines in different datasets. We observe that our method consistently outperforms the baseline methods significantly under all open set benchmarks. The same trend is

Table 2. AUROC for our methods and flow-based baselines. Results are averaged over 5 random class partitions. The best results are highlighted in **bold**.

Method	MNIST	SVHN	CIFAR10	CIFAR+10	CIFAR+50	TinyImageNet
DIGLM	0.643	0.559	0.583	0.590	0.594	0.520
DIGLM + OE	0.721	0.643	0.655	0.670	0.671	0.596
OpenHybrid (ours)	**0.995**	**0.947**	**0.950**	**0.962**	**0.955**	**0.793**

Table 3. AUROC for cross-dataset out-of-distribution detection between CIFAR-10 and CIFAR-100.

Train → Test (OOD)	OE [10]	Ours
CIFAR10 → CIFAR100	0.933	**0.951**
CIFAR100 → CIFAR10	0.757	**0.856**

Table 4. F-scores (For readers who are interested in classification accuracy: Our approach achieves overall accuracy 0.947 in MNIST, 0.929 in SVHN and 0.868 in CIFAR10. However, we believe F-score is a better measurement which considers data imbalance.) of the proposed OpenHybrid models using pretrained encoder and joint training.

Method	MNIST	SVHN	CIFAR10
Pretrained encoder	0.847	0.842	0.791
Joint training	**0.942**	**0.912**	**0.865**

observed for the f-score metric, e.g., we achieved 0.865 in CIFAR10 while DIGLM achieves only 0.673 and DIGLM+OE achieves 0.701) (Table 3).

Cross-Dataset OOD Settings. We further evaluate our approach on two cross-dataset settings: training on CIFAR10 and testing on CIFAR100 and vice versa. We compare the AUROC of our method directly with the numbers reported in [10]. The results suggest that our approach is still competitive in such settings. It is worth noting that training on CIFAR100 and testing on CIFAR10 is a harder task, probably due to the higher number of training classes. Our approach achieves higher gains (+10%) in this setting.

4.3 Discussion

The Benefit of Joint Training. We further compare the end-to-end trained OpenHybrid with a different training strategy based on alternative training. The framework is still the same. However, during training, the encoder and classifier are pretrained first on the training data. The flow-based model was then trained separately with both encoder and classifier being frozen. Table 4 shows a comparison between the two methods using F-score. The slack parameter s is chosen to be 80 for all datasets. We observe that joint training consistently outperforms OpenHybrid with a fixed pretrained encoder.

A Study on the Parameters. Our loss function contains a trade-off parameter λ. We varied this value among 0.5, 1 and 2 in the MNIST dataset and observed AUROC scores 0.993, 0.995, and 0.998, respectively. The model seems not sensitive to this variable but it is a parameter that can be tuned to further improve

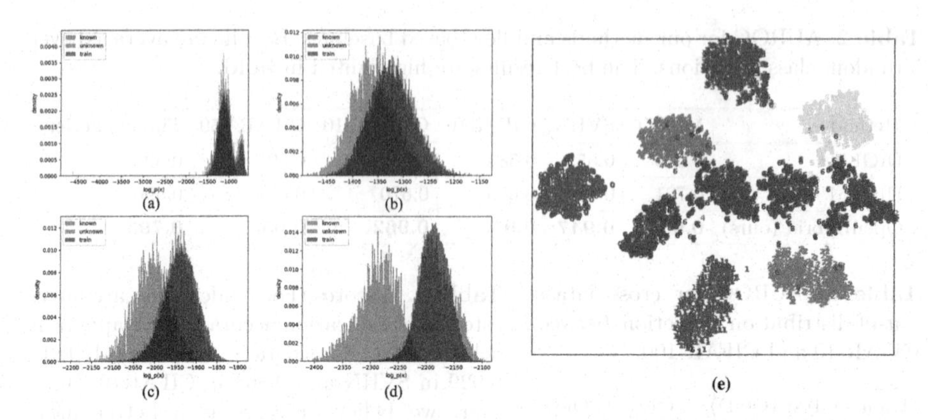

Fig. 3. Left: Histograms of log-likelihoods for MNIST (0–5 as known classes and 6–9 as unknown classes) made by (a) DIGLM, (b) DIGLM + OE, (c) OpenHybrid with pretrained encoder and end-to-end OpenHybrid. The blue color indicates training samples, the pink indicates known samples in the test, and the green is unknown samples. Right (e): t-SNE visualization of the latent space by end-to-end OpenHybrid. Different colors represent different classes. Brown color represents the unkown digits (6–9) (Color figure online).

performance. Another important parameter is the number of residual blocks in the flow model. We varied this value among 4, 8, 10 and 16 in CIFAR10 benchmark. Surprisingly, we still observe stable AUROC results (0.945, 0.948, 0.950, 0.958). So, for practitioners who may have resource constraints, it is advised to consider a smaller flow-based network when using the OpenHybrid framework.

A Visualization of the Estimated Density. Figure 3 (left) shows the histograms of log-likelihoods for MNIST (0–5 as known classes and 6–9 as unknown classes) made by DIGLM, OE, OpenHybrid with pretrained encoder and OpenHybrid with joint training. For DIGLM (a), the three histograms almost overlap so it is impossible to detect the unknown class by setting a threshold. The density estimation is improved with the help of OE (b), however, there is still a large area of overlap. The distribution overlap becomes further smaller but still not ideal when using OpenHybrid with pretrained encoder (c). In contrast, we observed the end-to-end OpenHybrid (d) produces the histogram of unknown samples well separated from those of known samples.

A Visualization of the Latent Space. Figure 3(e) shows a t-SNE [34] plot of the latent space learned by end-to-end OpenHybrid. The brown color represents the unkonwn classes (digit 6, 7, 8, 9) which is well separated from other color (known classes from 0 to 5). Interestingly, the model also learns to separate digits 6–9, which is in an unsupervised fashion. Although the MNIST dataset is simple compared to other real datasets, this result shows the potential of representation learning using hybrid models as a promising research direction.

A Disappeared Issue of Flow-Based Models. Nalisnick et al. [19] raised the issue that the flow-based model trained on CIFAR10 will assign a higher

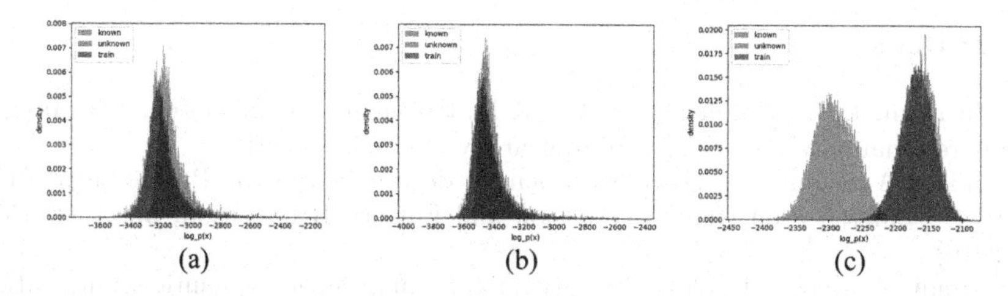

Fig. 4. Histograms of log-likelihoods for CIFAR10 (known samples) and SVHN (unknown samples) made by (a) DIGLM, (b) DIGLM + OE and (c) the proposed OpenHybrid. The blue color indicates training samples, the pink indicates known samples in the test, and the green is unknown samples (Color figure online).

log-likelihood value to SVHN. So we further conduct an experiment on this setting, where we use the full 10 classes of the CIFAR10 as known classes, and the SVHN as an unknown class. Our approach achieves 0.998 AUROC on this setting. Figure 4 shows the histograms of log-likelihoods under this setting. Similar to the observation made by [19], in Fig. 4(a), the histogram of unknown samples (green) is shifted more to the right than that of known samples (blue and pink), *i.e.*, unknown samples are assigned a larger log-likelihood value than known samples. In Fig. 4(b), OE seems to help but it does not fully address the problem as well. Our method is shown in Fig. 4(c) which clearly distinguish the two distributions. The histogram of unknown samples is almost entirely to the left of known samples. We believe a potential reason is that the original input space is a multimodal distribution and our method projects the input data into a latent space which is probably more suitable to the unimodal assumption of flow-based models. While we are unable to prove this theoretically, we hope our results could inspire future works on deeper understanding of flow-based models.

5 Conclusion

We presented the OpenHybrid framework for open set recognition. Our approach is built upon a flow-based model for density estimation and a discriminative classifier, with a shared latent space. Extensive experiments show that our approach achieves the state of the art. A common issue of flow-based models is that they often assign larger likelihood to out-of-distribution samples. We empirically observe on various datasets that this issue disappear by learning a joint feature space. Ablation study also suggests that joint training is another key contributing factor to the superior open set recognition performance.

Acknowledgement. We would like to thank Balaji Lakshminarayanan and Olaf Ronneberger for meaningful discussions. This research was supported by the National Science Foundation of China under Grants 61772257 and the Fundamental Research Funds for the Central Universities 020914380080.

References

1. Behrmann, J., Grathwohl, W., Chen, R.T., Duvenaud, D., Jacobsen, J.H.: Invertible residual networks. arXiv preprint arXiv:1811.00995 (2018)
2. Bendale, A., Boult, T.E.: Towards open set deep networks. In: Proceedings of the IEEE Conference on Computer Vision and Pattern Recognition, pp. 1563–1572 (2016)
3. Bevandić, P., Krešo, I., Oršić, M., Šegvić, S.: Simultaneous semantic segmentation and outlier detection in presence of domain shift. In: Fink, G.A., Frintrop, S., Jiang, X. (eds.) DAGM GCPR 2019. LNCS, vol. 11824, pp. 33–47. Springer, Cham (2019). https://doi.org/10.1007/978-3-030-33676-9_3
4. Chen, T.Q., Behrmann, J., Duvenaud, D.K., Jacobsen, J.H.: Residual flows for invertible generative modeling. In: Advances in Neural Information Processing Systems, pp. 9913–9923 (2019)
5. Dinh, L., Krueger, D., Bengio, Y.: NICE: non-linear independent components estimation. arXiv preprint arXiv:1410.8516 (2014)
6. Dinh, L., Sohl-Dickstein, J., Bengio, S.: Density estimation using real NVP. arXiv preprint arXiv:1605.08803 (2016)
7. Ge, Z., Demyanov, S., Chen, Z., Garnavi, R.: Generative openmax for multi-class open set classification. arXiv preprint arXiv:1707.07418 (2017)
8. Geng, C., Huang, S., Chen, S.: Recent advances in open set recognition: a survey. arXiv preprint arXiv:1811.08581 (2018)
9. Goodfellow, I., et al.: Generative adversarial nets. In: Advances in Neural Information Processing Systems, pp. 2672–2680 (2014)
10. Hendrycks, D., Mazeika, M., Dietterich, T.: Deep anomaly detection with outlier exposure. arXiv preprint arXiv:1812.04606 (2018)
11. Mendes Júnior, P.R., et al.: Nearest neighbors distance ratio open-set classifier. Mach. Learn. **106**(3), 359–386 (2016). https://doi.org/10.1007/s10994-016-5610-8
12. Kamoi, R., Kobayashi, K.: Likelihood assignment for out-of-distribution inputs in deep generative models is sensitive to prior distribution choice. arXiv preprint arXiv:1911.06515 (2019)
13. Kingma, D.P., Dhariwal, P.: Glow: generative flow with invertible 1x1 convolutions. In: Advances in Neural Information Processing Systems, pp. 10215–10224 (2018)
14. Krizhevsky, A., Hinton, G., et al.: Learning multiple layers of features from tiny images (2009)
15. Krizhevsky, A., Sutskever, I., Hinton, G.E.: Imagenet classification with deep convolutional neural networks. In: Advances in Neural Information Processing Systems, pp. 1097–1105 (2012)
16. Le, Y., Yang, X.: Tiny imagenet visual recognition challenge. CS 231N (2015)
17. LeCun, Y., Cortes, C., Burges, C.: MNIST handwritten digit database (2010)
18. Lee, K., Lee, H., Lee, K., Shin, J.: Training confidence-calibrated classifiers for detecting out-of-distribution samples (2017)
19. Nalisnick, E., Matsukawa, A., Teh, Y.W., Gorur, D., Lakshminarayanan, B.: Do deep generative models know what they don't know? arXiv preprint arXiv:1810.09136 (2018)
20. Nalisnick, E., Matsukawa, A., Teh, Y.W., Gorur, D., Lakshminarayanan, B.: Hybrid models with deep and invertible features. arXiv preprint arXiv:1902.02767 (2019)
21. Nalisnick, E., Matsukawa, A., Teh, Y.W., Lakshminarayanan, B.: Detecting out-of-distribution inputs to deep generative models using typicality. arXiv preprint arXiv:1906.02994 (2019)

22. Neal, L., Olson, M., Fern, X., Wong, W.-K., Li, F.: Open set learning with counterfactual images. In: Ferrari, V., Hebert, M., Sminchisescu, C., Weiss, Y. (eds.) ECCV 2018. LNCS, vol. 11210, pp. 620–635. Springer, Cham (2018). https://doi.org/10.1007/978-3-030-01231-1_38
23. Netzer, Y., Wang, T., Coates, A., Bissacco, A., Wu, B., Ng, A.Y.: Reading digits in natural images with unsupervised feature learning (2011)
24. Oza, P., Patel, V.M.: C2AE: class conditioned auto-encoder for open-set recognition. In: Proceedings of the IEEE Conference on Computer Vision and Pattern Recognition, pp. 2307–2316 (2019)
25. Perera, P., Nallapati, R., Xiang, B.: OCGAN: one-class novelty detection using GANs with constrained latent representations. In: Proceedings of the IEEE Conference on Computer Vision and Pattern Recognition, pp. 2898–2906 (2019)
26. Ren, J., et al.: Likelihood ratios for out-of-distribution detection. In: Advances in Neural Information Processing Systems, pp. 14680–14691 (2019)
27. Rozsa, A., Günther, M., Boult, T.E.: Adversarial robustness: softmax versus openmax. arXiv preprint arXiv:1708.01697 (2017)
28. Ruff, L., et al.: Deep one-class classification. In: International Conference on Machine Learning, pp. 4393–4402 (2018)
29. Saito, K., Yamamoto, S., Ushiku, Y., Harada, T.: Open set domain adaptation by backpropagation. In: Ferrari, V., Hebert, M., Sminchisescu, C., Weiss, Y. (eds.) ECCV 2018. LNCS, vol. 11209, pp. 156–171. Springer, Cham (2018). https://doi.org/10.1007/978-3-030-01228-1_10
30. Scheirer, W.J., Jain, L.P., Boult, T.E.: Probability models for open set recognition. IEEE Trans. Pattern Anal. Mach. Intell. **36**(11), 2317–2324 (2014)
31. Schölkopf, B., Williamson, R.C., Smola, A.J., Shawe-Taylor, J., Platt, J.C.: Support vector method for novelty detection. In: Advances in Neural Information Processing Systems, pp. 582–588 (2000)
32. Serrà, J., Álvarez, D., Gómez, V., Slizovskaia, O., Núñez, J.F., Luque, J.: Input complexity and out-of-distribution detection with likelihood-based generative models. CoRR abs/1909.11480 (2019). http://dblp.uni-trier.de/db/journals/corr/corr1909.html#abs-1909-11480
33. Shu, L., Xu, H., Liu, B.: DOC: deep open classification of text documents. arXiv preprint arXiv:1709.08716 (2017)
34. Van Der Maaten, L.: Accelerating t-SNE using tree-based algorithms. J. Mach. Learn. Res. **15**(1), 3221–45 (2014)
35. Venkataram, V.M.: Open set text classification using neural networks. Ph.D. thesis, Kraemer Family Library, University of Colorado Colorado Springs (2018)
36. Vernekar, S., Gaurav, A., Abdelzad, V., Denouden, T., Salay, R., Czarnecki, K.: Out-of-distribution detection in classifiers via generation. arXiv preprint arXiv:1910.04241 (2019)
37. Yoshihashi, R., Shao, W., Kawakami, R., You, S., Iida, M., Naemura, T.: Classification-reconstruction learning for open-set recognition. In: Proceedings of the IEEE Conference on Computer Vision and Pattern Recognition, pp. 4016–4025 (2019)
38. Zhang, H., Patel, V.M.: Sparse representation-based open set recognition. IEEE Trans. Pattern Anal. Mach. Intell. **39**(8), 1690–1696 (2016)
39. Zhang, H., Li, A., Han, X., Chen, Z., Zhang, Y., Guo, Y.: Improving open set domain adaptation using image-to-image translation. In: 2019 IEEE International Conference on Multimedia and Expo (ICME), pp. 1258–1263. IEEE (2019)

TopoGAN: A Topology-Aware Generative Adversarial Network

Fan Wang(✉), Huidong Liu, Dimitris Samaras, and Chao Chen

Stony Brook University, Stony Brook, NY 11794, USA
{fanwang1,huidliu,samaras}@cs.stonybrook.edu, chao.chen.1@stonybrook.edu

Abstract. Existing generative adversarial networks (GANs) focus on generating realistic images based on CNN-derived image features, but fail to preserve the structural properties of real images. This can be fatal in applications where the underlying structure (e.g.., neurons, vessels, membranes, and road networks) of the image carries crucial semantic meaning. In this paper, we propose a novel GAN model that learns the topology of real images, i.e., connectedness and loopy-ness. In particular, we introduce a new loss that bridges the gap between synthetic image distribution and real image distribution in the topological feature space. By optimizing this loss, the generator produces images with the same structural topology as real images. We also propose new GAN evaluation metrics that measure the topological realism of the synthetic images. We show in experiments that our method generates synthetic images with realistic topology. We also highlight the increased performance that our method brings to downstream tasks such as segmentation.

Keywords: Topology · Persistent homology · Generative Adversarial Network

1 Introduction

Generative adversarial networks (GANs) [20] have been very successful in generating realistic images. GANs train a generator to synthesize images that are similar to real images, and at the same time, a discriminator to distinguish these fake images from real ones. Through a minimax game, the generator converges to a network that generates synthetic images sampled from a distribution that matches the distribution of the real images.

When designing GANs, a key question is how to bridge the gap between the synthetic and real image distributions not only in appearance, but also in semantics. As shown [38,44], widely-used GANs [5,20,22,34,47,56] only match the first order moments of the distributions within a CNN-based image feature space. Newer methods match the synthetic/real image distributions using higher

Electronic supplementary material The online version of this chapter (https://doi.org/10.1007/978-3-030-58580-8_8) contains supplementary material, which is available to authorized users.

A. Vedaldi et al. (Eds.): ECCV 2020, LNCS 12348, pp. 118–136, 2020.
https://doi.org/10.1007/978-3-030-58580-8_8

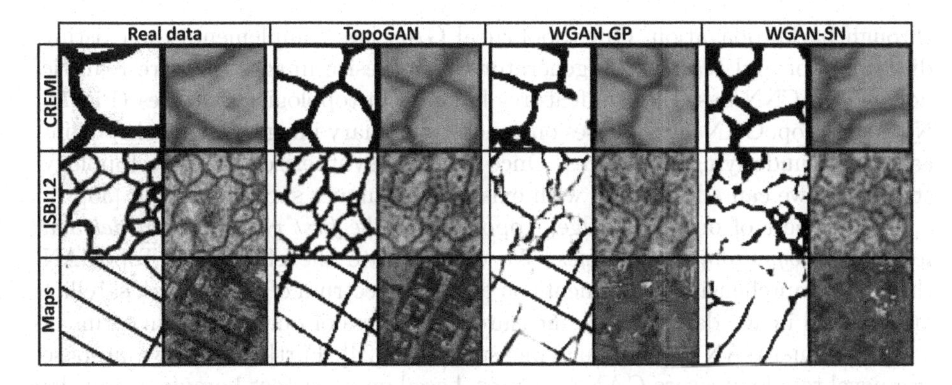

Fig. 1. Sample images in which the structures are neuron membranes and road networks from satellite images. From top to bottom: neuron images (CREMI [15]), neuron images (ISBI12 [4]) and satellite images (Google Maps [30]). From left to right: real images, images synthesized by TopoGAN, WGAN-GP and WGAN-SN. Each real/synthetic mask is paired with a textured image. For synthetic images, texture is added by a separately-trained pix2pix [30] network.

order statistics, e.g. second order statistics of the image features [43,44]. Kossaifi et al. [32] explicitly add a statistical shape prior for face images into the generator. The intuition is that the more high order information a generator can learn, the more semantically realistic the synthesized images will be.

In this paper, we pay attention to the structural information of an image. In many applications, images contain structures with rich topology, e.g., biomedical images with membrane, neuron or vessel structures, and satellite images with road maps (Fig. 1). These structures and their topology, i.e., connectivity and loopy-ness, carry important semantic/functionality information. Structural fidelity becomes crucial if we want to use the synthetic images to train downstream methods that hinge on the structural information, e.g., diagnosis algorithms based on the structural richness of retinal vessels, navigation systems based on road network topology, or neuron classifiers based on neuron morphology and connectivity.

In this paper, we propose TopoGAN, the first GAN model that learns topology from real data. Topology directly measures structural complexity, such as the numbers of connected components and holes. This information is very difficult to learn, due to its global nature. The conventional GAN discriminator distinguishes synthetic and real images in terms of CNN-based features, but is agnostic to topological dissimilarity. Thus, the generator cannot learn real image topology. In Fig. 1, structures synthesized by conventional GANs (WGAN-GP and WGAN-SN) tend to be broken and disconnected.

Our main technical contribution is a novel *topological GAN loss* that explicitly matches the synthetic and real image distributions in terms of their topology. Based on persistent homology theory [16], we map both synthetic and real images into a topological feature space, where their topological dissimilarity can be measured as a loss. We show that our loss is differentiable and can be minimized through backpropagation. Our topological GAN loss complements the existing discriminator and teaches the generator to synthesize images that are realistic not only in CNN-based image features but also in topological features (Fig. 1). Note that TopoGAN only focuses on generating binary images (i.e. masks) delineating the underlying structures. Once we have synthesized realistic topology structures, we can add texture with existing techniques such as pix2pix [30].

To the best of our knowledge, TopoGAN is *the first generative model that learns topology from real images.* We demonstrate the efficacy of TopoGAN through comprehensive experiments on a broad spectrum of biomedical, satellite and natural image datasets. We measure the success of our method in terms of a conventional GAN performance measure, FID [24]. Furthermore, we propose two novel topology-aware GAN measures, based on persistent homology and the Betti number. We show that TopoGAN outperforms baseline GAN models by a large margin in these topology-aware measures. Finally, we show that synthesized images with learnt topology can improve performance in downstream tasks such as image segmentation. In summary, our contributions are three-fold:

- We propose a topological GAN loss that measures the distance between synthetic and real image distributions in the space of topological features. Compared to previous topological loss that is applied to individual instances [28], our loss is the first to enforce topological similarity between distributions.
- We show that this loss is differentiable and incorporate it into GAN training.
- We propose novel topology-aware measures to evaluate generator performance in topological feature space.

2 Related Work

Generative Adversarial Nets (GANs) [20] are very popular for modeling data distributions. However, GAN training is very unstable. WGAN [5], WGAN-GP [22], WGAN-TS [37], WGAN-QC [36] and others, use the Wasserstein distance to train GANs. Different gradient penalty strategies [22,36,40,60] can stabilize GAN training effectively. Apart from the gradient penalty, Spectral Normalization (SN) [41] is also widely used for GAN training [7,65]. PatchGAN [30] applies a GAN to local patches instead of the whole image in order to capture high frequency signals. Such local/high frequency signals are very useful in various generative models, such as Pix2pix [30,63], CycleGAN [68] and SinGAN [57].

Several geometry-related GANs exploit geometric information on images. The geometricGAN [35] adopts the large margin idea from SVMs [14] to learn the discriminator and generator. The Localized GAN (LGAN) [50] uses local coordinates to parameterize the local geometry of the data manifold. The Geometry-Aware GAN (GAGAN) [32] is tailored for generating facial images using face shape priors. The Geometry-Consistent GAN (GcGAN) [18] uses a geometry-consistency constraint to preserve the image's semantic structure. Geometric transformations are restricted to image flipping and rotation.

We note that high-order structural information has been used in adversarial networks for semantic segmentation. Existing methods [19,29,39] use adversarial losses in the semantic segmentation space as they encode high-order structural information. However, these methods do not explicitly preserve topology.

Topological Information for Image Analysis. Many methods have been proposed to directly use persistent homology as a feature extraction mechanism. The extracted topological feature can be vectorized [3], and used as input to kernel machines [9,33,54] or deep neural networks [26]. For fully supervised image segmentation tasks, topological information has been explicitly used as a constraint/loss to improve segmentation quality [11,28,64]. Mosinska et al. [42] model topology implicitly with feature maps from pretrained VGG networks [58], but the method does not generalize to structures of unseen geometries. We also refer to methods developed for retinal vessels [23] and lung airways [51]. These methods only focus on connectivity (0-dimensional topology) and cannot generalize to high-dimensional topology. In machine learning, topological information has been used to analyze data manifold topology [10,25,46,53] and to leverage advanced structural information for graph learning [66,67].

In generative models, Khrulkov and Oseledets [31] use data manifold topology to compare synthetic and real data distributions as a qualitative measure of generative models. However, their measure still focuses on the standard image feature space, and cannot really evaluate whether the generator has learned the real image topology. Brüel-Gabrielsson et al. [8] use a loss to enforce the connectivity constraint in the generated images. However, enforcing hand-crafted topological constraints (e.g., connectedness) does not help the generator to learn the true topological distribution from real data. TopoGAN is the first generative model that automatically learns topological properties from real images.

3 Method

Our TopoGAN matches synthetic and real image distributions for both image and topology features. For this purpose, in addition to the conventional discriminator and generator losses, we introduce a new loss term for the generator, $L_{topo}(P_{data}, G)$. This loss term, called the *topological GAN loss*, measures how close the images generated by G are to the real images in terms of topology. Minimizing it forces the synthetic images to have similar topology as the real images. The discriminator loss is shown in Eq. (1). The generator loss (Eq. (2)) is a sum of the conventional generator loss and the new loss. Formally, we have

$$\arg\max_D \left[\mathbb{E}_{x\sim P_{data}} \log D(x) + \mathbb{E}_{z\sim P_z} \log(1 - D(G(z)))\right], \tag{1}$$

$$\arg\min_G \Big[\underbrace{\mathbb{E}_{z\sim P_z} \log(1 - D(G(z)))}_{\text{conventional generator loss}} + \lambda \underbrace{L_{topo}(P_{data}, G)}_{\text{topological GAN loss}}\Big], \tag{2}$$

where λ controls the weight of the topological GAN loss.

We focus on generating binary images, i.e., masks delineating structures such as vessels, neuron membranes, road networks, etc. The generator outputs a real-valued grey-scale image as the synthetic mask. The discriminator treats the input

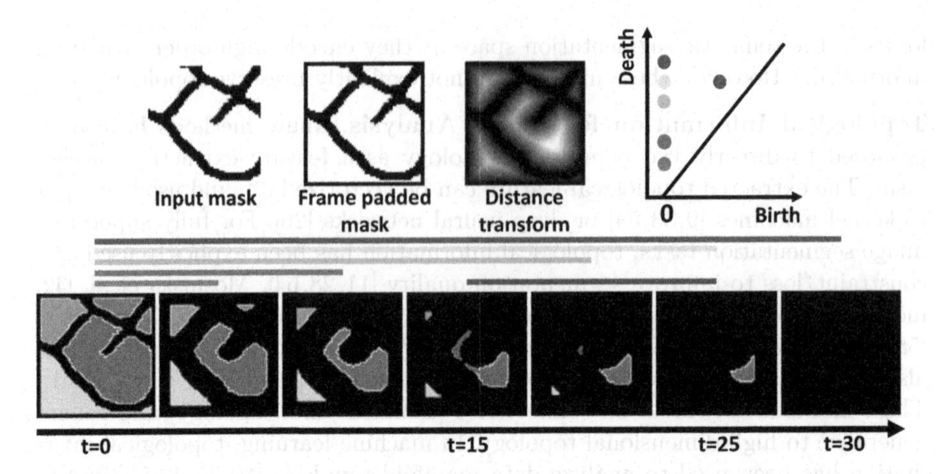

Fig. 2. Illustration of persistent homology. Top row from left to right: the input mask, padded with a frame (so that all branches form holes), the distance transform and the output persistence diagram. Bottom row: the sequence of sublevel sets with different threshold values. Different holes are born and filled. The original holes are all born at $t = 0$. The almost-hole (red region, red bar, red dot) is born at a later time ($t = 15$). (Color figure online)

image (real or synthetic) as a real-valued grey-scale image ranging between 0 and 1. After mask synthesis, a separately-trained pix2pix [30] network fills in the textures based on each mask.

The rest of this section describes how to define and optimize the topological GAN loss. In Sect. 3.1, we explain how to extract the topological feature (called persistence diagram) of an input mask using the theory of persistent homology. In Sect. 3.2 and 3.3, we formalize the topological GAN loss by comparing the distributions of persistence diagrams computed from synthetic and real images respectively. Minimizing this loss practically moves a synthetic persistence diagram toward its matched real persistence diagram. This diagram modification effectively grows the structure/mask to complete almost-loops. This teaches the generator to synthesize images without incomplete loops.

As a separate technical contribution, we propose two new topology-aware metrics to compare the distributions in the topological feature space in Sect. 3.4.

3.1 Persistent Homology: From Images to Topological Features

We explain how to extract the topological feature of an input mask using the theory of persistent homology. We compute a *persistence diagram* capturing not only holes/loops, but also almost-holes/almost-loops (structures that almost form a hole or a loop) (Fig. 2). We describe the basic concepts, leaving technical details to supplemental material and a classic topological data analysis reference [16].

Given a topological space, $y \subseteq \mathbb{R}^2$, the holes and connected components are its 1- and 0-dimensional topological structures respectively. We mainly focus on 1-dimensional topology in this paper. The number of holes is the *Betti number*,

β_y. In Fig. 2, we show a sample mask from the CREMI dataset, delineating a neuron membrane structure. We add a frame around the patch so that all structures are accounted for via 1-dimensional homology. In algebraic topology [45], we are effectively computing the relative homology.

We observe 5 holes (Betti number $\beta_y = 5$) in the figure and the Betti number is only able to capture the complete holes. The dangling branch in the middle of the image almost creates a new hole. But this almost-hole is not captured by the Betti number. To effectively account for these almost-holes in our computations, we leverage the distance transform and the theory of persistent homology [16]. We review the distance transform:

Definition 1. *The Distance Transform (DT) [17] generates a map D for each pixel p on a binary image I:* $D(p) = \min_{q \in \Omega}\{||p - q|| \mid I(q) = 0\}$, *in which* Ω *is the image domain.*

Instead of only looking at the original function, we apply a distance transform to the mask and get a non-negative scalar function defined on the whole image domain, $f_y : \Omega \to \mathbb{R}^+$. We define the *sublevel set* of f_y as the domain thresholded by a particular threshold t, formally, $\Omega^t_{f_y} = \{x \in \Omega | f_y(x) \le t\}$. We notice that one can take different sublevel sets with different thresholds. For certain threshold values, the almost-hole becomes a complete hole. The sequence of all possible sublevel sets, formally called the *filtration* induced by f_y, essentially captures the growing process of the initial mask.

Persistent homology takes the whole filtration and inspects its topological structures (holes, connected components, and higher dimensional topological structures). Each topological structure lives during an interval of threshold values. In Fig. 2, the five original holes are born at $t = 0$ and filled at different times, when they are filled up by the growing mask. The almost-hole (in red) is born at $t = 15$, when the purple hole is split into two. It dies at $t = 25$. All the holes (with life spans drawn as horizontal bars) are recorded as a 2D point set called a *persistence diagram*. The birth time and death time of each hole become the two coordinates of its corresponding point. In this diagram, we have 5 points with $birth = 0$ and a red point with a non-zero birth time, for the almost-hole[1].

3.2 Distance Between Diagrams and Topological GAN Loss

In this section, we formalize our topological GAN loss. Using the distance transform and persistent homology, we transform each input binary image y into its corresponding persistence diagram, $\mathrm{dgm}(f_y)$, which we call the topological feature of y. We first introduce the distance between any two persistence diagrams, which measures the topological dissimilarity between two images. Next, we define our topological GAN loss as the distance between two sets of diagrams, computed from synthetic masks and real masks respectively. We use optimal transport [62] to match the two sets of diagrams, and then define the loss as the total distance

[1] The persistence diagram definition does not require the input to be a distance transform. It can be an arbitrary scalar function defined on a topological space.

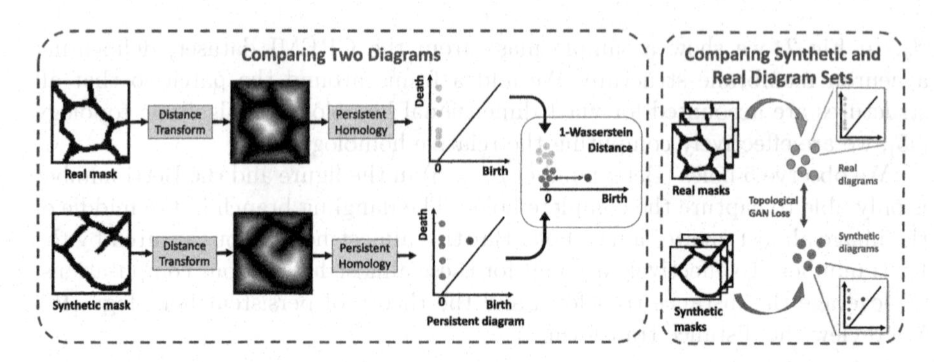

Fig. 3. Our topology-processing component. The input is a batch of real masks and synthetic masks. Each real or synthetic mask goes through the distance transform and persistent homology computation. We get its persistent diagram, a set of 2D points. We compare the two diagrams using the 1-Wasserstein distance only on birth times. The loss is defined as the matching distance between the two sets of diagrams (synthetic and real), computed using optimal transport.

between the matched diagram pairs. An illustration of the topological GAN loss can be found in Fig. 3.

The distance between persistence diagrams has been well studied. One can treat two diagrams as two point sets on a 2D plane and measure their p-Wasserstein distance. This distance is well-behaved [12,13].

In this paper, we use a modified version of the classic p-Wasserstein distance between diagrams. In particular, we only focus on the birth time, and drop their death time. The reason is that we are mainly focusing on the gaps one needs to close to complete an almost-hole (depending on birth time) and not particularly concerned with the size of the hole (corresponding to death time). Formally, we project all points of the two diagrams to the birth axis and compute their 1-Wasserstein distance, i.e., the optimal matching distance between the two point sets within the birth axis, as illustrated in Fig. 3. We note that points in the diagrams of the synthetic and real images are mostly paired with nearby points. The only exception is the red point corresponding to the almost-hole. The matching distance essentially measures the gap of the almost-hole. The diagram distance measures how easy it is to fix the synthetic image so it has the same number of holes as the real one. Formally, the distance between two diagrams dgm_1 and dgm_2 is

$$\mathcal{W}_1(\mathrm{dgm}_1, \mathrm{dgm}_2) = \min_{\sigma \in \Sigma} \sum_{x \in \mathrm{dgm}_1} |b_x - b_{\sigma(x)}| = \sum_{x \in \mathrm{dgm}_1} |b_x - b_{\sigma^*(x)}|, \quad (3)$$

in which Σ is the set of all possible one-to-one correspondences between the two diagrams, and σ^* is the optimal matching one can choose. Here b_x denotes the birth time of a point x in dgm_1. Similarly, $b_{\sigma(x)}$ and $b_{\sigma^*(x)}$ are the birth times of x's match $\sigma(x)$ and optimal match $\sigma^*(x)$ in dgm_2. The matching may not exist when there are different numbers of points from the two diagrams.

To this end, we can add infinitely many points to the diagonal line ($b = d$) so the unmatched points can be matched to the diagonal line.[2] In practice, our algorithm for matching computation is very similar to the sliced Wasserstein distance [6] for persistence diagrams, except that we only use one of the infinitely many slices, i.e., $d = 0$.

Topological GAN Loss Defined via Matching Persistence Diagrams. Next, we define our loss, which measures the difference between two diagram distributions. The loss should be (1) simple to compute; and (2) efficient in matching the two distributions. Due to these constraints, it is not straightforward to use other approaches such as the kernel mean embedding (which is used in Sect. 3.4 to define GAN metrics). See supplemental material for discussion.

We propose a loss that is easy to compute and can be efficiently optimized. We find a pairwise matching between synthetic and real diagrams and sum up the diagram distance between all matched pairs as the loss. Let $\mathcal{D}_{syn}$ and $\mathcal{D}_{real}$ be the two sets of persistence diagrams generated from synthetic and real images. Suppose we have an optimal matching between the two diagram sets, π^*. Our loss is the total matching distance between all matched synthetic-real diagram pairs. Recall $\mathcal{W}_1$ is the diagram distance (Eq. (3)). We have

$$L_{topo} = \sum\nolimits_{\text{dgm}_i \in \mathcal{D}_{syn}} \mathcal{W}_1\big(\text{dgm}_i, \pi^*(\text{dgm}_i)\big). \tag{4}$$

To find the optimal matching π^* between synthetic and real diagram sets, we use the optimal transport technique. Denote $\text{dgm}_i^s \in \mathcal{D}_{syn}$ and $\text{dgm}_j^r \in \mathcal{D}_{real}$. Let n_{syn} and n_{real} be the size of $\mathcal{D}_{syn}$ and $\mathcal{D}_{real}$. We solve Monge-Kantorovich's primal problem [62] to find the optimal transport plan:

$$\gamma^* = \min_{\gamma \in \Gamma} \sum_{i=1}^{n_{syn}} \sum_{j=1}^{n_{real}} \mathcal{W}_1(\text{dgm}_i^s, \text{dgm}_j^r) \cdot \gamma_{ij} \tag{5}$$

where $\Gamma = \{\gamma \in \mathbb{R}_+^{n_{syn} \times n_{real}} | \gamma \mathbf{1}_{n_{real}} = 1/n_{syn} \cdot \mathbf{1}_{n_{syn}}, \gamma^\intercal \mathbf{1}_{n_{syn}} = 1/n_{real} \cdot \mathbf{1}_{n_{real}}\}$. $\mathbf{1}_n$ is an n-dimensional vector of all ones. Denote by γ^* the optimal solution to Eq. (5). We compute the optimal matching (π^*) by mapping the i-th synthetic dgm_i^s to the best matched real diagram w.r.t. the optimal transportation plan, i.e., $\text{dgm}_{h(i)}^r$ such that $h(i) = \arg\max_j \gamma_{ij}^*$ [49]. Formally, $\pi^*(\text{dgm}_i^s) = \text{dgm}_{h(i)}^r$.

3.3 Gradient of the Loss

We derive the gradient of the topological GAN loss (Eq. (4)). The loss can be decomposed into the sum of the loss terms for individual synthetic diagrams, $L_{topo} = \sum_i L_{topo}^i$, in which the i-th loss term $L_{topo}^i = \mathcal{W}_1(\text{dgm}_i, \pi^*(\text{dgm}_i))$. Here the i-th synthetic diagram is generated from the distance transform of the i-th synthetic mask, y_i, $\text{dgm}_i = \text{dgm}(f_{y_i})$. Meanwhile, y_i is a binary mask computed

[2] There are more technical reasons for adding the diagonal line into the diagram, related to the stability of the metric. See [12].

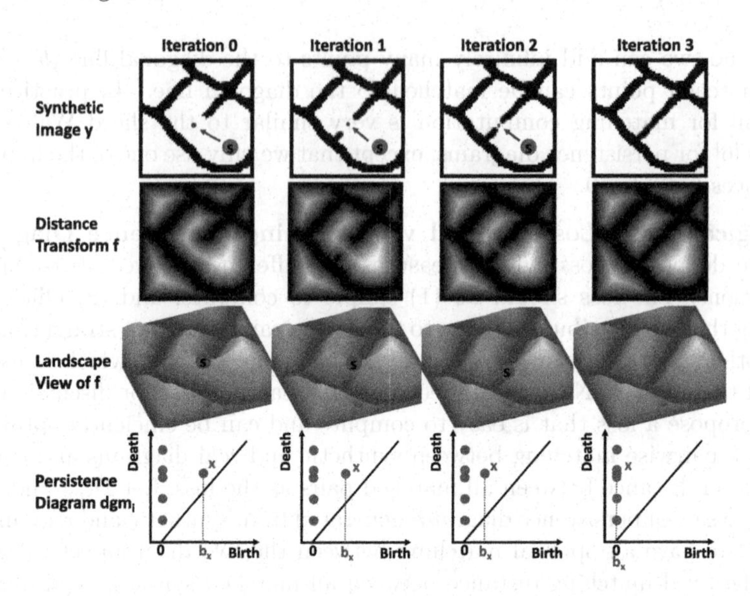

Fig. 4. From top to bottom: the same synthetic image being fixed at different iterations, their distance transforms, landscape views of distance transforms, and persistence diagrams. The red marker s is the saddle point, whose function value is the birth time of the almost-hole x. (Color figure online)

by thresholding the generated image $G(z_i)$. It suffices to calculate the gradient of L^i_{topo} with regard to the generator G.

Before deriving the gradient, we illustrate the intuition of the gradient descent in Fig. 4. For a particular synthetic image, we show how the mask is modified at different iterations and how the persistence diagram changes accordingly. As the gradient descent continues, the almost-hole in the synthetic image is slowly closed up to form a complete hole. At each iteration, the mask y grows toward the saddle point s of the distance transform f. The distance transform function value at the saddle s, $f(s)$, decreases toward zero. In the persistence diagram, the corresponding dot, x, moves toward left because its birth time $b_x = f(s)$ decreases. This reduces the 1-Wasserstein distance between the synthetic diagram and its matched real diagram.

Formally, by chain rule, we have $\frac{\partial L^i_{topo}}{\partial G} = \frac{\partial L^i_{topo}}{\partial \operatorname{dgm}_i} \cdot \frac{\partial \operatorname{dgm}_i}{\partial f_{y_i}} \cdot \frac{\partial f_{y_i}}{\partial G(z_i)} \cdot \frac{\partial G(z_i)}{\partial G}$. Next, we calculate each of the multiplicands.

Derivative of the Loss w.r.t. Persistence Diagrams. Recall that by Eqs. (3) and (4), we can rewrite the i-th loss term as $L^i_{topo} = \sum_{x \in \operatorname{dgm}_i} |b_x - b_{\sigma^*(x)}| = \sum_{x \in \operatorname{dgm}_i} \operatorname{sign}(b_x - b_{\sigma^*(x)})(b_x - b_{\sigma^*(x)})$. The equation depends on two optimal matchings, π^* and σ^*. The first one, $\pi^* : \mathcal{D}_{syn} \rightarrow \mathcal{D}_{real}$, is calculated by optimal transport between two sets of diagrams, $\mathcal{D}_{syn}$ and $\mathcal{D}_{real}$. The second optimal matching, $\sigma^* : \operatorname{dgm}_i \rightarrow \pi^*(\operatorname{dgm}_i)$, is calculated by 1D optimal transport between points of the two matched diagrams. Without loss of generality, we assume for all $x \in \operatorname{dgm}_i$ and $x' \in \pi^*(\operatorname{dgm}_i)$, their birth time differences $(b_x - b_{x'})$'s are distinct nonzero values.

While the optimal transport plan (γ^* in Eq. (5)) changes continuously as we change the input synthetic diagrams, the matchings π^* and σ^* only change at singularities (a measure-zero set). Within a small neighborhood of the input, we can assume constant optimal mappings π^* and σ^*, and constant $\text{sign}(b_x - b_{\sigma^*(x)})$ and $b_{\sigma^*(x)}$ as well. The gradient can be formally written as the partial derivative of the loss with regard to the birth and death times of each point $x \in \text{dgm}_i$: $\frac{\partial L^i_{topo}}{\partial b_x} = \text{sign}\left(b_x - b_{\sigma^*(x)}\right)$, $\frac{\partial L^i_{topo}}{\partial d_x} = 0$.

Intuitively, the negative gradient direction $-\frac{\partial L^i_{topo}}{\partial \text{dgm}_i}\frac{\partial \text{dgm}_i}{\partial G}$ moves each point x in the synthetic diagram dgm_i toward its matched point in the matched real diagram, $\sigma^*(x)$, horizontally (but not vertically). See Fig. 4 for an illustration.

Derivative of the Persistence Diagram w.r.t. the Distance Transform. The derivative of the loss w.r.t. death time is zero. Therefore, we only need to care about the derivative of the birth time b_x w.r.t. the distance transform f_{y_i}, $\frac{\partial b_x}{\partial f_{y_i}}$. An important observation is that the birth time of any almost-hole in a filtration is the function value of the saddle point of f_{y_i} sitting right in the middle of the gap, denoted as s_x. Formally, $b_x = \langle \delta_{s_x}, f_{y_i} \rangle$, in which δ_{s_x} is the Dirac delta function at the saddle point s_x. Taking the gradient, we have $\frac{\partial b_x}{\partial f_{y_i}} = \delta_{s_x}$.

Intuitively, $-\frac{\partial L^i_{topo}}{\partial b_x}\frac{\partial b_x}{\partial f_{y_i}}\frac{\partial f_{y_x}}{\partial G}$, the negative gradient w.r.t. the b_x of the diagram, moves the saddle point function value $b_x = f_{y_i}(s_x)$ up or down so it gets closer to the matched real diagram point's birth time, $b_{\sigma^*(x)}$. See Fig. 4.

Derivative of the Distance Transform w.r.t. the Synthetic Image $G(z_i)$. Finally, we compute the derivative of f_{y_i} with regard to the i-th synthetic image $G(z_i)$. Intuitively, focusing on the saddle point s_x, to increase or decrease its distance transform $f_{y_i}(s_x)$, the gradient needs to grow the mask y_i at its nearest boundary point to s_x, called r. This is achieved by changing the synthetic image values of the few pixels near r. As seen in Fig. 4, as we proceed, the mask grows toward the saddle point. More derivation details are in supplemental material.

3.4 Topology-Aware Metrics for GAN Evaluation

We introduce two novel metrics that can evaluate GAN performance in terms of topology. Conventionally, generator quality has been evaluated by comparing synthetic and real image distributions in the space of CNN-based image features. For example, both the Inception score (IS) [56] and the Fréchet Inception distance (FID) [24] use an Inception network pre-trained on ImageNet to map images into a feature space. The topological properties of images are not guaranteed to be preserved in such CNN-based image feature space.

In this paper, for the first time, we propose metrics that directly measure the topological difference between synthetic and real image distributions. The first metric, called the *Betti score*, is directly based on the topology of the mask, measured by the Betti number. Recall the Betti number counts the number of holes in a given synthetic or real mask. A Betti score computes a histogram for all synthetic masks and another histogram for all real masks. Then it compares the two histograms using their χ^2 distance. The definition can easily extend to zero-dimensional topology, i.e., counting the number of connected components.

Our second score is based on persistence diagrams which account for both holes and almost holes. We use the kernel mean embedding method [21]. Assume a given kernel for persistence diagrams, we can define an implicit function, Φ, mapping all synthetic/real persistence diagrams into a Hilbert space, $\mathcal{H}$. In such space, it becomes easy to compute the mean of each diagram set, $\Phi(\mathcal{D}_{syn}) := \frac{1}{n_{syn}} \sum_{i=1}^{n_{syn}} \Phi(\text{dgm}_i^s)$ and $\Phi(\mathcal{D}_{real}) := \frac{1}{n_{real}} \sum_{i=1}^{n_{real}} \Phi(\text{dgm}_i^r)$. We measure the difference between the synthetic and real diagram sample sets using *maximum mean discrepancy (MMD)*,

$$\text{MMD}(\mathcal{D}_{syn}, \mathcal{D}_{real}) := \| \Phi(\mathcal{D}_{syn}) - \Phi(\mathcal{D}_{real}) \|_{\mathcal{H}} .$$

It was proven that this sample-based MMD will converge to its continuous analog. We propose to use the unbiased MMD [21] (details are in the supplemental material). In terms of the kernel for persistence diagrams, there are many options [9,33]. Here we use the Gaussian kernel based on the 1-Wasserstein distance between diagrams, $k_{\mathcal{W}_1}(\text{dgm}_i, \text{dgm}_j) = \exp\left(-\frac{\mathcal{W}_1(\text{dgm}_i, \text{dgm}_j)}{\sigma^2}\right)$.

Our two metrics are generally useful to evaluate GAN results w.r.t. topology. We will evaluate TopoGAN using FID, unbiased MMD and Betti score.

4 Experiments

TopoGAN is built on top of WGAN-GP with deep convolutional generative adversarial networks [52] (DCGANs) as backbone network architectures. Details of TopoGAN's implementation, training, and computation cost are in Sec. B of the supplemental material. We compare TopoGAN against two baseline GANs: Wasserstein GAN with gradient penalty (WGAN-GP) and Wasserstein GAN with Spectral Normalization (WGAN-SN). These methods are best known for stabilizing GAN training and avoiding mode collapse. To demonstrate the potential of TopoGAN in practice, we showcase it in a downstream task: segmentation.

Datasets. TopoGAN is evaluated on five datasets: **CREMI** [15], **ISBI12** [4], **Google Maps** scraped by [30], **CMP Facade Database** [61], and **Retina** dataset. The first two are neuron image segmentation datasets and we randomly sample 7500 and 1500 patches of size 64 × 64 respectively from their segmentation masks. **Google Maps** (aerial photos ↔ maps) and **CMP Facade Database** (facades ↔ labels) consist of paired RGB images. THe RGB images of maps and labels are converted into grayscale images. We extract 4915 patches of size 64 × 64 from the converted maps and resize all 606 facade labels to 128 × 128. The **Retina** dataset consists of 98 retina segmentations we collected from 4 datasets: **IOSTAR** (40) [1,2], **DRIVE** (20) [59], **STARE** (20) [27], and **CHASE_DB1** (28) [48]. All retina images are cropped and resized to 128 × 128 resolution.

Quantitative and Qualitative Results. In Table 1, we report the performance of TopoGAN and two baselines w.r.t. three metrics: FID, unbiased MMD,

Table 1. Comparisons against baseline GANs on FID, unbiased MMD, and Betti score across five datasets. The standard deviations are based on 3 runs. We omit reporting unbiased MMD and Betti score of WGAN-SN on Retina as WGAN-SN fails to produce reasonable results.

	CREMI	ISBI12	Retina	Maps	Facade
	FID				
WGAN-GP	21.64±0.138	83.90±0.718	179.69±19.008	72.00±0.469	122.13±0.822
WGAN-SN	34.15±0.153	78.61±0.411	269.12±2.276	175.52±0.217	126.10±1.901
TopoGAN	**20.96**±0.195	**31.90**±0.248	**169.21**±21.976	**60.48**±0.467	**119.11**±0.874
	Unbiased MMD				
WGAN-GP	0.142±0.014	0.558±0.010	1.735±0.050	0.482±0.007	0.137±0.004
WGAN-SN	0.326±0.016	0.602±0.006	–	0.724±0.005	0.166±0.005
TopoGAN	**0.134**±0.019	**0.405**±0.003	**1.602**±0.114	**0.471**±0.010	**0.080**±0.002
	Betti score				
WGAN-GP	0.236±0.003	0.908±0.104	0.541±0.188	0.223±0.010	0.176±0.006
WGAN-SN	0.125±0.002	1.775±0.039	–	0.255±0.020	0.142±0.017
TopoGAN	**0.015**±0.001	**0.802**±0.058	**0.457**±0.144	**0.177**±0.004	**0.124**±0.002

and Betti score. TopoGAN outperforms the two baselines significantly in the two topology-aware metrics proposed in Sect. 3.4: unbiased MMD and Betti score. The superior performance of TopoGAN proves that the topological GAN loss successfully enforced the structural/topological faithfulness of the generated images, as desired. Further comparisons of the topological quality of the synthesized images at different training epochs can be found in the supplemental material. Meanwhile, we observe that TopoGAN is also better in FID. This suggests that topological integrity could serve as an important visual cue when deciding image quality by human standards.

Qualitative results are in Fig. 5. For fair comparison, we use the same set of noise inputs to generate data for each GAN method. We observe that the masks produced by TopoGAN have more clear boundaries and complete cycles. They are topologically more similar to the real data (i.e., having similar Betti numbers). TopoGAN also shows better performance in texture images (details of how these textures are generated will be explained later). On the contrary, baselines WGAN-GP and WGAN-SN tend to generate broken structures. The **Retina** dataset is challenging for all GAN models. This is due to the small training set (98) and the heterogeneity of the dataset; its images are from multiple datasets with different geometry, resolutions, aspect ratios, and contrasts.

Segmentation Application. We demonstrate that TopoGAN improves performance in a downstream binary segmentation task. For each dataset, we train a segmentation network with real training data, synthetic data, and real data augmented with synthetic data. The networks trained with synthetic data from TopoGAN are compared against networks trained with data from baseline GANs and with real training data. The segmentation networks are evaluated on test data with three segmentation metrics: (1) pixel accuracy, (2) Dice score, and (3) Adapted Rand Index (ARI). We report the results on dice score in Table 2, and leave the results on other scores to the supplemental material.

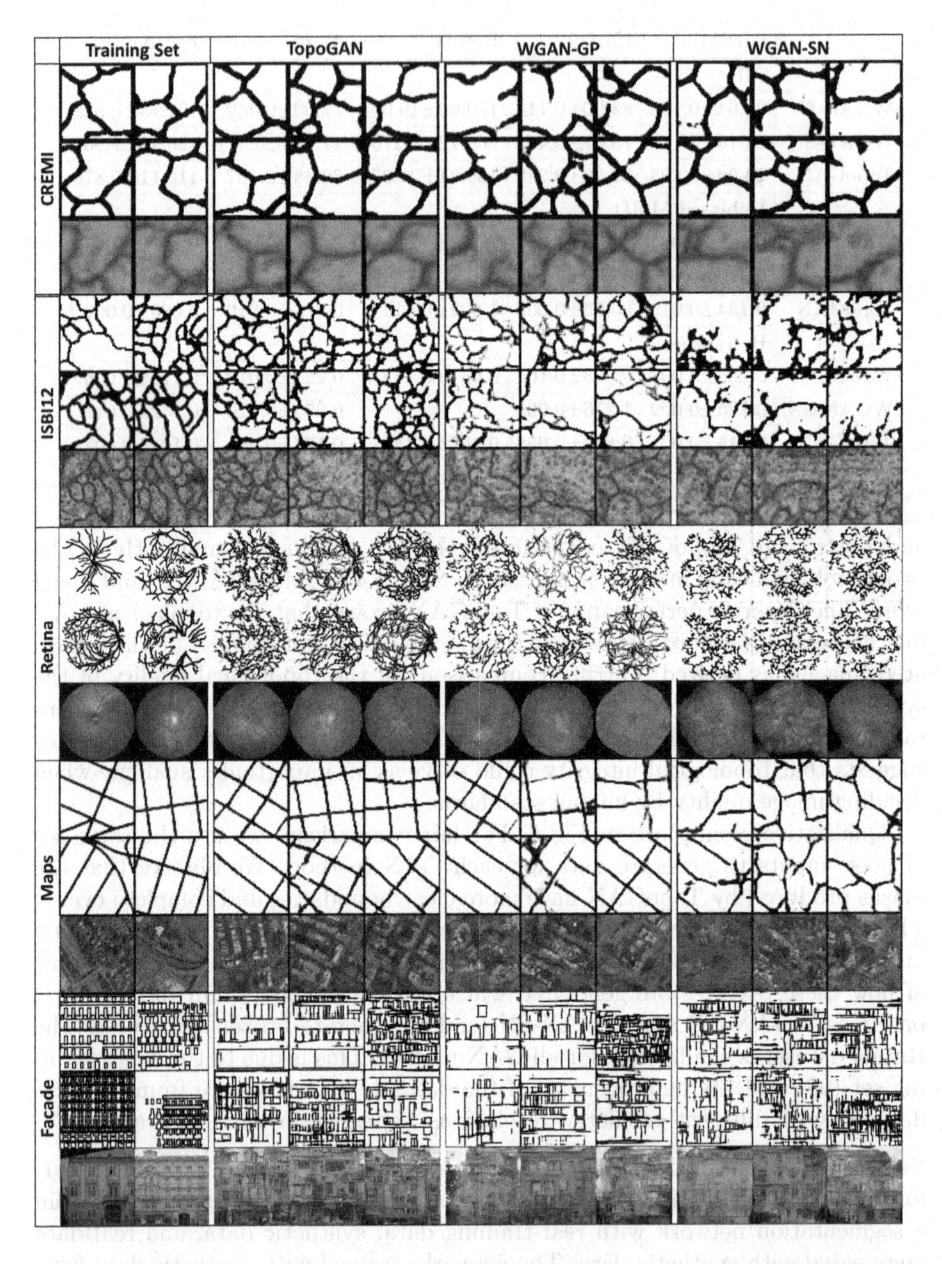

Fig. 5. Qualitative comparisons of TopoGAN to WGAN-GP and WGAN-SN on 5 datasets. From left to right: real masks from training set, generated masks from TopoGAN, WGAN-GP, and WGAN-SN. For each dataset, the third row shows texture images corresponding to the masks on the second row.

Table 2. Dice score of segmentation networks on real test data. For each dataset, we train a total of 21 segmentation networks with real training data, synthetic data from TopoGAN and two baselines, and real data augmented with synthetic data. We report mean and standard deviation of a 3-fold cross validation.

	CREMI	ISBI12	Retina
Real data	0.896±0.004	0.932±0.011	0.883±0.010
WGAN-GP	0.820±0.018	0.927±0.005	0.891±0.012
WGAN-SN	0.827±0.019	0.902±0.008	–
TopoGAN	0.851±0.011	0.933±0.006	0.892±0.013
WGAN-GP+real data	0.897±0.008	0.943±0.007	0.899±0.010
WGAN-SN+real data	0.900±0.004	0.905±0.054	–
TopoGAN+real data	**0.902**±0.006	**0.944**±0.008	**0.906**±0.014

To produce synthetic pairs (fake masks ↔ textured masks), a pix2pix [30] network is first trained with real data pairs. The trained pix2pix network takes as inputs the GAN-generated masks and produces textured masks on which a segmentation network can be trained on. We use U-Net [55] as our segmentation network. We use a three-fold cross validation and report both the mean and standard deviation of the Dice score for all datasets. Note that only we only segment **CREMI**, **ISBI12**, and **Retina**, as the other two datasets are not segmentation datasets and have no ground truth training data.

Segmentation results are summarized in Table 2. TopoGAN with pure synthetic data achieves comparable results to segmentation networks trained with real data on dataset **ISBI12** and **Retina**. Segmentations augmented with synthetic data always perform better than real data or synthetic data alone. TopoGAN plus real data produces the best results followed closely by WGAN-GP plus real data. Details of evaluation metrics, segmentation networks training procedure and full result table can be found in supplementary.

5 Conclusion

This paper proposed TopoGAN, the first GAN method explicitly learning image topology of the image from real data. We proposed a topological GAN loss and showed that this loss is differentiable and can be easily incorporated into GAN training. In addition, we proposed novel metrics to measure topological differences between synthesized and real images. Empirically, we have shown that TopoGAN generates images with better topological features than state-of-the-art GANs both quantitatively and qualitatively.

Acknowledgement. Fan Wang and Chao Chen's research was partially supported by NSF IIS-1855759, CCF-1855760 and IIS-1909038. Huidong Liu and Dimitris Samaras were partially supported from the Partner University Fund, the SUNY2020 Infrastructure Transportation Security Center, and a gift from Adobe.

References

1. Abbasi-Sureshjani, S., Smit-Ockeloen, I., Bekkers, E., Dashtbozorg, B., ter Haar Romeny, B.: Automatic detection of vascular bifurcations and crossings in retinal images using orientation scores. In: 2016 IEEE 13th International Symposium on Biomedical Imaging (ISBI), pp. 189–192, April 2016. https://doi.org/10.1109/ISBI.2016.7493241
2. Abbasi-Sureshjani, S., Smit-Ockeloen, I., Zhang, J., Ter Haar Romeny, B.: Biologically-inspired supervised vasculature segmentation in SLO retinal fundus images. In: Kamel, M., Campilho, A. (eds.) Image Analysis and Recognition, pp. 325–334. Springer International Publishing, Cham (2015)
3. Adams, H., et al.: Persistence images: a stable vector representation of persistent homology. J. Mach. Learn. Res. **18**(1), 218–252 (2017)
4. Arganda-Carreras, I., et al.: Crowdsourcing the creation of image segmentation algorithms for connectomics. Front. Neuroanat. **9**, 142 (2015). https://doi.org/10.3389/fnana.2015.00142. https://www.frontiersin.org/article/10.3389/fnana.2015.00142
5. Arjovsky, M., Chintala, S., Bottou, L.: Wasserstein generative adversarial networks. In: International Conference on Machine Learning, pp. 214–223 (2017)
6. Bonneel, N., Rabin, J., Peyré, G., Pfister, H.: Sliced and radon Wasserstein Barycenters of measures. J. Math. Imaging Vis. **51**(1), 22–45 (2015)
7. Brock, A., Donahue, J., Simonyan, K.: Large scale GAN training for high fidelity natural image synthesis. In: International Conference on Learning Representations (2018)
8. Brüel-Gabrielsson, R., Nelson, B.J., Dwaraknath, A., Skraba, P., Guibas, L.J., Carlsson, G.: A topology layer for machine learning. arXiv preprint arXiv:1905.12200 (2019)
9. Carriere, M., Cuturi, M., Oudot, S.: Sliced Wasserstein kernel for persistence diagrams. In: Proceedings of the 34th International Conference on Machine Learning, vol. 70, pp. 664–673. JMLR. org (2017)
10. Chen, C., Ni, X., Bai, Q., Wang, Y.: A topological regularizer for classifiers via persistent homology. In: The 22nd International Conference on Artificial Intelligence and Statistics, pp. 2573–2582 (2019)
11. Clough, J.R., Oksuz, I., Byrne, N., Schnabel, J.A., King, A.P.: Explicit topological priors for deep-learning based image segmentation using persistent homology. In: Chung, A.C.S., Gee, J.C., Yushkevich, P.A., Bao, S. (eds.) IPMI 2019. LNCS, vol. 11492, pp. 16–28. Springer, Cham (2019). https://doi.org/10.1007/978-3-030-20351-1_2
12. Cohen-Steiner, D., Edelsbrunner, H., Harer, J.: Stability of persistence diagrams. Discrete Comput. Geom. **37**(1), 103–120 (2007)
13. Cohen-Steiner, D., Edelsbrunner, H., Harer, J., Mileyko, Y.: Lipschitz functions have l p-stable persistence. Found. Comput. Math. **10**(2), 127–139 (2010)
14. Cortes, C., Vapnik, V.: Support-vector networks. Mach. Learn. **20**(3), 273–297 (1995)
15. Miccai challenge on circuit reconstruction from electron microscopy images. https://cremi.org/
16. Edelsbrunner, H., Harer, J.: Computational Topology: An Introduction. American Mathematical Society, Providence (2010)
17. Fabbri, R., Costa, L.D.F., Torelli, J.C., Bruno, O.M.: 2D Euclidean distance transform algorithms: a comparative survey. ACM Comput. Surv. (CSUR) **40**(1), 1–44 (2008)

18. Fu, H., Gong, M., Wang, C., Batmanghelich, K., Zhang, K., Tao, D.: Geometry-consistent generative adversarial networks for one-sided unsupervised domain mapping. In: Proceedings of the IEEE Conference on Computer Vision and Pattern Recognition, pp. 2427–2436 (2019)
19. Ghafoorian, M., Nugteren, C., Baka, N., Booij, O., Hofmann, M.: EL-GAN: embedding loss driven generative adversarial networks for lane detection. In: Leal-Taixé, L., Roth, S. (eds.) ECCV 2018. LNCS, vol. 11129, pp. 256–272. Springer, Cham (2019). https://doi.org/10.1007/978-3-030-11009-3_15
20. Goodfellow, I., et al.: Generative adversarial nets. In: Advances in Neural Information Processing Systems, pp. 2672–2680 (2014)
21. Gretton, A., Borgwardt, K., Rasch, M., Schölkopf, B., Smola, A.: A kernel two-sample test. J. Mach. Learn. Res. **13**, 723–773 (2012)
22. Gulrajani, I., Ahmed, F., Arjovsky, M., Dumoulin, V., Courville, A.C.: Improved training of Wasserstein GANs. In: Advances in Neural Information Processing Systems, pp. 5767–5777 (2017)
23. He, Y., et al.: Fully convolutional boundary regression for retina OCT segmentation. In: Shen, D., et al. (eds.) MICCAI 2019. LNCS, vol. 11764, pp. 120–128. Springer, Cham (2019). https://doi.org/10.1007/978-3-030-32239-7_14
24. Heusel, M., Ramsauer, H., Unterthiner, T., Nessler, B., Hochreiter, S.: GANs trained by a two time-scale update rule converge to a local nash equilibrium. In: Guyon, I., et al., (eds.) Advances in Neural Information Processing Systems, vol. 30, pp. 6626–6637. Curran Associates, Inc. (2017). http://papers.nips.cc/paper/7240-gans-trained-by-a-two-time-scale-update-rule-converge-to-a-local-nash-equilibrium.pdf
25. Hofer, C., Kwitt, R., Niethammer, M., Dixit, M.: Connectivity-optimized representation learning via persistent homology. In: International Conference on Machine Learning, pp. 2751–2760 (2019)
26. Hofer, C., Kwitt, R., Niethammer, M., Uhl, A.: Deep learning with topological signatures. In: Advances in Neural Information Processing Systems, pp. 1634–1644 (2017)
27. Hoover, A.D., Kouznetsova, V., Goldbaum, M.: Locating blood vessels in retinal images by piecewise threshold probing of a matched filter response. IEEE Trans. Med. Imaging **19**(3), 203–210 (2000). https://doi.org/10.1109/42.845178
28. Hu, X., Li, F., Samaras, D., Chen, C.: Topology-preserving deep image segmentation. In: Advances in Neural Information Processing Systems, pp. 5658–5669 (2019)
29. Hwang, J.J., Ke, T.W., Shi, J., Yu, S.X.: Adversarial structure matching for structured prediction tasks. In: Proceedings of the IEEE Conference on Computer Vision and Pattern Recognition, pp. 4056–4065 (2019)
30. Isola, P., Zhu, J.Y., Zhou, T., Efros, A.A.: Image-to-image translation with conditional adversarial networks. In: CVPR (2017)
31. Khrulkov, V., Oseledets, I.: Geometry score: a method for comparing generative adversarial networks. In: Dy, J., Krause, A. (eds.) Proceedings of the 35th International Conference on Machine Learning. Proceedings of Machine Learning Research, vol. 80, pp. 2621–2629. PMLR, Stockholmsmässan, Stockholm Sweden, 10–15 July 2018
32. Kossaifi, J., Tran, L., Panagakis, Y., Pantic, M.: GAGAN: geometry-aware generative adversarial networks. In: Proceedings of the IEEE Conference on Computer Vision and Pattern Recognition, pp. 878–887 (2018)

33. Kusano, G., Hiraoka, Y., Fukumizu, K.: Persistence weighted Gaussian kernel for topological data analysis. In: International Conference on Machine Learning, pp. 2004–2013 (2016)
34. Li, C.L., Chang, W.C., Cheng, Y., Yang, Y., Póczos, B.: MMD GAN: towards deeper understanding of moment matching network. In: Advances in Neural Information Processing Systems, pp. 2203–2213 (2017)
35. Lim, J.H., Ye, J.C.: Geometric GAN. arXiv preprint arXiv:1705.02894 (2017)
36. Liu, H., Gu, X., Samaras, D.: Wasserstein GAN with quadratic transport cost. In: The IEEE International Conference on Computer Vision (ICCV), October 2019
37. Liu, H., Xianfeng, G., Samaras, D.: A two-step computation of the exact GAN Wasserstein distance. In: International Conference on Machine Learning, pp. 3165–3174 (2018)
38. Liu, S., Bousquet, O., Chaudhuri, K.: Approximation and convergence properties of generative adversarial learning. In: Advances in Neural Information Processing Systems, pp. 5545–5553 (2017)
39. Luc, P., Couprie, C., Chintala, S., Verbeek, J.: Semantic segmentation using adversarial networks. arXiv preprint arXiv:1611.08408 (2016)
40. Mescheder, L., Nowozin, S., Geiger, A.: Which training methods for GANs do actually converge? In: International Conference on Machine Learning (2018)
41. Miyato, T., Kataoka, T., Koyama, M., Yoshida, Y.: Spectral normalization for generative adversarial networks. In: International Conference on Machine Learning (2018)
42. Mosinska, A., Márquez-Neila, P., Koziński, M., Fua, P.: Beyond the pixel-wise loss for topology-aware delineation. In: Proceedings of the IEEE Conference on Computer Vision and Pattern Recognition (CVPR), June 2018
43. Mroueh, Y., Sercu, T.: Fisher GAN. In: Advances in Neural Information Processing Systems, pp. 2513–2523 (2017)
44. Mroueh, Y., Sercu, T., Goel, V.: McGan: mean and covariance feature matching GAN. In: Precup, D., Teh, Y.W. (eds.) Proceedings of the 34th International Conference on Machine Learning. Proceedings of Machine Learning Research, vol. 70, pp. 2527–2535. PMLR, International Convention Centre, Sydney, Australia, 06–11 August 2017
45. Munkres, J.R.: Elements of Algebraic Topology. CRC Press, Boca Raton (2018)
46. Ni, X., Quadrianto, N., Wang, Y., Chen, C.: Composing tree graphical models with persistent homology features for clustering mixed-type data. In: Proceedings of the 34th International Conference on Machine Learning, vol. 70, pp. 2622–2631. JMLR. org (2017)
47. Nowozin, S., Cseke, B., Tomioka, R.: F-GAN: training generative neural samplers using variational divergence minimization. In: Advances in Neural Information Processing Systems, pp. 271–279 (2016)
48. Owen, C.G., et al.: Measuring retinal vessel tortuosity in 10-year-old children: validation of the computer-assisted image analysis of the retina (CAIAR) program. Invest. Ophthalmol. Vis. Sci. **50**(5), 2004–2010 (2009). https://doi.org/10.1167/iovs.08-3018
49. Peyré, G., Cuturi, M.: Computational optimal transport foundations and trends. Mach. Learn. **11**(2019), 355 (1803)
50. Qi, G.J., Zhang, L., Hu, H., Edraki, M., Wang, J., Hua, X.S.: Global versus localized generative adversarial nets. In: Proceedings of the IEEE Conference on Computer Vision and Pattern Recognition, pp. 1517–1525 (2018)

51. Qin, Y., et al.: AirwayNet: a voxel-connectivity aware approach for accurate airway segmentation using convolutional neural networks. In: Shen, D., et al. (eds.) MICCAI 2019. LNCS, vol. 11769, pp. 212–220. Springer, Cham (2019). https://doi.org/10.1007/978-3-030-32226-7_24
52. Radford, A., Metz, L., Chintala, S.: Unsupervised representation learning with deep convolutional generative adversarial networks. arXiv preprint arXiv:1511.06434 (2015)
53. Ramamurthy, K.N., Varshney, K., Mody, K.: Topological data analysis of decision boundaries with application to model selection. In: International Conference on Machine Learning, pp. 5351–5360 (2019)
54. Reininghaus, J., Huber, S., Bauer, U., Kwitt, R.: A stable multi-scale kernel for topological machine learning. In: Proceedings of the IEEE Conference on Computer Vision and Pattern Recognition, pp. 4741–4748 (2015)
55. Ronneberger, O., Fischer, P., Brox, T.: U-Net: convolutional networks for biomedical image segmentation. In: Navab, N., Hornegger, J., Wells, W.M., Frangi, A.F. (eds.) MICCAI 2015. LNCS, vol. 9351, pp. 234–241. Springer, Cham (2015). https://doi.org/10.1007/978-3-319-24574-4_28
56. Salimans, T., Goodfellow, I., Zaremba, W., Cheung, V., Radford, A., Chen, X.: Improved techniques for training GANs. In: Advances in Neural Information Processing Systems, pp. 2234–2242 (2016)
57. Shaham, T.R., Dekel, T., Michaeli, T.: Singan: learning a generative model from a single natural image. In: Proceedings of the IEEE International Conference on Computer Vision, pp. 4570–4580 (2019)
58. Simonyan, K., Zisserman, A.: Very deep convolutional networks for large-scale image recognition. In: International Conference on Learning Representations (2015)
59. Staal, J., Abràmoff, M.D., Niemeijer, M., Viergever, M.A., Van Ginneken, B.: Ridge-based vessel segmentation in color images of the retina. IEEE Trans. Med. Imaging **23**(4), 501–509 (2004)
60. Thanh-Tung, H., Tran, T., Venkatesh, S.: Improving generalization and stability of generative adversarial networks. In: International Conference on Learning Representations (2019)
61. Tyleček, R., Šára, R.: Spatial pattern templates for recognition of objects with regular structure. In: Weickert, J., Hein, M., Schiele, B. (eds.) GCPR 2013. LNCS, vol. 8142, pp. 364–374. Springer, Heidelberg (2013). https://doi.org/10.1007/978-3-642-40602-7_39
62. Villani, C.: Optimal Transport: Old and New, vol. 338. Springer Science & Business Media, Berlin Heidelberg (2008). https://doi.org/10.1007/978-3-540-71050-9
63. Wang, T.C., Liu, M.Y., Zhu, J.Y., Tao, A., Kautz, J., Catanzaro, B.: High-resolution image synthesis and semantic manipulation with conditional GANs. In: Proceedings of the IEEE Conference on Computer Vision and Pattern Recognition, pp. 8798–8807 (2018)
64. Wu, P., et al.: Optimal topological cycles and their application in cardiac trabeculae restoration. In: Niethammer, M., et al. (eds.) IPMI 2017. LNCS, vol. 10265, pp. 80–92. Springer, Cham (2017). https://doi.org/10.1007/978-3-319-59050-9_7
65. Zhang, H., Goodfellow, I., Metaxas, D., Odena, A.: Self-attention generative adversarial networks. In: International Conference on Machine Learning, pp. 7354–7363 (2019)
66. Zhao, Q., Wang, Y.: Learning metrics for persistence-based summaries and applications for graph classification. In: Advances in Neural Information Processing Systems, pp. 9859–9870 (2019)

67. Zhao, Q., Ye, Z., Chen, C., Wang, Y.: Persistence enhanced graph neural network. In: International Conference on Artificial Intelligence and Statistics, pp. 2896–2906 (2020)
68. Zhu, J.Y., Park, T., Isola, P., Efros, A.A.: Unpaired image-to-image translation using cycle-consistent adversarial networks. In: 2017 IEEE International Conference on Computer Vision (ICCV) (2017)

Learning to Localize Actions from Moments

Fuchen Long[1], Ting Yao[2(✉)], Zhaofan Qiu[1], Xinmei Tian[1], Jiebo Luo[3], and Tao Mei[2]

[1] University of Science and Technology of China, Hefei, China
longfc.ustc@gmail.com, zhaofanqiu@gmail.com, xinmei@ustc.edu.cn
[2] JD AI Research, Beijing, China
tingyao.ustc@gmail.com, tmei@jd.com
[3] University of Rochester, Rochester, NY, USA
jluo@cs.rochester.edu

Abstract. With the knowledge of action moments (i.e., trimmed video clips that each contains an action instance), humans could routinely localize an action temporally in an untrimmed video. Nevertheless, most practical methods still require all training videos to be labeled with temporal annotations (action category and temporal boundary) and develop the models in a fully-supervised manner, despite expensive labeling efforts and inapplicable to new categories. In this paper, we introduce a new design of transfer learning type to learn action localization for a large set of action categories, but only on action moments from the categories of interest and temporal annotations of untrimmed videos from a small set of action classes. Specifically, we present Action Herald Networks (AherNet) that integrate such design into an one-stage action localization framework. Technically, a weight transfer function is uniquely devised to build the transformation between classification of action moments or foreground video segments and action localization in synthetic contextual moments or untrimmed videos. The context of each moment is learnt through the adversarial mechanism to differentiate the generated features from those of background in untrimmed videos. Extensive experiments are conducted on the learning both across the splits of ActivityNet v1.3 and from THUMOS14 to ActivityNet v1.3. Our AherNet demonstrates the superiority even comparing to most fully-supervised action localization methods. More remarkably, we train AherNet to localize actions from 600 categories on the leverage of action moments in Kinetics-600 and temporal annotations from 200 classes in ActivityNet v1.3.

This work was performed at JD AI Research.

Electronic supplementary material The online version of this chapter (https://doi.org/10.1007/978-3-030-58580-8_9) contains supplementary material, which is available to authorized users.

A. Vedaldi et al. (Eds.): ECCV 2020, LNCS 12348, pp. 137–154, 2020.
https://doi.org/10.1007/978-3-030-58580-8_9

1 Introduction

With the tremendous increase in Internet bandwidth and the power of the cloud, video data is growing explosively and video-based intelligent services are becoming gradually accessible to ordinary users. This trend encourages the development of recent technological advances, which facilitates a variety of video understanding applications [3,25,26,34]. In between, one of the most fundamental challenges is the process of temporal action localization [7,11,24,29,46,59], which is to predict the temporal boundary of each action in an untrimmed video and categorize each action according to visual content as well. Most existing action localization systems still perform "intensive manual labeling" to collect temporal annotations (action category and temporal boundary) of actions in untrimmed videos and then train localization models in a fully-supervised manner. Such paradigm requires strong supervision, which is expensive to annotate for new categories and thus limits the number of action categories. In the meantime, there are various datasets (e.g., Kinetics [12]) which include expert labeled data of trimmed action moments for action recognition. A valid question then emerges as is it possible to achieve action localization for a large set of categories, with only trimmed action moments from these categories and temporal annotations from a small set of action classes? If possible, it is readily to adapt state-of-the-art action localization methods to support thousands of action categories in real-world deployment.

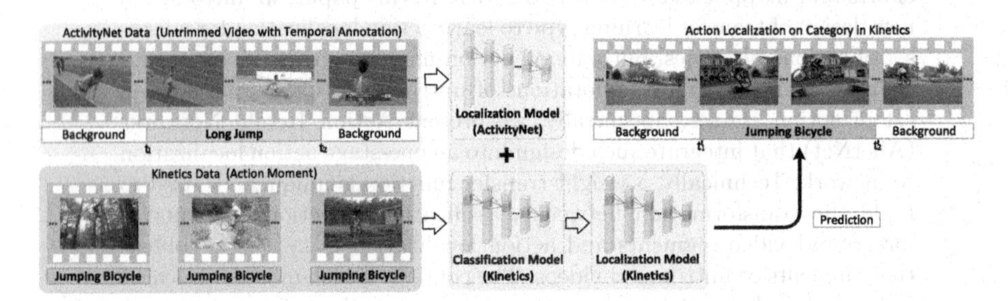

Fig. 1. Action localization modeling for a large set of categories based on only action moments of these categories (e.g., Kinetics [12]) and untrimmed videos from a small set of categories with temporal annotations (e.g., ActivityNet [17]).

With this motivation, Fig. 1 conceptually depicts the pipeline of action localization in our work. Given a large set of categories which have only trimmed action moments (e.g., Kinetics [12]) and a small set of classes which have fully temporal annotations on untrimmed videos (e.g., ActivityNet [17]), we aim for a model that enables to temporally localize and recognize actions from the large set of categories. Note that the categories in the two sets could be completely different. The main difficulties inherently originate from two aspects: 1) how to build the connection between classification and localization? 2) how to hallucinate the context or background of an action moment in training? We propose to mitigate the first issue through the design of weight transfer. In view that action localization generally consists of temporal action proposal and temporal action classification, the network weights for temporal action classification could be derived from those for action recognition of trimmed videos. In our case, the trimmed videos are either foreground video segments in untrimmed videos or action moments. As such, the weight transfer is considered as a bridge between classification and localization. We utilize the recipe of adversarial learning to alleviate the second issue. A discriminator is devised to differentiate the generated context features from those of background in untrimmed videos.

By consolidating the idea of learning action localization models on a mixture of action moments and fully temporal annotations, we present a new Action Herald Networks (AherNet) in an one-stage localization framework. AherNet mainly includes two modules, i.e., weight transfer between classification and localization on untrimmed videos with temporal annotations, and localization modeling on action moments with synthetic contexts. On one hand, the first module naturally constructs a correspondence between action localization in an untrimmed video and action classification of "action moment", i.e., the foreground video segment extracted from the untrimmed video. Technically, we learn a weight transfer function which transforms network parameters for foreground segment classification to those for temporal action classification in localization on untrimmed videos. On the other hand, to simulate action localization on action moments data, we hallucinate the features of context or background of an action moment via adversarial learning. The connection between action moment classification and localization of the action from the context is also built by the weight transfer function, whose parameters are shared. The whole AherNet is end-to-end optimized by minimizing proposal loss, classification loss and adversarial loss.

The main contribution of this work is a new paradigm between supervised and weakly-supervised training, that enables action localization models to support thousands of action categories, with only trimmed action moments from these categories and temporal annotations from a small set of classes. This also leads to the elegant view of how to bridge the task of classification and localization, and how to produce the context of action moments to simulate localization in training, which are problems not yet fully understood.

2 Related Work

Temporal Action Localization. We briefly group the temporal action localization into two categories: two-stage and one-stage action localization. Two-stage action localization approaches [16,43,45,55,58,59] first detect temporal action proposals [3,6,9,10,29,33,39,49,57] and then classify [40,41] the proposals into known action classes. For instance, Buch *et al.* [3] develop a recurrent GRU-based action proposal model followed by a S-CNN [46] classifier for localization. To further facilitate action localization by uniting separate optimization of two stages, there have been several one-stage techniques [2,4,27,32,56] being proposed. All these methods require the training data with fully temporal annotations. Instead, our AherNet models action localization for a large set of categories based on only action moments of these categories and untrimmed videos from a small set of categories with temporal annotations.

Parameter Prediction. Parameter prediction in neural networks is capable of building the connections between the related tasks. Several weight adaptation methods [18,23,50] learn specific matrix to adapt the image classification weights for object detection. Most recently, Hu *et al.* [19] explore the direction of parameter transferring from object detection to instance segmentation by a general function, which enables the transformed Mask R-CNN [15] to segment 3000 visual concepts. In our work, we utilize the parameter prediction to bridge the task of classification and localization.

Adversarial Learning. Inspired by the Generative Adversarial Networks (GAN) [14], the adversarial learning has been widely used in various vision tasks, e.g., image translation [20] and domain adaptation [8,51]. The training processing of GAN [14] corresponds to a minimax two-player game to make the distribution of fake data close to the real data distribution. In the context of our work, we simulate action localization on action moments with generated action contexts. Through adversarial learning, the generated contextual features become indiscriminative from real background features of untrimmed video.

Weakly-Supervised Action Localization. The weakly-supervised action localization approaches [31,37,38,42,44,52] only utilize the category supervision of untrimmed videos for localization, whose setting and scenario are different from our paradigm. Most of them build an attention mechanism to detect actions.

In short, our work mainly focuses on a new learning paradigm of scaling action localization to a large set of categories. The proposal of AherNet contributes by studying not only bridging action classification and localization through weight transfer, but also how the generated context of action moments should be better leveraged to support action localization learning.

3 Action Herald Networks

In this section we present the proposed Action Herald Networks (AherNet) in detail. Figure 2 illustrates an overview of our architecture. It consists of two modules, i.e., weight transfer between classification and localization, and localization modeling on action moments. Given an untrimmed video, the foreground video segment is extracted as the "action moment." A 3D ConvNet is exploited as the base network to extract a sequence of clip-level features for the untrimmed video and foreground segment, respectively. Each feature sequence is concatenated into a feature map, followed by a cascaded of 1D temporal convolutional layers to output feature maps on different scales. For action classification of foreground segment, global pooling is employed on the features of all the cells in each feature map to produce the features on each scale, which are projected via a matrix for segment-level classification. Such matrix is adapted by a weight transfer function to that used in action localization for the untrimmed video. In that case, we perform the adapted matrix on each feature map to obtain the projection of the features of every cell (anchor) in that map for temporal action classification. Similar processes are implemented on action moments and the extensions with

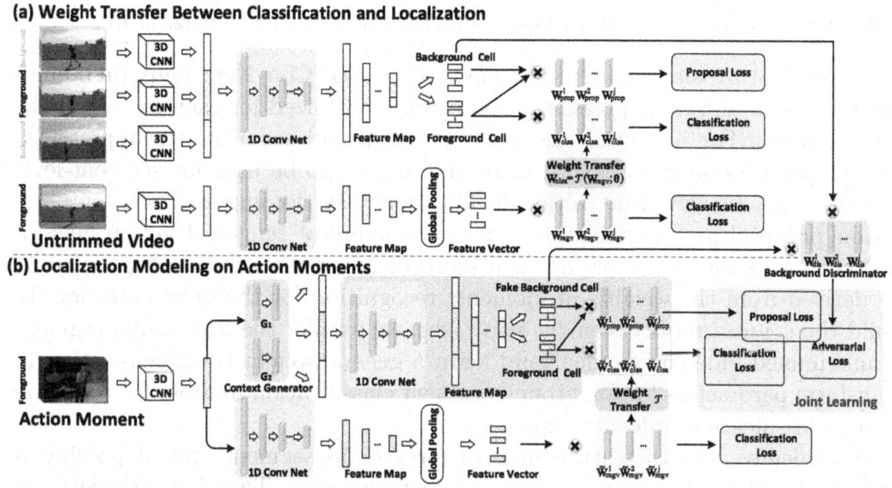

Fig. 2. An overview of our Action Herald Networks (AherNet) architecture. The foreground segments of untrimmed videos are first extracted as "action moments." The input untrimmed video and foreground segment is encoded into a series of clip-level features via a 3D ConvNet, which are sequentially concatenated as a feature map, respectively. A cascaded of 1D convolutional layers is applied to generate multiple feature maps on different scales. For classification of foreground segment, global pooling is exploited on all cells of feature map to produce the features on each scale, which are projected via a matrix for segment-level classification. The matrix is adapted by a weight transfer function (orange box) to that used in action localization for untrimmed video. In localization, the adapted matrix is performed on each cell in the feature map to obtain the projection for temporal action classification. Similar process are implemented on action moments and the extensions with generated context. The synthetic contexts of moments are confused with the background of untrimmed videos via adversarial learning (green box) and the parameters of weight transfer function are shared. Our AherNet is jointly optimized with proposal loss, classification loss and adversarial loss. In the inference stage, only the localization part (blue box) learnt on the moments with contexts is utilized to predict action instances (Color figure online).

contexts. The features of contexts are hallucinated through adversarial learning and the parameters of the weight transfer function are shared. The network is jointly optimized with proposal loss, classification loss and adversarial loss.

3.1 Base Backbone

We build our action localization model on a weight-sharing 1D convolutional networks. Given an input untrimmed video or action moment, a sequence of clip-level features are extracted from a 3D ConvNet. We concatenate all the features into one feature map and then feed the map into a cascaded of 1D convolutional layers (anchor layers) to generate multiple feature maps on eight temporal scales. These feature maps are further exploited for action classification of the action moment or temporal action localization of the untrimmed video.

3.2 Weight Transfer Between Classification and Localization

Given the feature maps of an untrimmed video in 1D ConvNet, temporal boundary regression and action classification can be optimized for each anchor in the feature maps. For an action moment or foreground segment, the representation of global pooling on each feature map is able to be used for segment-level classification. In view that action localization task decomposes into temporal action proposal and classification, the parameters of temporal action classification in localization to predict the score of a specific action category could be derived from the weights of moments recognition for the same category. To build the connection between the two tasks, we extract the foreground segment of untrimmed video as moment and learn a generic weight transfer function to transform parameters for foreground segment classification to those for temporal action classification in localization.

Specifically, in j-th feature map of foreground segment, global pooling is first employed on that map to produce a feature vector. Then a matrix $\mathbf{W}^j_{regv,c}$ is utilized to project the feature vector into the probability of category c for segment-level classification. As for localization on untrimmed video, we adopt a 1D convolutional layer with stride of 1 to obtain the score of each cell (anchor) in that map for anchor-level classification. The parameters in that 1D convolutional layer to predict score of category c are denoted as $\mathbf{W}^j_{clsa,c}$. To bridge classification and localization for the specific category c, a generic weight transfer function $\mathcal{T}$ is introduced to predict $\mathbf{W}^j_{clsa,c}$ from $\mathbf{W}^j_{regv,c}$ as follows:

$$\mathbf{W}^j_{clsa,c} = \mathcal{T}(\mathbf{W}^j_{regv,c}; \theta^j), \tag{1}$$

where θ^j are the learnt parameters irrespective of action category. $\mathcal{T}$ can be implemented with one or two fully-connected layers activated by different functions. Through sharing θ^j with the transfer module in j-th anchor layer between classification and localization on moments, $\mathcal{T}$ is generalized to the categories of action moments. The weights of segment-level classification for those categories can be transferred to the weights of anchor-level classification. As such, the weight transfer function is considered as a bridge to leverage the knowledge encoded in the action classification weights for action localization learning.

3.3 Localization Modeling on Action Moments

With the obtained anchor-level classification weights predicted by weight transfer function on action moments, we still can not perform action localization training since there is no background for optimizing temporal action proposal. To leverage action moments data for training localization model, a natural way is to hallucinate the background of moment to synthetize a complete action video. We therefore propose to generate action moment contextual features for localization modeling in an adversarial manner.

Figure 3 illustrates the process of action context generation for action moments. We denote the concatenated feature map of action moment and

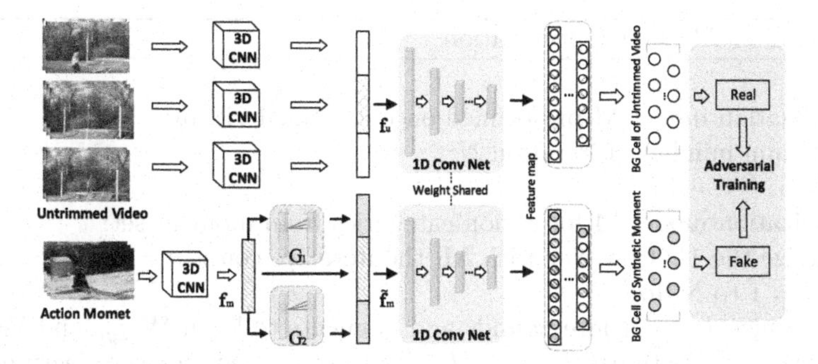

Fig. 3. Action context generation through adversarial learning. (BG: background)

untrimmed video extracted by 3D ConvNet as $\mathbf{f}_m$ and $\mathbf{f}_u$. Taking $\mathbf{f}_m$ as prior knowledge, two generators (G_1 and G_2) with the structure of two 1D convolutional layers are followed to synthesize the starting and ending contextual feature, respectively. The synthetic moments feature $\tilde{\mathbf{f}}_m$ is generated by concatenating $\mathbf{f}_m$ with the two generated contextual features as follows:

$$\tilde{\mathbf{f}}_m = A(G_1(\mathbf{f}_m), \mathbf{f}_m, G_2(\mathbf{f}_m)), \tag{2}$$

where A denotes the concatenation operation. By feeding the synthetic feature $\tilde{\mathbf{f}}_m$ and the original feature $\mathbf{f}_u$ of untrimmed videos into the 1D convolutional networks of localization model, multiple feature maps are produced on different scales. Each cell (anchor) in the j-th feature map reflects an action proposal, and the default temporal boundary of the t-th cell is defined as:

$$m_c = (t + 0.5)/T^j, \quad m_w = r_d/T^j, \tag{3}$$

where m_c and m_w are the center location and width. T^j and r_d represents the temporal length and scale ratio, respectively. For each cell, we denote the intersection over union (IoU) between the corresponding proposal and it's closest ground truth as g_{iou}. If the g_{iou} is larger than 0.8, we regard the cell as foreground cell. If the g_{iou} is lower than 0.3, it will be set as background cell. In each feature map, a discriminator is introduced to differentiate the background cells of synthetic moments from those of untrimmed videos. The simulation of action localization is employed on the concatenated synthetic feature.

Through adversarial learning, the contextual features of synthetic moments tend to be real through the guidance from those of untrimmed videos. Meanwhile, the anchor-level classification loss in localization modeling serves as a conditional constraint for adversarial training. The loss alleviates the generation of trivial background features and regularizes the generated context of each moment to preserve semantic information of action category.

3.4 Network Optimization

Given the global pooling feature vector f_p^j of j-th feature map, the segment-level classification loss (L_{reg}) for foreground segment or action moment is formulated

Algorithm 1. AherNet Optimization

Input:
Localization model $\mathbf{M}$ pre-trained on untrimmed videos;
Maximum number of iteration N;
Output:
Localization model $\tilde{\mathbf{M}}$ for action categories from moment set;
1: Initialize the 1D ConvNet with $\mathbf{M}$, the iterative count $n = 1$;
2: **for** n = 1 to N **do**
3: Optimize L_{reg} for foregrounds and moments to learn $\mathbf{W}_{regv}$ and $\tilde{\mathbf{W}}_{regv}$;
4: Fix $\mathbf{W}_{regv}$, optimize L_{cls} and L_{prop} of untrimmed videos to learn θ;
5: Apply θ to $\tilde{\mathbf{W}}_{regv}$ and obtain $\tilde{\mathbf{W}}_{clsa}$ for synthetic moments classification;
6: Fix 1D ConvNet, optimize context generators through L_{ad_G}, L_{cls} and L_{prop} of synthetic moments. Then fix context generators, optimize 1D ConvNet through L_{ad_D}, L_{cls} and L_{prop} of synthetic moments;
7: **end for**
8: **return** $\tilde{\mathbf{M}}$

via softmax loss:

$$L_{reg} = -\sum_{n=0}^{C-1} I_{n=c} \log(p_n^j), \tag{4}$$

where C represents the total number of action categories in untrimmed video set or moment set. The indicator function $I_{n=c} = 1$ if n equals to ground truth label c, otherwise $I_{n=c} = 0$. The probability p_n^j is projected by $\mathbf{W}_{regv}^j$ on f_p^j.

For the optimization of action localization, three 1D-conv layers are utilized on each feature map of untrimmed video or synthetic moment to predict anchor-level classification scores, offset parameters and overlap parameter for each cell (anchor). The anchor-level classification scores are predicted by transformed weights $\mathbf{W}_{clsa}$ and the formulation of loss function L_{cls} is the same with Eq. (4). The offset parameters $(\Delta c, \Delta w)$ denote temporal offsets relative to default center location m_c and width m_w, which are leveraged to adjust temporal coordinate:

$$\varphi_c = m_c + \alpha_1 m_w \Delta c \quad \text{and} \quad \varphi_w = m_w \exp(\alpha_2 \Delta w), \tag{5}$$

where φ_c, φ_w are refined center location and width of the corresponding proposal. α_1 and α_2 are used to balance the impact of temporal offsets. The offset loss is devised as Smooth L1 loss [13] (S_{L1}) between the foreground proposal and the closest ground truth, which is computed by

$$L_{of} = S_{L1}(\varphi_c - g_c) + S_{L1}(\varphi_w - g_w), \tag{6}$$

where g_c and g_w represents the center location and width of the proposal's closest ground truth instance, respectively. Furthermore, we define an overlap parameter y_{ov} to regress IoU between the proposal and it's closest ground truth for proposal re-ranking in localization. The mean square error (MSE) loss is adopted to optimize it as follows:

$$L_{ov} = (y_{ov} - g_{iou})^2. \tag{7}$$

Since both of the offset loss (L_{of}) and overlap loss (L_{ov}) are optimized for temporal action proposal, the sum of the two is regarded as the proposal loss (L_{prop}).

In the moment context generation stage, we define G as context generators of action moments, while D represents the discriminator of background cell on the feature map. We denote $\mathcal{F}_u$ and $\mathcal{F}_m$ as the set of extracted feature maps of untrimmed video and moment set, respectively. After producing the background cells b_u and b_m of each set, the adversarial loss is formulated as

$$\begin{aligned} L_{ad_D} &= -E_{\mathbf{f}_u \sim \mathcal{F}_u}[\log(D(b_u; \mathbf{f}_u))] - E_{\mathbf{f}_m \sim \mathcal{F}_m}[\log(1 - D(b_m; G(\mathbf{f}_m)))], \\ L_{ad_G} &= -E_{\mathbf{f}_m \sim \mathcal{F}_m}[\log(D(b_m; G(\mathbf{f}_m)))]. \end{aligned} \tag{8}$$

The overall training objective of our AherNet is formulated as a multi-task loss by integrating classification loss in segment-level (L_{reg}) and anchor-level (L_{cls}), proposal loss (L_{prop}) and adversarial loss (L_{ad}). The weight-sharing 1D convolutional networks of localization model are first pre-trained on untrimmed videos for initialization. Then we propose an alternating training strategy in each iteration to optimize the whole networks in an end-to-end manner. Algorithm 1 details the optimization strategy of our AherNet.

3.5 Inference and Post-processing

During prediction of action localization on action moment set, the context generators have been removed. The final ranking score s_f of each candidate action proposal is calculated by anchor-level classification scores $\mathbf{p} = [p_0, p_1, ..., p_{C-1}]$ and overlap parameter y_{ov} with $s_f = \max(\mathbf{p}) \cdot y_{ov}$. Given the predicted action instance $\phi = \{\varphi_c, \varphi_w, C_a, s_f\}$ with refined boundary (φ_c, φ_w), predicted action label C_a, and ranking score s_f, we employ the non-maximum suppression (NMS) for post-processing.

4 Experiments

We empirically verify the merit of our AherNet by conducting the experiments of temporal action localization across three different settings with three popular video benchmarks: ActivityNet v1.3 [17], THUMOS14 [21] and Kinetics-600 [12].

4.1 Datasets

The **ActivityNet v1.3** dataset contains 19,994 videos in 200 classes collected from YouTube. The dataset is divided into three disjoint subsets: training, validation and testing, by 2:1:1. All the videos in the dataset have temporal annotations. The labels of testing set are not publicly available and the performances of action localization on ActivityNet dataset are reported on validation set. The **THUMOS14** dataset includes 1,010 videos for validation and 1,574 videos for testing from 20 classes. Among all the videos, there are 220 and 212 videos with temporal annotations in validation and testing set, respectively. The **Kinetics-600** is a large-scale action recognition dataset which consists of around 480K videos from 600 action categories. The 480K videos are divided into 390K, 30K, 60K for training, validation and test sets, respectively. Each video in the dataset is a 10-s clip of action moment annotated from raw YouTube video.

4.2 Experimental Settings

Data Splits. For each setting, our AherNet involves two datasets, untrimmed video set with temporal annotations and action moment set with only category labels. In the first setting, we split the classes of ActivityNet v1.3 into two parts according to the dataset taxonomy. The untrimmed video set (ANet-UN) contains 87 classes and the action moment set (ANet-AM) consists of the remaining 113 classes. We extract the foreground segments of training videos from 113 classes as the training data and take the original videos in the validation set from 113 classes as the validation data. In view that we aim to transfer action localization capability on the categories in ANet-UN to those in ANet-AM, this setting is named as ANet-UN→ANet-AM. The second setting treats all the 220 validation videos in THUMOS14 (TH14) as untrimmed video set and the foreground segments of all the training videos in ActivityNet v1.3 as action moment set (ANet-FG). All the validation videos in ActivityNet v1.3 are exploited as the validation data. Similarly, we name this setting as TH14→ANet-FG. In the third setting, we utilize ActivityNet v1.3 (ANet) and Kinetics-600 (K600) as untrimmed video set and action moment set, respectively. To verify action localization on 600 categories in Kinetics-600, we crawled at least 10 raw YouTube videos of action moments in validation set for each class. In total, the validation data contains 6,459 videos. This setting is namely ANet→K600 for short.

Implementations. We utilize Pseudo-3D [40] network as our 3D ConvNet for clip-level feature extraction. The network input is a 16-frame clip and the sample rate of frames is set as 8. The 2,048-way outputs from pool5 layer are extracted as clip-level features. During training, we choose three temporal scale ratios $\{r_d\}_{d=1}^{3} = [2^0, 2^{1/3}, 2^{2/3}]$ derived from [30]. The parameter α_1 and α_2 are set as 1.0 by cross validation. The threshold of NMS is set as 0.90. We implement our AherNet on Tensorflow [1] platform. In all the experiments, our networks are trained by utilizing adaptive moment estimation optimizer (Adam) [22]. The initial learning rate is set as 0.0001, and decreased by 10% after every $5k$ on first two data split settings and $15k$ on the final setting. The mini-batch size is 16.

Evaluation Metrics. On all the three settings, we employ the mean average precision (mAP) values with IoU thresholds between 0.5 and 0.95 (inclusive) with a step size 0.05 as the metric for comparison.

4.3 Evaluation on Weight Transfer

We first examine the module of weight transfer between classification and localization in our AherNet. We compare several implementations of the weight transfer function $\mathcal{T}$, e.g., different number of fully-connected layers plus various activation functions (ReLU, LeakyReLU [35] and ELU [5]), and three baseline approaches of AherNet^0, AherNet$^-$ and AherNet*. AherNet^0 is a purely classification-based model which learns a snippet-level classifier to predict the action score sequentially and splits action instances with multi-threshold strategy on the score sequence. As such, AherNet^0 is regarded as the lower bound. AherNet$^-$ deploys a "proposal+classification" scheme without weight transfer module. The action proposal model in AherNet$^-$ is learnt on untrimmed video set and directly performed on validation videos to output temporal action proposals. The classifier trained on action moment set is employed to predict the category of each action proposal. AherNet* is an oracle run that exhaustively exploits the original videos of moment and trains a localization model in a fully-supervised manner. From this view, AherNet* is considered as the upper bound.

Table 1 summarizes the average mAP performances over all IoU thresholds of different methods on the first two settings. AherNet with weight transfer function of two fully-connected layers plus ELU activation consistently exhibits better performance than other implementations across the two settings. As expected, AherNet^0 performs worst since the method solely capitalizes on classification for localization problem without any knowledge of temporal action proposal. With the use of action proposal model learnt on untrimmed video set, AherNet$^-$ surpasses AherNet^0 by 2.6% and 1.1% on the settings of ANet-UN→ANet-AM and TH14→ANet-FG. AherNet further boosts up the average mAP from 12.8% and 10.4% of AherNet$^-$ to 17.2% and 24.3%, respectively. The results verify the merit of weight transfer in AherNet for bridging classification and localization, and

Table 1. Exploration of different implementations of the weight transfer function in our AherNet. (fc means fully-connected layer).

Approach	ANet-UN → ANet-AM	TH14 → ANet-FG
AherNet^0	10.2	9.3
AherNet$^-$	12.8	10.4
AherNet,1-fc,none	16.1	23.2
AherNet,1-fc, ReLU	16.4	23.4
AherNet,1-fc, LeakyReLU	16.5	23.5
AherNet,1-fc,ELU	16.7	23.9
AherNet,2-fc, LeakyReLU	16.9	24.2
AherNet,2-fc,ELU	**17.2**	**24.3**
AherNet,3-fc,ELU	16.8	24.1
AherNet*	22.6	28.9

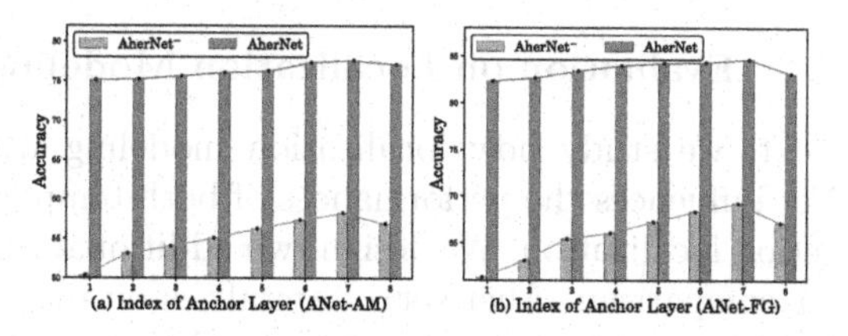

Fig. 4. Action classification accuracy of AherNet$^-$ and AherNet in different anchor layer on (a) ANet-UN→ANet-AM and (b) TH14→ANet-FG.

Table 2. The evaluations of localization modeling of AherNet.

Approach	ANet-UN→ ANet-AM		TH14 → ANet-FG	
	AUC	mAP	AUC	mAP
AherNet^0	41.8	10.2	11.2	9.3
AherNet$^-$	52.6	12.8	16.4	10.4
AherNet$_M$	53.5	13.2	49.7	17.3
AherNet$_{A-}$	54.6	14.7	51.0	19.1
AherNet	**58.3**	**17.2**	**55.5**	**24.3**
AherNet*	61.1	22.6	63.4	28.9

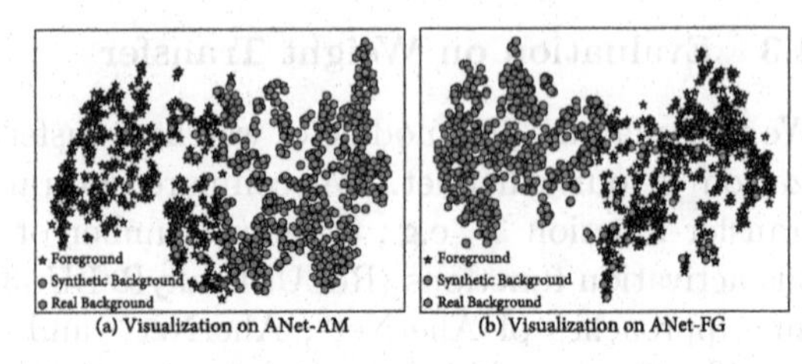

Fig. 5. Feature visualization of AherNet: (a) ANet-UN→ANet-AM and (b)TH14→ANet-FG.

scaling action localization to a large set of categories with only action moments. In practice, AherNet has great potential to support localization for thousands of categories. More importantly, when evaluating action localization model on the categories with full temporal annotation in the training, AherNet slightly outperforms AherNet*, e.g., 25.4% vs. 25.2% and 27.7% vs. 26.9% on the actions in ANet-UN and TH14. This also demonstrates the advantage of leveraging action moments data in AherNet training to enhance action localization model.

Figure 4 further details the average classification accuracy over all proposals in each anchor layer. Specifically, we feed the same proposals generated by AherNet into the classifier of AherNet$^-$ for accuracy computation on the same scale. Because most proposals in ActivityNet range over about 40% of the whole videos and such receptive field is nicely characterized by each anchor in the 7th layer, it is not a surprise that both AherNet and AherNet$^-$ achieve the highest accuracy on that layer. Benefited from the capture of contexts in joint optimization with temporal action proposal, AherNet leads to better and more stable performances than AherNet$^-$. The results again validate the weight transfer module.

4.4 Evaluation on Localization Modeling

Next, we study how localization modeling with context generation in AherNet influences the performances of both temporal action proposal and temporal action localization. We design two additional runs of AherNet$_M$ and AherNet$_{A-}$ for comparison. AherNet$_M$ capitalizes on only action moment set and directly learns an anchor-based action localization network by considering the starting/ending points of each moment as the time stamps of the action. AherNet$_{A-}$ is a variant of AherNet by removing adversarial learning. The context generator is pre-trained on untrimmed video set through minimizing $L2$ loss between the converted background from foreground and the real background.

Table 2 shows the measure of area under Average Recall vs. Average Number of proposals per video curves (AUC) for action proposal and mAP performances for action localization. Overall, AherNet$_M$ leads to a performance boost against AherNet$^-$ on both settings. In particular, AherNet$_M$ improves the AUC value from 16.4% to 49.7% on TH14→ANet-FG. Such results basically indicate that AherNet$_M$ is a practical choice for learning action localization directly

Table 3. Temporal action detection performances on ActivityNet v1.3, measured by mAP at different IoU thresholds α.

ActivityNet v1.3, mAP@α				
Approach	0.5	0.75	0.95	Average
Fully-supervised localization				
Wang *et al.* [53]	45.11	4.11	0.05	16.41
Singh *et al.* [47]	26.01	15.22	2.61	14.62
Singh *et al.* [48]	22.71	10.82	0.33	11.31
CDC [43]	45.30	26.00	0.20	23.80
TAG-D [54]	39.12	23.48	5.49	23.98
Lin *et al.* [28]	48.99	32.91	7.87	32.26
BSN [29]	52.50	33.53	8.85	33.72
Weakly-supervised localization				
STPN [37]	29.30	16.90	2.60	-
Nguyen *et al.* [38]	36.40	19.20	2.90	-
Partially-supervised localization				
AherNet	**40.33**	**25.04**	**3.92**	**24.31**

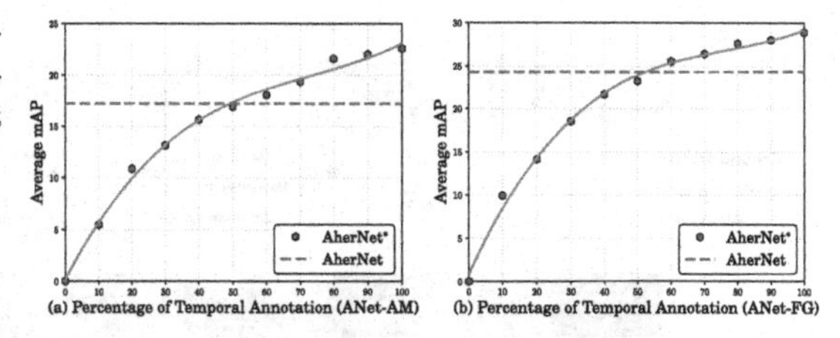

Fig. 6. Average mAP comparisons of AherNet* learnt with different ratio of temporal annotation and AherNet, on (a) ANet-UN→ANet-AM and (b) TH14→ANet-FG.

on moment data. AherNet$_{A-}$ is benefited from context generation for action moment set and the gain of mAP over AherNet$_M$ is 1.5% and 1.8%, respectively. Moreover, the upgrade of context generator from pre-training solely on untrimmed videos in AherNet$_{A-}$ to adversarial learning across the two video sets in AherNet contributes a mAP increase of 2.5% and 5.2%.

To examine the generated features of background, we further visualize the features of foreground, synthetic and real background for action moments by using t-SNE [36]. Specifically, we randomly select 500 anchors of foreground, synthetic and real background from 200 moments and the original videos in validation data, respectively. The first 256 principal components of the features of each anchor are extracted by PCA and projected into 2D space using t-SNE as shown in Fig. 5. It is clear that the generated features of background by AherNet are indistinguishable from those of real background on both ANet-AM and ANet-FG sets, that confirms the effectiveness of context generation.

4.5 Evaluation on Model Capacity of AherNet

We discuss our AherNet with several state-of-the-art fully-supervised and weakly-supervised action localization methods. Table 3 lists the mAP performances under different IoU thresholds on ActivityNet v1.3 and such evaluation corresponds to the second setting of TH14→ANet-FG for AherNet. The goal of weakly-supervised methods is to train action localization models for a set of

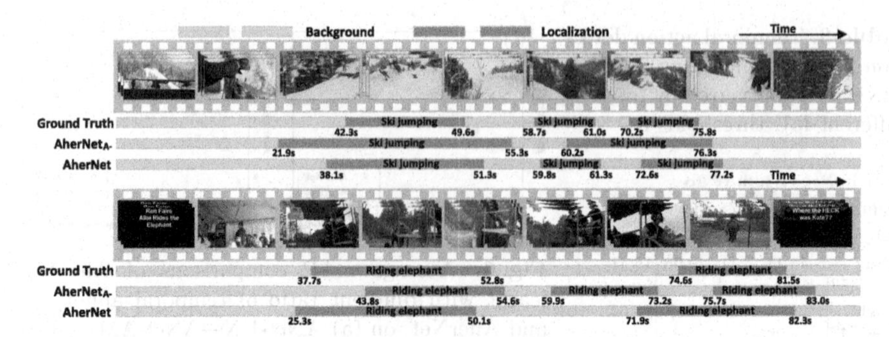

Fig. 7. Example of two action localization results on Kinetics-600.

Table 4. Performance comparisons of temporal action localization on Kinetics-600, measured by mAP at different IoU thresholds α.

ANet→K600, **Kinetics-600, mAP@α**				
Approach	0.5	0.75	0.95	Average
AherNet^0	19.26	16.72	0.88	14.18
AherNet$^-$	21.76	17.85	1.71	15.96
AherNet$_M$	28.98	20.71	2.95	19.05
AherNet$_{A-}$	32.71	23.04	5.08	21.68
AherNet	**36.19**	**26.96**	**6.55**	**24.43**

categories which have untrimmed videos with only video-level labels. Instead, our AherNet enables the training of localization model for the categories of interest with action moments from these categories (e.g., ANet-FG) and temporal annotations from a small set of classes (e.g., TH14). Compared to the most recent advance [38] in weakly-supervised localization, AherNet leads to a mAP boost of 3.9% and 5.8% under the IoU of 0.5 and 0.75, respectively. AherNet is also comparable or even superior to several fully-supervised localization models, e.g., [43,54], which rely on full temporal annotations for all the categories. More importantly, the partially-supervised learning paradigm of our AherNet extends action localization to potentially thousands of categories in a more deployable way.

To further quantitatively analyze the capability of AherNet, we compare AherNet with the fully-supervised version of AherNet* trained on different proportions of temporal annotations as shown in Fig. 6. As expected, the average mAP performances of AherNet* constantly improve with respect to the increase of temporal annotations in training on both datasets. The results are desirable in the way that AherNet* starts to surplus the performance of AherNet till more than 50% temporal annotations are leveraged.

4.6 Large-Scale Action Localization of AherNet

We finally take a step further to learn action localization model for 600 actions in Kinetics-600 dataset, which refers to the third setting of ANet→K600. Since the temporal annotations are not available for the validation videos of Kinetics-600, we collected the raw YouTube videos of action moments in our validation set and invited ten evaluators to label annotations. Table 4 summarizes the mAP at different IoU thresholds on the setting of ANet→K600. The performance trends are similar with those on the first two settings. AherNet boosts up the average mAP from 14.18% to 24.43%, indicating the impact of AherNet on the generalization of action localization for a large set of categories. Figure 7 showcases localization results of two videos from Kinetics-600, showing that AherNet nicely models the temporal dynamics and predicts accurate temporal boundaries.

5 Conclusions

We have presented Action Herald Networks (AherNet) which scale action localization to a large set of categories. Particularly, we study the problem from a new learning paradigm of training localization model with only trimmed action moments from the large set of categories plus temporal annotations on untrimmed videos from a small set of action classes. To materialize our idea, we have devised an one-stage action localization framework which consists of two key modules: weight transfer between classification and localization, and localization modeling on action moments. The former extracts foreground segments from untrimmed videos as action moments, and learns a weight transfer function between foreground segment classification and temporal action classification in localization. The latter simulates action localization on action moments data by hallucinating the background features of an action moment via adversarial learning. Experiments conducted on two settings, i.e., across the splits of ActivityNet v1.3 and from THUMOS14 to ActivityNet v1.3, validate our proposal. More remarkably, we build a large-scale localization model for 600 categories in Kinetics-600.

Acknowledgements. This work is partially supported by Beijing Academy of Artificial Intelligence (BAAI) and the National Key R&D Program of China under contract No. 2017YFB1002203.

References

1. Abadi, M., et al.: TensorFlow: a system for large-scale machine learning. In: OSDI (2016)
2. Buch, S., Escorcia, V., Ghanem, B., Fei-Fei, L., Niebles, J.C.: End-to-end, single-stream temporal action detection in untrimmed videos. In: BMVC (2017)
3. Buch, S., Escorcia, V., Shen, C., Ghanem, B., Niebles, J.C.: SST: single-stream temporal action proposals. In: CVPR (2017)

4. Chao, Y.W., Vijayanarasimhan, S., Seybold, B., Ross, D.A., Deng, J., Sukthankar, R.: Rethinking the faster R-CNN architecture for temporal action localization. In: CVPR (2018)
5. Clevert, D.A., Unterthiner, T., Hochreiter, S.: Fast and accurate deep network learning by exponential linear units (ELUs). In: ICLR (2016)
6. Escorcia, V., Caba Heilbron, F., Niebles, J.C., Ghanem, B.: DAPs: deep action proposals for action understanding. In: Leibe, B., Matas, J., Sebe, N., Welling, M. (eds.) ECCV 2016. LNCS, vol. 9907, pp. 768–784. Springer, Cham (2016). https://doi.org/10.1007/978-3-319-46487-9_47
7. Gaidon, A., Harchaoui, Z., Schmid, C.: Temporal localization of actions with actoms. IEEE Trans. PAMI **35**(11), 2782–2795 (2013)
8. Ganin, Y., Lempitsky, V.: Unsupervised domain adaptation by backpropagation. In: ICML (2015)
9. Gao, J., Chen, K., Nevatia, R.: CTAP: complementary temporal action proposal generation. In: Ferrari, V., Hebert, M., Sminchisescu, C., Weiss, Y. (eds.) ECCV 2018. LNCS, vol. 11206, pp. 70–85. Springer, Cham (2018). https://doi.org/10.1007/978-3-030-01216-8_5
10. Gao, J., Yang, Z., Sun, C., Chen, K., Nevatia, R.: TURN TAP: temporal unit regression network for temporal action proposals. In: ICCV (2017)
11. De Geest, R., Gavves, E., Ghodrati, A., Li, Z., Snoek, C., Tuytelaars, T.: Online action detection. In: Leibe, B., Matas, J., Sebe, N., Welling, M. (eds.) ECCV 2016. LNCS, vol. 9909, pp. 269–284. Springer, Cham (2016). https://doi.org/10.1007/978-3-319-46454-1_17
12. Ghanem, B., et al.: The ActivityNet large-scale activity recognition challenge 2018 summary. arXiv preprint arXiv:1808.03766 (2018)
13. Girshick, R.: Fast R-CNN. In: ICCV (2015)
14. Goodfellow, I.J., et al.: Generative Adversarial Nets. In: NIPS (2014)
15. He, K., Gkioxari, G., Dollar, P., Girshick, R.: Mask R-CNN. In: ICCV (2017)
16. Heilbron, F.C., Barrios, W., Escorica, V., Ghanem, B.: SCC: semantic context cascade for efficient action detection. In: CVPR (2017)
17. Heilbron, F.C., Escorcia, V., Ghanem, B., Niebles, J.C.: ActivityNet: a large-scale video benchmark for human activity understanding. In: CVPR (2015)
18. Hoffman, J., Guadarrama, S., Tzeng, E., Hu, R., Donahue, J.: LSDA: large scale detection through adaptation. In: NIPS (2014)
19. Hu, R., Dollar, P., He, K., Darell, T., Girshick, R.: Learning to segment every thing. In: CVPR (2018)
20. Isola, P., Zhu, J.Y., Zhou, T., Efros, A.A.: Image-to-image translation with conditional adversarial networks. In: CVPR (2017)
21. Jiang, Y.G., Liu, J., Zamir, A.R., Toderici, G.: THUMOS challenge: action recognition with a large number of classes (2014). http://crcv.ucf.edu/THUMOS14
22. Kingma, D.P., Ba, J.: Adam: a method for stochastic optimization. In: ICLR (2015)
23. Kuen, J., Perazzi, F., Lin, Z., Zhang, J., Tan, Y.P.: Scaling object detection by transferring classification weights. In: ICCV (2019)
24. Lea, C., Michael D. Flynn, R.V., Reiter, A., Hager, G.D.: Temporal convolutional network for action segmentation and detection. In: CVPR (2017)
25. Li, D., Qiu, Z., Dai, Q., Yao, T., Mei, T.: Recurrent tubelet proposal and recognition networks for action detection. In: Ferrari, V., Hebert, M., Sminchisescu, C., Weiss, Y. (eds.) ECCV 2018. LNCS, vol. 11210, pp. 306–322. Springer, Cham (2018). https://doi.org/10.1007/978-3-030-01231-1_19
26. Li, D., Yao, T., Qiu, Z., Li, H., Mei, T.: Long short-term relation networks for video action detection. In: ACM MM (2019)

27. Lin, T., Zhao, X., Shou, Z.: Single shot temporal action detection. In: ACM MM (2017)
28. Lin, T., Zhao, X., Shou, Z.: Temporal convolution based action proposal: submission to activitynet 2017. arXiv preprint arXiv:1707.06750 (2017)
29. Lin, T., Zhao, X., Su, H., Wang, C., Yang, M.: BSN: boundary sensitive network for temporal action proposal generation. In: Ferrari, V., Hebert, M., Sminchisescu, C., Weiss, Y. (eds.) ECCV 2018. LNCS, vol. 11208, pp. 3–21. Springer, Cham (2018). https://doi.org/10.1007/978-3-030-01225-0_1
30. Lin, T.Y., Goyal, P., Girshick, R., He, K., Dollar, P.: Focal loss for dense object detection. In: ICCV (2017)
31. Liu, D., Jiang, T., Wang, Y.: Completeness modeling and context separation for weakly supervised temporal action localization. In: CVPR (2019)
32. Long, F., Yao, T., Qiu, Z., Tian, X., Luo, J., Mei, T.: Gaussian temporal awareness networks for action localization. In: CVPR (2019)
33. Long, F., Yao, T., Qiu, Z., Tian, X., Mei, T., Luo, J.: Coarse-to-fine localization of temporal action proposals. IEEE Trans. Multimed. **22**(6), 1577–1590 (2020)
34. Lu, S., Wang, Z., Mei, T., Guan, G., Feng, D.D.: A bag-of-importance model with locality-constrained coding based feature learning for video summarization. IEEE Trans. Multimed. **16**(6), 1497–1509 (2014)
35. Maas, A.L., Hannun, A.Y., Ng, A.Y.: Rectifier nonlinearities improve neural network acoustic models. In: ICML (2013)
36. van der Maaten, L., Hinton, G.: Visualizing data using t-SNE. JMLR (2008)
37. Nguyen, P., Liu, T., Prasad, G., Han, B.: Weakly supervised action localization by sparse temporal pooling network. In: CVPR (2018)
38. Nguyen, P.X., Ramanan, D., Fowlkes, C.C.: Weakly-supervised action localization with background modeling. In: ICCV (2019)
39. Oneata, D., Verbeek, J., Schmid, C.: Action and event recognition with fisher vectors on a compact feature set. In: ICCV (2013)
40. Qiu, Z., Yao, T., Mei, T.: Learning spatio-temporal representation with pseudo-3D residual networks. In: ICCV (2017)
41. Qiu, Z., Yao, T., Ngo, C.W., Tian, X., Mei, T.: Learning spatio-temporal representation with local and global diffusion. In: CVPR (2019)
42. Shi, B., Dai, Q., Mu, Y., Wang, J.: Weakly-supervised action localization by generative attention modeling. In: CVPR (2020)
43. Shou, Z., Chan, J., Zareian, A., Miyazawa, K., Chang, S.F.: CDC: convolutional-de-convolutional network for precise temporal action localization in untrimmed videos. In: CVPR (2017)
44. Shou, Z., Gao, H., Zhang, L., Miyazawa, K., Chang, S.-F.: AutoLoc: weakly-supervised temporal action localization in untrimmed videos. In: Ferrari, V., Hebert, M., Sminchisescu, C., Weiss, Y. (eds.) ECCV 2018. LNCS, vol. 11220, pp. 162–179. Springer, Cham (2018). https://doi.org/10.1007/978-3-030-01270-0_10
45. Shou, Z., et al.: Online detection of action start in untrimmed, streaming videos. In: Ferrari, V., Hebert, M., Sminchisescu, C., Weiss, Y. (eds.) ECCV 2018. LNCS, vol. 11207, pp. 551–568. Springer, Cham (2018). https://doi.org/10.1007/978-3-030-01219-9_33
46. Shou, Z., Wang, D., Chang, S.F.: Temporal action localization in untrimmed videos via multi-stage CNNs. In: CVPR (2016)
47. Singh, B., Marks, T.K., Jones, M., Tuzel, O., Shao, M.: A multi-stream bi-directional recurrent neural network for fine-grained action detection. In: CVPR (2016)

48. Singh, G., Cuzzolin, F.: Untrimmed video classification for activity detection: submission to ActivityNet challenge. arXiv preprint arXiv:1607.01979 (2016)
49. Tang, K., Yao, B., Fei-Fei, L., Koller, D.: Combining the right features for complex event recognition. In: ICCV (2013)
50. Tang, Y., Wang, J., Gao, B., Dellandrea, E., Gaizauskas, R., Chen, L.: Large scale semi-supervised object detection using visual and semantic knowledge transfer. In: CVPR (2016)
51. Tzeng, E., Hoffman, J., Saenko, K., Darrell, T.: Adversarial discriminative domain adaption. In: CVPR (2017)
52. Wang, L., Xiong, Y., Lin, D., Gool, L.V.: UntrimmedNets for weakly supervised action recognition and detection. In: CVPR (2017)
53. Wang, R., Tao, D.: UTS at activitynet 2016. In: CVPR ActivityNet Challenge Workshop (2016)
54. Xiong, Y., Zhao, Y., Wang, L., Lin, D., Tang, X.: A pursuit of temporal accuracy in general activity detection. arXiv preprint arXiv:1703.02716 (2017)
55. Xu, H., Das, A., Saenko, K.: R-C3D: region convolutional 3D network for temporal activity detection. In: ICCV (2017)
56. Yeung, S., Russakovsky, O., Mori, G., Fei-Fei, L.: End-to-end learning of action detection from frame glimpses in videos. In: CVPR (2016)
57. Yuan, J., Ni, B., Yang, X., Kassim, A.A.: Temporal action localization with pyramid of score distribution features. In: CVPR (2016)
58. Zeng, R., et al.: Graph convolutional networks for temporal action localization. In: ICCV (2019)
59. Zhao, Y., Xiong, Y., Wang, L., Wu, Z., Tang, X., Lin, D.: Temporal action detection with structured segment networks. In: ICCV (2017)

ForkGAN: Seeing into the Rainy Night

Ziqiang Zheng[1], Yang Wu[2(✉)], Xinran Han[3], and Jianbo Shi[3]

[1] UISEE Technology (Beijing) Co., Ltd., Beijing, China
zhengziqiang1@gmail.com
[2] Kyoto University, Kyoto, Japan
wu.yang.8c@kyoto-u.ac.jp
[3] University of Pennsylvania, Philadelphia, USA
{hxinran,jshi}@seas.upenn.edu

Abstract. We present a ForkGAN for task-agnostic image translation that can boost multiple vision tasks in adverse weather conditions. Three tasks of image localization/retrieval, semantic image segmentation, and object detection are evaluated. The key challenge is achieving high-quality image translation without any explicit supervision, or task awareness. Our innovation is a fork-shape generator with one encoder and two decoders that disentangles the domain-specific and domain-invariant information. We force the cyclic translation between the weather conditions to go through a common encoding space, and make sure the encoding features reveal no information about the domains. Experimental results show our algorithm produces state-of-the-art image synthesis results and boost three vision tasks' performances in adverse weathers.

Keywords: Light illumination · Image-to-image translation · Image synthesis · Generative adversarial networks

1 Introduction

Data bias is a well-known challenge for deep learning methods. An AI algorithm trained on one dataset often has to pay a performance deficit in a different dataset. Take an example of image recognition on a rainy night. An object detector trained on a day time dataset could suffer 30–50% accuracy drop on rainy night images. One solution is simply collecting more labeled data in those adverse weather conditions [7,8,10,19,24]. This is expensive and more fundamentally does not address the data bias issue.

Domain adaptation [13,23,28] is a general solution to this data bias problem. Our work is related to a sub-branch of this approach focusing on *image-to-image translation* techniques to explicitly synthesize images in uncommon domains. In the context of day and night domain change, two strategies have been explored

Electronic supplementary material The online version of this chapter (https://doi.org/10.1007/978-3-030-58580-8_10) contains supplementary material, which is available to authorized users.

A. Vedaldi et al. (Eds.): ECCV 2020, LNCS 12348, pp. 155–170, 2020.
https://doi.org/10.1007/978-3-030-58580-8_10

Fig. 1. Our ForkGAN can boost performance for multiple vision tasks with night to day image translation: localization (by SIFT point matching between daytime and nighttime images), semantic segmentation and object detection (by data augmentation) in autonomous driving (results are all shown on the nighttime images).

recently: one is day-to-night approach [22], which transfers annotated daytime data to nighttime so that the annotations can be reused through data augmentation; the other [1] uses a night-to-day translator to generate images suitable for existing models trained on daytime data. The two strategies both demonstrated that the precise domain translation methods can boost the other vision tasks. In this paper, we look into a far more challenging case of a rainy night. Our experiments show that existing approaches perform poorly in this case, particularly when we have no supervised data annotation on the rainy night images.

The fundamental challenge is that what makes an image looks good to a human might not improve computer vision algorithms. A computer vision algorithm can handle certain types of lighting change surprisingly well, while minor artifacts invisible to human could be harmful to vision algorithms.

A straightforward method is to introduce task-specific supervision on the new domain to ensure the image translation is task aware. We believe task-aware approaches only shift the data bias problem to a task bias problem. Instead, we ask if we can create a *task agnostic* image-to-image translation algorithm that improves computer vision algorithms without any supervision or task information. Figure 1 shows our solution can achieve this goal on three untrained tasks: image localization, semantic segmentation, and object detection.

Problem Analysis. Domain translation between adverse conditions (*e.g.* nighttime) and standard conditions (*e.g.* daytime) is inherently a challenging unsupervised or weakly-supervised learning problem, as it is impossible to get precisely aligned ground-truth image pairs captured at a different time for dynamic driving scenes where a lot of moving objects exist. Many objects (*e.g.* the vehicles and street lamps) look totally differently across different weather conditions. There are global scene level texture differences such as raindrops, as well as regional changes such as cars' reflection on the wet road. There is a common semantic and geometrical level similarity between the adverse and normal domain, as well as vast differences. Precisely disentangling the invariant and variant features, without any supervision or task knowledge, is our key objective.

Proposed Solution. An ideal task-agnostic image translation preserves the image contents at all scale levels: scene level layout to object details such as letters on a traffic sign, while automatically adjusting to the illumination and weather conditions. For CycleGAN-based models that mainly rely on cycle-consistency losses, altering the global conditions can be done effectively, but faithfully maintaining the informative content details is not guaranteed. We first 'tie' the two encoding space of the CycleGAN together, to make sure we have kept only the necessary invariant information in both domains. We further explicitly check this encoding is domain agnostic: by looking at the encoded features, we cannot tell the domain they come from. This step can potentially remove much invariant information. We add a 'Fork' branch to check if we have encoded sufficient information to reconstruct the original image data in both domains. The model is called ForkGAN. It has the following main contributions.

- We propose a Fork-shaped Cyclic generative module that can decouple domain-invariant content and domain-specific style during domain translation. We force both encoders to go through a common encoding space and explicitly use an anti-contrastic loss to ensure necessary invariant information is produced in the disentanglement.
- We introduce a Fork-branch on each generator stage, to ensure sufficient information is kept for image recognition tasks in both domains.
- We boost the performance of localization, semantic segmentation and object detection in adverse conditions using our ForkGAN.

2 Related Work

2.1 Unpaired Image-to-Image Translation

Many models have been proposed for unpaired image-to-image translation task, which aims to translate images from source domain to the corresponding desired images in target domain without corresponding image pairs for training. Introduced by Zhu et al. [28], CycleGAN is a classical and elegant solution for unpaired image-to-image translation. The cycle-consistency loss provides a natural and nice way to regularize the image translation, and it has become a widely used base. However, it does not enforce the translated image to share the same semantic space as the source image, and therefore its disentanglement ability is rather weak. UNIT [17] added a shared latent space assumption and enforced weight sharing between the two generators. However, weight sharing does not always guarantee that the network will learn to disentangle the images from different domains. To improve the diversity of generated results, models such as MUNIT [12], DRIT [16] were proposed to better decompose visual information into domain-invariant content and domain-specific style. To handle the translation among multiple domains, StarGAN [5] was developed by combining an additional classification loss. One drawback of those models is that the user will need to specify a Etyle code Eor label to sample from. For application in autonomous driving, we want the model to translate the image in adverse weathers to an appropriate condition without any human guidance during inference time.

2.2 Low-Light Image Enhancement

Besides translating images from adverse conditions (*e.g.* night domain) to standard conditions, another possible approach to tackle the lack of visibility at night is to use low-light image enhancement models. Those models aim to improve the visual quality of underexposed photos by manipulating the color, brightness and contrast of the image. Recently, more deep learning based models have been proposed to solve underexposure problem. EnlightenGAN [14] can perform low-light enhancement without paired training data. The model increases the luminosity of an image while preserving the texture and structure of objects. However, without emphasis on foreground objects, EnlightenGAN provides limited details that are helpful for driving purposes. Different from these low-light image enhancement methods, we target to translate the whole image to day time and enhance the weak object signals in the dark.

2.3 Bad Weather Vision Tasks

Adverse weathers and undesirable illumination conditions pose challenges to common vision tasks such as localization, semantic segmentation and object detection. Visual localization and navigation allow the vehicle or robot to estimate its location and orientation in the real world. One efficient approach for this task is to use image retrieval techniques [1] and feature matching methods [2,18]. However, these methods suffer from performance degradation when the query image is sampled from different illumination and weather conditions as compared with the labelled database. ToDayGAN [1] modified the image translation model to improve image retrieval performance for localisation task. Porav et al. [21] proposed a system that translates input images to a desired domain to optimize feature-matching results.

For semantic segmentation, Porav et al. [20] proposed a system that uses light-weight adapters to transform images of different weather and lighting conditions to an ideal condition for training off-the-shelf computer vision models. To train the adapters, they chose a sequence of reference images under ideal condition, and use CycleGAN [28] to synthesize images in different weathers while preserving the geometry and structure of the reference images. They then trained adapters to transform images from specific domains so using the new images can achieve better performance on related vision tasks.

Object detection, despite its importance, has received less attention in recent works on driving in adverse weathers. A related work in this direction is from He et al. [8], where the authors developed a multi-adversarial Faster R-CNN framework for domain-adaptive object detection in driving scenario. Their source and

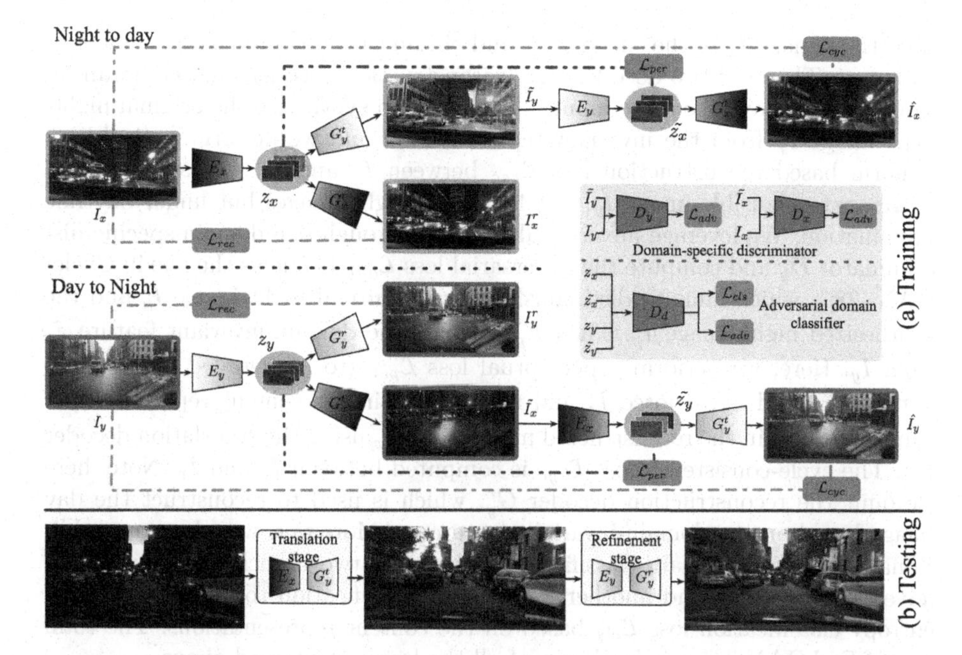

Fig. 2. The framework of ForkGAN. (a) shows the training stage while (b) represents the testing inference. I_x and I_y denote random image from night domain $\mathbb{X}$ and day domain $\mathbb{Y}$, separately. E_x and E_y are the encoders to encode the night and day images, separately. G_x^t and G_y^t are responsible to achieve domain translation based on the domain-invariant representations z and $\tilde{z}$. G_x^r and G_y^r aim to reconstruct the input images based on the representations. D_x and D_y are the discriminators of $\mathbb{X}$ and $\mathbb{Y}$, while D_d is the domain adversarial classifier.

target domain pairs involve regular and foggy Cityscapes, synthetic and real data from two different driving datasets with similar weather conditions. AugGAN [11] aims to combine an image parsing network to enhance object detection performance in nighttime images through day-to-night translation on synthetic datasets. However, it requires paired auxiliary annotations (*e.g.* semantic segmentation maps), which are sometimes expensive or hard to acquire, to regularize the image parsing network. Our ForkGAN addresses object detection under more challenging weather conditions - driving scenes at nighttime with reflections and noise from rain and even storms, without any auxiliary annotations.

3 Proposed Method

3.1 ForkGAN Overall Framework

Our ForkGAN performs image translation with unpaired data using a novel fork-shape architecture. The fork-shape module contains one encoder and two decoders. Take night-to-day translation in Fig. 2 as an example, first we feed a nighttime image I_x to the encoder E_x and obtain the domain-invariant representation z_x. Then the two decoders G_x^r (reconstruction decoder) and G_x^t (translation decoder) have the same input z_x. G_x^r aims to synthesize the original nighttime image I_x^r from the invariant representation and we perform a pixel-level l_1-norm based reconstruction loss $\mathcal{L}_{rec}$ between I_x^r and I_x. G_x^t is responsible to generate plausible image $\tilde{I}_y$ that looks like night images but under daytime illumination. We leverage adversarial training through one domain-specific discriminator D_y and compute the adversarial loss $\mathcal{L}_{adv}$ (same as the one in Cycle-GAN [28]), which aims to distinguish the random real night image I_y and the synthesized night image $\tilde{I}_y$. Then E_y extracts the domain-invariant feature $\tilde{z_x}$ from $\tilde{I}_y$. Here, we perform a perceptual loss $\mathcal{L}_{per}$ (to be detailed in Sect. 3.2) between $\tilde{z_x}$ and z_x to force I_x and $\tilde{I}_y$ to have similar content representation. Finally we obtain the reconstructed night image $\hat{I_x}$ using the translation decoder G_x^t. The cycle-consistency loss $\mathcal{L}_{cyc}$ is computed between $\hat{I_x}$ and I_x. Note, here we omit the reconstruction decoder G_y^r, which is used to reconstruct the day image based on the domain-invariant feature z_y. Moreover, we adopt one additional adversarial domain classifier D_d, which has two branch outputs: one for adversarial training and another for domain classification to obtain the cross-entropy classification loss $\mathcal{L}_{cls}$ based on the content representations. The total loss of ForkGAN is a weighted sum of all the losses mentioned above:

$$\mathcal{L}(E, G^r, G^t) = \mathcal{L}_{adv} + \mathcal{L}_{cls} + \mathcal{L}_{per} + \gamma\mathcal{L}_{cyc} + \epsilon\mathcal{L}_{rec}, \tag{1}$$

and we set $\gamma = \epsilon = 10$ in our experiments. With the total loss, the three components E, G^r, and G^t are optimized together so that the learned model is unbiased and can disentangle the domain-invariant content and domain-specific style. During inference time, our ForkGAN provides a two-stage translation procedure as shown in Fig. 2. Take night-to-day translation as an example, the input night image is translated to a daytime image using E_x and G_y^t, and the output is regarded as input of the refinement stage. E_y and G_y^r synthesize more precise translation output, which gives the final output of our ForkGAN.

3.2 ForkGAN - Disentanglement Stage

Previous Cycle-GAN based methods target to preserve the appearance of input images through an indirect pixel-level cycle-consistency loss and generate plausible translated image by leveraging adversarial loss to the translated images. However, some weak but informative domain-invariant characteristics are usually ignored during translation stage. Sometimes, the generator $\mathcal{F} : \mathbb{X} \longrightarrow \mathbb{Y}$ can fool the discriminator and minimize the adversarial loss by changing the global conditions that dominate more pixels, while ignoring local features such as cars and pedestrians. This leads to a trivial translation solution that throws away some informative signals. In the opposite direction, $\mathcal{G} : \mathbb{Y} \longrightarrow \mathbb{X}$ has a strong ability to remap the translated image to the original domain under a strong pixel-level cycle-consistency loss. In the two stages, there is no guarantee that the domain-invariant and domain-specific feature can be disentangled. Our fork-shape generator can achieve better disentanglement because:

(1) E aims to extract the domain-invariant content and discard the domain-specific style. G^r and G^t target to reload the style representation of the source and target domain separately. The three components have same parameter quantity, which ensures comparable network capacity so that there is no explicit bias or dependency among the three components. If E is too weak and fails to extract the informative content, the reconstruction loss $\mathcal{L}_{rec}$ is large. If the domain-invariant representation z_x is still mixed with the domain-specific information, the translation decoder will fail to generate reasonable translated output.

(2) We impose a perceptual loss as

$$\mathcal{L}_{per} = \tau(\sum_{n=1}^{N} \lambda_n ||\Phi_n(\tilde{z_x})) - \Phi_n(z_x)||_1), \tag{2}$$

which makes $\tilde{z_x}$ perceptually similar to z_x (designed according to the perceptual loss in [4]). Here, Φ_n denotes the feature extractor at the n_{th} level of the pre-trained VGG-19 network on ImageNet. The hyper-parameter λ_n controls the influence of perceptual loss at different levels and here we set λ_n all 1. Different from the way perceptual loss is typically used (feeding image data to the VGG network), we rearrange the feature maps of z_x and $\tilde{z_x}$ through bilinear interpolation to fit into only the last three layers of VGG. Such a modification enables an effective perceptual consistency check between z_x and $\tilde{z_x}$ at the feature level. If E_x and E_y fail to eliminate the domain-specific information completely, the perceptual loss between z_x and $\tilde{z_x}$ will be large. The perceptual loss can also help preserve the content information during translation stage.

(3) The adversarial domain classifier targets to distinguish the real/fake distribution and classify the content representation. We aim to match the distribution of z and $\tilde{z}$ through adversarial training. Specifically, we assign an opposite label to z and $\tilde{z}$ to implement an classification training to obtain $\mathcal{L}_{cls}$. We perform the classification loss using both z and $\tilde{z}$. If the classifier could not distinguish which domain the content representation is from, it indicates that the extracted representation does not carry any domain-specific style information.

Based on above reasons, the design of our model and training objectives can provide strong constraints to achieve disentanglement.

3.3 ForkGAN - Refinement Stage

In the fork-shape module, the generator has two branches: the translation branch and the reconstruction branch. We apply an additional refinement stage to the translated output using autoencoder E_y and reconstructions decoder G_y^r. This pair is trained during disentanglement stage and therefore the refinement does not introduce new parameters. During this stage, the reconstruction branch G_y^r

can refine the fake outputs ($\tilde{I_x}$ and $\tilde{I_y}$) by knowledge learned from reconstructing the real images, thus generating more realistic images and strengthening weak signals. We adopt additional pixel-wise Gaussian noise disturbance to the domain-invariant content representation z to improve the robustness of the reconstruction branch and make it less input-sensitive. We also hope that the reconstructed decoder can generate complementary information from additional noise even if some domain-invariant content feature is missed. In this way, the two-stage translation shown in Fig. 2 can obtain better translation performance even in adverse environment.

3.4 Dilated Convolution and Multi-scale Discriminator Architecture

Considering the occlusion and reflection of images captured in adverse weathers, it is difficult to recognize the objects essential to the task of navigation (*e.g.* traffic signs, lanes and other vehicles). A possible solution is to adopt a large receptive field to alleviate the occlusion issue. To do that, we use dilated residual networks [26] for the generator with fewer parameters. The dilated convolution can help the three components of our generator understand the relationship of different parts. To achieve high-resolution image-to-image translation, we adopt the multi-scale discriminator architecture [12,16,25] to improve the ability to distinguish the fake images and real images. The proposed architecture could fuse the information from multiple scales and generate more realistic outputs.

4 Experiments

4.1 Datasets and Evaluation Metrics

Datasets: Alderley is originally proposed for the SeqSLAM algorithm [19], which collected the images for the same route twice: once on a sunny day and another time on a stormy rainy night. Every frame in the dataset is GPS-tagged, and thus each nighttime frame has a corresponding daytime frame. The images collected at nighttime are blurry with a lot of reflections, which render the front vehicles, lanes and traffic signs difficult to be recognized. For this dataset, we use the first consecutive four fifths for training and others for evaluation. Since this dataset has day-night correspondences, we use it for quantitative evaluation on image localization task. Unfortunately, it doesn't provide ground-truth annotations for semantic segmentation and object detection, so we use another dataset instead for those two tasks. **BDD100K** [27] is a large scale high-resolution autonomous driving dataset, which collected 100,000 video clips in multiple cities and under various conditions. For each video, it selects a key frame to provide detailed annotations (such as the bounding box of various objects, the dense pixel annotation, the daytime annotation and so on). We reorganized this dataset according to the annotation, and obtained 27,971 night images for training and 3,929 night images for evaluation. We obtained 36728/5258 train/val split for day images. We inherit the data split from the BDD100K dataset. We perform semantic segmentation and object detection on this dataset.

Image Quality Metric: FID [9] evaluates the distance between the real sample distribution and the generated sample distribution. Lower FID score indicates higher image generation quality.

Vision Task Metrics: For **Localization:** SIFT [18] is good measure to find the feature matching points between two images. We measure the localization performance by the SIFT interesting points matching. **Semantic Segmentation:** Intersection-over-Union(IoU) is a commonly used metric for semantic segmentation. For each object class, the IoU is the overlap between predicted segmentation map and the ground truth, divided by their union. In the case of multiple classes, we take the average IoU of all classes (*i.e.*, mIoU) to indicate the overall performance of the model. **Object Detection:** We use mean average precision (mAP) to evaluate the performance and also report the average precision scores for individual classes to have a more thorough evaluation.

4.2 Experiment Settings and Implementation Details

We compare our proposed method with other state-of-the-art image translation methods such as UNIT [17], CycleGAN [28], MUNIT [12], DRIT [16], UGATIT [15] and StarGAN [1]. Additionally, we also compare with low-light enhancement methods such as EnlightenGAN [14] and ToDayGAN [1]. We follow the instructions of those methods and make a fair setting for comparison.

The encoder E contains 3 Conv-Ins-ReLU modules and 4 dilated residual blocks, while both reconstructed decoder G^r and the translated decoder G^t have 4 dilated residual blocks and 3 Deconv-Ins-ReLU modules followed by a Tanh activity function. All the domain-specific discriminators adopt the multi-scale discriminator architecture and we set the number of scales as 2. For the adversarial domain classifier, the backbone has 4 Conv-Ins-ReLU blocks, the adversarial branch has one additional convolution layer to get adversarial output, while the classification branch has one more fully-connected layer to obtain a domain classification output. We adopt Adam optimizer and set learning rate to 0.0002.

4.3 Localization by SIFT Point Matching

We aim to perform translation at an extremely difficult setting on Alderley dataset. Figure 3 shows the qualitative translation result comparisons. UNIT and MUNIT fail to perform reasonable translation and generate plausible objects. DRIT has lost the detailed information and missed some objects after domain translation. The result of EnlightenGAN fails to provide meaningful visual information and it only changed the illumination slightly. ToDayGAN and UGATIT obtain better translation results and have captured the visual objects in the darkness. But they cannot preserve the visual objects (*e.g.*, traffic signs and cars) well. In contrast, our method has stronger ability to capture these weak signals and preserve them better. For this dataset, we perform experiments using

512 * 256 resolution. We compute the number of SIFT matching points between the translated daytime images and the corresponding natural daytime images. Table 1 reports the quantitative comparison. Our ForkGAN obtains the best SIFT result through precise night-to-day image translation. It has also achieved the best image generation quality with lowest FID score. By improving the ability to maintain and enhance the SIFT matching, it can benefit place recognition and visual localization.

Ablation Studies. Several experiments are designed for ablation studies. Firstly, we remove the Fork-shape architecture (denoted as w/o Fork-shape) of the generator, and follow the setting of Cycle-GAN methods to optimize the model. The result generated by the vanilla generator has artifacts on cars as there is no guarantee on the disentanglement between the domain-invariant and domain-specific information. Then we investigate the effectiveness of the Fork-shape generator itself only (with a name of "Fork-shape"). Note, we do not compute $\mathcal{L}_{per}$ and the adversarial domain classification loss $\mathcal{L}_{cls}$ in this setting. Due to the

Fig. 3. The visual/qualitative translation result comparison of different methods. Please zoom in to check more details on the content and quality. The parts covered by red and green boxes show the enlarged cropped region in the corresponding image. (Color figure online)

Table 1. Evaluation metric results of different methods for night → day translation task on Alderley dataset [19]. The original denotes the scores of the original real night images. FID reports the visual image quality (lower is better) while SIFT reports the localization performance (higher is better).

Method	EnlightenGAN [14]	UNIT [17]	MUNIT [12]	CycleGAN [28]	DRIT [16]
FID/SIFT	249/2.00	155/2.68	138/2.75	167/3.36	145/3.71
Method	StarGAN [5]	UGATIT [15]	ToDayGAN [1]	ForkGAN	Original
FID/SIFT	117/3.28	170/2.51	104/4.14	**61.2/12.1**	210/3.12

Table 2. Quantitative comparisons for ablation studies on Alderley dataset [19].

Method	w/o Fork-shape	Fork-shape	Fork-shape+DRB
FID/SIFT	146/4.26	131/7.12	113/8.12
Fork-shape+DRB+MSD	w/o refinement	w/ shared decoders	ForkGAN
95.3/9.14	70.7/11.5	73.8/9.29	**61.2/12.1**

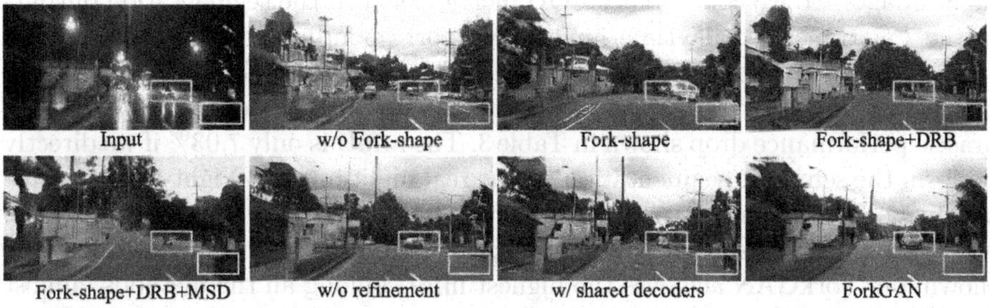

Fig. 4. Visual results for ablation studies on Alderley. Red boxes highlight some details. (Color figure online)

reconstruction loss, the synthesized images have fewer artifacts having all but the Fork-shape. Based on this, we aim to explore the improvement from the dilated residual blocks (DRB for abbreviation) by evaluating "Fork-shape+DRB". A larger receptive field can help the generator to better capture the objects in the dark. Then we adopt a multi-scale discriminator architecture (MSD for abbreviation), here we set $n = 2$. As reported in Table 2, the MSD architecture can also lead to the improvement on FID and SIFT matching (as shown by the results of "Fork-shape+DRB+MSD"). In the next experiment, we use everything we have covered for training ForkGAN, and just exclude the refinement stage when we use it for testing, which is denoted by "w/o refinement". As shown in Fig. 4 and Table 2, adding $\mathcal{L}_{per}$ and $\mathcal{L}_{cls}$ to "Fork-shape+DRB+MSD" can achieve better disentanglement, which leads to significantly better translation outputs. Finally, we apply the "ForkGAN" with the refinement stage at the testing stage and observe that the refinement can greatly improve the detailed part generation. Last but not the least, we also evaluate a twisted version of ForkGAN by letting the translation decoder and reconstruction decoder of the same domain (*e.g.*, G_y^t and G_y^r) share the same parameters (basically using the same decoder instead of two different ones), and have it denoted by "w/ shared decoders". As showed, it results in a significant performance drop when compared with "ForkGAN" which doesn't have shared decoders. The two decoders for the same domain look similar, but they are constrained by different losses and thus have different duties, which complement each other. Putting the loads on one single decoder makes it much harder to achieve the goal and leads to inferior model. All the quantitative comparisons of above different settings are listed in Table 2 and qualitative ones are given in Fig. 4.

4.4 Semantic Segmentation

Moreover, we perform high-resolution (1024×512) night-to-day image translation to boost the semantic segmentation performance. Figure 5 presents the translated results and corresponding segmentation outputs of various methods. For semantic segmentation, we use a pre-trained Deeplab-v3 model[1] on Cityscapes dataset [6]. The BDD100K dataset provides segmentation ground truth of 137 night images. So we compute the IOU metric between the segmentation outputs of the 137 translated daytime images and corresponding segmentation ground truth. The quantitative comparison is listed in Table 3. Since there is no night image on Cityscapes dataset, the segmentation performance of night image has a drastic performance drop shown in Table 3. The mIoU is only 7.03% if we directly perform the semantic segmentation on the real night images. Night-to-day translation model provides a powerful tool to improve segmentation performance, where stronger translation model should lead to larger performance boost. As shown, our ForkGAN achieves the highest mIoU among all the methods, almost doubling the original night image segmentation result. We also observe that the synthesized daytime images produced by some comparative translation methods obtain worse segmentation performance than the original night images. MUNIT and DRIT methods both fail to synthesize plausible outputs from challenging night images and thus obtain poor mIoU scores. ToDayGAN, while achieving reasonable night-to-day translation, obtains higher mIoU score than original night images. Our ForkGAN preserves detailed information during night-to-day image translation, especially the small traffic signs and pedestrians. So our method can boost the segmentation performance by preserving and enhancing the crucial

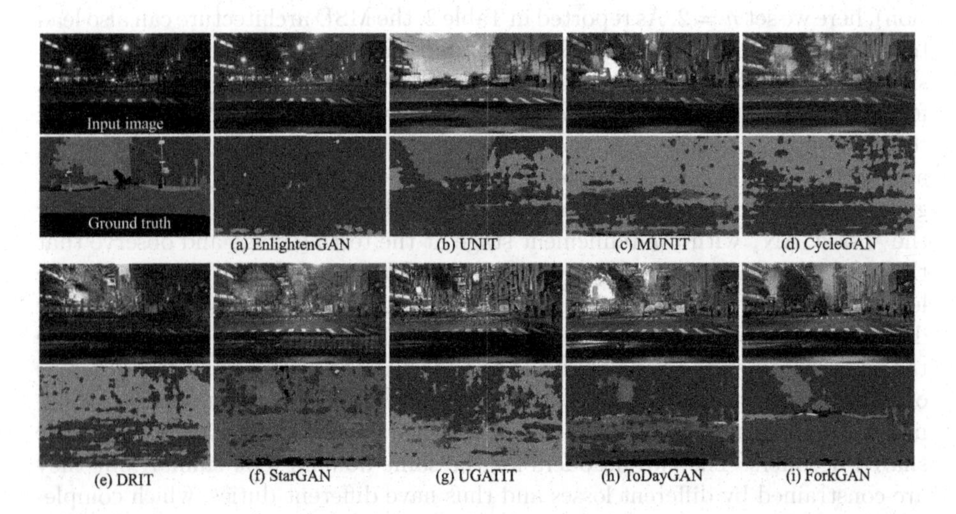

Fig. 5. The visual translation (the first row) and segmentation performance (the second row) comparison of different methods, with models pre-trained on Cityscapes [6].

[1] https://github.com/srihari-humbarwadi/DeepLabV3_Plus-Tensorflow2.0.

detailed information. To quantitatively compare the translation quality, we also compute the FID score to measure the distance between the generated sample distribution and the real image distribution in Table 3.

Table 3. Quantitative comparison of different methods for night → day translation task on BDD100K dataset [27]. The original denotes the outputs of the original real night images. FID reports the visual image quality (lower is better) while mIoU (percentage) reports the segmentation performance (higher is better).

Method	EnlightenGAN [14]	UNIT [17]	MUNIT [12]	CycleGAN [28]	DRIT [16]
FID/mIoU	90.3/6.03	62.1/2.47	61.1/2.44	51.7/1.88	53.1/2.45
Method	StarGAN [5]	UGATIT [15]	ToDayGAN [1]	ForkGAN	Original
FID/mIoU	68.3/6.63	72.2/3.83	43.8/8.19	**37.6/14.4**	101/7.03

4.5 Object Detection with Data Augmentation

In autonomous driving, it is laborious and sometimes difficult to collect abundant data with annotations in a wide variety of weather and illumination conditions for object detection. Most of available datasets contain images mostly from daytime driving. Models trained on those datasets are subject to performance degradation once they are tested on a different domain such as nighttime. One possible solution is to augment nighttime data with annotated daytime images through domain translation such that we can make the most use of available annotations. Our ForkGAN can also perform day-to-night translation to aid off-the-shelf detection model to adapt to different domains. We compare our ForkGAN with the most related ToDayGAN on BDD100K dataset in two settings. In both settings, we have unlabelled images from both day and night domains for training image translation network, as well as bounding box annotations for daytime images for training detection network, either with real or translated images:

(1) **Day Labels Only** - No nighttime labeled image is available at training time: We use ForkGAN to translate daytime images to night images and preserve the corresponding bounding boxes. Then we train an object detection network on those translated nighttime images. For comparison, we also train two separate detection networks using raw daytime images (*Day Real*) and translated nighttime images by ToDayGAN. The quantitative results are shown in Table 4. We observe that ForkGAN can improve the detection performance on night images. Visualization of detection results are shown in Fig. 6. The ability to detect small traffic signs in dark has been improved through domain adaptation.

(2) **Day + Night Labels** - Both nighttime and daytime labeled images are available for training: We again apply ForkGAN to translate the daytime images to night images for data augmentation. The detection

network is trained on both real and translated night images. We also report the performance of the detection network trained only on real night images '(*Night Real*) and night image augmentation with ToDayGAN (*Night+ToDayGAN*). Figure 6 and Table 4 show the visual and quantitative comparison. By combining with translated night images, the detection performance has been improved, which indicates the detection task can benefit from domain translation.

Table 4. Comparisons for object detection on 3,929 validation nighttime images. The first three rows show the results from setting (1), while the rest are from setting (2). We apply faster-rcnn-r50-fpn-1x detector based on MMDetection [3] in all the experiments.

Method	mAP	Person	Rider	Car	Bus	Truck	Bike	Motor	Traffic light	Traffic sign
Day(Real)	22.1	26.1	**14.3**	37.5	29.8	30.7	**18.5**	**16.3**	14.6	33.1
+ToDayGAN	19.5	23.5	10.4	35.9	32.5	29.4	16.0	11.0	9.0	26.7
+ForkGAN	**22.9**	**26.3**	13.0	**41.2**	**33.3**	**32.1**	16.4	15.9	**16.2**	**34.5**
Night(Real)	23.9	26.6	13.0	42.0	33.8	35.0	16.7	16.9	18.2	36.0
Night+ToDayGAN	24.2	26.9	14.1	42.3	36.5	36.8	20.2	19.1	17.6	35.7
Night+ForkGAN	**26.2**	**28.1**	**16.1**	**42.5**	**37.8**	**38.7**	**22.1**	**21.9**	**18.3**	**36.2**

Fig. 6. Visual comparison of detection results on BDD100K, where "Day"/"Night" denotes training with daytime/nighttime images. Areas pointed by yellow arrows are worth attention. ForkGAN can improve the detection of small objects. We show all the results of person, rider, car, bus, truck, bike, motor, traffic light and traffic signs. (Color figure online)

5 Conclusion and Future Work

We propose a novel framework ForkGAN to achieve unbiased image translation, which is beneficial to multiple vision tasks: localization/retrieval, semantic segmentation and object detection in adverse conditions. It disentangles domain-invariant and domain-specific information through a fork-shape module, enhanced by an adversarial domain classifier and an across-translation perceptual loss. Extensive experiments have demonstrated its superiority and effectiveness. Possible future works include designing a multi-task learning network to share the backbone of different vision tasks and performing object detection in the domain-invariant content space, which can be more compact and more efficient.

Acknowledgement. This work was supported by a MSRA Collaborative Research 2019 Grant.

References

1. Anoosheh, A., Sattler, T., Timofte, R., Pollefeys, M., Van Gool, L.: Night-to-day image translation for retrieval-based localization. In: 2019 International Conference on Robotics and Automation (ICRA), pp. 5958–5964. IEEE (2019)
2. Bay, H., Ess, A., Tuytelaars, T., Van Gool, L.: Speeded-up robust features (SURF). Comput. Vis. Image Underst. **110**(3), 346–359 (2008)
3. Chen, K., et al.: MMDetection: open MMLab detection toolbox and benchmark. arXiv preprint arXiv:1906.07155 (2019)
4. Chen, Q., Koltun, V.: Photographic image synthesis with cascaded refinement networks. In: IEEE International Conference on Computer Vision, pp. 1511–1520 (2017)
5. Choi, Y., Choi, M., Kim, M., Ha, J.W., Kim, S., Choo, J.: StarGAN: unified generative adversarial networks for multi-domain image-to-image translation. In: CVPR, pp. 8789–8797 (2018)
6. Cordts, M., et al.: The cityscapes dataset for semantic urban scene understanding. In: Proceedings of the IEEE Conference on Computer Vision and Pattern Recognition (CVPR) (2016)
7. Halder, S.S., Lalonde, J.F., de Charette, R.: Physics-based rendering for improving robustness to rain. In: IEEE/CVF International Conference on Computer Vision (2019)
8. He, Z., Zhang, L.: Multi-adversarial Faster-RCNN for unrestricted object detection. In: Proceedings of the IEEE International Conference on Computer Vision, pp. 6668–6677 (2019)
9. Heusel, M., Ramsauer, H., Unterthiner, T., Nessler, B., Hochreiter, S.: GANs trained by a two time-scale update rule converge to a local Nash equilibrium. In: NIPSs, pp. 6626–6637 (2017)
10. Hu, X., Fu, C.W., Zhu, L., Heng, P.A.: Depth-attentional features for single-image rain removal. In: Proceedings of the IEEE Conference on Computer Vision and Pattern Recognition, pp. 8022–8031 (2019)
11. Huang, S.-W., Lin, C.-T., Chen, S.-P., Wu, Y.-Y., Hsu, P.-H., Lai, S.-H.: AugGAN: cross domain adaptation with GAN-based data augmentation. In: Ferrari, V., Hebert, M., Sminchisescu, C., Weiss, Y. (eds.) ECCV 2018. LNCS, vol. 11213, pp. 731–744. Springer, Cham (2018). https://doi.org/10.1007/978-3-030-01240-3_44
12. Huang, X., Liu, M.-Y., Belongie, S., Kautz, J.: Multimodal unsupervised image-to-image translation. In: Ferrari, V., Hebert, M., Sminchisescu, C., Weiss, Y. (eds.) ECCV 2018. LNCS, vol. 11207, pp. 179–196. Springer, Cham (2018). https://doi.org/10.1007/978-3-030-01219-9_11

13. Isola, P., Zhu, J.Y., Zhou, T., Efros, A.A.: Image-to-image translation with conditional adversarial networks. In: CVPR, pp. 5967–5976 (2017)
14. Jiang, Y., et al.: EnlightenGAN: deep light enhancement without paired supervision. arXiv preprint arXiv:1906.06972 (2019)
15. Kim, J., Kim, M., Kang, H., Lee, K.: U-GAT-IT: unsupervised generative attentional networks with adaptive layer-instance normalization for image-to-image translation. arXiv preprint arXiv:1907.10830 (2019)
16. Lee, H.-Y., Tseng, H.-Y., Huang, J.-B., Singh, M., Yang, M.-H.: Diverse image-to-image translation via disentangled representations. In: Ferrari, V., Hebert, M., Sminchisescu, C., Weiss, Y. (eds.) ECCV 2018. LNCS, vol. 11205, pp. 36–52. Springer, Cham (2018). https://doi.org/10.1007/978-3-030-01246-5_3
17. Liu, M.Y., Breuel, T., Kautz, J.: Unsupervised image-to-image translation networks. In: Advances in Neural Information Processing Systems, pp. 700–708 (2017)
18. Lowe, D.G., et al.: Object recognition from local scale-invariant features. In: ICCV, vol. 99, pp. 1150–1157 (1999)
19. Milford, M.J., Wyeth, G.F.: SeqSLAM: visual route-based navigation for sunny summer days and stormy winter nights. In: 2012 IEEE International Conference on Robotics and Automation, pp. 1643–1649. IEEE (2012)
20. Porav, H., Bruls, T., Newman, P.: Don't worry about the weather: unsupervised condition-dependent domain adaptation. In: 2019 IEEE Intelligent Transportation Systems Conference (ITSC), pp. 33–40. IEEE (2019)
21. Porav, H., Maddern, W., Newman, P.: Adversarial training for adverse conditions: robust metric localisation using appearance transfer. In: 2018 IEEE International Conference on Robotics and Automation (ICRA), pp. 1011–1018. IEEE (2018)
22. Romera, E., Bergasa, L.M., Yang, K., Alvarez, J.M., Barea, R.: Bridging the day and night domain gap for semantic segmentation. In: 2019 IEEE Intelligent Vehicles Symposium (IV), pp. 1312–1318. IEEE (2019)
23. Ros, G., Alvarez, J.M.: Unsupervised image transformation for outdoor semantic labelling. In: 2015 IEEE Intelligent Vehicles Symposium (IV), pp. 537–542. IEEE (2015)
24. Sakaridis, C., Dai, D., Van Gool, L.: Semantic foggy scene understanding with synthetic data. Int. J. Comput. Vis. **126**(9), 973–992 (2018)
25. Wang, T.C., Liu, M.Y., Zhu, J.Y., Tao, A., Kautz, J., Catanzaro, B.: High-resolution image synthesis and semantic manipulation with conditional GANs. In: Proceedings of the IEEE Conference on Computer Vision and Pattern Recognition, pp. 8798–8807 (2018)
26. Yu, F., Koltun, V., Funkhouser, T.: Dilated residual networks. In: Proceedings of the IEEE Conference on Computer Vision and Pattern Recognition, pp. 472–480 (2017)
27. Yu, F., et al.: BDD100K: a diverse driving video database with scalable annotation tooling. arXiv preprint arXiv:1805.04687 (2018)
28. Zhu, J.Y., Park, T., Isola, P., Efros, A.A.: Unpaired image-to-image translation using cycle-consistent adversarial networks. In: International Conference on Computer Vision, pp. 2223–2232 (2017)

TCGM: An Information-Theoretic Framework for Semi-supervised Multi-modality Learning

Xinwei Sun[1], Yilun Xu[2], Peng Cao[2], Yuqing Kong[2](✉), Lingjing Hu[3], Shanghang Zhang[4](✉), and Yizhou Wang[2,5]

[1] Microsoft Research-Asia, Beijing, China
xinsun@microsoft.com
[2] Center on Frontiers of Computing Studies, Advanced Institute of Information Technology, Department of Computer Science, Peking University, Beijing, China
{xuyilun,caopeng2016,yuqing.kong,Yizhou.Wang}@pku.edu.cn
[3] Yanjing Medical College, Capital Medical University, Beijing, China
hulj@ccmu.edu.cn
[4] UC Berkeley, Berkeley, USA
shz@eecs.berkeley.edu
[5] Deepwise AI Lab, Beijing, China

Abstract. Fusing data from multiple modalities provides more information to train machine learning systems. However, it is prohibitively expensive and time-consuming to label each modality with a large amount of data, which leads to a crucial problem of semi-supervised multi-modal learning. Existing methods suffer from either ineffective fusion across modalities or lack of theoretical guarantees under proper assumptions. In this paper, we propose a novel information-theoretic approach - namely, **T**otal **C**orrelation **G**ain **M**aximization (TCGM) – for semi-supervised multi-modal learning, which is endowed with promising properties: (i) it can utilize effectively the information across different modalities of unlabeled data points to facilitate training classifiers of each modality (ii) it has theoretical guarantee to identify Bayesian classifiers, i.e., the ground truth posteriors of all modalities. Specifically, by maximizing TC-induced loss (namely TC gain) over classifiers of all modalities, these classifiers can cooperatively discover the equivalent class of ground-truth classifiers; and identify the unique ones by leveraging limited percentage of labeled data. We apply our method to various tasks and achieve state-of-the-art results, including the news classification (Newsgroup dataset), emotion recognition (IEMOCAP and MOSI datasets), and disease prediction (Alzheimer's Disease Neuroimaging Initiative dataset).

X. Sun and Y. Xu—Equal Contribution.

Electronic supplementary material The online version of this chapter (https://doi.org/10.1007/978-3-030-58580-8_11) contains supplementary material, which is available to authorized users.

A. Vedaldi et al. (Eds.): ECCV 2020, LNCS 12348, pp. 171–188, 2020.
https://doi.org/10.1007/978-3-030-58580-8_11

Keywords: Total Correlation · Semi-supervised · Multi-modality · Conditional independence · Information intersection

1 Introduction

Learning with data from multiple modalities has the advantage to facilitate information fusion from different perspectives and induce more robust models, compared with only using a single modality. For example, as shown in Fig. 1, to diagnose whether a patient has a certain disease or not, we can consult to its X-ray images, look into its medical records, or get results from clinical pathology. However, in many real applications, especially in some difficult ones (e.g. medical diagnosis), annotating such large-scale training data is prohibitively expensive and time-consuming. As a consequence, each modality of data may only contain a small proportion of labeled data from professional annotators, leaving a large proportion of unlabeled. This leads to an essential and challenging problem of semi-supervised multi-modality learning: *how to effectively train accurate classifiers by aggregating unlabeled data of all modalities?*

To achieve this goal, many methods have been proposed in the literature, which can be roughly categorized into two branches: (i) co-training strategy [6]; and (ii) learning joint representation across modalities in an unsupervised way [25,27]. These methods suffer from either too strong assumptions or loss of information during fusing. Specifically, the co-training strategy relies largely on the "compatible" assumption that the conditional distributions of the data point labels in each modality are the same, which may not be satisfied in the real settings, as self-claimed in [6]; while the latter branch of methods fails to capture the higher-order dependency among modalities, hence may end up in learning a trivial solution that maps all the data points to the same representation.

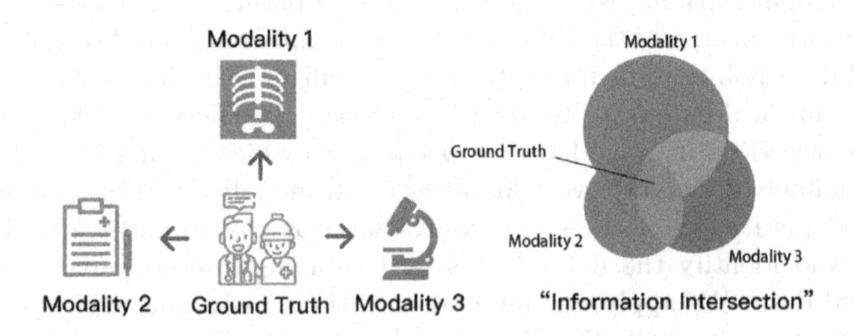

Fig. 1. Multiple modalities are independent conditioning on the ground truth; Ground truth is the "information intersection" of all of the modalities.

A common belief in multi-modality learning [6,13,20,22] is that conditioning on ground truth label Y, these modalities are conditionally independent, as illustrated in Fig. 1. For example, to diagnose if one suffers from a certain disease, an efficient way is to leverage as many as modalities that are related to the disease, e.g., X-ray image, medical records and the clinical pathology. Since each modality captures the characteristics of the disease from different aspects, the information extracted from these modalities, in addition to the label, are not necessarily correlated with each other. This suggests that the ground truth label can be regarded as the "information intersection" across all the modalities, i.e., the amount of agreement shared by all the modalities.

Inspired by such an assumption and the fact that the *Total Correlation* [28] can measure the amount of information shared by M ($M \geq 2$) variables, in this paper, we propose **T**otal **C**orrelation **G**ain (TCG), which is a function of classifiers of all the modalities, as a surrogate goal for maximization of mutual information, in order to infer the ground-truth labels (i.e., information intersection among these modalities). Based on the proposed TCG, we devise an information-theoretic framework called ***Total Correlation Gain Maximization*** (TCGM) for semi-supervised multi-modal learning. By maximizing TCG among all the modalities, the classifiers for different modalities cooperatively discover the information intersection across all the modalities. It can be proved that the optimal classifiers for such a Total Correlation Gain are equivalent to the Bayesian posterior classifiers given each modality under some permutation function. With further leverage of labeled data, we can identify the Bayesian posterior classifiers. Furthermore, we devise an aggregator that employs all the modalities to forecast the labels of data. A simulated experiment is conducted to verify this theoretical result.

We apply TCGM on various tasks: (i) News classification with three preprocessing steps as different modalities, (ii) Emotion recognition with videos, audios, and texts as three modalities and (iii) disease prediction on medical imaging with the Structural magnetic resonance imaging (sMRI) and Positron emission tomography (PET) modalities. On these tasks, our method consistently outperforms the baseline methods especially when a limited percentage of labeled data are provided. To validate the benefit of jointly learning, we visualize that some cases of Alzheimer's Disease whose label are difficult to be predicted via supervised learning with single modality; while our jointly learned single modal classifier is able to correctly classify such hard samples.

The contributions can be summarized as follows: (i) We propose a novel information-theoretic approach TCGM for semi-supervised multi-modality learning, which can effectively utilize information across all modalities. By maximizing the total correlation gain among all the modalities, the classifiers for different modalities cooperatively discover the information intersection across all the modalities - the ground truth. (ii) To the best of our knowledge, **TCGM** is the first in the literature that can be theoretically proved that, under the conditional independence assumption, it can identify the ground-truth Bayesian classifier given each modality. Further, by aggregating these classifiers, our method can learn the Bayesian classifier given all modalities. (iii) We achieve the state-of-the-art results on various semi-supervised multi-modality tasks including news classification, emotion recognition and disease prediction of medical imaging.

2 Related Work

Semi-supervised Multi-modal Learning. It is commonly believed in the literature that information of label is shared across all modalities. Existing work, which can be roughly categorized into two branches, suffers from either stronger but not reasonable assumptions or failure to capture the information (i.e., label) shared by all modalities. The first branch applies the co-training algorithm proposed by Blum et al. [6]. [3,11,12,17,21] use weak classifiers trained by the labeled data from each modality to bootstrap each other by generating labels for the unlabeled data. However, the underlying compatible condition of such a method, which assumes the same conditional distributions for data point labels in each modality, may not be consistent with the real settings.

The second branch of work [9,10,18,25,27,31] centers on learning joint representations that project unimodal representations all together into a multi-modal space in an unsupervised way and then using the labeled data from each modality to train a classifier to predict the label of the learned joint representation. A representative of such a framework is the soft-Hirschfeld-Gebelein-Rényi (HGR) framework [31], which proposed to maximize the correlation among non-linear representations of each modality. However, HGR only measures the linear dependence between pair modalities, since it follows the principle of maximizing the correlation between features of different modalities. In contrast, our framework, i.e., Total Correlation Gain Maximization can pursue information about higher-order dependence. Due to the above reasons, both branches can not avoid learning a naive solution that classifies all data points into the same class.

To overcome these limitations, we propose an information-theoretic loss function based on Total Correlation which can not only require the assumption in the first branch of work but also can be able to identify the ground-truth label, which is the information intersection among these modalities. Therefore, our method can avoid the trivial solution and can learn the optimal, i.e., the Bayesian Posterior classifiers of each modality.

Total Correlation/Mutual Information Maximization. Total Correlation [28], as an extension of Mutual Information, measures the amount of information shared by M ($M \geq 2$) variables. There are several works in the literature that have combined Mutual Information ($M = 2$) with deep learning algorithms and have shown superior performance on various tasks. Belghazi et al. [4] presents a mutual information neural estimator, which is utilized in a handful of applications based on the mutual information maximization (e.g., unsupervised learning of representations [14], learning node representations within graph-structured data [30]). Kong and Schoenebeck [19] provide another mutual information estimator in the co-training framework for the peer prediction mechanism, which has been combined with deep neural networks for crowdsourcing [8]. Xu et al. [32] proposes an alternative definition of information, which is more effective for structure learning. However, those three estimators can only be applied to two-view settings. To the best of our knowledge, there are no similar studies that focus on a general number of modalities, which is very often in real applications. In this paper, we propose to leverage Total Correlation to fill in such a gap.

3 Preliminaries

Notations. Given a random variable X, $\mathcal{X}$ denotes its realization space and $x \in \mathcal{X}$ denotes an instance. The $\mathrm{P}_{\mathcal{X}}$ denotes the probability distribution function over $\mathcal{X}$ and $p(x) := d\mathrm{P}_{\mathcal{X}}(x)$ denotes the density function w.r.t. the Lebesgue measure. Further, given a finite set $\mathcal{X}$, $\Delta_{\mathcal{X}}$ denotes the set of all distributions over $\mathcal{X}$. For every integer M, $[M]$ denotes the set $\{1, 2, \ldots, M\}$. For a vector $\mathbf{v}$, v_i denotes its i-th element.

Total Correlation. The Total Correlation (TC), as an extension of mutual information, measures the "amount of information" shared by M ($\geq$2) random variables:

$$\mathrm{TC}(X^1, \ldots, X^m) = \sum_{i=1}^{M} \mathrm{H}(X^i) - \mathrm{H}(X^1, \ldots, X^M), \tag{1}$$

where H is the Shannon entropy. As defined, the TC degenerates to mutual information when $M = 2$. The $\sum_{i=1}^{M} \mathrm{H}(X^i)$ measures the total amount of information when treating $X^1, \ldots, X^M$ independently; while the $\mathrm{H}(X^1, \ldots, X^M)$ measures the counterpart when treating these M variables as a whole. Therefore, the difference between them implies the redundant information, i.e., the information shared by these M variables.

Similar to mutual information, the TC is equivalent to the Kullback-Leibler (KL)-divergence between $\mathrm{P}_{\mathcal{X}^1 \times \ldots \times \mathcal{X}^M}$ and product of marginal distribution $\otimes_{m=1}^{M} \mathrm{P}_{\mathcal{X}^m}$:

$$\mathrm{TC}(X^1, \ldots, X^M) = \mathrm{D_{KL}}\left(d\mathrm{P}_{\mathcal{X}^1 \times \ldots \times \mathcal{X}^M} \mid\mid d \otimes_{m=1}^{M} \mathrm{P}_{\mathcal{X}^m}\right) = \mathbb{E}_{\mathrm{P}_{\mathcal{X}^1 \times \ldots \times \mathcal{X}^M}} \log \frac{d\mathrm{P}_{\mathcal{X}^1 \times \ldots \times \mathcal{X}^M}}{d \otimes_{m=1}^{M} \mathrm{P}_{\mathcal{X}^m}}, \tag{2}$$

where $\mathrm{D_{KL}}(\mathcal{P} \mid\mid \mathcal{Q}) = \mathbb{E}_{\mathcal{P}} \log \frac{d\mathcal{P}}{d\mathcal{Q}}$. Intuitively, larger KL divergence between joint and marginal distribution indicates more dependence among these M variables. To better characterize such a property, we give a formal definition of "Point-wise Total Correlation" (PTC):

Definition 1 (Point-wise Total Correlation). *Given M random variables $X^1, \ldots, X^M$, the Point-wise Total Correlation on $(x^1, \ldots, x^m) \in \mathcal{X}^1 \times \ldots \times \mathcal{X}^M$, i.e., $\mathrm{PTC}(x^1, \ldots, x^M)$ is defined as $\mathrm{PTC}(x^1, \ldots, x^M) = \log \frac{p(x^1, \ldots, x^M)}{p(x^1) \ldots p(x^M)}$. Further, the $R(x^1, \ldots, x^M) := \frac{p(x^1, \ldots, x^M)}{p(x^1) \ldots p(x^M)}$ is denoted as the joint-margin ratio.*

Remark 1. The Point-wise Total Correlation can be understood as the point-wise distance between joint distribution and the marginal distribution. In more details, as noted from [15], by applying first-order Taylor-expansion, we have $\log \frac{p(x)}{q(x)} \approx \log 1 + \frac{p(x) - q(x)}{q(x)} = \frac{p(x) - q(x)}{q(x)}$. Therefore, the expected value of $\mathrm{PTC}(\cdot)$ can well measure the amount of information shared among these variables, which will be shown later in detailed.

For simplicity, we denote $p^{[M]}(x) := p(x^1, \ldots, x^M)$ and $q^{[M]}(x) := \prod_{i=1}^M p(x^i)$. According to dual representation in [26], we have the following lower bound for KL divergence between p and q, and hence TC.

Lemma 1 (Dual version of f-divergence [26])

$$\mathrm{D_{KL}}\left(p^{[M]} \,\|\, q^{[M]}\right) \geq \sup_{g \in \mathcal{G}} \mathbb{E}_{x \sim p^{[M]}}[g(x)] - \mathbb{E}_{x \sim q^{[M]}}\left[e^{g(x)-1}\right] \tag{3}$$

where $\mathcal{G}$ is the set of functions that maps $\mathcal{X}^1 \times \mathcal{X}^2 \times \cdots \times \mathcal{X}^M$ to $\mathbb{R}$. The equality holds if and only if $g(x^1, x^2, \ldots, x^M) = 1 + \mathrm{PTC}(x^1, \ldots, x^M)$, and the supremum is $\mathrm{D_{KL}}\left(p^{[M]} \,\|\, q^{[M]}\right) = \mathbb{E}_{p^{[M]}}(\mathrm{PTC})$.

The Lemma 1 is commonly utilized for estimation of Mutual information [4] or optimization as variational lower bound in the machine learning literature. Besides, it also informs that the PTC is the optimal function to describe the amount of information shared by these M variables. Indeed, such shared information is the information intersection among these variables, i.e., *conditional on such information, these M variables are independent of each other.* To quantitatively describe this, we first introduce the *conditional total correlation* (CTC). Similar to TC, CTC measures the amount of information shared by these M variables conditioning on some variable Z:

Definition 2 (Conditional Total Correlation (CTC)). *Given $M+1$ random variables $X^1, \ldots, X^M, Z$, the Conditional Total Correlation ($\mathrm{CTC}(X^1, \ldots, X^M|Z)$) is defined as* $\mathrm{CTC}(X^1, \ldots, X^M|Z) = \sum_{i=1}^M \mathrm{H}(X^i|Z) - \mathrm{H}(X^1, \ldots, X^M|Z)$.

4 Method

Problem Statement. In the semi-supervised multi-modal learning scenario, we have access to an unlabeled dataset $\mathcal{D}_u = \{x_i^{[M]}\}_i$ and a labeled dataset $\mathcal{D}_l = \{(x_i^{[M]}, y_i)\}_i$. Each label $y_i \in \mathcal{C}$, where $\mathcal{C}$ denotes the set of classes. Each datapoint $x_i^{[M]} := \{x_i^1, x_i^2, \ldots, x_i^M | x_i^m \in \mathcal{X}^m\}$ consists of M modalities, where $\mathcal{X}^m$ denotes the domain of the m-th modality. Datapoints and labels in $\mathcal{D}_l$ are i.i.d. samples drawn from the joint distribution $U_{X^{[M]},Y}(x^1, x^2, \ldots, x^M, y) := \Pr(X^1 = x^1, X^2 = x^2, \ldots, X^M = x^M, Y = y)$. Data points in $\mathcal{D}_u$ are i.i.d. samples drawn from joint distribution $U_{X^{[M]}}(x^1, x^2, \ldots, x^M) := \sum_{c \in \mathcal{C}} U_{X^{[M]},Y}(x^1, x^2, \ldots, x^M, y = c)$. Denote the prior of the ground truth labels by $\mathbf{p}^* = (\Pr(Y = c))_c$. Upon the labeled and unlabeled datasets, our goal is to train M classifiers $h^{[M]} := \{h^1, h^2, \ldots, h^M\}$ and an aggregator ζ such that $\forall m, h^m : \mathcal{X}^m \to \Delta_{\mathcal{C}}$ predicts the ground truth y based on a m-th modality x^m and $\zeta : \mathcal{X}^1 \times \mathcal{X}^2 \times \cdots \times \mathcal{X}^M \to \Delta_{\mathcal{C}}$ predicts the ground truth y based on all of the modalities $x^{[M]}$.

Outline. We will first introduce the assumptions regarding the ground truth label Y and prior distribution on $(X^1,\dots,X^M,Y)$ in Sect. 4.1. In Sect. 4.2, we will present our method, i.e., maximize the total correlation gain on unlabeled dataset $\mathcal{D}_u$. Finally, we will introduce our algorithm for optimization in Sect. 4.3.

4.1 Assumptions for Identification of *Y*

In this section, we first introduce two basic assumptions to ensure that the ground-truth label can be identified. According to Proposition 2.1 in [1], the label Y can be viewed as the generating factor of data X. Such a result can be extended to multiple variables (please refer supplementary for details), which implies that Y is the common generating factor of $X^1,\dots,X^M$. Motivated by this, it is natural to assume that the ground truth label Y is the "information intersection" among $X^1,\dots,X^M$, i.e., all of the modalities are independent conditioning on the ground-truth:

Assumption 1 (Conditional Independence). *Conditioning on* Y, $X^1,X^2,\dots,X^M$ *are independent, i.e.,* $\forall x^1,\dots,x^M$, $\Pr(X^{[M]}=x^{[M]}|Y=c)=\prod_m \Pr(X^m=x^m|Y=c)$, for any $c\in\mathcal{C}$.

On the basis of this assumption, one can immediately get the conditional total correlation gain $\mathrm{CTC}(X^1,\dots,X^M|Y)=0$. In other words, conditioning on Y, there is no extra information shared by these M modalities, which is commonly assumed in the literature of semi-supervised learning [6,13,20,22]. However, the Y may not be the unique information intersection among these M modalities. Specifically, the following lemma establishes the rules for such information intersection to hold:

Lemma 2. *Given Assumption 1,* $R(x^1,\dots,x^M)$ *(Joint-marginal ratio Definition 1) has* $R(x^1,\dots,x^M)=\sum_{c\in\mathcal{C}}\frac{\prod_m \Pr(Y=c|X^m=x^m)}{(\Pr(Y=c))^{M-1}}$. *Further, the optimal* g *in Lemma 1 satisfies* $g(x^1,\dots,x^M)=1+\log\sum_{c\in\mathcal{C}}\frac{\prod_m \Pr(Y=c|x^m)}{(\Pr(Y=c))^{M-1}}$.

In other words, in addition to $\{\Pr(Y=c|X^m=x^m)\}_m,\mathbf{p}_c^*$, there are other solutions $\{a_{x^1},\dots,a_{x^M}\},r$ with $a_{x^i}\in\Delta_{\mathcal{C}}$ (for $i\in\mathcal{C}$) and $r\in\Delta_{\mathcal{C}}$ that can make the g optimal, as long as its joint-marginal ratio is equal to the ground-truth one:

$$\sum_{c\in\mathcal{C}}\frac{\prod_m a_{x^m}}{(r_c^a)^{M-1}}=R(x^1,\dots,x^M) \tag{4}$$

To make $\{\Pr(Y=c|X^m=x^m)\}_m,\mathbf{p}_c^*$ identifiable w.r.t. a trivial permutation, we make the following trivial assumption on $\Pr(X^1,\dots,X^M,Y)$.

Assumption 2 (Well-defined Prior). *The solutions* $\{a_{x^1},\dots,a_{x^M}\},r^a$ *and* $\{b_{x^1},\dots,b_{x^M}\},r^b$ *for Eq. (4) are equivalent under the permutation* $\prod$ *on* $\mathcal{C}$: $a_{x^m}=\prod b_{x^m}$, $r^a=\prod r^b$.

4.2 Total Correlation Gain Maximization (TCGM)

Assumption 1 indicates that the label Y is the generating factor of all modalities, and Assumption 2 further ensures its uniqueness under permutation. Our goal is to learn the ground-truth label Y which is the information intersection among M modalities. In this section, we propose a novel framework, namely ***Total Correlation Gain Maximization*** (**TCGM**) to capture such an information intersection, which is illustrated in Fig. 2. To the best of our knowledge, we are the first to theoretically prove the identification of ground truth classifiers on semi-supervised multi-modality data, by generalizing [8,19] that can only handle two views in multi-view scenario. The high-level spirit is designing TC-induced loss over classifiers of every modality. By maximizing such a loss, these classifiers can converge to Bayesian posterior, which is the optimal solution of TC as expectation of the loss. First, we introduce the basic building blocks for our method.

Classifiers $h^{[M]}$. In order to leverage the powerful representation ability of deep neural network (DNN), each classifier $h^m(x^m; \Theta^m)$ is modeled by a DNN with parameters Θ^m. For each modality m, we denote the set of all such classifiers by H^m and $H^{[M]} := \{H^1, H^2, \ldots, H^M\}$.

Modality Classifiers-Aggregator ζ. Given M classifiers for each modality $h^{[M]}$ and a distribution $\mathbf{p} = (p_c)_c \in \Delta_{\mathcal{C}}$, the aggregator ζ which predicts the ground-truth label by aggregating classifiers of all modalities, is constructed by $\zeta(x^{[M]}; h^{[M]}, \mathbf{p}) = \text{Normalize}\left(\left(\frac{\prod_m h^m(x_i^m)_c}{(p_c)^{M-1}}\right)_c\right)$, where $\text{Normalize}(\mathbf{v}) := \frac{\mathbf{v}}{\sum_c v_c}$ for all $\mathbf{v} \in \Delta_{\mathcal{C}}$.

Reward Function $\mathcal{R}$. We define a reward function $\mathcal{R} : \overbrace{\Delta_{\mathcal{C}} \times \ldots \times \Delta_{\mathcal{C}}}^{M} \to \mathbb{R}$ that measures the "amount of agreement" among these classifiers. Note that the desired classifiers should satisfy Eq. (4).

Inspired by Lemma 1, we can take the empirical total correlation gain of N samples, i.e., the lower bound of Total Correlation as our maximization function. Specifically, given a reward function $\mathcal{R}$, the empirical total correlation with respect to classifiers $h^{[M]}$, a prior $\mathbf{p} \in \Delta_{\mathcal{C}}$ measures the empirical "amount of agreement" for these M classifiers at the same sample $(x_i^1, \ldots, x_i^M) \in \mathcal{D}_u$, with additionally a punishment of them on different samples: $x_{i_1}^1 \in \mathcal{X}^1, \ldots, x_{i_M}^M \in \mathcal{X}^M$ with $i_1 \neq i_2 \neq \ldots \neq i_M$:

$$TCg[\mathcal{R}](\{x_i^{[M]}\}_{i=1}^N; h^{[M]}, p) := \frac{1}{N} \sum_i \mathcal{R}(h^1(x_i^1), \ldots, h^M(x_i^M)) \\ - \frac{1}{N!/(N-M)!} \sum_{i_1 \neq i_2 \neq \cdots \neq i_M} e^{\mathcal{R}(h^1(x_{i_1}^1), \ldots, h^M(x_{i_M}^M)) - 1} \tag{5}$$

for simplicity we denote $TCg[\mathcal{R}](\{x_i^{[M]}\}_{i=1}^N; h^{[M]}, p)$ as $TCg^{(N)}$. Intuitively, we expect our classifiers to be consistent on the same sample; on the other hand, to disagree on different samples to avoid learning a trivial solution that classifies all data points into the same class.

Definition 3 (Bayesian posterior classifiers/aggregator). *The* $h_*^{[M]}$ *and* ζ_* *are called* Bayesian posterior classifiers *and* Bayesian posterior aggregator *if they satisfy*

$$\forall m, h_*^m(x^m)_c = \Pr(Y = c|X^m = x^m);\ \zeta_*(x^{[M]})_c = \Pr(Y = c|X^{[M]} = x^{[M]}).$$

Note that from Eq. (5) that our maximization goal, i.e., $TCg^{(N)}$ relies on the form of reward function $\mathcal{R}$. The following Lemma tells us the form of optimal reward function, with which we can finally give an explicit form of $TCg^{(N)}$.

Lemma 3. *The* $\mathcal{R}_*$ *that maximizes the expectation of* $TCg^{(N)}$ *can be represented as the Point-wise Total Correlation function, which is the function of Bayesian classifiers and the prior of ground truth labels* $(\mathbf{p}_c^*)_c$*:* $\mathcal{R}_*(h^1(x^1), \ldots, h^M(x^M)) = 1 + \mathrm{PTC}(x^1, \ldots, x^M) = 1 + \log \sum_{c \in \mathcal{C}} \frac{\prod_m h_*^m(x^m)_c}{(\mathbf{p}_c^*)^{M-1}}$.

Total Correlation Gain. Bring $\mathcal{R}_*$ to Eq. (5), we have:

$$\begin{aligned} TCg(\{x_i^{[M]}\}_{i=1}^N; h^{[M]}, p) :=& 1 + \frac{1}{N} \sum_i \log \sum_{c \in \mathcal{C}} \frac{\prod_m h^m(x_i^m)_c}{(p_c)^{M-1}} \\ &- \frac{1}{N!/(N-M)!} \sum_{i_1 \neq i_2 \neq \cdots \neq i_M} \sum_{c \in \mathcal{C}} \frac{\prod_m h^m(x_{i_m}^m)_c}{(p_c)^{M-1}}. \end{aligned} \quad (6)$$

As inspired by Lemma 3, we have that these Bayesian posterior classifiers are maximizers of the expected total correlation gain. Therefore, we can identify the equivalent class of Bayesian posteriors by minimizing $-TCg^{(N)}$ on unlabeled dataset $\mathcal{D}_u$. By additionally minimize expected cross entropy (CE) loss on $\mathcal{D}_l$, we can identify the unique Bayesian classifiers since they are respectively the minimizers of CE loss.

Theorem 1 (Main theorem). *Define the expected total correlation gain* $eTCg(h^{[M]}, p)$*:*

$$eTCg(h^1, \ldots, h^M, p) := \mathbb{E}_{x_i^{[M]} \sim U_{X^{[M]}}} \left(TCg(x_i^{[M]}; h^{[M]}, p) \right)$$

Given the conditional independence Assumption 1 and well-defined prior Assumption 2, we have that the maximum value of eTCg is Total Correlation of M modalities, i.e., $\mathrm{TC}(X^1, \ldots, X^M)$*. Besides,*

Ground-truth $\rightarrow$ ***Maximizer***$(h_*^{[M]}, \mathbf{p}^*)$ *is a maximizer of* $eTCg(h^{[M]}, p)$*. In other words,* $\forall h^{[M]} \in H^{[M]}, p \in \Delta_C$, $eTCg(h_*^1, \ldots, h_*^M, \mathbf{p}^*) \geq eTCg(h^1, \ldots, h^M, p)$*. The corresponding optimal aggregator is* $\zeta_\star$*, i.e.,* $\zeta_\star(x^{[M]})_c = \Pr(Y = c|X^{[M]} = x^{[M]})$*.*

Maximizer $\rightarrow$ ***(Permuted) Ground-truth.*** *If the prior is well defined, then for any maximizer of eTCg,* $(\tilde{h}^{[M]}, \tilde{\mathbf{p}})$*, there is a permutation* $\tilde{\prod} : \mathcal{C} \rightarrow \mathcal{C}$ *such that:*

$$\tilde{h}^m(x^m)_c = \mathrm{P}(Y = \tilde{\prod}(c)|X^m = x^m),\ \tilde{\mathbf{p}}_c = \mathrm{P}(Y = \tilde{\prod}(c)) \quad (7)$$

The proof is in Appendix A. Note from our main theorem that by maximizing the $eTCg$, we can get the total correlation of M modalities, which is the ground-truth label Y, and also the equivalent class of Bayesian posterior classifier under permutation function. In order to identify the Bayesian posterior classifiers, we can further minimize cross-entropy loss on labeled data $\mathcal{D}_l$ since the Bayesian posterior classifiers are the only minimizers of the expected cross-entropy loss. On the other hand, compared with only using $\mathcal{D}_l$ to train classifiers, our method can leverage more information from $\mathcal{D}_u$, which can be shown in the experimental result later.

4.3 Optimization

Since $eTCg$ is intractable, we alternatively maximize the empirical total correlation gain, i.e., $TCg^{(N)}$ to learn the optimal classifiers. To identify the unique Bayesian posteriors, we should further utilize labeled dataset $\mathcal{D}_l$ in a supervised way. Our whole optimization process is shown in Appendix, which adopts iteratively optimization strategy that is roughly contains two steps in each round: (i) We train the M classifiers using the classic cross entropy loss on the labeled dataset $\mathcal{D}_l$ and (ii) using our information-theoretic loss function $\mathcal{L}_{\text{TC}}$ on the unlabeled dataset $\mathcal{D}_u$. To learn the Bayesian posterior classifiers more accurately, the (ii) can help to learn the equivalent class of Bayesian Posterior Classifiers and (i) is to learn the correct and unique classifiers. As shown in Fig. 2, by optimizing $\mathcal{L}_{\text{TC}}^{(B)}$ (Eq. (5) with B denoting the number of samples in each batch), we reward the M classifiers for their agreements on the same data point and punish the M classifiers for their agreements on different data points.

Loss Function $\mathcal{L}_{\text{CE}}$ for Labeled Data. We use the classic cross entropy loss for labeled data. Formally, for a batch of data points $\{x_i^{[M]}\}_{i=1}^B$ drawn from labeled data $\mathcal{D}_l$, the cross entropy loss $\mathcal{L}_{\text{CE}}$ for each classifier h^m is defined as $\mathcal{L}_{\text{CE}}(\{(x_i^{[M]}, y_i)\}_{i=1}^B; h^m) := \frac{1}{B}\sum_i -\log(h^m(x_i^m)_{y_i})$.

Loss Function $\mathcal{L}_{\text{TC}}^{(B)}$ for Unlabeled Data. For a batch of data points $\{x_i^{[M]}\}_{i=1}^B$ drawn from unlabeled data $\mathcal{D}_u$, our loss function $\mathcal{L}_{\text{TC}}^{(B)} := -TCg^{(B)}$ that is defined in Eq. (6) with N replaced by number of batch size B. When N is large, we only sample a fixed number of samples from product of marginal distribution to estimate the second term in Eq. (6), which makes training more amenable.

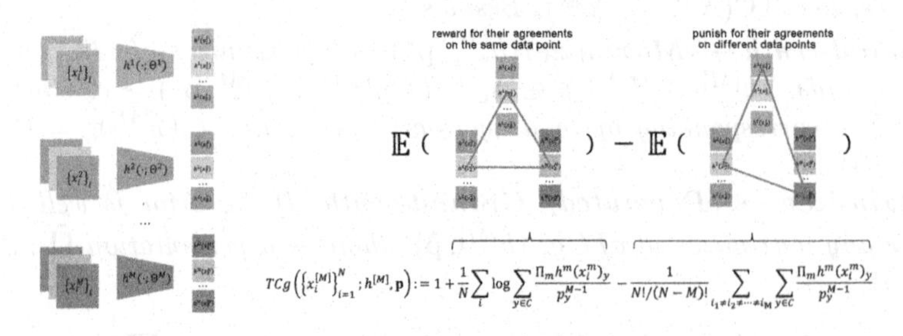

Fig. 2. Empirical Total Correlation Gain $TCg^{(N)}$

Prediction. After optimization, we can get the classifiers $\{h_m\}_m$. The prior $\mathbf{p}_c$ can be estimated from data, i.e., $\mathbf{p}_c = \frac{\sum_{i\in[N]} \mathbf{1}(y_i=c)}{N}$. Then based on Eq. (7), we can get the aggregator classifier ζ for prediction. Specifically, given a new sample $\tilde{x}^{[M]}$, the predicted label is $\tilde{y} := \arg\max_c \zeta(\tilde{x}^{[M]})_c$.

Time Complexity. The overall time complexity for optimizing our TCGM is linear scale to the number of modalities, i.e., $\mathcal{O}(M)$. Please refer to appendix for detailed analysis.

5 Preliminary Experiments

We first conduct a simulated experiment on synthetic data to validate our theoretical result of **TCGM**. Specifically, we will show the effectiveness of Total Correlation Gain TCg for unsupervised clustering of data. Further, with few labeled data, our **TCGM** can give accurate classification. In more detail, we synthesize the data of three modalities from a specific Gaussian distribution $P(X^i|y)$ $(i = 1, 2, 3)$. The clustering accuracy is calculated as classification accuracy by assuming the label is known. As shown in Fig. 3, our method **TCGM** has competitive performance compared to well established clustering algorithms K-means++ [2] and spectral clustering [24]. Based on the promising unsupervised learning result, as shown by the light blue line (the top line) in Fig. 3, our method can accurately classify the data even with only a small portion of labeled data. In contrast, **HGR** [31] degrades since it can only capture the linear dependence and fail when higher-order dependency exists.

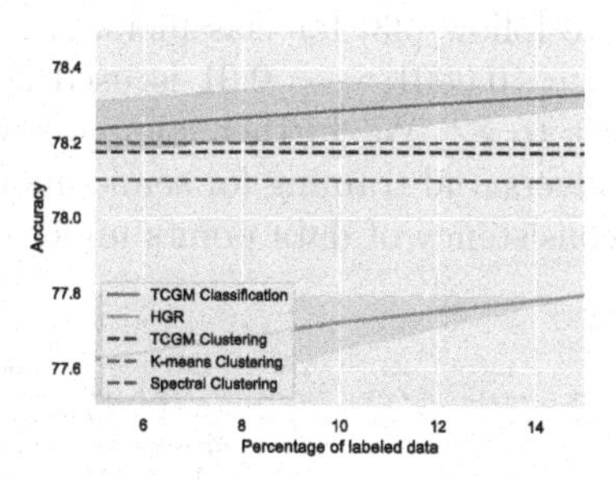

Fig. 3. Clustering and classification accuracy.

6 Applications

In this section, we evaluate our method on various multi-modal classification tasks: (i) News classification (Newsgroup) (ii) Emotion recognition: **IEMOCAP**, **MOSI** and (iii) Disease prediction of Alzheimer's Disease on 3D medical Imaging: **Alzheimer's Disease Neuroimaging Initiative (ADNI)**. Our method **TCGM** is compared with: **CE** separately trains classifiers for each modality by minimizing cross entropy loss of only labeled data; **HGR** [31] learns representation by maximizing correlation of different modalities; and **LMF** [23] performs multimodal fusion using low-rank tensors. The optimal hyperparameters are selected according to validation accuracy, among which the learning rate is optimized from $\{0.1, 0.01, 0.001, 0.0001\}$. All experiments are repeated five times with different random seeds. The mean test accuracies and standard deviations of single classifiers the aggregator (ζ) are reported.

6.1 News Classification

Dataset. **Newsgroup** [5][1] is a group of news classification datasets. Following [16], each data point has three modalities, PAM, SMI and UMI, collected from three different preprocessing steps[2]. We evaluate **TCGM** and the baseline methods on 3 datasets from **Newsgroup**: News-M2, News-M5, News-M10. They contain 500, 500, 1000 data points with 2, 5, 10 categories respectively. Following [33], we use 60% for training, 20% for validation and 20% for testing for all of these three datasets.

Implementation Details. We synthesize two different label rates (the percentage of labeled data points in each modality): {10%, 30%} for each dataset. We follow [33] for classifiers. Adam with default parameters and learning rate $\gamma_u = 0.0001, \gamma_l = 0.01$ is used as the optimizer during training. Batch size is set to 32. We further compare with two additional baselines: **VAT** [29] uses adversarial training for semi-supervised learning; **PVCC** [33] that considers the consistency of data points under different modalities.

	PAM	SMI	LMI	All
CE	78.20 ± 4.21	84.20 ± 7.01	77.40 ± 5.03	92.00 ± 2.24
VAT	80.60 ± 6.02	84.00 ± 4.74	75.80 ± 2.77	90.80 ± 2.05
PVCC	80.80 ± 2.77	80.60 ± 2.97	78.40 ± 7.64	92.20 ± 3.35
HGR	–	–	–	92.40 ± 2.19
TCGM	**93.60 ± 1.14**	**93.80 ± 2.17**	**93.00 ± 1.58**	**98.20 ± 1.10**

News-M2, Label Rate 10%

	PAM	SMI	LMI	All
CE	64.20 ± 3.90	67.80 ± 8.20	53.40 ± 4.72	78.00 ± 6.89
VAT	65.60 ± 1.82	66.80 ± 8.38	52.40 ± 3.29	79.40 ± 5.94
PVCC	67.80 ± 8.38	67.00 ± 1.22	54.60 ± 6.02	79.00 ± 6.28
HGR	–	–	–	79.80 ± 7.40
TCGM	**89.00 ± 2.24**	**90.60 ± 5.27**	**76.00 ± 3.54**	**97.20 ± 1.10**

News-M5, Label Rate 10%

	PAM	SMI	LMI	All
CE	27.80 ± 4.40	31.70 ± 4.86	31.00 ± 3.74	38.10 ± 5.47
VAT	30.70 ± 1.25	30.60 ± 4.02	27.40 ± 3.05	37.20 ± 5.20
PVCC	31.80 ± 2.89	34.10 ± 3.11	31.10 ± 4.14	46.40 ± 5.08
HGR	–	–	–	38.20 ± 3.72
TCGM	**45.50 ± 6.29**	**47.60 ± 3.73**	**52.70 ± 5.12**	**64.10 ± 2.16**

News-M10, Label Rate 10%

	PAM	SMI	LMI	All
CE	86.40 ± 2.30	91.20 ± 1.92	85.40 ± 1.95	96.40 ± 1.82
VAT	89.00 ± 3.87	91.20 ± 3.56	84.40 ± 2.70	96.20 ± 1.92
PVCC	83.40 ± 4.16	82.60 ± 5.22	80.00 ± 4.90	93.20 ± 2.49
HGR	–	–	–	95.20 ± 3.27
TCGM	**94.20 ± 2.17**	**93.20 ± 3.27**	**92.40 ± 1.34**	**98.40 ± 1.14**

News-M2, Label Rate 30%

	PAM	SMI	LMI	All
CE	85.80 ± 3.63	81.80 ± 2.86	63.80 ± 5.40	96.20 ± 3.11
VAT	85.40 ± 2.88	83.00 ± 2.55	60.20 ± 8.98	97.00 ± 2.83
PVCC	81.80 ± 5.17	72.20 ± 6.46	57.00 ± 9.59	88.80 ± 6.91
HGR	–	–	–	89.80 ± 4.02
TCGM	**89.20 ± 2.28**	**87.20 ± 4.32**	**73.60 ± 4.28**	**98.20 ± 1.10**

News-M5, Label Rate 30%

	PAM	SMI	LMI	All
CE	49.00 ± 4.40	50.80 ± 2.49	51.20 ± 2.51	69.50 ± 4.57
VAT	46.80 ± 4.31	49.80 ± 4.58	50.30 ± 3.49	69.90 ± 5.10
PVCC	52.60 ± 4.02	48.60 ± 4.68	46.80 ± 2.02	66.90 ± 3.49
HGR	–	–	–	61.20 ± 5.38
TCGM	**63.60 ± 4.60**	**61.30 ± 3.40**	**64.70 ± 1.75**	**86.70 ± 2.97**

News-M10, Label Rate 30%

Fig. 4. Test accuracies (mean ± std. dev.) on newsgroups datasets

As shown in Fig. 4, **TCGM** achieves the best classification accuracy for both single classifier and aggregators, especially when the label rate is small. This shows the efficacy of utilizing the cross-modal information during training as compared to others that are unable to utilize the cross-modal information. Moreover, we can achieve further improvement by aggregating classifiers on all modalities, which shows the benefit of aggregating knowledge from different modalities.

6.2 Emotion Recognition

Dataset. We evaluate our methods on two multi-modal emotion recognition datasets: **IEMOCAP** dataset [7] and **MOSI** dataset [34]. The goal for both datasets is to identify speaker emotions based on the collected videos, audios

[1] http://qwone.com/~jason/20Newsgroups/.

[2] PAM (Partitioning Around Medoïds preprocessing), SMI (Supervised Mutual Information preprocessing) and UMI (Unsupervised Mutual Information preprocessing).

and texts. The IEMOCAP consists of 151 sessions of recorded dialogues, with 2 speakers per session for a total of 302 videos across the dataset. The MOSI is composed of 93 opinion videos from YouTube movie reviews. We follow the settings in [23] for the data splits of training, validation and test set. For IEMOCAP, we conduct experiments on three different emotions: happy, angry and neutral emotions; for MOSI dataset we consider the binary classification of emotions: positive and negative.

Implementation Details. We synthesize three label rates for each dataset (the percentage of labeled data points in each modality): $\{0.5\%, 1\%, 1.5\%\}$ for IEMOCAP and $\{1\%, 2\%, 3\%\}$ for MOSI. For a fair comparison, we follow architecture setting in [23]. We adopt the modality encoder architectures in [23] as the single modality classifiers for CE and TCGM, while adopting the aggregator on the top of modality encoders for LMF and HGR. Adam with default parameters and learning rate $\gamma_u = 0.0001, \gamma_l = 0.001$ is used as the optimizer. The batch size is set to 32.

We report the AUC (Area under ROC curve) for the aggregators on all the modalities and single modality classifiers by different methods. We only report the AUC of LMF and HGR on all modalities since they do not have single modality classifiers. For single modality classifiers, we show results on the text modality on happy emotion (d), audio modality on neutral emotion (e) the video modality on angry emotion on IEMOCAP; and (h) the video modality (i) the audio modality on MOSI. Please refer to supplementary material for complete experimental result. As shown in Fig. 5, aggregators trained by TCGM outperform all the baselines given only tiny fractions of labeled data. **TCGM** improves the AUC of the single modality classifiers significantly, which shows the efficacy of utilizing the cross-modal information during the training of our method. As label rates continue to grow, the advantage of our method over CE decreases since more information is provided for CE to learn the ground-truth label.

Our method also outperforms other methods in terms of the prediction based on all the modalities, especially when the label rate is small. This shows the superiority of our method when dealing with a limited amount of annotations.

6.3 Disease Prediction of Alzheimer's Disease

Dataset. Early prediction of Alzheimer's Disease (AD) is attracting increasing attention since it is irreversible and very challenging. Besides, due to privacy issues and high collecting costs, an efficient classifier with limited labeled data is desired. To validate the effectiveness of our method on this challenging task, we only keep labels of a limited percentage of data, which is obtained from the most popular **Alzheimer's Disease Neuroimaging Initiative (ADNI)** dataset[3], with 3D images sMRI and PET. DARTEL VBM pipeline [2] is implemented to pre-process the sMRI data, and then images of PET were reoriented into

[3] www.loni.ucla.edu/ADNI.

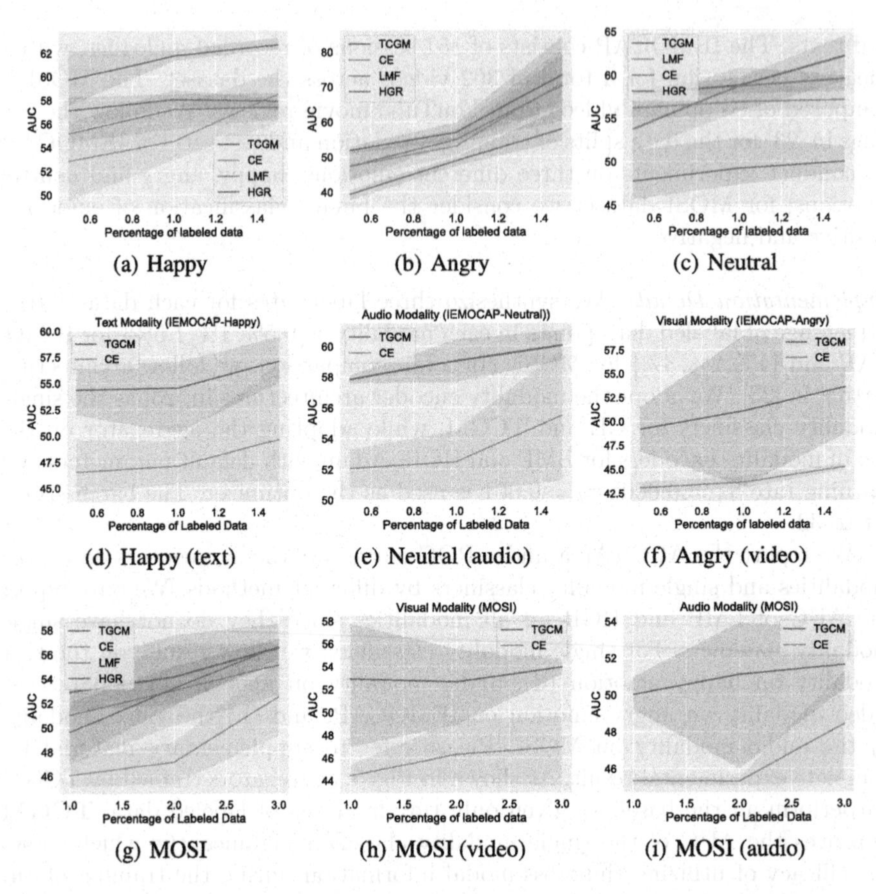

(a) Happy (b) Angry (c) Neutral

(d) Happy (text) (e) Neutral (audio) (f) Angry (video)

(g) MOSI (h) MOSI (video) (i) MOSI (audio)

Fig. 5. (a, b, c) AUC of Aggregators on happy, angry and neutral emotion recognition on IEMOCAP data: (d, e, f) AUC on text, audio and video modality classifiers on IEMOCAP: AUC of other composition of (modality, emotion) are listed in supplementary material. **(g, h, i) AUC on MOSI data**: AUC of (g) Aggregators on all modalities and single classifiers on (h) Video modality (i) Audio modality.

a standard 91 × 109 × 91 voxel image grid in MNI152 space, which is same with sMRIs'. To limit the size of images, only the hippocampus on both sides are extracted as input in the experiments. We denote subjects that convert to Alzheimer's disease (MCI_c) as AD, and subjects that remain stable (MCI_s) as NC (Normal Control). Our dataset contains 300 samples in total, with 144 AD and 156 NC. We randomly choose 60% for training, 20% for validation and 20% for testing stage.

Implementation Details. We synthesize two different label rates (the percentage of labeled data points): $\{10\%, 50\%\}$. DenseNet is used as the classifier. Two 3D convolutional layers with the kernel size 3 × 3 × 3 are adopted to replace the first 2D convolutional layers with the kernel size 7×7. We use four dense blocks with the size of $(6, 12, 24, 16)$. To preserve more low-level local information, we discard the first max-pooling layer that follows after the first convolution layer. Adam with default parameters and learning rate $\gamma_u = \gamma_l = 0.001$ are used as the optimizer during training. We set Batch Size as only 12 due to the large memory usage of 3D images. Random crop of $64 \times 64 \times 64$, random flip and random transpose are applied as data augmentation.

	sMRI	PET	All
CE	54.48 ± 8.76	58.62 ± 3.23	62.07 ± 2.72
HGR	–	–	60.34 ± 5.59
TCGM	**63.44 ± 2.83**	**62.41 ± 3.53**	**65.17 ± 1.44**

ADNI, Label Rate 10%

	sMRI	PET	All
CE	63.79 ± 3.66	60.69 ± 2.83	65.52 ± 3.66
HGR	–	–	66.22 ± 4.16
TCGM	**65.86 ± 3.09**	**62.76 ± 1.97**	**67.24 ± 2.10**

ADNI, Label Rate 50%

Fig. 6. Test accuracies (mean ± std.) on ADNI dataset

Figure 6 shows the accuracy of classifiers for each modality and the aggregator. Our method **TCGM** outperforms the baseline methods in all settings especially when the label rate is small, which is desired since it is costly to label data.

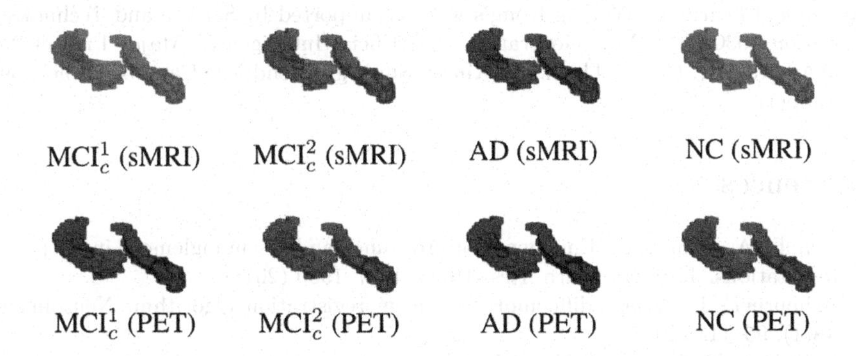

Fig. 7. Volume (sMRI, top line) and SUV (PET, bottom line) of MCI_c^1, MCI_c^2, AD and NC. Darker color implies smaller volume and SUV, i.e., more probability of being AD.

To further illustrate the advantage of our model over others in terms of leveraging the knowledge of another modality, we visualize two MCI_cs, denoted as MCI_c^1 and MCI_c^2, which are mistakenly classified as NC by **CE**'s classifier for sMRI and PET modality, respectively. The volume and standardized uptake value (SUV) (a measurement of the degree of metabolism), whose information are respectively contained by sMRI and PET data, are linearly mapped to the darkness of the red and blue. Darker color implies smaller volume and SUV, i.e., more probability of being AD. As shown in Fig. 7, the volume (SUV) of MCI_c^1 (MCI_c^2) is similar to NC, hence it is reasonable for **CE** to mistakenly classify it by only using the information of volume (SUV). In contrast, **TCGM** for each modality can correctly classify both cases as AD, which shows the better learning of the information intersection (i.e., the ground truth) during training, facilitated by the leverage of knowledge from another modality.

7 Conclusion

In this paper, we propose an information-theoretic framework on multi-modal data, Total Correlation Gain Maximization (TCGM), in the scenario of semi-supervised learning. Specifically, we learn to infer the ground truth labels shared by all modalities by maximizing the total correlation gain. Conditioning on a common assumption that all modalities are independent given the ground truth label, it can be theoretically proved our method can learn the Bayesian posterior classifier for each modality and the Bayesian posterior aggregator for all modalities. Extensive experiments on Newsgroup, IEMOCAP, MOSI and ADNI datasets are conducted and achieve promising results, which demonstrates the benefit and utility of our framework.

Acknowledgement. Yizhou Wang's work is supported by MOST-2018AAA0102004, NSFC-61625201, DFG TRR169/NSFC Major International Collaboration Project "Crossmodal Learning". Yuqing Kong's work is supported by Science and Technology Innovation 2030 "The New Generation of Artificial Intelligence" Major Project No. 2018AAA0100901, China. Thanks to Xinwei Sun's girlfriend Yue Cao, for all her love and support.

References

1. Achille, A., Soatto, S.: Emergence of invariance and disentanglement in deep representations. J. Mach. Learn. Res. **19**(1), 1947–1980 (2018)
2. Ashburner, J.: A fast diffeomorphic image registration algorithm. Neuroimage **38**(1), 95–113 (2007)
3. Balcan, M.F., Blum, A., Yang, K.: Co-training and expansion: towards bridging theory and practice. In: Advances in Neural Information Processing Systems, pp. 89–96 (2005)
4. Belghazi, M.I., et al.: MINE: mutual information neural estimation. arXiv preprint arXiv:1801.04062 (2018)
5. Bisson, G., Grimal, C.: Co-clustering of multi-view datasets: a parallelizable approach. In: 2012 IEEE 12th International Conference on Data Mining, pp. 828–833. IEEE (2012)
6. Blum, A., Mitchell, T.: Combining labeled and unlabeled data with co-training. In: Proceedings of the Eleventh Annual Conference on Computational Learning Theory, pp. 92–100. ACM (1998)
7. Busso, C., et al.: IEMOCAP: interactive emotional dyadic motion capture database. Lang. Resour. Eval. **42**, 335–359 (2008)

8. Cao, P., Xu, Y., Kong, Y., Wang, Y.: Max-MIG: an information theoretic approach for joint learning from crowds. In: ICLR 2019 (2018)
9. Chandar, S., Khapra, M.M., Larochelle, H., Ravindran, B.: Correlational neural networks. Neural Comput. **28**(2), 257–285 (2016)
10. Chang, X., Xiang, T., Hospedales, T.M.: Scalable and effective deep CCA via soft decorrelation. In: Proceedings of the IEEE Conference on Computer Vision and Pattern Recognition, pp. 1488–1497 (2018)
11. Cheng, Y., Zhao, X., Huang, K., Tan, T.: Semi-supervised learning and feature evaluation for RGB-D object recognition. Comput. Vis. Image Underst. **139**, 149–160 (2015)
12. Christoudias, C.M., Saenko, K., Morency, L.P., Darrell, T.: Co-adaptation of audio-visual speech and gesture classifiers. In: Proceedings of the 8th International Conference on Multimodal Interfaces, pp. 84–91. ACM (2006)
13. Dasgupta, S., Littman, M.L., McAllester, D.A.: PAC generalization bounds for co-training. In: Advances in Neural Information Processing Systems, pp. 375–382 (2002)
14. Hjelm, R.D., Fedorov, A., Lavoie-Marchildon, S., Grewal, K., Trischler, A., Bengio, Y.: Learning deep representations by mutual information estimation and maximization. arXiv preprint arXiv:1808.06670 (2018)
15. Huang, S.L., Suh, C., Zheng, L.: Euclidean information theory of networks. IEEE Trans. Inf. Theory **61**(12), 6795–6814 (2015)
16. Hussain, S.F., Grimal, C., Bisson, G.: An improved co-similarity measure for document clustering. In: International Conference on Machine Learning and Applications (2010)
17. Jones, R.: Learning to extract entities from labeled and unlabeled text. Ph.D. thesis, Citeseer (2005)
18. Kim, D.H., Lee, M.K., Choi, D.Y., Song, B.C.: Multi-modal emotion recognition using semi-supervised learning and multiple neural networks in the wild. In: Proceedings of the 19th ACM International Conference on Multimodal Interaction, pp. 529–535. ACM (2017)
19. Kong, Y., Schoenebeck, G.: Water from two rocks: maximizing the mutual information. In: Proceedings of the 2018 ACM Conference on Economics and Computation, pp. 177–194. ACM (2018)
20. Leskes, B.: The value of agreement, a new boosting algorithm. In: Auer, P., Meir, R. (eds.) COLT 2005. LNCS (LNAI), vol. 3559, pp. 95–110. Springer, Heidelberg (2005). https://doi.org/10.1007/11503415_7
21. Levin, A., Viola, P.A., Freund, Y.: Unsupervised improvement of visual detectors using co-training. In: ICCV, vol. 1, p. 2 (2003)
22. Lewis, D.D.: Naive (Bayes) at forty: the independence assumption in information retrieval. In: Nédellec, C., Rouveirol, C. (eds.) ECML 1998. LNCS, vol. 1398, pp. 4–15. Springer, Heidelberg (1998). https://doi.org/10.1007/BFb0026666
23. Liu, Z., Shen, Y., Lakshminarasimhan, V.B., Liang, P.P., Zadeh, A., Morency, L.P.: Efficient low-rank multimodal fusion with modality-specific factors. In: ACL (2018)
24. Ng, A.Y., Jordan, M.I., Weiss, Y.: On spectral clustering: analysis and an algorithm. In: NIPS (2001)
25. Ngiam, J., Khosla, A., Kim, M., Nam, J., Lee, H., Ng, A.Y.: Multimodal deep learning. In: Proceedings of the 28th International Conference on Machine Learning (ICML 2011), pp. 689–696 (2011)
26. Nguyen, X., Wainwright, M.J., Jordan, M.I.: Estimating divergence functionals and the likelihood ratio by convex risk minimization. IEEE Trans. Inf. Theory **56**(11), 5847–5861 (2010)

27. Sohn, K., Shang, W., Lee, H.: Improved multimodal deep learning with variation of information. In: Advances in Neural Information Processing Systems, pp. 2141–2149 (2014)
28. Studený, M., Vejnarová, J.: The multiinformation function as a tool for measuring stochastic dependence. In: Jordan, M.I. (ed.) Learning in Graphical Models. ASID, vol. 89, pp. 261–297. Springer, Dordrecht (1998). https://doi.org/10.1007/978-94-011-5014-9_10
29. Miyato, T., Maeda, S., Koyama, M., Ishii, S.: Virtual adversarial training: a regularization method for supervised and semi-supervised learning. IEEE Trans. Pattern Anal. Mach. Intell. **41**, 1979–1993 (2018)
30. Veličković, P., Fedus, W., Hamilton, W.L., Liò, P., Bengio, Y., Hjelm, R.D.: Deep graph infomax. arXiv preprint arXiv:1809.10341 (2018)
31. Wang, L., et al.: An efficient approach to informative feature extraction from multimodal data. arXiv preprint arXiv:1811.08979 (2018)
32. Xu, Y., Zhao, S., Song, J., Stewart, R.J., Ermon, S.: A theory of usable information under computational constraints. ArXiv abs/2002.10689 (2020)
33. Yang, Y., Zhan, D.C., Sheng, X.R., Jiang, Y.: Semi-supervised multi-modal learning with incomplete modalities. In: IJCAI, pp. 2998–3004 (2018)
34. Zadeh, A., Zellers, R., Pincus, E., Morency, L.P.: MOSI: multimodal corpus of sentiment intensity and subjectivity analysis in online opinion videos. ArXiv abs/1606.06259 (2016)

ExchNet: A Unified Hashing Network for Large-Scale Fine-Grained Image Retrieval

Quan Cui[1,3], Qing-Yuan Jiang[2], Xiu-Shen Wei[3(✉)], Wu-Jun Li[2], and Osamu Yoshie[1]

[1] Graduate School of IPS, Waseda University, Fukuoka, Japan
cui-quan@toki.waseda.jp, yoshie@waseda.jp
[2] National Key Laboratory for Novel Software Technology, Department of Computer Science and Technology, Nanjing University, Nanjing, China
qyjiang24@gmail.com, liwujun@nju.edu.cn
[3] Megvii Research Nanjing, Megvii Technology, Nanjing, China
weixs.gm@gmail.com

Abstract. Retrieving content relevant images from a large-scale fine-grained dataset could suffer from intolerably slow query speed and highly redundant storage cost, due to high-dimensional real-valued embeddings which aim to distinguish subtle visual differences of fine-grained objects. In this paper, we study the novel fine-grained hashing topic to generate compact binary codes for fine-grained images, leveraging the search and storage efficiency of hash learning to alleviate the aforementioned problems. Specifically, we propose a unified end-to-end trainable network, termed as ExchNet. Based on attention mechanisms and proposed attention constraints, ExchNet can firstly obtain both local and global features to represent object parts and the whole fine-grained objects, respectively. Furthermore, to ensure the discriminative ability and semantic meaning's consistency of these part-level features across images, we design a local feature alignment approach by performing a feature exchanging operation. Later, an alternating learning algorithm is employed to optimize the whole ExchNet and then generate the final binary hash codes. Validated by extensive experiments, our ExchNet consistently outperforms state-of-the-art generic hashing methods on five fine-grained datasets. Moreover, compared with other approximate nearest neighbor methods, ExchNet achieves the best speed-up and storage reduction, revealing its efficiency and practicality.

Keywords: Fine-Grained Image Retrieval · Learning to hash · Feature alignment · Large-scale image search

Q. Cui, Q.-Y. Jiang—Equal contribution.

Electronic supplementary material The online version of this chapter (https://doi.org/10.1007/978-3-030-58580-8_12) contains supplementary material, which is available to authorized users.

A. Vedaldi et al. (Eds.): ECCV 2020, LNCS 12348, pp. 189–205, 2020.
https://doi.org/10.1007/978-3-030-58580-8_12

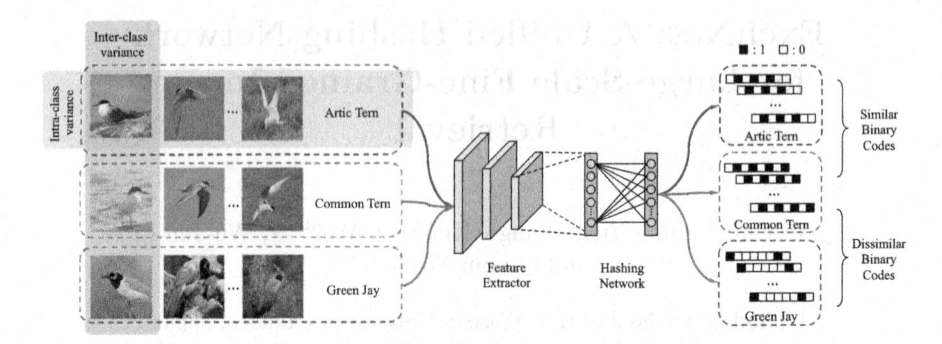

Fig. 1. Illustration of the fine-grained hashing task. Fine-grained images could share large intra-class variances but small inter-class variances. Fine-grained hashing aims to generate compact binary codes with tiny Hamming distances for images of the same sub-category, as well as distinct codes for images from different sub-categories.

1 Introduction

Fine-Grained Image Retrieval (FGIR) [19,26,31,36,41,42] is a practical but challenging computer vision task. It aims to retrieve images belonging to various sub-categories of a certain meta-category (e.g., birds, cars and aircrafts) and return images with the same sub-category as the query image. In real FGIR applications, previous methods could suffer from slow query speed and redundant storage costs due to both the explosive growth of massive fine-grained data and high-dimensional real-valued features.

Learning to hash [3,6,7,10,14,16,17,21,22,34,35] has proven to be a promising solution for large-scale image retrieval because it can greatly reduce the storage cost and increase the query speed. As a representative research area of approximate nearest neighbor (ANN) search [1,6,13], hashing aims to embed data points as similarity-preserving binary codes. Recently, hashing has been successfully applied in a wide range of image retrieval tasks, e.g., face image retrieval [18], person re-identification [5,43], etc. We hereby explore the effectiveness of hashing for *fine-grained* image retrieval.

To the best of our knowledge, this is the first work to study the fine-grained hashing problem, which refers to the problem of designing hashing for fine-grained objects. As shown in Fig. 1, the task is desirable to generate compact binary codes for fine-grained images sharing both large intra-class variances and small inter-class variances. To deal with the challenging task, we propose a unified end-to-end trainable network ExchNet to first learn fine-grained tailored features and then generate the final binary hash codes.

In concretely, our ExchNet consists of three main modules, including representation learning, local feature alignment and hash code learning, as shown in Fig. 2. In the representation learning module, beyond obtaining the holistic image representation (i.e., global features), we also employ the attention mechanism to capture the part-level features (i.e., local features) for representing fine-grained objects' parts. Localizing parts and embedding part-level cues are

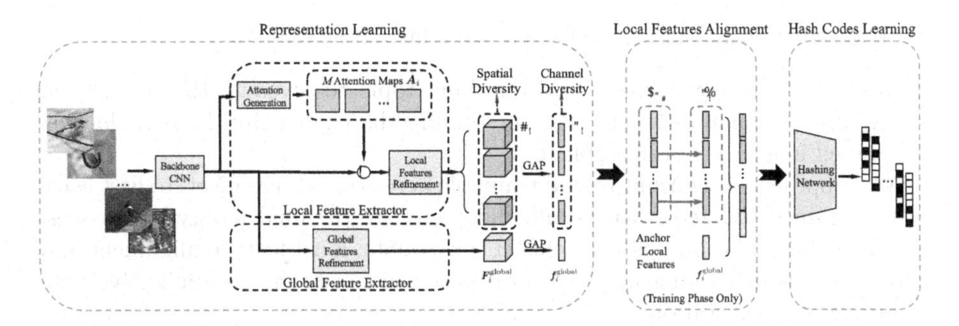

Fig. 2. Framework of our proposed ExchNet, which consists of three modules. (1) The representation learning module, as well as the attention mechanism with spatial and channel diversity learning constraints, is designed to obtain both local and global features of fine-grained objects. (2) The local feature alignment module is used to align obtained local features w.r.t. object parts across different fine-grained images. (3) The hash codes learning module is performed to generate the compact binary codes.

crucial for fine-grained tasks, since these discriminative but subtle parts (e.g., bird heads or tails) play a major role to distinguish different sub-categories. Moreover, we also develop two kinds of attention constraints, i.e., spatial and channel constraints, to collaboratively work together for further improving the discriminative ability of these local features. In the following, to ensure that these part-level features can correspond to their own corresponding parts across different fine-grained images, we design an anchor based feature alignment approach to align these local features. Specifically, in the local feature alignment module, we treat the anchored local features as the "prototype" w.r.t. its sub-category by averaging all the local features of that part across images. Once local features are well aligned for their own parts, even if we exchange one specific part's local feature of an input image with the same part's local feature of the prototype, the image meanings derived from the image representations and also the final hash codes should be both extremely similar. Inspired by this motivation, we perform a feature exchanging operation upon the anchored local features and other learned local features, which is illustrated in Fig. 3. After that, for effectively training the network with our feature alignment fashion, we utilize an alternating algorithm to solve the hashing learning problem and update anchor features simultaneously.

To quantitatively prove both effectiveness and efficiency of our ExchNet, we conduct comprehensive experiments on five fine-grained benchmark datasets, including the large-scale ones, i.e., *NABirds* [11], *VegFru* [12] and *Food101* [23]. Particularly, compared with competing approximate nearest neighbor methods, our ExchNet achieves up to hundreds times speedup for large-scale fine-grained image retrieval without significant accuracy drops. Meanwhile, compared with state-of-the-art generic hashing methods, ExchNet could consistently outperform these methods by a large margin on all the fine-grained datasets. Additionally, ablation studies and visualization results justify the effectiveness of our tailored model designs like local feature alignment and proposed attention approach.

The contributions of this paper are summarized as follows:

- We study the novel fine-grained hashing topic to leverage the search and storage efficiency of hash codes for solving the challenging large-scale fine-grained image retrieval problem.
- We propose a unified end-to-end trainable network, i.e., ExchNet, to first learn fine-grained tailored features and then generate the final binary hash codes. Particularly, the proposed attention constraints, local feature alignment and anchor-based learning fashion contribute well to obtain discriminative fine-grained representations.
- We conduct extensive experiments on five fine-grained datasets to validate both effectiveness and efficiency of our proposed ExchNet. Especially for the results on large-scale datasets, ExchNet exhibits its outperforming retrieval performance on either speedup, memory usages and retrieval accuracy.

2 Related Work

Fine-Grained Image Retrieval. Fine-Grained Image Retrieval (FGIR) is an active research topic emerged in recent years, where the database and query images could share small inter-class variance but large intra-class variance. In previous works [36], handcrafted features were initially utilized to tackle the FGIR problem. Powered by deep learning techniques, more and more deep learning based FGIR methods [19,26,31–33,36,41,42] were proposed. These deep methods can be roughly divided into two parts, i.e., supervised and unsupervised methods. In supervised methods, FGIR is defined as a metric learning problem. Zheng et al. [41] designed a novel ranking loss and a weakly-supervised attractive feature extraction strategy to facilitate the retrieval performance. Zheng et al. [42] improved their former work [41] with a normalize-scale layer and decorrelated ranking loss. As to unsupervised methods, Selective Convolutional Descriptor Aggregation (SCDA) [31] was proposed to localize the main object in fine-grained images firstly, and then discard the noisy background and keep useful deep descriptors for fine-grained image retrieval.

Deep Hashing. Hashing methods can be divided into two categories, i.e., data-independent methods [6] and data-dependent methods [10,17], based on whether training points are used to learn hash functions. Generally speaking, data-dependent methods, also named as Learning to Hash (L2H) methods, can achieve better retrieval performance with the help of the learning on training data. With the rise of deep learning, some L2H methods integrate deep feature learning into hash frameworks and achieve promising performance. As previous work, many deep hashing methods [2,3,7,14,16,17,21,22,30,35,38,39] for large-scale image retrieval have been proposed. Compared with deep unsupervised hashing methods [7,14,21], deep supervised hashing methods [14,16,17,35] can achieve superior retrieval accuracy as they can fully explore the semantic information. Specifically, the previous work [35] was essentially a two-stage method

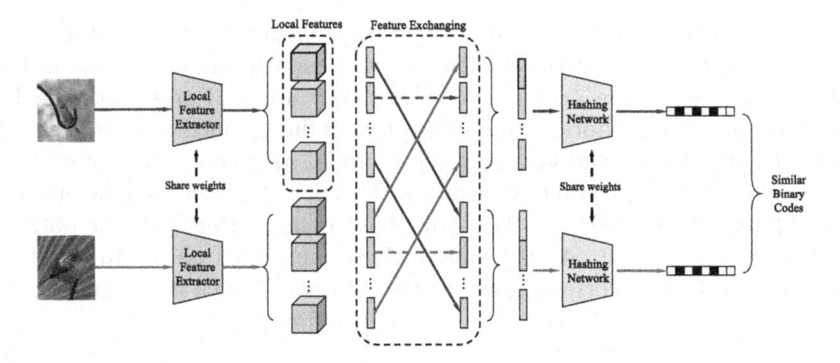

Fig. 3. Key idea of our local feature alignment approach: given an image pair of a fine-grained category, exchanging their local features of the same object parts should not change their corresponding hash codes, i.e., these hash codes should be the same as those generated without local feature exchanging and their Hamming distance should be still close also.

which tried to learn binary codes in the first stage and employed feature learning guided by the learned binary codes in the second stage. Then, there appeared numerous one-stage deep supervised hashing methods, including Deep Pairwise Supervised Hashing (DPSH) [17], Deep Supervised Hashing (DSH) [22], and Deep Cauchy Hashing (DCH) [3], which aimed to integrate feature learning and hash code learning into an end-to-end framework.

3 Methodology

The framework of our ExchNet is presented in Fig. 2, which contains three key modules, i.e., the representation learning module, local feature alignment module, and hash code learning module.

3.1 Representation Learning

The learning of discriminative and meaningful local features is mutually correlated with fine-grained tasks [9,15,20,37,40], since these local features can greatly benefit the distinguishing of sub-categories with subtle visual differences deriving from the discriminative fine-grained parts (e.g., bird heads or tails). In consequence, as shown in Fig. 2, beyond the global feature extractor, we also introduce a local feature extractor in the representation learning module. Specifically, by considering model efficiency, we hereby propose to learn local features with the attention mechanism, rather than other fine-grained techniques with tremendous computation cost, e.g., second-order representations [15,20] or complicated network architectures [9,37,40].

Given an input image $\boldsymbol{x}_i$, a backbone CNN is utilized to extract a holistic deep feature $\boldsymbol{E}_i \in \mathbb{R}^{H \times W \times C}$, which serves as the appetizer for both the local feature extractor and the global feature extractor.

It is worth mentioning that the attention is engaged in the middle of the feature extractor. Since, in the shallow layers of deep neural networks, low-level context information (e.g., colors and edges, etc.) are well preserved, which is crucial for distinguish subtle visual differences of fine-grained objects. Then, by feeding $\boldsymbol{E}_i$ into the attention generation module, M pieces of attention maps $\boldsymbol{A}_i \in \mathbb{R}^{M \times H \times W}$ are generated and we use $\boldsymbol{A}_i^j \in \mathbb{R}^{H \times W}$ to denote the attentive region of the j-th ($j \in \{1, \ldots, M\}$) part cues for $\boldsymbol{x}_i$. After that, the obtained part-level attention map $\boldsymbol{A}_i^j$ is element-wisely multiplied on $\boldsymbol{E}_i$ to select the attentive local feature corresponding to the j-th part, which is formulated as:

$$\hat{\boldsymbol{E}}_i^j = \boldsymbol{E}_i \otimes \boldsymbol{A}_i^j, \tag{1}$$

where $\hat{\boldsymbol{E}}_i^j \in \mathbb{R}^{H \times W \times C}$ represents the j-th attentive local feature of $\boldsymbol{x}_i$, and "$\otimes$" denotes the Hadamard product on each channel. For simplification, we use $\hat{\mathcal{E}}_i = \{\hat{\boldsymbol{E}}_i^1, \ldots, \hat{\boldsymbol{E}}_i^M\}$ to denote a set of local features and, subsequently, $\hat{\mathcal{E}}_i$ is fed into the later Local Features Refinement (LFR) network composed of a stack of convolution layers to embed these attentive local features into higher-level semantic meanings:

$$\mathcal{F}_i = f_{\texttt{LFR}}(\hat{\mathcal{E}}_i), \tag{2}$$

where the output of the network is denoted as $\mathcal{F}_i = \{\boldsymbol{F}_i^1, \ldots, \boldsymbol{F}_i^M\}$, which represents the final local feature maps w.r.t. high-level semantics. We denote $\boldsymbol{f}_i^j \in \mathbb{R}^{C'}$ as the local feature vector after applying global average pooling (GAP) on $\boldsymbol{F}_i^j \in \mathbb{R}^{H' \times W' \times C'}$ as:

$$\boldsymbol{f}_i^j = f_{\texttt{GAP}}(\boldsymbol{F}_i^j). \tag{3}$$

On the other side, as to the global feature extractor, for $\boldsymbol{x}_i$, we directly adopt a Global Features Refinement (GFR) network composed of conventional convolutional operations to embed $\boldsymbol{E}_i$, which is presented by:

$$\boldsymbol{F}_i^{\text{global}} = f_{\texttt{GFR}}(\boldsymbol{E}_i). \tag{4}$$

We use $\boldsymbol{F}_i^{\text{global}} \in \mathbb{R}^{H' \times W' \times C'}$ and $\boldsymbol{f}_i^{\text{global}} \in \mathbb{R}^{C'}$ to denote the learned global feature and the corresponding holistic feature vector after GAP, respectively.

Furthermore, to facilitate the learning of localizing local feature cues (i.e., capturing fine-grained parts), we impose the spatial diversity and channel diversity constraints over the local features in $\mathcal{F}_i$.

Specifically, it is a natural choice to increase the diversity of local features by differentiating the distributions of attention maps [40]. However, it might cause a problem that the holistic feature can not be activated in some spatial positions, while the attention map has large activation values on them due to over-applied constraints upon the learned attention maps. Instead, in our method, we design and apply constraints on the local features. In concretely, for the local feature $\boldsymbol{F}_i^j$, we obtain its "aggregation map" $\hat{\boldsymbol{A}}_i^j \in \mathbb{R}^{H' \times W'}$ by adding all C' feature maps through the channel dimension and apply the softmax function on it for converting it into a valid distribution, then flat it into a vector $\hat{\boldsymbol{a}}_i^j$. Based on the Hellinger distance, we propose a spatial diversity induced loss as:

$$\mathcal{L}_{\mathtt{sp}}(\boldsymbol{x}_i) = 1 - \frac{1}{\sqrt{2}\binom{M}{2}} \sum_{l,k=1}^{M} \left\| \sqrt{\hat{\boldsymbol{a}}_i^l} - \sqrt{\hat{\boldsymbol{a}}_i^k} \right\|_2 , \tag{5}$$

where $\binom{M}{2}$ is used to denote the combinatorial number of ways to pick 2 unordered outcomes from M possibilities. The spatial diversity constraint drives the aggregation maps to be activated in spatial positions as diverse as possible. As to the channel diversity constraint, we first convert the local feature vector $\boldsymbol{f}_i^j$ into a valid distribution, which can be formulated by

$$\boldsymbol{p}_i^j = \mathtt{softmax}(\boldsymbol{f}_i^j), \ \forall j \in \{1, \ldots, M\}. \tag{6}$$

Subsequently, we propose a constraint loss over $\{\boldsymbol{p}_i^j\}_{j=1}^M$ as:

$$\mathcal{L}_{\mathtt{cp}}(\boldsymbol{x}_i) = \left[t - \frac{1}{\sqrt{2}\binom{M}{2}} \sum_{l,k=1}^{M} \left\| \sqrt{\boldsymbol{p}_i^l} - \sqrt{\boldsymbol{p}_i^k} \right\|_2 \right]_+ , \tag{7}$$

where $t \in [0, 1]$ is a hyper-parameter to adjust the diversity and $[\cdot]_+$ denotes $max(\cdot, 0)$. Equipping with the channel diversity constraint could benefit the network to depress redundancies in features through channel dimensions. Overall, our spatial diversity and channel diversity constraints can work in a collaborative way to obtain discriminative local features.

3.2 Learning to Align by Local Feature Exchanging

Upon the representation learning module, the alignment on local features is necessary for confirming that they represent and more importantly correspond to common fine-grained parts across images, which are essential to fine-grained tasks. Hence, we propose an anchor-based local features alignment approach assisted with our feature exchanging operation.

Intuitively, local features from the same object part (e.g., bird heads of a bird species) should be embedded with almost the same semantic meaning. As illustrated by Fig. 3, our key idea is that, if local features were well aligned, exchanging the features of identical parts for two input images belonging to the same sub-category should not change the generated hash codes. Inspired by that, we propose a local feature alignment strategy by leveraging the feature exchanging operation, which happens between learned local features and anchored local features. As a foundation for feature exchanging, a set of dynamic anchored local features $\mathcal{C}_{y_i} = \{\boldsymbol{c}_{y_i}^1, \ldots, \boldsymbol{c}_{y_i}^M\}$ for class y_i should be maintained, in which the j-th anchored local feature $\boldsymbol{c}_{y_i}^j$ is obtained by averaging all j-th part's local features of training samples from class y_i. At the end of each training epoch, anchored local features will be recalculated and updated. Subsequently, as shown in Fig. 4,

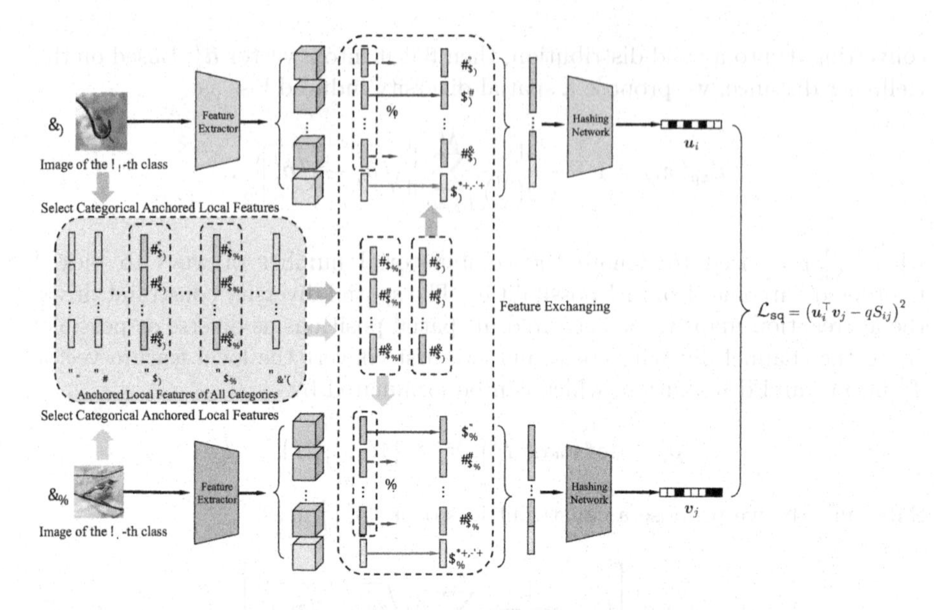

Fig. 4. Our feature exchanging and hash codes learning in the training phase. According to the class indices (i.e., y_i and y_j), we first select categorical anchor features $\mathcal{C}_{y_i}$ and $\mathcal{C}_{y_j}$ for samples $\boldsymbol{x}_i$ and $\boldsymbol{x}_j$, respectively. Then, for each input image, the feature exchanging operation is conducted between its learned and anchored local features. After that, hash codes are generated with exchanged features and the learning is driven by preserving pairwise similarities of hash codes $\boldsymbol{u}_i$ and $\boldsymbol{v}_j$.

for a sample $\boldsymbol{x}_i$ whose category is y_i, we exchange a half of the learned local features in $\mathcal{G}_i = \{\boldsymbol{f}_i^1, \ldots, \boldsymbol{f}_i^M\}$ with its corresponding anchored local features in $\mathcal{C}_{y_i} = \{\boldsymbol{c}_{y_i}^1, \ldots, \boldsymbol{c}_{y_i}^M\}$. The exchanging process can be formulated as:

$$\forall j \in \{1, \ldots, M\}, \hat{\boldsymbol{f}}_i^j \triangleq \begin{cases} \boldsymbol{f}_i^j, & \text{if } \xi_j \geq 0.5, \\ \boldsymbol{c}_{y_i}^j, & \text{otherwise,} \end{cases} \tag{8}$$

where $\xi_j \sim \mathcal{B}(0.5)$ is a random variable following the Bernoulli distribution for the j-th part. The local features after exchanging are denoted as $\hat{\mathcal{G}}_i = \{\hat{\boldsymbol{f}}_i^1, \ldots, \hat{\boldsymbol{f}}_i^M\}$ and fed into the hashing learning module for generating binary codes and computing similarity preservation losses.

3.3 Hash Code Learning

After obtaining both global features and local features, we concatenate them together and feed them into the hashing learning module. Specifically, the hashing network contains a fully connected layer and a $\texttt{sign}(\cdot)$ activation function layer. In our method, we choose an asymmetric hashing for ExchNet for its flexibility [25]. Concretely, we utilize two hash functions, defined as $g(\cdot)$ and $h(\cdot)$, to learn two different binary codes for the same training sample. The learning procedure is as follows:

$$\boldsymbol{u}_i = g([\hat{\mathcal{G}}_i; \boldsymbol{f}_i^{\text{global}}]_{\texttt{cat}}) = \texttt{sign}(\boldsymbol{W}^{(g)}[\hat{\mathcal{G}}_i; \boldsymbol{f}_i^{\text{global}}]_{\texttt{cat}}), \tag{9}$$

$$\boldsymbol{v}_i = h([\hat{\mathcal{G}}_i; \boldsymbol{f}_i^{\text{global}}]_{\texttt{cat}}) = \texttt{sign}(\boldsymbol{W}^{(h)}[\hat{\mathcal{G}}_i; \boldsymbol{f}_i^{\text{global}}]_{\texttt{cat}}), \tag{10}$$

where $[;]_{\text{cat}}$ denotes the concatenation operator, and $\boldsymbol{u}_i, \boldsymbol{v}_i \in \{-1, +1\}^q$ denote the two different binary codes of the i-th sample. q represents the code length. $\boldsymbol{W}^{(g)}$ and $\boldsymbol{W}^{(h)}$ present the parameters of hash functions $g(\cdot)$ and $h(\cdot)$[1], respectively. We denote $\boldsymbol{U} = \{\boldsymbol{u}_i\}_{i=1}^n$ and $\boldsymbol{V} = \{\boldsymbol{v}_i\}_{i=1}^n$ as learned binary codes. Inspired by [14], we only keep binary codes $\boldsymbol{v}_i$ and set hash function $h(\cdot)$ implicitly. Hence, we can perform feature learning and binary codes learning simultaneously.

To preserve the pairwise similarity, we adopt the squared loss and define the following objective function:

$$\mathcal{L}_{\texttt{sq}}(\boldsymbol{u}_i, \boldsymbol{v}_j, \boldsymbol{\mathcal{C}}) = \left(\boldsymbol{u}_i^\top \boldsymbol{v}_j - qS_{ij}\right)^2, \tag{11}$$

where $\boldsymbol{u}_i = g([\hat{\mathcal{G}}_i; \boldsymbol{f}_i^{\text{global}}]_{\texttt{cat}})$, S_{ij} is the pairwise similarity label and $\boldsymbol{\mathcal{C}} = \{\mathcal{C}_i\}_{i=1}^M$. We use Θ to denote the parameters of deep neural network and hash layer. The aforementioned process is generally illustrated by Fig. 4.

Due to the zero-gradient problem caused by the $\texttt{sign}(\cdot)$ function, $L_{\texttt{sq}}(\cdot,\cdot,\cdot)$ becomes intractable to optimize. In this paper, we relax $g(\cdot) = \texttt{sign}(\cdot)$ into $\hat{g}(\cdot) = \texttt{tanh}(\cdot)$ to alleviate this problem. Then, we can derive the following loss function:

$$\hat{\mathcal{L}}_{\texttt{sq}}(\hat{\boldsymbol{u}}_i, \boldsymbol{v}_j, \boldsymbol{\mathcal{C}}) = \left(\hat{\boldsymbol{u}}_i^\top \boldsymbol{v}_j - qS_{ij}\right)^2, \tag{12}$$

where $\hat{\boldsymbol{u}}_i = \hat{g}([\hat{\mathcal{G}}_i; \boldsymbol{f}_i^{\text{global}}]_{\texttt{cat}})$ and $\boldsymbol{U}$ is relaxed as $\hat{\boldsymbol{U}} = \{\hat{\boldsymbol{u}}_i\}_{i=1}^n$.

Then, given a set of image samples $\mathcal{X} = \{\boldsymbol{x}_1, \ldots, \boldsymbol{x}_n\}$ and their pairwise labels $\boldsymbol{S} = \{S_{ij}\}_{i,j=1}^n$, we can get the following objective function by combining Eqs. (5), (7) and (12):

$$\min_{\boldsymbol{V},\Theta,\boldsymbol{\mathcal{C}}} \mathcal{L}(\mathcal{X}) = \sum_{i,j=1}^{n} \hat{\mathcal{L}}_{\texttt{sq}}(\hat{\boldsymbol{u}}_i, \boldsymbol{v}_j; S_{ij}) + \lambda \sum_{i=1}^{n} \mathcal{L}_{\texttt{sp}}(\boldsymbol{x}_i) + \gamma \sum_{i=1}^{n} \mathcal{L}_{\texttt{cp}}(\boldsymbol{x}_i) \tag{13}$$

$$\texttt{s.t.} \forall i \in \{1, \ldots, n\}, \hat{\boldsymbol{u}}_i = \hat{g}([\hat{\mathcal{G}}_i; \boldsymbol{f}_i^{\text{global}}]_{\texttt{cat}}), \boldsymbol{v}_j \in \{-1, +1\}^q,$$

where S_{ij} represents the similarity between the i-th and j-th samples, q denotes the code length, λ and γ are hyper-parameters.

3.4 Learning Algorithm

To solve the optimization problem in Eq. (13), we design an alternating algorithm to learn $\boldsymbol{V}$, Θ, and $\boldsymbol{\mathcal{C}}$. Specifically, we learn one parameter with the others fixed.

[1] We omit the bias term for simplicity.

Learn Θ with $\boldsymbol{V}$ and $\mathcal{C}$ Fixed. When $\boldsymbol{V}, \mathcal{C}$ fixed, we use back-propagation (BP) to update the parameters Θ. In particular, for input sample $\boldsymbol{x}_i$, we first calculate the following gradient:

$$\nabla_{\Theta}\mathcal{L}(\boldsymbol{X}) = \sum_{i,j=1}^{n} \nabla_{\Theta}\mathcal{L}_{\text{sq}}(\hat{\boldsymbol{u}}_i, \boldsymbol{v}_j) + \lambda \sum_{i=1}^{n} \nabla_{\Theta}\mathcal{L}_{\text{sp}}(\boldsymbol{x}_i) + \gamma \sum_{i=1}^{n} \nabla_{\Theta}\mathcal{L}_{\text{cp}}(\boldsymbol{x}_i). \quad (14)$$

Then, we use the back-propagation algorithm to update Θ.

Learn $\boldsymbol{V}$ with Θ and $\mathcal{C}$ Fixed. When Θ, $\mathcal{C}$ are fixed, we rewrite $\mathcal{L}(\boldsymbol{V})$ as follows:

$$\mathcal{L}(\boldsymbol{V}) = \sum_{i,j=1}^{n} \left(\hat{\boldsymbol{u}}_i^{\top}\boldsymbol{v}_j - qS_{ij}\right)^2 = \|\widetilde{\boldsymbol{U}}\boldsymbol{V}^{\top} - q\boldsymbol{S}\|_F^2 \quad (15)$$

$$= \|\widetilde{\boldsymbol{U}}\boldsymbol{V}^{\top}\|_F^2 - 2q\text{tr}(\boldsymbol{S}^{\top}\widetilde{\boldsymbol{U}}\boldsymbol{V}^{\top}) + \texttt{const}. \quad (16)$$

Because $\boldsymbol{V}$ is defined over $\{-1,+1\}^{n\times q}$, we learn $\boldsymbol{V}$ column by column as that in ADSH [14]. Specifically, we can get the closed-form solution for the k-th column $\boldsymbol{V}_{*k}$ as follows:

$$\boldsymbol{V}_{*k} = \texttt{sign}(\boldsymbol{V}_{/k}\widetilde{\boldsymbol{U}}_{/k}^{\top}\widetilde{\boldsymbol{U}}_{*k} - q\boldsymbol{Q}_{*k}), \quad (17)$$

where $\boldsymbol{Q} = \boldsymbol{S}^{\top}\widetilde{\boldsymbol{U}}$ and $\boldsymbol{V}_{/k}$ denotes the matrix excluding the k-th column.

Learn $\mathcal{C}$ with $\boldsymbol{V}$ and Θ Fixed. When Θ, $\boldsymbol{V}$ fixed, we use the following equation to update each $\mathcal{C}_i \in \mathcal{C}$:

$$\forall k, \boldsymbol{c}_i^k = \frac{1}{n_i}\sum_{i=1}^{n_i} \boldsymbol{f}_i^k, \quad (18)$$

where n_i denotes the number of samples in class y_i.

3.5 Out-of-Sample Extension

When we finish the training phase, we can generate the binary code for the sample $\boldsymbol{x}_i$ by $\boldsymbol{u}_i = \texttt{sign}(\boldsymbol{W}^{(g)}[\mathcal{G}_i; \boldsymbol{f}_i^{\text{global}}]_{\texttt{cat}})$.

4 Experiments

4.1 Datasets

For comparisons, we select two widely used fine-grained datasets, i.e., *CUB* [29] and *Aircraft* [24], as well as three popular large-scale fine-grained datasets, i.e., *NABirds* [11], *VegFru* [12], and *Food101* [23], to conduct experiments.

Specifically, *CUB* is a bird classification benchmark dataset containing 11,788 images from 200 bird species. It is officially split into 5,994 for training and 5,794 for test. *Aircraft* contains 10,000 images from 100 kinds of aircraft model variants with 6667 for training and 3333 for test. Moreover, for large-scale datasets, *NABirds* has 555 common species of birds in North America with 23,929 training images and 24,633 test images. *VegFru* is a large-scale fine-grained dataset covering vegetables and fruits from 292 categories with 29,200 for training and 116,931 for test. *Food101* contains 101 kinds of foods with 101,000 images. For each class, 250 test images are manually reviewed for correctness while 750 training images still contain some amount of noises.

4.2 Baselines and Implementation Details

Baselines. For comparisons with other ANN algorithms, we select two tree-based ANN methods, i.e., BallTree [8] and KDTree [1], and one production quantization based ANN method, i.e., Product Quantization (PQ) [13]. The linear scan means that we directly perform exhaustive search based on the learned real-valued features. For comparisons with other hashing baselines, we choose eight state-of-the-art generic hashing methods. They are LSH [6], SH [34], ITQ [10], SDH [28], DPSH [17], DSH [22], HashNet [4], and ADSH [14]. Among these methods, DPSH, DSH, HashNet and ADSH are based on deep learning and others are not.

Implementation Details. For comparisons with other ANN algorithms, we carry out experiments on *Food101* in which the database is the largest. We first utilize the triplet loss [27] to learn 512-D and 1024-D feature embeddings for its frequent usages in fine-grained retrieval tasks. Then, the performance of linear scan is tested on the learned features. More experimental settings about BallTree [8], KDTree [1] and PQ [13] can be found in the supplementary materials. For our ExchNet, the retrieval procedure is divided into coarse ranking to select top N as candidates and re-ranking to return top K ($K < N$) from top N candidates. We adopt the real-valued features learned with the triplet loss directly. As presented in Table 1, we report results including precision at top K (P@K), wall clock time (WC time), speed up ratio, and memory cost.

Our backbone employs the first three stages of ResNet50 and the attention generation module is the fourth stage of ResNet50 without downsample convolutions. The LFR and GFR of ExchNet are independent networks, sharing the same architecture with the fourth stage of ResNet50. The optimizer is standard mini-batch stochastic gradient descent with weight decay 1×10^{-4}. The mini-batch size M is set to 64 and the iteration times T_{max} is 100. Learning rate is set to 0.001, which is divided by 10 at the 60-th and 80-th iteration, respectively. The hyper-parameter t is set to 0.4. The number of training epochs is 20. For efficient training, we randomly sample a subset of the training set in each epoch. Specifically, for *CUB*, *Aircraft*, *Food101*, we sample 2,000 samples per epoch, while 4,000 samples are randomly selected for other datasets. To provide reliable local features for our local feature alignment strategy, in the first 50 iterations, since both local and global features are not well learned, the part-level feature exchanging operation is disabled for avoiding aligning meaningless local features.

Table 1. Retrieval performance comparisons on the *Food101* dataset.

Method	512-dim				1024-dim			
	P@10 (↑)	WCtime (↓)	Speedup (↑)	Memory (↓)	P@10 (↑)	WCtime (↓)	Speedup (↑)	Memory (↓)
Linear	80.05%	9,481.03	1×	207.2 MB	80.28%	22,377.96	1×	414.1 MB
BallTree	77.22%	236.23	40.13×	28.1 MB	77.74%	213.88	104.62×	28.1 MB
KDTree	77.42%	70.16	135.13×	28.8 MB	77.73%	73.57	304.14×	28.7 MB
PQ	77.12%	43.49	217.99×	524.5 KB	77.18%	72.47	308.74×	1.0 MB
Ours	**77.69%**	**40.54**	**233.85×**	**404.0 KB**	**78.06%**	**56.57**	**395.53×**	**404.0 KB**

Table 2. Comparisons of retrieval accuracy (MAP) on all the fine-grained datasets.

Method	#Bits	LSH	SH	ITQ	SDH	DPSH	DSH	HashNet	ADSH	**Ours**
CUB	12bits	2.26%	5.55%	6.80%	10.52%	8.68%	4.48%	12.03%	20.03%	**25.14%**
	24bits	3.59%	6.72%	9.42%	16.95%	12.51%	7.97%	17.77%	50.33%	**58.98%**
	32bits	5.01%	7.63%	11.19%	20.43%	12.74%	7.72%	19.93%	61.68%	**67.74%**
	48bits	6.16%	8.32%	12.45%	22.23%	15.58%	11.81%	22.13%	65.43%	**71.05%**
Aircraft	12bits	1.69%	3.28%	4.38%	4.89%	8.74%	8.14%	14.91%	15.54%	**33.27%**
	24bits	2.19%	3.85%	5.28%	6.36%	10.87%	10.66%	17.75%	23.09%	**45.83%**
	32bits	2.38%	4.04%	5.82%	6.90%	13.54%	12.21%	19.42%	30.37%	**51.83%**
	48bits	2.82%	4.28%	6.05%	7.65%	13.94%	14.45%	20.32%	50.65%	**59.05%**
NABirds	12bits	0.90%	2.12%	2.53%	3.10%	2.17%	1.56%	2.34%	2.53%	**5.22%**
	24bits	1.68%	3.14%	4.22%	6.72%	4.08%	2.33%	3.29%	8.23%	**15.69%**
	32bits	2.43%	3.71%	5.38%	8.86%	3.61%	2.44%	4.52%	14.71%	**21.94%**
	48bits	3.09%	4.05%	6.10%	10.38%	3.20%	3.42%	4.97%	25.34%	**34.81%**
VegFru	12bits	1.28%	2.36%	3.05%	5.92%	6.33%	4.60%	3.70%	8.24%	**23.55%**
	24bits	2.21%	4.04%	5.51%	11.55%	9.05%	8.91%	6.24%	24.90%	**35.93%**
	32bits	3.39%	5.65%	7.48%	14.55%	10.28%	11.23%	7.83%	36.53%	**48.27%**
	48bits	4.51%	6.56%	8.74%	16.45%	9.11%	17.12%	10.29%	55.15%	**69.30%**
Food101	12bits	1.57%	4.51%	6.46%	10.21%	11.82%	6.51%	24.42%	35.64%	**45.63%**
	24bits	2.48%	5.79%	8.20%	11.44%	13.05%	8.97%	34.48%	40.93%	**55.48%**
	32bits	2.64%	5.91%	9.70%	13.36%	16.41%	13.10%	35.90%	42.89%	**56.39%**
	48bits	3.07%	6.63%	10.07%	15.55%	20.06%	17.18%	39.65%	48.81%	**64.19%**

4.3 Comparisons with Other ANN Methods

To prove the practicality and effectiveness of our proposed method, comparisons with other ANN methods are presented in this section. All experiments are conducted based on hash codes of 32bits generated by our model.

In Table 1, we present the retrieval performance on the *Food101* dataset. Specifically, we present the P@10, WC time, speedup, and memory cost for all methods. We can observe that, compared with the linear search, our method can achieve up to 233× and 395× acceleration on features of 512-D and 1024-D, respectively. The memory cost of our method is also much less than tree-based methods. The best speed-up and the lowest storage usage prove the practicality of our proposed method. Meanwhile, our method can achieve state-of-the-art retrieval accuracies, which demonstrates that our ExchNet is the most effective one compared with other ANN methods. Above results illustrate our ExchNet deserves to be the optimal choice for fine-grained image retrieval.

4.4 Comparisons with State-of-the-Art Hashing Methods

In Table 2, we present the mean average precision (MAP) results for comparisons with state-of-the-art hashing methods on all datasets. From Table 2, we can observe that our method can achieve the best retrieval performance in all cases. On fine-grained datasets (*CUB* and *Aircraft*) of relatively small size, almost all the generic hashing methods (except for ADSH) can not achieve a satisfactory performance, i.e., a relatively low MAP. Also, our ExchNet outperforms the most powerful baseline ADSH considerably. It can verify that given limited training data, our proposed method could still perform well. As to large-scale fine-grained datasets, the improvements become more significant. Particularly, comparing with the most powerful baselines, we achieve 12% and 14% MAP improvements on the 32 bits and 48 bits evaluation experiments of the large-scale *VegFru* dataset. Meanwhile, we achieve 14% and 16% MAP improvements on the 32 bits and 48 bits experiments of the *Food101* dataset. It shows that, with sufficient training data, we can get better retrieval results with our ExchNet on large-scale fine-grained datasets.

4.5 Ablation Studies

Effectiveness of the Exchanging-Based Feature Alignment. We verify the effectiveness of the local feature alignment approach (cf. Sect. 3.2) in this section. The retrieval accuracy are present in Fig. 5, where "Ours w/o Exchange" means that we do not perform the feature exchanging operation (i.e., the local feature alignment) during training. Note that "Ours w/o Exchange" is degenerated to the ADSH [14] learned with our proposed representation learning architecture instead of ResNet50. Hence, we also present the results of ADSH.

It can be observed that our method can achieve the best accuracy thanks to the feature exchanging operation. Specifically, on *CUB* and *Aircraft* datasets, our proposed method with the exchanging operation performs considerably better than that without exchanging. The performance improvement on the large-scale fine-grained datasets (e.g., *Food101*) becomes more significant. Above results illustrate that our proposed local features alignment strategy is effective, especially on large-scale datasets. Moreover, even if bits of hash codes are limited, our feature alignment strategy could still benefit fine-grained retrieval greatly.

Sensitivity to Hyper Parameter M. In our ExchNet, we use M to denote the number of local features, which is also the number of attention maps. In this section, we present the influence of the hyper-parameter M by ablation studies.

As presented in Fig. 6, we vary M as 2, 4 and 6. From that figure, it is observed that satisfactory retrieval accuracies are achieved regardless of different M values, and the best fine-grained retrieval accuracy is obtained when $M = 4$. As analyzed, redundant local features (i.e., overmuch object parts when M is large) might cause redundancies in local feature representations, while the lack of local features (i.e., scant object parts when M is small) may result in that fine-grained images are under-represented for distinguishing subtle visual differences.

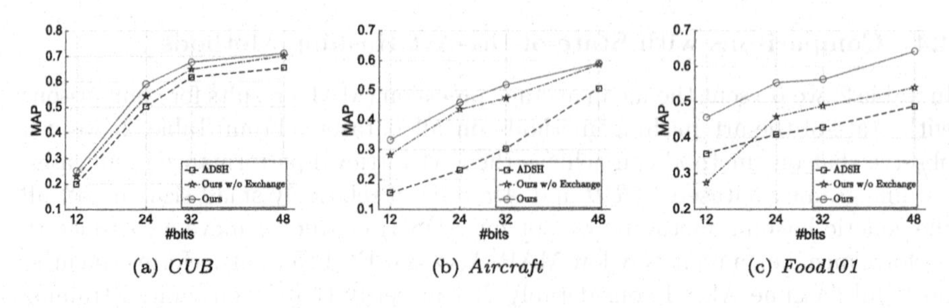

(a) *CUB* (b) *Aircraft* (c) *Food101*

Fig. 5. Effectiveness of our feature exchanging operation.

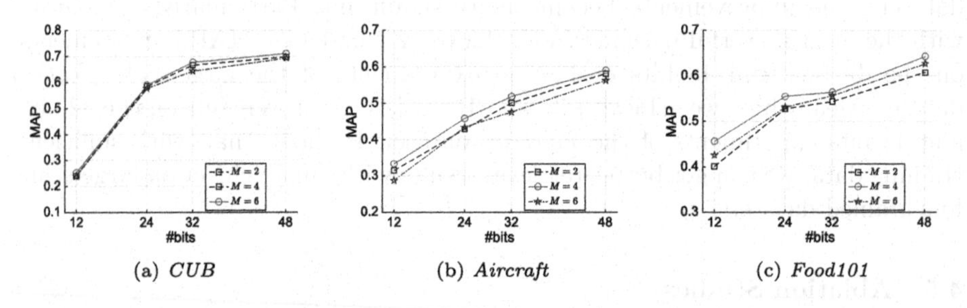

(a) *CUB* (b) *Aircraft* (c) *Food101*

Fig. 6. Influence of hyper parameter M which denotes the number of local features.

Those might be the reasons why M is too small or large will cause slightly accuracy drops. Moreover, comparable retrieval results of different M values show that our ExchNet is not sensitive to M.

5 Conclusions

In this paper, we studied the practical but challenging fine-grained hashing task, which aims to solve large-scale FGIR problems by leveraging the search and storage efficiency of compact hash codes. Specifically, we proposed a unified network ExchNet to obtain representative fine-grained local and global features by performing our attention approach equipped with the tailored attention constraints. Then, ExchNet utilized its local feature alignment to align these local features to their corresponding object parts across images. Later, an alternating learning algorithm was employed to return the final fine-grained binary codes. Compared with ANN methods and competing generic hash methods, experiments validated both effectiveness and efficiency of our ExchNet. In the future, we would like to explore a more challenging unsupervised fine-grained hashing topic.

Acknowledgements. Quan Cui's contribution was made when he was an intern at Megvii Research Nanjing. This research was supported by the National Key Research and Development Program of China under Grant 2017YFA0700800 and "111" Program B13022. Qing-Yuan Jiang and Wu-Jun Li were supported by the NSFC-NRF Joint Research Project (No. 61861146001).

References

1. Bentley, J.L.: Multidimensional binary search trees used for associative searching. ACM Commun. **18**(9), 509–517 (1975)
2. Cakir, F., He, K., Sclaroff, S.: Hashing with binary matrix pursuit. In: Ferrari, V., Hebert, M., Sminchisescu, C., Weiss, Y. (eds.) ECCV 2018. LNCS, vol. 11209, pp. 344–361. Springer, Cham (2018). https://doi.org/10.1007/978-3-030-01228-1_21
3. Cao, Y., Long, M., Liu, B., Wang, J.: Deep cauchy hashing for hamming space retrieval. In: CVPR, pp. 1229–1237 (2018)
4. Cao, Z., Long, M., Wang, J., Yu, P.S.: HashNet: deep learning to hash by continuation. In: ICCV, pp. 5609–5618 (2017)
5. Chen, J., Wang, Y., Qin, J., Liu, L., Shao, L.: Fast person re-identification via cross-camera semantic binary transformation. In: CVPR, pp. 5330–5339 (2017)
6. Datar, M., Immorlica, N., Indyk, P., Mirrokni, V.S.: Locality-sensitive hashing scheme based on p-stable distributions. In: SoCG, pp. 253–262 (2004)
7. Dizaji, K.G., Zheng, F., Sadoughi, N., Yang, Y., Deng, C., Huang, H.: Unsupervised deep generative adversarial hashing network. In: CVPR, pp. 3664–3673 (2018)
8. Dolatshah, M., Hadian, A., Minaei-Bidgoli, B.: Ball*-tree: efficient spatial indexing for constrained nearest-neighbor search in metric spaces. CoRR abs/1511.00628 (2015)
9. Fu, J., Zheng, H., Mei, T.: Look closer to see better: recurrent attention convolutional neural network for fine-grained image recognition. In: CVPR, pp. 4438–4446 (2017)
10. Gong, Y., Lazebnik, S.: Iterative quantization: a procrustean approach to learning binary codes. In: CVPR, pp. 817–824 (2011)
11. Horn, G.V., et al.: Building a bird recognition app and large scale dataset with citizen scientists: the fine print in fine-grained dataset collection. In: CVPR, pp. 595–604 (2015)
12. Hou, S., Feng, Y., Wang, Z.: VegFru: a domain-specific dataset for fine-grained visual categorization. In: ICCV, pp. 541–549 (2017)
13. Jégou, H., Douze, M., Schmid, C.: Product quantization for nearest neighbor search. IEEE TPAMI **33**(1), 117–128 (2011)
14. Jiang, Q.Y., Li, W.J.: Asymmetric deep supervised hashing. In: AAAI, pp. 3342–3349 (2018)
15. Li, P., Xie, J., Wang, Q., Gao, Z.: Towards faster training of global covariance pooling networks by iterative matrix square root normalization. In: CVPR, pp. 947–955 (2018)
16. Li, Q., Sun, Z., He, R., Tan, T.: Deep supervised discrete hashing. In: NeurIPS, pp. 2482–2491 (2017)
17. Li, W.J., Wang, S., Kang, W.C.: Feature learning based deep supervised hashing with pairwise labels. In: IJCAI, pp. 1711–1717 (2016)
18. Lin, J., Li, Z., Tang, J.: Discriminative deep hashing for scalable face image retrieval. In: IJCAI, pp. 2266–2272 (2017)
19. Lin, K., Yang, F., Wang, Q., Piramuthu, R.: Adversarial learning for fine-grained image search. In: ICME, pp. 490–495 (2019)
20. Lin, T.Y., RoyChowdhury, A., Maji, S.: Bilinear CNN models for fine-grained visual recognition. In: CVPR, pp. 1449–1457 (2015)
21. Liong, V.E., Lu, J., Wang, G., Moulin, P., Zhou, J.: Deep hashing for compact binary codes learning. In: CVPR, pp. 2475–2483 (2015)

22. Liu, H., Wang, R., Shan, S., Chen, X.: Deep supervised hashing for fast image retrieval. In: CVPR, pp. 2064–2072 (2016)
23. Bossard, L., Guillaumin, M., Van Gool, L.: Food-101 – mining discriminative components with random forests. In: Fleet, D., Pajdla, T., Schiele, B., Tuytelaars, T. (eds.) ECCV 2014. LNCS, vol. 8694, pp. 446–461. Springer, Cham (2014). https://doi.org/10.1007/978-3-319-10599-4_29
24. Maji, S., Rahtu, E., Kannala, J., Blaschko, M.B., Vedaldi, A.: Fine-grained visual classification of aircraft. CoRR abs/1306.5151 (2013)
25. Neyshabur, B., Srebro, N., Salakhutdinov, R.R., Makarychev, Y., Yadollahpour, P.: The power of asymmetry in binary hashing. In: NeurIPS, pp. 2823–2831 (2013)
26. Pang, C., Li, H., Cherian, A., Yao, H.: Part-based fine-grained bird image retrieval respecting species correlation. In: ICIP, pp. 2896–2900 (2017)
27. Schroff, F., Kalenichenko, D., Philbin, J.: FaceNet: a unified embedding for face recognition and clustering. In: CVPR, pp. 815–823 (2015)
28. Shen, F., Shen, C., Liu, W., Shen, H.T.: Supervised discrete hashing. In: CVPR, pp. 37–45 (2015)
29. Wah, C., Branson, S., Welinder, P., Perona, P., Belongie, S.: The caltech-UCSD birds-200-2011 dataset (2011)
30. Wang, G., Hu, Q., Cheng, J., Hou, Z.: Semi-supervised generative adversarial hashing for image retrieval. In: Ferrari, V., Hebert, M., Sminchisescu, C., Weiss, Y. (eds.) ECCV 2018. LNCS, vol. 11219, pp. 491–507. Springer, Cham (2018). https://doi.org/10.1007/978-3-030-01267-0_29
31. Wei, X.S., Luo, J.H., Wu, J., Zhou, Z.H.: Selective convolutional descriptor aggregation for fine-grained image retrieval. IEEE TIP **26**(6), 2868–2881 (2017)
32. Wei, X.S., Wang, P., Liu, L., Shen, C., Wu, J.: Piecewise classifier mappings: learning fine-grained learners for novel categories with few examples. IEEE TIP **28**(12), 6116–6125 (2019)
33. Wei, X.S., Xie, C.W., Wu, J., Shen, C.: Mask-CNN: localizing parts and selecting descriptors for fine-grained bird species categorization. Pattern Recogn. **76**, 704–714 (2018)
34. Weiss, Y., Torralba, A., Fergus, R.: Spectral hashing. In: NeurIPS, pp. 1753–1760 (2008)
35. Xia, R., Pan, Y., Lai, H., Liu, C., Yan, S.: Supervised hashing for image retrieval via image representation learning. In: AAAI, pp. 2156–2162 (2014)
36. Xie, L., Wang, J., Zhang, B., Tian, Q.: Fine-grained image search. IEEE Trans. Multimed. **17**(5), 636–647 (2015)
37. Yang, Z., Luo, T., Wang, D., Hu, Z., Gao, J., Wang, L.: Learning to navigate for fine-grained classification. In: Ferrari, V., Hebert, M., Sminchisescu, C., Weiss, Y. (eds.) Computer Vision – ECCV 2018. LNCS, vol. 11218, pp. 438–454. Springer, Cham (2018). https://doi.org/10.1007/978-3-030-01264-9_26
38. Yuan, X., Ren, L., Lu, J., Zhou, J.: Relaxation-free deep hashing via policy gradient. In: Ferrari, V., Hebert, M., Sminchisescu, C., Weiss, Y. (eds.) ECCV 2018. LNCS, vol. 11208, pp. 141–157. Springer, Cham (2018). https://doi.org/10.1007/978-3-030-01225-0_9
39. Zhang, J., et al.: Generative domain-migration hashing for sketch-to-image retrieval. In: Ferrari, V., Hebert, M., Sminchisescu, C., Weiss, Y. (eds.) ECCV 2018. LNCS, vol. 11206, pp. 304–321. Springer, Cham (2018). https://doi.org/10.1007/978-3-030-01216-8_19
40. Zheng, H., Fu, J., Mei, T., Luo, J.: Learning multi-attention convolutional neural network for fine-grained image recognition. In: CVPR, pp. 5209–5217 (2017)

41. Zheng, X., Ji, R., Sun, X., Wu, Y., Huang, F., Yang, Y.: Centralized ranking loss with weakly supervised localization for fine-grained object retrieval. In: IJCAI, pp. 1226–1233 (2018)
42. Zheng, X., Ji, R., Sun, X., Zhang, B., Wu, Y., Huang, F.: Towards optimal fine grained retrieval via decorrelated centralized loss with normalize-scale layer. In: AAAI, vol. 33, pp. 9291–9298 (2019)
43. Zhu, F., Kong, X., Zheng, L., Fu, H., Tian, Q.: Part-based deep hashing for large-scale person re-identification. IEEE TIP **26**(10), 4806–4817 (2017)

TSIT: A Simple and Versatile Framework for Image-to-Image Translation

Liming Jiang[1], Changxu Zhang[2], Mingyang Huang[3], Chunxiao Liu[3], Jianping Shi[3], and Chen Change Loy[1(✉)]

[1] Nanyang Technological University, Singapore, Singapore
{liming002,ccloy}@ntu.edu.sg
[2] University of California, Berkeley, CA, USA
zhangcx@berkeley.edu
[3] SenseTime Research, Beijing, China
{huangmingyang,liuchunxiao,shijianping}@sensetime.com

Abstract. We introduce a simple and versatile framework for image-to-image translation. We unearth the importance of normalization layers, and provide a carefully designed two-stream generative model with newly proposed feature transformations in a coarse-to-fine fashion. This allows multi-scale semantic structure information and style representation to be effectively captured and fused by the network, permitting our method to scale to various tasks in both unsupervised and supervised settings. No additional constraints (*e.g.*, cycle consistency) are needed, contributing to a very clean and simple method. Multi-modal image synthesis with arbitrary style control is made possible. A systematic study compares the proposed method with several state-of-the-art task-specific baselines, verifying its effectiveness in both perceptual quality and quantitative evaluations. GitHub: https://github.com/EndlessSora/TSIT.

1 Introduction

Image-to-image translation [16] aims at translating one image representation to another. Recent advances [10,21,22,30,32], especially Generative Adversarial Networks (GANs) [10], have made remarkable success in various image-to-image translation tasks. Previous studies usually present specialized solutions for a specific form of application, ranging from arbitrary style transfer [13,14,24,27, 44,49,53] in the unsupervised setting, to semantic image synthesis [4,16,28,33, 34,41] in the supervised setting.

In this study, we are interested in devising a general and unified framework that is applicable to different image-to-image translation tasks without degradation in synthesis quality. This is *non-trivial* given the different natures of different tasks. For instance, in certain conditional image synthesis tasks (*e.g.*, arbitrary

Electronic supplementary material The online version of this chapter (https://doi.org/10.1007/978-3-030-58580-8_13) contains supplementary material, which is available to authorized users.

A. Vedaldi et al. (Eds.): ECCV 2020, LNCS 12348, pp. 206–222, 2020.
https://doi.org/10.1007/978-3-030-58580-8_13

Fig. 1. Our framework is simple and versatile for various image-to-image translation tasks. For unsupervised arbitrary style transfer, diverse scenarios (*e.g.*, natural images, real-world scenes, artistic paintings) can be handled. For supervised semantic image synthesis, our method is robust to different scenes (*e.g.*, outdoor, street scene, indoor). Multi-modal image synthesis is feasible by a *single* model with controllable styles.

style transfer), paired data are usually not available. Under this unsupervised setting, translation task demands additional constraints on cycle consistency [19,27,44,53], semantic features [39], pixel gradients [1], or pixel values [36]. In semantic image synthesis (*i.e.*, translation from segmentation labels to images), training pairs are available. This task is more data-dependent and typically needs losses to minimize per-pixel distance between the generated sample and ground truth. In addition, specialized structures [4,28,33,41] are required to maintain spatial coherence and resolution. Due to the different needs, existing methods exploit their own specially designed components. It is difficult to cross-use these components or integrate them into a unified framework.

To address the aforementioned challenges, we propose a Two-Stream Image-to-image Translation (TSIT) framework, which is *versatile* for various image-to-image translation tasks (see Fig. 1). The framework is simple as it is based purely on feature transformation. Unlike previous approaches [13,33] that only consider either semantic structure or style representation, we factorize *both* the structure and style in multi-scale *feature levels* via a symmetrical *two-stream* network. The two streams jointly influence the new image generation in a coarse-to-fine manner via a consistent feature transformation scheme. Specifically, the content spatial structure is preserved by an element-wise feature adaptive denormalization (FADE) from the content stream, while the style information is exerted by feature adaptive instance normalization (FAdaIN) from the style stream. Standard loss functions such as adversarial loss and perceptual loss are used, without additional constraints like cycle consistency. The pipeline is applicable to both unsupervised and supervised settings, easing the preparation of data.

The **contributions** of our work are summarized as follows. We propose TSIT, a simple and versatile framework, which is effective for various image-to-image translation tasks. Despite the succinct design, our network is readily adaptable to various tasks and achieves compelling results. The good performance is achieved by (1) *multi-scale* feature normalization (FADE and FAdaIN) scheme that captures *coarse-to-fine* structure and style information, and (2) a *two-stream* network design that integrates *both* content and style effectively, reducing artifacts and making multi-modal image synthesis possible (see Fig. 1). In comparison to several state-of-the-art task-specific baselines [4,14,28,33,34,41,49], our method achieves comparable or even better results in both perceptual quality and quantitative evaluations.

2 Related Work

Image-to-Image Translation. Existing methods can be classified into two categories: unsupervised and supervised. With only unpaired data, unsupervised image-to-image translation problem is inherently ill-posed. Additional constraints are needed on *e.g.*, cycle consistency [19,27,44,53], semantic features [39], pixel gradients [1], or pixel values [36]. In contrast, supervised methods, such as `pix2pix` [16], are more data-dependent, requiring well-annotated paired training samples. Subsequent approaches [4,28,33,34,41] extend the supervised problem for generating high-resolution images or keeping effective semantic meaning.

Limited by learning only one-to-one mapping between two domains, some of the GAN-based methods [19,27,44,53] suffer from generating images with low diversity. Recent studies explore more deeply into both multi-domain translation [6,26] and multi-modal translation [14,24,48], significantly increasing generation diversity. MUNIT [14] is a representative method that disentangles domain-invariant content and domain-specific style representation, enriching the synthesized images. Multi-mapping translation is defined in a very recent work, DMIT [49], which is designed to capture multi-modal image nature in each domain.

Existing image-to-image translation methods lack the scalability to adapt to different tasks under diverse difficult settings. Different demands of unsupervised and supervised settings oblige previous methods to exploit customized modules. Cross-using these components will be suboptimal due to either degradation in quality or introduction of additional constraints. It is non-trivial to integrate them into a single framework and improve robustness. In this study, we design a two-stream network with newly proposed feature transformations inspired by [33] and [13]. Our method is succinct yet able to link various tasks.

Arbitrary Style Transfer. Style transfer is closely relevant to image-to-image translation in the unsupervised setting. Style transfer aims at retaining the content structure of an image, while manipulating its style representation adopted from other images. Classical methods [3,8,9,17] gradually improve this task from optimization-based to real-time, allowing multiple style transfer during inference. Huang *et al.* introduce AdaIN [13], an effective normalization strategy for arbitrary style transfer. Several studies [5,23,29,38,43,45,51] improve stylization via wavelet transforms [45], graph cuts [51], or iterative error-correction [38]. Besides, most collection-guided [14] style transfer methods are GAN-based [14,24,27,44,49,53], showing impressive results.

Previous works usually consider either content or style information. In contrast, our framework succeeds in seeking a balance between content and style, and adaptively fuses them well. The proposed method achieves user-controllable multi-modal style manipulation by only a *single* model. Compared to customized style transfer methods, our approach achieves better synthesis quality in many scenarios including natural images, real-world scenes, and artistic paintings.

Semantic Image Synthesis. We define semantic image synthesis as in [33], aiming at synthesizing a photorealistic image from a semantic segmentation mask. Semantic image synthesis is a special form of supervised image-to-image translation. The domain gap of this task is large. Therefore, keeping effective semantic information to enhance fidelity without losing diversity is challenging.

`Pix2pix` [16] first adopts conditional GAN [30] in the semantic image synthesis task. Pix2pixHD [41] contains a multi-scale generator and multi-scale discriminators to generate high-resolution images. SPADE [33] takes a noise map as input, and resizes the semantic label map for modulating the activations in normalization layers by a learned affine transformation. CC-FPSE [28] employs a weight prediction network for generator. A semantics-embedding discriminator is used to enhance fine details and semantic alignments between the generated samples and the input semantic layouts. In addition to these GAN-based methods, CRN [4] applies a cascaded refinement network with regression loss as the supervision. SIMS [34] is a semi-parametric method, retrieving fragments from a memory bank and refining the canvas by a refinement network.

Different from prior works, we design a symmetrical two-stream framework. The network learns feature-level semantic structure information and style representation instead of directly resizing the input mask like SPADE [33]. Coarse-to-fine feature representations are learned by neural networks, adaptively keeping high fidelity without diminishing diversity.

3 Methodology

We consider three key requirements in formulating a robust and scalable method to link various tasks: (1) *Both* semantic structure information and style representation should be considered and fused adaptively. (2) The content and style information should be learned by networks in *feature level* instead of in image level to fit the nature of diverse semantic tasks. (3) The network structure and loss functions should be *simple* for easy training without additional constraints.

3.1 Network Structure

Based on the aforementioned considerations, we design a Two-Stream Image-to-image Translation (TSIT) framework, as illustrated in Fig. 2. TSIT consists of four components: content stream, style stream, generator, and discriminators (omitted in Fig. 2). The first three main components are fully convolutional and symmetrically designed. The details of the submodules, including content/style residual block, FADE residual block, FADE module in the FADE residual block, are as shown in Fig. 3. We will discuss them separately in this section.

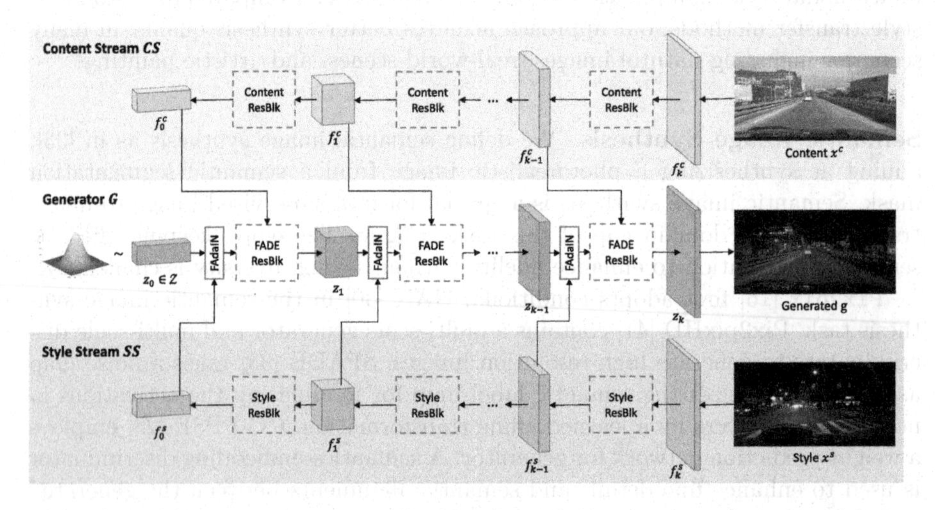

Fig. 2. The proposed Two-Stream Image-to-image Translation (TSIT) framework. The multi-scale patch-based discriminators are omitted. A Gaussian noise map is taken as the latent input for the generator. The feature representations of the content and style images are extracted by the corresponding streams for multi-scale feature transformations. The symmetrical networks fuses semantic structure and style representation in an end-to-end training. Submodules of our network are shown in Fig. 3.

Content/Style Stream. Unlike the traditional conditional GAN [30], we place the two-stream networks, *i.e.*, content stream and style stream, on each side of the generator (see Fig. 2). These two streams are symmetrical with the same network structure, aiming at extracting corresponding feature representations in different levels. We construct content/style stream based on standard residual blocks [11]. We call them content/style residual blocks. As shown in Fig. 3(a), each block has three convolutional layers, one of which is designed for the learned skip connection. The activation function is Leaky ReLU. The function of content/style stream is to extract features and feed them to the corresponding feature transformation layers in the generator. Multi-scale content/style representation in *feature levels* can be learned by content/style stream, adaptively fitting different feature transformations.

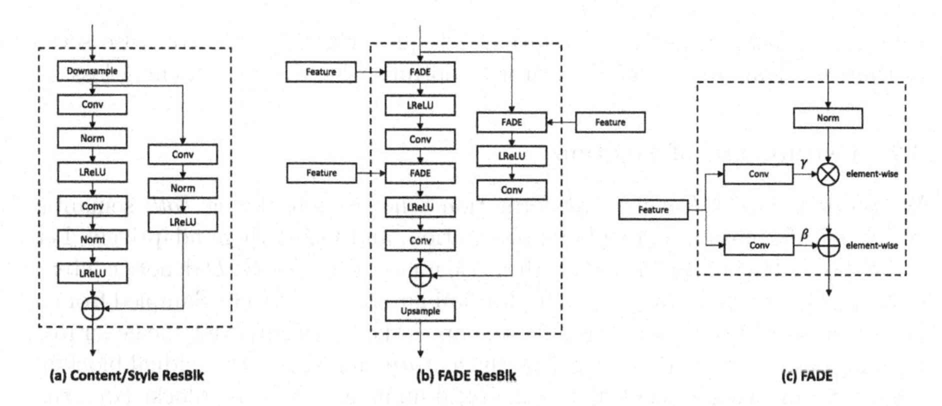

Fig. 3. Submodules of our framework. (a) is a content/style residual block in the symmetrical content/style streams. (b) is a FADE residual block in the generator. (c) is a FADE module in the FADE residual block. It performs *element-wise* denormalization by modulating the normalized activation using a learned affine transformation defined by the modulation parameters γ and β.

Generator. The generator has a completely inverse structure *w.r.t.* the content/style stream. This is intentionally designed to consistently match the level of semantic abstraction at different feature scales. A noise map is sampled from a Gaussian distribution as the latent input, and the feature maps from corresponding layers in content/style stream are taken as multi-scale feature inputs. The proposed feature transformations are implemented by a FADE residual block (Fig. 3(b)) and a FAdaIN module. In the FADE residual block, we use an inverse architecture *w.r.t.* the content/style residual block and replace the batch normalization [15] layer with the FADE module (Fig. 3(c)). The FADE module performs *element-wise* denormalization by modulating the normalized activation using a learned affine transformation defined by the modulation parameters γ and β. The FAdaIN module is used to exert style information through feature adaptive instance normalization. More discussions are given in Sect. 3.2.

The entire image generation process is performed in a coarse-to-fine manner. In particular, multi-scale content/style features are injected to refine the generated image constantly from high-level latent code to low-level image representation. Semantic structure and style information are learnable and effectively fused in an end-to-end training.

Discriminators. We exploit the standard multi-scale patch-based discriminators (omitted in Fig. 2) in [33,41]. Three regular discriminators with an identical architecture are included to discriminate images at different scales. Despite the same structure, patch-based training allows the discriminator operating at the coarsest scale to have the largest receptive field, capturing global information of the image. Whereas the one operating at the finest scale has the smallest receptive field, making the generator produce better details. Multi-scale patch-based discriminators further improve the robustness of our method for image-to-image translation tasks in different resolutions. Besides, the discriminators also serve as feature extractors for the generator to optimize the feature matching loss.

3.2 Feature Transformation

We propose a new feature transformation scheme, considering *both* semantic structure information and style representation, and fusing them adaptively. Let x^c be the content image and x^s be the style image. CS, SS, G, D denote content stream, style stream, generator, and discriminators, respectively. Sampled from a Gaussian distribution, $z_0 \in \mathbb{Z}$ is a noise map as the latent input for the generator (Fig. 2). Let $z_i \in \{z_0, z_1, z_2, ..., z_k\}$ be the feature map after i-th residual block in the generator, with k denoting the the total number of residual blocks (*i.e.*, the upsampling times in the generator). Let $f_i^c \in \{f_0^c, f_1^c, f_2^c, ..., f_k^c\}$ represent the corresponding feature representations extracted by the content stream (Fig. 2), $f_i^s \in \{f_0^s, f_1^s, f_2^s, ..., f_k^s\}$ with the similar meaning in the style stream.

Feature Adaptive Denormalization (FADE). Our method is inspired by spatially adaptive denormalization (SPADE) [33]. Different from SPADE that resizes a semantic mask as its input, we generalize the input to multi-scale *feature representation* f_i^c of the content image x^c. In this way, we fully exploit semantic information captured by the content stream CS.

Formally, we define N as the batch size, L_i as the number of feature map channels in each layer. H_i and W_i are height and width, respectively. We first apply batch normalization [15] to normalize the generator feature map z_i in a channel-wise manner. Then, we modulate the normalized feature by using the learned parameters scale γ_i and bias β_i. The denormalized activation ($n \in N$, $l \in L_i$, $h \in H_i$, $w \in W_i$) is:

$$\gamma_i^{l,h,w} \cdot \frac{z_i^{n,l,h,w} - \mu_i^l}{\sigma_i^l} + \beta_i^{l,h,w}, \tag{1}$$

where μ_i^l and σ_i^l are the mean and standard deviation, respectively, of the generator feature map z_i before the batch normalization [15] in channel l:

$$\mu_i^l = \frac{1}{NH_iW_i} \sum_{n,h,w} z_i^{n,l,h,w}, \tag{2}$$

$$\sigma_i^l = \sqrt{\frac{1}{NH_iW_i} \sum_{n,h,w} \left(z_i^{n,l,h,w}\right)^2 - (\mu_i^l)^2}. \tag{3}$$

The denormalization operation is *element-wise*, and the parameters $\gamma_i^{l,h,w}$ and $\beta_i^{l,h,w}$ are learned by one-layer convolutions from f_i^c in the FADE module (see Fig. 3(c)). Compared to previous conditional normalization methods [8,13,33], FADE experiences more perceptible influence from coarse-to-fine feature representations, thus it can better preserve semantic structure information.

Feature Adaptive Instance Normalization (FAdaIN). To better fuse style representation, we introduce another feature transformation, named feature adaptive instance normalization (FAdaIN). This method is inspired by adaptive instance normalization (AdaIN) [13], with a generalization to enable the style stream SS to learn multi-scale *feature-level* style representation f_i^s of the style image x^s more effectively.

We use the same notation z_i to represent the feature map after i-th FADE residual block in the generator. FAdaIN adaptively computes the affine parameters from the corresponding style feature f_i^s with the same scale from SS:

$$\text{FAdaIN}\left(z_i, f_i^s\right) = \sigma\left(f_i^s\right)\left(\frac{z_i - \mu\left(z_i\right)}{\sigma\left(z_i\right)}\right) + \mu\left(f_i^s\right), \tag{4}$$

where $\mu\left(z_i\right)$ and $\sigma\left(z_i\right)$ are the mean and standard deviation, respectively, of z_i.

Exploiting FAdaIN, coarse-to-fine style features at different layers can be fused adaptively with the corresponding semantic structure features learned by FADE, allowing our framework to be trained end-to-end and versatile to different tasks. Furthermore, owing to the effectiveness of FAdaIN in capturing multi-scale style feature representations, multi-modal image synthesis is made possible with arbitrary style control.

3.3 Objective

We use standard losses in our objective function. Following [28,33], we adopt a hinge loss term [25,31,50] as our adversarial loss. For the generator, we apply hinge-based adversarial loss, perceptual loss [17], and feature matching loss [41]. For the multi-scale discriminators, only hinge-based adversarial loss is used to distinguish whether the image is real or fake. The generator and discriminator are trained alternately to play a min-max game. The generator loss $\mathcal{L}_G$ and the discriminator loss $\mathcal{L}_D$ can be written as:

$$\mathcal{L}_G = -\mathbb{E}\left[D\left(g\right)\right] + \lambda_P \mathcal{L}_P\left(g, x^c\right) + \lambda_{FM} \mathcal{L}_{FM}\left(g, x^s\right), \tag{5}$$

$$\mathcal{L}_D = -\mathbb{E}\left[\min\left(-1 + D\left(x^s\right), 0\right)\right] - \mathbb{E}\left[\min\left(-1 - D\left(g\right), 0\right)\right], \tag{6}$$

where $g = G\left(z_0, x^c, x^s\right)$ denotes the generated image, z_0, x^c, x^s denote the input noise map in latent space, the content image, and the style image, respectively. $\mathcal{L}_P$ is the perceptual loss [17] that minimizes the difference between the feature representations extracted by VGG-19 [17] network. $\mathcal{L}_{FM}$ is the feature matching loss [41] that matches the intermediate features at different layers of multi-scale discriminators. λ_P and λ_{FM} are the corresponding weights.

The simple objective functions make our framework stable and easy to train. Thanks to the two-stream network, the typical KL loss [21,28,33,49] for multi-modal image synthesis becomes optional. Despite the simplicity, TSIT is a highly versatile tool, readily adaptable to various image-to-image translation tasks.

4 Settings

Implementation Details. We use Adam [20] optimizer and set $\beta_1 = 0$, $\beta_2 = 0.9$. Two time-scale update rule [12] is applied, where the learning rates for the generator (including two streams) and the discriminators are 0.0001 and 0.0004, respectively. We exploit Spectral Norm [31] for all layers in our network. We adopt SyncBN and IN [40] for the generator and the multi-scale discriminators, respectively. For the perceptual loss [17], we use the feature maps of `relu1_1`, `relu2_1`, `relu3_1`, `relu4_1`, `relu5_1` layers from a pretrained VGG-19 [37] model, with the weights [1/32, 1/16, 1/8, 1/4, 1]. For the feature matching loss [41], we select features of three layers from the discriminator at each scale. All the experiments are conducted on NVIDIA Tesla V100 GPUs. Please refer to our *supplementary material* for additional implementation details.

Applications. The proposed framework is versatile for various image-to-image translation tasks. We consider three representative applications of conditional image synthesis: arbitrary style transfer (unsupervised), semantic image synthesis (supervised), and multi-modal image synthesis (enriching generation diversity). Please refer to our *supplementary material* for details of our application exploration.

Datasets. For arbitrary style transfer, we consider diverse scenarios. We use Yosemite summer → winter dataset (natural images) provided by [53]. We classify BDD100K [47] (real-world scenes) into different times and perform day → night translation. Besides, we use Photo → art dataset (artistic paintings) in [53]. For semantic image synthesis, we select several challenging datasets (*i.e.*, Cityscapes [7] and ADE20K [52]). For multi-modal image synthesis, we further classify BDD100K [47] into different time and weather conditions, and perform controllable time and weather translation. The details of the datasets can be found in the *supplementary material.*

Evaluation Metrics. Besides comparing perceptual quality, we employ the standard evaluation protocol in prior works [2,14,18,28,33] for quantitative evaluation. For arbitrary style transfer, we apply Fréchet Inception Distance (FID, evaluating similarity of distribution between the generated images and the real images, lower is better) [12] and Inception Score (IS, considering clarity and diversity, higher is better) [35]. For semantic image synthesis, we strictly follow [28,33], adopting FID [12] and segmentation accuracy (mean Intersection-over-Union (mIoU) and pixel accuracy (accu)). The segmentation models are: DRN-D-105 [46] for Cityscapes [7], and UperNet101 [42] for ADE20K [52].

Baselines. We compare our method with several state-of-the-art task-specific baselines. For a fair comparison, we mainly employ GAN-based methods. In the unsupervised setting, MUNIT [14] and DMIT [49] are included, with the strong ability to capture the multi-modal nature of images while keeping quality. In the supervised setting, we compare against CRN [4], SIMS [34], pix2pixHD [41], SPADE [33], and CC-FPSE [28].

5 Results

Arbitrary Style Transfer. The results of *Yosemite summer → winter season transfer* are shown in Fig. 4. Baselines [14,49] tend to impose the color of the style image (winter) to the whole content image (summer). Besides, MUNIT sometimes introduces unnecessary artistic effects, and DMIT generates some grid-like artifacts. In comparison, our generated results are clearer and more semantics-aware spatially. The results of *BDD100K day → night time translation* are shown in Fig. 5. Some objects (*e.g.*, road sign, car) generated by MUNIT are too dark, and the whole image tends to have some unnatural colors. DMIT introduces obvious artifacts to the car or sky. In contrast, our method produces more photorealistic samples in this task. In *photo → art style transfer*, we choose some hard cases to make a clear comparison (see Fig. 6) due to the very strong ability of all the methods in this task. Our method can transfer the styles well while effectively keeping the content structure. MUNIT tends to impose a homogeneous color to the image. Although DMIT achieves slightly better stylization than our method in certain cases (in Row 3 of Fig. 6), it also brings some grid-like distortions.

Fig. 4. Yosemite summer → winter season transfer results compared to baselines.

The quantitative evaluation results are shown in Table 1. Our approach achieves better performance than baselines [14,49] in all the tasks. We also note

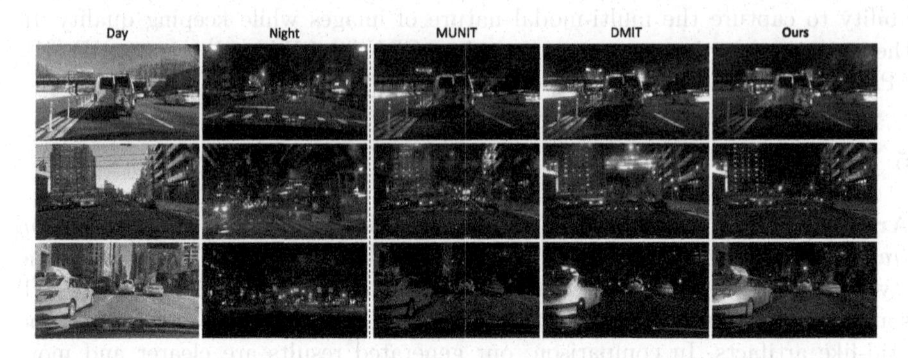

Fig. 5. BDD100K day → night time translation results compared to baselines.

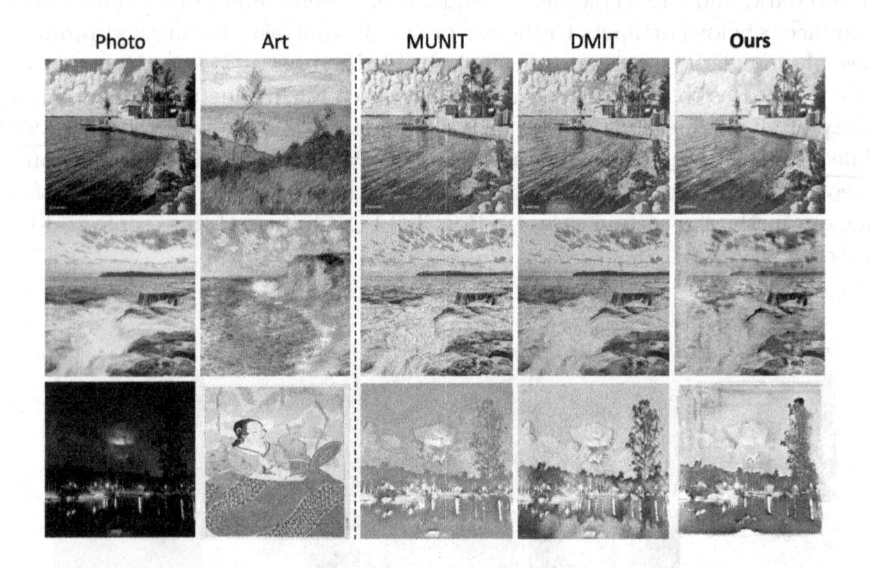

Fig. 6. Photo → art style transfer results compared to baselines.

Table 1. The FID and IS scores of our method compared to state-of-the-art methods in arbitrary style transfer tasks. A lower FID and a higher IS indicate better performance.

Methods	Summer → winter		Day → night		Photo → art	
	FID ↓	IS ↑	FID ↓	IS ↑	FID ↓	IS ↑
MUNIT [14]	118.225	2.537	110.011	2.185	167.314	3.961
DMIT [49]	87.969	2.884	83.898	2.156	166.933	3.871
Ours	**80.138**	**2.996**	**79.697**	**2.203**	**165.561**	**4.020**

that the gap is relatively small in photo → art style transfer, in line with the close qualitative performance in this task (see Fig. 6).

Semantic Image Synthesis. We choose two state-of-the-art baselines, SPADE [33] and CC-FPSE [28], to show some qualitative comparison results of semantic image synthesis (Fig. 7). Our method demonstrates better perceptual quality than these task-specific baselines. In street scene (Column 1), our method generates better details on key objects (car, pedestrian). In road scene (Column 2), SPADE generates atypical colors on the roads, while CC-FPSE produces unnatural edges on the cars, hardly fitting the background (road). For outdoor natural images (Column 3), all the methods share a similar generation quality. Our method is slightly better due to less distortions on the grass. In indoor scene (Column 4 and 5), SPADE and CC-FPSE produce obvious artifacts in some cases (Column 5). In contrast, our method is more robust to diverse scenarios.

The quantitative evaluation results are shown in Table 2 (the values used for comparison are taken from [28,33]). The proposed method achieves comparable performance with the very strong specialized methods [4,28,33,34,41] for semantic image synthesis. Note that SIMS [34] yields the best FID score but poor segmentation performance on Cityscapes, because it stitches image patches from

Fig. 7. Semantic image synthesis results compared to baselines.

Table 2. The mIoU, pixel accuracy (accu) and FID scores of our method compared to state-of-the-art methods in semantic image synthesis tasks. A higher mIoU, a higher pixel accuracy (accu) and a lower FID indicate better performance.

Methods	Cityscapes			ADE20K		
	mIoU ↑	accu ↑	FID ↓	mIoU ↑	accu ↑	FID ↓
CRN [4]	52.4	77.1	104.7	22.4	68.8	73.3
SIMS [34]	47.2	75.5	**49.7**	N/A	N/A	N/A
pix2pixHD [41]	58.3	81.4	95.0	20.3	69.2	81.8
SPADE [33]	62.3	81.9	71.8	38.5	79.9	33.9
CC-FPSE [28]	65.5	82.3	54.3	**43.7**	**82.9**	31.7
Ours	**65.9**	**94.4**	59.2	38.6	80.8	**31.6**

a memory bank of training set while not keeping the exactly consistent position in the synthesized image. Our approach achieves state-of-the-art segmentation performance on Cityscapes and the best FID score on ADE20K, suggesting its robustness to fit the nature of different image-to-image translation tasks.

Multi-modal Image Synthesis. We perform multi-modal image synthesis for time and weather image-to-image translation (see Fig. 8) on BDD100K [47]. Training only a *single* model, we translate the images of weather *sunny* to different times and weathers (*i.e.*, *night*, *snowy*, *cloudy*, *rainy*). Our method effectively adapts to different style control and keeps photorealistic generation quality. Although the weather *snowy* is not very obvious in BDD100K [47], our approach successfully introduces some snow-like effects on trees and grounds (Column 2).

Fig. 8. BDD100K multi-modal image synthesis for different time and weather translation results by a *single* model.

Fig. 9. Cross validation of ineffectiveness of task-specific methods in inverse settings.

Cross Validation. We also conduct experiments to evaluate the performance of existing specialized methods in inverse settings (*i.e.*, using unsupervised methods to do semantic image synthesis/using supervised methods to perform arbitrary style transfer). We selected two representative methods, MUNIT [14] and SPADE [33]. Without modifying the architecture, we tuned the loss weights and tried to get the best generation results. To ensure a fair comparison, we also tried to compute perceptual loss with the content (day) image for SPADE to match the setting of TSIT. Representative results of cross validation are shown in Fig. 9. The proposed method shows much better results than baseline methods. MUNIT fails to adapt to semantic image synthesis. SPADE loses details of key objects and introduces very strong artifacts despite translating the color correctly.

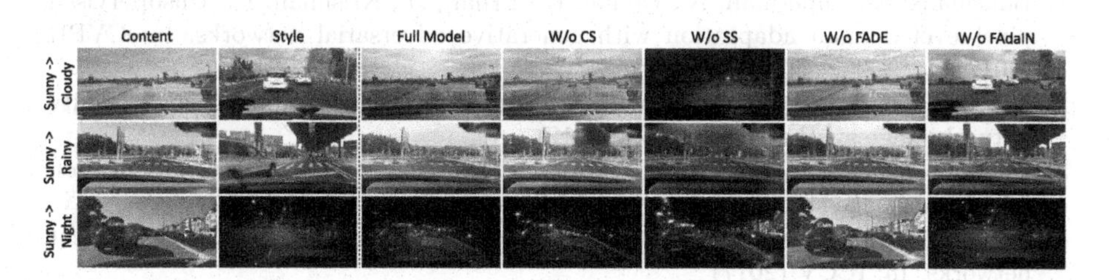

Fig. 10. Ablation studies of key modules (*i.e.*, content stream (CS), style stream(SS)) and feature transformations in multi-modal image synthesis task.

Ablation Studies. We present ablation studies of key modules (*i.e.*, content stream (CS), style stream (SS)) and the proposed feature transformations (see Fig. 10. More ablation study results can be found in the *supplementary material*). We perform multi-modal image synthesis to show the effectiveness of different components. Our full model generates high-quality results (Column 3). When we directly inject the resized content image without CS, the semantic structure information becomes weak, leading to several artifacts in the sky (Column 4).

Without SS, the model cannot perform multi-modal image synthesis at all (Column 5). The style representation is dominated by the night style. When we concatenate the feature maps of CS with the ones of the generator instead of using FADE, the concatenation introduces too much content information, leading to several failure cases (*e.g.*, *sunny* → *night* in Column 6). If we discard FAdaIN by concatenating the feature maps of SS with the ones of the generator, the style becomes too strong, causing serious style regionalization problem (Column 7).

6 Conclusion

We propose TSIT, a simple and versatile framework for image-to-image translation. The proposed symmetrical two-stream network allows the image generation to be effectively conditioned on the multi-scale feature-level semantic structure information and style representation via feature transformations. A systematic study verifies the effectiveness of our method in diverse tasks compared to state-of-the-art task-specific baselines. We believe that designing a unified and versatile framework for more tasks is an important direction in the image generation area. Incorporating unconditional image synthesis tasks and introducing more variability into the two streams/latent space can be interesting future works.

Acknowledgements. This work is supported by the SenseTime-NTU Collaboration Project, Singapore MOE AcRF Tier 1 (2018-T1-002-056), and NTU NAP.

References

1. Bousmalis, K., Silberman, N., Dohan, D., Erhan, D., Krishnan, D.: Unsupervised pixel-level domain adaptation with generative adversarial networks. In: CVPR (2017)
2. Brock, A., Donahue, J., Simonyan, K.: Large scale GAN training for high fidelity natural image synthesis. In: ICLR (2018)
3. Chen, D., Yuan, L., Liao, J., Yu, N., Hua, G.: StyleBank: an explicit representation for neural image style transfer. In: CVPR (2017)
4. Chen, Q., Koltun, V.: Photographic image synthesis with cascaded refinement networks. In: ICCV (2017)
5. Chiu, T.Y.: Understanding generalized whitening and coloring transform for universal style transfer. In: ICCV (2019)
6. Choi, Y., Choi, M., Kim, M., Ha, J.W., Kim, S., Choo, J.: StarGAN: unified generative adversarial networks for multi-domain image-to-image translation. In: CVPR (2018)
7. Cordts, M., et al.: The cityscapes dataset for semantic urban scene understanding. In: CVPR (2016)
8. Dumoulin, V., Shlens, J., Kudlur, M.: A learned representation for artistic style. arXiv preprint arXiv:1610.07629 (2016)
9. Gatys, L.A., Ecker, A.S., Bethge, M.: A neural algorithm of artistic style. arXiv preprint arXiv:1508.06576 (2015)

10. Goodfellow, I.,et al.: Generative adversarial nets. In: NeurIPS (2014)
11. He, K., Zhang, X., Ren, S., Sun, J.: Deep residual learning for image recognition. In: CVPR (2016)
12. Heusel, M., Ramsauer, H., Unterthiner, T., Nessler, B., Hochreiter, S.: GANs trained by a two time-scale update rule converge to a local nash equilibrium. In: NeurIPS (2017)
13. Huang, X., Belongie, S.: Arbitrary style transfer in real-time with adaptive instance normalization. In: ICCV (2017)
14. Huang, X., Liu, M.-Y., Belongie, S., Kautz, J.: Multimodal unsupervised image-to-image translation. In: Ferrari, V., Hebert, M., Sminchisescu, C., Weiss, Y. (eds.) ECCV 2018. LNCS, vol. 11207, pp. 179–196. Springer, Cham (2018). https://doi.org/10.1007/978-3-030-01219-9_11
15. Ioffe, S., Szegedy, C.: Batch normalization: accelerating deep network training by reducing internal covariate shift. In: ICML (2015)
16. Isola, P., Zhu, J.Y., Zhou, T., Efros, A.A.: Image-to-image translation with conditional adversarial networks. In: CVPR (2017)
17. Johnson, J., Alahi, A., Fei-Fei, L.: Perceptual losses for real-time style transfer and super-resolution. In: Leibe, B., Matas, J., Sebe, N., Welling, M. (eds.) ECCV 2016. LNCS, vol. 9906, pp. 694–711. Springer, Cham (2016). https://doi.org/10.1007/978-3-319-46475-6_43
18. Karras, T., Laine, S., Aila, T.: A style-based generator architecture for generative adversarial networks. In: CVPR (2019)
19. Kim, T., Cha, M., Kim, H., Lee, J.K., Kim, J.: Learning to discover cross-domain relations with generative adversarial networks. In: ICML (2017)
20. Kingma, D.P., Ba, J.: Adam: a method for stochastic optimization. arXiv preprint arXiv:1412.6980 (2014)
21. Kingma, D.P., Welling, M.: Auto-encoding variational bayes. arXiv preprint arXiv:1312.6114 (2013)
22. Kingma, D.P., Mohamed, S., Rezende, D.J., Welling, M.: Semi-supervised learning with deep generative models. In: NeurIPS (2014)
23. Kotovenko, D., Sanakoyeu, A., Lang, S., Ommer, B.: Content and style disentanglement for artistic style transfer. In: ICCV (2019)
24. Lee, H.-Y., Tseng, H.-Y., Huang, J.-B., Singh, M., Yang, M.-H.: Diverse image-to-image translation via disentangled representations. In: Ferrari, V., Hebert, M., Sminchisescu, C., Weiss, Y. (eds.) ECCV 2018. LNCS, vol. 11205, pp. 36–52. Springer, Cham (2018). https://doi.org/10.1007/978-3-030-01246-5_3
25. Lim, J.H., Ye, J.C.: Geometric GAN. arXiv preprint arXiv:1705.02894 (2017)
26. Liu, A.H., Liu, Y.C., Yeh, Y.Y., Wang, Y.C.F.: A unified feature disentangler for multi-domain image translation and manipulation. In: NeurIPS (2018)
27. Liu, M.Y., Breuel, T., Kautz, J.: Unsupervised image-to-image translation networks. In: NeurIPS (2017)
28. Liu, X., Yin, G., Shao, J., Wang, X., et al.: Learning to predict layout-to-image conditional convolutions for semantic image synthesis. In: NeurIPS (2019)
29. Lu, M., Zhao, H., Yao, A., Chen, Y., Xu, F., Zhang, L.: A closed-form solution to universal style transfer. In: ICCV (2019)
30. Mirza, M., Osindero, S.: Conditional generative adversarial nets. arXiv preprint arXiv:1411.1784 (2014)
31. Miyato, T., Kataoka, T., Koyama, M., Yoshida, Y.: Spectral normalization for generative adversarial networks. arXiv preprint arXiv:1802.05957 (2018)
32. Van den Oord, A., Kalchbrenner, N., Espeholt, L., Vinyals, O., Graves, A., et al.: Conditional image generation with PixelCNN decoders. In: NeurIPS (2016)

33. Park, T., Liu, M.Y., Wang, T.C., Zhu, J.Y.: Semantic image synthesis with spatially-adaptive normalization. In: CVPR (2019)
34. Qi, X., Chen, Q., Jia, J., Koltun, V.: Semi-parametric image synthesis. In: CVPR (2018)
35. Salimans, T., Goodfellow, I., Zaremba, W., Cheung, V., Radford, A., Chen, X.: Improved techniques for training GANs. In: NeurIPS (2016)
36. Shrivastava, A., Pfister, T., Tuzel, O., Susskind, J., Wang, W., Webb, R.: Learning from simulated and unsupervised images through adversarial training. In: CVPR (2017)
37. Simonyan, K., Zisserman, A.: Very deep convolutional networks for large-scale image recognition. arXiv preprint arXiv:1409.1556 (2014)
38. Song, C., Wu, Z., Zhou, Y., Gong, M., Huang, H.: ETNet: error transition network for arbitrary style transfer. In: NeurIPS (2019)
39. Taigman, Y., Polyak, A., Wolf, L.: Unsupervised cross-domain image generation. arXiv preprint arXiv:1611.02200 (2016)
40. Ulyanov, D., Vedaldi, A., Lempitsky, V.: Instance normalization: the missing ingredient for fast stylization. arXiv preprint arXiv:1607.08022 (2016)
41. Wang, T.C., Liu, M.Y., Zhu, J.Y., Tao, A., Kautz, J., Catanzaro, B.: High-resolution image synthesis and semantic manipulation with conditional GANs. In: CVPR (2018)
42. Xiao, T., Liu, Y., Zhou, B., Jiang, Y., Sun, J.: Unified perceptual parsing for scene understanding. In: Ferrari, V., Hebert, M., Sminchisescu, C., Weiss, Y. (eds.) ECCV 2018. LNCS, vol. 11209, pp. 432–448. Springer, Cham (2018). https://doi.org/10.1007/978-3-030-01228-1_26
43. Yang, S., Wang, Z., Wang, Z., Xu, N., Liu, J., Guo, Z.: Controllable artistic text style transfer via shape-matching GAN. In: ICCV (2019)
44. Yi, Z., Zhang, H., Tan, P., Gong, M.: DualGAN: unsupervised dual learning for image-to-image translation. In: ICCV (2017)
45. Yoo, J., Uh, Y., Chun, S., Kang, B., Ha, J.W.: Photorealistic style transfer via wavelet transforms. In: ICCV (2019)
46. Yu, F., Koltun, V., Funkhouser, T.: Dilated residual networks. In: CVPR (2017)
47. Yu, F., et al.: BDD100K: a diverse driving video database with scalable annotation tooling. arXiv preprint arXiv:1805.04687 (2018)
48. Yu, X., Cai, X., Ying, Z., Li, T., Li, G.: SingleGAN: image-to-image translation by a single-generator network using multiple generative adversarial learning. In: Jawahar, C.V., Li, H., Mori, G., Schindler, K. (eds.) ACCV 2018. LNCS, vol. 11365, pp. 341–356. Springer, Cham (2019). https://doi.org/10.1007/978-3-030-20873-8_22
49. Yu, X., Chen, Y., Liu, S., Li, T., Li, G.: Multi-mapping image-to-image translation via learning disentanglement. In: NeurIPS (2019)
50. Zhang, H., Goodfellow, I., Metaxas, D., Odena, A.: Self-attention generative adversarial networks. arXiv preprint arXiv:1805.08318 (2018)
51. Zhang, Y., et al.: Multimodal style transfer via graph cuts. In: ICCV (2019)
52. Zhou, B., Zhao, H., Puig, X., Fidler, S., Barriuso, A., Torralba, A.: Scene parsing through ADE20K dataset. In: CVPR (2017)
53. Zhu, J.Y., Park, T., Isola, P., Efros, A.A.: Unpaired image-to-image translation using cycle-consistent adversarial networks. In: ICCV (2017)

ProxyBNN: Learning Binarized Neural Networks via Proxy Matrices

Xiangyu He[1,2], Zitao Mo[1], Ke Cheng[1,2], Weixiang Xu[1,2], Qinghao Hu[1], Peisong Wang[1], Qingshan Liu[4], and Jian Cheng[1,2,3](✉)

[1] NLPR, Institute of Automation, Chinese Academy of Sciences, Beijing, China
{xiangyu.he,qinghao.hu,peisong.wang,jcheng}@nlpr.ia.ac.cn

[2] School of Artificial Intelligence, University of Chinese Academy of Sciences, Beijing, China

[3] Center for Excellence in Brain Science and Intelligence Technology, Beijing, China

[4] Nanjing University of Information Science and Technology, Nanjing, China

Abstract. Training Binarized Neural Networks (BNNs) is challenging due to the discreteness. In order to efficiently optimize BNNs through backward propagations, real-valued auxiliary variables are commonly used to accumulate gradient updates. Those auxiliary variables are then directly quantized to binary weights in the forward pass, which brings about large quantization errors. In this paper, by introducing an appropriate proxy matrix, we reduce the weights quantization error while circumventing explicit binary regularizations on the full-precision auxiliary variables. Specifically, we regard pre-binarization weights as a linear combination of the basis vectors. The matrix composed of basis vectors is referred to as the proxy matrix, and auxiliary variables serve as the coefficients of this linear combination. We are the first to empirically identify and study the effectiveness of learning both basis and coefficients to construct the pre-binarization weights. This new proxy learning contributes to new leading performances on benchmark datasets.

Keywords: Binarized Neural Networks · Proxy matrix

1 Introduction

Binary embedding is a fundamental technique in machine learning applications, such as retrival [12,16], clustering [3,19], matching [8,38] and classification [9, 20]. The popular signum function quantizes data points to ± 1, which enables compact storage (*i.e.*, $32\times$ compression than floating point) and efficient bitwise operations (*i.e.*, replacing time-consuming inner-product with *xnor-popcnt*) [32]. However, $sgn(\cdot)$ is non-smooth with derivative 0 everywhere except at 0, which makes gradient-driven optimizations incapable, especially for training BNNs.

Electronic supplementary material The online version of this chapter (https://doi.org/10.1007/978-3-030-58580-8_14) contains supplementary material, which is available to authorized users.

A. Vedaldi et al. (Eds.): ECCV 2020, LNCS 12348, pp. 223–241, 2020.
https://doi.org/10.1007/978-3-030-58580-8_14

Pioneer works present constructive training algorithms according to the sense of growth of networks [30] and verify the information capacity of binary weights [21]. Variational Bayes methods [43,44] propose to train discrete multilayer neural networks using Expectation Propagation (EP). Recent gradient-based methods with Straight-Through-Estimator (STE) show that a linear backprop function for the non-linear activation surprisingly leads to promising results on CIFAR-10 [9,20,49]. XNOR-Net [37] further introduces a scaling factor to relax the binary constraint and show notable improvements on ImageNet dataset. Regardless of their differences, a real-valued auxiliary variable is commonly used to accumulate gradient updates and then binarized to ±1 at inference time [6,20,28,37]. To minimize the weights binarization error, recent BNNs impose explicit binary regularizations on the auxiliary variables that lead to the bimodal distribution [10,13,14,45]. Though bimodality that encourages auxiliary variables to be around binary values may facilitate binarization intuitively, it can be hard to change positive auxiliary variables to negative by small gradient steps and vice versa (Note that large gradient steps can be risky for BNNs training since there are no accurate gradients for binary weights but approximations).

In this paper, we try to reduce weights quantization errors while avoiding the explicit constraint that forces the full-precision auxiliary variables to be around ±1. To this end, we investigate the following question: is there a latent parameter space which can serve our goal, to bridge full-precision auxiliary variables and binary weights? We introduce proxy matrix R as a basis of the latent parameter space. Every filter before binarization can be written as a linear combination of basis vectors. The coefficients of this linear combination are referred to as the auxiliary variables. Since the basis can be the key component in proxy learning, we conduct empirical studies on the construction of R, based on the view of minimizing both weights quantization errors and the global cost function. It is shown that a well-designed proxy matrix leads to smooth optimization landscapes with superior performances. Exhaustive experiments show that our proxy learning strategy notably outperforms the state-of-the-art on ImageNet dataset.

2 Related Works

Binarized neural network has been a long-standing topic in machine learning community [32,33]. Due to its high memory and computing efficiency, it becomes an ideal solution to the deployment of computation-intensive deep convolutional neural networks on low-power devices [4,51]. Previous literatures prove that the manual-designed backpropagation of binarization/ternarization still performs well on small datasets, not only for weights compression but activations quantization [9,20,26,49]. DoReFa [52] further presents low-bit weights, activations and gradients to accelerate both training and inference on customized devices.

To narrow the gap between BNNs and full-precision networks on the challenging ImageNet, XNOR-Net [37] proposes scaling factors for both weights and activation functions to minimize the quantization error. The following works further develop various regularization functions that encourage training weights around

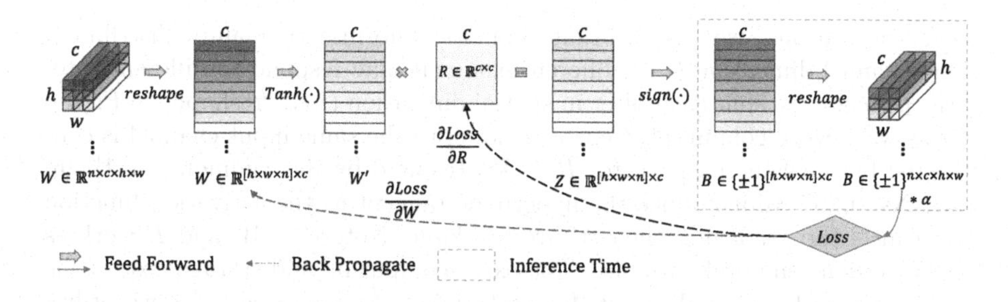

Fig. 1. Overview of the proxy learning for 3×3 binary weights

binary values [10,13,14] and controls the range of activations [11]. In light of the success of scaling factors, XNOR++ [6] improves the performances by learned both spatial and channel-wise scaling factors. To compensate for the information loss of binarization, Bi-Real [28] proposes double residual connections with full-precision downsampling layers and [6] replaces ReLU by PReLU. Due to the gradient mismatch, [10,28,48] formulate quantization forward/backward as differentiable non-linear mapping functions. More recently, probabilistic training methods [35,41] circumvent the need to approximate the gradient of $sign()$ by sampling from the weight distribution. Since BNN training is not well-founded, there are still tremendous efforts on the study of BNNs' optimizations [1,5,17,29] and how to explain the effectiveness of BNNs [2]. All those methods pave the way for a better understanding of binarized neural networks.

3 Methodology

3.1 Formulation

We quickly revisit the popular gradient-based method proposed in BinaryConnect [9], which maintains real-valued latent variables W for gradient updates. In the forward pass, W are binarized to ± 1 by

$$W_b = sgn(W) \tag{1}$$

to perform binary convolutions $W_b \otimes sgn(X)$, where X is the input feature map.

Given a basis R of the latent parameter space, we decompose the previous W into R and coordinates (or components) W'. Thus, we present a new pattern of learning binary weights

$$sgn(Z) = sgn(W'R), \quad W' = \phi(W) \tag{2}$$

where $W \in \mathbb{R}^{[h \times w \times n] \times c}$, $R \in \mathbb{R}^{c \times c1}$ and $\phi(\cdot)$ is a nonlinear mapping. As illustrated in Fig. 1, during gradient descent ProxyBNN learns coordinate representations W' and updates the manual-designed basis R simultaneously. For

[1] h, w, n and c are kernel height, width, kernel number and input channel number, respectively. For 1×1 convolutions and FC layers, $[h \times w \times n] \times c$ degrades into $n \times c$.

binary activations, we assume that semantic information mainly distributed along channel dimension (*i.e.*, different channels may respond to different categories). Hence, we split each filter in spatial dimension (*i.e.*, "reshape" in Fig.1). In this way, every column of Z corresponding to the same input channel is constructed by the same basis vector $R_i \in \mathbb{R}^c$. Inspired by the common $[-1, 1]$-clip in BNNs [9,20], we introduce the hyperbolic tangent as the activation function ϕ to cancel the gradients when W are too large. Note that W and R work as high-precision temporal variables. The extra computing and storage cost of the basis and coordinates only exist during training. At inference time, we utilize the well-trained B which is the same as previous BNNs.

3.2 Proxy Learning Procedure

To optimize the global objective of deep neural networks with binary constraints, we formulate the n layers BNN training to a constrained optimization problem

$$\min_{Z} \ \ell(Z), \quad s.t.\ Z_i = \alpha_i B_i,\ B_i \in \{+1, -1\}^{[h\times w]\times c}, \quad i = 1, \cdots, n \tag{3}$$

where $\alpha_i \in \mathbb{R}$ is a real-valued scaling factor to relax the binary constraint on Z_i [37] and $\ell(\cdot)$ is cross-entropy loss. Note that we introduce α_i and B_i as independent variables, which will be used in binary convolutions after training. If the first equation constraint is brought to the objective via a regularization parameter γ, we show that the resulting form can be solved by updating B_i, α_i and Z_i iteratively,

$$\mathcal{L}_\gamma = \min_{\alpha,Z,B} \ \ell(\psi(Z)) + \gamma \sum_{i=1}^{n} ||Z_i - \alpha_i B_i||_F^2, \quad s.t.\ B_i \in \{+1, -1\}^{[h\times w]\times c}, \tag{4}$$

where $\psi(\cdot)$ is a binary mapping that relaxes Z to $\mathbb{R}^{[h\times w\times n]\times c}$ and guarantees binary weights in the forward pass.

Fix Z_i, α_i, update B_i. In this step, we treat Z_i and α_i as constants and update B_i to minimize $\mathcal{L}_\gamma$. Since B_i only exists in the second term, we have

$$B_i^{t+1} = \arg\min ||Z_i^t - \alpha_i^t B_i||_F^2 = \arg\max\ tr(\alpha_i^t B_i^T Z_i^t) \tag{5}$$

where $tr(\alpha B^T Z) = \sum_{m=1}^{c} \sum_{n=1}^{h\cdot w} = \alpha B_{n,m} Z_{n,m}$. Given the binary constraint on B_i, the solution is simply $B_i^{t+1} = sgn(\alpha_i^t Z_i^t)$.

Fix Z_i, B_i, update α_i. Here we use the updated B_i^{t+1} and minimize $\mathcal{L}_\gamma$ in terms of α_i. Since Z_i^t and B_i^{t+1} are fixed in this step, problem (4) becomes independent subtasks

$$\min_{\alpha_i} ||Z_i^t - \alpha_i B_i^{t+1}||_F^2 = \min_{\alpha_i} \ (hwc)\alpha_i^2 - 2tr({B_i^{t+1}}^T Z_i^t)\alpha_i + const. \tag{6}$$

Note that α_i is a full-precision scalar and (6) is quadratic, the optimum can be easily obtained as $\alpha_i^{t+1} = \frac{tr({B_i^{t+1}}^T Z_i^t)}{hwc}$.

Fix α_i, B_i, update Z_i. To update the latent variable Z_i, we perform a gradient descent step since the objective function $\ell(\cdot)$ for BNN is differentiable and the second term in (4) is a quadratic regularization term, which is differentiable and convex. Following the rule of SGD, the derivative of Z_i^t is calculated as follow

$$\frac{\partial \mathcal{L}_\gamma}{\partial Z_i^t} = \frac{\partial \ell}{\partial \psi(Z_i^t)} \frac{\partial \psi(Z_i^t)}{\partial Z_i^t} + 2\gamma(Z_i^t - \alpha_i^{t+1} B_i^{t+1}). \tag{7}$$

Given the optimal solution of α_i and B_i at each step, we obtain the binary mapping $\psi(Z_i) = \frac{||Z_i||_1}{c \times h \times w} sgn(Z_i)$ in vector form (*i.e.*, $Z_i \in \mathbb{R}^{[h \times w \times c] \times 1}$). Then, the gradient with respect to the k-th element in Z_i is defined as[2]

$$\frac{\partial \ell}{\partial Z_{i,k}} := \frac{sgn(Z_{i,k})}{h \cdot w \cdot c} \sum_{j=1}^{hwc} \frac{\partial \ell}{\partial \psi(Z_i)_j} sgn(Z_{i,j}) + \frac{\partial \ell}{\partial \psi(Z_i)_k}. \tag{8}$$

Combining Eq. (7) and Eq. (8), we obtain the derivative to W', R as

$$\frac{\partial \mathcal{L}_\gamma}{\partial R} = \frac{\partial \mathcal{L}_\gamma}{\partial Z}^T W', \quad \frac{\partial \mathcal{L}_\gamma}{\partial W'} = \frac{\partial \mathcal{L}_\gamma}{\partial Z} R^T. \tag{9}$$

Following the standard gradient update step in [22], $W^{t+1} \leftarrow W^t - \beta_1 \nabla_W \mathcal{L}_\gamma$ and $R^{t+1} \leftarrow R^t - \beta_2 \nabla_R \mathcal{L}_\gamma$ where β_1 and β_2 are the learning rates, we have the updated $Z^{t+1} = \phi(W^{t+1}) R^{t+1}$.

3.3 The Construction of Basis

Although the basis R can be trained end-to-end as shown in the previous section, we empirically prove that the construction of the initial basis matters in ProxyBNN training.

Random Matrix. The most intuitive choice is a random initialization where every element $R_{i,j} \sim \mathcal{N}(0, 1)$. We include it as a baseline scheme to conduct fair comparisons.

Minimizing Square Error (MSE) Matrix. In light of the empirical success of minimizing weights quantization error [10,13,14,24,37], we consider the following square object

$$\min_R ||W'R - sgn(W'R)||_F^2. \tag{10}$$

Beginning with the identity matrix initialization of R, we adopt an iterative optimization procedure to find a local minimum of (10). In each iteration, $W'R^t$ is first assigned to the binary codewords, and then R^{t+1} is updated to minimize the square error, *i.e.*, calculating the Moore-Penrose inverse of W' then multiplied by $sgn(W'R^t)$. Since the pseudo-inverse relies on Singular Value Decomposition (SVD), which is time-consuming for large matrix, we conduct MSE construction only once and notice no accuracy improvement (even result in worse performance) with more re-construction during training.

[2] Further details in appendix 1.

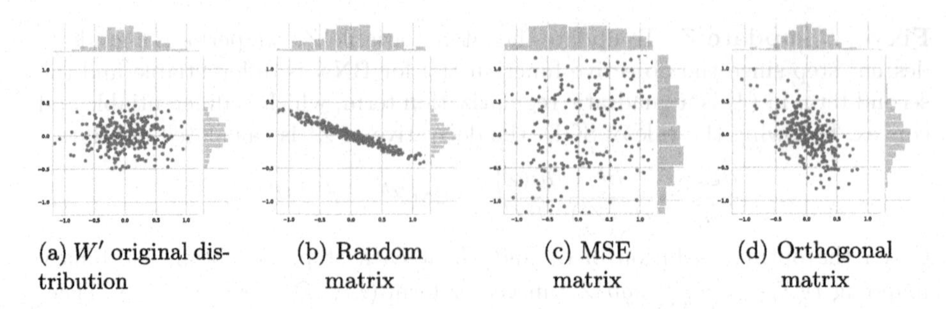

(a) W' original distribution (b) Random matrix (c) MSE matrix (d) Orthogonal matrix

Fig. 2. Toy examples of the effects of different proxy matrices. (a) shows the original distribution of W'. (b-d) illustrate the distributions of $W'R$, *i.e.*, Z

Orthogonal Matrix. The main idea of introducing orthogonal matrix is simply that: *similar* coordinates $w'_i, w'_j \in \mathbb{R}^{1\times c}$ may correspond to *similar* representations $z_i, z_j \in \mathbb{R}^{1\times c}$ in Euclidean space, given $z_i = w'_i R$. That is, we try to preserve the similarity relationship (locality structure) between coordinates while minimizing the quantization error. In this case, an orthogonal matrix R with $||w'_i - w'_j||_2 = ||w'_i R - w'_j R||_2$ becomes an ideal solution. Then, we reformulate problem (10) as

$$\min_R \; ||W'R - sgn(W'R)||_F^2, \quad s.t. \; R^T R = I. \tag{11}$$

The rows of coordinate matrix $W' \in \mathbb{R}^{[h\times w\times n]\times c}$ can be seen as a set of $h \times w \times n$ data points $\{w'_1, w'_2, \cdots, w'_{h\cdot w\cdot n}\}, w'_i \in \mathbb{R}^{1\times c}$, and (11) forms the classical hashing problem. Here we use *ITQ* proposed in [12] for solving hashing codes to obtain the optimal R. The alternating update is similar to MSE. We first binarize $W'R$ in each step, then the objective function corresponds to the classic Orthogonal Procrustes problem [40],

$$U\Sigma V^T = \text{svd}(sgn(W'R)^T W'R), \quad R = VU^T. \tag{12}$$

Before alternating optimization, we use a random orthogonal matrix to initialize R and train 10 epochs to warm up W'. We only conduct the construction once and then update R with small gradient steps.

3.4 The Effect of Proxy Learning

Toy Example. To better understand the proposed proxy learning, we first show a 2D toy example then analyze the experimental phenomenon in real networks. As shown in Fig. (2b, 2c), both random matrix and MSE matrix change the original data structure, especially MSE minimizes quantization errors at the cost of ruining the 2-dimensional Gaussian distribution, which approximates uniform distributions. Figure 2d shows the orthogonal matrix serves as a similarity-preserving rotation, which not only quantizes weights with small errors but maintains the structure of W'.

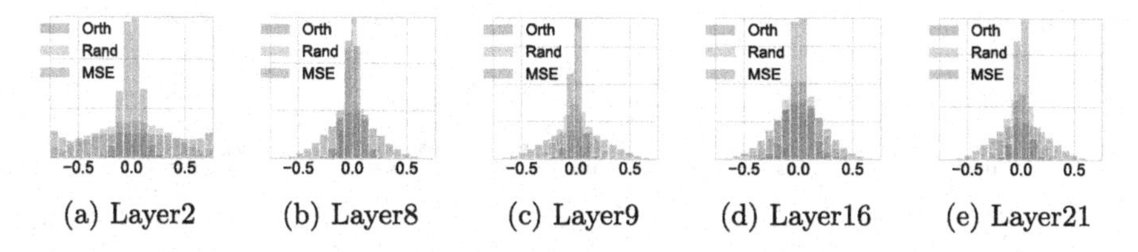

Fig. 3. Histograms of W' of WRN22 on CIFAR-100 (best viewed in color) (Color figure online)

Besides, the variance of W' in each direction is different. Directly quantizing both low-variance directions and high-variance directions (with more information) to 1-bit can be suboptimal. An orthogonal matrix balances the variance of different directions (*e.g.*, different channels in real networks), which facilitates the binary encoding.

Similar phenomena exist in practical WRN22 network. As shown in Fig. 4, MSE matrix leads to the largest variance in the channel dimension among three candidates, which is consistent with Fig. 2. For the random matrix, it has a wider distribution interval of variances than the orthogonal matrix, which reflects imbalanced variances across different channels, as shown in Fig. 2b (*e.g.*, high variance in x-dimension and low variance in y-dimension).

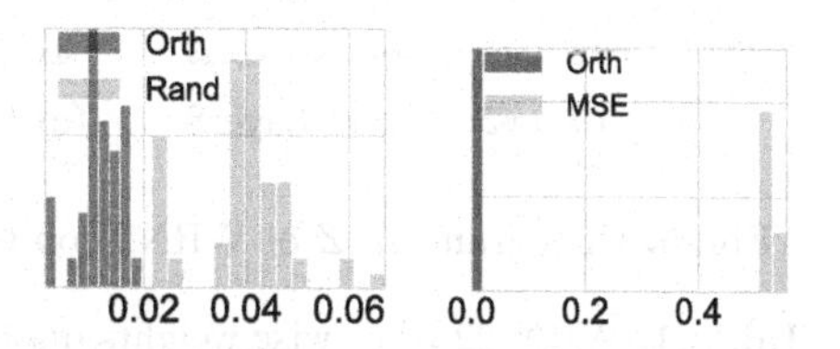

Fig. 4. Histograms of the variance of $Z \in \mathbb{R}^{h \cdot w \cdot n \cdot c}$ in the channel dimension (To be specific, we compute the variance of $Z_i \in \mathbb{R}^{h \cdot w \cdot n}$, $i = 1, \cdots, c$ then visualize the distribution of c samples. The more concentrated distribution indicates the more balanced variance.)

Weights Distribution. To clearly verify the effectiveness of the proxy matrix R, we visualize the distributions of W' and $W'R$. Figure 3 illustrates that all schemes' W' are approximate Gaussian distributions similar to weights in full-precision counterparts. We further demonstrate Z in Fig. 5. The baseline random matrix (*i.e.*, the first row) illustrates a bimodal distribution, which is a sensible result for pre-binarization weights to minimize quantization error. Since MSE matrix is based on $\min ||W'R - sgn(W'R)||_F$, the initial MSE basis naturally makes two peaks move towards ± 1, as shown in the second row of Fig. 5. However, it seems counterintuitive, the orthogonal scheme still generates a unimodal distribution. Here is the question: *Does either the unimodal distribution or the bimodal distribution contribute to "accurate" binary networks?*

Quantization Error v.s. Classification Error. Table 1 details the trade-off between layer-wise quantization error and the final accuracy. Here we define the quantization error as: $Q(Z, \alpha, B) = \frac{1}{h \cdot w \cdot n \cdot c} \sum_{i=1}^{n} ||Z_i - \alpha_i B_i||_F^2$. All proxy learning schemes obtain smaller average quantization errors compared with baseline

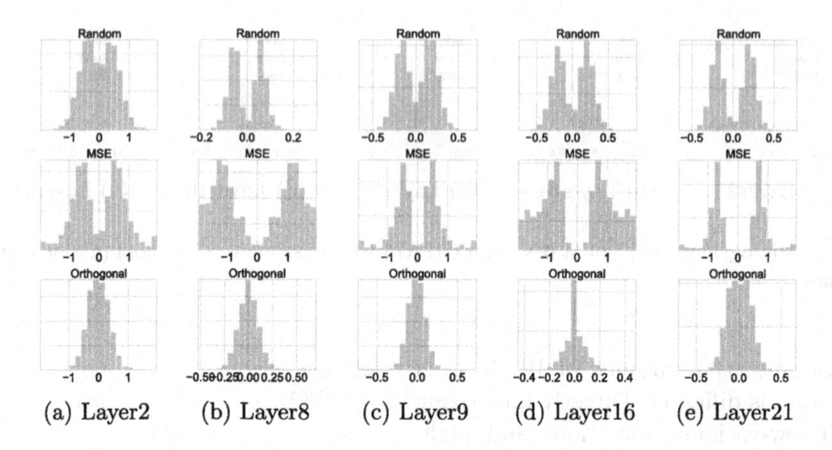

Fig. 5. Histograms of Z of WRN22 on CIFAR-100, *i.e.*, the distributions of $W'R$

Table 1. WRN-22 layer-wise weights quantization error and final accuracy. "Average" refers to the mean quantization error, averaged across all elements

Layer	ProxyBNN Orthogonal	ProxyBNN Random	ProxyBNN MSE	Bi-Real [28]
Conv2	**0.0473**	0.0834	0.0695	0.1797
Conv3	0.0108	0.0248	**0.0072**	0.1022
Conv5	0.0081	0.0139	**0.0021**	0.0664
Conv6	0.0071	0.0114	**0.0014**	0.0086
Conv7	0.0043	0.0031	**0.0011**	0.0003
Conv8	0.0043	**0.0006**	0.0009	0.0001
Conv9	**0.0043**	0.0070	0.0073	0.0670
Conv10	0.0065	0.0036	**0.0029**	0.0390
Conv12	0.0039	**0.0007**	0.0018	0.0029
Conv13	0.0023	**0.0005**	0.0010	0.0002
Conv14	0.0014	**0.0004**	0.0007	0.0001
Conv15	0.0017	**0.0003**	0.0017	0.0001
Conv16	**0.0031**	0.0089	0.0079	0.0397
Conv17	0.0068	0.0076	**0.0037**	0.0374
Conv19	0.0060	0.0069	**0.0035**	0.0373
Conv20	0.0057	0.0068	**0.0029**	0.0331
Conv21	0.0055	0.0058	**0.0035**	0.0277
Conv22	0.0056	0.0055	**0.0026**	0.0235
Average	0.0054	0.0061	**0.0035**	0.0294
Acc. (%)	**71.61**	69.10	59.32	69.73

Bi-Real-Net [28]. To be specific, ProxyBNN minimizes the binarization loss in the first and last few layers, which may facilitate feature extraction and semantic analysis. Note that MSE focuses on how to quantize weights locally, which generates over 8× smaller average loss than Bi-Real, yet results in poor per-

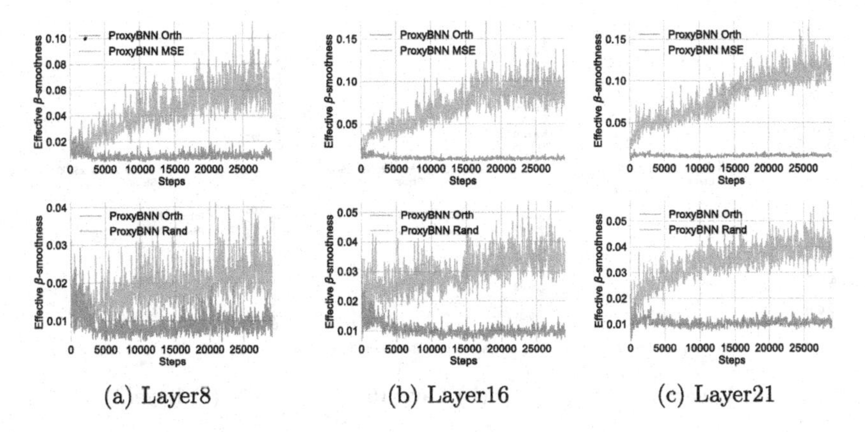

Fig. 6. Analysis of the "effective" β-smoothness [39] of WRN22 network. For a layer we measure the maximum ℓ_2-norm difference in gradient. The lower the values indicate the smoother loss landscape (best viewed in color) (Color figure online)

formance. The orthogonal scheme presents a better trade-off between weights binarization loss and the global cost function, and achieves the highest performance. It is shown that unimodal weights distributions (*i.e.*, the third row in Fig.5) can be another group of solutions to minimizing quantization error, when jointly optimized with cross-entropy loss $\ell(\cdot)$.

Optimization Landscape. If pre-binarization variables are close to zero, a small gradient step can change binary weights from positive to negative and vice versa, which may make the training easier. Motivated by this hypothesis, we analyze the optimization landscape of different bases and observe the superiority of the orthogonal scheme. Following [39], we measure the stability and smoothness of the landscape by Lipschitzness and "effective" β-smoothness of the loss function. As shown in Fig. 6, we observe consistent differences between these schemes. The improved Lipschitzness encourages us to take a step in the direction of a computed gradient, which provides a fairly accurate estimate of the real gradient [39]. Figure 7 also demonstrates the effect of different bases on the stability/Lipschitzness of the gradients. No matter how weights quantization loss changes (Conv8/16/21 correspond to three cases in Table 1), the orthogonal scheme still outperforms other candidates.

4 Experiments

To verify the effectiveness of the proposed approach, in this section, we introduce three benchmark datasets: CIFAR-10, CIFAR-100, and ImageNet. We comprehensively evaluate our method on the mainstream deep CNN architectures, including AlexNet [23], VGG [42], ResNet [15] and Wide ResNet [50].

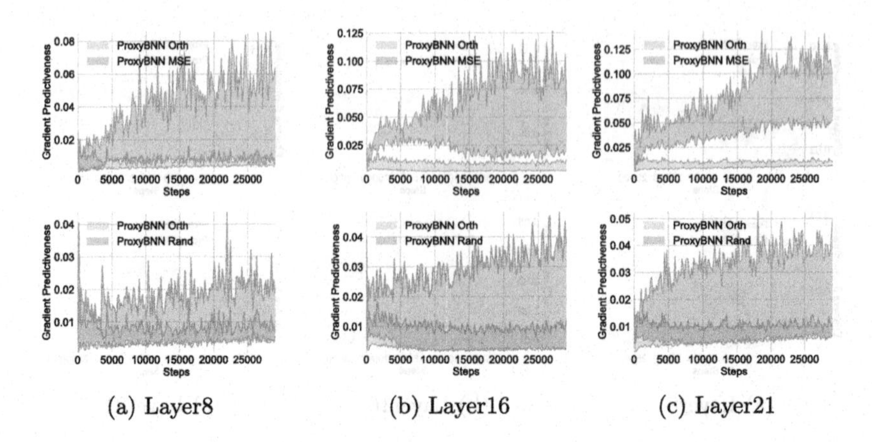

Fig. 7. Analysis of the gradient predictiveness [39] of WRN22 network. The shaded region corresponds to the variation in ℓ_2-norm changes in gradient over the distance. The thinner shade in plots show the smoother loss landscape and thus less training difficulty (best viewed in color) (Color figure online)

4.1 Experimental Setup

Network Structure. Since modified network structures can be the game-changer for training BNNs, we follow the same settings as prior works to make fair comparisons. For AlexNet, we use the same architecture from XNOR-Net [37] where batch normalization layers are added before activations and LRN layers are omitted. ResNet-18/34 refer to the original structure introduced in [15], unless specified. In binary weights experiments, we simply replace full-precision convolution layers with binary weights counterparts without any bells and whistles. When both activations and weights are quantized to 1-bit (including 1×1 downsample layers), we use batch-normalization before each activation function [10,37]. The modified ResNet/WRN [13,14,27,28] consist of double skip connections [28], PReLU activations [7] and real-valued downsampling layers [28]. The operations are reordered as Batch-Normalization → Binarization → Binary-Convolution → Activation, as proposed in XNOR-Net [37]. VGG9 is a VGG-like structure with six convolutional layers and three fully-connected layers, first described in BinaryConnect [9]. We use the same modification as [37,46]. As in almost all previous works, the first and last layers in all experiments are kept real.

Activation Binarization. There have been tremendous efforts on exploiting binary activations [10,28,37,46,48]. To verify the robustness of ProxyBNN, we consider two simple settings in our experiments: the signum function proposed in BinaryNet [20] and *round*(*clip*(x)) introduced in DoReFa [52]. We conduct the first setting in CIFAR experiments then we apply the second technique to the ImageNet networks.

Table 2. Performances of ProxyBNN trained with different bases. Top-1 accuracies on benchmark datasets are reported (single stage, trained from scratch)

Model	#Param	Dataset	Orthogonal	Random	MSE
ResNet18	2.80M	Cifar10	**91.87** (±0.36)	88.36 (±0.49)	67.88 (±0.38)
	2.82M	Cifar100	**67.17** (±0.73)	53.58 (±1.08)	30.69 (±0.84)
WRN22	4.30M	Cifar10	**92.96** (±0.11)	91.24 (±0.22)	86.77 (±0.72)
	4.33M	Cifar100	**71.57** (±0.14)	68.93 (±0.29)	58.00 (±0.57)
ResNet-18	11.70M	ImageNet	**58.7**	53.7	36.8

Table 3. Error rates (%) on CIFAR-100 using WRN22

(a) Ablation studies on penalty factor γ

γ	Error (%)
0.001	30.28 (±0.11)
0.0001	28.51 (±0.10)
$1e^{-5}$	**28.43** (±0.14)
$1e^{-6}$	29.94 (±0.50)
$1e^{-7}$	31.33 (±0.47)

(b) Effect of using different initial learning rates for R

init. lr_R	Error (%)
lr_w	30.77 (±0.29)
$lr_w \times 0.1$	**28.43** (±0.14)
$lr_w \times 0.01$	30.35 (±0.05)
$lr_w \times 0.001$	31.44 (±0.30)

Ablation Study. In this section, we first evaluate the effects of the penalty weight γ and different learning rates for the proxy matrix. Table 3a indicates that a proper γ matters in the balance between cross-entropy loss and the penalty term. We also observe that the basis should be updated a little slower than the coordinates, as shown in Table 3b. To further verify the superiority of the orthogonal scheme, we evaluate different bases on benchmarks (Table 2). The performance gap is consistent with that in Table 1. Besides, Fig. 8 shows that the property of the orthogonal basis roughly remains after training, *i.e.*, $R_i^T R_j \approx 0\ \forall i \neq j$ (for clarity, we normalize the max value to 1). Based on Table 3, we apply the best settings to the following experiments without finetuning.

4.2 Results

CIFAR-10/100. The CIFAR-10/100 dataset consist of 50,000 train images and a test set of 10,000 across 10/100 classes. Unless specified, the images are padded by 4 pixels on each side then randomly cropped to 32×32 [13,14,27]. We use a batch size of 128 for training, optimized by Adam [22] with cosine scheduler. The initial learning rate for W is set to 0.005 and the weight decay is $1e^{-6}$ (same for both R and W). All networks are trained for 310 epochs.

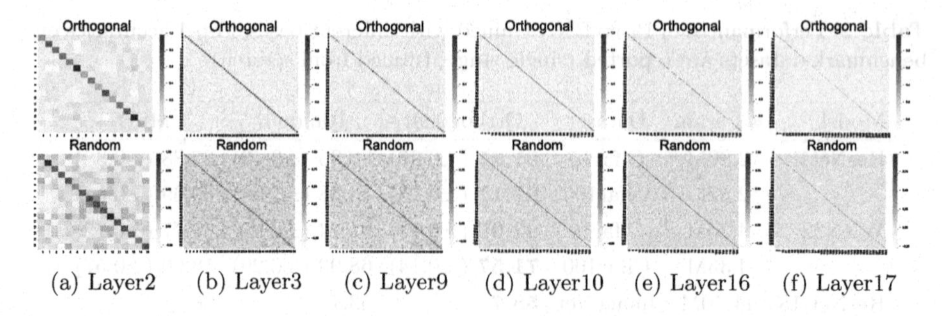

(a) Layer2 (b) Layer3 (c) Layer9 (d) Layer10 (e) Layer16 (f) Layer17

Fig. 8. Visualizations of $R^T R$, where R is the learned proxy matrix

Table 4. Test accuracies on CIFAR-10/100, comparison with different 1-bit methods. † indicates modified architectures [28] (details in Sect. 4.1). "MS" refers to the multi-stage training strategy, *e.g.*, binarizing the activations first, and then using the model as initialization to train fully binarized networks. "Center" means using center loss [47] during stage-2. "FP32" is the full-precision baseline. "Aug" indicates 32×32 random cropping with 4 pixels padding

Model	Kernel-Stage	Method	Cifar10 (%)	Cifar100 (%)	MS	Center	Aug
ResNet18†	16-16-32-64	PCNN [13]	78.93	41.41	✓	–	✓
		GBCN [27]	81.22	47.96	✓	✓	✓
		ProxyBNN	**84.53**	**52.07**	–	–	✓
		FP32	90.77	65.15	–	–	✓
WRN22†	64-64-128-256	PCNN [13]	91.37	69.98	✓	–	✓
		GBCN [27]	92.72	**71.85**	✓	✓	✓
		BONN [14]	92.36	–	✓	✓	✓
		ProxyBNN	**92.96**	71.57	–	–	✓
		FP32	95.75	77.34	–	–	✓
VGG9	128-256-512	BNN [20]	89.9	–	–	–	–
		XNOR [37]	89.8	–	–	–	–
		SiBNN [46]	90.2	–	–	–	–
		ProxyBNN	**90.5**	63.23	–	–	–
		FP32	91.7	67.01	–	–	–
ResNet18†	32-64-128-256	PCNN [13]	87.76	60.29	✓	–	✓
		GBCN [27]	87.69	62.01	✓	✓	✓
		ProxyBNN	**91.87**	**67.17**	–	–	✓
		FP32	93.88	72.51	–	–	✓

We compare our results with prior state-of-the-arts, as shown in Table 4. Both over-parameterized architectures, such as VGG/WRN with the kernel stage of 64-64-128-256, and compact ResNet-18 are considered. Our method in the worst case is still competitive with recent works (results reported in the original papers), without other techniques.

Table 5. Comparison with state-of-the-art methods on ResNets. "MS" refers to the multi-stage training strategy, *e.g.*, binarizing the activations first, and then using the model as initialization to train fully binarized networks. For binary weights experiments, "MS" is fine-tuning from full-precision weights. † indicates modified architectures [28] (details in Sect. 4.1)

Model	Method	Weight	Activation	Top-1 (%)	Top-5 (%)	MS
ResNet-18	XNOR [37]	1	1	51.2	73.2	–
	BNN+ [10]	1	1	53.0	72.6	✓
	QNet [48]	1	1	53.6	75.3	✓
	XNOR++ [6]	1	1	57.1	79.9	–
	ProxyBNN	1	1	**58.7**	**81.2**	–
	BWN [37]	1	32	60.8	83.0	–
	BWHN [18]	1	32	64.3	85.9	✓
	ADMM [24]	1	32	64.8	86.2	–
	IR-Net [36]	1	32	66.5	86.8	–
	ProxyBNN	1	32	**67.3**	**87.2**	–
	FP32	32	32	69.3	89.2	–
ResNet-18†	Bi-Real [28]	1	1	56.4	79.5	✓
	PCNN [13]	1	1	57.3	80.0	✓
	GBCN [27]	1	1	57.8	80.9	✓
	IR-Net [36]	1	1	58.1	80.0	–
	BONN [14]	1	1	59.3	81.6	✓
	SiBNN [46]	1	1	59.7	81.8	–
	ProxyBNN	1	1	63.3	84.3	–
	ProxyBNN	1	1	**63.7**	**84.8**	✓
	PCNN [13]	1	32	63.5	85.1	–
	ProxyBNN	1	32	**67.7**	**87.7**	–
	FP32	32	32	68.5	88.3	–
ResNet-34	ProxyBNN	1	32	**70.7**	**89.6**	–
	FP32	32	32	73.3	91.3	–
ResNet-34†	ABC [25]	1	1	52.4	76.5	–
	WRPN [31]	1	1	60.5	–	–
	Bi-Real [28]	1	1	62.2	83.9	✓
	IR-Net [36]	1	1	62.9	84.1	–
	SiBNN [46]	1	1	63.3	84.4	–
	ProxyBNN	1	1	**66.3**	**86.5**	–
	FP32	32	32	70.4	89.3	–

ImageNet. ImageNet (ILSVRC2012) is one of the most challenging image classification benchmarks with over 1.2 million training images and 50K validation

Table 6. Comparison with state-of-the-art methods on AlexNet. "MS" refers to the multi-stage training strategy. For binary weights experiments, "MS" is fine-tuning from full-precision weights, otherwise, training from scratch

Model	Method	Weight	Activation	Top-1 (%)	Top-5 (%)	MS
AlexNet	DoReFa [52]	1	1	43.6	–	–
	XNOR [37]	1	1	44.2	69.2	–
	RAD [11]	1	1	47.8	71.5	–
	QNet [48]	1	1	47.9	72.5	✓
	SiBNN [46]	1	1	50.5	74.6	–
	ProxyBNN	1	1	**51.4**	**75.5**	–
	DoReFa [52]	1	32	53.9	76.3	–
	BWN [37]	1	32	56.8	79.4	–
	ADMM [24]	1	32	57.0	79.7	–
	QNet [48]	1	32	58.8	**81.7**	✓
	ProxyBNN	1	32	**59.3**	81.3	–
	FP32	32	32	61.8	83.5	–

images, that cover 1000 object classes. As in [13,14,27,28], we conduct the standard PyTorch [34] data preprocessing for both training and inference, *i.e.*, random resized 224×224 (227×227 for AlexNet) crop with the standard horizontal flip. We follow the settings in CIFAR experiments, except that the initial learning rate is set to 0.001 and the training time is 110 epochs.

For binary weights and further activation binarization, we compare the proposed algorithm with the state-of-the-art approaches. Table 5 shows the performance gap between binary weights networks and full-precision counterparts have been narrowed to less than three points. The performance improvement in Table 6 is consistent with ResNet. When comparing to multi-bit methods such as 5 bases ABC-ResNet18 [25] with 85.9% Top-5 accuracy, our approach achieves $25\times$ less computing cost, yet suffers only -1.1% accuracy loss.

5 Conclusions

In this paper, we present a new technique for training binarized neural networks, that decomposes pre-binarization weights into the basis and coordinates. We consider different construction schemes for the basis and empirically analyze the superiority of the orthogonal scheme. When jointly optimized by weights quantization error and cross-entropy loss, the orthogonal scheme preserves the unimodal distribution while minimizing the binarization error. Our experiments demonstrate that ProxyBNN has a better generalization capacity than previous methods on benchmark datasets. These results show that mainstream architectures can generally benefit from the proposed proxy learning, which enables the deployment of deep binarized neural networks on low-power devices.

Acknowledgement. This work was supported in part by National Natural Science Foundation of China (No.61972396, 61876182, 61906193), National Key Research and Development Program of China (No. 2019AAA0103402), the Strategic Priority Research Program of Chinese Academy of Science(No.XDB32050200), the Advance Research Program (No. 31511130301), and Jiangsu Frontier Technology Basic Research Project (No. BK20192004).

References

1. Alizadeh, M., Fernández-Marqués, J., Lane, N.D., Gal, Y.: A systematic study of binary neural networks' optimisation. In: International Conference on Learning Representations (2019). https://openreview.net/forum?id=rJfUCoR5KX
2. Anderson, A.G., Berg, C.P.: The high-dimensional geometry of binary neural networks. In: International Conference on Learning Representations (2018). https://openreview.net/forum?id=B1IDRdeCW
3. Andoni, A., Indyk, P.: Near-optimal hashing algorithms for approximate nearest neighbor in high dimensions. Commun. ACM **51**(1), 117–122 (2008). http://doi.acm.org/10.1145/1327452.1327494
4. Bahou, A.A., Karunaratne, G., Andri, R., Cavigelli, L., Benini, L.: XNORBIN: A 95 top/s/w hardware accelerator for binary convolutional neural networks. In: 2018 IEEE Symposium in Low-Power and High-Speed Chips, COOL CHIPS 2018, Yokohama, Japan, 18–20 Apr 2018, pp. 1–3. IEEE Computer Society (2018). https://doi.org/10.1109/CoolChips.2018.8373076
5. Bethge, J., Yang, H., Bornstein, M., Meinel, C.: Back to simplicity: How to train accurate BNNs from scratch? CoRR abs/1906.08637 (2019). http://arxiv.org/abs/1906.08637
6. Bulat, A., Tzimiropoulos, G.: XNOR-Net++: Improved binary neural networks. In: British Machine Vision Conference, BMVC 2019 (2019)
7. Bulat, A., Tzimiropoulos, G., Kossaifi, J., Pantic, M.: Improved training of binary networks for human pose estimation and image recognition. CoRR abs/1904.05868 (2019). http://arxiv.org/abs/1904.05868
8. Cheng, J., Leng, C., Wu, J., Cui, H., Lu, H.: Fast and accurate image matching with cascade hashing for 3d reconstruction. In: 2014 IEEE Conference on Computer Vision and Pattern Recognition, CVPR 2014, Columbus, OH, USA, 23–28 Jun 2014, pp. 1–8 (2014). https://doi.org/10.1109/CVPR.2014.8
9. Courbariaux, M., Bengio, Y., David, J.: Binaryconnect: Training deep neural networks with binary weights during propagations. In: Advances in Neural Information Processing Systems 28: Annual Conference on Neural Information Processing Systems 2015, 7–12 Dec 2015, Montreal, Quebec, Canada, pp. 3123–3131 (2015). http://papers.nips.cc/paper/5647-binaryconnect-training-deep-neural-networks-with-binary-weights-during-propagations
10. Darabi, S., Belbahri, M., Courbariaux, M., Nia, V.P.: BNN+: improved binary network training. CoRR abs/1812.11800 (2018). http://arxiv.org/abs/1812.11800
11. Ding, R., Chin, T., Liu, Z., Marculescu, D.: Regularizing activation distribution for training binarized deep networks. In: IEEE Conference on Computer Vision and Pattern Recognition, CVPR 2019, Long Beach, CA, USA, 16–20 June 2019, pp. 11408–11417 (2019)

12. Gong, Y., Lazebnik, S., Gordo, A., Perronnin, F.: Iterative quantization: A procrustean approach to learning binary codes for large-scale image retrieval. IEEE Trans. Pattern Anal. Mach. Intell. **35**(12), 2916–2929 (2013). https://doi.org/10.1109/TPAMI.2012.193
13. Gu, J., et al.: Projection convolutional neural networks for 1-bit CNNs via discrete back propagation. In: The Thirty-Third AAAI Conference on Artificial Intelligence, AAAI 2019, The Thirty-First Innovative Applications of Artificial Intelligence Conference, IAAI 2019, The Ninth AAAI Symposium on Educational Advances in Artificial Intelligence, EAAI 2019, Honolulu, Hawaii, USA, 27 Jan–1 Feb 2019, pp. 8344–8351 (2019). https://doi.org/10.1609/aaai.v33i01.33018344
14. Gu, J., et al.: Bayesian optimized 1-bit CNNs. In: IEEE Proceedings of the IEEE International Conference on Computer Vision ICCV 2019, Seoul, South Korea (2019)
15. He, K., Zhang, X., Ren, S., Sun, J.: Deep residual learning for image recognition. In: 2016 IEEE Conference on Computer Vision and Pattern Recognition, CVPR 2016, Las Vegas, NV, USA, 27–30 Jun 2016, pp. 770–778 (2016). https://doi.org/10.1109/CVPR.2016.90
16. He, X., Wang, P., Cheng, J.: K-nearest neighbors hashing. In: IEEE Conference on Computer Vision and Pattern Recognition, CVPR 2019, Long Beach, CA, USA, 16–20 Jun 2019, pp. 2839–2848 (2019). http://openaccess.thecvf.com/content_CVPR_2019/html/He_K-Nearest_Neighbors_Hashing_CVPR_2019_paper.html
17. Helwegen, K., Widdicombe, J., Geiger, L., Liu, Z., Cheng, K., Nusselder, R.: Latent weights do not exist: Rethinking binarized neural network optimization. In: Wallach, H.M., Larochelle, H., Beygelzimer, A., d'Alché-Buc, F., Fox, E.B., Garnett, R. (eds.) Advances in Neural Information Processing Systems 32: Annual Conference on Neural Information Processing Systems 2019, NeurIPS 2019, 8–14 Dec 2019, Vancouver, BC, Canada, pp. 7531–7542 (2019). http://papers.nips.cc/paper/8971-latent-weights-do-not-exist-rethinking-binarized-neural-network-optimization
18. Hu, Q., Wang, P., Cheng, J.: From hashing to CNNs: Training binary weight networks via hashing. In: Proceedings of the Thirty-Second AAAI Conference on Artificial Intelligence, (AAAI-18), the 30th innovative Applications of Artificial Intelligence (IAAI-18), and the 8th AAAI Symposium on Educational Advances in Artificial Intelligence (EAAI-18), New Orleans, Louisiana, USA, 2–7 Feb 2018, pp. 3247–3254 (2018). https://www.aaai.org/ocs/index.php/AAAI/AAAI18/paper/view/16466
19. Hu, Q., Wu, J., Bai, L., Zhang, Y., Cheng, J.: Fast k-means for large scale clustering. In: Proceedings of the 2017 ACM on Conference on Information and Knowledge Management, CIKM 2017, Singapore, 06–10 Nov 2017, pp. 2099–2102 (2017). https://doi.org/10.1145/3132847.3133091
20. Hubara, I., Courbariaux, M., Soudry, D., El-Yaniv, R., Bengio, Y.: Binarized neural networks. In: Advances in Neural Information Processing Systems 29: Annual Conference on Neural Information Processing Systems 2016, 5–10 Dec 2016, Barcelona, Spain, pp. 4107–4115 (2016). http://papers.nips.cc/paper/6573-binarized-neural-networks
21. Ji, C., Psaltis, D.: Capacity of two-layer feedforward neural networks with binary weights. IEEE Trans. Inf. Theory **44**(1), 256–268 (1998). https://doi.org/10.1109/18.651033

22. Kingma, D.P., Ba, J.: Adam: A method for stochastic optimization. In: Bengio, Y., LeCun, Y. (eds.) 3rd International Conference on Learning Representations, ICLR 2015, San Diego, CA, USA, 7–9 May 2015, Conference Track Proceedings (2015). http://arxiv.org/abs/1412.6980
23. Krizhevsky, A., Sutskever, I., Hinton, G.E.: ImageNet classification with deep convolutional neural networks. Commun. ACM **60**(6), 84–90 (2017). http://doi.acm.org/10.1145/3065386
24. Leng, C., Dou, Z., Li, H., Zhu, S., Jin, R.: Extremely low bit neural network: Squeeze the last bit out with ADMM. In: Proceedings of the Thirty-Second AAAI Conference on Artificial Intelligence, (AAAI-18), the 30th innovative Applications of Artificial Intelligence (IAAI-18), and the 8th AAAI Symposium on Educational Advances in Artificial Intelligence (EAAI-18), New Orleans, Louisiana, USA, 2–7 Feb 2018, pp. 3466–3473 (2018). https://www.aaai.org/ocs/index.php/AAAI/AAAI18/paper/view/16767
25. Lin, X., Zhao, C., Pan, W.: Towards accurate binary convolutional neural network. In: Advances in Neural Information Processing Systems 30: Annual Conference on Neural Information Processing Systems 2017, 4–9 Dec 2017, Long Beach, CA, USA, pp. 345–353 (2017). http://papers.nips.cc/paper/6638-towards-accurate-binary-convolutional-neural-network
26. Lin, Z., Courbariaux, M., Memisevic, R., Bengio, Y.: Neural networks with few multiplications. In: 4th International Conference on Learning Representations, ICLR 2016, San Juan, Puerto Rico, 2–4 May 2016, Conference Track Proceedings (2016). http://arxiv.org/abs/1510.03009
27. Liu, C., Ding, W., Hu, Y., Zhang, B., Liu, J., Guo, G.: GBCNs: Genetic binary convolutional networks for enhancing the performance of 1-bit DCNNs. In: The Thirty-Fourth AAAI Conference on Artificial Intelligence (AAAI-20) (February 2020)
28. Liu, Z., Wu, B., Luo, W., Yang, X., Liu, W., Cheng, K.: Bi-real net: Enhancing the performance of 1-bit CNNs with improved representational capability and advanced training algorithm. In: Computer Vision - ECCV 2018–15th European Conference, Munich, Germany, 8–14 Sept 2018, Proceedings, Part XV, pp. 747–763 (2018). https://doi.org/10.1007/978-3-030-01267-0_44
29. Martinez, B., Yang, J., Bulat, A., Tzimiropoulos, G.: Training binary neural networks with real-to-binary convolutions. In: International Conference on Learning Representations (2020). https://openreview.net/forum?id=BJg4NgBKvH
30. Mayoraz, E., Aviolat, F.: Constructive training methods for feedforward neural networks with binary weights. Int. J. Neural Syst. **7**(2), 149–66 (1995)
31. Mishra, A.K., Nurvitadhi, E., Cook, J.J., Marr, D.: WRPN: wide reduced-precision networks. In: 6th International Conference on Learning Representations, ICLR 2018, Vancouver, BC, Canada, 30 Apr–3 May 2018, Conference Track Proceedings (2018). https://openreview.net/forum?id=B1ZvaaeAZ
32. Oliveira, A.L., Sangiovanni-Vincentelli, A.L.: Learning complex Boolean functions: Algorithms and applications. In: Advances in Neural Information Processing Systems 6, [7th NIPS Conference, Denver, Colorado, USA, 1993], pp. 911–918 (1993). http://papers.nips.cc/paper/857-learning-complex-boolean-functions-algorithms-and-applications

33. Pagallo, G., Haussler, D.: Boolean feature discovery in empirical learning. Mach. Learn. **5**, 71–99 (1990). https://doi.org/10.1007/BF00115895
34. Paszke, A., et al.: Pytorch: An imperative style, high-performance deep learning library. In: Advances in Neural Information Processing Systems 32: Annual Conference on Neural Information Processing Systems 2019, NeurIPS 2019, 8–14 Dec 2019, Vancouver, BC, Canada, pp. 8024–8035 (2019). http://papers.nips.cc/paper/9015-pytorch-an-imperative-style-high-performance-deep-learning-library
35. Peters, J.W., Genewein, T., Welling, M.: Probabilistic binary neural networks (2019). https://openreview.net/forum?id=B1fysiAqK7
36. Qin, H., Gong, R., Liu, X., Wei, Z., Yu, F., Song, J.: IR-Net: Forward and backward information retention for highly accurate binary neural networks. In: The IEEE Conference on Computer Vision and Pattern Recognition (CVPR), Seattle Wastington, USA (June 2020)
37. Rastegari, M., Ordonez, V., Redmon, J., Farhadi, A.: XNOR-Net: Imagenet classification using binary convolutional neural networks. In: Computer Vision - ECCV 2016–14th European Conference, Amsterdam, The Netherlands, 11–14 Oct 2016, Proceedings, Part IV, pp. 525–542 (2016). https://doi.org/10.1007/978-3-319-46493-0_32
38. Rublee, E., Rabaud, V., Konolige, K., Bradski, G.R.: ORB: An efficient alternative to SIFT or SURF. In: IEEE International Conference on Computer Vision, ICCV 2011, Barcelona, Spain, 6–13 Nov 2011, pp. 2564–2571 (2011). https://doi.org/10.1109/ICCV.2011.6126544
39. Santurkar, S., Tsipras, D., Ilyas, A., Madry, A.: How does batch normalization help optimization? In: Advances in Neural Information Processing Systems 31: Annual Conference on Neural Information Processing Systems 2018, NeurIPS 2018, 3–8 Dec 2018, Montréal, Canada, pp. 2488–2498 (2018). http://papers.nips.cc/paper/7515-how-does-batch-normalization-help-optimization
40. Schonemann, P.H.: A generalized solution of the orthogonal procrustes problem. Psychometrika **31**(1), 1–10 (1966)
41. Shayer, O., Levi, D., Fetaya, E.: Learning discrete weights using the local reparameterization trick. In: 6th International Conference on Learning Representations, ICLR 2018, Vancouver, BC, Canada, 30 Apr–3 May 2018, Conference Track Proceedings. OpenReview.net (2018). https://openreview.net/forum?id=BySRH6CpW
42. Simonyan, K., Zisserman, A.: Very deep convolutional networks for large-scale image recognition. In: 3rd International Conference on Learning Representations, ICLR 2015, San Diego, CA, USA, 7–9 May 2015, Conference Track Proceedings (2015). http://arxiv.org/abs/1409.1556
43. Soudry, D., Hubara, I., Meir, R.: Expectation backpropagation: Parameter-free training of multilayer neural networks with continuous or discrete weights. In: Advances in Neural Information Processing Systems 27: Annual Conference on Neural Information Processing Systems 2014, 8–13 Dec 2014, Montreal, Quebec, Canada, pp. 963–971 (2014). http://papers.nips.cc/paper/5269-expectation-backpropagation-parameter-free-training-of-multilayer-neural-networks-with-continuous-or-discrete-weights
44. Soudry, D., Meir, R.: Mean field Bayes backpropagation: scalable training of multilayer neural networks with binary weights (2013)
45. Tang, W., Hua, G., Wang, L.: How to train a compact binary neural network with high accuracy? In: Proceedings of the Thirty-First AAAI Conference on Artificial Intelligence, 4–9 Feb 2017, San Francisco, California, USA, pp. 2625–2631 (2017). http://aaai.org/ocs/index.php/AAAI/AAAI17/paper/view/14619

46. Wang, P., He, X., Li, G., Zhao, T., Cheng, J.: Sparsity-inducing binarized neural networks. In: The Thirty-Fourth AAAI Conference on Artificial Intelligence (AAAI-20) (February 2020)
47. Wen, Y., Zhang, K., Li, Z., Qiao, Y.: A discriminative feature learning approach for deep face recognition. In: Computer Vision-ECCV 2016–14th European Conference, Amsterdam, The Netherlands, 11–14 Oct 2016, Proceedings, Part VII, pp. 499–515 (2016). https://doi.org/10.1007/978-3-319-46478-7_31
48. Yang, J., et al.: Quantization networks. In: The IEEE Conference on Computer Vision and Pattern Recognition (CVPR), Long Beach CA, USA (June 2019)
49. Yazdani, M.: Linear backprop in non-linear networks. In: Compact Deep Neural Network Representation with Industrial Applications Workshop, Advances in Neural Information Processing Systems 31: Annual Conference on Neural Information Processing Systems 2018, NeurIPS 2018, 3–8 Dec 2018, Montréal, Canada (2018)
50. Zagoruyko, S., Komodakis, N.: Wide residual networks. In: Proceedings of the British Machine Vision Conference 2016, BMVC 2016, York, UK, 19–22 Sept 2016 (2016). http://www.bmva.org/bmvc/2016/papers/paper087/index.html
51. Zhao, T., He, X., Cheng, J., Hu, J.: Bitstream: Efficient computing architecture for real-time low-power inference of binary neural networks on CPUs. In: Boll, S., Lee, K.M., Luo, J., Zhu, W., Byun, H., Chen, C.W., Lienhart, R., Mei, T. (eds.) 2018 ACM Multimedia Conference on Multimedia Conference, MM 2018, Seoul, Republic of Korea, 22–26 Oct 2018, pp. 1545–1552. ACM (2018). https://doi.org/10.1145/3240508.3240673
52. Zhou, S., Ni, Z., Zhou, X., Wen, H., Wu, Y., Zou, Y.: DoReFa-Net: Training low bitwidth convolutional neural networks with low bitwidth gradients. CoRR abs/1606.06160 (2016). http://arxiv.org/abs/1606.06160

HMOR: Hierarchical Multi-person Ordinal Relations for Monocular Multi-person 3D Pose Estimation

Can Wang[2], Jiefeng Li[1], Wentao Liu[2], Chen Qian[2], and Cewu Lu[1](✉)

[1] Shanghai Jiao Tong University, Shanghai, China
{ljf_likit,lucewu}@sjtu.edu.cn
[2] SenseTime Research, Beijing, China
{wangcan,liuwentao,qianchen}@sensetime.com

Abstract. Remarkable progress has been made in 3D human pose estimation from a monocular RGB camera. However, only a few studies explored 3D multi-person cases. In this paper, we attempt to address the lack of a global perspective of the top-down approaches by introducing a novel form of supervision - *Hierarchical Multi-person Ordinal Relations (HMOR)*. The HMOR encodes interaction information as the ordinal relations of depths and angles hierarchically, which captures the *body-part* and *joint* level semantic and maintains global consistency at the same time. In our approach, an integrated top-down model is designed to leverage these ordinal relations in the learning process. The integrated model estimates human bounding boxes, human depths, and root-relative 3D poses simultaneously, with a coarse-to-fine architecture to improve the accuracy of depth estimation. The proposed method significantly outperforms state-of-the-art methods on publicly available multi-person 3D pose datasets. In addition to superior performance, our method costs lower computation complexity and fewer model parameters.

Keywords: 3D human pose · Ordinal relations · Integrated model

1 Introduction

Estimating 3D human poses from a monocular RGB camera is fundamental and challenging. It has found applications in robotics [13,72], activity recognition [32,50], human-object interaction detection [15,28,29,51], and content creation for graphics [1,4]. With deep neural networks [19,43,44,57] and large scale

C. Wang and J. Li are equal contribution, order determined by coin flip.
C. Lu is the member of Qing Yuan Research Institute and MoE Key Lab of Artificial Intelligence, AI Institute, Shanghai Jiao Tong University, China.

Electronic supplementary material The online version of this chapter (https://doi.org/10.1007/978-3-030-58580-8_15) contains supplementary material, which is available to authorized users.

A. Vedaldi et al. (Eds.): ECCV 2020, LNCS 12348, pp. 242–259, 2020.
https://doi.org/10.1007/978-3-030-58580-8_15

publicly available datasets [3,21,23,31,33,36,38,56], significant improvement has been achieved in the field of 3D pose estimation. Most of the works [9,17,35,47, 59,60,64,71] focus on estimating the single-person pose. Recently, some methods [38,39,48,52,53,66,67] start to deal with multi-person cases. However, recovering absolute 3D poses in the camera-centered coordinate system is quite a challenge. Since multi-person activities take place in cluttered scenes, inherent depth ambiguity and occlusions make it still difficult to estimate the absolute position of multiple instances.

Recently, top-down approaches [39,52,53] achieve noticeable improvements in estimating multi-person 3D poses. These approaches first perform human detection and estimate the 3D pose of each person by a single-person pose estimator. However, the pose estimator is applied to each bounding box separately, which raises the doubt that the top-down models are not able to understand multi-person relationships and handle complex scenes. Without a broad view of the input scenario, it is challenging to get rid of inherent depth ambiguity and occlusion problems. In this paper, the relationship among multiple persons is fully considered to address this limitation of top-down approaches.

We propose a novel form of supervision for 3D pose estimation - *Hierarchical Multi-person Ordinal Relations (HMOR)*. HMOR explicitly encodes the interaction information as ordinal relations, supervising the networks to output 3D poses in the correct order. Different from previous works [46,54,61] that only use relative depth information, HMOR considers both depths and angles relations and expresses the ordinal information hierarchically, i.e., *instance* → *part* → *joint*, which makes up for the lack of a global perspective of the top-down approaches.

Further, we propose an integrated top-down model to learn this knowledge by encoding it into the learning process. The integrated model can be end-to-end trained with back-propagation and performs *human detection*, *pose estimation*, and *human-depth estimation* simultaneously. Since metric depth from a single image is fundamentally ambiguous, estimating absolute 3D pose suffers from inaccurate human-depth estimation. To improve the accuracy, we take a coarse-to-fine approach to estimate human depth: i) initializes a global depth map, and ii) finetunes the human depths by estimating the correction residual.

We evaluate our method on two multi-person [23,38] and one single-person 3D pose datasets [21]. Our method significantly outperforms previous multi-person 3D pose estimation methods [26,37–39,52,67] by **12.3** PCK_{abs} improvement on the MuPoTS-3D [38] dataset, and **20.5** mm improvement on CMU Panoptic [23] dataset, with lower computation complexity and fewer model parameters. Compared to state-of-the-art single-person methods [17,59,60,68], our method does not need ground-truth bounding-box in the inference phase and still achieves comparable performance. Additionally, our proposed method is compatible with 2D pose annotations, which allows the 2D-3D mixed training strategy.

The contributions of this paper can be summarized as follows:

- We propose HMOR, a novel form of supervision, to explicitly leverage the relationship among multiple persons for pose estimation. HMOR divides human relations into three levels: *instance*, *part* and *joint*. This hierarchical

manner ensures both the global consistency and the fine-grained accuracy of the predicted results.
- An integrated end-to-end top-down model is proposed for multi-person 3D pose estimation from a monocular RGB input. We design a coarse-to-fine architecture to improve the accuracy of human-depth estimation. Our model jointly performs human detection, human-depth estimation, and 2D/3D pose estimation.

2 Related Work

Multi-person 2D Pose Estimation. Most of the multi-person 2D pose estimation methods can be divided into two categories: bottom-up and top-down approaches. Bottom-up approaches localize the body joints and group them into different persons. Traditional top-down approaches first detect human bounding boxes in the image and then estimate single-person 2D poses separately.

Representative works [5,22,25,42] of the bottom-up approaches are reviewed. Cao et al. [5] propose part affinity fields (PAFs) to model human bones. Complete skeletons are assembled by detected joints with PAFs. Newell et al. [42] introduce a pixel-wise tag to assign joints to a specific person. Kocabas et al. [25] assign joints to detected persons by a pose residual network.

Top-down approaches [10,16,18,27,58,62,63] achieve impressive accuracy in multi-person 2D pose estimation. Mask R-CNN [18] is an end-to-end model to estimate multiple human poses but still process multiple persons separately. Fang et al. [16] propose a two-stage framework (RMPE) to reduce the effect of the inaccurate human detector. Sun et al. [58] propose the HRNet that maintains high-resolution representations through the whole process.

Single-person 3D Pose Estimation. There are two approaches to the problem of single-person 3D pose estimation from monocular RGB: single-stage and two-stage approaches.

Single-stage approaches [24,36,47,59,60] directly locate 3D human joints from the input image. For example, Pavlakos et al. [47] propose a coarse-to-fine approach to estimate a 3D heatmap for pose estimation. Kanazawa et al. [24] recover 3D pose and body mesh by minimizing the reprojection loss. Sun et al. [60] operate an integral operation as soft-argmax to obtain 3D pose coordinates in a differentiable manner.

Two-stage approaches [2,7,17,35,40,45,64,65,71] first estimate 2D pose or utilize the off-the-shelf accurate 2D pose estimator, and then lift them to the 3D space. Martinez et al. [35] propose a simple baseline to regress 3D pose from 2D coordinates directly. Moreno-Noguer [40] obtains more precise pose estimation by the distance matrix representation. Yang et al. [64] utilize a multi-source discriminator to generate anthropometrically valid poses.

Multi-person 3D Pose Estimation. A few works explore the problem of multi-person 3D pose estimation from a monocular RGB. Rogez et al. [52,53]

propose LCR-Net and LCR-Net++. They locate human bounding boxes and classify those boxes into a set of K anchor-poses. A regression module is proposed to refine the anchor-pose to the final prediction. Instead of using a learning-based manner, they obtain the human depth by minimizing the distance between the projected 3D pose and the estimated 2D pose. Mehta et al. [38] propose a bottom-up method. Their proposed occlusion-robust pose-map (ORPM) enables full body pose inference even under strong partial occlusions. Zanfir et al. [67] propose MubyNet, a bottom-up model. MubyNet integrates a limb scoring model and formulates the person grouping problem as an integer program. Moon et al. [39] propose a top-down two-stage model. They utilize the off-the-shelf human detection model and then perform single-person 3D pose estimation and root-joint localization. Those top-down approaches are not able to utilize multi-person relations since they estimate individual 3D pose separately. The bottom-up approaches are still suffering from limited accuracy. Our method combines the advantages of both approaches and boosts multi-person absolute 3D pose estimation by leveraging the multi-person relations in the integrated end-to-end top-down model.

Ordinal Relations. In the context of computer vision, several works learn ordinal apparent depth [8,73] or reflectance [41,69] relationship as weak supervision. They motivated by the fact that ordinal relations are easier for humans to annotate. In the case of single-person 3D pose estimation, [46,54,55] use depth relations of body joints to generate 3D pose from 2D pose.

3 Method

We propose a novel representation, *Hierarchical Multi-person Ordinal Relation (HMOR)*, to explicitly leverage ordinal relations among multiple persons and improve the performance of 3D pose estimation. Compared with previous works [46,54,61] that use ordinal relation in 3D pose estimation, HMOR extends this idea in three dimensions: i) single-person to multi-persons, ii) *joint* level to hierarchical *instance-part-joint* levels, iii) depth relations to angle relations. Further, we develop an integrated model to aggregate HMOR into the end-to-end training process. In this section, we first describe the unified representation of the absolute multi-person 3D pose recovery under the top-down framework (Sect. 3.1). Then we detail the encoding and training schemes of the proposed HMOR (Sect. 3.2). Finally, the integrated model with a coarse-to-fine depth estimation design is elaborated (Sect. 3.3).

3.1 Representation

Our task is to recover multiple absolute 3D human poses $\mathcal{P} = \{\mathbf{P}_m^{abs}\}_{m=1}^N$ in the camera-centered coordinate system, where N denotes the number of persons in the input RGB image. We assume that there are J joints in a single 3D pose skeleton. The mth absolute 3D pose can be formulated as:

$$\mathbf{P}_m^{abs} = \{\mathbf{k}_{m,j} : (x_{m,j}^{abs}, y_{m,j}^{abs}, z_{m,j}^{abs})^\mathsf{T}\}_{j=1}^J, \quad (1)$$

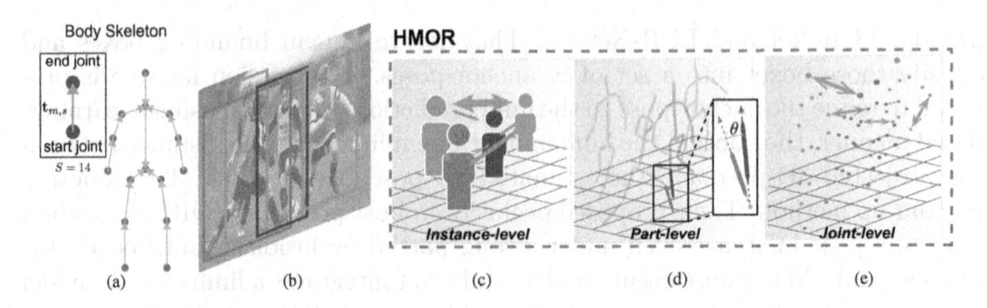

Fig. 1. Illustration of the proposed HMOR. (a) Definition of skeletal parts. (b) Monocular input image. (c–e) Hierarchical Multi-person Ordinal Relations. HMOR supervises the ordinal relations among multiple persons

where $\mathbf{k}_{m,j}$ is the jth joint position of the mth absolute pose.

Human bounding boxes $\{\hat{\mathbf{B}}_m\}_{m=1}^N$, root-relative 3D poses $\{\hat{\mathbf{P}}_m^{rel}\}_{m=1}^N$, and absolute depth of the root-joint $\{\hat{z}_{m,R}^{abs}\}_{m=1}^N$ are needed to estimate the absolute 3D poses. We term root-joint's absolute depth as human depth, corresponding to the pelvis bone position (the Rth joint of the body skeleton). We use $\hat{}$ to denote the predicted values. The mth human bounding box $\hat{\mathbf{B}}_m$ and root-relative 3D pose $\hat{\mathbf{P}}_m^{rel}$ are formulated as:

$$\hat{\mathbf{B}}_m = (\hat{u}_m^{top}, \hat{v}_m^{top}, \hat{w}_m, \hat{h}_m)^\intercal, \tag{2}$$

$$\hat{\mathbf{P}}_m^{rel} = \{(\hat{u}_{m,j}, \hat{v}_{m,j}, \hat{z}_{m,j}^{rel})^\intercal\}_{j=1}^J, \tag{3}$$

where $\hat{u}_{m,j}$ and $\hat{v}_{m,j}$ represent pixel coordinates of the estimated body joint with respect to the bounding box. $\hat{z}_{m,j}^{rel}$ denotes the estimated depth of joint j relative to the root-joint. $\hat{u}_m^{top}$, $\hat{v}_m^{top}$, $\hat{w}_m$, and $\hat{h}_m$ are the top left corner coordinates, the width, and the height of the predicted bounding box, respectively. With the intrinsic matrix $\mathbf{M}$, the final absolute 3D pose $\hat{\mathbf{P}}_m^{abs}$ can be obtained via back-projection, where each joint is calculated by:

$$\begin{pmatrix} \hat{x}_{m,j}^{abs} \\ \hat{y}_{m,j}^{abs} \\ \hat{z}_{m,j}^{abs} \end{pmatrix} = (\hat{z}_{m,j}^{rel} + \hat{z}_{m,R}^{abs})\mathbf{M}^{-1} \begin{pmatrix} \hat{u}_{m,j} + \hat{u}_m^{top} \\ \hat{v}_{m,j} + \hat{v}_m^{top} \\ 1 \end{pmatrix}. \tag{4}$$

3.2 Hierarchical Multi-person Ordinal Relations

Our initial goal is to leverage multi-person interaction relations to improve the performance of 3D pose estimation. Traditional top-down methods [39,52,53] lack a global perspective because they estimate single human poses in each bounding box separately. Therefore, they are vulnerable to truncation, self-occlusions, and inter-person occlusions. Here, we develop a novel form of supervision named *Hierarchical Multi-person Ordinal Relations (HMOR)* to model human relations explicitly. Basically, given an image of human activities, we divide the relationship into three levels: i) instance-level depth relations, ii) part-level angle relations, iii) joint-level depth relations. In each level, HMOR formulates pair-wise ordinal relations and punishes the wrong-order pairs. In the following, we detail our HMOR formulations that reflect interpretable relations of human activities.

Instance-Level Depth Relations. In a given camera view, for two persons $(\mathbf{p}_1, \mathbf{p}_2)$, we denote the instance depth-relation function as $R_{ins}(\mathbf{p}_1, \mathbf{p}_2; \mathbf{n}_\perp)$, taking the value:

- $+1$, if $\mathbf{p}_1$ is closer than $\mathbf{p}_2$ in the $\mathbf{n}_\perp$ direction,
- -1, if $\mathbf{p}_2$ is closer than $\mathbf{p}_1$ in the $\mathbf{n}_\perp$ direction,
- 0, if the depths of two person are equal,

where $\mathbf{n}_\perp$ is the camera normal vector. We define the position of a person as the arithmetic mean of its body joints, i.e. $\mathbf{p}_m = \frac{1}{J}\sum_j^J \hat{\mathbf{k}}_{m,j}$. The ordinal error of a pair of instances is denoted as:

$$err_{ins}(\hat{\mathbf{p}}_1, \hat{\mathbf{p}}_2) = \log(1 + \max(0, R_{ins}(\hat{\mathbf{p}}_1, \hat{\mathbf{p}}_2; \mathbf{n}_\perp) * [(\hat{\mathbf{p}}_1 - \hat{\mathbf{p}}_2) \cdot \mathbf{n}_\perp])). \quad (5)$$

This differentiable instance ranking expression will punish the wrong-order instance pairs and ignore the correct results. For example, if $\mathbf{p}_1$ is closer than $\mathbf{p}_2$, and the prediction relation is correct, i.e., $(\hat{\mathbf{p}}_1 - \hat{\mathbf{p}}_2) \cdot \mathbf{n}_\perp < 0$, the multiplication result will be smaller than 0 and ignored by the maximum operation.

Supervising the instance-level depth relations is to help the network build a global understanding of the input scenario. Ablative study in Sect. 4.4 reveals that the accuracy of human-depth estimation benefits a lot from instance-level depth relations.

Part-Level Angle Relations. As shown in Fig. 1(a), we divide the body skeleton into $S = 14$ parts according to the kinematically connected joints. Each part $\mathbf{t}$ is a vector defined by start-joint $\mathbf{k}_{start}$ and end-joint $\mathbf{k}_{end}$, i.e., $\mathbf{t} = \mathbf{k}_{end} - \mathbf{k}_{start}$. Since body-parts are a set of 3D vectors with direction and length values, we can not directly compare their depths. Here, we utilize a unique attribute of body-part – direction, and compare their angle relations. To simplify the ordinal relation of angles, we first project the body-part vector $\mathbf{t}_{m,s}$ onto the camera plane:

$$\mathbf{t}_{m,s}^{\mathbf{n}_\perp} = \mathbf{t}_{m,s} - (\mathbf{t}_{m,s} \cdot \mathbf{n}_\perp)\mathbf{n}_\perp, \quad (6)$$

where m is the person index, and s is the body-part index. In a given camera view, for a pair of body parts $(\mathbf{t}_{m_1,s_1}, \mathbf{t}_{m_2,s_2})$, we denote the angle-relation function as $R_{arg}(\mathbf{t}_{m_1,s_1}, \mathbf{t}_{m_2,s_2}; \mathbf{n}_\perp)$, taking the value:

- $+1$, if $\mathrm{Arg}(\mathbf{t}_{m_1,s_1}^{\mathbf{n}_\perp}) < \mathrm{Arg}(\mathbf{t}_{m_2,s_2}^{\mathbf{n}_\perp})$,
- -1, if $\mathrm{Arg}(\mathbf{t}_{m_1,s_1}^{\mathbf{n}_\perp}) > \mathrm{Arg}(\mathbf{t}_{m_2,s_2}^{\mathbf{n}_\perp})$,
- 0, if $\mathrm{Arg}(\mathbf{t}_{m_1,s_1}^{\mathbf{n}_\perp}) = \mathrm{Arg}(\mathbf{t}_{m_2,s_2}^{\mathbf{n}_\perp})$,

where $\mathrm{Arg}(\mathbf{t}^{\mathbf{n}_\perp})$ computes the principal value of the argument of the projection vector. The ordinal error of a pair of body-parts is:

$$err_{part}(\hat{\mathbf{t}}_{m_1,s_1}, \hat{\mathbf{t}}_{m_2,s_2}) = [R_{arg}(\hat{\mathbf{t}}_{m_1,s_1}, \hat{\mathbf{t}}_{m_2,s_2}; \mathbf{n}_\perp) * [(\hat{\mathbf{t}}_{m_1,s_1} \times \hat{\mathbf{t}}_{m_2,s_2}) \cdot \mathbf{n}_\perp]]_+. \quad (7)$$

With the cross-product operation $\times$, we supervise the direction of the angle between a pair of body-parts. If the angle between $\hat{\mathbf{t}}_{m_1,s_1}$ and $\hat{\mathbf{t}}_{m_2,s_2}$ is in the correct direction, the projection of the cross-product $(\hat{\mathbf{t}}_{m_1,s_1} \times \hat{\mathbf{t}}_{m_2,s_2}) \cdot \mathbf{n}_\perp$ will have an opposite sign of $R_{arg}(\cdot)$. Therefore, the negative multiplication results will be ignored by the $[\cdot]_+$ operation.

Another intuitive way is to express body-parts as particles and supervise their depth relations, using the average position of its two endpoints. To compare vector and particle representations, we conduct ablative experiments and find out vector is superior to particle representation. We suspect this is because the depth relations have been fully utilized in the other two levels, supervising depths of body-part is redundant. More experimental details are reported in Sect. 4.4.

Joint-Level Depth Relations. The definition of body joint depth-relation function $R_{jt}(\mathbf{k}_{m_1,s_1}, \mathbf{k}_{m_2,s_2}; \mathbf{n}_\perp)$ is similar to R_{ins}:

- $+1$, if $\mathbf{k}_{m_1,s_1}$ is closer than $\mathbf{k}_{m_2,s_2}$ in the $\mathbf{n}_\perp$ direction,
- -1, if $\mathbf{k}_{m_2,s_2}$ is closer than $\mathbf{k}_{m_1,s_1}$ in the $\mathbf{n}_\perp$ direction,
- 0, if the depths of two joints are equal.

The ordinal error of a pair of joints is denoted as:

$$err_{jt}(\hat{\mathbf{k}}_{m_1,s_1}, \hat{\mathbf{k}}_{m_2,s_2}) = \log(1+[R_{jt}(\hat{\mathbf{k}}_{m_1,s_1}, \hat{\mathbf{k}}_{m_2,s_2}; \mathbf{n}_\perp]_+ * [(\hat{\mathbf{k}}_{m_1,s_1} - \hat{\mathbf{k}}_{m_2,s_2}) \cdot \mathbf{n}_\perp])). \quad (8)$$

Denoting the set of estimated *persons*, *body-parts*, and *joints* pairs as $\mathcal{I}_{ins}$, $\mathcal{I}_{part}$, and $\mathcal{I}_{jt}$, respectively, the HMOR loss is computed as follows:

$$\mathcal{L}_{HMOR} = \frac{1}{|\mathcal{I}_{ins}|} \sum_{\hat{\mathbf{p}}_1, \hat{\mathbf{p}}_2} err_{ins} + \frac{1}{|\mathcal{I}_{part}|} \sum_{\hat{\mathbf{t}}_1, \hat{\mathbf{t}}_2} err_{part} + \frac{1}{|\mathcal{I}_{jt}|} \sum_{\hat{\mathbf{k}}_1, \hat{\mathbf{k}}_2} err_{jt} \quad (9)$$

Augmented Training Scheme. As mentioned before, HMOR computes the ordinal relations with respect to a vector $\mathbf{n}_\perp$. Initially, this vector is set as the camera normal vector. However, we notice that annotations from 3D human pose datasets (Human3.6M, MuPoTS-3D, and CMU Panoptic) are mostly captured in an laboratory environment, limited to the fixed viewing angle. To alleviate camera restrictions, we sample virtual views to improve the generalization ability.

In the training phase, we generate a virtual view vector $\mathbf{n}_v$ by rotating the camera normal vector $\mathbf{n}_\perp$ randomly. We adapt the uniform sphere sampling strategy from Marsaglia et al. [34]:

$$\mathbf{n}_v = (\sqrt{1-u^2}\cos\theta, \sqrt{1-u^2}\sin\theta, u)^\mathsf{T}, \quad (10)$$

where $\theta \sim U[0, 2\pi)$ and $u \sim U[0, 1]$. In this way, HMOR can calculate the ordinal relations with respect to an arbitrary viewing angle. The effectiveness of the sampled view is validated in Sect. 4.4.

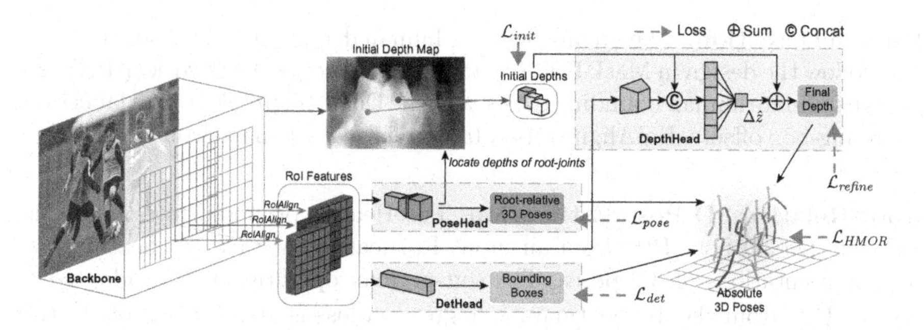

Fig. 2. Architecture of the integrated model. The ResNet-50 based backbone network extract RoI features and initial depth map. PoseHead and DetHead perform root-relative 3D pose estimation and human detection, respectively. DepthHead retrieves initial depths from the depth map and predicts refined human depths by correction residual $\Delta\hat{z}$. This architecture allows the 2D-3D mixed training strategy

Additionally, a mixed datasets training strategy is utilized for a fair comparison with previous methods in experiments. HMOR is compatible with 2D pose datasets and single-person 3D pose datasets. Given an image only with 2D pose annotations, we can define the part-level angle relations, since the 2D pose skeletons are the projections of body-parts with respect to $\mathbf{n}_{\perp}$. As for single-person cases, HMOR only supervises the *joint* and *body-part* relations of an individual person and ignore instance-level relations.

3.3 Integrated End-to-end Model

In our approach, an integrated end-to-end top-down model is designed to aggregate HMOR into the end-to-end training process. Although the disjoint model [39] can use different strong networks for different tasks (e.g., human detection, pose estimation, depth estimation), an integrated model has three advantages over the disjoint learning model: 1) Fewer model parameters. 2) Only an integrated model can leverage the multi-person relations since the disjoint learning methods train their model with single person annotations separately. 3) The multi-task training strategy of the integrated model can benefit each task. In our experiments, the integrated model is found to have much better performance than the disjoint learning methods.

The overall architecture of our model is summarized in Fig. 2. Our model consists of two stages. In the first stage, the backbone network extracts RoIs and the initial depth map. PoseHead and DetHead estimate root-relative 3D poses and human bounding boxes from RoIs, respectively. In the second stage, we retrieve the initial depths of root-joints from the depth map. The DepthHead takes the RoI features and initial depth as input and outputs the correction residual $\Delta\hat{z}$. The residual is added to the initial depth to obtain refined human depths. Aggregating the outputs from DetHead, PoseHead, and DepthHead, the absolute 3D poses $\hat{\mathbf{P}}^{abs}$ are estimated via back-projection as Eq. 4.

Human Detection. The architecture for human detection and the loss function L_{det} follow the design in Mask R-CNN [18]. Region Proposal Network (RPN) proposes candidate human bounding boxes, and the DetHead predicts class labels and bounding-box offsets. RoiAlign is used to extract feature maps from each RoI.

Root-Relative 3D Pose Estimation. PoseHead is proposed to estimate the root-relative 3D pose $\hat{\mathbf{P}}^{rel}$ from an input RoI feature. We use 3D heatmaps as the representations of 3D poses. The soft-argmax operation [60] is adopted to extract $\hat{\mathbf{P}}^{rel}$ from the 3D heatmap. ℓ_1 regression loss is applied to root-relative coordinates $\hat{\mathbf{P}}^{rel}_m$:

$$\mathcal{L}_{pose} = \frac{1}{N}\frac{1}{J}\sum_{m}^{N} \|\hat{\mathbf{P}}^{rel}_m - \mathbf{P}^{rel}_m\|_1. \tag{11}$$

RoIAlign extracts 14×14 RoI features, which are fed into the PoseHead subsequently. We adopt a simple network as PoseHead, including three residual blocks for feature extraction, a transposed convolution [14] for upsampling, a batch normalization layers [20], a ReLU activation function, and a 1×1 convolution. The size of an output heatmap is $28 \times 28 \times 28$.

Human Depth Estimation. Direct human-depth regression from an input RoI is challenging. Part of the challenges comes from the variety of camera parameters and human body shapes. Furthermore, the inputs of DepthHead are fixed-size RoI features, which erase the information of projected body shapes and sizes. Inspired by the idea of iterative error feedback (IEF) from previous works [6,12,24], we design a coarse-to-fine estimation approach to enhance the accuracy of human-depth regression. The model will first predict an initial depth of root-joint $\hat{z}^{init}$. Then the DepthHead takes the RoI features and the initial depth $\hat{z}^{init}$ as an input and outputs the residual Δz. Ideally, the refined depth is updated by adding this residual to the initial estimate $\hat{z}^{refine} = \hat{z}^{init} + \Delta z$.

Depth Initialize. To estimate the initial depths of root-joints, we directly regress an initial depth map. During training, we first normalize the absolute depth value by focal lengths and then calculate the loss $\mathcal{L}_{init}$ between the ground truth and the initial depth map in the area around the root-joint's 2D pixel location:

$$z^{norm}_R = z^{abs}_R / \sqrt{f_x \cdot f_y}. \tag{12}$$

$$\mathcal{L}_{init} = \frac{1}{N}\sum_{m}^{N} \|z^{norm}_{m,R} - \hat{z}^{init}_{m,R}\|_1, \tag{13}$$

Depth Refinement. In the refinement step, we retrieve the initial-depth values of root-joints from the depth map according to their 2D pixel locations. Because the input features are resized by RoIAlign, we first need to transfer the original depth to the equivalent depth of the resized person. According to the pinhole camera model:

$$z_R^{eq,norm} = z_R^{norm} \cdot \sqrt{\frac{A_{Box}}{A_{RoI}}}, \tag{14}$$

$$\hat{z}_R^{eq,init} = \hat{z}_R^{init} \cdot \sqrt{\frac{A_{Box}}{A_{RoI}}}, \tag{15}$$

where A_{Box} denotes the area of the bounding box, and A_{RoI} denotes the area of RoI. DepthHead extracts 1D features from RoIs. Then the equivalent initial-depth values $\hat{z}_{m,R}^{eq,init}$ are concatenated with the extracted features and fed into an *fc* layer to predict the residual $\Delta\hat{z}$. The loss function of the refinement step $\mathcal{L}_{refine}$ is defined as:

$$\mathcal{L}_{refine} = \frac{1}{N}\sum_{m}^{N} \|z_{m,R}^{eq,norm} - \hat{z}_{m,R}^{eq,init} - \Delta\hat{z}_m\|_1. \tag{16}$$

In the testing phase, we can recover the absolute depth of root-joint $\hat{\mathbf{z}}_{m,R}^{abs}$ as:

$$\hat{\mathbf{z}}_{m,R}^{abs} = (\Delta\hat{z}_m + \hat{\mathbf{z}}_{m,R}^{eq,init}) \cdot \sqrt{\frac{f_x \cdot f_y \cdot A_{RoI}}{A_{Box}}}. \tag{17}$$

The DepthHead uses three residual blocks (following ResNet [19]) and an average pooling layer to extract 1D features. The FC layer contains 512 neurons.

The end-to-end training loss is formulated as:

$$\mathcal{L} = \mathcal{L}_{det} + \mathcal{L}_{pose} + \mathcal{L}_{init} + \mathcal{L}_{refine} + \mathcal{L}_{HMOR}. \tag{18}$$

4 Experiment

In this section, we first introduce the datasets employed for quantitative evaluation and elaborate implementation details. Then we report our results and compare the proposed method with state-of-the-art methods. Finally, ablation experiments are conducted to evaluate our contributions and show how each choice contributes to our state-of-the-art performance.

4.1 Datasets

MuCo-3DHP and MuPoTS-3D: MuCo-3DHP is a multi-person composited 3D human pose training dataset. MuPoTS-3D is the real-world scenes test set. Following [38,39], 400K composited frames are utilized for training.

CMU Panoptic: CMU Panoptic [23] is a multi-person 3D pose dataset captured in an indoor dome with multiple cameras. Here we follow the evaluation protocol of [66,67].

3DPW: 3D Poses in the Wild (3DPW) [33] is a recent challenging dataset, captured mostly in outdoor conditions. It contains 60 video sequences (24 train, 24 test, and 12 validation).

Human3.6M: Human3.6M [21] is an indoor benchmark for single-person 3D pose estimation. A total of 11 professional actors (6 male, 5 female) perform 15 activities in a laboratory environment.

Table 1. Quantitative comparisons with state-of-the-art methods on the MuPoTS-3D dataset. "-" shows the results that are not available

Method	$AUC_{rel}\uparrow$	$3DPCK_{rel}\uparrow$	$3DPCK_{abs}\uparrow$
LCRNet [52]	-	53.8	-
Single Shot [38]	-	66.0	-
LCRNet++ [53]	-	70.6	-
Xnect [37]	-	70.4	-
Moon et al. [39]	39.8	81.8	31.5
Ours	**43.5**	**82.0**	**43.8**

Table 2. Quantitative comparisons of MPJPE on the CMU Panoptic dataset

Method	Haggling	Mafia	Ultimatum	Pizza	Mean$\downarrow$
Popa [49]	217.9	187.3	193.6	221.3	203.4
Zanfir [66]	140.0	165.9	150.7	156.0	153.4
Zanfir [67]	72.4	78.8	66.8	94.3	72.1
Ours	**50.9**	**50.5**	**50.7**	**68.2**	**51.6**

4.2 Implementation Details

Our method is implemented in PyTorch. We adopt a ResNet-50 [19] based FPN [30] as our model backbone. The backbone is initialized with the ImageNet [11] pre-trained model. The settings of each network head are reported in Sect. 3.3. We resize the image to 1333×800 and feed into the network. SGD is used for optimization, with a mini-batch size of 32. All tasks are trained simultaneously. We adopt the linear learning rate warm-up policy. The learning rate is set to 0.2/3 at first and gradually increases to 0.2 after 2.5 k iterations. We reduce the learning rate by a factor of 10 at the 10th and 20th epochs. In each experiment, our model is trained for 30 epochs with 16 NVIDIA 1080 Ti GPUs. We perform data augmentations including horizontal flip and multi-scale training. Additional COCO [31] 2D pose estimation data are used in the training phase. For evaluation, we report the flip-test results. All reported numbers have been obtained with a single model without ensembling.

4.3 Compare with Prior Art

MuPoTS-3D. We compare our method against state-of-the-art methods under three protocols. PCK_{abs} is used to evaluate absolute camera-centered coordinates of 3D poses. Additionally, PCK_{rel} and AUC_{rel} are used to evaluate root-relative 3D poses after root alignment. Quantitative results are reported in Table 1. Without bells and whistles, our method surpasses state-of-the-art methods by 12.3 PCK_{abs} (**39.0**% relative improvement). Our method demonstrates a clear advantage for handling multi-person 3D poses.

Table 3. Quantitative comparisons on the Human3.6M dataset

	Single-Person							Multi-Person		
Method	Moreno [40]	Zhou [70]	Martinez [35]	Sun [59]	Fang [17]	Sun [60]	Zhou [68]	Rogez [53]	Moon [39]	**Ours**
PA MPJPE↓	76.5	55.3	47.7	48.3	45.7	40.6	**27.9**	42.7	35.2	**30.5**
	Single-Person					Multi-Person				
Method	Martinez [35]	Fang [17]	Sun [59]	Sun [60]	Zhou [68]	Rogez [52]	Metha [38]	Rogez [53]	Moon [39]	**Ours**
MPJPE↓	62.9	60.4	59.1	49.6	**39.9**	87.7	69.9	63.5	54.4	**48.6**

As for root-relative results, our method achieves 82.0 PCK_{rel} and 43.5 AUC_{rel}. Note that the PCK result relies on the threshold value. AUC can reflect a more reliable result since it computes the area under the PCK curve from various thresholds. Our method outperforms the previous methods by 3.7 AUC_{rel}.

CMU Panoptic. Following previous works [66,67], we evaluate our method under MPJPE after root alignment. Table 2 provides experimental results. In this dataset, the activities take place in a small room. Thus, the scenarios are severely affected by the occlusion problem. Our method effectively reduces the interference of occlusion and outperforms state-of-the-art methods by 20.5 mm MPJPE (**28.4**% relative improvement).

Human3.6M. We conduct experiments on Human3.6M dataset to evaluate the performance of root-relative 3D pose estimation. Two experimental protocols are widely used. *Protocol 1* uses *PA MPJPE* and *Protocol 2* uses *MPJPE* as evaluation metrics. As most of the previous methods use the ground-truth bounding box, our method does not require any ground-truth information at inference time. Quantitative results are reported in Table 3. Our method achieves comparable performance with single-person methods and outperforms previous multi-person 3D pose estimation methods.

4.4 Ablation Study

In this study, we evaluate the effectiveness of the proposed HMOR and integrated model. We evaluate on 3DPW dataset that contains in-the-wild complex scenes to demonstrate the strength of our model. We further propose ABS-MPJPE to evaluate the absolute 3D pose estimation results without root alignment.

Effect of Hierarchical Multi-person Ordinal Relations. In this experiment, we study the effectiveness of using HMOR supervision. We first implement

Table 4. Ablative study on the effects of HMOR

	Settings	3DPW		
		MPJPE↓	PA- ↓	ABS- ↓
(a)	baseline	95.7	63.6	169.3
	+ $\mathcal{L}_{abs}$	94.6	61.1	158.2
	+ *jt*	89.9	59.7	132.8
	+ *part*	90.2	60.3	143.2
	+ *instance*	93.3	61.2	128.3
	+ *jt* + *part*	89.1	58.3	125.9
	+ *jt* + *instance*	89.2	58.5	122.3
	+ *part* + *instance*	89.5	59.5	123.6
	+ *jt* + *part* + *instance*	88.3	57.8	119.6
(b)	+ *jt* + *particle-part* + *instance*	89.0	58.2	119.5
(c)	+ *jt* + *part* + *instance* + *sample views* (Final)	**87.7**	**57.4**	**118.5**
(d)	w/o refine depth	88.4	58.1	133.6

a vanilla baseline without HMOR supervision. Moreover, we implement another baseline by directly supervising the predicted absolute 3D poses with an ℓ_1 loss $\mathcal{L}_{abs}$. Intuitively, since the human poses are evaluated in the camera coordinate system, the local optimum for $\mathcal{L}_{abs}$ is consistent with the evaluation metrics.

The experimental results are shown in Table 4(a). The model trained with $\mathcal{L}_{abs}$ supervision has better performance than the vanilla baseline, but is still inferior to HMOR supervision. HMOR supervision brings 7.4 mm MPJPE improvement. By removing three types of relations separately, we can observe that *instance* relation affects the absolute pose accuracy (ABS-MPJPE) most, while *part* and *joint* relations mainly affect the root-relative pose accuracy.

Variants of HMOR. In this experiment, we examine a variant of HMOR. When handling the part relations, we represent a body part as a particle rather than a vector. The position of a body part is defined as the average of its two endpoints. Similar to *joint* and *instance*, we supervise the depth relations of the particle body-parts. The experimental results are shown in Table 4(b). The particle representation produces inferior performance than the vector representation.

Effect of Sampled Views. Table 4(c) reports the result of training with sampled views. Compare with the results in Table 4(a) that only use the original camera normal vector $\mathbf{n}_\perp$, sampled views provide 0.6 mm MPJPE improvement.

Effect of Coarse-to-Fine Depth Surpervision. In this experiment, we study the effectiveness of the coarse-to-fine design for human depth estimation. We remove the refinement step and output the initial value directly. The experimental results are shown in Table 4(d). We observe that the coarse-to-fine design is necessary to produce accurate human depth.

Fig. 3. Qualitative results of our proposed method on COCO validation set (left) and MuPoTS-3D test set (right)

Table 5. Ablative study on computation complexity and model parameters

Method	#Params↓	GFLOPs↓	AUC_{rel} ↑	PCK_{rel} ↑	PCK_{abs} ↑
Moon [39]	167.7M	547.8	39.8	81.8	31.5
Ours	**45.0M**	**320.2**	**43.5**	**82.0**	**43.8**

Computation Complexity. The experimental results of computation complexity and model parameters are listed in Table 5. We compare our method with Moon et al. [39], which is the only open-source multi-person 3D pose estimation method. Our approach obtains superior results to the state-of-the-art 3D pose estimation method (both absolute pose and root-relative pose), with significantly lower computation complexity and fewer model parameters.

5 Conclusion

In this paper, we proposed a novel form of supervision - HMOR, to learn multi-person 3D poses from a monocular RGB image. HMOR supervises the multi-person ordinal relations in a hierarchical manner, which captures fine-grained semantics and maintains global consistency at the same time. To end-to-end learn the ordinal relations, we further proposed an integrated model with a coarse-to-fine depth-estimation architecture. We demonstrate the effectiveness of our proposed method on standard benchmarks. The proposed method surpasses state-of-the-art multi-person 3D pose estimation methods, with lower computation complexity and fewer model parameters. We believe the idea of leveraging multi-person relations can be further explored to improve 3D pose estimation, e.g., exploit the relations via network design.

Acknowledgements. This work is supported in part by the National Key R&D Program of China, No. 2017YFA0700800, National Natural Science Foundation of China under Grants 61772332Shanghai Qi Zhi Institute.

References

1. Aitpayev, K., Gaber, J.: Creation of 3D human avatar using kinect. ATFECM **01**, 1–3 (2012)
2. Akhter, I., Black, M.J.: Pose-conditioned joint angle limits for 3D human pose reconstruction. In: CVPR (2015)
3. Andriluka, M., Pishchulin, L., Gehler, P., Schiele, B.: 2D human pose estimation: new benchmark and state of the art analysis. In: CVPR (2014)
4. Boulic, R., Bécheiraz, P., Emering, L., Thalmann, D.: Integration of motion control techniques for virtual human and avatar real-time animation. In: VRST (1997)
5. Cao, Z., Simon, T., Wei, S.E., Sheikh, Y.: Realtime multi-person 2D pose estimation using part affinity fields. In: CVPR (2017)
6. Carreira, J., Agrawal, P., Fragkiadaki, K., Malik, J.: Human pose estimation with iterative error feedback. In: CVPR (2016)
7. Chen, C.H., Ramanan, D.: 3D human pose estimation = 2D pose estimation + matching. In: CVPR (2017)
8. Chen, W., Fu, Z., Yang, D., Deng, J.: Single-image depth perception in the wild. In: NeurIPS (2016)
9. Chen, X., Lin, K.Y., Liu, W., Qian, C., Lin, L.: Weakly-supervised discovery of geometry-aware representation for 3D human pose estimation. In: CVPR (2019)
10. Chen, Y., Wang, Z., Peng, Y., Zhang, Z., Yu, G., Sun, J.: Cascaded pyramid network for multi-person pose estimation. In: CVPR (2018)

11. Deng, J., Dong, W., Socher, R., Li, L.J., Li, K., Fei-Fei, L.: ImageNet: a large-scale hierarchical image database. In: CVPR (2009)
12. Dollár, P., Welinder, P., Perona, P.: Cascaded pose regression. In: CVPR (2010)
13. Du, G., Zhang, P.: Markerless human-robot interface for dual robot manipulators using kinect sensor. Robot Comput. Int. Manuf. **30**(2), 150–159 (2014)
14. Dumoulin, V., Visin, F.: A guide to convolution arithmetic for deep learning (2016). arXiv:1603.07285
15. Fang, H.S., Cao, J., Tai, Y.W., Lu, C.: Pairwise body-part attention for recognizing human-object interactions. In: ECCV (2018)
16. Fang, H.S., Xie, S., Tai, Y.W., Lu, C.: RMPE: Regional multi-person pose estimation. In: ICCV (2017)
17. Fang, H.S., Xu, Y., Wang, W., Liu, X., Zhu, S.C.: Learning pose grammar to encode human body configuration for 3D pose estimation. In: AAAI (2018)
18. He, K., Gkioxari, G., Dollár, P., Girshick, R.: Mask R-CNN. In: ICCV (2017)
19. He, K., Zhang, X., Ren, S., Sun, J.: Deep residual learning for image recognition. In: CVPR (2016)
20. Ioffe, S., Szegedy, C.: Batch normalization: accelerating deep network training by reducing internal covariate shift. In: ICML (2015)
21. Ionescu, C., Papava, D., Olaru, V., Sminchisescu, C.: Human3.6m: large scale datasets and predictive methods for 3D human sensing in natural environments. TPAMI **36**, 1325–1339 (2014)
22. Jin, S., Liu, W., Ouyang, W., Qian, C.: Multi-person articulated tracking with spatial and temporal embeddings. In: CVPR (2019)
23. Joo, H., et al.: Panoptic studio: a massively multiview system for social motion capture. In: ICCV (2015)
24. Kanazawa, A., Black, M.J., Jacobs, D.W., Malik, J.: End-to-end recovery of human shape and pose. In: CVPR (2018)
25. Kocabas, M., Karagoz, S., Akbas, E.: MultiPoseNet: fast multi-person pose estimation using pose residual network. In: ECCV (2018)
26. Kolotouros, N., Pavlakos, G., Black, M.J., Daniilidis, K.: Learning to reconstruct 3d human pose and shape via model-fitting in the loop. In: ICCV (2019)
27. Li, J., Wang, C., Zhu, H., Mao, Y., Fang, H.S., Lu, C.: Crowdpose: efficient crowded scenes pose estimation and a new benchmark. In: CVPR (2019)
28. Li, Y.L., et al.: Detailed 2d–3d joint representation for human-object interaction. In: CVPR (2020)
29. Li, Y.L., et al.: Pastanet: toward human activity knowledge engine. In: CVPR (2020)
30. Lin, T.Y., Dollár, P., Girshick, R., He, K., Hariharan, B., Belongie, S.: Feature pyramid networks for object detection. In: CVPR (2017)
31. Lin, T.Y., et al.: Microsoft COCO: common objects in context. In: ECCV (2014)
32. Luvizon, D.C., Tabia, H., Picard, D.: Learning features combination for human action recognition from skeleton sequences. Pattern Recogn. Lett. **99**, 13–20 (2017)
33. von Marcard, T., Henschel, R., Black, M., Rosenhahn, B., Pons-Moll, G.: Recovering accurate 3D human pose in the wild using IMUs and a moving camera. In: ECCV (2018)
34. Marsaglia, G., et al.: Choosing a point from the surface of a sphere. Ann. Math. Stat. **43**, 645–646 (1972)
35. Martinez, J., Hossain, R., Romero, J., Little, J.J.: A simple yet effective baseline for 3D human pose estimation. In: ICCV (2017)
36. Mehta, D., et al.: Monocular 3D human pose estimation in the wild using improved CNN supervision. In: 3DV (2017)

37. Mehta, D., et al.: Xnect: Real-time multi-person 3D human pose estimation with a single RGB camera (2019). arXiv preprint arXiv:1907.00837
38. Mehta, D., et al.: Single-shot multi-person 3D pose estimation from monocular RGB. In: 3DV (2018)
39. Moon, G., Chang, J.Y., Lee, K.M.: Camera distance-aware top-down approach for 3D multi-person pose estimation from a single RGB image. In: ICCV (2019)
40. Moreno-Noguer, F.: 3D human pose estimation from a single image via distance matrix regression. In: CVPR (2017)
41. Narihira, T., Maire, M., Yu, S.X.: Learning lightness from human judgement on relative reflectance. In: CVPR (2015)
42. Newell, A., Huang, Z., Deng, J.: Associative embedding: End-to-end learning for joint detection and grouping. In: NeurIPS (2017)
43. Newell, A., Yang, K., Deng, J.: Stacked hourglass networks for human pose estimation. In: ECCV (2016)
44. Pang, B., Zha, K., Cao, H., Shi, C., Lu, C.: Deep RNN framework for visual sequential applications. In: CVPR (2019)
45. Park, S., Hwang, J., Kwak, N.: 3D human pose estimation using convolutional neural networks with 2D pose information. In: ECCV (2016)
46. Pavlakos, G., Zhou, X., Daniilidis, K.: Ordinal depth supervision for 3D human pose estimation. In: CVPR (2018)
47. Pavlakos, G., Zhou, X., Derpanis, K.G., Daniilidis, K.: Coarse-to-fine volumetric prediction for single-image 3D human pose. In: CVPR (2017)
48. Pirinen, A., Gärtner, E., Sminchisescu, C.: Domes to drones: self-supervised active triangulation for 3D human pose reconstruction. In: NeurIPS (2019)
49. Popa, A.I., Zanfir, M., Sminchisescu, C.: Deep multitask architecture for integrated 2D and 3D human sensing. In: CVPR (2017)
50. Presti, L.L., La Cascia, M.: 3D skeleton-based human action classification: a survey. Pattern Recogn. **53**, 130–147 (2016)
51. Qi, S., Wang, W., Jia, B., Shen, J., Zhu, S.C.: Learning human-object interactions by graph parsing neural networks. In: ECCV (2018)
52. Rogez, G., Weinzaepfel, P., Schmid, C.: LCR-Net: localization-classification-regression for human pose. In: CVPR (2017)
53. Rogez, G., Weinzaepfel, P., Schmid, C.: LCR-Net++: multi-person 2D and 3D pose detection in natural images. TPAMI (2019)
54. Ronchi, M.R., Mac Aodha, O., Eng, R., Perona, P.: It's all relative: monocular 3D human pose estimation from weakly supervised data. In: BMVC (2018)
55. Sharma, S., Varigonda, P.T., Bindal, P., Sharma, A., Jain, A.: Monocular 3D human pose estimation by generation and ordinal ranking. In: ICCV (2019)
56. Sigal, L., Balan, A.O., Black, M.J.: Humaneva: synchronized video and motion capture dataset and baseline algorithm for evaluation of articulated human motion. IJCV **87**, 1–24 (2010)
57. Simonyan, K., Zisserman, A.: Very deep convolutional networks for large-scale image recognition (2014). arXiv:1409.1556
58. Sun, K., Xiao, B., Liu, D., Wang, J.: Deep high-resolution representation learning for human pose estimation. In: CVPR (2019)
59. Sun, X., Shang, J., Liang, S., Wei, Y.: Compositional human pose regression. In: ICCV (2017)
60. Sun, X., Xiao, B., Wei, F., Liang, S., Wei, Y.: Integral human pose regression. In: ECCV (2018)
61. Wang, M., Chen, X., Liu, W., Qian, C., Lin, L., Ma, L.: Drpose3d: depth ranking in 3D human pose estimation. IJCAI (2018)

62. Xiao, B., Wu, H., Wei, Y.: Simple baselines for human pose estimation and tracking. In: ECCV (2018)
63. Xiu, Y., Li, J., Wang, H., Fang, Y., Lu, C.: Pose flow: efficient online pose tracking (2018). arXiv preprint arXiv:1802.00977
64. Yang, W., Ouyang, W., Wang, X., Ren, J., Li, H., Wang, X.: 3D human pose estimation in the wild by adversarial learning. In: CVPR (2018)
65. Yasin, H., Iqbal, U., Kruger, B., Weber, A., Gall, J.: A dual-source approach for 3D pose estimation from a single image. In: CVPR (2016)
66. Zanfir, A., Marinoiu, E., Sminchisescu, C.: Monocular 3D pose and shape estimation of multiple people in natural scenes-the importance of multiple scene constraints. In: CVPR (2018)
67. Zanfir, A., Marinoiu, E., Zanfir, M., Popa, A.I., Sminchisescu, C.: Deep network for the integrated 3D sensing of multiple people in natural images. In: NeurIPS (2018)
68. Zhou, K., Han, X., Jiang, N., Jia, K., Lu, J.: Hemlets pose: learning part-centric heatmap triplets for accurate 3d human pose estimation. In: ICCV (2019)
69. Zhou, T., Krahenbuhl, P., Efros, A.A.: Learning data-driven reflectance priors for intrinsic image decomposition. In: ICCV (2015)
70. Zhou, X., Zhu, M., Pavlakos, G., Leonardos, S., Derpanis, K.G., Daniilidis, K.: Monocap: monocular human motion capture using a CNN coupled with a geometric prior. TPAMI (2018)
71. Zhou, X., Huang, Q., Sun, X., Xue, X., Wei, Y.: Towards 3D human pose estimation in the wild: a weakly-supervised approach. In: ICCV (2017)
72. Zimmermann, C., Welschehold, T., Dornhege, C., Burgard, W., Brox, T.: 3D human pose estimation in RGBD images for robotic task learning. In: ICRA (2018)
73. Zoran, D., Isola, P., Krishnan, D., Freeman, W.T.: Learning ordinal relationships for mid-level vision. In: ICCV (2015)

Mask2CAD: 3D Shape Prediction by Learning to Segment and Retrieve

Weicheng Kuo[1,2(✉)], Anelia Angelova[1,2], Tsung-Yi Lin[1,2], and Angela Dai[3]

[1] Google AI, Mountain View, USA
weicheng@google.com, anelia@google.com, tsungyi@google.com
[2] Robotics at Google, Munich, Germany
[3] Technical University of Munich, Munich, Germany
angela.dai@tum.de

Abstract. Object recognition has seen significant progress in the image domain, with focus primarily on 2D perception. We propose to leverage existing large-scale datasets of 3D models to understand the underlying 3D structure of objects seen in an image by constructing a CAD-based representation of the objects and their poses. We present Mask2CAD, which jointly detects objects in real-world images and for each detected object, optimizes for the most similar CAD model and its pose. We construct a joint embedding space between the detected regions of an image corresponding to an object and 3D CAD models, enabling retrieval of CAD models for an input RGB image. This produces a clean, lightweight representation of the objects in an image; this CAD-based representation ensures a valid, efficient shape representation for applications such as content creation or interactive scenarios, and makes a step towards understanding the transformation of real-world imagery to a synthetic domain. Experiments on real-world images from Pix3D demonstrate the advantage of our approach in comparison to state of the art. To facilitate future research, we additionally propose a new image-to-3D baseline on ScanNet which features larger shape diversity, real-world occlusions, and challenging image views.

1 Introduction

Object recognition and localization in images has been a core task of computer vision with a well-studied history. Recent years have shown incredible progress in identifying objects in RGB images by predicting their bounding boxes or segmentation masks [10,17,27]. Although these advances are very promising, recognizing 3D attributes of objects such as shape and pose is crucial to many real-world applications. In fact, 3D perception is fundamental towards human understanding of imagery and real-world environments – from a single RGB

Electronic supplementary material The online version of this chapter (https://doi.org/10.1007/978-3-030-58580-8_16) contains supplementary material, which is available to authorized users.

A. Vedaldi et al. (Eds.): ECCV 2020, LNCS 12348, pp. 260–277, 2020.
https://doi.org/10.1007/978-3-030-58580-8_16

image a human can easily perceive geometric structure, and is paramount for enabling higher-level scene understanding such as inter-object relationships, or interaction with an environment by exploration or manipulation of objects.

At the same time, we are now seeing a variety of advances in understanding the shape of a single object from image view(s), driven by exploration of various geometric representations: voxels [7,41,44], points [11,45], meshes [9,15,42], and implicit surfaces [31,33]. While these generative approaches have shown significant promise in inferring the geometry of single objects, these approaches tend to generate geometry that may not necessarily represent a valid shape, with tendency towards noise or oversmoothing, and excessive tessellation. Such limitations render these results unsuitable for many applications, for instance content creation, real-time robotics scenarios, or interaction in mixed reality environments. In addition, the ability to digitize the objects of real world images to CAD models opens up new possibilities in helping to bridge the real-synthetic domain gap by transforming real-world images to a synthetic representation where far more training data is available.

In contrast, we propose Mask2CAD to join together the capabilities of 2D recognition and 3D reconstruction by leveraging CAD model representations of objects. Such CAD models are now readily available [4] and represent valid real-world object shapes, in a clean, compact representation – a representation widely used by existing production applications. Thus, we aim to infer from a single RGB image object detection in the image as well as 3D representations of each detected object as CAD models aligned to the image view. This provides a geometrically clean, compact reconstruction of the objects in an image, and a lightweight representation for downstream applications.

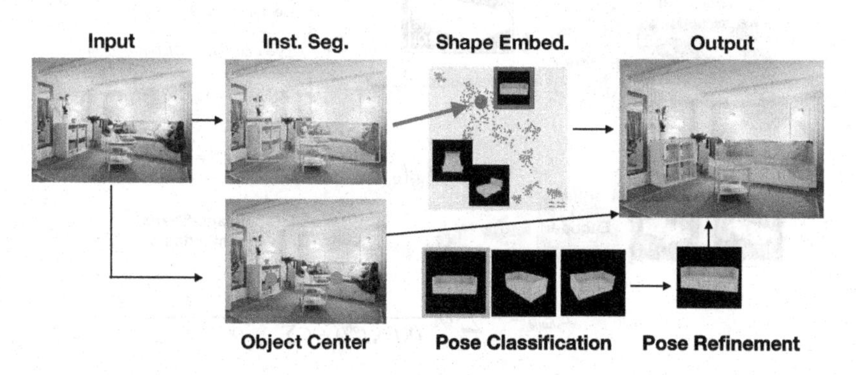

Fig. 1. Mask2CAD aims to predict object mask, and 3D shape in the scene. We achieve this by formulating an image-shape embedding learning problem. Combined with pose and object center prediction, Mask2CAD outputs realistic 3D shapes of objects from a single RGB image input. The entire system is differentiable and learned end-to-end.

Our Mask2CAD approach jointly detects object regions in an image and learns to map these image regions and CAD models to a shared embedding space

(See Fig. 1). At train time we learn a joint embedding which brings together corresponding image-CAD pairs, and pushes apart other pairs. At test time, we retrieve shapes by their renderings from the embedding space. To align the shapes to the image, we develop a pose prediction branch to classify and refine the shape alignment. We train our approach on the Pix3D dataset [39], achieving more accurate reconstructions than state of the art. Importantly, our retrieval-based approach allows adaptation to new domain by simply adding CAD models to the CAD model set without any re-training. Experiments on unseen shapes of the Pix3D dataset [39] show notable improvement when we have access to all CAD models at test time (but no access to corresponding RGB images of the unseen models and no re-training). By leveraging CAD models as shape representation, we are able to predict multiple distinct 3D objects per image efficiently (approximately 60 ms per image).

In addition to Pix3D, we also apply Mask2CAD on ScanNet and propose the first single-image to 3D object reconstruction baseline. Compared to Pix3D, this dataset contains 25K images, an order of magnitude more 3D shapes, complex real-world occlusions, diverse views and lighting conditions. Despite these challenges, Mask2CAD still manages to place appropriate CAD models that match the image observation (see Fig. 5). We hope Mask2CAD could serve as a benchmark for future retrieval methods and reference for generative methods.

Mask2CAD opens up possibilities for object-based 3D understanding of images for content creation and interactive scenarios, and provides an initial step towards transforming real images to a synthetic representation (Fig. 2).

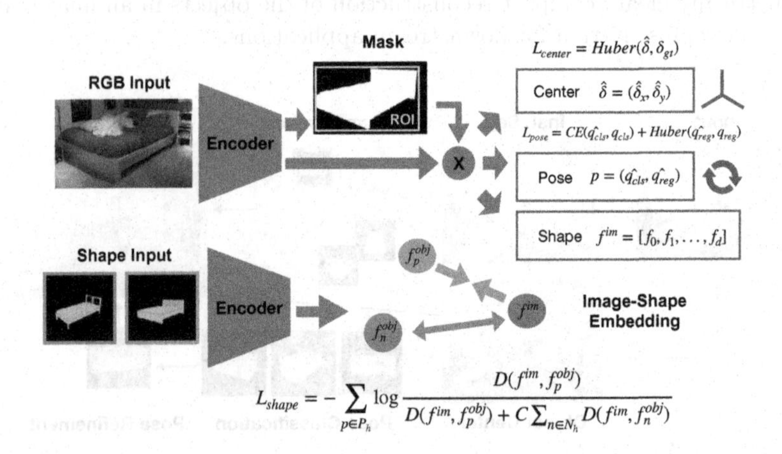

Fig. 2. Overview of our Mask2CAD approach for joint object segmentation and shape retrieval from a single RGB image. At train time, object detection is performed on an RGB image to produce a bounding box, segmentation mask, and feature descriptor for each detected object. The object feature descriptor is then used to train for an image-CAD embedding space for shape retrieval, as well as pose regression for the object rotation and center regression for its location. The embedding space is constructed through a triplet loss with corresponding and non-corresponding shapes to a detected object region of an image.

2 Related Work

Object Recognition in Images. Our work draws inspiration from the success of 2D object detection and segmentation in the image domain, where myriad methods have been developed to predict 2D object bounding boxes and class labels from a single image input [12,27,29,34,35]. Recent approaches have focused on additionally predicting instance masks for each object [17,23]. We build from this 2D object detection and segmentation to jointly learn to predict shape as well.

Single-View Object Reconstruction. In recent years, a variety of approaches have been developed to infer 3D shape from a single RGB image observation, largely focusing on the single object scenario and exploring a variety of shape representations in the context of learning-based methods. Regular voxel grids are a natural extension of the 2D image domain, and have been shown to effectively predict global shape structures [7,44], but remain limited by computational and memory constraints when scaling to high resolutions, as well as uneconomical in densely representing free space. Thus methods which focus representation power on surface geometry have been developed, including hierarchical approaches on voxels such as octrees [36,41], or point-based representations [11,45]. More recently, methods have been developed to predict triangle meshes, largely based on strong topological assumptions such as deforming an existing template mesh [42] while free-form generative approaches tend to remain limited by computational complexity to very small numbers of vertices [9]. Implicit reconstruction approaches have also shown impressive shape reconstruction results at relatively high resolution by predicting the occupancy [31] or signed distance field value [33] for point sampled locations.

Mesh R-CNN [13] pioneered an approach to extend such single object reconstruction to jointly detect and reconstruct shape geometry for each detected object in an RGB image. Mesh R-CNN extends upon the object recognition pipeline of Mask R-CNN [17] to predict initial voxel-based occupancy of an object, which is then refined by a graph convolutional network to produce an output mesh for each object.

While these approaches for object reconstruction have shown significant promise in predicting general, structural shape properties, due to the low-level nature of the reconstruction approaches (generating on a per-voxel/per-point basis), the reconstructed objects tend to be noisy or oversmoothed, may not represent valid real-world shapes (e.g., disconnected in thin regions, missing object symmetries), and inefficiently represented in geometry (e.g., over-tessellated to achieve higher resolutions). In contrast, our Mask2CAD approach leverages existing CAD models of objects to jointly segment and retrieve the 3D shape for each object in an image, producing both an accurate reconstruction and clean, compact geometric representation.

CAD Model Priors for 3D Reconstruction. Leveraging geometric model-based priors for visual understanding has been established near the inception of computer vision and robotic understanding [3,6,37], although constrained by the geometric models available. With increasing availability of larger-scale CAD model datasets [4,39], we have seen a rejuvenation in understanding the objects of a scene by retrieving and aligning similar CAD models. Most methods focus on aligning CAD models to RGB-D scan, point cloud geometry, or 2D-3D surface mapping though various geometric feature matching techniques [1,2,14,20,22, 25,38]. Izadinia and Seitz [21] and Huang et al. [18] propose to optimize for both scene layouts and CAD models of objects from image input, leveraging analysis-by-synthesis approaches; these methods involve costly optimization for each input image (minutes to hours).

From single image views of a object, Li et al. [26] propose a method to construct a joint embedding space between RGB images and CAD models by first constructing the space based on handcrafted shape similarity descriptors, and then optimizing for the image embeddings into the shape space. Our approach also optimizes for a joint embedding space between image views and CAD models in order to perform retrieval; however, we construct our space by jointly learning from both image and CAD in an end-to-end fashion without any explicit encoding of shape similarity.

3 Mask2CAD

3.1 Overview

From a single RGB image, Mask2CAD detects and localizes objects by recognizing similar 3D models from a candidate set, and inferring their pose alignment to the image. We focus on real-world imagery and jointly learn the 2D-3D relations in an end-to-end fashion. This produces an object-based reconstruction and understanding of the image, where each object by nature is characterized by a valid, complete 3D model with a clean, efficient geometric representation.

Specifically, from an input image, we first detect all objects in the image domain by predicting their bounding boxes, class labels, and segmentation masks. From these detected image regions, we then learn to construct a shared embedding space between these image regions and 3D CAD models of objects, which enables retrieving a geometrically similar model for the image observation. We simultaneously predict the object alignment to the image as 5 dof pose optimization (z-depth translation given), yielding a 3D understanding of the objects in the image.

Object Detection. For object detection, we build upon ShapeMask [23], a state-of-the-art instance segmentation approach. ShapeMask takes as input an RGB image and outputs detected objects characterized by their bounding boxes, class labels, and segmentation masks. The one-stage detection approach of RetinaNet [27] is leveraged to generate object bounding box detections, which are then refined into instance masks by a learned set of shape priors. We modify it to leverage the learned features for our 3D shape prediction. Each bounding box detection is used to crop features from the corresponding level of feature pyramid to produce a feature descriptor F_i for the instance mask prediction M_i

(e.g. 32 × 32) of object i; we then take the product $M_i \circ F_i$ as the representative feature map for object i as seen in the image. This is then used to inform the following CAD model retrieval and pose alignment.

3.2 Joint Embedding Space for Image-CAD Retrieval

The core of our approach lies in learning to seamlessly map between image views of an object and 3D CAD models, giving an association between image and 3D geometry. The CAD models represent an explicit prior on object geometry, providing an inherently clean, complete, and compact 3D representation of an object. We learn this mapping between image-CAD by constructing a shared embedding space – importantly, as we show in Sect. 4, our approach to jointly learn this embedding space constructs a space that is robust to new, unseen CAD models at test time.

Constructing a joint embedding space between image regions and 3D object geometry requires mapping across two very different domains, where in contrast to a geometric CAD model, an image is view-dependent and composed of the interaction of scene geometry with lighting and material. To facilitate the construction of this shared space between, we thus represent each object similar to a light field descriptor [5], rendering a set of k views $I_0^i, ..., I_k^i$ for an object O_i. For all our experiments, we use $k = 16$; the set of canonical views for each object is determined by K-medoid clustering of the views seen of the object category in the training set. In addition, we augment the pool of anchor-positive pairs by using slightly jittered groundtruth views of the objects.

The embedding space is then established between the image region features $M_i \circ F_i$ from the detection, and the 3D object renderings $I_0^j, ..., I_k^j$. Representative features for the image region descriptions and object renderings are extracted by a series of convolutional layers applied to each input. The convolutional networks to extract these features are structured symmetrically, although we do not share weights due to the different input domains (See Sect. 3.4 for more details). We refer to the resulting extracted feature descriptors as f^{im} and f^{obj} for the image regions and object views, respectively. The f^{im} come from the regions of interest (ROI) shared with the 2D detection and segmentation branch. More specifically, the encoder backbone is a ResNet feature pyramid network and the decoder is a stack of 3 × 3 convolution layers on the ROI features.

We guide the construction of the embedding space with a noise contrastive estimation loss [32] for f^{im} describing a detected image region

$$L_c = -\sum_{p \in P_h} \log \frac{D(f^{im}, f_p^{obj})}{D(f^{im}, f_p^{obj}) + C \sum_{n \in N_h} D(f^{im}, f_n^{obj})} \tag{1}$$

where f_p^{obj} represents the feature descriptor of a corresponding 3D object to the image region, f_n^{obj} the feature descriptor of a non-corresponding object, C a weighting parameter, and D the cosine distance function:

$$D(x, y) := \frac{1}{\tau}\left(\frac{x}{||x||}\right)^T\left(\frac{y}{||y||}\right) \tag{2}$$

where τ is the temperature. P_h and N_h denote the set of hard positive and negative examples for the image region. Details of hard-example mining are provided in the next section. The positively corresponding objects are determined by the CAD annotations to the images, and negatively corresponding objects are the non-corresponding CAD renderings in the training batch.

Since the object detection already provides class of the object, the negatives are only sampled from the shapes under the same class; that is, our embedding spaces are constructed for each class category although the weights are shared among them.

Empirically, we find it important to place more sampling weights on the rare classes because the number of valid pairs scale *quadratically* with the number of same-class examples in the batch. We apply the inverse square root method as in [16] to enhance the rare class examples with a threshold $t = 0.1$, which leads to improved performance on rare classes without compromising dominant classes.

Hard Example Mining. Hard example mining is known to be crucial for embedding learning, as most examples are easy and do not contain much information to improve the model. We employ both hard positive and hard negative mining in Mask2CAD as follows. For each image region (anchor), we sample top-P_h positive object views and top-N_h negative object views by their distances to anchor. Similar to [23], we sample Q objects for each image during the training ($Q = 8$). Since the number of objects in a batch scales linearly with Q, we set $P_h = 4Q = 32$ and $N_h = 16Q = 128$. Summation over hard examples allows the loss to focus on difficult cases and perform better.

Shape Retrieval. Once this embedding space is constructed, we can then perform shape retrieval at test time to provide a 3D understanding of the objects in an image. An input image at test time is processed by the 2D detection to provide a bounding box, segmentation mask, and feature descriptor for each detected object. We then use a nearest neighbor retrieval into the embedding space with $N_k = 1$ based on cosine distance to find the most similar CAD model for each detected object. We have tried larger N_k values and majority vote schemes but did not see any performance gain.

3.3 Pose Prediction

We additionally aim to predict the pose of the retrieved 3D object such that it aligns best to the input image. We thus propose a pose prediction branch which outputs the rotation and translation of the object. Starting with the $M_i \circ F_i$ feature map for a detected object, the object translation is directly regressed with a Huber loss [19] as follows:

$$L_\delta(x) = \begin{cases} \frac{1}{2}x^2 & \text{for } |x| \leq \delta, \\ \delta(|x| - \frac{1}{2}\delta), & \text{otherwise.} \end{cases} \tag{3}$$

The object rotation is simultaneously predicted; the rotation is first classified to a set of K discretized rotation bins using cross entropy loss; this coarse estimate is then refined through a regression step using a Huber loss. This coarse-to-fine approach helps to navigate the non-euclidean rotation space, and enables continuous rotation predictions.

For rotation prediction, we represent the rotation as a quaternion, and compute the set of rotation bins by K-medoid clustering based on train object rotations. To further refine this coarse prediction, we then predict a refined rotation by estimating the delta from the classified bin using a Huber loss. The delta is represented as a R^4 quaternion. We initialize the bias of the last layer with $(0.95, 0, 0, 0)$ such that the quaternion is close to identity transform at the beginning. Note that during training, we only train the refinement for classified rotations within θ to avoid regressing to dissimilar targets.

To obtain full prediction in the camera space, we need to predict the translation of the object in addition to the shape and rotation. A naive approach is to use the bounding box center as the object center in 2D and cast a ray through the center to intersects with the given groundtruth z-plane. Unfortunately, the bounding boxes tend to be unstable against the rotation and their centers can end up far from the actual object center.

We thus regress the object center as a bounding regression problem. More specifically, for each ROI, we task the network with predicting $(\delta_x,\ \delta_y)$, where the δs are the shift between bounding box center and actual object center as a ratio of object width and height. At train time, we optimize $(\delta_x,\ \delta_y)$ with the aforementioned Huber loss. At test time, we apply $(\delta_x,\ \delta_y)$ to the box center to obtain the object 3D translations (assuming groundtruth depth is given [13]).

3.4 Implementation Details

We use ShapeMask [23] as the instance segmentation backbone. The model backbone is initialized from COCO-pretrained checkpoint and uses ResNet-50 architecture so as to be comparable to Mesh R-CNN in our experiments. The shape rendering branch uses a lightweight ResNet-18 backbone initialized randomly.

We freeze the weights of the backbone ResNet-50 layers after initialization and optimize both branches jointly for 48K iterations until convergence (about 1000 epochs for Pix3D), which takes approximately 13 h. The learning rate is set to be 0.08 and decays by 0.1 at 32K and 40K iterations. The losses for the retrieval and pose estimation are weighted with 0.5, 0.25, and 5.0 for the embedding loss, pose classification loss, and pose regression loss. We use $C = 1.5$ and $\tau = 0.15$ in our contrastive loss, and Huber loss margin of $\delta = 0.15$ for the pose and center regression. For pose prediction, we set $K = 16$ bins, and $\theta = \pi/6$.

For each example, we randomly sample 3 out of $k = 16$ canonical view renderings and one jittered groundtruth view rendering to add to the contrastive learning pool. Similar to ShapeMask, we apply ROI jittering to the image region for training the segmentation, embedding, and pose estimation branches. The noise is set to 0.025 following ShapeMask. We also apply data augmentation by horizontal image flips with 50% probability. For such image flips, the pose labels were also adjusted accordingly.

4 Experiments

We evaluate our approach on the Pix3D dataset [39], which comprises 10,069 images annotated with corresponding 3D models of the objects in the images. We aim to jointly detect and predict the 3D shapes for the objects in the images. We evaluate on the train/test split used by Mesh R-CNN [13] for the same task. Additionally, we propose the first single-image 3D object reconstruction baseline on the ScanNet dataset [8], which tends to contain more cluttered, in-the-wild views of objects.

Evaluation Metric. We adopt the popular metrics from 2D object recognition, and similar to Mesh R-CNN [13], evaluate AP^{box} and AP^{mask} on the 2D detections, and AP^{mesh} on the 3D shape predictions for the objects. Similar to Mesh R-CNN, we evaluate AP^{mesh} using the precision-recall for $F1^{0.3}$. However, note that while Mesh R-CNN evaluate these metrics at IoU 0.5 (AP50), we adopt the standard COCO object detection protocol of AP50-AP95 (denoted as AP), averaging over 10 IoU thresholds of 0.5 : 0.05 : 0.95 [28]. This enables characterization of high-accuracy shape reconstructions captured at more strict IoU thresholds, demonstrating a more comprehensive description of the accuracy of the shape predictions. In addition to AP, we also report individual AP^{mesh} scores for IoU thresholds of 0.5 and 0.75 following Mask R-CNN [17]. For better reproducibility, we report every metric as an average of 5 independent runs throughout this paper.

Comparison to State of the Art. We compare our Mask2CAD approach for 3D object understanding from RGB images by joint segmentation and retrieval to Mesh R-CNN [13], who first propose this task on Pix3D [39]. Table 1 shows our shape prediction results in comparison to Mesh-RCNN on their $\mathcal{S}_1$ split of the Pix3D dataset. We evaluate AP^{mesh}, averaged over all class categories, as well as per-category. In contrast to Mesh R-CNN, whose results show effective coarse predictions but suffer significantly at AP75, our shape and pose predictions maintain high-accuracy reconstructions. We show qualitative comparisons in Fig. 3.

Table 1. Performance on Pix3D [39] $\mathcal{S}_1$. We report mean AP^{mesh} as well as per category AP^{mesh}. AP is averaged from AP50-AP95 following the COCO detection protocol. All AP performances are in %. We outperform the state-of-the-art approach on all AP metrics. This improvement mostly derives from maintaining more robust performance in the high AP regime above AP50. Additionally, we observe that Mask2CAD performs well on furniture categories and not so well on tools and miscellaneous objects which exhibit highly irregular shapes.

Pix3D $\mathcal{S}_1$	AP	AP50	AP75	*chair*	*sofa*	*table*	*bed*	*desk*	*bkcs*	*wrdrb*	*tool*	*misc*
Mesh R-CNN [13]	17.2	51.2	7.4	17.6	30.0	11.0	20.0	21.0	10.1	14.3	**6.5**	**24.5**
Mask2CAD	**33.2**	**54.9**	**30.8**	**19.6**	**55.8**	**29.2**	**39.4**	**31.6**	**42.4**	**60.3**	4.2	15.9

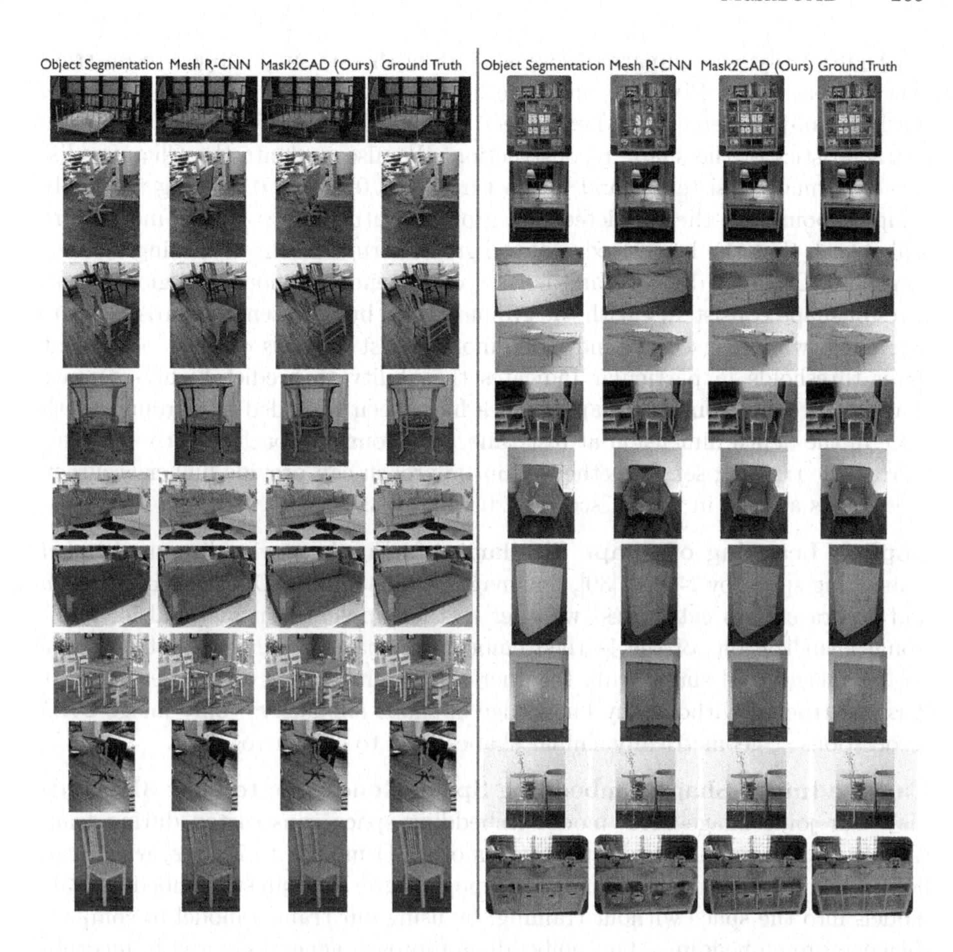

Fig. 3. Mask2CAD predictions on Pix3D [39]. The detected object is highlighted on the lefthand side of each column, with shape predictions denoted in purple, and ground truth in green. In contrast to Mesh R-CNN [13], our approach can achieve more accurate shape predictions with geometry in a clean, efficient representation. (Color figure online)

Table 2. Performance on Pix3D [39] with ground truth object bounding boxes given. We report Chamfer distance, Normal consistency and F1 scores. Note that for these experiments, the Mesh R-CNN-based approaches are additionally provided the ground truth scale in the depth dimension of the object.

Pix3D $\mathcal{S}_1$ gt	Chamfer ↓	Normal ↑	$\text{F1}^{0.1}$ ↑	$\text{F1}^{0.3}$ ↑	$\text{F1}^{0.5}$ ↑
Mask R-CNN + Pixel2Mesh [13]	1.30	0.70	16.4	51.0	68.4
Mesh R-CNN (Voxel-Only) [13]	1.28	0.57	9.9	37.3	56.1
Mesh R-CNN (Sphere-Init) [13]	1.30	0.69	16.8	51.4	68.8
Mesh R-CNN [13]	1.11	0.71	18.7	56.4	73.5
Mask2CAD (Ours)	**0.99**	**0.74**	**25.6**	**66.4**	**79.3**

Additionally, we compare to several state-of-the-art single object reconstruction approaches on Pix3D $\mathcal{S}_1$ in Table 2; for each approach we provide ground truth 2D object detections, i.e. perfect bounding boxes. We evaluate various characteristics of the shape reconstruction. We also evaluate the Chamfer distance, normal consistency, and F1 at thresholds 0.1, 0.3, 0.5, using randomly sampled points on the predicted and ground truth meshes, where meshes are scaled such that the longest edge of the ground-truth mesh's bounding box has length 10. Chamfer distance and normal consistency provide more global measures of shape consistency with the ground truth, but can tend towards favoring averaging, while F1 scores tend to be more robust towards outliers, and F1 at lower thresholds in particular indicates the ability to predict highly-accurate shapes. Note the competing approaches have been provided the ground truth scale in the depth dimension at test time, while our approach directly retrieves it from the training set. Nonetheless, our approach can provide higher-accuracy predictions as seen in the F1 scores at 0.1 and 0.3.

Implicit Learning of Shape Similarity. In Fig. 4, we visualize our learned embedding space by t-SNE [30], for image regions and CAD models of the *sofa* and *bookcase* class categories (we refer to the supplemental material for additional visualizations of the learned embedding spaces). We find that not only do the images and shapes mix together in this embedding space, despite that it is constructed without any knowledge of shape similarity – only image-CAD associations –, geometrically similar shapes tend to cluster together.

Can the Image-Shape Embedding Space Generalize to New 3D Models? Our joint image-CAD model embedding space constructed during train time leverages ground truth annotations of CAD models to images, which can be costly to acquire. During inference time, however, we can still embed new 3D models into the space without training, by using our trained model to compute their feature embeddings. Our embedding approach generalizes well in incorporating these new models.

We demonstrate this on the $\mathcal{S}_2$ split of Pix3D, training on a subset of the 3D CAD models, with test images comprising views of a disjoint set of objects than those in the training set. Generalization under this regime is difficult, particularly for a retrieval-based approach. However, in Table 3 we show clear improvements when using all available CAD models at test time in comparison to only the CAD models in the train set, despite not having seen any of the new objects nor their corresponding image views.

To help the model generalize better, we apply more data augmentation than the $\mathcal{S}_1$ split, including HSV-space jittering, random crop and resize of the renderings, and augmenting the box and image jittering magnitude as used in ShapeMask [23].

Comparison with ShapeNet Reconstruction Methods. In Table 4, we compare the Pix3D $\mathcal{S}_1$ model on the validation set with the other methods that train on ShapeNet with real data augmentation. The evaluation protocol and implementation follows [39]. Mask2CAD results are reported on 1165 chairs in

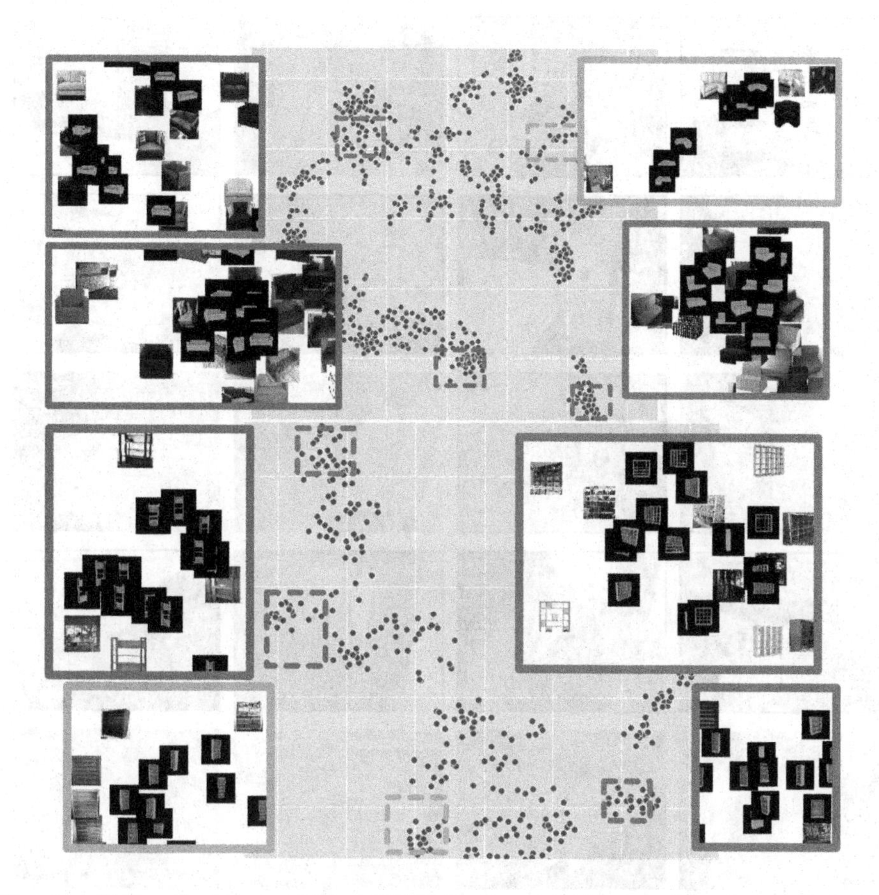

Fig. 4. t-SNE embeddings of Mask2CAD for the sofa (top) and bookcase (bottom) classes. More visualizations can be found in the Supp. Materials. Red points correspond to images, and blue to shapes. Both images and shapes mix well together in the embedding space. Note that despite lack of shape similarity information during training, clusters tend to form together in geometric similarity, e.g., L-shaped sofas (yellow), single seat sofas without armrests (orange), single seat sofas with armrests (blue), double seat sofas (green). This stands in contrast to the embedding space construction of [26] which explicitly enforces shape similarity in its light field descriptors. (Color figure online)

Table 3. Test-time generalization on Pix3D [39] $\mathcal{S}_2$. The performance improves on all categories with the addition 139 of CAD models at test time without re-training.

Pix3D $\mathcal{S}_2$	AP	AP50	AP75	*chair*	*sofa*	*table*	*bed*	*desk*	*bkcs*	*wrdrb*	*tool*	*misc*
Mask2CAD (Ours)	6.5	17.3	3.8	3.2	35.4	1.2	14.0	0.2	2.2	1.6	0.6	0.0
Mask2CAD (Ours) + CAD	**8.2**	**20.7**	**4.8**	**4.5**	**37.8**	**3.6**	**16.9**	**2.7**	**2.2**	**5.3**	**0.9**	**0.1**

Fig. 5. Example Mask2CAD predictions on ScanNet [8] images. Our approach shows encouraging results in its application to the more diverse set of image views, lighting, occlusions, and object categories of ScanNet.

the $\mathcal{S}_1$ test split of Mesh R-CNN [13], as an average over 5 independent runs. Surprisingly, Mask2CAD achieves significantly better shape predictions than the state-of-the-art methods (0.288 IoU, and 0.013 Chamfer Distance), showing the capability of retrieval.

Table 4. Mask2CAD on Pix3D [39] $\mathcal{S}_1$ test split in comparison with other methods that train on ShapeNet models with real data augmentation.

	Mask2CAD (Ours)	FroDO [24]	Sun et al. [39]	MarrNet [43]	3D-R2N2 [7]
IoU	**0.613**	0.325	0.282	0.231	0.136
Chamfer	**0.086**	0.099	0.119	0.144	0.239

Baseline on ScanNet Dataset. We additionally apply Mask2CAD to real-world images from the ScanNet dataset [8], which contains RGB-D video data of 1513 indoor scenes. We use the 25K frame subset provided by the dataset for training and validation. The train/val split contains 19387/5436 images respectively, and the images come from separate scenes with distinct objects. Compared to Pix3D, this dataset has an order of magnitude more shapes, as well as many more occlusions, diverse image views, lighting conditions, and importantly, metric 3D groundtruth of the scene. We believe this could be a suitable benchmark for object 3D prediction from a real single image.

We use the CAD labels from Scan2CAD [1] by projecting the CAD models to each image view and use the amodal box, mask, pose, and shape for training. We additionally remove the objects whose centers are out of frame from training and evaluation. We also remove the object categories that appear less than 1000 times in the training split, resulting in eight categories: bed, sofa, chair (inc. toilet), bin, cabinet (inc. fridge), display, table, and bookcase. Regarding shape similarity, we adopt F score = 0.5 as the threshold for Mesh AP computation, because the Scan2CAD annotations come from ShapeNet and do not provide exact matches to the images. As Scan2CAD provides 9-DoF annotation for each object, we apply the groundtruth z depth and (x, y, z) scale to the predicted shape before computing the shape similarity metrics. We trained Mask2CAD for 72000 iterations with HSV-color, ROI, and image scale jittering using the same learning rate schedule as Pix3D. The quantitative results are reported in Table 5. Despite the complexity of ScanNet data, Mask2CAD manages to recognize the object shapes in these images, as shown in Fig. 5. Our CAD model retrieval and alignment shows promising results and a potential for facilitating content creation pipelines.

Runtime. At test time Mask2CAD is efficient and runs at approximately 60 ms per 640 by 640 image on Pix3D, including 2D detection and segmentation as well as shape retrieval and pose estimation.

Table 5. Performance on ScanNet [8]. We report mean AP^{mesh} as well as per category AP^{mesh} following Pix3D protocol.

ScanNet 25K	AP	AP50	AP75	*bed*	*sofa*	*chair*	*cab*	*bin*	*disp*	*table*	*bkcs*
Mask2CAD(Ours)	8.4	23.1	4.9	14.2	13.0	13.2	7.5	7.8	5.9	2.9	3.1

Limitations. While Mask2CAD shows promising progress in attaining 3D understanding of the objects from a single image, we believe there are many avenues for further development. For instance, our retrieval-based approach can suffer in the case of objects that differ too strongly from the existing CAD model database, and we believe that mesh-based approaches to deform and refine geometry [40,42] have significant potential to complement our approach. Additionally, we believe a holistic 3D scene understanding characterizing not only the objects in an environment but all elements in the scene is a promising direction for 3D perception and semantic understanding.

5 Conclusion

We propose Mask2CAD, a novel approach for 3D perception from 2D images. Our method leverages a CAD model representation, and jointly detects objects for an input image and retrieves and aligns a similar CAD model to the detected region. We show that our approach produces accurate shape reconstructions and is capable of generalizing to unseen 3D objects at test time. The final output of Mask2CAD is a CAD-based object understanding of the image, where each object is represented in a clean, lightweight fashion. We believe that this makes a promising step in 3D perception from images as well as transforming real-world imagery to a synthetic representation, opening up new possibilities for digitization of real-world environments for applications such as content creation or domain transfer.

Acknowledgements. We would like to thank Georgia Gkioxari for her advice on Mesh R-CNN and the support of the ZD.B (Zentrum Digitalisierung.Bayern) for Angela Dai.

References

1. Avetisyan, A., Dahnert, M., Dai, A., Savva, M., Chang, A.X., Nießner, M.: Scan2CAD: learning cad model alignment in RGB-D scans. In: CVPR (2019)
2. Bansal, A., Russell, B., Gupta, A.: Marr revisited: 2D–3D alignment via surface normal prediction. In: Proceedings of the IEEE Conference on Computer Vision and Pattern Recognition, pp. 5965–5974 (2016)
3. Binford, T.O.: Survey of model-based image analysis systems. Int. J. Robot. Res. **1**(1), 18–64 (1982)
4. Chang, A.X., et al.: ShapeNet: an information-rich 3D model repository. Technical report. arXiv:1512.03012 [cs.GR], Stanford University – Princeton University – Toyota Technological Institute at Chicago (2015)
5. Chen, D.Y., Tian, X.P., Shen, Y.T., Ouhyoung, M.: On visual similarity based 3D model retrieval. In: Computer Graphics Forum, vol. 22, pp. 223–232. Wiley Online Library (2003)
6. Chin, R.T., Dyer, C.R.: Model-based recognition in robot vision. ACM Comput. Surv. (CSUR) **18**(1), 67–108 (1986)

7. Choy, C.B., Xu, D., Gwak, J.Y., Chen, K., Savarese, S.: 3D-R2N2: a unified approach for single and multi-view 3D object reconstruction. In: Leibe, B., Matas, J., Sebe, N., Welling, M. (eds.) ECCV 2016. LNCS, vol. 9912, pp. 628–644. Springer, Cham (2016). https://doi.org/10.1007/978-3-319-46484-8_38
8. Dai, A., Chang, A.X., Savva, M., Halber, M., Funkhouser, T., Nießner, M.: ScanNet: richly-annotated 3D reconstructions of indoor scenes. In: Proceedings of the Computer Vision and Pattern Recognition (CVPR). IEEE (2017)
9. Dai, A., Nießner, M.: Scan2Mesh: from unstructured range scans to 3D meshes. In: Proceedings of the IEEE Conference on Computer Vision and Pattern Recognition, pp. 5574–5583 (2019)
10. Duan, K., Bai, S., Xie, L., Qi, H., Huang, Q., Tian, Q.: CenterNet: keypoint triplets for object detection. In: Proceedings of the IEEE International Conference on Computer Vision, pp. 6569–6578 (2019)
11. Fan, H., Su, H., Guibas, L.J.: A point set generation network for 3D object reconstruction from a single image. In: Proceedings of the IEEE Conference on Computer Vision and Pattern Recognition, pp. 605–613 (2017)
12. Girshick, R.: Fast R-CNN. In: Proceedings of the IEEE International Conference on Computer Vision, pp. 1440–1448 (2015)
13. Gkioxari, G., Malik, J., Johnson, J.: Mesh R-cnn. arXiv preprint arXiv:1906.02739 (2019)
14. Grabner, A., Roth, P.M., Lepetit, V.: Location field descriptors: single image 3D model retrieval in the wild. In: 2019 International Conference on 3D Vision (3DV), pp. 583–593. IEEE (2019)
15. Groueix, T., Fisher, M., Kim, V.G., Russell, B.C., Aubry, M.: A papier-mâché approach to learning 3D surface generation. In: Proceedings of the IEEE Conference on Computer Vision and Pattern Recognition, pp. 216–224 (2018)
16. Gupta, A., Dollar, P., Girshick, R.: LVIS: a dataset for large vocabulary instance segmentation. In: Proceedings of the IEEE Conference on Computer Vision and Pattern Recognition, pp. 5356–5364 (2019)
17. He, K., Gkioxari, G., Dollár, P., Girshick, R.: Mask R-CNN. In: 2017 IEEE International Conference on Computer Vision (ICCV), pp. 2980–2988. IEEE (2017)
18. Huang, S., Qi, S., Zhu, Y., Xiao, Y., Xu, Y., Zhu, S.-C.: Holistic 3D scene parsing and reconstruction from a single RGB image. In: Ferrari, V., Hebert, M., Sminchisescu, C., Weiss, Y. (eds.) ECCV 2018. LNCS, vol. 11211, pp. 194–211. Springer, Cham (2018). https://doi.org/10.1007/978-3-030-01234-2_12
19. Huber, P.J.: Robust estimation of a location parameter. In: Kotz, S., Johnson, N.L. (eds.) Breakthroughs in Statistics. Springer Series in Statistics (Perspectives in Statistics), pp. 492–518. Springer, New York (1992). https://doi.org/10.1007/978-1-4612-4380-9_35
20. Izadinia, H., Seitz, S.M.: Scene recomposition by learning-based ICP. In: CVPR (2020)
21. Izadinia, H., Shan, Q., Seitz, S.M.: IM2CAD. In: Proceedings of the IEEE Conference on Computer Vision and Pattern Recognition, pp. 5134–5143 (2017)
22. Kim, Y.M., Mitra, N.J., Huang, Q., Guibas, L.: Guided real-time scanning of indoor objects. In: Computer Graphics Forum, vol. 32, pp. 177–186. Wiley Online Library (2013)
23. Kuo, W., Angelova, A., Malik, J., Lin, T.Y.: Shapemask: learning to segment novel objects by refining shape priors. In: Proceedings of the IEEE International Conference on Computer Vision, pp. 9207–9216 (2019)
24. Li, K., et al.: FroDO: from detections to 3D objects. arXiv preprint arXiv:2005.05125 (2020)

25. Li, Y., Dai, A., Guibas, L., Nießner, M.: Database-assisted object retrieval for real-time 3D reconstruction. In: Computer Graphics Forum, vol. 34, pp. 435–446. Wiley Online Library (2015)
26. Li, Y., Su, H., Qi, C.R., Fish, N., Cohen-Or, D., Guibas, L.J.: Joint embeddings of shapes and images via CNN image purification. ACM Trans. Graph. (TOG) **34**(6), 1–12 (2015)
27. Lin, T.Y., Goyal, P., Girshick, R., He, K., Dollár, P.: Focal loss for dense object detection. In: Proceedings of the IEEE International Conference on Computer Vision, pp. 2980–2988 (2017)
28. Lin, T.-Y., et al.: Microsoft COCO: common objects in context. In: Fleet, D., Pajdla, T., Schiele, B., Tuytelaars, T. (eds.) ECCV 2014. LNCS, vol. 8693, pp. 740–755. Springer, Cham (2014). https://doi.org/10.1007/978-3-319-10602-1_48
29. Liu, W., et al.: SSD: single shot multibox detector. In: Leibe, B., Matas, J., Sebe, N., Welling, M. (eds.) ECCV 2016. LNCS, vol. 9905, pp. 21–37. Springer, Cham (2016). https://doi.org/10.1007/978-3-319-46448-0_2
30. Maaten, L.v.d., Hinton, G.: Visualizing data using t-SNE. J. Mach. Learn. Res. **9**, 2579–2605 (2008)
31. Mescheder, L., Oechsle, M., Niemeyer, M., Nowozin, S., Geiger, A.: Occupancy networks: learning 3D reconstruction in function space. In: Proceedings of the IEEE Conference on Computer Vision and Pattern Recognition, pp. 4460–4470 (2019)
32. Oord, A.v.d., Li, Y., Vinyals, O.: Representation learning with contrastive predictive coding. arXiv preprint arXiv:1807.03748 (2018)
33. Park, J.J., Florence, P., Straub, J., Newcombe, R., Lovegrove, S.: DeepSDF: learning continuous signed distance functions for shape representation. In: Proceedings of the IEEE Conference on Computer Vision and Pattern Recognition, pp. 165–174 (2019)
34. Redmon, J., Divvala, S., Girshick, R., Farhadi, A.: You only look once: unified, real-time object detection. In: Proceedings of the IEEE Conference on Computer Vision and Pattern Recognition, pp. 779–788 (2016)
35. Ren, S., He, K., Girshick, R., Sun, J.: Faster R-CNN: towards real-time object detection with region proposal networks. In: Advances in Neural Information Processing Systems, pp. 91–99 (2015)
36. Riegler, G., Osman Ulusoy, A., Geiger, A.: OctNet: Learning deep 3D representations at high resolutions. In: Proceedings of the IEEE Conference on Computer Vision and Pattern Recognition, pp. 3577–3586 (2017)
37. Roberts, L.G.: Machine perception of three-dimensional solids. Ph.D. thesis, Massachusetts Institute of Technology (1963)
38. Shao, T., Xu, W., Zhou, K., Wang, J., Li, D., Guo, B.: An interactive approach to semantic modeling of indoor scenes with an RGBD camera. ACM Trans. Graph. (TOG) **31**(6), 1–11 (2012)
39. Sun, X., et al.: Pix3D: dataset and methods for single-image 3D shape modeling. In: Proceedings of the IEEE Conference on Computer Vision and Pattern Recognition, pp. 2974–2983 (2018)
40. Tang, J., Han, X., Pan, J., Jia, K., Tong, X.: A skeleton-bridged deep learning approach for generating meshes of complex topologies from single RGB images. In: Proceedings of the IEEE Conference on Computer Vision and Pattern Recognition, pp. 4541–4550 (2019)
41. Tatarchenko, M., Dosovitskiy, A., Brox, T.: Octree generating networks: efficient convolutional architectures for high-resolution 3D outputs. In: Proceedings of the IEEE International Conference on Computer Vision, pp. 2088–2096 (2017)

42. Wang, N., Zhang, Y., Li, Z., Fu, Y., Liu, W., Jiang, Y.-G.: Pixel2Mesh: generating 3D mesh models from single RGB images. In: Ferrari, V., Hebert, M., Sminchisescu, C., Weiss, Y. (eds.) ECCV 2018. LNCS, vol. 11215, pp. 55–71. Springer, Cham (2018). https://doi.org/10.1007/978-3-030-01252-6_4
43. Wu, J., Wang, Y., Xue, T., Sun, X., Freeman, B., Tenenbaum, J.: MarrNet: 3D shape reconstruction via 2.5 D sketches. In: Advances in Neural Information Processing Systems, pp. 540–550 (2017)
44. Wu, J., Zhang, C., Xue, T., Freeman, B., Tenenbaum, J.: Learning a probabilistic latent space of object shapes via 3D generative-adversarial modeling. In: Advances in Neural Information Processing Systems, pp. 82–90 (2016)
45. Yang, G., Huang, X., Hao, Z., Liu, M.Y., Belongie, S., Hariharan, B.: PointFlow: 3D point cloud generation with continuous normalizing flows. In: Proceedings of the IEEE International Conference on Computer Vision, pp. 4541–4550 (2019)

A Unified Framework of Surrogate Loss by Refactoring and Interpolation

Lanlan Liu[1,2], Mingzhe Wang[2], and Jia Deng[2(✉)]

[1] University of Michigan, Ann Arbor, MI 48105, USA
llanlan@umich.edu

[2] Princeton University, Princeton, NJ 08544, USA
{mingzhew,jiadeng}@cs.princeton.edu

Abstract. We introduce UniLoss, a unified framework to generate surrogate losses for training deep networks with gradient descent, reducing the amount of manual design of task-specific surrogate losses. Our key observation is that in many cases, evaluating a model with a performance metric on a batch of examples can be refactored into four steps: from input to real-valued scores, from scores to comparisons of pairs of scores, from comparisons to binary variables, and from binary variables to the final performance metric. Using this refactoring we generate differentiable approximations for each non-differentiable step through interpolation. Using UniLoss, we can optimize for different tasks and metrics using one unified framework, achieving comparable performance compared with task-specific losses. We validate the effectiveness of UniLoss on three tasks and four datasets. Code is available at https://github.com/princeton-vl/uniloss.

Keywords: Loss design · Image classification · Pose estimation

1 Introduction

Many supervised learning tasks involve designing and optimizing a loss function that is often different from the actual performance metric for evaluating models. For example, cross-entropy is a popular loss function for training a multi-class classifier, but when it comes to comparing models on a test set, classification error is used instead.

Why not optimize the performance metric directly? Because many metrics or output decoders are non-differentiable and cannot be optimized with gradient-based methods such as stochastic gradient descent. Non-differentiability occurs when the output of the task is discrete (e.g. class labels), or when the output is continuous but the performance metric has discrete operations (e.g. percentage of real-valued predictions within a certain range of the ground truth).

Electronic supplementary material The online version of this chapter (https://doi.org/10.1007/978-3-030-58580-8_17) contains supplementary material, which is available to authorized users.

A. Vedaldi et al. (Eds.): ECCV 2020, LNCS 12348, pp. 278–293, 2020.
https://doi.org/10.1007/978-3-030-58580-8_17

To address this issue, designing a differentiable loss that serves as a surrogate to the original metric is standard practice. For standard tasks with simple output and metrics, there exist well-studied surrogate losses. For example, cross-entropy or hinge loss for classification, both of which have proven to work well in practice.

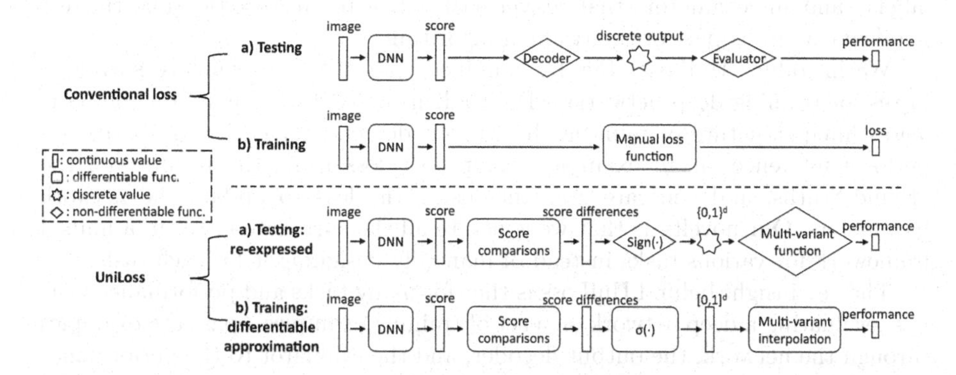

Fig. 1. Computation graphs for conventional losses and UniLoss. Top: (a) testing for conventional losses. The decoder and evaluator are usually non-differentiable. (b) training for conventional losses. To avoid the non-differentiability, conventional methods optimize a manually-designed differentiable loss function instead during training. Bottom: (a) refactored testing in UniLoss. We refactor the testing so that the non-differentiability exists only in Sign(·) and the multi-variant function. (b) training in UniLoss with the differentiable approximation of refactored testing. $\sigma(\cdot)$ is the sigmoid function. We approximate the non-differentiable components in the refactored testing pipeline with interpolation methods.

However, designing surrogate losses can sometimes incur substantial manual effort, including a large amount of trial and error and hyper-parameter tuning. For example, a standard evaluation of single-person human pose estimation—predicting the 2D locations of a set of body joints for a single person in an image—involves computing the percentage of predicted body joints that are within a given radius of the ground truth. This performance metric is non-differentiable. Existing work instead trains a deep network to predict a heatmap for each type of body joints, minimizing the difference between the predicted heatmap and a "ground truth" heatmap consisting of a Gaussian bump at the ground truth location [17,28]. The decision for what error function to use for comparing heatmaps and how to design the "ground truth" heatmaps are manually tuned to optimize performance.

This manual effort in conventional losses is tedious but necessary, because a poorly designed loss can be misaligned with the final performance metric and lead to ineffective training. As we show in the experiment section, without carefully-tuned loss hyper-parameters, conventional manual losses can work poorly.

In this paper, we seek to reduce the efforts of manual design of surrogate losses by introducing a unified surrogate loss framework applicable to a wide range of tasks. We provide a unified framework to mechanically generate a surrogate loss given a performance metric in the context of deep learning. This means that we only need to specify the performance metric (e.g. classification error) and the inference algorithm—the network architecture, a "decoder" that converts the network output (e.g. continuous scores) to the final output (e.g. discrete class labels), and an "evaluator" that converts the labels to final metric—and the rest is taken care of as part of the training algorithm.

We introduce UniLoss (Fig. 1), a unified framework to generate surrogate losses for training deep networks with gradient descent. We maintain the basic algorithmic structure of mini-batch gradient descent: for each mini-batch, we perform inference on all examples, compute a loss using the results and the ground truths, and generate gradients using the loss to update the network parameters. Our novelty is that we generate all the surrogate losses in a unified framework for various tasks instead of manually designing it for each task.

The key insight behind UniLoss is that for many tasks and performance metrics, evaluating a deep network on a set of training examples—pass the examples through the network, the output decoder, and the evaluator to the performance metric—can be refactored into a sequence of four transformations: the training examples are first transformed to a set of real scores, then to some real numbers representing comparisons (through subtractions) of certain pairs of the real-valued scores, then to a set of binary values representing the signs of the comparisons, and finally to a single real number. Note that the four transforms do not necessarily correspond to running the network inference, the decoder, and the evaluator.

Take multi-class classification as an example, the training examples are first transformed to a set of scores (one per class per example), and then to pairwise comparisons (subtractions) between the scores for each example (i.e. the argmax operation), and then to a set of binary values, and finally to a classification accuracy.

The final performance metric is non-differentiable with respect to network weights because the decoder and the evaluator are non-differentiable. But this refactoring allows us to generate a differentiable approximation of each non-differentiable transformation through interpolation.

Specifically, the transformation from comparisons to binary variables is non-differentiable, we can approximate it by using the sigmoid function to interpolate the sign function. And the transformation from binary variables to final metric may be nondifferentiable, we can approximate it by multivariate interpolation.

The proposed UniLoss framework is general and can be applied to various tasks and performance metrics. Given any performance metric involving discrete operations, to the best of our knowledge, the discrete operations can always be refactored to step functions that first make some differentiable real-number comparisons non-differentiable, and any following operations, which fit in our framework. Example tasks include classification scenarios such as accuracy in image classification, precision and recall in object detection; ranking scenarios such as average precision in binary classification, area under curve in image retrieval; pixel-wise prediction scenarios such as mean IOU in segmentation, PCKh in pose estimation.

To validate its effectiveness, we perform experiments on three representative tasks from three different scenarios. We show that UniLoss performs well on a classic classification setting, multi-class classification, compared with the well-established conventional losses. We also demonstrate UniLoss's ability in a ranking scenario that evolves ranking multiple images in an evaluation set: average precision (area under the precision-recall curve) in unbalanced binary classification. In addition, we experiment with pose estimation where the output is structured as pixel-wise predictions.

Our main contributions in this work are:

- We present a new perspective of finding surrogate losses: evaluation can be refactored as a sequence of four transformations, where each nondifferentiable transformation can be tackled individually.
- We propose a new method: a unified framework to generate losses for various tasks reducing task-specific manual design.
- We validate the new perspective and the new method on three tasks and four datasets, achieving comparable performance with conventional losses.

2 Related Work

2.1 Direct Loss Minimization

The line of direct loss minimization works is related to UniLoss because we share a similar idea of finding a good approximation of the performance metric. There have been many efforts to directly minimize specific classes of tasks and metrics.

For example, [26] optimized ranking metrics such as Normalized Discounted Cumulative Gain by smoothing them with an assumed probabilistic distribution of documents. [11] directly optimized mean average precision in object detection by computing "pseudo partial derivatives" for various continuous variables. [18] explored to optimize the 0–1 loss in binary classification by search-based methods including branch and bound search, combinatorial search, and also coordinate descent on the relaxations of 0–1 losses. [16] proposed to improve the conventional cross-entropy loss by multiplying a preset constant with the angle in the inner product of the `softmax` function to encourage large margins between classes. [7] proposed an end-to-end optimization approach for speech enhancement by directly optimizing short-time objective intelligibility (STOI) which is a differentiable performance metric.

In addition to the large algorithmic differences, these works also differ from ours in that they are tightly coupled with specific tasks and applications.

[9] and [24] proved that under mild conditions, optimizing a max-margin structured-output loss is asymptotically equivalent to directly optimizing the performance metrics. Specifically, assume a model in the form of a differentiable scoring function $S(x, y; w) : \mathcal{X} \times \mathcal{Y} \to \mathbf{R}$ that evaluates the compatibility of output y with input x. During inference, they predict the y_w with the best score:

$$y_w = \underset{y}{\text{argmax}}\ S(x, y; w). \tag{1}$$

During training, in addition to this regular inference, they also perform the *loss-augmented inference* [9,29]:

$$y^{\dagger} = \underset{y}{\text{argmax}}\ S(x, y; w) + \epsilon\xi(y_w, y), \quad (2)$$

where ξ is the final performance metric (in terms of error), and ϵ is a small time-varying weight. [24] generalized this result from linear scoring functions to arbitrary scoring functions, and developed an efficient loss-augmented inference algorithm to directly optimize average precision in ranking tasks.

While above max-margin losses can ideally work with many different performance metrics ξ, its main limitation in practical use is that it can be highly nontrivial to design an efficient algorithm for the loss-augmented inference, as it often requires some clever factorization of the performance metric ξ over the components of the structured output y. In fact, for many metrics the loss-augmented inference is NP-hard and one must resort to designing approximate algorithms, which further increases the difficulty of practical use.

In contrast, our method does not demand the same level of human ingenuity. The main human effort involves refactoring the inference code and evaluation code to a particular format, which may be further eliminated by automatic code analysis. There is no need to design a new inference algorithm over discrete outputs and analyze its efficiency. The difficulty of designing loss-augmented inference algorithms for each individual task makes it impractical to compare fairly with max-margin methods on diverse tasks, because it is unclear how to design the inference algorithms.

Recently, some prior works propose to directly optimize the performance metric by learning a parametric surrogate loss [6,8,13,22,30]. During training, the model is updated to minimize the current surrogate loss while the parametric surrogate loss is also updated to align with the performance metric.

Compared to these methods, UniLoss does not involve any learnable parameters in the loss. As a result, UniLoss can be applied universally across different settings without any training, and the parametric surrogate loss has to be trained separately for different tasks and datasets.

Reinforcement Learning inspired algorithms have been used to optimize performance metrics for structured output problems, especially those that can be formulated as taking a sequence of actions [4,15,21,31,33]. For example, [15] use policy gradients [25] to optimize metrics for image captioning.

We differ from these approaches in two key aspects. First, we do not need to formulate a task as a sequential decision problem, which is natural for certain tasks such as text generation, but unnatural for others such as human pose estimation. Second, these methods treat performance metrics as black boxes, whereas we assume access to the code of the performance metrics, which is a valid assumption in most cases. This access allows us to reason about the code and generate better gradients.

2.2 Surrogate Losses

There has been a large body of literature studying surrogate losses, for tasks including multi-class classification [1,3,5,19,20,27,32], binary classification [3,19,20,32] and pose estimation [28]. Compared to them, UniLoss reduces the manual effort to design task-specific losses. UniLoss, as a general loss framework, can be applied to all these tasks and achieve comparable performance.

3 UniLoss

3.1 Overview

UniLoss provides a unified way to generate a surrogate loss for training deep networks with mini-batch gradient descent without task-specific design. In our general framework, we first re-formulate the evaluation process and then approximate the non-differentiable functions using interpolation.

Original Formulation. Formally, let $\mathbf{x} = (x_1, x_2, \ldots, x_n) \in \mathcal{X}^n$ be a set of n training examples and $\mathbf{y} = (y_1, y_2, \ldots, y_n) \in \mathcal{Y}^n$ be the ground truth. Let $\phi(\cdot; w) : \mathcal{X} \to \mathbf{R}^d$ be a deep network parameterized by weights w that outputs a d-dimensional vector; let $\delta : \mathbf{R}^d \to \mathcal{Y}$ be a decoder that decodes the network output to a possibly discrete final output; let $\xi : \mathcal{Y}^n \times \mathcal{Y}^n \to \mathbf{R}$ be an evaluator. ϕ and δ are applied in a mini-batch fashion on $\mathbf{x} = (x_1, x_2, \ldots, x_n)$; the performance e of the deep network is then

$$e = \xi(\delta(\phi(\mathbf{x}; w)), \mathbf{y}). \tag{3}$$

Refactored Formulation. Our approach seeks to find a surrogate loss to minimize e, with the novel observation that in many cases e can be refactored as

$$e = g(h(f(\phi(\mathbf{x}; w), \mathbf{y}))), \tag{4}$$

where $\phi(\cdot; w)$ is the same as in Eq. 3, representing a deep neural network, $f : \mathbf{R}^{n \times d} \times \mathcal{Y}^n \to \mathbf{R}^l$ is differentiable and maps outputted real numbers and the ground truth to l comparisons each representing the difference between certain pair of real numbers, $h : \mathbf{R}^l \to \{0, 1\}^l$ maps the l score differences to l binary variables, and $g : \{0, 1\}^l \to \mathbf{R}$ computes the performance metric from binary variables. Note that h has a restricted form that always maps continuous values to binary values through sign function, whereas g can be arbitrary computation that maps binary values to a real number.

We give intermediate outputs some notations:

- Training examples $\mathbf{x}, \mathbf{y}$ are transformed to scores $\mathbf{s} = (s_1, s_2, \ldots, s_{nd})$, where $\mathbf{s} = \phi(\mathbf{x}; w)$.
- $\mathbf{s}$ is converted to comparisons (differences of two scores) $\mathbf{c} = (c_1, c_2, \ldots, c_l)$, where $\mathbf{c} = f(\mathbf{s}, \mathbf{y})$.

- $\mathbf{c}$ is converted to binary variables $\mathbf{b} = (b_1, b_2, \ldots, b_l)$ representing the binary outcome of the comparisons, where $\mathbf{b} = h(\mathbf{c})$.
- The binary variables are transformed to a single real number by $e = g(\mathbf{b})$.

This new refactoring of a performance metric allows us to decompose the metric e with g, h, f and ϕ, where ϕ and f are differentiable functions but h and g are often non-differentiable. The non-differentiability of h and g causes e to be non-differentiable with respect to network weights w.

Differentiable Approximation. Our UniLoss generates differentiable approximations of the non-differentiable h and g through interpolation, thus making the metric e optimizable with gradient descent. Formally, UniLoss gives a differentiable approximation $\tilde{e}$

$$\tilde{e} = \tilde{g}(\tilde{h}(f(\phi(\mathbf{x}; w), \mathbf{y}))), \tag{5}$$

where f and ϕ are the same as in Eq. 4, and $\tilde{h}$ and $\tilde{g}$ are the differentiable approximation of h and g. We explain a concrete example of multi-class classification and introduce the refactoring and interpolation in detail based on this example in the following sections.

3.2 Example: Multi-class Classification

We take multi-class classification as an example to show how refactoring works. First, we give formal definitions of multi-class classification and the performance metric: prediction accuracy.

Input is a mini-batch of images $\mathbf{x} = (x_1, x_2, \ldots, x_n)$ and their corresponding ground truth labels are $\mathbf{y} = (y_1, y_2, \ldots, y_n)$ where n is the batch size. $y_i \in \{1, 2, \ldots, p\}$ and p is the number of classes, which happens to be the same value as d in Sect. 3.1. A network $\phi(\cdot; w)$ outputs a score matrix $\mathbf{s} = [s_{i,j}]_{n \times p}$ and $s_{i,j}$ represents the score for the i-th image belongs to the class j.

The decoder $\delta(\mathbf{s})$ decodes $\mathbf{s}$ into the discrete outputs $\tilde{\mathbf{y}} = (\tilde{y}_1, \tilde{y}_2, \ldots, \tilde{y}_n)$ by

$$\tilde{y}_i = \underset{1 \le j \le p}{\operatorname{argmax}}\, s_{i,j}, \tag{6}$$

and $\tilde{y}_i$ represents the predicted label of the i-th image for $i = 1, 2, \ldots, n$.

The evaluator $\xi(\tilde{\mathbf{y}}, \mathbf{y})$ evaluates the accuracy e from $\tilde{\mathbf{y}}$ and $\mathbf{y}$ by

$$e = \frac{1}{n} \sum_{i=1}^{n} [y_i = \tilde{y}_i], \tag{7}$$

where $[\cdot]$ is the Iverson bracket.

Considering above together, the predicted label for an image is correct if and only if the score of its ground truth class is higher than the score of every other class:

$$[y_i = \tilde{y}_i] = \bigwedge_{\substack{1 \le j \le p \\ j \ne y_i}} [s_{i,y_i} - s_{i,j} > 0], \text{for all} 1 \le i \le n, \tag{8}$$

where $\wedge$ is logical and.

We thus refactor the decoding and evaluation process as a sequence of $f(\cdot)$ that transforms $\mathbf{s}$ to comparisons—$s_{i,y_i} - s_{i,j}$for all$1 \leq i \leq n, 1 \leq j \leq p,$ and $j \neq y_i$ ($n \times (p-1)$ comparisons in total), $h(\cdot)$ that transforms comparisons to binary values using $[\cdot > 0]$, and $g(\cdot)$ that transforms binary values to e using logical and. Next, we introduce how to refactor the above procedure into our formulation and approximate g and h.

3.3 Refactoring

Given a performance metric, we refactor it in the form of Eq. 4. We first transform the training images into scores $\mathbf{s} = (s_1, s_2, \ldots, s_{nd})$. We then get the score comparisons (differences of pairs of scores) $\mathbf{c} = (c_1, c_2, \ldots, c_l)$ using $\mathbf{c} = f(\mathbf{s}, \mathbf{y})$. Each comparison is $c_i = s_{k_i^1} - s_{k_i^2}, 1 \leq i \leq l, 1 \leq k_i^1, k_i^2 \leq nd$. The function h then transforms the comparisons to binary values by $\mathbf{b} = h(\mathbf{c})$. h is the sign function, i.e. $b_i = [c_i > 0], 1 \leq i \leq l$. The function g then computes e by $e = g(\mathbf{b})$, where g can be arbitrary computation that converts binary values to a real number. In practice, g can be complex and vary significantly across tasks and metrics.

Given any performance metrics involving discrete operations in function ξ and δ in Eq. 3 (otherwise the metric e is differentiable and trivial to be handled), the computation of function $\xi(\delta(\cdot))$ can be refactored as a sequence of continuous operations (which is optional), discrete operations that make some differentiable real numbers non-differentiable, and any following operations. The discrete operations always occur when there are step functions, which can be expressed as comparing two numbers, to the best of our knowledge.

This refactoring is usually straightforward to obtain from the specification of the decoding and evaluating procedures. The only manual effort is in identifying the discrete comparisons (binary variables). Then we simply write the discrete comparisons as function f and h, and represent its following operations as function g.

In later sections we will show how to identify the binary variables for three commonly-used metrics in three scenarios, which can be easily extended to other performance metrics. On the other hand, this process is largely a mechanical exercise, as it is equivalent to rewriting some existing code in an alternative rigid format.

3.4 Interpolation

The two usually non-differentiable functions h and g are approximated by interpolation methods individually.

Scores to Binaries: *h*. In $\mathbf{b} = h(\mathbf{c})$, each element $b_i = [c_i > 0]$. We approximate the step function $[\cdot]$ using the `sigmoid` function. That is, $\tilde{\mathbf{b}} = \tilde{h}(\mathbf{c}) = (\tilde{b}_1, \tilde{b}_2, \ldots, \tilde{b}_l)$, and each element

$$\tilde{b}_i = \texttt{sigmoid}(c_i), \tag{9}$$

where $1 \leq i \leq l$. We now have $\tilde{h}$ as the differentiable approximation of h.

Binaries to Performance: g. We approximate $g(\cdot)$ in $e = g(\mathbf{b})$ by multivariate interpolation over the input $\mathbf{b} \in \{0,1\}^l$. More specifically, we first sample a set of configurations as "anchors" $\mathbf{a} = (\mathbf{a}_1, \mathbf{a}_2, \ldots, \mathbf{a}_t)$, where $\mathbf{a}_i$ is a configuration of $\mathbf{b}$, and compute the output values $g(\mathbf{a}_1), g(\mathbf{a}_2), \ldots, g(\mathbf{a}_t)$, where $g(\mathbf{a}_i)$ is the actual performance metric value and t is the number of anchors sampled.

We then get an interpolated function over the anchors as $\tilde{g}(\cdot; \mathbf{a})$. We finally get $\tilde{e} = \tilde{g}(\tilde{\mathbf{b}}; \mathbf{a})$, where $\tilde{\mathbf{b}}$ is computed from $\tilde{h}$, f and ϕ.

By choosing a differentiable interpolation method, the $\tilde{g}$ function becomes trainable using gradient-based methods. We use a common yet effective interpolation method: inverse distance weighting (IDW) [23]:

$$\tilde{g}(\mathbf{u}; \mathbf{a}) = \begin{cases} \frac{\sum_{i=1}^{t} \frac{1}{d(\mathbf{u},\mathbf{a}_i)} g(\mathbf{a}_i)}{\sum_{i=1}^{t} \frac{1}{d(\mathbf{u},\mathbf{a}_i)}}, & d(\mathbf{u}, \mathbf{a}_i) \neq 0 \text{for} 1 \leq i \leq t; \\ g(\mathbf{a}_i), & d(\mathbf{u}, \mathbf{a}_i) = 0 \text{for some} i. \end{cases} \tag{10}$$

where $\mathbf{u}$ represents the input to $\tilde{g}$ and $d(\mathbf{u}, \mathbf{a}_i)$ is the Euclidean distance between $\mathbf{u}$ and $\mathbf{a}_i$.

We select a subset of anchors based on the current training examples. We use a mix of three types of anchors—good anchors with high performance values globally, bad anchors with low performance values globally, and nearby anchors that are close to the current configuration, which is computed from the current training examples and network weights. By using both the global information from the good and bad anchors and the local information from the nearby anchors, we are able to get an informative interpolation surface. On the contrast, a naive random anchor sampling strategy does not give informative interpolation surface and cannot train the network at all in our experiments.

More specifically, we adopt a straightforward anchor sampling strategy for all tasks and metrics: we obtain good anchors by flipping some bits from the best anchor, which is the ground truth. The bad anchors are generated by randomly sampling binary values. The nearby anchors are obtained by flipping some bits from the current configuration.

4 Experimental Results

To use our general framework UniLoss on each task, we refactor the evaluation process of the task into the format in Eq. 4, and then approximate the non-differentiable functions h and g using the interpolation method in Sect. 3.

We validate the effectiveness of the UniLoss framework in three representative tasks in different scenarios: a ranking-related task using a set-based metric—average precision, a pixel-wise prediction task, and a common classification task. For each task, we demonstrate how to formulate the evaluation process to our refactoring and compare our UniLoss with interpolation to the conventional task-specific loss. More implementation details and analysis can be found in the material.

4.1 Tasks and Metrics

Average Precision for Unbalanced Binary Classification. Binary classification is to classify an example from two classes—positives and negatives. Potential applications include face classification and image retrieval. It has unbalanced number of positives and negatives in most cases, which results in that a typical classification metric such as accuracy as in regular classification cannot demonstrate how good is a model properly. For example, when the positives to negatives is 1:9, predicting all examples as negatives gets 90% accuracy.

On this unbalanced binary classification, other metrics such as precision, recall and average precision (AP), i.e. area under the precision-recall curve, are more descriptive metrics. We use AP as our target metric in this task.

It is notable that AP is fundamentally different from accuracy because it is a set-based metric. It can only be evaluated on a set of images, and involves not only the correctness of each image but also the ranking of multiple images. This task and metric is chosen to demonstrate that UniLoss can effectively optimize for a set-based performance metric that requires ranking of the images.

PCKh for Single-Person Pose Estimation. Single-person pose estimation predicts the localization of human joints. More specifically, given an image, it predicts the location of the joints. It is usually formulated as a pixel-wise prediction problem, where the neural network outputs a score for each pixel indicating how likely is the location can be the joint.

Following prior work, we use PCKh (Percentage of Correct Keypoints wrt to head size) as the performance metric. It computes the percentage of the predicted joints that are within a given radius r of the ground truth. The radius is half of the head segment length. This task and metric is chosen to validate the effectiveness of UniLoss in optimizing for a pixel-wise prediction problem.

Accuracy for Multi-class Classification. Multi-class classification is a common task that has a well-established conventional loss—cross-entropy loss.

We use accuracy (the percentage of correctly classified images) as our metric following the common practice. This task and metric is chosen to demonstrate that for a most common classification setting, UniLoss still performs similarly effectively as the well-established conventional loss.

4.2 Average Precision for Unbalanced Binary Classification

Dataset and Baseline. We augment the handwritten digit dataset MNIST to be a binary classification task, predicting zeros or non-zeros. Given an image containing a single number from 0 to 9, we classify it into the zero (positive) class or the non-zero (negative) class. The positive-negative ratio of 1:9. We create a validation split by reserving 6k images from the original training set.

We use a 3-layer fully-connected neural network with 500 and 300 neurons in each hidden layer respectively. Our baseline model is trained with a 2-class cross-entropy loss. We train both baseline and UniLoss with a fixed learning rate of 0.01 for 30 epochs. We sample 16 anchors for each anchor type in our anchor interpolation for all of our experiments except in ablation studies.

Formulation and Refactoring. The evaluation process is essentially ranking images using pair-wise comparisons and compute the area under curve based on the ranking. It is determined by whether positive images are ranked higher than negative images. Given that the output of a mini-batch of n images is $\mathbf{s} = (s_1, s_2, \ldots, s_n)$, where s_i represents the predicted score of i-th image to be positive. The binary variables are $\mathbf{b} = \{b_{i,j} = [c_{i,j} > 0] = [s_i - s_j > 0]\}$, where i belongs to positives and j belongs to negatives.

Results. UniLoss achieves an AP of 0.9988, similarly as the baseline cross-entropy loss (0.9989). This demonstrates that UniLoss can effectively optimize for a performance metric (AP) that is complicated to compute and involves a batch of images.

4.3 PCKh for Single-Person Pose Estimation

Dataset and Baseline. We use MPII [2] which has around 22K images for training and 3K images for testing. For simplicity, we perform experiments on the joints of head only, but our method could be applied to an arbitrary number of human joints without any modification.

We adopt the Stacked Hourglass [17] as our model. The baseline loss is the Mean Squared Error (MSE) between the predicted heatmaps and the manually-designed "ground truth" heatmaps. We train a single-stack hourglass network for both UniLoss and MSE using RMSProp [12] with an initial learning rate 2.5e−4 for 30 epochs and then drop it by 4 for every 10 epochs until 50 epochs.

Formulation and Refactoring. Assume the network generates a mini-batch of heatmaps $\mathbf{s} = (\mathbf{s}^1, \mathbf{s}^2, \ldots, \mathbf{s}^n) \in \mathbf{R}^{n \times m}$, where n is the batch size, m is the number of pixels in each image. The pixel with the highest score in each heatmap is predicted as a key point during evaluation. We note the pixels within the radius r around the ground truth as positive pixels, and other pixels as negative and each heatmap $\mathbf{s}^k$ can be flatted as $(s^k_{pos,1}, s^k_{pos,2}, \ldots, s^k_{pos,m_k}, s^k_{neg,1}, \ldots, s^k_{neg,m-m_k})$, where m_k is the number of positive pixels in the k-th heatmap and $s^k_{pos,j}$ ($s^k_{neg,j}$) is the score of the j-th positive (negative) pixel in the k-th heatmap.

PCKh requires to find out if a positive pixel has the highest score among others. Therefore, we need to compare each pair of positive and negative pixels and this leads to the binary variables $\mathbf{b} = (b_{k,i,j})$for$1 \leq k \leq n$, $1 \leq i \leq m_k$, $1 \leq j \leq m - m_k$, where $b_{k,i,j} = [s^k_{pos,i} - s^k_{neg,j} > 0]$, i.e. the comparison between the i-th positive pixel and the j-th negative pixel in the k-th heatmap.

Table 1. PCKh of Stacked Hourglass with MSE and UniLoss on the MPII validation.

Loss	MSE $\sigma = 0.1$	$\sigma = 0.5$	$\sigma = 0.7$	$\sigma = 1$	$\sigma = 3$	$\sigma = 5$	$\sigma = 10$	UniLoss
PCKh	91.31	95.13	93.06	95.71	95.74	94.99	92.25	**95.77**

Table 2. Accuracy of ResNet-20 with CE loss and UniLoss on the CIFAR-10 and CIFAR-100 test set.

Loss	CIFAR-10	CIFAR-100
CE (cross-entropy) loss	91.49	65.90
UniLoss	91.64	65.92

Table 3. Ablation study for mini-batch sizes on CIFAR-10.

Batch size	8	16	32	64	128	256	512	1024
Accuracy	87.89	90.05	90.82	91.23	**91.64**	90.94	89.10	87.20

Results. It is notable that the manual design of the target heatmaps is a part of the MSE loss function for pose estimation. It heavily relies on the careful design of the ground truth heatmaps. If we intuitively set the pixels at the exact joints to be 1 and the rest of pixels as 0 in the heatmaps, the training diverges.

Luckily, [28] proposed to design target heatmaps as a 2D Gaussian bump centered on the ground truth joints, whose shape is controlled by its variance σ and the bump size. The success of the MSE loss function relies on the choices of σ and the bump size. UniLoss, on the other hand, requires no such design.

As shown in Table 1, our UniLoss achieves a 95.77 PCKh which is comparable as the 95.74 PCKh for MSE with the best σ. This validates the effectiveness of UniLoss in optimizing for a pixel-wise prediction problem.

We further observe that the baseline is sensitive to the shape of 2D Gaussian, as in Table 1. Smaller σ makes target heatmaps concentrated on ground truth joints and makes the optimization to be unstable. Larger σ generates vague training targets and decreases the performance. This demonstrates that conventional losses require dedicated manual design while UniLoss can be applied directly.

4.4 Accuracy for Multi-class Classification

Dataset and Baseline. We use CIFAR-10 and CIFAR-100 [14], with 32×32 images and 10/100 classes. They each have 50k training images and 10k test images. Following prior work [10], we split the training set into a 45k-5k train-validation split.

We use the ResNet-20 architecture [10]. Our baselines are trained with cross-entropy (CE) loss. All experiments are trained following the same augmentation and pre-processing techniques as in prior work [10]. We use an initial learning

Table 4. Ablation study for number of anchors on the three tasks.

Task	#Anchors in each type	Performance
Bin. classification (AP)	5	0.9988
	10	0.9987
	16	0.9988
	5	91.55%
Classification (Acc)	10	91.54%
	16	91.64%
	5	94.79%
Pose estimation (PCKh)	10	94.95%
	16	95.77%

rate of 0.1, divided by 10 and 100 at the 140th epoch and the 160th epoch, with a total of 200 epochs trained for both baseline and UniLoss on CIFAR-10. On CIFAR-100, we train baseline with the same training schedule and UniLoss with 5x training schedule because we only train 20% binary variables at each step. For a fair comparison, we also train baseline with the 5x training schedule but observe no improvement.

Formulation and Refactoring. As shown in Sect. 3.2, given the output of a mini-batch of n images $\mathbf{s} = (s_{1,1}, s_{1,2}.., s_{n,p})$, we compare the score of the ground truth class and the scores of other $p-1$ classes for each image. That is, for the i-th image with the ground truth label y_i, $b_{i,j} = [s_{i,y_i} - s_{i,j} > 0]$, where $1 \leq j \leq p$, $j \neq y_i$, and $1 \leq i \leq n$. For tasks with many binary variables such as CIFAR-100, we train a portion of binary variables in each update to accelerate training.

Results. Our implementation of the baseline method obtains a slightly better accuracy (91.49%) than that was reported in [10]—91.25% on CIFAR-10 and obtains 65.9% on CIFAR-100. UniLoss performs similarly (91.64% and 65.92%) as baselines on both datasets (Table 2), which shows that even when the conventional loss is well-established for the particular task and metric, UniLoss still matches the conventional loss.

4.5 Discussion of Hyper-parameters

Mini-Batch Sizes. We also use a mini-batch of images for updates with UniLoss. Intuitively, as long as the batch size is not extremely small or large, it should be able to approximate the distribution of the whole dataset.

We explore different batch sizes on the CIFAR-10 multi-class classification task, as shown in Table 3. The results match with our hypothesis—as long as the batch size is not extreme, the performance is similar. A batch size of 128 gives the best performance.

Number of Anchors. We explore different number of anchors in the three tasks. We experiment with 5, 10, 16 as the number of anchors for each type of the good, bad and nearby anchors. That is, we have 15, 30, 48 anchors in total respectively. Table 4 shows that binary classification and classification are less sensitive to the number of anchors, while in pose estimation, fewer anchors lead to slightly worse performance. It is related to the number of binary variables in each task: pose estimation has scores for each pixel, thus has much more comparisons than binary classification and classification. With more binary variables, more anchors tend to be more beneficial.

5 Conclusion and Limitations

We have introduced UniLoss, a framework for generating surrogate losses in a unified way, reducing the amount of manual design of task-specific surrogate losses. The proposed framework is based on the observation that there exists a common refactoring of the evaluation computation for many tasks and performance metrics. Using this refactoring we generate a unified differentiable approximation of the evaluation computation, through interpolation. We demonstrate that using UniLoss, we can optimize for various tasks and performance metrics, achieving comparable performance as task-specific losses.

We now discuss some limitations of UniLoss. One limitation is that the interpolation methods are not yet fully explored. We adopt the most straightforward yet effective way in this paper, such as the sigmoid function and IDW interpolation for simplicity and an easy generalization across different tasks. But there are potentially other sophisticated choices for the interpolation methods and for the sampling strategy for anchors.

The second limitation is that proposed anchor sampling strategy is biased towards the optimal configuration that corresponds to the ground truth when there are multiple configurations that can lead to the optimal performance.

The third limitation is that ranking-based metrics may result in a quadratic number of binary variables if pairwise comparison is needed for every pair of scores. Fortunately in many cases such as the ones discussed in this paper, the number of binary variables is not quadratic because many comparisons does not contribute to the performance metric.

The fourth limitation is that currently UniLoss still requires some amount of manual effort (although less than designing a loss from scratch) to analyze the given code of the decoder and the evaluator for the refactoring. Combining automatic code analysis with our framework can further reduce manual efforts in loss design.

References

1. Allwein, E.L., Schapire, R.E., Singer, Y.: Reducing multiclass to binary: a unifying approach for margin classifiers. In: ICML (2000)

2. Andriluka, M., Pishchulin, L., Gehler, P., Schiele, B.: 2D human pose estimation: new benchmark and state of the art analysis. In: CVPR (2014)
3. Bartlett, P.L., Jordan, M.I., McAuliffe, J.D.: Convexity, classification, and risk bounds. J. Am. Stat. Assoc. **101**(473), 138–156 (2006)
4. Caicedo, J.C., Lazebnik, S.: Active object localization with deep reinforcement learning. In: ICCV (2015)
5. Crammer, K., Singer, Y.: On the algorithmic implementation of multiclass kernel-based vector machines. In: ICML (2001)
6. Engilberge, M., Chevallier, L., Pérez, P., Cord, M.: SoDeep: a sorting deep net to learn ranking loss surrogates. In: CVPR (2019)
7. Fu, S.W., Wang, T.W., Tsao, Y., Lu, X., Kawai, H.: End-to-end waveform utterance enhancement for direct evaluation metrics optimization by fully convolutional neural networks. IEEE/ACM Trans. Audio Speech Lang. Process. **26**(9), 1570–1584 (2018)
8. Grabocka, J., Scholz, R., Schmidt-Thieme, L.: Learning surrogate losses. arXiv preprint arXiv:1905.10108 (2019)
9. Hazan, T., Keshet, J., McAllester, D.A.: Direct loss minimization for structured prediction. In: NeurIPS (2010)
10. He, K., Zhang, X., Ren, S., Sun, J.: Deep residual learning for image recognition. In: CVPR (2016)
11. Henderson, P., Ferrari, V.: End-to-end training of object class detectors for mean average precision. In: Lai, S.-H., Lepetit, V., Nishino, K., Sato, Y. (eds.) ACCV 2016. LNCS, vol. 10115, pp. 198–213. Springer, Cham (2017). https://doi.org/10.1007/978-3-319-54193-8_13
12. Hinton, G., Srivastava, N., Swersky, K.: Lecture 6A overview of mini-batch gradient descent. Coursera Lecture slides (2012). https://class.coursera.org/neuralnets-2012-001/lecture
13. Huang, C., et al.: Addressing the loss-metric mismatch with adaptive loss alignment. arXiv preprint arXiv:1905.05895 (2019)
14. Krizhevsky, A., Hinton, G.: Learning multiple layers of features from tiny images (2009)
15. Liu, S., Zhu, Z., Ye, N., Guadarrama, S., Murphy, K.: Improved image captioning via policy gradient optimization of spider. In: ICCV (2017)
16. Liu, W., Wen, Y., Yu, Z., Yang, M.: Large-margin softmax loss for convolutional neural networks. In: ICML (2016)
17. Newell, A., Yang, K., Deng, J.: Stacked hourglass networks for human pose estimation. In: Leibe, B., Matas, J., Sebe, N., Welling, M. (eds.) ECCV 2016. LNCS, vol. 9912, pp. 483–499. Springer, Cham (2016). https://doi.org/10.1007/978-3-319-46484-8_29
18. Nguyen, T., Sanner, S.: Algorithms for direct 0–1 loss optimization in binary classification. In: ICML (2013)
19. Ramaswamy, H.G., Agarwal, S., Tewari, A.: Convex calibrated surrogates for low-rank loss matrices with applications to subset ranking losses. In: NeurIPS (2013)
20. Ramaswamy, H.G., Babu, B.S., Agarwal, S., Williamson, R.C.: On the consistency of output code based learning algorithms for multiclass learning problems. In: Conference on Learning Theory, pp. 885–902 (2014)
21. Ranzato, M., Chopra, S., Auli, M., Zaremba, W.: Sequence level training with recurrent neural networks. In: ICLR (2016)
22. Santos, C.N.d., Wadhawan, K., Zhou, B.: Learning loss functions for semi-supervised learning via discriminative adversarial networks. arXiv preprint arXiv:1707.02198 (2017)

23. Shepard, D.: A two-dimensional interpolation function for irregularly-spaced data. In: Proceedings of the 23rd ACM National Conference. ACM (1968)
24. Song, Y., Schwing, A., Urtasun, R., et al.: Training deep neural networks via direct loss minimization. In: ICML (2016)
25. Sutton, R.S., McAllester, D.A., Singh, S.P., Mansour, Y.: Policy gradient methods for reinforcement learning with function approximation. In: NeurIPS (2000)
26. Taylor, M., Guiver, J., Robertson, S., Minka, T.: SoftRank: optimizing non-smooth rank metrics. In: WSDM (2008)
27. Tewari, A., Bartlett, P.L.: On the consistency of multiclass classification methods. In: ICML (2007)
28. Tompson, J.J., Jain, A., LeCun, Y., Bregler, C.: Joint training of a convolutional network and a graphical model for human pose estimation. In: NeurIPS (2014)
29. Tsochantaridis, I., Joachims, T., Hofmann, T., Altun, Y.: Large margin methods for structured and interdependent output variables. In: ICML (2005)
30. Wu, L., Tian, F., Xia, Y., Fan, Y., Qin, T., Jian-Huang, L., Liu, T.Y.: Learning to teach with dynamic loss functions. In: Advances in Neural Information Processing Systems, pp. 6466–6477 (2018)
31. Yeung, S., Russakovsky, O., Mori, G., Fei-Fei, L.: End-to-end learning of action detection from frame glimpses in videos. In: CVPR (2016)
32. Zhang, T.: Statistical behavior and consistency of classification methods based on convex risk minimization. Ann. Stat. (2004)
33. Zhou, Y., Xiong, C., Socher, R.: Improving end-to-end speech recognition with policy learning. In: 2018 IEEE International Conference on Acoustics, Speech and Signal Processing (ICASSP), pp. 5819–5823. IEEE (2018)

Deep Reflectance Volumes: Relightable Reconstructions from Multi-view Photometric Images

Sai Bi[1(✉)], Zexiang Xu[1,2], Kalyan Sunkavalli[2], Miloš Hašan[2], Yannick Hold-Geoffroy[2], David Kriegman[1], and Ravi Ramamoorthi[1]

[1] University of California, San Diego, USA
bisai@cs.ucsd.edu
[2] Adobe Research, San Jose, USA

Abstract. We present a deep learning approach to reconstruct scene appearance from unstructured images captured under collocated point lighting. At the heart of Deep Reflectance Volumes is a novel volumetric scene representation consisting of opacity, surface normal and reflectance voxel grids. We present a novel physically-based differentiable volume ray marching framework to render these scene volumes under arbitrary viewpoint and lighting. This allows us to optimize the scene volumes to minimize the error between their rendered images and the captured images. Our method is able to reconstruct real scenes with challenging non-Lambertian reflectance and complex geometry with occlusions and shadowing. Moreover, it accurately generalizes to *novel* viewpoints and lighting, including non-collocated lighting, rendering photorealistic images that are significantly better than state-of-the-art mesh-based methods. We also show that our learned reflectance volumes are editable, allowing for modifying the materials of the captured scenes.

Keywords: View synthesis · Relighting · Appearance acquisition · Neural rendering

1 Introduction

Capturing a real scene and re-rendering it under novel lighting conditions and viewpoints is one of the core challenges in computer vision and graphics. This is classically done by reconstructing the 3D scene geometry, typically in the form of a mesh, and computing per-vertex colors or reflectance parameters, to support arbitrary re-rendering. However, 3D reconstruction methods like multi-view stereo are prone to errors in textureless and non-Lambertian regions [37,47], and accurate reflectance acquisition usually requires dense, calibrated capture using sophisticated devices [5,55].

Electronic supplementary material The online version of this chapter (https://doi.org/10.1007/978-3-030-58580-8_18) contains supplementary material, which is available to authorized users.

A. Vedaldi et al. (Eds.): ECCV 2020, LNCS 12348, pp. 294–311, 2020.
https://doi.org/10.1007/978-3-030-58580-8_18

Fig. 1. Given a set of images taken using a mobile phone with flashlight (sampled images are shown in (a)), our method learns a volume representation of the captured object by estimating the opacity volume, normal volume (b) and reflectance volumes such as albedo (c) and roughness (d). Our volume representation enables free navigation of the object under arbitrary viewpoints and novel lighting conditions (e).

Recent works have proposed learning-based approaches to capture scene appearance. One class of methods use surface-based representations [15,20] but are restricted to specific scene categories and cannot synthesize photo-realistic images. Other methods bypass explicit reconstruction, instead focusing on relighting [58] or view synthesis sub-problems [31,56].

Our goal is to make high-quality scene acquisition and rendering practical with off-the-shelf devices under mildly controlled conditions. We use a set of unstructured images captured around a scene by a single mobile phone camera with flash illumination in a dark room. This practical setup acquires multi-view images under collocated viewing and lighting directions—referred to as photometric images [56]. While the high-frequency appearance variation in these images (due to sharp specular highlights and shadows) can result in low-quality mesh reconstruction from state-of-the-art methods (see Fig. 3), we show that our method can accurately model the scene and realistically reproduce complex appearance information like specularities and occlusions.

At the heart of our method is a novel, physically-based neural volume rendering framework. We train a deep neural network that simultaneously learns the geometry and *reflectance* of a scene as volumes. We leverage a decoder-like network architecture, where an encoding vector together with the corresponding network parameters are learned during a per-scene optimization (training) process. Our network decodes a volumetric scene representation consisting of opacity, normal, diffuse color and roughness volumes, which model the global geometry, local surface orientations and spatially-varying reflectance parameters of the scene, respectively. These volumes are supplied to a differentiable rendering module to render images with collocated light-view settings at training time, and arbitrary light-view settings at inference time (see Fig. 2).

We base our differentiable rendering module on classical volume ray marching approaches with opacity (alpha) accumulation and compositing [24,52]. In particular, we compute point-wise shading using local normal and reflectance properties, and accumulate the shaded colors with opacities along each marching ray of sight. Unlike the opacity used in previous view synthesis work [31,62] that is only accumulated along view directions, we propose to learn global scene opacity that can be accumulated from both view and light directions. As shown in Fig. 1, we demonstrate that our scene opacity can be effectively learned and used to compute accurate hard shadows under *novel lighting*, despite the fact that the training process never observed images with shadows that are taken under non-collocated view-light setups. Moreover, different from previous volume-based works [31,62] that learn a single color at each voxel, we reconstruct per-voxel reflectance and handle complex materials with high glossiness. Our neural rendering framework thus enables rendering with complex view-dependent and light-dependent shading effects including specularities, occlusions and shadows. We compare against a state-of-the-art mesh-based method [37], and demonstrate that our method is able to achieve more accurate reconstructions and renderings (see Fig. 3). We also show that our approach supports scene material editing by modifying the reconstructed reflectance volumes (see Fig. 8). To summarize, our contributions are:

- A practical neural rendering framework that reproduces high-quality geometry and appearance from unstructured mobile phone flash images and enables view synthesis, relighting, and scene editing.
- A novel scene appearance representation using opacity, normal and reflectance volumes.
- A physically-based differentiable volume rendering approach based on deep priors that can effectively reconstruct the volumes from input flash images.

2 Related Works

Geometry Reconstruction. There is a long history in reconstructing 3D geometry from images using traditional structure from motion and multi-view stereo (MVS) pipelines [13,25,47]. Recently deep learning techniques have also been applied to 3D reconstruction with various representations, including volumes [18,45], point clouds [1,42,51], depth maps [16,59] and implicit functions [10,35,40]. We aim to model scene geometry for realistic image synthesis, for which mesh-based reconstruction [23,32,38] is the most common way in many applications [6,37,44,61]. However, it remains challenging to reconstruct accurate meshes for challenging scenes where there are textureless regions and thin structures, and it is hard to incorporate a mesh into a deep learning framework [26,30]; the few mesh-based deep learning works [15,20] are limited to category-specific reconstruction and cannot produce photo-realistic results. Instead, we leverage a physically-based opacity volume representation that can be easily embedded in a deep learning system to express scene geometry of arbitrary shapes.

Reflectance Acquisition. Reflectance of real materials is classically measured using sophisticated devices to densely acquire light-view samples [12,33], which is impractical for common users. Recent works have improved the practicality with fewer samples [39,57] and more practical devices (mobile phones) [2,3,17,28]; however, most of them focus on flat planar objects. A few single-view techniques based on photometric stereo [4,14] or deep learning [29] are able to handle arbitrary shape, but they merely recover limited single-view scene content. To recover complete shape with spatially varying BRDF from multi-view inputs, previous works usually rely on a pre-reconstructed initial mesh and images captured under complex controlled setups to reconstruct per-vertex BRDFs [7,21,53,55,63]. While a recent work [37] uses a mobile phone for practical acquisition like ours, it still requires MVS-based mesh reconstruction, which is ineffective for challenging scenes with textureless, specular and thin-structure regions. In contrast, we reconstruct spatially varying volumetric reflectance via deep network based optimization; we avoid using any initial geometry and propose to jointly reconstruct geometry and reflectance in a holistic framework.

Relighting and View Synthesis. Image-based techniques have been extensively explored in graphics and vision to synthesize images under novel lighting and viewpoint without explicit complete reconstruction [8,11,27,43]. Recently, deep learning has been applied to view synthesis and most methods leverage either view-dependent volumes [49,56,62] or canonical world-space volumes [31,48] for geometric-aware appearance inference. We extend them to a more general physically-based volumetric representation which explicitly expresses both geometry and reflectance, and enables relighting with view synthesis. On the other hand, learning-based relighting techniques have also been developed. Purely image-based methods are able to relight scenes with realistic specularities and soft shadows from sparse inputs, but unable to reproduce accurate hard shadows [19,50,58,60]; some other methods [9,44] propose geometry-aware networks and make use of pre-acquired meshes for relighting and view synthesis, and their performance is limited by the mesh reconstruction quality. A work [36] concurrent to ours models scene geometry and appearance by reconstructing a continuous radiance field for pure view synthesis. In contrast, Deep Reflectance Volumes explicitly express scene geometry and reflectance, and reproduce accurate high-frequency specularities and hard shadows. Ours is the first comprehensive neural rendering framework that enables both relighting and view synthesis with complex shading effects.

3 Rendering with Deep Reflectance Volumes

Unlike a mesh that is comprised of points with complex connectivity, a volume is a regular 3D grid, suitable for convolutional operations. Volumes have been widely used in deep learning frameworks for 3D applications [54,59]. However, previous neural volumetric representations have only represented pixel colors; this can be used for view synthesis [31,62], but does not support relighting or scene editing. Instead, we propose to jointly learn geometry and reflectance (i.e.

material parameters) volumes to enable broader rendering applications including view synthesis, relighting and material editing in a comprehensive framework. Deep Reflectance Volumes are learned from a deep network and used to render images in a fully differentiable end-to-end process as shown in Fig. 2. This is made possible by a new differentiable volume ray marching module, which is motivated by physically-based volume rendering. In this section, we introduce our volume rendering method and volumetric scene representation. We discuss how we learn these volumes from unstructured images in Sect. 4.

3.1 Volume Rendering Overview

In general, volume rendering is governed by the physically-based volume rendering equation (radiative transfer equation) that describes the radiance that arrives at a camera [34,41]:

$$L(\boldsymbol{c}, \boldsymbol{\omega}_o) = \int_0^\infty \tau(\boldsymbol{c}, \boldsymbol{x})[L_e(\boldsymbol{x}, \boldsymbol{\omega}_o) + L_s(\boldsymbol{x}, \boldsymbol{\omega}_o)]dx, \tag{1}$$

This equation integrates emitted, L_e, and in-scattered, L_s, light contributions along the ray starting at camera position $\boldsymbol{c}$ in the direction $-\boldsymbol{\omega}_o$. Here, x represents distance along the ray, and $\boldsymbol{x} = \boldsymbol{c} - x\boldsymbol{\omega}_o$ is the corresponding 3D point. $\tau(\boldsymbol{c}, \boldsymbol{x})$ is the transmittance factor that governs the loss of light along the line segment between $\boldsymbol{c}$ and $\boldsymbol{x}$:

$$\tau(\boldsymbol{c}, \boldsymbol{x}) = e^{-\int_0^x \sigma_t(z)dz}, \tag{2}$$

where $\sigma_t(z)$ is the extinction coefficient at location z on the segment. The in-scattered contribution is defined as:

$$L_s(\boldsymbol{x}, \boldsymbol{\omega}_o) = \int_{\mathcal{S}} f_p(\boldsymbol{x}, \boldsymbol{\omega}_o, \boldsymbol{\omega}_i) L_i(\boldsymbol{x}, \boldsymbol{\omega}_i) d\boldsymbol{\omega}_i, \tag{3}$$

in which $\mathcal{S}$ is a unit sphere, $f_p(\boldsymbol{x}, \boldsymbol{\omega}_o, \boldsymbol{\omega}_i)$ is a generalized (unnormalized) phase function that expresses how light scatters at a point in the volume, and $L_i(\boldsymbol{x}, \boldsymbol{\omega}_i)$ is the incoming radiance that arrives at $\boldsymbol{x}$ from direction $\boldsymbol{\omega}_i$.

In theory, fully computing $L(\boldsymbol{c}, \boldsymbol{\omega}_o)$ requires multiple-scattering computation using Monte Carlo methods [41], which is computationally expensive and unsuitable for deep learning techniques. We consider a simplified case with a single point light, single scattering and no volumetric emission. The transmittance between the scattering location and the point light is handled the same way as between the scattering location and camera. The generalized phase function $f_p(\boldsymbol{x}, \boldsymbol{\omega}_o, \boldsymbol{\omega}_i)$ becomes a reflectance function $f_r(\boldsymbol{\omega}_o, \boldsymbol{\omega}_i, \boldsymbol{n}(\boldsymbol{x}), R(\boldsymbol{x}))$ which computes reflected radiance at $\boldsymbol{x}$ using its local surface normal $\boldsymbol{n}(\boldsymbol{x})$ and the reflectance parameters $R(\boldsymbol{x})$ of a given surface reflectance model. Therefore, Eq. 1 and Eq. 3 can be simplified and written concisely as [24,34]:

$$L(\boldsymbol{c}, \boldsymbol{\omega}_o) = \int_0^\infty \tau(\boldsymbol{c}, \boldsymbol{x}) \tau(\boldsymbol{x}, \boldsymbol{l}) f_r(\boldsymbol{\omega}_o, \boldsymbol{\omega}_i, \boldsymbol{n}(\boldsymbol{x}), R(\boldsymbol{x})) L_l(\boldsymbol{x}, \boldsymbol{\omega}_i) dx, \tag{4}$$

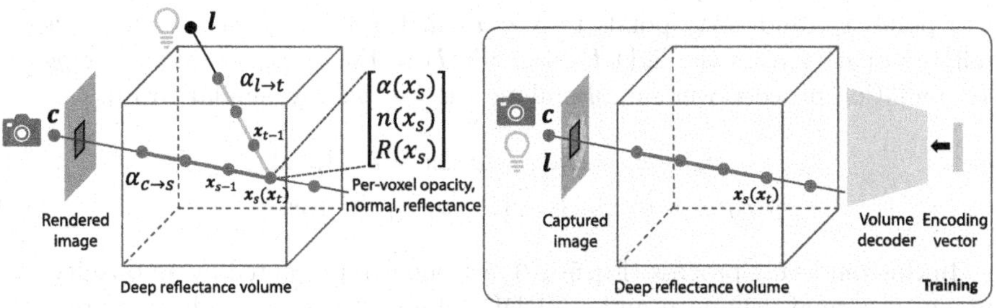

Fig. 2. We propose Deep Reflectance Volume representation to capture scene geometry and appearance, where each voxel consists of opacity α, normal n and reflectance (material coefficients) R. During rendering, we perform ray marching through each pixel and accumulate contributions from each point $\boldsymbol{x}_s$ along the ray. Each contribution is calculated using the local normal, reflectance and lighting information. We accumulate opacity from both the camera $\alpha_{c\to s}$ and the light $\alpha_{l\to t}$ to model the light transport loss in both occlusions and shadows. To predict such a volume, we start from an encoding vector, and decode it into a volume using a 3D convolutional neural network; thus the combination of the encoding vector and network weights is the unknown variable being optimized (trained). We train on images captured with collocated camera and light by enforcing a loss function between rendered images and training images.

where $\boldsymbol{l}$ is the light position, $\boldsymbol{\omega}_i$ corresponds to the direction from $\boldsymbol{x}$ to $\boldsymbol{l}$, $\tau(\boldsymbol{c}, \boldsymbol{x})$ still represents the transmittance from the scattering point $\boldsymbol{x}$ to the camera $\boldsymbol{c}$, the term $\tau(\boldsymbol{x}, \boldsymbol{l})$ (that was implicitly involved in Eq. 3) is the transmittance from the light $\boldsymbol{l}$ to $\boldsymbol{x}$ and expresses light extinction before scattering, and $L_l(\boldsymbol{x}, \boldsymbol{\omega}_i)$ represents the light intensity arriving at $\boldsymbol{x}$ without considering light extinction.

3.2 A Discretized, Differentiable Volume Rendering Module

To make volume rendering practical in a learning framework, we further approximate Eq. 4 by turning it into a discretized version, which can be evaluated by ray marching [24,34,52]. This is classically expressed using opacity compositing, where opacity α is used to represent the transmittance with fixed ray marching step size Δx. Points are sequentially sampled along a given ray, $\boldsymbol{\omega}_o$ from the camera position, $\boldsymbol{c}$ as:

$$\boldsymbol{x}_s = \boldsymbol{x}_{s-1} - \boldsymbol{\omega}_o \Delta x = \boldsymbol{c} - s\boldsymbol{\omega}_o \Delta x. \tag{5}$$

The radiance L_s and opacity $\alpha_{c\to s}$ along this path, $c \to s$, are recursively accumulated until $\boldsymbol{x}_s$ exits the volume as:

$$L_s = L_{s-1} + [1 - \alpha_{c\to(s-1)}][1 - \alpha_{l\to(t-1)}]\alpha(\boldsymbol{x}_s)L(\boldsymbol{x}_s), \tag{6}$$

$$\alpha_{c\to s} = \alpha_{c\to(s-1)} + [1 - \alpha_{c\to(s-1)}]\alpha(\boldsymbol{x}_s), \tag{7}$$

$$L(\boldsymbol{x}_s) = f_r(\boldsymbol{\omega}_o, \boldsymbol{\omega}_i, \boldsymbol{n}(\boldsymbol{x}_s), R(\boldsymbol{x}_s))L_l(\boldsymbol{x}_s, \boldsymbol{\omega}_i). \tag{8}$$

Here, $L(\boldsymbol{x}_s)$ computes the reflected radiance from the reflectance function and the incoming light, $\alpha_{c\to s}$ represents the accumulated opacity from the camera

$\boldsymbol{c}$ to point $\boldsymbol{x}_s$, and corresponds to $\tau(\boldsymbol{c}, \boldsymbol{x})$ in Eq. 4. $\alpha_{l \to t}$ represents the accumulated opacity from the light $\boldsymbol{l}$—i.e., $\tau(\boldsymbol{x}, \boldsymbol{l})$ in Eq. 4—and requires a *separate* accumulation process over samples along the $\boldsymbol{l} \to \boldsymbol{x}_s$ ray, similar to Eq. 7:

$$\boldsymbol{x}_s = \boldsymbol{x}_t = \boldsymbol{x}_{t-1} - \boldsymbol{\omega}_i \Delta x = \boldsymbol{l} - t\boldsymbol{\omega}_i \Delta x, \tag{9}$$

$$\alpha_{l \to t} = \alpha_{l \to (t-1)} + [1 - \alpha_{l \to (t-1)}]\alpha(\boldsymbol{x}_t). \tag{10}$$

In this rendering process (Eq. 5–10), a scene is represented by an opacity volume α, a normal volume $\boldsymbol{n}$ and a BRDF volume R; together, these express the geometry and reflectance of the scene, and we refer to them as *Deep Reflectance Volumes*. The simplified opacity volume α is essentially one minus the transmission τ (depending on the physical extinction coefficient σ_t) over a ray segment of a fixed step size Δx; this means that α is dependent on Δx.

Our physically-based ray marching is fully differentiable, so it can be easily incorporated in a deep learning framework and backpropagated through. With this rendering module, we present a neural rendering framework that simultaneously learns scene geometry and reflectance from captured images.

We support any differentiable reflectance model f_r and, in practice, use the simplified Disney BRDF model [22] that is parameterized by diffuse albedo and specular roughness (please refer to the supplementary materials for more details). Our opacity volume is a general geometry representation, accounting for both occlusions (view opacity accumulation in Eq. 7) and shadows (light opacity accumulation in Eq. 10). We illustrate our neural rendering with ray marching in Fig. 2. Note that, because our acquisition setup has collocated camera and lighting, $\alpha_{l \to t}$ becomes equivalent to $\alpha_{c \to s}$ during training, thus requiring only one-pass opacity accumulation from the camera. However, the learned opacity can still be used for re-rendering under any *non-collocated lighting* with two-pass opacity accumulation.

Note that while alpha compositing-based rendering functions have been used in previous work on view synthesis, their formulations are not physically-based [31] and are simplified versions that don't model lighting [49,62]. In contrast, our framework is physically-based and models single-bounce light transport with complex reflectance, occlusions and shadows.

4 Learning Deep Reflectance Volumes

4.1 Overview

Given a set of images of a real scene captured under multiple known viewpoints with collocated lighting, we propose to use a neural network to reconstruct a Deep Reflectance Volume representation of a real scene. Similar to Lombardi et al. [31], our network starts from a 512-channel deep encoding vector that encodes scene appearance; in contrast to their work, where this volume only represents RGB colors, we decode a vector to an opacity volume α, normal volume $\boldsymbol{n}$ and reflectance volume R for rendering. Moreover, our scene encoding vector is not predicted by any network encoder; instead, we jointly optimize for a scene encoding vector and scene-dependent decoder network.

Our network infers the geometry and reflectance volumes in a transformed 3D space with a learned warping function W. During training, our network learns the warping function W, and the geometry and reflectance volumes α_w, $\boldsymbol{n}_w$, R_w, where the subscript w refers to a volume in the warped space. The corresponding world-space scene representation is expressed by $V(\boldsymbol{x}) = V_w(W(\boldsymbol{x}))$, where V is α, $\boldsymbol{n}$ or R. In particular, we use bilinear interpolation to fetch a corresponding value at an arbitrary position $\boldsymbol{x}$ in the space from the discrete voxel values. We propose a decoder-like network, which learns to decode the warping function and the volumes from the deep scene encoding vector. We use a rendering loss between rendered and captured images as well as two regularizing terms.

4.2 Network Architecture

Geometry and Reflectance. To decode the geometry and reflectance volumes (α_w, $\boldsymbol{n}_w$, R_w), we use upsampling 3D convolutional operations to 3D-upsample the deep scene encoding vector to a multi-channel volume that contains the opacity, normal and reflectance. In particular, we use multiple transposed convolutional layers with stride 2 to upsample the volume, each of which is followed by a LeakyRelu activation layer. The network regresses an 8-channel $128 \times 128 \times 128$ volume that includes α_w, $\boldsymbol{n}_w$ and R_w—one channel for opacity α_w, three channels for normal $\boldsymbol{n}_w$, and four channels for reflectance R_w (three for albedo and one for roughness). These volumes express the scene geometry and reflectance in a transformed space, which can be warped to the world space for ray marching.

Warping Function. To increase the effective resolution of the volume, we learn an affine-based warping function similar to [31]. The warping comprises a global warping and a spatially-varying warping. The global warping is represented by an affine transformation matrix W_g. The spatially varying warping is modeled in the inverse transformation space, which is represented by six basis affine matrices $\{W_j\}_{j=1}^{16}$ and a $32 \times 32 \times 32$ 16-channel volume B that contains spatially-varying linear weights of the 16 basis matrices. Specifically, given a world-space position $\boldsymbol{x}$, the complete warping function W maps it into a transformed space by:

$$W(\boldsymbol{x}) = [\sum_{j=1}^{16} B_j(\boldsymbol{x})W_j]^{-1} W_g \boldsymbol{x}, \tag{11}$$

where $B_j(\boldsymbol{x})$ represents the normalized weight of the jth warping basis at $\boldsymbol{x}$. Here, each global or local basis affine transformation matrix W_* is composed of rotation, translation and scale parameters, which are optimized during the training process. Our network decodes the weight volume B from the deep encoding vector using a multi-layer perceptron network with fully connected layers.

4.3 Loss Function and Training Details

Loss Function. Our network learns the scene volumes using a rendering loss computed using the differentiable ray marching process discussed in Sect. 3. During training, we randomly sample pixels from the captured images and do the ray marching (using known camera calibration) to get the rendered pixel colors L_k of pixel k; we supervise them with the ground truth colors $\tilde{L}_k$ in the captured images using a L_2 loss. In addition, we also apply regularization terms from additional priors similar to [31]. We only consider opaque objects in this work and enforce the accumulated opacity along any camera ray $\alpha_{c_k \to s'}$ (see Eq. 7, here k denotes a pixel and s' reflects the final step that exits the volume) to be either 0 or 1, corresponding to a background or foreground pixel, respectively. We also regularize the per-voxel opacity to be sparse over the space by minimizing the spatial gradients of the logarithmic opacity. Our total loss function is given by:

$$\sum_k \|L_k - \tilde{L}_k\|^2 + \beta_1 \sum_k [\log(\alpha_{c_k \to s'}) + \log(1 - \alpha_{c_k \to s'})] + \beta_2 \sum \|\nabla_{\boldsymbol{x}} \log \alpha(\boldsymbol{x})\| \tag{12}$$

Here, the first part reflects the data term, the second regularizes the accumulated α and the third regularizes the spatial sparsity.

Training Details. We build our volume as a cube located at $[-1, 1]^3$. During training, we randomly sample 128×128 pixels from 8 captured images for each training batch, and perform ray marching through the volume using a step size of 1/64. Initially, we set $\beta_1 = \beta_2 = 0.01$; we increase these weights to $\beta_1 = 1.0$, $\beta_2 = 0.1$ after 300000 iterations, which helps remove the artifacts in the background and recover sharp boundaries.

5 Results

In this section we show our results on real captured scenes. We first introduce our acquisition setup and data pre-processing. Then we compare against the state-of-the-art mesh-based appearance acquisition method, followed by a detailed analysis of the experiments. We also demonstrate material editing results with our approach. Please refer to the supplementary materials for video results.

Data Acquisition. Our approach learns the volume representation in a scene dependent way from images with collocated view and light; this requires adequately dense input images well distributed around a target scene to learn complete appearance. Such data can be practically acquired by shooting a video using a handheld cellphone; we show one result using this practical handheld setup in Fig. 4. For other results, we use a robotic arm to automatically capture more uniformly distributed images around scenes for convenience and thorough evaluations; this allows us to evaluate the performance of our method with different numbers of input images that are roughly uniformly distributed as shown in Fig. 5. In the robotic arm setups, we mount a Samsung Galaxy Note 8 cellphone to the robotic arm and capture about 480 images using its camera and the built-in flashlight in a dark room; we leave out a subset of 100 images for validation purposes and use the others for training. We use the same phone to capture a 4-min video of the object in CAPTAIN and select one image for training for every 20 frames, which effectively gives us 310 training images.

Data Pre-processing. Our captured objects are roughly located around the center of the images. We select one fixed rectangular region around the center that covers the object across all frames and use it to crop the images as input for training. The resolution of the cropped training images fed to our network ranges from 400×500 to 1100×1100. Note that we do not use a foreground mask for the object. Our method leverages the regularization terms in training (see Sect. 4.3), which automatically recovers a clean background. We calibrate the captured images using structure from motion (SfM) in COLMAP [46] to get the camera intrinsic and extrinsic parameters. Since SfM may fail to register certain views, the actual number of training images varies from 300 to 385 in different scenes. We estimate the center and bounding box of the captured object with the sparse reconstructions from SfM. We translate the center of the object to the origin and scale it to fit into the $[-1, 1]^3$ cube.

Implementation and Timing. We implement our system (both neural network and differentiable volume rendering components) using PyTorch. We train our network using four NVIDIA 2080Ti RTX GPUs for about two days (about 450000 iterations; though 200000 iterations for 1 day typically already converges to good results, see Fig. 7). At inference time, we directly render the scene from the reconstructed volumes without the network. It takes about 0.8 s to render a 700×700 image under collocated view and light. For non-collocated view and light, the rendering requires connecting each shading point to the light source with additional light-dependent opacity accumulation, which is very expensive if done naively. To facilitate this process, we perform ray marching from the light's point of view and precompute the accumulated opacity at each spatial position of the volume. During rendering, the accumulated opacity for the light ray can be directly sampled from the precomputed volume. By doing so, our final rendering under arbitrary light and view takes about 2.3 s.

Comparisons with Mesh-Based Reconstruction. We use a practical acquisition setup where we capture unstructured images using a mobile phone with its built-in flashlight on in a dark room. Such a mildly controlled acquisition setup is rarely supported by previous works [7,21,55,56,58,63]. Therefore, we compare with the state-of-the-art method proposed by Nam et al. [37] for mesh-based geometry and reflectance reconstruction, that uses the same cellphone setup as ours to reconstruct a mesh with per-vertex BRDFs, and supports both relighting and view synthesis. Figure 3 shows comparisons on renderings under both collocated and non-collocated view-light conditions. The comparison results are generated from the same set of input images, and we requested the authors of [37] run their code on our data and compared on the rendered images provided by the authors. Please refer to the supplementary materials for video comparisons.

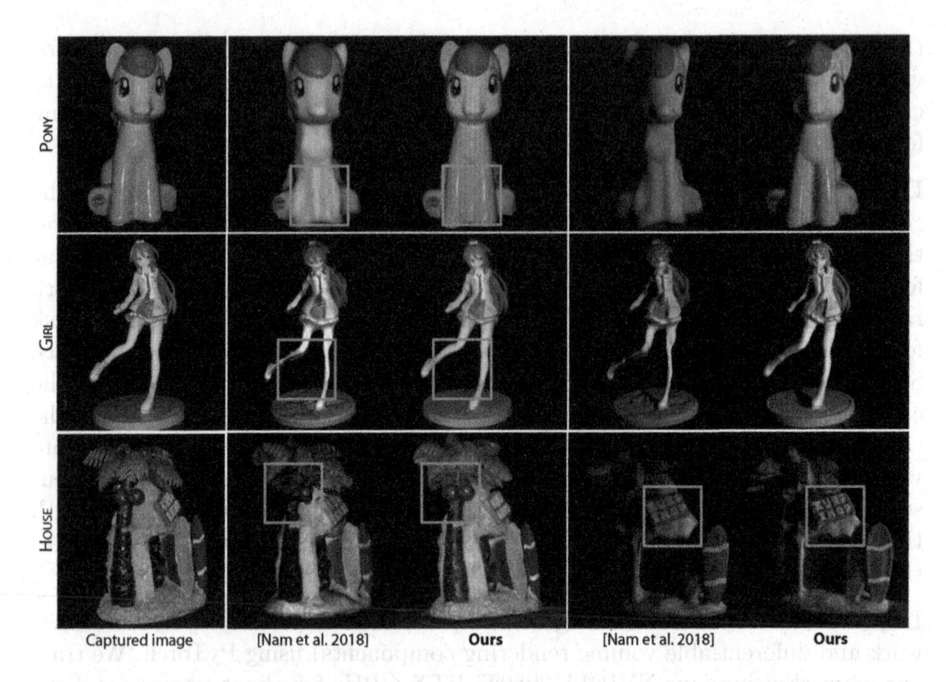

Fig. 3. Comparisons with mesh-based reconstruction. We show renderings of the captured object under both collocated (column 2, 3) and non-collocated (column 4, 5) camera and light. We compare our volume-based neural reconstruction against a state-of-the-art method [37] that reconstructs mesh and per-vertex BRDFs. Nam et al. [37] fails to handle such challenging cases and recovers inaccurate geometry and appearance. In contrast our method produces photo-realistic results.

As shown in Fig. 3, our results are significantly better than the mesh-based method in terms of both geometry and reflectance. Note that, Nam et al. [37] leverage a state-of-the-art MVS method [47] to reconstruct the initial mesh from captured images and performs an optimization to further refine the geometry; this however still fails to recover the accurate geometry in texture-less, specular and thin-structured regions in those challenging scenes, which leads to seriously distorted shapes in Pony, over-smoothness and undesired structures in House, and degraded geometry in Girl. Our learning-based volumetric representation avoids these mesh-based issues and models the scene geometry accurately with many details. Moreover, it is also very difficult for the classical per-vertex BRDF optimization in [37] to recover high-frequency specularities, which leads to over-diffuse appearance in most of the scenes; this is caused by the lack of constraints for the high-frequency specular effects, which appear in very few pixels in limited input views. In contrast, our optimization is driven by our novel neural rendering framework with deep network priors, which effectively correlates the sparse specularities in different regions through network connections and recovers realistic specularities and other appearance effects.

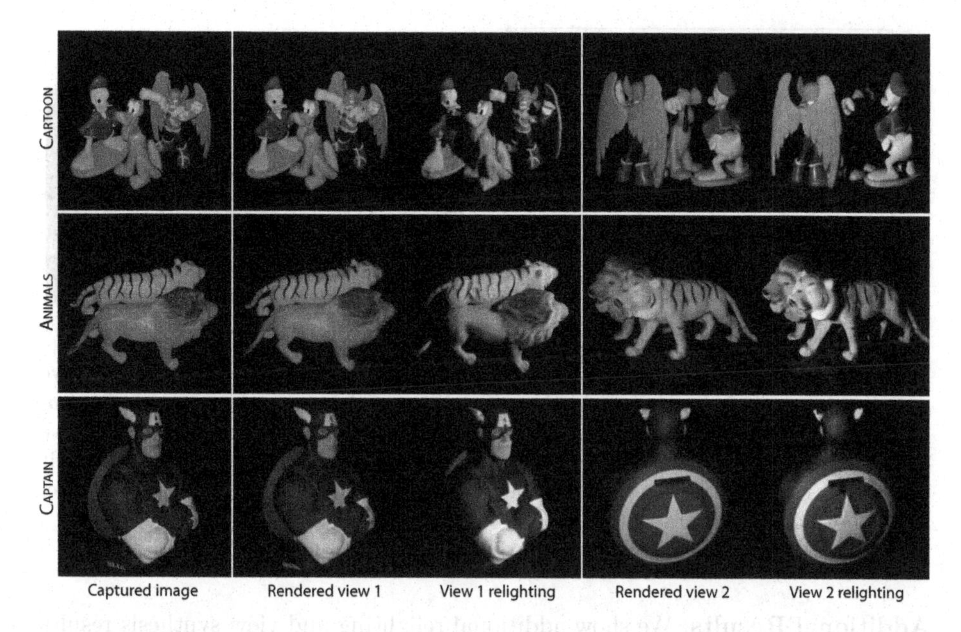

Fig. 4. Additional results on real scenes. We show renderings under novel view and lighting conditions. Our method is able to handle scenes with multiple objects (top two rows) and model the complex occlusions between them. Our method can also generate high-quality results from casual handheld video captures (third row), which demonstrates the practicability of our approach.

	25	50	100	200	385
PSNR	25.33	26.36	26.95	27.85	28.13
SSIM	0.70	0.73	0.75	0.80	0.81

Fig. 5. We evaluate the performance of our method on the House scene with different numbers of training images. Although we use all 385 images in our final experiments, our method is able to achieve comparable performance with as few as 200 images for this challenging scene.

	House	Cartoon
[48]	0.786/25.81	0.532/16.34
Ours	**0.896/30.44**	**0.911/29.14**

Fig. 6. We compare against DeepVoxels on synthesizing novel views under collocated lights and report the PSNR/SSIM scores. The results show that our method generates more accurate renderings. Note that we retrain our model with a resolution of 512 $\times$ 512 for a fair comparison.

Comparison on Synthesizing Novel Views. We also make a comparison on synthesizing novel views under collocated lights against a view synthesis method DeepVoxels [48], which encodes view-dependent appearance in a learnt 3D-aware neural representation. Note that DeepVoxels does not support relighting. As shown in Fig. 6, our method is able to generate renderings of higher quality with higher PSNR/SSIM scores. In contrast, DeepVoxels fails to reason about the complex geometry in our real scenes, thus resulting in degraded image quality. Please refer to the supplementary materials for visual comparison results.

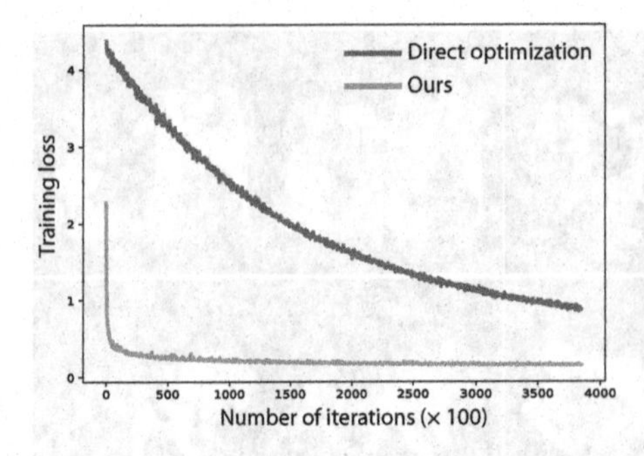

Fig. 7. We compare our deep prior based optimization against direct optimization of the volume and warping function without using networks. Direct optimization converges significantly slower than our method, which demonstrates the effectiveness of regularization by the networks.

Fig. 8. Our approach supports intuitive editing of the material properties of a captured object. In this example we decrease the roughness of the object to make it look like glossy marble instead of plastic.

Additional Results. We show additional relighting and view synthesis results of complex real scenes in Fig. 4. Our method is able to handle scenes with multiple objects, as shown in scene CARTOON and ANIMALS. Our volumetric representation can accurately model complex occlusions between objects and reproduce realistic cast shadows under novel lighting, which are never observed by our network during the training process. In the CAPTAIN scene, we show the result generated from *handheld* mobile phone captures. We select frames from the video at fixed intervals as training data. Despite the potential existence of motion blur and non-uniform coverage, our method is able to generate high-quality results, which demonstrates the robustness and practicality of our approach. Please refer to the supplementary materials for video results.

Evaluation of the Number of Inputs. Our method relies on an optimization over adequate input images that capture the scene appearance across different view/light directions. We evaluate how our reconstruction degrades with the decrease of training images on the HOUSE scene. We uniformly select a subset of views from the full training images and train our model on them. We evaluate the trained model on the test images, and report the SSIMs and PSNRs in Fig. 5. As we can see from the results, there is an obvious performance drop when there are fewer than 100 training images due to insufficient constraints. On the other hand, while we use the full 385 images for our final results, our method in fact achieves comparable performance with only 200 for this scene, as reflected by their close PSNRs and SSIMs.

Comparison with Direct Optimization. Our neural rendering leverages a "deep volume prior" to drive the volumetric optimization process. To justify the effectiveness of this design, we compare with a naive method that directly optimizes the parameters in each voxel and the warping parameters using the same loss function. We show the optimization progress in Fig. 7. Note that, the naive method converges significantly slower than ours, where the independent voxel-wise optimization without considering across-voxel correlations cannot properly disentangle the ambiguous information in the captured images; yet, our deep optimization is able to correlate appearance information across the voxels with deep convolutions, which effectively minimizes the reconstruction loss.

Material Editing. Our method learns explicit volumes with physical meaning to represent the reflectance of real scenes. This enables broad image synthesis applications like editing the materials of captured scenes. We show one example in Fig. 8, where we successfully make the scene glossier by decreasing the learned roughness in the volume. Note that, the geometry and colors are still preserved in the scene, while novel specularities are introduced which are not part of the material appearance in the scene. This example illustrates that our network disentangles the geometry and reflectance of the scene in a reasonable way, thereby enabling sub-scene component editing without influencing other components.

Limitations. We reconstruct the deep reflectance volumes with a resolution of 128^3, which is restricted by available GPU memory. While we have applied a warping function to increase the actual utilization of the volume space, and demonstrated that it is able to generate compelling results on complex real scenes, it may fail to fully reproduce the geometry and appearance of scenes with highly complex surface normal variations and texture details. Increasing the volume resolution may resolve this issue. In the future, it would also be interesting to investigate how to efficiently apply sparse representations such as octrees in our framework to increase the capacity of our volume representation. The current reflectance model we are using is most appropriate for opaque surfaces. Extensions to other materials like hair, fur or glass could be potentially addressed by applying other reflectance models in our neural rendering framework.

6 Conclusion

We have presented a novel approach to learn a volume representation that models both geometry and reflectance of complex real scenes. We predict per-voxel opacity, normal, and reflectance from unstructured multi-view mobile phone captures with the flashlight. We also introduce a physically-based differentiable rendering module to enable renderings of the volume under arbitrary viewing and lighting directions. Our method is practical, and supports novel view synthesis, relighting and material editing, which has significant potential benefits in scenarios such as 3D visualization and VR/AR applications.

Acknowledgements. We thank Giljoo Nam for help with the comparisons. This work was supported in part by ONR grants N000141712687, N000141912293, N000142012529, NSF grant 1617234, Adobe, the Ronald L. Graham Chair and the UC San Diego Center for Visual Computing.

References

1. Achlioptas, P., Diamanti, O., Mitliagkas, I., Guibas, L.: Learning representations and generative models for 3D point clouds. In: ICML, pp. 40–49 (2018)
2. Aittala, M., Aila, T., Lehtinen, J.: Reflectance modeling by neural texture synthesis. ACM Trans. Graph. **35**(4), 65:1–65:13 (2016)
3. Aittala, M., Weyrich, T., Lehtinen, J.: Two-shot SVBRDF capture for stationary materials. ACM Trans. Graph. **34**(4), 110:1–110:13 (2015)
4. Alldrin, N., Zickler, T., Kriegman, D.: Photometric stereo with non-parametric and spatially-varying reflectance. In: CVPR, pp. 1–8. IEEE (2008)
5. Baek, S.H., Jeon, D.S., Tong, X., Kim, M.H.: Simultaneous acquisition of polarimetric SVBRDF and normals. ACM Trans. Graph. **37**(6), 268-1 (2018)
6. Bi, S., Kalantari, N.K., Ramamoorthi, R.: Patch-based optimization for image-based texture mapping. ACM Trans. Graph. **36**(4), 106-1 (2017)
7. Bi, S., Xu, Z., Sunkavalli, K., Kriegman, D., Ramamoorthi, R.: Deep 3D capture: geometry and reflectance from sparse multi-view images. In: CVPR, pp. 5960–5969 (2020)
8. Buehler, C., Bosse, M., McMillan, L., Gortler, S., Cohen, M.: Unstructured lumigraph rendering. In: SIGGRAPH, pp. 425–432. ACM (2001)
9. Chen, Z., Chen, A., Zhang, G., Wang, C., Ji, Y., Kutulakos, K.N., Yu, J.: A neural rendering framework for free-viewpoint relighting. In: CVPR, June 2020
10. Chen, Z., Zhang, H.: Learning implicit fields for generative shape modeling. arXiv preprint arXiv:1812.02822 (2018)
11. Debevec, P., Hawkins, T., Tchou, C., Duiker, H.P., Sarokin, W., Sagar, M.: Acquiring the reflectance field of a human face. In: Proceedings of the 27th Annual Conference on Computer Graphics and Interactive Techniques, pp. 145–156. ACM Press/Addison-Wesley Publishing Co. (2000)
12. Foo, S.C.: A gonioreflectometer for measuring the bidirectional reflectance of material for use in illumination computation. Ph.D. thesis, Citeseer (1997)
13. Furukawa, Y., Ponce, J.: Accurate, dense, and robust multiview stereopsis. IEEE Trans. Pattern Anal. Mach. Intell. **32**(8), 1362–1376 (2009)
14. Goldman, D.B., Curless, B., Hertzmann, A., Seitz, S.M.: Shape and spatially-varying BRDFs from photometric stereo. IEEE Trans. Pattern Anal. Mach. Intell. **32**(6), 1060–1071 (2009)
15. Groueix, T., Fisher, M., Kim, V.G., Russell, B.C., Aubry, M.: A papier-mâché approach to learning 3D surface generation. In: CVPR, pp. 216–224 (2018)
16. Huang, P.H., Matzen, K., Kopf, J., Ahuja, N., Huang, J.B.: DeepMVS: learning multi-view stereopsis. In: CVPR, pp. 2821–2830 (2018)
17. Hui, Z., Sunkavalli, K., Lee, J.Y., Hadap, S., Wang, J., Sankaranarayanan, A.C.: Reflectance capture using univariate sampling of BRDFs. In: ICCV, pp. 5362–5370 (2017)
18. Ji, M., Gall, J., Zheng, H., Liu, Y., Fang, L.: SurfaceNet: an end-to-end 3D neural network for multiview stereopsis. In: ICCV, pp. 2307–2315 (2017)
19. Kanamori, Y., Endo, Y.: Relighting humans: occlusion-aware inverse rendering for full-body human images. ACM Trans. Graph. **37**(6), 1–11 (2018)
20. Kanazawa, A., Tulsiani, S., Efros, A.A., Malik, J.: Learning category-specific mesh reconstruction from image collections. In: Ferrari, V., Hebert, M., Sminchisescu, C., Weiss, Y. (eds.) ECCV 2018. LNCS, vol. 11219, pp. 386–402. Springer, Cham (2018). https://doi.org/10.1007/978-3-030-01267-0_23

21. Kang, K., et al.: Learning efficient illumination multiplexing for joint capture of reflectance and shape (2019)
22. Karis, B., Games, E.: Real shading in unreal engine 4 (2013)
23. Kazhdan, M., Bolitho, M., Hoppe, H.: Poisson surface reconstruction. In: Proceedings of the Fourth Eurographics Symposium on Geometry Processing, vol. 7 (2006)
24. Kniss, J., Premoze, S., Hansen, C., Shirley, P., McPherson, A.: A model for volume lighting and modeling. IEEE Trans. Vis. Comput. Graph. **9**(2), 150–162 (2003)
25. Kutulakos, K.N., Seitz, S.M.: A theory of shape by space carving. ICCV **38**(3), 199–218 (2000). https://doi.org/10.1023/A:1008191222954
26. Ladicky, L., Saurer, O., Jeong, S., Maninchedda, F., Pollefeys, M.: From point clouds to mesh using regression. In: ICCV, pp. 3893–3902 (2017)
27. Levoy, M., Hanrahan, P.: Light field rendering. In: Proceedings of the 23rd Annual Conference on Computer Graphics and Interactive Techniques, pp. 31–42. ACM (1996)
28. Li, Z., Sunkavalli, K., Chandraker, M.: Materials for masses: SVBRDF acquisition with a single mobile phone image. In: Ferrari, V., Hebert, M., Sminchisescu, C., Weiss, Y. (eds.) ECCV 2018. LNCS, vol. 11207, pp. 74–90. Springer, Cham (2018). https://doi.org/10.1007/978-3-030-01219-9_5
29. Li, Z., Xu, Z., Ramamoorthi, R., Sunkavalli, K., Chandraker, M.: Learning to reconstruct shape and spatially-varying reflectance from a single image. In: SIGGRAPH Asia 2018, p. 269. ACM (2018)
30. Liao, Y., Donne, S., Geiger, A.: Deep marching cubes: learning explicit surface representations. In: CVPR, pp. 2916–2925 (2018)
31. Lombardi, S., Simon, T., Saragih, J., Schwartz, G., Lehrmann, A., Sheikh, Y.: Neural volumes: learning dynamic renderable volumes from images. ACM Trans. Graph. **38**(4), 65 (2019)
32. Lorensen, W.E., Cline, H.E.: Marching cubes: a high resolution 3D surface construction algorithm. ACM SIGGRAPH Comput. Graph. **21**(4), 163–169 (1987)
33. Matusik, W., Pfister, H., Brand, M., McMillan, L.: A data-driven reflectance model. ACM Trans. Graph. **22**(3), 759–769 (2003)
34. Max, N.: Optical models for direct volume rendering. IEEE Trans. Vis. Comput. Graph. **1**(2), 99–108 (1995)
35. Mescheder, L., Oechsle, M., Niemeyer, M., Nowozin, S., Geiger, A.: Occupancy networks: learning 3D reconstruction in function space. arXiv preprint arXiv:1812.03828 (2018)
36. Mildenhall, B., Srinivasan, P.P., Tancik, M., Barron, J.T., Ramamoorthi, R., Ng, R.: NeRF: representing scenes as neural radiance fields for view synthesis (2020)
37. Nam, G., Lee, J.H., Gutierrez, D., Kim, M.H.: Practical SVBRDF acquisition of 3D objects with unstructured flash photography. In: SIGGRAPH Asia 2018, p. 267. ACM (2018)
38. Newcombe, R.A., et al.: KinectFusion: real-time dense surface mapping and tracking. In: Proceedings of the 2011 10th IEEE International Symposium on Mixed and Augmented Reality, ISMAR 2011, pp. 127–136. IEEE Computer Society, Washington, DC, USA (2011)
39. Nielsen, J.B., Jensen, H.W., Ramamoorthi, R.: On optimal, minimal BRDF sampling for reflectance acquisition. ACM Trans. Graph. **34**(6), 1–11 (2015)
40. Niemeyer, M., Mescheder, L., Oechsle, M., Geiger, A.: Differentiable volumetric rendering: learning implicit 3D representations without 3D supervision. In: CVPR, pp. 3504–3515 (2020)

41. Novák, J., Georgiev, I., Hanika, J., Jarosz, W.: Monte Carlo methods for volumetric light transport simulation. In: Computer Graphics Forum, vol. 37, pp. 551–576. Wiley Online Library (2018)
42. Paschalidou, D., Ulusoy, O., Schmitt, C., Van Gool, L., Geiger, A.: RayNet: learning volumetric 3D reconstruction with ray potentials. In: CVPR, pp. 3897–3906 (2018)
43. Peers, P., et al.: Compressive light transport sensing. ACM Trans. Graph. **28**(1), 3 (2009)
44. Philip, J., Gharbi, M., Zhou, T., Efros, A.A., Drettakis, G.: Multi-view relighting using a geometry-aware network. ACM Trans. Graph. **38**(4), 1–14 (2019)
45. Richter, S.R., Roth, S.: Matryoshka networks: predicting 3D geometry via nested shape layers. In: CVPR, pp. 1936–1944 (2018)
46. Schönberger, J.L., Frahm, J.M.: Structure-from-motion revisited. In: CVPR (2016)
47. Schönberger, J.L., Zheng, E., Frahm, J.-M., Pollefeys, M.: Pixelwise view selection for unstructured multi-view stereo. In: Leibe, B., Matas, J., Sebe, N., Welling, M. (eds.) ECCV 2016. LNCS, vol. 9907, pp. 501–518. Springer, Cham (2016). https://doi.org/10.1007/978-3-319-46487-9_31
48. Sitzmann, V., Thies, J., Heide, F., Nießner, M., Wetzstein, G., Zollhofer, M.: DeepVoxels: learning persistent 3D feature embeddings. In: CVPR, pp. 2437–2446 (2019)
49. Srinivasan, P.P., Tucker, R., Barron, J.T., Ramamoorthi, R., Ng, R., Snavely, N.: Pushing the boundaries of view extrapolation with multiplane images. In: CVPR, pp. 175–184 (2019)
50. Sun, T., et al.: Single image portrait relighting. ACM Trans. Graph. (Proceedings SIGGRAPH) (2019)
51. Wang, J., Sun, B., Lu, Y.: MVPNet: multi-view point regression networks for 3D object reconstruction from a single image. arXiv preprint arXiv:1811.09410 (2018)
52. Wittenbrink, C.M., Malzbender, T., Goss, M.E.: Opacity-weighted color interpolation, for volume sampling. In: Proceedings of the 1998 IEEE Symposium on Volume Visualization, pp. 135–142 (1998)
53. Wu, H., Wang, Z., Zhou, K.: Simultaneous localization and appearance estimation with a consumer RGB-D camera. IEEE Trans. Vis. Comput. Graph. **22**(8), 2012–2023 (2015)
54. Wu, Z., et al.: 3D ShapeNets: a deep representation for volumetric shapes. In: CVPR, pp. 1912–1920 (2015)
55. Xia, R., Dong, Y., Peers, P., Tong, X.: Recovering shape and spatially-varying surface reflectance under unknown illumination. ACM Trans. Graph. **35**(6), 187 (2016)
56. Xu, Z., Bi, S., Sunkavalli, K., Hadap, S., Su, H., Ramamoorthi, R.: Deep view synthesis from sparse photometric images. ACM Trans. Graph. **38**(4), 76 (2019)
57. Xu, Z., Nielsen, J.B., Yu, J., Jensen, H.W., Ramamoorthi, R.: Minimal BRDF sampling for two-shot near-field reflectance acquisition. ACM Trans. Graph. **35**(6), 188 (2016)
58. Xu, Z., Sunkavalli, K., Hadap, S., Ramamoorthi, R.: Deep image-based relighting from optimal sparse samples. ACM Trans. Graph. **37**(4), 126 (2018)
59. Yao, Y., Luo, Z., Li, S., Fang, T., Quan, L.: MVSNet: depth inference for unstructured multi-view stereo. In: Ferrari, V., Hebert, M., Sminchisescu, C., Weiss, Y. (eds.) ECCV 2018. LNCS, vol. 11212, pp. 785–801. Springer, Cham (2018). https://doi.org/10.1007/978-3-030-01237-3_47
60. Zhou, H., Hadap, S., Sunkavalli, K., Jacobs, D.W.: Deep single-image portrait relighting. In: ICCV, pp. 7194–7202 (2019)

61. Zhou, Q.Y., Koltun, V.: Color map optimization for 3D reconstruction with consumer depth cameras. ACM Trans. Graph. **33**(4), 155 (2014)
62. Zhou, T., Tucker, R., Flynn, J., Fyffe, G., Snavely, N.: Stereo magnification: learning view synthesis using multiplane images. ACM Trans. Graph. **37**(4), 1–12 (2018)
63. Zhou, Z., et al.: Sparse-as-possible SVBRDF acquisition. ACM Trans. Graph. **35**(6), 189 (2016)

Memory-Augmented Dense Predictive Coding for Video Representation Learning

Tengda Han(✉), Weidi Xie, and Andrew Zisserman

Visual Geometry Group, Department of Engineering Science, University of Oxford, Oxford, UK
{htd,weidi,az}@robots.ox.ac.uk

Abstract. The objective of this paper is self-supervised learning from video, in particular for representations for action recognition. We make the following contributions: (i) We propose a new architecture and learning framework *Memory-augmented Dense Predictive Coding* (MemDPC) for the task. It is trained with a *predictive attention mechanism* over the set of *compressed memories*, such that any future states can always be constructed by a convex combination of the condensed representations, allowing to make multiple hypotheses efficiently. (ii) We investigate visual-only self-supervised video representation learning from RGB frames, or from unsupervised optical flow, or both. (iii) We thoroughly evaluate the quality of the learnt representation on four different downstream tasks: action recognition, video retrieval, learning with scarce annotations, and unintentional action classification. In all cases, we demonstrate state-of-the-art or comparable performance over other approaches with orders of magnitude fewer training data.

1 Introduction

Recent advances in self-supervised representation learning for images have yielded impressive results, *e.g.* [11,26–28,34,50,57,70], with performance matching or exceeding that of supervised representation learning on downstream tasks. However, in the case of videos, although there have been similar gains for *multi-modal* self-supervised representation learning, *e.g.* [2,4,39,47,52,56], progress on learning *only* from the video stream (without additional audio or text streams) is lagging behind. The objective of this paper is to improve the performance of video only self-supervised learning.[1]

Compared to still images, videos should be a more suitable source for self-supervised representation learning as they naturally provide various augmentation, such as object out of plane rotations and deformations. In addition,

[1] Code is available at http://www.robots.ox.ac.uk/~vgg/research/DPC.

Electronic supplementary material The online version of this chapter (https://doi.org/10.1007/978-3-030-58580-8_19) contains supplementary material, which is available to authorized users.

A. Vedaldi et al. (Eds.): ECCV 2020, LNCS 12348, pp. 312–329, 2020.
https://doi.org/10.1007/978-3-030-58580-8_19

videos contain additional temporal information that can be used to disambiguate actions *e.g.* open vs. close. The temporal information can also act as a free supervisory signal to train a model to predict the future states from the past either passively by watching videos [24,45,59] or actively in an interactive environment [16], and thereby learn a video representation.

Fig. 1. Can you predict the next frame? Future prediction naturally involves challenges from multiple hypotheses, *e.g.* the motion of each leaf, reflections on the water, hands and the golf club can be in many possible positions

Unfortunately, the exact future is indeterministic (a problem long discussed in the history of science, and known as "Laplace's Demon"). As shown in Fig. 1, this problem is directly apparent in the stochastic variability of scenes, *e.g.* trying to predict the exact motion of each leaf on a tree when the wind blows, or the changing reflections on the water. More concretely, consider the action of 'playing golf' – once the action starts, a future frame could have the hands and golf club in many possible positions, depending on the person who is playing. Learning visual representation by predicting the future therefore requires designing specific training schemes that simultaneously circumvents the unpredictable details in exact frames, and also handles multiple hypotheses and incomplete data – in particular only one possible future is exposed by the frames of one video.

Various approaches have been developed to deal with the multiple possible futures for an action. Vondrick *et al.* [59] explicitly generates multiple hypotheses, and only the hypothesis that is closest to the true observation is chosen during optimization, however, this approach limits the number of possible future states. Another line of work [24,50] circumvents this difficulty by using contrastive learning – the model is only asked to predict *one* future state that assigns higher similarity to the true observation than to any distractor observation (from different videos or from elsewhere in the same video). Recalling the 'playing golf' example, the embedding must capture the hand movement for this action, but not necessarily the precise position and velocity, only sufficiently to disambiguate future frames.

In this paper, we continue the idea of contrastive learning, but improve it by the addition of a *Compressive Memory*, which maps "lifelong" experience to a set of compressed memories and helps to better anticipate the future. We make the following four contributions: *First*, we propose a novel Memory-augmented Dense Predictive Coding (`MemDPC`) architecture. It is trained with a *predictive attention mechanism* over the set of *compressed memories*, such that any future states can always be constructed by a convex combination of the condensed representations, allowing it to make multiple hypotheses efficiently. *Second*, we investigate visual only self-supervised video representation learning from RGB frames, or from unsupervised optical flow, or both. *Third*, we argue that, in addition to the standard linear probes and fine-tuning [56,69], that have been used for evaluating representation from self-supervised learning, a non-linear probe should also be used, and demonstrate the difference that this probe makes. *Finally*, we evaluate the quality of learnt feature representation on four different downstream tasks: action recognition, learning under low-data regime (scarce annotations), video retrieval, and unintentional action classification; and demonstrate state-of-the-art performance over other approaches with similar settings on *all* tasks.

2 Related Work

Self-supervised Learning for Images has undergone rapid progress in visual representation learning recently [11,26–28,34,50,57,70]. Generally speaking, the success can be attributed to one specific training paradigm, namely contrastive learning [12,23], *i.e.* contrast the positive and negative sample pairs.

Self-supervised Learning for Videos has explored various ideas to learn representations by exploiting spatio-temporal information [1,2,8,15,20,21,30–33,35,37,42–45,48,59,60,62–64,66]. Of more relevance here is the line of research using contrastive learning, *e.g.* [2,4,5,39,51,52] learn from visual-audio correspondence, [47] learns from video and narrations, and our previous work [24] learns video representations by predicting future states.

Memory Models have been considered as one of the fundamental building blocks towards intelligence. In the deep learning literature, two different lines of research have received extensive attention, one is to build networks that involve an internal memory which can be implicitly updated in a recurrent manner, *e.g.* LSTM [29] and GRU [13]. The other line of research focuses on augmenting feed-forward models with an explicit memory that can be read or written to with an attention-based procedure [6,14,22,41,55,58,61,67]. In this work, our compressive memory falls in the latter line, *i.e.* an external memory module.

3 Methodology

The proposed Memory-augmented Dense Predictive Coding (`MemDPC`), is a conceptually simple model for learning a video representation with contrastive predictive coding. The key novelty is to augment the previous DPC model with a *Compressive Memory*. This provides a mechanism for handling the multiple future hypotheses required in learning due to the problem that only one possible future is exposed by a particular video.

The architecture is shown in Fig. 2. As in the case of DPC, the video is partitioned into 8 blocks with 5 frames each, and an encoder network f generates an embedding z for each block. For inference, these embeddings are aggregated over time by a function g into a video level embedding c. During training, the future block embeddings $\hat{z}$ are predicted and used to select the true embedding in the dense predictive coding manner. In MemDPC, the prediction of $\hat{z}$ is from a convex combination of memory elements (rather from c directly as in DPC), and it is this restriction that also enables the network to handle multiple hypotheses, as will be explained below.

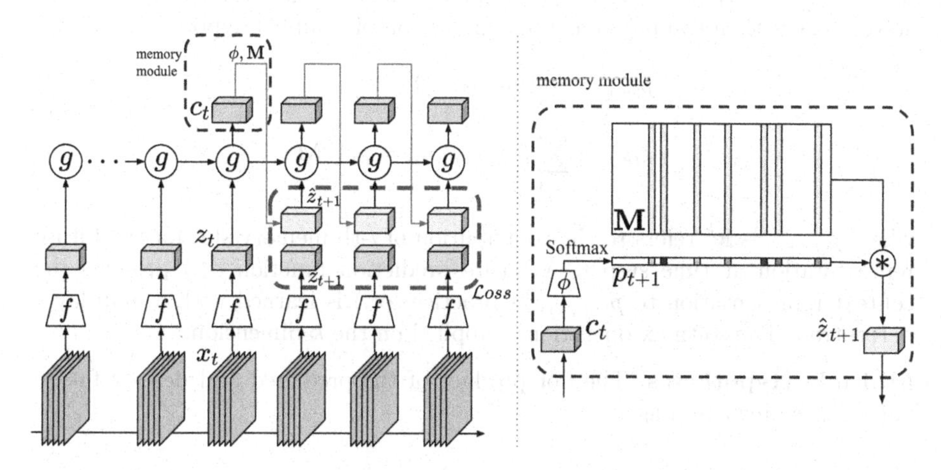

Fig. 2. Architecture of the Memory-augmented Dense Predictive Coding (MemDPC). Note, the memory module is only used during the self-supervised training. The c_t embedding is used for downstream tasks

3.1 Memory-Augmented Dense Predictive Coding (MemDPC)

Video Block Encoder. As shown in Fig. 2, we partition the input video sequence into multiple blocks $x_1, ..., x_t, x_{t+1}, ...$, where each block is composed of multiple video frames. Then a shared feature extractor $f(.)$ (architecture details are given in the supplementary material) extracts the video features z_i from each video block x_i:

$$z_i = f(x_i) \tag{1}$$

Temporal Aggregation. After acquiring block representations, multiple block embeddings are aggregated to obtain a context feature c_t, summarizing the information over a longer temporal window. Specifically,

$$c_t = g(z_1, z_2, ..., z_t) \tag{2}$$

in our case, we simply adopt Recurrent Neural Networks (RNNs) for $g(.)$, but other auto-regressive model should also be feasible for temporal aggregation.

Compressive Memory. In order to enable efficient multi-hypotheses estimation, we augment the predictive models with an external common compressive memory. This external memory bank is shared for the entire dataset during training, and is accessed by a *predictive addressing mechanism* that infers a probability distribution over the memory entries, where each memory entry acts as a potential hypothesis.

In detail, the compressed memory bank is written $\mathbf{M} = [m_1, m_2, ..., m_k]^\top \in \mathbb{R}^{k \times C}$, where k is the memory size and C is the dimension of each compressed memory. During training, a predictive memory addressing mechanism (Eq. 3) is used to draw a hypothesis from the compressed memory, and the predicted future states $\hat{z}_{t+1}$ is then computed as the expectation of sampled hypotheses (Eq. 4):

$$p_{t+1} = \text{Softmax}(\phi(c_t)) \tag{3}$$

$$\hat{z}_{t+1} = \sum_{i=1}^{k} p_{(i,t+1)} \cdot m_i = p_{t+1}\mathbf{M} \tag{4}$$

where $p_{(i,t+1)} \in \mathbb{R}^k$ refers to the contribution of i-th memory slot for the future representation at time step t. A future prediction function $\phi(.)$ projects the context representation to $p_{(i,t+1)}$, in practice, $\phi(.)$ is learned with a multilayer perceptron. The softmax operation is applied on the k dimension.

Multiple Hypotheses. The dot product of the predicted and desired future pairs can be rewritten as:

$$\psi(\hat{z}^\top, z) = \left(\sum_{i=1}^{k} p_i \cdot m_i^\top\right) z = \sum_{i=1}^{k} p_i \cdot (m_i^\top z) \tag{5}$$

where $m_i^\top z$ refers to the dot product (*i.e.* similarity) between a single memory slot and the feature states from the observation. The objective of $\phi(.)$ is to predict a probability distribution over k hypotheses in the memory bank, such that the expectation of $m_i^\top z$ for a positive pair is larger than that of negative pairs. Since the future is uncertain, the desired future feature z might be similar to one of the multiple hypotheses in the memory bank, say either m_p or m_q, for instance. To handle this uncertainty, the future prediction function $\phi(.)$ just needs to put a higher probability on both the p and q slots, such that Eq. 5 is always large no matter which state the future is. In this way, the burden of modelling the future uncertainly is allocated to the memory bank $\mathbf{M}$ and future prediction function $\phi(.)$, thus the backbone encoder $f(.)$ and $g(.)$ can save capacity and capture the high-level action trajectory.

Memory Mechanism Discussion. Note, in contrast to the memory mechanism in Wu *et al.* [65] and MoCo [26], which has the goal of storing more data samples to increase the number of negative samples during contrastive learning, our Compressive Memory has the goal of aiding learning by compressing all the potential hypotheses within the entire dataset, and allowing access through the predictive addressing mechanism. The memory mechanism shares similarity with NetVLAD [3], which represents a feature distribution with "trainable centroids". However, in NetVLAD the goal is for compact and discriminative feature aggregation, and it encodes a weighted sum of *residuals* between feature vectors and the centroids. In contrast, our goal with $\phi(.)$ is to predict attention weights for the entries in the memory bank $\mathbf{M}$, in order to construct the future state as a convex combination these entries. The model can also sequentially predict further into the future with the *same* memory bank.

3.2 Contrastive Learning

Contrastive Learning generally refers to the training paradigm that forces the similarity scores of positive pairs to be higher than those of negatives.

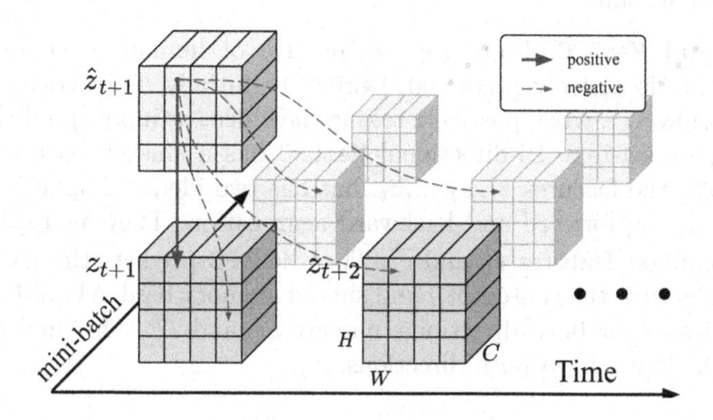

Fig. 3. Details of the contrastive loss. Contrastive loss is computed densely, *i.e.* over both spatial and temporal dimension of the feature

Specifically, in MemDPC, we predict the future states recursively, ending up with a sequence of predicted features $\hat{z}_{t+1}, \hat{z}_{t+2}, \ldots, \hat{z}_{end}$ and the video feature from the true observations $z_{t+1}, z_{t+2}, \ldots, z_{end}$. As shown in Fig. 3, each predicted $\hat{z}$ in practise is a dense feature map. To simplify the notation, we denote temporal index with i and denote other indexes including spatial index and batch-wise index as k, where batch-wise index means the index in the current mini-batch, $k \in \{(1,1,1),(1,1,2),...,(B,H,W)\}$. The objective function to minimize becomes:

$$\mathcal{L} = -\mathbb{E}\left[\sum_{i,k} \log \frac{e^{\psi(\hat{z}_{i,k}^{\top}, z_{i,k})}}{e^{\psi(\hat{z}_{i,k}^{\top}, z_{i,k})} + \sum_{(j,m)\neq(i,k)} e^{\psi(\hat{z}_{i,k}^{\top}, z_{j,m})}}\right] \tag{6}$$

where $\psi(\cdot)$ is acting as a *critic* function, in our case, we simply use dot product between the two vectors (we also experiment with L2-normalization, and find it gives similar performance on downstream tasks). Essentially, the objective function acts as a multi-way classifier, and the goal of optimization is to learn the video block encoder that assigns the highest values for $(\hat{z}_{i,k}, z_{i,k})$ *i.e.* higher similarity between the predicted future states and that from true observations originating from the same video and spatial-temporal aligned position.

3.3 Improving Performance by Extensions

As MemDPC is a general self-supervised learning framework, it can be combined with other 'modules' like two-stream networks and bi-directional RNN to improve the quality of the visual representations.

Two-Stream Architecture. We represent dense optical flow as images by stacking the x and y displacement fields and another zero-valued array to make them 3-channel images. There is no need to modify the MemDPC framework, and it can be directly applied to optical flow inputs by simply replacing the input x_t from RGB frames to optical flow frames. We use late fusion like [19,53] to combine both streams.

Bi-directional MemDPC. From the perspective of human perception, where only the future is actively predicted, MemDPC is initially designed to be single-directional. However, when passively taking the videos as input, predicting backwards becomes feasible. Bi-directional MemDPC has a shared feature extractor $f(.)$ to extract the features $z_1, z_2, ..., z_t$, but has two identical aggregators $g^f(.)$ and $g^b(.)$ denoting forward and backward aggregation. They aggregate the bi-directional context features c_t^f and c_t^b. Then MemDPC predicts the past and the future features with the shared $\phi(.)$ and shared memory bank $\mathbf{M}$, and constructs contrastive losses for both directions, namely $\mathcal{L}^f$ and $\mathcal{L}^b$. The final loss is the average of the losses from both directions.

4 How to Evaluate Self-supervised Learning?

The standard way to evaluate the quality of the learned representation is to assess the performance on downstream tasks using two protocols: (i) a linear probe – freezing the network and only train a linear head for the downstream task; or (ii) fine-tuning the entire network for the downstream task. For example, in (i) if the downstream task is classification, *e.g.* of UCF101, then a linear classifier is trained on top of the frozen base network. In (ii) the self-supervised training of the base network only provides the initialization. However, there is no particular reason why self-supervision should lead to features that are linearly separable, even if the representation has encoded semantic information. Consequently, in addition to the two protocols mentioned above, we also evaluate the frozen features with non-linear probing, *e.g.* in the case of a classification downstream task, a non-linear MLP head is trained as the final classifier. In the experiments we evaluate the representation on four different downstream tasks.

Action Classification is a common evaluation task for self-supervised learning on videos and it allows us to compare against other methods. After self-supervised training, our MemDPC can be evaluated on this task under two settings: (i) linear and non-linear probing with a fixed network (here the entire backbone network, namely $f(.)$, $g(.)$); and (ii) fine-tuning the entire network end-to-end with supervised learning. For the embedding, as shown in Fig. 4, we take the input video blocks $x_1, x_2, ..., x_t$ in the same way as MemDPC and extract the context feature c_t using the feature extractor $f(.)$ for each block and temporal aggregator $g(.)$; then we spatially pool the context feature c_t to obtain the embedding. We describe the training details in Sect. 5.3. The detailed experiment can be found in Sect. 5.4.

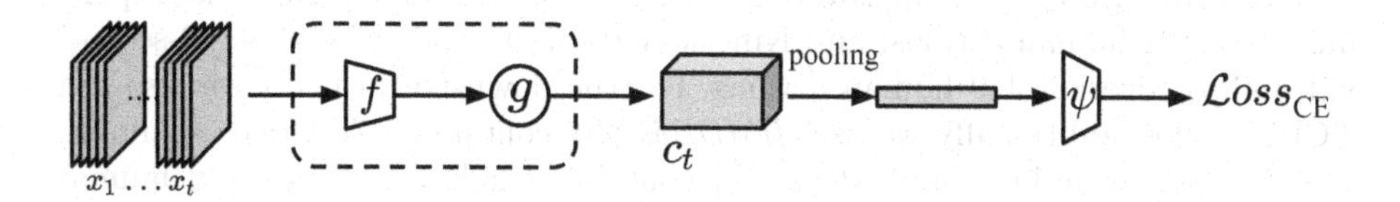

Fig. 4. Architecture of the action classification framework

Data Efficiency and Generalizability are reflected by the effectiveness of the representation under a scarce-annotation regime. For this task, we take the `MemDPC` representation and finetune it for action classification task, but limit the model to only use 10%, 20% and 50% of the labelled training samples, then we report the accuracy on the same testing set. The classifier has the identical training pipeline as shown in Fig. 4, and training details are explained in Sect. 5.3. The detailed experiment can be found in Sect. 5.5.

Video Action Retrieval directly evaluates the quality of the representation without any further training, aiming to provide a straightforward understanding on the quality of the learnt representation. Here, we use the simplest non-parametric classifier, *i.e.* k-nearest neighbours, to determine whether semantically similar actions are close in the high-dimensional space. Referring to Fig. 4, for each video, we truncate it into blocks $x_1, x_2, ..., x_t$ and extract the context feature c_t with the $f(.)$ and $g(.)$ trained with `MemDPC`. We spatially pool c_t to get a context feature vector, which is directly used as a query vector for measuring the similarity with other videos in the dataset. The detailed experiment can be found in Sect. 5.6.

Unintentional Actions is a straightforward application for a predictive framework like `MemDPC`. We evaluate our representation on the task of unintentional event classification that is proposed in the recent Oops dataset [17]. The core of unintentional events detection in video is a problem of anomaly detection. Usually, one of the predicted hypotheses tends to match true future relatively well for most of the videos. The discrepancy between them yields a measurement of future predictability, or 'surprise' level. A big surprise or a mismatched prediction can be used to locate the failing moment. In detail, we design the model as follows: first, we compute both the predicted feature $\hat{z}_i$ and the corresponding true feature z_i, and let a function $\xi(.)$ to measure their discrepancy. We train the model with two settings: (i) freezing the representation and only train the classifier $\xi(.)$; (ii) finetuning the entire network. The detailed structure for the classification task can be found in Sect. 5.7.

5 Experiments

5.1 Datasets

For the self-supervised training, two video action recognition datasets are used, but labels are dropped during training: *UCF101* [54], containing 13k videos spanning over 101 human actions; and Kinetics400 (*K400*) [36] with 306k 10-second video clips covering 400 human actions. For the downstream tasks we also use UCF101, and additionally we use: *HMDB51* [40] containing 7k videos spanning over 51 human actions; and *Oops* [17] containing 20k videos of daily human activities with unexpected failed moments, among them 14k videos have the time stamps of the failed moments manually labelled.

5.2 Self-supervised Training

In our experiment, we use a (2+3D)-ResNet, following [18,24], as the encoder $f(.)$, where the first two residual blocks res2 and res3 have 2D convolutional kernels, and only res4 and res5 have 3D kernels. Specifically, (2+3D)-ResNet18 and (2+3D)-ResNet34 are used in our experiments, denoted as R18 and R34 below. For the temporal aggregation, $g(\cdot)$, we use an one-layer GRU with kernel size 1×1, with the weights shared among all spatial positions on the feature map. The future prediction function, $\phi(.)$, is a two-layer convolutional network. We choose the size of the memory bank $\mathbf{M}$ to be 1024 based on experiments in Table 1. Network architecture are given in the supplementary material.

For the data, raw videos are decoded at a frame rate 24–30 fps, and each data sample consists of 40 consecutive frames, sampled with a temporal stride of 3 from the raw video. As input to `MemDPC`, they are divided into 8 video clips – so that each encoder $f(.)$ inputs 5 frames, covering around 0.5 s, and the 40 frames around 4 s. For optical flow, in order to eliminate extra supervisory signals in the self-supervised training stage, we use the **un-supervised** TV-L1 algorithm [68], and follow the same pre-processing procedures as [10], *i.e.* truncating large motions with more than ± 20 in both channels, appending a third 0s channel, and transforming the values from $[-20, 20]$ to $[0, 255]$. For data augmentation, we apply clip-wise random crop and horizontal flip, and frame-wise color jittering and random greyscale, for both the RGB and optical flow streams. We experiment with both 128×128 and 224×224 input resolution. The original video resolution is 256×256 and it is firstly cropped to 224×224 then rescaled if needed. Self-supervised training uses the Adam [38] optimizer with initial learning rate 10^{-3}. The learning rate is decayed once to 10^{-4} when the validation loss plateaus. We use a batch size of 16 samples per GPU.

5.3 Supervised Classification

For all action classification downstream tasks, the input follows the same frame sampling procedure as when the model is trained with self-supervised learning, and then we train the classifier with cross-entropy loss as shown in Fig. 4. A dropout of 0.9 is applied on the final layer. For data augmentation, we use clip-wise random crop, random horizontal flip, and random color jittering. The classifier is trained with Adam with a 10^{-3} initial learning rate, and decayed once to 10^{-4} when the validation loss plateaus. During testing, we follow the standard pipeline, *i.e.* ten-crop (center and four corner crops, w/o horizontal flip), take the same sequence length as training from the video, and average the prediction from the sampling temporal moving window.

5.4 Evaluation: Action Classification

We conduct two sets of experiments: (i) ablation studies on the effectiveness of the different modules in the `MemDPC`, by self-supervised learning on UCF101, (ii) to compare with other state-of-the-art approaches, we run `MemDPC` on K400 with self-supervised learning. For both settings, the representation quality is evaluated on UCF101 and HMDB51 with linear probing, non-linear probing, and end-to-end finetuning.

Ablations on UCF101. In this section, we conduct extensive experiments to validate the effectiveness of compressive memory, bidirectional aggregation, and self-supervised learning on optical flow. Note that, in each experiment, we keep the settings identical, and only vary one variable at a time.

As shown in Table 1, the following phenomena can be observed: *First*, comparing experiment $\mathcal{C}2$ against $\mathcal{B}1$ (68.2 vs. 61.8), networks initialized with self-supervised `MemDPC` clearly present better generalization than a randomly initialized network; *Second*, comparing with a strong baseline ($\mathcal{A}$), the proposed compressive memory boost the learnt representation by around 5% (68.2 vs. 63.6), and the optimal memory size for UCF101 is 1024; *Third*, `MemDPC` acts as a general learning framework that can also help to boost the generalizability of motion representations, a 7.3% boost can be seen from $\mathcal{D}1$ vs. $\mathcal{B}2$ (81.9 vs. 74.6); *Fourth*, the bidirectional aggregation provides a small boost to the accuracy by about 1% ($\mathcal{E}1$ vs. $\mathcal{C}2$, $\mathcal{E}2$ vs. $\mathcal{D}1$, $\mathcal{E}3$ vs. $\mathcal{D}2$). *Lastly*, after fusing both streams, $\mathcal{D}2$ achieves 84% classification accuracy, confirming our claim that self-supervised learning with only the video stream (without additional audio or text streams) can also end up with strong action recognition models.

Table 1. Ablation studies. We train `MemDPC` on UCF101 and evaluate on UCF101 action classification by finetuning the entire network. We group rows for clarity: $\mathcal{A}$ is the reimplementation of DPC, $\mathcal{B}$ are random initialization baselines, $\mathcal{C}$ for different memory size, $\mathcal{D}$ incorporates optical flow, $\mathcal{E}$ incorporates a bi-directional RNN

#	Network	Self-sup.				Sup. (top1)
		Dataset	Input	Resolution	Memory size	UCF101(ft)
$\mathcal{A}$	R18	UCF101	RGB	128×128	- (DPC [24])	63.6
$\mathcal{B}$1	R18	-(rand. init.)	RGB	128×128	–	61.8
$\mathcal{B}$2	R18	-(rand. init.)	Flow	128×128	–	74.6
$\mathcal{B}$3	R18×2	-(rand. init.)	RGB+F	128×128	–	78.7
$\mathcal{C}$1	R18	UCF101	RGB	128×128	512	65.3
$\mathcal{C}$2	R18	UCF101	RGB	128×128	1024	68.2
$\mathcal{C}$3	R18	UCF101	RGB	128×128	2048	68.0
$\mathcal{D}$1	R18	UCF101	Flow	128×128	1024	81.9
$\mathcal{D}$2	R18×2	UCF101	RGB+F	128×128	1024	84.0
$\mathcal{E}$1	R18-bd	UCF101	RGB	128×128	1024	69.2
$\mathcal{E}$2	R18-bd	UCF101	Flow	128×128	1024	82.3
$\mathcal{E}$3	R18-bd×2	UCF101	RGB+F	128×128	1024	84.3

Comparison with Others. In this section, we train `MemDPC` on K400 and evaluate the action classification performance on UCF101 and HMDB51. Specifically, we evaluate three settings: (1) finetuning the entire network (denoted as Freeze = ✗); (2) freeze the backbone and only train a linear classifier, *i.e.* linear probe (denoted as Freeze = ✓); (3) freeze the backbone and only train a **non-linear** classifier, *i.e.* non-linear probe (denoted as 'n.l.').

As shown in Table 2, for the same amount of data (K400) and visual-only input, `MemDPC` surpasses all previous state-of-the-art self-supervised methods on both UCF101 and HMDB51 (although there exist small differences in architecture, *e.g.* for 3DRotNet, ST-Puzzle, DPC, SpeedNet). When freezing the representation, it can be seen that a non-linear probe gives better results than a linear probe, and in practice a non-linear classifier is still very cheap to train.

Other self-supervised training methods on the same benchmarks are not directly comparable, even ignoring the architecture differences, due to the duration of videos used or to the number of modalities used. For example, CBT [56] uses a longer version of K600 (referred to as K600+ in the table), the size is about 9 times that of the standard K400 that we use, and CBT requires RotNet [35] initialization while `MemDPC` can be trained from scratch. Nevertheless, our performance exceeds that of CBT. Other works use additional modalities for pre-text tasks like audio [2,39,51,52], or narrations [47], and train on larger datasets. Despite these disadvantages, we demonstrate that `MemDPC` trained with only visual inputs, can achieve competitive results on the finetuning protocol.

Table 2. Comparison with state-of-the-art approaches. In the left columns, we show the pre-training setting, *e.g.* dataset, resolution, architectures with encoder depth, modality. In the right columns, the top-1 accuracy is reported on the downstream action classification task for different datasets, *e.g.* UCF, HMDB, K400. The dataset parenthesis shows the total video duration in time (**d** for day, **y** for year). 'Frozen ✗' means the network is end-to-end finetuned from the pretrained representation, shown in the top half of the table; 'Frozen ✓' means the pretrained representation is fixed and classified with a linear layer, 'n.l.' denotes a non-linear classifier. For input, 'V' refers to visual only (colored with blue), 'A' is audio, 'T' is text narration. MemDPC models with † refer to the two-stream networks, where the predictions from RGB and Flow networks are averaged

Method	Date	Dataset (duration)	Res.	Arch.	Depth	Modality	Frozen	UCF	HMDB
CBT [56]	2019	K600+ (273d)	112	S3D	23	V	✓	54.0	29.5
MIL-NCE [47]	2020	HTM (15y)	224	S3D	23	V+T	✓	82.7	53.1
MIL-NCE [47]	2020	HTM (15y)	224	I3D	22	V+T	✓	83.4	54.8
XDC [2]	2019	IG65M (21y)	224	R(2+1)D	26	V+A	✓	85.3	56.0
ELO [52]	2020	Youtube8M- (8y)	224	R(2+1)D	65	V+A	✓	–	64.5
MemDPC†		K400 (28d)	224	R-2D3D	33	V	✓	54.1	30.5
MemDPC†		K400 (28d)	224	R-2D3D	33	V	✓n.l.	58.5	33.6
OPN [44]	2017	UCF (1d)	227	VGG	14	V	✗	59.6	23.8
3D-RotNet [35]	2018	K400 (28d)	112	R3D	17	V	✗	62.9	33.7
ST-Puzzle [37]	2019	K400 (28d)	224	R3D	17	V	✗	63.9	33.7
VCOP [66]	2019	UCF (1d)	112	R(2+1)D	26	V	✗	72.4	30.9
DPC [24]	2019	K400 (28d)	224	R-2D3D	33	V	✗	75.7	35.7
CBT [56]	2019	K600+ (273d)	112	S3D	23	V	✗	79.5	44.6
DynamoNet [15]	2019	Youtube8M-1 (1.9y)	112	STCNet	133	V	✗	88.1	59.9
SpeedNet [7]	2020	K400 (28d)	224	S3D-G	23	V	✗	81.1	48.8
AVTS [39]	2018	K400 (28d)	224	I3D	22	V+A	✗	83.7	53.0
AVTS [39]	2018	AudioSet (240d)	224	MC3	17	V+A	✗	89.0	61.6
XDC [2]	2019	K400 (28d)	224	R(2+1)D	26	V+A	✗	84.2	47.1
XDC [2]	2019	IG65M (21y)	224	R(2+1)D	26	V+A	✗	94.2	67.4
GDT [51]	2020	K400 (28d)	112	R(2+1)D	26	V+A	✗	88.7	57.8
MIL-NCE [47]	2020	HTM (15y)	224	S3D-G	23	V+T	✗	91.3	61.0
ELO [52]	2020	Youtube8M-2 (13y)	224	R(2+1)D	65	V+A	✗	93.8	67.4
MemDPC		K400 (28d)	224	R-2D3D	33	V	✗	78.1	41.2
MemDPC†		K400 (28d)	224	R-2D3D	33	V	✗	86.1	54.5
Supervised [25]		K400 (28d)	224	R3D	33	V	✗	87.7	59.1

5.5 Evaluation: Data Efficiency

In Fig. 5, we show the data efficiency of `MemDPC` on both RGB input and optical flow with action recognition on the UCF101 dataset. As we reduce the labelled training samples, action classifier trained on `MemDPC` representation generalize significantly better than the classifier trained from scratch. Also, to match the performance of a random initialized classifier trained on 100% labelled data, a classifier trained on `MemDPC` initialization only requires less than 50% labelled data for both RGB and optical flow input.

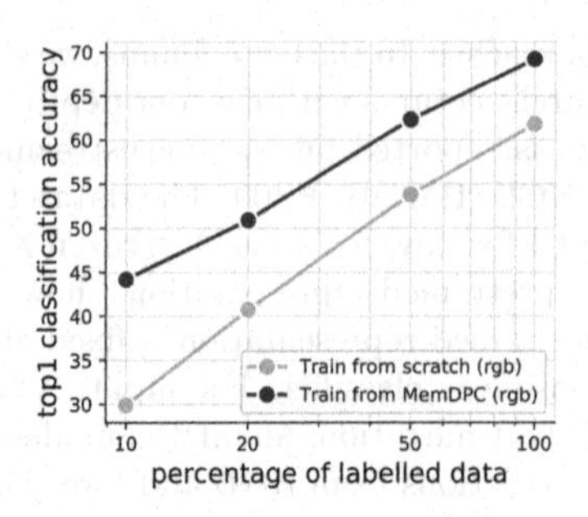

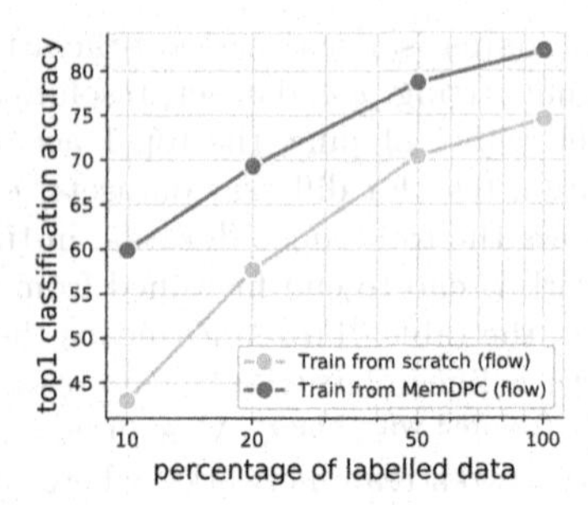

Fig. 5. Data efficiency of MemDPC representations. Left is RGB input and right is optical flow input. The MemDPC is trained on UCF101 and it is evaluated on action classification (finetuning protocol) on UCF101 with a reduced number of labels

5.6 Evaluation: Video Retrieval

In this protocol, we evaluate our representation with nearest-neighbour video retrieval, features are extracted from the model, which is only trained with self-supervised learning, no further finetuning is allowed.

Experiments are shown on two datasets: UCF101 and HMDB51. For both datasets, within the training set or within the testing set, multiple clips could be from the same source video, hence they are visually similar and make the retrieval task trivial. We follow the practice of [46,66], and use each clip in the test set to query the k nearest clips in the training set.

For each clip, we sample multiple 8 video blocks with a sliding window, and extract the context representation c_t for each window. We spatial-pool each c_t and take the average over all the windows. For distance measurement, we use cosine distance. We report Recall at k (R@k) as the evaluation metric. That is, as long as one clip of the same class is retrieved in the top k nearest neighbours, a correct retrieval is counted.

Table 3. Comparison with others on Nearest-Neighbour video retrieval on UCF101 and HMDB51. Testing set clips are used to retrieve training set videos and R@k is reported, where $k \in [1, 5, 10, 20]$. Note that all the models reported were only pretrained on UCF101 with self-supervised learning except SpeedNet

Method	Date	Dataset	UCF				HMDB			
			R@1	R@5	R@10	R@20	R@1	R@5	R@10	R@20
Jigsaw [49]	2016	UCF	19.7	28.5	33.5	40.0	–	–	–	–
OPN [44]	2017	UCF	19.9	28.7	34.0	40.6	–	–	–	–
Buchler [9]	2018	UCF	25.7	36.2	42.2	49.2	–	–	–	–
VCOP [66]	2019	UCF	14.1	30.3	40.4	51.1	7.6	22.9	34.4	48.8
VCP [46]	2020	UCF	18.6	33.6	42.5	53.5	7.6	24.4	36.3	53.6
SpeedNet [7]	2020	K400	13.0	28.1	37.5	49.5	–	–	–	–
MemDPC-RGB		UCF	20.2	**40.4**	**52.4**	**64.7**	**7.7**	**25.7**	**40.6**	**57.7**
MemDPC-Flow		UCF	**40.2**	**63.2**	**71.9**	**78.6**	**15.6**	**37.6**	**52.0**	**65.3**

In Table 3, we show the retrieval performance on UCF101 and HMDB51. Note that the MemDPC benchmarked here is only trained on UCF101, the same as [46,66]. For fair comparison, MemDPC in this experiment uses a R18 backbone, which has the same depth but less parameters than the 3D-ResNet used in [46,66]. With RGB inputs, our MemDPC gets state-of-the-art performance on all the metrics except R@1 in UCF101, where the method from Buchler *et al.* [9] specializes well on R@1. While for Flow inputs, MemDPC significantly outperforms all previous methods by a large margin. We also qualitatively show video retrieval results in the supplementary material.

5.7 Evaluation: Unintentional Actions

We evaluate MemDPC on the Oops dataset on unintentional action classification. In Oops, there is one failure moment in the middle of each video. When cutting the video into short clips, the clip overlapping the failure moment is defined as a 'transitioning' action, the clips before are 'intentional' actions, and the clips afterwards are 'unintentional' actions. The core task is therefore to classify each short video clip into one of three categories,

In this experiment, we use a R18 based MemDPC model that takes 128×128 resolution video frames as input. After MemDPC is trained on K400 and the Oops training set videos with self-supervised learning, we further train it for unintentional action classification with a linear probe, and end-to-end finetuning (as shown in Table 4). The training details are given in the supplementary material. State-of-the-art performance is demonstrated by our MemDPC on this unintentional action classification task, even outperforming the model pretrained on K700 with full supervision with finetuning.

Table 4. MemDPC on unintentional action classification tasks. Note that our backbone 2+3D-ResNet18 has the same depth as 3D-ResNet18 used in [17] but with less parameters. MemDPC model is trained on K400 and the OOPS training set without using labels, and the network is then finetuned with supervision from the OOPS training set

Task	Method	Backbone	Freeze	Finetune
Classification	K700 supervision	3D-ResNet18	**53.6**	64.0
	Video Speed [17]	3D-ResNet18	53.4	61.6
	MemDPC	R18	53.0	**64.4**

6 Conclusion

In this paper, we propose a new architecture and learning framework (MemDPC) for self-supervised learning from video, in particular for representations for action recognition. With the novel compressive memory, the model can efficiently handle the nature of multiple hypotheses in the self-supervised predictive learning procedure. In order to thoroughly evaluate the quality of the learnt representation, we conduct experiments on four different downstream tasks, namely action recognition, video retrieval, learning with scarce annotations, and unintentional action classification. In all cases, we demonstrate state-of-the-art or competitive performance over other approaches that use orders of magnitude more training data. Above all, for the first time, we show that it is possible to learn high-quality video representations with self-supervised learning, from the visual stream alone (without additional audio or text streams).

Acknowledgements. Funding for this research is provided by a Google-DeepMind Graduate Scholarship, and by the EPSRC Programme Grant Seebibyte EP/M013774/1. We would like to thank João F. Henriques, Samuel Albanie and Triantafyllos Afouras for helpful discussions.

References

1. Agrawal, P., Carreira, J., Malik, J.: Learning to see by moving. In: Proceedings of the ICCV, pp. 37–45. IEEE (2015)
2. Alwassel, H., Mahajan, D., Torresani, L., Ghanem, B., Tran, D.: Self-supervised learning by cross-modal audio-video clustering. arXiv preprint arXiv:1911.12667 (2019)
3. Arandjelović, R., Gronat, P., Torii, A., Pajdla, T., Sivic, J.: NetVLAD: CNN architecture for weakly supervised place recognition. In: Proceedings of the CVPR (2016)
4. Arandjelović, R., Zisserman, A.: Look, listen and learn. In: Proceedings of the ICCV (2017)
5. Arandjelović, R., Zisserman, A.: Objects that sound. In: Ferrari, V., Hebert, M., Sminchisescu, C., Weiss, Y. (eds.) ECCV 2018. LNCS, vol. 11205, pp. 451–466. Springer, Cham (2018). https://doi.org/10.1007/978-3-030-01246-5_27
6. Bahdanau, D., Cho, K., Bengio, Y.: Neural machine translation by jointly learning to align and translate. In: Proceedings of the ICLR (2015)
7. Benaim, S., et al.: SpeedNet: learning the speediness in videos. In: Proceedings of the CVPR (2020)
8. Brabandere, B.D., Jia, X., Tuytelaars, T., Gool, L.V.: Dynamic filter networks. In: NeurIPS (2016)
9. Büchler, U., Brattoli, B., Ommer, B.: Improving spatiotemporal self-supervision by deep reinforcement learning. In: Ferrari, V., Hebert, M., Sminchisescu, C., Weiss, Y. (eds.) ECCV 2018. LNCS, vol. 11219, pp. 797–814. Springer, Cham (2018). https://doi.org/10.1007/978-3-030-01267-0_47
10. Carreira, J., Zisserman, A.: Quo vadis, action recognition? A new model and the kinetics dataset. In: Proceedings of the CVPR (2017)
11. Chen, T., Kornblith, S., Norouzi, M., Hinton, G.: A simple framework for contrastive learning of visual representations. In: Proceedings of the ICML (2020)
12. Chopra, S., Hadsell, R., LeCun, Y.: Learning a similarity metric discriminatively, with application to face verification. In: Proceedings of the CVPR (2005)
13. Chung, J., Gulcehre, C., Cho, K., Bengio, Y.: Empirical evaluation of gated recurrent neural networks on sequence modeling. arXiv preprint arXiv:1412.3555 (2014)

14. Devlin, J., Chang, M., Lee, K., Toutanova, K.: BERT: pre-training of deep bidirectional transformers for language understanding. arXiv preprint arXiv:1810.04805 (2018)
15. Diba, A., Sharma, V., Gool, L.V., Stiefelhagen, R.: DynamoNet: dynamic action and motion network. In: Proceedings of the ICCV (2019)
16. Dosovitskiy, A., Koltun, V.: Learning to act by predicting the future. In: Proceedings of the ICLR (2017)
17. Epstein, D., Chen, B., Vondrick, C.: Oops! Predicting unintentional action in video. In: Proceedings of the CVPR (2020)
18. Feichtenhofer, C., Pinz, A., Wildes, R.P., Zisserman, A.: What have we learned from deep representations for action recognition? In: Proceedings of the CVPR (2018)
19. Feichtenhofer, C., Pinz, A., Zisserman, A.: Convolutional two-stream network fusion for video action recognition. In: Proceedings of the CVPR (2016)
20. Fernando, B., Bilen, H., Gavves, E., Gould, S.: Self-supervised video representation learning with odd-one-out networks. In: Proceedings of the ICCV (2017)
21. Gan, C., Gong, B., Liu, K., Su, H., Guibas, L.J.: Geometry guided convolutional neural networks for self-supervised video representation learning. In: Proceedings of the CVPR (2018)
22. Graves, A., Wayne, G., Danihelka, I.: Neural turing machines. arXiv preprint arXiv:1410.5401 (2014)
23. Gutmann, M.U., Hyvärinen, A.: Noise-contrastive estimation: a new estimation principle for unnormalized statistical models. In: AISTATS (2010)
24. Han, T., Xie, W., Zisserman, A.: Video representation learning by dense predictive coding. In: Workshop on Large Scale Holistic Video Understanding, ICCV (2019)
25. Hara, K., Kataoka, H., Satoh, Y.: Can spatiotemporal 3D CNNs retrace the history of 2D CNNs and ImageNet? In: CVPR (2018)
26. He, K., Fan, H., Wu, A., Xie, S., Girshick, R.: Momentum contrast for unsupervised visual representation learning. In: Proceedings of the CVPR (2020)
27. Hénaff, O.J., Razavi, A., Doersch, C., Eslami, S.M.A., van den Oord, A.: Data-efficient image recognition with contrastive predictive coding. arXiv preprint arXiv:1905.09272 (2019)
28. Hjelm, R.D., et al.: Learning deep representations by mutual information estimation and maximization. In: Proceedings of the ICLR (2019)
29. Hochreiter, S., Schmidhuber, J.: Long short-term memory. Neural Comput. **9**(8), 1735–1780 (1997)
30. Isola, P., Zoran, D., Krishnan, D., Adelson, E.H.: Learning visual groups from co-occurrences in space and time. In: Proceedings of the ICLR (2015)
31. Jakab, T., Gupta, A., Bilen, H., Vedaldi, A.: Unsupervised learning of object landmarks through conditional image generation. In: NeurIPS (2018)
32. Jayaraman, D., Grauman, K.: Learning image representations tied to ego-motion. In: Proceedings of the ICCV (2015)
33. Jayaraman, D., Grauman, K.: Slow and steady feature analysis: higher order temporal coherence in video. In: Proceedings of the CVPR (2016)
34. Ji, X., Henriques, J.F., Vedaldi, A.: Invariant information clustering for unsupervised image classification and segmentation. In: Proceedings of the ICCV (2019)
35. Jing, L., Tian, Y.: Self-supervised spatiotemporal feature learning by video geometric transformations. arXiv preprint arXiv:1811.11387 (2018)
36. Kay, W., ET AL.: The kinetics human action video dataset. arXiv preprint arXiv:1705.06950 (2017)

37. Kim, D., Cho, D., Kweon, I.S.: Self-supervised video representation learning with space-time cubic puzzles. In: AAAI (2019)
38. Kingma, D.P., Ba, J.: Adam: A method for stochastic optimization. arXiv preprint arXiv:1412.6980 (2014)
39. Korbar, B., Tran, D., Torresani, L.: Cooperative learning of audio and video models from self-supervised synchronization. In: NeurIPS (2018)
40. Kuehne, H., Jhuang, H., Garrote, E., Poggio, T., Serre, T.: HMDB: a large video database for human motion recognition. In: Proceedings of the ICCV, pp. 2556–2563 (2011)
41. Kumar, A., et al.: Ask me anything: dynamic memory networks for natural language processing. In: Proceedings of the ICML (2016)
42. Lai, Z., Lu, E., Xie, W.: MAST: A memory-augmented self-supervised tracker. In: Proceedings of the CVPR (2020)
43. Lai, Z., Xie, W.: Self-supervised learning for video correspondence flow. In: Proceedings of the BMVC (2019)
44. Lee, H., Huang, J., Singh, M., Yang, M.: Unsupervised representation learning by sorting sequence. In: Proceedings of the ICCV (2017)
45. Lotter, W., Kreiman, G., Cox, D.: Deep predictive coding networks for video prediction and unsupervised learning. In: Proceedings of the ICLR (2017)
46. Luo, D., Liu, C., Zhou, Y., Yang, D., Ma, C., Ye, Q., Wang, W.: Video cloze procedure for self-supervised spatio-temporal learning. In: AAAI (2020)
47. Miech, A., Alayrac, J.B., Smaira, L., Laptev, I., Sivic, J., Zisserman, A.: End-to-end learning of visual representations from uncurated instructional videos. In: Proceedings of the CVPR (2020)
48. Misra, I., Zitnick, C.L., Hebert, M.: Shuffle and learn: unsupervised learning using temporal order verification. In: Leibe, B., Matas, J., Sebe, N., Welling, M. (eds.) ECCV 2016. LNCS, vol. 9905, pp. 527–544. Springer, Cham (2016). https://doi.org/10.1007/978-3-319-46448-0_32
49. Noroozi, M., Favaro, P.: Unsupervised learning of visual representations by solving Jigsaw puzzles. In: Leibe, B., Matas, J., Sebe, N., Welling, M. (eds.) ECCV 2016. LNCS, vol. 9910, pp. 69–84. Springer, Cham (2016). https://doi.org/10.1007/978-3-319-46466-4_5
50. van den Oord, A., Li, Y., Vinyals, O.: Representation learning with contrastive predictive coding. arXiv preprint arXiv:1807.03748 (2018)
51. Patrick, M., Asano, Y.M., Fong, R., Henriques, J.F., Zweig, G., Vedaldi, A.: Multi-modal self-supervision from generalized data transformations. arXiv preprint arXiv:2003.04298 (2020)
52. Piergiovanni, A., Angelova, A., Ryoo, M.S.: Evolving losses for unsupervised video representation learning. In: Proceedings of the CVPR (2020)
53. Simonyan, K., Zisserman, A.: Two-stream convolutional networks for action recognition in videos. In: NeurIPS (2014)
54. Soomro, K., Zamir, A.R., Shah, M.: UCF101: A dataset of 101 human actions classes from videos in the wild. arXiv preprint arXiv:1212.0402 (2012)
55. Sukhbaatar, S., Szlam, A., Weston, J., Fergus, R.: End-to-end memory networks. In: NeurIPS (2015)
56. Sun, C., Baradel, F., Murphy, K., Schmid, C.: Contrastive bidirectional transformer for temporal representation learning. arXiv preprint arXiv:1906.05743 (2019)
57. Tian, Y., Krishnan, D., Isola, P.: Contrastive multiview coding. arXiv preprint arXiv:1906.05849 (2019)
58. Vaswani, A., et al.: Attention is all you need. In: NIPS (2017)

59. Vondrick, C., Pirsiavash, H., Torralba, A.: Anticipating visual representations from unlabelled video. In: CVPR (2016)
60. Vondrick, C., Shrivastava, A., Fathi, A., Guadarrama, S., Murphy, K.: Tracking emerges by colorizing videos. In: Ferrari, V., Hebert, M., Sminchisescu, C., Weiss, Y. (eds.) ECCV 2018. LNCS, vol. 11217, pp. 402–419. Springer, Cham (2018). https://doi.org/10.1007/978-3-030-01261-8_24
61. Wang, X., Girshick, R., Gupta, A., He, K.: Non-local neural networks. In: Proceedings of the CVPR (2018)
62. Wang, X., Gupta, A.: Unsupervised learning of visual representations using videos. In: Proceedings of the ICCV (2015)
63. Wang, X., Jabri, A., Efros, A.A.: Learning correspondence from the cycle-consistency of time. In: CVPR (2019)
64. Wiles, O., Koepke, A.S., Zisserman, A.: Self-supervised learning of a facial attribute embedding from video. In: Proceedings of the BMVC (2018)
65. Wu, Z., Xiong, Y., Yu, S., Lin, D.: Unsupervised feature learning via non-parametric instance-level discrimination. In: Proceedings of the CVPR, vol. abs/1805.01978 (2018)
66. Xu, D., Xiao, J., Zhao, Z., Shao, J., Xie, D., Zhuang, Y.: Self-supervised spatiotemporal learning via video clip order prediction. In: CVPR (2019)
67. Xu, K., et al.: Show, attend and tell: neural image caption generation with visual attention. In: Proceedings of the ICML (2015)
68. Zach, C., Pock, T., Bischof, H.: A duality based approach for realtime TV-L^1 optical flow. In: Hamprecht, F.A., Schnörr, C., Jähne, B. (eds.) DAGM 2007. LNCS, vol. 4713, pp. 214–223. Springer, Heidelberg (2007). https://doi.org/10.1007/978-3-540-74936-3_22
69. Zhang, R., Isola, P., Efros, A.A.: Colorful image colorization. In: Leibe, B., Matas, J., Sebe, N., Welling, M. (eds.) ECCV 2016. LNCS, vol. 9907, pp. 649–666. Springer, Cham (2016). https://doi.org/10.1007/978-3-319-46487-9_40
70. Zhuang, C., Zhai, A.L., Yamins, D.: Local aggregation for unsupervised learning of visual embeddings. In: ICCV (2019)

PointMixup: Augmentation for Point Clouds

Yunlu Chen[1(✉)], Vincent Tao Hu[1(✉)], Efstratios Gavves[1], Thomas Mensink[1,2], Pascal Mettes[1], Pengwan Yang[1,3], and Cees G. M. Snoek[1]

[1] University of Amsterdam, Amsterdam, The Netherlands
{y.chen3,t.hu}@uva.nl
[2] Google Research, Amsterdam, The Netherlands
[3] Peking University, Beijing, China

Abstract. This paper introduces data augmentation for point clouds by interpolation between examples. Data augmentation by interpolation has shown to be a simple and effective approach in the image domain. Such a mixup is however not directly transferable to point clouds, as we do not have a one-to-one correspondence between the points of two different objects. In this paper, we define data augmentation between point clouds as a shortest path linear interpolation. To that end, we introduce PointMixup, an interpolation method that generates new examples through an optimal assignment of the path function between two point clouds. We prove that our PointMixup finds the shortest path between two point clouds and that the interpolation is assignment invariant and linear. With the definition of interpolation, PointMixup allows to introduce strong interpolation-based regularizers such as mixup and manifold mixup to the point cloud domain. Experimentally, we show the potential of PointMixup for point cloud classification, especially when examples are scarce, as well as increased robustness to noise and geometric transformations to points. The code for PointMixup and the experimental details are publicly available (Code is available at: https://github.com/yunlu-chen/PointMixup/).

Keywords: Interpolation · Point cloud classification · Data augmentation

1 Introduction

The goal of this paper is to classify a cloud of points into their semantic category, be it an airplane, a bathtub or a chair. Point cloud classification is challenging, as they are sets and hence invariant to point permutations. Building

Y. Chen V. T. Hu—Equal contribution.

Electronic supplementary material The online version of this chapter (https://doi.org/10.1007/978-3-030-58580-8_20) contains supplementary material, which is available to authorized users.

A. Vedaldi et al. (Eds.): ECCV 2020, LNCS 12348, pp. 330–345, 2020.
https://doi.org/10.1007/978-3-030-58580-8_20

on the pioneering PointNet by Qi *et al.* [15], multiple works have proposed deep learning solutions to point cloud classification [12,16,23,29,30,36]. Given the progress in point cloud network architectures, as well as the importance of data augmentation in improving classification accuracy and robustness, we study how could data augmentation be naturally extended to support also point cloud data, especially considering the often smaller size of point clouds datasets (*e.g.*ModelNet40 [31]). In this work, we propose point cloud data augmentation by interpolation of existing training point clouds.

Fig. 1. Interpolation between point clouds. We show the interpolation between examples from different classes (airplane/chair, and monitor/bathtub) with multiple ratios λ. The interpolants are learned to be classified as $(1-\lambda)$ the first class and λ the second class. The interpolation is not obtained by learning, but induced by solving the optimal bijective correspondence which allows the minimum overall distance that each point in one point cloud moves to the assigned point in the other point cloud.

To perform data augmentation by interpolation, we take inspiration from augmentation in the image domain. Several works have shown that generating new training examples, by interpolating images and their corresponding labels, leads to improved network regularization and generalization, *e.g.*, [8,24,26,34]. Such a mixup is feasible in the image domain, due to the regular structure of images and one-to-one correspondences between pixels. However, this setup does not generalize to the point cloud domain, since there is no one-to-one correspondence and ordering between points. To that end, we seek to find a method to enable interpolation between permutation invariant point sets.

In this work, we make three contributions. First, we introduce data augmentation for point clouds through interpolation and we define the augmentation as a shortest path interpolation. Second, we propose PointMixup, an interpolation between point clouds that computes the optimal assignment as a path function between two point clouds, or the latent representations in terms of point cloud. The proposed interpolation strategy therefore allows usage of successful regularizers of Mixup and Manifold Mixup [26] on point cloud. We prove that (*i*) our PointMixup indeed finds the shortest path between two point clouds; (*ii*) the assignment does not change for any pairs of the mixed point clouds for any interpolation ratio; and (*iii*) our PointMixup is a linear interpolation, an important property since labels are also linearly interpolated. Figure 1 shows two pairs of point clouds, along with our interpolations. Third, we show the empirical benefits of our data augmentation across various tasks, including classification, few-shot learning, and semi-supervised learning. We furthermore show that our approach is agnostic to the network used for classification, while we also become more robust to noise and geometric transformations to the points.

2 Related Work

Deep Learning for Point Clouds. Point clouds are unordered sets and hence early works focus on analyzing equivalent symmetric functions which ensures permutation invariance. [15,17,33]. The pioneering PointNet work by Qi *et al.* [15] presented the first deep network that operates directly on unordered point sets. It learns the global feature with shared multi-layer perceptions and a max pooling operation to ensure permutation invariance. PointNet++ [16] extends this idea further with hierarchical structure by relying on a heuristic method of farthest point sampling and grouping to build the hierarchy. Likewise, other recent methods follow to learn hierarchical local features either by grouping points in various manners [10,12,23,29,30,32,36]. Li *et al.* [12] propose to learn a transformation from the input points to simultaneously solve the weighting of input point features and permutation of points into a latent and potentially canonical order. Xu *et al.* [32] extends 2D convolution to 3D point clouds by parameterizing a family of convolution filters. Wang *et al.* [29] proposed to leverage neighborhood structures in both point and feature spaces.

In this work, we aim to improve point cloud classification for any point-based approach. To that end, we propose a new model-agnostic data augmentation. We propose a Mixup regularization for point clouds and show that it can build on various architectures to obtain better classification results by reducing the generalization error in classification tasks. A very recent work by Li *et al.* [11] also considers improving point cloud classification by augmentation. They rely on auto-augmentation and a complicated adversarial training procedure, whereas in this work we propose to augment point clouds by interpolation.

Interpolation-Based Regularization. Employing regularization approaches for training deep neural networks to improve their generalization performances have become standard practice in deep learning. Recent works consider a regularization by interpolating the example and label pairs, commonly known as Mixup [8,24,34]. Manifold Mixup [26] extends Mixup by interpolating the hidden representations at multiple layers. Recently, an effort has been made on applying Mixup to various tasks such as object detection [35] and segmentation [7]. Different from existing works, which are predominantly employed in the image domain, we propose a new optimal assignment Mixup paradigm for point clouds, in order to deal with their permutation-invariant nature.

Recently, Mixup [34] has also been investigated from a semi-supervised learning perspective [2,3,27]. Mixmatch [3] guesses low-entropy labels for unlabelled data-augmented examples and mixes labelled and unlabelled data using

Mixup [34]. Interpolation Consistency Training [27] utilizes the consistency constraint between the interpolation of unlabelled points with the interpolation of the predictions at those points. In this work, we show that our PointMixup can be integrated in such frameworks to enable semi-supervised learning for point clouds.

3 Point Cloud Augmentation by Interpolation

3.1 Problem Setting

In our setting, we are given a training set $\{(S_m, c_m)\}_{m=1}^M$ consisting of M point clouds. $S_m = \{p_n^m\}_{n=1}^N \in \mathcal{S}$ is a point cloud consisting of N points, $p_n^m \in \mathbb{R}^3$ is the 3D point, $\mathcal{S}$ is the set of such 3D point clouds with N elements. $c_m \in \{0,1\}^C$ is the one-hot class label for a total of C classes. The goal is to train a function $h : \mathcal{S} \mapsto [0,1]^C$ that learns to map a point cloud to a semantic label distribution. Throughout our work, we remain agnostic to the type of function h used for the mapping and we focus on data augmentation to generate new examples.

Data augmentation is an integral part of training deep neural networks, especially when the size of the training data is limited compared to the size of the model parameters. A popular data augmentation strategy is Mixup [34]. Mixup performs augmentation in the image domain by linearly interpolating pixels, as well as labels. Specifically, let $I_1 \in \mathbb{R}^{W\times H\times 3}$ and $I_2 \in \mathbb{R}^{W\times H\times 3}$ denote two images. Then a new image and its label are generated as:

$$I_{\text{mix}}(\lambda) = (1-\lambda)\cdot I_1 + \lambda \cdot I_2, \quad (1)$$

$$c_{\text{mix}}(\lambda) = (1-\lambda)\cdot c_1 + \lambda \cdot c_2, \quad (2)$$

where $\lambda \in [0,1]$ denotes the mixup ratio. Usually λ is sampled from a beta distribution $\lambda \sim \text{Beta}(\gamma, \gamma)$. Such a direct interpolation is feasible for images as the data is aligned. In point clouds, however, linear interpolation is not straightforward. The reason is that point clouds are sets of points in which the point elements are orderless and permutation-invariant. We must, therefore, seek a definition of interpolation on unordered sets.

3.2 Interpolation Between Point Clouds

Let $S_1 \in \mathcal{S}$ and $S_2 \in \mathcal{S}$ denote two training examples on which we seek to perform interpolation with ratio λ to generate new training examples. Given a pair of source examples S_1 and S_2, an interpolation function, $f_{S_1\to S_2} : [0,1] \mapsto \mathcal{S}$ can be *any continuous function*, which forms a curve that joins S_1 and S_2 in a metric space $(\mathcal{S}, d)$ with a proper distance function d. This means that it is up to us to define what makes an interpolation good. We define the concept of shortest-path interpolation in the context of point cloud:

Definition 1 (Shortest-path interpolation). *In a metric space $(\mathcal{S}, d)$, a shortest-path interpolation $f^*_{S_1\to S_2} : [0,1] \mapsto \mathcal{S}$ is an interpolation between the* *given pair of source examples $S_1 \in \mathcal{S}$ and $S_2 \in \mathcal{S}$, such that for any $\lambda \in [0,1]$, $d(S_1, S^{(\lambda)}) + d(S^{(\lambda)}, S_2)) = d(S_1, S_2)$ holds for $S^{(\lambda)} = f^*_{S_1 \to S_2}(\lambda)$ being the interpolant.*

We say that Definition 1 ensures the shortest path property because the triangle inequality holds for any properly defined distance d : $d(S_1, S^{(\lambda)}) + d(S^{(\lambda)}, S_2)) \geq d(S_1, S_2)$. The intuition behind this definition is that the shortest path property ensures the uniqueness of the label distribution on the interpolated data. To put it otherwise, when computing interpolants from different sources, the interpolants generated by the shortest-path interpolation is more likely to be discriminative than the ones generated by a non-shortest-path interpolation (Fig. 2).

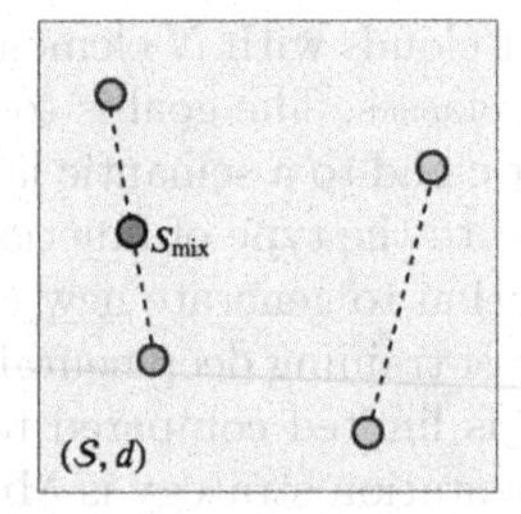

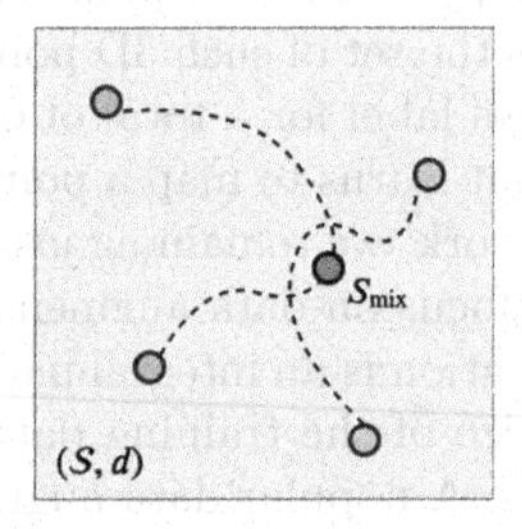

Fig. 2. Intuition of shortest-path interpolation. The examples lives on a metric space $(\mathcal{S}, d)$ as dots in the figure. The dashed lines are the interpolation paths between different pairs of examples. When the shortest-path property is ensured (left), the interpolation paths from different pairs of source examples are likely to be not intersect in a complicated metric space. While in non-shortest path interpolation (right), the paths can intertwine with each other with a much higher probability, making it hard to tell which pair of source examples does the mixed data come from.

To define an interpolation for point clouds, therefore, we must first select a reasonable distance metric. Then, we opt for the shorterst-path interpolation function based on the selected distance metric. For point clouds a proper distance metric is the Earth Mover's Distance (EMD), as it captures well not only the geometry between two point clouds, but also local details as well as density distributions [1,5,13]. EMD measures the least amount of total displacement required for each of the points in the first point cloud, $x_i \in S_1$, to match a corresponding point in the second point cloud, $y_j \in S_2$. Formally, the EMD for point clouds solves the following assignment problem:

$$\phi^* = arg\,min_{\phi \in \mathbf{\Phi}} \sum_i \|x_i - y_{\phi(i)}\|_2, \tag{3}$$

where $\mathbf{\Phi} = \{\{1, \ldots, N\} \mapsto \{1, \ldots, N\}\}$ is the set of possible bijective assignments, which give one-to-one correspondences between points in the two point clouds. Given the optimal assignment ϕ^*, the EMD is then defined as the average effort to move S_1 points to S_2:

$$d_{\text{EMD}} = \frac{1}{N} \sum_i \|x_i - y_{\phi^*(i)}\|_2. \tag{4}$$

3.3 PointMixup: Optimal Assignment Interpolation for Point Clouds

We propose an interpolation strategy, which can be used for augmentation that is analogous of Mixup [34] but for point clouds. We refer to this proposed PointMixup as *Optimal Assignment (OA) Interpolation*, as it relies on the optimal assignment on the basis of the EMD to define the interpolation between clouds. Given the source pair of point clouds $S_1 = \{x_i\}_{i=1}^N$ and $S_2 = \{y_j\}_{j=1}^N$, the Optimal Assignment (OA) interpolation is a path function $f^*_{S_1 \to S_2} : [0,1] \mapsto \mathcal{S}$. With $\lambda \in [0,1]$,

$$f^*_{S_1 \to S_2}(\lambda) = \{u_i\}_{i=1}^N, \quad \text{where} \tag{5}$$

$$u_i = (1-\lambda) \cdot x_i + \lambda \cdot y_{\phi^*(i)}, \tag{6}$$

in which ϕ^* is the optimal assignment from S_1 to S_2 defined by Eq. 3. Then the interpolant $S_{\mathbf{OA}}^{S_1 \to S_2, (\lambda)}$ (or $S_{\mathbf{OA}}^{(\lambda)}$ when there is no confusion) generated by the OA interpolation path function $f^*_{S_1 \to S_2}(\lambda)$ is the required augmented data for point cloud Mixup.

$$S_{\mathbf{OA}}^{(\lambda)} = \{(1-\lambda) \cdot x_i + \lambda \cdot y_{\phi^*(i)}\}_{i=1}^N. \tag{7}$$

Under the view of $f^*_{S_1 \to S_2}$ being a path function in the metric space $(\mathcal{S}, d_{\text{EMD}})$, f is expected to be the shortest path joining S_1 and S_2 since the definition of the interpolation is induced from the EMD.

3.4 Analysis

Intuitively we expect that PointMixup is a shortest path linear interpolation. That is, the interpolation lies on the shortest path joining the source pairs, and the interpolation is linear with regard to λ in $(\mathcal{S}, d_{\text{EMD}})$, since the definition of the interpolation is derived from the EMD. However, it is non-trivial to show the optimal assignment interpolation abides to a shortest path linear interpolation, because the optimal assignment between the mixed point cloud and either of the source point cloud is unknown. It is, therefore, not obvious that we can ensure whether there exists a shorter path between the mixed examples and the source examples. To this end, we need to provide an in-depth analysis.

To ensure the uniqueness of the label distribution from the mixed data, we need to show that the *shortest path property* w.r.t. the EMD is fulfilled. Moreover, we need to show that the proposed interpolation is *linear* w.r.t the EMD, in order to ensure that the input interpolation has the same ratio as the label interpolation. Besides, we evaluate the *assignment invariance property* as a prerequisite knowledge for the proof for the linearity. This property implies that there exists no shorter path between interpolants with different λ, *i.e.*, the shortest path between the interpolants is a part of the shortest path between the source examples. Due to space limitation, we sketch the proof for each property. The complete proofs are available in the supplementary material.

We start with the shortest path property. Since the EMD for point cloud is a metric, the triangle inequality $d_{EMD}(A,B) + d_{EMD}(B,C) \geq d_{EMD}(A,C)$ holds (for which a formal proof can be found in [19]). Thus we formalize the shortest path property into the following proposition:

Property 1 (shortest path). *Given the source examples S_1 and S_2, $\forall \lambda \in [0,1]$, $d_{EMD}(S_1, S_{OA}^{(\lambda)}) + d_{EMD}(S_{OA}^{(\lambda)}, S_2) = d_{EMD}(S_1, S_2)$.*

Sketch of Proof. From the definition of the EMD we can derive $d_{\text{EMD}}(S_1, S_{\mathbf{OA}}^{(\lambda)}) + d_{\text{EMD}}(S_2, S_{\mathbf{OA}}^{(\lambda)}) \leq d_{\text{EMD}}(S_1, S_2)$. Then from the triangle inequity of the EMD, only the equality remains. □

We then introduce the assignment invariance property of the OA Mixup as an intermediate step for the proof of the linearity of OA Mixup. The property shows that the assignment does not change for any pairs of the mixed point clouds with different λ. Moreover, the assignment invariance property is important to imply that the shortest path between the any two mixed point clouds is part of the shortest path between the two source point clouds.

Property 2 (assignment invariance). *$S_{OA}^{(\lambda_1)}$ and $S_{OA}^{(\lambda_2)}$ are two mixed point clouds from the same given source pair of examples S_1 and S_2 as well as the mix ratios λ_1 and λ_2 such that $0 \leq \lambda_1 < \lambda_2 \leq 1$. Let the points in $S_{OA}^{(\lambda_1)}$ and $S_{OA}^{(\lambda_2)}$ be $u_i = (1-\lambda_1) \cdot x_i + \lambda_1 \cdot y_{\phi^*(i)}$ and $v_k = (1-\lambda_2) \cdot x_k + \lambda_2 \cdot y_{\phi^*(k)}$, where ϕ^* is the optimal assignment from S_1 to S_2. Then the identical assignment ϕ_I is the optimal assignment from $S_{OA}^{(\lambda_1)}$ to $S_{OA}^{(\lambda_2)}$.*

Sketch of Proof. We first prove that the identical mapping is the optimal assignment from S_1 to $S_{\mathbf{OA}}^{(\lambda_1)}$ from the definition of the EMD. Then we prove that ϕ^* is the optimal assignment from $S_{\mathbf{OA}}^{(\lambda_1)}$ to S_2. Finally we prove that the identical mapping is the optimal assignment from $S_{\mathbf{OA}}^{(\lambda_1)}$ to $S_{\mathbf{OA}}^{(\lambda_2)}$ similarly as the proof for the first intermediate argument. □

Given the property of assignment invariance, the linearity follows:

Property 3 (linearity). *For any mix ratios λ_1 and λ_2 such that $0 \leq \lambda_1 < \lambda_2 \leq 1$, the mixed point clouds $S_{OA}^{(\lambda_1)}$ and $S_{OA}^{(\lambda_2)}$ satisfies that $d_{EMD}(S_{OA}^{(\lambda_1)}, S_{OA}^{(\lambda_2)}) = (\lambda_2 - \lambda_1) \cdot d_{EMD}(S_1, S_2)$.*

Sketch of Proof. The proof can be directly derived from the fact that the identical mapping is the optimal assignment between $S_{\mathbf{OA}}^{(\lambda_1)}$ and $S_{\mathbf{OA}}^{(\lambda_2)}$. □

The linear property of our interpolation is important, as we jointly interpolate the point clouds and the labels. By ensuring that the point cloud interpolation is linear, we ensure that the input interpolation has the same ratio as the label interpolation.

On the basis of the properties, we find that PointMixup is a shortest path linear interpolation between point clouds in $(\mathcal{S}, d_{\text{EMD}})$.

3.5 Manifold PointMixup: Interpolate Between Latent Point Features

In standard PointMixup, only the inputs, *i.e.*, the XYZ point cloud coordinates are mixed. The input XYZs are low-level geometry information and sensitive to disturbances and transformations, which in turn limits the robustness of the PointMixup. Inspired by Manifold Mixup [26], we can also use the proposed interpolation solution to mix the latent representations in the hidden layers of point cloud networks, which are trained to capture salient and high-level information that is less sensitive to transformations. PointMixup can be applied for the purpose of Manifold Mixup to mix both at the XYZs and different levels of latent point cloud features and maintain their respective advantages, which is expected to be a stronger regularizer for improved performance and robustness.

We describe how to mix the latent representations. Following [26], at each batch we randomly select a layer l to perform PointMixup from a set of layers L, which includes the input layer. In a point cloud network, the intermediate latent representation at layer l (before the global aggregation stage such as the max pooling aggregation in PointNet [15] and PointNet++ [16]) is $Z_{(l)} = \{(x_i, z_i^{(x)})\}_{i=1}^{N_z}$, in which x_i is 3D point coordinate and $z_i^{(x)}$ is the corresponding high-dimensional feature. For the mixed latent representation, given the latent representation of two source examples are $Z_{(l),1} = \{(x_i, z_i^{(x)})\}_{i=1}^{N_z}$ and $Z_{(l),2} = \{(y_i, z_i^{(y)})\}_{i=1}^{N_z}$, the optimal assignment ϕ^* is obtained by the 3D point coordinates x_i, and the mixed latent representation then becomes

$$\begin{aligned}
Z_{(l),\mathbf{OA}}^{(\lambda)} &= \{(x_i^{\text{mix}}, z_i^{\text{mix}})\}, \qquad \text{where} \\
x_i^{\text{mix}} &= (1-\lambda)\cdot x_i + \lambda \cdot y_{\phi^*(i)}, \\
z_i^{\text{mix}} &= (1-\lambda)\cdot z_i^{(x)} + \lambda \cdot z_{\phi^*(i)}^{(y)}.
\end{aligned}$$

Specifically in PointNet++, three layers of representations are randomly selected to perform Manifold Mixup: the input, and the representations after the first and the second SA modules (See appendix of [16]).

4 Experiments

4.1 Setup

Datasets. We focus in our experiments on the **ModelNet40** dataset [31]. This dataset contains 12,311 CAD models from 40 man-made object categories, split into 9,843 for training and 2,468 for testing. We furthermore perform experiments on the **ScanObjectNN** dataset [25]. This dataset consists of real-world point cloud objects, rather than sampled virtual point clouds. The dataset consists of 2,902 objects and 15 categories. We report on two variants of the dataset, a standard variant OBJ_ONLY and one with heavy permutations from rigid transformations PB_T50_RS [25].

Following [12], we discriminate between settings where each dataset is pre-aligned and unaligned with horizontal rotation on training and test point cloud examples. For the unaligned settings, we randomly rotate the training point cloud along the up-axis. Then, before solving the optimal assignment, we perform a simple additional alignment step to fit and align the symmetry axes between the two point clouds to be mixed. Through this way, the point clouds are better aligned and we obtain more reasonable point correspondences. Last, we also perform experiments using only 20% of the training data (Fig. 3).

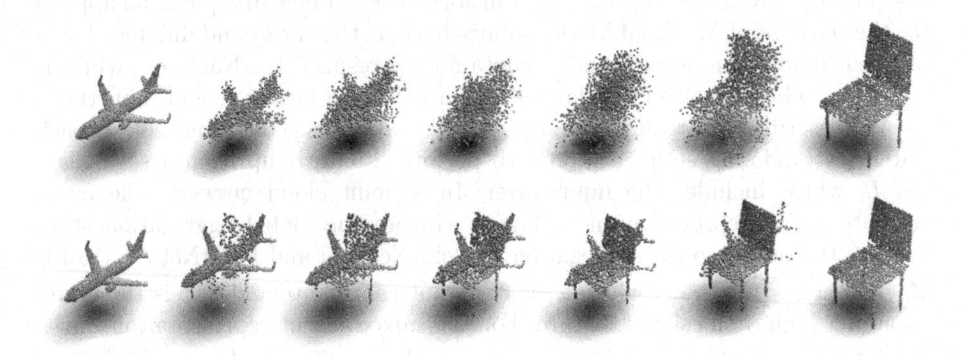

Fig. 3. Baseline interpolation variants. Top: point cloud interpolation through random assignment. Bottom: interpolation through sampling.

Network Architectures. The main network architecture used throughout the paper is PointNet++ [16]. We also report results with PointNet [15] and DGCNN [29], to show that our approach is agnostic to the architecture that is employed. PointNet learns a permutation-invariant set function, which does not capture local structures induced by the metric space the points live in. PointNet++ is a hierarchical structure, which segments a point cloud into smaller clusters and applies PointNet locally. DGCNN performs hierarchical operations by selecting a local neighbor in the feature space instead of the point space, resulting in each point having different neighborhoods in different layers.

Experimental Details. We uniformly sample 1,024 points on the mesh faces according to the face area and normalize them to be contained in a unit sphere, which is a standard setting [12,15,16]. In case of mixing clouds with different number of points, we can simply replicate random elements from the each point set to reach the same cardinality. During training, we augment the point clouds on-the-fly with random jitter for each point using Gaussian noise with zero mean and 0.02 standard deviation. We implement our approach in PyTorch [14]. For network optimization, we use the Adam optimizer with an initial learning rate of 10^{-3}. The model is trained for 300 epochs with a batch size of 16. We follow previous work [26,34] and draw λ from a beta distribution $\lambda \sim \text{Beta}(\gamma, \gamma)$. We also perform Manifold Mixup [26] in our approach, through interpolation on the transformed and pooled points in intermediate network layers. In this work, we opt to use the efficient algorithm and adapt the open-source implementation from [13] to solve the optimal assignment approximation. Training for 300 epochs takes around 17 h without augmentation and around 19 h with PointMixup or Manifold PointMixup on a single NVIDIA GTX 1080 ti.

Baseline Interpolations. For our comparisons to baseline point cloud augmentations, we compare to two variants. The first variant is random assignment interpolation, where a random assignment $\phi^{\mathbf{RA}}$ is used, to connect points from both sets, yielding:

$$S_{\mathbf{RA}}^{(\lambda)} = \{(1-\lambda) \cdot x_i + \lambda \cdot y_{\phi^{\mathbf{RA}}(i)}\}.$$

The second variant is point sampling interpolation, where random draws without replacement of points from each set are made according to the sampling frequency λ:

$$S_{\mathbf{PS}}^{(\lambda)} = S_1^{(1-\lambda)} \cup S_2^{(\lambda)},$$

where $S_2^{(\lambda)}$ denotes a randomly sampled subset of S_2, with $\lfloor \lambda N \rfloor$ elements. ($\lfloor \cdot \rfloor$ is the floor function.) And similar for S_1 with $N - \lfloor \lambda N \rfloor$ elements, such that $S_{\mathbf{PS}}^{(\lambda)}$ contains exactly N points. The intuition of the point sampling variant is that for point clouds as unordered sets, one can move one point cloud to another through a set operation such that it removes several random elements from set S_1 and replace them with same amount of elements from S_2.

4.2 Point Cloud Classification Ablations

We perform four ablation studies to show the workings of our approach with respect to the interpolation ratio, comparison to baseline interpolations and other regularizations, as well robustness to noise.

Effect of Interpolation Ratio. The first ablation study focuses on the effect of the interpolation ratio in the data augmentation for point cloud classification. We perform this study on ModelNet40 using the PointNet++ architecture. The results are shown in Fig. 4 for the pre-aligned setting. We find that regardless of the interpolation ratio used, our approach provides a boost over the setting without augmentation by interpolation. PointMixup positively influences point cloud classification. The inclusion of manifold mixup adds a further boost to the scores. Throughout further experiments, we use $\gamma = 0.4$ for input mixup and $\gamma = 1.5$ for manifold mixup in unaligned setting, and $\gamma = 1.0$ for input mixup and $\gamma = 2.0$ for manifold mixup in pre-aligned setting.

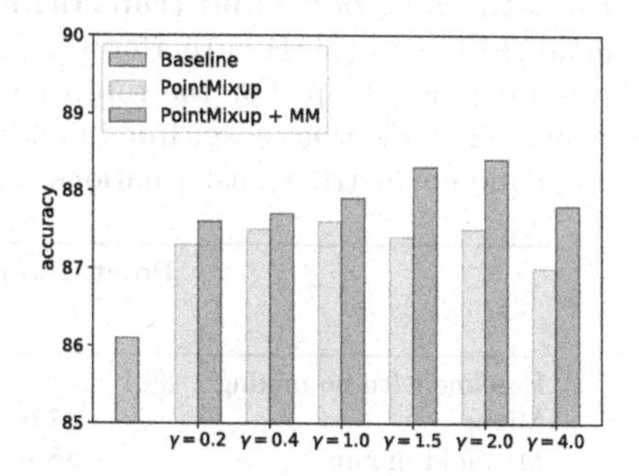

Fig. 4. Effect of interpolation ratios. MM denotes Manifold Mixup.

Comparison to Baseline Interpolations. In the second ablation study, we investigate the effectiveness of our PointMixup compared to the two interpolation baselines. We again use ModelNet40 and PointNet++. We perform the evaluation on both the pre-aligned and unaligned dataset variants, where for both we also report results with a reduced training set. The results are shown in Table 1. Across both the alignment variants and dataset sizes, our PointMixup obtains favorable results. This result highlights the effectiveness of our approach, which abides to the shortest path linear interpolation definition, while the baselines do not.

Table 1. Comparison of PointMixup to baseline interpolations on ModelNet40 using PointNet++. PointMixup compares favorable to excluding interpolation and to the baselines, highlighting the benefits of our shortest path interpolation solution.

Manifold mixup	No mixup	Random assignment		Point sampling		PointMixup	
	×	×	✓	×	✓	×	✓
Full dataset							
Pre-aligned	91.9	91.6	91.9	92.2	92.5	92.3	**92.7**
Unaligned	90.7	90.8	91.1	90.9	91.4	91.3	**91.7**
Reduced dataset							
Pre-aligned	86.1	85.5	87.3	87.2	87.6	87.6	**88.6**
Unaligned	84.4	84.8	85.4	85.7	86.5	86.1	**86.6**

Table 2. Evaluating our approach to other data augmentations (left) **and its robustness to noise and transformations** (right). We find that our approach with manifold mixup (MM) outperforms augmentations such as label smoothing and other variations of mixup. For the robustness evaluation, we find that our approach with strong regularization power from manifold mixup provides more robustness to random noise and geometric transformations.

	PointMixup	
	×	✓
Baseline with no mixing	86.1	–
Mixup	–	87.6
Manifold mixup	–	**88.6**
Mix input, not labels	–	86.6
Mix input from same class	–	86.4
Mixup latent (layer 1)	–	86.9
Mixup latent (layer 2)	–	86.8
Label smoothing (0.1)	87.2	–
Label smoothing (0.2)	87.3	–

Transforms	**PointMixup**		
	×	w/o MM	w/ MM
Noise $\sigma^2 = 0.01$	91.3	91.9	**92.3**
Noise $\sigma^2 = 0.05$	35.1	51.5	**56.5**
Noise $\sigma^2 = 0.1$	4.03	4.27	**7.42**
Z-rotation [-30,30]	74.3	70.9	**77.8**
X-rotation [-30,30]	73.2	70.8	**76.8**
Y-rotation [-30,30]	87.6	87.9	**88.7**
Scale (0.6)	85.8	84.5	**86.3**
Scale (2.0)	59.2	67.7	**72.9**
DropPoint (0.2)	84.9	78.1	**90.9**

PointMixup with Other Regularizers. Third, we evaluate how well PointMixup works by comparing to multiple existing data regularizers and mixup variants, again on ModelNet40 and PointNet++. We investigate the following augmentations: (*i*) Mixup [34], (*ii*) Manifold Mixup [26], (*iii*) mix input only without target mixup, (*iv*) mix latent representation at a fixed layer (manifold mixup does so at random layers), and (*v*) label smoothing [22]. Training is performed on the reduced dataset to better highlight their differences. We show the results in Table 2 on the left. Our approach with manifold mixup obtains the highest scores. The label smoothing regularizer is outperformed, while we also obtain better scores than the mixup variants. We conclude that PointMixup is forms an effective data augmentation for point clouds.

Robustness to Noise. By adding additional augmented training examples, we enrich the dataset. This enrichment comes with additional robustness with respect to noise in the point clouds. We evaluate the robustness by adding random noise perturbations on point location, scale, translation and different rotations. Note that for evaluation of robustness against up-axis rotation, we use the models which are trained with the pre-aligned setting, in order to test also the performance against rotation along the up-axis as a novel transform. The results are in Table 2 on the right. Overall, our approach including manifold mixup provides more stability across all perturbations. For example, with additional noise ($\sigma = 0.05$), we obtain an accuracy of 56.5, compared to 35.1 for the baseline. We similar trends for scaling (with a factor of two), with an accuracy of 72.9 versus 59.2. We conclude that PointMixup makes point cloud networks such as PointNet++ more stable to noise and rigid transformations.

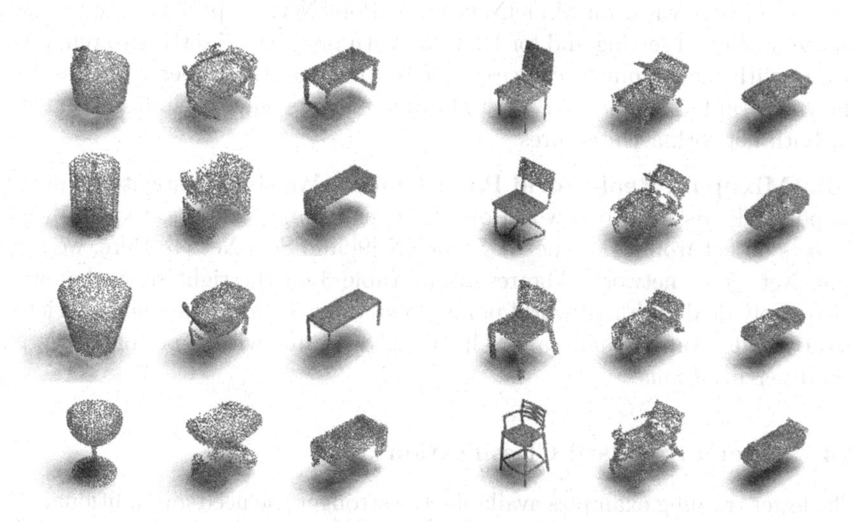

Fig. 5. Qualitative examples of PointMixup. We provide eight visualizations of our interpolation. The four examples on the left show interpolations for different configurations of cups and tables. The four examples on the right show interpolations for different chairs and cars.

Qualitative Analysis. In Fig. 5, we show eight examples of PointMix for point cloud interpolation; four interpolations of cups and tables, four interpolations of chairs and cars. Through our shortest path interpolation, we end up at new training examples that exhibit characteristics of both classes, making for sensible point clouds and mixed labels, which in turn indicate why PointMixup is beneficial for point cloud classification.

4.3 Evaluation on Other Networks and Datasets

With PointMixup, new point clouds are generated by interpolating existing point clouds. As such, we are agnostic to the type of network or dataset. To highlight this ability, we perform additional experiments on extra networks and an additional point cloud dataset.

Table 3. PointMixup on other networks (left) **and another dataset** (right). We find our approach is beneficial regardless the network or dataset.

	PointNet		DGCNN		
	×	w/o MM	×	w/o MM	w/ MM
Full	89.2	**89.9**	92.7	92.9	**93.1**
Reduced	81.3	**83.4**	88.2	88.8	**89.0**

	ScanObjectNN		
	×	w/o MM	w/ MM
Standard	86.6	87.6	**88.5**
Perturbed	79.3	80.2	**80.6**

PointMixup on Other Network Architectures. We show the effect of PointMixup to two other networks, namely PointNet [15] and DGCNN [29]. The experiments are performed on ModelNet40. For PointNet, we perform the evaluation on the unaligned setting and for DGCNN with pre-aligned setting to remain consistent with the alignment choices made in the respective papers. The results are shown in Table 3 on the left. We find improvements when including PointMixup for both network architectures.

PointMixup on Real-World Point Clouds. We also investigate PointMixup on point clouds from real-world object scans, using ScanObjectNN [25], which collects object from 3D scenes in SceneNN [9] and ScanNet [4]. Here, we rely on PointNet++ as network. The results in Table 3 on the right show that we can adequately deal with real-world point cloud scans, hence we are not restricted to point clouds from virtual scans. This result is in line with experiments on point cloud perturbations.

4.4 Beyond Standard Classification

The fewer training examples available, the stronger the need for additional examples through augmentation. Hence, we train PointNet++ on ModelNet40 in both a few-shot and semi-supervised setting.

Semi-supervised Learning. Semi-supervised learning learns from a dataset where only a small portion of data is labeled. Here, we show how PointMixup directly enables semi-supervised learning for point clouds. We start from Interpolation Consistency Training [27], a state-of-the-art semi-supervised approach, which utilizes Mixup between unlabeled points. Here, we use our Mixup for point clouds within their semi-supervised approach. We evaluate on ModelNet40 using 400, 600, and 800 labeled point clouds. The result of semi-supervised learning are illustrated in Table 4 on the left. Compared to the supervised baseline, which only uses the available labelled examples, our mixup enables the use of additional unlabelled training examples, resulting in a clear boost in scores. With 800 labelled examples, the accuracy increases from 73.5% to 82.0%, highlighting the effectiveness of PointMixup in a semi-supervised setting.

Table 4. Evaluating PointMixup in the context of semi-supervised (left) **and few-shot learning** (right). When examples are scarce, as is the case for both settings, using our approach provides a boost to the scores.

	Semi-supervised classification	
	Supervised	[27]+PointMixup
400 examples	69.4	**76.7**
600 examples	72.6	**80.8**
800 examples	73.5	**82.0**

	Few-shot classification	
	[20]	+ PointMixup
5-way 1-shot	72.3	**77.2**
5-way 3-shot	80.2	**82.2**
5-way 5-shot	84.2	**85.9**

Few-Shot Learning. Few-shot classification aims to learn a classifier to recognize unseen classes during training with limited examples. We follow [6,18,20,21,28] to regard few-shot learning a typical meta-learning method, which learns how to learn from limited labeled data through training from a collection of tasks, *i.e.*, episodes. In an N-way K-shot setting, in each task, N classes are selected and K examples for each class are given as a support set, and the query set consists of the examples to be predicted. We perform few-shot classification on ModelNet40, from which we select 20 classes for training, 10 for validation, and 10 for testing. We utilize PointMixup within ProtoNet [20] by constructing mixed examples from the support set and update the model with the mixed examples before making predictions on the query set. We refer to the supplementary material for the details of our method and the settings. The results in Table 4 on the right show that incorporating our data augmentation provides a boost in scores, especially in the one-shot setting, where the accuracy increases from 72.3% to 77.2%.

5 Conclusion

This work proposes PointMixup for data augmentation on point clouds. Given the lack of data augmentation by interpolation on point clouds, we start by defining it as a shortest path linear interpolation. We show how to obtain PointMixup between two point clouds by means of an optimal assignment interpolation between their point sets. As such, we arrive at a Mixup for point clouds, or latent point cloud representations in the sense of Manifold Mixup, that can handle permutation invariant nature. We first prove that PointMixup abides to our shortest path linear interpolation definition. Then, we show through various experiments that PointMixup matters for point cloud classification. We show that our approach outperforms baseline interpolations and regularizers. Moreover, we highlight increased robustness to noise and geometric transformations, as well as its general applicability to point-based networks and datasets. Lastly, we show the potential of our approach in both semi-supervised and few-shot settings. The generic nature of PointMixup allows for a comprehensive embedding in point cloud classification.

Acknowledgment. This research was supported in part by the SAVI/MediFor and the NWO VENI What & Where projects.

References

1. Achlioptas, P., Diamanti, O., Mitliagkas, I., Guibas, L.: Learning representations and generative models for 3D point clouds. In: ICML (2018)
2. Berthelot, D., et al.: ReMixMatch: semi-supervised learning with distribution alignment and augmentation anchoring. CoRR (2019)
3. Berthelot, D., Carlini, N., Goodfellow, I., Papernot, N., Oliver, A., Raffel, C.A.: MixMatch: a holistic approach to semi-supervised learning. In: NeurIPS (2019)
4. Dai, A., Chang, A.X., Savva, M., Halber, M., Funkhouser, T., Nießner, M.: ScanNet: richly-annotated 3D reconstructions of indoor scenes. In: CVPR (2017)
5. Fan, H., Su, H., Guibas, L.J.: A point set generation network for 3D object reconstruction from a single image. In: CVPR (2017)
6. Finn, C., Abbeel, P., Levine, S.: Model-agnostic meta-learning for fast adaptation of deep networks. In: ICML (2017)
7. French, G., Aila, T., Laine, S., Mackiewicz, M., Finlayson, G.: Consistency regularization and CutMix for semi-supervised semantic segmentation. CoRR (2019)
8. Guo, H., Mao, Y., Zhang, R.: MixUp as locally linear out-of-manifold regularization. In: AAAI (2019)
9. Hua, B.S., Pham, Q.H., Nguyen, D.T., Tran, M.K., Yu, L.F., Yeung, S.K.: SceneNN: a scene meshes dataset with annotations. In: 3DV (2016)
10. Li, J., Chen, B.M., Hee Lee, G.: SO-Net: self-organizing network for point cloud analysis. In: CVPR, pp. 9397–9406 (2018)
11. Li, R., Li, X., Heng, P.A., Fu, C.W.: PointAugment: an auto-augmentation framework for point cloud classification. In: CVPR, pp. 6378–6387 (2020)
12. Li, Y., Bu, R., Sun, M., Wu, W., Di, X., Chen, B.: PointCNN: convolution on X-transformed points. In: NeurIPS (2018)
13. Liu, M., Sheng, L., Yang, S., Shao, J., Hu, S.M.: Morphing and sampling network for dense point cloud completion. CoRR (2019)
14. Paszke, A., et al.: PyTorch: an imperative style, high-performance deep learning library. In: NeurIPS (2019)
15. Qi, C.R., Su, H., Mo, K., Guibas, L.J.: PointNet: deep learning on point sets for 3D classification and segmentation. In: CVPR (2017)
16. Qi, C.R., Yi, L., Su, H., Guibas, L.J.: PointNet++: deep hierarchical feature learning on point sets in a metric space. In: NeurIPS (2017)
17. Ravanbakhsh, S., Schneider, J., Poczos, B.: Deep learning with sets and point clouds. arXiv preprint arXiv:1611.04500 (2016)

18. Ravi, S., Larochelle, H.: Optimization as a model for few-shot learning. In: ICLR (2016)
19. Rubner, Y., Tomasi, C., Guibas, L.J.: The earth mover's distance as a metric for image retrieval. IJCV (2000)
20. Snell, J., Swersky, K., Zemel, R.: Prototypical networks for few-shot learning. In: NeurIPS (2017)
21. Sung, F., Yang, Y., Zhang, L., Xiang, T., Torr, P.H., Hospedales, T.M.: Learning to compare: relation network for few-shot learning. In: CVPR (2018)
22. Szegedy, C., Vanhoucke, V., Ioffe, S., Shlens, J., Wojna, Z.: Rethinking the inception architecture for computer vision. In: CVPR (2016)
23. Thomas, H., Qi, C.R., Deschaud, J.E., Marcotegui, B., Goulette, F., Guibas, L.J.: KPConv: flexible and deformable convolution for point clouds. In: ICCV, pp. 6411–6420 (2019)
24. Tokozume, Y., Ushiku, Y., Harada, T.: Between-class learning for image classification. In: CVPR (2018)
25. Uy, M.A., Pham, Q.H., Hua, B.S., Nguyen, T., Yeung, S.K.: Revisiting point cloud classification: a new benchmark dataset and classification model on real-world data. In: ICCV (2019)
26. Verma, V., et al.: Manifold mixup: better representations by interpolating hidden states. In: ICML (2019)
27. Verma, V., Lamb, A., Kannala, J., Bengio, Y., Lopez-Paz, D.: Interpolation consistency training for semi-supervised learning. IJCAI (2019)
28. Vinyals, O., Blundell, C., Lillicrap, T., Wierstra, D., et al.: Matching networks for one shot learning. In: NeurIPS (2016)
29. Wang, Y., Sun, Y., Liu, Z., Sarma, S.E., Bronstein, M.M., Solomon, J.M.: Dynamic graph CNN for learning on point clouds. TOG (2019)
30. Wu, W., Qi, Z., Fuxin, L.: PointConv: deep convolutional networks on 3D point clouds. In: CVPR (2019)
31. Wu, Z., et al.: 3D ShapeNets: a deep representation for volumetric shapes. In: CVPR (2015)
32. Xu, Y., Fan, T., Xu, M., Zeng, L., Qiao, Yu.: SpiderCNN: deep learning on point sets with parameterized convolutional filters. In: Ferrari, V., Hebert, M., Sminchisescu, C., Weiss, Y. (eds.) ECCV 2018. LNCS, vol. 11212, pp. 90–105. Springer, Cham (2018). https://doi.org/10.1007/978-3-030-01237-3_6
33. Zaheer, M., Kottur, S., Ravanbakhsh, S., Poczos, B., Salakhutdinov, R.R., Smola, A.J.: Deep sets. In: NeurIPS, pp. 3391–3401 (2017)
34. Zhang, H., Cisse, M., Dauphin, Y.N., Lopez-Paz, D.: mixup: Beyond empirical risk minimization. In: ICLR (2017)
35. Zhang, Z., He, T., Zhang, H., Zhang, Z., Xie, J., Li, M.: Bag of freebies for training object detection neural networks. In: CoRR (2019)
36. Zhang, Z., Hua, B.S., Yeung, S.K.: ShellNet: efficient point cloud convolutional neural networks using concentric shells statistics. In: ICCV (2019)

Identity-Guided Human Semantic Parsing for Person Re-identification

Kuan Zhu[1,2(✉)], Haiyun Guo[1], Zhiwei Liu[1,2], Ming Tang[1,3], and Jinqiao Wang[1,2]

[1] National Laboratory of Pattern Recognition, Institute of Automation, Chinese Academy of Sciences, Beijing, China
{kuan.zhu,haiyun.guo,zhiwei.liu,tangm,jqwang}@nlpr.ia.ac.cn
[2] School of Artificial Intelligence, University of Chinese Academy of Sciences, Beijing, China
[3] Shenzhen Infinova Limited, Shenzhen, China

Abstract. Existing alignment-based methods have to employ the pre-trained human parsing models to achieve the pixel-level alignment, and cannot identify the personal belongings (e.g., backpacks and reticule) which are crucial to person re-ID. In this paper, we propose the identity-guided human semantic parsing approach (ISP) to locate both the human body parts and personal belongings at pixel-level for aligned person re-ID only with person identity labels. We design the cascaded clustering on feature maps to generate the pseudo-labels of human parts. Specifically, for the pixels of all images of a person, we first group them to foreground or background and then group the foreground pixels to human parts. The cluster assignments are subsequently used as pseudo-labels of human parts to supervise the part estimation and ISP iteratively learns the feature maps and groups them. Finally, local features of both human body parts and personal belongings are obtained according to the self-learned part estimation, and only features of visible parts are utilized for the retrieval. Extensive experiments on three widely used datasets validate the superiority of ISP over lots of state-of-the-art methods. Our code is available at https://github.com/CASIA-IVA-Lab/ISP-reID.

Keywords: Person re-ID · Weakly-supervised human parsing · Aligned representation learning

1 Introduction

Person re-identification (re-ID), which aims to associate the person images captured by different cameras from various viewpoints, has attracted increasing attention from both the academia and the industry. However, the task of person re-ID is inherently challenging because of the ubiquitous misalignment issue, which is commonly caused by part occlusions, inaccurate person detection, human pose variations or camera viewpoints changing. All these factors can significantly change the visual appearance of a person in images and greatly increase the difficulty of this retrieval problem.

A. Vedaldi et al. (Eds.): ECCV 2020, LNCS 12348, pp. 346–363, 2020.
https://doi.org/10.1007/978-3-030-58580-8_21

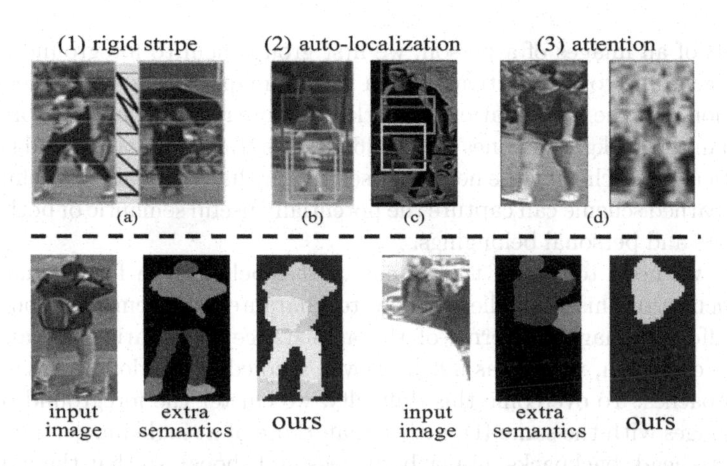

Fig. 1. The alignment-based methods. From (a) to (d): AlignedReID [50], MSCAN [20], DPL [49], MHN [3]. The extra semantic in the second row is predicted by the pre-trained parsing model [21], which exclude the personal belongings and are error-prone when one person is occluded by another. Our method is the first extra semantic free method which can locate both the human body parts and personal belongings at pixel-level, and explicitly identify the visible parts in an occluded image

In recent years, plenty of efforts have been made to alleviate the misalignment problem. The extra semantic free methods try to address the misalignment issue through a self-learned style. However, they can only achieve coarse alignment at region-level. These methods could be roughly summarized to the following streams: (1) The rigid stripe based methods, which directly partition the person image into fixed horizontal stripes [36,41,50,53]. (2) The auto-localization based methods, which try to locate human parts through the learned grids [20,49,52]. (3) The attention based methods, which construct the part alignment through enhancing the discriminative regions and suppressing the background [3,23,48, 56]. Most of the above methods are coarse with much background noise in their located parts and do not consider the situation that some human parts disappear in an image due to occlusion. The first row of Fig. 1 illustrates these streams.

The extra semantic based methods inject extra semantic in terms of part/pose to achieve the part alignment at pixel-level [18,24,31,51]. Their success heavily counts on the accuracy of the extra pre-trained human parsing models or pose estimators. Most importantly, the identifiable personal belongings (e.g., backpacks and reticule), which are the potentially useful contextual cues for identifying a person, cannot be recognized by these pre-trained models and discarded as background. The failure cases of the extra semantic based methods are shown in the second row of Fig. 1.

In this paper, we propose an extra semantic free method, Identity-guided Semantic Parsing (ISP), which can locate both human body parts and potential personal belongings at pixel-level only with the person identity labels. Specifically, we design the cascaded clustering on feature maps and regard the cluster assignments as the pseudo-labels of human parts to supervise the part estimation. For the pixels of all images of a person, we first group them to foreground or background according to their activations on feature maps, basing on the reasonable assumption that the classification networks are more responsive to the foreground pixels than the background ones [40,45,46]. In this stage, the foreground parts are automatically searched by the network itself rather than manually predefined, and the self-learned scheme can capture the potentially useful semantic of both human body parts and personal belongings.

Next, we need to assign the human part labels to the foreground pixels. The difficulty of this stage lies in how to guarantee the semantic consistency across different images in terms of the appearance/pose variations, and especially the occlusion, which has not been well studied in previous extra semantic free approaches. To overcome this difficulty, we cluster the foreground pixels of all the images with the same ID, rather than those of a single image, into human parts (e.g., head, backpacks, upper-body, legs and shoes), so that the number of assigned semantic parts of a single image can adaptively vary when the instance is occluded. Consequently, our scheme is robust to the occlusion and the assigned pseudo-labels of human parts across different images are ensured to be semantically consistent. The second row of Fig. 1 shows the assigned pseudo-labels.

We iteratively cluster the pixels of feature maps and employ the cluster assignments as pseudo-labels of human parts to learn the part representations. In this iterative mechanism, the generated pseudo-labels become finer and finer, resulting in the more and more accurate part estimation. The predicted probability maps of part estimation are then used to conduct the part pooling for partial representations of both human body parts and personal belongings. During matching, we only consider local features of the shared-visible parts between probe and gallery images. Besides, ISP is a generally applicable and backbone-agnostic approach, which can be readily applied in popular networks.

We summarize the contributions of this work as follows:

- In this paper, we propose the identity-guided human semantic parsing approach (ISP) for aligned person re-ID, which can locate both the human body parts and personal belongings (e.g., backpacks and reticule) at pixel-level only with the image-level supervision of person identities.
- To the best of our knowledge, ISP is the first extra semantic free method that can explicitly identify the visible parts from the occluded images. The occluded parts are excluded and only features of the shared-visible parts between probe and gallery images are considered during the feature matching.
- We set the new state-of-the-art performance on three person re-ID datasets, Market-1501 [55], DukeMTMC-reID [57] and CUHK03-NP [22,58].

2 Related Work

2.1 Semantic Learning with Image-Level Supervision

To the best of our knowledge, there is no previous work to learning human semantic parsing with image-level supervision but only weakly-supervised methods for semantic segmentation [11,12,16,19,32,42,45,60], which aim to locate objects like person, horse or dog at pixel-level with image-level supervision. However, all these methods cannot be used for the weakly-supervised human parsing task because they focus on different levels. Besides, their complex network structures and objective functions are not suitable for the end-to-end learning of person re-ID. Therefore, we draw little inspiration from these methods.

2.2 Alignment-Based Person Re-ID

The alignment-baed methods can be roughly summarized to the four streams:

Rigid Stripe Based Approaches. Some researchers directly partition the person image into rigid horizontal stripes to learn local features [36,41,50,53]. Wang et al. [41] design a multiple granularity network, which contains horizontal stripes of different granularities. Zhang et al. [50] introduce a shortest path loss to align rigidly divided local stripes. However, the stripe-based partition is too coarse to well align the human parts and introduces lots of background noise.

Auto-Localization Based Approaches. A few works have been proposed to automatically locate the discriminative parts by incorporating a regional selection sub-network [20,49]. Li et al. [20] exploit the STN [17] for locating latent parts and subsequently extract aligned part features. However, the located grids of latent parts are still coarse and with much overlap. Besides, they produce a fixed number of latent parts, which cannot handle the occluded images.

Attention Based Approaches. Attention mechanism constructs alignment by suppressing background noise and enhancing the discriminative regions [3,23,30,39,48,56]. However, these methods cannot explicitly locate the semantic parts and the consistency of focus area between images is not guaranteed.

Extra Semantic Based Approaches. Many works employ extra semantic in terms of part/pose to locate body parts [6,26,28,29,31,33,38,44,47,54] and try to achieve the pixel-level alignment. Kalayeh et al. [18] propose to employ pre-trained human parsing model to provide extra semantic. Zhang et al. [51] further adopt DensePose [1] to get densely semantic of 24 regions for a person. However, the requiring of extra semantic limits the utility and robustness of these methods. First, the off-the-shelf models can make mistakes in semantic estimation and these methods cannot recorrect the mistakes throughout the training. Second, the identifiable personal belongings like backpacks and reticule, which are crucial for person re-ID, cannot be recognized and ignored as background.

In this paper, we adopt the clustering to learn the human semantic parsing only with person identity labels, which can locate both human body parts and personal belongings at pixel-level. Clustering is a classical unsupervised learning method that groups similar features, while its capability has not been fully explored in the end-to-end training of deep neural networks. Recently, Mathilde et al. [2] adopt clustering to the end-to-end unsupervised learning of image classification. Lin et al. [25] also use clustering to solve the unsupervised person re-ID task. Different from them, we go further by grouping pixels to human parts to

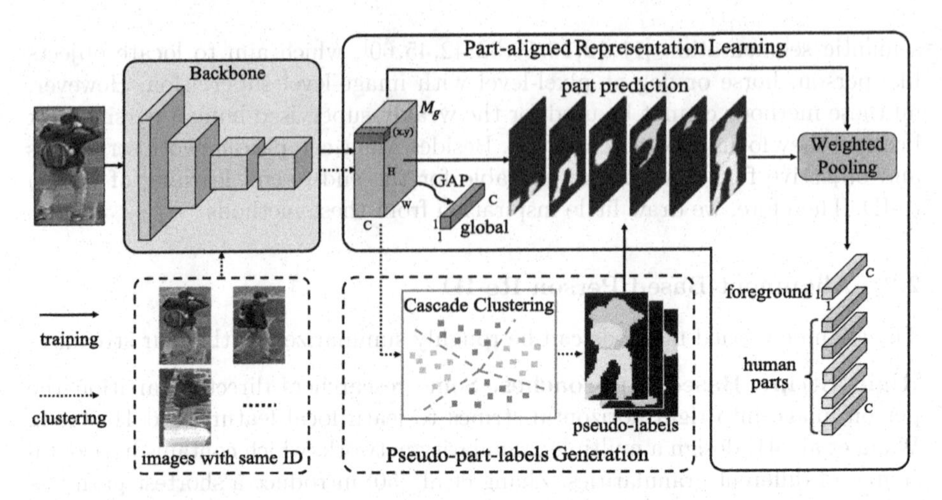

Fig. 2. The overview of ISP. The *solid line* represents the training phase and the *dotted line* represents the clustering phase. The two stages are iteratively done until the network converges. ISP is a generally applicable and backbone-agnostic approach.

generate the pseudo-part-labels at pixel-level, which is more challenging due to various noises. Moreover, the results of clustering must guarantee the semantic consistency across images.

3 Methodology

The overview of ISP is shown in Fig. 2. There are mainly two processes in our approach, i.e., pseudo-part-labels generation and part-aligned representation learning. We repeat the above two processes until the network converges.

3.1 Pixel-Level Part-Aligned Representation Learning

Given n training person images $\{X_i\}_{i=1}^n$ from n_{id} distinct people and their identity labels $\{y_i\}_{i=1}^n$ (where $y_i \in \{1, ..., n_{id}\}$), we could learn the human semantic parsing to obtain the pixel-level part-aligned representations for person re-ID. For image x_i, the backbone mapping function (defined as f_θ) will output the global feature map:

$$M_g^{c \times h \times w} = f_\theta(x_i) \tag{1}$$

where θ is the parameters of backbone, and c, h, w is the channel, height and width. For clear exposition, we omit the channel dimension and denote by $M_g(x, y)$ the feature at spatial position (x, y), which is a vector of c-dim.

The main idea of our pixel-level part-aligned representations is to represent human parts with the representations of pixels belonging to that part, which is the aggregation of the pixel-wise representations weighted by a set of confidence maps. Each confidence map is used to surrogate a human part. Assuming there are $K-1$ human parts and one background part in total, we need to estimate K confidence maps of different semantic parts for every person image. It should be noted that we treat the personal belongings as one category of human parts. The K confidence maps is defined as $P_0, P_1, ..., P_{K-1}$, where each confidence map P_k is associated with a semantic part. We denote by $P_k(x, y)$ the confidence of pixel (x, y) belonging to semantic part k. Then the feature map of part k can be extracted from the global feature map by:

$$M_k = P_k \circ M_g \tag{2}$$

where $k \in \{0, ..., K-1\}$ and $\circ$ is the element-wise product. Adding M_k from $k=1$ to $k=K-1$ in element-wise will get the foreground feature map M_f. Ideally, for the occluded part k in an occluded person image, $\forall_{(x,y)} P_k(x, y) = 0$ should be satisfied, which is reasonable that the network should not produce representations for the invisible parts.

3.2 Cascaded Clustering for Pseudo-Part-Labels Generation

Existing studies integrate human parsing results to help capture the human body parts at pixel-level [18,24,31]. However, there are still many useful contextual cues like backpacks and reticule that do not fall into the scope of manually predefined human body parts. We design the cascaded clustering on feature maps M_g to generate the pseudo-labels of human parts, which includes both human body parts and personal belongings.

Specifically, in the first stage, for all M_g of the same person, we group their pixels into the foreground or background according to the activation, basing on the conception that the foreground pixels have a higher response than background ones [40,45,46]. In this stage, the discriminative foreground parts are automatically searched by the network and the self-learned scheme could apply both the human body parts and the potential useful personal belongings with high response. We regard the l_2-norm of $M_g(x, y)$ as the activation of pixel (x, y). For all pixels of a M_g, we normalize their activations with their maximum:

$$a(x, y) = \frac{||M_g(x, y)||_2}{\max_{(i,j)} ||M_g(i, j)||_2} \tag{3}$$

where (i, j) is the positions in the M_g and the maximum of $a(x, y)$ equals to 1.

In the second stage, we cluster all the foreground pixels assigned by the first clustering step into $K-1$ semantic parts. The number of semantic parts for a single image could be less than $K-1$ when the person is occluded because the cluster samples are foreground pixels of all M_g from the images of the same person, rather than M_g of a single image. Therefore, the clustering is robust to the occlusion and the part assignments across different images are ensured to be semantically consistent. In this stage, we focus on the similarities and differences between pixels rather than activation thus l_2-normalization is used:

$$D(x, y) = \frac{M_g(x, y)}{||M_g(x, y)||_2} \tag{4}$$

The cluster assignments are then used as the pseudo-labels of human parts, which contain the personal belongings as one foreground part, to supervise the learning of human semantic parsing. We assign label 0 to background and the body parts are assigned to label $\{1, ..., K-1\}$ according to the average position from top to down. ISP iteratively does the cascade clustering on feature maps and uses the assignments as pseudo-part-labels to learn the partial representations. In this iterative mechanism, the generated pseudo-labels become finer and finer, resulting in more and more accurate part estimation for aligned person re-ID.

Optimization. For part prediction, we use a linear layer followed by softmax activation as the classifier, which is formulated as:

$$P_k(x,y) = softmax(W_k^T M_g(x,y)) = \frac{\exp(W_k^T M_g(x,y))}{\sum_{i=0}^{K-1} \exp(W_i^T M_g(x,y))} \tag{5}$$

where $k \in \{0, ..., K-1\}$ and W is the parameters of linear layer.

We assign the probability $P_k(x,y)$ as the confidence of pixel (x,y) belonging to the semantic part k and employ cross-entropy loss to optimize the classifier:

$$\mathcal{L}_{parsing} = \sum_{x,y} -\log P_{k_i}(x,y) \tag{6}$$

where k_i is the generated pseudo-label of human parts for pixel (x,y).

3.3 Objective Function

The representation for semantic part k is obtained by $F_k = GAP(M_k)$, where GAP means global average pooling. We concatenate all F_k except $k = 0$ and regard the outcome as a whole representation of local parts for training. Besides, the representations for foreground and global image are directly obtained by $F_f = GAP(M_f)$, $F_g = GAP(M_g)$. In fact, the probability map product together with the GAP is the operation of weighted pooling as indicated in Fig. 2.

In the training phase, we employ three groups of basic losses for the representations of local part, foreground and global image separately, which are denoted as $\mathcal{L}_p$, $\mathcal{L}_f$ and $\mathcal{L}_g$. For each basic loss group, we follow [27] to combine the triplet loss [10] and cross-entropy loss with label smoothing [37]. Therefore, the overall objective function is:

$$\mathcal{L}_{reid} = \mathcal{L}_p + \mathcal{L}_f + \mathcal{L}_g + \alpha \mathcal{L}_{parsing} \tag{7}$$

where α is the balanced weight and is set to 0.1 in our experiments.

3.4 Aligned Representation Matching

As illustrated in Fig. 3, the final distance between query and gallery images consists of two parts. One is the distance of global and foreground features, which always exist. The other is the distance of the partial features between the shared-visible human parts. The matching strategy is inspired by [28], but [28]

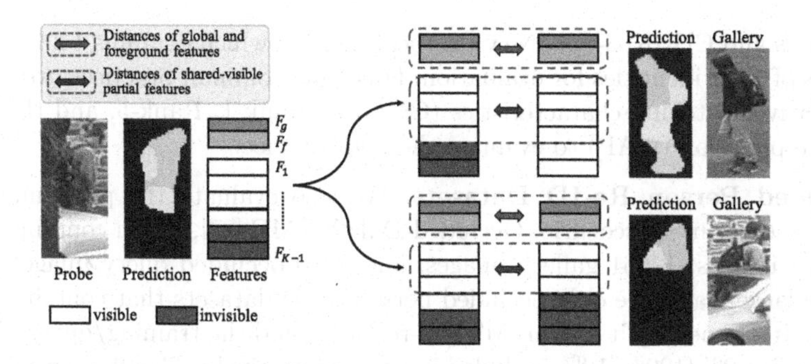

Fig. 3. The matching strategy of ISP. The distance between probe and gallery images are measured by features of the global image and foreground part, which always exist, and the features of shared-visible parts.

utilizes extra pose information and only achieves stripe-level alignment, while we do not require any extra semantic and could identify the visible parts at pixel-level. As the $\text{argmax}_i P_i(x, y)$ indicates the semantic part of pixel (x, y) belonging to, we could easily obtain the label of whether part k is visible $l_k \in \{0, 1\}$ by:

$$l_k = \begin{cases} 1, & \text{if } \exists (x, y) \in \{(x, y) | \text{argmax}_i P_i(x, y) = k\} \\ 0, & \text{else} \end{cases} \quad (i = 0, ..., K-1) \tag{8}$$

Now the distance d_k of the kth part between query and gallery images is:

$$d_k = D(F_k^q, F_k^g) \quad (k = 1, ..., K-1) \tag{9}$$

where $D()$ denotes the distance metric, which is cosine distance in this paper. F_k^q, F_k^g denote the kth partial feature of the query and gallery image, respectively. Similarly, the measure distance between global and foreground features are formulated as: $d_g = D(F_g^q, F_g^g)$, $d_f = D(F_f^q, F_f^g)$. Then, the final distance d could be obtained by:

$$d = \frac{\sum_{k=1}^{K-1} (l_k^q \cdot l_k^g) d_k + (d_g + d_f)}{\sum_{k=1}^{K-1} (l_k^q \cdot l_k^g) + 2} \tag{10}$$

If the kth parts of both the query and gallery images are visible, $l_k^q \cdot l_k^g = 1$. Else, $l_k^q \cdot l_k^g = 0$. To the best of our knowledge, ISP is the first extra semantic free method that explicitly addresses the occlusion problem.

4 Experiments

4.1 Datasets and Evaluation Metrics

Holistic Person Re-ID Datasets. We select three widely used holistic person re-ID benchmarks, Market-1501 [55] which contains 32668 person images of 1501 identities, DukeMTMC-reID [57] which contains 36411 person images of 1402 identities and CUHK03-NP (New Protocol) [22,58] which contains 14097 person images of 1467 identities for evaluation. Following common practices, we use the cumulative matching characteristics (CMC) at Rank-1, Rank-5, and the mean average precision (mAP) to evaluate the performance.

Occluded Person Re-ID Datasets. We also evaluate the performance of ISP in the occlusion scenario. Occluded-DukeMTMC [28], which contains 15618 training images, 17661 gallery images, and 2210 occluded query images, is by far the largest and the only occluded person re-ID datasets that contains training set. It is a new split of DukeMTMC-reID [57] and the training/query/gallery set contains 9%/100%/10% occluded images, respectively. Therefore, we demonstrate the effectiveness of ISP in occluded scenario on this dataset.

4.2 Implementation Details

Data Preprocessing. The input images are resized to 256 × 128 and the global feature map M_g is 1/4 of the input size. As for data augmentation, we adopt the commonly used random cropping [43], horizontal flipping and random erasing [39,43,59] (with a probability of 0.5) in both the baseline and our schemes.

Optimization. The backbone network is initialized with the pre-trained parameters on ImageNet [4]. We warm up the model for 10 epochs with a linearly growing learning rate from 3.5×10^{-5} to 3.5×10^{-4}. Then, the learning rate is decreased by a factor of 0.1 at 40th and 70th epoch. We observe that 120 epochs are enough for model converging. The batch size is set to 64 and adam method is adopted to optimize the model. All our methods are implemented on PyTorch.

Clustering for Reassignment. We adopt k-means as our clustering algorithm and reassign the clusters every n epochs, which is a tradeoff between the parameter updating and the pseudo-label generation. We find out that simply setting $n = 1$ is nearly optimal. We do not define any initial pseudo-labels for person images and the first clustering is directly conducted on the feature maps output by the initialized backbone. As for the time consumption, to train a model, the overall clustering time is about 6.3 h/5.4 h/3.1 h for datasets of DukeMTMC-reID/Market1501/CUHK03-NP through multi-processes with one NVIDIA TITAN X GPU. Most importantly, the testing time is not increased at all.

4.3 Comparison with State-of-the-Art Methods

We compare our method with the state-of-the-art methods for holistic and occluded person re-ID in Table 1 and Table 2, respectively.

DukeMTMC-reID. ISP achieves the best performance and outperforms others by at least 0.5%/1.6% in Rank-1/mAP. On this dataset, the semantic extracted by pre-trained model is error-prone [51], which leads to significant performance degradation for extra semantic based methods. This also proves the learned semantic parts are superior to the outside ones in robustness.

Table 1. Comparison with state-of-the-art methods of the holistic re-ID problem. The $1^{st}/2^{nd}$ results are shown in italic/bold, respectively. The methods in the 1st group are rigid stripe based. The methods in the 2nd group are auto-localization based. The 3rd group is attention based methods. The methods in the 4th group are extra semantic based. The last line is our method

Methods	Ref	DukeMTMC			Market1501			CUHK03-NP			
		R-1	R-5	mAP	R-1	R-5	mAP	R-1	mAP	R-1	mAP
AlignedReID [50]	Arxiv18	–	–	–	91.8	97.1	79.3	–	–	–	–
PCB+RPP [36]	ECCV18	83.3	–	69.2	93.8	97.5	81.6	–	–	63.7	57.5
MGN [41]	MM18	88.7	–	**78.4**	*95.7*	–	86.9	68.0	67.4	66.8	66.0
MSCAN [20]	CVPR17	–	–	–	80.8	–	57.5	–	–	–	–
PAR [52]	ICCV17	–	–	–	81.0	92.0	63.4	–	–	–	–
DuATM [30]	CVPR18	81.8	90.2	64.6	91.4	97.1	76.6	–	–	–	–
Mancs [39]	ECCV18	84.9	–	71.8	93.1	–	82.3	69.0	63.9	65.5	60.5
IANet [13]	CVPR19	87.1	–	73.4	94.4	–	83.1	–	–	–	–
CASN+PCB [56]	CVPR19	87.7	–	73.7	94.4	–	82.8	73.7	68.0	71.5	64.4
CAMA [48]	CVPR19	85.8	–	72.9	94.7	98.1	84.5	70.1	66.5	66.6	64.2
MHN-6 [3]	ICCV19	**89.1**	**94.6**	77.2	95.1	98.1	85.0	77.2	72.4	71.7	65.4
SPReID [18]	CVPR18	84.4	–	71.0	92.5	–	81.3	–	–	–	–
PABR [34]	ECCV18	84.4	92.2	69.3	91.7	96.9	79.6	–	–	–	–
AANet [38]	CVPR19	87.7	–	74.3	93.9	–	83.4	–	–	–	–
DSA-reID [51]	CVPR19	86.2	–	74.3	*95.7*	-	**87.6**	*78.9*	*75.2*	*78.2*	*73.1*
P^2-Net [6]	ICCV19	86.5	93.1	73.1	95.2	**98.2**	85.6	**78.3**	73.6	74.9	68.9
PGFA [28]	ICCV19	82.6	–	65.5	91.2	–	76.8	–	–	–	–
ISP (ours)	ECCV20	*89.6*	*95.5*	*80.0*	**95.3**	*98.6*	*88.6*	76.5	**74.1**	**75.2**	**71.4**

Market1501. ISP achieves the best performance on mAP accuracy and the second best on Rank-1. We further find that the improvement of mAP score brought by ISP is larger than that of Rank-1, which indicates ISP effectively advances the ranking positions of misaligned person images as mAP is a comprehensive index that considers all the ranking positions of the target images.

CUHK03-NP. ISP achieves the second best results. In CUHK03-NP, a great many of images contain incomplete person bodies and some human parts even disappear from all the images of a person. But ISP requires at least every semantic part appears once for a person to guarantee a high consistency on semantic. Even so, ISP still outperforms all other approaches except DSA-reID [51] which employs extra supervision by a pre-trained DensePose model [1], while our method learns the pixel-level semantic without any extra supervision.

Occluded-DukeMTMC. ISP sets a new state-of-the-art performance and outperforms others by a large margin, at least 11.2%/14.3% in Rank-1/mAP. ISP could explicitly identify the visible parts at pixel-level from the occluded images and only the shared-visible parts between query and gallery images are considered during the feature matching, which greatly improves the performance. As shown in Table 2, the aligned representation matching strategy brings considerable improvement, e.g., 3.3% of Rank-1 and 0.9% of mAP.

Table 2. Comparison with state-of-the-art methods of the occluded re-ID problem on Occluded-DukeMTMC. Methods in the 1st group are for the holistic re-ID problem. Methods in the 2nd group utilize extra pose information for occluded re-ID problem. The methods in the 3rd group do not adopt extra semantic. The last line is our method

Methods	Rank-1	Rank-5	Rank-10	mAP
HACNN [23]	34.4	51.9	59.4	26.0
Adver Occluded [15]	44.5	–	–	32.2
PCB [36]	42.6	57.1	62.9	33.7
Part Bilinear [34]	36.9	–	–	–
FD-GAN [5]	40.8	–	–	–
PGFA [28]	51.4	68.6	74.9	37.3
DSR [8]	40.8	58.2	65.2	30.4
SFR [9]	42.3	60.3	67.3	32.0
$\text{ISP}_{w/oarm}$	59.5	73.5	78.0	51.4
ISP (ours)	**62.8**	**78.1**	**82.9**	**52.3**

4.4 The Performance of the Learned Human Semantic Parsing

As there are no manual part labels for the person re-ID datasets, we adopt the state-of-the-art parsing model SCHP [21] pre-trained on Look into Person (LIP) [24] to create the "ground-truth" of four parts.[1] Then we adopt segmentation Intersection over Union (IoU) to evaluate both the accuracy of the pseudo-part-labels of the training set and that of the semantic estimation on the testing set, which are detailed in Table 3, and the results are high enough for the IoU evaluation metric. We also find an interesting phenomenon that the accuracy of the predicted semantic estimation on testing set is mostly higher than that of the pseudo-part-labels for the training set, which indicates that, with the person re-ID supervision, our part estimator is robust to the false pseudo-labels and obtains an enhanced generalization capability.

Table 3. The human semantic parsing performance (%) of ISP ($K = 6$)

IoU	Datasets	Foreground	Head	Legs	Shoes
Pseudo-labels accuracy	DukeMTMC	65.66	68.17	61.83	58.89
	Market1501	65.45	54.74	67.02	55.25
	CUHK03-NP-Labeled	51.26	68.21	52.24	57.60
Prediction accuracy	DukeMTMC	66.94	71.35	68.02	62.60
	Market1501	63.44	55.78	69.10	56.32
	CUHK03-NP-Labeled	53.51	59.96	50.14	59.08

[1] The parts of hat, hair, sunglass, face in LIP is aggregated as the "ground-truth" for Head; the parts of left-leg, right-leg, socks, pants are aggregated as Legs; the parts of left-shoe and right-shoe are aggregated as Shoes.

We further conduct three visualization experiments to show the effect of ISP. First, we visualize our pseudo-part-labels under different K and show the comparison with SCHP [21] in Fig. 4, which validates that ISP can recognize the personal belongings (e.g. backpacks and reticule) as one human part while the pre-trained parsing model cannot. The first row of Fig. 4 shows our capability of explicitly locating the visible parts in occluded images. Moreover, Fig. 4 also validates the semantic consistency in ISP.

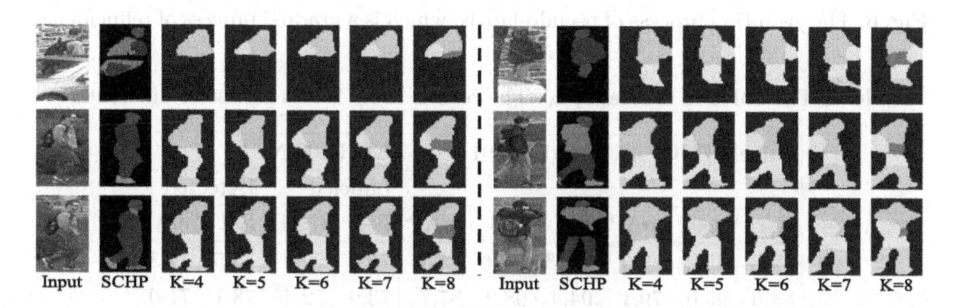

Fig. 4. The assigned pseudo-labels of human parts. From left to right: input images, semantic estimation by SCHP [21], the assigned pseudo-labels with different K.

Second, to validate the necessity of the cascaded clustering, we compare the pseudo-part-labels by cascaded clustering with Variant 1 which directly clusters the semantic in one step, and Variant 2 which removes the l_2 normalization. Figure 5 indicates that the alignment of Variant 1 is coarse and error-prone, and Variant 2 assigns an unreasonable semantic part with sub-response surround human bodies, which indicates the clustering is influenced by the activation.

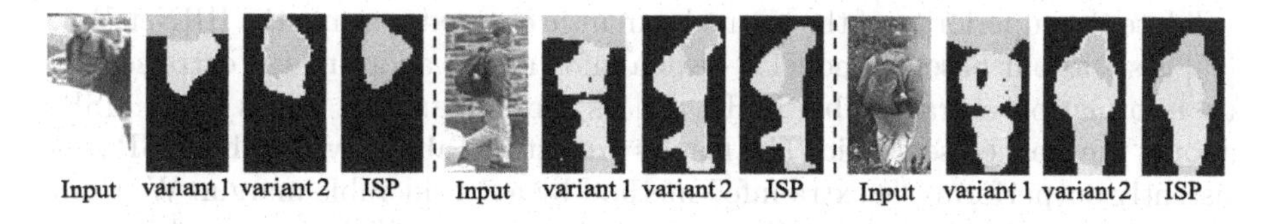

Fig. 5. The necessity of the operations of cascaded clustering ($K = 6$).

Finally, Fig. 6 shows the evolution process of the pseudo-part-labels ($K = 6$), which presents a clear process of coarse-to-fine. The first clustering is directly conducted on feature maps output by the initialized network.

4.5 Ablation Studies on Re-ID Performance

Choice of K Clustering Categories. Intuitively, the number of clustering centers K determines the granularity of the aligned parts. We perform the quantitative ablation studies to clearly find the most suitable K. As detailed in Table 4,

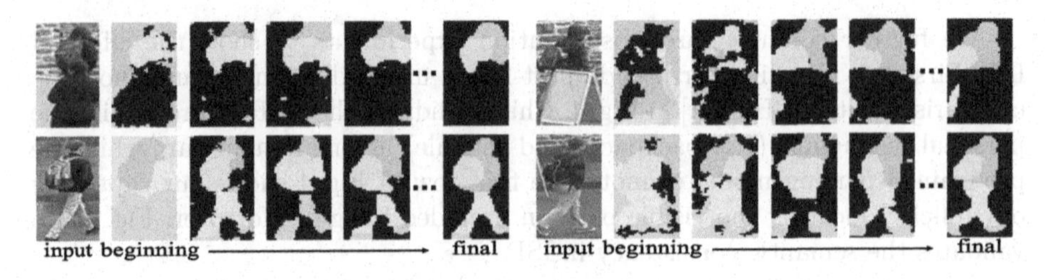

Fig. 6. The evolution process of pseudo-labels, which is a gradual process of refinement.

Table 4. The ablation studies of K. The results show ISP is robust to different K

K	DukeMTMC			Market1501			CUHK03-NP			
							Labeled		Detected	
	R-1	R-5	mAP	R-1	R-5	mAP	R-1	mAP	R-1	mAP
4	89.6	95.2	79.2	**95.3**	**98.6**	**88.6**	75.9	72.9	73.5	71.3
5	88.6	95.0	79.1	94.4	98.2	87.7	73.9	72.4	73.1	71.0
6	89.0	95.1	78.9	95.2	98.4	88.4	75.9	73.8	**75.2**	**71.4**
7	**89.6**	**95.5**	**80.0**	95.0	98.2	88.4	**76.5**	**74.1**	73.6	70.8
8	88.9	94.7	78.4	94.9	98.5	88.6	75.9	73.1	74.0	71.4

the performance of ISP is robust to different K. Besides, we also find that $K = 5$ is always the worst, which is consistent with its lowest accuracy of semantic parsing. For example, $K = 5$ only obtains the pseudo-labels accuracy (IoU) of 64.25%, 53.23% and 54.83% for foreground, legs and shoes on DukeMTMC-reID.

Learned Semantic vs. Extra Semantic. We further conduct experiments to validate the superiority of the learned semantic over extra semantic. HRNet-W32 [35] is set as our baseline model. "+extra info" means adopting the extra semantic information extracted by SCHP [21] as the human part labels while "ISP" adopts the learned semantic. The results are list in Table 5, which show ISP consistently outperforms "+extra info" method by a considerable margin. We think this is mainly because: (1) ISP can recognize the identifiable personal belongs while "+extra info" cannot. (2) "+extra info" cannot recorrect the semantic estimation errors throughout the training while ISP can recorrect its mistakes every epoch, thus ISP is less likely to miss the key clues.

Choice of Backbone Architecture. As ISP is a backbone-agnostic approach, we show the effectiveness of ISP with different backbones including ResNet [7], SeResNet [14] and HRNet [35]. A bilinear upsample layer is added to scale up the final feature maps of ResNet50 and SeResNet50 to the same size of HRNet-W32. As Table 6 shows, HRNet-W32 obtains the highest performance. We think it is because HRNet maintains high-resolution representations throughout the network, which could contain more semantic information.

Table 5. The comparison of learned semantic and extra semantic

Model	DukeMTMC			Market1501			CUHK03-NP			
							Labeled		Detected	
	R-1	R-5	mAP	R-1	R-5	mAP	R-1	mAP	R-1	mAP
Baseline	87.7	94.3	77.2	94.0	97.9	85.9	71.9	68.5	67.6	64.7
+extra info	88.6	94.7	79.1	94.8	98.4	87.7	73.3	71.9	72.2	69.6
ISP	89.6	95.5	80.0	95.3	98.6	88.6	76.5	74.1	75.2	71.4

Table 6. The ablation studies of different backbone networks on DukeMTMC-reID.

Backbone	#params	R-1	R-5	mAP
HRNet-W32	28.5M	89.6	95.5	80.0
ResNet50	25.6M	88.7	94.9	78.9
SeResNet50	28.1M	88.8	95.2	79.2

The Matching Results. We compare the ranking results of baseline and ISP in Fig. 7, which indicates ISP can well overcome the misalignment problem including part occlusions, inaccurate person detection, and human pose variations. Besides, we can also observe the benefit of identifying the personal belongings.

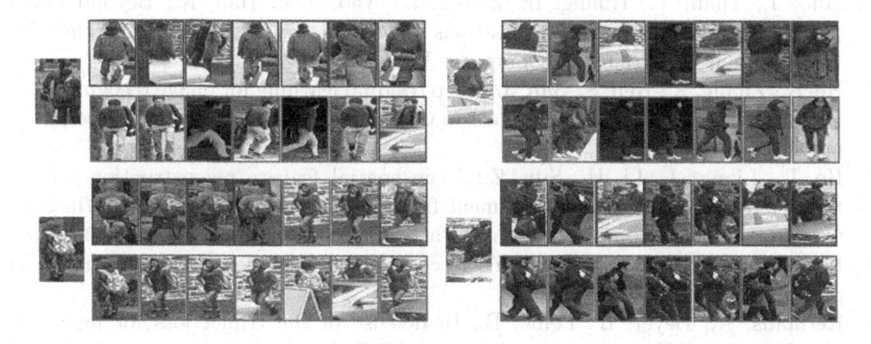

Fig. 7. The ranking results of baseline (the first row) and ISP (the second row).

5 Conclusion

In this paper, we propose the identity-guided human semantic parsing method for aligned person re-identification, which can locate both human body parts and personal belongings at pixel-level only with image-level supervision of person identities. Extensive experiments validate the superiority of our method.

Acknowledgement. This work was supported by Key-Area Research and Development Program of Guangdong Province (No. 2020B010165001), National Natural Science Foundation of China (No. 61772527, 61976210, 61702510), China Postdoctoral Science Foundation No. 2019M660859, Open Project of Key Laboratory of Ministry of Public Security for Road Traffic Safety (No. 2020ZDSYSKFKT04).

References

1. Alp Güler, R., Neverova, N., Kokkinos, I.: DensePose: dense human pose estimation in the wild. In: Proceedings of the IEEE Conference on Computer Vision and Pattern Recognition, pp. 7297–7306 (2018)
2. Caron, M., Bojanowski, P., Joulin, A., Douze, M.: Deep clustering for unsupervised learning of visual features. In: Ferrari, V., Hebert, M., Sminchisescu, C., Weiss, Y. (eds.) Computer Vision – ECCV 2018. LNCS, vol. 11218, pp. 139–156. Springer, Cham (2018). https://doi.org/10.1007/978-3-030-01264-9_9
3. Chen, B., Deng, W., Hu, J.: Mixed high-order attention network for person re-identification. In: Proceedings of the IEEE International Conference on Computer Vision (2019)
4. Deng, J., et al.: ImageNet: a large-scale hierarchical image database. In: 2009 IEEE Conference on Computer Vision and Pattern Recognition, pp. 248–255. IEEE (2009)
5. Ge, Y., Li, Z., Zhao, H., Yin, G., Yi, S., Wang, X., et al.: FD-GAN: pose-guided feature distilling GAN for robust person re-identification. In: Advances in Neural Information Processing Systems, pp. 1222–1233 (2018)
6. Guo, J., Yuan, Y., Huang, L., Zhang, C., Yao, J.G., Han, K.: Beyond human parts: dual part-aligned representations for person re-identification. In: The IEEE International Conference on Computer Vision (ICCV), October 2019
7. He, K., Zhang, X., Ren, S., Sun, J.: Deep residual learning for image recognition. In: Proceedings of the IEEE Conference on Computer Vision and Pattern Recognition, pp. 770–778 (2016)
8. He, L., Liang, J., Li, H., Sun, Z.: Deep spatial feature reconstruction for partial person re-identification: alignment-free approach. In: Proceedings of the IEEE Conference on Computer Vision and Pattern Recognition, pp. 7073–7082 (2018)
9. He, L., Sun, Z., Zhu, Y., Wang, Y.: Recognizing partial biometric patterns. arXiv preprint arXiv:1810.07399 (2018)
10. Hermans, A., Beyer, L., Leibe, B.: In defense of the triplet loss for person re-identification. arXiv preprint arXiv:1703.07737 (2017)
11. Hong, S., Noh, H., Han, B.: Decoupled deep neural network for semi-supervised semantic segmentation. In: Advances in Neural Information Processing Systems, pp. 1495–1503 (2015)
12. Hong, S., Yeo, D., Kwak, S., Lee, H., Han, B.: Weakly supervised semantic segmentation using web-crawled videos. In: Proceedings of the IEEE Conference on Computer Vision and Pattern Recognition, pp. 7322–7330 (2017)
13. Hou, R., Ma, B., Chang, H., Gu, X., Shan, S., Chen, X.: Interaction-and-aggregation network for person re-identification. In: The IEEE Conference on Computer Vision and Pattern Recognition (CVPR), June 2019
14. Hu, J., Shen, L., Sun, G.: Squeeze-and-excitation networks. In: Proceedings of the IEEE Conference on Computer Vision and Pattern Recognition, pp. 7132–7141 (2018)

15. Huang, H., Li, D., Zhang, Z., Chen, X., Huang, K.: Adversarially occluded samples for person re-identification. In: Proceedings of the IEEE Conference on Computer Vision and Pattern Recognition, pp. 5098–5107 (2018)
16. Huang, Z., Wang, X., Wang, J., Liu, W., Wang, J.: Weakly-supervised semantic segmentation network with deep seeded region growing. In: Proceedings of the IEEE Conference on Computer Vision and Pattern Recognition, pp. 7014–7023 (2018)
17. Jaderberg, M., Simonyan, K., Zisserman, A., et al.: Spatial transformer networks. In: Advances in Neural Information Processing Systems, pp. 2017–2025 (2015)
18. Kalayeh, M.M., Basaran, E., Gökmen, M., Kamasak, M.E., Shah, M.: Human semantic parsing for person re-identification. In: Proceedings of the IEEE Conference on Computer Vision and Pattern Recognition, pp. 1062–1071 (2018)
19. Lee, J., Kim, E., Lee, S., Lee, J., Yoon, S.: FickleNet: weakly and semi-supervised semantic image segmentation using stochastic inference. In: Proceedings of the IEEE Conference on Computer Vision and Pattern Recognition, pp. 5267–5276 (2019)
20. Li, D., Chen, X., Zhang, Z., Huang, K.: Learning deep context-aware features over body and latent parts for person re-identification. In: Proceedings of the IEEE Conference on Computer Vision and Pattern Recognition, pp. 384–393 (2017)
21. Li, P., Xu, Y., Wei, Y., Yang, Y.: Self-correction for human parsing. arXiv preprint arXiv:1910.09777 (2019)
22. Li, W., Zhao, R., Xiao, T., Wang, X.: DeepReId: deep filter pairing neural network for person re-identification. In: Proceedings of the IEEE Conference on Computer Vision and Pattern Recognition, pp. 152–159 (2014)
23. Li, W., Zhu, X., Gong, S.: Harmonious attention network for person re-identification. In: Proceedings of the IEEE Conference on Computer Vision and Pattern Recognition, pp. 2285–2294 (2018)
24. Liang, X., Gong, K., Shen, X., Lin, L.: Look into person: joint body parsing & pose estimation network and a new benchmark. IEEE Trans. Pattern Anal. Mach. Intell. **41**(4), 871–885 (2018)
25. Lin, Y., Dong, X., Zheng, L., Yan, Y., Yang, Y.: A bottom-up clustering approach to unsupervised person re-identification. In: Proceedings of the AAAI Conference on Artificial Intelligence, vol. 33, pp. 8738–8745 (2019)
26. Liu, J., Ni, B., Yan, Y., Zhou, P., Cheng, S., Hu, J.: Pose transferrable person re-identification. In: Proceedings of the IEEE Conference on Computer Vision and Pattern Recognition, pp. 4099–4108 (2018)
27. Luo, H., Gu, Y., Liao, X., Lai, S., Jiang, W.: Bag of tricks and a strong baseline for deep person re-identification. In: Proceedings of the IEEE Conference on Computer Vision and Pattern Recognition Workshops (2019)
28. Miao, J., Wu, Y., Liu, P., Ding, Y., Yang, Y.: Pose-guided feature alignment for occluded person re-identification. In: Proceedings of the IEEE International Conference on Computer Vision, pp. 542–551 (2019)
29. Saquib Sarfraz, M., Schumann, A., Eberle, A., Stiefelhagen, R.: A pose-sensitive embedding for person re-identification with expanded cross neighborhood re-ranking. In: Proceedings of the IEEE Conference on Computer Vision and Pattern Recognition, pp. 420–429 (2018)
30. Si, J., Zhang, H., Li, C.G., Kuen, J., Kong, X., Kot, A.C., Wang, G.: Dual attention matching network for context-aware feature sequence based person re-identification. In: Proceedings of the IEEE Conference on Computer Vision and Pattern Recognition, pp. 5363–5372 (2018)

31. Song, C., Huang, Y., Ouyang, W., Wang, L.: Mask-guided contrastive attention model for person re-identification. In: Proceedings of the IEEE Conference on Computer Vision and Pattern Recognition, pp. 1179–1188 (2018)
32. Souly, N., Spampinato, C., Shah, M.: Semi and weakly supervised semantic segmentation using generative adversarial network. arXiv preprint arXiv:1703.09695 (2017)
33. Su, C., Li, J., Zhang, S., Xing, J., Gao, W., Tian, Q.: Pose-driven deep convolutional model for person re-identification. In: Proceedings of the IEEE International Conference on Computer Vision, pp. 3960–3969 (2017)
34. Suh, Y., Wang, J., Tang, S., Mei, T., Lee, K.M.: Part-aligned bilinear representations for person re-identification. In: Ferrari, V., Hebert, M., Sminchisescu, C., Weiss, Y. (eds.) Computer Vision – ECCV 2018. LNCS, vol. 11218, pp. 418–437. Springer, Cham (2018). https://doi.org/10.1007/978-3-030-01264-9_25
35. Sun, K., Xiao, B., Liu, D., Wang, J.: Deep high-resolution representation learning for human pose estimation. In: CVPR (2019)
36. Sun, Y., Zheng, L., Yang, Y., Tian, Q., Wang, S.: Beyond part models: person retrieval with refined part pooling (and a strong convolutional baseline). In: Ferrari, V., Hebert, M., Sminchisescu, C., Weiss, Y. (eds.) ECCV 2018. LNCS, vol. 11208, pp. 501–518. Springer, Cham (2018). https://doi.org/10.1007/978-3-030-01225-0_30
37. Szegedy, C., Vanhoucke, V., Ioffe, S., Shlens, J., Wojna, Z.: Rethinking the inception architecture for computer vision. In: Proceedings of the IEEE Conference on Computer Vision and Pattern Recognition, pp. 2818–2826 (2016)
38. Tay, C.P., Roy, S., Yap, K.H.: AANet: attribute attention network for person re-identifications. In: The IEEE Conference on Computer Vision and Pattern Recognition (CVPR), June 2019
39. Wang, C., Zhang, Q., Huang, C., Liu, W., Wang, X.: Mancs: a multi-task attentional network with curriculum sampling for person re-identification. In: Ferrari, V., Hebert, M., Sminchisescu, C., Weiss, Y. (eds.) ECCV 2018. LNCS, vol. 11208, pp. 384–400. Springer, Cham (2018). https://doi.org/10.1007/978-3-030-01225-0_23
40. Wang, F., Jiang, M., Qian, C., Yang, S., Li, C., Zhang, H., Wang, X., Tang, X.: Residual attention network for image classification. In: Proceedings of the IEEE Conference on Computer Vision and Pattern Recognition, pp. 3156–3164 (2017)
41. Wang, G., Yuan, Y., Chen, X., Li, J., Zhou, X.: Learning discriminative features with multiple granularities for person re-identification. In: 2018 ACM Multimedia Conference on Multimedia Conference, pp. 274–282. ACM (2018)
42. Wang, X., You, S., Li, X., Ma, H.: Weakly-supervised semantic segmentation by iteratively mining common object features. In: Proceedings of the IEEE Conference on Computer Vision and Pattern Recognition, pp. 1354–1362 (2018)
43. Wang, Y., et al.: Resource aware person re-identification across multiple resolutions. In: Proceedings of the IEEE Conference on Computer Vision and Pattern Recognition, pp. 8042–8051 (2018)
44. Wei, L., Zhang, S., Yao, H., Gao, W., Tian, Q.: GLAD: global-local-alignment descriptor for pedestrian retrieval. In: Proceedings of the 25th ACM International Conference on Multimedia, pp. 420–428. ACM (2017)
45. Wei, Y., Feng, J., Liang, X., Cheng, M.M., Zhao, Y., Yan, S.: Object region mining with adversarial erasing: a simple classification to semantic segmentation approach. In: Proceedings of the IEEE Conference on Computer Vision and Pattern Recognition, pp. 1568–1576 (2017)

46. Woo, S., Park, J., Lee, J.-Y., Kweon, I.S.: CBAM: convolutional block attention module. In: Ferrari, V., Hebert, M., Sminchisescu, C., Weiss, Y. (eds.) ECCV 2018. LNCS, vol. 11211, pp. 3–19. Springer, Cham (2018). https://doi.org/10.1007/978-3-030-01234-2_1
47. Xu, J., Zhao, R., Zhu, F., Wang, H., Ouyang, W.: Attention-aware compositional network for person re-identification. In: Proceedings of the IEEE Conference on Computer Vision and Pattern Recognition, pp. 2119–2128 (2018)
48. Yang, W., Huang, H., Zhang, Z., Chen, X., Huang, K., Zhang, S.: Towards rich feature discovery with class activation maps augmentation for person re-identification. In: Proceedings of the IEEE Conference on Computer Vision and Pattern Recognition, pp. 1389–1398 (2019)
49. Yao, H., Zhang, S., Hong, R., Zhang, Y., Xu, C., Tian, Q.: Deep representation learning with part loss for person re-identification. IEEE Trans. Image Process. **28**(6), 2860–2871 (2019)
50. Zhang, X., et al.: AlignedReID: surpassing human-level performance in person re-identification. arXiv preprint arXiv:1711.08184v2 (2018)
51. Zhang, Z., Lan, C., Zeng, W., Chen, Z.: Densely semantically aligned person re-identification. In: Proceedings of the IEEE Conference on Computer Vision and Pattern Recognition, pp. 667–676 (2019)
52. Zhao, L., Li, X., Zhuang, Y., Wang, J.: Deeply-learned part-aligned representations for person re-identification. In: Proceedings of the IEEE International Conference on Computer Vision, pp. 3219–3228 (2017)
53. Zheng, F., et al.: Pyramidal person re-identification via multi-loss dynamic training. In: The IEEE Conference on Computer Vision and Pattern Recognition (CVPR), June 2019
54. Zheng, L., Huang, Y., Lu, H., Yang, Y.: Pose invariant embedding for deep person re-identification. IEEE Trans. Image Process. (2019)
55. Zheng, L., Shen, L., Tian, L., Wang, S., Wang, J., Tian, Q.: Scalable person re-identification: a benchmark. In: Computer Vision, IEEE International Conference on Computer Vision, pp. 1116–1124 (2015)
56. Zheng, M., Karanam, S., Wu, Z., Radke, R.J.: Re-identification with consistent attentive siamese networks. In: Proceedings of the IEEE Conference on Computer Vision and Pattern Recognition, pp. 5735–5744 (2019)
57. Zheng, Z., Zheng, L., Yang, Y.: Unlabeled samples generated by GAN improve the person re-identification baseline in vitro. In: Proceedings of the IEEE International Conference on Computer Vision, pp. 3754–3762 (2017)
58. Zhong, Z., Zheng, L., Cao, D., Li, S.: Re-ranking person re-identification with k-reciprocal encoding. In: Proceedings of the IEEE Conference on Computer Vision and Pattern Recognition, pp. 1318–1327 (2017)
59. Zhong, Z., Zheng, L., Kang, G., Li, S., Yang, Y.: Random erasing data augmentation. arXiv preprint arXiv:1708.04896 (2017)
60. Zhou, Y., Zhu, Y., Ye, Q., Qiu, Q., Jiao, J.: Weakly supervised instance segmentation using class peak response. In: Proceedings of the IEEE Conference on Computer Vision and Pattern Recognition, pp. 3791–3800 (2018)

Learning Gradient Fields for Shape Generation

Ruojin Cai[✉], Guandao Yang, Hadar Averbuch-Elor, Zekun Hao, Serge Belongie, Noah Snavely, and Bharath Hariharan

Cornell University, Ithaca, USA
rc844@cornell.edu

Abstract. In this work, we propose a novel technique to generate shapes from point cloud data. A point cloud can be viewed as samples from a distribution of 3D points whose density is concentrated near the surface of the shape. Point cloud generation thus amounts to moving randomly sampled points to high-density areas. We generate point clouds by performing stochastic gradient ascent on an unnormalized probability density, thereby moving sampled points toward the high-likelihood regions. Our model directly predicts the gradient of the log density field and can be trained with a simple objective adapted from score-based generative models. We show that our method can reach state-of-the-art performance for point cloud auto-encoding and generation, while also allowing for extraction of a high-quality implicit surface. Code is available at https://github.com/RuojinCai/ShapeGF.

Keywords: 3D generation · Generative models

1 Introduction

Point clouds are becoming increasingly popular for modeling shapes, as many modern 3D scanning devices process and output point clouds. As such, an increasing number of applications rely on the recognition, manipulation, and synthesis of point clouds. For example, an autonomous vehicle might need to detect cars in sparse LiDAR point clouds. An augmented reality application might need to scan in the environment. Artists may want to further manipulate scanned objects to create new objects and designs. A *prior* for point clouds would be useful for these applications as it can densify LiDAR clouds, create additional training data for recognition, complete scanned objects or synthesize new ones. Such a prior requires a powerful generative model for point clouds.

R. Cai and G. Yang—Equal contribution.

Electronic supplementary material The online version of this chapter (https://doi.org/10.1007/978-3-030-58580-8_22) contains supplementary material, which is available to authorized users.

A. Vedaldi et al. (Eds.): ECCV 2020, LNCS 12348, pp. 364–381, 2020.
https://doi.org/10.1007/978-3-030-58580-8_22

Fig. 1. To generate shapes, we sample points from an arbitrary prior (depicting the letters "E", "C", "C", "V" in the examples above) and move them stochastically along a learned gradient field, ultimately reaching the shape's surface. Our learned fields also enable extracting the surface of the shape, as demonstrated on the right.

In this work, we are interested in learning a generative model that can sample shapes represented as point clouds. A key challenge here is that point clouds are sets of arbitrary size. Prior work often generates a fixed number of points instead [1,16,17,48,64]. This number, however, may be insufficient for some applications and shapes, or too computationally expensive for others. Instead, following recent works [31,52,60], we consider a point cloud as a set of *samples* from an underlying distribution of 3D points. This new perspective not only allows one to generate an arbitrary number of points from a shape, but also makes it possible to model shapes with varying topologies. However, it is not clear how to best parameterize such a distribution of points, and how to learn it using *only* a limited number of sampled points.

Prior research has explored modeling the distribution of points that represent the shape using generative adversarial networks (GANs) [31], flow-based models [60], and autoregressive models [52]. While substantial progress has been made, these methods have some inherent limitations for modeling the distribution representing a 3D shape. The training procedure can be unstable for GANs or prohibitively slow for invertible models, while autoregressive models assume an ordering, restricting their flexibility for point cloud generation. Implicit representations such as DeepSDF [44] and OccupancyNet [36] can be viewed as modeling this probability density of the 3D points directly, but these models require ground truth signed distance fields or occupancy fields, which are difficult to obtain from point cloud data alone without corresponding meshes.

In this paper, we take a different approach and focus on the end goal – being able to draw an arbitrary number of samples from the distribution of points.

Working backward from this goal, we observe that the sampling procedure can be viewed as moving points from a generic prior distribution towards high likelihood regions of the shape (i.e., the surface of the shape). One way to achieve that is to move points gradually, following the gradient direction, which indicates where the density grows the most [57]. To perform such sampling, one only needs to model the gradient of log-density (known as the *Stein score function* [34]). In this paper, we propose to model a shape by learning the gradient field of its log-density. To learn such a gradient field from a set of sampled points from the shape, we build upon a *denoising score matching* framework [27,50]. Once we learn a model that outputs the gradient field, the sampling procedure can be done using a variant of stochastic gradient ascent (i.e. *Langevin dynamics* [50,57]).

Our method offers several advantages. First, our model is trained using a simple L_2 loss between the predicted and a "ground-truth" gradient field estimated from the input point cloud. This objective is much simpler to optimize than adversarial losses used in GAN-based techniques. Second, because it models the gradient directly and does not need to produce a normalized distribution, it imposes minimal restrictions on the model architecture in comparison to flow-based or autoregressive models. This allows us to leverage more expressive networks to model complicated distributions. Because the partition function need not be estimated, our model is also much *faster* to train. Finally, our model is able to furnish an implicit surface of the shape, as shown in Fig. 1, without requiring ground truth surfaces during training. We demonstrate that our technique can achieve state-of-the-art performance in both point cloud auto-encoding and generation. Moreover, our method can retain the same performance when trained with much sparser point clouds.

Our key contributions can be summarized as follows:

- We propose a novel point cloud generation method by extending score-based generative models to learn conditional distributions.
- We propose a novel algorithm to extract high-quality implicit surfaces from the learned model without the supervision from ground truth meshes.
- We show that our model can achieve state-of-the-art performance for point cloud auto-encoding and generation.

2 Related Work

Point Cloud Generative Modeling. Point clouds are widely used for representing and generating 3D shapes due to their simplicity and direct relation to common data acquisition techniques (LiDARs, depth cameras, etc.). Earlier generative models either treat point clouds as a fixed-dimensional matrix (i.e. $N \times 3$ where N is predefined) [1,16,17,48,52,55,63,64], or relies on heuristic set distance functions such as Chamfer distance and Earth Mover Distance [5,11,17,23,61]. As pointed out in Yang *et al.* [60] and Sect. 1, both of these approaches lead to several drawbacks. Alternatively, we can model the point cloud as samples from a distribution of 3D points. Toward this end, Sun *et al.* [52] applies an autoregressive model to model the distribution of points, but it requires assuming an ordering while generating points. Li *et al.* [31] applies a GAN [3,24] on both this distribution of 3D points as well as the distribution of shapes. PointFlow [60] applies normalizing flow [43] to model such distribution, so sampling points amounts to moving them to the surface according to a learned vector field. In addition to modeling the movement of points, PointFlow also tracks the change of volume in order to normalize the learned distribution, which is computationally expensive [8]. While our work applies a GAN to learn the distribution of latent code similar to Li *et al.* and Achilioptas *et al.*, we take a different approach to model the distribution of 3D points. Specifically, we predict the gradient of log-density field to model the non-normalized probability density, thus circumventing the need to compute the partition function and achieves faster training time with a simple L2 loss.

Generating Other 3D Representations. Common representations emerged for deep generative 3D modeling include voxel-based [21,59], mesh-based [2,18,25, 33,45,53], and assembly-based techniques [32,38]. Recently, implicit representations are gaining increasing popularity, as they are capable of representing shapes with high level of detail [10,36,37,44]. They also allow for learning a structured decomposition of shapes, representing local regions with Gaussian functions [19,20] or other primitives [26,49,54]. In order to reconstruct the mesh surface from the learned implicit field, these methods require finding the zero iso-surface of the learned occupancy field (e.g. using the Marching Cubes algorithm [35]). Our learned gradient field also allows for high-quality surface reconstruction using similar methods. However, we do not require prior information on the shape (e.g., signed distance values) for training, which typically requires a watertight input mesh. Recently, SAL [4] learns a signed distance field using only point cloud as supervision. Different from SAL, our model directly outputs the gradients of the log-density instead field of the signed distance, which allows our model to use arbitrary network architecture without any constraints. As a result, our method can scale to more difficult settings such as train on larger dataset (e.g. ShapeNet [6]) or train with sparse scanned point clouds.

Energy-Based Modeling. In contrast to flow-based models [8,12,22,28,46,60] and auto-regressive models [40–42,52], energy-based models learn a non-normalized probability distribution [29], thus avoid computation to estimate the partition function. It has been successfully applied to tasks such as image segmentation [14,15], where a normalized probability density function is hard to define. Score matching was first proposed for modeling energy-based models in [27] and deals with "matching" the model and the observed data log-density gradients, by minimizing the squared distance between them. To improve its performance and scalability, various extensions have been proposed, including denoising score matching [56] and sliced score matching [51]. Most recently, Song and Ermon [50] introduced data perturbation and annealed Langevin dynamics to the original denoising score matching method, providing an effective way to model data embedded on a low dimensional manifold. Their method was applied to the image generation task, achieving performance comparable to GANs. In this work, we extend this method to model conditional distributions and demonstrate its suitability to the task of point cloud generation, viewing point clouds as samples from the 2D manifold (shape surface) in 3D space.

3 Method

In this work, we are interested in learning a generative model that can sample shapes represented as point clouds. Therefore, we need to model two distributions. First, we need to model the distribution of shapes, which encode how shapes vary across an entire collection of shapes. Once we can sample a particular shape of interest, then we need a mechanism to sample a point clouds from its surface. As previously discussed, a point cloud is best viewed as samples from a distribution of 3D (or 2D) points, which encode a particular shape. To sample point clouds of arbitrary size for this shape, we also need to model this distribution of points.

Specifically, we assume a set of shapes $\mathcal{X} = \{X^{(i)}\}_{i=1}^{N}$ are provided as input. Each shape in $\mathcal{X}$ is represented as a point cloud sampled from its surface, defined by $X^{(i)} = \{x_j^i\}_{j=1}^{M_i}$. Our goal is to learn both the distribution of shapes and the distribution of points, conditioned on a particular shape from the data. To achieve that, we first propose a model to learn the distribution of points encoding a shape from a set of points $X^{(i)}$ (Sect. 3.1– 3.5). Then we describe how to model the distribution of shapes from the set of point clouds (i.e. $\mathcal{X}$) in Sect. 3.6.

3.1 Shapes as a Distribution of 3D Points

We would like to define a distribution of 3D points $P(x)$ such that sampling from this distribution will provide us with a surface point cloud of the object. Thus, the probability density encoding the shape should concentrate on the shape surface. Let S be the set of points on the surface and $P_S(x)$ be the uniform distribution over the surface. Sampling from $P_S(x)$ will create a point cloud uniformly sampled from the surface of interest. However, this distribution is hard to work with: for all points that are not in the surface $x \notin S$, $P_S(x) = 0$. As a result, $P_S(x)$ is discontinuous and has usually zero support over its ambient space (i.e. $\mathbb{R}^3$), which impose challenges in learning and modeling. Instead, we approximate $P_S(x)$ by smoothing the distribution with a Gaussian kernel:

$$Q_{\sigma,S}(x) = \int_{s\in\mathbb{R}^3} P_S(s)\mathcal{N}(x; s, \sigma^2 I)ds. \tag{1}$$

As long as the standard deviation σ is small enough, $Q_{\sigma,S}(x)$ will approximate the true data distribution $P_S(x)$ whose density concentrates near the surface. Therefore, sampling from $Q_{\sigma,S}(x)$ will yield points near the surface S.

As discussed in Sect. 1, instead of modeling $Q_{\sigma,S}$ directly, we will model the gradient of the logarithmic density (i.e. $\nabla_x \log Q_{\sigma,S}(x)$). Sampling can then be performed by starting from a prior distribution and performing gradient ascent on this field, thus moving points to high probability regions.

In particular, we will model the gradient of the log-density using a neural network $g_\theta(x, \sigma)$, where x is a location in 3D (or 2D) space. We will first analyze several properties of this gradient field $\nabla_x \log Q_{\sigma,S}(x)$. Then we describe how we train this neural network and how we sample points using the trained network.

3.2 Analyzing the Gradient Field

In this section we provide an interpretation of how $\nabla_x \log Q_{\sigma,S}(x)$ behaves with different σ's. Computing a Monte Carlo approximation of $Q_{\sigma,S}(x)$ using a set of observations $\{x_i\}_{i=1}^m$, we obtain a mixture of Gaussians with modes centered at x_i and radially-symmetric kernels:

$$Q_{\sigma,S}(x) = \mathbb{E}_{s\sim P_S}\left[\mathcal{N}(x; s, \sigma^2 I)\right] \approx \frac{1}{m}\sum_{i=1}^{m}\mathcal{N}(x; x_i, \sigma^2 I) \triangleq A_\sigma(x, \{x_i\}_{i=1}^m).$$

The gradient field can thus be approximated by the gradient of the logarithmic of this Gaussian mixture:

$$\nabla_x \log A_\sigma(x, \{x_i\}_{i=1}^m) = \frac{1}{\sigma^2}\left(-x + \sum_{i=1}^{m} x_i w_i(x, \sigma)\right), \tag{2}$$

where the weight $w_{ij}(x, \sigma)$ is computed from a softmax with temperature $2\sigma^2$:

$$w_i(x, \sigma) = \frac{\exp\left(-\frac{1}{2\sigma^2}\|x - x_i\|^2\right)}{\sum_{j=1}^{m}\exp\left(-\frac{1}{2\sigma^2}\|x - x_j\|^2\right)}. \tag{3}$$

Since $\sum_i w_i(x, \sigma) = 1$, $\sum_i x_i w_i(x, \sigma)$ falls within the convex hull created by the sampled surface points $\{x_i\}_{i=1}^m$. Therefore, the direction of this gradient of the logarithmic density field points from the sampled location towards a point inside the convex hull of the shape. When the temperature is high (i.e. σ is large), then the weights $w_i(x, \sigma)$ will be roughly the same and $\sum_i x_i w_i(x, \sigma)$ behaves like averaging all the x_i's. Therefore, the gradient field will point to a coarse shape that resembles an average of the surface points. When the temperature is low (i.e. σ is small), then $w_i(x, \sigma)$ will be close to 0 except when x_i is the closest to x. As a result, $\sum_i x_i w_i(x, \sigma)$ will behave like an $\text{argmin}_{x_i} \|x_i - x\|$. The gradient direction will thus point to the nearest point on the surface. In this case, the norm of the gradient field approximates a distance field of the surface up to a constant σ^{-2}. This allows the gradient field to encode fine details of the shape and move points to the shape surface more precisely. Figure 2 shows a visualization of the field in the 2D case for a series of different σ's.

3.3 Training Objective

As mentioned in Sect. 3.1, we would like to train a deep neural network $g_\theta(x, \sigma)$ to model the gradient of log-density: $\nabla_x \log Q_{\sigma,S}(x)$. One simple objective achieving this is minimizing the L2 loss between them [27]:

$$\ell_{\mathsf{direct}}(\sigma, S) = \mathbb{E}_{x\sim Q_{\sigma,S}(x)}\left[\frac{1}{2}\|g_\theta(x, \sigma) - \nabla_x \log Q_{\sigma,S}(x)\|_2^2\right]. \tag{4}$$

However, optimizing such an objective is difficult as it is generally hard to compute $\nabla_x \log Q_{\sigma,S}(x)$ from a finite set of observations.

Inspired by *denoising score matching* methods [50,56], we can write $Q_{\sigma,S}(x)$ as a perturbation of the data distribution $P_S(x)$, produced with a Gaussian noise with standard deviation σ. Specifically, $Q_{\sigma,S}(x) = \int P_S(s)q_\sigma(\tilde{x}|s)dx$, where $q_\sigma(\tilde{x}|s) = \mathcal{N}(\tilde{x}|s, \sigma^2 I)$. As such, optimizing the objective in Eq. 4 can be shown to be equivalent to optimizing the following [56]:

$$\ell_{\mathsf{denoising}}(\sigma, S) = \mathbb{E}_{s\sim P_S, \tilde{x}\sim q_\sigma(\tilde{x}|s)} \left[\frac{1}{2}\|g_\theta(\tilde{x}, \sigma) - \nabla_{\tilde{x}} \log q_\sigma(\tilde{x}|s)\|_2^2\right]. \tag{5}$$

Since $\nabla_{\tilde{x}} \log q_\sigma(\tilde{x}|s) = \frac{s-\tilde{x}}{\sigma^2}$, this loss can be easily computed using the observed point cloud $X = \{x_j\}_{j=1}^m$ as following:

$$\ell(\sigma, X) = \frac{1}{|X|}\sum_{x_i \in X} \|g_\theta(\tilde{x}_i, \sigma) - \frac{x_i - \tilde{x}_i}{\sigma^2}\|_2^2, \quad \tilde{x}_i \sim \mathcal{N}(x_i, \sigma^2 I). \tag{6}$$

Multiple Noise Levels. One problem with the abovementioned objective is that most $\tilde{x}_i$ will concentrate near the surface if σ is small. Thus, points far away from the surface will not be supervised. This can adversely affect the sampling quality, especially when the prior distribution puts points to be far away from the surface. To alleviate this issue, we follow Song and Ermon [50] and train g_θ for multiple σ's, with $\sigma_1 \geq \cdots \geq \sigma_k$. Our final model is trained by jointly optimizing $\ell(\sigma_i, X)$ for all σ_i. The final objective is computed empirically as:

$$\mathcal{L}(\{\sigma_i\}_{i=1}^k, X) = \sum_{i=1}^k \lambda(\sigma_i)\ell(\sigma_i, X), \tag{7}$$

where $\lambda(\sigma_i)$ are parameters weighing the losses $\ell(\sigma_i, X)$. $\lambda(\sigma_i)$ is chosen so that the weighted losses roughly have the same magnitude during training.

3.4 Point Cloud Sampling

Sampling a point cloud from the distribution is equivalent to moving points from a prior distribution to the surface (i.e. the high-density region). Therefore, we can perform stochastic gradient ascent on the logarithmic density field. Since $g_\theta(x, \sigma)$ approximates the gradient of the log-density field (i.e. $\nabla_x \log Q_{\sigma,S}(x)$), we could thus use $g_\theta(x, \sigma)$ to update the point location x. In order for the points to reach all the local maxima, we also need to inject random noise into this process. This amounts to using Langevin dynamics to perform sampling [57].

Specifically, we first sample a point x_0 from a prior distribution π. The prior is usually chosen to be simple distribution such as a uniform or a Gaussian distribution. We empirically demonstrate that the sampling performance won't be affected as long as the prior points are sampled from places where the perturbed points would reach during training. We then perform the following recursive update with step size $\alpha > 0$:

$$x_{t+1} = x_t + \frac{\alpha}{2} g_\theta(x_t, \sigma) + \sqrt{\alpha}\epsilon_t, \quad \epsilon_t \sim \mathcal{N}(0, I). \tag{8}$$

Under mild conditions, $p(x_T)$ converges to the data distribution $Q_{\sigma,S}(x)$ as $T \to \infty$ and $\epsilon \to 0$ [57]. Even when such conditions fail to hold, the error in Eq. 8 is usually negligible when α is small and T is large [9,13,39,50].

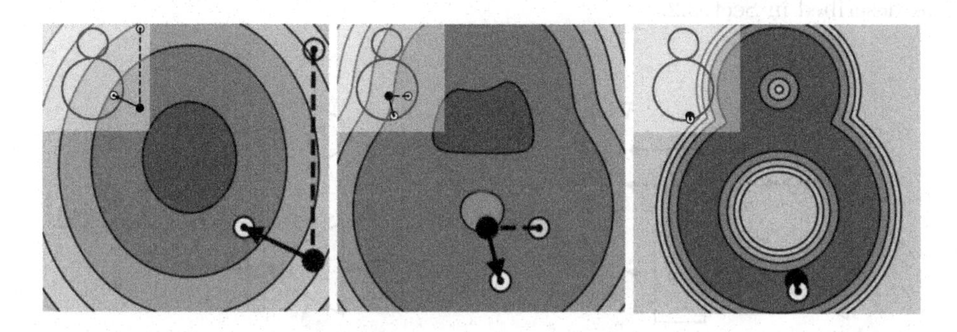

Fig. 2. Log density field with different σ (biggest to smallest) and a Langevin Dynamic point update step with that σ. Deeper color indicates higher density. The ground truth shape is shown in the upper left corner. Dotted line indicated Gaussian noise and solid arrows indicates gradient step. As sigma decreases, the log-density field changes from coarse to fine, and points are moved closer to the surface.

Prior works have observed that a main challenge for using Langevin dynamics is its slow mixing time [50,58]. To alleviate this issue, Song and Ermon [50] propose an annealed version of Langevin dynamics, which gradually anneals the noise for the score function. Specifically, we first define a list of σ_i with $\sigma_1 \geq \cdots \geq \sigma_k$, then train one single denoising score matching model that could approximate q_{σ_i} for all i. Then, annealed Langevin dynamics will recursively compute the x_t while gradually decreasing σ_i:

$$x'_{t+1} = x_t + \frac{\sqrt{\alpha}\sigma_i\epsilon_t}{\sigma_k}, \quad \epsilon_t \sim \mathcal{N}(0, I), \tag{9}$$

$$x_{t+1} = x'_{t+1} + \frac{\alpha\sigma_i^2}{2\sigma_k^2} g_\theta(x'_{t+1}, \sigma_i). \tag{10}$$

Figure 2 demonstrates the sampling across the annealing process in a 2D point cloud. As discussed in Sect. 3.3, larger σ's correspond to coarse shapes while smaller σ's correspond to fine shape. Thus, this annealed Langevin dynamics can be thought of as a coarse-to-fine refinement of the shape. Note that we make the noise perturbation step before the gradient update step, which leads to cleaner point clouds. The supplementary material contains detailed hyperparameters.

3.5 Implicit Surface Extraction

Next we show that our learned gradient field (e.g. g_θ) also allows for obtaining an implicit surface. The key insight here is that the surface is defined as the set of points that reach the maximum density in the data distribution $P_S(x)$, and thus these points have *zero* gradient. Another interpretation is that when σ is small enough (i.e. $Q_{\sigma,S}(x)$ approximates the true data distribution $p(x)$), the gradient for points near the surface will point to its nearest point on the surface, as described in Sect. 3.2:

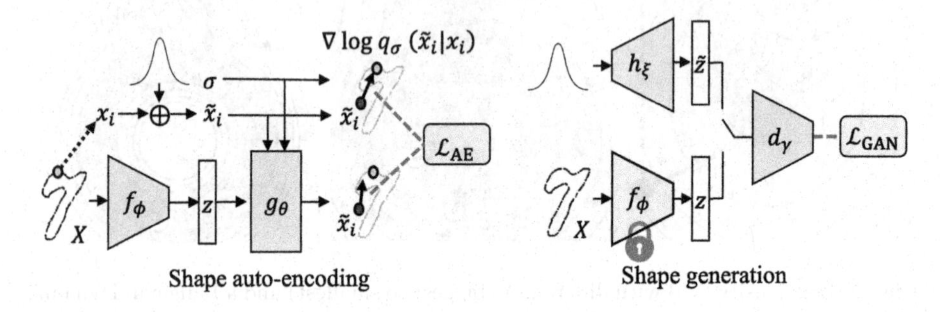

Fig. 3. Illustration of training pipe for shape auto-encoding and generation.

$$g_\theta(x, \sigma) \approx \frac{1}{\sigma^2}\left(-x + \text{argmin}_{s \in S} \|x - s\|\right). \tag{11}$$

Thus, for a point near the surface, its norm equals *zero* if and only if $x \in S$ (provided the arg min is unique). Therefore, the shape can be approximated by the zero iso-surface of the gradient norm:

$$S \approx \{x \mid \|g_\theta(x, \sigma)\| = \delta\}, \tag{12}$$

for some $\delta > 0$ that is sufficiently small. One caveat is that points for which the arg min in Eq. 11 is not unique may also have a zero gradient. These correspond to *local minimas* of the likelihood. In practice, this is seldom a problem for surface extraction, and it is possible to discard these regions by conducting the second partial derivative test.

Also as mentioned in Sect. 3.2, when the σ is small, the norm of the gradient field approximates a distance field of the surface, scaled by a constant σ^{-2}. This allows us to retrieval the surface S efficiently using an off-the-shelf ray-casting technique [47] (see Figs. 1, 4 and 5).

3.6 Generating Multiple Shapes

In the previous sections, we focused on learning the distribution of points that represent a single shape. Our next goal is to model the distribution of shapes. We, therefore, introduce a latent code z to encode which specific shape we want to sample point clouds from. Furthermore, we adapt our gradient decoder to be conditional on the latent code z (in addition to σ and the sampled point).

As illustrated in Fig. 3, the training is conducted in two stages. We first train an auto-encoder with an encoder f_ϕ that takes a point cloud and outputs the latent code z. The gradient decoder is provided with z as input and produces a gradient field with noise level σ. The auto-encoding loss is thus:

$$\mathcal{L}_{AE}(\mathcal{X}) = \mathop{\mathbb{E}}_{X \sim \mathcal{X}} \left[\frac{1}{2|X|} \sum_{x \in X, \sigma_i} \lambda(\sigma_i) \left\| g_\theta(\tilde{x}, f_\phi(X), \sigma_i) - \frac{x - \tilde{x}}{\sigma_i^2} \right\|_2^2 \right], \tag{13}$$

where each $\tilde{x_j}$ is drawn from a $\mathcal{N}(x_j, \sigma_i^2 I)$ for a corresponding σ_i. This first stage provides us with a network that can model the distribution of points representing the shape encoded in the latent variable z. Once the auto-encoder is fully trained, we apply a latent-GAN [1] to learn the distribution of the latent code $p(z) = p(f_\phi(X))$, where X is a point cloud sampled from the data distribution. Doing so provides us with a generator h_ξ that can sample a latent code from $p(z)$, allowing us control over which shape will be generated. To sample a novel shape, we first sample a latent code $\tilde{z}$ using h_ξ. We can then use the trained gradient decoder g_θ to sample point clouds or extract an implicit surface from the shape represented as z. For more details about hyperparameters and model architecture, please refer to the supplementary material.

4 Experiments

In this section, we will evaluate our model's performance in point cloud auto-encoding (Sect. 4.1), up-sampling (Sect. 4.1), and generation (Sect. 4.2) tasks. Finally, we present an ablation study examining our model design choices (Sect. 4.3). Implementation details will be shown in the supplementary materials.

Datasets. Our experiments focus mainly on two datasets: MNIST-CP and ShapeNet. MNIST-CP was recently proposed by Yifan *et al.* [62] and consists of 2D contour points extracted from the MNIST [30] dataset, which contains 50K and 10K training and testing images. Each point cloud contains 800 points. The ShapeNet [7] dataset contains 35708 shapes in training set and 5158 shapes in test set, capturing 55 categories. For our method, we normalize all point clouds by centering their bounding boxes to the origin and scaling them by a constant such that all points range within the cube $[-1,1]^3$ (or the square in the 2D case).

Evaluation Metrics. Following prior works [1,23,60], we use the symmetric Chamfer Distance (CD) and the Earth Mover's Distance (EMD) to evaluate the reconstruction quality of the point clouds. To evaluate the generation quality, we use metrics in Yang *et al.* [60] and Achlioptas *et al.* [1]. Specifically, Achilioptas *et al.* [1] suggest using Minimum Matching Distance (MMD) to measure fidelity of the generated point cloud and Coverage (COV) to measure whether the set of generated samples cover all the modes of the data distribution. Yang *et al.* [60]

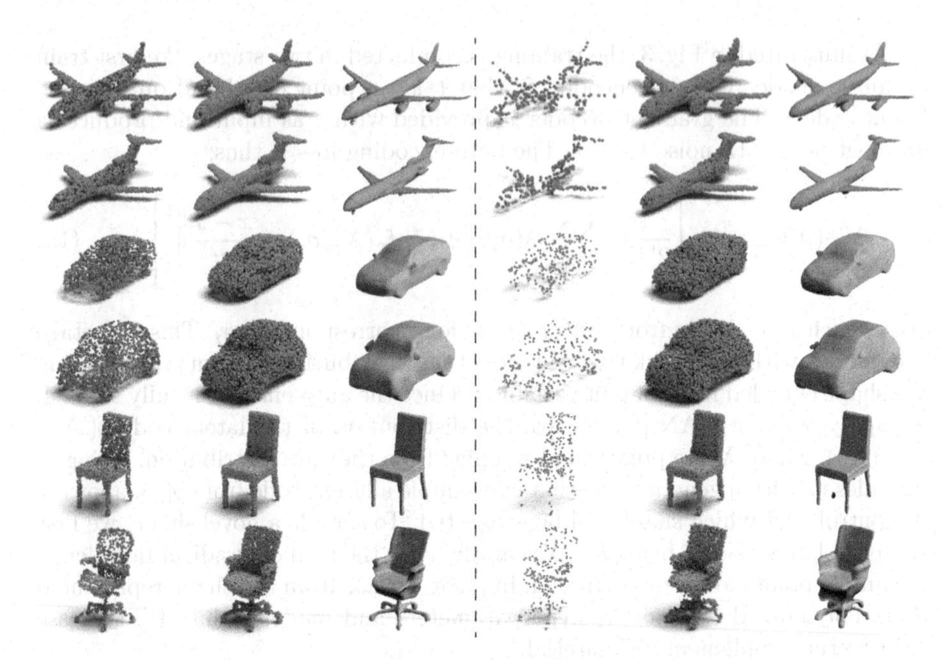

Fig. 4. Shape auto-encoding test results. Our model can accurately reconstruct shapes given 2048 points (left) or only 256 points (right) describing the shape. Output point clouds are illustrated in the center and implicit surfaces on the left.

propose to use the accuracy of a k-NN classifier performing two-sample tests. The idea is that if the sampled shapes seem to be drawn from the actual data distribution, then the classifier will perform like a random guess (i.e. results in 50% accuracy). To evaluate our results, we first conduct per-shape normalization to center the bounding box of the shape and scale its longest length to be 2, which allows the metric to focus on the geometry of the shape and not the scale.

4.1 Shape Auto-encoding

In this section, we evaluate how well our model can learn the underlying distribution of points by asking it to auto-encode a point cloud. We conduct the auto-encoding task for five settings: all 2D point clouds in MNIST-CP, 3D point clouds on the whole ShapeNet, and three categories in ShapeNet (Airplane, Car, Chair). In this experiment, our method is compared with the current state-of-the-art AtlasNet [23] with patches and with sphere. Furthermore, we also compare against Achilioptas *et al.* [1] which predicts point clouds as a fixed-dimensional array, and PointFlow [60] which uses a flow-based model to represent the distribution. We follow the experiment set-up in PointFlow to report performance in both CD and EMD in Table 1. Since these two metrics depend on the scale of the point clouds, we also report the upper bound in the "oracle" column. The upper

Table 1. Shape auto-encoding on the MNIST-CP and ShapeNet datasets. The best results are highlighted in bold. CD is multiplied by 10^4 and EMD is multiplied by 10^2.

Dataset	Metric	l-GAN [1]		AtlasNet [23]		PF [60]	Ours	Oracle
		CD	EMD	Sphere	Patches			
MNIST-CP	CD	8.204	–	7.274	4.926	17.894	**2.669**	1.012
	EMD	40.610	–	19.920	15.970	8.705	**7.341**	4.875
Airplane	CD	1.020	1.196	1.002	0.969	1.208	**0.96**	0.837
	EMD	4.089	2.577	2.672	2.612	2.757	**2.562**	2.062
Chair	CD	9.279	11.21	6.564	6.693	10.120	**5.599**	3.201
	EMD	8.235	6.053	5.790	5.509	6.434	**4.917**	3.297
Car	CD	5.802	6.486	5.392	5.441	6.531	**5.328**	3.904
	EMD	5.790	4.780	4.587	4.570	5.138	**4.409**	3.251
ShapeNet	CD	7.120	8.850	5.301	**5.121**	7.551	5.154	3.031
	EMD	7.950	5.260	5.553	5.493	5.176	**4.603**	3.103

bound is produced by computing the error between two different point clouds with the same number of points sampled from the same underlying meshes.

Our method consistently outperforms all other methods on the EMD metric, which suggests that our point samples follow the distribution or they are more uniformly distributed among the surface. Note that our method outperforms PointFlow in both CD and EMD for all datasets, but requires much less time to train. Our training for the Airplane category can be completed in about less than 10 h, yet reproducing the results for PointFlow's pretrained model takes at least two days. Our method can even sometimes outperform Achilioptas *et al.*and AtlasNet in CD, which is the loss they are directly optimizing at.

Table 2. Auto-encoding sparse point clouds. We randomly sample N points from each shape (in the Airplane dataset). During training, the model is provided with M points (the columns). CD is multiplied by 10^4 and EMD is multiplied by 10^2.

N	CD					EMD				
	2048	1024	512	256	128	2048	1024	512	256	128
$10K$	0.993	1.057	0.999	1.136	1.688	2.463	2.608	2.589	3.042	3.715
$3K$	1.080	1.059	1.003	1.142	1.753	2.533	2.586	2.557	2.997	3.878
$1K$	–	–	1.021	1.149	1.691	–	–	2.565	2.943	3.633

Point Cloud Upsampling. We conduct a set of experiments on subsampled ShapeNet point clouds. These experiments are primarily focused on showing that

(i) our model can learn from sparser datasets, and that (ii) we can infer a dense shape from a sparse input. In the regular configuration (reported above), we learn from $N = 10K$ points which are uniformly sampled from each shape mesh model. During training, we sample $M = 2048$ points (from the $10K$ available in total) to be the input point cloud. To evaluate our model, we perform the Langevin dynamic procedure (described in Sect. 3.4) over 2048 points sampled from the prior distribution and compare these to 2048 points from the reference set.

To evaluate whether our model can effectively upsample point clouds and learn from a sparse input, we train models with $N = [1K, 3K, 10K]$ and $M = [128, 256, 512, 1024, 2048]$ on the Airplane dataset. To allow for a fair comparison, we evaluate all models using the same number of output points (i.e. 2048 points are sampled from the prior distribution in all cases). As illustrated in Table 2, we obtain comparable auto-encoding performance while training with significantly sparser shapes. Interestingly, the number of points available from the model (i.e. N) does not seem to affect performance, suggesting that we can indeed learn from sparser datasets. Several qualitative examples auto-encoding shapes from the regular and sparse configurations are shown in Fig. 4. We also demonstrate that our model can also provide a smooth iso-surface, even when only a sparse point cloud (i.e. 256 points) is provided as input.

Fig. 5. Generation results. We shown results from r-GAN, GCN, TreeGAN (Tree), and PointFlow (PF) are illustrated on the left for comparison. Generated point clouds are illustrated alongside the corresponding implicit surfaces.

4.2 Shape Generation

We quantitatively compare our method's performance on shape generation with r-GAN [1], GCN-GAN [55], TreeGAN [48], and PointFlow [60]. We use the same experiment setup as PointFlow except for the data normalization before the evaluation. The generation results are reported in Table 3. Though our model requires a two-stage training, the training can be done within one day with a 1080 Ti GPU, while reproducing PointFlow's results requires training for at least two days on the same hardware. Despite using much less training time, our model achieves comparable performance to PointFlow, the current state-of-the-art. As demonstrated in Fig. 5, our generated shapes are also visually cleaner.

Table 3. Shape generation results. ↑ means the higher the better, ↓ means the lower the better. MMD-CD is multiplied by 10^3 and MMD-EMD is multiplied by 10^2.

Category	Model	MMD (↓)		COV (%, ↑)		1-NNA (%, ↓)	
		CD	EMD	CD	EMD	CD	EMD
Airplane	r-GAN [1]	1.657	13.287	38.52	19.75	95.80	100.00
	GCN [55]	2.623	15.535	9.38	5.93	95.16	99.12
	Tree [48]	1.466	16.662	44.69	6.91	95.06	100.00
	PF [60]	1.408	7.576	39.51	**41.98**	**83.21**	**82.22**
	Ours	**1.285**	**7.364**	**47.65**	**41.98**	85.06	83.46
	Train	1.288	7.036	45.43	45.43	72.10	69.38
Chair	r-GAN [1]	18.187	32.688	19.49	8.31	84.82	99.92
	GCN [55]	23.098	25.781	6.95	6.34	86.52	96.48
	Tree [48]	16.147	36.545	40.33	8.76	74.55	99.92
	PF [60]	15.027	19.190	40.94	44.41	67.60	72.28
	Ours	**14.818**	**18.791**	**46.37**	**46.22**	**66.16**	**59.82**
	Train	15.893	18.472	50.45	52.11	53.93	54.15

Table 4. Ablation study comparing auto-encoding performance on the Airplane dataset. CD is multiplied by 10^4 and EMD is multiplied by 10^2.

Metric	Single noise level			Prior distribution		
	0.1	0.05	0.01	Uniform	Fixed	Gaussian
CD	2.545	1.573	1009.357	0.993	0.993	0.996
EMD	4.400	8.455	36.715	2.463	2.476	2.475

4.3 Ablation Study

We conduct an ablation study quantifying the importance of learning with multiple noise levels. As detailed in Sects. 3.3–3.4, we train s_θ for multiple σ's. During inference, we sample point clouds using an annealed Langevin dynamics procedure, using the same σ's seen during training. In Table 4 we show results for models trained with a single noise level and tested without annealing. As illustrated in the table, the model does not perform as well when learning using a single noise level only. This is especially noticeable for the model trained on the smallest noise level in our model ($\sigma = 0.01$), as large regions in space are left unsupervised, resulting in significant errors.

We also demonstrate that our model is insensitive to the choice of the prior distribution. We repeat the inference procedure for our auto-encoding experiment, initializing the prior points with a Gaussian distribution or in a fixed location (using the same trained model). Results are reported on the right side of Table 4. Different prior configurations don't affect the performance, which is expected due to the stochastic nature of our solution. We further demonstrate our model's robustness to the prior distribution in Fig. 1, where the prior depicts 3D letters.

5 Conclusions

In this work, we propose a generative model for point clouds which learns the gradient field of the logarithmic density function encoding a shape. Our method not only allows sampling of high-quality point clouds, but also enables extraction of the underlying surface of the shape. We demonstrate the effectiveness of our model on point cloud auto-encoding, generation, and super-resolution. Future work includes extending our work to model texture, appearance, and scenes.

Acknowledgment. This work was supported in part by grants from Magic Leap and Facebook AI, and the Zuckerman STEM leadership program.

References

1. Achlioptas, P., Diamanti, O., Mitliagkas, I., Guibas, L.: Learning representations and generative models for 3D point clouds. In: ICML (2018)
2. Anguelov, D., Srinivasan, P., Koller, D., Thrun, S., Rodgers, J., Davis, J.: Scape: shape completion and animation of people. In: ACM SIGGRAPH 2005 Papers, pp. 408–416 (2005)
3. Arjovsky, M., Chintala, S., Bottou, L.: Wasserstein generative adversarial networks. In: ICML (2017)
4. Atzmon, M., Lipman, Y.: SAL: sign agnostic learning of shapes from raw data. In: Proceedings of the IEEE/CVF Conference on Computer Vision and Pattern Recognition, pp. 2565–2574 (2020)
5. Ben-Hamu, H., Maron, H., Kezurer, I., Avineri, G., Lipman, Y.: Multi-chart generative surface modeling. ACM Trans. Graph. (TOG) **37**(6), 1–15 (2018)
6. Chang, A.X., et al.: ShapeNet: an information-rich 3D model repository. Technical report. arXiv:1512.03012 [cs.GR], Stanford University — Princeton University — Toyota Technological Institute at Chicago (2015)
7. Chang, A.X., et al.: ShapeNet: an information-rich 3D model repository. arXiv preprint arXiv:1512.03012 (2015)
8. Chen, T.Q., Rubanova, Y., Bettencourt, J., Duvenaud, D.K.: Neural ordinary differential equations. In: NeurIPS (2018)
9. Chen, T., Fox, E., Guestrin, C.: Stochastic gradient Hamiltonian Monte Carlo. In: International Conference on Machine Learning, pp. 1683–1691 (2014)
10. Chen, Z., Zhang, H.: Learning implicit fields for generative shape modeling. In: Proceedings of the IEEE Conference on Computer Vision and Pattern Recognition, pp. 5939–5948 (2019)
11. Deprelle, T., Groueix, T., Fisher, M., Kim, V., Russell, B., Aubry, M.: Learning elementary structures for 3D shape generation and matching. In: Advances in Neural Information Processing Systems, pp. 7433–7443 (2019)
12. Dinh, L., Krueger, D., Bengio, Y.: Nice: non-linear independent components estimation. CoRR abs/1410.8516 (2014)
13. Du, Y., Mordatch, I.: Implicit generation and generalization in energy-based models. arXiv preprint arXiv:1903.08689 (2019)

14. Fan, A., Fisher III, J.W., Kane, J., Willsky, A.S.: MCMC curve sampling and geometric conditional simulation. In: Computational Imaging VI, vol. 6814, p. 681407. International Society for Optics and Photonics (2008)
15. Fan, A.C., Fisher, J.W., Wells, W.M., Levitt, J.J., Willsky, A.S.: MCMC curve sampling for image segmentation. In: Ayache, N., Ourselin, S., Maeder, A. (eds.) MICCAI 2007. LNCS, vol. 4792, pp. 477–485. Springer, Heidelberg (2007). https://doi.org/10.1007/978-3-540-75759-7_58
16. Fan, H., Su, H., Guibas, L.J.: A point set generation network for 3D object reconstruction from a single image. In: CVPR (2017)
17. Gadelha, M., Wang, R., Maji, S.: Multiresolution tree networks for 3D point cloud processing. In: Ferrari, V., Hebert, M., Sminchisescu, C., Weiss, Y. (eds.) ECCV 2018. LNCS, vol. 11211, pp. 105–122. Springer, Cham (2018). https://doi.org/10.1007/978-3-030-01234-2_7
18. Gao, L., Yang, J., Wu, T., Yuan, Y.J., Fu, H., Lai, Y.K., Zhang, H.: SDM-NET: deep generative network for structured deformable mesh. ACM Trans. Graph. (TOG) **38**(6), 1–15 (2019)
19. Genova, K., Cole, F., Sud, A., Sarna, A., Funkhouser, T.: Deep structured implicit functions. arXiv preprint arXiv:1912.06126 (2019)
20. Genova, K., Cole, F., Vlasic, D., Sarna, A., Freeman, W.T., Funkhouser, T.: Learning shape templates with structured implicit functions. In: Proceedings of the IEEE International Conference on Computer Vision, pp. 7154–7164 (2019)
21. Girdhar, R., Fouhey, D.F., Rodriguez, M., Gupta, A.: Learning a predictable and generative vector representation for objects. In: Leibe, B., Matas, J., Sebe, N., Welling, M. (eds.) ECCV 2016. LNCS, vol. 9910, pp. 484–499. Springer, Cham (2016). https://doi.org/10.1007/978-3-319-46466-4_29
22. Grathwohl, W., Chen, R.T.Q., Bettencourt, J., Sutskever, I., Duvenaud, D.: FFJORD: free-form continuous dynamics for scalable reversible generative models. In: ICLR (2019)
23. Groueix, T., Fisher, M., Kim, V.G., Russell, B., Aubry, M.: AtlasNet: a Papier-Mâché approach to learning 3D surface generation. In: CVPR (2018)
24. Gulrajani, I., Ahmed, F., Arjovsky, M., Dumoulin, V., Courville, A.C.: Improved training of Wasserstein GANs. In: NeurIPS (2017)
25. Hanocka, R., Hertz, A., Fish, N., Giryes, R., Fleishman, S., Cohen-Or, D.: MeshCNN: a network with an edge. ACM Trans. Graph. (TOG) **38**(4), 1–12 (2019)
26. Hao, Z., Averbuch-Elor, H., Snavely, N., Belongie, S.: DualSDF: semantic shape manipulation using a two-level representation. In: Proceedings of the IEEE/CVF Conference on Computer Vision and Pattern Recognition, pp. 7631–7641 (2020)
27. Hyvärinen, A.: Estimation of non-normalized statistical models by score matching. J. Mach. Learn. Res. **6**, 695–709 (2005)
28. Kingma, D.P., Dhariwal, P.: Glow: generative flow with invertible 1x1 convolutions. In: NeurIPS (2018)
29. LeCun, Y., Chopra, S., Hadsell, R., Ranzato, M., Huang, F.: A tutorial on energy-based learning. Predict. Struct. Data **1** (2006)
30. LeCun, Y., Cortes, C., Burges, C.: MNIST handwritten digit database (2010)
31. Li, C.L., Zaheer, M., Zhang, Y., Poczos, B., Salakhutdinov, R.: Point cloud GAN. arXiv preprint arXiv:1810.05795 (2018)
32. Li, J., Xu, K., Chaudhuri, S., Yumer, E., Zhang, H., Guibas, L.: GRASS: generative recursive autoencoders for shape structures. ACM Trans. Graph. (TOG) **36**(4), 1–14 (2017)

33. Litany, O., Bronstein, A., Bronstein, M., Makadia, A.: Deformable shape completion with graph convolutional autoencoders. In: Proceedings of the IEEE Conference on Computer Vision and Pattern Recognition, pp. 1886–1895 (2018)
34. Liu, Q., Lee, J., Jordan, M.: A kernelized stein discrepancy for goodness-of-fit tests. In: International Conference on Machine Learning, pp. 276–284 (2016)
35. Lorensen, W.E., Cline, H.E.: Marching cubes: a high resolution 3D surface construction algorithm. ACM SIGGRAPH Comput. Graph. **21**(4), 163–169 (1987)
36. Mescheder, L., Oechsle, M., Niemeyer, M., Nowozin, S., Geiger, A.: Occupancy networks: learning 3D reconstruction in function space. In: Proceedings of the IEEE Conference on Computer Vision and Pattern Recognition, pp. 4460–4470 (2019)
37. Michalkiewicz, M., Pontes, J.K., Jack, D., Baktashmotlagh, M., Eriksson, A.: Deep level sets: implicit surface representations for 3D shape inference. arXiv preprint arXiv:1901.06802 (2019)
38. Mo, K., Guerrero, P., Yi, L., Su, H., Wonka, P., Mitra, N., Guibas, L.J.: StructureNet: hierarchical graph networks for 3D shape generation. arXiv preprint arXiv:1908.00575 (2019)
39. Nijkamp, E., Hill, M., Han, T., Zhu, S.C., Wu, Y.N.: On the anatomy of MCMC-based maximum likelihood learning of energy-based models. arXiv preprint arXiv:1903.12370 (2019)
40. Van den Oord, A., Kalchbrenner, N., Espeholt, L., Vinyals, O., Graves, A., et al.: Conditional image generation with pixelcnn decoders. In: NeurIPS (2016)
41. Oord, A.v.d., et al.: WaveNet: a generative model for raw audio. arXiv preprint arXiv:1609.03499 (2016)
42. Oord, A.v.d., Kalchbrenner, N., Kavukcuoglu, K.: Pixel recurrent neural networks. In: ICML (2016)
43. Papamakarios, G., Nalisnick, E., Rezende, D.J., Mohamed, S., Lakshminarayanan, B.: Normalizing flows for probabilistic modeling and inference. arXiv preprint arXiv:1912.02762 (2019)
44. Park, J.J., Florence, P., Straub, J., Newcombe, R., Lovegrove, S.: DeepSDF: learning continuous signed distance functions for shape representation. In: Proceedings of the IEEE Conference on Computer Vision and Pattern Recognition, pp. 165–174 (2019)
45. Pons-Moll, G., Romero, J., Mahmood, N., Black, M.J.: Dyna: a model of dynamic human shape in motion. ACM Trans. Graph. (TOG) **34**(4), 1–14 (2015)
46. Rezende, D.J., Mohamed, S.: Variational inference with normalizing flows. In: ICML (2015)
47. Roth, S.D.: Ray casting for modeling solids. Comput. Graph. Image Process. **18**(2), 109–144 (1982)
48. Shu, D.W., Park, S.W., Kwon, J.: 3D point cloud generative adversarial network based on tree structured graph convolutions. In: Proceedings of the IEEE International Conference on Computer Vision, pp. 3859–3868 (2019)
49. Smirnov, D., Fisher, M., Kim, V.G., Zhang, R., Solomon, J.: Deep parametric shape predictions using distance fields. arXiv preprint arXiv:1904.08921 (2019)
50. Song, Y., Ermon, S.: Generative modeling by estimating gradients of the data distribution. In: Advances in Neural Information Processing Systems, pp. 11895–11907 (2019)
51. Song, Y., Garg, S., Shi, J., Ermon, S.: Sliced score matching: a scalable approach to density and score estimation. arXiv preprint arXiv:1905.07088 (2019)

52. Sun, Y., Wang, Y., Liu, Z., Siegel, J.E., Sarma, S.E.: PointGrow: autoregressively learned point cloud generation with self-attention. arXiv preprint arXiv:1810.05591 (2018)
53. Tan, Q., Gao, L., Lai, Y.K., Xia, S.: Variational autoencoders for deforming 3D mesh models. In: Proceedings of the IEEE Conference on Computer Vision and Pattern Recognition, pp. 5841–5850 (2018)
54. Tulsiani, S., Su, H., Guibas, L.J., Efros, A.A., Malik, J.: Learning shape abstractions by assembling volumetric primitives. In: Proceedings of the IEEE Conference on Computer Vision and Pattern Recognition, pp. 2635–2643 (2017)
55. Valsesia, D., Fracastoro, G., Magli, E.: Learning localized generative models for 3d point clouds via graph convolution (2018)
56. Vincent, P.: A connection between score matching and denoising autoencoders. Neural Comput. **23**(7), 1661–1674 (2011)
57. Welling, M., Teh, Y.W.: Bayesian learning via stochastic gradient Langevin dynamics. In: Proceedings of the 28th International Conference on Machine Learning (ICML-11), pp. 681–688 (2011)
58. Wenliang, L., Sutherland, D., Strathmann, H., Gretton, A.: Learning deep kernels for exponential family densities. arXiv preprint arXiv:1811.08357 (2018)
59. Wu, J., Zhang, C., Xue, T., Freeman, B., Tenenbaum, J.: Learning a probabilistic latent space of object shapes via 3D generative-adversarial modeling. In: Advances in Neural Information Processing Systems, pp. 82–90 (2016)
60. Yang, G., Huang, X., Hao, Z., Liu, M.Y., Belongie, S., Hariharan, B.: PointFlow: 3D point cloud generation with continuous normalizing flows. In: Proceedings of the IEEE International Conference on Computer Vision, pp. 4541–4550 (2019)
61. Yang, Y., Feng, C., Shen, Y., Tian, D.: FoldingNet: point cloud auto-encoder via deep grid deformation. In: CVPR (2018)
62. Yifan, W., Wu, S., Huang, H., Cohen-Or, D., Sorkine-Hornung, O.: Patch-based progressive 3D point set upsampling. In: Proceedings of the IEEE Conference on Computer Vision and Pattern Recognition, pp. 5958–5967 (2019)
63. Zamorski, M., Zieba, M., Nowak, R., Stokowiec, W., Trzcinski, T.: Adversarial autoencoders for generating 3D point clouds. arXiv preprint arXiv:1811.07605 2 (2018)
64. Zamorski, M., Zieba, M., Nowak, R., Stokowiec, W., Trzciński, T.: Adversarial autoencoders for generating 3D point clouds. arXiv preprint arXiv:1811.07605 (2018)

COCO-FUNIT: Few-Shot Unsupervised Image Translation with a Content Conditioned Style Encoder

Kuniaki Saito[1,2](✉), Kate Saenko[1], and Ming-Yu Liu[2]

[1] Boston University, Boston, USA
{keisaito,saenko}@bu.edu
[2] NVIDIA, Santa Clara, USA
mingyul@nvidia.com

Abstract. Unsupervised image-to-image translation intends to learn a mapping of an image in a given domain to an analogous image in a different domain, without explicit supervision of the mapping. Few-shot unsupervised image-to-image translation further attempts to generalize the model to an unseen domain by leveraging example images of the unseen domain provided at inference time. While remarkably successful, existing few-shot image-to-image translation models find it difficult to preserve the structure of the input image while emulating the appearance of the unseen domain, which we refer to as the *content loss* problem. This is particularly severe when the poses of the objects in the input and example images are very different. To address the issue, we propose a new few-shot image translation model, COCO-FUNIT, which computes the style embedding of the example images conditioned on the input image and a new module called the constant style bias. Through extensive experimental validations with comparison to the state-of-the-art, our model shows effectiveness in addressing the *content loss* problem. Code and pretrained models are available at https://nvlabs.github.io/COCO-FUNIT/.

Keywords: Image-to-image translation · Generative Adversarial Networks

1 Introduction

Image-to-Image translation [18,44] concerns learning a mapping that can translate an input image in one domain into an analogous image in a different domain. Unsupervised image-to-image translation [5,21,25,26,28,39,47] attempts to learn such a mapping without paired data. Thanks to the introduction of novel

---

Electronic supplementary material The online version of this chapter (https://doi.org/10.1007/978-3-030-58580-8_23) contains supplementary material, which is available to authorized users.

A. Vedaldi et al. (Eds.): ECCV 2020, LNCS 12348, pp. 382–398, 2020.
https://doi.org/10.1007/978-3-030-58580-8_23

Fig. 1. Given as few as one style example image from an object class unseen during training, our model can generate a photorealistic translation of the input content image in the unseen domain.

network architectures and learning objective terms, the state-of-the-art has advanced significantly in the past few years. However, while existing unsupervised image-to-image translation models can generate realistic translations, they still have several drawbacks. First, they require a large amount of images from the source and target domains for training. Second, they cannot be used to generate images in unseen domains. These limitations are addressed in the few-shot *unsupervised* image-to-image translation framework [27]. By leveraging example-guided episodic training, the few-shot image translation framework [27] learns to extract the domain-specific style information from a few example images in the unseen domain during test time, mixes it with the domain-invariant content information extracted from the input image, and generates a few-shot translation output as illustrated in Fig. 2 (Fig. 1).

However, despite showing encouraging results on relatively simple tasks such as animal face and flower translation, the few-shot translation framework [27] frequently generates unsatisfactory translation outputs when the model is applied to objects with diverse appearance, such as animals with very different body poses. Often, the translation output is not well-aligned with the input image. The domain invariant content that is supposed to remain unchanged disappears after translation, as shown in Fig. 3. We will call this issue the *content loss* problem. We hypothesize that solving the content loss problem would produce more faithful and photorealistic few-shot image translation results.

But why does the content loss problem occur? To learn the translation in an unsupervised manner, Liu *et al.* [27] rely on inductive bias injected by the network design and adversarial training [10] to transfer the appearance from the example images in the unseen domain to the input image. However, as there is no supervision, it is difficult to control what to be transferred precisely. Ideally, the

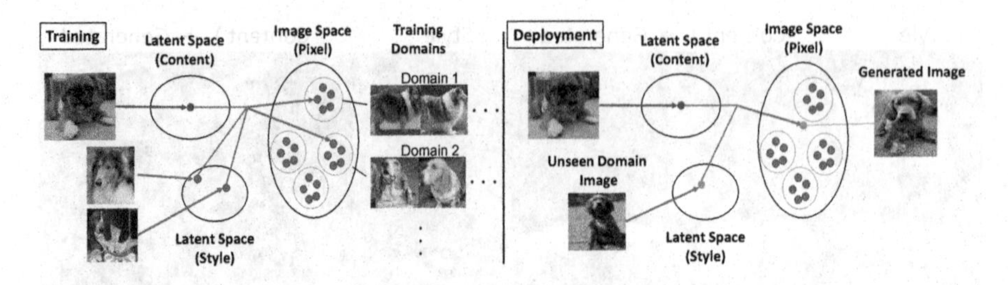

Fig. 2. Few-shot image-to-image translation. **Training.** The training set consists of many domains. We train a model to translate images between these domains. **Deployment.** We apply the trained model to perform few-shot image translation. Given a few examples from a test domain, we aim to translate a content image into an image analogous to the test class.

transferred appearance should contain just the style. In reality, it often contains other information, such as the object pose.

In this paper, we propose a novel network architecture to counter the content loss problem. We design a style encoder called the *content-conditioned style encoder*, to hinder the transmission of task-irrelevant appearance information to the image translation process. In contrast to the existing style encoders, our style code is computed by conditioning on the input content image. We use a new architecture design to limit the variance of the style code. We conduct an extensive experimental validation with a comparison to the state-of-the-art method using several newly collected and challenging few-shot image translation datasets. Experimental results, including both automatic performance metrics and user studies, verify the effectiveness of the proposed method in dealing with the content loss problem.

2 Related Works

Image-to-Image Translation. Most of the existing models are based on the Generative Adversarial Network (GAN) [10] framework. Unlike unconditional GANs [10,12,19,20,30], which learn to map random vectors to images, existing image-to-image translation models are mostly based on conditional GANs where they learn to generate a corresponding image in the target domain conditioned on the input image in the source domain. Depending on the availability of paired input and output images as supervision in the training dataset, image-to-image translation models can be divided into supervised [4,18,29,33,35,40,41,43,44,46,48,49] or unsupervised [1,3,5,9,16,21,23,25,26,28,34,37,39,47]. Our work falls in the category of unsupervised image-to-image translation. However, instead of learning a mapping between two specific domains, we aim at learning a flexible mapping that can be used to generate images in many unseen domains. Specifically, the mapping is only determined at test time, via example images. When using example images from a different unseen domain, the same model can generate images in the new unseen domain.

Fig. 3. Illustration of the *content loss* problem. The images generated by the baseline [27] fail to preserve domain invariant appearance information in the content image. The animals' bodies are sometimes merged with the background (column 3, & 4), scales of the generated body parts are sometimes inconsistent with the input (column 5), and some body parts absent in the content image show up (column 1 & 2). Our proposed method solves this "content loss" problem.

Multi-domain Image Translation. Several works [2,5,6,17] extend the unsupervised image translation to multiple domains. They learn a mapping between multiple seen domains, simultaneously. Our work differs from the multi-domain image translation works in that we aim to translate images to *unseen* domains.

Few-Shot Image Translation. Several few-shot methods are proposed to generate human images [13,38,41,42], scenes [41], or human faces [11,42,45] given a few instances and semantic layouts in a test time. These methods operate in the supervised setting. During training, they assume access to paired input (layout) and output data. Our work is most akin to the FUNIT work [27] as we aim to learn to generalize the translation to unseen domain without paired input and output data. We build on top of the FUNIT work where we first identify the *content loss* problem and then address it with a novel content-conditioned style encoder architecture.

Example-Guided Image Translation refers to methods that generate a translation of an input conditioning on some example images. Existing works in this space [16,27,33] use a style encoder to extract style information from the example images. Our work is also an example-guided image translation method. However, unlike the prior works where the style code is computed independent of the input image, our style code is computed by conditioning on the input image, where we normalize the style code using the content to prevent over-transmission of the style information to the output.

Neural Style Transfer studies approaches to transfer textures from a painting to a real photo. While existing neural style transfer methods [8,15,24] can generalize to unseen textures, they cannot generalize to unseen shapes, necessary for image-to-image translation. Our work is inspired by these works, but we focus on generalizing the generation of both unseen shapes and textures, which is essential to few-shot unsupervised image-to-image translation.

3 Method

In this section, we start with a brief explanation of the problem setup, introduce the basic architecture, and then describe our proposed architecture. Throughout the paper, the two words, "class" and "domain", are used interchangeably since we treat each object class as a domain.

Problem Setting. Figure 2 provides an overview of the few-shot image translation problem [27]. Let X be a training set consists of images from K different domains. For each image in X, the class label is known. Note that we operate in the unsupervised setting where corresponding images between domains are *unavailable.* The few-shot image-to-image translation model learns to map a "content" image in one domain to an analogous image in the domain of the input "style" examples. In the test phase, the model sees a few example images from an unseen domain and performs the translation.

During training, a pair of content and style images x_c, x_k is randomly sampled. Let x_k denote a style image in domain k. The content image x_c can be from any domains in K. The generator G translates x_c into an image of class k ($\bar{\boldsymbol{x}}_k$) while preserving the content information of x_c.

$$\bar{\boldsymbol{x}}_k = G(x_c, x_k) \tag{1}$$

In the test phase, the generator takes style images from a domain unseen during training, which we call the target domain. The target domain can be any related domain, not included in K.

FUNIT Baseline. FUNIT uses an example-guided conditional generator architecture as illustrated in the top-left of Fig. 4. It consists of three modules, 1) content encoder E_c, 2) style encoder E_s, and 3) image decoder F. E_c takes content image x_c as input and outputs content embedding z_c. E_s takes style image x_s as input and output style embedding z_s. Then, F generates an image using z_c and z_s, where z_s is used to generate the mean and scale parameters of adaptive instance normalization (AdaIN) layers [15] in F. The AdaIN design is based on

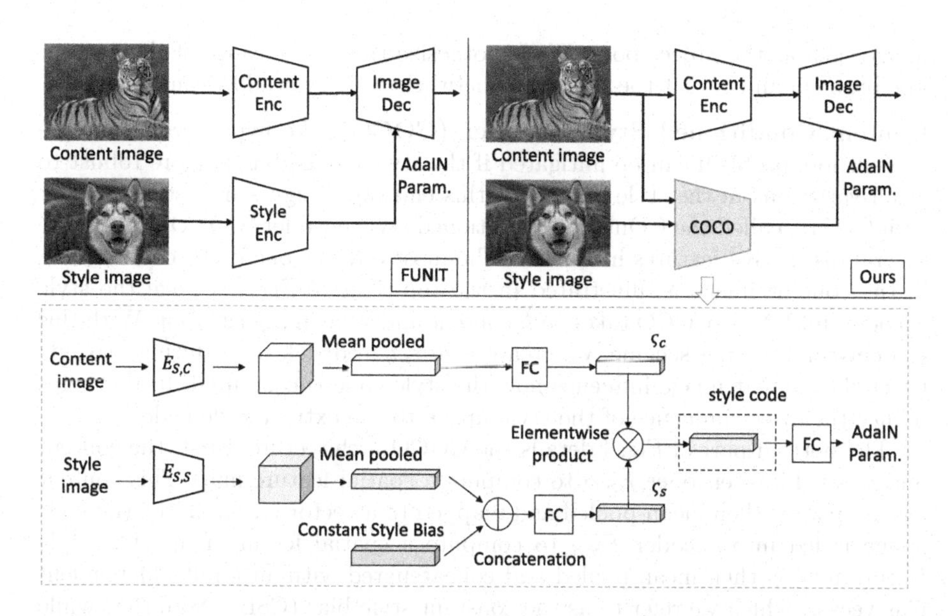

Fig. 4. Top. The FUNIT baseline [27] vs. our COCO-FUNIT. To highlight, we use a novel style encoder called the content-conditioned style encoder where the content image is also used in computing the style code for few-shot unsupervised image-to-image translation. **Bottom.** Detail design of the content-conditioned style encoder. Please refer to the main text for more details.

the assumption that the domain-specific information can be governed by the first and second order statistics of the activation and has been used in several GAN frameworks [16,20,27]. We further note that when multiple example/style images are present. FUNIT extracts a style code from each image and uses the average style code as the final input to F. To sum up, in FUNIT the image translation is formalized as follows,

$$z_c = E_c(x_c), \quad z_s = E_s(x_s), \quad \bar{x} = F(z_c, z_s). \tag{2}$$

Content Loss. As illustrated in Fig. 3, the FUNIT method suffers from the content loss problem—the translation result is not well-aligned with the input image. While a direct theoretical analysis is likely elusive, we conduct an empirical study, aiming at identify the cause of the content loss problem. As shown in Fig. 5, we compute different translation results of a content image based on a different style image where each of the style images is cropped from the same original style image. In the plot, we show variations of the deviation of the extracted style code due to different crops. Ideally, the plot should be constant as long as the crop covers sufficient appearance signature of the target class since that should be all required to generate a translation in the unseen domain. However, the FUNIT style encoder produces very different style codes as using different crops. Clearly, the style code contains other information about the style image such as the object pose. We hypothesize this is the cause of the content loss problem and revisit the translator network design for addressing it.

Content-Conditioned Style Encoder (COCO). We hypothesize that the content loss problem can be mitigated if the style embedding is more robust to small variations in the style image. To this end, we design a new style encoder architecture, called the COntent-COnditioned style encoder (COCO). There are several distinctive features in COCO. The most obvious one is the conditioning in the content image as illustrated in the top-right of Fig. 4. Unlike the style encoder in FUNIT, COCO takes *both* content and style image as input. With this content-conditioning scheme, we create a *direct* feedback path during learning to let the content image influence how the style code is computed. It also helps reduce the direct influence of the style image to the extract style code.

The bottom part of Fig. 4 details the COCO architecture. First, the content image is fed into encoder $E_{S,C}$ to compute a spatial feature map. This content feature map is then mean-pooled and mapped to a vector ζ_c. Similarly, the style image is fed into encoder $E_{S,S}$ to compute a spatial feature map. The style feature map is then mean-pooled and concatenated with an input-independent bias vector, which we refer to as the constant style bias (CSB). Note that while the regular bias in deep networks is added to the activations, in CSB, the bias is concatenated with the activations. The CSB provides a fixed input to the style encoder, which helps compute a style code that is less sensitive to the variations in the style image. In the experiment section, we show that the CSB can also be used to control the type of appearance information that is transmitted from the style image. When the CSB is activated, mostly texture-based appearance information is transferred. Note that the dimension of the CSB is set to 1024 through the paper.

The concatenation of the style vector and the CSB is mapped to a vector ζ_s via a fully connected layer. We then perform an element-wise product operation to ζ_c and ζ_s, which is our final style code. The style code is then mapped to produce the AdaIN parameters for generating the translation. Through this element-wise product operation, the resulting style code is heavily influenced by the content image. One way to look at this mechanism is that it produces a customized style code for the input content image.

We use the COCO as a drop-in replacement for the style encoder in FUNIT. Let ϕ denote the COCO mapping. The translation output is then computed via

$$z_c = E_c(x_c),\ z_s = \phi(E_{s,s}(x_s), E_{s,c}(x_c)),\ \bar{\boldsymbol{x}} = F(z_c, z_s). \tag{3}$$

As shown in Fig. 5, the style code extracted by the COCO is more robust to variations in the style image. Note that we set $E_{S,C} \equiv E_C$ to keep the number of parameters in our model similar to that in FUNIT.

We note that the proposed COCO architecture shows only one way to generate the style code conditioned on the content and to utilize the CSB. Certainly, there exist other design choices that could potentially lead to better translation performance. However, since this is the first time these two components are used for the few-shot image-to-image translation task, we focus on analyzing their contribution in one specific design, i.e., our design. An exhaustive exploration is beyond the scope of the paper and is left for future work.

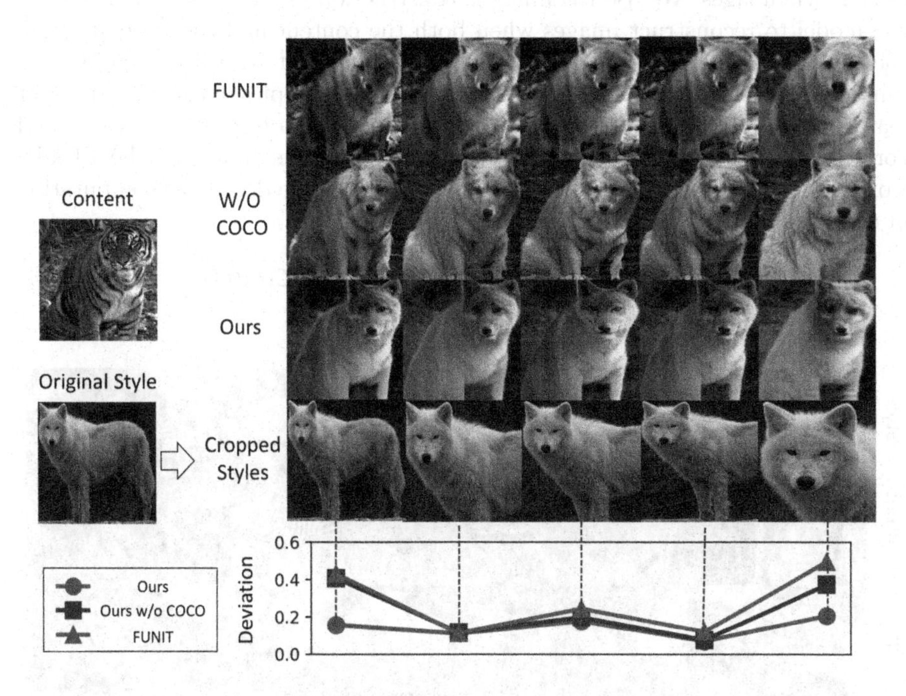

Fig. 5. We compare variations of the computed style codes due to variations in the style images for different methods. Note that for a fair comparison, in addition to the original FUNIT baseline [27], we create an improved FUNIT method by using our improved design for the content encoder, image decoder, and discriminator, which is termed "Ours w/o COCO". "Ours" is our full algorithm where we use COCO as a drop-in replacement for the style encoder in the FUNIT framework. In the bottom part of the figure, we plot the variations of the style code due to using different crops of a style image. Specifically, the style code for each style image is first extracted for each method. We then compute the mean of the style codes for each method. The magnitudes of the deviations from the mean style code are then plotted. Note that to calibrate the network weights in different methods, all the style codes are first normalized by the mean extracted from 500 style images for each method. As shown in the figure, "Ours" produces more consistent translation outputs, which is a direct consequence of a more consistent style code extraction mechanism.

In addition to the COCO, we also improve the design of the content encoder, image decoder, and discriminator in the FUNIT work [27]. For the content encoder and image decoder, we find that replacing the vanilla convolutional layers in the original design with residual blocks [14] improves the performance so does replacing the multi-task adversarial discriminator with the project-based discriminator [32]. In Appendix D of our full technical report [36], we report their individual contribution to the few-shot image translation performance.

Learning. We train our model using three objective terms. We use the GAN loss ($\mathcal{L}_{\text{GAN}}(D, G)$) to ensure the realism of the generated images given the class of the style images. We use the image reconstruction loss ($\mathcal{L}_{\text{R}}(G)$) to encourage the model to reconstruct images when both the content and the style are from the same domain. We use the discriminator feature matching loss ($\mathcal{L}_{\text{FM}}(G)$) to minimize the feature distance between real and fake samples in the discriminator feature space, which has the effect of stabilizing the adversarial training and contributes to generating better translation outputs as shown in the FUNIT work. In Appendix B of our full technical report [36], we detail the computation of each loss. Overall the objective is

$$\min_{D} \max_{G} \mathcal{L}_{\text{GAN}}(D, G) + \lambda_{\text{R}} \mathcal{L}_{\text{R}}(G) + \lambda_{\text{F}} \mathcal{L}_{\text{FM}}(G), \tag{4}$$

Fig. 6. Results on one-shot image-to-image translation. Column 1 & 2 are from the Carnivores dataset. Column 3 & 4 are from the Birds dataset. Column 5 & 6 are from the Mammals dataset. Column 7 & 8 are from the Motorbikes dataset.

Table 1. Results on the benchmark datasets.

Dataset	Method	mFID ↓	PAcc ↑	mIoU ↑	User style Preference ↑	User content Preference ↑
Carnivores	FUNIT	147.8	59.8	44.6	16.5	11.9
	Ours	**107.8**	**66.5**	**52.1**	**83.5**	**88.1**
Mammals	FUNIT	245.8	35.3	23.3	23.6	27.8
	Ours	**109.3**	**48.8**	**35.5**	**76.4**	**72.2**
Birds	FUNIT	89.2	52.4	37.2	38.5	37.5
	Ours	**74.6**	**53.3**	**38.3**	**61.5**	**62.5**
Motorbikes	FUNIT	275.0	85.6	73.8	17.8	17,4
	Ours	**56.2**	**94.6**	**90.3**	**82.2**	**82.6**

where λ_R and λ_F denote trade-off parameters for two losses. We set λ_R 0.1 and λ_F 1.0 in all of the experiments.

Fig. 7. Two-shot image translation results on the Carnivores dataset.

Fig. 8. Two-shot image translation results on the Birds dataset.

4 Experiments

We evaluate our method on challenging datasets that contain large pose variations, part variations, and category variations. Unlike existing few-shot image-to-image translation works, which focus on translations between reasonably-aligned images or simple objects, our interest is in the translations between likely misaligned images of highly articulate objects. Throughout the experiments, we use 256×256 as our default image resolution for both inputs and outputs.

Fig. 9. Two-shot image translation results on the Mammals dataset.

Implementation. We use Adam [22] with $lr = 0.0001$, $\beta_1 = 0.0$, and $\beta_2 = 0.999$ for all methods. Spectral normalization [31] is applied to the discriminator. The final generator is a historical average version of the intermediate generators [19] where the update weight is 0.001. We train the model for 150,000 iterations in total. For every competing model, we compute the scores every 10,000 iterations and report the scores of the iteration that achieves the smallest mFID. Each training batch consists of 64 content images, which are evenly distributed on a DGX machine with 8 V100 GPUs, each with 32 GB RAM.

Datasets. We benchmark our method using 4 datasets. Each of the dataset contains objects with diverse poses, parts, and appearances.

- *Carnivores.* We build the dataset using images from the ImageNet dataset[7]. We pick up images from the 149 carnivorous animals and used 119 as the source/seen classes and 30 as the target/unseen classes.

Table 2. Ablation study on the Carnivores and Birds dataset. "Ours w/o CC" represents a baseline where the content conditioning part in COCO is removed. "Ours w/o CSB" represents a baseline where the CSB is removed. Detailed architecture of these baselines are given in Appendix A of our full technical report [36].

Method	Carnivores			Birds		
	mFID↓	PAcc↑	mIou ↑	mFID↓	PAcc↑	mIou ↑
Ours w/o COCO	**99.6**	62.5	47.8	**68.8**	52.8	37.9
Ours w/o CSB	107.1	61.8	46.9	74.1	52.5	37.7
Ours w/o CC	110.0	**66.7**	**52.1**	75.3	52.8	37.9
Ours	107.8	66.5	**52.1**	74.6	**53.3**	**38.3**

- *Mammals.* We collect 152 classes of herbivore animal images using Google image search and combine them with the Carnivores dataset to build the Mammals dataset. Out of the 301 classes, 236 classes are used for the source/seen and the rest is used for the target/unseen.
- *Birds.* We collect 205 classes of bird images using Google image search. 172 classes are used for training and the rest is used for the testing.
- *Motorbikes.* We also collected 109 classes of motorbike images in the same way. 92 classes are used as the source and the rest is used for the target.

Evaluation Protocol. For each dataset, we train a model using the source classes mentioned above and test the performance on the target classes for each competing methods. In the test phase, we randomly sample 25,000 content images and pair each of them with a few style images from a target class to compute the translation. Unless specified otherwise, we use the one-shot setting for performance evaluation as it is the most challenging few-shot setting. We evaluate the quality of the translated images using various metrics as explained below.

Performance Metrics. Ideally, a translated image should keep the structure of the input content image, such as the pose or scale of body parts, unchanged when emulating the appearances of the unseen domain. Existing work mainly focused on the style transfer evaluation because the experiments are performed on well-aligned images or images of simple objects. To consider both the style translation and content preservation, we employ the following metrics. First, we evaluate the style transfer by measuring distance between the distribution of the translated images and the distribution of the real images in the unseen domain. Second, the content preservation is evaluated by measuring correspondence between a content and a translated image. Third, we conduct a user study to compute human preference scores on both the style transfer and content preservation of the translation results. The details of the performance metrics are given in Appendix C of our technical report [36].

Baseline. We compare our method with the FUNIT method because it outperforms many baselines with a large margin as described in Liu *et al.* [27].

Therefore, a direct comparison with this baseline can verify the effectiveness of the proposed method for the few-shot image-to-image translation task.

Main Results. The comparison results is summarized in Table 1. As shown, our method outperforms FUNIT by a large margin in all the datasets on both automatic metrics and human preference scores. This validates the effectiveness of our method for few-shot unsupervised image-to-image translation. Figure 6 and 3 compare the one-shot translation results computed by the FUNIT method and our approach. We find images generated by the FUNIT method contain many artifacts while our method can generate photorealistic and faithful translation outputs. In Fig. 7, 8, and 9, we further visualize two-shot translation results. More visualization results are provided in the supplementary materials (Fig. 10).

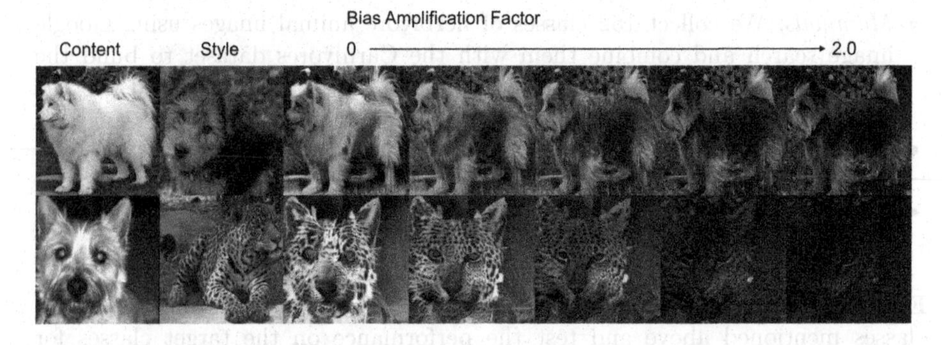

Fig. 10. By changing the amplification factor λ of the CSB, our model generates different translation outputs for the same pair of content and style images.

Fig. 11. We interpolate the style codes from two example images from two different unseen domains. Our model can generate photorealistic results using these interpolated style codes. More results are in the supplementary materials.

Ablation Study. In Table 2, we ablate modules in our architecture and measure their impact on the few-shot translation performance using the Carnivores and Birds datasets. Now, let us walk through the results. First, we find using the CSB improve content preservation scores ("Ours w/o CSB" vs "Ours"), reflected by the better PAcc and mIoU scores achieved. Second, using content conditioning improves style transferring ("Ours w/o CC" vs "Ours"), reflected by the better mFID scores achieved. We also note that despite "Ours w/o COCO" achieves a better mFID, it is in the expense of large content loss.

Effect of the CSB. We conduct an experiment to understand how the CSB designed added to our COCO influences the translation results. Specifically, during testing, we multiply the CSB with a scalar λ. We then change the λ value to visualize its effect as shown in Fig. reffig:csbspsmanipulation. Interestingly, different values of λ generate different translation results. When the value is small, the model mostly changes the texture of the content image. With a large λ value, both the shape and texture are changed.

Unseen Style Blending. Here, we show an application where we combine two style images from two unseen domains to create a new unseen domain. Specifically, we first extract two style codes from two images from two different unseen domains. We then mix their styles by linear interpolating the style codes. The results are shown in Fig. 11 where the leftmost image is the content and row indicated by s1 and s2 are the two style images. We find the intermediate style codes render plausible translation results.

Failure Cases. While our approach effectively addresses the content loss problem, it still have several failure modes. We discuss these failure modes in the supplementary materials.

5 Conclusion

We introduced the COCO-FUNIT architecture, a new style encoder for few-shot image-to-image translation that extracts the style code from the example images from the unseen domain conditioning on the input content image and uses a constant style bias design. We showed that the COCO-FUNIT can effectively address the content loss problem, proven challenging for few-shot image-to-image-translation.

Acknowledgements. We would like to thank Jan Kautz for his insightful feedback on our project. Kuniaki Saito and Kate Saenko were supported by Honda, DARPA and NSF Award No. 1535797. Kuniaki Saito contributed to this work during his internship at NVIDIA.

References

1. AlBahar, B., Huang, J.B.: Guided image-to-image translation with bi-directional feature transformation. In: IEEE Conference on Computer Vision and Pattern Recognition (CVPR) (2019)

2. Anoosheh, A., Agustsson, E., Timofte, R., Van Gool, L.: ComboGAN: unrestrained scalability for image domain translation. arXiv preprint arXiv:1712.06909 (2017)
3. Benaim, S., Wolf, L.: One-shot unsupervised cross domain translation. In: Advances in Neural Information Processing Systems (NeurIPS) (2018)
4. Chen, Q., Koltun, V.: Photographic image synthesis with cascaded refinement networks. In: IEEE International Conference on Computer Vision (ICCV) (2017)
5. Choi, Y., Choi, M., Kim, M., Ha, J.W., Kim, S., Choo, J.: StarGAN: unified generative adversarial networks for multi-domain image-to-image translation. In: IEEE Conference on Computer Vision and Pattern Recognition (CVPR) (2018)
6. Choi, Y., Uh, Y., Yoo, J., Ha, J.W.: StarGAN v2: diverse image synthesis for multiple domains. In: IEEE Conference on Computer Vision and Pattern Recognition (CVPR) (2020)
7. Deng, J., Dong, W., Socher, R., Li, L.J., Li, K., Fei-Fei, L.: ImageNet: a large-scale hierarchical image database. In: IEEE Conference on Computer Vision and Pattern Recognition (CVPR) (2009)
8. Gatys, L.A., Ecker, A.S., Bethge, M.: Texture synthesis using convolutional neural networks. In: Advances in Neural Information Processing Systems (NeurIPS) (2015)
9. Gokaslan, A., Ramanujan, V., Ritchie, D., Kim, K.I., Tompkin, J.: Improving shape deformation in unsupervised image-to-image translation. In: Ferrari, V., Hebert, M., Sminchisescu, C., Weiss, Y. (eds.) ECCV 2018. LNCS, vol. 11216, pp. 662–678. Springer, Cham (2018). https://doi.org/10.1007/978-3-030-01258-8_40
10. Goodfellow, I., et al: Generative adversarial networks. In: Advances in Neural Information Processing Systems (NeurIPS) (2014)
11. Gu, Q., Wang, G., Chiu, M.T., Tai, Y.W., Tang, C.K.: LADN: local adversarial disentangling network for facial makeup and de-makeup. In: IEEE International Conference on Computer Vision (ICCV) (2019)
12. Gulrajani, I., Ahmed, F., Arjovsky, M., Dumoulin, V., Courville, A.C.: Improved training of Wasserstein GANs. In: Advances in Neural Information Processing Systems (NeurIPS) (2017)
13. Han, X., Wu, Z., Wu, Z., Yu, R., Davis, L.S.: VITON: an image-based virtual try-on network. In: IEEE Conference on Computer Vision and Pattern Recognition (CVPR) (2018)
14. He, K., Zhang, X., Ren, S., Sun, J.: Deep residual learning for image recognition. In: IEEE Conference on Computer Vision and Pattern Recognition (CVPR) (2016)
15. Huang, X., Belongie, S.: Arbitrary style transfer in real-time with adaptive instance normalization. In: IEEE International Conference on Computer Vision (ICCV) (2017)
16. Huang, X., Liu, M.-Y., Belongie, S., Kautz, J.: Multimodal unsupervised image-to-image translation. In: Ferrari, V., Hebert, M., Sminchisescu, C., Weiss, Y. (eds.) ECCV 2018. LNCS, vol. 11207, pp. 179–196. Springer, Cham (2018). https://doi.org/10.1007/978-3-030-01219-9_11
17. Hui, L., Li, X., Chen, J., He, H., Yang, J.: Unsupervised multi-domain image translation with domain-specific encoders/decoders. arXiv preprint arXiv:1712.02050 (2017)
18. Isola, P., Zhu, J.Y., Zhou, T., Efros, A.A.: Image-to-image translation with conditional adversarial networks. In: IEEE Conference on Computer Vision and Pattern Recognition (CVPR) (2017)
19. Karras, T., Aila, T., Laine, S., Lehtinen, J.: Progressive growing of GANs for improved quality, stability, and variation. In: International Conference on Learning Representations (ICLR) (2018)

20. Karras, T., Laine, S., Aila, T.: A style-based generator architecture for generative adversarial networks. In: IEEE Conference on Computer Vision and Pattern Recognition (CVPR) (2019)
21. Kim, T., Cha, M., Kim, H., Lee, J.K., Kim, J.: Learning to discover cross-domain relations with generative adversarial networks. In: International Conference on Machine Learning (ICML) (2017)
22. Kingma, D., Ba, J.: Adam: A method for stochastic optimization. In: International Conference on Learning Representations (ICLR) (2015)
23. Lee, H.-Y., Tseng, H.-Y., Huang, J.-B., Singh, M., Yang, M.-H.: Diverse image-to-image translation via disentangled representations. In: Ferrari, V., Hebert, M., Sminchisescu, C., Weiss, Y. (eds.) ECCV 2018. LNCS, vol. 11205, pp. 36–52. Springer, Cham (2018). https://doi.org/10.1007/978-3-030-01246-5_3
24. Li, Y., Fang, C., Yang, J., Wang, Z., Lu, X., Yang, M.H.: Universal style transfer via feature transforms. In: Advances in Neural Information Processing Systems (NeurIPS) (2017)
25. Liang, X., Lee, L., Dai, W., Xing, E.P.: Dual motion GAN for future-flow embedded video prediction. In: Advances in Neural Information Processing Systems (NeurIPS) (2017)
26. Liu, M.Y., Breuel, T., Kautz, J.: Unsupervised image-to-image translation networks. In: Advances in Neural Information Processing Systems (NeurIPS) (2017)
27. Liu, M.Y., Huang, X., Mallya, A., Karras, T., Aila, T., Lehtinen, J., Kautz, J.: Few-shot unsupervised image-to-image translation. In: IEEE International Conference on Computer Vision (ICCV) (2019)
28. Liu, M.Y., Tuzel, O.: Coupled generative adversarial networks. In: Advances in Neural Information Processing Systems (NeurIPS) (2016)
29. Liu, X., Yin, G., Shao, J., Wang, X., et al.: Learning to predict layout-to-image conditional convolutions for semantic image synthesis. In: Advances in Neural Information Processing Systems (NeurIPS) (2019)
30. Mao, X., Li, Q., Xie, H., Lau, R.Y., Wang, Z., Smolley, S.P.: Least squares generative adversarial networks. In: IEEE International Conference on Computer Vision (ICCV) (2017)
31. Miyato, T., Kataoka, T., Koyama, M., Yoshida, Y.: Spectral normalization for generative adversarial networks. In: International Conference on Learning Representations (ICLR) (2018)
32. Miyato, T., Koyama, M.: cGANs with projection discriminator. In: International Conference on Learning Representations (ICLR) (2018)
33. Park, T., Liu, M.Y., Wang, T.C., Zhu, J.Y.: Semantic image synthesis with spatially-adaptive normalization. In: IEEE Conference on Computer Vision and Pattern Recognition (CVPR) (2019)
34. Pumarola, A., Agudo, A., Martinez, A.M., Sanfeliu, A., Moreno-Noguer, F.: GANimation: anatomically-aware facial animation from a single image. In: Ferrari, V., Hebert, M., Sminchisescu, C., Weiss, Y. (eds.) ECCV 2018. LNCS, vol. 11214, pp. 835–851. Springer, Cham (2018). https://doi.org/10.1007/978-3-030-01249-6_50
35. Qi, X., Chen, Q., Jia, J., Koltun, V.: Semi-parametric image synthesis. In: IEEE Conference on Computer Vision and Pattern Recognition (CVPR) (2018)
36. Saito, K., Saenko, K., Liu, M.Y.: COCO-FUNIT: few-shot unsupervised image translation with a content conditioned style encoder. arXiv preprint arXiv:2007.07431 (2020)
37. Shen, Z., Huang, M., Shi, J., Xue, X., Huang, T.S.: Towards instance-level image-to-image translation. In: IEEE Conference on Computer Vision and Pattern Recognition (CVPR) (2019)

38. Siarohin, A., Lathuilière, S., Tulyakov, S., Ricci, E., Sebe, N.: Animating arbitrary objects via deep motion transfer. In: IEEE Conference on Computer Vision and Pattern Recognition (CVPR) (2019)
39. Taigman, Y., Polyak, A., Wolf, L.: Unsupervised cross-domain image generation. In: International Conference on Learning Representations (ICLR) (2017)
40. Wang, C., Zheng, H., Yu, Z., Zheng, Z., Gu, Z., Zheng, B.: Discriminative region proposal adversarial networks for high-quality image-to-image translation. In: Ferrari, V., Hebert, M., Sminchisescu, C., Weiss, Y. (eds.) ECCV 2018. LNCS, vol. 11205, pp. 796–812. Springer, Cham (2018). https://doi.org/10.1007/978-3-030-01246-5_47
41. Wang, M., Yang, G.Y., Li, R., Liang, R.Z., Zhang, S.H., Hall, P.M., Hu, S.M.: Example-guided style-consistent image synthesis from semantic labeling. In: IEEE Conference on Computer Vision and Pattern Recognition (CVPR) (2019)
42. Wang, T.C., Liu, M.Y., Tao, A., Liu, G., Kautz, J., Catanzaro, B.: Few-shot video-to-video synthesis. In: Advances in Neural Information Processing Systems (NeurIPS) (2019)
43. Wang, T.C., Liu, M.Y., Zhu, J.Y., Liu, G., Tao, A., Kautz, J., Catanzaro, B.: Video-to-video synthesis. In: Advances in Neural Information Processing Systems (NeurIPS) (2018)
44. Wang, T.C., Liu, M.Y., Zhu, J.Y., Tao, A., Kautz, J., Catanzaro, B.: High-resolution image synthesis and semantic manipulation with conditional GANs. In: IEEE Conference on Computer Vision and Pattern Recognition (CVPR) (2018)
45. Zakharov, E., Shysheya, A., Burkov, E., Lempitsky, V.: Few-shot adversarial learning of realistic neural talking head models. In: IEEE International Conference on Computer Vision (ICCV) (2019)
46. Zhao, B., Meng, L., Yin, W., Sigal, L.: Image generation from layout. In: IEEE Conference on Computer Vision and Pattern Recognition (CVPR) (2019)
47. Zhu, J.Y., Park, T., Isola, P., Efros, A.A.: Unpaired image-to-image translation using cycle-consistent adversarial networks. In: IEEE International Conference on Computer Vision (ICCV) (2017)
48. Zhu, J.Y., Zhang, R., Pathak, D., Darrell, T., Efros, A.A., Wang, O., Shechtman, E.: Toward multimodal image-to-image translation. In: Advances in Neural Information Processing Systems (NeurIPS) (2017)
49. Zhu, S., Urtasun, R., Fidler, S., Lin, D., Change Loy, C.: Be your own prada: fashion synthesis with structural coherence. In: IEEE International Conference on Computer Vision (ICCV) (2017)

Corner Proposal Network for Anchor-Free, Two-Stage Object Detection

Kaiwen Duan[1], Lingxi Xie[2], Honggang Qi[1], Song Bai[3], Qingming Huang[1,4](✉), and Qi Tian[2](✉)

[1] University of Chinese Academy of Sciences, Beijing, China
duankaiwen17@mails.ucas.ac.cn, {hgqi,qmhuang}@ucas.ac.cn
[2] Huawei Inc., Shenzhen, China
198808xc@gmail.com, tian.qi1@huawei.com
[3] Huazhong University of Science and Technology, Wuhan, China
songbai.site@gmail.com
[4] Peng Cheng Laboratory, Shenzhen, China

Abstract. The goal of object detection is to determine the class and location of objects in an image. This paper proposes a novel anchor-free, two-stage framework which first extracts a number of object proposals by finding potential corner keypoint combinations and then assigns a class label to each proposal by a standalone classification stage. We demonstrate that these two stages are effective solutions for improving recall and precision, respectively, and they can be integrated into an end-to-end network. Our approach, dubbed Corner Proposal Network (CPN), enjoys the ability to detect objects of various scales and also avoids being confused by a large number of false-positive proposals. On the MS-COCO dataset, CPN achieves an AP of 49.2% which is competitive among state-of-the-art object detection methods. CPN also fits the scenario of computational efficiency, which achieves an AP of 41.6%/39.7% at 26.2/43.3 FPS, surpassing most competitors with the same inference speed. Code is available at https://github.com/Duankaiwen/CPNDet.

Keywords: Object detection · Anchor-free detector · Two-stage detector · Corner keypoints · Object proposals

1 Introduction

Powered by the rapid development of deep learning [21], in particular deep convolutional neural networks [13,18,35], researchers have designed effective algorithms for object detection [11]. This is a challenging task since objects can appear in any scale, shape, and position in a natural image, yet the appearance of objects of the same class can be very different.

The two keys to a detection approach are to find objects with different kinds of geometry (*i.e.*, high recall) as well as to assign an accurate label to each detected object (*i.e.*, high precision). Existing object detection approaches

A. Vedaldi et al. (Eds.): ECCV 2020, LNCS 12348, pp. 399–416, 2020.
https://doi.org/10.1007/978-3-030-58580-8_24

are roughly categorized according to how objects are located and how their classes are determined. For the first issue, early research efforts are mostly **anchor-based** [3,6,10,24,27,34,46], which involved placing a number of size-fixed bounding boxes on the image plane, while this methodology was later challenged by **anchor-free** [8,17,19,39,42,50] methods which suggested depicting each object with one or few keypoints and the geometry. These potential objects are named proposals, and for each of them, the class label is either inherited from a previous output or verified by an individual classifier trained for this purpose. This brings a debate between the so-called **two-stage** and **one-stage** approaches, in which people tend to believe that the former works slower but produces higher detection accuracy.

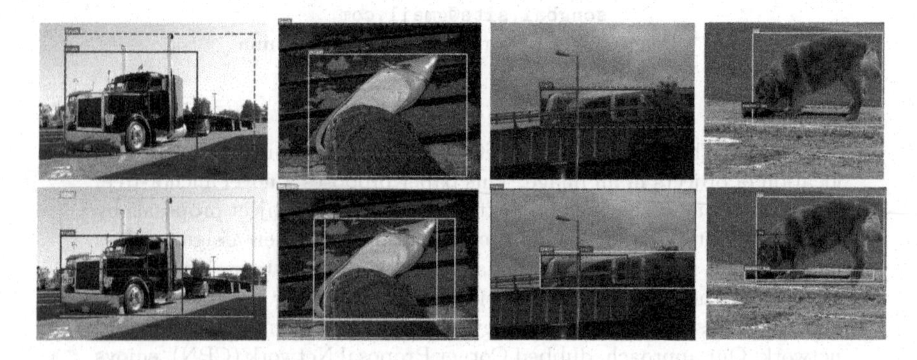

Fig. 1. Typical errors by existing object detection approaches (best viewed in color). **Top**: an anchor-based method (*e.g.*, Faster R-CNN [34]) may have difficulty in finding objects with a peculiar shape (*e.g.*, with a very large size or an extreme aspect ratio). **Bottom**: an anchor-free method (*e.g.*, CornerNet [19]) may mistakenly group irrelevant keypoints into an object. Green, blue and red bounding-boxes indicate true positives, false positives and false negatives, respectively. (Color figure online)

This paper provides an alternative opinion on the design of object detection approaches. There are two arguments. **First**, the recall of a detection approach is determined by its ability to locate objects of different geometries, especially those with rare shapes, and anchor-free methods (in particular, the methods based on locating the border of objects) are potentially better in this task. **Second**, anchor-free methods often incur a large number of false positives, and thus an individual classifier is strongly required to improve the precision of detection, see Fig. 1. Therefore, we inherit the merits of both anchor-free and two-stage object detectors and design an efficient, end-to-end implementation.

Our approach is named **Corner Proposal Network** (CPN). It detects an object by locating the top-left and bottom-right corners of it and then assigning a class label to it. We make use of the keypoint detection method of CornerNet [19] but, instead of grouping the keypoints with keypoint feature embedding, enumerate all valid corner combinations as potential objects. This leads to a large number of proposals, most of which are false positives. We then train a classifier to discriminate real objects from incorrectly paired keypoints based on the corresponding regional features. There are two steps for classification, with the first one, a binary classifier, filtering out a large part of proposals (*i.e.*, that do not correspond to objects), and the second one, with stronger features, re-ranking the survived proposals with multi-class classification scores.

The effectiveness of CPN is verified on the MS-COCO dataset [25], one of the most challenging object detection benchmarks. Using a 104-layer stacked Hourglass network [30] as the backbone, CPN reports an AP of 49.2%, which outperforms the previously best anchor-free detectors, CenterNet [8], by a significant margin of 2.2%. In particular, CPN enjoys even larger accuracy gain in detecting objects with peculiar shapes (*e.g.*, very large or small areas or extreme aspect ratios), demonstrating the advantage of using anchor-free methods for proposal extraction. Last but not least, CPN can also fit scenarios that desire for network efficiency. Working on a lighter backbone, DLA-34 [43], and switching off image flip in inference, CPN achieves an AP of 41.6% at 26.2 FPS or 39.7% at 43.3 FPS, surpassing most competitors with the same inference speed.

2 Related Work

Object detection is an important yet challenging task in computer vision. It aims to obtain a tight bounding-box as well as a class label for each object in an image. In recent years, with the rapid development of deep learning, most powerful object detection methods are based on training deep neural networks [10,11]. According to the way of localizing objects, existing detection approaches can be roughly categorized into anchor-based and anchor-free methods.

An **anchor-based approach** starts with placing a large number of anchors, which are regional proposals with different but fixed scales and shapes, and are uniformly distributed on the image plane. These anchors are then considered as object proposals and an individual classifier is trained to determine the objectness as well as the class of each proposal [34]. Beyond this framework, researchers made efforts in two aspects, namely, improving the basic quality of regional features extracted from the proposal, and arriving at a better alignment between the proposals and features. For the first type of efforts, typical examples include using more powerful network backbones [13,14,36] and using hierarchical features to represent a region [23,27,34]. Regarding the second type, there exist methods to align anchors to features [46,47], align features to anchors [5,7], and adjust the anchors after classification has been done [3,27,34].

Alternatively, an **anchor-free approach** does not assume the objects to come from uniformly distributed anchors. Early efforts including DenseBox [15] and UnitBox [44] proved that the detectors can achieve the detection task without anchors. Recently, anchor-free approaches have been greatly promoted by the development of keypoint detection [29] and the assist of the focal loss [24]. The fundamental of anchor-free approaches is usually one or few keypoints. Depending on how keypoints are used for object depiction, anchor-free approaches can be roughly categorized into point-grouping detectors and point-vector detectors. Point-grouping detectors, including CornerNet [19], CenterNet [8], ExtremeNet [49], *etc.*, group more than one keypoints into an object, while the point-vector detectors such as FCOS [39], CenterNet [48], FoveaBox [17], SAPD [50], *etc.*, use a keypoint and a vector of object geometry (*e.g.*, the width, height, or its distance to the borders) to determine the shape of objects.

Based on the object proposals, it remains to determine whether each proposal is an object and what class of object it is. There is also discussion on using **two-stage** and **one-stage** detectors for object detection. A two-stage detector [3,12, 22,31,34] refers to an individual classifier is trained for this purpose, while a one-stage detector [5,24,27,33,46] mostly uses classification cues from the previous stage. Two-stage detectors are often more accurate but slower, compared to one-stage detectors. To accelerate it, an efficient method is to partition classification into two steps [3,6,12,23,31], with the first step filtering out most easy judged false positives, and the second step using heavier computation to assign each survived proposal a class label.

3 Our Approach

Object detection starts with an image $\mathbf{I}$ denoted by raw pixels, on which a few rectangles, often referred to as bounding-boxes, that tightly covers the objects are labeled with a class label. Denote a ground-truth bounding-box as $\mathbf{b}_n^\star$, $n = 1, 2, \ldots, N$, the corresponding class label as $c_n^\star$, and $\mathbf{I}_n^\star$ represents the image region within the corresponding bounding-box. The goal is to locate a few bounding-boxes, $\mathbf{b}_m$, $m = 1, 2, \ldots, M$, and assign each one with a class label, c_m, so that the sets of $\{\mathbf{b}_n^\star, c_n^\star\}_{n=1}^N$ and $\{\mathbf{b}_m, c_m\}_{m=1}^M$ are as close as possible.

3.1 Anchor-Based or Anchor-Free? One-Stage or Two-Stage?

We focus on two important choices of object detection, namely, whether to use anchor-based or anchor-free methods for proposal extraction, and whether to use one-stage or two-stage methods for determining the class of proposals. Based on these discussions, we present a novel framework in the next subsection.

We first investigate anchor-based *vs.* anchor-free methods. Anchor-based methods first place a number of anchors on the image as object proposals and then use an individual classifier to judge the objectness and class of each proposal. Most often, each anchor is associated with a specific position on the image and its size is fixed, although the following process named bounding-box regression can slightly change its geometry. Anchor-free methods do not assume the objects to come from anchors of relatively fixed geometry, and instead, locate one or few keypoints of an object and determine its geometry and/or class afterward.

Our core opinion is that **anchor-free methods have better flexibility of locating objects with arbitrary geometry, and thus a higher recall.** This is mainly due to the design nature of anchors, which is mostly empirical (*e.g.*, to reduce the number of anchors and improve efficiency, only common object sizes and shapes are considered), the detection algorithm is potential of lower flexibility and objects with a peculiar shape can be missing. Typical examples are shown in Fig. 1, and a quantitative study is provided in Table 1. We evaluate four object proposal extraction methods as well as our work on the MS-COCO validation dataset and show that anchor-free methods often have a higher overall recall, which is mainly due to their advantages in two scenarios. **First**, when the object is very large, *e.g.*, larger than 400^2 pixels, Faster R-CNN, an anchor-based approach, does not report a much higher recall. This is not expected because large objects should be easier to detect, as the other three anchor-free methods suggest. **Second**, Faster R-CNN suffers a very low recall when the aspect ratio of the object becomes peculiar, *e.g.*, 5:1 and 8:1, in which cases the recalls are significantly lower than CornerNet [19] and CenterNet [8], because no pre-defined anchors (also used in other variants [3,27,31]) can fit these objects. A similar phenomenon is also observed in FCOS, an anchor-free approach which represents an object by a keypoint and the distance to the border, because it is difficult to predict an accurate distance when the border is far from the center. Since CornerNet and CenterNet group corner (and center) keypoints into an object, they somewhat get rid of this trouble. Therefore, we choose anchor-free methods, in particular, **point-grouping** methods (CornerNet and CenterNet), to improve the recall of object detection. Moreover, we report the corresponding results of CPN, the method proposed in this paper, which demonstrates that CPN inherits the merits of CenterNet and CornerNet and has better flexibility of locating objects, especially with peculiar shapes.

Table 1. Comparison among the average recall (AR) of anchor-based and anchor-free detection methods. Here, the average recall is recorded for targets of different aspect ratios and different sizes. To explore the limit of the average recall for each method, we exclude the impacts of bounding-box categories and sorts on recall, and compute it by allowing at most 1000 object proposals. AR_{1+}, AR_{2+}, AR_{3+} and AR_{4+} denote box area in the ranges of $(96^2, 200^2]$, $(200^2, 300^2]$, $(300^2, 400^2]$, and $(400^2, +\infty)$, respectively. 'X' and 'HG' stand for ResNeXt and Hourglass, respectively.

Method	Backbone	AR	AR_{1+}	AR_{2+}	AR_{3+}	AR_{4+}	$AR_{5:1}$	$AR_{6:1}$	$AR_{7:1}$	$AR_{8:1}$
Faster R-CNN [34]	X-101-64 × 4d	57.6	73.8	77.5	79.2	86.2	43.8	43.0	34.3	23.2
FCOS [39]	X-101-64 × 4d	64.9	82.3	87.9	89.8	95.0	45.5	40.8	34.1	23.4
CornerNet [19]	HG-104	66.8	85.8	92.6	95.5	98.5	50.1	48.3	40.4	36.5
CenterNet [8]	HG-104	66.8	87.1	93.2	95.2	96.9	50.7	45.6	40.1	32.3
CPN (this work)	HG-104	68.8	88.2	93.7	95.8	99.1	54.4	50.6	46.2	35.4

However, anchor-free methods free the constraints of finding object proposals, it encounters a major difficulty of building a close relationship between keypoints and objects, since the latter often requires richer semantic information. As shown in Fig. 1, lacking semantics can incur a large number of false positives and thus harms the precision of detection. We take CornerNet [19] and CenterNet [8] with potentially high recalls as examples. As shown in Table 2, the CornerNets with 52-layer and 104-layer Hourglass networks achieved APs of 37.6% and 41.0% on

Table 2. Anchor-free detection methods such as CornerNet and CenterNet suffer a large number of false positives and can benefit from incorporating richer semantics for judgment. Here, $\mathrm{AP}_{\mathrm{original}}$, $\mathrm{AP}_{\mathrm{refined}}$, and $\mathrm{AP}_{\mathrm{correct}}$ indicate the AP of the original output, after non-object proposals are removed, and after the correct label is assigned to each survived proposal. Both $\mathrm{AP}_{\mathrm{refined}}$ and $\mathrm{AP}_{\mathrm{correct}}$ require ground-truth labels.

Method	Backbone	$\mathrm{AP}_{\mathrm{original}}$	$\mathrm{AP}_{\mathrm{refined}}$	$\mathrm{AP}_{\mathrm{correct}}$	$\mathrm{AR}^{100}_{\mathrm{original}}$	$\mathrm{AR}^{100}_{\mathrm{refined}}$	$\mathrm{AR}^{100}_{\mathrm{correct}}$
CornerNet [19]	HG-52	37.6	53.8	60.3	56.7	65.3	69.2
CornerNet [19]	HG-104	41.0	56.6	61.6	59.1	67.0	70.4
CenterNet [8]	HG-52	41.3	51.9	56.6	59.5	61.1	64.2
CenterNet [8]	HG-104	44.8	55.3	59.9	62.2	65.1	68.4

the MS-COCO validation dataset, while many of the detected 'objects' are false positives. Either when we remove the non-object proposals or assign each preserved proposal with a correct label, the detection accuracy goes up significantly. This observation also holds on CenterNet [8], which added a center point to filter out false positives but obviously did not remove them all. To further alleviate this problem, we need to inherit the merits of two-stage methods, which extract the features within proposals and train a classifier to filter out false positives.

3.2 The Framework of Corner Proposal Network

Motivated by the above, the goal of our approach is to integrate the advantages of anchor-free methods and alleviate their drawbacks by leveraging the mechanism of discrimination from two-stage methods. We present a new framework named **Corner-Proposal-Network** (CPN). It uses an anchor-free method to extract object proposals followed by efficient regional feature computation and classification to filter out false positives. Figure 2 shows the overall framework which contains two stages, and details of the two stages are elaborated as follows.

- **Stage 1: Anchor-free Proposals with Corner Keypoints**

The first stage is an anchor-free proposal extraction process, in which we assume that each object is located by two keypoints determining its top-left and bottom-right corners. We follow CornerNet [19] to locate an object with a pair of keypoints located in its top-left and bottom-right corners, respectively. For each class, we compute two heatmaps (*i.e.*, the top-left heatmap and the bottom-right heatmap, each value on a heatmap indicates the probability that a corner keypoint occurs in the corresponding position) with a 4×-reduced resolution compared to the original image. The heatmaps are equipped with two loss terms, namely, a focal loss $\mathcal{L}^{\mathrm{corner}}_{\mathrm{det}}$ to locate the keypoint on the heatmap and a offset loss $\mathcal{L}^{\mathrm{corner}}_{\mathrm{offset}}$ to learn its offset to the accurate corner position. Then, a fixed number of keypoints (K top-left and K bottom-right) are extracted from all heatmaps. This implies that each corner keypoint is equipped with a class label.

Next, each valid pair of keypoints defines an object proposal. Here by valid we mean that two keypoints belong to the same class (*i.e.*, extracted from the top-left heatmap and the bottom-right heatmap of the same class), and the x and y coordinates of the top-left point are smaller than that of the bottom-right

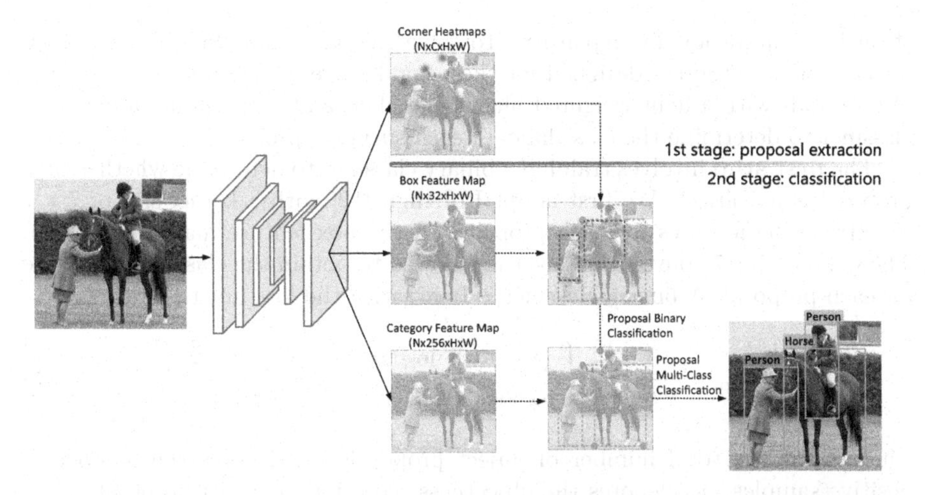

Fig. 2. The overall framework of proposed CPN. It first predicts corners to compose a number of object proposals, and then applies two-step classification to filter out false positives and assign a class label for each survived proposal.

point, respectively. This leads to a large number of false positives (incorrectly paired corner keypoints) on each image, and we leave the task of discriminating and classifying these proposals in the second stage.

As a side comment, we emphasize that although we extract object proposals based on CornerNet, the follow-up mechanism of determining objectness and class is quite different. CornerNet generates objects by projecting the keypoints to a one-dimensional space, and grouping keypoints with closely embedded numbers into the same instance. We argue that the embedding process, while necessary under the assumption that no additional computation can be used, can incur significant errors in pairing keypoints. In particular, there is no guarantee that the embedding function (assigning a number to each object) is learnable, and more importantly, the loss function only works in each training image to force the embedded numbers of different objects to be separated, but this mechanism often fails to generalize to unseen scenarios, *e.g.*, even when multiple training images are simply concatenated together, the embedding function that works well on separate images can fail dramatically. Differently, our method determines object instances using an individual classifier, which makes full use of the internal features to improve accuracy. Please refer to Table 6 for the advantage of an individual classifier over instance embedding.

- **Stage 2: Two-step Classification for Filtering Proposals**

Thanks to the high resolution of the keypoint heatmap and a flexible mechanism of grouping keypoints, the detected objects can be of an arbitrary size, and the upper-bound of recall is largely improved. However, this strategy increases the number of proposals. For example, we follow CenterNet [8] to set K to be 70, which results in an average of 2,500 object proposals on each image. Most of these proposals are false positives, and individually validating and classifying each of them is computationally expensive. To solve this issue, an efficient, **two-step** classification method is designed for the second stage, which first removes 80% of proposals with a light-weighted binary classifier, and then applies a fine-level classifier to determine the class label of each survived proposal.

The **first step** involves training a binary classifier to determine whether each proposal is an object. We first adopt RoIAlign [12] with a kernel size of 7×7 to extract the features for each proposal on the box feature map (see Fig. 2). Then, a $32 \times 7 \times 7$ convolution layer is followed to obtain the classification score for each proposal. A binary classifier is built, with the loss function being:

$$\mathcal{L}_{\text{prop}} = -\frac{1}{N} \sum_{m=1}^{M} \begin{cases} (1 - p_m)^{\alpha} \log(p_m), & \text{if IoU}_m \geqslant \tau \\ p_m^{\alpha} \log(1 - p_m), & \text{otherwise} \end{cases}, \tag{1}$$

where M is the total number of object proposals, N denotes the number of positive samples, p_m denotes the objectness score for the m-th proposal, $p_m \in [0, 1]$, and IoU_m denotes the maximum IoU value between the m-th proposal and all the ground-truth bounding-boxes. τ is the IoU threshold, set to be 0.7 throughout this paper. This is to sample a few positive examples to avoid training data imbalance [24]. $\alpha = 2$ is a hyper-parameter that smoothes the loss function. According to [24], we use $\pi = 0.1$, so the value of the biases is around -2.19.

The **second step** follows to assign a class label for each survived proposal. This step is very important, since the class labels associated to the corner key-points are not always reliable. Although we rely on the corner classes to reject invalid corner pairs, the consensus between them may be incorrect due to the lack of information from the ROI regions, so we need a more powerful classifier that incorporates the ROI features to make the final decision. To this end, we train another classifier with C outputs where C is the number of classes in the dataset. This classifier is also built upon the RoIAlign-features extracted in the first step, but instead extract the features from the category feature map (see Fig. 2) to preserve more information and a C dimensional vector is obtained using a $256 \times 7 \times 7$ convolution layer, for each of the survived proposals. Then, a C-way classifier is built with a similar loss function considering the class label:

$$\mathcal{L}_{\text{class}} = -\frac{1}{\hat{N}} \sum_{m=1}^{\hat{M}} \sum_{c=1}^{C} \begin{cases} (1 - q_{m,c})^{\beta} \log(q_{m,c}), & \text{if IoU}_{m,c} \geqslant \tau \\ q_{m,c}^{\beta} \log(1 - q_{m,c}), & \text{otherwise} \end{cases}, \tag{2}$$

where $\hat{M}$ and $\hat{N}$ denote the number of survived proposals and the number of positive samples within them, respectively. $\text{IoU}_{m,c}$ denotes the maximum IoU value between the m-th proposal and all the ground-truth bounding-boxes of the c-th class, and the IoU threshold, τ, remains unchanged. $q_{m,c}$ is the classification score for the c-th class of the m-th object, and β plays a similar role as α, and we also fix it to be 2 in this paper.

Here we emphasize the differences between DeNet [40] and our method, although they are similar in the idea level. First, we equip each corner keypoint with a multi-class label rather than a binary label, thus we can make use of the class labels to reject the unnecessary invalid corner pairs to save the computational costs of the overall framework. Second, we use an extra lightweight binary classification network to first reduce the number of proposals to be processed by the classification network, which improves the efficiency of our approach, while DeNet used one-step classification. Finally, we design a novel variant of the focal loss for the two classifiers, which is different from the maximum likelihood function in DeNet. This is mainly designed to solve the significant imbalance between the positive and negative proposals during the training process.

3.3 The Inference Process

The inference process simply repeats the training process but uses thresholds to filter out clearly low-quality proposals. Note that even with augmented positive training data, the predicted scores, p_m and $q_{m,c}$, are biased towards 0. So, in the inference stage, we use a relatively low threshold (0.2 in this paper) in the first step to allow more proposals to survive. For each proposal, provided the RoIAlign-features, the computational cost of the first classifier is about 10% of that of the second one. Under the threshold of 0.2, the average fraction of survived proposals is around 20%, making the overheads of these two stages comparable.

For each proposal survived to the second step, we assign it with up to 2 class labels, corresponding to the dominant class of the corner keypoints and that of the proposal (two classes may be identical, if not, the proposal becomes two proposals with potentially different scores). For each candidate class, we denote s_1 as the corner classification score (the average of two corner keypoints, in the range of $(0, 1)$), and s_2 as the regional classification score (the probability of the proposal class label, predicted by the multi-class classifier, also in the range of $(0, 1)$). We assume that both scores contribute to the final score, and a positive evidence should be added if either score is larger than 0.5. Therefore, we compute the score by $s_c = (s_1 + 0.5)(s_2 + 0.5)$, then we will apply normalization to rescale this score into the $[0, 1]$. We finally preserve 100 proposals with highest scores into evaluation. In Table 4, we will show that two classifiers provide complementary information and boost the detection accuracy.

4 Experiments

4.1 Dataset and Evaluation Metrics

We evaluate our approach on the MS-COCO dataset [25], one of the most popular object detection benchmarks. It contains a total of 120K images with more than 1.5 million bounding boxes covering 80 object categories, making it a very challenging dataset. Following the common practice [23,24], we train our model using the 'trainval35k', which is the union set of 80K training images and 35K (a subset of) validation images. We report evaluation results on the standard test-dev set, which has no public annotations, by uploading the results to the evaluation server. For the ablation studies and visualization experiments, we use the 5K validation images remained in the validation set.

We use the average precision (AP) metric defined in MS-COCO to characterize the performance of our approach as well as other competitors. AP computes the average precision over ten IoU thresholds (*i.e.*, 0.5:0.05:0.95), for all categories. Meanwhile, we follow the convention to report some other important metrics, *e.g.*, AP_{50} and AP_{75} are computed at single IoU thresholds of 0.50 and 0.75 [9], and $\mathrm{AP}_{\mathrm{small}}$, $\mathrm{AP}_{\mathrm{medium}}$, and $\mathrm{AP}_{\mathrm{large}}$ are computed under different object scales (*i.e.*, small for area $< 32^2$, medium for $32^2 <$ area $< 96^2$, and large for area $> 96^2$), respectively. All metrics are computed by allowing at most 100 proposals preserved on each testing image.

4.2 Implementation Details

We implement our method using Pytorch [32], and refer to some codes from CornerNet [19], mmdetection [4] and CenterNet [48]. We use CornerNet [19] and CenterNet [8] as our baselines. The stacked Hourglass networks [30] with 52 and 104 layers are trained for keypoint extraction, with all modifications made by CornerNet preserved. In addition, we experiment another backbone named DLA-34 [43]. We follow the modifications made by CenterNet [48], but replace the deformable convolutional layers with normal layers.

Training. All networks are trained from scratch, except the DLA-34, which is initialized with ImageNet pretrain. Cascade corner pooling [8] is used to help the network better detect corners. The input image is resized into 511 × 511, and the output resolutions for the four feature maps (the top-left and bottom-right heatmaps, the proposal and class feature maps) are all 128×128. The data augmentation strategy presented in [19] is used. The overall loss function is

$$\mathcal{L} = \mathcal{L}_{\mathrm{det}}^{\mathrm{corner}} + \mathcal{L}_{\mathrm{offset}}^{\mathrm{corner}} + \mathcal{L}_{\mathrm{prop}} + \mathcal{L}_{\mathrm{class}}, \tag{3}$$

which we use an Adam [16] optimizer to train our model. On eight NVIDIA Tesla-V100 (32 GB) GPUs, we use a batch size of 48 (6 samples on each card) and train the model for 200K iterations with a base learning rate of 2.5×10^{-4} followed by another 50K iterations with a reduced learning rate of 2.5×10^{-5}. The training lasts about 9 days, 5 days and 3 days for the backbones of Hourglass-104, Hourglass-52 and DLA-34, respectively.

Inference. Following [19], both single-scale and multi-scale detection processes are performed. For single-scale testing, we feed the image with the original resolution into the network, while for multi-scale testing, the image is resized into different resolutions (0.6×, 1×, 1.2×, 1.5×, and 1.8×) and then fed into the network. Flip argumentation is added to both single-scale or multi-scale evaluation. For multi-scale evaluation, the predictions for all scales (including the flipped variants) are fused into the final result. we use soft-NMS [2] to suppress the redundant bounding-boxes, and preserve 100 top-scored boxes for evaluation.

4.3 Comparisons with State-of-the-Art Detectors

We report the inference accuracy of CPN on the MS-COCO test-dev set and compare with the state-of-the-art detectors, as shown in Table 3. CPN obtains a significant improvement compared to CornerNet [19] and CenterNet [8], two direct baselines. Specifically, CPN-52 (indicating that the backbone is Hourglass-52) reports a single-scale testing AP of 43.9% and a multi-scale testing AP of 45.8%, which surpasses CornerNet-104, with a deeper backbone, by 3.4% and 3.7%, respectively. The best performance of CPN reaches an AP of 49.2%, surpassing all published anchor-free approaches to the best of our knowledge. Meanwhile, CPN is also competitive among anchor-based detectors, *e.g.*, CPN-52 reports a single-scale testing AP of 43.9% which is comparable to 44.1% of AlignDet [5], and CPN-104 reports a single-scale testing AP of 47.0% which is comparable 47.4% of PANet [26].

CPN also takes the lead in other metrics. For example, AP_{50} and AP_{75} reflect the accuracy of proposal localization and class prediction. Compared to CenterNet, CPN reports higher AP scores especially for AP_{75} (*e.g.*, CPN-104 reports a single-scale testing AP_{75} of 51.0%, claiming an improvement of 2.9% over CenterNet). This suggests that some inaccurate bounding boxes with IoU value between 0.5 and 0.7 are difficult for CenterNet to filter out with merely center information incorporated. AP_{S}, AP_{M} and AP_{L} reflect the detection accuracy for objects with different scales. CPN improves more for AP_{M} and AP_{L} than AP_{S} (*e.g.*, CPN-104 reports single-scale testing AP_{S}, AP_{M} and AP_{L} of 26.5%, 50.2% and 60.7%, which improves by 0.9%, 2.8% and 3.3% from CenterNet, respectively). This is because medium and large objects require richer semantic information to be extracted from the proposal, which is not likely to be handled well with a center keypoint.

4.4 Classification Improves Precision

We investigate the improvement of precision brought by the classification stage. Note that there are two classifiers, with the first one (binary) determines the objectness of each proposal, and the second one (multi-class) providing complementary information of the class label. We perform ablation study in Table 4 to analyze the contribution of individual classifiers – in the scenarios that the multi-class classifier is missing, we directly use the class label of the corner keypoints as the final prediction. On the one hand, both classifiers can improve the AP of the basic model (one-stage corner keypoint grouping) significantly (3.4% and 3.8% absolute gains). On the other hand, two classifiers provide complementary information, so that combining them can further improve the AP by more than 1%. This indicates that the semantic information required for determining the objectness and class label of a proposal is slightly different.

We have discussed the importance of the incorrect bounding box suppression for the improvement of detection accuracy and recall in Sect. 3.1. For a more intuitive analysis of false positives, we adopt the average false discovery rate

Table 3. Inference accuracy (%) of CPN and state-of-the-art detectors on the COCO *test-dev* set. CPN ranks among the top of state-of-the-art detectors. 'R', 'X', 'HG', 'DCN' and '†' denote ResNet, ResNeXt, Hourglass, Deformable Convolution Network [7], and multi-scale training or testing, respectively.

Method	Backbone	Input Size	AP	AP_{50}	AP_{75}	AP_S	AP_M	AP_L
Anchor-based:								
Faster R-CNN [23]	R-101	600	36.2	59.1	39.0	18.2	39.0	48.2
RetinaNet [24]	R-101	800	39.1	59.1	42.3	21.8	42.7	50.2
Cascade R-CNN [3]	R-101	800	42.8	62.1	46.3	23.7	45.5	55.2
Libra R-CNN [31]	X-101-64x4d	800	43.0	64.0	47.0	25.3	45.6	54.6
Grid R-CNN [28]	X-101-64x4d	800	43.2	63.0	46.6	25.1	46.5	55.2
YOLOv4 [1]	CSPDarknet-53	608	43.5	65.7	47.3	26.7	46.7	53.3
AlignDet [5]	X-101-32x8d	800	44.1	64.7	48.9	26.9	47.0	54.7
AB+FSAF [51] †	X-101-64x4d	800	44.6	65.2	48.6	29.7	47.1	54.6
FreeAnchor [47] †	X-101-32x8d	≤1280	47.3	66.3	51.5	30.6	50.4	59.0
PANet [26] †	X-101-64x4d	840	47.4	67.2	51.8	30.1	51.7	60.0
TridentNet [22] †	R-101-DCN	800	48.4	69.7	53.5	31.8	51.3	60.3
ATSS [45] †	X-101-64x4d-DCN	800	50.7	68.9	56.3	33.2	52.9	62.4
EfficientDet [38]	EfficientNet [37]	1536	**53.7**	**72.4**	**58.4**	–	–	–
Anchor-free:								
GA-Faster-RCNN [41]	R-50	800	39.8	59.2	43.5	21.8	42.6	50.7
FoveaBox [17]	R-101	800	42.1	61.9	45.2	24.9	46.8	55.6
ExtremeNet [49] †	HG-104	≤1.5×	43.2	59.8	46.4	24.1	46.0	57.1
FCOS [39] w/ imprv	X-101-64x4d	800	44.7	64.1	48.4	27.6	47.5	55.6
CenterNet [48] †	HG-104	≤1.5×	45.1	63.9	49.3	26.6	47.1	57.7
RPDet [42] †	R-101-DCN	800	46.5	67.4	50.9	30.3	49.7	57.1
SAPD [50]	X-101-64x4d	800	45.4	65.6	48.9	27.3	48.7	56.8
SAPD [50]	X-101-64x4d-DCN	800	47.4	67.4	51.1	28.1	50.3	61.5
CornerNet [19]	HG-104	ori.	40.5	56.5	43.1	19.4	42.7	53.9
CornerNet [19] †	HG-104	≤1.5×	42.1	57.8	45.3	20.8	44.8	56.7
CenterNet [8]	HG-104	ori.	44.9	62.4	48.1	25.6	47.4	57.4
CenterNet [8] †	HG-104	≤1.8×	47.0	64.5	50.7	28.9	49.9	58.9
CPN	DLA-34	ori.	41.7	58.9	44.9	20.2	44.1	56.4
CPN	HG-52	ori.	43.9	61.6	47.5	23.9	46.3	57.1
CPN	HG-104	ori.	47.0	65.0	51.0	26.5	50.2	60.7
CPN †	DLA-34	≤1.8×	44.5	62.3	48.3	25.2	46.7	58.2
CPN †	HG-52	≤1.8×	45.8	63.9	49.7	26.8	48.4	59.4
CPN †	HG-104	≤1.8×	**49.2**	**67.4**	**53.7**	**31.0**	**51.9**	**62.4**

Table 4. The detection performance (%) of different classification options on CPN.

Backbone	B-Classifier	M-Classifier	AP	AP_{50}	AP_{75}	AP_S	AP_M	AP_L	AR_{100}
HG-52			38.6	54.5	41.4	19.0	40.6	54.4	57.4
	✓		42.0	59.7	45.1	24.2	45.0	56.6	60.5
		✓	42.4	60.2	45.7	23.3	44.7	58.6	59.0
	✓	✓	**43.8**	**61.4**	**47.4**	**24.7**	**46.5**	**60.0**	**60.9**

Table 5. We report the average false discovery rates (%, lower is better) for CornerNet, CenterNet and CPN on the MS-COCO validation dataset. The results show that our approach generates fewer false positives. Under the same corner keypoint extractor, this is the key to outperform the baselines in the AP metrics.

Method	Backbone	AF	AF_5	AF_{25}	AF_{50}	AF_S	AF_M	AF_L
CornerNet [19]	HG-52	40.4	35.2	39.4	46.7	62.5	36.9	28.0
CenterNet [8]	HG-52	35.1	30.7	34.2	40.8	53.0	31.3	24.4
CPN	HG-52	**33.4**	**29.5**	**32.5**	**38.6**	**52.0**	**29.2**	**21.0**
CornerNet [19]	HG-104	37.8	32.7	36.8	43.8	60.3	33.2	25.1
CenterNet [8]	HG-104	32.4	28.2	31.6	37.5	50.7	27.1	23.0
CPN	HG-104	**30.6**	**26.9**	**29.7**	**35.5**	**48.8**	**25.7**	**19.2**

(AF) metric[1] [8] to quantify the fraction of incorrectly grouped proposals for different detectors. Results are shown in Table 5. CPN-52 and CPN-104 report AF of 33.4% and 30.6%, respectively, which are lower than the direct baselines, CornerNet and CenterNet.

Table 6. The detection performance (%) of using different ways (instance embedding and binary classification) to determine the validity of a proposal.

Method	Backbone	Objectness	AP	AP_{50}	AP_{75}	AP_S	AP_M	AP_L	AR_{100}
CornerNet [19]	HG-52	Embedding	38.3	54.2	40.5	18.5	39.6	52.2	56.7
		B-Classifier	**42.3**	**59.5**	**45.4**	**24.6**	**45.4**	**57.6**	**59.9**
CenterNet [8]	HG-52	Embedding	41.3	59.2	43.9	23.6	43.6	53.6	59.0
		B-Classifier	**42.6**	**59.8**	**45.8**	**25.1**	**45.7**	**57.7**	**60.1**

Note that these three methods share a similar way of extracting corner keypoints, but CornerNet suffers large AF values due to the lack of validation beyond the proposals. CenterNet, by forcing a center keypoint to be detected,

[1] AF $= 1 - \widetilde{\mathrm{AP}}$, where $\widetilde{\mathrm{AP}}$ is computed over IoU thresholds of $[0.05 : 0.05 : 0.5]$ for all categories. $\mathrm{AF}_\tau = 1 - \widetilde{\mathrm{AP}}_\tau$, where $\widetilde{\mathrm{AP}}_\tau$ is computed at the IoU threshold of $\tau\%$, $\mathrm{AF}_{\mathrm{scale}} = 1 - \widetilde{\mathrm{AP}}_{\mathrm{scale}}$, where scale $\in$ {small, medium, large} indicates the object size.

was believed effective in filtering out false positives, and our approach, by reevaluating the proposal based on regional features, performs better than CenterNet. More importantly, by inserting an individual classification stage, CPN alleviates the false positives produced by instance embedding, as shown in Table 6.

4.5 Inference Speed

To show that CPN can generate high-quality bounding boxes with small computational costs, we report the inference speed for CPN on the MS-COCO validation dataset under different settings and compare the results to state-of-the-art efficient detectors, as shown in Table 7. For fair comparison, we test the inference speed for all detectors on an NVIDIA Tesla-V100 GPU. CPN-104 reports an FPS/AP of 7.3/46.8%, which is both faster and better than CenterNet-104 (5.1/44.8%) under the same setting. With a lighter backbone of Hourglass-52, CPN-52 reports an FPS/AP of 9.9/43.8%, which outperforms both CornerNet-52 and CenterNet-52. This indicates that two-stage detectors are not necessarily slow – our solution, by sharing a large amount of computation between the first (for keypoint extraction) and the second (for feature extraction) stage, achieves a good trade-off between inference speed and accuracy.

In the scenarios that require faster inference speed, CPN can be further accelerated by replacing with a lighter backbone and not using flip augmentation at the inference stage. For example, with the backbone of DLA-34 [43], CPN reports FPS/AP of 43.3/39.7% and 26.2/41.6% with and without flip augmentation, surpassing other competitors with the similar computational cost.

Table 7. Inference speed of CPN under different conditions *vs.* other detectors on the MS-COCO validation dataset. FPS is measured on the on an NVIDIA Tesla-V100 GPU. CPN achieves a good trade-off between accuracy and speed.

Method	Backbone	Input Size	Flip	AP	FPS
FCOS [39]	X-101-64 × 4d	800	×	42.6	8.1
Faster R-CNN [34]	X-101-64 × 4d	800	×	41.1	8.2
CornerNet-lite [20]	HG-Squeeze	ori	✓	36.5	22.0
CornerNet [19]	HG-104	ori	✓	41.0	5.8
CenterNet [8]	HG-104	ori	✓	44.8	5.1
CPN	HG-104	ori	✓	46.8	7.3
CPN	HG-52	ori	✓	43.8	9.9
CPN	HG-104	0.7×ori	×	40.5	17.9
CPN	DLA-34	ori	✓	41.6	26.2
CPN	DLA-34	ori	×	39.7	43.3

5 Conclusions

In this paper, we present an anchor-free, two-stage object detection framework. It starts with extracting corner keypoints and composing them into object proposals, and applies two-step classification to filter out false positives. With the above two stages, the recall and precision of detection are significantly improved, and the final result ranks among the top of existing object detection methods. We have also achieved a satisfying trade-off between accuracy and complexity.

The most important take-away is that anchor-free methods are more flexible in proposal extraction, while an individual discrimination stage is required to improve precision. When implemented properly, such a two-stage framework can be efficient in evaluation. Therefore, the debate on using one-stage or two-stage detectors seems not critical, but the importance of accurately localizing and recognizing instances becomes more significant.

Acknowledgments. This research was supported in part by National Natural Science Foundation of China: 61620106009, 61931008, U1636214 and 61836002, and Key Research Program of Frontier Sciences, CAS: QYZDJ-SSWSYS013. We thank Dr. Wei Zhang and Jianzhong He for instructive discussions.

References

1. Bochkovskiy, A., Wang, C.Y., Liao, H.Y.M.: YOLOv4: optimal speed and accuracy of object detection. arXiv preprint arXiv:2004.10934 (2020)
2. Bodla, N., Singh, B., Chellappa, R., Davis, L.S.: Soft-NMS-improving object detection with one line of code. In: Proceedings of the IEEE International Conference on Computer Vision, pp. 5561–5569 (2017)
3. Cai, Z., Vasconcelos, N.: Cascade R-CNN: delving into high quality object detection. In: Proceedings of the IEEE Conference on Computer Vision and Pattern Recognition, pp. 6154–6162 (2018)
4. Chen, K., et al.: MMDetection: open MMLab detection toolbox and benchmark. arXiv preprint arXiv:1906.07155 (2019)
5. Chen, Y., Han, C., Wang, N., Zhang, Z.: Revisiting feature alignment for one-stage object detection. arXiv preprint arXiv:1908.01570 (2019)
6. Dai, J., Li, Y., He, K., Sun, J.: R-FCN: object detection via region-based fully convolutional networks. In: Advances in Neural Information Processing Systems, pp. 379–387 (2016)
7. Dai, J., et al.: Deformable convolutional networks. In: Proceedings of the IEEE International Conference on Computer Vision, pp. 764–773 (2017)
8. Duan, K., Bai, S., Xie, L., Qi, H., Huang, Q., Tian, Q.: CenterNet: keypoint triplets for object detection. In: Proceedings of the IEEE International Conference on Computer Vision, pp. 6569–6578 (2019)
9. Everingham, M., Van Gool, L., Williams, C.K., Winn, J., Zisserman, A.: The pascal visual object classes (VOC) challenge. Int. J. Comput. Vis. **88**(2), 303–338 (2010)
10. Girshick, R.: Fast R-CNN. In: Proceedings of the IEEE International Conference on Computer Vision, pp. 1440–1448 (2015)
11. Girshick, R., Donahue, J., Darrell, T., Malik, J.: Rich feature hierarchies for accurate object detection and semantic segmentation. In: Proceedings of the IEEE Conference on Computer Vision and Pattern Recognition, pp. 580–587 (2014)

12. He, K., Gkioxari, G., Dollár, P., Girshick, R.: Mask R-CNN. In: Proceedings of the IEEE international conference on computer vision, pp. 2961–2969 (2017)
13. He, K., Zhang, X., Ren, S., Sun, J.: Deep residual learning for image recognition. In: Proceedings of the IEEE Conference on Computer Vision and Pattern Recognition, pp. 770–778 (2016)
14. Huang, G., Liu, Z., Van Der Maaten, L., Weinberger, K.Q.: Densely connected convolutional networks. In: Proceedings of the IEEE Conference on Computer vision and Pattern Recognition, pp. 4700–4708 (2017)
15. Huang, L., Yang, Y., Deng, Y., Yu, Y.: DenseBox: unifying landmark localization with end to end object detection. arXiv preprint arXiv:1509.04874 (2015)
16. Kingma, D.P., Ba, J.: Adam: a method for stochastic optimization. Comput. Sci. (2014)
17. Kong, T., Sun, F., Liu, H., Jiang, Y., Li, L., Shi, J.: FoveaBox: beyound anchor-based object detection. IEEE Trans. Image Process. **29**, 7389–7398 (2020)
18. Krizhevsky, A., Sutskever, I., Hinton, G.E.: ImageNet classification with deep convolutional neural networks. In: Advances in Neural Information Processing Systems, pp. 1097–1105 (2012)
19. Law, H., Deng, J.: CornerNet: detecting objects as paired keypoints. In: European Conference on Computer Vision, pp. 734–750 (2018)
20. Law, H., Teng, Y., Russakovsky, O., Deng, J.: CornerNet-lite: efficient keypoint based object detection. arXiv preprint arXiv:1904.08900 (2019)
21. LeCun, Y., Bengio, Y., Hinton, G.: Deep learning. Nature **521**(7553), 436–444 (2015)
22. Li, Y., Chen, Y., Wang, N., Zhang, Z.: Scale-aware trident networks for object detection. In: Proceedings of the IEEE International Conference on Computer Vision, pp. 6054–6063 (2019)
23. Lin, T.Y., Dollár, P., Girshick, R., He, K., Hariharan, B., Belongie, S.: Feature pyramid networks for object detection. In: Proceedings of the IEEE Conference on Computer Vision and Pattern Recognition, pp. 2117–2125 (2017)
24. Lin, T.Y., Goyal, P., Girshick, R., He, K., Dollár, P.: Focal loss for dense object detection. In: Proceedings of the IEEE International Conference on Computer Vision, pp. 2980–2988 (2017)
25. Lin, T.-Y., et al.: Microsoft COCO: common objects in context. In: Fleet, D., Pajdla, T., Schiele, B., Tuytelaars, T. (eds.) ECCV 2014. LNCS, vol. 8693, pp. 740–755. Springer, Cham (2014). https://doi.org/10.1007/978-3-319-10602-1_48
26. Liu, S., Qi, L., Qin, H., Shi, J., Jia, J.: Path aggregation network for instance segmentation. In: Proceedings of the IEEE Conference on Computer Vision and Pattern Recognition, pp. 8759–8768 (2018)
27. Liu, W., Anguelov, D., Erhan, D., Szegedy, C., Reed, S., Fu, C.-Y., Berg, A.C.: SSD: Single Shot MultiBox Detector. In: Leibe, B., Matas, J., Sebe, N., Welling, M. (eds.) ECCV 2016. LNCS, vol. 9905, pp. 21–37. Springer, Cham (2016). https://doi.org/10.1007/978-3-319-46448-0_2
28. Lu, X., Li, B., Yue, Y., Li, Q., Yan, J.: Grid R-CNN. In: Proceedings of the IEEE Conference on Computer Vision and Pattern Recognition, pp. 7363–7372 (2019)
29. Newell, A., Huang, Z., Deng, J.: Associative embedding: end-to-end learning for joint detection and grouping. In: Advances in Neural Information Processing Systems, pp. 2277–2287 (2017)
30. Newell, A., Yang, K., Deng, J.: Stacked hourglass networks for human pose estimation. In: Leibe, B., Matas, J., Sebe, N., Welling, M. (eds.) ECCV 2016. LNCS, vol. 9912, pp. 483–499. Springer, Cham (2016). https://doi.org/10.1007/978-3-319-46484-8_29

31. Pang, J., Chen, K., Shi, J., Feng, H., Ouyang, W., Lin, D.: Libra R-CNN: towards balanced learning for object detection. In: Proceedings of the IEEE Conference on Computer Vision and Pattern Recognition, pp. 821–830 (2019)
32. Paszke, A., et al.: Automatic differentiation in PyTorch (2017)
33. Redmon, J., Divvala, S., Girshick, R., Farhadi, A.: You only look once: unified, real-time object detection. In: Proceedings of the IEEE Conference on Computer Vision and Pattern Recognition, pp. 779–788 (2016)
34. Ren, S., He, K., Girshick, R., Sun, J.: Faster R-CNN: towards real-time object detection with region proposal networks. In: Advances in Neural Information Processing Systems, pp. 91–99 (2015)
35. Simonyan, K., Zisserman, A.: Very deep convolutional networks for large-scale image recognition. arXiv preprint arXiv:1409.1556 (2014)
36. Sun, K., Xiao, B., Liu, D., Wang, J.: Deep high-resolution representation learning for human pose estimation. In: Proceedings of the IEEE Conference on Computer Vision and Pattern Recognition, pp. 5693–5703 (2019)
37. Tan, M., Le, Q.V.: EfficientNet: rethinking model scaling for convolutional neural networks. arXiv preprint arXiv:1905.11946 (2019)
38. Tan, M., Pang, R., Le, Q.V.: EfficientDet: scalable and efficient object detection. In: Proceedings of the IEEE Conference on Computer Vision and Pattern Recognition, pp. 10781–10790 (2020)
39. Tian, Z., Shen, C., Chen, H., He, T.: FCOS: fully convolutional one-stage object detection. In: Proceedings of the IEEE International Conference on Computer Vision, pp. 9627–9636 (2019)
40. Tychsen-Smith, L., Petersson, L.: DeNet: scalable real-time object detection with directed sparse sampling. In: Proceedings of the IEEE International Conference on Computer Vision, pp. 428–436 (2017)
41. Wang, J., Chen, K., Yang, S., Loy, C.C., Lin, D.: Region proposal by guided anchoring. In: Proceedings of the IEEE Conference on Computer Vision and Pattern Recognition, pp. 2965–2974 (2019)
42. Yang, Z., Liu, S., Hu, H., Wang, L., Lin, S.: RepPoints: point set representation for object detection. In: Proceedings of the IEEE International Conference on Computer Vision, pp. 9657–9666 (2019)
43. Yu, F., Wang, D., Shelhamer, E., Darrell, T.: Deep layer aggregation. In: Proceedings of the IEEE Conference on Computer Vision and Pattern Recognition, pp. 2403–2412 (2018)
44. Yu, J., Jiang, Y., Wang, Z., Cao, Z., Huang, T.: UnitBox: an advanced object detection network. In: Proceedings of the 24th ACM International Conference on Multimedia, pp. 516–520 (2016)
45. Zhang, S., Chi, C., Yao, Y., Lei, Z., Li, S.Z.: Bridging the gap between anchor-based and anchor-free detection via adaptive training sample selection. In: Proceedings of the IEEE/CVF Conference on Computer Vision and Pattern Recognition, pp. 9759–9768 (2020)
46. Zhang, S., Wen, L., Bian, X., Lei, Z., Li, S.Z.: Single-shot refinement neural network for object detection. In: Proceedings of the IEEE Conference on Computer Vision and Pattern Recognition, pp. 4203–4212 (2018)
47. Zhang, X., Wan, F., Liu, C., Ji, R., Ye, Q.: FreeAnchor: learning to match anchors for visual object detection. In: Advances in Neural Information Processing Systems, pp. 147–155 (2019)
48. Zhou, X., Wang, D., Krähenbühl, P.: Objects as points. arXiv preprint arXiv:1904.07850 (2019)

49. Zhou, X., Zhuo, J., Krahenbuhl, P.: Bottom-up object detection by grouping extreme and center points. In: Proceedings of the IEEE Conference on Computer Vision and Pattern Recognition, pp. 850–859 (2019)
50. Zhu, C., Chen, F., Shen, Z., Savvides, M.: Soft anchor-point object detection. arXiv preprint arXiv:1911.12448 (2019)
51. Zhu, C., He, Y., Savvides, M.: Feature selective anchor-free module for single-shot object detection. In: Proceedings of the IEEE Conference on Computer Vision and Pattern Recognition, pp. 840–849 (2019)

PhraseClick: Toward Achieving Flexible Interactive Segmentation by Phrase and Click

Henghui Ding[1(✉)], Scott Cohen[2], Brian Price[2], and Xudong Jiang[1]

[1] Nanyang Technological University, Singapore, Singapore
{ding0093,exdjiang}@ntu.edu.sg
[2] Adobe Research, San Jose, USA
{scohen,bprice}@adobe.com

Abstract. Existing interactive object segmentation methods mainly take spatial interactions such as bounding boxes or clicks as input. However, these interactions do not contain information about explicit attributes of the target-of-interest and thus cannot quickly specify what the selected object exactly is, especially when there are diverse scales of candidate objects or the target-of-interest contains multiple objects. Therefore, excessive user interactions are often required to reach desirable results. On the other hand, in existing approaches attribute information of objects is often not well utilized in interactive segmentation. We propose to employ phrase expressions as another interaction input to infer the attributes of target object. In this way, we can 1) leverage spatial clicks to locate the target object and 2) utilize semantic phrases to qualify the attributes of the target object. Specifically, the phrase expressions focus on "what" the target object is and the spatial clicks are in charge of "where" the target object is, which together help to accurately segment the target-of-interest with smaller number of interactions. Moreover, the proposed approach is flexible in terms of interaction modes and can efficiently handle complex scenarios by leveraging the strengths of each type of input. Our multi-modal phrase+click approach achieves new state-of-the-art performance on interactive segmentation. To the best of our knowledge, this is the first work to leverage both clicks and phrases for interactive segmentation.

Keywords: Interactive segmentation · Click · Phrase · Flexible · Attribute

1 Introduction

Interactive object segmentation (or interactive object selection) aims to accurately segment the image into foreground and background given a minimal amount of user interactive inputs. It allows user to gradually refine the prediction with further interaction inputs if any mistakes are made in prediction.

A. Vedaldi et al. (Eds.): ECCV 2020, LNCS 12348, pp. 417–435, 2020.
https://doi.org/10.1007/978-3-030-58580-8_25

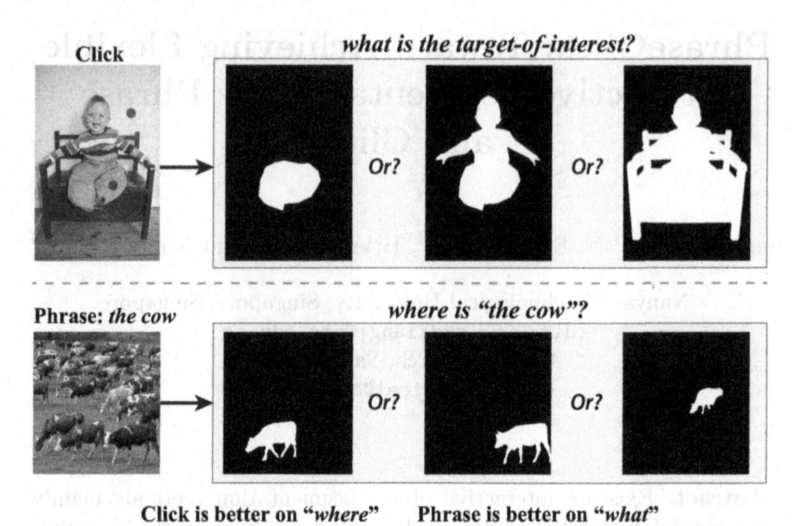

Fig. 1. (Best viewed in color) As interaction inputs, both the click and phrase have their advantages and disadvantages. Here we show ambiguities that exist when using only click or phrase interactions. We propose to utilize both interaction types to make selection flexible and robust to handle complex scenarios.

These inputs usually come in the form of user clicks/strokes [5,28,34,41,51,58] or bounding boxes [16,29,49,57]. This form of input gives hard constraints regarding the location of the object of interest. With the recent advances in deep learning such methods can now often select familiar objects with a small amount of input. Alternately, systems have been proposed that instead use language-based input to drive the selection [24,35,50,61]. Natural language phrases can be used by a neural network to infer high-level attribute information about what the object of interest looks like that can then be used to select the objects.

While great strides have been made in interactive selection, each of these interaction approaches may still fall short and require additional and excessive user interaction. For example, click-based methods are required to infer the target object given only spatial constraints and usually are trained to select entire objects. However, the region of user interest may instead be an object part or a combination of multiple objects. For example, the first row in Fig. 1, the click on the boy may indicate the trousers, the boy, or the whole foreground (boy and chair). This leads to ambiguities that must be overcame with additional user inputs, which directly runs counter to the goal of minimizing user interaction. Click-based methods also generally assume accurate input, but with mobile devices it can be difficult for users to accurately click on objects, especially given that the user's finger is occluding the object of interest. It remains a significant challenge to accurately segment a target-of-interest with a few clicks.

On the other hand, language-driven segmentation methods learn the overall appearance of objects and must infer their location. A language phrase naturally and easily overcomes ambiguities such as whether the target is an object, object part, or collection of objects. A phrase can also provide rough spatial information ("the woman on the left"). Besides, for mobile devices like smartphones, speech is a natural and desired interface and easier than precise touching on a small phone screen. However, in many cases an object name and rough location is not sufficient to produce a desired result. For example, in images where there are multiple objects with similar appearance (an image of dozens of cows, see Fig. 1) it can be very difficult to articulate using language a single target instance. In such cases, directly clicking on the object is much easier for a user. It can also be difficult to verbally articulate some required corrections that do not correspond to an entire semantically-meaningful region. Further, due to the long tail distribution of objects in images, it is difficult to obtain labels and training data for all possible objects.

The strengths and weaknesses of click-based and phrase-based inputs are complementary with clicks giving hard spatial constraints and phrases giving high-level attribute information. An effective combination of these inputs may reduce the amount of user interactions needed for accurately selecting objects of interest. Given this observation, we propose to build a versatile interactive segmentation network that accepts both clicks and phrases as interaction input. We use a convolutional neural network (CNN) to process the input image and clicks, and employ the bi-directional LSTM (bi-LSTM) to encode the phrases and infer language features. To bridge click-based spatial constraints with phrase-based attribute constraints, we introduce a novel attribute guided feature attention module to effectively integrate language and vision features together. Our approach can better handle complex scenarios via utilizing advantages of these two interactions.

The main contributions of this paper can be summarized as follows: 1) To the best of our knowledge, this is the first work to leverage both clicks and phrases for interactive segmentation. The proposed approach allows the user the flexibility of using the interaction method that is most suitable for a given task, making the system more practical than past approaches. 2) We propose an attribute guided feature attention module to bring clicks and phrases information together. It extracts discriminative attribute clues from interactions and integrate the vision and language attribute clues. 3) Extensive experimental results have demonstrated that phrase expressions are indeed effective at boosting the performance of interactive segmentation, especially in some complex scenarios and with few clicks.

2 Related Work

2.1 Interactive Object Segmentation

Many interactive segmentation methods have been proposed in the past decades using many different interaction types such as bounding boxes [16,29,49,57], contours [1,6,26,45], strokes [5,31,51,53] and clicks [25,28,34,40,41,58]. There are also some language-based segmentation methods [24,35,61], but they only provide an initial result and cannot further refine the result to correct mistakes.

Early methods rely on low-level features, such as color similarity or boundary properties [26,45]. For example, [5,31,49] adopt graphical models, [18] employs random walker and [3,11,47] are based on geodesic approaches. However, low-level features are not robust and hence excessive user interactions are required for these methods to achieve desirable segmentation results. Recently, thanks to the great success of convolution neural networks (CNN) [36,44,55,56,63–70], CNN-based interactive segmentation methods have achieved exciting progress. For example, Xu et al. [58] concatenate Euclidean distance maps to user foreground and background clicks with the image as input to a fully convolutional network (FCN) [39]. Hu et al. [25] employ a two-stream network to deal with image and user clicks separately to make the user interaction impact the result more directly. Le et al. [28] introduce boundary clicks to perform object selection. Besides the clicks, bounding boxes have also been used in CNN-based methods [57]. As a variant to using bounding boxes, Papadopoulos et al. [46] propose extreme points and Maninis et al. [41] use extreme points to generate Gaussian heatmap and crop the image to achieve instance segmentation. Agustsson et al. [2] further segment all regions jointly with extreme points and scribbles. As these methods receive only spatial constraints, they cannot provide high-level attribute information to the method and thus may require additional user input to overcome this drawback in challenging cases. Different from these methods, our approach receives both spatial constraints and high-level attribute information.

2.2 Semantic/Instance Segmentation

Semantic segmentation and instance segmentation are closely related with interactive segmentation. Driven by the significant success of CNN, many deep-learning-based works have been proposed for semantic segmentation [8,12–15,22,52,71] and instance segmentation [7,20,23,37]. For example, Long et al. [39] propose the FCN to train the segmentation network end-to-end. He et al. [20] propose to add an instance-level segmentation mask branch on the top of Faster R-CNN [48]. However, it is not reasonable to directly transform semantic/instance segmentation to interactive segmentation [58]. Interactive object segmentation methods respond to user's inputs instead of predefined labels, thus the ability to segment any unseen objects is required and this is impractical in current segmentation approaches. In this work, we modify the semantic segmentation framework [9] to have it accept interaction inputs, and re-train the modified framework on interactive object segmentation datasets.

2.3 Referring Expression Comprehension

Referring expression comprehension methods [27,35,38,54,60–62] detect a specific object in an image given a referring expression. Some comprehension methods can be used to segment the referential object [24,35,59,61]. However, these methods only compute an initial segmentation and cannot further correct the

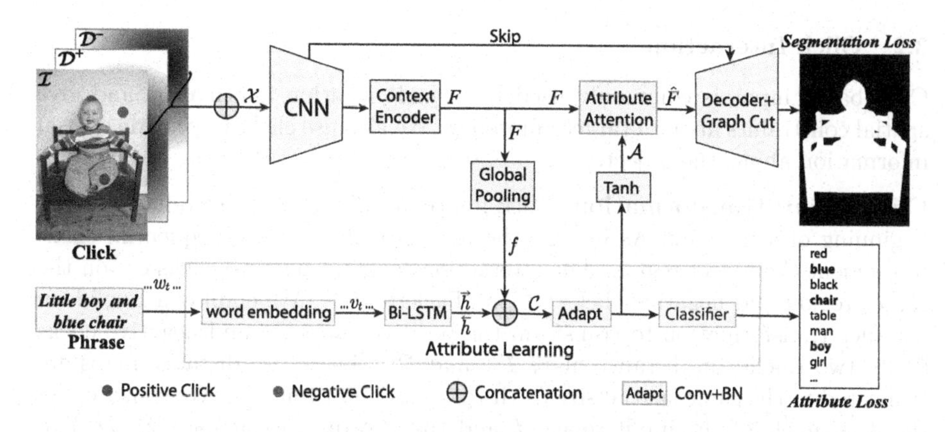

Fig. 2. The proposed approach accepts clicks and phrases as interaction input, but it is *not* necessary to enter both types of interactions at the same time.

segmentation mask, which is required for interactive segmentation if any mistakes are made. For example, MAttNet [61] depends on the instance segmentation results of Mask R-CNN [20]. MAttNet first scores and ranks the similarity between phrase embedding and instance objects generated by Mask R-CNN, and then outputs the mask of best matched object. These methods cannot further improve the segmentation mask even with new additional phrase input. Thus, they cannot meet the requirement of practical application for interactive segmentation. In contrast, our approach enables users to add interactive information until the segmentation results meets the users' requirements.

3 Approach

In this work, we propose to build a versatile interactive segmentation network that can take both clicks and phrases as interaction input. Compared with previous approaches that accept only one type of interaction, the proposed approach 1) is more flexible in terms of interaction ways and 2) can better handle complex scenarios via utilizing advantages of these two interactions.

3.1 Network Architecture

The overall architecture of our approach is shown in Fig. 2, the proposed approach accepts both clicks and phrases as interaction input. We use the ResNet-101 [21] based DeepLabv3+ [9] as the backbone of vision part. The clicks are transformed to distance maps and concatenated with original image to form a 5-channel input for the CNN as in [58]. For the language part, the phrase is processed by a word-to-vector model and then a bi-directional LSTM to extract the language clues. During testing, we employ graph cut [5] as a post-processing tool to refine the final segmentation mask.

3.2 Click Interaction

Click-based input is commonly used to provide location information and give spatial constraints about a target of interest. We also use clicks to provide spatial information about the object.

Click Input Transformation. Click information is usually introduced at the beginning of a network. As in [58], the user provides a set of sequential clicks to segment the target region. The interactions contain positive clicks $\mathcal{C}^+$ on the target region and negative clicks $\mathcal{C}^-$ on "background". We employ a Euclidean distance transformation to transform the positive clicks $\mathcal{C}^+$ and negative clicks $\mathcal{C}^-$ to two Euclidean distance maps $\mathcal{D}^+$ and $\mathcal{D}^-$. These two distance maps are truncated to the same spatial size as the original input image $\mathcal{I}$. We concatenate the 3 channels of the input image $\mathcal{I}$ and the two distance maps $(\mathcal{D}^+, \mathcal{D}^-)$ to form the 5 channel input:

$$\mathcal{X} = \mathcal{I} \oplus \mathcal{D}^+ \oplus \mathcal{D}^- \tag{1}$$

where $\oplus$ denotes concatenation and $\mathcal{X}$ is the 5-channel input to the network.

Click Simulation Protocol. We follow the simulation strategies proposed in [58] to form a set of synthetic positive and negative clicks. Furthermore, to encourage the model to learn to generate correct prediction from ambiguous clicks, we follow [33] and introduce another sampling strategy that samples clicks on the overlapped foreground and overlapped background of different object masks, for example, the positive and negative clicks in Fig. 2.

3.3 Phrase Interaction

While clicks as a kind of spatial interaction can indicate the position of target object, such spatial interactions cannot explicitly express the semantic attributes of a target object. A language phrase naturally and easily expresses attributes of objects and is commonly used in referring segmentation [10,24,27,54,60–62]. However, referring segmentation methods only compute an initial segmentation mask and cannot further improve the segmentation mask even with new additional phrase input, which does not agree with the goal of interactive segmentation.

Bringing Click and Phrase Together. In this work, we explore using phrase expressions as another interaction input to express the attributes of a target-of-interest and quickly narrow the range of candidates, which can assist the click interaction process and decrease the user interaction times. For regions that are difficult to articulate, such as small mistakes that need correction or objects that do not have easily identifiable attributes that will separate them from surrounding objects, clicks provide a strong spatial constraint to supplement the attribute information.

Phrase Expression Annotation. The interactive segmentation datasets have no phrase annotation. To train the proposed network, corresponding phrase annotation for each segmentation mask is required. Therefore, we annotate phrase expressions for every image in Grabcut [49] and Berkeley [43], some examples are shown in Fig. 3.

3.4 Attribute Feature Attention Module

To uniformly integrate click and phrase interactions in a single network, we propose an attribute attention module. CNN features contain rich attributes information like color or shape, which are preserved in different feature channels [10]. Based on this observation, we explore to infer attribute clues of target-of-interest from vision features of $\mathcal{X}$ (in Eq. 1) and language features of phrases. The attribute clues of the target object are changed with user's interaction, *i.e.*phrases and clicks in this work. We employ the attribute clues to compute channel-wise attribute attention, and then leverage this attribute attention to emphasize some specific channels and suppress others in order to help the network to determine the object of interest.

Fig. 3. Examples of our annotated phrase-segmentation pairs for Grabcut [49] and Berkeley [43].

First, to process language input, we use a word-to-vector model to embed each word $w_t \in \{w_1, ..., w_T\}$, $v_t = word2vec(w_t)$. Next, we employ the bi-directional LSTM (bi-LSTM) to encode the holistic context of phrase expression:

$$\overrightarrow{h}_t = \overrightarrow{LSTM}(v_t, \overrightarrow{h}_{t-1}), \quad \overleftarrow{h}_t = \overleftarrow{LSTM}(v_t, \overleftarrow{h}_{t+1}) \tag{2}$$

Besides the context from phrase expression, we also extract visual context vector from CNN features, $f = pooling(F)$, where *pooling* is a global average pooling operation, F with size of $H \times W \times \#C$ is the high-level feature of CNN, f with size of $1 \times 1 \times \#C$ is visual context vector that preserves visual clues in channels. Then we concatenate the visual context vector and the bi-directional context vectors of phrase to generate final attribute context representation:

$$\mathcal{C} = \overrightarrow{h} \oplus \overleftarrow{h} \oplus f \tag{3}$$

where $\overrightarrow{h} = (\overrightarrow{h}_1, ..., \overrightarrow{h}_t, ..., \overrightarrow{h}_T)$ and $\overleftarrow{h} = (\overleftarrow{h}_1, ..., \overleftarrow{h}_t, ..., \overleftarrow{h}_T)$, and $\mathcal{C}$ is the attribute context vector. Next we adapt the attribute context vector to make it have the same number of channels ($\#C = 512$ in our experiments) with features in CNN, and normalize the values to range of $[-1, 1]$ by:

$$\mathcal{A} = tanh(Adapt(\mathcal{C}, \Theta)) \tag{4}$$

where $tanh(x) = (e^x - e^{-x})/(e^x + e^{-x})$, *Adapt* is `Conv+BN` and Θ is its learnable parameters. Now we get the attribute attention weight $\mathcal{A}$, in which different attribute channels have different attention weights in range of $[-1, 1]$. Finally, as shown in Fig. 4, we apply the attribute attention weights on feature maps in CNN:

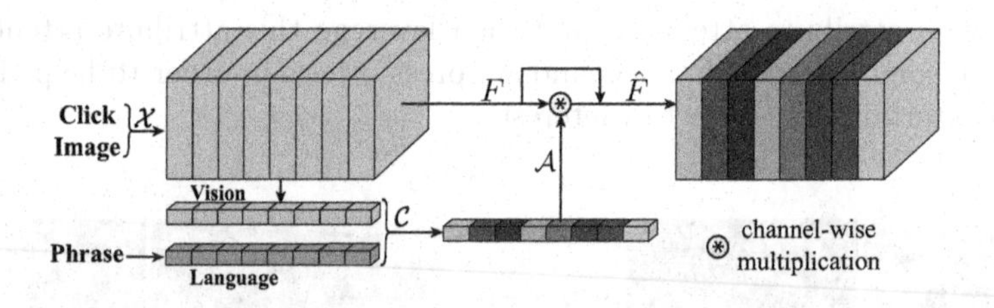

Fig. 4. Attribute Attention: It emphasizes feature channels that have larger response in semantic attribute learning, which is based on the phrase interaction, click interaction and visual patterns.

$$\hat{F} = \beta \mathcal{A} * F + F \tag{5}$$

where β is a learnable weight and $*$ denotes channel-wise multiplication. To have our approach more flexible in terms of interaction methods, the phrase interaction is not compulsory. When there is no phrase interaction input, the $\overrightarrow{h}$ and $\overleftarrow{h}$ are set to 0.

The attribute attention module helps to bridge the clicks and phrases information together. These two interaction inputs work together to reduce the interaction times for improved performance.

3.5 Training Loss

The proposed approach is a unified framework that brings vision part and language part together. To train such a network end-to-end and balance the effect of two different interactions, we carefully design its loss function. First, we use the binary cross-entropy loss for the segmentation branch training. Given the segmentation ground truth t and the segmentation branch output s, the segmentation loss is:

$$\mathcal{L}_m = -tlog(\sigma(s)) - (1 - t)log(1 - \sigma(s)) \tag{6}$$

where $\sigma(x) = 1/(1 + e^{-x})$ is sigmoid function.

Table 1. Ablation study of the proposed approach on testA of RefeCOCO.

Interaction inputs	Phrase	Click	Graph cut	IoU
Phrase-only	✓			50.98
Click-only		✓		77.93
PhraseClick	✓	✓		83.94
PhraseClick	✓	✓	✓	85.02

Additionally, in order to help the network optimize the attribute learning process and catch meaningful visual and phrasal attributes, we introduce an attribute loss for the attribute learning module. We first generate the attribute label $\{a_i\}_{i=1}^N$ following [27], $a_i \in \{0,1\}$ indicates whether the *i-th* attribute word exists in the input phrase. Denoting the output of attribute learning module as p_i, the objective function for attribute module is defined as:

$$\mathcal{L}_a = \sum_{i=1}^{N} w_i(a_i log(\sigma(p_i)) + (1 - a_i) log(1 - \sigma(p_i)) \tag{7}$$

where w_i is a weighting factor to address the unbalance of different attributes, and a_i indicates whether the *i-th* attribute word exists in the input phrase. The final loss for the proposed network is $\mathcal{L} = \omega_m \mathcal{L}_m + \omega_a \mathcal{L}_a$, where ω_m and ω_a are used to balance the contributions of the segmentation and attribute loss.

4 Experiments

4.1 Implementation Details

The network and all the experiments are implemented based on the public Pytorch platform. We use ResNet-101 [21] based DeepLabv3+ [9] as the backbone of the segmentation branch. The first convolution layer is modified to $5 \times 64 \times 7 \times 7$ to deal with the 5-channel input. The proposed attribute attention module is placed after ASPP. The parameters of newly added layers are randomly initialized from a Gaussian distribution with standard variance of 10^{-2}. The network is trained by SGD with batch size set to 10. For batch processing, we resize the inputs to 512×512 pixels during training. We adopt random horizontal flipping to augment the training data. The learning rate is set to 1×10^{-8} and the parameters of the new layers are trained with a higher learning rate 1×10^{-7}. Momentum and weight decay are fixed to 0.99 and 1×10^{-4} respectively. We empirically set ω_m and ω_a to 1 and 10. The network is evaluated with Intersection-over-Union (**IoU**) [39]. To jointly train the attribute learning branch and segmentation branch, we train our network on RefCOCO [27] first, which contains referring phrase expressions and segmentation masks for every instance objects in each of 19994 images. Then, for interactive segmentation datasets like PASCAL VOC, we take the name of categories as the phrase input, like person and car. We annotate phrase expressions for the Grabcut and Berkeley datasets, as shown in Fig. 3.

4.2 Ablation Studies

We conduct the ablation studies on RefCOCO [27], as shown in Table 1. First, to verify the effectiveness of the proposed attribute attention module, we discard the click interaction and only take the phrase as input. With *Phrase-only* input, our network can achieve an acceptable IoU of 50.98%. We also display some visualized segmentation score maps in Fig. 6. Both the quantitative and qualitative results demonstrate that the attribute attention module does help the network to select the target-of-interest according to referential phrase expression.

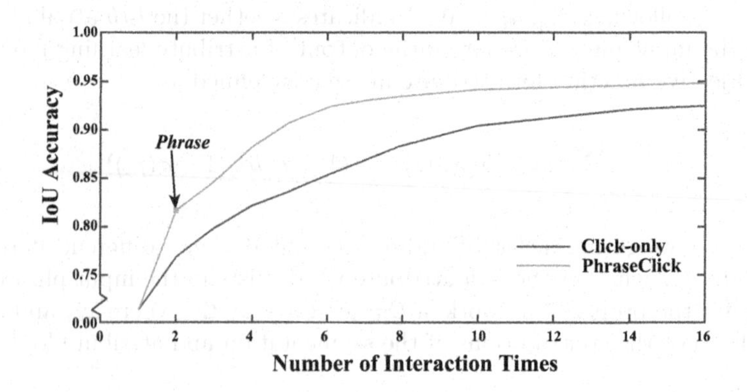

Fig. 5. Segmentation accuracy with different number of interaction times.

Fig. 6. Examples of segmentation score maps (soft value from 0 to 1) of *Phrase-only.*

We also conduct a *Click-only* experiment in which the $\overrightarrow{h}$ and $\overleftarrow{h}$ in Eq. (3) are set to 0. This achieves 77.93% IoU by 3 clicks, which is much higher than phrase only. It shows the advantage of click interaction for interactive object segmentation. Next, we verify whether the phrase interaction can speed up the click-based interactive segmentation process and improve the segmentation performance. To this end, we first compare the segmentation performance of *Click-only*

Table 2. Comparison with previous state-of-the-art interactive object segmentation methods. We demonstrate the number of interactions that each method needed to reach a certain segmentation performance.

Methods	PASCAL VOC (85% IoU)	GrabCut (90% IoU)	Berkeley (90% IoU)
GraphCut [5]	15.06	11.10	14.33
Geodesic Matting [4]	14.75	12.44	15.96
Random Walker [18]	11.37	12.30	14.02
Geodesic Convexity [19]	11.73	8.38	12.57
iFCN [58]	6.88	6.04	8.65
RIS-Net [34]	5.12	5.00	6.03
ITIS [40]	3.80	5.60	–
LDN [32]	–	4.79	–
FCTSFN [25]	4.58	3.76	6.49
PhraseClick (Ours)	**3.12**	**2.06**	**3.26**

vs. PhraseClick. As shown in Table 1, compared with *Click-only* without phrase, *PhraseClick* achieves 6.01% better performance on IoU. This demonstrates that the phrase expression and attribute attention module visibly enhances the performance of click-based interactive segmentation. Note, in Table 1, the interaction inputs for *PhraseClick* are one phrase and two clicks (one positive and one negative). For fair comparison, there is one more positive click for *Click-only* than *PhraseClick* in Table 1, *i.e.*the number of interaction inputs for *Click-only* and *PhraseClick* is the same. Next, we show the segmentation performance with different number of interaction times in Fig. 5. We start with an initial positive click, then gradually add a phrase/click as the next interaction according to the current segmentation prediction. As shown in Fig. 5, to reach a specific IoU performance, the network taking the *PhraseClick* as interaction inputs is faster than the network that only takes *Click* input. This shows the phrase expression helps to speed up the interaction process and improve the segmentation performance. During the inference, we employ the graph cut [5] as post-processing method to refine the segmentation mask, which improve the performance by 1.08% IoU. The performance gain is brought by the refinement in boundary regions.

4.3 Results on Benchmarks

Interactive Object Segmentation

We first compare the proposed approach with previous state-of-the-art interactive segmentation methods. Interactive object segmentation enables users to add interactive information until the desired selection is reached. We conduct the interactive segmentation experiments on three public benchmarks with instance object annotations: PASCAL VOC [17], Grabcut [49], Berkeley [43].

To evaluate the performance of interactive object segmentation, the standard process is: 1) starting with an initial positive click at the center of the target-of-interest, the network generates an initial segmentation mask for the target object; 2) then a succeeding positive/negative click is added at the center of

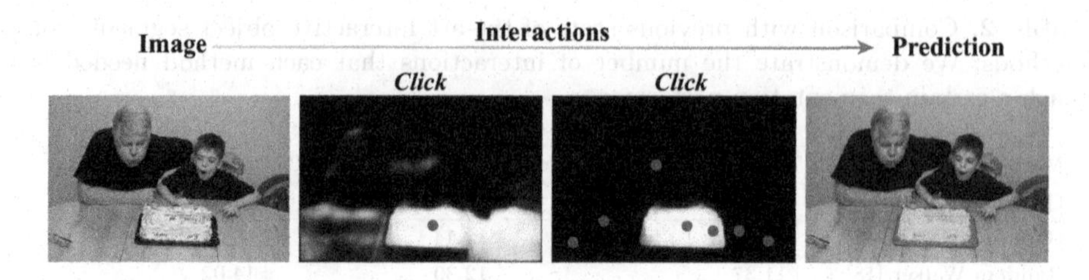

Fig. 7. Existing click-based methods' interaction process: click first, then gradually refine with more clicks.

Table 3. User study. *PhraseClick* requires less interactions than *Click-only.*

Methods	User#1	User#2	User#3	User#4	User#5	User#6	User#7	Average
Click-only	4.38	4.41	4.65	4.52	4.33	4.59	4.57	4.49
PhraseClick	3.02	3.13	3.17	3.14	3.00	3.07	3.12	**3.09**

the largest wrongly predicted piece to add/remove this piece; 3) the 2nd step is repeated again and again until the desirable accuracy is reached or the maximum number (set to 20 as [58]) of clicks is reached. Different from existing interactive segmentation methods that only use clicks as input, the proposed approach accepts both the clicks and phrases as interaction input. Therefore, to evaluate our approach and compare with existing state-of-the-art works, we input the phrase for the target-of-interest after the initial click in the first step, which together are counted as two interaction times. Then the subsequent interactions only use clicks. Table 2 shows the number of interactions that each approach requires to reach a certain IoU. We achieve new state-of-the-art performance on the three benchmarks, which shows the advantage of the proposed approach.

Furthermore, comparing to existing methods that accept a single type of interaction (as shown in Fig. 7), the proposed approach is more flexible in terms of interaction and can better handle complex scenarios via utilizing advantages of both clicks and phrases. Although we fix the order of phrase input to be the second interaction after the initial positive click in Table 2, this is just for convenience of evaluation. In practice, users can input the phrase at any time and in any order, *e.g.*phrase first and then refine with clicks. We show some interaction examples in Fig. 8. In the first row in Fig. 8, we start with a phrase to get the initial segmentation prediction, then refine the initial prediction with clicks. It demonstrates the advantages of clicks at locating "where" the object is when there are multiple objects with the similar appearance. In the second row in Fig. 8, phrases are employed to disambiguate the confused results of clicks, which shows the advantages of phrases at describing "what" the object of interest is. The first two rows show how the integration of clicks and phrases help to improve the performance of interactive segmentation and better handle complex scenarios. Our model can also take only clicks similar to previous click-based segmentation methods, as shown in the last row in Fig. 8, which shows the abilities of clicks to correct mistakes and refine details.

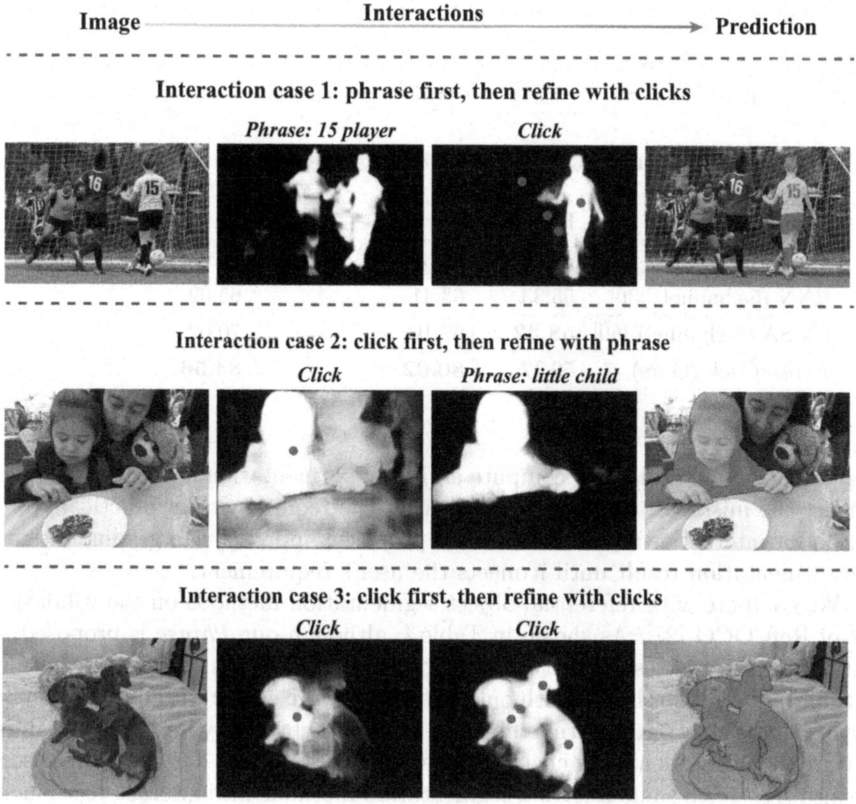

Fig. 8. Our interaction process is more flexible. The user can choose either click or phrase in each interaction step, making the system more practical than past approaches. Furthermore, our approach can better handle complex scenarios via utilizing advantages of these two interactions.

To justify the robustness and effectiveness of the proposed PhraseClick with different real users, we conduct a user study in Table 3. We randomly choose 50 images from PASCAL VOC, RefCOCO, and Berkeley for this testing. Seven users participated in this study. We record the number of interactions they required to reach 85% IoU segmentation performance. As shown in Table 3, our *PhraseClick* requires less interactions than *Click-only*. In practical applications, the amount of time to input a phrase depends greatly on implementation and the input device. Given a strong voice-to-text algorithm, the time to input a phrase could be close to click. In our specific case, we do not have a sophisticated voice-to-text system, so we allow the user to type in the desired phrase. This will be slower than using voice and likely slower than clicks.

Referential Object Segmentation
Although the proposed network is designed for interactive object segmentation, it can also be used for referential object segmentation. Existing referential object

Table 4. Comparison with state-of-the-art referring segmentation methods.

Methods	RMI [35]	DMN [42]	RNN [30]	MAttNet [61]	CMSA [60]	*Phrase*	*PhraseClick*
IoU	45.18	49.78	55.33	56.51	58.32	50.42	**80.02**

Table 5. Comparison with our modified referring segmentation methods.

# of interactions	Phrase*1	Phrase*1 + Click*1	Phrase*1 + Click*2
DMN (5-channel) [42]	49.78	60.32	62.08
RNN (5-channel) [30]	55.33	63.41	65.02
CMSA (5-channel) [60]	**58.32**	67.10	70.02
PhraseClick (Ours)	50.42	**80.02**	**84.56**

segmentation methods only compute an initial segmentation and cannot further correct the mistakes, thus they cannot meet the requirement of practical application for interactive segmentation. The proposed approach can gradually refine the segmentation result until it meets the user's requirement.

We compare with referential object segmentation methods on the validation set of RefCOCO [27]. As shown in Table 4, although our *Phrase* is proposed as part of a network design that includes clicking, it achieves competitive results with phrase-only methods. With only one click (*PhraseClick*), the proposed approach significantly outperforms the state-of-the-arts by a very large margin, illustrating the utility of adding clicks to phrase-based selection. For fair comparison, we also transform some referring segmentation methods into interactive segmentation by replacing their RGB input with the same five-channel input our method uses (Sect. 3.2) and compare their performance with different number of interactions, as shown in Table 5. Our approach significantly outperform others as interaction times increase, showing that a naive transformation of referential object segmentation methods is not as effective as our model that integrates the phrase information in an attribute guided feature attention module.

5 Conclusion

In this work, we propose an interactive segmentation network that can take either clicks or phrases or both as interaction input, which utilizes the complementary merits of these two interactions. Besides the commonly used spatial constraints like clicks, we introduce phrase expression as another interaction input to infer semantic attributes and propose an attribute attention module to integrate such attributes information into the network. Specifically, the language phrases focuses on "what" the target-of-interest is and the spatial clicks are in charge of "where" the target-of-interest is. Extensive experimental results have shown that the proposed approach is flexible in terms of interaction and can handle complex scenarios well.

References

1. Acuna, D., Ling, H., Kar, A., Fidler, S.: Efficient interactive annotation of segmentation datasets with Polygon-RNN++. In: Proceedings of the IEEE Conference on Computer Vision and Pattern Recognition, pp. 859–868 (2018)
2. Agustsson, E., Uijlings, J.R., Ferrari, V.: Interactive full image segmentation by considering all regions jointly. In: IEEE Conference on Computer Vision and Pattern Recognition, pp. 11622–11631 (2019)
3. Bai, X., Sapiro, G.: A geodesic framework for fast interactive image and video segmentation and matting. In: 2007 IEEE 11th International Conference on Computer Vision, pp. 1–8. IEEE (2007)
4. Bai, X., Sapiro, G.: Geodesic matting: a framework for fast interactive image and video segmentation and matting. Int. J. Comput. Vis. **82**(2), 113–132 (2009)
5. Boykov, Y.Y., Jolly, M.P.: Interactive graph cuts for optimal boundary & region segmentation of objects in nd images. In: IEEE International Conference on Computer Vision, vol. 1, pp. 105–112. IEEE (2001)
6. Castrejon, L., Kundu, K., Urtasun, R., Fidler, S.: Annotating object instances with a Polygon-RNN. In: Proceedings of the IEEE Conference on Computer Vision and Pattern Recognition, pp. 5230–5238 (2017)
7. Chen, L.C., Hermans, A., Papandreou, G., Schroff, F., Wang, P., Adam, H.: MaskLab: instance segmentation by refining object detection with semantic and direction features. In: Proceedings of the IEEE Conference on Computer Vision and Pattern Recognition, pp. 4013–4022 (2018)
8. Chen, L.C., Papandreou, G., Kokkinos, I., Murphy, K., Yuille, A.L.: DeepLab: semantic image segmentation with deep convolutional nets, atrous convolution, and fully connected CRFs. arXiv:1606.00915 (2016)
9. Chen, L.-C., Zhu, Y., Papandreou, G., Schroff, F., Adam, H.: Encoder-decoder with atrous separable convolution for semantic image segmentation. In: Ferrari, V., Hebert, M., Sminchisescu, C., Weiss, Y. (eds.) ECCV 2018. LNCS, vol. 11211, pp. 833–851. Springer, Cham (2018). https://doi.org/10.1007/978-3-030-01234-2_49
10. Chen, Y.W., Tsai, Y.H., Wang, T., Lin, Y.Y., Yang, M.H.: Referring expression object segmentation with caption-aware consistency. arXiv preprint arXiv:1910.04748 (2019)
11. Criminisi, A., Sharp, T., Blake, A.: GeoS: geodesic image segmentation. In: Forsyth, D., Torr, P., Zisserman, A. (eds.) ECCV 2008. LNCS, vol. 5302, pp. 99–112. Springer, Heidelberg (2008). https://doi.org/10.1007/978-3-540-88682-2_9
12. Ding, H., Jiang, X., Liu, A.Q., Thalmann, N.M., Wang, G.: Boundary-aware feature propagation for scene segmentation. In: Proceedings of the IEEE International Conference on Computer Vision, pp. 6819–6829 (2019)
13. Ding, H., Jiang, X., Shuai, B., Liu, A.Q., Wang, G.: Context contrasted feature and gated multi-scale aggregation for scene segmentation. In: Proceedings of the IEEE Conference on Computer Vision and Pattern Recognition (CVPR), pp. 2393–2402, June 2018
14. Ding, H., Jiang, X., Shuai, B., Liu, A.Q., Wang, G.: Semantic correlation promoted shape-variant context for segmentation. In: Proceedings of the IEEE Conference on Computer Vision and Pattern Recognition (CVPR), pp. 8885–8894, June 2019
15. Ding, H., Jiang, X., Shuai, B., Liu, A.Q., Wang, G.: Semantic segmentation with context encoding and multi-path decoding. IEEE Trans. Image Process. **29**, 3520–3533 (2020)

16. Dutt Jain, S., Grauman, K.: Predicting sufficient annotation strength for interactive foreground segmentation. In: Proceedings of the IEEE International Conference on Computer Vision (2013)
17. Everingham, M., Van Gool, L., Williams, C.K., Winn, J., Zisserman, A.: The pascal visual object classes (VOC) challenge. Int. J. Comput. Vis. **88**(2) (2010)
18. Grady, L.: Random walks for image segmentation. IEEE Trans. Pattern Anal. Mach. Intell. **28**(11), 1768–1783 (2006)
19. Gulshan, V., Rother, C., Criminisi, A., Blake, A., Zisserman, A.: Geodesic star convexity for interactive image segmentation. In: IEEE Computer Society Conference on Computer Vision and Pattern Recognition, pp. 3129–3136. IEEE (2010)
20. He, K., Gkioxari, G., Dollár, P., Girshick, R.: Mask R-CNN. In: Proceedings of the IEEE International Conference on Computer Vision, pp. 2961–2969 (2017)
21. He, K., Zhang, X., Ren, S., Sun, J.: Deep residual learning for image recognition. In: The IEEE Conference on Computer Vision and Pattern Recognition (2016)
22. Hu, P., Caba, F., Wang, O., Lin, Z., Sclaroff, S., Perazzi, F.: Temporally distributed networks for fast video semantic segmentation. In: Proceedings of the IEEE Conference on Computer Vision and Pattern Recognition, pp. 8818–8827 (2020)
23. Hu, R., Dollár, P., He, K., Darrell, T., Girshick, R.: Learning to segment every thing. In: Proceedings of the IEEE Conference on Computer Vision and Pattern Recognition, pp. 4233–4241 (2018)
24. Hu, R., Rohrbach, M., Darrell, T.: Segmentation from natural language expressions. In: Leibe, B., Matas, J., Sebe, N., Welling, M. (eds.) ECCV 2016. LNCS, vol. 9905, pp. 108–124. Springer, Cham (2016). https://doi.org/10.1007/978-3-319-46448-0_7
25. Hu, Y., Soltoggio, A., Lock, R., Carter, S.: A fully convolutional two-stream fusion network for interactive image segmentation. Neural Netw. **109** (2019)
26. Kass, M., Witkin, A., Terzopoulos, D.: Snakes: active contour models. Int. J. Comput. Vis. **1**(4) (1988)
27. Kazemzadeh, S., Ordonez, V., Matten, M., Berg, T.: ReferitGame: referring to objects in photographs of natural scenes. In: Proceedings of the 2014 Conference on Empirical Methods In Natural Language Processing (EMNLP), pp. 787–798 (2014)
28. Le, H., Mai, L., Price, B., Cohen, S., Jin, H., Liu, F.: Interactive boundary prediction for object selection. In: Ferrari, V., Hebert, M., Sminchisescu, C., Weiss, Y. (eds.) Computer Vision – ECCV 2018. LNCS, vol. 11218, pp. 20–36. Springer, Cham (2018). https://doi.org/10.1007/978-3-030-01264-9_2
29. Lempitsky, V.S., Kohli, P., Rother, C., Sharp, T.: Image segmentation with a bounding box prior. In: ICCV, vol. 76 (2009)
30. Li, R., Li, K., Kuo, Y.C., Shu, M., Qi, X., Shen, X., Jia, J.: Referring image segmentation via recurrent refinement networks. In: Proceedings of the IEEE Conference on Computer Vision and Pattern Recognition, pp. 5745–5753 (2018)
31. Li, Y., Sun, J., Tang, C.K., Shum, H.Y.: Lazy snapping. ACM Trans. Graph. (ToG) (2004)
32. Li, Z., Chen, Q., Koltun, V.: Interactive image segmentation with latent diversity. In: Proceedings of the IEEE Conference on Computer Vision and Pattern Recognition, pp. 577–585 (2018)
33. Liew, J.H., Cohen, S., Price, B., Mai, L., Ong, S.H., Feng, J.: MultiSeg: semantically meaningful, scale-diverse segmentations from minimal user input. In: The IEEE International Conference on Computer Vision (2019)

34. Liew, J., Wei, Y., Xiong, W., Ong, S.H., Feng, J.: Regional interactive image segmentation networks. In: 2017 IEEE International Conference on Computer Vision (ICCV), pp. 2746–2754. IEEE (2017)
35. Liu, C., Lin, Z., Shen, X., Yang, J., Lu, X., Yuille, A.: Recurrent multimodal interaction for referring image segmentation. In: Proceedings of the IEEE International Conference on Computer Vision, pp. 1271–1280 (2017)
36. Liu, J., et al.: Feature boosting network for 3D pose estimation. IEEE Trans. Pattern Anal. Mach. Intell. **42**(2), 494–501 (2020)
37. Liu, S., Qi, L., Qin, H., Shi, J., Jia, J.: Path aggregation network for instance segmentation. In: Proceedings of the IEEE Conference on Computer Vision and Pattern Recognition, pp. 8759–8768 (2018)
38. Liu, X., Wang, Z., Shao, J., Wang, X., Li, H.: Improving referring expression grounding with cross-modal attention-guided erasing. In: Proceedings of the IEEE Conference on Computer Vision and Pattern Recognition (2019)
39. Long, J., Shelhamer, E., Darrell, T.: Fully convolutional networks for semantic segmentation. In: Proceedings of the IEEE Conference on Computer Vision and Pattern Recognition, pp. 3431–3440 (2015)
40. Mahadevan, S., Voigtlaender, P., Leibe, B.: Iteratively trained interactive segmentation. In: BMVC (2018)
41. Maninis, K.K., Caelles, S., Pont-Tuset, J., Van Gool, L.: Deep extreme cut: from extreme points to object segmentation. In: Proceedings of the IEEE Conference on Computer Vision and Pattern Recognition, pp. 616–625 (2018)
42. Margffoy-Tuay, E., Pérez, J.C., Botero, E., Arbeláez, P.: Dynamic multimodal instance segmentation guided by natural language queries. In: Ferrari, V., Hebert, M., Sminchisescu, C., Weiss, Y. (eds.) ECCV 2018. LNCS, vol. 11215, pp. 656–672. Springer, Cham (2018). https://doi.org/10.1007/978-3-030-01252-6_39
43. McGuinness, K., O'connor, N.E.: A comparative evaluation of interactive segmentation algorithms. Pattern Recognit. **43**(2), 434–444 (2010)
44. Mei, J., Wu, Z., Chen, X., Qiao, Y., Ding, H., Jiang, X.: DeepdeBlur: text image recovery from blur to sharp. Multimed. Tools Appl. **78**(13), 18869–18885 (2019)
45. Mortensen, E.N., Barrett, W.A.: Intelligent scissors for image composition. In: Proceedings of the 22nd Annual Conference on Computer Graphics and Interactive Techniques. ACM (1995)
46. Papadopoulos, D.P., Uijlings, J.R., Keller, F., Ferrari, V.: Extreme clicking for efficient object annotation. In: IEEE International Conference on Computer Vision, pp. 4930–4939 (2017)
47. Price, B.L., Morse, B., Cohen, S.: Geodesic graph cut for interactive image segmentation. In: IEEE Conference on Computer Vision and Pattern Recognition, pp. 3161–3168. IEEE (2010)
48. Ren, S., He, K., Girshick, R., Sun, J.: Faster R-CNN: towards real-time object detection with region proposal networks. In: Advances in Neural Information Processing Systems, pp. 91–99 (2015)
49. Rother, C., Kolmogorov, V., Blake, A.: "GrabCut": interactive foreground extraction using iterated graph cuts. ACM Trans. Graph. (TOG) **23**(3), 309–314 (2004)
50. Rupprecht, C., Laina, I., Navab, N., Hager, G.D., Tombari, F.: Guide me: interacting with deep networks. In: Proceedings of the IEEE Conference on Computer Vision and Pattern Recognition, pp. 8551–8561 (2018)
51. Shi, J., Malik, J.: Normalized cuts and image segmentation. Departmental Papers (CIS), p. 107 (2000)

52. Shuai, B., Ding, H., Liu, T., Wang, G., Jiang, X.: Toward achieving robust low-level and high-level scene parsing. IEEE Trans. Image Process. **28**(3), 1378–1390 (2018)
53. Vezhnevets, V., Konouchine, V.: GrowCut: interactive multi-label nd image segmentation by cellular automata. In: Proceedings of Graphicon, vol. 1, pp. 150–156. Citeseer (2005)
54. Wang, P., Wu, Q., Cao, J., Shen, C., Gao, L., Hengel, A.v.d.: Neighbourhood watch: referring expression comprehension via language-guided graph attention networks. In: Proceedings of the IEEE Conference on Computer Vision and Pattern Recognition (2019)
55. Wang, X., Ding, H., Jiang, X.: Dermoscopic image segmentation through the enhanced high-level parsing and class weighted loss. In: 2019 IEEE International Conference on Image Processing (ICIP), pp. 245–249. IEEE (2019)
56. Wang, X., Jiang, X., Ding, H., Liu, J.: Bi-directional dermoscopic feature learning and multi-scale consistent decision fusion for skin lesion segmentation. IEEE Trans. Image Process. **29**, 3039–3051 (2019)
57. Xu, N., Price, B., Cohen, S., Yang, J., Huang, T.: Deep GrabCut for object selection. In: BMVC (2017)
58. Xu, N., Price, B., Cohen, S., Yang, J., Huang, T.S.: Deep interactive object selection. In: Proceedings of the IEEE Conference on Computer Vision and Pattern Recognition, pp. 373–381 (2016)
59. Ye, L., Liu, Z., Wang, Y.: Dual convolutional LSTM network for referring image segmentation. IEEE Trans. Multimed. (2020)
60. Ye, L., Rochan, M., Liu, Z., Wang, Y.: Cross-modal self-attention network for referring image segmentation. In: Proceedings of the IEEE Conference on Computer Vision and Pattern Recognition, pp. 10502–10511 (2019)
61. Yu, L., Lin, Z., Shen, X., Yang, J., Lu, X., Bansal, M., Berg, T.L.: MAttNet: modular attention network for referring expression comprehension. In: Proceedings of the IEEE Conference on Computer Vision and Pattern Recognition, pp. 1307–1315 (2018)
62. Yu, L., Poirson, P., Yang, S., Berg, A.C., Berg, T.L.: Modeling context in referring expressions. In: Leibe, B., Matas, J., Sebe, N., Welling, M. (eds.) ECCV 2016. LNCS, vol. 9906, pp. 69–85. Springer, Cham (2016). https://doi.org/10.1007/978-3-319-46475-6_5
63. Zeng, Y., Lin, Z., Yang, J., Zhang, J., Shechtman, E., Lu, H.: High-resolution image inpainting with iterative confidence feedback and guided upsampling. In: European Conference on Computer Vision. Springer (2020)
64. Zeng, Y., Lu, H., Zhang, L., Feng, M., Borji, A.: Learning to promote saliency detectors. In: IEEE Conference on Computer Vision and Pattern Recognition (2018)
65. Zeng, Y., Zhuge, Y., Lu, H., Zhang, L.: Joint learning of saliency detection and weakly supervised semantic segmentation. In: IEEE International Conference on Computer Vision (2019)
66. Zeng, Y., Zhuge, Y., Lu, H., Zhang, L., Qian, M., Yu, Y.: Multi-source weak supervision for saliency detection. In: IEEE Conference on Computer Vision and Pattern Recognition (2019)
67. Zhang, L., Dai, J., Lu, H., He, Y.: A bi-directional message passing model for salient object detection. In: CVPR (2018)
68. Zhang, L., Lin, Z., Zhang, J., Lu, H., He, Y.: Fast video object segmentation via dynamic targeting network. In: ICCV (2019)

69. Zhang, L., Zhang, J., Lin, Z., Lu, H., He, Y.: Capsal: Leveraging captioning to boost semantics for salient object detection. In: CVPR (2019)
70. Zhang, L., Zhang, J., Lin, Z., Mech, R., Lu, H., He, Y.: Unsupervised video object segmentation with joint hotspot tracking. In: ECCV (2020)
71. Zhao, H., Shi, J., Qi, X., Wang, X., Jia, J.: Pyramid scene parsing network. In: The IEEE Conference on Computer Vision and Pattern Recognition (CVPR) (2017)

Unified Multisensory Perception: Weakly-Supervised Audio-Visual Video Parsing

Yapeng Tian[1(✉)], Dingzeyu Li[2], and Chenliang Xu[1]

[1] University of Rochester, Rochester, USA
{yapengtian,chenliang.xu}@rochester.edu
[2] Adobe Research, Seattle, USA
dinli@adobe.com

Abstract. In this paper, we introduce a new problem, named audio-visual video parsing, which aims to parse a video into temporal event segments and label them as either audible, visible, or both. Such a problem is essential for a complete understanding of the scene depicted inside a video. To facilitate exploration, we collect a *Look, Listen, and Parse* (LLP) dataset to investigate audio-visual video parsing in a weakly-supervised manner. This task can be naturally formulated as a Multimodal Multiple Instance Learning (MMIL) problem. Concretely, we propose a novel hybrid attention network to explore unimodal and cross-modal temporal contexts simultaneously. We develop an attentive MMIL pooling method to adaptively explore useful audio and visual content from different temporal extent and modalities. Furthermore, we discover and mitigate modality bias and noisy label issues with an individual-guided learning mechanism and label smoothing technique, respectively. Experimental results show that the challenging audio-visual video parsing can be achieved even with only video-level weak labels. Our proposed framework can effectively leverage unimodal and cross-modal temporal contexts and alleviate modality bias and noisy labels problems.

Keywords: Audio-visual video parsing · Weakly-supervised · LLP dataset

1 Introduction

Human perception involves complex analyses of visual, auditory, tactile, gustatory, olfactory, and other sensory data. Numerous psychological and brain cognitive studies [3,20,46,51] show that combining different sensory data is crucial for human perception. However, the vast majority of work [9,26,48,64] in

---

Electronic supplementary material The online version of this chapter (https://doi.org/10.1007/978-3-030-58580-8_26) contains supplementary material, which is available to authorized users.

A. Vedaldi et al. (Eds.): ECCV 2020, LNCS 12348, pp. 436–454, 2020.
https://doi.org/10.1007/978-3-030-58580-8_26

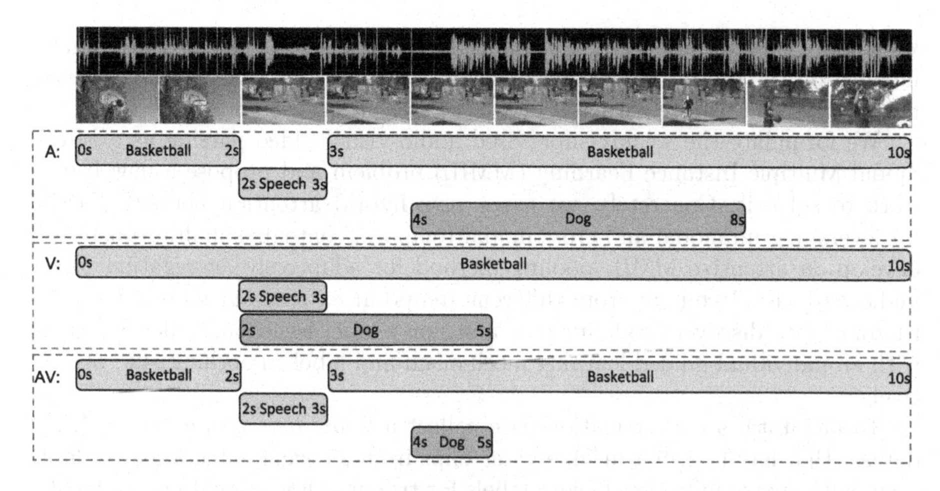

Fig. 1. Our audio-visual video parsing model aims to parse a video into different audio (audible), visual (visible), and audio-visual (audi-visible) events with correct categories and boundaries. A dog in the video visually appears from 2nd second to 5th second and make barking sounds from 4th second to 8th second. So, we have audio event (4 s–8 s), visual event (2 s–5 s), and audio-visual event (4 s–5 s) for the *Dog* event category.

scene understanding, an essential perception task, focuses on visual-only methods ignoring other sensory modalities. They are inherently limited. For example, when the object of interest is outside of the field-of-view (FoV), one would rely on audio cues for localization. While there is little data on tactile, gustatory, or olfactory signals, we do have an abundance of multimodal audiovisual data, e.g., YouTube videos.

Utilizing and learning from both auditory and visual modalities is an emerging research topic. Recent years have seen progress in learning representations [1, 2,19,23,37,38], separating visually indicated sounds [8,10–13,65,66,70], spatially localizing visible sound sources [37,45,55], and temporally localizing audio-visual synchronized segments [27,55,63]. However, past approaches usually assume audio and visual data are always correlated or even temporally aligned. In practice, when we analyze the video scene, many videos have audible sounds, which originate outside of the FoV, leaving no visual correspondences, but still contribute to the overall understanding, such as out-of-screen running cars and a narrating person. Such examples are ubiquitous, which leads us to some basic questions: what video events are audible, visible, and "audi-visible," where and when are these events inside of a video, and how can we effectively detect them?

To answer the above questions, we pose and try to tackle a fundamental problem: *audio-visual video parsing* that recognizes event categories bind to sensory modalities, and meanwhile, finds temporal boundaries of when such an event starts and ends (see Fig. 1). However, learning a fully supervised audio-visual video parsing model requires densely annotated event modality and category labels with corresponding event onsets and offsets, which will make the labeling process extremely expensive and time-consuming. To avoid tedious labeling, we explore weakly-supervised learning for the task, which only requires sparse labeling on the presence or absence of video events. The weak labels are easier to annotate and can be gathered in a large scale from web videos.

We formulate the weakly-supervised audio-visual video parsing as a Multimodal Multiple Instance Learning (MMIL) problem and propose a new framework to solve it. Concretely, we use a new hybrid attention network (HAN) for leveraging unimodal and cross-modal temporal contexts simultaneously. We develop an attentive MMIL pooling method for adaptively aggregating useful audio and visual content from different temporal extent and modalities. Furthermore, we discover modality bias and noisy label issues and alleviate them with an individual-guided learning mechanism and label smoothing [42], respectively.

To facilitate our investigations, we collect a *Look, listen, and Parse* (LLP) dataset that has 11, 849 YouTube video clips from 25 event categories. We label them with sparse video-level event labels for training. For evaluation, we label a set of precise labels, including event modalities, event categories, and their temporal boundaries. Experimental results show that it is tractable to learn audio-visual video parsing even with video-level weak labels. Our proposed HAN model can effectively leverage multimodal temporal contexts. Furthermore, modality bias and noisy label problems can be addressed with the proposed individual learning strategy and label smoothing, respectively. Besides, we make a discussion on the potential applications enabled by audio-visual video parsing.

The contributions of our work include: (1) a new audio-visual video parsing task towards a unified multisensory perception; (2) a novel hybrid attention network to leverage unimodal and cross-modal temporal contexts simultaneously; (3) an effective attentive MMIL pooling to aggregate multimodal information adaptively; (4) a new individual guided learning approach to mitigate the modality bias in the MMIL problem and label smoothing to alleviate noisy labels; and (5) a newly collected large-scale video dataset, named LLP, for audio-visual video parsing. Dataset, code, and pre-trained models are publicly available in https://github.com/YapengTian/AVVP-ECCV20.

2 Related Work

In this section, we discuss some related work on temporal action localization, sound event detection, and audio-visual learning.

Temporal Action Localization. Temporal action localization (TAL) methods usually use sliding windows as action candidates and address TAL as a classification problem [9,25,29,47,48,67] learning from full supervisions. Recently, weakly-supervised approaches are proposed to solve the TAL. Wang *et al.* [60] present an UntrimmedNet with a classification module and a selection module to learn the action models and reason about the temporal duration of action instances, respectively. Hide-and-seek [49] randomly hides certain sequences while training to force the model to explore more discriminative content. Paul *et al.* [40] introduce a co-activity similarity loss to enforce instances in the same

Fig. 2. Some examples from the LLP dataset.

class to be similar in the feature space. Inspired by the class activation map method [68], Nguyen *et al.* [36] propose a sparse temporal pooling network (STPN). Liu *et al.* [28] incorporate both action completeness modeling and action-context separation into a weakly-supervised TAL framework. Unlike actions in TAL, video events in audio-visual video parsing might contain motionless or even out-of-screen sound sources and the events can be perceived by either audio or visual modalities. Even though, we extend two recent weakly-supervised TAL methods: STPN [36] and CMCS [28] to address visual event parsing and compare them with our model in Sect. 6.2.

Sound Event Detection. Sound event detection (SED) is a task of recognizing and locating audio events in acoustic environments. Early supervised approaches rely on some machine learning models, such as support vector machines [7], Gaussian mixture models [17] and recurrent neural networks [39]. To bypass strongly labeled data, weakly-supervised SED methods are developed [6,22,31,62]. These methods only focus on audio events from constrained domains, such as urban sounds [44] and domestic environments [32] and visual information is ignored. However, our audio-visual video parsing will exploit both modalities to parse not only event categories and boundaries but also event perceiving modalities towards a unified multisensory perception for unconstrained videos.

Audio-Visual Learning. Benefiting from the natural synchronization between auditory and visual modalities, audio-visual learning has enabled a set of new problems and applications including representation learning [1,2,19,23,35, 37,38], audio-visual sound separation [8,10–13,65,66,70], vision-infused audio inpainting [69], sound source spatial localization [37,45,55], sound-assisted action recognition [14,21,24], audio-visual video captioning [41,53,54,61], and audio-visual event localization [27,55,56,63]. Most previous work assumes that temporally synchronized audio and visual content are always matched conveying the same semantic meanings. However, unconstrained videos can be very noisy: sound sources might not be visible (*e.g.*, an out-of-screen running car and a narrating person) and not all visible objects are audible (*e.g.*, a static motorcycle and people dancing with music). Different from previous methods, we pose and seek to tackle a fundamental but unexplored problem: audio-visual video parsing for parsing unconstrained videos into a set of video events associated with event categories, boundaries, and modalities. Since the existing methods cannot directly address our problem, we modify the recent weakly-supervised audio-visual event localization methods: AVE [55] and AVSDN [27] adding additional audio and visual parsing branches as baselines.

3 LLP: The Look, Listen and Parse Dataset

To the best of our knowledge, there is no existing dataset that is suitable for our research. Thus, we introduce a *Look, Listen, and Parse* dataset for audio-visual video scene parsing, which contains 11,849 YouTube video clips spanning over 25 categories for a total of 32.9 h collected from AudioSet [15]. A wide range of video events (*e.g.*, human speaking, singing, baby crying, dog barking, violin playing, and car running, and vacuum cleaning *etc.*) from diverse domains (*e.g.*, human activities, animal activities, music performances, vehicle sounds, and domestic environments) are included in the dataset. Some examples in the LLP dataset are shown in Fig. 2.

Videos in the LLP have 11,849 video-level event annotations on the presence or absence of different video events for facilitating weakly-supervised learning. Each video is 10 s long and has at least 1 s audio or visual events. There are 7,202 videos that contain events from more than one event categories and per video has averaged 1.64 different event categories. To evaluate audio-visual scene parsing performance, we annotate individual audio and visual events with second-wise temporal boundaries for randomly selected 1,849 videos from the LLP dataset. Note that the audio-visual event labels can be derived from the audio and visual event labels. Finally, we have totally 6,626 event annotations, including 4,131 audio events and 2,495 visual events for the 1,849 videos. Merging the individual audio and visual labels, we obtain 2,488 audio-visual event annotations. To do validation and testing, we split the subset into a validation set with 649 videos and a testing set with 1,200 videos. Our weakly-supervised audio-visual video parsing network will be trained using the 10,000 videos with weak labels and the trained models are developed and tested on the validation and testing sets with fully annotated labels, respectively.

4 Audio-Visual Video Parsing with Weak Labels

We define the *Audio-Visual Video Parsing* as a task to group video segments and parse a video into different temporal audio, visual, and audio-visual events associated with semantic labels. Since event boundary in the LLP dataset was annotated at second-level, video events will be parsed at scene-level not object/instance level in our experimental setting. Concretely, given a video sequence containing both audio and visual tracks, we divide it into T non-overlapping audio and visual snippet pairs $\{V_t, A_t\}_{t=1}^{T}$, where each snippet is $1s$ long and V_t and A_t denote visual and audio content in the same video snippet, respectively. Let $\boldsymbol{y}_t = \{(y_t^a, y_t^v, y_t^{av})|[y_a^t]_c, [y_v^t]_c, [y_{av}^t]_c \in \{0,1\}, c = 1, ..., C\}$ be the event label set for the video snippet $\{V_t, A_t\}$, where c refers to the c-th event category and y_t^a, y_t^v, and y_t^{av} denote audio, visual, and audio-visual event labels, respectively. Here, we have a relation: $y_t^{av} = y_t^a * y_t^v$, which means that

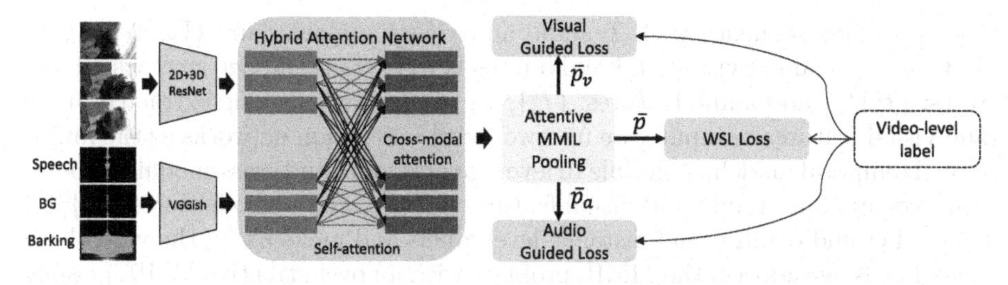

Fig. 3. The proposed audio-visual video parsing framework. It uses pre-trained CNNs to extract snippet-level audio and visual features and leverages multimodal temporal contexts with the proposed hybrid attention network (HAN). For each snippet, we will predict both audio and visual event labels from the aggregated features by the HAN. Attentive MMIL pooling is utilized to adaptively predict video-level event labels for weakly-supervised learning (WSL) and individual guided learning is devised to mitigate the modality bias issue.

audio-visual events occur only when there exists both audio and visual events at the same time and from the same event categories.

In this work, we explore the audio-visual video parsing in a weakly-supervised manner. We only have video-level labels for training, but will predict precise event label sets for all video snippets during testing, which makes the weakly-supervised audio-visual video parsing be a multi-modal multiple instance learning (MMIL) problem. Let a video sequence with T audio and visual snippet pairs be a bag. Unlike the previous audio-visual event localization [55] that is formulated as a MIL problem [30] where an audio-visual snippet pair is regarded as an instance, each audio snippet and the corresponding visual snippet occurred at the same time denote two individual instances in our MMIL problem. So, a positive bag containing video events will have at least one positive video snippet; meanwhile at least one modality has video events in the positive video snippet. During training, we can only access bag labels. During inference, we need to know not only which video snippets have video events but also which sensory modalities perceive the events. The temporal and multi-modal uncertainty in this MMIL problem makes it very challenging.

5 Method

First, we present the overall framework that formulates the weakly-supervised audio-visual video parsing as an MMIL problem in Sect. 5.1. Built upon this framework, we propose a new multimodal temporal model: hybrid attention network in Sect. 5.2; attentive MMIL pooling in Sect. 5.3; addressing modality bias and noisy label issues in Sect. 5.4.

5.1 Audio-Visual Video Parsing Framework

Our framework, as illustrated in Fig. 3, has three main modules: audio and visual feature extraction, multimodal temporal modeling, and attentive MMIL pooling.

Given a video sequence with T audio and visual snippet pairs $\{V_t, A_t\}_{t=1}^T$, we first use pre-trained visual and audio models to extract snippet-level visual features: $\{f_v^t\}_{t=1}^T$ and audio features: $\{f_a^t\}_{t=1}^T$, respectively. Taking extracted audio and visual features as inputs, we use two hybrid attention networks as the multimodal temporal modeling module to leverage unimodal and cross-modal temporal contexts and obtain updated visual features $\{\hat{f}_v^t\}_{t=1}^T$ and audio features $\{\hat{f}_a^t\}_{t=1}^T$. To predict audio and visual instance-level labels and make use of the video-level weak labels, we address the MMIL problem with a novel attentive MMIL pooling module outputting video-level labels.

5.2 Hybrid Attention Network

Natural videos tend to contain continuous and repetitive rather than isolated audio and visual content. In particular, audio or visual events in a video usually redundantly recur many times inside the video, both within the same modality (unimodal temporal recurrence [34,43]), as well as across different modalities (audio-visual temporal synchronization [23] and asynchrony [59]). The observation suggests us to jointly model the temporal recurrence, co-occurrence, and asynchrony in a unified approach. However, existing audio-visual learning methods [27,55,63] usually ignore the audio-visual temporal asynchrony and explore unimodal temporal recurrence using temporal models (*e.g.*, LSTM [18] and Transformer [58]) and audio-visual temporal synchronization using multimodal fusion (*e.g.*, feature fusion [55] and prediction ensemble [21]) in a isolated way. To simultaneously capture multimodal temporal contexts, we propose a new temporal model: Hybrid Attention Network (HAN), which uses a self-attention network and a cross-attention network to adaptively learn which bimodal and cross-modal snippets to look for each audio or visual snippet, respectively.

At each time step t, a hybrid attention function g in HAN will be learned from audio and visual features: $\{f_a^t, f_v^t\}_{t=1}^T$ to update f_a^t and f_v^t, respectively. The updated audio feature $\hat{f}_a^t$ and visual feature $\hat{f}_v^t$ can be computed as:

$$\hat{f}_a^t = g(f_a^t, \boldsymbol{f}_a, \boldsymbol{f}_v) = f_a^t + g_{sa}(f_a^t, \boldsymbol{f}_a) + g_{ca}(f_a^t, \boldsymbol{f}_v) \ , \tag{1}$$

$$\hat{f}_v^t = g(f_v^t, \boldsymbol{f}_a, \boldsymbol{f}_v) = f_v^t + g_{sa}(f_v^t, \boldsymbol{f}_v) + g_{ca}(f_v^t, \boldsymbol{f}_a) \ , \tag{2}$$

where $\boldsymbol{f}_a = [f_a^1; ...; f_a^T]$ and $\boldsymbol{f}_v = [f_v^1; ...; f_v^T]$; g_{sa} and g_{ca} are self-attention and cross-modal attention functions, respectively; skip-connections can help preserve the identity information from the input sequences. The two attention functions are formulated with the same computation mechanism. With $g_{sa}(f_a^t, \boldsymbol{f}_a)$ and $g_{ca}(f_a^t, \boldsymbol{f}_v)$ as examples, they are defined as:

$$g_{sa}(f_a^t, \boldsymbol{f}_a) = \sum_{t=1}^{T} w_t^{sa} f_a^t = softmax(\frac{f_a^t \boldsymbol{f}_a^{'}}{\sqrt{d}}) \boldsymbol{f}_a \ , \tag{3}$$

$$g_{ca}(f_a^t, \boldsymbol{f}_v) = \sum_{t=1}^{T} w_t^{ca} f_v^t = softmax(\frac{f_a^t \boldsymbol{f}_v^{'}}{\sqrt{d}}) \boldsymbol{f}_v \ , \tag{4}$$

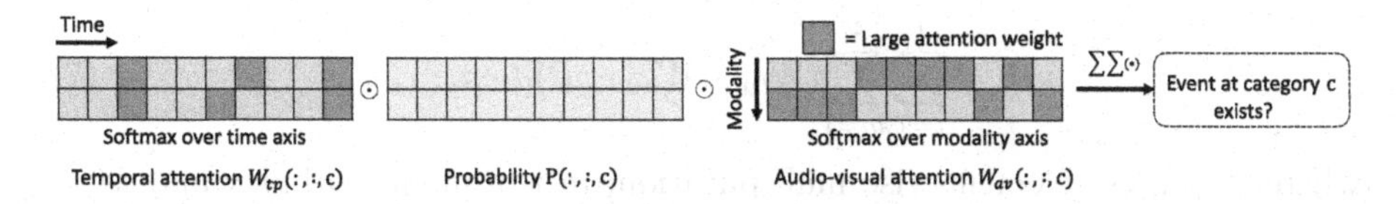

Fig. 4. Attentive MMIL Pooling. For event category c, temporal and audio-visual attention mechanisms will adaptively select informative event predictions crossing temporal and modality axes, respectively, for predicting whether there is an event at the category.

where the scaling factor d is equal to the audio/visual feature dimension and $(\cdot)'$ denotes the transpose operator. Clearly, the self-attention and cross-modal attention functions in HAN will assign large weights to snippets, which are similar to the query snippet containing the same video events within the same modality and cross different modalities. The experimental results show that the HAN modeling unimodal temporal recurrence, multimodal temporal co-occurrence, and audio-visual temporal asynchrony can well capture unimodal and cross-modal temporal contexts and improves audio-visual video parsing performance.

5.3 Attentive MMIL Pooling

To achieve audio-visual video parsing, we predict all event labels for audio and visual snippets from temporal aggregated features: $\{\hat{f}_a^t, \hat{f}_v^t\}_{t=1}^T$. We use a shared fully-connected layer to project audio and visual features to different event label space and adopt a sigmoid function to output probability for each event category:

$$p_a^t = sigmoid(FC(\hat{f}_a^t)) \ , \tag{5}$$

$$p_v^t = sigmoid(FC(\hat{f}_v^t)) \ , \tag{6}$$

where p_a^t and p_v^t are predicted audio and visual event probabilities at timestep t, respectively. Here, the shared FC layer can implicitly enforce audio and visual features into a similar latent space. The reason to use sigmoid to output an event probability for each event category rather than softmax to predict a probability distribution over all categories is that a single snippet may have multiple labels rather than only a single event as assumed in Tian *et al.* [55].

Since audio-visual events only occur when sound sources are visible and their sounds are audible, the audio-visual event probability p_{av}^t can be derived from individual audio and visual predictions: $p_{av}^t = p_a^t * p_v^t$. If we have direct supervisions for all audio and visual snippets from different time steps, we can simply learn the audio-visual video parsing network in a fully-supervised manner. However, in this MMIL problem, we can only access a video-level weak label $\bar{\boldsymbol{y}}$ for all audio and visual snippets: $\{A_t, V_t\}_{t=1}^T$ from a video. To learn our network with weak labels, as illustrated in Fig. 4, we propose a attentive MMIL pooling method to predict video-level event probability: $\bar{\boldsymbol{p}}$ from $\{p_a^t, p_v^t\}_{t=1}^T$. Concretely, the $\bar{\boldsymbol{p}}$ is computed by:

$$\bar{\boldsymbol{p}} = \sum_{t=1}^{T}\sum_{m=1}^{M}(W_{tp} \odot W_{av} \odot P)[t, m, :] \ , \tag{7}$$

where $\odot$ denotes element-wise multiplication; m is a modality index and $M = 2$ refers to audio and visual modalities; W_{tp} and W_{av} are temporal attention and audio-visual attention tensors predicted from $\{\hat{f}_a^t, \hat{f}_v^t\}_{t=1}^T$, respectively, and P is the probability tensor built by $\{p_a^t, p_v^t\}_{t=1}^T$ and we have $P(t, 0, :) = p_a^t$ and $P(t, 1, :) = p_v^t$. To compute the two attention tensors, we first compose an input feature tensor F, where $F(t, 0, :) = \hat{f}_a^t$ and $F(t, 1, :) = \hat{f}_v^t$. Then, two different FC layers are used to transform the F into two tensors: F_{tp} and F_{av}, which has the same size as P. To adpatively select most informative snippets for predicting probabilities of different event categories, we assign different weights to snippets at different time steps with a temporal attention mechanism:

$$W_{tp}[:, m, c] = softmax(F_{tp}[:, m, c]) \ , \tag{8}$$

where $m = 1, 2$ and $c = 1, \ldots, C$. Accordingly, we can adaptively select most informative modalities with the audio-visual attention tensor:

$$W_{av}[t, :, c] = softmax(F_{av}[t, :, c]) \ , \tag{9}$$

where $t = 1, \ldots, T$ and $c = 1, \ldots, C$. The snippets within a video from different temporal steps and different modalities may have different video events. The proposed attentive MMIL pooling can well model this observation with the tensorized temporal and multimodal attention mechanisms.

With the predicted video-level event probability $\bar{\boldsymbol{p}}$ and the ground truth label $\bar{\boldsymbol{y}}$, we can optimize the proposed weakly-supervised learning model with a binary cross-entropy loss function: $\mathcal{L}_{wsl} = CE(\bar{\boldsymbol{p}}, \bar{\boldsymbol{y}}) = -\sum_{c=1}^{C} \bar{\boldsymbol{y}}[c]log(\bar{\boldsymbol{p}}[c])$.

5.4 Alleviating Modality Bias and Label Noise

The weakly supervised audio-visual video parsing framework only uses less detailed annotations without requiring expensive densely labeled audio and visual events for all snippets. This advantage makes this weakly supervised learning framework appealing. However, it usually enforces models to only identify discriminative patterns in the training data, which was observed in previous weakly-supervised MIL problems [49,50,68]. In our MMIL problem, the issue becomes even more complicated since there are multiple modalities and different modalities might not contain equally discriminative information. With weakly-supervised learning, the model tends to only use information from the most discriminative modality but ignore another modality, which can probably achieve good video classification performance but terrible video parsing performance on the events from ignored modality and audio-visual events. Since a video-level label contains all event categories from audio and visual content within the video, to alleviate the modality bias in the MMIL, we propose to use explicit supervisions to both modalities with a guided loss:

$$\mathcal{L}_g = CE(\bar{\boldsymbol{p}}_a, \bar{\boldsymbol{y}}_a) + CE(\bar{\boldsymbol{p}}_v, \bar{\boldsymbol{y}}_v) \ , \tag{10}$$

where $\bar{\boldsymbol{y}}_a = \bar{\boldsymbol{y}}_v = \bar{\boldsymbol{y}}$, and $\bar{\boldsymbol{p}}_a = \sum_{t=1}^{T}(W_{tp} \odot P)[t, 0, :]$ and $\bar{\boldsymbol{p}}_v = \sum_{t=1}^{T}(W_{tp} \odot P)[t, 1, :]$ are video-level audio and visual event probabilities, respectively.

However, not all video events are audio-visual events, which means that an event occurred in one modality might not occur in another modality and then the corresponding event label will be label noise for one of the two modalities. Thus, the guided loss: $\mathcal{L}_g$ suffers from noisy label issue. For the example shown in Fig. 3, the video-level label is {*Speech*, *Dog*} and the video-level visual event label is only {*Dog*}. The {*Speech*} will be a noisy label for the visual guided loss. To handle the problem, we use label smoothing [52] to lower the confidence of positive labels with smoothing $\bar{\boldsymbol{y}}$ and generate smoothed labels: $\bar{\boldsymbol{y}}_a$ and $\bar{\boldsymbol{y}}_v$. They are formulated as: $\bar{\boldsymbol{y}}_a = (1-\epsilon_a)\bar{\boldsymbol{y}} + \frac{\epsilon_a}{K}$ and $\bar{\boldsymbol{y}}_v = (1-\epsilon_v)\bar{\boldsymbol{y}} + \frac{\epsilon_v}{K}$, where $\epsilon_a, \epsilon_v \in [0, 1)$ are two confidence parameters to balance the event probability distribution and a uniform distribution: $u = \frac{1}{K}$ $(K > 1)$. For a noisy label at event category c, when $\bar{\boldsymbol{y}}[c] = 1$ and real $\bar{\boldsymbol{y}}_a[c] = 0$, we have $\bar{\boldsymbol{y}}[c] = (1-\epsilon_a)\bar{\boldsymbol{y}}[c] + \epsilon_a > (1-\epsilon_a)\bar{\boldsymbol{y}} + \frac{\epsilon_a}{K} = \bar{\boldsymbol{y}}_a[c]$ and the smoothed label will become more reliable. Label smoothing technique is commonly adopted in a lot of tasks, such as image classification [52], speech recognition [5], and machine translation [58] to reduce over-fitting and improve generalization capability of deep models. Different from the past methods, we use smoothed labels to mitigate label noise occurred in the individual guided learning. Our final model is optimized with the two loss terms: $\mathcal{L} = \mathcal{L}_{wsl} + \mathcal{L}_g$.

6 Experiments

6.1 Experimental Settings

Implementation Details. For a 10-second-long video, we first sample video frames at 8 fps and each video is divided into non-overlapping snippets of the same length with 8 frames in 1 s. Given a visual snippet, we extract a 512-D snippet-level feature with fused features extracted from ResNet152 [16] and 3D ResNet [57]. In our experiments, batch size and number of epochs are set as 16 and 40, respectively. The initial learning rate is 3*e*-4 and will drop by multiplying 0.1 after every 10 epochs. Our models optimized by Adam can be trained using one NVIDIA 1080Ti GPU.

Baselines. Since there are no existing methods to address the audio-visual video parsing, we design several baselines based on previous state-of-the-art weakly-supervised sound detection [22,62], temporal action localization [28,36], and audio-visual event localization [27,55] methods to validate the proposed framework. To make [27,55] possible to address audio-visual scene parsing, we add additional audio and visual branches to predict audio and visual event probabilities supervised with an additional guided loss as defined in Sect. 5.4. For fair comparisons, the compared approaches use the same audio and visual features as our method.

Evaluation Metrics. To comprehensively measure the performance of different methods, we evaluate them on parsing all types of events (individual audio,

Table 1. Audio-visual video parsing accuracy (%) of different methods on the LLP test dataset. These methods all use the same audio and visual features as inputs for a fair comparison. The top-1 results in each line are highlighted.

Event type	Methods	Segment-level	Event-level
Audio	Kong *et al.* 2018 [22]	39.6	29.1
	TALNet [62]	50.0	41.7
	AVE [55]	47.2	40.4
	AVSDN [27]	47.8	34.1
	Ours	**60.1**	**51.3**
Visual	STPN [36]	46.5	41.5
	CMCS [28]	48.1	45.1
	AVE [55]	37.1	34.7
	AVSDN [27]	52.0	46.3
	Ours	**52.9**	**48.9**
Audio-Visual	AVE [55]	35.4	31.6
	AVSDN [27]	37.1	26.5
	Ours	**48.9**	**43.0**
Type@AV	AVE [55]	39.9	35.5
	AVSDN [27]	45.7	35.6
	Ours	**54.0**	**47.7**
Event@AV	AVE [55]	41.6	36.5
	AVSDN [27]	50.8	37.7
	Ours	**55.4**	**48.0**

visual, and audio-visual events) under both segment-level and event-level metrics. To evaluate overall audio-visual scene parsing performance, we also compute aggregated results, where Type@AV computes averaged audio, visual, and audio-visual event evaluation results and Event@AV computes the F-score considering all audio and visual events for each sample rather than directly averaging results from different event types as the Type@AV. We use both segment-level and event-level F-scores [33] as metrics. The segment-level metric can evaluate snippet-wise event labeling performance. For computing event-level F-score results, we extract events with concatenating positive consecutive snippets in the same event categories and compute the event-level F-score based on mIoU = 0.5 as the threshold.

6.2 Experimental Comparison

To validate the effectiveness of the proposed audio-visual video parsing network, we compare it with weakly-supervised sound event detection methods: Kong *et al.* 2018 [22] and TALNet [62] on audio event parsing, weakly-supervised action localization methods: STPN [36] and CMCS [28] on visual event parsing, and modified audio-visual event localization methods: AVE [55] and AVSD [27] on

Table 2. Ablation study on learning mechanism, attentive MMIL pooling, hybrid attention network, and handling noisy labels. Segment-level audio-visual video parsing results are shown. The best results for each ablation study are highlighted.

Loss	MMIL Pooling	Temporal Net	Handle Noisy Label	Audio	Visual	Audio-Visual	Type@AV	Event@AV
$\mathcal{L}_{wsl}$	Attentive	×	×	**56.9**	16.4	17.2	30.2	43.3
$\mathcal{L}_g$	Attentive	×	×	42.3	43.9	34.5	40.3	42.0
$\mathcal{L}_{wsl} + \mathcal{L}_g$	Attentive	×	×	45.1	**51.7**	**35.0**	**44.0**	**48.9**
$\mathcal{L}_{wsl} + \mathcal{L}_g$	Max	×	×	31.6	43.6	22.5	32.6	39.1
$\mathcal{L}_{wsl} + \mathcal{L}_g$	Mean	×	×	40.2	43.2	**35.0**	39.5	39.7
$\mathcal{L}_{wsl} + \mathcal{L}_g$	Attentive	×	×	**45.1**	**51.7**	**35.0**	**44.0**	**48.9**
$\mathcal{L}_{wsl} + \mathcal{L}_g$	Attentive	×	×	45.1	51.7	35.0	44.0	48.9
$\mathcal{L}_{wsl} + \mathcal{L}_g$	Attentive	GRU [4]	×	52.0	49.4	39.0	46.8	51.0
$\mathcal{L}_{wsl} + \mathcal{L}_g$	Attentive	Transformer [58]	×	53.4	**53.8**	41.8	49.7	53.3
$\mathcal{L}_{wsl} + \mathcal{L}_g$	Attentive	HAN	×	**58.4**	52.8	**48.4**	**53.2**	**54.5**
$\mathcal{L}_{wsl}$	Attentive	HAN	×	39.6	40.5	20.1	33.4	44.9
$\mathcal{L}_g$	Attentive	HAN	×	57.5	52.5	47.4	52.5	53.8
$\mathcal{L}_{wsl} + \mathcal{L}_g$	Attentive	HAN	×	**58.4**	**52.8**	**48.4**	**53.2**	**54.5**
$\mathcal{L}_{wsl} + \mathcal{L}_g$	Max	HAN	×	55.7	52.0	**48.6**	52.1	51.8
$\mathcal{L}_{wsl} + \mathcal{L}_g$	Mean	HAN	×	56.0	51.9	46.3	51.4	52.9
$\mathcal{L}_{wsl} + \mathcal{L}_g$	Attentive	HAN	×	**58.4**	**52.8**	48.4	**53.2**	**54.5**
$\mathcal{L}_{wsl} + \mathcal{L}_g$	Attentive	HAN	×	58.4	52.8	48.4	53.2	54.5
$\mathcal{L}_{wsl} + \mathcal{L}_g$	Attentive	HAN	Bootstrap [42]	59.0	52.6	47.8	53.1	55.2
$\mathcal{L}_{wsl} + \mathcal{L}_g$	Attentive	HAN	Label Smoothing [52]	**60.1**	**52.9**	**48.9**	**54.0**	**55.4**

audio, visual, and audio-visual event parsing. The quantitative results are shown in Table 1. We can see that our method outperforms compared approaches on all audio-visual video parsing subtasks under both the segment-level and event-level metrics, which demonstrates that our network can predict more accurate snippet-wise event categories with more precise event onsets and offsets for testing videos.

Individual Guided Learning. From Table 2, we observe that the model without individual guided learning can achieve pretty good performance on audio event parsing but incredibly bad visual parsing results leading to terrible audio-visual event parsing; w/ only $\mathcal{L}_g$ model can achieve both reasonable audio and visual event parsing results; our model trained with both $\mathcal{L}_{wsl}$ and $\mathcal{L}_g$ outperforms model train without and with only $\mathcal{L}_g$. The results indicate that the model trained only $\mathcal{L}_{wsl}$ find discriminative information from mostly sounds and visual information is not well-explored during training and the individual learning can effectively handle the modality bias issue. In addition, when the network is trained with only $\mathcal{L}_g$, it actually models audio and visual event parsing as two individual MIL problems in which only noisy labels are used. Our MMIL framework can learn from clean weak labels with $\mathcal{L}_{wsl}$ and handle the modality bias with $\mathcal{L}_g$ achieves the best overall audio-visual video parsing performance. Moreover, we would like to note that the modality bias issue is from audio and visual data unbalance in training videos, which are originally from an audio-oriented dataset: AudioSet. Since the issue occurred after just 1 epoch training, it is not over-fitting.

Attentive MMIL Pooling. To validate the proposed Attentive MMIL Pooling, we compare it with two commonly used methods: Max pooling and Mean pooling. Our Attentive MMIL Pooling (see Table 2) is superior over the both compared methods. The Max MMIL pooling only selects the most discriminative snippet for each training video, thus it cannot make full use of informative audio and visual content. The Mean pooling does not distinguish the importance of different audio and visual snippets and equally aggregates instance scores in a bad way, which can obtain good audio-visual event parsing but poor individual audio and visual event parsing since a lot of audio-only and visual-only events are incorrectly parsed as audio-visual events. Our attentive MMIL pooling allows assigning different weights to audio and visual snippets within a video bag for each event category, thus can adaptively discover useful snippets and modalities.

Hybrid Attention Network. We compare our HAN with two popular temporal networks: GRU and Transformer and a base model without temporal modeling in Table 2. The models with GRU and Transformer are better than the base model and our HAN outperforms the GRU and Transformer. The results demonstrate that temporal aggregation with exploiting temporal recurrence is important for audio-visual video parsing and our HAN with jointly exploring unimodal temporal recurrence, multimodal temporal co-occurrence, and audio-visual temporal asynchrony is more effective in leveraging the multimodal temporal contexts. Another surprising finding of the HAN is that it actually tends to alleviate the modality bias by enforcing cross-modal modeling.

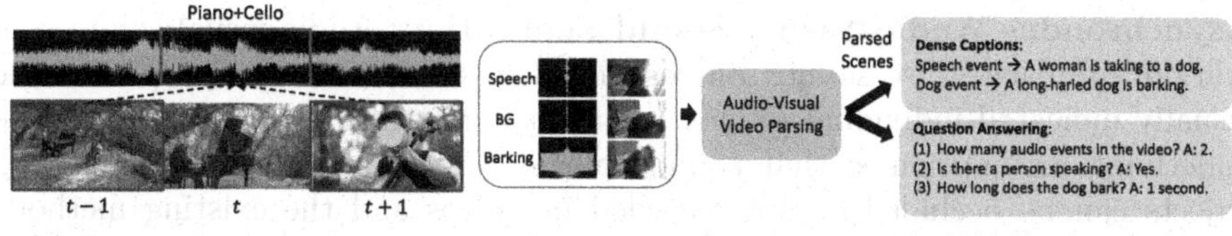

(a) Asynchronous Separation (b) Scene-Aware Video Understanding

Fig. 5. Potential applications of audio-visual video parsing. (a) Temporally asynchronous visual events detected by audio-visual video parsing highlighted in blue boxes can provide related visual information to separate *Cello* sound from the audio mixture in the red box. (b) Parsed scenes can provide important cues for audio-visual scene-aware video dense captioning and question answering.

Noisy Label. Table 2 also shows results of our model without handling the noisy label, with Bootstrap [42] and label smoothing-based method. We can find that Bootstrap updating labels using event predictions even decreases performance due to error propagation. Label smoothing-based method with reducing confidence for potential false positive labels can help to learn a more robust model with improved audio-visual video parsing results.

7 Limitation

To mitigate the modality bias issue, the guided loss is introduced to enforce that each modality should also be able to make the correct prediction on its own. Then, a new problem appears: the guide loss is not theoretically correct because some of the events only appear in one modality, so the labels are wrong. Finally, the label smoothing is used to alleviate the label noise. Although the proposed methods work at each step, they also introduce new problems. It is worth to design a one-pass approach. One possible solution is to introduce a new learning strategy to address the modality bias problem rather than using the guided loss. For example, we could perform modality dropout to enforce the model to explore both audio and visual information during training.

8 Conclusion and Future Work

In this work, we investigate a fundamental audio-visual research problem: audio-visual video parsing in a weakly-supervised manner. We introduce baselines and propose novel algorithms to address the problem. Extensive experiments on the newly collected LLP dataset support our findings that the audio-visual video parsing is tractable even learning from cheap weak labels, and the proposed model is capable of leveraging multimodal temporal contexts, dealing with modality bias, and mitigating label noise. Accurate audio-visual video parsing opens the door to a wide spectrum of potential applications, as discussed below.

Asynchronous Audio-Visual Sound Separation. Audio-visual sound separation approaches use sound sources in videos as conditions to separate the visually indicated individual sounds from sound mixtures [8,11–13,65,66]. The underlying assumption is that sound sources are visible. However, sounding objects can be occluded or not recorded in videos and the existing methods will fail to handle these cases. Our audio-visual video parsing model can find temporally asynchronous cross-modal events, which can help to alleviate the problem. For the example in Fig. 5(a), the existing audio-visual separation models will fail to separate the *Cello* sound from the audio mixture at the time step t, since the sound source *Cello* is not visible in the segment. However, our model can help to find temporally asynchronous visual events with the same semantic label as the audio event *Cello* for separating the sound. In this way, we can improve the robustness of audio-visual sound separation by leveraging temporally asynchronous visual content identified by our audio-visual video parsing models.

Audio-Visual Scene-Aware Video Understanding. The current video understanding community usually focuses on the visual modality and regards information from sounds as a bonus assuming that audio content should be associated with the corresponding visual content. However, we want to argue that auditory and visual modalities are equally important and most natural videos contain numerous audio, visual, and audio-visual events rather than only visual and audio-visual events. Our audio-visual scene parsing can achieve a unified multisensory perception, therefore it has the potential to help us build an audio-visual scene-aware video understanding system regarding all audio and visual events in videos(see Fig. 5(b)).

Acknowledgment. We thank the anonymous reviewers for the constructive feedback. This work was supported in part by NSF 1741472, 1813709, and 1909912. The article solely reflects the opinions and conclusions of its authors but not the funding agents.

References

1. Arandjelovic, R., Zisserman, A.: Look, listen and learn. In: ICCV (2017)
2. Aytar, Y., Vondrick, C., Torralba, A.: SoundNet: Learning sound representations from unlabeled video. In: Advances in Neural Information Processing Systems, pp. 892–900 (2016)
3. Bulkin, D.A., Groh, J.M.: Seeing sounds: visual and auditory interactions in the brain. Current Opinion Neurobiol. **16**(4), 415–419 (2006)
4. Cho, K., et al.: Learning phrase representations using RNN encoder-decoder for statistical machine translation. arXiv preprint arXiv:1406.1078 (2014)
5. Chorowski, J., Jaitly, N.: Towards better decoding and language model integration in sequence to sequence models. arXiv preprint arXiv:1612.02695 (2016)
6. Chou, S.Y., Jang, J.S.R., Yang, Y.H.: Learning to recognize transient sound events using attentional supervision. In: IJCAI, pp. 3336–3342 (2018)
7. Elizalde, B., et al.: Experiments on the dcase challenge 2016: acoustic scene classification and sound event detection in real life recording. arXiv preprint arXiv:1607.06706 (2016)

8. Ephrat, A., et al.: Looking to listen at the cocktail party: a speaker-independent audio-visual model for speech separation. arXiv preprint arXiv:1804.03619 (2018)
9. Gaidon, A., Harchaoui, Z., Schmid, C.: Temporal localization of actions with actoms. IEEE Trans. Pattern Anal. Mach. Intell. **35**(11), 2782–2795 (2013)
10. Gan, C., Huang, D., Zhao, H., Tenenbaum, J.B., Torralba, A.: Music gesture for visual sound separation. In: Proceedings of the IEEE/CVF Conference on Computer Vision and Pattern Recognition, pp. 10478–10487 (2020)
11. Gao, R., Feris, R., Grauman, K.: Learning to separate object sounds by watching unlabeled video. In: Proceedings of the European Conference on Computer Vision (ECCV), pp. 35–53 (2018)
12. Gao, R., Grauman, K.: 2.5 D visual sound. In: Proceedings of the IEEE Conference on Computer Vision and Pattern Recognition, pp. 324–333 (2019)
13. Gao, R., Grauman, K.: Co-separating sounds of visual objects. In: Proceedings of the IEEE International Conference on Computer Vision, pp. 3879–3888 (2019)
14. Gao, R., Oh, T.H., Grauman, K., Torresani, L.: Listen to look: action recognition by previewing audio. arXiv preprint arXiv:1912.04487 (2019)
15. Gemmeke, J.F., et al.: Audio set: an ontology and human-labeled dataset for audio events. In: IEEE International Conference on Acoustics, Speech and Signal Processing (ICASSP), pp. 776–780. IEEE (2017)
16. He, K., Zhang, X., Ren, S., Sun, J.: Deep residual learning for image recognition. In: Proceedings of the IEEE Conference on Computer Vision and Pattern Recognition, pp. 770–778 (2016)
17. Heittola, T., Mesaros, A., Eronen, A., Virtanen, T.: Context-dependent sound event detection. EURASIP J. Audio Speech Music Process. **2013**(1), 1 (2013)
18. Hochreiter, S., Schmidhuber, J.: Long short-term memory. Neural Comput. **9**(8), 1735–1780 (1997)
19. Hu, D., Nie, F., Li, X.: Deep multimodal clustering for unsupervised audiovisual learning. In: Proceedings of the IEEE Conference on Computer Vision and Pattern Recognition, pp. 9248–9257 (2019)
20. Jacobs, R.A., Xu, C.: Can multisensory training aid visual learning? A computational investigation. J. Vis. **19**(11), 1–1 (2019)
21. Kazakos, E., Nagrani, A., Zisserman, A., Damen, D.: Epic-fusion: audio-visual temporal binding for egocentric action recognition. In: Proceedings of the IEEE International Conference on Computer Vision, pp. 5492–5501 (2019)
22. Kong, Q., Xu, Y., Wang, W., Plumbley, M.D.: Audio set classification with attention model: a probabilistic perspective. In: IEEE International Conference on Acoustics, Speech and Signal Processing (ICASSP), pp. 316–320. IEEE (2018)
23. Korbar, B., Tran, D., Torresani, L.: Cooperative learning of audio and video models from self-supervised synchronization. In: Advances in Neural Information Processing Systems, pp. 7763–7774 (2018)
24. Korbar, B., Tran, D., Torresani, L.: Scsampler: sampling salient clips from video for efficient action recognition. In: Proceedings of the IEEE International Conference on Computer Vision, pp. 6232–6242 (2019)
25. Lin, T., Zhao, X., Su, H., Wang, C., Yang, M.: BSN: boundary sensitive network for temporal action proposal generation. In: Proceedings of the European Conference on Computer Vision (ECCV), pp. 3–19 (2018)
26. Lin, T.-Y., et al.: Microsoft COCO: common objects in context. In: Fleet, D., Pajdla, T., Schiele, B., Tuytelaars, T. (eds.) ECCV 2014. LNCS, vol. 8693, pp. 740–755. Springer, Cham (2014). https://doi.org/10.1007/978-3-319-10602-1_48

27. Lin, Y.B., Li, Y.J., Wang, Y.C.F.: Dual-modality seq2seq network for audio-visual event localization. In: IEEE International Conference on Acoustics, Speech and Signal Processing (ICASSP), ICASSP 2019, pp. 2002–2006. IEEE (2019)
28. Liu, D., Jiang, T., Wang, Y.: Completeness modeling and context separation for weakly supervised temporal action localization. In: Proceedings of the IEEE Conference on Computer Vision and Pattern Recognition, pp. 1298–1307 (2019)
29. Long, F., Yao, T., Qiu, Z., Tian, X., Luo, J., Mei, T.: Gaussian temporal awareness networks for action localization. In: Proceedings of the IEEE Conference on Computer Vision and Pattern Recognition, pp. 344–353 (2019)
30. Maron, O., Lozano-Pérez, T.: A framework for multiple-instance learning. In: Advances in Neural Information Processing Systems, pp. 570–576 (1998)
31. McFee, B., Salamon, J., Bello, J.P.: Adaptive pooling operators for weakly labeled sound event detection. IEEE/ACM Trans. Audio Speech Lang. Process. **26**(11), 2180–2193 (2018)
32. Mesaros, A., et al.: DCASE 2017 challenge setup: tasks, datasets and baseline system (2017)
33. Mesaros, A., Heittola, T., Virtanen, T.: Metrics for polyphonic sound event detection. Appl. Sci. **6**(6), 162 (2016)
34. Naphade, M.R., Huang, T.S.: Discovering recurrent events in video using unsupervised methods. In: Proceedings of the International Conference on Image Processing, vol. 2, pp. II-II. IEEE (2002)
35. Ngiam, J., Khosla, A., Kim, M., Nam, J., Lee, H., Ng, A.Y.: Multimodal deep learning. In: ICML (2011)
36. Nguyen, P., Liu, T., Prasad, G., Han, B.: Weakly supervised action localization by sparse temporal pooling network. In: Proceedings of the IEEE Conference on Computer Vision and Pattern Recognition, pp. 6752–6761 (2018)
37. Owens, A., Efros, A.A.: Audio-visual scene analysis with self-supervised multisensory features. In: Ferrari, V., Hebert, M., Sminchisescu, C., Weiss, Y. (eds.) ECCV 2018. LNCS, vol. 11210, pp. 639–658. Springer, Cham (2018). https://doi.org/10.1007/978-3-030-01231-1_39
38. Owens, A., Wu, J., McDermott, J.H., Freeman, W.T., Torralba, A.: Ambient sound provides supervision for visual learning. In: Leibe, B., Matas, J., Sebe, N., Welling, M. (eds.) ECCV 2016. LNCS, vol. 9905, pp. 801–816. Springer, Cham (2016). https://doi.org/10.1007/978-3-319-46448-0_48
39. Parascandolo, G., Huttunen, H., Virtanen, T.: Recurrent neural networks for polyphonic sound event detection in real life recordings. In: IEEE International Conference on Acoustics, Speech and Signal Processing (ICASSP), pp. 6440–6444. IEEE (2016)
40. Paul, S., Roy, S., Roy-Chowdhury, A.K.: W-TALC: weakly-supervised temporal activity localization and classification. In: Proceedings of the European Conference on Computer Vision (ECCV), pp. 563–579 (2018)
41. Rahman, T., Xu, B., Sigal, L.: Watch, listen and tell: multi-modal weakly supervised dense event captioning. In: Proceedings of the IEEE International Conference on Computer Vision, pp. 8908–8917 (2019)
42. Reed, S., Lee, H., Anguelov, D., Szegedy, C., Erhan, D., Rabinovich, A.: Training deep neural networks on noisy labels with bootstrapping. arXiv preprint arXiv:1412.6596 (2014)
43. Roma, G., Nogueira, W., Herrera, P., de Boronat, R.: Recurrence quantification analysis features for auditory scene classification. IEEE AASP challenge on detection and classification of acoustic scenes and events 2 (2013)

44. Salamon, J., Jacoby, C., Bello, J.P.: A dataset and taxonomy for urban sound research. In: Proceedings of the 22nd ACM International Conference on Multimedia, pp. 1041–1044 (2014)
45. Senocak, A., Oh, T.H., Kim, J., Yang, M.H., So Kweon, I.: Learning to localize sound source in visual scenes. In: Proceedings of the IEEE Conference on Computer Vision and Pattern Recognition, pp. 4358–4366 (2018)
46. Shams, L., Seitz, A.R.: Benefits of multisensory learning. Trends Cogn. Sci. **12**(11), 411–417 (2008)
47. Shou, Z., Chan, J., Zareian, A., Miyazawa, K., Chang, S.F.: CDC: convolutional-de-convolutional networks for precise temporal action localization in untrimmed videos. In: Proceedings of the IEEE Conference on Computer Vision and Pattern Recognition, pp. 5734–5743 (2017)
48. Shou, Z., Wang, D., Chang, S.F.: Temporal action localization in untrimmed videos via multi-stage CNNs. In: Proceedings of the IEEE Conference on Computer Vision and Pattern Recognition, pp. 1049–1058 (2016)
49. Singh, K.K., Lee, Y.J.: Hide-and-seek: forcing a network to be meticulous for weakly-supervised object and action localization. In: IEEE International Conference on Computer Vision (ICCV), pp. 3544–3553. IEEE (2017)
50. Song, H.O., Lee, Y.J., Jegelka, S., Darrell, T.: Weakly-supervised discovery of visual pattern configurations. In: Advances in Neural Information Processing Systems, pp. 1637–1645 (2014)
51. Spence, C., Squire, S.: Multisensory integration: maintaining the perception of synchrony. Curr. Biol. **13**(13), R519–R521 (2003)
52. Szegedy, C., Vanhoucke, V., Ioffe, S., Shlens, J., Wojna, Z.: Rethinking the inception architecture for computer vision. In: Proceedings of the IEEE Conference on Computer Vision and Pattern Recognition, pp. 2818–2826 (2016)
53. Tian, Y., Guan, C., Goodman, J., Moore, M., Xu, C.: An attempt towards interpretable audio-visual video captioning. arXiv preprint arXiv:1812.02872 (2018)
54. Tian, Y., Guan, C., Justin, G., Moore, M., Xu, C.: Audio-visual interpretable and controllable video captioning. In: The IEEE Conference on Computer Vision and Pattern Recognition (CVPR) Workshops, June 2019
55. Tian, Y., Shi, J., Li, B., Duan, Z., Xu, C.: Audio-visual event localization in unconstrained videos. In: Ferrari, V., Hebert, M., Sminchisescu, C., Weiss, Y. (eds.) ECCV 2018. LNCS, vol. 11206, pp. 252–268. Springer, Cham (2018). https://doi.org/10.1007/978-3-030-01216-8_16
56. Tian, Y., Shi, J., Li, B., Duan, Z., Xu, C.: Audio-visual event localization in the wild. In: IEEE Computer Society Conference on Computer Vision and Pattern Recognition Workshops (2019)
57. Tran, D., Wang, H., Torresani, L., Ray, J., LeCun, Y., Paluri, M.: A closer look at spatiotemporal convolutions for action recognition. In: Proceedings of the IEEE Conference on Computer Vision and Pattern Recognition, pp. 6450–6459 (2018)
58. Vaswani, A., et al.: Attention is all you need. In: Advances in Neural Information Processing Systems, pp. 5998–6008 (2017)
59. Vroomen, J., Keetels, M., De Gelder, B., Bertelson, P.: Recalibration of temporal order perception by exposure to audio-visual asynchrony. Cogn. Brain Res. **22**(1), 32–35 (2004)
60. Wang, L., Xiong, Y., Lin, D., Van Gool, L.: Untrimmednets for weakly supervised action recognition and detection. In: Proceedings of the IEEE Conference on Computer Vision and Pattern Recognition, pp. 4325–4334 (2017)

61. Wang, X., Wang, Y.F., Wang, W.Y.: Watch, listen, and describe: globally and locally aligned cross-modal attentions for video captioning. arXiv preprint arXiv:1804.05448 (2018)
62. Wang, Y., Li, J., Metze, F.: A comparison of five multiple instance learning pooling functions for sound event detection with weak labeling. In: IEEE International Conference on Acoustics, Speech and Signal Processing (ICASSP), ICASSP 2019, pp. 31–35. IEEE (2019)
63. Wu, Y., Zhu, L., Yan, Y., Yang, Y.: Dual attention matching for audio-visual event localization. In: Proceedings of the IEEE International Conference on Computer Vision (ICCV) (2019)
64. Yang, Y., Liu, J., Shah, M.: Video scene understanding using multi-scale analysis. In: IEEE 12th International Conference on Computer Vision, pp. 1669–1676. IEEE (2009)
65. Zhao, H., Gan, C., Ma, W.C., Torralba, A.: The sound of motions. In: Proceedings of the IEEE International Conference on Computer Vision, pp. 1735–1744 (2019)
66. Zhao, H., Gan, C., Rouditchenko, A., Vondrick, C., McDermott, J., Torralba, A.: The sound of pixels. In: Proceedings of the European Conference on Computer Vision (ECCV), pp. 570–586 (2018)
67. Zhao, Y., Xiong, Y., Wang, L., Wu, Z., Tang, X., Lin, D.: Temporal action detection with structured segment networks. In: Proceedings of the IEEE International Conference on Computer Vision, pp. 2914–2923 (2017)
68. Zhou, B., Khosla, A., Lapedriza, A., Oliva, A., Torralba, A.: Learning deep features for discriminative localization. In: Proceedings of the IEEE Conference on Computer Vision and Pattern Recognition, pp. 2921–2929 (2016)
69. Zhou, H., Liu, Z., Xu, X., Luo, P., Wang, X.: Vision-infused deep audio inpainting. In: Proceedings of the IEEE International Conference on Computer Vision, pp. 283–292 (2019)
70. Zhou, H., Xu, X., Lin, D., Wang, X., Liu, Z.: Sep-stereo: visually guided stereophonic audio generation by associating source separation. In: Vedaldi, A., Bischof, H., Brox, T., Frahm, J.-M. (eds.) ECCV 2020. LNCS, vol. 12357, pp. 52–69. Springer, Cham (2020). https://doi.org/10.1007/978-3-030-58610-2_4

Learning Delicate Local Representations for Multi-person Pose Estimation

Yuanhao Cai[1,2], Zhicheng Wang[1(✉)], Zhengxiong Luo[1,3], Binyi Yin[1,4], Angang Du[1,5], Haoqian Wang[2], Xiangyu Zhang[1], Xinyu Zhou[1], Erjin Zhou[1], and Jian Sun[1]

[1] Megvii Inc., Beijing, China
{caiyuanhao,wangzhicheng,zxy,zej,zhangxiangyu,sunjian}@megvii.com
[2] Tsinghua University, Beijing, China
wanghaoqian@tsinghua.edu.cn
[3] Chinese Academy of Sciences, Beijing, China
[4] Beihang University, Beijing, China
[5] Ocean University of China, Qingdao, China

Abstract. In this paper, we propose a novel method called Residual Steps Network (RSN). RSN aggregates features with the same spatial size (Intra-level features) efficiently to obtain delicate local representations, which retain rich low-level spatial information and result in precise keypoint localization. Additionally, we observe the output features contribute differently to final performance. To tackle this problem, we propose an efficient attention mechanism - Pose Refine Machine (PRM) to make a trade-off between local and global representations in output features and further refine the keypoint locations. Our approach won the 1st place of COCO Keypoint Challenge 2019 and achieves state-of-the-art results on both COCO and MPII benchmarks, without using extra training data and pretrained model. Our single model achieves 78.6 on COCO test-dev, 93.0 on MPII test dataset. Ensembled models achieve 79.2 on COCO test-dev, 77.1 on COCO test-challenge dataset. The source code is publicly available for further research at https://github.com/caiyuanhao1998/RSN/.

Keywords: Human pose estimation · COCO · MPII · Feature aggregation · Attention mechanism

1 Introduction

The goal of multi-person pose estimation is to locate keypoints of all persons in a single image. It is a fundamental task for human motion recognition, kinematics analysis, human-computer interaction, animation etc. For years, human pose estimation was based on handcraft features. Recently, It has made great progress with the development of deep convolutional neural network. The task

Y. Cai and Z. Wang—The first two authors contribute equally to this work. This work is done when Yuanhao Cai, Zhengxiong Luo, Binyi Yin and Angang Du are interns at Megvii Research.

A. Vedaldi et al. (Eds.): ECCV 2020, LNCS 12348, pp. 455–472, 2020.
https://doi.org/10.1007/978-3-030-58580-8_27

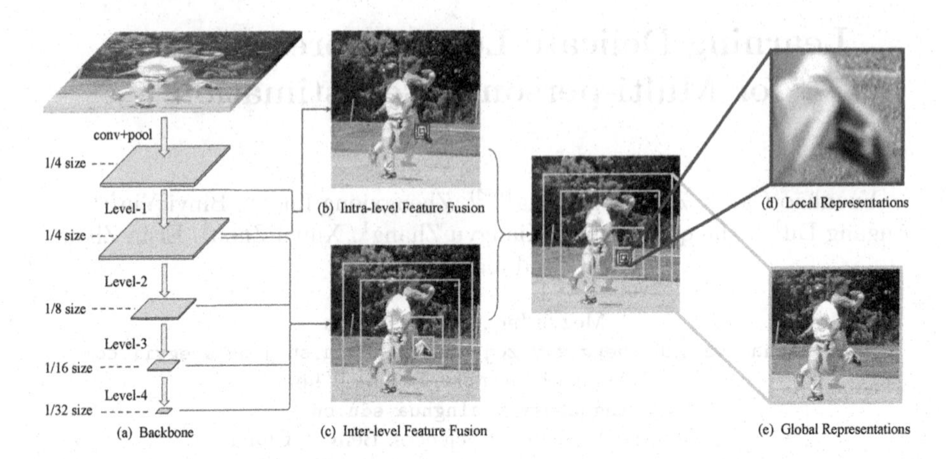

Fig. 1. Comparison of intra-level feature fusion and inter-level feature fusion. (a) Backbone. "1/4 size" means 1/4 size of input image. (b) Intra-level feature fusion of level 1. (c) Inter-level feature fusion. (d) Local Representations. (e) Global representations.

of human pose estimation concerns both keypoint localization and classification. Spatial information benefits the localization task, while semantic information is good for the classification task. To extract these two kinds of information, current methods mainly focus on aggregating inter-level features. For instance, HRNet [23] maintains spatial information in high-resolution sub-network and gradually adds semantic information to it from low-resolution sub-networks. In this way, features of different levels are fully aggregated. In CPN [2], features of four different spatial levels are extracted by the backbone, and they are combined by a head network. Although these methods are different in the ways of feature fusion, the features to be aggregated are always from different levels. On the contrast, the feature fusion within the same level stays less explored in the task of human pose estimation.

The comparison of intra-level feature fusion (level 1) and inter-level feature fusion is illustrated in Fig. 1. The feature maps are continuously downsampled to 1/4, 1/8, 1/16, 1/32 size of input image in Fig. 1(a). We define consecutive feature maps with the same spatial size as one level. As Fig. 1(c) depicts, there is a big gap between the receptive fields of features from different levels, which are indicated by light blue bounding boxes. As a result, representations learned by inter-level feature fusion are relatively coarse, which impede the localization of human pose from precise. As Fig. 1(b) shows, the gap between the receptive fields of intra-level features which are indicated by red bounding boxes is relatively small. As shown in Fig. 1(d), fusing intra-level features can extract much more delicate local representations retaining more precise spatial information, which is critical to keypoint localization.

To learn better local representations, we propose a novel network architecture - Residual Steps Network (RSN). The Residual Steps Block (RSB) of RSN fuses features inside each level using dense element-wise sum operations, which is shown in Fig. 2(c). The inner structure of RSB is deeply connected and motivated by DenseNet [10], which has a good performance for human pose estimation owing to retaining rich low-level features by deep connections. However, deep connections bring about explosion of the network capacity as it goes deeper. Thus, DenseNet performs poorly when the network becomes large. RSN is motivated by DenseNet but is quite different in that RSN uses element-wise sum rather than concatenation to circumvent network capacity explosion. RSN is modestly less dense connected in the block than DenseNet, which further promotes the efficiency. Additionally, we observe that the output features containing both global and local representations contribute differently to final performance. In light of this observation, we propose an attention module - Pose Refine Machine (PRM) to rebalance the output features of the network. The architecture of PRM is illustrated in Fig. 3 and analyzed in Sect. 3.3. To better illustrate the advantages of our approach, we analyze the differences between RSN and current methods in Sect. 2.2.

In conclusion, our contributions can be summarized as three points:

1. We propose a novel network - RSN, which aims to learn delicate local representations by efficient intra-level feature fusion.
2. We propose an attention mechanism - PRM, which goes further to make a trade-off between local and global representations, and benefits the final performance.
3. Comprehensive experiments demonstrate that Our approach outperforms the state-of-the-art methods on both COCO and MPII datasets without using extra training data and pretrained model. Moreover, the proposed approach is much faster than HRNet with comparable performance on both GPU and CPU platforms.

2 Related Work

2.1 Multi-person Pose Estimation

Current methods of human pose estimation fall into two categories: top-down methods [2,5,8,11,17,19–23,28] and bottom-up methods [1,14,18,27]. Top-down methods first detect the positions of all persons, then estimate the pose of each person. Bottom-up methods first detect all the human keypoints in an image and then assemble these points into groups to form different individuals. Since this paper mainly concentrates on feature fusion strategies, we discuss these methods in terms of feature fusion.

2.2 Feature Fusion

Recently, many methods [2,17,23,28,30] of human pose estimation use inter-level feature fusion to extract more spatial and semantic information. Newell et al. [17] propose a U-shape convolutional neural network (CNN) named Hourglass. In a single stage of hourglass, high-level features are added to low-level features

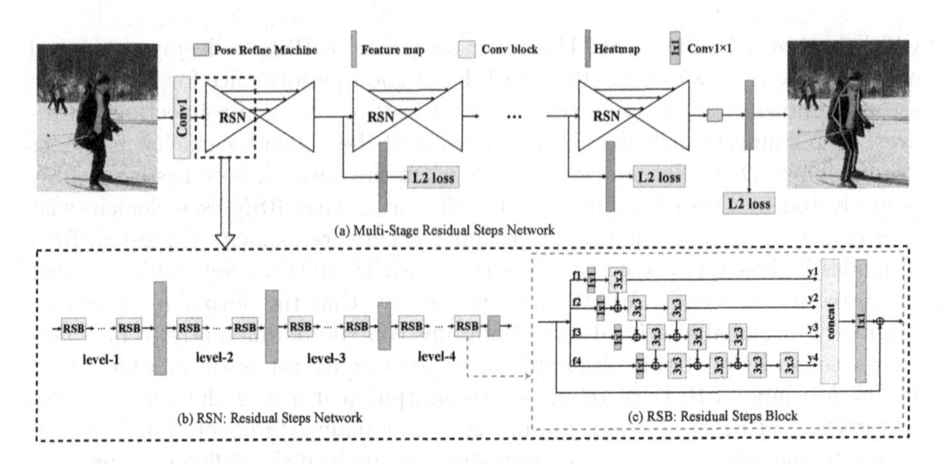

Fig. 2. Our pipeline. (a) is the multi-stage network architecture. It is cascaded by several Residual Steps Networks (RSNs). (b) is the backbone of RSN. (c) is the structure of Residual Steps Block (RSB), which is the basic block of RSN. RSB is designed for learning delicate local representations through dense element-wise sum connections. A Pose Refine Machine (PRM) is used in the last stage and it is analyzed in Sect. 3.4.

after upsampling. Later works such as Yang et al. [30] show great performance of using inter-level feature fusion. Chen et al. [2] combines inter-level features using a RefineNet. Sun et al. [23] set up four parallel sub-networks. The features of these four sub-networks aggregate with each other through high-to-low or low-to-high way.

Though many methods have validates the effectiveness of inter-level feature fusion, intra-level feature fusion is rarely explored in human pose estimation. However, it has extensive applications in other tasks such as semantic segmentation and image classification [4,7,10,24,29,33]. In a block of Inception [24], features pass through several convolutional layers with different kernels separately and then added up. DenseNet [10] fuses intra-level features using continuous concatenating operations. This implementation retains low-level features to improve the performance. However, when the network goes deeper, the capacity increases sharply and much redundant information appears in the network, resulting in poor efficiency. Different from DenseNet, RSN uses element-wise sum rather than concatenation to circumvent network capacity explosion. In addition, RSN is modestly less densely connected in the constituent unit, which further promotes the efficiency.

Res2Net [7] and OSNet [33] focus on multi-scale representations. Both of them lack dense connections between adjacent branches. The dense connections contribute sufficient gradients and make low-level features better supervised. Therefore, lack of dense connections between adjacent branches results in less precise spatial information, which is essential to keypoint localization. Suffering from this limitations, both Res2Net and OSNet are inferior to RSN in the task of human pose estimation. In Sect. 4.1, we validate the efficiency of DenseNet, Res2Net and RSN.

2.3 Attention Mechanism

Attention mechanism [6,9,13,22,26,31,32] is almost used in all areas of computer vision. Current methods of attention mechanism mainly fall into two categories: channel attention [9,22,26,31] and spatial attention [6,13,22,26,32]. Woo et al. [26] propose a channel attention module with global average pooling and max pooling. Kligvasser et al. [13] propose a spatial activation function with depth-wise separable convolution. Other works such as Hu et al. [9] show the advantages of using attention mechanism. However, most prior attention modules are lack of representing capacity and focus on optimizing the backbone. We design PRM to make a trade-off between local and global representations in output features by using powerful while computation-economical operations.

3 Proposed Method

The overall pipeline of our method is illustrated in Fig. 2. The multi-stage network architecture is cascaded by several single-stage modules - Residual Steps Network (RSN), shown in Fig. 2(a). As Fig. 2(b) shows, RSN differs from ResNet in the architecture of constituent unit. RSN consists of Residual Steps Blocks (RSBs) while ResNet is comprised of "bottleneck" blocks. Figure 2(c) illustrates the structure of RSB. A Pose Refine Machine (PRM) is used in the last stage and it is analyzed in Sect. 3.3.

3.1 Delicate Local Representations Learning

Residual Steps Network is designed for learning delicate local representations by repeatedly enhancing efficient intra-level feature fusion inside RSB, which is the constituent unit of RSN. As shown in Fig. 2(c), RSB firstly divides the features into four splits f_i (i = 1, 2, 3, 4), then implements a conv1 × 1 (convolutional layer with kernel size 1 × 1) separately. Each feature output from conv1 × 1 undergoes incremental numbers of conv3 × 3. The output features y_i (i = 1, 2, 3, 4) are then concatenated to go through a conv1 × 1. An identity connection is employed as the ResNet bottleneck. Because the incremental numbers of conv3 × 3 look like steps, the network is therefore named Residual Steps Network.

The receptive fields of RSB range across several values, and the max one is 15. Compared with a single receptive field in ResNet bottleneck as shown in Table 1, RSB indicates more delicate information viewed in the network. In addition, it is deeply connected inside RSB. On the ith branch, the front $i-1$ conv3 × 3 receive the features output from the $(i-1)^{th}$ branch. The ith conv3 × 3 is then designed to refine the fusion of the features output from the $(i-1)^{th}$ conv3 × 3. Benefit from the dense connection structure, small-gap receptive fields of features are fully fused resulting in delicate local representations, which retain precise spatial and semantic information. Additionally, during the training process, the deeply connected structure contributes sufficient gradients, so the low-level features are better supervised, which benefits the keypoint localization task. We investigate how the branch number of RSB influences the prediction results in Sect. 4.1. Four-branch architecture has the best performance.

Table 1. The receptive field comparison between RSB and other methods.

Architecture	y1	y2	y3	y4
ResNet	3	3	3	3
OSNet	3	5	7	9
Res2Net	1	3	3,5	3,5,7
RSN	3	5,7	7,9,11	9,11,13,15

3.2 Receptive Field Analysis

In this part, we analyze the receptive fields in RSB and other methods. Firstly, the formula for calculating the receptive field of the kth convolutional layer is written as Eq. 1

$$l_k = l_{k-1} + [(f_k - 1) * \prod_{i=1}^{k-1} s_i] \quad (1)$$

l_k denotes the size of the receptive field corresponding to the kth layer, f_k denotes the kernel size of the kth layer and s_i denotes the stride of the ith layer. In this part, we only focus on the change of relative receptive fields in a block. Every f_k is 3 and s_i is 1. Thus, Eq. 1 can be simplified to Eq. 2

$$l_k = l_{k-1} + 2 \quad (2)$$

Using this formula, we calculate the relative receptive fields of RSB and other methods, as shown in Table 1. It indicates that RSN has a wider range of scales than ResNet, Res2Net and OSNet. The scale of each human joint varies a lot. For instance, the scale of eye is small while that of hip is large. For this reason, architecture with wider range of receptive fields is more fit for extracting features relating to different joints. Also, wider range of receptive fields helps to learn more discriminant semantic representations, which benefits the keypoint classification task. More importantly, RSN builds dense connections between the features with small-gap receptive fields inside RSB. The deeply connected architecture contributes to learning delicate local representations, which are essential to precise human pose estimation.

3.3 Pose Refine Machine

In the last module of multi-stage network (Fig. 2(a)), an attention mechanism - Pose Refine Machine (PRM) is used to reweight the output features, as shown in Fig. 3. The first component of PRM is a conv3 × 3, then the features are input into three paths. The top path is an identity connection. The middle one, motivated by SENet [9], passes through a global pooling, two conv1 × 1 and a sigmoid activation to get a weight vector α. The bottom path passes through a conv1 × 1, a depth-wise separable conv9 × 9 and a sigmoid activation to get

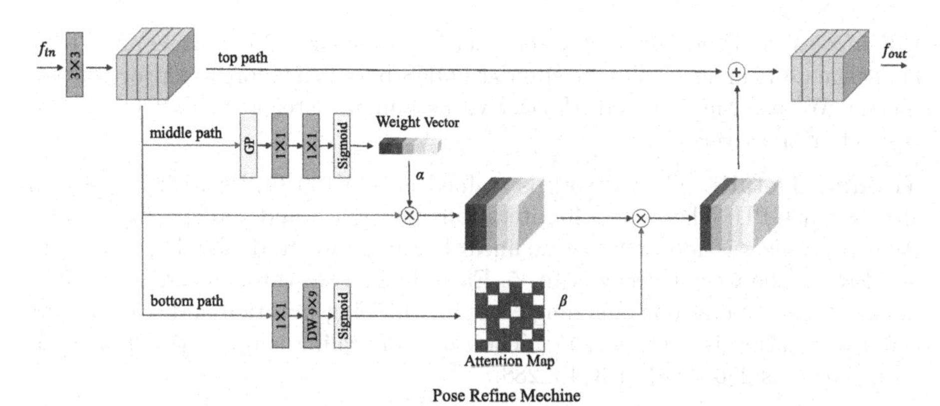

Fig. 3. Architecture of Pose Refine Machine (PRM). GP denotes global pooling. DW denotes depth-wise separable convolution. α denotes the weight vector. β denots the attention map. The top path is an identity connection, the middle path is designed to reweight features in channel wise and the bottom path is proposed for spatial attention.

an attention map β. Element-wise sum and multiplication are conducted among the three paths to get the output features. Define the input features of PRM as f_{in}, the output features as f_{out}, the first conv3×3 as $K(\cdot)$ and element-wise multiplication as $\odot$. Then PRM can be formulated as Eq. 3.

$$f_{out} = K(f_{in}) \odot (1 + \beta \odot \alpha) \tag{3}$$

As for the output of RSN, features after intra-level and inter-level aggregation are mixed together containing both low-level precise spatial information and high-level discriminant semantic information. Spatial information is good for keypoint localization while semantic information benefits keypoints classification. These features contribute differently to the final prediction. Therefore, to tackle this imbalance problem, PRM is designed to make a trade-off between local and global representations in output features of RSN. Compared to prior work of attention mechanism, we use powerful while computation-economical operations, e.g. conv3×3, conv1×1 and DW conv9×9. The top identity mapping in PRM is good for retaining local features which benefits precise keypoint localization. The middle path is designed to reweight the features in channel wise and the bottom path is proposed for spatial attention.

4 Experiments

4.1 COCO Keypoints Detection

Datasets, Evaluation Metric, Human Detection. COCO dataset [16] includes over 200K images and 250K person instances labeled with 17 joints. We use only COCO train2017 dataset for training (including about 57K images and 150K person instances). We evaluate our method on COCO minival dataset

(5K images) and the testing datasets including test-dev (20K images) and test-challenge (20K images). We use standard OKS-based AP score as the evaluation metric. We use MegDet and MegDet-v2 as human detecor on COCO val and test sets respectively.

Training Details. The network is trained on 8 V100 GPUs with mini-batch size 48 per GPU. There are 140k iterations per epoch and 200 epochs in total. Adam optimizer is adopted and the linear learning rate gradually decreases from 5e−4 to 0. The weight decay is 1e−5. Each image goes through a series of data augmentation operations including cropping, flipping, rotation and scaling. The range of rotation is $-45^\circ \sim +45^\circ$. The range of scaling is 0.7–1.35. The size of input image is 256×192 or 384×288.

Testing Details. We apply a post-Gaussian filter to the estimated heatmaps. Following the strategy of hourglass [17], we average the predicted heatmaps of original image with the results of corresponding flipped image. Then we implement a quarter offset from the highest response to the second highest one to get the locations of keypoints. The same with CPN [2], the pose score is the multiplication of the average score of keypoints and the bounding box score.

Table 2. Results of ResNet, Res2Net, Baseline1,2 and RSN on COCO val set

Backbone	Input size	AP	Δ	GFLOPs
ResNet-18	256 × 192	70.7	0	2.3
Res2Net-18	256 × 192	71.3	+0.6	2.2
Baseline1-18	256 × 192	72.9	+2.1	2.5
Baseline2-18	256 × 192	72.1	+1.4	2.5
RSN-18	**256** × 192	**73.6**	**+2.9**	**2.5**
ResNet-50	256 × 192	72.2	0	4.6
Res2Net-50	256 × 192	72.8	+0.6	4.5
Baseline1-50	256 × 192	73.7	+1.5	6.4
Baseline2-50	256 × 192	72.7	+0.5	6.4
RSN-50	**256** × 192	**74.7**	**+2.5**	**6.4**
ResNet-101	256 × 192	73.2	0	7.5
Res2Net-101	256 × 192	73.9	+0.7	7.5
RSN-101	**256** × 192	**75.8**	**+2.5**	**11.5**
4 × ResNet-50	256 × 192	76.8	0	20.6
4 × Res2Net-50	256 × 192	77.0	+0.2	20.1
4 × **RSN-50**	**256** × 192	**78.6**	**+1.8**	**27.5**
4 × ResNet-50	384 × 288	77.5	0	46.4
4 × Res2Net-50	384 × 288	77.6	+0.1	45.2
4 × **RSN-50**	**384** × 288	**79.2**	**+1.7**	**61.9**

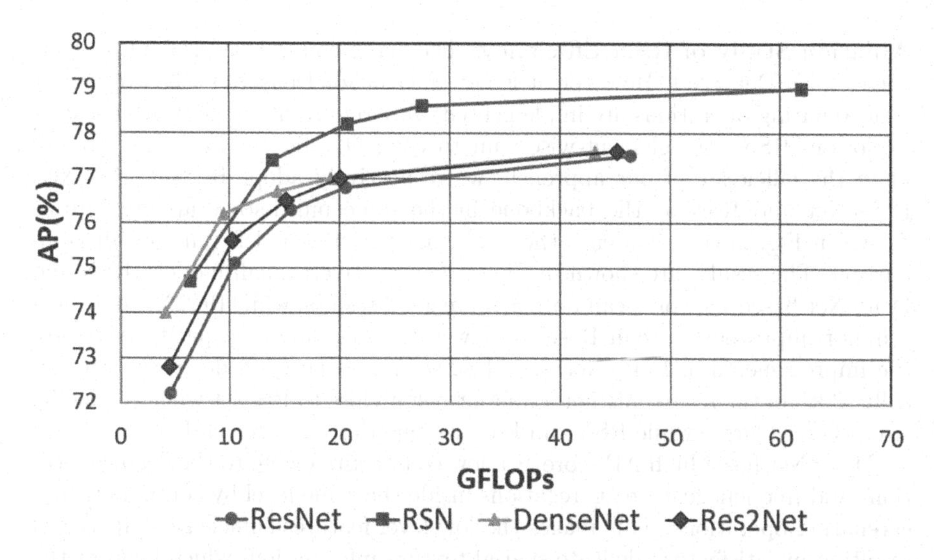

Fig. 4. Illustrating how the performances of ResNet, Res2Net, DenseNet and RSN are affected by GFLOPs. The results are reported on COCO minival dataset.

Ablation Study of RSN Improvement. In Sect. 2.2, we analyze the differences between RSN and current methods. In this part, we validate the effectiveness of intra-level feature fusion method in RSN. Since it is known dividing the network into branches, e.g., Inception and ResNetXt [29], can improve the recognition performance, we add two baselines into comparison. Baseline 1: remove the intra-level fusion (i.e., the vertical arrows) from Fig. 2(c). This baseline can reveal whether the proposed intra-level fusion is important. Baseline 2: replace f1-f4 in Baseline 1 with 4 f3's respectively, which is more like the conventional branching strategy. We keep the same GFLOPs of Baseline1, Baseline2 with RSN by adapting channels. Ablation experiments are implemented on ResNet, Res2Net, Bseline1, Baseline2, and RSN based networks. PRM is left out for more strong comparison. The results on COCO val are reported in Table 2.

As Table 2 shows, RSN boosts the performance by relatively larger extent with acceptable computation cost addition, while Res2Net can only obtain limited gain. For instance, RSN-18 is 2.9 points AP higher than ResNet-18 adding only 0.2 GFLOPs and 2.3 points AP higher than Res2Net-18 adding only 0.3 GFLOPs. However, Res2Net-18 obtains only 0.6 AP gain than ResNet-18. RSN always works much better than ResNet and Res2Net with comparable GFLOPs. In addition, it is worth noting that when model complexity is relatively low, RSN still has a remarkable performance, which indicates that RSN is more compact and efficient. For instance, compared with ResNet-101 and Res2Net-101, RSN-18 has a similar AP, however, with only a third of computation cost. On the other hand, RSN achieves higher AP than Baseline1 and Baseline2 with the same GFLOPs, e.g., RSN-50 is 1 AP higher than Baseline1-50 and 2 AP higher than Baseline2-50. This observation strongly demonstrate the superiority of the intra-level feature fusion mode of RSN.

Ablation Study of RSN Efficiency. The dense connection principle of RSN comes form DenseNet. However, it is not efficient for DenseNet when too many concatenating operations are implemented. To circumvent the network capacity explosion, RSN uses element-wise sum to connect adjacent branches. To validate the efficiency of our approach, we respectively adopt ResNet, Res2Net, DenseNet and RSN as the backbone in the same multi-stage architecture as shown in Fig. 2(a) to compare the performance. PRM is left out for fair comparison. The results are shown in Fig. 4. For relatively small models, RSN and DenseNet based networks can both achieve good results, while Res2Net only gets a minor improvement than ResNet. However, as the model capacity increases, the improvements of DenseNet and Res2Net based network decrease dramatically. Both of them can only get a inferior result close to ResNet when the model size becomes large, while RSN can keep its superiority to the end.

DenseNet has a high AP score at a low complexity owing to the deep connections and frequent feature aggregations inside the same level by continuous concatenating operations. This makes the low-level features sufficiently supervised resulting in satisfactory delicate spatial texture information, which benefits the keypoint localization. However, as the computation cost raises, the concatenating operations of DenseNet become redundant. It combines quite a large range of less utilized information. As for Res2Net, narrower range of receptive fields and lack of efficient intra-level feature fusion to extract delicate local representations make it much inferior than RSN.

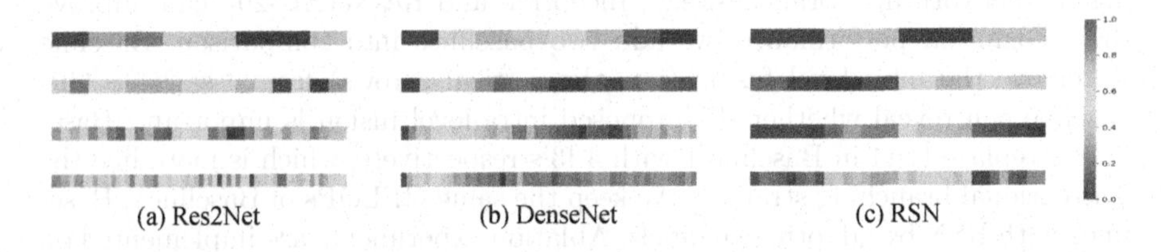

Fig. 5. The average absolute filter weights of the last conv1 × 1 layers of each level in trained Res2Net-50 (a), DenseNet-169(b) and RSN-50(c). Larger weights means higher utilization. The weights of Res2Net are smaller than those of RSN. Most weights in DenseNet have values close to zero. While RSN can utilize most channels better.

In order to embody the differences of Res2Net, DenseNet and RSN more essentially, we show the average absolute filter weights of the last conv1 × 1 layers of each level in trained Res2Net-50, DenseNet-169 and RSN-50 in Fig. 5. The highly used weights become less from level 1 to level 4 in DenseNet. The overall useful weights of DenseNet are less than those of RSN, which can be deduced from Fig. 5 (b) and (c) that the red area in each level of DenseNet is much smaller than that of RSN. According to the analysis in Sect. 3.2, RSN can enhance the efficient fusion of intra-level features with dense element-wise sum connections. There are not accumulative concatenating operations like DenseNet. Thus, RSN is less occupied by the redundant features with low utilization. On the other

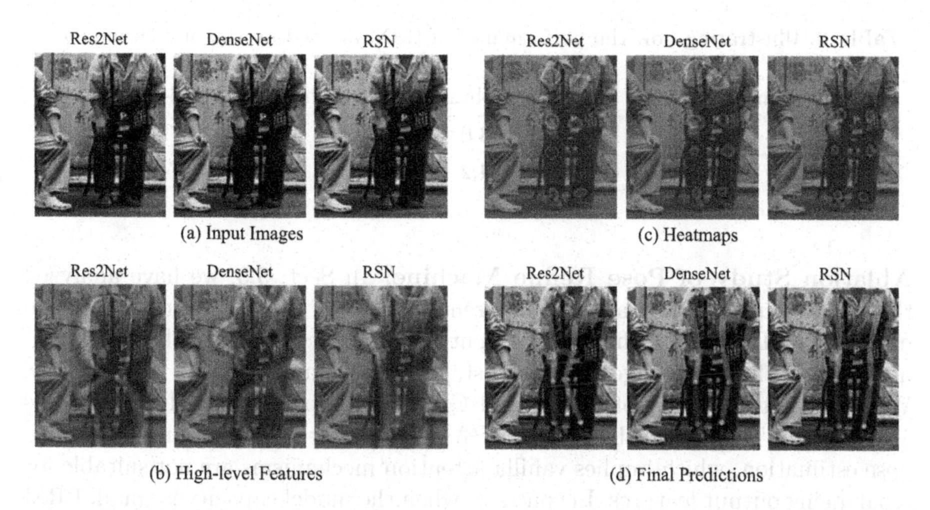

Fig. 6. Visual analysis of Res2Net-50, DenseNet-169 and RSN-50. (a) Input images, (b) High-level feature maps (level 4), (c) Low-level heatmaps (level 1), (d) Final predictions.

hand, compared with Res2Net, the more densely connected architecture and wider range of receptive fields make the intra-level feature fusion of RSN more effective, that is why the red area of RSN is much larger than that of Res2Net and the weights of RSN are more fully used, just as shown in Fig. 5(a) and (c). As a result, the RSN model can keep its high efficiency and considerable improvement from the beginning to the end, just as shown in Fig. 4.

Additionally, to highlight the advantages of RSN more intuitively, we conduct visual analysis of Res2Net-50, DenseNet-169 and RSN-50, as shown in Fig. 6. In Fig. 6(b), the high-level response to human body of Res2Net and DenseNet either covers incomplete body area or too large area of background. Only RSN has a relatively complete and appropriate response area to the human body. As a result, in final prediction, Res2Net is easily misled by the background information, DenseNet ignores some keypoints such as shoulders, while RSN can locate the keypoints better and reduce the interference of background information. As Fig. 6(c) shows, the heatmaps of RSN are much clearer and the locations of the responses are much more accurate.

Ablation Study of RSN Architecture. When designing RSN, we firstly deploy the dense connection principle of DenseNet. Then, for a break-down ablation, we set different branch number as variants to discuss the designing of RSN and explore a best trade-off between branch representing capacity and the degree of intra-level fusion. experiments are done on RSN-18 and RSN-50. We increase branch numbers from 2 to 6 while keeping the model capacity unchanged by adapting channels. As Table 3 shows, the performance firstly becomes better and attains its peak when there are 4 branches. However, when the branch number continues growing, the results get worse. Thus, 4 is the best choice.

Table 3. Illustrating how the performance of RSN affected by the branch number.

Backbone	Input size	2-branch	3-branch	4-branch	5-branch	6-branch
RSN-18	256×192	73.1	73.0	**73.6**	73.2	72.9
RSN-50	256×192	73.9	74.2	**74.7**	74.3	74.0

Ablation Study of Pose Refine Machine. In Sect. 3.3, we have analyzed the architecture of PRM and the differences between PRM and prior attention mechanism. To validate the improvement of PRM, we perform ablation experiments on both single-stage and multi-stage network architecture. Additionally, We validate the impact of SENet and CBAM by replacing PRM. The results are shown in Table 4. SE-block and CBAM decrease the performance of human pose estimation, which implies vanilla attention mechanisms are not suitable for rebalancing output features. In contrast, when the model capacity is small, PRM has a considerable improvement. As for relatively high AP baseline, PRM still obtains 0.4 AP gain. These observations demonstrate the robustness of PRM.

Results on COCO Test-Dev and Test-Challenge. We validate our approach on COCO test-dev and test-challenge sets. The results are shown in Table 5 and Table 6. For fair comparison, we pay attention to the performances of single models with comparable GFLOPs, without using extra training data. In Table 5,

Table 4. Ablation experiments of Pose Refine Machine on COCO minival dataset.

Backbone	Attention	Input size	AP	Δ	GFLOPs
ResNet-18	None	256×192	70.7	0	2.3
ResNet-18	SE-block	256×192	70.5	−0.2	2.3
ResNet-18	CBAM	256×192	69.9	−0.8	2.3
ResNet-18	PRM	256×192	**72.2**	**+1.5**	**4.1**
ResNet-50	None	256×192	72.2	0	4.6
ResNet-50	SE-block	256×192	72.1	−0.1	4.6
ResNet-50	CBAM	256×192	71.1	−1.1	4.6
ResNet-50	PRM	256×192	**73.4**	**+1.2**	**6.4**
$4 \times$ ResNet-50	None	256×192	76.8	0	20.6
$4 \times$ ResNet-50	SE-block	256×192	76.6	−0.2	20.6
$4 \times$ ResNet-50	CBAM	256×192	76.1	−0.7	20.6
$4 \times$ ResNet-50	PRM	256×192	**77.2**	**+0.4**	**22.4**
$4 \times$ RSN-50	None	256×192	78.6	0	27.5
$4 \times$ RSN-50	SE-block	256×192	78.6	0	27.5
$4 \times$ RSN-50	CBAM	256×192	78.0	−0.6	27.5
$4 \times$ RSN-50	PRM	256×192	**79.0**	**+0.4**	**29.3**

Table 5. Results on COCO test-dev dataset. "*" means using ensembled models. Pretrain = pretrain the backbone on the ImageNet classification task.

Method	Extra data	Pretrain	Backbone	Input Size	Params	GFLOPs	AP	AP.5	AP.75	AP(M)	AP(L)	AR
CMUpose [1]	×	-	-	-	-	-	61.8	84.9	67.5	57.1	68.2	66.5
G-MRI [20]	×	-	ResNet-101	353 × 257	42.6M	57.0	64.9	85.5	71.3	62.3	70.0	69.7
G-RMI [20]	√	-	ResNet-101	353 × 257	42.6M	57.0	68.5	87.1	75.5	65.8	73.3	73.3
CPN [2]	×	√	ResNet-Inception	384 × 288	58.8M	29.2	72.1	91.4	80.0	68.7	77.2	78.5
CPN* [2]	×	√	ResNet-Inception	384 × 288	-	-	73.0	91.7	80.9	69.5	78.1	79.0
SimpleBase [28]	×	√	ResNet-152	384 × 288	68.6M	35.6	73.7	91.9	81.1	70.3	80.0	79.0
HRNet-W32 [23]	×	√	HRNet-W32	384 × 288	28.5M	16.0	74.9	92.5	82.8	71.3	80.9	80.1
HRNet-W48 [23]	×	√	HRNet-W48	384 × 288	63.6M	32.9	75.5	92.5	83.3	71.9	81.5	80.5
SimpleBase* [28]	√	√	ResNet-152	384 × 288	-	-	76.5	92.4	84.0	73.0	82.7	81.5
HRNet-W48 [23]	√	√	HRNet-W48	384 × 288	63.6M	32.9	77.0	92.7	84.5	73.4	**83.1**	82.0
MSPN [15]	√	×	4×ResNet-50	384 × 288	71.9M	58.7	77.1	93.8	84.6	73.4	82.3	82.3
Ours(RSN)	×	×	RSN-18	256 × 192	12.5M	2.5	71.6	92.6	80.3	68.8	75.8	77.7
Ours(RSN)	×	×	RSN-50	256 × 192	25.7M	6.4	72.5	93.0	81.3	69.9	76.5	78.8
Ours(RSN)	×	×	2 × RSN-50	256 × 192	54.0M	13.9	75.5	93.6	84.0	73.0	79.6	81.3
Ours(RSN)	×	×	**4 × RSN-50**	256 × 192	111.8M	**29.3**	**78.0**	**94.2**	**86.5**	**75.3**	82.2	**83.4**
Ours(RSN)	×	×	**4 × RSN-50**	384 × 288	111.8M	**65.9**	**78.6**	**94.3**	**86.6**	**75.5**	**83.3**	**83.8**
Ours(RSN*)	×	×	**4 × RSN-50**	-	-	-	**79.2**	**94.4**	**87.1**	**76.1**	**83.8**	**84.1**

Table 6. Results on COCO test-challenge dataset. "*" means using ensembled models.

Method	Extra data	Pretrain	Backbone	Input Size	Params	GFLOPs	AP	AP.5	AP.75	AP(M)	AP(L)	AR
G-RMI [20]	√	-	ResNet-101	353 × 257	42.6M	57.0	69.1	85.9	75.2	66.0	74.5	75.1
CPN* [2]	×	√	ResNet-Inception	384 × 288	-	-	72.1	90.5	78.9	67.9	78.1	78.7
Sea Monsters*	√	-	-	-	-	-	74.1	90.6	80.4	68.5	82.1	79.5
SimpleBase* [28]	√	√	ResNet-152	384 × 288	-	-	74.5	90.9	80.8	69.5	82.9	80.5
MSPN* [15]	√	×	4 × ResNet-50	384 × 288	-	-	76.4	92.9	82.6	71.4	83.2	82.2
Ours(RSN*)	×	×	**4 × RSN-50**	-	-	-	**77.1**	**93.3**	**83.6**	**72.2**	**83.6**	**82.6**

our method outperforms HRNet by 2.5 AP (78.0 v.s. 75.5), and outperforms SimpleBaseline by 4.3 AP on COCO test-dev dataset. Additionally, as Table 6 shows, our approach outperforms MSPN (winner of COCO kps Challenge 2018) by 0.7 AP on test-challenge set. Note that we don't even use pretrained model.

Inference Speed. Current methods of human pose estimation mainly focus on promoting the performance while deploying resource-intensive networks with large depth and width. This leads to inefficient inference. Interestingly, we observe RSN can make a better trade-off between accuracy and inference speed than prior work. For fair comparison, we train RSN and HRNet under the same settings in Sect. 4, with 256×192 input size. Both use MegDet as human detector when testing. We use pps to measure inference speed, i.e., Persons inferred Per Second. On the same GPU (RTX 2080ti), results of COCO val are reported, HRNet-w16 with 1.9 G and 7.2 M achieves 71.9 AP and 31.8 pps, RSN-18 with 2.5G and 12.5 M achieves 73.6 AP and 64.9 pps, HRNet-w32 with 7.1 G and 28.5M achieves 74.6 AP and 26.5 pps, RSN-50 with 6.4 G and 25.7 M achieves 74.7 AP and 42.6 pps. HRNet-w48 with 14.6G and 63.6M achieves 75.5 AP and 24.7 pps, $2 \times$ RSN-50 with 13.9 G and 54.0 M achieves 77.4 AP and 20.2 pps. In addition, the inference speed on CPU (Intel(R) Xeon(R) Gold6013@2.1GHZ) also shows, RSNs with higher performances are faster than HRNet by all sizes. These results suggest that RSN is more accurate, compact and efficient.

Effect of Human Detection. We use MegDet as human detector in ablation study, which achieves 49.4 AP on COCO val. For test sets, we use MegDet-v2, which has 59.8 AP on COCO val. As human detection has an influence on the final performance of top-down approach, We perform ablations to investigate the impact of human detector on COCO test-dev. $4 \times$ RSN-50 at input size of 256×192 achieves 77.3 AP using MegDet, and 78.0 using MegDet-v2. $4 \times$ RSN-50 at input size of 384×288 achieves 77.9 using MegDet, and 78.6 using MegDet-v2.

4.2 MPII Human Pose Estimation

We validate RSN on MPII test set, a single-person pose estimation benchmark. As shown in Table 7, RSN boosts the SOTA performance by 0.7 in PCKh@0.5, which demonstrates the superiority and generalization ability of our method.

Table 7. PCKh@0.5 results on MPII test dataset.

Method	Hea	Sho	Elb	Wri	Hip	Kne	Ank	Mean
Chen et al. [3]	98.1	96.5	92.5	88.5	90.2	89.6	86.0	91.9
Yang et al. [30]	98.5	96.7	92.5	88.7	91.1	88.6	86.0	92.0
Ke et al. [12]	98.5	96.8	92.7	88.4	90.6	89.3	86.3	92.1
Tang et al. [25]	98.4	96.9	92.6	88.7	91.8	89.4	86.2	92.3
Sun et al. [23]	**98.6**	96.9	92.8	89.0	91.5	89.0	85.7	92.3
ours(4 × RSN-50)	98.5	**97.3**	**93.9**	**89.9**	**92.0**	**90.6**	**86.8**	**93.0**

Fig. 7. Prediction results on COCO (top line) and MPII (bottom line) val sets.

5 Conclusion

In this paper, we propose a novel method, Residual Steps Network, which aims to learn delicate local representations by efficient intra-level feature fusion. To make a better trade-off between local and global representations in output features, we design Pose Refine Machine. Our method yields the best results on two benchmarks, COCO and MPII. Some prediction results are visualized in Fig. 7.

Acknowledgments. This work was supported in part by the National Key Research and Development Program of China under Grant 2017YFA0700800.

References

1. Cao, Z., Simon, T., Wei, S.E., Sheikh, Y.: Realtime multi-person 2d pose estimation using part affinity fields. In: The IEEE Conference on Computer Vision and Pattern Recognition (CVPR) (2017)
2. Chen, Y., Wang, Z., Peng, Y., Zhang, Z., Yu, G., Sun, J.: Cascaded pyramid network for multi-person pose estimation. In: The IEEE Conference on Computer Vision and Pattern Recognition (CVPR) (2018)

3. Chen, Y., Shen, C., Wei, X.S., Liu, L., Yang, J.: Adversarial PoseNet: a structure-aware convolutional network for human pose estimation. In: The IEEE International Conference on Computer Vision (ICCV) (2017)
4. Chen, Y., Li, J., Xiao, H., Jin, X., Yan, S., Feng, J.: Dual path networks. In: Guyon, I., et al. (eds.) Advances in Neural Information Processing Systems, vol. 30, pp. 4467–4475. Curran Associates, Inc. (2017). http://papers.nips.cc/paper/7033-dual-path-networks.pdf
5. Fang, H.S., Xie, S., Tai, Y.W., Lu, C.: RMPE: regional multi-person pose estimation. In: The IEEE International Conference on Computer Vision (ICCV) (2017)
6. Fu, J., et al.: Dual attention network for scene segmentation. In: The IEEE Conference on Computer Vision and Pattern Recognition (CVPR) (2019)
7. Gao, S.H., Cheng, M.M., Zhao, K., Zhang, X.Y., Yang, M.H., Torr, P.: Res2Net: a new multi-scale backbone architecture. In: IEEE TPAMI (2020)
8. He, K., Gkioxari, G., Dollar, P., Girshick, R.: Mask R-CNN. In: The IEEE International Conference on Computer Vision (ICCV) (2017)
9. Hu, J., Shen, L., Sun, G.: Squeeze-and-excitation networks. In: The IEEE Conference on Computer Vision and Pattern Recognition (CVPR) (2018)
10. Huang, G., Liu, Z., van der Maaten, L., Weinberger, K.Q.: Densely connected convolutional networks. In: The IEEE Conference on Computer Vision and Pattern Recognition (CVPR) (2017)
11. Huang, S., Gong, M., Tao, D.: A coarse-fine network for keypoint localization. In: The IEEE International Conference on Computer Vision (ICCV) (2017)
12. Ke, L., Chang, M.C., Qi, H., Lyu, S.: Multi-scale structure-aware network for human pose estimation. In: The European Conference on Computer Vision (ECCV) (2018)
13. Kligvasser, I., Rott Shaham, T., Michaeli, T.: xunit: learning a spatial activation function for efficient image restoration. In: The IEEE Conference on Computer Vision and Pattern Recognition (CVPR) (2018)
14. Kocabas, M., Karagoz, S., Akbas, E.: MultiPoseNet: fast multi-person pose estimation using pose residual network. In: Ferrari, V., Hebert, M., Sminchisescu, C., Weiss, Y. (eds.) ECCV 2018. LNCS, vol. 11215, pp. 437–453. Springer, Cham (2018). https://doi.org/10.1007/978-3-030-01252-6_26
15. Li, W., et al.: Rethinking on multi-stage networks for human pose estimation. arXiv preprint arXiv:1901.00148 (2019)
16. Lin, T.-Y., et al.: Microsoft COCO: common objects in context. In: Fleet, D., Pajdla, T., Schiele, B., Tuytelaars, T. (eds.) ECCV 2014. LNCS, vol. 8693, pp. 740–755. Springer, Cham (2014). https://doi.org/10.1007/978-3-319-10602-1_48
17. Newell, A., Yang, K., Deng, J.: Stacked hourglass networks for human pose estimation. In: Leibe, B., Matas, J., Sebe, N., Welling, M. (eds.) ECCV 2016. LNCS, vol. 9912, pp. 483–499. Springer, Cham (2016). https://doi.org/10.1007/978-3-319-46484-8_29
18. Newell, A., Huang, Z., Deng, J.: Associative embedding: end-to-end learning for joint detection and grouping. In: Advances in Neural Information Processing Systems, vol. 30, pp. 2277–2287 (2017)
19. Insafutdinov, E., Pishchulin, L., Andres, B., Andriluka, M., Schiele, B.: DeeperCut: a deeper, stronger, and faster multi-person pose estimation model. In: Leibe, B., Matas, J., Sebe, N., Welling, M. (eds.) ECCV 2016. LNCS, vol. 9910, pp. 34–50. Springer, Cham (2016). https://doi.org/10.1007/978-3-319-46466-4_3
20. Papandreou, G., et al.: Towards accurate multi-person pose estimation in the wild. In: The IEEE Conference on Computer Vision and Pattern Recognition (CVPR) (2017)

21. Pishchulin, L., Insafutdinov, E., Tang, S., Andres, B., Andriluka, M., Gehler, P.V., Schiele, B.: Deepcut: Joint subset partition and labeling for multi person pose estimation. In: The IEEE Conference on Computer Vision and Pattern Recognition (CVPR) (2016)
22. Su, K., Yu, D., Xu, Z., Geng, X., Wang, C.: Multi-person pose estimation with enhanced channel-wise and spatial information. In: The IEEE Conference on Computer Vision and Pattern Recognition (CVPR) (2019)
23. Sun, K., Xiao, B., Liu, D., Wang, J.: Deep high-resolution representation learning for human pose estimation. In: CVPR (2019)
24. Szegedy, C., et al.: Going deeper with convolutions. In: The IEEE Conference on Computer Vision and Pattern Recognition (CVPR) (2015)
25. Tang, W., Yu, P., Wu, Y.: Deeply learned compositional models for human pose estimation. In: Ferrari, V., Hebert, M., Sminchisescu, C., Weiss, Y. (eds.) ECCV 2018. LNCS, vol. 11207, pp. 197–214. Springer, Cham (2018). https://doi.org/10.1007/978-3-030-01219-9_12
26. Woo, S., Park, J., Lee, J.-Y., Kweon, I.S.: CBAM: convolutional block attention module. In: Ferrari, V., Hebert, M., Sminchisescu, C., Weiss, Y. (eds.) ECCV 2018. LNCS, vol. 11211, pp. 3–19. Springer, Cham (2018). https://doi.org/10.1007/978-3-030-01234-2_1
27. Xia, F., Wang, P., Chen, X., Yuille, A.L.: Joint multi-person pose estimation and semantic part segmentation. In: The IEEE Conference on Computer Vision and Pattern Recognition (CVPR) (2017)
28. Xiao, B., Wu, H., Wei, Y.: Simple baselines for human pose estimation and tracking. In: Ferrari, V., Hebert, M., Sminchisescu, C., Weiss, Y. (eds.) ECCV 2018. LNCS, vol. 11210, pp. 472–487. Springer, Cham (2018). https://doi.org/10.1007/978-3-030-01231-1_29
29. Xie, S., Girshick, R., Dollar, P., Tu, Z., He, K.: Aggregated residual transformations for deep neural networks. In: The IEEE Conference on Computer Vision and Pattern Recognition (CVPR) (2017)
30. Yang, W., Li, S., Ouyang, W., Li, H., Wang, X.: Learning feature pyramids for human pose estimation. In: The IEEE International Conference on Computer Vision (ICCV) (2017)
31. Yu, C., Wang, J., Peng, C., Gao, C., Yu, G., Sang, N.: Learning a discriminative feature network for semantic segmentation. In: The IEEE Conference on Computer Vision and Pattern Recognition (CVPR) (2018)
32. Zhao, H., et al.: PSANet: point-wise spatial attention network for scene parsing. In: Ferrari, V., Hebert, M., Sminchisescu, C., Weiss, Y. (eds.) ECCV 2018. LNCS, vol. 11213, pp. 270–286. Springer, Cham (2018). https://doi.org/10.1007/978-3-030-01240-3_17
33. Zhou, K., Yang, Y., Cavallaro, A., Xiang, T.: Omni-scale feature learning for person re-identification. In: ICCV (2019)

Learning to Plan with Uncertain Topological Maps

Edward Beeching[1(✉)], Jilles Dibangoye[1], Olivier Simonin[1], and Christian Wolf[2]

[1] INRIA Chroma team, CITI Lab. INSA Lyon, Villeurbanne, France
{edward.beeching,jilles.dibangoye,olivier.simonin}@insa-lyon.fr
[2] Université de Lyon, INSA-Lyon, LIRIS, CNRS, Lyon, France
christian.wolf@insa-lyon.fr
https://team.inria.fr/chroma/en/

Abstract. We train an agent to navigate in 3D environments using a hierarchical strategy including a high-level graph based planner and a local policy. Our main contribution is a data driven learning based approach for planning under uncertainty in topological maps, requiring an estimate of shortest paths in valued graphs with a probabilistic structure. Whereas classical symbolic algorithms achieve optimal results on noise-less topologies, or optimal results in a probabilistic sense on graphs with probabilistic structure, we aim to show that machine learning can overcome missing information in the graph by taking into account rich high-dimensional node features, for instance visual information available at each location of the map. Compared to purely learned neural white box algorithms, we structure our neural model with an inductive bias for dynamic programming based shortest path algorithms, and we show that a particular parameterization of our neural model corresponds to the Bellman-Ford algorithm. By performing an empirical analysis of our method in simulated photo-realistic 3D environments, we demonstrate that the inclusion of visual features in the learned neural planner outperforms classical symbolic solutions for graph based planning.

Keywords: Visual navigation · Topological maps · Graph neural networks

1 Introduction

A critical part of intelligence is navigation, memory and planning. An animal that is able to store and recall pertinent information about their environment is likely to exceed the performance of an animal whose behavior is purely reactive.

Project page https://edbeeching.github.io/papers/learning_to_plan.

Electronic supplementary material The online version of this chapter (https://doi.org/10.1007/978-3-030-58580-8_28) contains supplementary material, which is available to authorized users.

A. Vedaldi et al. (Eds.): ECCV 2020, LNCS 12348, pp. 473–490, 2020.
https://doi.org/10.1007/978-3-030-58580-8_28

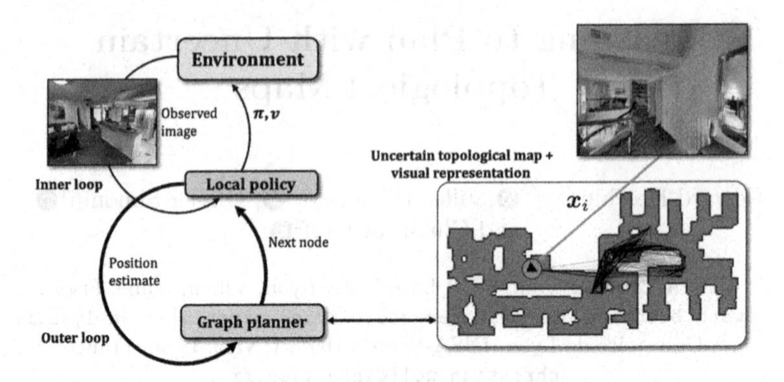

Fig. 1. An agent navigates to a goal location with a hierarchical planner. A high-level planner proposes target nodes in a topological map, which are used as an objective for a local point-goal policy. The graph is estimated from an explorative rollout and, as such, uncertain: the opacities of the edges correspond to estimations of connectivity between nodes (darker lines = higher confidence). We observe a low probability of connection between the node at the agent's position and its nearest neighbor, whereas from the visual observation associated to the node we can see there is a traversable space between the two nodes.

Many control and navigation problems in partially observed 3D environments involve long term dependencies and planning. It has been shown that humans and other animals navigate through the use of waypoints combined with a local locomotion policy [22,55]. In this work, we mimic this strategy by proposing a hierarchical planner, which performs high-level long term planning using an uncertain topological map (a valued graph including visual features) combined with a local RL-based policy navigating between high-level waypoints proposed by the graph planner. Our main contribution is a way to combine symbolic planning with machine learning, and we look to structure a neural network architecture to incorporate landmark based planning in unseen 3D environments.

When solving visual navigation tasks, biological or artificial agents require an internal representation of the environment if they want to solve more complex tasks than random exploration. We target a scenario where an agent is trained on a large-scale set of 3D environments to learn to reason on planning and navigation. When faced with a previously unseen environment, the agent is given the opportunity to build a representation by doing an explorative rollout from a previously learned explorative policy. It can then exploit this internal representation in subsequent visual navigation tasks. This corresponds to many realistic situations, where robots are deployed to indoor environments and are allowed to familiarize themselves before performing their tasks [43].

Our agent constructs an imperfect topological map of its environment, where nodes correspond to places and valued edges to connections. Edges are assigned two different values, spatial distances and probabilities indicating whether it is possible to navigate between the two nodes. Nodes are also assigned rich visual features extracted from images taken at the corresponding places in the environment. After deployment, the agent faces visual navigation tasks requiring it to find a specific location in the environment provided by a set of images corresponding to different viewpoints, extending the task proposed in [63]. The objective is to identify the goal location in the internal representation, and to provide an estimate for the shortest path to it. The main difficulty we address here is the fact that this path is an estimate only, since the ground truth path is not available during testing.

Whilst planning in graphs with known connectivity has been solved for many decades [7,15], planning under uncertainty remains an ongoing area of research. Whereas optimal results in a probabilistic sense exist for graphs with probabilistic connectivity, we aim to show that machine learning can overcome missing information in the graph by taking into account rich high-dimensional node features extracted from image observations associated with specific nodes. We train a graph neural network in a fully supervised way to predict estimates of the shortest path, using vision to overcome uncertainty in the connectivity information. We present a new variant of graph neural networks imbued with specific inductive bias, and we show that this structure can be parameterized to fallback to the classical Bellman-Ford algorithm.

Figure 1 illustrates the hierarchical planner: a neural graph based planner runs an outer loop providing estimates for next way-point on a graph, which are used as target nodes for a local RL-based policy running an inner loop and providing feedback to high-level planner on reached locations. Both planners take into account visual features, either stored in the graph (graph based planner), or directly as observations provided by the environment (local policy). The two planners are trained separately—the graph based planner in a fully supervised way from ground truth graphs, the local policy with RL and a point-goal strategy. This work makes the following contributions:

- A hierarchical model combining high-level graph based planning with a local point goal policy for robot navigation;
- A neural planner that combines an uncertain topological map with node features to learn to estimate shortest paths in noisy and unknown environments.
- A variant of graph networks encoding inductive bias inspired by dynamic programming-based shortest path algorithms.
- We evaluate this method in challenging and visually realistic 3D environments and show it outperforms symbolic planning on noisy topological maps.

2 Related Work

Classical Planning and Graph Search—A large body of work is available on classical planning on graphs, notable references include [31,42]. In robotics, there have been a number of works applying classical planning in topological maps for indoor robot navigation, for instance [48,54].

Planning Under Imperfect Information—In many realistic robotic problems, the current state of the world is unknown. Though sensor observations

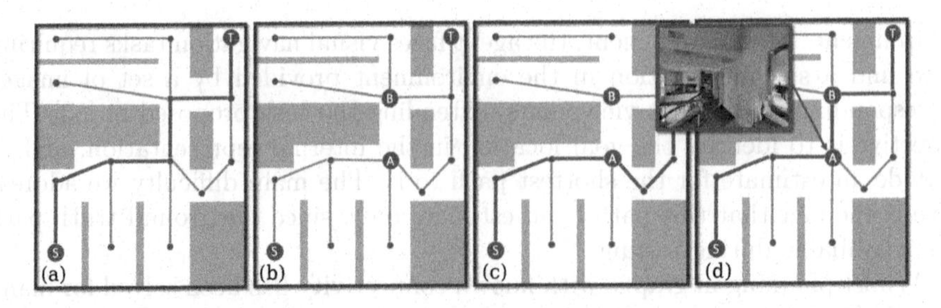

Fig. 2. Illustration of the different types of solutions to the high-level graph planning problem: (a) the ground truth graph (unavailable during testing) with the shortest path from node S to node T in red; (b) the uncertain graph available during test time. This graph is fully connected and for each edge a connection probability is available. For clarity we here show only edges where the connection probability is above a threshold. The edge from $A \rightarrow B$ is wrongly estimated as not connected; (c) an "optimal" path taking into account both probabilities and distances; (d) A learned shortest path, where the visual features at node A indicate passage to node B. We supervise a network to predict the GT path (a). (Color figure online)

provide measurements about the current state of the world, these measurements are usually incomplete or noisy because of disturbances that distort their values. Planning problems that face these issues are referred to as planning problems under imperfect information. Research on this topic has a long history, starting with the seminal work by [2] presenting the first non-trivial exact dynamic programming algorithm for partially observable Markov decision processes (POMDPs). While there are other models [31, chap 12], POMDPs emerged as the standard framework to formalize and solve (single-agent) sequential decision-making problems with imperfect information about the state of the world [26]. As the agent does not have access to the actual state of the world, it acts based solely on its entire history of actions and observations, or the corresponding belief state, *i.e.,* the posterior probability distribution over the states given the history [2,50]. Approaches for finding optimal solutions have been investigated in the 2000s, ranging from dynamic programming [26] to heuristic search methods [30,51]. Key to these approaches is the idea that one can recast the original problem into a continuous-state fully observable Markov decision process, where states are belief states or histories [2]. Doing so allows theory and algorithm that applies for MDPs to also apply to POMDPs, albeit in much larger (and possibly continuous) state space. Another significant result of this literature is proof that the optimal value function is a piece-wise linear and convex function of the belief states, which allows the design of algorithms with faster rates of convergence [50]. For a thorough discussion on existing solvers for POMDPs, the reader can refer to [47].

Deep Reinforcement Learning—The field of Deep Reinforcement Learning (RL) has gained attention with successes on board games [49] and Atari games [37]. Recent works have applied Deep RL for the control of an agent in 3D environments [24,36], exploring the use of auxiliary tasks such as depth prediction, loop detection and reward prediction to accelerate learning. Other recent work uses street-view scenes to train an agent to navigate in city environments [35]. To infer long term dependencies and store pertinent information about the partially observable environment, network architectures typically incorporate recurrent memory such as Gated Recurrent Units [13] or Long Short-Term Memory [23]. Extensions to memory based neural approaches began with Neural Turing Machines [19] and Differentiable Neural Computers [20], and have since been adapted to expand the capacity of Deep RL agents [56]. Spatially structured memory architectures have been shown to augment an agent's performance in 3D environments and are broadly split into two categories: metric maps which discretize the environment into a grid based structure and topological maps which produce node embeddings at key points in the environment. Research in learning to use a metric map is extensive and includes spatially structured memory [40], Neural SLAM based approaches [61] and approaches incorporating projective geometry and neural memory [8,21], these techniques are combined, extended and evaluated in [6]. Other works include that of Value Iteration Networks (VIN) [53] which approximate the value iteration algorithm with a CNN, applied planning in small fully observable state spaces (grid worlds). While VIN and our work structure planners, VINs use convolutions to approximate classical value iteration, while we use a graph representation and a novel GNN architecture with recurrent updates to approximate the Bellman-Ford algorithm. [27] plans under uncertainty in partially observable gridworld environments. Here uncertainty refers to POMPs, the classical QMDP algorithm is used as inductive bias for a neural network, whereas in our work uncertainty is over node connectivity in a graph constructed in a previously unseen environment. [52] is applied in observable state spaces to learn a forward model in a latent space to plan appropriate actions; they are not hierarchical, are not graph-based and do not appear to plan under uncertainty.

Research combining learning, navigation in 3D environments and topological representations has been limited in recent years with notable works being [43] who create graph a through random exploration in ViZDoom RL environment [28]. [16] performs planning in 3D environments on a graph-based structure created from randomly sampled observations, with node distances estimated with value estimates. The downside of these approaches is that to generalize to an unseen environment, many random samples must be taken to populate the graph.

Graph Neural Networks—Graph Neural Networks (GNN) are deep networks that operate on graphs directly. They have recently shown great promise in domains such as knowledge graphs [44], chemical analysis [18], protein interactions [17], physics simulations [3] and social network analysis [29]. These types of architectures enable learning from both node features and graph connectivity. Several review papers have covered graph neural networks in great detail [4,9,57,62]. GNNs have been applied to shortest path planning in travelling salesmen problems [25,33] and it has been reasoned that they can approximate optimal symbolic planning algorithms such as the Bellman-Ford algorithm [60]. This work applies a novel variant of GNN in order to solve approximate planning problems, where classical methods may struggle to deal with uncertainty.

3 Hierarchical Navigation with Uncertain Graphs

We train an agent to navigate in a 3D visual environment and to exploit an internal representation, which it is allowed to obtain from an explorative rollout before the episode. Our objective is image goal, i.e. target-driven navigation to a location which is provided through a (visual) image. We extend the task introduced in [63] by generalizing to unseen environment configurations without the need to retrain the agent for a novel environment.

From the explorative rollout obtained with an agent trained with RL, which is further described in Sect. 3.3, we create an uncertain topological map covering the environment, i.e. a valued graph $\mathcal{G} = \{\mathcal{V}, \boldsymbol{V}, \boldsymbol{E}, \boldsymbol{L}, \boldsymbol{D}\}$, where $\mathcal{V} = \{1, \ldots N\}$ is a set of nodes, $\boldsymbol{V}$ is a $K \times N$ matrix of rich visual node features of dimensions K, $\boldsymbol{E} \in [0,1]^{N \times N}$ is a set of edge probabilities where $\boldsymbol{E}_{i,j}$ is the probability of having an edge between nodes i and j, $\boldsymbol{L}$ is a matrix of node locations and $\boldsymbol{D}$ is a distance matrix, where $\boldsymbol{D}_{i,j}$ is a distance between nodes i and j. While $\boldsymbol{D}$ encodes a distance in a path planning sense, $\boldsymbol{E}$ encodes the probability of j being directly accessible from i with obstructions. The uncertainty encoded by this probability can be considered to be a combination of aleatory variability, i.e. uncertainty associated with natural randomness of the environment, as well as epistemic uncertainty, i.e. uncertainty associated with variability in computational models for estimating the graph, in our case the explorative policy trained with RL and taking into account visual observations.

Once the topological map is obtained, the objective of the agent at each episode is to navigate to a location given an image, which is provided as additional observation at each time step. The agent acts in 3D environments like Habitat[34] (see Sect. 5), receiving images of the environment as observations and predicting actions from a discrete space (*forward*, *turn left 10°*, *turn right 10°*). We propose a hierarchical planner performing actions at two different levels:

A high-level graph based planner that operates on longer time scale τ and iteratively proposes new point-goals nodes p_g^τ that are predicted, by a Graph Neural Network, to be on the shortest path from the agent to the estimated location of the target image.

A local policy that has been trained to navigate to a local point-goal p_g^τ, which has been provided by the high-level policy. The local policy operates for a maximum of m time-steps, where m is a hyper-parameter, set to 10. The agent has been trained with an additional *STOP* action, so that it can learn to terminate the local policy in the case that it reaches p_g^τ in under m steps.

The two planners communicate through estimated locations, the graph planner indicating the next waypoint to the local policy as a location, and the local policy providing an estimate of its location back to the high-level planner. The planner updates its current node estimate as the nearest node and planning continues.

3.1 High-Level Planning with Uncertain Graphs

The objective of the high-level planner is to estimate the shortest path from the current position $S \in \mathcal{V}$ in the graph to a terminal node $T \in \mathcal{V}$, whose identity is estimated as the node whose visual features are closest to the target image in cosine distance. Planning takes into account the distances between nodes encoded in $\boldsymbol{D}$ as well as estimated edge connectivity encoded in $\boldsymbol{E}$. As an edge (i, j) may have a large connection probability $\boldsymbol{E}_{i,j}$ but still be obstructed in reality, the goal is to learn a trainable planner parameterized by parameters θ, which takes into account visual features $\boldsymbol{V}$ to overcome the uncertainty in the graph connectivity. To this end, we assume the ground truth connectivity $\boldsymbol{E}^*$ available during training only. Figure 2 illustrates the different types of solutions this problem admits: the optimal shortest path is only available on ground truth data (Fig. 2a), the objective is to use the noisy uncertain graph (Fig. 2b) and provide an estimate of the optimal solution taking into account visual features (Fig. 2d). This is unlike the optimal solution in a probabilistic sense calculated from a symbolic algorithm (Fig. 2c).

We propose a trainable planner, which consists of a novel graph neural network architecture with dedicated inductive bias for planning. Akin to graph networks [5], the node embeddings are updated with messages over the edges, which propagate information over the full graph. While it has been shown that graph networks can be trained to perform planning [59], we aim to closely mimic the structure of the Bellman-Ford algorithm and we embue the planner with additional inductive bias and a supervised objective to explicitly learn to calculate shortest paths from data. To this end, each node i of the graph is assigned an embedding $\boldsymbol{x}_i = [\boldsymbol{v}_i, \boldsymbol{e}_i, t_i, \boldsymbol{d}_i, \boldsymbol{s}_i]$ where $\boldsymbol{v}_i$ are visual features from the memory matrix $\boldsymbol{V}$, t_i is a boolean value indicating if the node is the target, $\boldsymbol{e}_i$ are the edge connection probabilities from node i to all other nodes, $\boldsymbol{d}_i$ are the distances for node i to all other nodes, $\boldsymbol{s}_i$ is a one hot vector identifying the node.

We motivate our neural model with the following objective: the planner should be able to exploit information contained in the graph connectivity, but also in the visual features, to be able to find the shortest path from a given current node to a target node. As with classical planning algorithms, it will thus be required to keep for each node a latent representation of the bound d_i on the shortest distance as well as information on the identity of the outgoing edge to the neighbor on the shortest path, the predecessor function $\Pi(i)$. Known algorithms (Dijstra, Bellman-Ford) perform iterative updates of these variables (d_i, Π_i) by comparing them with neighboring nodes and their inter-node distances, updating the bound d_i and Π_i when a shorter path is found than the current one. This is usually done by iterating over the successors of a given node i.

In our trained model, these variables are not made explicit, but they are supposed to be learned as a unique vectorial latent representation for each node i in the form of an internal state $\boldsymbol{r}_i$, which generally holds current information on the reasoning of the agent. The input to each iteration of the graph network is, for each node i, the node embedding $\boldsymbol{x}_i$, and the node state $\boldsymbol{r}_i$, which we concatenate to form a single node vector $\boldsymbol{n}_i = [\boldsymbol{x}_i, \boldsymbol{r}_i] = [\boldsymbol{v}_i, \boldsymbol{e}_i, t_i, \boldsymbol{d}_i, \boldsymbol{s}_i, \boldsymbol{r}_i]$.

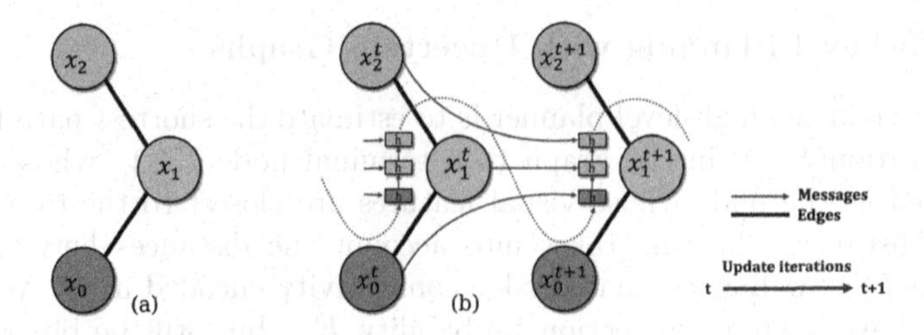

Fig. 3. (a) An example graph; (b) One iteration of the neural planner's message passing and bound update. Messages from neighbors are serialized and fed through a GRU, an inductive bias for learning minima necessary for bound updates.

As classically done in graph neural networks, this representation is updated iteratively by exchanging messages between nodes in the form of trainable functions. The messages and trainable functions of our model are given as follows, illustrated in Fig. 3, and will be motivated in detail further below.

$$\boldsymbol{m}_{i,j} = \boldsymbol{W}_1[\boldsymbol{n}_i, \boldsymbol{n}_j] \odot \sigma(\boldsymbol{W}_2[\boldsymbol{n}_i, \boldsymbol{n}_j]) \tag{1}$$

$$\boldsymbol{r}'_i = \phi^{r \leftarrow h}(\{\boldsymbol{m}_{i,j}\}_{\forall j}, \boldsymbol{h}_i) \tag{2}$$

Here, $\odot$ is the Hadamard product, $\boldsymbol{W}_.$ are weight matrices, and $\boldsymbol{r}'_i$ is the updated latent representation. The features $\boldsymbol{x}_i$ do not change during these operations.

Equation (1) is inspired from gated linear layers [14], and enables each node to identify whether it is the target, and update its representation of the bound. We use gated linear layers in order to provide the network with the capacity to update bound estimates for its neighbors.

Equation (2) integrates messages from all neighbors j of node i, updating its latent representation. Since planning requires this step to update internal bounds on shortest paths, akin to shortest path algorithms that rely on dynamic programming, we serialize the updates from different neighbors into a sequence of updates, which allows the network to learn to calculate minimum functions on bound estimates. In particular, we model this through a Gated Recurrent Unit [12], using a hidden state vector $\boldsymbol{h}_i$ associated to each node i. The step is structured to mimic the min operation of the Bellman-Ford algorithm (see Sect. 3.2 for details on this equivalence).

Equation (2) can thus be rewritten in more detail as follows: Going sequentially over the different neighbors j of node i, the hidden state $\boldsymbol{h}_i$ is updated as follows:

$$\boldsymbol{h}_i^{[j]} = \boldsymbol{W}_3\boldsymbol{m}_{i,j} + \boldsymbol{W}_4\boldsymbol{h}_i^{[j-1]} \tag{3}$$

For simplicity, we omitted the gating equations of GRUs and presented a single layer GRU. In practice we include all gating operations and use a stacked GRU with two layers. The output of the recurrent unit is a non-linear function of the last hidden state, providing the new latent value $\boldsymbol{r}'_i = MLP(\boldsymbol{h}_i^N)$.

The above messages are exchanged and accumulated for k steps where k is a hyper-parameter which should be at least the largest span of the graphs in the dataset. The action distribution is then estimated for each node as a linear mapping of the node embeddings followed by a softmax activation function.

3.2 Relations to Optimal Symbolic Planners

As mentioned before, our neural planner could in theory be instantiated with a specific set of network parameters such that it corresponds to a known symbolic planner calculating an optimal path in a certain sense. To illustrate the relationship of the network structure, in particular the recurrent nature of the graph updates, we will layout details for the case where the planner performs the estimation of a shortest path given the distance matrix and ignoring the uncertainty information—an adaptation to an optimal planner in the probabilistic sense can be done in a straightforward manner. To avoid misunderstandings, we insist that the reasoning developed in this sub section is for illustration and general understanding of the chosen inductive network bias only, the real network parameters are fully trained with supervised learning as explained in Sect. 4.

Handcrafting a parameterization requires imposing a structure on the node state $\boldsymbol{r}_i$, which otherwise is a learned representation. In our case, the node state will be composed of the bound b_i on the shortest path from the given node to the target node (a scalar), and the current estimate Π_i of the identity of predecessor node of node i w.r.t. the shortest path, which can be represented as a 1-in-K encoded vector indicating a distribution over nodes. Standard Bellman-Ford iteratively updates the bound for a given node i by examining all its neighbors j and checking whether a shorter path can be found passing through neighbor j. This can be written in a sequential form s.t. the bound gets updated iterating through the neighbors $j = 1 \ldots J_i$ of node i:

$$
\begin{aligned}
b_i^{[0]} &= b_i \\
b_i^{[j]} &= \min(b_i^{[j-1]}, b_j + d_{ij}) \\
b_i' &= b_i^{[J_i]}
\end{aligned}
\tag{4}
$$

where b_i and b_i' are the bounds before and after the round of updates for node i.

In our neural formulation, the message updates given in Eq. (2), further developed in (3), mimic the Bellman-Ford bound update given in Eqs. (4). This provided motivation for our choice of a recurrent neural network in the graph neural network, as we require the update of the recurrent state h_i^j in Eq. (3) to be able to perform a minimum operation and an arg min operation (or differentiable approximations of min and arg min).

3.3 Graph Creation from Explorative Rollouts

Graphs were generated during the initial rollout from an exploratory policy trained with Reinforcement Learning. During training, the agent interacts with training environments and receives RGB-D image observations calculated as a projection from the 3D environment. The agent is trained to explore the environment and to maximize coverage, similar to [10,11].

To learn to estimate the graph connectivity, we add an auxiliary loss to the agent's objective function, $f_{link}(\boldsymbol{o}_i, \boldsymbol{o}_j, \boldsymbol{h}_i)$ which is trained to classify whether two locations are in line of sight of each other, conditioned on the visual features $\boldsymbol{o}_i, \boldsymbol{o}_j$ from the two locations and the agent's hidden state $\boldsymbol{h}_i$. Node features were calculated with a CNN [32]. Ground truth line of sight measurements were computed by 2D ray tracing on an occupancy map of each environment. In order to limit the size of the graph to a maximum number of nodes k, we aim to maximize each node's coverage of the environment using a Gaussian kernel function. At each time step a new node is observed by the agent, previous node positions are compared with a Gaussian kernel function (Eq. 5) in order to identify the index of the most redundant node r, which is removed from the graph and replaced with the new node, node connectivities are then recomputed with $f_{link}(.)$, where $\boldsymbol{L}_i$ is the location of node i.

$$r = \arg\min_i \sum_j K(\boldsymbol{L}_i, \boldsymbol{L}_j), \qquad K(\boldsymbol{v}, \boldsymbol{v}') = \exp\left(\frac{-\|\boldsymbol{v} - \boldsymbol{v}'\|^2}{2\sigma^2}\right), \tag{5}$$

4 Training

The High-Level Graph Based Planner—is trained in a purely supervised way. We generate ground truth labels by running a symbolic algorithm (Dijkstra [15]) on a set of valued ground truth training graphs described with the method detailed in Sect. 3.3. In particular, the supervised training algorithm takes as input *uncertain/noisy* graphs, which include visual features, and is supervised to learn to produce paths, which are calculated from known ground truth graphs unavailable during test time. During training we treat path planning as a classification problem where for a given target, each node must learn to predict the subsequent node on the optimal path to the target. Formally for each node i we predict a distribution $\boldsymbol{A}_i$ and aim to match a ground-truth distribution $\boldsymbol{A}_i^*$, which is a one-hot vector, minimizing cross entropy loss $\mathcal{L}(\boldsymbol{A}, \boldsymbol{A}^*) = -\sum_{i=1}^{n} \boldsymbol{A}_i^* \log \boldsymbol{A}_i$.

We augment training with a novel version of mod-drop [39], an algorithm for multi-modal data, which drops modalities probabilistically during training. During training we extend the node connection probabilities with the ground truth node adjacencies and mask either the probabilities or the adjacencies with a probability of 50%, during training we linearly taper the masking probability from 50% to 100% over the first 250 epochs. Ensuring that the final model requires only connection probabilities, but the reasoning performed during message passing and recurrent updates can be bootstrapped from the ground truth adjacencies. Training curves on unseen validation data are shown in Fig. 5.

The Local Policy—is a recurrent version of AtariNet [38] with two output heads for the action distribution and value estimates. The network was trained with a reinforcement learning algorithm Proximal Policy Optimization (PPO) [45] to navigate with discrete actions to a local point-goal. Point-goals were generated to be within 5 m of the spawn location of agent. A dense reward was provided that corresponds to a decrease in geodesic distance to the target, a

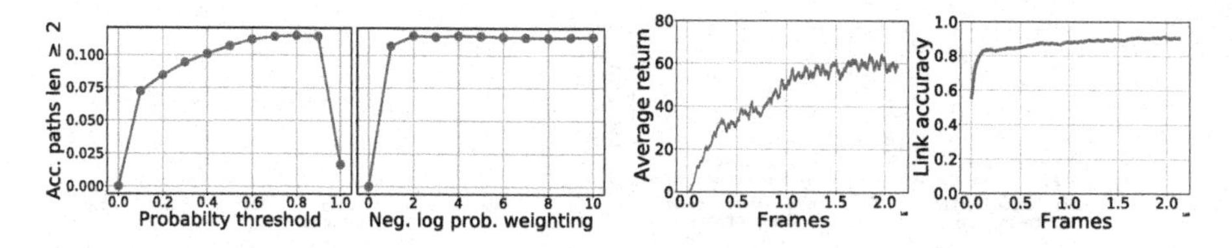

Fig. 4. Left: Symbolic baselines Dijkstra on thresholded probs. & cost function (5.1). Right: Average return & Acc. of line of sight predictions.

Table 1. H-SPL and acc. of the neural planner's predictions.

(a) Uncertain graphs

Method	Acc	H-SPL
Symbolic (threshold)	0.114	0.184
Symbolic (custom cost)	0.115	0.269
Neural (w/o visual)	0.251	0.468
Neural (w visual)	**0.262**	**0.501**

(b) Ground truth graphs

Method	Acc	H-SPL
Symbolic (GT)	**1.00**	**1.00**
Neural planner (GT)	0.921	0.983

large reward (10.0) was provided when the agent reached the target and the *STOP* action was used. The episode was terminated when either the *STOP* action was used or after 500 time-steps. A negative reward of −0.01 was given at each time-step .

The Explorative Policy for Graph Creation—is trained with PPO [46]. We aim to maximize coverage that is within the field of view of the agent. We create an occupancy grid of the environment with a grid spacing of 10 cm. The first time a cell is observed the agent receives a reward of 0.1. A cell is considered to observable if it is free space, within 3 m of the agent and in the field of view of the agent. Agent performance is shown in Fig. 4.

5 Experiments

We evaluated our method in simulated 3D environments in the Habitat [34] simulator with the visually realistic Gibson dataset [58]. During training, the agent interacts with 72 different environments, where each environment corresponds to a different home. We evaluate on a set of 16 held out environments.

5.1 High-Level Graph-Based Planner

The neural planner was implemented in PyTorch [41]. We compare two metrics, accuracy of prediction of the next way-point along the optimal path and the SPL metric [1], both for paths of length two or greater. As we evaluate SPL for both the high level planner and the hierarchical planner-controller, we refer to the high-level planner's SPL as H-SPL to avoid ambiguity.

Symbolic Baselines—We compare the neural planner to two symbolic baselines, which reason on the uncertain graph only, without taking into account

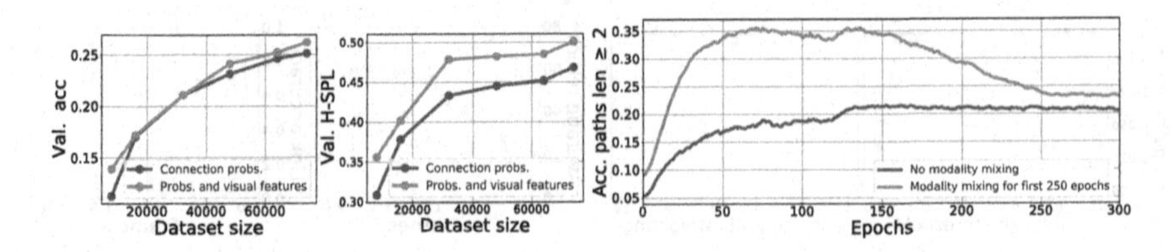

Fig. 5. Left & centre: Accuracy and H-SPL with increasing size of data when training with and without visual features. Right: Modality mixing.

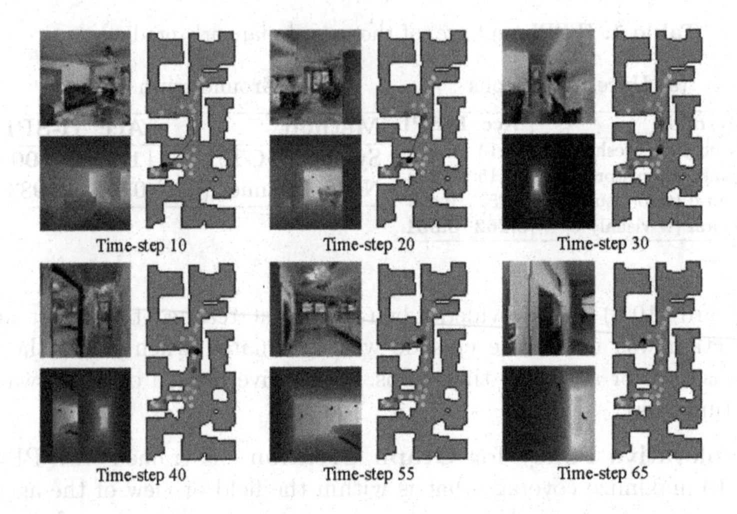

Fig. 6. Time-steps from a rollout of the hierarchical planner (graph+local). For each time-step: left – RGB-D obs., right – map of the environment (unseen) with graph nodes, source node, target node, agent position (black), agent's nearest neighbour, local point-goal provided by the planner and planned path. (Color figue online)

rich node features. While these baselines are *"optimal"* with respect to their respective objective functions, they are optimal with respect to the amount of information available to them, which is uncertain: *(i) Thresholding*—In order to generate non-probabilistic edge connections, we threshold the connection probabilities with values ranging from 0–1 in steps of 0.1. After threshholding the graph, path planning was performed with Dijkstra's algorithm; *(ii) A custom cost function* for Dijkstra's algorithm weighting distances and probabilities: $cost(i,j) = \boldsymbol{D}_{i,j} - \lambda \log(\boldsymbol{E}_{i,j})$. We vary λ in order to control the trade-off of distance and connection probability. When λ is 0, the graph is a fully connected graph, whereas high values of λ would lead to finding the most probable path.

Results of both symbolic baselines are shown in Fig. 4, we observe that they perform poorly under uncertainty. We evaluate the accuracy of their predictions with respect to the symbolic baseline on the ground truth graph. We report accuracy on source-target pairs separated by at least 2 steps.

Table 2. Performance of the hierarchical graph planner and local policy

Method: Planner + Local policy	Success rate	SPL
Graph oracle (optimal point-goals, not comparable)	0.963	0.882
Random	0.152	0.111
Recurrent image-goal agent	0.548	0.248
Symbolic (threshold)	0.621	0.527
Symbolic (custom cost)	0.707	0.585
Neural planner (sampling)	0.966	0.796
Neural planner (deterministic)	**0.983**	**0.877**

Image Driven Recurrent Baseline—We also compare to an end-to-end RL approach where the current obs. and target are provided to a CNN based RL agent. The architecture is a siamese CNN with a recurrent GRU. We train with a dense reward of improvement in geodesic dist. and provide a reward of 10 when the agent reaches the goal. We train with PPO for 200 M frames.

In Table 1, we compare the neural planner with the baselines. We can see, that even without visual features, the neural planner is able to outperform the "optimal" symbolic baselines. This can be explained with the fact, that the baselines optimize a fixed criterion, whereas the neural planner can learn to exploit patterns in the connection probability matrix $\boldsymbol{E}$ to infer valuable information on shortest ground truth path. The gap further increases when the neural planner can use visual features. Results of modality mixing (see Sect. 4) are shown in Fig. 5. As a sanity check, Table 1b compares the optimal symbolic planner against the neural planner trained with ground truth adjacencies provided as input. The results of the neural planner are close to optimum in this case.

We evaluated our approach on dataset sizes ranging from 8,000 graphs to 74,000 graphs (Fig. 5). One graph contains 32×32 possible source-target combinations, leading to a maximum amount of 75,000,000 training instances.

5.2 Hierarchical Planning and Control (Topological and Local Policy)

We evaluated the neural graph planner coupled with the local policy. For a given episode, the graph planner estimates the next node in the path to a target image and provides its location to the local policy, which executes for m time-steps. The planner then re-plans from the nearest neighbor to the agent's current position, this back and forth process of planning and navigating continues until either the agent reaches the target or 500 low-level time-steps have been conducted. We report accuracy as percentage of runs completed successfully and SPL in Table 2, albeit measured on low-level trajectories as opposed to graph space. We combine the local policy with various graph planners, and can see that the neural graph planners greatly outperform the symbolic baselines. We perform

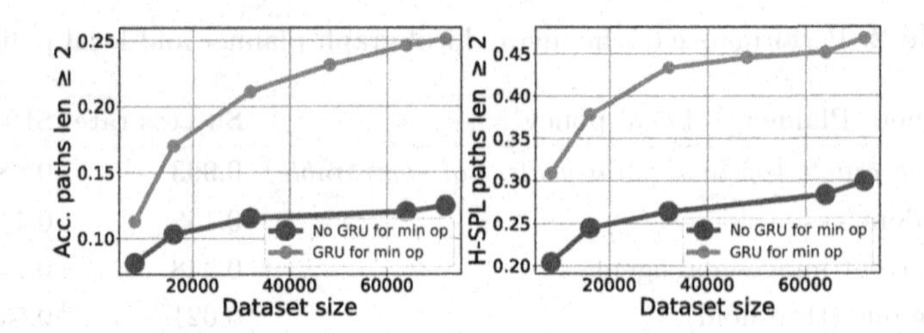

Fig. 7. Ablation of a GRU for the accumulation of incoming messages. The GRU was added to ensure that the model could represent the Bellman-Ford algorithm.

two evaluations of the neural planner; a deterministic evaluation where point-goals are chosen with the argmax of the $\boldsymbol{A}$ distribution and a non-deterministic one by sampling from $\boldsymbol{A}$. The motivation is that by sampling, the planner can escape from local minima and loops created by errors in approximation. This is confirmed when studying rollouts from the agents, and also quantitatively through the performances shown in Table 2. A visualization of steps from an episode is shown in Fig. 6, where in step 10 we can see that navigation is robust w.r.t. local errors in planning.

5.3 Ablation: Effect of Chosen Inductive Bias

As developed in Sects. 3.1 and 3.2, our graph based planner includes a particular inductive bias, which allows it to represent the Bellman-Ford algorithm for the calculation of shortest or best paths. This bias is implemented as a recurrent model (a GRU) running sequentially over the message passing procedures, as illustrated in Fig. 3. Figure 7 ablates the effect of this additional bias as a function of data sizes ranging from 8,000 to 74,000 example graphs, each one evaluated with $32^2 = 1{,}024$ different combinations of starting and end points.

The differences are substantial, we can see that our model is able to exploit increasing amounts of data and translate them into gains in performance, whereas standard graph convolutional networks do not—we conjecture that they lack in structure allowing them to pick up the required reasoning.

6 Conclusion

We demonstrated that path planning can be approximated with learning when structured in a manner that is akin to classical path planning algorithms. We have performed an empirical analysis of the proposed solution in photo-realistic 3D environments and have shown that in uncertain environments graph neural networks can outperform their symbolic counterparts by incorporating rich visual features. Our method can be used to augment a vision based agent with the ability to form long term plans under uncertainty in novel environments, without a priori knowledge of the particular environment. We have analysed the empirical performance of the neural planning algorithm with a variety of dataset sizes, shown that the high-level planner can be coupled with a low-level policy and evaluated the hierarchical performance on an image-goal task.

Acknowledgements. This work was funded by grant Deepvision (ANR-15-CE23-0029, STPGP479356-15), a joint French/Canadian call by ANR & NSERC; Compute was provided by the CNRS/IN2P3 Computing Center (Lyon, France), and by GENCI-IDRIS (Grant 2019-100964).

References

1. Anderson, P., et al.: On evaluation of embodied navigation agents (2018)
2. Åström, K.J.: Optimal control of Markov processes with incomplete state information. J. Math. Anal. Appl. **10**(1), 174–205 (1965)
3. Battaglia, P., Pascanu, R., Lai, M., Rezende, D.J., et al.: Interaction networks for learning about objects, relations and physics. In: Advances in Neural Information Processing Systems, pp. 4502–4510 (2016)
4. Battaglia, P.W., et al.: Relational inductive biases, deep learning, and graph networks. arXiv preprint arXiv:1806.01261 (2018)
5. Battaglia, P., et al.: Relational inductive biases, deep learning, and graph networks. arXiv preprint 1807.09244 (2018)
6. Beeching, E., Wolf, C., Dibangoye, J., Simonin, O.: EgoMap: projective mapping and structured egocentric memory for deep RL (2020)
7. Bellman, R.: On a routing problem. Q. Appl. Math. **16**(1), 87–90 (1958)
8. Bhatti, S., Desmaison, A., Miksik, O., Nardelli, N., Siddharth, N., Torr, P.H.S.: Playing doom with slam-augmented deep reinforcement learning. arxiv preprint 1612.00380 (2016)
9. Bronstein, M.M., Bruna, J., LeCun, Y., Szlam, A., Vandergheynst, P.: Geometric deep learning: going beyond Euclidean data. IEEE Sig. Process. Mag. **34**(4), 18–42 (2017)
10. Chaplot, D.S., Gandhi, D., Gupta, S., Gupta, A., Salakhutdinov, R.: Learning to explore using active neural slam. In: International Conference on Learning Representations (2020). https://openreview.net/forum?id=HklXn1BKDH
11. Chen, T., Gupta, S., Gupta, A.: Learning exploration policies for navigation. In: International Conference on Learning Representations (2019). https://openreview.net/forum?id=SyMWn05F7
12. Chung, J., Gulcehre, C., Cho, K., Bengio, Y.: Empirical evaluation of gated recurrent neural networks on sequence modeling. arXiv preprint arXiv:1412.3555 (2014)
13. Chung, J., Gulcehre, C., Cho, K., Bengio, Y.: Gated Feedback Recurrent Neural Networks. In: ICML (2015)
14. Dauphin, Y.N., Fan, A., Auli, M., Grangier, D.: Language modeling with gated convolutional networks. In: Proceedings of the 34th International Conference on Machine Learning, vol. 70, pp. 933–941. JMLR.org (2017)
15. Dijkstra, E.W.: A note on two problems in connexion with graphs. Numerische mathematik **1**(1), 269–271 (1959)
16. Eysenbach, B., Salakhutdinov, R.R., Levine, S.: Search on the replay buffer: bridging planning and reinforcement learning. In: Advances in Neural Information Processing Systems, vol. 32, pp. 15220–15231. Curran Associates, Inc. (2019)
17. Fout, A., Byrd, J., Shariat, B., Ben-Hur, A.: Protein interface prediction using graph convolutional networks. In: Advances in Neural Information Processing Systems, pp. 6530–6539 (2017)

18. Gilmer, J., Schoenholz, S.S., Riley, P.F., Vinyals, O., Dahl, G.E.: Neural message passing for quantum chemistry. In: Proceedings of the 34th International Conference on Machine Learning, vol. 70, pp. 1263–1272. JMLR.org (2017)
19. Graves, A., Wayne, G., Danihelka, I.: Neural turing machines. arXiv preprint arXiv:1410.5401 (2014)
20. Graves, A., et al.: Hybrid computing using a neural network with dynamic external memory. Nature **538**(7626), 471 (2016)
21. Gupta, S., Davidson, J., Levine, S., Sukthankar, R., Malik, J.: Cognitive mapping and planning for visual navigation. In: IEEE Conference on Computer Vision and Pattern Recognition (CVPR), pp. 7272–7281, July 2017. https://doi.org/10.1109/CVPR.2017.769
22. Gupta, S., Fouhey, D., Levine, S., Malik, J.: Unifying map and landmark based representations for visual navigation. arXiv preprint arXiv:1712.08125 (2017)
23. Hochreiter, S., Schmidhuber, J.: Long short-term memory. Neural Comput. **9**(8), 1735–1780 (1997)
24. Jaderberg, M., et al.: Reinforcement learning with unsupervised auxiliary tasks. In: ICLR (2017)
25. Joshi, C.K., Laurent, T., Bresson, X.: An efficient graph convolutional network technique for the travelling salesman problem. arXiv preprint arXiv:1906.01227 (2019)
26. Kaelbling, L.P., Littman, M.L., Cassandra, A.R.: Planning and acting in partially observable stochastic domains. Artif. Intell. **101**(1–2), 99–134 (1998)
27. Karkus, P., Hsu, D., Lee, W.S.: QMDP-net: deep learning for planning under partial observability (2017)
28. Kempka, M., Wydmuch, M., Runc, G., Toczek, J., Jaskowski, W.: ViZDoom: a doom-based AI research platform for visual reinforcement learning. In: IEEE Conference on Computatonal Intelligence and Games, CIG (2017). https://doi.org/10.1109/CIG.2016.7860433, https://arxiv.org/pdf/1605.02097.pdf
29. Kipf, T.N., Welling, M.: Semi-supervised classification with graph convolutional networks. In: International Conference on Learning Representations (2017)
30. Kurniawati, H.: SARSOP: efficient point-based POMDP planning by approximating optimally reachable belief spaces. In: Proceedings of the Robotics: Science and Systems (2008)
31. LaValle, S.M.: Planning Algorithms. Cambridge University Press, New York (2006)
32. Lecun, Y., Eon Bottou, L., Bengio, Y., Haaner, P.: Gradient-based learning applied to document recognition. Proc. IEEE **86**(11), 2278–2324 (1998)
33. Li, Z., Chen, Q., Koltun, V.: Combinatorial optimization with graph convolutional networks and guided tree search. In: Advances in Neural Information Processing Systems, pp. 539–548 (2018)
34. Savva, M., et al.: Habitat: a platform for embodied AI research. In: Proceedings of the IEEE/CVF International Conference on Computer Vision (ICCV) (2019)
35. Mirowski, P., et al.: Learning to navigate in cities without a map. arxiv pre-print 1804.00168v2 (2018)
36. Mirowski, P., et al.: Learning to navigate in complex environments. In: ICLR (2017)
37. Mnih, V., et al.: Human-level control through deep reinforcement learning. Nature **518**(7540), 529–533 (2015)
38. Mnih, V., et al.: Human-level control through deep reinforcement learning. Nature **518**, (2015). https://doi.org/10.1038/nature14236, https://storage.googleapis.com/deepmind-media/dqn/DQNNaturePaper.pdf

39. Neverova, N., Wolf, C., Taylor, G., Nebout, F.: ModDrop: adaptive multi-modal gesture recognition. IEEE Trans. Pattern Anal. Mach. Intell. **38**(8), 1692–1706 (2015)
40. Parisotto, E., Salakhutdinov, R.: Neural map: structured memory for deep reinforcement learning. In: ICLR (2018)
41. Paszke, A., et al.: Pytorch: an imperative style, high-performance deep learning library. In: Wallach, H., Larochelle, H., Beygelzimer, A., d'Alché Buc, F., Fox, E., Garnett, R. (eds.) Advances in Neural Information Processing Systems, vol. 32, pp. 8024–8035. Curran Associates, Inc. (2019)
42. Remolina, E., Kuipers, B.: Towards a general theory of topological maps. Artif. Intell. **152**, 47–104 (2004)
43. Savinov, N., Dosovitskiy, A., Koltun, V.: Semi-parametric topological memory for navigation. In: International Conference on Learning Representations (2018). https://openreview.net/forum?id=SygwwGbRW
44. Schlichtkrull, M., Kipf, T.N., Bloem, P., van den Berg, R., Titov, I., Welling, M.: Modeling relational data with graph convolutional networks. In: Gangemi, A., et al. (eds.) ESWC 2018. LNCS, vol. 10843, pp. 593–607. Springer, Cham (2018). https://doi.org/10.1007/978-3-319-93417-4_38
45. Schulman, J., Wolski, F., Dhariwal, P., Radford, A., Klimov, O.: Proximal policy optimization algorithms. arxiv pre-print 1707.06347 (2017)
46. Schulman, J., Wolski, F., Dhariwal, P., Radford, A., Klimov, O.: Proximal policy optimization algorithms. arXiv preprint arXiv:1707.06347 (2017)
47. Shani, G., Pineau, J., Kaplow, R.: A survey of point-based POMDP solvers. Auton. Agents Multi-Agent Syst. **27**(1), 1–51 (2013). https://doi.org/10.1007/s10458-012-9200-210.1007/s10458-012-9200-2
48. Shatkay, H., Kaelbling, L.P.: Learning topological maps with weak local odometric information. In: IJCAI, vol. 2, pp. 920–929 (1997)
49. Silver, D., et al.: A general reinforcement learning algorithm that masters chess, shogi, and go through self-play. Science **362**(6419), 1140–1144 (2018)
50. Smallwood, R.D., Sondik, E.J.: The optimal control of partially observable Markov processes over a finite horizon. Oper. Res. **21**(5), 1071–1088 (1973)
51. Smith, T., Simmons, R.: Heuristic search value iteration for POMDPs. In: Proceedings of the 20th Conference on Uncertainty in Artificial Intelligence, pp. 520–527 (2004)
52. Srinivas, A., Jabri, A., Abbeel, P., Levine, S., Finn, C.: Universal planning networks (2018)
53. Tamar, A., Wu, Y., Thomas, G., Levine, S., Abbeel, P.: Value iteration networks (2016)
54. Thrun, S.: Learning metric-topological maps for indoor mobile robot navigation. Artif. Intell. **99**(1), 21–71 (1998)
55. Wang, R.F., Spelke, E.S.: Human spatial representation: insights from animals. Trends Cogn. Sci. **6**(9), 376–382 (2002). https://doi.org/10.1016/s1364-6613(02)01961-7
56. Wayne, G., et al.: Unsupervised predictive memory in a goal-directed agent. arxiv preprint 1803.10760 (2018)
57. Wu, Z., Pan, S., Chen, F., Long, G., Zhang, C., Yu, P.S.: A comprehensive survey on graph neural networks. arXiv preprint arXiv:1901.00596 (2019)
58. Xia, F., R. Zamir, A., He, Z.Y., Sax, A., Malik, J., Savarese, S.: Gibson Env: real-world perception for embodied agents. In: IEEE Conference on Computer Vision and Pattern Recognition (CVPR), IEEE (2018)

59. Xu, K., Li, J., Zhang, M., Du, S., Kawarabayashi, K., Jegelka, S.: What can neural networks reason about? arxiv preprint 1905.13211 (2019)
60. Xu, K., Li, J., Zhang, M., Du, S.S., Kawarabayashi, K.i., Jegelka, S.: What can neural networks reason about? arXiv preprint arXiv:1905.13211 (2019)
61. Zhang, J., Tai, L., Boedecker, J., Burgard, W., Liu, M.: Neural SLAM. arxiv preprint 1706.09520 (2017)
62. Zhou, J., et al.: Graph neural networks: a review of methods and applications. arXiv preprint arXiv:1812.08434 (2018)
63. Zhu, Y., et al.: Target-driven visual navigation in indoor scenes using deep reinforcement learning. In: IEEE International Conference on Robotics and Automation (ICRA), pp. 3357–3364. IEEE (2017)

Neural Design Network: Graphic Layout Generation with Constraints

Hsin-Ying Lee[2], Lu Jiang[1], Irfan Essa[1,4], Phuong B. Le[1], Haifeng Gong[1], Ming-Hsuan Yang[1,2,3], and Weilong Yang[1](✉)

[1] Google Research, Mountain View, USA
[2] University of California, Merced, Merced, USA
[3] Yonsei University, Seoul, South Korea
[4] Georgia Institute of Technology, Atlanta, USA
weilongyang@google.com

Abstract. Graphic design is essential for visual communication with layouts being fundamental to composing attractive designs. Layout generation differs from pixel-level image synthesis and is unique in terms of the requirement of mutual relations among the desired components. We propose a method for design layout generation that can satisfy user-specified constraints. The proposed neural design network (NDN) consists of three modules. The first module predicts a graph with complete relations from a graph with user-specified relations. The second module generates a layout from the predicted graph. Finally, the third module fine-tunes the predicted layout. Quantitative and qualitative experiments demonstrate that the generated layouts are visually similar to real design layouts. We also construct real designs based on predicted layouts for a better understanding of the visual quality. Finally, we demonstrate a practical application on layout recommendation.

1 Introduction

Graphic design surrounds us on a daily basis, from image advertisements, movie posters, and book covers to more functional presentation slides, websites, and mobile applications. Graphic design is a process of using text, images, and symbols to visually convey messages. Even for experienced graphic designers, the design process is iterative and time-consuming with many false starts and dead ends. This is further exacerbated by the proliferation of platforms and users with significantly different visual requirements and desires.

In graphic design, layout – the placement and sizing of components (e.g., title, image, logo, banner, etc.) – plays a significant role in dictating the flow of the viewer's attention and, therefore, the order by which the information is received.

H.-Y. Lee—Work done during their internship at Google Research.

Electronic supplementary material The online version of this chapter (https://doi.org/10.1007/978-3-030-58580-8_29) contains supplementary material, which is available to authorized users.

A. Vedaldi et al. (Eds.): ECCV 2020, LNCS 12348, pp. 491–506, 2020.
https://doi.org/10.1007/978-3-030-58580-8_29

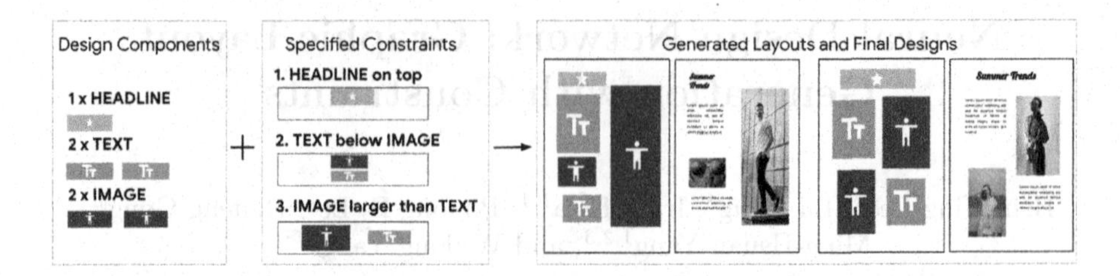

Fig. 1. Graphic layout generation with user constraints. We present realistic use cases of the proposed model. Given the desired components and partial user-specified constraints among them, our model can generate layouts following these constraints. We also present example designs constructed based on the generated layouts.

Creating an effective layout requires understanding and balancing the complex and interdependent relationships amongst all of the visible components. Variations in the layout change the hierarchy and narrative of the message.

In this work, we focus on the layout generation problem that places components based on the component attributes, relationships among components, and user-specified constraints. Figure 1 illustrates examples where users specify a collection of assets and constraints, then the model would generate a design layout that satisfies all input constraints, while remaining visually appealing. Generative models have seen a success in rendering realistic natural images [7,17,27]. However, learning-based graphic layout generation remains less explored. Existing studies tackle layout generation based on templates [3,12] or heuristic rules [25], and more recently using learning-based generation methods [16,22,33]. However, these approaches are limited in handling relationships among components. High-level concepts such as mutual relationships of components in a layout are less likely to be captured well with conventional generative models in pixel space. Moreover, the use of generative models to account for user preferences and constraints is non-trivial. Therefore, effective feature representations and learning approaches for graphic layout generation remain challenging.

In this work, we introduce neural design network (NDN), a new approach of synthesizing a graphic design layout given a set of components with user-specified attributes and constraints. We employ directional graphs as our feature representation for components and constraints since the attributes of components (node) and relations among components (edge) can be naturally encoded in a graph. NDN takes as inputs a graph constructed by desired components as well as user-specified constraints, and then outputs a layout where bounding boxes of all components are predicted. NDN consists of three modules. First, the *relation prediction* module takes as input a graph with partial edges, representing components and user-specified constraints, and infers a graph with complete relationships among components. Second, in the *layout generation* module, the model predicts bounding boxes for components in the complete graph in an iterative manner. Finally, in the *refinement* module, the model further fine-tunes the bounding boxes to improve the alignment and visual quality.

We evaluate the proposed method qualitatively and quantitatively on three datasets under various metrics to analyze the visual quality. The three experimental datasets are RICO [4,24], Magazine [33], and an image banner advertisement dataset collected in this work. These datasets reasonably cover several typical applications of layout design with common components such as images, texts, buttons, toolbars and relations such as above, larger, around, etc. We construct real designs based on the generated layouts to assess the quality. We also demonstrate the efficacy of the proposed model by introducing a practical layout recommendation application.

To summarize, we make the following contributions in this work:

- We propose a new approach that can generate high-quality design layouts for a set of desired components and user-specified constraints.
- We validate that our method performs favorably against existing models in terms of realism, alignment, and visual quality on three datasets.
- We demonstrate real use cases that construct designs from generated layouts and a layout recommendation application. Furthermore, we collect a real-world advertisement layout dataset to broaden the variety of existing layout benchmarks.

2 Related Work

Natural Scene Layout Generation. Layout is often used as the intermediate representation in the image generation task conditioned on text [9,11,31] or scene graph [15]. Instead of directly learning the mapping from the source domain (e.g., text and scene graph) to the image domain, these methods model the operation as a two-stage framework. They first predict layouts conditioned on the input sources, and then generate images based on the predicted layouts. Recently, Jyothi *et al.* propose the LayoutVAE [16], which is a generative framework that can synthesize scene layout given a set of labels. However, a graphic design layout has several fundamental differences to a natural scene layout. The demands for relationship and alignment among components are strict in graphic design. A few pixels offsets of components can either cause a difference in visual experience or even ruin the whole design. The graphic design layout does not only need to look realistic but also needs to consider the aesthetic perspective.

Graphic Design Layout Generation. Early work on design layout or document layout mostly relies on templates [3,12], exemplars [21], or heuristic design rules [25,30]. These methods rely on predefined templates and heuristic rules, for which professional knowledge is required. Therefore, they are limited in capturing complex design distributions. Other work leverages saliency maps [1] and attention mechanisms [26] to capture the visual importance of graphic designs and to trace the user's attention. Recently, generative models are applied to graphic design layout generation [22,33]. The LayoutGAN model [22] can generate layouts consisting of graphic elements like rectangles and triangles. However, the LayoutGAN model generates layout from input noises and fails to handle

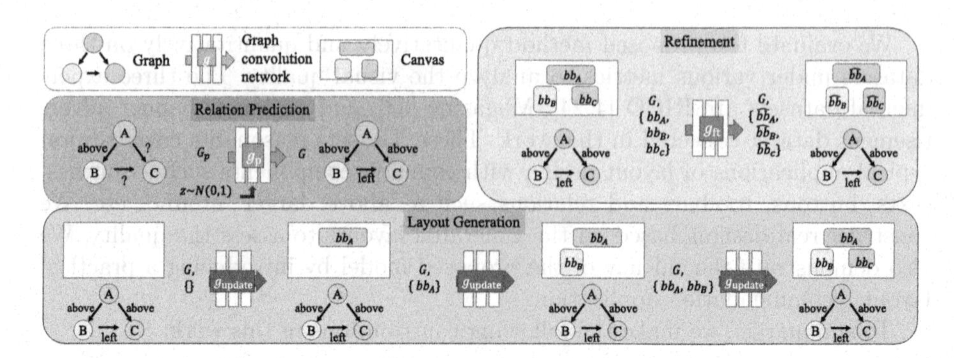

Fig. 2. Framework Illustration. Neural design network consists of three modules: relation prediction, bounding box prediction, and refinement. We illustrate the process with a three-component example. In **relation prediction** module, the model takes as inputs a graph with partial relations along with a latent vector (encoded from the graph with complete relations during training, sampled from prior during testing), and outputs a graph with complete relations. Only the graph with location relations is shown in the figure for brevity. In **layout generation** module, the model takes a graph with complete relations as inputs, and predicts the bounding boxes of components in an iterative manner. In **refinement** module, the model further fine-tune the layout.

layout generation given a set of components with specified attributes, which is the common setting in graphic design. The Layout Generative Network [33] is a content-aware layout generation framework that can render layouts conditioned on attributes of components. While the goals are similar, the conventional GAN-based framework cannot explicitly model relationships among components and user-specified constraints.

Graph Neural Networks in Vision. Graph Neural Networks (GNNs) [6,8,29] aim to model dependence among nodes in a graph via message passing. GNNs are useful for data that can be formulated in a graph data structure. Recently, GNNs and related models have been applied to classification [20], scene graph [2,15,23,32,34], motion modeling [13], and molecular property prediction [5,14], to name a few. In this work, we model a design layout as a graph and apply GNNs to capture the dependency among components.

3 Graphic Layout Generation

Our goal is to generate design layouts given a set of design components with user-specified constraints. For example, in image ads creation, the designers can input the constraints such as "logo at bottom-middle of canvas", "call-to-action button of size (100px, 500px)", "call-to-action-button is below logo", etc. The goal is to synthesize a set of design layouts that satisfy both the user-specified constraints as well as common rules in image ads layouts. Unlike layout templates, these layouts are dynamically created and can serve as inspirations for designers.

We introduce the neural design network using graph neural network and conditional variational auto-encoder (VAE) [19,28] with the goal of capturing better representations of design layouts. Figure 2 illustrates the process of generating a three-component design with the proposed neural design network. In the rest of this section, we first describe the problem overview in Sect. 3.1. Then we detail three modules in NDN: the relation prediction (Sect. 3.2) and layout generation modules (Sect. 3.3), and refinement module (Sect. 3.4).

3.1 Problem Overview

The inputs to our network are a set of design components and user-specified constraints. We model the inputs as a graph, where each design component is a node and their relationships are edges. In this paper, we study two common relationships between design components: *location* and *size*.

Define $G = \{G_{\text{loc}}, G_{\text{size}}\} = (O, E_{\text{loc}}, E_{\text{size}})$, where $O = \{o_0, o_1, ..., o_n\}$ is a set of n components with each $o_i \in \mathcal{C}$ coming from a set of categories $\mathcal{C}$. We use o_0 to denote the *canvas* that is fixed in both location and size, and o_i to denote other design components that need to be placed on the *canvas*, such as *logo*, *button*. $E_{\text{loc}} = \{l_1 ..., l_{m_l}\}$ and $E_{\text{size}} = \{s_1 ..., s_{m_s}\}$ are sets of directed edges with $l_k = (o_i, r_l, o_j)$ and $s_k = (o_i, r_s, o_j)$, where $r_l \in \mathcal{R}_{\text{loc}}$ and $r_s \in \mathcal{R}_{\text{size}}$. Here, $\mathcal{R}_{\text{size}}$ specifies the relative size of the component, such as *smaller* or *bigger*, and r_l can be *left*, *right*, *above*, *below*, *upper-left*, *lower-left*, etc. In addition, if anchoring on the *canvas* o_0, we extend the $\mathcal{R}_{\text{loc}}$ to capture the location that is relative to the *canvas*, *e.g.*, upper-left of the canvas.

Furthermore, in reality, designers often do not specify all the constraints. This results in an input graph with missing edges. Figure 2 shows an example of a three-component design with only one specified constraint "(A, above, B)" and several unknown relations "?". To this end, we augment $\mathcal{R}_{\text{loc}}$ and $\mathcal{R}_{\text{size}}$ to include an additional *unknown* category, and represent graphs that contain unknown size or location relations as $G^{\text{p}}_{\text{size}}$ and $G^{\text{p}}_{\text{loc}}$, respectively, to indicate they are the partial graphs. In Sect. 3.2, we describe how to predict the unknown relations in the partial graphs.

Finally, we denote the output layout of the neural design network as a set of bounding boxes $\{bb_1, ..., bb_{|O|}\}$, where $bb_i = \{x_i, y_i, w_i, h_i\}$ represents the location and shape.

In all modules, we apply the graph convolutional networks on graphs. The graph convolutional networks take as the input the features of nodes and edges, and outputs updated features. The input features can be one-hot vectors representing the categories or any embedded representations.

More implementation details can be found in the supplementary material.

3.2 Relation Prediction

In this module, we aim to infer the unknown relations in the user-specified constraints. Figure 2 shows an example where a three-component graph is given and we need to predict the missing relations between A, B, and C. For brevity, we denote the graphs with complete relations as G, and the graphs with partial relations as G^{p}, which can be either $G^{\mathrm{p}}_{\mathrm{size}}$ or $G^{\mathrm{p}}_{\mathrm{loc}}$. Note that since the augmented relations include the *unknown* category, both G^{p} and G are complete graphs in practice. We also use e_i to refer to either l_i or s_i depending on the context.

We model the prediction process as a paired graph-to-graph translation task: from G^{p} to G. Since the translation is multimodal, i.e., a graph with partial relations can be translated to many possible graphs with complete relations. Therefore, we adopt a similar framework to the multimodal image-to-image translation [35] and treat G^{p} as the source domain and G as the target domain. Similar to [35], the translation is a conditional generation process that maps the source graph, along with a latent code, to the target graph. The latent code is encoded from the corresponding target graph G to achieve reconstruction during training, and is sampled from a prior during testing. The conditional translation encoding process is modeled as:

$$\begin{aligned} z &= g_c(G) & z &\in \mathcal{Z}, \\ \{h_i\} &= g_p(G^{\mathrm{p}}, z) & i &= 1, ..., |\tilde{E}|, \\ \{\hat{e}_i\} &= h_{\mathrm{pred}}(\{h_i\}) & i &= 1, ..., |E|, \end{aligned} \tag{1}$$

where g_c and g_p are graph convolutional networks, and h_{pred} is a relation predictor. In addition, $\tilde{E}$ is the set of edges in the target graph. Note that $|\tilde{E}| = |E|$ since the graph is a complete graph.

The model is trained with the reconstruction loss $L_{\mathrm{cls}} = \mathrm{CE}(\{\hat{e}_i\}, \{e_i\})$ on the relation categories, where the CE indicates cross-entropy function, and a KL loss on the encoded latent vectors to facilitate sampling at inference time: $L_{\mathrm{KL}_1} = \mathbb{E}[D_{\mathrm{KL}}((z)\|\mathcal{N}(0,1))]$, where $D_{\mathrm{KL}}(p\|q) = -\int p(z) \log \frac{p(z)}{q(z)} \mathrm{d}z$. The objective of the relation prediction module is:

$$L_{\mathrm{rel}} = \lambda_{\mathrm{cls}} L_{\mathrm{cls}} + \lambda_{\mathrm{KL}_1} L_{\mathrm{KL}_1}. \tag{2}$$

The reconstruction loss captures the knowledge that the predicted relations should agree with the existing relations in G^{p}, and fill in any missing edge with the most likely relation discovered in the training data.

3.3 Layout Generation

Given a graph with complete relations, this module aims to generate the design layout by predicting the bounding boxes for all nodes in the graph.

Let G be the graph with complete relations constructed from G_{size} and G_{loc}, the output of the relation prediction module. We model the generation process using a graph-based iterative conditional VAE model. We first obtain the features of each component by

$$\{f_i\}_{i=1\sim|O|} = g_{\mathrm{enc}}(G), \tag{3}$$

where g_{enc} is a graph convolutional network. These features capture the relative relations among all components. We then predict bounding boxes in an iterative manner starting from an empty canvas (i.e., all bounding boxes are unknown). As shown in Fig. 2, the prediction of each bounding box is conditioned on the initial features as well as the current canvas, i.e., predicted bounding boxes from previous iterations. At iteration k, the condition can be modeled as:

$$\begin{aligned} t_k &= (\{f_i\}_{i=1\sim|O|}, \{bb_i\}_{i=1\sim k-1}), \\ c_k &= g_{\text{update}}(t_k), \end{aligned} \tag{4}$$

where g_{update} is another graph convolutional network. t_k is a tuple of features and current canvas at iteration k, and c_k is a vector. Then we apply conditional VAE on the current bounding box bb_k conditioned on c_k.

$$\begin{aligned} z &= h_{bb}^{\text{enc}}(bb_k, c_k), \\ \hat{bb}_k &= h_{bb}^{\text{dec}}(z, c_k), \end{aligned} \tag{5}$$

where h_{bb}^{enc} and h_{bb}^{dec} represent encoders and decoders consisting of fully connected layers. We train the model with conventional VAE loss: a reconstruction loss $L_{\text{recon}} = \sum_{i=1}^{|O|} \|\hat{bb}_i - bb_i\|_1$ and a KL loss $L_{\text{KL}_2} = \mathbb{E}[D_{\text{KL}}(p(z|c_k, bb_k)\|p(z|c_k))]$. The objective of the layout generation module is:

$$L_{\text{layout}} = \lambda_{\text{recon}} L_{\text{recon}} + \lambda_{\text{KL}_2} L_{\text{KL}_2}. \tag{6}$$

The model is trained with teacher forcing where the ground truth bounding box at step k will be used as the input for step $k+1$. At test time, the model will use the actual output boxes from previous steps. In addition, the latent vector z will be sampled from a conditional prior distribution $p(z|c_k)$, where p is a prior encoder.

Bounding Boxes with Predefined Shapes. In many design use cases, it is often required to constrain some design components to fixed size. For example, the logo size needs to be fixed in the ad design. To achieve this goal, we augment the original layout generation module with an additional VAE encoder $\bar{h}_{bb}^{enc}$ to ensure the encoded latent vectors z can be decoded to bounding boxes with desired widths and heights. Similar to 5, given a ground-truth bounding box $bb_k = (x_k, y_k, w_k, h_k)$, we obtain the reconstructed bounding box $\hat{bb}_k = (\hat{x}_k, \hat{y}_k, \hat{w}_k, \hat{h}_k)$ with $\bar{h}_{bb}^{enc}$ and h_{bb}^{dec}. Then, instead of applying reconstruction loss on whole bounding boxes tuples, we only enforce the reconstruction of width and height with

$$L_{\text{recon}}^{\text{size}} = \sum_{i=1}^{|O|} \|\hat{w}_i - w_i\|_1 + \|\hat{h}_i - h_i\|_1. \tag{7}$$

The objective of the augmented layout generation module is given by:

$$L'_{\text{layout}} = \lambda_{\text{recon}}^{\text{size}} L_{\text{recon}}^{\text{size}} + L_{\text{layout}}. \tag{8}$$

3.4 Layout Refinement

We predict bounding boxes in an iterative manner that requires to fix the predicted bounding boxes from the previous iteration. As a result, the overall bounding boxes might not be optimal, as shown in the layout generation module in Fig. 2. To tackle this issue, we fine-tune the bounding boxes for better alignment and visual quality in the final layout refinement module. Given a graph G with ground-truth bounding boxes $\{bb_i\}$, we simulate the misalignment by randomly apply offsets $\delta \sim U(-0.05, 0.05)$ on $\{bb_i\}$, where U is the uniform distribution. We obtain misaligned bounding boxes $\{\bar{bb}_i\} = \{bb_i + \delta_i\}$. We apply a graph convolutional network g_{ft} for finetuning:

$$\{\hat{bb}_i\} = g_{\mathrm{ft}}(G, \{\bar{bb}_i\}). \tag{9}$$

The model is trained with reconstruction loss $L_{\mathrm{ft}} = \sum_i \|\{\hat{bb}_i\} - \{bb_i\}\|_1$.

4 Experiments and Analysis

Datasets. We perform the evaluation on three datasets:

- **Magazine** [33]. The dataset contains $4k$ images of magazine pages and 6 categories (texts, images, headlines, over-image texts, over-image headlines, backgrounds).
- **RICO** [4,24]. The original dataset contains $91k$ images of the Android apps interface and 27 categories. We choose 13 most frequent categories (toolbars, images, texts, icons, buttons, inputs, list items, advertisements, pager indicators, web views, background images, drawers, modals) and filter the number of components within an image to be less than 10, totaling $21k$ images.
- **Image banner ads**. We collect 500 image banner ads of the size 300×250 via image search using keywords such as "car ads". We annotate bounding boxes of 6 categories: images, regions of interest, logos, brand names, texts, and buttons.

Evaluated Methods. We evaluate and compare the following algorithms:

- **sg2im** [15]. The model is proposed to generate a natural scene layout from a given scene graph. The sg2im method takes as inputs graphs with complete relations in the setting where all constraints are provided. When we compare with this method in the setting where no constraint is given, we simplify the input scene graph by removing all relations. We refer the simplified model as **sg2im-none**.
- **LayoutVAE** [16]. This model takes a label set as input, and predicts the number of components for each label as well as the locations of each component. We compare with the second stage of the LayoutVAE model (i.e., the bounding box prediction stage) by giving the number of components for each label. In addition, we refer to **LayoutVAE-loo** as the model that predicts the bounding box of a single component when all other components are provided and fixed (the leave-one-out setting).

- **Neural Design Network**. We refer to **NDN-none** when the input contains no prior constraint, **NDN-all** in the same setting as **sg2im** when all constraints are provided, and **NDN-loo** in the same setting as **LayoutVAE-loo**.

We do not compare our method with LayoutGAN [22] since LayoutGAN generates outputs in an unconditional manner (i.e., generation from sampled noise vectors). Even in the no-constraint setting, it is difficult to conduct fair comparisons as multiple times of resampling are required to generate the same combinations of components.

4.1 Implementation Details

In this work, h_{bb}^{enc}, h_{bb}^{dec}, and h_{pred} consists of 3 fully-connected layers. In addition, g_c, g_p, g_{enc}, and g_{update} consist of 3 graph convolution layers. The dimension of latent vectors z in the relation prediction and layout generation module is 32. The input features of nodes and edges are obtained from a dictionary mapping, which is trained along with the model. For training, we use the Adam optimizer [18] with batch size of 512, learning rate of 0.0001, and $(\beta_1, \beta_2) = (0.5, 0.999)$. In all experiments, we set the hyper-parameters as follows: $\lambda_{\text{cls}} = 1$, $\lambda_{\text{KL1}} = 0.005$, $\lambda_{\text{recon}} = \lambda_{\text{KL2}} = 1$, and $\lambda_{\text{recon}} = 10$. We use a predefined order of component sets in all experiments.

For the relation prediction module, the graphs with partial constraint are generated from the ground-truth graph with $0.2 - 0.9$ dropout rate. For the layout generation module, the input graphs with complete relations are constructed from the ground-truth layouts. The location and size relations are obtained by ground-truth bounding boxes. The corresponding outputs are the bounding boxes from the ground-truth layouts.

Since the location relations are discretized and mutually exclusive, there might be some ambiguity. For example, a component is both "above" and "right" of another component when it is in the upper-right direction to the other. To handle the ambiguity, we predefine the order when conflicts occur. Specifically, "above" and "below" have higher priority than "left of" and "right of".

More implementation details can be found in the supplementary material.

4.2 Quantitative Evaluation

Realism and Accuracy. We evaluate the visual quality following Fréchet Inception Distance (FID) [10] by measuring how close the distribution of generated layout is to the real ones. We train a binary layout classifier to discriminate between good and bad layouts. The bad layouts are generated by randomly moving component locations of good layouts. The classifier consists of four graph convolution layers and three fully connected layers. The binary classifier achieves classification accuracy of 94%, 90, and 95% on the Ads, Magazine, and RICO datasets, respectively. We extract the features of the second from the last fully connected layer to measure FID.

Table 1. Quantitative comparisons. We compare the proposed method to other works on three datasets using three settings: no-constraint setting that no prior constraint is provided (first row), all-constraint setting that all relations are provided (second row), and leave-one-out setting that aims to predict the bounding box of a component with ground-truth bounding boxes of other components provided. The FID metric measures the realism and diversity, the alignment metric measures the alignment among components, and the prediction error metric measures the prediction accuracy in the leave-one-out setting.

Datasets	Ads		Magazine		RICO	
	FID ↓	Align. ↓	FID ↓	Align. ↓	FID ↓	Align. ↓
sg2im-none	**116.63**	**0.63**	95.81	**0.97**	269.60	**0.14**
LayoutVAE	138.11 ± 38.91	1.21 ± 0.08	81.56 ± 36.78	$.314 \pm 0.11$	192.11 ± 29.97	1.19 ± 0.39
NDN-none	129.68 ± 32.12	0.91 ± 0.07	$\mathbf{69.43 \pm 32.92}$	2.51 ± 0.09	$\mathbf{143.51 \pm 22.36}$	0.91 ± 0.03
sg2im	230.44	0.0069	102.35	0.0178	190.68	0.007
NDN-all	$\mathbf{168.44 \pm 21.83}$	$\mathbf{0.61 \pm 0.05}$	$\mathbf{82.77 \pm 16.24}$	$\mathbf{1.51 \pm 0.09}$	$\mathbf{64.78 \pm 11.60}$	$\mathbf{0.32 \pm 0.02}$
	Pred. error ↓	Align. ↓	Pred. error ↓	Align. ↓	Pred. error ↓	Align. ↓
LayoutVAE-loo	0.071 ± 0.002	0.48 ± 0.01	0.059 ± 0.002	1.41 ± 0.02	0.045 ± 0.0021	0.39 ± 0.02
NDN-loo	$\mathbf{0.043 \pm 0.001}$	$\mathbf{0.36 \pm 0.01}$	$\mathbf{0.024 \pm 0.0002}$	$\mathbf{1.30 \pm 0.01}$	$\mathbf{0.018 \pm 0.002}$	$\mathbf{0.14 \pm 0.01}$
Real data	-	0.0034	-	0.0126	-	0.0012

We measure FID in two settings. First, a model predicts bounding boxes without any constraints. That is, only the number and the category of components are provided. We compare with LayoutVAE and sg2im-none in this setting. Second, a model predicts bounding boxes with all constraints provided. We compare with sg2im in this setting since LayoutVAE cannot take constraints as inputs. The first two rows in Table 1 present the results of these two settings. Since LayoutVAE and the proposed method are both stochastic models, we generate 100 samples for each testing design in each trial. The results are averaged over 5 trials. In both no-constraint and all-constraint settings, the proposed method performs favorably against the other schemes.

We also measure the prediction accuracy in the leave-one-out setting, i.e., predicting the bounding box of a component when bounding boxes of other components are provided. We measure the accuracy by the $L1$ error between the predicted and the ground-truth bounding boxes. The third row of Table 1 shows the comparison to the LayoutVAE-loo method in this setting. The proposed method gains better accuracy with statistical significance (≥95%), indicating that the graph-based framework encodes better relations among components.

Alignment. Alignment is an important principle in design creation. In most good designs, components need to be either in center alignment or edge alignment (e.g., left- or right-aligned). Therefore, in addition to realism, we explicitly measure the alignment among components using:

$$\frac{1}{N_D} \sum_d \sum_i \min_{j, i \neq j} \{\min(l(c_i^d, c_j^d), m(c_i^d, c_j^d), r(c_i^d, c_j^d)\}, \tag{10}$$

Table 2. Ablation on partial constraints and the refinement module. We measure the FID and alignment of the proposed method taking different percentages of prior constraints as inputs using the RICO dataset. We also show that the refinement module can further improve the visual quality as well as the alignment.

Unary size (%)	Binary size (%)	Unary location (%)	Binary location (%)	Refinement	FID ↓	Align. ↓
0	0	0	0	✓	143.51 ± 22.36	0.91 ± 0.03
20	20	0	0	✓	141.64 ± 20.01	0.87 ± 0.03
0	0	20	20	✓	129.92 ± 23.76	0.81 ± 0.03
20	20	20	20		126.18 ± 23.11	0.74 ± 0.02
20	20	20	20	✓	125.41 ± 21.68	0.70 ± 0.02
100	100	100	100		70.55 ± 12.68	0.36 ± 0.02
100	100	100	100	✓	64.78 ± 11.60	0.32 ± 0.02

Table 3. Components order. We compare the performance of our model using different strategies of deciding orders of components. We evaluate the FID score on the RICO dataset.

Order	Size	Occurrence	Random
FID	132.84	136.22	143.51
Pred. error	1.08 ± 0.04	1.02 ± 0.04	0.91 ± 0.03

Table 4. Constraint consistency. We measure the consistency of the relations among generated components and the user-specified constraints.

Dataset	Ads	Magazine	RICO
Constraint consistency (%)	96.8	95.9	98.2

where N_D is the number of generated layouts, c_k^d is the k_{th} component of the d_{th} layout. In addition, l, m, and r are alignment functions where the distances between the left, center, and right of components are measured, respectively.

Table 1 presents the results in the no-constraint, all-constraint, and leave-one-out settings. The results are also averaged over 5 trials. The proposed method performs favorably against other methods. The sg2im-none method gets better alignment score since it tends to predict bounding boxes in several fixed locations when no prior constraint is provided, which leads to worse FID. For similar reasons, the sg2im method gains a slightly higher alignment score on RICO.

Partial Constraints. Previous experiments are conducted under the settings of either no constraints or all constraints provided. Now, we demonstrate the efficacy of the proposed method on handling partial constraints. Table 2 shows the results of layout prediction with different percentages of prior constraints provided. We evaluate the partial constraints setting using the RICO dataset, which is the most difficult dataset in terms of diversity and complexity. Ideally, the FID and alignment scores should be similar regardless of the percentage of constraints given. However, in the challenging RICO dataset, the prior information of size and location still greatly improves the visual quality, as shown in Table 2, The location constraints contribute to more improvement since they explicitly provide guidance from the ground-truth layouts. As for the alignment score, layouts in all settings perform similarly. Furthermore, the refinement module can slightly improve the alignment score as well as FID.

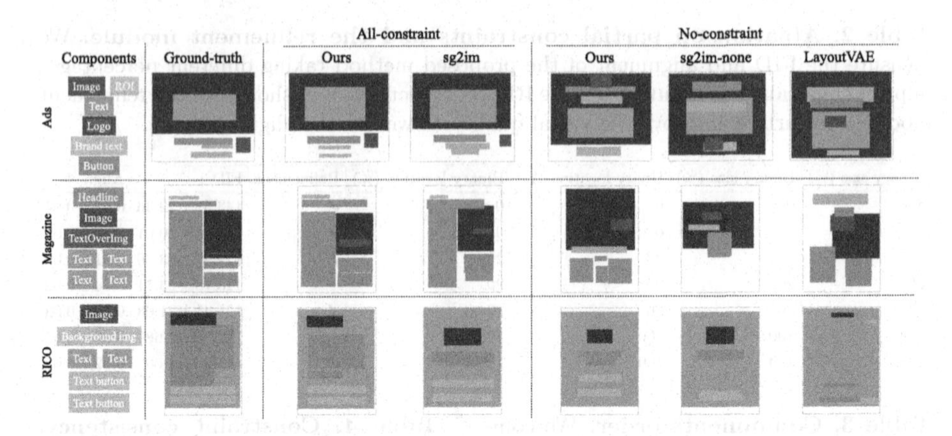

Fig. 3. Qualitative comparison. We evaluate the proposed method with the LayoutVAE and Sg2im methods in both no-constraint and all-constraint setting. The proposed method can better model the relations among components and generate layouts of better visual quality.

User Constraint Consistency. The major goal of the proposed model is to generate layouts according to user-specified constraints. Therefore, we explicitly measure the consistency between the relations among generated components and the original user-specified constraints. Table 4 shows that the generated layouts reasonably conform to the input constraints.

Order of Components. Since the proposed model predicts layouts in an iterative manner, the order of the components plays an important role. We evaluate our method using three different strategies of defining orders: ordered by size, ordered by occurrences, and random order. We show the comparisons in Table 3. We have a similar finding as in LayoutVAE that the order of components affects the generation results. However, we use the random order in all our experiments since our goal is not only to generate layouts, but also enable flexible user control. In user cases such as leave-one-out prediction and layout recommendation, using random order can better align the training and testing scenarios.

4.3 Qualitative Evaluation

We compare the proposed method with related work in Fig. 3. In the all-constraint setting, both the sg2im method and the proposed model can generate reasonable layouts similar to the ground-truth layouts. However, the proposed model can better tackle alignment and overlapping issues. In the no-constraint setting, the sg2im-none method tends to place components of the same categories at the same location, like the "text"s in the second row and the "text"s and "text button"s in the third row. The LayoutVAE method, on the other hand, cannot handle relations among components well without using graphs. The proposed method can generate layouts with good visual quality, even with no constraint provided.

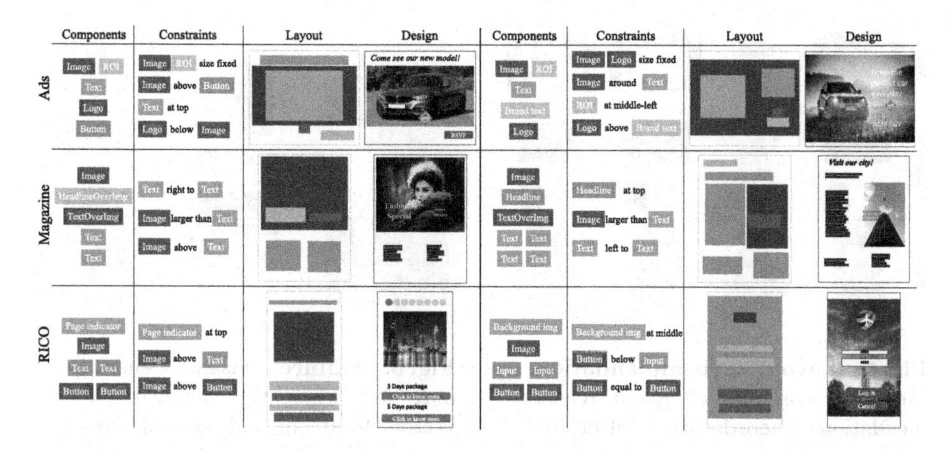

Fig. 4. Layout generation with partial user-specified constraints. We generate layouts according to different user-specified constraints on location and size. Furthermore, we construct designs with real assets based on the generated layouts to better visualize the quality of our model.

Partial Constraints. In Fig. 4, we present the results of layout generation given several randomly selected constraints on size and location. Our model generates design layouts that are both realistic and follows user-specified constraints. To better visualize the quality of the generated layouts, we present designs with real assets generated from the predicted layouts. Furthermore, we can constrain the size of specific components to desired shapes (e.g., we fix the *image* and *logo* size in the first row of Fig. 4.) using the augmented layout generation module.

Layout Recommendation. The proposed model can also help designers decide the best locations of a specific design component (e.g., *logo*, *button*, or *headline*) when a partial design layout is provided. This can be done by building graphs with partial location and size relations based on the current canvas and set the relations to target components as *unknown*. We then complete this graph using the relation prediction module. Finally, conditioned on the predicted graph as well as current canvas, we perform iterative bounding boxes prediction with the layout generation module. Figure 5 shows examples of layout recommendations.

Failure Cases. Several reasons may lead to undesirable generation. First, due to the limited amount of training data, the sampled latent vectors used for generation might locate in undersampled spaces that are not fully exploited during training. Second, the characteristic of the set of components is too different from the training data. For example, the lower-left image in Fig. 6 demonstrates a generation requiring three buttons and two logos, which are less likely to exist in real designs.

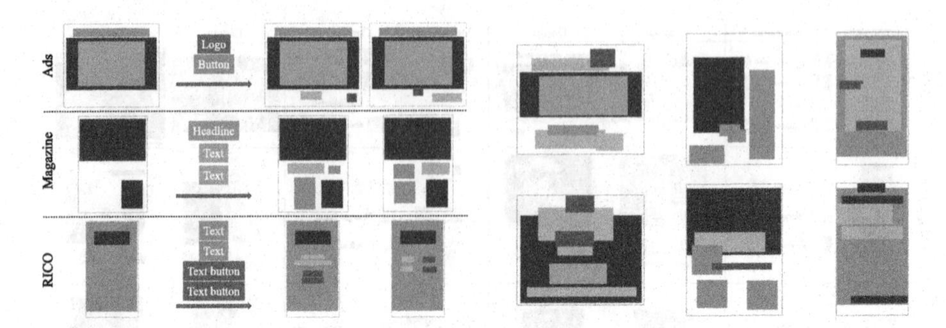

Fig. 5. Layout recommendation. We show examples of layout recommendations where locations of desired components are recommended given the current layouts.

Fig. 6. Failure cases. Generation may fail when the sampled latent vectors locate in under-sample spaces or the characteristics of inputs differ greatly from that in the training data.

5 Conclusion and Future Work

In this work, we propose a neural design network to handle design layout generation given user-specified constraints. The proposed method can generate layouts that are visually appealing and follow the constraints with a three-module framework, including a relation prediction module, a layout generation module, and a refinement module. Extensive quantitative and qualitative experiments demonstrate the efficacy of the proposed model. We also present examples of constructing real designs based on generated layouts, and an application of layout recommendation.

Visual design creation is an impactful but understudied topic in our community. It is extremely challenging. Our work is among one of the first works tackling graphic design in a well-defined setting that is reasonably close to the real use case. However, graphic design is a complicated process involving content attributes such as color, font, semantic labels, etc. Future directions may include content-aware graphic design or fine-grained layout generation beyond the bounding box.

Acknowledgements. This work is supported in part by the NSF CAREER Grant #1149783.

References

1. Bylinskii, Z., et al.: Learning visual importance for graphic designs and data visualizations. In: UIST (2017)
2. Cheng, Y.C., Lee, H.Y., Sun, M., Yang, M.H.: Controllable image synthesis via SegVAE. In: ECCV (2020)
3. Damera-Venkata, N., Bento, J., O'Brien-Strain, E.: Probabilistic document model for automated document composition. In: DocEng (2011)

4. Deka, B., et al.: Rico: a mobile app dataset for building data-driven design applications. In: UIST (2017)
5. Duvenaud, D.K., et al.: Convolutional networks on graphs for learning molecular fingerprints. In: NeurIPS (2015)
6. Goller, C., Kuchler, A.: Learning task-dependent distributed representations by backpropagation through structure. In: ICNN (1996)
7. Goodfellow, I., et al.: Generative adversarial nets. In: NeurIPS (2014)
8. Gori, M., Monfardini, G., Scarselli, F.: A new model for learning in graph domains. In: IJCNN (2005)
9. Gupta, T., Schwenk, D., Farhadi, A., Hoiem, D., Kembhavi, A.: Imagine this! scripts to compositions to videos. In: ECCV (2018)
10. Heusel, M., Ramsauer, H., Unterthiner, T., Nessler, B., Hochreiter, S.: GANs trained by a two time-scale update rule converge to a local Nash equilibrium. In: NeurIPS (2017)
11. Hong, S., Yang, D., Choi, J., Lee, H.: Inferring semantic layout for hierarchical text-to-image synthesis. In: CVPR (2018)
12. Hurst, N., Li, W., Marriott, K.: Review of automatic document formatting. In: DocEng (2009)
13. Jain, A., Zamir, A.R., Savarese, S., Saxena, A.: Structural-ENN: Deep learning on spatio-temporal graphs. In: CVPR (2016)
14. Jin, W., Yang, K., Barzilay, R., Jaakkola, T.: Learning multimodal graph-to-graph translation for molecular optimization. In: ICLR (2019)
15. Johnson, J., Gupta, A., Fei-Fei, L.: Image generation from scene graphs. In: CVPR (2018)
16. Jyothi, A.A., Durand, T., He, J., Sigal, L., Mori, G.: LayoutVAE: stochastic scene layout generation from a label set. In: ICCV (2019)
17. Karras, T., Laine, S., Aila, T.: A style-based generator architecture for generative adversarial networks. In: CVPR (2019)
18. Kingma, D., Ba, J.: Adam: a method for stochastic optimization. In: ICLR (2015)
19. Kingma, D.P., Welling, M.: Auto-encoding variational bayes. In: ICLR (2014)
20. Kipf, T.N., Welling, M.: Semi-supervised classification with graph convolutional networks. In: ICLR (2017)
21. Kumar, R., Talton, J.O., Ahmad, S., Klemmer, S.R.: Bricolage: example-based retargeting for web design. In: SIGCHI (2011)
22. Li, J., Yang, J., Hertzmann, A., Zhang, J., Xu, T.: LayoutGAN: generating graphic layouts with wireframe discriminators. In: ICLR (2019)
23. Li, Y., Jiang, L., Yang, M.H.: Controllable and progressive image extrapolation. arXiv preprint arXiv:1912.11711 (2019)
24. Liu, T.F., Craft, M., Situ, J., Yumer, E., Mech, R., Kumar, R.: Learning design semantics for mobile apps. In: UIST (2018)
25. O'Donovan, P., Agarwala, A., Hertzmann, A.: Learning layouts for single-pagegraphic designs. TVCG **20**, 1200–1213 (2014)
26. Pang, X., Cao, Y., Lau, R.W., Chan, A.B.: Directing user attention via visual flow on web designs. ACM TOG **35**, 1–11 (2016). (Proc. SIGGRAPH)
27. Razavi, A., van den Oord, A., Vinyals, O.: Generating diverse high-fidelity images with VQ-VAE-2. arXiv preprint arXiv:1906.00446 (2019)
28. Rezende, D.J., Mohamed, S., Wierstra, D.: Stochastic backpropagation and approximate inference in deep generative models. In: ICML (2014)
29. Scarselli, F., Gori, M., Tsoi, A.C., Hagenbuchner, M., Monfardini, G.: The graph neural network model. TNN **20**, 61–80 (2008)

30. Tabata, S., Yoshihara, H., Maeda, H., Yokoyama, K.: Automatic layout generation for graphical design magazines. In: SIGGRAPH (2019)
31. Tan, F., Feng, S., Ordonez, V.: Text2scene: Generating abstract scenes from textual descriptions. In: CVPR (2019)
32. Tseng, H.Y., Lee, H.Y., Jiang, L., Yang, W., Yang, M.H.: RetrieveGAN: image synthesis via differentiable patch retrieval. In: ECCV (2020)
33. Zheng, X., Qiao, X., Cao, Y., Lau, R.W.: Content-aware generative modeling of graphic design layouts. In: SIGGRAPH (2019)
34. Yang, J., Lu, J., Lee, S., Batra, D., Parikh, D.: Graph R-CNN for scene graph generation. In: ECCV (2018)
35. Zhu, J.Y., et al.: Toward multimodal image-to-image translation. In: NeurIPS (2017)

Learning Open Set Network with Discriminative Reciprocal Points

Guangyao Chen[1], Limeng Qiao[1], Yemin Shi[1], Peixi Peng[1(✉)], Jia Li[2,3], Tiejun Huang[1,3], Shiliang Pu[4], and Yonghong Tian[1,3(✉)]

[1] Department of Computer Science and Technology, Peking University, Beijing, China
{gy.chen,qiaolm,pxpeng,yhtian}@pku.edu.cn

[2] State Key Laboratory of Virtual Reality Technology and Systems, SCSE, Beihang University, Beijing, China

[3] Peng Cheng Laboratory, Shenzhen, China

[4] Hikvision Research Institute, Hangzhou, China

Abstract. Open set recognition is an emerging research area that aims to simultaneously classify samples from predefined classes and identify the rest as 'unknown'. In this process, one of the key challenges is to reduce the risk of generalizing the inherent characteristics of numerous unknown samples learned from a small amount of known data. In this paper, we propose a new concept, *Reciprocal Point*, which is the potential representation of the extra-class space corresponding to each known category. The sample can be classified to known or unknown by the otherness with reciprocal points. To tackle the open set problem, we offer a novel open space risk regularization term. Based on the bounded space constructed by reciprocal points, the risk of unknown is reduced through multi-category interaction. The novel learning framework called **R**eciprocal **P**oint **L**earning (RPL), which can indirectly introduce the unknown information into the learner with only known classes, so as to learn more compact and discriminative representations. Moreover, we further construct a new large-scale challenging aircraft dataset for open set recognition: **Air**craft 300 (Air-300). Extensive experiments on multiple benchmark datasets indicate that our framework is significantly superior to other existing approaches and achieves state-of-the-art performance on standard open set benchmarks.

1 Introduction

Recent error rate for classification task by Deep Convolutional Neural Network has surpassed the human-level performance [9]. Such a significant improvement promotes its attempts in real-world applications. However, a robust recognition system should not only be able to identify all test instances of the seen or known

Electronic supplementary material The online version of this chapter (https://doi.org/10.1007/978-3-030-58580-8_30) contains supplementary material, which is available to authorized users.

A. Vedaldi et al. (Eds.): ECCV 2020, LNCS 12348, pp. 507–522, 2020.
https://doi.org/10.1007/978-3-030-58580-8_30

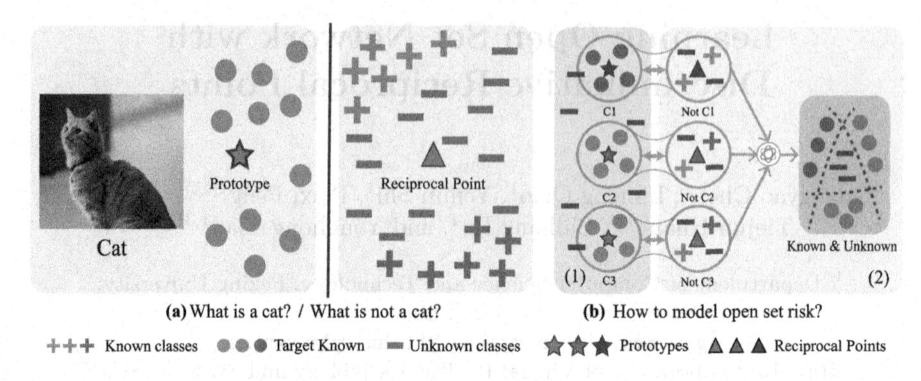

Fig. 1. (a): How to identify a cat in open set problem? Most methods focus on learning the representative potential feature of cats as prototypes. In contrast, *Reciprocal Points*, representative potential feature of *non-cat*, identify the cat by otherness. (b): (1) Focusing only on known information, it is hard to reduce open space risk without unknown information. (2) The *extra-class* space represented by *Reciprocal Points* can bring more connections with unknown.

classes, but also be able to handle unknown samples or novel events that have not been seen. Recent works on open set recognition [23] have formalized the process for performing recognition in the setting that labels the input as one of the known classes or as unknown. Apart from returning the most likely category, the robust recognition system must also support the rejection of unseen/unknown. An open set scenario has classes in testing phase that are not seen during training, which brings great difficulties to solve open set recognition.

In the majority of deep neural networks, the last layer for multiclass recognition is the SoftMax function which produces a probability distribution over the known classes. However, the SoftMax layer brings a significant difficulty to open set recognition because of its closed nature. Bendale *et al.* [2] improves SoftMax to OpenMax to estimate the probability of unknown classes but does not optimize for unknown classes during the training phase. Potential solutions to the open set recognition should optimize for unknown classes, as well as the known classes [23]. More knowledge of unknown information can facilitate reducing of open space risk. Many methods [8,19,21,26] utilize generative models for generating samples as unknown. However, Nalisnick *et al.* [18] finds the instability of the mainstream generation models to the open set problem. How to model open space risk with only known training data is still an urgent problem to be solved. For example, as shown in Fig. 1(a), how to identify a cat in open set problem. Most classification methods pay attention to *"what is a cat?"*. They look for a more representative feature of the cat. Yang *et al.* [25] proposes convolutional prototype learning (CPL) to solve the multiclass recognition problem in the open world. For neural networks, each known class k exists in its own way. Although focusing only on known samples can reduce the intra-class distance, it is still inevitable that it will bring open space risks because the dark information in the unknown space cannot be considered at all. In addition, for each known category k, most unknown samples obviously belong to the space of non-k and their features should be more similar to the representation of non-k correspondingly for neural networks, which means that the corresponding unknown information is more implicit in each non-k embedding space. Therefore, we further focus on *"what is not a cat?"* to learn representations in the form of one more latent vector of non-cat data. The new representation is called *Reciprocal Point*, which can bring more connections with unknown by exploring each extra-class space.

Based on the exploration of the extra-class space, we propose a novel framework, termed as **R**eciprocal **P**oint **L**earning (RPL), which can effectively reduce the open space risk by bounded space while learning reciprocal points of each known class, and the whole framework of the proposed method is shown in Fig. 2. Specifically, we extract the feature of each sample with a deep Convolutional Neural Network model. Since the reciprocal point is the learnable representation of the extra-class space, we classify the input by the otherness between the embedding feature and the reciprocal point. Then we extend the model to capture the risk of the unknown based on reciprocal points by a novel open space risk regularization term. The extra-class embedding space of each known class is limited to a bounded range through the reciprocal point and the learnable margin. When multiple classes interact with each other in the training stage, all known classes are not only pushed to the periphery of the space by the corresponding reciprocal point but also pulled in a certain bounded range by other reciprocal points. Finally, the embedding space of the network is limited to a bounded range. All known classes are distributed around the periphery of the embedding space, and the unknown samples are limited to the internal bounded space. The bounded domain by RPL can prevent the neural network from generating arbitrarily high confidence for unknown [10]. Although only known samples are available during the training stage, the interval between known and unknown classes is separated by reciprocal points indirectly. The advantage of RPL is that it can shrink unknown space while considering the known space classification and form a good bouned feature space distribution in which the magnitudes of unknown samples are lower than those of known samples.

Moreover, in order to facilitate the research of open set recognition in the real visual world, we further construct a new large-scale aircraft dataset from the Internet: **Air**craft 300 (Air-300), which contains 320,000 annotated colored images from 300 different classes in total. Note that each category contains 100 images at least, and a maximum of 10,000 images, which leads to the long tail distribution of our proposed dataset. Compared with the existing benchmark datasets, the tailored Air-300 dataset maintains a long tail distribution to simulate the real visual world, and can also be utilized to perform fine-grained classification and object recognition.

To summarize our contributions: (1) we propose a new concept, *Reciprocal Point*, which is the potential representation of the extra-class space corresponding to each known category. (2) Meanwhile, we offer a novel framework for learning open set network. Through introducing unknown information by reciprocal

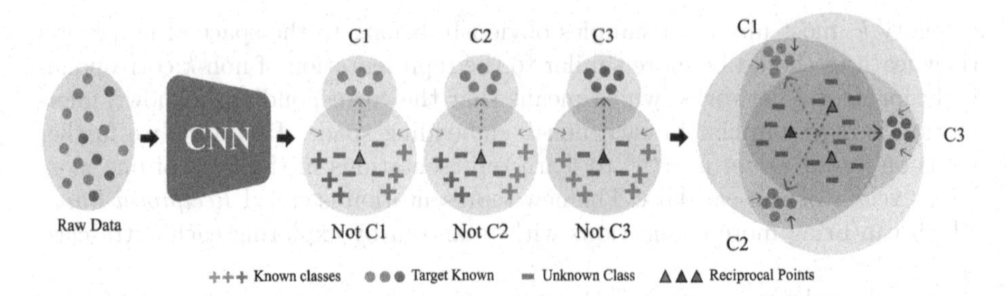

Fig. 2. An overview of the proposed Reciprocal Point Learning (RPL) approach to open set recognition. The extra-class embedding space of each known class is limited to a bounded range through the reciprocal point and the learnable margin. With interaction of multiple classes, the known classes are pushed to the periphery of the feature space, and the unknown classes are limited in the bounded space.

points, the neural network can learn more compact and robust feature space, and separate known space and unknown space effectively. (3) Then, we further introduce a new challenging large-scale dataset with long-tailed distribution, Air-300, which is a new dataset to tailor for open set recognition. (4) The results on multiple benchmark datasets show that our approach significantly outperforms other existing state-of-the-art deep open set classifiers.

2 Related Work

Open Set Recognition. Inspired by a classifier with rejection option [1,3,27], Scheirer *et al.* [23] define the open set recognition problem for the first time and propose a base framework to perform training and evaluation. The deep neural network has achieved great success in many areas and is introduced to open set recognition by Bendale *et al.* [2]. They prove threshold on SoftMax probability does not yield a robust model for open set recognition. Openmax [2] detects unknown classes by modeling distance of activation vectors, which are from the mean penultimate vector of each class with EVT. Ge *et al.* [8] propose G-Openmax, a direct extension of Openmax, which trains deep neural networks with unknown samples generated by generative models. However, this method does not extend to tasks that generate modeling difficulties. Yoshihashi *et al.* [26] addresses deep reconstruction-based representation learning in open set recognition by adding image reconstructed as a regularizer. However, this method introduces auxiliary decoder network for training, which brings additional overhead to training.

Out-of-Distribution Detection. Based on the concern for safety of AI systems, the detection of out-of-distribution (OOD) examples, one of the concerns regarding the distribution shift, is first introduced by Hendrycks *et al.* in [11]. OOD is the detection of samples that do not belong to the training set but can appear during testing [11]. Several works [11,12,15,16] seek to address these problems by giving deep neural network classifiers a means of assigning anomaly scores to inputs, which are used for detecting OOD. Hendrycks *et al.* [11] demonstrated that anomalous samples had a lower maximum softmax probability than in-distribution samples based on a deep, pre-trained classifier. Liang *et al.* [16] propose ODIN to allow more effective detection by using temperature scaling and adding small perturbations to the input. Similar with G-Openmax, De Vries *et al.* [12] utilize generative models for generating most effective samples from out-of-distribution and deriving a new OOD score form this branch. Hendrycks *et al.* [12] propose Outlier Exposure, which uses an auxiliary dataset in order to teach the network better representations for anomaly detection. Detecting out-of-distribution (OOD) is related to the rejection of unknown classes in open set recognition (OSR). They all are studying separating in-distribution (known) and out-of distribution (unkown) [11,23].

Prototype Learning. Prototypes are learnable representations in the form of one or more latent vector per class. Wen *et al.* [24] proposed a center loss to learn centers for deep features of each identity and used the centers to reduce intra-class variance. It can enhance the discrimination power of features. Yang *et al.* in [25] propose Convolutional Prototype Learning (CPL) for robust classification. Moreover, a Generalized CPL (GCPL) with prototype loss was proposed as a regularization to improve the intra-class compactness of the feature representation. However, center loss and GCPL only explicitly encourage intra-class compactness. Conversely, the reciprocal point limits the embedding space outside each known class to bounded space. Known and unknown classes are separated by maximizing otherness between known and corresponding reciprocal points.

3 Reciprocal Point Learning

3.1 Preliminaries

We first establish preliminaries related to open set learning, following which we formally formulate the proposed idea. Suppose we are given a set of n labeled samples $\mathcal{D_L} = \{(x_1, y_1), \ldots, (x_n, y_n)\}$ where $y_i \in \{1, \ldots, N\}$ is the label of x_i and a larger amount of test data $\mathcal{D_T} = \{t_1, \ldots, t_u\}$ where the label of t_i belong to $\{1, \ldots, N\} \cup \{N+1, \ldots, N+U\}$ and U is the number of unknown classes in actual scenarios. We then denote the embedding space of category k as $\mathcal{S}_k$ and its corresponding **open space** as $\mathcal{O}_k = \mathbb{R}^d - \mathcal{S}_k$, where $\mathbb{R}^d$ is the d-dimensional full space. In order to formalize and then manage open space risk finely, we further address the positive open space from other known classes as $\mathcal{O}_k^{pos}$ and the remaining infinite unknown space as negative open space $\mathcal{O}_k^{neg}$, namely $\mathcal{O}_k = \mathcal{O}_k^{pos} \cup \mathcal{O}_k^{neg}$.

The Open Set Recognition Problem. For simplicity, we will first introduce the open set recognition of a single class and then extend the entire learning process to a multiclass form. Given the labeled data $\mathcal{D_L}$ with N known classes, let samples from category k be positive training data $\mathcal{D}_{\mathcal{L}}^{k}$ (space $\mathcal{S}_k$), samples from other known classes be negative training data $\mathcal{D}_{\mathcal{L}}^{\neq k}$ (space $\mathcal{O}_k^{pos}$), and samples from $\mathbb{R}^d$ but except $\mathcal{D}_{\mathcal{L}}$ be potential unknown data $\mathcal{D}_{\mathcal{U}}$ (space $\mathcal{O}_k^{neg}$). Let $\psi^k : \mathbb{R}^d \mapsto \{0,1\}$ be a binary measurable prediction function, mapping embedding x to the label y. For 1-class open set recognition problem, the overall goal is to optimize a discriminant binary function ψ_k by minimizing the expected error $\mathcal{R}^k$ as:

$$\arg\min_{\psi_k}\{\mathcal{R}^k \mid \mathcal{R}_{\epsilon}(\psi_k, \mathcal{S}_k \cup \mathcal{O}_k^{pos}) + \alpha \cdot \mathcal{R}_o(\psi_k, \mathcal{O}_k^{neg})\} \tag{1}$$

where α is a positive regularization parameter, $\mathcal{R}_{\epsilon}$ is the empirical classification risk on known data and $\mathcal{R}_o$ is the **open space risk** function to measure the uncertainty of labeling the unknown samples as the class of interest or as unknown, with further formulating as a non-zero integral function on space $\mathcal{O}_k^{neg}$.

Afterwards we identify open set recognition problem by integrating multiple binary classification tasks (*one vs. rest*) into a multiclass recognition problem (see Fig. 2), by summarizing the expected risk defined in the Eq.(1) category by category, i.e. $\sum_{k=1}^{N} \mathcal{R}^k$, which leads to the following formulation as:

$$\arg\min_{f \in \mathcal{H}} \{\mathcal{R}_{\epsilon}(f, \mathcal{D}_L) + \alpha \cdot \sum_{k=1}^{N} \mathcal{R}_o(f, \mathcal{D}_U)\} \tag{2}$$

where $f : \mathbb{R}^d \mapsto \mathbb{N}$ is a measurable multiclass recognition function (see more details of derivation in supplementary material). According to the Eq. (2), solving the open set recognition problem is equivalent to minimize the combination of the empirical classification risk on labeled known data and the open space risk on potential unknown data simultaneously, over the space of allowable recognition functions, which leads to be more distinguishable between known and unknown spaces. Further, we propose reciprocal points for both closed set classification and reducing the open space risk on $\mathcal{D}_U$ during training.

3.2 Reciprocal Points for Classification

The **reciprocal points** for category k are denoted as $\mathcal{P}^k = \{p_i^k | i = 1, \ldots, M\}$, where M is the number of reciprocal points for each class and $\mathcal{P}^k$ can be regarded as the set of latent representations for the sub-dataset $\mathcal{D}_{\mathcal{L}}^{\neq k} \cup \mathcal{D}_{\mathcal{U}}$ (space $\mathcal{O}_k$). Unlike the learning driven by prototype or center loss, we agree the fact that most samples from the classes in $\mathcal{O}_k$ are more similar to the classificatory reciprocal points $\mathcal{P}^k$ than samples from $\mathcal{S}_k$, which can be formulated as:

$$\forall d \in \zeta(\mathcal{P}^k,\ \mathcal{D}_L^k),\ \max(\zeta(\mathcal{P}^k,\ \mathcal{D}_L^{\neq k} \cup \mathcal{D}_U))\ \leq\ d \tag{3}$$

where $\zeta(\cdot,\ \cdot)$ calculates a set of distances of all samples between two sets and $\max(\cdot)$ is a maximal function.

In fact, the most intuitive goal of open set learning is usually to separate known and unknown spaces as much as possible. To achieve this, for any category k, we propose to separate the two mutually exclusive space by maximizing the distance between the reciprocal points of the category and its corresponding known samples. Specifically, we optimize a set of d-dimensional representations $\mathcal{P}^k$, or reciprocal points, of each class through an embedding function f_θ with learnable parameters θ. Given the sample x and classificatory reciprocal points $\mathcal{P}_k$, we calculate the distance between them with the formulation as:

$$d(f_\theta(x), \mathcal{P}^k) = \frac{1}{M}\sum\nolimits_{i=1}^{M} ||f_\theta(x) - p_i^k||_2^2 \tag{4}$$

Afterwards, based on the distance we propose above, our framework estimates the otherness between the embedding feature $f_\theta(x)$ and the reciprocal points $\mathcal{P}^k$ of all known classes to determine which category it belongs to. Actually, following the nature of reciprocal points, the probability of sample x belongs to category k is proportional to the otherness between $f_\theta(x)$ and the reciprocal points of category k, i.e. $p(y = k|x) \propto d(f_\theta(x), \mathcal{P}^k)$, which means the greater the distance between $f_\theta(x)$ and $\mathcal{P}^k$, the greater the probability that x will be assigned with label k. According to the sum-to-one property, we embody the final probability as:

$$p(y = k|x, f_\theta, \mathcal{P}) = \frac{e^{\gamma d(f_\theta(x),\ \mathcal{P}^k)}}{\sum_{i=1}^{N} e^{\gamma d(f_\theta(x),\ \mathcal{P}^i)}} \tag{5}$$

where γ is a hyper-parameter that controls the hardness of probability assignment. Learning proceeds by minimizing the reciprocal points classification loss based negative log-probability of the true class k via SGD as:

$$\mathcal{L}_c(x; \theta, \mathcal{P}) = -\ log\ p(y = k|x, f_\theta, \mathcal{P}) \tag{6}$$

which can be further seen as integrating multiple binary classification tasks to solve the multiclass open set recognition problem. Through minimizing Eq. (6), on the one hand, maximizing the otherness between known data $\mathcal{D}_L$ and reciprocal points set $\mathcal{P}$ has a facilitating effect on maximizing the interval between closed space $\mathcal{S}_k$ and open space $\mathcal{O}_k$, which is consistent with our initial goal. On the other hand, corresponding to $\mathcal{R}_\epsilon(f, \mathcal{D}_L)$ in the Eq. 2, the reciprocal points classification loss reduces the empirical classification risk through the reciprocal points.

3.3 Reducing Open Space Risk

In solving the open set recognition problem, in addition to utilizing the reciprocal points loss to manage the empirical risk, we also further reduce the open space risk $\mathcal{R}_o(f, \mathcal{D}_U)$ in Eq. 2 with introducing a regularization term. According to the notion of open set in Sec. 3.1, each particular category k has positive open space $\mathcal{O}_k^{pos}$ and infinite negative open space $\mathcal{O}_k^{neg}$ respectively. For multiclass open-set recognition scenarios, we further union multiple class-wise open spaces into a global open space $\mathcal{O}_G$ as:

$$\mathcal{O}_G = \bigcap\nolimits_{k=1}^{N} (\mathcal{O}_k^{pos} \cup \mathcal{O}_k^{neg}) \tag{7}$$

Algorithm 1. The reciprocal point learning algorithm.

Input: Training data $\{x_i\}$. Initialized parameters θ in convolutional layers. Parameters $\mathcal{P}$ and R in loss layers, respectively. Hyperparameter λ, γ and learning rate μ. The number of iteration $t \leftarrow 0$.
Output: The parameters θ, $\mathcal{P}$ and R.
1: while *not converge* do
2: $t \leftarrow t + 1$.
3: Compute the joint loss by $\mathcal{L}^t = \mathcal{L}_c^t + \lambda \cdot \mathcal{L}_o^t$.
4: Compute the backpropagation error $\frac{\partial \mathcal{L}^t}{\partial x^t}$ for each i by $\frac{\partial \mathcal{L}^t}{\partial x^t} = \frac{\partial \mathcal{L}_c^t}{\partial x^t} + \lambda \cdot \frac{\partial \mathcal{L}_o^t}{\partial x^t}$.
5: Update the parameters $\mathcal{P}$ by $\mathcal{P}^{t+1} = \mathcal{P}^t - \mu^t \cdot \frac{\partial \mathcal{L}^t}{\partial \mathcal{P}^t} = \mathcal{P}^t - \mu^t \cdot (\frac{\partial \mathcal{L}_c^t}{\partial \mathcal{P}^t} + \lambda \cdot \frac{\partial \mathcal{L}_o^t}{\partial \mathcal{P}^t})$.
6: Update the parameters R by $R^{t+1} = R^t - \mu^t \cdot \frac{\partial \mathcal{L}^t}{\partial R^t} = R^t - \lambda \cdot \mu^t \cdot \frac{\partial \mathcal{L}_o^t}{\partial R^t}$.
7: Update the parameters θ by $\theta^{t+1} = \theta^t - \mu^t \cdot \frac{\partial \mathcal{L}^t}{\partial \theta^t} = \theta^t - \mu^t \cdot (\frac{\partial \mathcal{L}_c^t}{\partial \theta^t} + \lambda \cdot \frac{\partial \mathcal{L}_o^t}{\partial \theta^t})$.
8: end while

where we would restrict the total open space risk for all known classes. Moreover, based on the principle of maximum entropy, for an unknown sample x_u without any prior information, a well-trained closed-set discriminant function tends to assign known labels to x_u with equal probability, which leads to the deep neural networks usually embed unknown samples into the inside of known spaces rather than random positions in the full space. This phenomenon is also consistent with the observation of [6] and our visualization results shown in Fig. 4.

Obviously, the idea of which wants to manage open space risk by restricting the open space to a bounded space directly is almost impossible because the open space contains a large number of potentially unknown samples. However, considering space $\mathcal{S}_k$ and $\mathcal{O}_k$ are complementary with each other, we indirectly bound the open space risk by constraining the distance between samples from $\mathcal{S}_k$ and reciprocal points $\mathcal{P}^k$ to a certain extent as:

$$\mathcal{L}_o(x; \theta, \mathcal{P}^k, R^k) = \frac{1}{M} \cdot \sum_{i=1}^{M} ||d(f_\theta(x), p_i^k) - R^k||_2^2 \tag{8}$$

where R^k is the learnable margin in our framework. Specifically, minimizing the Eq. (8) is equivalent to making R^k and $\zeta(\mathcal{D}_L^k,\ \mathcal{P}^k)$ in Eq. (3) as close as possible, so that the following formula is established as:

$$\max(\zeta(\mathcal{D}_L^{\neq k} \cup \mathcal{D}_U,\ \mathcal{P}^k)) \leq\ R^k \tag{9}$$

Considering bounded space $\mathcal{B}(\mathcal{P}^k, R^k)$ with the reciprocal points $\mathcal{P}^k$ as centers and R^k as its corresponding intervals, in order to separate known and unknown space, we further utilize these bounded spaces to approximate the global unknown space $\mathcal{O}_G$ as much as possible. As a result, calculating the regularization loss with Eq. (8) can be viewed as managing open space risk $\mathcal{R}_o(f, \mathcal{D}_U)$ in Eq. (6).

3.4 Learning Open Set Network

Finally, the overall loss function of reciprocal points learning combines the empirical classification risk and reducing open space risk as:

$$\mathcal{L}(x;\theta,\mathcal{P},R) = \mathcal{L}_c(x;\theta,\mathcal{P}) + \lambda\mathcal{L}_o(x;\theta,\mathcal{P},R) \quad (10)$$

where λ is a hyper-parameter of controlling the weight of reducing open space risk module and $\theta, \mathcal{P}, R$ represent the learnable parameters of our framework. As shown in Fig. 2, learning open set recognition problem with Eq. (10) results in pushing the known spaces to the periphery of global open space and then separating the two spaces as much as possible. In Algorithm 1, we summarize the learning details of the open set network with joint supervision.

How to Detect Unknown Classes? Samples from a class on which the network was not trained would have probability scores distributed across all the known classes [6]. Similar to thresholding softmax scores in [11], we agree that unknown samples have a closer distance with all reciprocal points than samples of known training classes based on a deep classifier. Therefore, the probability that test instance x belongs to known classes is proportional to the distance between the test instance and the farthest reciprocal point corresponding to category k:

$$p(known|x) \propto \max_{k\in\{1,\ldots,N\}} d(f(x),\mathcal{P}^k) \quad (11)$$

The models for open set problem focus on two key questions, ***Q1****) what is a good score for open-set identification?* (i.e., identifying a class as known or unknown), and given a score, ***Q2****) what is a good operating threshold for the model?* [21]. Furthermore, since how rare or common samples from unknown space is not known in the actual scenario, the approaches to open set recognition which require an arbitrary threshold or sensitivity for comparison is unreasonable [19]. Thus, we utilize the difference between known and unknown distribution to measure the learned models' ability to detect unknown, which provides a calibration-free measure of detection performance.

4 Experiments

4.1 Experiments for Open Set Identification

Evaluation Metrics. The evaluation protocol defined in [19] is considered and Area Under the Receiver Operating Characteristic (AUROC) curve is used as evaluation metric. AUROC is a threshold-independent metric [4]. It can be interpreted as the probability that a positive example is assigned a higher detection score than a negative example [7]. Following the protocol in [19], we report the AUROC averaged over five randomized trials.

Network Architecture. Except for RPL-WRN, the encoder and decoder architecture for this experiment is same to the architecture used in [19]. For a fair

Table 1. The AUROC results of open set identification. The second column shows the backbone for each method, including Encoder (E) or Decoder (D). The architecture used to infer the unknown class are highlighted in bold. Values other than RPL are taken from [21]. RPL++ means using GCPL to assist RPL training. Best performing methods are highlighted in bold.

Method	Backbone	MNIST	SVHN	CIFAR10	CIFAR+10	CIFAR+50	TinyImageNet
Softmax	**E**	97.8	88.6	67.7	81.6	80.5	57.7
Openmax [2]	**E**	98.1	89.4	69.5	81.7	79.6	57.6
RPL	**E**	98.9	93.4	82.7	84.2	83.2	68.8
RPL++	**E**	**99.3**	**95.1**	**86.1**	**85.6**	**85.0**	**70.2**
G-OpenMax [8]	**E** + E + D	98.4	89.6	67.5	82.7	81.9	58.0
OSRCI [19]	**E** + E + D	98.8	91.0	69.9	83.8	82.7	58.6
C2AE [21]	**E** + D	98.9	92.2	89.5	95.5	93.7	74.8
RPL-WRN	**E**(WRN)	**99.6**	**96.8**	**90.1**	**97.6**	**96.8**	**80.9**

comparison, we trained RPL with a Wide-ResNet [28] with depth 40, width 4 and dropout 0, WRN-40-4. The parameters of WRN-40-4 are about $9M$, which is less than the networks of C2AE [21] and OSRCI [19]. λ for RPL is set to 0.1 in the training stage. More details in the supplementary material.

Datasets. Similar to [21], we provide a simple summary of these protocols for each dataset.

- **MNIST**, **SVHN**, **CIFAR10**. For MNIST [14], SVHN [20] and CIFAR10 [13], by randomly sampling 6 known classes and 4 unknown classes.
- **CIFAR+10**, **CIFAR+50**. For CIFAR+N experiments, 4 classes are sampled from CIFAR10 for training. N non-overlapping classes are used as unknown, which are sampled from the CIFAR100 dataset [13].
- **TinyImageNet**. For experiments with TinyImageNet [22], 20 known classes and 180 unknown classes are randomly sampled for evaluation.

Result Comparisons. As shown in Table 1, RPL outperforms other methods based on encoder in [19] in all datasets. Noted that RPL with only using known training samples and a single encoder outperforms methods based on encoder and decoder in MNIST and SVHN. With less information and simple training method, RPL achieves better performance than OSRCI [19] and G-OpenMax [8] using generation-based model on all datasets. Furthermore, we need fewer network parameters and a more simple strategy for training to identify unknown classes. RPL-WRN has similar parameters to C2AE model and performs significantly better than other recent state-of-the-art methods in SVHN, CIFAR, and TinyImageNet. It shows that open set network trained by RPL learns better scores for identifying known and unknown classes.

4.2 Experiments for Open Long-Tailed Recognition

Datasets. In real visual world, the frequency distribution of visual classes is long-tailed, with a few common classes and many more rare classes [17].

To simulate natural data distribution, we adopt ImageNet-LT [17] and a new image datasets, Air-300, to evaluate our algorithm for open set identification.

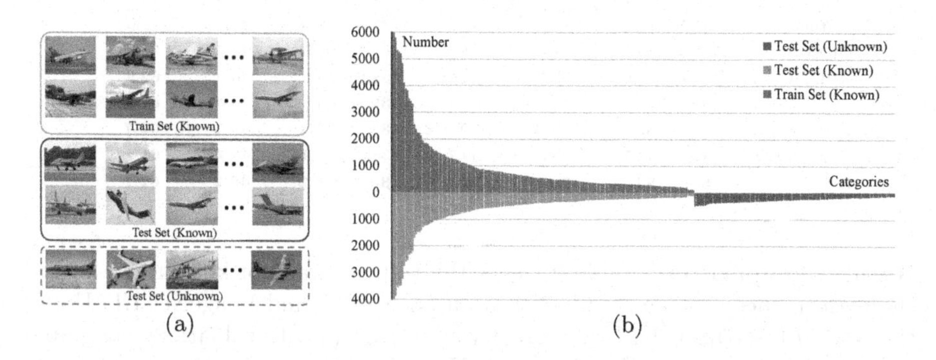

Fig. 3. (a) The overview of Air-300. (b) The number of images with classes.

ImageNet-LT. ImageNet-LT is a long-tailed version of the original ImageNet-2012 [5]. It has 115.8K images from 1000 categories, with maximally 1280 images per class and minimally 5 images per class. The test set contains the open set based on the additional classes of images in ImageNet-2010.

Aircraft 300 Dataset. To facilitate research on open set recognition and evaluate the performance of different approaches, we further construct a new large-scale dataset from the Internet, **Aircraft 300** (Air-300), which contains 320,000 annotated colour images from 300 different classes in total, see thumbnails in Fig. 3a briefly. Each category contains 100 images at least, and a maximum of 10,000 images, which leads to the long tail distribution as shown in Fig. 3b. According to the number of images in each category, we then divide all classes into two parts with 180 known classes for training and 120 novel unknown classes for testing respectively. Furthermore, we aggregate 300 different aircraft models into 10 super classes according to their roles, such as a bomber, fighter, helicopter and so on. As such, each image is annotated with a fine-grain label (the sub-class category) and a coarse-grain label (the superclasss category) by human. Note we also split these 10 superclasses into two parts for coarse-grain open set recognition. Compared with the existing benchmark datasets, the tailored Air-300 dataset is not only much larger and brings new challenges to open set recognition, but also can be utilized to perform fine-grain classification and object recognition respectively.

Evaluation Metrics. The AUROC and AUPR are adopted for evaluation, AUPR-Known and AUPR-Unknown denote the area under the precision-recall curve where known and unknown are specified as positives, respectively.

Network Architecture. All methods are trained based on WRN-40-4 for Air-300 and ResNet50 for ImageNet-LT [17]. We optimize GCPL performance to best by adjusting its parameters. More details in the supplementary material.

Table 2. Test accuracy and open set identification of different methods on Air-300.

Method		Air-300 (WRN-40-4)				ImageNet-LT (ResNet50)			
		Acc	AUROC	AUPR-K	AUPR-U	Acc	AUROC	AUPR-K	AUPR-U
Softmax		85.7	77.1	85.8	48.0	37.8	53.3	45.0	60.5
GCPL		84.5	79.1	88.0	61.3	37.1	54.5	45.7	61.9
RPL	$\lambda = 0$	84.6	64.6	78.5	43.6	38.7	55.0	46.0	62.5
	$\lambda = 0.01$	87.9	82.5	89.3	68.9	**39.0**	55.0	46.0	62.5
	$\lambda = 0.1$	**88.8**	**83.1**	**89.3**	**70.2**	**39.0**	**55.1**	**46.1**	**62.7**
	$\lambda = 1$	86.7	81.7	87.5	69.7	38.2	54.4	45.7	61.6
RPL++		**89.0**	**84.1**	**90.1**	**72.7**	**39.7**	**55.2**	**46.2**	**62.7**

Result Comparisons. Table 2 shows the test accuracy of known classes and the performance of open set identification on Air-300 and ImageNet-LT. Under the $\lambda = 0.1$ settings, RPL has greatly improved in all evaluations compared to softmax and GCPL. For air-300, RPL not only improves the classification accuracy of closed set, but also is about 6% higher than softmax in AUROC. ImageNet-LT has more categories and fewer training samples, which makes it very challenging for open set recognition. It is difficult to detect unknown when the accuracy of closed set is not so well, but RPL still improves the performance of classification and open set identification by nearly 2%. It shows that RPL is effective for the open set problem of long tail distribution.

4.3 Further Analysis

Analysis of Reducing Open Space Risk.

RPL vs. Softmax. Reciprocal points classification loss term (the first part of Eq. 2) defines constraints for reciprocal point. However, similar to softmax, the learned representation is still linear separable only with this classification loss. See Fig. 4a and 4b, reciprocal points without $\mathcal{L}_o$ are learned to the origin, while RPL with $\mathcal{L}_o$ can achieve better distribution (shown in Fig. 4d). For softmax and RPL without $\mathcal{L}_o$, there is a significant overlap between the known and the unknown classes, so as to the neural network views unknown samples as some known classes with high confidence. In contrast, the whole feature space learned by RPL with $\mathcal{L}_o$ is in a limited range (1–6 in abscissa for unknowns and 5–8 in abscissa for knowns in Fig. 4d), which prevents high confidence in unknown classes.

RPL vs. GCPL. As shown in Fig. 4c, GCPL used the prototypes to reduce intra-class variance. However, without considering the unknown, GCPL extends unknown classes to the whole feature space, resulting in a significant overlap with known. By introducing $\mathcal{L}_o$, different known categories are spread to the periphery of the space, while unknown categories are restricted to the interior. It can be seen that a clear gap is maintained between the two types of samples (known vs. unknown) in Fig. 4d. RPL improves the robustness of neural networks by preventing the misjudgment of the unknown class through the bounded restriction, thereby enhancing and stabilizing the classification of known categories.

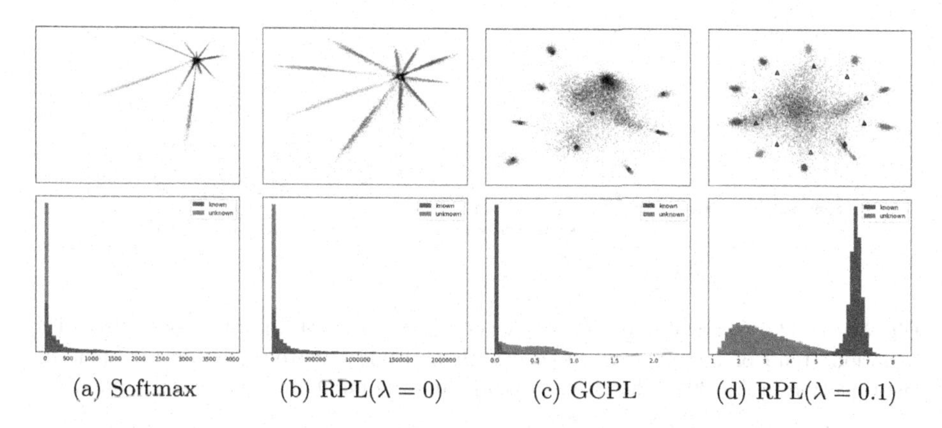

(a) Softmax (b) RPL($\lambda = 0$) (c) GCPL (d) RPL($\lambda = 0.1$)

Fig. 4. (1) The first row is visualization in the learned feature space of MNIST as known and FashionMNIST as unknown. Gray shapes are data of unknown, and circles in color are data of known samples. Different colors represent different classes. Colored stars and triangles represent the prototypes and the reciprocal points learned of different known categories, respectively. (2) The second row is the maximum distance distribution between features and prototypes or reciprocal points.

Margin and λ. See Table 2, the performance of RPL is related to the setting of λ. With the increase of λ, RPL has greatly improved on open set identification. However, when λ is too big, the accuracy of known classes will decrease slightly, which is related to $\mathcal{L}_o$ limiting the size of the whole feature space, while the learning margin in RPL can show the size of the whole feature space. Figure 5a shows the variation trend of margin with λ. With the increase of λ, the learned margin becomes smaller, and the limitation of $\mathcal{L}_o$ on feature space also tends larger. Different training data need different reasonable margins and space sizes to ensure that known and unknown can be classified correctly. Figure 5a proves that the margin increases with the number of known classes with fixed lambda. Reasonable λ can effectively control the interaction among known classes, by learning the more appropriate embedding space size.

Experiments with Multiple Reciprocal Points. Figure 5b shows the impact of increasing the number of reciprocal points on open set identification to detect rare unknown classes. We can observe that the performance of open set identification is relatively stable. The performance is slightly improved when employing 8 reciprocal points. Based on a good feature space distribution, increasing the number of reciprocal points for each known category can slightly improve the ability of the neural networks to distinguish unknown. However, too many reciprocal points will bring more training overhead and affect performance.

Combination of Reciprocal Point and Prototype. The reciprocal point can push known classes to the periphery in the limited feature space, thereby improving the performance of the open set identification. We deem that the reciprocal point and the prototype are opposite and complementary. Here we combine the reciprocal point and the prototype, as *RPL++*. In the training phase of RPL, we

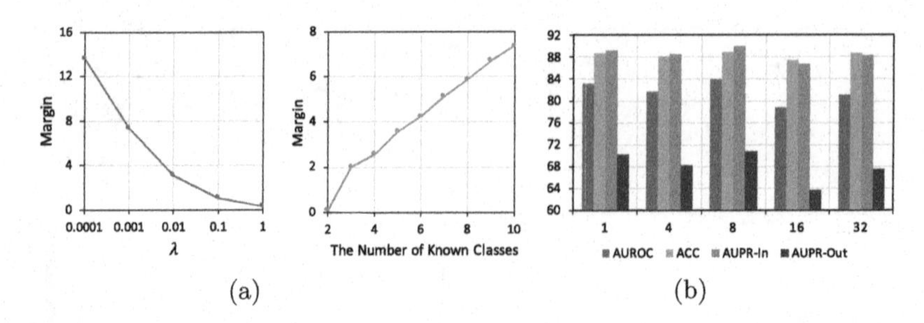

Fig. 5. (a) The trend of margin with λ and the number of known classes. (b) is the variation trend of testing performance with multiple reciprocal points for open set identification in Air-300. The abscissa represents the number of reciprocal points.

add prototypes and further train reciprocal points and prototypes by RPL and prototype loss [25]. As shown in Table 1, RPL++ achieves better performance than RPL. Moreover, the prototype further improves the performance of RPL on open long-tailed recognition in Table 2. The reciprocal points form an excellent embedding space structure, then the prototype further narrows intra-class variance, to further divide the known classes and the unknown classes.

5 Conclusion

This paper proposes a new concept, *Reciprocal Point*, which is the potential representation of the extra-class space corresponding to each known category. We also propose a novel learning framework towards open set learning. The approach introduces unknown information by reciprocal points, to optimize a better feature space to separate known and unknown. Comprehensive experiments conducted on multiple datasets demonstrate that our method outperforms previous state-of-the-art open set classifiers in all cases. We also publish a open long-tailed dataset, the Air-300, which is a challenging dataset to simulate natural data distribution for open set recognition and other visual tasks.

Acknowledgments. This work is partially supported by grants from the National Key R&D Program of China under grant 2017YFB1002400, the Key-Area Research and Development Program of Guangdong Province under Grant 2019B010153002, and the National Natural Science Foundation of China under contract No. 61825101, No. 61702515 and No. U1611461.

References

1. Bartlett, P.L., Wegkamp, M.H.: Classification with a reject option using a hinge loss. J. Mach. Learn. Res. **9**, 1823–1840 (2008)
2. Bendale, A., Boult, T.E.: Towards open set deep networks. In: Proceedings of the IEEE Conference on Computer Vision and Pattern Recognition, pp. 1563–1572 (2016)

3. Da, Q., Yu, Y., Zhou, Z.H.: Learning with augmented class by exploiting unlabeled data. In: Twenty-Eighth AAAI Conference on Artificial Intelligence (2014)
4. Davis, J., Goadrich, M.: The relationship between precision-recall and ROC curves. In: Proceedings of the 23rd International Conference on Machine Learning, pp. 233–240. ACM (2006)
5. Deng, J., Dong, W., Socher, R., Li, L.J., Li, K., Fei-Fei, L.: ImageNet: a large-scale hierarchical image database. In: IEEE Conference on Computer Vision and Pattern Recognition, pp. 248–255. IEEE (2009)
6. Dhamija, A.R., Günther, M., Boult, T.: Reducing network agnostophobia. In: Advances in Neural Information Processing Systems, pp. 9157–9168 (2018)
7. Fawcett, T.: An introduction to ROC analysis. Pattern Recogn. Lett. **27**(8), 861–874 (2006)
8. Ge, Z., Demyanov, S., Chen, Z., Garnavi, R.: Generative OpenMax for multi-class open set classification. arXiv preprint arXiv:1707.07418 (2017)
9. He, K., Zhang, X., Ren, S., Sun, J.: Delving deep into rectifiers: surpassing human-level performance on imagenet classification. In: Proceedings of the IEEE International Conference on Computer Vision, pp. 1026–1034 (2015)
10. Hein, M., Andriushchenko, M., Bitterwolf, J.: Why ReLU networks yield high-confidence predictions far away from the training data and how to mitigate the problem. In: Proceedings of the IEEE Conference on Computer Vision and Pattern Recognition, pp. 41–50 (2019)
11. Hendrycks, D., Gimpel, K.: A baseline for detecting misclassified and out-of-distribution examples in neural networks. arXiv preprint arXiv:1610.02136 (2016)
12. Hendrycks, D., Mazeika, M., Dietterich, T.G.: Deep anomaly detection with outlier exposure. arXiv preprint arXiv:1812.04606 (2018)
13. Krizhevsky, A., Hinton, G.: Learning multiple layers of features from tiny images. Technical report. Citeseer (2009)
14. LeCun, Y., Bottou, L., Bengio, Y., Haffner, P., et al.: Gradient-based learning applied to document recognition. Proc. IEEE **86**(11), 2278–2324 (1998)
15. Lee, K., Lee, H., Lee, K., Shin, J.: Training confidence-calibrated classifiers for detecting out-of-distribution samples. arXiv preprint arXiv:1711.09325 (2017)
16. Liang, S., Li, Y., Srikant, R.: Enhancing the reliability of out-of-distribution image detection in neural networks. arXiv preprint arXiv:1706.02690 (2017)
17. Liu, Z., Miao, Z., Zhan, X., Wang, J., Gong, B., Yu, S.X.: Large-scale long-tailed recognition in an open world. arXiv preprint arXiv:1904.05160 (2019)
18. Nalisnick, E., Matsukawa, A., Teh, Y.W., Gorur, D., Lakshminarayanan, B.: Do deep generative models know what they don't know? arXiv preprint arXiv:1810.09136 (2018)
19. Neal, L., Olson, M., Fern, X., Wong, W.K., Li, F.: Open set learning with counterfactual images. In: Proceedings of the European Conference on Computer Vision (ECCV), pp. 613–628 (2018)
20. Netzer, Y., Wang, T., Coates, A., Bissacco, A., Wu, B., Ng, A.Y.: Reading digits in natural images with unsupervised feature learning (2011)
21. Oza, P., Patel, V.M.: C2AE: class conditioned auto-encoder for open-set recognition. In: Proceedings of the IEEE Conference on Computer Vision and Pattern Recognition, pp. 2307–2316 (2019)
22. Russakovsky, O., et al.: ImageNet large scale visual recognition challenge. Int. J. Comput. Vis. **115**(3), 211–252 (2015)
23. Scheirer, W.J., de Rezende Rocha, A., Sapkota, A., Boult, T.E.: Toward open set recognition. IEEE Trans. Pattern Anal. Mach. Intell. **35**(7), 1757–1772 (2013)

24. Wen, Y., Zhang, K., Li, Z., Qiao, Yu.: A discriminative feature learning approach for deep face recognition. In: Leibe, B., Matas, J., Sebe, N., Welling, M. (eds.) ECCV 2016. LNCS, vol. 9911, pp. 499–515. Springer, Cham (2016). https://doi.org/10.1007/978-3-319-46478-7_31
25. Yang, H.M., Zhang, X.Y., Yin, F., Liu, C.L.: Robust classification with convolutional prototype learning. In: Proceedings of the IEEE Conference on Computer Vision and Pattern Recognition, pp. 3474–3482 (2018)
26. Yoshihashi, R., Shao, W., Kawakami, R., You, S., Iida, M., Naemura, T.: Classification-reconstruction learning for open-set recognition. arXiv preprint arXiv:1812.04246 (2018)
27. Yuan, M., Wegkamp, M.: Classification methods with reject option based on convex risk minimization. J. Mach. Learn. Res. **11**, 111–130 (2010)
28. Zagoruyko, S., Komodakis, N.: Wide residual networks. arXiv preprint arXiv:1605.07146 (2016)

Convolutional Occupancy Networks

Songyou Peng[1,2](✉), Michael Niemeyer[2,3], Lars Mescheder[2,4], Marc Pollefeys[1,5], and Andreas Geiger[2,3]

[1] ETH Zurich, Zurich, Switzerland
songyou.peng@inf.ethz.ch
[2] Max Planck Institute for Intelligent Systems, Tübingen, Germany
[3] University of Tübingen, Tübingen, Germany
[4] Amazon, Tübingen, Germany
[5] Microsoft, Zurich, Switzerland

Abstract. Recently, implicit neural representations have gained popularity for learning-based 3D reconstruction. While demonstrating promising results, most implicit approaches are limited to comparably simple geometry of single objects and do not scale to more complicated or large-scale scenes. The key limiting factor of implicit methods is their simple fully-connected network architecture which does not allow for integrating local information in the observations or incorporating inductive biases such as translational equivariance. In this paper, we propose Convolutional Occupancy Networks, a more flexible implicit representation for detailed reconstruction of objects and 3D scenes. By combining convolutional encoders with implicit occupancy decoders, our model incorporates inductive biases, enabling structured reasoning in 3D space. We investigate the effectiveness of the proposed representation by reconstructing complex geometry from noisy point clouds and low-resolution voxel representations. We empirically find that our method enables the fine-grained implicit 3D reconstruction of single objects, scales to large indoor scenes, and generalizes well from synthetic to real data.

1 Introduction

3D reconstruction is a fundamental problem in computer vision with numerous applications. An ideal representation of 3D geometry should have the following properties: a) encode complex geometries and arbitrary topologies, b) scale to large scenes, c) encapsulate local and global information, and d) be tractable in terms of memory and computation.

Unfortunately, current representations for 3D reconstruction do not satisfy all of these requirements. Volumetric representations [25] are limited in terms of resolution due to their large memory requirements. Point clouds [9] are lightweight

L. Mescheder—This work was done prior to joining Amazon.

Electronic supplementary material The online version of this chapter (https://doi.org/10.1007/978-3-030-58580-8_31) contains supplementary material, which is available to authorized users.

A. Vedaldi et al. (Eds.): ECCV 2020, LNCS 12348, pp. 523–540, 2020.
https://doi.org/10.1007/978-3-030-58580-8_31

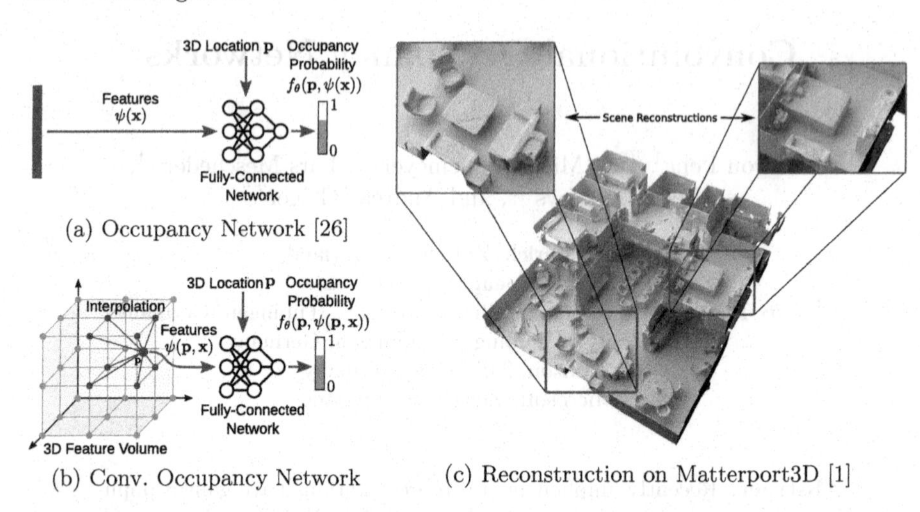

(a) Occupancy Network [26]

(b) Conv. Occupancy Network

(c) Reconstruction on Matterport3D [1]

Fig. 1. Convolutional Occupancy Networks. Traditional implicit models (a) are limited in their expressiveness due to their fully-connected network structure. We propose Convolutional Occupancy Networks (b) which exploit convolutions, resulting in scalable and equivariant implicit representations. We query the convolutional features at 3D locations $\mathbf{p} \in \mathbb{R}^3$ using linear interpolation. In contrast to Occupancy Networks (ONet) [26], the proposed feature representation $\psi(\mathbf{p}, \mathbf{x})$ therefore depends on *both* the input $\mathbf{x}$ and the 3D location $\mathbf{p}$. Figure (c) shows a reconstruction of a two-floor building from a noisy point cloud on the Matterport3D dataset [1].

3D representations but discard topological relations. Mesh-based representations [13] are often hard to predict using neural networks.

Recently, several works [3,26,27,31] have introduced deep implicit representations which represent 3D structures using learned occupancy or signed distance functions. In contrast to explicit representations, implicit methods do not discretize 3D space during training, thus resulting in continuous representations of 3D geometry without topology restrictions. While inspiring many follow-up works [10,11,23,24,28–30,41], all existing approaches are limited to single objects and do not scale to larger scenes. The key limiting factor of most implicit models is their simple fully-connected network architecture [26,31] which neither allows for integrating local information in the observations, nor for incorporating inductive biases such as translation equivariance into the model. This prevents these methods from performing *structured reasoning* as they only act globally and result in overly smooth surface reconstructions.

In contrast, translation equivariant convolutional neural networks (CNNs) have demonstrated great success across many 2D recognition tasks including object detection and image segmentation. Moreover, CNNs naturally encode information in a hierarchical manner in different network layers [50,51]. Exploiting these inductive biases is expected to not only benefit 2D but also 3D tasks, e.g., reconstructing 3D shapes of multiple similar chairs located in the same room. In this work, we seek to combine the complementary strengths of convolutional neural networks with those of implicit representations.

Towards this goal, we introduce *Convolutional Occupancy Networks*, a novel representation for accurate large-scale 3D reconstruction[1] with continuous implicit representations (Fig. 1). We demonstrate that this representation not only preserves fine geometric details, but also enables the reconstruction of complex indoor scenes at scale. Our key idea is to establish rich input features, incorporating inductive biases and integrating local as well as global information. More specifically, we exploit convolutional operations to obtain translation equivariance and exploit the local self-similarity of 3D structures. We systematically investigate multiple design choices, ranging from canonical planes to volumetric representations. Our contributions are summarized as follows:

- We identify major limitations of current implicit 3D reconstruction methods.
- We propose a flexible translation equivariant architecture which enables accurate 3D reconstruction from object to scene level.
- We demonstrate that our model enables generalization from synthetic to real scenes as well as to novel object categories and scenes.

Our code and data are provided at https://github.com/autonomousvision/convolutional_occupancy_networks.

2 Related Work

Learning-based 3D reconstruction methods can be broadly categorized by the output representation they use.

Voxels: Voxel representations are amongst the earliest representations for learning-based 3D reconstruction [5,46,47]. Due to the cubic memory requirements of voxel-based representations, several works proposed to operate on multiple scales or use octrees for efficient space partitioning [8,14,25,37,38,42]. However, even when using adaptive data structures, voxel-based techniques are still limited in terms of memory and computation.

Point Clouds: An alternative output representation for 3D reconstruction is 3D point clouds which have been used in [9,21,34,49]. However, point cloud-based representations are typically limited in terms of the number of points they can handle. Furthermore, they cannot represent topological relations.

Meshes: A popular alternative is to directly regress the vertices and faces of a mesh [12,13,17,20,22,44,45] using a neural network. While some of these works require deforming a template mesh of fixed topology, others result in non-watertight reconstructions with self-intersecting mesh faces.

Implicit Representations: More recent implicit occupancy [3,26] and distance field [27,31] models use a neural network to infer an occupancy probability

[1] With 3D reconstruction, we refer to 3D surface reconstruction throughout the paper.

or distance value given any 3D point as input. In contrast to the aforementioned explicit representations which require discretization (e.g., in terms of the number of voxels, points or vertices), implicit models represent shapes continuously and naturally handle complicated shape topologies. Implicit models have been adopted for learning implicit representations from images [23,24,29,41], for encoding texture information [30], for 4D reconstruction [28] as well as for primitive-based reconstruction [10,11,15,32]. Unfortunately, all these methods are limited to comparably simple 3D geometry of single objects and do not scale to more complicated or large-scale scenes. The key limiting factor is the simple fully-connected network architecture which does not allow for integrating local features or incorporating inductive biases such as translation equivariance.

Notable exceptions are PIFu [40] and DISN [48] which use pixel-aligned implicit representations to reconstruct people in clothing [40] or ShapeNet objects [48]. While these methods also exploit convolutions, all operations are performed in the *2D image domain*, restricting these models to image-based inputs and reconstruction of single objects. In contrast, in this work, we propose to aggregate features in *physical 3D space*, exploiting both 2D and 3D convolutions. Thus, our world-centric representation is independent of the camera viewpoint and input representation. Moreover, we demonstrate the feasibility of implicit 3D reconstruction at scene-level as illustrated in Fig. 1c.

In concurrent work, Chibane et al.[4] present a model similar to our convolutional volume decoder. In contrast to us, they only consider a single variant of convolutional feature embeddings (3D), use lossy discretization for the 3D point cloud encoding and only demonstrate results on single objects and humans, as opposed to full scenes. In another concurrent work, Jiang et al. [16] leverage shape priors for scene-level implicit 3D reconstruction. In contrast to us, they use 3D point normals as input and require optimization at inference time.

3 Method

Our goal is to make implicit 3D representations more expressive. An overview of our model is provided in Fig. 2. We first encode the input $\mathbf{x}$ (e.g., a point cloud) into a 2D or 3D feature grid (left). These features are processed using convolutional networks and decoded into occupancy probabilities via a fully-connected network. We investigate planar representations (a + c + d), volumetric representations (b + e) as well as combinations thereof in our experiments. In the following, we explain the encoder (Sect. 3.1), the decoder (Sect. 3.2), the occupancy prediction (Sect. 3.3) and the training procedure (Sect. 3.4) in more detail.

3.1 Encoder

While our method is independent of the input representation, we focus on 3D inputs to demonstrate the ability of our model in recovering fine details and scaling to large scenes. More specifically, we assume a noisy sparse point cloud (e.g., from structure-from-motion or laser scans) or a coarse occupancy grid as input $\mathbf{x}$.

We first process the input $\mathbf{x}$ with a task-specific neural network to obtain a feature encoding for every point or voxel. We use a one-layer 3D CNN for voxelized inputs, and a shallow PointNet [35] with local pooling for 3D point clouds. Given these features, we construct planar and volumetric feature representations in order to encapsulate local neighborhood information as follows.

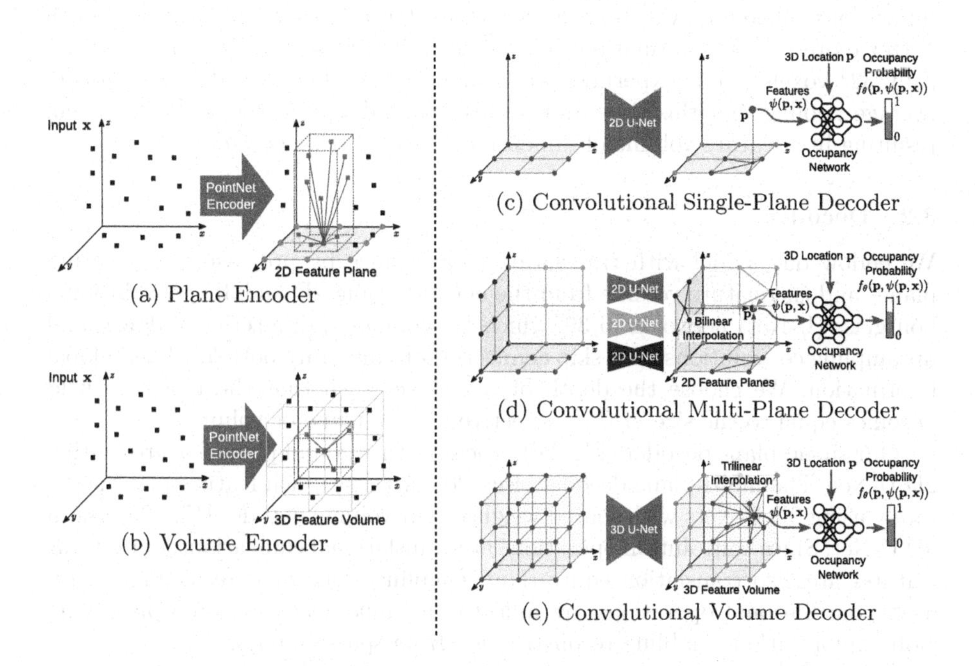

(a) Plane Encoder

(b) Volume Encoder

(c) Convolutional Single-Plane Decoder

(d) Convolutional Multi-Plane Decoder

(e) Convolutional Volume Decoder

Fig. 2. Model Overview. The **encoder (left)** first converts the 3D input $\mathbf{x}$ (e.g., noisy point clouds or coarse voxel grids) into features using task-specific neural networks. Next, the features are projected onto one or multiple planes (Fig. 2a) or into a volume (Fig. 2b) using average pooling. The **convolutional decoder (right)** processes the resulting feature planes/volume using 2D/3D U-Nets to aggregate local and global information. For a query point $\mathbf{p} \in \mathbb{R}^3$, the point-wise feature vector $\psi(\mathbf{x}, \mathbf{p})$ is obtained via bilinear (Fig. 2c and Fig. 2d) or trilinear (Fig. 2e) interpolation. Given feature vector $\psi(\mathbf{x}, \mathbf{p})$ at location $\mathbf{p}$, the occupancy probability is predicted using a fully-connected network $f_\theta(\mathbf{p}, \psi(\mathbf{p}, \mathbf{x}))$.

Plane Encoder. As illustrated in Fig. 2a, for each input point, we perform an orthographic projection onto a canonical plane (i.e., a plane aligned with the axes of the coordinate frame) which we discretize at a resolution of $H \times W$ pixel cells. For voxel inputs, we treat the voxel center as a point and project it to the plane. We aggregate features projecting onto the same pixel using average pooling, resulting in planar features with dimensionality $H \times W \times d$, where d is the feature dimension.

In our experiments, we analyze two variants of our model: one variant where features are projected onto the ground plane, and one variant where features are projected to all three canonical planes. While the former is computationally more efficient, the latter allows for recovering richer geometric structure in the z dimension.

Volume Encoder. While planar feature representations allow for encoding at large spatial resolution (128^2 pixels and beyond), they are restricted to two dimensions. Therefore, we also consider volumetric encodings (see Fig. 2b) which better represent 3D information, but are restricted to smaller resolutions (typically 32^3 voxels in our experiments). Similar to the plane encoder, we perform average pooling, but this time over all features falling into the same *voxel* cell, resulting in a feature volume of dimensionality $H \times W \times D \times d$.

3.2 Decoder

We endow our model with translation equivariance by processing the feature planes and the feature volume from the encoder using 2D and 3D convolutional hourglass (U-Net) networks [6,39] which are composed of a series of down- and upsampling convolutions with skip connections to integrate both local and global information. We choose the depth of the U-Net such that the receptive field becomes equal to the size of the respective feature plane or volume.

Our single-plane decoder (Fig. 2c) processes the ground plane features with a 2D U-Net. The multi-plane decoder (Fig. 2d) processes each feature plane separately using 2D U-Nets with shared weights. Our volume decoder (Fig. 2e) uses a 3D U-Net. Since convolution operations are translational equivariant, our output features are also translation equivariant, enabling structured reasoning. Moreover, convolutional operations are able to "inpaint" features while preserving global information, enabling reconstruction from sparse inputs.

3.3 Occupancy Prediction

Given the aggregated feature maps, our goal is to estimate the occupancy probability of any point $\mathbf{p}$ in 3D space. For the single-plane decoder, we project each point $\mathbf{p}$ orthographically onto the ground plane and query the feature value through bilinear interpolation (Fig. 2c). For the multi-plane decoder (Fig. 2d), we aggregate information from the 3 canonical planes by summing the features of all 3 planes. For the volume decoder, we use trilinear interpolation (Fig. 2e).

Denoting the feature vector for input $\mathbf{x}$ at point $\mathbf{p}$ as $\psi(\mathbf{p}, \mathbf{x})$, we predict the occupancy of $\mathbf{p}$ using a small fully-connected occupancy network:

$$f_\theta(\mathbf{p}, \psi(\mathbf{p}, \mathbf{x})) \rightarrow [0, 1] \tag{1}$$

The network comprises multiple ResNet blocks. We use the network architecture of [29], adding ψ to the input features of every ResNet block instead of the more memory intensive batch normalization operation proposed in earlier works [26]. In contrast to [29], we use a feature dimension of 32 for the hidden layers. Details about the network architecture can be found in the supplementary.

3.4 Training and Inference

At training time, we uniformly sample query points $\mathbf{p} \in \mathbb{R}^3$ within the volume of interest and predict their occupancy values. We apply the binary cross-entropy loss between the predicted $\hat{o}_{\mathbf{p}}$ and the true occupancy values $o_{\mathbf{p}}$:

$$\mathcal{L}(\hat{o}_{\mathbf{p}}, o_{\mathbf{p}}) = -[o_{\mathbf{p}} \cdot \log(\hat{o}_{\mathbf{p}}) + (1 - o_{\mathbf{p}}) \cdot \log(1 - \hat{o}_{\mathbf{p}})] \tag{2}$$

We implement all models in PyTorch [33] and use the Adam optimizer [19] with a learning rate of 10^{-4}. During inference, we apply *Multiresolution IsoSurface Extraction* (MISE) [26] to extract meshes given an input $\mathbf{x}$. As our model is fully-convolutional, we are able to reconstruct large scenes by applying it in a "sliding-window" fashion at inference time. We exploit this property to obtain reconstructions of entire apartments (see Fig. 1).

4 Experiments

We conduct three types of experiments to evaluate our method. First, we perform **object-level reconstruction** on ShapeNet [2] chairs, considering noisy point clouds and low-resolution occupancy grids as inputs. Next, we compare our approach against several baselines on the task of **scene-level reconstruction** using a synthetic indoor dataset of various objects. Finally, we demonstrate **synthetic-to-real generalization** by evaluating our model on real indoor scenes [1,7].

Datasets

ShapeNet [2]: We use all 13 classes of the ShapeNet subset, voxelizations, and train/val/test split from Choy et al.[5]. Per-class results can be found in supplementary.

Synthetic Indoor Scene Dataset: We create a synthetic dataset of 5000 scenes with multiple objects from ShapeNet (chair, sofa, lamp, cabinet, table). A scene consists of a ground plane with randomly sampled width-length ratio, multiple objects with random rotation and scale, and randomly sampled walls.

ScanNet v2 [7]: This dataset contains 1513 real-world rooms captured with an RGB-D camera. We sample point clouds from the provided meshes for testing.

Matterport3D [1]: Matterport3D contains 90 buildings with multiple rooms on different floors captured using a Matterport Pro Camera. Similar to ScanNet, we sample point clouds for evaluating our model on Matterport3D.

Baselines

ONet [26]: Occupancy Networks is a state-of-the-art implicit 3D reconstruction model. It uses a fully-connected network architecture and a global encoding of the input. We compare against this method in all of our experiments.

PointConv: We construct another simple baseline by extracting point-wise features using PointNet++ [36], interpolating them using Gaussian kernel regression and feeding them into the same fully-connected network used in our approach. While this baseline uses local information, it does not exploit convolutions.

SPSR [18]: Screened Poisson Surface Reconstruction (SPSR) is a traditional 3D reconstruction technique which operates on oriented point clouds as input. Note that in contrast to all other methods, SPSR requires additional surface normals which are often hard to obtain for real-world scenarios.

Table 1. Object-level 3D reconstruction from point clouds. Left: We report GPU memory, IoU, Chamfer-L_1 distance, normal consistency and F-Score for our approach (2D plane and 3D voxel grid dimensions in brackets), the baselines ONet [26] and PointConv on ShapeNet (mean over all 13 classes). Right: The training progression plot shows that our method converges faster than the baselines.

	GPU Memory	IoU	Chamfer-L_1	Normal C.	F-Score
PointConv	5.1G	0.689	0.126	0.858	0.644
ONet [26]	7.7G	0.761	0.087	0.891	0.785
Ours-2D (64^2)	1.6G	0.833	0.059	0.914	0.887
Ours-2D (3×64^2)	2.4G	**0.884**	**0.044**	**0.938**	**0.942**
Ours-3D (32^3)	5.9G	0.870	0.048	0.937	0.933

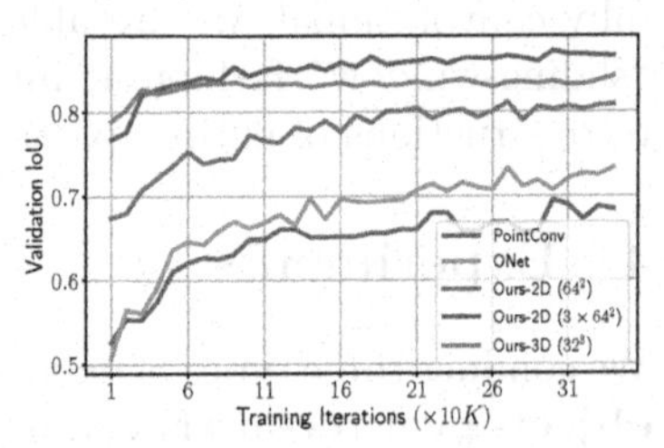

Metrics

Following [26], we consider Volumetric IoU, Chamfer Distance, Normal Consistency for evaluation. We further report F-Score [43] with the default threshold value of 1% unless otherwise specified. Details can be found in the supplementary.

4.1 Object-Level Reconstruction

We first evaluate our method on the single object reconstruction task on ShapeNet [2]. We consider two different types of 3D inputs: noisy point clouds and low-resolution voxels. For the former, we sample 3000 points from the mesh and apply Gaussian noise with zero mean and standard deviation 0.05. For the latter, we use the coarse 32^3 voxelizations from [26]. For the query points (i.e., for which supervision is provided), we follow [26] and uniformly sample 2048 and 1024 points for noisy point clouds and low-resolution voxels, respectively. Due to the different encoder architectures for these two tasks, we set the batch size to 32 and 64, respectively.

Reconstruction from Point Clouds. Table 1 and Fig. 3 show quantitative and qualitative results. Compared to the baselines, all variants of our method achieve equal or better results on all three metrics. As evidenced by the training progression plot on the right, our method reaches a high validation IoU after only few iterations. This verifies our hypothesis that leveraging convolutions and local features benefits 3D reconstruction in terms of both accuracy and efficiency. The results show that, in comparison to PointConv which directly aggregates features from point clouds, projecting point-features to planes or volumes followed by 2D/3D CNNs is more effective. In addition, decomposing 3D representations from volumes into three planes with higher resolution (64^2 vs. 32^3) improves

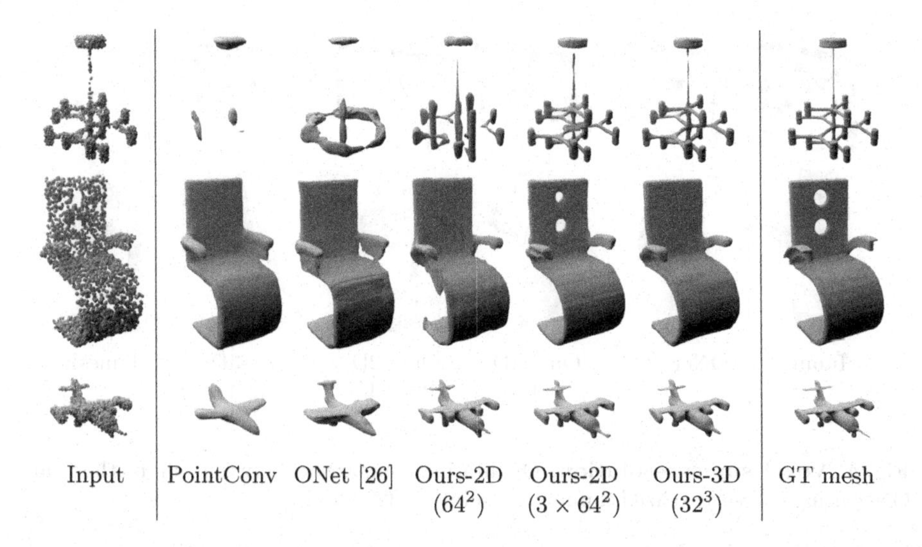

Fig. 3. Object-level 3D reconstruction from point clouds. Comparison of our convolutional representation to ONet and PointConv on ShapeNet.

Table 2. Voxel super-resolution. 3D reconstruction results from low resolution voxelized inputs (32^3 voxels) on the ShapeNet dataset (mean over 13 classes).

	GPU memory	IoU	Chamfer-L_1	Normal C.	F-Score
Input	–z	0.631	0.136	0.810	0.440
ONet [26]	4.8G	0.703	0.110	0.879	0.656
Ours-2D (64^2)	2.4G	0.652	0.145	0.861	0.592
Ours-2D (3×64^2)	4.0G	**0.752**	0.092	0.905	**0.735**
Ours-3D (32^3)	10.8G	**0.752**	**0.091**	**0.912**	0.729

performance while at the same time requiring less GPU memory. More results can be found in supplementary.

Voxel Super-Resolution: Besides noisy point clouds, we also evaluate on the task of voxel super-resolution. Here, the goal is to recover high-resolution details from coarse (32^3) voxelizations of the shape. Table 2 and Fig. 4 show that our method with three planes achieves comparable results over our volumetric method while requiring only 37% of the GPU memory. In contrast to reconstruction from point clouds, our single-plane approach fails on this task. We hypothesize that a single plane is not sufficient for resolving ambiguities in the coarse but regularly structured voxel input.

4.2 Scene-Level Reconstruction

To analyze whether our approach can scale to larger scenes, we now reconstruct 3D geometry from point clouds on our synthetic indoor scene dataset. Due to the

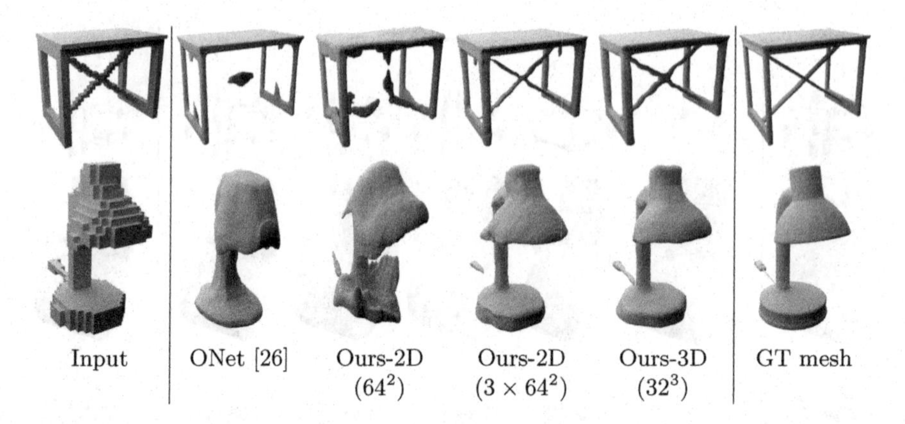

Fig. 4. Voxel super-resolution. Qualitative comparison between our method and ONet using coarse voxelized inputs at resolution 32^3 voxels.

Table 3. Scene-level reconstruction on synthetic rooms. Quantitative comparison for reconstruction from noisy point clouds. We do not report IoU for SPSR as SPSR generates only a single surface for walls and the ground plane. To ensure a fair comparison to SPSR, we compare all methods with only a single surface for walls/ground planes when calculating Chamfer-L_1 and F-Score.

	IoU	Chamfer-L_1	Normal consistency	F-Score
ONet [26]	0.475	0.203	0.783	0.541
PointConv	0.523	0.165	0.811	0.790
SPSR [18]	–	0.223	0.866	0.810
SPSR [18] (trimmed)	–	0.069	0.890	0.892
Ours-2D (128^2)	0.795	0.047	0.889	0.937
Ours-2D (3×128^2)	0.805	0.044	0.903	0.948
Ours-3D (32^3)	0.782	0.047	0.902	0.941
Ours-3D (64^3)	**0.849**	**0.042**	**0.915**	**0.964**
Ours-2D-3D ($3 \times 128^2 + 32^3$)	0.816	0.044	0.905	0.952

increasing complexity of the scene, we uniformly sample 10000 points as input point cloud and apply Gaussian noise with standard deviation of 0.05. During training, we sample 2048 query points, similar to object-level reconstruction. For our plane-based methods, we use a resolution to 128^2. For our volumetric approach, we investigate both 32^3 and 64^3 resolutions. Hypothesizing that the plane and volumetric features are complementary, we also test the combination of the multi-plane and volumetric variants.

Table 3 and Fig. 5 show our results. All variants of our method are able to reconstruct geometric details of the scenes and lead to smooth results. In contrast, ONet and PointConv suffer from low accuracy while SPSR leads to noisy surfaces. While high-resolution canonical plane features capture fine details they

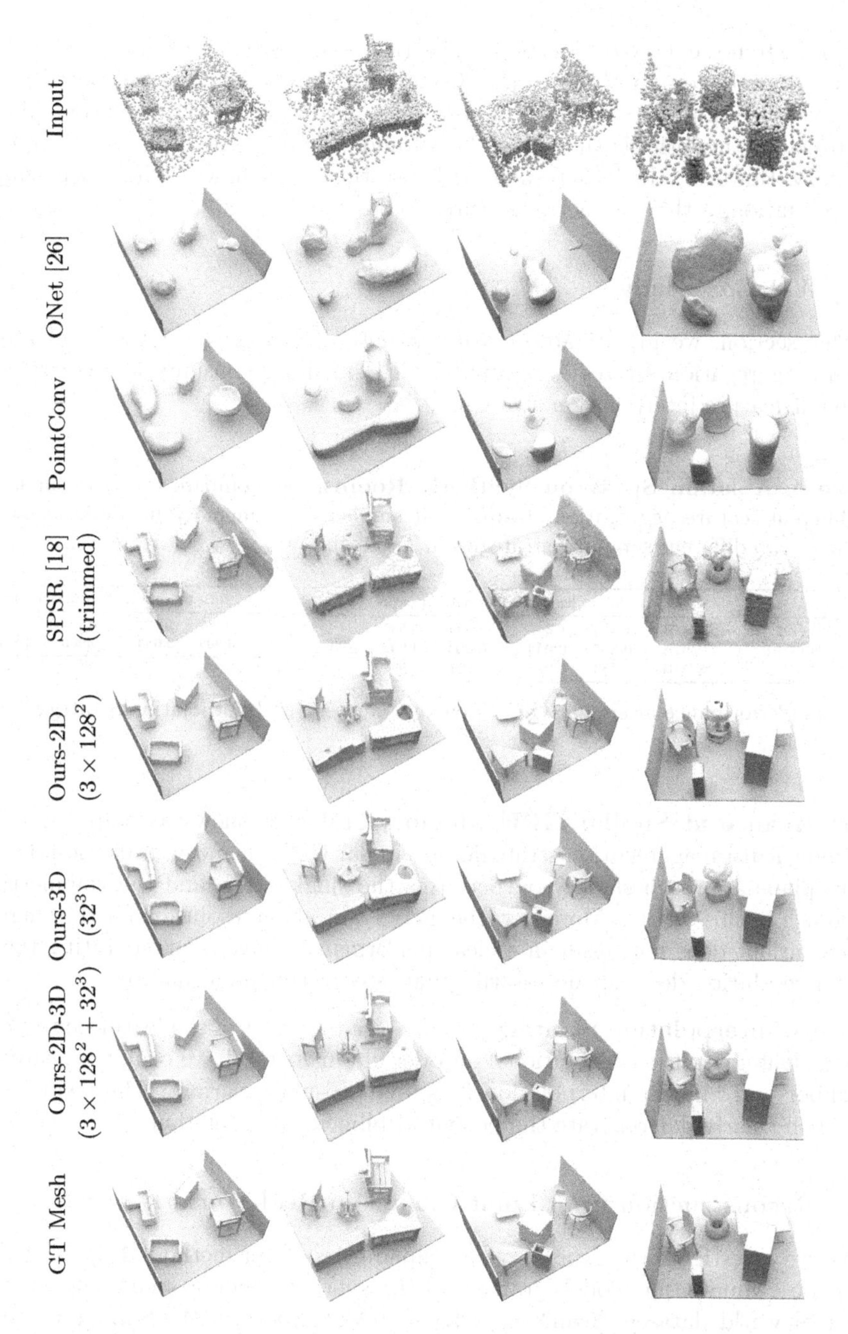

Fig. 5. Scene-level reconstruction on synthetic rooms. Qualitative comparison for point-cloud based reconstruction on the synthetic indoor scene dataset.

are prone to noise. Low-resolution volumetric features are instead more robust to noise, yet produce smoother surfaces. Combining complementary volumetric and plane features improves results compared to considering them in isolation. This confirms our hypothesis that plane-based and volumetric features are complementary. However, the best results in this setting are achieved when increasing the resolution of the volumetric features to 64^3.

4.3 Ablation Study

In this section, we investigate on our synthetic indoor scene dataset different feature aggregation strategies at similar GPU memory consumption as well as different feature interpolation strategies.

Table 4. Ablation Study on Synthetic Rooms. We compare the performance of different feature aggregation strategies at similar GPU memory in Table 4a and evaluate two different sampling strategies in Table 4b.

	GPU Memory	IoU	Chamfer-L_1	Normal C.	F-Score
Ours-2D (192^2)	9.5GB	0.773	0.047	0.889	0.937
Ours-2D (3×128^2)	9.3GB	**0.805**	**0.044**	**0.903**	**0.948**
Ours-3D (32^3)	8.5GB	0.782	0.047	0.902	0.941

(a) Performance at similar GPU Memory

	IoU	Chamfer-L_1	Normal C.	F-Score
Nearest Neighbor	0.766	0.052	0.885	0.920
Bilinear	**0.805**	**0.044**	**0.903**	**0.948**

(b) Interpolation Strategy

Performance at Similar GPU Memory: Table 4a shows a comparison of different feature aggregation strategies at similar GPU memory utilization. Our multi-plane approach slightly outperforms the single plane and the volumetric approach in this setting. Moreover, the increase in plane resolution for the single plane variant does not result in a clear performance boost, demonstrating that higher resolution does not necessarily guarantee better performance.

Feature Interpolation Strategy: To analyze the effect of the feature interpolation strategy in the convolutional decoder of our method, we compare nearest neighbor and bilinear interpolation for our multi-plane variant. The results in Table 4b clearly demonstrate the benefit of bilinear interpolation.

4.4 Reconstruction from Point Clouds on Real-World Datasets

Next, we investigate the generalization capabilities of our method. Towards this goal, we evaluate our models trained on the synthetic indoor scene dataset on the real world datasets ScanNet v2 [7] and Matterport3D [1]. Similar to our previous experiments, we use 10000 points sampled from the meshes as input.

ScanNet v2: Our results in Table 5 show that among all our variants, the volumetric-based models perform best, indicating that the plane-based approaches are more affected by the domain shift. We find that 3D CNNs are more robust to noise as they aggregate features from all neighbors which results in smooth outputs. Moreover, all variants outperform the learning-based baselines by a significant margin.

Table 5. Scene-level reconstruction on ScanNet. Evaluation of point-based reconstruction on the real-world ScanNet dataset. As ScanNet does not provide watertight meshes, we trained all methods on the synthetic indoor scene dataset. Remark: In ScanNet, walls/floors are only observed from one side. To not wrongly penalize methods for predicting walls and floors with thickness (0.01 in our training set), we chose a F-Score threshold of 1.5% for this experiment.

	Chamfer-L_1	F-Score
ONet [26]	0.398	0.390
PointConv	0.316	0.439
SPSR [18]	0.293	0.731
SPSR [18] (trimmed)	0.086	0.847

	Chamfer-L_1	F-Score
Ours-2D (128^2)	0.139	0.747
Ours-2D (3×128^2)	0.142	0.776
Ours-3D (32^3)	0.095	0.837
Ours-3D (64^3)	**0.077**	**0.886**
Ours-2D-3D ($3 \times 128^2 + 32^3$)	0.099	0.847

The qualitative comparison in Fig. 6 shows that our model is able to smoothly reconstruct scenes with geometric details at various scales. While Screened PSR [18] also produces reasonable reconstructions, it tends to close the resulting meshes and hence requires a carefully chosen trimming parameter. In contrast, our method does not require additional hyperparameters.

Matterport 3D Dataset. Finally, we investigate the scalability of our method to larger scenes which comprise multiple rooms and multiple floors. For this experiment, we exploit the Matterport3D dataset. Unlike before, we implement a fully convolutional version of our 3D model that can be scaled to any size by running on overlapping crops of the input point cloud in a sliding window fashion. The overlap is determined by the size of the receptive field to ensure correctness of the results. Figure 1 shows the resulting 3D reconstruction. Our method reconstructs details inside each room while adhering to the room layout. Note that the geometry and point distribution of the Matterport3D dataset differs significantly from the synthetic indoor scene dataset which our model is trained on. This demonstrates that our method is able to generalize not only to unseen classes, but also novel room layouts and sensor characteristics. More implementation details and results can be found in supplementary.

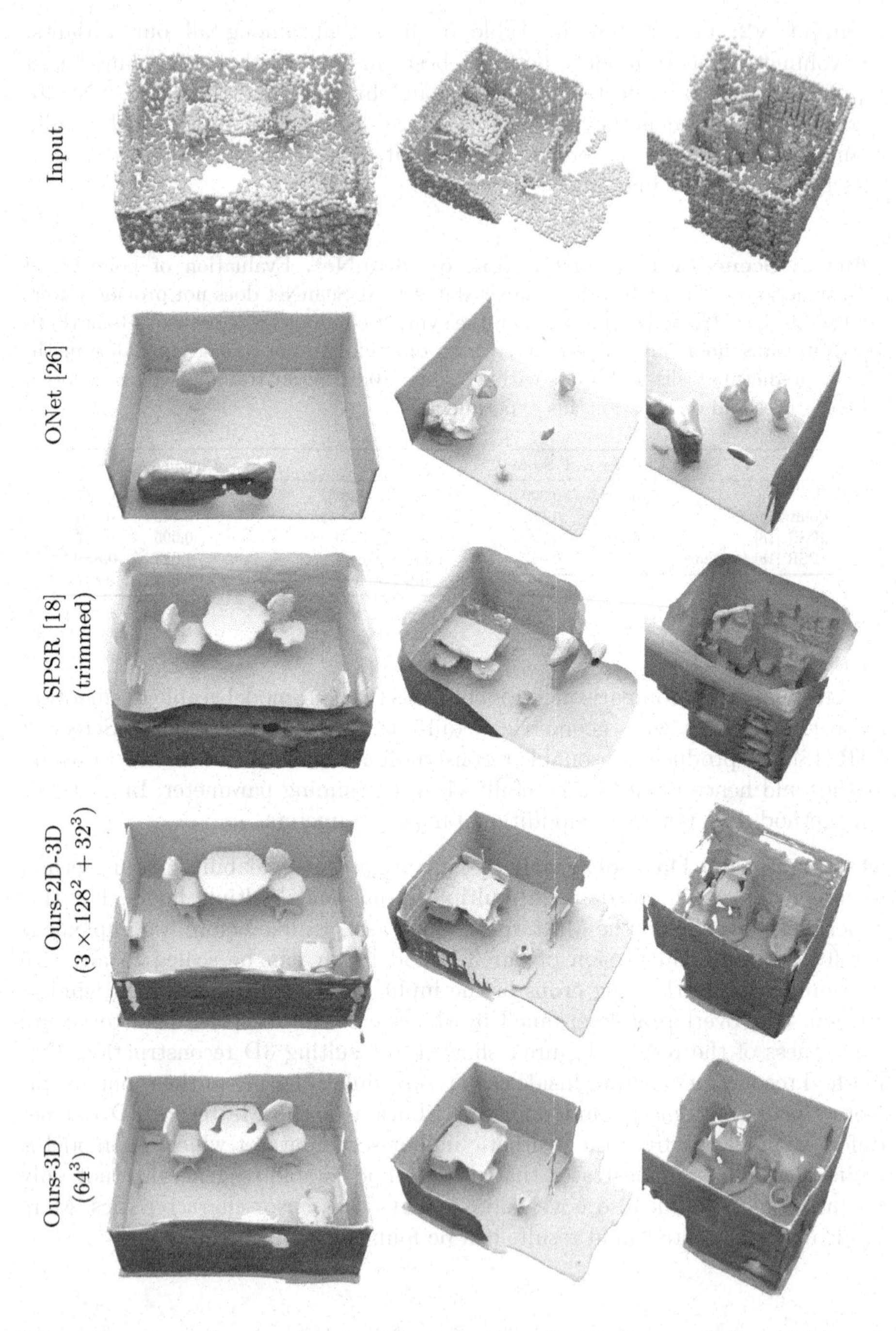

Fig. 6. Scene-level reconstruction on ScanNet. Qualitative results for point-based reconstruction on ScanNet [7]. All learning-based methods are trained on the synthetic room dataset and evaluated on ScanNet.

5 Conclusion

We introduced Convolutional Occupancy Networks, a novel shape representation which combines the expressiveness of convolutional neural networks with the advantages of implicit representations. We analyzed the tradeoffs between 2D and 3D feature representations and found that incorporating convolutional operations facilitates generalization to unseen classes, novel room layouts and large-scale indoor spaces. We find that our 3-plane model is memory efficient, works well on synthetic scenes and allows for larger feature resolutions. Our volumetric model, in contrast, outperforms other variants on real-world scenarios while consuming more memory.

Finally, we remark that our method is not rotation equivariant and only translation equivariant with respect to translations that are multiples of the defined voxel size. Moreover, there is still a performance gap between synthetic and real data. While the focus of this work was on learning-based 3D reconstruction, in future work, we plan to apply our novel representation to other domains such as implicit appearance modeling and 4D reconstruction.

Acknowledgements. This work was supported by an NVIDIA research gift. The authors thank Max Planck ETH Center for Learning Systems (CLS) for supporting Songyou Peng and the International Max Planck Research School for Intelligent Systems (IMPRS-IS) for supporting Michael Niemeyer.

References

1. Chang, A., et al.: Matterport3D: learning from RGB-D data in indoor environments. In: Proceedings of the International Conference on 3D Vision (3DV) (2017)
2. Chang, A.X., et al.: ShapeNet: an information-rich 3D model repository. arXiv.org 1512.03012 (2015)
3. Chen, Z., Zhang, H.: Learning implicit fields for generative shape modeling. In: Proceedings of the IEEE Conference on Computer Vision and Pattern Recognition (CVPR) (2019)
4. Chibane, J., Alldieck, T., Pons-Moll, G.: Implicit functions in feature space for 3D shape reconstruction and completion. In: Proceedings of the IEEE Conference on Computer Vision and Pattern Recognition (CVPR) (2020)
5. Choy, C.B., Xu, D., Gwak, J., Chen, K., Savarese, S.: 3D-R2N2: a unified approach for single and multi-view 3D object reconstruction. In: Proceedings of the European Conference on Computer Vision (ECCV) (2016)
6. Çiçek, Ö., Abdulkadir, A., Lienkamp, S.S., Brox, T., Ronneberger, O.: 3D U-Net: learning dense volumetric segmentation from sparse annotation. In: Ourselin, S., Joskowicz, L., Sabuncu, M.R., Unal, G., Wells, W. (eds.) MICCAI 2016. LNCS, vol. 9901, pp. 424–432. Springer, Cham (2016). https://doi.org/10.1007/978-3-319-46723-8_49
7. Dai, A., Chang, A.X., Savva, M., Halber, M., Funkhouser, T., Niessner, M.: Scannet: richly-annotated 3D reconstructions of indoor scenes. In: Proceedings of IEEE Conference on Computer Vision and Pattern Recognition (CVPR) (2017)

8. Dai, A., Qi, C.R., Nießner, M.: Shape completion using 3D-encoder-predictor CNNs and shape synthesis. In: Proceedings of IEEE Conference on Computer Vision and Pattern Recognition (CVPR) (2017)
9. Fan, H., Su, H., Guibas, L.J.: A point set generation network for 3D object reconstruction from a single image. In: Proceedings of the IEEE Conference on Computer Vision and Pattern Recognition (CVPR) (2017)
10. Genova, K., Cole, F., Sud, A., Sarna, A., Funkhouser, T.A.: Local deep implicit functions for 3D shape. In: Proceedings of the IEEE Conference on Computer Vision and Pattern Recognition (CVPR) (2020)
11. Genova, K., Cole, F., Vlasic, D., Sarna, A., Freeman, W.T., Funkhouser, T.: Learning shape templates with structured implicit functions. In: Proceedings of the IEEE International Conference on Computer Vision (ICCV) (2019)
12. Gkioxari, G., Malik, J., Johnson, J.: Mesh R-CNN. In: Proceedings of the IEEE International Conference on Computer Vision (ICCV) (2019)
13. Groueix, T., Fisher, M., Kim, V.G., Russell, B.C., Aubry, M.: AtlasNet: a papier-mâché approach to learning 3D surface generation. In: Proceedings of the IEEE Conference on Computer Vision and Pattern Recognition (CVPR) (2018)
14. Hane, C., Tulsiani, S., Malik, J.: Hierarchical surface prediction for 3D object reconstruction. In: Proceedings of the International Conference on 3D Vision (3DV) (2017)
15. Jeruzalski, T., Deng, B., Norouzi, M., Lewis, J.P., Hinton, G.E., Tagliasacchi, A.: NASA: neural articulated shape approximation. In: Proceedings of the European Conference on Computer Vision (ECCV) (2020)
16. Jiang, C., Sud, A., Makadia, A., Huang, J., Nießner, M., Funkhouser, T.: Local implicit grid representations for 3D scenes. In: Proceedings of the IEEE Conference on Computer Vision and Pattern Recognition (CVPR) (2020)
17. Kanazawa, A., Tulsiani, S., Efros, A.A., Malik, J.: Learning category-specific mesh reconstruction from image collections. In: Ferrari, V., Hebert, M., Sminchisescu, C., Weiss, Y. (eds.) ECCV 2018. LNCS, vol. 11219, pp. 386–402. Springer, Cham (2018). https://doi.org/10.1007/978-3-030-01267-0_23
18. Kazhdan, M.M., Hoppe, H.: Screened poisson surface reconstruction. ACM Trans. Graph. **32**(3), 29 (2013)
19. Kingma, D.P., Ba, J.: Adam: a method for stochastic optimization. In: Proceedings of the International Conference on Machine learning (ICML) (2015)
20. Liao, Y., Donne, S., Geiger, A.: Deep marching cubes: learning explicit surface representations. In: Proceedings of the IEEE Conference on Computer Vision and Pattern Recognition (CVPR) (2018)
21. Lin, C., Kong, C., Lucey, S.: Learning efficient point cloud generation for dense 3D object reconstruction. In: Proceedings of the Conference on Artificial Intelligence (AAAI) (2018)
22. Lin, C., et al.: Photometric mesh optimization for video-aligned 3D object reconstruction. In: Proceedings of the IEEE Conference on Computer Vision and Pattern Recognition (CVPR) (2019)
23. Liu, S., Zhang, Y., Peng, S., Shi, B., Pollefeys, M., Cui, Z.: DIST: rendering deep implicit signed distance function with differentiable sphere tracing. In: Proceedings of the IEEE Conference on Computer Vision and Pattern Recognition (CVPR) (2020)
24. Liu, S., Saito, S., Chen, W., Li, H.: Learning to infer implicit surfaces without 3D supervision. In: Advances in Neural Information Processing Systems (NeurIPS) (2019)

25. Maturana, D., Scherer, S.: Voxnet: A 3D convolutional neural network for real-time object recognition. In: Proceedings of the IEEE International Conference on Intelligent Robots and Systems (IROS) (2015)
26. Mescheder, L., Oechsle, M., Niemeyer, M., Nowozin, S., Geiger, A.: Occupancy networks: Learning 3D reconstruction in function space. In: Proceedings of the IEEE Conference on Computer Vision and Pattern Recognition (CVPR) (2019)
27. Michalkiewicz, M., Pontes, J.K., Jack, D., Baktashmotlagh, M., Eriksson, A.: Implicit surface representations as layers in neural networks. In: Proceedings of the IEEE International Conference on Computer Vision (ICCV) (2019)
28. Niemeyer, M., Mescheder, L., Oechsle, M., Geiger, A.: Occupancy flow: 4D reconstruction by learning particle dynamics. In: Proceedings of the IEEE International Conference on Computer Vision (ICCV) (2019)
29. Niemeyer, M., Mescheder, L.M., Oechsle, M., Geiger, A.: Differentiable volumetric rendering: learning implicit 3D representations without 3D supervision. In: Proceedings of the IEEE Conference on Computer Vision and Pattern Recognition (CVPR) (2020)
30. Oechsle, M., Mescheder, L., Niemeyer, M., Strauss, T., Geiger, A.: Texture fields: learning texture representations in function space. In: Proceedings of the IEEE International Conference on Computer Vision (ICCV) (2019)
31. Park, J.J., Florence, P., Straub, J., Newcombe, R.A., Lovegrove, S.: Deepsdf: Learning continuous signed distance functions for shape representation. In: Proceedings of the IEEE Conference on Computer Vision and Pattern Recognition (CVPR) (2019)
32. Paschalidou, D., van Gool, L., Geiger, A.: Learning unsupervised hierarchical part decomposition of 3D objects from a single RGB image. In: Proceedings of the IEEE Conference on Computer Vision and Pattern Recognition (CVPR) (2020)
33. Paszke, A., et al.: Pytorch: an imperative style, high-performance deep learning library. In: Advances in Neural Information Processing Systems (NeurIPS) (2019)
34. Prokudin, S., Lassner, C., Romero, J.: Efficient learning on point clouds with basis point sets. In: Proceedings of the IEEE International Conference on Computer Vision (ICCV) (2019)
35. Qi, C.R., Su, H., Mo, K., Guibas, L.J.: PointNet: deep learning on point sets for 3D classification and segmentation. In: Proceedings of the IEEE Conference on Computer Vision and Pattern Recognition (CVPR) (2017)
36. Qi, C.R., Yi, L., Su, H., Guibas, L.J.: Pointnet++: deep hierarchical feature learning on point sets in a metric space. In: Advances in Neural Information Processing Systems (NeurIPS) (2017)
37. Riegler, G., Ulusoy, A.O., Bischof, H., Geiger, A.: OctNetFusion: learning depth fusion from data. In: Proceedings of the International Conference on 3D Vision (3DV) (2017)
38. Riegler, G., Ulusoy, A.O., Geiger, A.: OctNet: learning deep 3D representations at high resolutions. In: Proceedings of the IEEE Conference on Computer Vision and Pattern Recognition (CVPR) (2017)
39. Ronneberger, O., Fischer, P., Brox, T.: U-Net: convolutional networks for biomedical image segmentation. In: Navab, N., Hornegger, J., Wells, W.M., Frangi, A.F. (eds.) MICCAI 2015. LNCS, vol. 9351, pp. 234–241. Springer, Cham (2015). https://doi.org/10.1007/978-3-319-24574-4_28
40. Saito, S., Huang, Z., Natsume, R., Morishima, S., Kanazawa, A., Li, H.: PIFu: pixel-aligned implicit function for high-resolution clothed human digitization. In: Proceedings of the IEEE International Conference on Computer Vision (ICCV) (2019)

41. Sitzmann, V., Zollhöfer, M., Wetzstein, G.: Scene representation networks: continuous 3D-structure-aware neural scene representations. In: Advances in Neural Information Processing Systems (NeurIPS) (2019)
42. Tatarchenko, M., Dosovitskiy, A., Brox, T.: Octree generating networks: efficient convolutional architectures for high-resolution 3D outputs. In: Proceedings of the IEEE International Conference on Computer Vision (ICCV) (2017)
43. Tatarchenko, M., Richter, S.R., Ranftl, R., Li, Z., Koltun, V., Brox, T.: What do single-view 3D reconstruction networks learn? In: Proceedings of the IEEE Conference on Computer Vision and Pattern Recognition (CVPR) (2019)
44. Wang, N., Zhang, Y., Li, Z., Fu, Y., Liu, W., Jiang, Y.-G.: Pixel2Mesh: generating 3D mesh models from single RGB images. In: Ferrari, V., Hebert, M., Sminchisescu, C., Weiss, Y. (eds.) ECCV 2018. LNCS, vol. 11215, pp. 55–71. Springer, Cham (2018). https://doi.org/10.1007/978-3-030-01252-6_4
45. Wen, C., Zhang, Y., Li, Z., Fu, Y.: Pixel2mesh++: multi-view 3D mesh generation via deformation. In: Proceedings of the IEEE International Conference on Computer Vision (ICCV) (2019)
46. Wu, J., Zhang, C., Xue, T., Freeman, B., Tenenbaum, J.: Learning a probabilistic latent space of object shapes via 3D generative-adversarial modeling. In: Advances in Neural Information Processing Systems (NeurIPS) (2016)
47. Wu, Z., et al.: 3D shapenets: a deep representation for volumetric shapes. In: Proceedings of the IEEE Conference on Computer Vision and Pattern Recognition (CVPR) (2015)
48. Xu, Q., Wang, W., Ceylan, D., Mech, R., Neumann, U.: DISN: deep implicit surface network for high-quality single-view 3D reconstruction. In: Advances in Neural Information Processing Systems (NeurIPS) (2019)
49. Yang, G., Huang, X., Hao, Z., Liu, M., Belongie, S.J., Hariharan, B.: PointFlow: 3D point cloud generation with continuous normalizing flows. In: Proceedings of the IEEE International Conference on Computer Vision (ICCV) (2019)
50. Zeiler, M.D., Fergus, R.: Visualizing and understanding convolutional networks. In: Fleet, D., Pajdla, T., Schiele, B., Tuytelaars, T. (eds.) ECCV 2014. LNCS, vol. 8689, pp. 818–833. Springer, Cham (2014). https://doi.org/10.1007/978-3-319-10590-1_53
51. Zhang, Q., Wu, Y.N., Zhu, S.: Interpretable convolutional neural networks. In: Proceedings of the IEEE Conference on Computer Vision and Pattern Recognition (CVPR) (2018)

Multi-person 3D Pose Estimation in Crowded Scenes Based on Multi-view Geometry

He Chen[1](✉), Pengfei Guo[1], Pengfei Li[1], Gim Hee Lee[2], and Gregory Chirikjian[1,2]

[1] The Johns Hopkins University, Baltimore, USA
{hchen136, pguo4, pli32, gchirik1}@jhu.edu
[2] National University of Singapore, Singapore, Singapore
gimhee.lee@comp.nus.edu.sg, mpegre@nus.edu.sg

Abstract. Epipolar constraints are at the core of feature matching and depth estimation in current multi-person multi-camera 3D human pose estimation methods. Despite the satisfactory performance of this formulation in sparser crowd scenes, its effectiveness is frequently challenged under denser crowd circumstances mainly due to two sources of ambiguity. The first is the mismatch of human joints resulting from the simple cues provided by the Euclidean distances between joints and epipolar lines. The second is the lack of robustness from the naive formulation of the problem as a least squares minimization. In this paper, we depart from the multi-person 3D pose estimation formulation, and instead reformulate it as crowd pose estimation. Our method consists of two key components: a graph model for fast cross-view matching, and a maximum a posteriori (MAP) estimator for the reconstruction of the 3D human poses. We demonstrate the effectiveness and superiority of our proposed method on four benchmark datasets. Our code is available at: https://github.com/HeCraneChen/3D-Crowd-Pose-Estimation-Based-on-MVG.

Keywords: 3D pose estimation · Occlusion · Correspondence problem

1 Introduction

Fast 3D human pose estimation for crowded scenes is an important component in many computer vision applications such as autonomous driving, surveillance, and robotics [12,17,26,29–31,41,43,47]. However, recovering 3D human pose from

H. Chen and P. Guo—Equal first author contribution.
G. H. Lee and G. Chirikjian—Jointly supervised this work.

A. Vedaldi et al. (Eds.): ECCV 2020, LNCS 12348, pp. 541–557, 2020.
https://doi.org/10.1007/978-3-030-58580-8_32

crowded real-world setting is a challenging endeavor due to the inherent depth ambiguity caused by 2D to 3D backprojections, self-occlusions, and occlusions by other people in crowded scenes [1,25,38]. A three-step process is commonly used in the multi-person multi-camera 3D pose estimation problem: 1) Detecting human body keypoints or parts in separate 2D views; 2) Matching people across different views; 3) Reconstructing 3D pose by triangulation. Unfortunately, the critical second step of matching people across different views is non-trivial. Well-known matching algorithms such as the Harris corner detector [19] and the Scale Invariant Feature Transform (SIFT) [35] give mostly wrong matches even after robust estimation with RANSAC [18]. The problem is further aggravated in the third step when these unreliable matches are used in a vanilla triangulation algorithm to recover the 3D points.

With the rapid development of deep learning, features are extracted more precisely and significant improvements are made for appearance-based feature matching across different viewpoints on the spatial level or different frames on the temporal level [34,40,48]. Despite the improvements, these methods are sub-optimal for the task of people matching across multiple views in crowded scenarios. The reasons are threefold. Firstly, intra-class variation of human body appearance is relatively smaller than objects such as architectural features or graffiti paintings, and thus more outliers can result if the aforementioned methods are deployed directly. Secondly, dense feature matching across whole images is usually computationally inefficient for applications such as autonomous driving, where real-time is one of the primary concerns. Thirdly, appearance-based matching has a lower correctness criterion than people-based matching across multiple views. On the other hand, it is interesting to note that the level of occlusion in the same object can differ drastically among different views. Therefore, it is reasonable to trust the slightly occluded views more than the highly occluded views in the process of triangulation.

In this paper, we propose a 3D crowd human pose estimation method based on multi-view geometry. Specifically, we focus on overcoming the bottlenecks of multi-person 3D pose estimation and pushing it further to dense crowd 3D pose estimation. To this end, we propose the matching of feet across multiple views to improve the accuracy of body joint correspondences. We first modify a 2D pose estimation network, i.e. the joint-candidates single person pose estimation (SPPE) [28] to include additional joints for the feet. Subsequently, we find the best matches of the feet across multiple views, and then extend the correspondences to the other joints using the kinematic chain of the human body. We cast the matching problem as a binary linear program and solve it efficiently with the Jonker-Volgenant algorithm [22]. Finally, we improve the robustness of triangulation by formulating the problem as a maximum a posteriori (MAP) estimation that weighs the likelihood term with the uncertainty of the 2D joint observation and enforces a prior on the average bone lengths of the estimated 3D human poses. We evaluate our proposed method on four challenging benchmark datasets. Experimental results show that our method outperforms all existing algorithms on these datasets.

Our main contributions in this work are summarized as follows:

- Design a simple and efficient people matching mechanism based on feet assignment across different views, which is applicable for dense crowds.
- Propose a more robust triangulation for 3D crowd reconstruction using MAP estimation that accounts for the uncertainty of 2D joint detection and enforces the average 3D bone lengths.
- Define a problem of crowd 3D human pose estimation, and argue its existence as a separate problem from multi-person multi-camera 3D human pose estimation.

2 Related Work

Single-Person Human Pose Estimation. A large amount of literature exists in this field due to the advancement of deep learning. We briefly summarize those for 3D human pose which are more closely related to this work. State-of-the-art methods can be divided into two categories, direct regression methods [10,21,36] and indirect regression methods based on heat maps [20,27,37]. In [37], a coarse-to-fine prediction scheme was developed by analyzing 3D human pose in a volumetric representation. Integral pose [42] unifies the heat map representation and joint regression by replacing the non-differentiable *argmax* with integral operation. Regardless of the good performance, learning 3D pose from a single image is still an ill-posed problem. Instead of finding one exact solution, [27] developed a multimodal mixture density network, so that multiple feasible solutions are found before refining into one solution. The authors of [20] proposed a volumetric aggregation from intermediate 2D backbone feature maps and combines 3D information from multiple 2D views. The aforementioned methods obtained state of the art performance for single person 3D pose estimation, but unfortunately in the multi-person scenario, additional ambiguity makes these methods suboptimal.

Multi-person Human Pose Estimation. Several recent works have focused on multi-person scenarios in problem formulation either based on monocular setting [44] or multi-view setting [2–5,13,24]. Results obtained from the multi-view setting are generally more precise due to the additional information. However, bottlenecks still exist in these multi-view based methods, *i.e.*how to cope with the correspondence problem and how to make the triangulation of depth information sufficiently robust against noise. In [24], epipolar constraints are directly applied for people assignment among different views. This worked perfectly when people in the scene stand far away from each other. However, this constraint is likely to fail when the scenario gets crowded. For instance, if some epipolar line of a particular joint happens to pass through several other people, it is hard to make sure that no other joint is closer to the line than the correct matching joint. The authors of [13] incorporated appearance cues by fusing re-identification with epipolar constraints. However, the two kinds of constraints are still independently considered. The 3D pictorial structure model [2,3] resolves ambiguities of mixed parts, occlusion, and false positives by building multi-view unary potentials, while at the same time integrating prior model by pairwise and ternary potential functions. This motivates our work in using MAP as a formulation to cope with measurement noise in triangulation process.

Previous 'multi-person' methods work on relatively sparse crowds. In [28], crowd pose estimation is firstly defined as a separate research field, but the problem is defined in 2D. When extending to 3D, more uncertainties are introduced. This encourages us to define the crowd pose estimation problem in 3D and explore a potential solution in this paper.

Feature Matching and Correspondence Problem. Feature correspondence in general raises stricter demand than feature matching due to the fact that both appearance and location need to be taken into consideration. In [6], a globally-optimal inlier set cardinality maximization approach is proposed to jointly estimate optimal camera pose and optimal correspondences. [46] solves the correspondence problem between two images by defining energy function measuring data consistency and spatial regularity. In [14], Point-Line Minimal Problems are thoroughly defined and analyzed. This provides a theoretical guidance to solve the specific problem of point line matching for the people assignment task.

3 Our Method

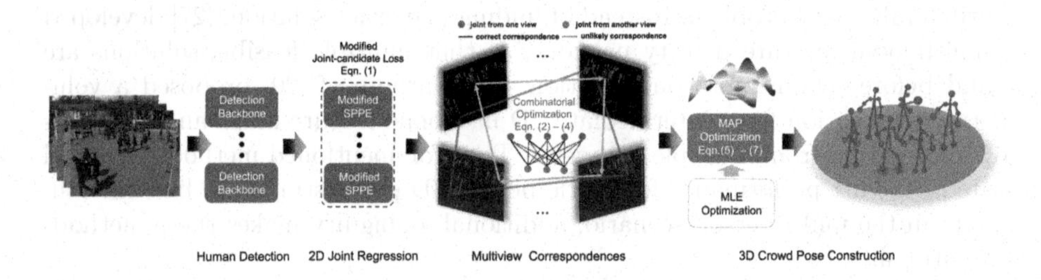

Fig. 1. The pipeline of our proposed approach. See text for more detail.

Figure 1 shows an overview of our approach. Human bounding box proposals are first obtained by an off-the-shelf detection network, and then fed into a modified SPPE network (Sect. 3.1) to estimate the 2D joints. Subsequently, we get the multi-view joint correspondences by solving a combinatorial optimization problem via graph matching (Sect. 3.2). Finally, the 3D crowd poses are reconstructed using a MAP formulation (Sect. 3.3) solved by the trust region method [11].

3.1 2D Pose Estimation

We leverage on the recently proposed CrowdPose network [28] trained on the CrowdPose Dataset [28] for 2D pose detection on the input images. The CrowdPose network follows a top-down framework. It first detects the bounding boxes of individual persons using YOLOv3 [39], and then performs joint-candidate SPPE and a global maximum joints association algorithm to estimate the 2D joints. Similar to other 2D pose estimation methods, the accuracy of the joint detection drops as it moves farther away from the center of a person (i.e. the 'hip' joint) despite the state-of-the-art performance of [28] on the benchmark datasets. As a result, detection of the 'ankle' joints, which are usually used to represent feet, are especially noisy. To mitigate this problem, we follow [7] in adding 6 additional joints on the feet (3 on each foot) and modify the loss function of the network into the weighted sum of the mean square error (i.e. MSE[.,.]) from the body joints and the feet joints as follows:

$$\mathcal{L} = \frac{1}{I + 6\lambda} \left\{ \sum_{i=1}^{I} \text{MSE}\left[\mathbf{P}_h^i, \mathbf{T}_h^i + \mu\mathbf{C}_h^i\right] + \lambda \sum_{i=I+1}^{I+7} \text{MSE}\left[\mathbf{P}_h^i, \mathbf{T}_h^i + \mu\mathbf{C}_h^i\right] \right\}. \quad (1)$$

I stands for the number of joints of the body part excluding the 6 joints representing the feet (e.g. $I = 17$ for MSCOCO [32]). $\mathbf{P}_h^i$ and $\mathbf{T}_h^i$ represents the output heatmap and the heatmap of the target joints, respectively, for the ith joint of the hth person. $\mathbf{C}_h^i$ represents detections of the same joint type from other persons that might be within the bounding box of the hth person. We include $\mathbf{C}_h^i$ into the loss function to learn a multi-modal heatmap $\mathbf{P}_h^i$. μ is the attention factor in the range of $[0, 1]$ to control the extent of the contribution of $\mathbf{C}_h^i$, which we set to 0.5 in all our implementations. We set $\lambda > 1$, so that the 6 additional joints on the feet receive more attention during training. Our network is trained on the Human Foot Keypoint Dataset [7].

3.2 Multi-view Correspondence with Graph Matching

Previous methods [13,24] apply epipolar constraints to all joints in order to solve the correspondence problem. We argue that this can give a suboptimal solution when the crowd becomes denser. This is because the epipolar line that corresponds to a joint in one view is likely to pass through multiple joints in the other view for a crowded scene. Consequently, this ambiguity renders the Euclidean distance between the epipolar line and joints to be a less ideal metric. We circumvent this challenge by casting the joint correspondence problem into a feet assignment problem. Specifically, we first establish the feet that belong to a same person across the multiple views, and then grow the joint correspondences from the feet using the kinematic chain of the human body.

Feet Assignment. We propose to use feet assignment to realize people matching as shown in Fig. 2(a). The core intuition is that prior information, appearance constraints, location constraints are naturally fused in such setting. We use the fact that at least one foot is on the ground when a person is walking as the prior information. The detected joints of the feet as described in Sect. 3.1 are used as the appearance information. To incorporate location constraints, we use the homographies between all view pairs to rectify the ground planes among different views into a common reference. We denote the homography between the ground planes of view j and k as $H_{j,k}$. Consequently, we can directly compare the joints of the feet across different views. We get ≥4 point correspondences between the ground plane of each pair of view j and k to compute $H_{j,k}$. It is interesting to note that applying the homography to all pixels in the image, we might get a twisted image which appear to be strange at first glance. This is based on the prior that this 'the world is 3D'. However, if we change the prior into 'the world is 2D', and treat everything as chalk art drawn on the ground, then everything in the rectified image starts to look reasonable. In this light, the problem of joint matching boils down to feet assignment.

Graph Building. A naive search for the optimal feet assignment is intractable due to the large combinatorial search space. To improve the efficiency of the search, we build a complete bipartite graph from the feet across two views and solve it as a linear assignment problem. Let $\mathcal{V}_j = \{v_{ij} : \forall i \in \{1, \ldots, a_j\}\}$ denote the set of pair-of-feet in view j. v_{ij} is the detected pair-of-feet with index i in view j, and a_j is the total number of detected pair-of-feet in view j. We further denote the set of edges in the complete bipartite graph for the pair of views j and k as $\mathcal{E}_{j,k} = \{e_{l,m} : \forall l \in \{1, \ldots, a_j\}, m \in \{1, \ldots, a_k\}\}$. The complete bipartite graph for each pair of views can then be formally written as:

$$\mathcal{K}_{a_j,a_k} = ((\mathcal{V}_j, \mathcal{V}_k), \mathcal{E}_{j,k}), \tag{2}$$

Optimal Cross-view Matching. Based on this construction, our goal becomes finding a subgraph $\mathcal{G} \subset \mathcal{K}_{a_j,a_k}$ by eliminating edges in the graph that represent the unlikely correspondences. We solve this edge elimination problem as a binary linear program that minimizes the total edge costs subjected to a set of linear constraints, i.e.

$$\begin{aligned} \min_{\mathbf{d}} \quad & \sum_{l=1}^{a_j} \sum_{m=1}^{a_k} c_{l,m} \cdot d_{l,m} \\ \text{s.t.} \quad & \sum_{l=1}^{a_j} c_{l,m} \leq 1, \quad \sum_{m=1}^{a_k} c_{l,m} \leq 1, \\ & \sum_{l=1}^{a_j} d_{l,m} = 1, \quad \sum_{m=1}^{a_k} d_{l,m} = 1, \quad \mathbf{d} \in \{0,1\}^{a_j \times a_k}. \end{aligned} \tag{3}$$

$d_{l,m} \in \mathbf{d}$ is a binary variable that represents the selection of the edge $e_{l,m}$ when it is equals to 1. $c_{l,m}$ is the cost of selecting the edge $e_{l,m}$, which we define as:

$$c_{l,m} = k_1 \cdot |p_l - H_{j,k} \cdot p_m| + k_2 \cdot ||\mathbf{v}_l|| - ||\mathbf{v}_m|| + k_3 \cdot \left(\frac{\mathbf{v}_l \times \mathbf{v}_m}{|\mathbf{v}_l| \cdot |\mathbf{v}_m|} \right), \tag{4}$$

where p_l and p_m respectively represents the location of two pairs of feet, $H_{j,k}$ represents the homography matrix between the two views j and k, $\mathbf{v}_l$ and $\mathbf{v}_m$ represent vectors of strides. k_1, k_2, k_3 are hyper parameters to adjust the importance between the foot location, stride size, and stride direction. The metric is visualized in Fig. 2(b).

Solver. We use the Jonker-Volgenant algorithm [22] as the solver to find the solution to the two-view feet assignment problem formulated in Eq. 3. We ensure consistency of the assignment across multiple views by resolving the conflict in the correspondences with priority given to edges with lower edge cost as defined in Eq. 4. A directed graph where the skeleton is a spanning union of disjoint cycles is obtained when the matching across n views is successful. Our matching algorithm has a time complexity of $\mathcal{O}((2N)^3) = \mathcal{O}(8N^3)$, where N is the number of persons per image. In contrast, the $O(n^4)$ implementation of Hungarian algorithm has a total time complexity of $\mathcal{O}((17N)^4)$ on 17 joints. Although the constant term is usually considered unimportant for time complexity analysis, it cannot be neglected in this study since $N < 30$ usually holds. Thus, our method is significantly faster.

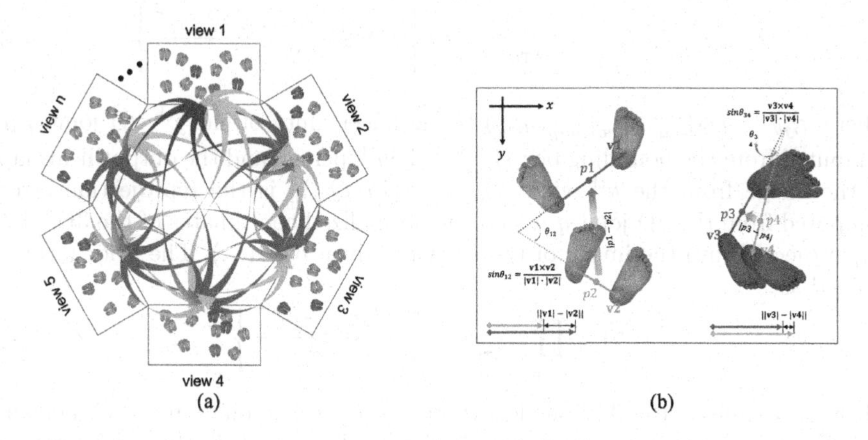

Fig. 2. People matching using feet assignment. (a) The matching process across n views, and (b) visualization of edge cost defined in Eq. 4.

3.3 3D Crowd Pose Reconstruction

Under the assumption that the camera parameters are known, we can reconstruct the 3D human poses by triangulation of the joint correspondences across the multiple views obtained from the previous section. One naive method of triangulation is to directly minimize the squared sum of perpendicular distances between the epipolar line and the detected joint. We refer to this naive method as the vanilla triangulation method. This is a classical method that works well in single person scenarios. However, in occluded scenes, the 2D joints are noisy and might have shifts of a few pixels. Consequently, this breaks the correspondence across multiple views and causes the 3D reconstructed points to be unreliable. We formulate a MAP optimization to mitigate the problem from the unreliable correspondences, where we model the likelihood with the 2D measurement uncertainty and use the prior term to constrain the bone lengths of the estimated body poses.

MAP Optimization. The ultimate goal of the proposed method is to estimate 3D coordinates of human joints. We formulate this as a MAP over the latent 3D poses $\mathbf{Q}$, i.e.

$$\mathbf{Q}_{\text{MAP}} = \underset{\mathbf{Q}}{\operatorname{argmax}} \prod_{i=1}^{N} P(Q_i) \prod_{j=1}^{M} \prod_{k=1}^{O} P(q_{ijk} \mid \mathcal{P}_k, Q_{ij}), \tag{5}$$

where N is the total number of persons in the scene, M is the number of joints per person, and O is the total number of camera views. q_{ijk} is the jth 2D joint of the ith person in the kth camera view. $Q_{ij} \in Q_i$ is the jth 3D joint from the 3D pose $Q_i \in \mathbf{Q}$ of the ith person in the scene. $\mathcal{P}_k$ is the projection matrix of the kth camera. The likelihood term is given by the following Gaussian distribution:

$$P(q_{ijk} \mid \mathcal{P}_k, Q_{ij}) = \frac{1}{2\pi\sigma_{ijk}} \exp\left\{ -\frac{\|q_{ijk} - \alpha(\mathcal{P}_k, Q_{ij})\|^2}{2{\sigma_{ijk}}^2} \right\}, \tag{6}$$

where $\sigma_{ijk} = f(s^i_{bbox}, s^k_{heatmap}, q_{ijk})$ is the uncertainty of the jth 2D joint q_{ijk} computed from the bounding box s^i_{bbox} of the ith person and the output heatmap of the image from the kth view. $\|q_{ijk} - \alpha(\mathcal{P}_k, Q_{ij})\|$ is the reprojection error computed from the 2D joint q_{ijk} and the normalized coordinates of the 3D joint Q_{ij} projected into the image of the kth view given by $\alpha(.,.)$. The prior term is defined as:

$$P(Q_i) = \prod_{l=1}^{L} \frac{1}{2\pi\sigma_l} \exp\left\{ -\frac{\|b^l_{\text{ref}} - b^l_i\|^2}{2\sigma_l^2} \right\}, \tag{7}$$

where b^l_i represents the lth bone length between two 3D joints in the ith person, and b^l_{ref} represents the average length of the lth bone. L is the total number of bones in the human body representation. σ_l is the standard deviation in the length of the lth bone. Intuitively, the prior term enforces the bone lengths of the estimated 3D human pose to be close to the average lengths.

Initialization and Solver. We initialize the iterative MAP optimization with the vanilla triangulation. Subsequently, we use the trust region method [11] as a solver for the MAP optimization. In addition, we empirically observe that performing the maximum likelihood estimation (MLE) with the initialized values as an intermediate step before MAP improves the final estimation of the 3d human poses.

4 Experiments

We evaluate our proposed method on four public datasets. These datasets consist of scenarios that include autonomous driving and surveillance with challenging situations such as moving camera and heavy occlusions.

4.1 Datasets

LOEWENPLATZ [15]. This is a dataset of driving recorder scenario captured in Zurich with two calibrated cameras. The dataset represents common scenarios that autonomous driving cars are likely to experience everyday.

Chariot Mk I [16]. This is a dataset captured by hand-held cameras. The cameras are moving and shaking, which resemble real-life scenarios from the perspective of the pedestrians.

Wildtrack [9]: This dataset emulates surveillance scenarios with the set-up of 7 fixed cameras. All cameras are fully calibrated, i.e. known intrinsics and extrinsics camera parameters. Occlusion is severe in each view of this dataset.

CMU Panoptic Dataset [23]: This dataset is captured in a studio and provides precise 3D ground truth in MSCOCO [32] format. In this paper, we evaluate the performance of our method quantitatively on the 'Ultimatum' sequences with complete 3D human pose annotations. This sequence consists of relatively more active and complicated social scenarios for human pose estimation than other sequences.

Table 1. Quantitative results for the Chariot Mk I, LOEWENPLATZ, and Wildtrack datasets using the evaluation metrics from MSCOCO [32].

Chariot Mk I	**AP**	AP^{50}	AP^{75}	AP^{M}	AP^{L}	**AR**	AR^{50}	AR^{75}	AR^{M}	AR^{L}
Belagiannis *et al.* [3]	48.1	64.8	59.3	63.7	64.6	58.1	62.7	55.9	54.4	61.9
Dong *et al.* [13]	69.3	87.4	73.6	77.5	75.4	71.9	87.5	81.7	78.1	80.0
Ours w/ Vanilla Trigulation	60.0	90.8	72.2	65.4	77.6	72.3	95.3	83.0	76.6	81.8
Ours w/ Proposed MAP	**89.8**	**98.9**	**92.7**	**91.7**	**99.5**	**93.9**	**99.8**	**96.0**	**95.4**	**99.6**
LOEWENPLATZ	**AP**	AP^{50}	AP^{75}	AP^{M}	AP^{L}	**AR**	AR^{50}	AR^{75}	AR^{M}	AR^{L}
Belagiannis *et al.* [3]	49.3	63.7	58.2	63.2	56.9	61.9	84.3	64.3	73.7	55.3
Dong *et al.* [13]	62.1	88.3	63.5	61.3	72.5	80.3	87.2	77.9	81.7	84.6
Ours w/ Vanilla Trigulation	66.7	93.8	73.1	71.6	84.4	78.2	96.7	84.5	80.1	88.9
Ours w/ Proposed Optimization	**81.8**	**97.1**	**88.7**	**83.3**	**90.8**	**88.9**	**98.5**	**93.5**	**90.0**	**94.4**
Wildtrack	**AP**	AP^{50}	AP^{75}	AP^{M}	AP^{L}	**AR**	AR^{50}	AR^{75}	AR^{M}	AR^{L}
Belagiannis *et al.* [3]	44.1	53.4	46.0	19.4	47.8	64.1	79.1	61.4	20.9	55.4
Dong *et al.* [13]	55.6	78.4	53.1	34.9	60.0	73.4	87.8	68.1	38.1	77.6
Ours w/ Vanilla Trigulation	55.3	79.6	50.6	33.2	60.1	77.3	88.7	72.9	38.6	78.4
Ours w/ Proposed MAP	**70.0**	**90.2**	**73.6**	**44.7**	**76.4**	**78.3**	**93.6**	**82.4**	**55.5**	**83.7**

4.2 Results

Quantitative Results. We adopt the key point evaluation metrics of MSCOCO [32], i.e. the average precision (AP), average recall (AR) and their variants. Specifically, the variants of AP and AR are specified by the Object Keypoint Similarity (OKS) that plays the same role as the Intersection over Union (IoU) in object detection. It measures the scale of the object, and the distance between predicted joints and ground truth points. The AP at OKS = .50:.05:.95 (primary challenge metric in MSCOCO [32] competitions) is used to measure the reprojection errors. Table 1 shows that our method outperforms the state-of-the-art algorithms on the Chariot Mk I, LOEWENPLATZ, and Wildtrack datasets using the evalution metrics from MSCOCO [32]. Table 2 shows the comparative performance for 2D key point detection of our modified body+foot candidate-joint SPPE network on the MSCOCO dataset [32]. Our method achieves a comparable performance with the best performing [8]. Furthermore, our method outperforms on AP@0.5:0.95 for medium objects, which is more valuable for our framework with the feet detection, matching and optimization stages. CMU Panoptic Dataset provides the 3D ground truth. Therefore, we use two metrics, i.e. mean per joint position error (MPJPE) and percentage of correct parts (PCP) instead of the reprojection error for direct evaluation. The results are shown in Table 3 and Table 4.

Table 2. Quantitative comparison of key point detection experiments on COCO body + foot validation set [7].

Method	**AP**	AP^{50}	AP^{75}	AP^{M}	AP^{L}
GT Bbox + CPM [45]	62.7	86.0	69.3	58.5	70.6
SSD [33] + CPM [45]	52.7	71.1	57.2	47.0	64.2
Cao *et al.* [8]	65.3	**85.2**	71.3	62.2	**70.7**
Ours	**65.3**	80.1	**72.2**	**74.1**	68.3

Table 3. Quantitative results for the proposed method on different joints of human body in CMU Panoptic Dataset (Ultimatum sequences, four cameras) using MPJPE (mm).

Metric	**Average**	Head	Shoulder	Elbow	Wrist	Hip	Knee	Foot
MPJPE	50.0	45.1	43.6	55.6	60.7	25.3	53.2	66.0

Table 4. Quantitative results for the proposed method on different body parts in CMU Panoptic Dataset (Ultimatum sequences, four cameras) using the PCP metric.

Metric	**PCP**	Head	Torso	Upper arms	Lower arms	Upper legs	Lower legs
Percentage	91.3	74.5	100.0	93.8	80.0	100.0	99.3

Table 5. Ablation study of MLE as an intermediate step on WildTrack dataset.

Method	ave	min	max	var
Ours w/o MLE	64.75	18.06	316.7	50.69
Ours w/ MLE	**38.55**	**2.18**	**219.29**	**27.52**

Qualitative Results. Figure 3, 4, and 5 show the qualitative results on the Wildtrack [9], CMU Panoptic [23], and LOEWENPLATZ [15] datasets, respectively. In Fig. 3, our approach gives good quality 3D reconstructions of the human poses even when heavy occlusion happens in the crowded scene. To validate effectiveness of the proposed method, we choose crowded scenes with at least 5 people appearing in each frame as shown Fig. 4. We further show the qualitative visualizations of the estimated 3D human pose of several single persons from our method with the ground truth. Location information is used to match estimated pose with ground truth of each individual person. Orange represents estimated skeleton and blue represents ground truth. We zoom in each skeleton to clearly show details. As can be observed, the blue skeleton and orange skeleton has a slight offset. Nonetheless, this offset is in a tolerable range. In Fig. 5, we evaluate our method under the setting of autonomous driving. The car went straight, turned left, and stopped at a crosswalk. We can see that our proposed method gives good 3D human pose estimations in different road scenes from a moving camera.

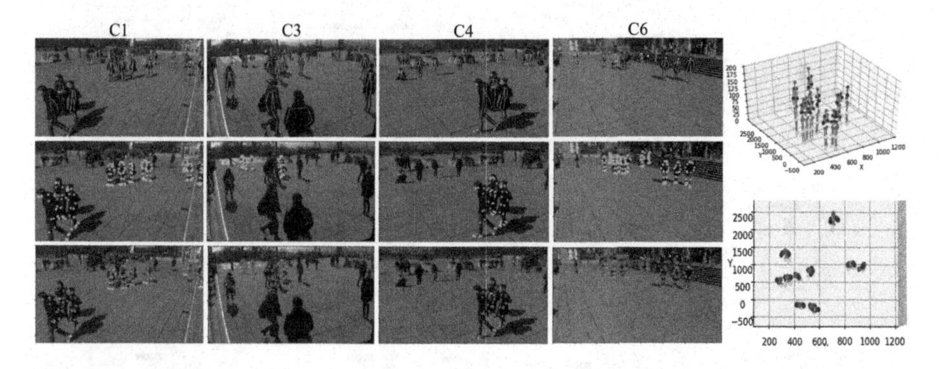

Fig. 3. Qualitative results on Wildtrack dataset. (First four columns) First row shows results of our modified candidate joint SPPE with attention on the feet; Second row shows the ground truth 2D joints (blue dots); Third row shows the reprojection of our estimated 3D joints (orange dots) overlaid on the ground truths (blue dots). The last column shows the (top) estimated 3D crowd human poses and its (bottom) top view. (Color figure online)

Table 6. Foot keypoint analysis on COCO foot validation set.

Method	**AP**	**AR**	AP^{75}	AR^{75}
Cao et al. [8]	77.9	82.5	82.1	85.6
Our	80.1	82.0	85.5	87.4

Table 7. Evaluation of correspondence process on CMU Panoptic Dataset.

Dataset	RANSAC	EC	Ours
Precision	46.0	86.5	93.7
Time complexity	NA	$O((17N)^4)$	$O((2N)^3)$

Ablation Study. We perform ablation studies to show the effectiveness of our proposed loss function Eq. 1 for 2D pose estimation, and the MLE as an intermediate step. We define an error distance between the reprojection of a 3D

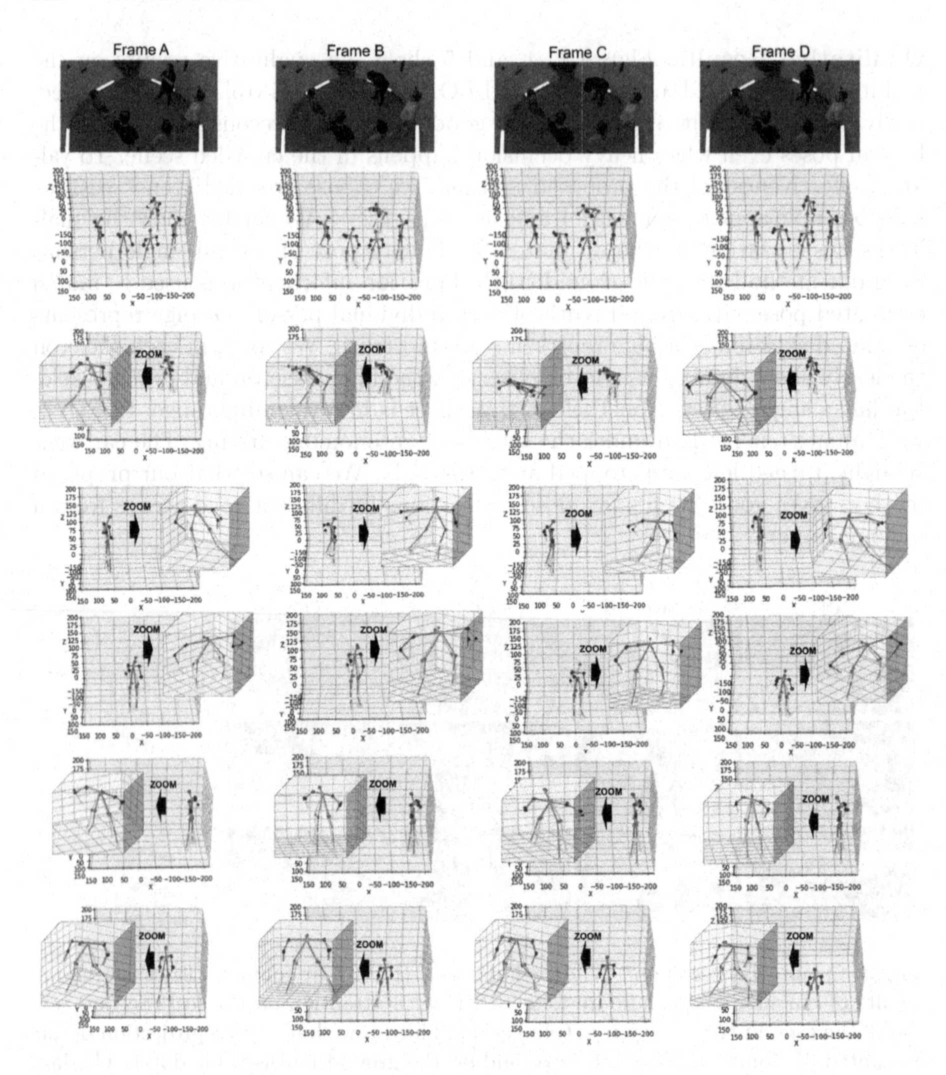

Fig. 4. Qualitative results on CMU Panoptic dataset. The first row shows images from the 4 cameras in the setup. The second row shows 3D crowd pose. The third to seventh row visualize the estimated 3D pose of each person (orange skeleton) and its corresponding ground truth (blue skeleton). (Color figure online)

point and its corresponding 2D ground truth for quantitative evaluation. Comparison is carried out between the results from MAP with and without MLE as an intermediate step on the WildTrack dataset. In Table 5, we show the average, minimum, maximum, and variance of the reprojection error distance. Figure 7 shows the histogram of error distribution in pixel unit. We can see that the smaller errors of the estimated 3D poses are obtained with the MLE as an intermediate step. Figure 6 demonstrates the effectiveness of our proposed loss

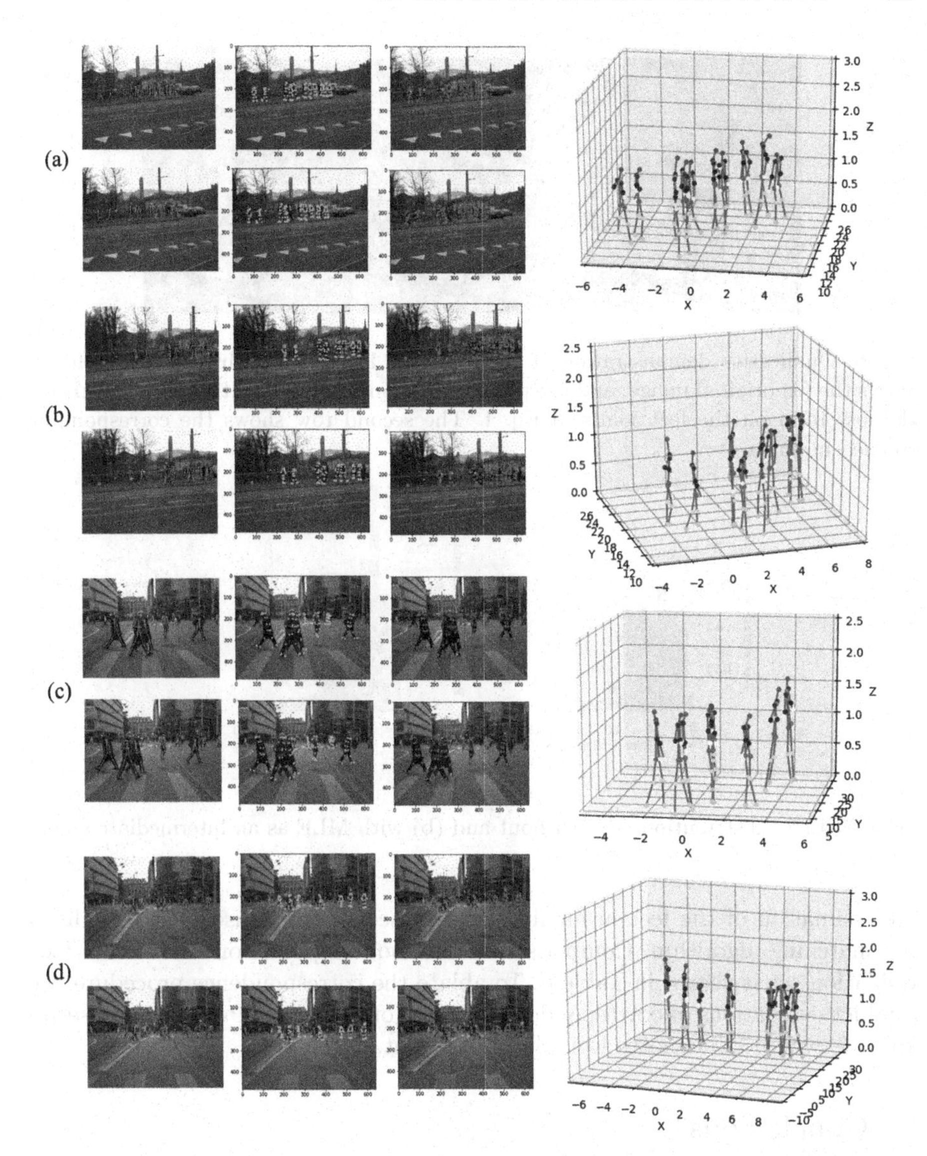

Fig. 5. Qualitative results on the LOEWENPLATZ dataset. The right most column shows the estimated 3D poses of scene (a)-(d). The first column shows the 2D skeletons detected by our modified SPPE network, the second column shows the ground truths of the 2D joints (blue dots), and the third column shows the reprojection of our estimated 3D joints (orange dots) and overlaid on the ground truths (blue dots). (Color figure online)

function Eq. 1 for 2D pose estimation. As can be seen in the figure, our network detects the 'big toe', 'small toe' and 'heel' instead of the usual 'ankle' for the representation of a foot. The increased attention of the feet joints improves

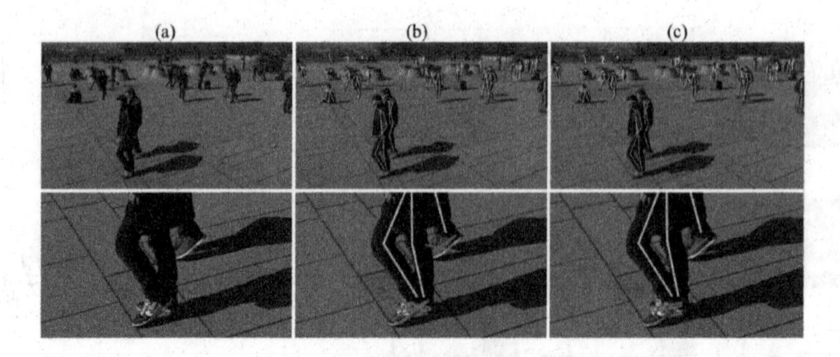

Fig. 6. Qualitative demonstration of our proposed loss function in Eq. 1. The figure shows the (a) original image, and the pose estimation result (b) **with** and (c) **without** the loss term on the feet joints in Eq. 1. The second row shows the corresponding zoomed-in images.

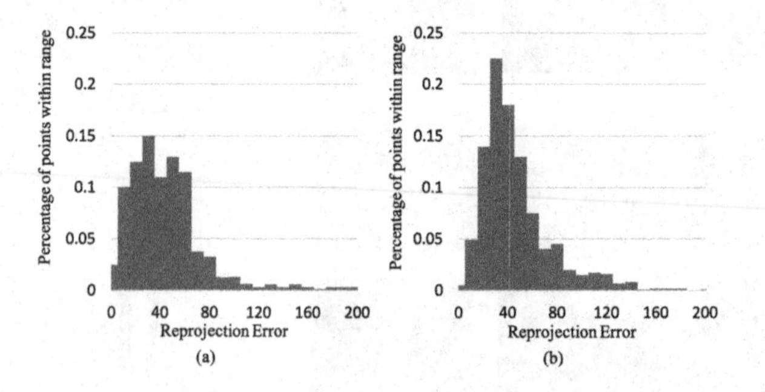

Fig. 7. Error distributions: (a) without and (b) with MLE as an intermediate step.

the estimation of the feet in highly occluded scene, and consequently facilities our matching algorithm. Comparison of the foot keypoints on the COCO foot validation set is shown in Table 6. To ablate the correspondence procedure, we conduct evaluations of correspondence process on the CMU Panoptic dataset in Table 7, where EC denotes Epipolar Constraint.

5 Conclusions

In this work, we propose a simple and effective approach for multi-person 3D pose estimation applicable to dense crowds. Matching of feet across multiple views improves the accuracy of body joint correspondences. A graph model is used for fast cross-view matching based on accurate estimation of foot joints. We cast the bipartite matching problem as a binary linear program and solve it efficiently with the Jonker-Volgenant algorithm. The robustness of triangulation is improved by using a MAP estimation that weighs the likelihood term with the uncertainty of the 2D joint observation and enforces a prior on the average bone lengths of the estimated 3D human poses. Experimental results show that our method outperforms all existing algorithms on four public datasets.

Acknowledgements. The authors would like to thank Yawei Li and Weixiao Liu for useful discussion. This work is supported in parts by the Office of Naval Research Award N00014-17-1-2142 and the Singapore MOE Tier 1 grant R-252-000-A65-114.

References

1. Baqué, P., Fleuret, F., Fua, P.: Deep occlusion reasoning for multi-camera multi-target detection. In: Proceedings of the ICCV, pp. 271–279 (2017)
2. Belagiannis, V., Amin, S., Andriluka, M., Schiele, B., Navab, N., Ilic, S.: 3D pictorial structures for multiple human pose estimation. In: Proceedings of the CVPR, pp. 1669–1676 (2014)
3. Belagiannis, V., Amin, S., Andriluka, M., Schiele, B., Navab, N., Ilic, S.: 3D pictorial structures revisited: Multiple human pose estimation. IEEE Trans. PAMI **38**(10), 1929–1942 (2015)
4. Belagiannis, V., Wang, X., Schiele, B., Fua, P., Ilic, S., Navab, N.: Multiple human pose estimation with temporally consistent 3D pictorial structures. In: Agapito, L., Bronstein, M.M., Rother, C. (eds.) ECCV 2014. LNCS, vol. 8925, pp. 742–754. Springer, Cham (2015). https://doi.org/10.1007/978-3-319-16178-5_52
5. Bogo, F., Kanazawa, A., Lassner, C., Gehler, P., Romero, J., Black, M.J.: Keep It SMPL: automatic estimation of 3D human pose and shape from a single image. In: Leibe, B., Matas, J., Sebe, N., Welling, M. (eds.) ECCV 2016. LNCS, vol. 9909, pp. 561–578. Springer, Cham (2016). https://doi.org/10.1007/978-3-319-46454-1_34
6. Campbell, D., Petersson, L., Kneip, L., Li, H.: Globally-optimal inlier set maximisation for simultaneous camera pose and feature correspondence. In: Proceedings of ICCV, pp. 1–10 (2017)
7. Cao, Z., Hidalgo, G., Simon, T., Wei, S.E., Sheikh, Y.: OpenPose: realtime multi-person 2D pose estimation using part affinity fields. In: arXiv preprint arXiv:1812.08008 (2018)
8. Cao, Z., Simon, T., Wei, S.E., Sheikh, Y.: Realtime multi-person 2D pose estimation using part affinity fields. In: Proceedings of the CVPR, pp. 7291–7299 (2017)
9. Chavdarova, T., et al.: WILDTRACK: a multi-camera HD dataset for dense unscripted pedestrian detection. In: Proceedings of the CVPR, pp. 5030–5039 (2018)
10. Chen, C.H., Ramanan, D.: 3D human pose estimation = 2D pose estimation + matching. In: Proceedings of the CVPR, pp. 7035–7043 (2017)
11. Conn, A.R., Gould, N.I., Toint, P.L.: Trust Region Methods, vol. 1. SIAM, Philadelphia (2000)
12. Dinesh Reddy, N., Vo, M., Narasimhan, S.G.: CarFusion: combining point tracking and part detection for dynamic 3D reconstruction of vehicles. In: Proceedings CVPR, pp. 1906–1915 (2018)
13. Dong, J., Jiang, W., Huang, Q., Bao, H., Zhou, X.: Fast and robust multi-person 3D pose estimation from multiple views. In: Proceedings of the CVPR, pp. 7792–7801 (2019)
14. Duff, T., Kohn, K., Leykin, A., Pajdla, T.: PLMP-point-line minimal problems in complete multi-view visibility. In: Proceedings of the ICCV, pp. 1675–1684 (2019)
15. Ess, A., Leibe, B., Schindler, K., Gool, L.V.: Robust multiperson tracking from a mobile platform. IEEE Trans. PAMI **31**, 1831–1846 (2009)
16. Ess, A., Leibe, B., Schindler, K., Van Gool, L.: A mobile vision system for robust multi-person tracking. In: Proceedings of the CVPR, pp. 1–8. IEEE (2008)

17. Fernando, T., Denman, S., Sridharan, S., Fookes, C.: Tracking by prediction: a deep generative model for mutli-person localisation and tracking. In: Proceedings of the WACV, pp. 1122–1132. IEEE (2018)
18. Fischler, M.A., Bolles, R.C.: Random sample consensus: a paradigm for model fitting with applications to image analysis and automated cartography. Commun. ACM **24**(6), 381–395 (1981)
19. Harris, C.G., Stephens, M., et al.: A combined corner and edge detector. In: Alvey Vision Conference, vol. 15, pp. 10–5244. Citeseer (1988)
20. Iskakov, K., Burkov, E., Lempitsky, V., Malkov, Y.: Learnable triangulation of human pose. In: Proceedings of the ICCV, pp. 7718–7727 (2019)
21. Jahangiri, E., Yuille, A.L.: Generating multiple diverse hypotheses for human 3D pose consistent with 2D joint detections. In: Proceedings of the ICCVW, pp. 805–814 (2017)
22. Jonker, R., Volgenant, A.: A shortest augmenting path algorithm for dense and sparse linear assignment problems. Computing **38**(4), 325–340 (1987)
23. Joo, H., et al.: Panoptic studio: a massively multiview system for social motion capture. In: Proceedings of the ICCV (2015)
24. Kadkhodamohammadi, A., Padoy, N.: A generalizable approach for multi-view 3D human pose regression. arXiv preprint arXiv:1804.10462 (2018)
25. Korman, S., Milam, M., Soatto, S.: OATM: occlusion aware template matching by consensus set maximization. In: Proceedings of the CVPR, pp. 2675–2683 (2018)
26. Kubo, H., Jayasuriya, S., Iwaguchi, T., Funatomi, T., Mukaigawa, Y., Narasimhan, S.G.: Programmable non-epipolar indirect light transport: Capture and analysis. IEEE Trans. VCG (2019)
27. Li, C., Lee, G.H.: Generating multiple hypotheses for 3D human pose estimation with mixture density network. In: Proceedings of the CVPR, pp. 9887–9895 (2019)
28. Li, J., Wang, C., Zhu, H., Mao, Y., Fang, H.S., Lu, C.: Crowdpose: efficient crowded scenes pose estimation and a new benchmark. In: Proceedings of the CVPR, pp. 10863–10872 (2019)
29. Li, Y., Agustsson, E., Gu, S., Timofte, R., Van Gool, L.: CARN: convolutional anchored regression network for fast and accurate single image super-resolution. In: Proceedings of the ECCV, p. 0 (2018)
30. Li, Y., Gu, S., Mayer, C., Van Gool, L., Timofte, R.: Group sparsity: the hinge between filter pruning and decomposition for network compression. In: Proceedings of CVPR (2020)
31. Li, Y., Tsiminaki, V., Timofte, R., Pollefeys, M., Van Gool, L.: 3D appearance super-resolution with deep learning. In: Proceedings of the CVPR, pp. 9671–9680 (2019)
32. Lin, T.-Y., et al.: Microsoft COCO: common objects in context. In: Fleet, D., Pajdla, T., Schiele, B., Tuytelaars, T. (eds.) ECCV 2014. LNCS, vol. 8693, pp. 740–755. Springer, Cham (2014). https://doi.org/10.1007/978-3-319-10602-1_48
33. Liu, W., et al.: SSD: single shot multibox detector. In: Leibe, B., Matas, J., Sebe, N., Welling, M. (eds.) ECCV 2016. LNCS, vol. 9905, pp. 21–37. Springer, Cham (2016). https://doi.org/10.1007/978-3-319-46448-0_2
34. Liu, X., et al.: Extremely dense point correspondences using a learned feature descriptor. In: Proceedings of the CVPR (2020)
35. Lowe, D.G.: Object recognition from local scale-invariant features. In: Proceedings of the ICCV, vol. 2, pp. 1150–1157 (1999)
36. Mahendran, S., Ali, H., Vidal, R.: 3D pose regression using convolutional neural networks. In: Proceedings of the ICCVW, pp. 2174–2182 (2017)

37. Pavlakos, G., Zhou, X., Derpanis, K.G., Daniilidis, K.: Coarse-to-fine volumetric prediction for single-image 3D human pose. In: Proceedings of the CVPR (2017)
38. Reddy, N.D., Vo, M., Narasimhan, S.G.: Occlusion-Net: 2D/3D occluded keypoint localization using graph networks. In: Proceedings of the CVPR, pp. 7326–7335 (2019)
39. Redmon, J., Farhadi, A.: YOLOv3: an incremental improvement. arXiv preprint arXiv:1804.02767 (2018)
40. Sindagi, V.A., Patel, V.M.: Multi-level bottom-top and top-bottom feature fusion for crowd counting. In: Proceedings of the ICCV, pp. 1002–1012 (2019)
41. Stoll, C., Hasler, N., Gall, J., Seidel, H.P., Theobalt, C.: Fast articulated motion tracking using a sums of Gaussians body model. In: Proceedings of the ICCV, pp. 951–958. IEEE (2011)
42. Sun, X., Xiao, B., Wei, F., Liang, S., Wei, Y.: Integral human pose regression. In: Ferrari, V., Hebert, M., Sminchisescu, C., Weiss, Y. (eds.) ECCV 2018. LNCS, vol. 11210, pp. 536–553. Springer, Cham (2018). https://doi.org/10.1007/978-3-030-01231-1_33
43. Vo, M., Yumer, E., Sunkavalli, K., Hadap, S., Sheikh, Y., Narasimhan, S.G.: Self-supervised multi-view person association and its applications. IEEE Trans. PAMI (2020)
44. Wang, C., Wang, Y., Lin, Z., Yuille, A.L.: Robust 3D human pose estimation from single images or video sequences. IEEE Trans. PAMI **41**(5), 1227–1241 (2018)
45. Wei, S.E., Ramakrishna, V., Kanade, T., Sheikh, Y.: Convolutional pose machines. In: Proceedings of the CVPR, pp. 4724–4732 (2016)
46. Windheuser, T., Cremers, D.: A convex solution to spatially-regularized correspondence problems. In: Leibe, B., Matas, J., Sebe, N., Welling, M. (eds.) ECCV 2016. LNCS, vol. 9906, pp. 853–868. Springer, Cham (2016). https://doi.org/10.1007/978-3-319-46475-6_52
47. Xin, S., Nousias, S., Kutulakos, K.N., Sankaranarayanan, A.C., Narasimhan, S.G., Gkioulekas, I.: A theory of fermat paths for non-line-of-sight shape reconstruction. In: Proceedings of the CVPR, pp. 6800–6809 (2019)
48. Theobald, S., Schmitt, A., Diebold, P.: Comparing scaling agile frameworks based on underlying practices. In: Hoda, R. (ed.) XP 2019. LNBIP, vol. 364, pp. 88–96. Springer, Cham (2019). https://doi.org/10.1007/978-3-030-30126-2_11

TIDE: A General Toolbox for Identifying Object Detection Errors

Daniel Bolya(✉), Sean Foley, James Hays, and Judy Hoffman

Georgia Institute of Technology, Atlanta, USA
dbolya@gatech.edu

Abstract. We introduce TIDE, a framework and associated toolbox (https://dbolya.github.io/tide/) for analyzing the sources of error in object detection and instance segmentation algorithms. Importantly, our framework is applicable across datasets and can be applied directly to output prediction files without required knowledge of the underlying prediction system. Thus, our framework can be used as a drop-in replacement for the standard mAP computation while providing a comprehensive analysis of each model's strengths and weaknesses. We segment errors into six types and, crucially, are the first to introduce a technique for measuring the contribution of each error in a way that isolates its effect on overall performance. We show that such a representation is critical for drawing accurate, comprehensive conclusions through in-depth analysis across 4 datasets and 7 recognition models.

Keywords: Error diagnosis · Object detection · Instance segmentation

1 Introduction

Object detection and instance segmentation are fundamental tasks in computer vision, with applications ranging from self-driving cars [6] to tumor detection [9]. Recently, the field of object detection has rapidly progressed, thanks in part to competition on challenging benchmarks, such as CalTech Pedestrians [8], Pascal [10], COCO [20], Cityscapes [6], and LVIS [12]. Typically, performance on these benchmarks is summarized by one number: mean Average Precision (mAP).

However, mAP suffers from several shortcomings, not the least of which is its complexity. It is defined as the area under the precision-recall curve for detections at a specific intersection-over-union (IoU) threshold with a correctly classified ground truth (GT), averaged over all classes. Starting with COCO [20], it became standard to average mAP over 10 IoU thresholds (interval of 0.05) to get a final $mAP^{0.5:0.95}$. The complexity of this metric poses a particular challenge when we wish to analyze errors in our detectors, as error types become intertwined, making it difficult to gauge how much each error type affects mAP.

Electronic supplementary material The online version of this chapter (https://doi.org/10.1007/978-3-030-58580-8_33) contains supplementary material, which is available to authorized users.

A. Vedaldi et al. (Eds.): ECCV 2020, LNCS 12348, pp. 558–573, 2020.
https://doi.org/10.1007/978-3-030-58580-8_33

Table 1. Comparison to Other Toolkits. We compare our desired features between existing toolkits and ours. ✔ indicates a toolkit has the feature, ✱ indicates that it partially does, and ✗ indicates that it doesn't.

Feature	Hoiem [14]	COCO [1]	UAP [4]	TIDE (Ours)
Compact summary of error types	✱	✗	✔	✔
Isolates error contribution	✱	✗	✗	✔
Dataset agnostic	✗	✗	✔	✔
Uses all detections	✗	✔	✔	✔
Allows for deeper analysis	✔	✔	✔	✔

Moreover, by optimizing for mAP alone, we may be inadvertently leaving out the relative importance of error types that can vary between applications. For instance, in tumor detection, correct classification arguably matters more than box localization; the existence of the tumor is essential, but the precise location may be manually corrected. In contrast, precise localization may be critical for robotic grasping where even slight mislocalizations can lead to faulty manipulation. Understanding how these sources of error relate to overall mAP is crucial to designing new models and choosing the proper model for a given task (Table 1).

Thus we introduce TIDE, a general Toolkit for Identifying Detection and segmentation Errors, in order to address these concerns. We argue that a complete toolkit should: 1.) compactly summarize error types, so comparisons can be made at a glance; 2.) fully isolate the contribution of each error type, such that there are no confounding variables that can affect conclusions; 3.) not require dataset-specific annotations, to allow for comparisons across datasets; 4.) incorporate all the predictions of a model, since considering only a subset hides information; 5.) allow for finer analysis as desired, so that the sources of errors can be isolated.

Why We Need a New Analysis Toolkit. Many works exist to analyze the errors in object detection and instance segmentation [7,15,17,22,24], but only a few provide a useful summary of all the errors in a model [1,4,14], and none have all the desirable properties listed above.

Hoiem et al. introduced the foundational work for summarizing errors in object detection [14], however their summary only explains false positives (with false negatives requiring separate analysis), and it depends on a hyperparameter N to control how many errors to consider, thus not fulfilling (4). Moreover, to use this summary effectively, this N needs to be swept over which creates 2d plots that are difficult to interpret (see error analysis in [11,21]), and thus in practice only partially addresses (1). Their approach also doesn't fulfill (3) because their error types require manually defined superclasses which are not only subjective, but difficult to meaningfully define for datasets like LVIS [12] with over 1200 classes. Finally, it only partially fulfills (2) since the classification errors are defined such that if the detection is both mislocalized and misclassified it will be considered as misclassified, limiting the effectiveness of conclusions drawn from classification and localization error.

The COCO evaluation toolkit [1] attempts to update Hoiem et al.'s work by representing errors in terms of their effect on the precision-recall curve (thus tying them closer to mAP). This allows them to use all detections at once (4), since the precision recall curve implicitly weights each error based on its confidence. However, the COCO toolkit generates 372 2d plots, each with 7 precision-recall curves, which requires a significant amount of time to digest and thus makes it difficult to compactly compare models (1). Yet, perhaps the most critical issue is that the COCO eval toolkit computes errors progressively which we show drastically misrepresents the contribution of each error (2), potentially leading to incorrect conclusions (see Sect. 2.3). Finally, the toolkit requires manual annotations that exist for COCO but not necessarily for other datasets (3).

As concurrent work, [4] attempts to find an upper bound for AP on these datasets and in the process addresses certain issues with the COCO toolkit. However, this work still bases their error reporting on the same progressive scheme that the COCO toolkit uses, which leads them to the dubious conclusion that background error is significantly more important all other types (see Fig. 2). As will be described in detail later, to draw reliable conclusions, it is essential that our toolkit work towards isolating the contribution of each error type (2).

Contributions. In our work, we address all 5 goals and provide a compact, yet detailed *summary* of the errors in object detection and instance segmentation. Each error type can be represented as a single meaningful number (1), making it compact enough to fit in ablation tables (see Table 2), incorporates all detections (4), and doesn't require any extra annotations (3). We also weight our errors based on their effect on overall performance while carefully avoiding the confounding factors present in mAP (2). And while we prioritize ease of interpretation, our approach is modular enough that the same set of errors can be used for more fine-grained analysis (5). The end result is a compact, meaningful, and expressive set of errors that is applicable across models, datasets, and even tasks.

We demonstrate the value of our approach by comparing several recent CNN-based object detectors and instance segmenters across several datasets. We explain how to incorporate the summary into ablation studies to quantitatively justify design choices. We also provide an example of how to use the summary of errors to guide more fine-grained analysis in order to identify specific strengths or weaknesses of a model.

We hope that this toolkit can form the basis of analysis for future work, lead model designers to better understand weaknesses in their current approach, and allow future authors to quantitatively and compactly justify their design choices. To this end, full toolkit code is released at https://dbolya.github.io/tide/ and opened to the community for future development.

2 The Tools

Object detection and instance segmentation primarily use one metric to judge performance: mean Average Precision (mAP). While mAP succinctly summarizes the performance of a model in one number, disentangling errors in object detection and instance segmentation from mAP is difficult: a false positive can be a duplicate detection, misclassification, mislocalization, confusion with background, or even both a misclassification and mislocalization. Likewise, a false negative could be a completely missed ground truth, or the potentially correct prediction could have just been misclassified or mislocalized. These error types can have hugely varying effects on mAP, making it tricky to diagnose problems with a model off of mAP alone.

We could categorize all these types of errors, but it's not entirely clear how to weight their relative importance. Hoiem et al. [14] weight false positives by their prevalence in the top N most confident errors and consider false negatives separately. However, this ignores the effect many low scoring detections could have (so effective use of it requires a sweep over N), and it doesn't allow comparison between false positives and false negatives.

There is one easy way to determine the importance of a given error to overall mAP, however: simply fix that error and observe the resulting change in mAP. Hoiem et al. briefly explored this method for certain false positives but didn't base their analysis off of it. This is also similar to how the COCO eval toolkit [1] plots errors, with one key difference: the COCO implementation computes the errors *progressively*. That is, it observes the change in mAP after fixing one error, but keep those errors fixed for the next error. This is nice because at the end result is trivially 100 mAP, but we find that fixing errors progressively in this manner is misleading and may lead to false conclusions (see Sect. 2.3).

So instead, we define errors in such a way that fixing all errors will still result in 100 mAP, but we weight each error individually starting from the original model's performance. This retains the nice property of including confidence and false negatives in the calculation, while keeping the magnitudes of each error type comparable.

2.1 Computing mAP

Before defining error types, we focus our attention on the definition of mAP to understand what may cause it to degrade. To compute mAP, we are first given a list of predictions for each image by the detector. Each ground truth in the image is then matched to at most one detection. To qualify as a positive match, the detection must have the same class as the ground truth and an IoU overlap greater than some threshold, t_f, which we will consider as 0.5 unless otherwise specified. If multiple detections are eligible, the one with the highest overlap is chosen to be true positive while all remaining are considered false positives.

Once each detection has matched with a ground truth (true positive) or not (false positive), all detections are collected from every image in the dataset and are sorted by descending confidence. Then the cumulative precision and recall over all detections is computed as:

$$P_c = \frac{TP_c}{TP_c + FP_c} \qquad R_c = \frac{TP_c}{N_{GT}} \tag{1}$$

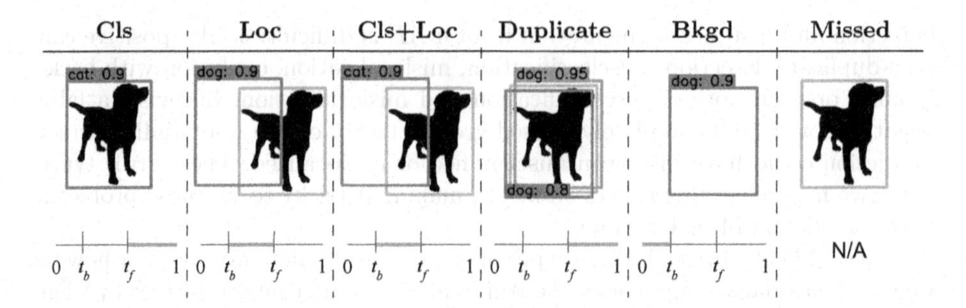

Fig. 1. Error Type Definitions. We define 6 error types, illustrated in the *top row*, where box colors are defined as: ■ = false positive detection; ■ = ground truth; ■ = true positive detection. The IoU with ground truth for each error type is indicated by an orange highlight and shown in the *bottom row.*

where for all detections with confidence $\geq c$, P_c denotes the precision, R_c recall, TP_c the number of true positives, and FP_c the number of false positives. N_{GT} denotes the number of GT examples in the current class.

Then, precision is interpolated such that P_c decreases monotonically, and AP is computed as a integral under the precision recall curve (approximated by a fixed-length Riemann sum). Finally, mAP is defined as the average AP over all classes. In the case of COCO [20], mAP is averaged over all IoU thresholds between 0.50 and 0.95 with a step size of 0.05 to obtain $mAP^{0.5:0.95}$.

2.2 Defining Error Types

Examining this computation, there are 3 places our detector can affect mAP: outputting false positives during the matching step, not outputting true positives (i.e., false negatives) for computing recall, and having incorrect calibration (i.e., outputting a higher confidence for a false positive then a true positive).

Main Error Types. In order to create a meaningful distribution of errors that captures the components of mAP, we bin all false positives and false negatives in the model into one of 6 types (see Fig. 1). Note that for some error types (classification and localization), a false positive can be paired with a false negative. We will use $IoU_{\max}$ to denote a false positive's maximum IoU overlap with a ground truth of the given category. The foreground IoU threshold is denoted as t_f and the background threshold is denoted as t_b, which are set to 0.5 and 0.1 (as in [14]) unless otherwise noted.

1. **Classification Error**: $IoU_{\max} \geq t_{\mathrm{f}}$ for GT of the *incorrect* class (i.e., localized correctly but classified incorrectly).
2. **Localization Error**: $t_{\mathrm{b}} \leq IoU_{\max} \leq t_{\mathrm{f}}$ for GT of the *correct* class (i.e., classified correctly but localized incorrectly).
3. **Both Cls and Loc Error**: $t_{\mathrm{b}} \leq IoU_{\max} \leq t_{\mathrm{f}}$ for GT of the *incorrect* class (i.e., classified incorrectly *and* localized incorrectly).

4. **Duplicate Detection Error**: $IoU_{\text{max}} \geq t_{\text{f}}$ for GT of the *correct* class but another higher-scoring detection already matched that GT (i.e., would be correct if not for a higher scoring detection).
5. **Background Error**: $IoU_{\text{max}} \leq t_{\text{b}}$ for all GT (i.e., detected background as foreground).
6. **Missed GT Error**: All undetected ground truth (false negatives) not already covered by classification or localization error.

This differs from [14] in a few important ways. First, we combine both `sim` and `other` errors into one classification error, since Hoiem et al.'s `sim` and `other` require manual annotations that not all datasets have and analysis of the distinction can be done separately. Then, both classification errors in [14] are defined for all detections with $IoU_{\text{max}} \geq t_b$, even if $IoU_{\text{max}} < t_f$. This confounds localization and classification errors, since using that definition, detections that are both mislocalized and misclassified are considered class errors. Thus, we separate these detections into their own category.

Weighting the Errors. Just counting the number of errors in each bin is not enough to be able to make direct comparisons between error types, since a false positive with a lower score has less effect on overall performance than one with a higher score. Hoiem et al. [14] attempt to address this by considering the top N highest scoring errors, but in practice N needed to be swept over to get the full picture, creating 2d plots that are hard to interpret (see the analysis in [11,21]).

Ideally, we'd like one comprehensive number that represents how each error type affects overall performance of the model. In other words, for each error type we'd like to ask the question, how much is this category of errors holding back the performance of my model? In order to answer that question, we can consider what performance of the model would be if it didn't make that error and use how that changed mAP.

To do this, for each error we need to define a corresponding "oracle" that fixes that error. For instance, if an oracle $o \in \mathcal{O}$ described how to change some false positives into true positives, we could call the AP computed after applying the oracle as AP_o and then compare that to the vanilla AP to obtain that oracle's (and corresponding error's) effect on performance:

$$\Delta AP_o = AP_o - AP \tag{2}$$

We know that we've covered all errors in the model if applying all the oracles together results in 100 mAP. In other words, given oracles $\mathcal{O} = \{o_1, \dots, o_n\}$:

$$AP_{o_1,\dots,o_n} = 100 \qquad AP + \Delta AP_{o_1,\dots,o_n} = 100 \tag{3}$$

Referring back to the definition of AP in Sect. 2.1, to satisfy Eq. 3 the oracles used together must fix all false positives and false negatives.

Considering this, we define the following oracles for each of the main error types described above:

1. **Classification Oracle**: Correct the class of the detection (thereby making it a true positive). If a duplicate detection would be made this way, suppress the lower scoring detection.
2. **Localization Oracle**: Set the localization of the detection to the GT's localization (thereby making it a true positive). Again, if a duplicated detection would be made this way, suppress the lower scoring detection.
3. **Both Cls and Loc Oracle**: Since we cannot be sure of which GT the detector was attempting to match to, just suppress the false positive detection.
4. **Duplicate Detection Oracle**: Suppress the duplicate detection.
5. **Background Oracle**: Suppress the hallucinated background detection.
6. **Missed GT Oracle**: Reduce the number of GT (N_{GT}) in the mAP calculation by the number of missed ground truth. This has the effect of stretching the precision-recall curve over a higher recall, essentially acting as if the detector was equally as precise on the missing GT. The alternative to this would be to add new detections, but it's not clear what the score should be for that new detection such that it doesn't introduce confounding variables. We discuss this choice further in the Appendix.

Other Error Types. While the previously defined types fully account for all error in the model, how the errors are defined doesn't clearly delineate false positive and negative errors (since cls, loc, and missed errors can all capture false negatives). There are cases where a clear split would be useful, so for those cases we define two separate error types by the oracle that would address each:

1. **False Positive Oracle**: Suppress all false positive detections.
2. **False Negative Oracle**: Set N_{GT} to the number of true positive detections.

Both of these oracles together account for 100 mAP like the previous 6 oracles do, but they bin the errors in a different way.

2.3 Limitations of Computing Errors Progressively

Note that we are careful to compute errors *individually* (i.e., each ΔAP starts from the vanilla AP with no errors fixed). Other approaches [1,4] compute their errors *progressively* (i.e., each ΔAP starts with the last error fixed, such that fixing the last error results in 100 AP). While we ensure that applying all oracles together also results in 100 AP, we find that a progressive ΔAP misrepresents the weight of each error type and is strongly biased toward error types *fixed last.*

To make this concrete, we can define progressive error $\Delta AP_{a|b}$ to be the change in AP from applying oracle a given that you've already applied oracle b:

$$\Delta AP_{a|b} = AP_{a,b} - AP_b \tag{4}$$

Then, computing errors progressively amounts to setting the importance of error i to $\Delta AP_{o_i|o_1,\ldots,o_{i-1}}$. This is problematic for two reasons: the definition of precision includes false positives in the *denominator*, meaning that if you start with fewer false positives (as would be the case when having fixed most false positives already), the change in precision will be *much higher*. Furthermore, any changes in recall (e.g., by fixing localization or classification errors) amplifies the effect of precision on mAP, since the integral now has more area.

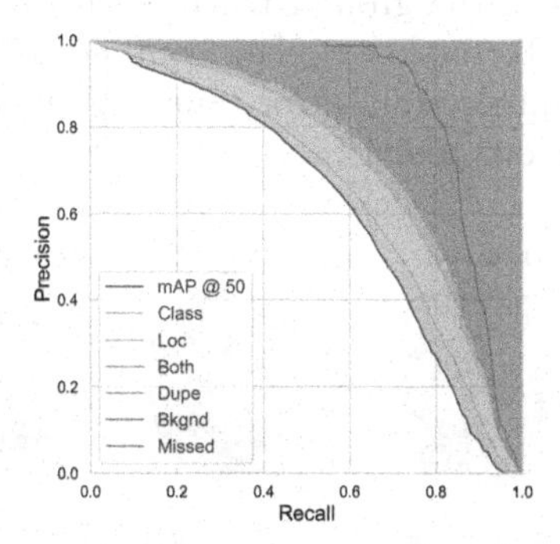

(a) Default COCO eval style error curves.

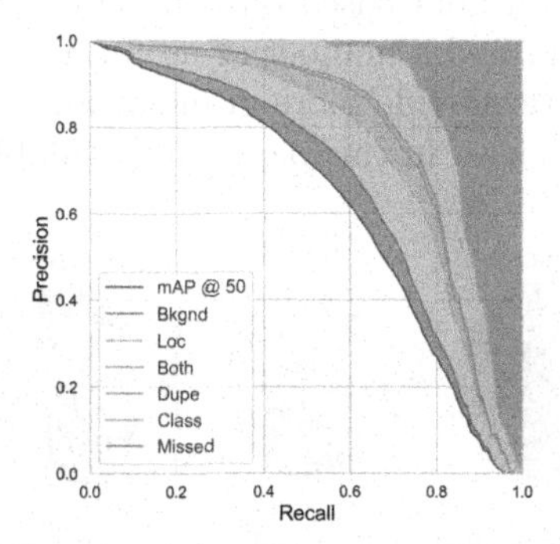

(b) Swapping the order of errors changes magnitudes drastically.

Fig. 2. The problem with computing errors progressively. The COCO eval `analyze` function [1] computes errors progressively, which we show for Mask R-CNN [13] detections on mAP_{50}. On the right, we swap the order of applying the classification and background oracles. The quantity of each error remains the same, but the perceived contribution from background error (purple region) significantly decreases, while it increases for all other errors. Because COCO computes background error second to last, this instills a belief that it's more important than other errors, which does not reflect reality (see Sect. 2.3). (Color figure online)

We show this empirically in Fig. 2, where Fig. 2a displays the original COCO eval style PR curves, while Fig. 2b simply swaps the order that background and classification error are computed. Just computing background first leads to an incredible decrease in the prevalence of its contribution (given by the area of the shaded region), meaning that the true weight of background error is likely much less than COCO eval reports. This makes it difficult to draw factual conclusions from analysis done this way.

Moreover, computing errors progressively doesn't make intuitive sense. When using these errors, you'd be attempting to address them individually, one at a time. There will never be an opportunity to correct *all* localization errors, and then start addressing the classification errors—there will always be some amount of error in each category left over after improving the method, so observing $AP_{a|b}$ isn't useful, because there is no state where you're starting with AP_b.

For these reasons, we entirely avoid computing errors progressively.

3 Analysis

In this section we demonstrate the generality and usefulness of our analysis toolbox by providing detailed analysis across various object detection and instance segmentation models and across different data and annotation sets. We also compare errors based on general qualities of the ground truth, such as object size, and find a number of useful insights. To further explain complicated error cases, we provide more granular analysis into certain error types. All modes of analysis used in this paper are available in our toolkit.

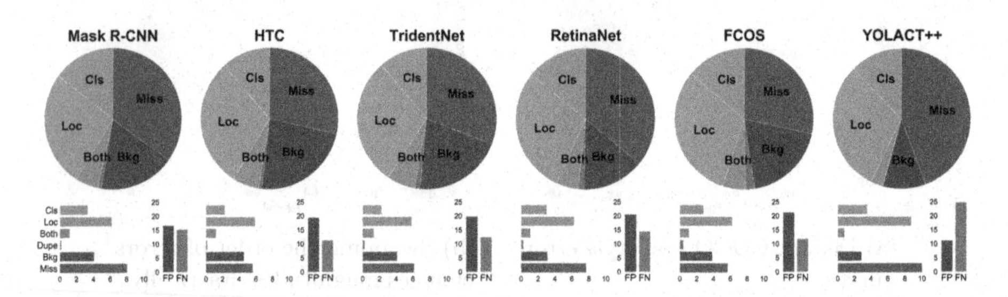

Fig. 3. Summary of errors on COCO Detection. Our model specific error analysis applied to various object detectors on COCO. The pie chart shows the relative contribution of each error, while the bar plots show their absolute contribution. For instance segmentation results, see the Appendix.

Models. We choose various object detectors and instance segmenters based on their ubiquity and/or unique qualities which allows us to study the performance trade-offs between different approaches and draw several insights. We use Mask R-CNN [13] as our baseline, as many other approaches build on top of the standard R-CNN framework. We additionally include three such models: Hybrid Task Cascades (HTC) [5], TridentNet [18], and Mask Scoring R-CNN (MS-RCNN) [13]. We include HTC due to its strong performance, being the 2018 COCO challenge winner. We include TridentNet [18] as it specifically focuses on increasing scale-invariance. Finally, we include MS R-CNN as a method which specifically focuses on fixing calibration based error. Distinct from the two-stage R-CNN style approaches, we also include three single-stage approaches, YOLACT/YOLACT++ [2,3] to represent real-time models, RetinaNet [19] as a strong anchor-based model, Fully Convolutional One-Stage Object Detection (FCOS) [23] as a non anchor-based approach. Where available, we use the ResNet101 versions of each model. Exact models are indicated in the Appendix.

Datasets. We present our core cross-model analysis on MS-COCO [20], a widely used and active benchmark. In addition, we seek to showcase the power of our toolbox to perform cross-dataset analysis by including three additional datasets: Pascal VOC [10] as a relatively simple object detection dataset, Cityscapes [6] providing high-res, densely annotation images with many small objects, and LVIS [12] using the same images at COCO but with a massive diversity of annotated objects with 1200+ mostly-rare class.

3.1 Validating Design Choices

The authors of each new object detector or instance segmenter make design choices they claim to affect their model's performance in different ways. While the goal is almost always to increase overall mAP, there remains the question: does the intuitive justification for a design choice hold up? In Fig. 3 we present the distribution of errors for all object detectors and instance segmenters we consider on COCO [20], and in this section we'll analyze the distribution of errors for each detector to see whether our errors line up with the intuitive justifications.

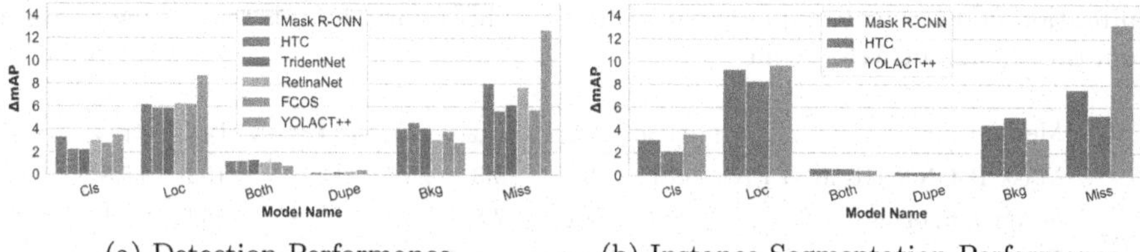

(a) Detection Performance. (b) Instance Segmentation Performance.

Fig. 4. Comparison across models on COCO. Weight of each error type compared across models. This has the same data as Fig. 3.

R-CNN Based Methods. First, HTC [5] makes two main improvements over Mask R-CNN: 1.) it iteratively refines predictions (i.e., a cascade) and passes information between all parts of the model each time, and 2.) it introduces a module specifically for improved detection of foreground examples that look like background. Intuitively, (1) would improve classification and localization significantly, as the prediction and the features used for the prediction are being refined 3 times. And indeed, the classification and localization errors for HTC are the lowest of the models we consider in Fig. 4 for both instance segmentation and detection. Then, (2) should have the effect of eliciting higher recall while potentially adding false positives where something in the background was misclassified as an object. And this is exactly what our errors reveal: HTC has the lowest missed GT error while having the highest background error (not counting YOLACT++, whose distribution of errors is quite unique).

Next, TridentNet [18] attempts to create scale-invariant features by having a separate pipeline for small, medium, and large objects that all share weights. Ideally this would improve classification and localization performance for objects of different scales. Both HTC and TridentNet end up having the same classification and localization performance, so we test this hypothesis further in Sect. 3.2. Because HTC and TridentNet make mostly orthogonal design choices, they would likely compliment each other well.

One-Stage Methods. RetinaNet [19] introduces focal loss that down-weights confident examples in order to be able to train on all background anchor boxes (rather than the standard 3 negative to 1 positive ratio). Training on all negatives by itself should cause the model to output fewer background false positives, but at the cost of significantly lower recall (since the detector would be biased toward predicting background). The goal of focal loss then is to train on all negatives without causing extra missed detections. We observe this is successful as RetinaNet has one of the lowest background errors across models in Fig. 4a, while retaining slightly less missed GT error than Mask R-CNN.

Table 2. Mask Rescoring. An ablation of MS-RCNN [13] and YOLACT++ [2] mask performance using the errors defined in this paper. ΔmAP_{50} is denoted as E for brevity, and only errors that changed are included. Mask scoring better calibrates localization, leading to decrease in localization error. However, by scoring based on localization, the calibration of other error types suffer. Note that this information is impossible to glean from the change in AP_{50} alone.

Method	$AP_{50}\uparrow$	$E_{cls}\downarrow$	$E_{loc}\downarrow$	$E_{bkg}\downarrow$	$E_{miss}\downarrow$	$E_{FP}\downarrow$	$E_{FN}\downarrow$
Mask R-CNN (R-101-FPN)	58.1	3.1	9.3	4.5	7.5	15.9	17.8
+ Mask Scoring	58.3	3.6	7.8	5.1	7.8	15.9	18.1
Improvement	+0.2	+0.4	**−1.5**	+0.7	+0.3	+0.0	+0.3
YOLACT++ (R-50-FPN)	51.8	3.3	10.4	3.2	13.0	10.7	27.7
+ Mask Scoring	52.3	3.6	9.7	3.2	13.2	10.1	28.2
Improvement	+0.5	+0.3	**−0.7**	+0.0	+0.2	−0.5	+0.6

Then FCOS [23] serves as a departure from traditional anchor-based models, predicting a newly defined box at each location in the image instead of regressing an existing prior. While the primary motivation for this design choice was simplicity, getting rid of anchor boxes has other tangible benefits. For instance, an anchor-based detector is at the mercy of its priors: if there is no applicable prior for a given object, then the detector is likely to completely miss it. FCOS on the other hand doesn't impose any prior-based restriction on its detections, leading to it having one of the lowest missed detection errors of all the models we consider (Fig. 4a). Note that it also has the highest duplication error because it uses an NMS threshold of 0.6 instead of the usual 0.5.

Real-Time Methods. YOLACT [3] is a real-time instance segmentation method that uses a modified version of RetinaNet as its backbone detector without focal loss. YOLACT++ [2] iterates on the former and additionally includes mask scoring (discussed in Table 2). Observing the distribution of errors in Fig. 3, it appears that design choices employed to speed up the model result in a completely different distribution of errors w.r.t. RetinaNet. Observing the raw magnitudes in Fig. 4a, this is largely due to YOLACT having much higher localization and missed detection error. However, the story changes when we look at instance segmentation, where it localizes almost as well as Mask R-CNN despite the bad performance of its detector (see Appendix). This substantiates their claim that YOLACT is more conducive to high quality masks and that its performance is likely limited by a poor detector.

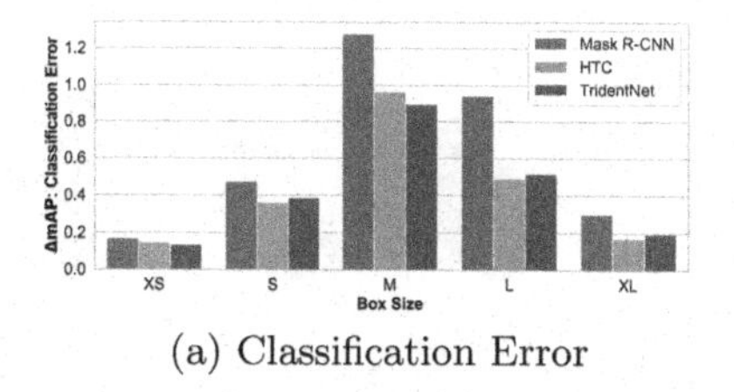

(a) Classification Error

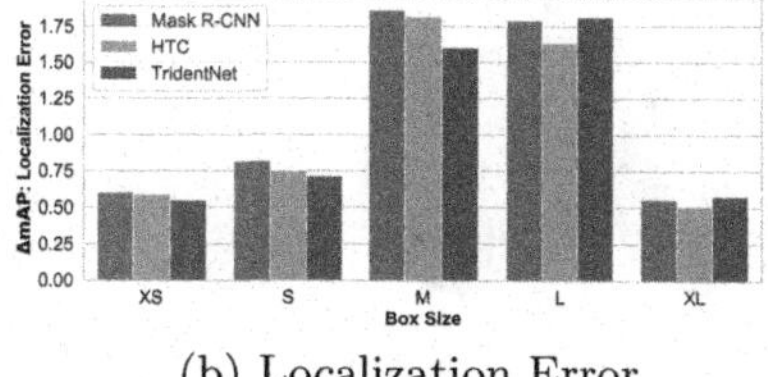

(b) Localization Error

Fig. 5. Comparison of Scales between HTC and TridentNet. Both HTC and TridentNet have the same classification and localization error on COCO detection. Using fine analysis, we can isolate the cause of these errors further.

A Note on Ablations. To demonstrate the potential usefulness of this toolkit for isolating error contribution and debugging, we showcase how an ablation over error types instead of only over mAP provides meaningful insights while still being compact. As an example, consider the trend of rescoring a mask's confidence based on its predicted IoU with a ground truth, as in Mask Scoring R-CNN [16] and YOLACT++ [2]. This modification is intended to increase the score of good quality masks and decrease the score of poor quality masks, which intuitively should result in better localization. In order to justify their claims, the authors of both papers provide qualitative examples where this is the case, but limit quantitative support to the change to an observed increase in mAP. Unfortunately, a change in mAP alone does not illuminate the cause of that change, and some ablations may show little change in mAP despite the method working. By adding the error types that were affected by the change to ablation tables (e.g., see Table 2) we not only provide quantitative evidence for the design choice, but also reveal side effects (such that classification calibration error went up), which were previously hidden by the raw increase in mAP.

3.2 Comparing Object Attributes for Fine Analysis

In order to compare performance across object attributes such as scale or aspect ratio, the typical approach is to compute mAP on a subset of the detections and ground truth that have the specified attributes (with effective comparison requiring normalized mAP [14]). While we offer this mode of analysis in our toolkit, this doesn't describe the effect of that attribute on overall performance, just how well a model performs on that attribute. Thus, we propose an additional approach based on the tools we defined earlier for summarizing error's affect on overall performance: simply fix errors and observe ΔmAP as before, but only those whose associated prediction or ground truth have the desired attribute.

Comparing Across Scale. As an example of using this approach across different scales of objects, we return to the case of TridentNet vs. HTC discussed in Sect. 3.1. Both models have the same classification and localization error and we would like to understand where the difference, if any, lies. Since TridentNet focuses specifically on scale-invariance, we turn our attention to performance

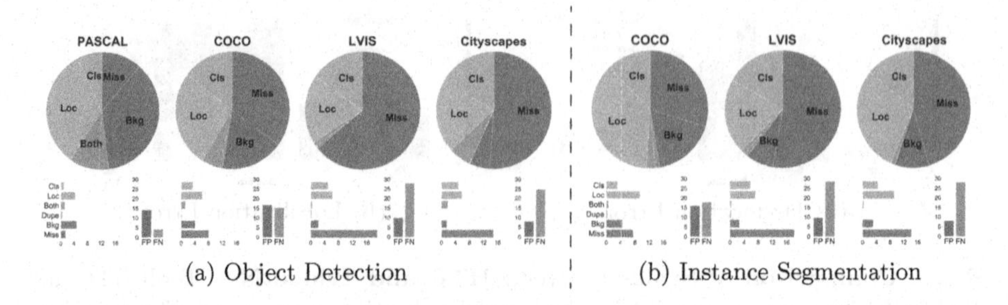

(a) Object Detection (b) Instance Segmentation

Fig. 6. Performance of Mask R-CNN Across Datasets. Because our toolkit is dataset agnostic, we can fix a detection architecture and compare performance across datasets to gain valuable insights into properties of each dataset.

across scales. We define objects with pixel areas of between 0 and 16^2 as extra small (XS), 16^2 to 32^2 as small (S), 32^2 to 96^2 as medium (M), 96^2 to 288^2 as (L), and 288^2 and above as extra large (XL). In Fig. 5 we apply our approach across HTC and TridentNet (with Mask R-CNN detections included for reference). This comparison reveals that TridentNet localizes and classifies medium sized objects better than HTC, while HTC is better at large objects. This could potentially be why the authors of TridentNet find that they can achieve nearly the same performance by only evaluating their branch for medium sized objects [18]. Other comparisons between subsets of detections such as across aspect ratios, anchor boxes, FPN layers, etc. are possible with the same approach.

3.3 Comparing Performance Across Datasets

Our toolkit is dataset agnostic, allowing us to compare the same model across several datasets, as in Fig. 6, where we compare Mask R-CNN (Faster R-CNN for Pascal) across Pascal VOC [10], COCO [20], Cityscapes [6], and LVIS [12].

In this comparison, the first immediately clear pattern is that Background error decreases both in overall prevalence (pie charts) and absolute magnitude (bar charts) with increasing density of annotations. Faster R-CNN on Pascal is dominated by background error, but of reduced concern on COCO. Both LVIS and Cityscapes, which are very densely annotated, have almost no background error at all. This potentially indicates that much of the background error in Pascal and COCO are simply due to unannotated objects (see Sect. 3.4).

As expected, missed ground truths are a large issue for densely annotated datasets like LVIS and Cityscapes. The core challenge on Cityscapes is the presence of many small objects, which are well known to be difficult to detect with modern algorithms. On the other hand, LVIS's challenge is to deal with the vast number of possible objects that the detector has to recognize. We can see from the relatively normal classification error on LVIS that the model isn't particularly suffering directly from misclassifying rare objects, but instead completely failing to detect them when they appear. This is also reflected in the false positive and false negative error distributions (vertical bars). Overall, Pascal is heavily biased toward false positives, COCO is mixed, and LVIS and Cityscapes are both biased toward false negatives.

On COCO, Mask R-CNN has a harder time localizing masks (Fig. 6b) than boxes (Fig. 6a), but the opposite is true for LVIS, possibly because of its higher quality masks, which are verified with expert studies [12]. Again, this potentially indicates that a lot of the error in instance segmentation may be derived by mis-annotations.

Fig. 7. Examples of Poor Annotations. In modern detectors, highly confident detections classified as both mislocalized and misclassified or background errors are likely to be mislabeled examples on COCO. In the first two images, the ground truth should have been labeled as crowds. In the third, some of the donuts simply weren't labeled. ■ = ground truth, ■ = predictions.

3.4 Unavoidable Errors

We find in Sect. 3.3 that a lot of the background and localization error may simply be due to mis- or unannotated ground truth. Examining the top errors more closely, we find that indeed (at least in COCO), many of the most confident errors are actually misannotated or ambiguously misannotated ground truth (see Fig. 7). For instance, 30 of the top 100 most confident localization errors in Mask R-CNN detections are due to bad annotations, while the number soars to 50 out of 100 for background error. These misannotations are simple mistakes like making the box too big or forgetting to mark a box as a crowd annotation. More examples are ambiguous: should a mannequin or action figure be annotated as a person? Should a sculpture of a cat be annotated as a cat? Should a reflection of an object be annotated as that object? Highly confident mistakes result in large changes in overall mAP, so misannotated ground truth considerably lower the maximum mAP a reasonable model can achieve.

This begs the question, what is the upper bound for mAP on these datasets? Existing analyses into the potential upper bound in object detection such as [4] don't seem to account for the rampant number of mislabeled examples. The state-of-the-art on the COCO challenge are slowly stagnating, so perhaps we are nearing the "reasonable" upper bound for these detectors. We leave this for future work to analyze.

4 Conclusion

In this work, we define meaningful error types and a way of tying these error types to overall performance such that it minimizes any confounding variables. We then apply the resulting framework to evaluate design decisions, compare performance on object attributes, and reveal properties of several datasets, including the prevalence of misannotated ground truth in COCO. We hope that our toolkit can not only serve as method to isolate and improve on errors in detection, but also lead to more interpretability in design decisions and clearer descriptions of the strengths and weaknesses of a model.

References

1. COCO Analysis Toolkit. http://cocodataset.org/#detection-eval. Accessed 01 Mar 2020
2. Bolya, D., Zhou, C., Xiao, F., Lee, Y.J.: YOLACT++: better real-time instance segmentation. arXiv:1912.06218 (2019)
3. Bolya, D., Zhou, C., Xiao, F., Lee, Y.J.: YOLACT: real-time instance segmentation. In: ICCV (2019)
4. Borji, A., Iranmanesh, S.M.: Empirical upper-bound in object detection and more. arXiv:1911.12451 (2019)
5. Chen, K., et al.: Hybrid task cascade for instance segmentation. In: CVPR (2019)
6. Cordts, M., et al.: The cityscapes dataset for semantic urban scene understanding. In: CVPR (2016)
7. Divvala, S.K., Hoiem, D., Hays, J.H., Efros, A.A., Hebert, M.: An empirical study of context in object detection. In: CVPR (2009)
8. Dollár, P., Wojek, C., Schiele, B., Perona, P.: Pedestrian detection: a benchmark. In: CVPR (2009)
9. Dong, H., Yang, G., Liu, F., Mo, Y., Guo, Y.: Automatic brain tumor detection and segmentation using U-Net based fully convolutional networks. In: MIUA (2017)
10. Everingham, M., Van Gool, L., Williams, C.K., Winn, J., Zisserman, A.: The pascal visual object classes (VOC) challenge. IJCV (2010)
11. Girshick, R., Donahue, J., Darrell, T., Malik, J.: Rich feature hierarchies for accurate object detection and semantic segmentation. In: CVPR (2014)
12. Gupta, A., Dollar, P., Girshick, R.: LVIS: a dataset for large vocabulary instance segmentation. In: CVPR (2019)
13. He, K., Gkioxari, G., Dollár, P., Girshick, R.: Mask R-CNN. In: ICCV (2017)
14. Hoiem, D., Chodpathumwan, Y., Dai, Q.: Diagnosing error in object detectors. In: Fitzgibbon, A., Lazebnik, S., Perona, P., Sato, Y., Schmid, C. (eds.) ECCV 2012. LNCS, vol. 7574, pp. 340–353. Springer, Heidelberg (2012). https://doi.org/10.1007/978-3-642-33712-3_25
15. Hosang, J., Benenson, R., Schiele, B.: How good are detection proposals, really? In: BMVC (2014)
16. Huang, Z., Huang, L., Gong, Y., Huang, C., Wang, X.: Mask scoring R-CNN. In: CVPR (2019)
17. Kabra, M., Robie, A., Branson, K.: Understanding classifier errors by examining influential neighbors. In: CVPR (2015)
18. Li, Y., Chen, Y., Wang, N., Zhang, Z.: Scale-aware trident networks for object detection. In: ICCV (2019)

19. Lin, T.Y., Goyal, P., Girshick, R., He, K., Dollár, P.: Focal loss for dense object detection. In: CVPR (2017)
20. Lin, T.-Y., et al.: Microsoft COCO: common objects in context. In: Fleet, D., Pajdla, T., Schiele, B., Tuytelaars, T. (eds.) ECCV 2014. LNCS, vol. 8693, pp. 740–755. Springer, Cham (2014). https://doi.org/10.1007/978-3-319-10602-1_48
21. Liu, W., et al.: SSD: single shot multibox detector. In: Leibe, B., Matas, J., Sebe, N., Welling, M. (eds.) ECCV 2016. LNCS, vol. 9905, pp. 21–37. Springer, Cham (2016). https://doi.org/10.1007/978-3-319-46448-0_2
22. Pepik, B., Benenson, R., Ritschel, T., Schiele, B.: What is holding back convnets for detection? In: Gall, J., Gehler, P., Leibe, B. (eds.) GCPR 2015. LNCS, vol. 9358, pp. 517–528. Springer, Cham (2015). https://doi.org/10.1007/978-3-319-24947-6_43
23. Tian, Z., Shen, C., Chen, H., He, T.: FCOS: fully convolutional one-stage object detection. In: ICCV (2019)
24. Zhu, H., Lu, S., Cai, J., Lee, Q.: Diagnosing state-of-the-art object proposal methods. arXiv:1507.04512 (2015)

PointContrast: Unsupervised Pre-training for 3D Point Cloud Understanding

Saining Xie[1](✉), Jiatao Gu[1], Demi Guo[1], Charles R. Qi[1], Leonidas Guibas[2], and Or Litany[2]

[1] Facebook AI Research, Menlo Park, USA
xiesaining@gmail.com
[2] Stanford University, Stanford, USA

Abstract. Arguably one of the top success stories of deep learning is transfer learning. The finding that pre-training a network on a rich source set (*e.g.*, ImageNet) can help boost performance once fine-tuned on a usually much smaller target set, has been instrumental to many applications in language and vision. Yet, very little is known about its usefulness in 3D point cloud understanding. We see this as an opportunity considering the effort required for annotating data in 3D. In this work, we aim at facilitating research on 3D representation learning. Different from previous works, we focus on high-level scene understanding tasks. To this end, we select a suit of diverse datasets and tasks to measure the effect of unsupervised pre-training on a large source set of 3D scenes. Our findings are extremely encouraging: using a unified triplet of architecture, source dataset, and contrastive loss for pre-training, we achieve improvement over recent best results in segmentation and detection across 6 different benchmarks for indoor and outdoor, real and synthetic datasets – demonstrating that the learned representation can generalize across domains. Furthermore, the improvement was similar to supervised pre-training, suggesting that future efforts should favor scaling data collection over more detailed annotation. We hope these findings will encourage more research on unsupervised pretext task design for 3D deep learning.

Keywords: Unsupervised learning · Point cloud recognition · Representation learning · 3D scene understanding

1 Introduction

Representation learning is one of the main driving forces of deep learning research. In 2D vision, the finding that pre-training a network on a rich source

D. Guo, C. R. Qi, L. Guibas and O. Litany—Work done while at Facebook AI Research.

Electronic supplementary material The online version of this chapter (https://doi.org/10.1007/978-3-030-58580-8_34) contains supplementary material, which is available to authorized users.

A. Vedaldi et al. (Eds.): ECCV 2020, LNCS 12348, pp. 574–591, 2020.
https://doi.org/10.1007/978-3-030-58580-8_34

set (*e.g.* ImageNet classification) can help boost performance once fine-tuned on the usually much smaller target set, has been key to the success of many applications. A particularly important setting, is when the pre-training stage is unsupervised, as this opens up the possibility to utilize a practically infinite train set size. Unsupervised pre-training has been remarkably successful in natural language processing [13,47], and has recently attracted increasing attention in 2D vision [3,3,8,8,26,26,27,38,40,40,64,81].

In the past few years, the field of 3D deep learning has witnessed much progress with an ever-increasing number of 3D representation learning schemes [1,9,12,15,16,21,22,34,62,69,75]. However, it still falls behind compared to its 2D counterpart as evidently, in all 3D scene understanding tasks, ad-hoc training *from scratch* on the target data is still the dominant approach. Notably, all existing representation learning schemes are tested either on single objects or low-level tasks (e.g. registration). This status quo can be attributed to multiple reasons: 1) Lack of large-scale and high-quality data: compared to 2D images, 3D data is harder to collect, more expensive to label, and the variety of sensing devices may introduce drastic domain gaps; 2) Lack of unified backbone architectures: in contrast to 2D vision where architectures such as ResNets were proven successful as backbone networks for pre-training and fine-tuning, point cloud network architecture designs are still evolving; 3) Lack of a comprehensive set of datasets and high-level tasks for evaluation.

The purpose of this work is to move the needle by initiating research on *unsupervised pre-training* with *supervised fine-tuning* in deep learning for 3D scene understanding. To do so, we cover four important ingredients: 1) Selecting a large dataset to be used at pre-training; 2) identifying a backbone architecture that can be shared across many different tasks; 3) evaluating two unsupervised objectives for pre-training the backbone network; and 4) defining an evaluation protocol on a set of diverse downstream datasets and tasks.

Specifically, we choose ScanNet [11] as our source set on which the pre-training takes place, and utilize a sparse residual U-Net [9,49] as the backbone architecture in all our experiments and focus on the point cloud representation of 3D data. For the pre-training objective, we evaluate two different contrastive losses: Hardest-contrastive loss [10], and PointInfoNCE – an extension of InfoNCE loss [40] used for pre-training in 2D vision. Next, we choose a broad set of target datasets and downstream tasks that includes: semantic segmentation on S3DIS [2], ScanNetV2 [11], ShapeNetPart [71] and Synthia 4D [50]; and object detection on SUN RGB-D [31,53,55,65] and ScanNetV2. Remarkably, our results indicate improved performance across all datasets and tasks (See Table 1 for a summary of the results). In addition, we found a relatively small advantage to pre-training with supervision. This implies that future efforts in collecting data for pre-training should favor scale over precise annotations.

Our contributions can be summarized as follows:

- We evaluate, for the first time, the transferability of learned representation in 3D point clouds to high-level scene understanding.

- Our results indicate that *unsupervised pre-training* improves performance across downstream tasks and datasets, while using a single unified architecture, source set and objective function.
- Powered by unsupervised pre-training, we achieve a new state-of-the-art performance on 6 different benchmarks.
- We believe these findings would encourage a change of paradigm on how we tackle 3D recognition and drive more research on 3D representation learning.

2 Related Work

Representation Learning in 3D. Deep neural networks are notoriously data hungry. This renders the ability to transfer learned representations between datasets and tasks extremely powerful. In 2D vision it has led to a surge of interest in finding optimal pretext unsupervised tasks [3,5,8,10,14,18,26,27,38–41,64,77,78,81]. We note that while many of these tasks are *low-level* (*e.g.* pixel or patch level reconstruction), they are evaluated based on their transferability to *high-level* tasks such as object detection. Being much harder to annotate, 3D tasks are potentially the biggest beneficiaries of unsupervised- and transfer-learning. This was shown in several works on single object tasks like reconstruction, classification and part segmentation [1,16,21,22,34,51,62,69]. Yet, generally much less attention has been devoted to representation learning in 3D that extends beyond the single-object level. Further, in the few cases that did study it, the focus was on low-level tasks like registration [12,15,75]. In contrast, here we wish to push forward research in 3D representation learning by focusing on transferability to more high-level tasks on more complex scenes.

Deep Architectures for Point Cloud Processing. In this work we focus on learning useful representation for point cloud data. Inspired by the success in 2D domain, we conjecture that an important ingredient in enabling such progress is the evident standardization of neural architectures. Canonical examples include VGGNet [54] and ResNet/ResNeXt [25,66]. In contrast, point cloud neural network design is much less mature, as is apparent by the abundance of new architectures that have been recently proposed. This has multiple reasons. First, is the challenge of processing unordered sets [37,45,48,74]. Second, is the choice of neighborhood aggregation mechanism which could either be hierarchical [16,32,33,46,76], spatial CNN-like [29,35,57,68,79], spectral [58,60,72] or graph-based [52,59,63,67]. Finally, since the points are discrete samples of an underlying surface, continuous convolutions have also been considered [4,61,70]. Recently Choy et al. proposed the Minkowski Engine [9], an extension of submanifold sparse convolutional networks [20] to higher dimensions. In particular, sparse convolutional networks facilitate the adoption of common deep architectures from 2D vision, which in turn can help standardize deep learning for point cloud. In this work, we use a unified UNet [49] architecture built with Minkowski Engine as the backbone network in all experiments and show it can gracefully transfer between tasks and datasets.

3 PointContrast Pre-training

In this section, we introduce our unsupervised pre-training pipeline. First, to motivate the necessity of a new pre-training scheme, we conduct a pilot study to understand the limitations of existing practice (pre-training on ShapeNet) in 3D deep learning (Sect. 3.1). After briefly reviewing an inspirational local feature learning work *Fully Convolutional Geometric Features (FCGF)* (Sect. 3.2), we introduce our unsupervised pre-training solution, *PointContrast*, in terms of pretext task (Sect. 3.3), loss function (Sect. 3.4), network architecture (Sect. 3.5) and pre-training dataset (Sect. 3.6).

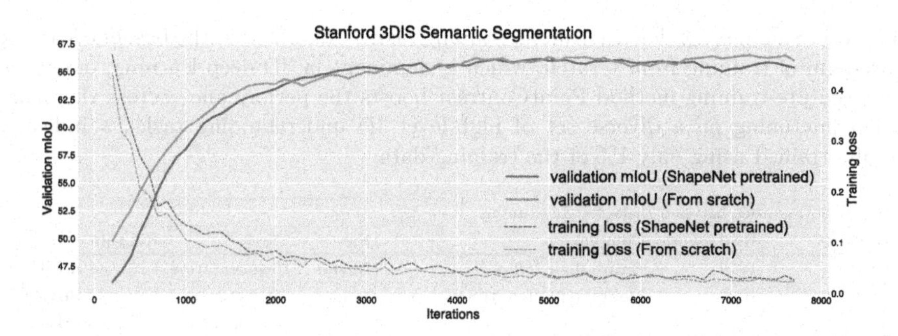

Fig. 1. Training from scratch *vs.* fine-tuning with ShapeNet pretrained weights.

3.1 Pilot Study: Is Pre-training on ShapeNet Useful?

Previous works on unsupervised 3D representation learning [1,16,21,22,34,62, 69] mainly focused on ShapeNet [7], a dataset of single-object CAD models. One underlying assumption is that by adopting ShapeNet as the ImageNet counterpart in 3D, features learned on *synthetic single objects* could transfer to other real-world applications. Here we take a step back and reassess this assumption by studying a straightforward supervised pre-training setup: we simply pre-train an encoder network on ShapeNet with *full supervision*, and fine-tune it with a U-Net on a downstream task (S3DIS semantic segmentation). Based on results in 2D representation learning, we use full supervision here as an upper bound to what could be gained from pre-training. We train a sparse ResNet-34 model (details to follow in Sect. 3.5) for 200 epochs. The model achieves a high validation accuracy of 85.4% on ShapeNet classification task. In Fig. 1, we show the downstream task training curves for (a) training from scratch and (b) fine-tuning with ShapeNet pretrained weights. Critically, one can observe that ShapeNet pre-training, even in the supervised fashion, *hampers* downstream task learning. Among many potential explanations, we highlight two major concerns:

- **Domain gap between source and target data:** Objects in ShapeNet are synthetic, normalized in scale, aligned in pose, and lack scene context. This makes pre-training and fine-tuning data distributions drastically different.

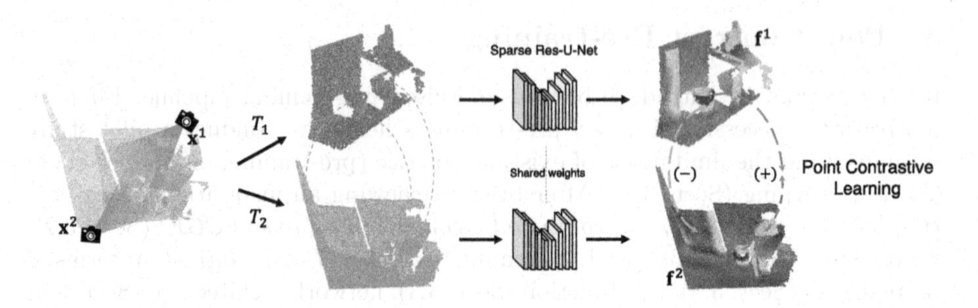

Fig. 2. PointContrast: pretext task for 3D pre-training.

Table 1. Summary of downstream fine-tuning tasks. Compared to the baseline learning paradigm of training from scratch, which is dominant in 3D deep learning, our unsupervised pre-training method PointContrast boosts the performance across the board when finetuning on a diverse set of high-level 3D understanding tasks. $*$ indicates results trained using only 1% of the training data.

PointContrast: downstream tasks for fine-tuning					
Datasets	Real/Synth.	Complexity	Env.	Task	Rel.gain
S3DIS	Real	Entire floor, office	Indoor	Segmentation	(+2.7%) mIoU
SUN RGB-D	Real	Medium-sized cluttered rooms	Indoor	Detection	(+3.1%) mAP0.5
ScanNetV2	Real	Large rooms	Indoor	Segmentation	(+1.9%) mIoU
				Detection	(+2.6%) mAP0.5
ShapeNet	Synth.	Single objects	Indoor & outdoor	Classification	(+4.0%) Acc.*
ShapeNetPart	Synth.	Object parts	Indoor & outdoor	Segmentation	(+2.2%) mIoU*
Synthia 4D	Synth.	Street scenes, driving envs.	Outdoor	Segmentation	(+3.3%) mIoU

- **Point-level representation matters:** In 3D deep learning, the local geometric features, *e.g.* those encoded by a point and its neighbors, have proven to be discriminative and critical for 3D tasks [45,46]. Directly training on *object instances* to obtain a global representation might be insufficient.

This led us to rethink the problem: if the goal of pre-training is to boost performance across many real world tasks, exploring pre-training strategies on single objects might offer limited potential. (1) To address the domain gap concern, it might be beneficial to directly pre-train the network on complex scenes with multiple objects, to better match the target distributions; (2) to capture point-level information, we need to design a pretext task and corresponding network architecture that is not only based on instance-level/global representations, but instead can capture dense/local features at the point level.

3.2 Revisiting Fully Convolutional Geometric Features (FCGF)

Here we revisit a previous approach FCGF [10] designed to learn geometric features for *low-level* tasks (*e.g.* registration) as our work is mainly inspired by FCGF. FCGF is a deep learning based algorithm that learns local feature descriptors on correspondence datasets via metric learning. FCGF has two major ingredients that help it stand out and achieve impressive results in registration recall: (1) **a fully-convolutional design** and (2) **point-level metric learning**. With a fully-convolutional network (FCN) [36] design, FCGF operates on the entire input point cloud (*e.g.* full indoor or outdoor scenes) without having to crop the scene into patches as done in previous works; this way the local descriptors can aggregate information from a large number of neighboring points (up to the extent of receptive field size). As a result, point-level metric learning becomes natural. FCGF uses a U-Net architecture that has full-resolution output (*i.e.* for N points, the network outputs N associated feature vectors), and positive/negative pairs for metric learning are defined at the point level.

Despite having a fundamentally different goal in mind, FCGF offers inspirations that might address the pretext task design challenges: A fully-convolutional design will allow us to pre-train on the target data distributions that involve complex scenes with a large number of points, and we could define the pretext task directly on points. Under this perspective, we pose the question: *Can we repurpose FCGF as the pretext task for high-level 3D understanding?*

Algorithm 1 General Framework of PointContrast

Input: Backbone architecture NN; Dataset $X = \{\mathbf{x}_i \in \mathbb{R}^{N \times 3}\}$; Point feature dimension D;
Output: Pre-trained weights for NN.
for *each point cloud* $\mathbf{x}$ *in* X **do**
- From $\mathbf{x}$, generate two views $\mathbf{x}^1$ and $\mathbf{x}^2$.
- Compute correspondence mapping (matches) M between points in $\mathbf{x}^1$ and $\mathbf{x}^2$.
- Sample two transformations $\mathbf{T}_1$ and $\mathbf{T}_2$.
- Compute point features $\mathbf{f}^1, \mathbf{f}^2 \in \mathbb{R}^{N \times D}$ by $\mathbf{f}^1 = \text{NN}(\mathbf{T}_1(\mathbf{x}^1))$ and $\mathbf{f}^2 = \text{NN}(\mathbf{T}_2(\mathbf{x}^2))$.
- Backprop. to update NN with contrastive loss $\mathcal{L}_c(\mathbf{f}^1, \mathbf{f}^2)$ on the matched points.

3.3 PointContrast as a Pretext Task

FCGF focuses on local descriptor learning for low-level tasks only. In contrast, a good pretext task for pre-training aims to learn *network weights* that are universally applicable and useful to many high-level 3D understanding tasks. To take the inspiration of FCGF and create such pretext tasks, there are several design choices that need to be revisited. In terms of *architecture*, since inference speed is a major concern in registration tasks, the network used in FCGF is very light-weight; Contrarily, the success of pre-training relies on over-parameterized networks, as clearly evidenced in other domains [8,13]. In terms of *dataset*, FCGF uses domain-specific registration datasets such as 3DMatch [75] and KITTI odometry [17], which lack both scale and generality. Finally, in terms of *loss design*, contrastive losses explored in FCGF are tailored for registration and it is interesting to explore other alternatives.

In Algorithm 1, we summarize the overall pretext task framework explored in this work. We name the framework *PointContrast*, since the high-level strategy of this pretext task is, contrasting—at the point level—between two transformed point clouds. Conceptually, given a point cloud $\mathbf{x}$ sampled from a certain distribution, we first generate two views $\mathbf{x}^1$ and $\mathbf{x}^2$ that are aligned in the same world coordinates. We then compute the correspondence mapping M between these two views. If $(i, j) \in M$ then point $\mathbf{x}_i^1$ and point $\mathbf{x}_j^2$ are a pair of matched points across two views. We then sample two random geometric transformations T_1 and T_2 to further transform the point clouds in two views. The transformation is what could make the pretext task challenging as the network needs to learn certain *equivariance* with respect to the geometric transformation imposed. In this work, we mainly consider rigid transformation including rotation, translation and scaling. Further details are provided in Appendix. Finally, a contrastive loss is defined over points in two views: we minimize the distance for matched points and maximize the distance of unmatched points. This framework, though coming from a very different motivation (metric learning for geometric local descriptors), shares a strikingly similar pipeline with recent contrastive-based methods for 2D unsupervised visual representation learning [8,23,64]. The key difference is that most work for 2D focuses on contrasting between instances/images, while in our work the contrastive learning is done densely at the point level.

3.4 Contrastive Learning Loss Design

Hardest-Contrastive Loss. The first loss function, hardest-contrastive loss we try, is borrowed from the best-performing loss design proposed in FCGF [10], which adopts a hard negative mining scheme in traditional margin-based contrastive learning formulation,

$$\mathcal{L}_c = \sum_{(i,j)\in\mathcal{P}} \left\{ [d(\mathbf{f}_i, \mathbf{f}_j) - m_p]_+^2 / |\mathcal{P}| + 0.5[m_n - \min_{k\in\mathcal{N}} d(\mathbf{f}_i, \mathbf{f}_k)]_+^2 / |\mathcal{N}_i| + 0.5[m_n - \min_{k\in\mathcal{N}} d(\mathbf{f}_j, \mathbf{f}_k)]_+^2 / |\mathcal{N}_j| \right\}$$

Here $\mathcal{P}$ is a set of matched (positive) pairs of points $\mathbf{x}_i^1$ and $\mathbf{x}_j^2$ from two views $\mathbf{x}^1$ and $\mathbf{x}^2$, and $\mathbf{f}_i^1$ and $\mathbf{f}_j^2$ are associated point features for the matched pair. $\mathcal{N}$ is a randomly sampled set of non-matched (negative) points which is used for the hardest negative mining, where the hardest sample is defined as the closest point in the $\mathcal{L}_2$ normalized feature space to a positive pair. $[x]_+$ denotes function $\max(0, x)$. $m_p = 0.1$ and $m_n = 1.4$ are margins for positive and negative pairs.

PointInfoNCE Loss. Here we propose an alternative loss design for PointContrast. InfoNCE proposed in [40] is widely used in recent unsupervised representation learning approaches for 2D visual understanding. By modeling the contrastive learning framework as a dictionary look-up process [23], InfoNCE poses contrastive learning as a classification problem and is implemented with a Softmax loss. Specifically, the loss encourages a query q to be similar to its positive key k^+ and dissimilar to, typically many, negative keys k^-. One challenge in 2D is to scale the number of negative keys [23].

However, in the domain of 3D, we have a different problem: usually the real-world 3D datasets are much smaller in terms of instance count, but the number of points for each instance (*e.g.* a indoor or outdoor scene) can be huge, *i.e.* 100K+ points even from one RGB-D frame. This unique property of 3D data property, together with the original motivation to modelling point level information, inspire us to propose the following PointInfoNCE loss:

$$\mathcal{L}_c = -\sum_{(i,j)\in\mathcal{P}} \log \frac{\exp(\mathbf{f}_i \cdot \mathbf{f}_j/\tau)}{\sum_{(\cdot,k)\in\mathcal{P}} \exp(\mathbf{f}_i \cdot \mathbf{f}_k/\tau)}$$

Here $\mathcal{P}$ is the set of all the positive matches from two views. In this formulation, we only consider points that have at least one match and do not use additional non-matched points as negatives. For a matched pair $(i,j) \in \mathcal{P}$, point feature $\mathbf{f}_i^1$ will serve as the query and $\mathbf{f}_j^2$ will serve as the positive key k^+. We use point feature $\mathbf{f}_k^2$ where $(\cdot,k) \in \mathcal{P}$ and $k \neq j$ as the set of negative keys. In practice, we sample a subset of 4096 matched pairs from $\mathcal{P}$ for faster training.

Compared to hardest-contrastive loss, the PointInfoNCE loss has a simpler formulation with less hyperparatmers. Perhaps more importantly, due to the large number of negative distractors, it is more robust against *mode collapsing* (features collapsed to a single vector) than the hardest-contrastive loss. In our experiments, we find that hard-contrastive loss is unstable and hard to train: the representation often collapses with extended training epochs (which is also observed in FCGF [10]).

3.5 A Sparse Residual UNet as Shared Backbone

We use a Sparse Residual UNet (SR-UNet) architecture in this work. It is a 34-layer UNet [49] architecture that has an encoder network of 21 convolution layers and a decoder network of 13 convolution/deconvolution layers. It follows the 2D ResNet basic block design and each conv/deconv layer in the network are followed by Batch Normalization (BN) [30] and ReLU activation. The overall UNet architecture has 37.85M parameters. We provide more information and a visualization of the network in Appendix. The SR-UNet architecture was originally designed in [9] that achieved significant improvement over prior methods on the challenging ScanNet semantic segmentation benchmark. In this work we explore if we can use this architecture as a unified design for both the pre-training task and a diverse set of fine-tuning tasks.

3.6 Dataset for Pre-training

For local geometric feature learning approaches, including FCGF [10], training and evaluation are typically conducted on domain and task specific datasets such as KITTI odometry [17] or 3DMatch [75]. Common registration datasets are typically constrained in either scale (training samples collected from just dozens of scenes), or generality (focusing on one specific application scenario, *e.g.* indoor scenes or LiDAR scans for self-driving car), or both. To facilitate future research on 3D unsupervised representation learning, in our work we utilize the ScanNet dataset for pre-training, aiming to address the scale issue. ScanNet is a collection of ~1500 indoor scenes. Created with a light-weight RGB-D scanning procedure, ScanNet is currently the largest of its kind.[1]

Here we create a point cloud pair dataset on top of ScanNet for the pretraining framework shown in Fig. 2. Given a scene $\mathbf{x}$, we extract pairs of partial scans $\mathbf{x}^1$ and $\mathbf{x}^2$ from different views. More precisely, for each scene, we first sub-sample RGB-D scans from the raw ScanNet videos every 25 frames, and align the 3D point clouds in the same world coordinates (by utilizing estimated camera poses for each frame). Then we collect point cloud pairs from the sampled frames and require that two point clouds in a pair have at least 30% overlap. We sample a total number of 870K point cloud pairs. Since the partial views are aligned in ScanNet scenes, it is straight-forward to compute the correspondence mapping M between two views with nearest neighbor search.

Although ScanNet only captures indoor data distributions, as we will see in Section 4.4, surprisingly it can generalize to other target distributions. We provide additional visualizations for the pre-training dataset in Appendix.

4 Fine-Tuning on Downstream Tasks

The most important motivation for representation learning is to learn features that can transfer well to different downstream tasks. There could be different evaluation protocols to measure the usefulness of the learned representation. For example, probing with a linear classifier [19], or evaluating in a semi-supervised setup [26]. The *supervised fine-tuning* strategy, where the pre-trained weights are used as the initialization and are further refined on the target downstream task, is arguably the most practically meaningful way of evaluating feature transferability. With this setup, good features could directly lead to performance gains in downstream tasks.

Under this perspective, in this section we perform extensive evaluations of the effectiveness of PointContrast framework by fine-tuning the pre-trained weights on multiple downstream tasks and datasets. We aim to cover a diverse suit of high-level 3D understanding tasks of different natures such as semantic segmentation, object detection and classification. In all experiments we use the same backbone network, pre-trained on the proposed ScanNet pair dataset (Sect. 3.6) using both PointInfoNCE and Hardest-Constrastive objectives.

4.1 ShapeNet: Classification and Part Segmentation

Setup. In Sect. 3.1 we have observed that weights learned on supervised ShapeNet classification are not able to transfer well to scene-level tasks. Here we explore the opposite direction: Are PointContrast features learned on ScanNet

[1] Admittedly, ScanNet is still much smaller in scale compared to 2D datasets.

Table 2. ShapeNet classification. Top: classification accuracy with limited labeled training data for finetuning. Bottom: classification accuracy on the least represented classes in the data (tail-classes). In all cases, PointContrast boosts performance. Relative improvement increases with scarcer training data and on less frequent classes.

Evaluating on all 55 classes	1% data	10% data	100% data
Trained from scratch	62.2	77.9	85.1
PointConstrast (Hardest-Contrastive)	**66.2** (+4.0)	**79.0** (+1.1)	**85.7** (+0.6)
PointConstrast (PointInfoNCE)	**65.8** (+3.6)	**78.8** (+0.9)	**85.7** (+0.6)
Using 100% training data	10 tail classes	30 tail classes	All 55 classes
Train from scratch	65.0	70.9	85.1
PointConstrast (Hardest-Contrastive)	**70.9** (+5.9)	**72.9** (+2.0)	**85.7** (+0.6)
PointConstrast (PointInfoNCE)	**67.8** (+2.8)	**72.0** (+1.1)	**85.7** (+0.6)

Table 3. ShapeNet part segmentation. Replacing the backbone architecture with SR-UNet already boosts performance. PointContrast pre-training further adds a significant gain, and outshines where labels are most limited.

Methods	IoU (1% data)	IoU (5% data)	IoU (100% data)
SO-Net [34]	64.0	69.0	–
PointCapsNet [80]	67.0	70.0	–
Multitask Unsupervised [22]	68.2	77.7	–
Train from scratch	71.8	79.3	84.7
PointConstrast (Hardest-Contrastive)	**74.0** (+2.2)	**79.9** (+0.6)	**85.1** (+0.4)
PointConstrast (PointInfoNCE)	**73.1** (+1.3)	**79.9** (+0.6)	**85.1** (+0.4)

useful for tasks on ShapeNet? To recap, ShapeNet [7] is a dataset of synthetic 3D objects of 55 common categories. It was curated by collecting CAD models from online open-sourced 3D repositories. In [71], part annotations were added to a subset of ShapeNet models segmenting them into 2–5 parts. In order to provide a comparison with existing approaches, here we utilize the ShapeNetCore dataset (SHREC 15 split) for classification, and the ShapeNet part dataset for part segmentation, respectively. We uniformly sample point clouds of 1024 points from each model for classification and 2048 points for part segmentation. Albeit containing overlapping indoor object categories with ScanNet, this dataset is substantially different as it is synthetic, and contains only single objects. We also follow recent works on 3D unsupervised representation learning [22] to explore a more challenging setup: using a very small percentage (*e.g.* 1%–10%) of training data to fine-tune the pre-trained model.

Results. As shown in Table 2 and Table 3, for both datasets, the effectiveness of pre-training are correlated with the availability of training data. In the ShapeNet classification task (Table 2), pre-training helps most where less training data is available, achieving a 4.0% improvement over the training-from-scratch baseline with the hardest-negative objective. We also note that ShapeNet is a class-imbalanced dataset and the minority (tail) classes are very infrequent. When using 100% of the training data, pre-training provides a class-balancing effect, as it boosts performance more on underrepresented (tail) classes. Table 3 shows a similar effects of pre-training on part segmentation performance. Notably, using SR-UNet backbone architecture already boosts performance; yet, pre-training is able to provide further gains, especially when training data is scarce.

4.2 S3DIS Segmentation

Setup. Stanford Large-Scale 3D Indoor Spaces (S3DIS) [2] dataset comprises 3D scans of 6 large-scale indoor areas collected from 3 office buildings. The scans are represented as point clouds and annotated with semantic labels of 13 object categories. Among the datasets used here for evaluation S3DIS is probably the most similar to ScanNet. Transferring features to S3DIS represents a typical scenario for fine-tuning: the downstream task dataset is similar yet much smaller than the pre-training dataset. For the commonly used benchmark split ("Area 5 test"), there are only about 240 samples in the training set. We follow [9] for pre-processing, and use standard data augmentations. See Appendix for details.

Results. Results are summarized in Table 4. Again, merely switching the SR-UNet architecture, training from scratch already improves upon prior art. Yet, fine-tuning the features learned by PointContrast achieves markedly better segmentation results in mIoU and mAcc. Notably, the effect persists across both loss types, achieving a 2.7% mIoU gain using Hardest-Contrastive loss and an on-par improvement of 2.1% mIoU for the PointInfoNCE variant.

Table 4. Stanford Area 5 Test (Fold 1) (S3DIS). Replacing the backbone network with SR-UNet improves upon prior art. Using PointContrast adds further significant boost with a mild preference for Hardest-contrastive over the PointInfoNCE objective. See Appendix for more methods in comparison.

Methods	mIoU	mAcc
PointNet [45]	41.1	49.0
PointCNN [35]	57.3	63.9
MinkowskiNet32 [9]	65.4	71.7
Train from scratch	68.2	75.5
PointConstrast (Hardest-Contrastive)	**70.9**	**77.0**
PointConstrast (PointInfoNCE)	70.3	76.9

4.3 SUN RGB-D Detection

Setup. We now attend to a different high-level 3D understanding task: object detection. Compared to segmentation tasks that estimate point labels, 3D object detection predicts 3D bounding boxes (localization) and their corresponding object labels (recognition). This calls for an architectural modification as the SR-UNet architecture does not directly output bounding box coordinates. Among many different choices [28,42,44,73], we identify the recently proposed VoteNet [43] as a good candidate for three main reasons. First, VoteNet is designed to work directly on point clouds with no additional input (e.g. images). Second, VoteNet originally uses PointNet++ [46] as the backbone architecture for feature extraction. Replacing this with a SR-UNet requires a minimal modification, keeping the proposal pipeline intact. In particular, we reuse the same hyperparameters. Third, VoteNet is the current state-of-the-art method that uses geometric features only, rendering an improvement markedly useful. We evaluate the detection performance on the SUN RGB-D dataset [55], a collection of single view RGB-D images. The train set contains 5K images annotated with amodal, 3D oriented bounding boxes for objects from 37 categories.

Results. We summarize the results in Table 5. We find that by simply switching in the backbone network, our baseline results (training from scratch) with the SR-UNet architecture achieves worse results (-1.4% mAP@0.25). This may be attributed to the fact that VoteNet design and hyperparamter settings were tailored to its PointNet++ backbone. However, PointContrast gracefully closes the gap by showing a +3.1% gain on mAP@0.5, which also sets a new state-of-the-art in this metric. The performance gain with harder evaluation metric (mAP@0.5) suggests that the PointContrast pre-training can greatly help localization.

Table 5. SUN RGB-D detection results. PointContrast demonstrates a substantial boost compared to training from scratch. We observe a larger improvement in localization as manifested by the ΔmAP being larger for @0.5 than @0.25.

Methods	Input	mAP@0.5	mAP@0.25
VoteNet [43]	Geo	–	57.0
VoteNet [43]	Geo+Height	32.9	57.7
Train from scratch	Geo	31.7	55.6
PointContrast(Hardest-Contrastive)	Geo	34.5	**57.5**
PointContrast(PointInfoNCE)	Geo	**34.8**	**57.5**

Table 6. Segmentation results on the 4D Synthia test set. All networks here are SR-UNet with 3D kernels, trained on individual 3D frames without temporal modeling.

Methods	mIoU	mAcc
MinkowskiNet32 [9]	78.7	91.5
Train from scratch	79.8	91.5
PointContrast (Hardest-contrastive)	82.6	**93.7**
PointContrast (PointInfoNCE)	**83.1**	**93.7**

4.4 Synthia4D Segmentation

Setup. Synthia4D [50] is a large synthetic dataset designed to facilitate the training of deep neural networks for visual inference in driving scenarios. Photo-realistic renderings are generated from a virtual city, allowing dense and precise annotations of 13 semantic classes, together with pixel-accurate depth. We follow the train/val/test split as prescribed by [9] in the clean setting. In the context of this work, Synthia4D is especially interesting since it is probably the most distant from our pre-training set (outdoor v.s. indoor, synthetic v.s. real). We test the segmentation performance using 3D SR-UNet on a per-frame basis.

Results. PointContrast pre-training brings substantial improvement over the baseline model trained from scratch (+2.3% mIoU) as seen in Table 6. PointInfoNCE performs noticeably better than the hardest-contrastive loss. With unsupervised pre-training, the overall results are much better than the previous state-of-the-art reported in [9]. Note that in [9] it has been shown that adding the temporal learning (*i.e.* using a 4D network instead of a 3D one) brings additional benefit. To use 3D pre-trained weights for a 4D network with an additional temporal dimension, we can simply inflate the convolutional kernels, following the standard practice in 2D video recognition [6]. We leave it as a future work.

Table 7. Segmentation results on ScanNet validation set. PointContrast boosts performance on the "in-domain" transfer task where the pre-training and fine-tuning datasets come from a common source, showing the usefulness of pre-training even when labels are available.

Methods	mIoU	mAcc
Train from scratch	72.2	80.7
PointContrast(Hardest-Contrastive)	73.3	81.0
PointContrast(PointInfoNCE)	**74.1**	**81.6**

Table 8. 3D object detection results on ScanNet validation set. Similarly to in-domain *segmentation* task, here as well PointContrast boost performance on *detection*, setting a new best result over prior art. See Appendix for more methods in comparison.

Methods	Input	mAP@0.5	mAP@0.25
DSS [28,56]	Geo+RGB	6.8	15.2
3D-SIS [28]	Geo+RGB (5 Views)	22.5	40.2
VoteNet [43]	Geo+Height	33.5	58.6
Train from scratch	Geo	35.4	56.7
PointContrast(Hardest-Contrastive)	Geo	37.3	**59.2**
PointContrast(PointInfoNCE)	Geo	**38.0**	58.5

4.5 ScanNet: Segmentation and Detection

Setup. Although typically the source dataset for pre-training and the target dataset for fine-tuning are different, because of the specific multi-view contrastive learning pipeline for pre-training, it is likely that PointContrast can learn different representations (*e.g.* invariance/equivariance to rigid transformations or robustness to noise) compared to directly training with supervision. Thus it is interesting to see whether the pre-trained weights can further improve the results on ScanNet itself. We use ScanNet semantic segmentation and object detection tasks to test our hypothesis. For the segmentation experiment, we use the SR-UNet architecture to directly predict point labels. For the detection experiment, we again follow VoteNet [43] and simply switch the original backbone network with the SR-UNet without other modifications to the detection head (See Appendix for details).

Results. Results are summarized in Table 7 and Table 8. Remarkably, on both detection and segmentation benchmark, models pre-trained with PointContrast outperform those trained from scratch. Notably, PointInfoNCE objective performs better than the Hardest-contrastive one, achieving a relative improvement of +1.9% in terms of segmentation mIoU and 2.6%+ in terms of detection mAP@0.5. Similar to SUN RGB-D detection, here we also observe that PointContrast features help most for localization as indicated by the larger margin of improvement for mAP@0.5 than mAP@0.25.

4.6 Analysis Experiments and Discussions

In this section we show additional experiments to provide more insights on our pre-training framework. We use S3DIS segmentation for the experiments below.

Supervised Pre-training. While the focus of this work is unsupervised pre-training, a natural baseline is to compare against supervised pre-training. To this end, we use the training-from-scratch baseline for the segmentation task on ScanNetV2 and finetune the network on S3DIS. This yields an mIoU of 71.2%, which is only 0.3% better than PointContrast unsupervised pre-training. We deem this a very *encouraging signal* that suggests that the gap between supervised and unsupervised representation learning in 3D has been mostly closed (*cf.* years of effort in 2D). One might argue that this is due to the limited quality and scale of ScanNet, but even at this scale the amount of labor involved in annotating thousands of rooms is large. The outcome of this, complements the conclusion we had so far: not only should we put resources into creating large-scale 3D datasets for pre-training; but if facing a trade-off between scaling the data size and annotating it, we should favor the former.

Fine-Tuning vs From-Scratch Under Longer Training Schedule. Recent study in 2D vision [24] suggests that simply by training from scratch for more epochs might close the gap from ImageNet pre-training. We conduct additional experiment to train the network from scratch with 2× and 3× schedules on S3DIS, relative to the 1× schedule of our default setup (10K iterations with batch size 48). We found that validation mIoU does not improve with longer training. In fact, the model exhibits overfitting due to the small dataset size, achieving 66.7% and 66.1% mIoU at 20K and 30K iteration, respectively. This suggests that potentially many of the 3D datasets could fall into the "breakdown regime" [24] where network pre-training is essential for good performance.

Holistic Scene as a Single View for PointContrast. To show that the multi-view design in PointContrast is important, we try a different variant where instead of having partial views $\mathbf{x}^1$ and $\mathbf{x}^2$, we directly use the reconstructed point cloud $\mathbf{x}$ (a full scene in ScanNet) PointContrast. We still apply independent transformations T_1 and T_2 to the same $\mathbf{x}$. We tried different variants and augmentations such as random cropping, point jittering, and dropout. We also tried different transformations for T_1 and T_2 of different degrees of freedom. However, with the best configuration we can get a validation mIoU on S3DIS of 68.35, which is just slightly better than the training from scratch baseline of 68.17. This suggests that the multi-view setup in PointContrast is critical. Potential reasons include: much more abundant and diverse training samples; natural noise due to camera instability as good regularization, as also observed in [75].

Acknowledgements. O.L. and L.G. were supported in part by NSF grant IIS-1763268, a Vannevar Bush Faculty Fellowship, and a grant from the SAIL-Toyota Center for AI Research.

References

1. Achlioptas, P., Diamanti, O., Mitliagkas, I., Guibas, L.: Learning representations and generative models for 3D point clouds. arXiv preprint arXiv:1707.02392 (2017)
2. Armeni, I., et al.: 3D semantic parsing of large-scale indoor spaces. In: ICCV (2016)
3. Bachman, P., Hjelm, R.D., Buchwalter, W.: Learning representations by maximizing mutual information across views. In: NeurIPS (2019)
4. Boulch, A.: ConvPoint: continuous convolutions for point cloud processing. Comput. Graph. **88**, 24–34 (2020)
5. Caron, M., Bojanowski, P., Joulin, A., Douze, M.: Deep clustering for unsupervised learning of visual features. In: ECCV (2018)
6. Carreira, J., Zisserman, A.: Quo vadis, action recognition? a new model and the kinetics dataset. In: CVPR (2017)
7. Chang, A.X., et al.: ShapeNet: an information-rich 3D model repository. arXiv preprint arXiv:1512.03012 (2015)
8. Chen, T., Kornblith, S., Norouzi, M., Hinton, G.: A simple framework for contrastive learning of visual representations. arXiv preprint arXiv:2002.05709 (2020)
9. Choy, C., Gwak, J., Savarese, S.: 4D spatio-temporal convnets: Minkowski convolutional neural networks. In: CVPR (2019)
10. Choy, C., Park, J., Koltun, V.: Fully convolutional geometric features. In: ICCV (2019)

11. Dai, A., Chang, A.X., Savva, M., Halber, M., Funkhouser, T., Nießner, M.: ScanNet: Richly-annotated 3d reconstructions of indoor scenes. In: CVPR (2017)
12. Deng, H., Birdal, T., Ilic, S.: PPF-FoldNet: unsupervised learning of rotation invariant 3D local descriptors. In: ECCV (2018)
13. Devlin, J., Chang, M.W., Lee, K., Toutanova, K.: BERT: pre-training of deep bidirectional transformers for language understanding. In: NAACL (2019)
14. Doersch, C., Gupta, A., Efros, A.A.: Unsupervised visual representation learning by context prediction. In: ICCV (2015)
15. Elbaz, G., Avraham, T., Fischer, A.: 3D point cloud registration for localization using a deep neural network auto-encoder. In: CVPR (2017)
16. Gadelha, M., Wang, R., Maji, S.: Multiresolution tree networks for 3D point cloud processing. In: ECCV (2018)
17. Geiger, A., Lenz, P., Stiller, C., Urtasun, R.: Vision meets robotics: the KITTI dataset. Int. J. Robot. Res. **32**(11), 1231–1237 (2013)
18. Gidaris, S., Singh, P., Komodakis, N.: Unsupervised representation learning by predicting image rotations. In: ICLR (2018)
19. Goyal, P., Mahajan, D., Gupta, A., Misra, I.: Scaling and benchmarking self-supervised visual representation learning. In: ICCV (2019)
20. Graham, B., Engelcke, M., van der Maaten, L.: 3D semantic segmentation with submanifold sparse convolutional networks. In: CVPR (2018)
21. Groueix, T., Fisher, M., Kim, V.G., Russell, B.C., Aubry, M.: A papier-mâché approach to learning 3D surface generation. In: CVPR (2018)
22. Hassani, K., Haley, M.: Unsupervised multi-task feature learning on point clouds. In: ICCV (2019)
23. He, K., Fan, H., Wu, Y., Xie, S., Girshick, R.: Momentum contrast for unsupervised visual representation learning. In: CVPR (2020)
24. He, K., Girshick, R., Dollár, P.: Rethinking ImageNet pre-training. In: ICCV (2019)
25. He, K., Zhang, X., Ren, S., Sun, J.: Deep residual learning for image recognition. In: CVPR (2016)
26. Hénaff, O.J., Razavi, A., Doersch, C., Eslami, S., Oord, A.V.D.: Data-efficient image recognition with contrastive predictive coding. arXiv preprint arXiv:1905.09272 (2019)
27. Hjelm, R.D., et al.: Learning deep representations by mutual information estimation and maximization. In: ICLR (2019)
28. Hou, J., Dai, A., Nießner, M.: 3D-SIS: 3D semantic instance segmentation of RGB-D scans. In: CVPR (2019)
29. Hua, B.S., Tran, M.K., Yeung, S.K.: Pointwise convolutional neural networks. In: CVPR (2018)
30. Ioffe, S., Szegedy, C.: Batch normalization: accelerating deep network training by reducing internal covariate shift. In: ICML (2015)
31. Janoch, A. et al.: A category-level 3d object dataset: putting the kinect to work. In: Fossati, A., Gall, J., Grabner, H., Ren, X., Konolige, K. (eds.) Consumer Depth Cameras for Computer Vision. Advances in Computer Vision and Pattern Recognition. Springer, London (2013). https://doi.org/10.1007/978-1-4471-4640-7_8
32. Klokov, R., Lempitsky, V.: Escape from cells: deep Kd-networks for the recognition of 3D point cloud models. In: ICCV (2017)
33. Lei, H., Akhtar, N., Mian, A.: Spherical convolutional neural network for 3D point clouds. arXiv preprint arXiv:1805.07872 (2018)
34. Li, J., Chen, B.M., Hee Lee, G.: SO-Net: self-organizing network for point cloud analysis. In: CVPR (2018)

35. Li, Y., Bu, R., Sun, M., Wu, W., Di, X., Chen, B.: PointCNN: convolution on X-transformed points. In: NeurIPS (2018)
36. Long, J., Shelhamer, E., Darrell, T.: Fully convolutional networks for semantic segmentation. In: CVPR (2015)
37. Maron, H., Litany, O., Chechik, G., Fetaya, E.: On learning sets of symmetric elements. arXiv preprint arXiv:2002.08599 (2020)
38. Misra, I., van der Maaten, L.: Self-supervised learning of pretext-invariant representations. In: CVPR (2020)
39. Noroozi, M., Favaro, P.: Unsupervised learning of visual representations by solving jigsaw puzzles. In: Leibe, B., Matas, J., Sebe, N., Welling, M. (eds.) ECCV 2016. LNCS, vol. 9910, pp. 69–84. Springer, Cham (2016). https://doi.org/10.1007/978-3-319-46466-4_5
40. Oord, A.V.D., Li, Y., Vinyals, O.: Representation learning with contrastive predictive coding. arXiv preprint arXiv:1807.03748 (2018)
41. Pathak, D., Krahenbuhl, P., Donahue, J., Darrell, T., Efros, A.A.: Context encoders: feature learning by inpainting. In: CVPR (2016)
42. Qi, C.R., Chen, X., Litany, O., Guibas, L.J.: ImVoteNet: boosting 3D object detection in point clouds with image votes. In: CVPR (2020)
43. Qi, C.R., Litany, O., He, K., Guibas, L.J.: Deep hough voting for 3D object detection in point clouds. In: ICCV (2019)
44. Qi, C.R., Liu, W., Wu, C., Su, H., Guibas, L.J.: Frustum PointNets for 3D object detection from RGB-D data. In: CVPR (2018)
45. Qi, C.R., Su, H., Mo, K., Guibas, L.J.: PointNet: deep learning on point sets for 3D classification and segmentation. In: CVPR (2017)
46. Qi, C.R., Yi, L., Su, H., Guibas, L.J.: PointNet++: deep hierarchical feature learning on point sets in a metric space. In: NeurIPS (2017)
47. Radford, A., Wu, J., Child, R., Luan, D., Amodei, D., Sutskever, I.: Language models are unsupervised multitask learners. OpenAI Blog **1**(8), 9 (2019)
48. Ravanbakhsh, S., Schneider, J., Poczos, B.: Deep learning with sets and point clouds. arXiv preprint arXiv:1611.04500 (2016)
49. Ronneberger, O., Fischer, P., Brox, T.: U-Net: convolutional networks for biomedical image segmentation. In: Navab, N., Hornegger, J., Wells, W.M., Frangi, A.F. (eds.) MICCAI 2015. LNCS, vol. 9351, pp. 234–241. Springer, Cham (2015). https://doi.org/10.1007/978-3-319-24574-4_28
50. Ros, G., Sellart, L., Materzynska, J., Vazquez, D., Lopez, A.M.: The synthia dataset: a large collection of synthetic images for semantic segmentation of urban scenes. In: CVPR (2016)
51. Sauder, J., Sievers, B.: Self-supervised deep learning on point clouds by reconstructing space. In: NeurIPS (2019)
52. Shen, Y., Feng, C., Yang, Y., Tian, D.: Mining point cloud local structures by kernel correlation and graph pooling. In: CVPR (2018)
53. Silberman, N., Hoiem, D., Kohli, P., Fergus, R.: Indoor segmentation and support inference from RGBD images. In: Fitzgibbon, A., Lazebnik, S., Perona, P., Sato, Y., Schmid, C. (eds.) ECCV 2012. LNCS, vol. 7576, pp. 746–760. Springer, Heidelberg (2012). https://doi.org/10.1007/978-3-642-33715-4_54
54. Simonyan, K., Zisserman, A.: Very deep convolutional networks for large-scale image recognition. arXiv preprint arXiv:1409.1556 (2014)
55. Song, S., Lichtenberg, S.P., Xiao, J.: Sun RGB-D: a RGB-D scene understanding benchmark suite. In: CVPR (2015)
56. Song, S., Xiao, J.: Deep sliding shapes for amodal 3D object detection in RGB-D images. In: CVPR (2016)

57. Su, H., et al.: SPLATNet: sparse lattice networks for point cloud processing. In: CVPR, pp. 2530–2539 (2018)
58. Te, G., Hu, W., Zheng, A., Guo, Z.: RGCNN: regularized graph CNN for point cloud segmentation. In: ACM Multimedia (2018)
59. Verma, N., Boyer, E., Verbeek, J.: FeastNet: feature-steered graph convolutions for 3D shape analysis. In: CVPR (2018)
60. Wang, C., Samari, B., Siddiqi, K.: Local spectral graph convolution for point set feature learning. In: ECCV (2018)
61. Wang, S., Suo, S., Ma, W.C., Pokrovsky, A., Urtasun, R.: Deep parametric continuous convolutional neural networks. In: CVPR (2018)
62. Wang, Y., Solomon, J.M.: Deep closest point: learning representations for point cloud registration. In: ICCV (2019)
63. Wang, Y., Sun, Y., Liu, Z., Sarma, S.E., Bronstein, M.M., Solomon, J.M.: Dynamic graph CNN for learning on point clouds. ACM TOG **38**(5), 1–12 (2019)
64. Wu, Z., Xiong, Y., Yu, S.X., Lin, D.: Unsupervised feature learning via non-parametric instance discrimination. In: CVPR (2018)
65. Xiao, J., Owens, A., Torralba, A.: SUN3D: a database of big spaces reconstructed using SFM and object labels. In: ICCV (2013)
66. Xie, S., Girshick, R., Dollár, P., Tu, Z., He, K.: Aggregated residual transformations for deep neural networks. In: CVPR (2017)
67. Xie, S., Liu, S., Chen, Z., Tu, Z.: Attentional ShapeContextNet for point cloud recognition. In: CVPR, pp. 4606–4615 (2018)
68. Xu, Y., Fan, T., Xu, M., Zeng, L., Qiao, Y.: SpiderCNN: deep learning on point sets with parameterized convolutional filters. In: ECCV (2018)
69. Yang, Y., Feng, C., Shen, Y., Tian, D.: FoldingNet: point cloud auto-encoder via deep grid deformation. In: CVPR (2018)
70. Yang, Z., Litany, O., Birdal, T., Sridhar, S., Guibas, L.: Continuous geodesic convolutions for learning on 3D shapes. In: arXiv preprint arXiv:2002.02506 (2020)
71. Yi, L., et al.: A scalable active framework for region annotation in 3D shape collections. In: SIGGRAPH Asia (2016)
72. Yi, L., Su, H., Guo, X., Guibas, L.J.: SyncSpecCNN: synchronized spectral CNN for 3D shape segmentation. In: CVPR (2017)
73. Yi, L., Zhao, W., Wang, H., Sung, M., Guibas, L.: GSPN: generative shape proposal network for 3d instance segmentation in point cloud. In: CVPR (2019)
74. Zaheer, M., Kottur, S., Ravanbakhsh, S., Poczos, B., Salakhutdinov, R.R., Smola, A.J.: Deep sets. In: NeurIPS (2017)
75. Zeng, A., Song, S., Nießner, M., Fisher, M., Xiao, J., Funkhouser, T.: 3DMatch: learning local geometric descriptors from RGB-D reconstructions. In: CVPR (2017)
76. Zeng, W., Gevers, T.: 3DContextNet: KD tree guided hierarchical learning of point clouds using local and global contextual cues. In: ECCV (2018)
77. Zhang, R., Isola, P., Efros, A.A.: Colorful image colorization. In: ECCV (2016)
78. Zhang, R., Isola, P., Efros, A.A.: Split-brain autoencoders: unsupervised learning by cross-channel prediction. In: CVPR (2017)
79. Zhang, Z., Hua, B.S., Yeung, S.K.: ShellNet: efficient point cloud convolutional neural networks using concentric shells statistics. In: ICCV (2019)
80. Zhao, Y., Birdal, T., Deng, H., Tombari, F.: 3D point capsule networks. In: CVPR (2019)
81. Zhuang, C., Zhai, A.L., Yamins, D.: Local aggregation for unsupervised learning of visual embeddings. In: ICCV (2019)

DSA: More Efficient Budgeted Pruning via Differentiable Sparsity Allocation

Xuefei Ning[1], Tianchen Zhao[2], Wenshuo Li[1], Peng Lei[2], Yu Wang[1](✉), and Huazhong Yang[2]

[1] Department of Electronic Engineering, Tsinghua University, Beijing, China
foxdoraame@gmail.com, yu-wang@tsinghua.edu.cn
[2] Department of Electronic Engineering, Beihang University, Beijing, China
ztc16@buaa.edu.cn

Abstract. Budgeted pruning is the problem of pruning under resource constraints. In budgeted pruning, how to distribute the resources across layers (i.e., sparsity allocation) is the key problem. Traditional methods solve it by discretely searching for the layer-wise pruning ratios, which lacks efficiency. In this paper, we propose Differentiable Sparsity Allocation (DSA), an efficient end-to-end budgeted pruning flow. Utilizing a novel *differentiable pruning process*, DSA finds the layer-wise pruning ratios with *gradient-based optimization*. It allocates sparsity in continuous space, which is more efficient than methods based on discrete evaluation and search. Furthermore, DSA could work in a *pruning-from-scratch* manner, whereas traditional budgeted pruning methods are applied to pre-trained models. Experimental results on CIFAR-10 and ImageNet show that DSA could achieve superior performance than current iterative budgeted pruning methods, and shorten the time cost of the overall pruning process by at least 1.5× in the meantime.

Keywords: Budgeted pruning · Structured pruning · Model compression

1 Introduction

Convolutional Neural Networks (CNNs) have demonstrated superior performances in computer vision tasks. However, CNNs are computational and storage intensive, which poses significant challenges on the NN deployments under resource constrained scenarios. Model compression techniques [16,20] are proposed to reduce the computational cost of CNNs. Moreover, there are situations (e.g., deploying the model onto certain hardware, meeting real-time constraints)

X. Ning and T. Zhao—Contributed equally to this work.

Electronic supplementary material The online version of this chapter (https://doi.org/10.1007/978-3-030-58580-8_35) contains supplementary material, which is available to authorized users.

A. Vedaldi et al. (Eds.): ECCV 2020, LNCS 12348, pp. 592–607, 2020.
https://doi.org/10.1007/978-3-030-58580-8_35

under which the resources (e.g., latency, energy) of the compressed models must be restricted under certain budgets. Budgeted pruning is introduced for handling these situations.

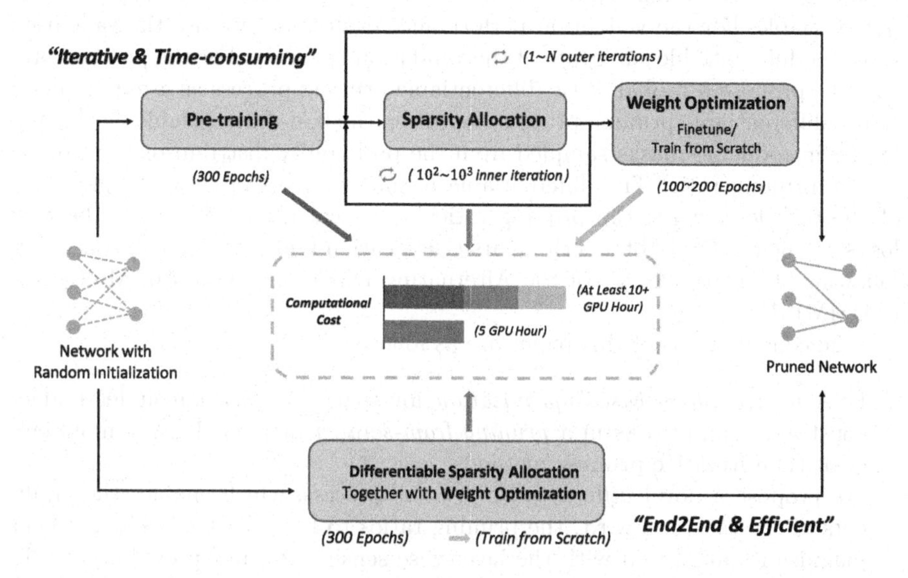

Fig. 1. Workflow comparison of the iterative pruning methods [7,14,15,22] and DSA

As shown in Fig. 1, the budgeted pruning problem could be divided into two sub-tasks: to decide how many channels to keep for each layer (i.e., sparsity allocation) and to acquire proper weights (i.e., weight optimization). Recent work [17] observes that once the pruned structure is acquired, the compressed model can achieve similar accuracies no matter it is trained from scratch or finetuned from the weights inherited from the original model. Therefore, sparsity allocation is the key problem for budgeted pruning.

To solve the sparsity allocation problem, the majority of methods [7,14,15,22] adopt an "iterative pruning flow" scheme. The workflow of these methods involves three stages: pre-training, sparsity allocation, and finetuning, as shown in Fig. 1. These methods conduct the sparsity allocation through a discrete search, which contains hundreds of search-evaluation iterations. For each candidate allocation, a time-consuming approximate evaluation is needed. Also, the discrete search in the large search space lacks sample efficiency. Moreover, these methods need to be applied to the pre-trained models, and model pre-training costs much computational effort. As a result, the overall iterative pruning flow is not efficient.

In order to improve the efficiency of budgeted pruning, we propose DSA, an end-to-end pruning flow. Firstly, DSA can work in a "pruning-from-scratch" manner, thus eliminating the cumbersome pre-training process (see Sect. 5.3). Secondly, DSA optimizes the sparsity allocation in continuous space with a gradient-based method, which is more efficient than methods based on discrete evaluation and search.

For applying the gradient-based optimization for allocating sparsity, we should make the evaluation of the validation accuracy and the pruning process differentiable. For the validation performance evaluation, we use the validation loss as a differentiable surrogate of the validation accuracy. For the pruning process, we propose a probabilistic differentiable pruning process as a replacement. In the differentiable pruning process, we soften the non-differentiable hard pruning by introducing masks sampled from the probability distributions controlled by the pruning ratio. The differentiable pruning process enables the gradients of the task loss, w.r.t. the pruning ratios to be calculated. Utilizing the task loss's gradients, DSA obtains the sparsity allocation under the budget constraint following the methodology of the Alternating Direction Method of Multipliers (ADMM) [1].

The contributions of this paper are as follows.

- DSA uses *gradient-based optimization* for sparsity allocation under budget constraints, and works in a *pruning-from-scratch* manner. DSA is more efficient than iterative pruning methods.
- We propose a novel *differentiable pruning process*, which enables the gradients of the task loss w.r.t. the pruning ratios to be calculated. The gradient magnitudes align well with the layer-wise sensitivity, thus providing an efficient way of measuring the sensitivity (See Sect. 5.3). Due to this property of the gradients, DSA can attribute higher keep ratios to more sensitive layers.
- We give a topological grouping procedure to handle the topological constraints that are introduced by skip connections and so on, thus the resulting model keeps the original connectivity.
- Experimental results on CIFAR-10 and ImageNet demonstrate the effectiveness of DSA. DSA consistently outperforms other iterative budgeted pruning methods with at least 1.5× speedup.

2 Related Work

2.1 Structured Pruning

Structured pruning intends to introduce structured sparsity into the NN models. SSL [20] removes structured subsets of weights by applying group lasso regularization and magnitude pruning. Some studies [4,16] add regularization on the batch normalization (BN) scaling parameters γ instead of the convolution weights. These methods focus on seeking the trade-off between model sparsity and performance via designing regularization terms. Since these methods allocate sparsity through regularizing the weights, the results are sensitive to the hyperparameters.

There are also studies that target at choosing which filters to prune, given the layer-wise pruning ratios. ThiNet [18] utilizes the information from the next layer to select filters. FPGM [6] exploits the distances to the geometric median as the importance criterion. These methods focus on exploring intra-layer filter importance criteria, instead of handling the inter-layer sparsity allocation.

2.2 Budgeted Pruning

To amend the regularization based methods for budgeted pruning, MorphNet [4] alternates between training with L_1 regularization and applying a channel multiplier. However, the uniform expansion of width neglects the different sensitivity of different layers and might fail to find the optimal resource allocation strategy under budget. To explicitly control the pruning ratios, ECC [21] updates the pruning ratios according to the energy consumption model. The pruning process is modeled as discrete constraints tying the weights and pruning ratios, and this constrained optimization is handled using a proximal projection. In our work, the pruning process is relaxed into a probabilistic and differentiable process, which enables the pruning ratios to be directly instructed by the task loss.

Other methods view the budgeted pruning problem as a discrete search problem, in which the sparsity allocation and finetuning are alternatively conducted for multiple stages. In each stage, a search-evaluation loop is needed to decide the pruning ratios. For the approximate evaluation, a hard pruning procedure and a walk-through of the validation dataset are usually required. As for the search strategy, NetAdapt [22] empirically adjusts the pruning ratios, while ensuring a certain resource reduction is achieved; AMC [7] employs reinforcement learning to instruct the learning of a controller, and uses the controller to sample the pruning ratios; AutoCompress [14] uses simulated annealing to explore in the search space; MetaPruning [15] improves the sensitivity analysis by introducing a meta network to generate weights for pruned networks. Apart from these methods that search for the layer-wise pruning ratio, LeGR [2] searches for appropriate affine transformation coefficients to calculate the global importance scores of the filters. These methods can guarantee that the resulting models meet the budget constraint, but the discrete search process is inefficient and requires a pre-trained model.

In contrast, DSA (Differentiable Sparsity Allocation) is an end-to-end pruning flow that allocates the inter-layer sparsity with a gradient-based method, which yields better performance and higher efficiency. Moreover, DSA works in a "pruning from scratch" manner, saving the cost of pre-training the model. The comparison of various properties across pruning methods is summarized in Table 1.

3 Problem Definition

For budgeted pruning, denoting the budget as $B_{\mathcal{F}}$, the weight as W, the optimization problem of the keep ratios $\mathcal{A} = \{\alpha^{(k)}\}_{k=1,\cdots,K}$ (1 - pruning ratio) of K layers can be written as:

$$\begin{aligned} \mathcal{A}^* &= \mathrm{argmax}_{\mathcal{A}} \mathrm{Acc}_v(W^*(\mathcal{A}), \mathcal{A}) \\ \text{s.t.} \quad W^*(\mathcal{A}) &= \mathrm{argmin}_W L_t(W, \mathcal{A}) \\ \mathcal{F}(\mathcal{A}) &\le B_{\mathcal{F}}, \quad 0 \le \mathcal{A} \le 1 \end{aligned} \tag{1}$$

Table 1. Comparison of structured pruning methods. **Headers**: The "budget control" column indicates whether the method could ensure the resulting model to satisfy the budget constraint; The "from scratch" column indicates whether the method could be applied to random initialized models rather than pre-trained ones; The "performance instruction" column describes how the task performance instructs the sparsity allocation, "indirect" means that the task performance instructs the sparsity allocation only indirectly through weights (e.g., magnitude pruning); The "gen. perf." column indicates whether the generalization performance guides the pruning process

Methods	Budget control	From scratch	End-to-End	Performance instruction	Gen. perf.
SSL [20]	**No**	Yes	Yes	Indirect	No
NetAdapt [22]	Yes	No	No	Discrete evaluation	Yes
AMC [7]	Yes	No	No	Discrete evaluation	Yes
MetaPruning [15]	Yes	No	No	Discrete evaluation	Yes
ECC [21]	Yes	No	Yes	Indirect	No
Ours	Yes	Yes	Yes	Differentiable	Yes

where Acc_v is the validation accuracy, and L_t is the training loss, and $\mathcal{F}(\mathcal{A})$ is the consumed resource corresponding to the keep ratios $\mathcal{A}$.

To solve this optimization problem, existing iterative methods [7,14,22] conduct sensitive analysis in each stage, and use discrete search methods to adjust $\mathcal{A}$. The common assumption adopted is that in each stage, $\mathrm{Acc}_v(\hat{W}^*(\mathcal{A}), \mathcal{A})$ should be correlated to $\mathrm{Acc}_v(W^*(\mathcal{A}), \mathcal{A})$, in which $\hat{W}^*(\mathcal{A})$ is approximated using the current weights (e.g., threshold-based pruning, local layer-wise least-square fitting), instead of finding $W^*(\mathcal{A})$ by finetuning.

4 Method

Since the validation accuracy Acc_v in Eq. 1 is not differentiable, we use the validation loss L_v as a differentiable surrogate of Acc_v. Then, the objective function becomes $\mathcal{A}^* = \mathrm{argmin}_{\mathcal{A}} L_v(W^*(\mathcal{A}), \mathcal{A})$ in the differentiable relaxation of the budgeted pruning problem in Eq. 1. As for the inner optimization of $W^*(\mathcal{A})$, we adapt the weights to the changes of the structure by adopting similar bi-level optimization as in DARTS [13]. The high-order gradients $\frac{\partial W^*(\mathcal{A})}{\partial \mathcal{A}}$ are ignored, thus $\frac{\partial L_v(W^*(\mathcal{A}),\mathcal{A})}{\partial \mathcal{A}} \approx \frac{\partial L_v(W,\mathcal{A})}{\partial \mathcal{A}}$.

The workflow of DSA is shown in Algorithm 1 and Fig. 2. First, DSA groups the network layers according to the topological constraints (Sect. 4.2), and assigned a keep ratio for each group of layers. The sparsity allocation problem is to decide the proper keep ratios $\mathcal{A}$ for the K groups. The optimization of the keep ratios $\mathcal{A}$ is conducted with gradient-based method in the *continuous* space. We apply an ADMM-inspired optimization method (Sect. 4.3) to utilize the gradients of both the task and budget loss to find a good sparsity allocation α^* that meets the budget constraint. Note that to enable the task loss's gradients to flow back to the keep ratios α, we design a novel differentiable pruning process (Sec. 4.1).

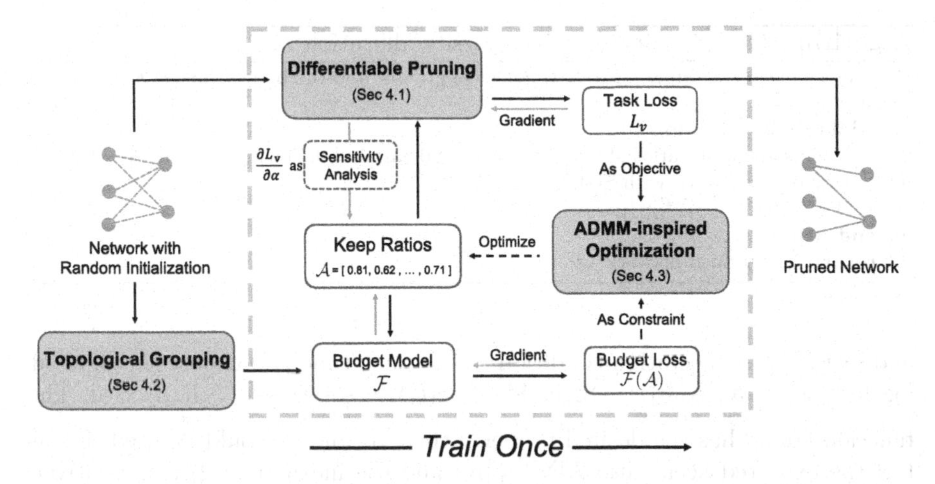

Fig. 2. DSA workflow. For feasible deployment, we first group the network layers according to the topological constraints (Sect. 4.2), and a keep ratio will be attributed to each topological group. The budget model $\mathcal{F}$ is also generated for measuring the budget loss $\mathcal{F}(\mathcal{A})$. Then, in the sparsity allocation procedure, the weights are updated using the task loss on the training dataset. And the keep ratios decision (i.e., inter-layer sparsity allocation) is conducted on the validation dataset using gradient-based steps (Secy. 4.3). Note that the differentiable characteristic of the pruning process (Sect. 4.1) enables the task loss's gradients to flow back to the keep ratios $\mathcal{A}$

4.1 Differentiable Pruning

Pruning Process Relaxation. In the traditional pruning process of a particular layer, given the keep ratio α, a subset of channels are chosen according to the channel-wise importance criteria $b_i \in \mathbb{R}^+, i = 1, \cdots, C$ (e.g., the L1 norm of the convolutional weights). In contrast, we use a probabilistic relaxation of the "hard" pruning process where channel i has the probability p_i to be kept. Then, channel-wise masks are sampled from the Bernoulli distribution of p_i: $m_i \sim \text{Bernoulli}(p_i), i = 1, \cdots, C$. The pruning process is illustrated in Fig. 3.

The channel-wise keep probability $p_i = f(\alpha, b_i)$ is computed using α. Due to the probabilistic characteristics of the pruning process, the proportion of the actual kept channels might deviate from α. We should make the expectation of the actual kept channels $E[\sum_{i=1}^{C} m_i] = \sum_{i=1}^{C} p_i = \sum_{i=1}^{C} f(\alpha, b_i)$ equal to αC, which we denote as the "expectation condition" requirement.

What is more, as we need a "hard" pruning scheme eventually, this probabilistic pruning process should become deterministic in the end. Defining the inexactness as $\mathcal{E} = E[|\sum_i m_i - \alpha C|^2] = \sum_i \text{Var}[m_i] = \sum_i p_i(1 - p_i)$, a proper control on the inexactness is desired such that the inexactness $\mathcal{E}$ can reach 0 at the end of pruning.

The choice of f is important to the stability and controllability of the pruning process. An S-shaped function family w.r.t. b, $f(b; \beta) : \mathbb{R}^+ \to (0, 1)$, parametrized by at least two parameters is required, so that we can control the inexactness

Algorithm 1. DSA: Differentiable sparsity allocation

1: Run topological grouping, get K topological groups, and the budget model $\mathcal{F}$
2: $\mathcal{A} = \mathbf{1}_K$
3: **while** $\mathcal{F}(\mathcal{A}) > B_{\mathcal{F}}$ **do**
4: Update the keep ratios $\mathcal{A}$, auxiliary and dual variables following Eq. 8 in Sec. 4.3
5: Update weights W with SGD:
$W_T = W_{T-1} - \eta_w \frac{\partial L_t}{\partial W}|_{W_T}$
6: **end while**
7: **return** the pruned network, $\mathcal{A}$

and satisfy the expectation condition at the same time. We choose the sigmoid-log function $f(b_i, \beta_1, \beta_2) = \text{Sigmoid} \circ \text{Log}(b_i) = \frac{1}{1+(\frac{b_i}{\beta_1})^{-\beta_2}}, \quad \beta_1, \beta_2 > 0$. This function family has the desired property that, β_1 and β_2 could be used to control the expected keep ratio $E[\sum_{i=1}^{C} m_i]$ and the inexactness $\mathcal{E}$ in a relatively independent manner. 1) In our work, β_2 is a parameter that follows an increasing schedule. As β_2 approaches infinity, the inexactness $\mathcal{E}$ approaches 0, and β_1 becomes the hard pruning threshold of this layer. 2) $\beta_1 = \beta_1(\alpha)$ is a function of α. It has the interpretation of the soft threshold for the base importance score. It is calculated by solving the implicit equation of expectation condition:

$$g(\beta_1) = \frac{1}{C} E[\sum_{i=1}^{C} m_i] - \alpha = \frac{1}{C} \sum_{i=1}^{C} f(b_i, \beta_1, \beta_2) - \alpha = 0 \tag{2}$$

Since $g(\beta_1)$ is monotonically decreasing, the root $\beta_1(\alpha)$ could be simply and efficiently found (e.g., with bisection or Newton methods).

To summarize, utilizing the differentiable pruning process, the forward process of the k-th layer can be written as

$$\begin{aligned} & y^{(k)}(w, y^{(k-1)}; \alpha) = m \odot \text{Conv-BN-ReLU}(w, y^{(k-1)}) \\ & \text{s.t.} \quad m_i \sim \text{Bernoulli}(p_i), \quad \sum_{i=1}^{C} p_i = \sum_{i=1}^{C} f(\alpha, b_i) = \alpha C, \quad i = 1, \cdots, C \end{aligned} \tag{3}$$

where $y^{(k-1)}, w, y^{(k)}$ denote the inputs, weights, outputs of this layer, and the superscript k is omitted for simplicity.

Differentiable Instruction by the Task Loss. The task loss L can be written as

$$L(\mathcal{A}, W) = E_{x \sim D}[E_{m_i^{(k)} \sim Ber(m; p_i^{(k)})}[\text{CE}(x, y; M, W)]] \tag{4}$$

where D is the training dataset, W denotes the weights, M is the set of masks $\{\{m_i^{(k)}\}_{i=1,\cdots,C^{(k)}}\}_{k=1,\cdots,K}$, $\mathcal{A}$ is the set of keep ratios $\{\alpha^{(k)}\}_{k=1,\cdots,K}$.

To enable differentiable tuning of the layer-wise keep ratios $\mathcal{A}$ instructed by both the task loss and the budget constraint, the major challenge is to derive

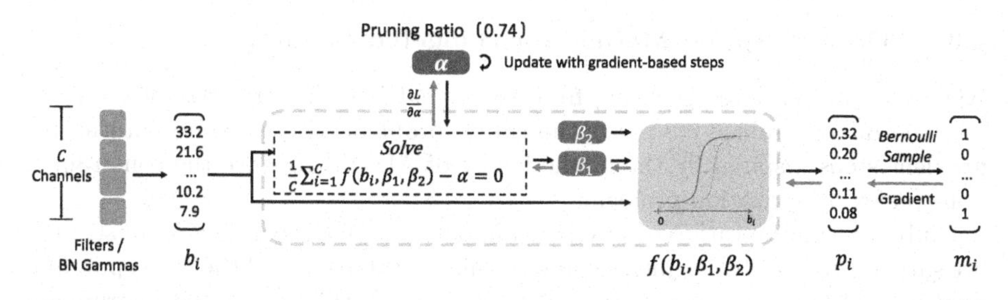

Fig. 3. The illustration of the differentiable pruning process of one layer. Given the base importances b_i and the keep ratio α, the process outputs channel-wise keep probabilities $p_i = f(b_i, \beta_1, \beta_2)$, which satisfy that the expectation condition $\sum_i^C p_i = \alpha C$. Then, the channel-wise masks m_i are sampled

the task loss's gradients w.r.t. $\alpha^{(k)}$: $\frac{\partial L}{\partial \alpha^{(k)}}$. First, we can calculate the implicit gradient $\frac{\partial \beta_1(\alpha)}{\partial \alpha}$ as:

$$\begin{aligned} \frac{1}{C}\sum_{i=1}^{C} \frac{\partial f(b_i, \beta_1, \beta_2)}{\partial \beta_1}\frac{\partial \beta_1}{\partial \alpha} - 1 = 0 \\ \frac{\partial \beta_1(\alpha)}{\partial \alpha} = \frac{C}{\sum_{i=1}^{C} f'(b_i, \beta_1, \beta_2)} \end{aligned} \tag{5}$$

Then, $\frac{\partial L}{\partial \alpha^{(k)}}$ could be calculated as:

$$\frac{\partial L}{\partial \alpha^{(k)}} = \frac{\partial \beta_1}{\partial \alpha}\sum_{i=1}^{C} \frac{\partial L}{\partial p_i}\frac{\partial p_i}{\partial \beta_1} = C\sum_{i=1}^{C} \frac{\partial L}{\partial p_i}\hat{f}_i'; \quad \hat{f}_i' = \frac{f_i'}{\sum f_i'} \tag{6}$$

where the superscript k is omitted for simplicity and $\frac{\partial L}{\partial p_i}$ could be approximated using Monte-Carlo samples of the reparametrization gradients.

Equation 6 could be interpreted as: The update of $\alpha^{(k)}$ is instructed using a weighted aggregation of the gradients of the task loss L w.r.t. the keep probabilities of channels $\frac{\partial L}{\partial p_i^{(k)}}$, and the aggregation weights are $f_i', i = 1, \cdots, C^{(k)}$.

4.2 Topological Grouping and Budget Modeling

For plain CNN, we can choose the layer-wise keep ratios $\alpha^{(k)}, k = 1, \cdots, K$ independently. However, for networks with shortcuts (e.g., ResNet), the naive scheme can lead to irregular computation patterns. Our grouping procedure for handling the topological constraints is described in Alg. 1 in the appendix. An example of grouping convolutions in two residual blocks is also shown in the appendix.

The $\mathcal{F}(\mathcal{A})$ function models relationship of the keep ratios $\mathcal{A}$ and the resources. Taking FLOPs as an example, $\mathcal{F}(\mathcal{A})$ could be represented as $\mathcal{F}(\mathcal{A}) = \mathcal{A}^T \mathcal{F}_A \mathcal{A} + \mathcal{F}_B^T \mathcal{A}$. For completeness, we summarize the calculation procedure of $\mathcal{F}_A$ and $\mathcal{F}_B$ in Alg. 1 in the appendix. Under budget constraints for resources other than FLOPs, regression models can be fitted to get the corresponding $\mathcal{F}$ model.

4.3 ADMM-Inspired Method for Budgeted Pruning

ADMM is an iterative algorithm framework that is widely used to solve unconstrained or constrained convex optimization problems. Here, we use alternative gradient steps inspired by the methodology of ADMM to solve the constrained non-convex optimization problem.

Substituting the variable $\mathcal{A}$ by $\Theta = \text{Sigmoid}^{-1}(\mathcal{A})$, the $0 \leq \mathcal{A} \leq 1$ constraints are satisfied naturally. By introducing auxiliary variable z and the corresponding dual variable u_2 for the equality constraint $z = \Theta$, the augmented Lagrangian is:

$$\mathcal{L}(\Theta, z, u_2) = L_v(\Theta) + I(\mathcal{F}(z) \leq B_{\mathcal{F}}) + u_2^T(\Theta - z) + \frac{\rho_2}{2}||\Theta - z||^2 \tag{7}$$

We then minimize the dual lower bound $\max_{u_2} \mathcal{L}(\Theta, z, u_2)$ of $L_v(\Theta)$. Equation 8 shows the 3 alternative steps in one iteration. The variables with the superscript "′" denote the values at the previous time step.

$$\begin{aligned} \Theta &= \text{argmin}_{\Theta} L_v(\Theta) + u_2^T(\Theta - z') + \frac{\rho_2}{2}||\Theta - z'||^2 \\ z &= \text{argmin}_z u_2^T(\Theta - z) + \frac{\rho_2}{2}||\Theta - z||^2 \quad \text{s.t.} \quad \mathcal{F}(z) \leq B_{\mathcal{F}} \\ u_2 &= u_2' + \rho_2(\Theta - z) \end{aligned} \tag{8}$$

The unconstrained sub-problem for Θ is hard to solve, since $L_v(\Theta)$ is a stochastic objective and W can only be regarded as being constant in a local region. Therefore, in each iteration, we only do one stochastic gradient step on one validation batch for Θ.

To solve the inner problem for the auxiliary variable z with an inequality constraint, we use the standard trick of converting $\mathcal{F}(z) \leq B_{\mathcal{F}}$ to $[\mathcal{F}(z) - B_{\mathcal{F}}]_+ = 0$, and then use gradient descent to solve the min-max optimization of the augmented lagrangian $\mathcal{L}^{(z)}(z, u_1)$ in Eq. 9. In each iteration of the inner optimization, one gradient descent step is applied to z: $z = z' - \eta_z \nabla_z \mathcal{L}^{(z)}(z, u_1)$, and one dual ascent step is applied to u_1: $u_1 = u_1' + \rho_1[\mathcal{F}(\Theta) - B_{\mathcal{F}}]_+$. This optimization is efficient since only z and u_1 need to be updated.

$$\mathcal{L}^{(z)}(z, u_1) = u_1[\mathcal{F}(z) - B_{\mathcal{F}}]_+ \frac{\rho_1}{2}[\mathcal{F}(z) - B_{\mathcal{F}}]^2 + u_2^T(\Theta - z) + \frac{\rho_2}{2}||\Theta - z||^2 \tag{9}$$

The dual variables u_1, u_2 can be interpreted as the regularization coefficients, which are dynamically adjusted according to the constraint violations.

Table 2. Pruning results of ResNet-20 and ResNet-56 on CIFAR-10. SSL and MorphNet are re-implemented with topological grouping. Accuracy drops for the referred results are calculated based on the reported baseline in their papers. **Headers**: "TG" stands for Topological Grouping; "FLOPs Budget" stands for the percentage of the pruned models' FLOPs compared to the full model

FLOPs budget	Method	TG	ResNet-20		ResNet-56	
			FLOPs ratio	Acc. (Acc. Drop)	FLOPs ratio	Acc. (Acc. Drop)
Baseline	Ours		100 %	92.17 %	100 %	93.12 %
75%	SSL [20]	✓	73.8%	91.08% (−1.09%)	69.0%	92.06% (−1.06%)
	Variational[b] [23]		83.5%	91.66% (−0.41%)	79.7%	92.26% (−0.78%)
	PFEC[b] [10]	✓	–	–	74.4%	91.31% (−1.75%)
	MorphNet [4]	✓	74.9%	90.64% (−1.53%)	69.2%	91.71% (−1.41%)
	DSA (Ours)	✓	74.0%	**92.10% (−0.07%)**	70.7%	**93.08% (−0.04%)**
50% (×2)	SSL [20]	✓	51.8%[a]	89.78 % (−2.39%)	45.5%	91.22% (−1.90%)
	MorphNet [4]	✓	47.1%	90.1% (−2.07%)	51.9%[a]	91.55% (−1.57%)
	AMC[b] [7]	✓	–	–	50%	91.9% (−0.9%)
	CP[b] [8]		–	–	50%	91.8% (−1.0%)
	Rethink[b] [17]		60.0%	91.07% (−1.34%)	50%	93.07% (−0.73%)
	SFP[b] [5]		57.8%	90.83% (−1.37%)	47.4%	92.26% (−1.33%)
	FPGM[b] [6]		57.8%	91.09% (−1.11%)	47.4%	**92.93%** (−0.66%)
	LCCL[b] [3]		64.0%	**91.68%** (−1.06%)	62.1%	92.81% (−1.54%)
	DSA (Ours)	✓	49.7%	91.38% **(−0.79%)**	47.8%	92.91% **(−0.22%)**
33.3% (×3)	SSL [20]	✓	34.6%[a]	89.06% (−3.11%)	38.1%[a]	91.32% (−1.80%)
	MorphNet [4]	✓	30.5%	88.72% (−3.45%)	39.7%[a]	91.21% (−1.91%)
	DSA (Ours)	✓	32.5%	**90.24% (−1.93%)**	32.6%	**92.20% (−0.92%)**

[a] These pruned models' FLOPs of SSL and MorphNet are higher than the budget constraints, since these regularization based methods lack explicit control of the resource consumption and even with carefully tuned hyperparameters, the resulting model might still violate the budget constraint.
[b] These methods' results are directly taken from their paper, and their accuracy drops are calculated based on their reported baseline accuracies.

5 Experiments

5.1 Setup

We conduct the experiments on CIFAR-10 and ImageNet. For CIFAR-10, the batch size is 128, and an SGD optimizer with momentum 0.9, weight decay 4e−5 is used to train the model for 300 epochs. The learning rate is initialized to 0.05 and decayed by 10 at epochs 120, 180, and 240. The differentiable sparsity allocation is conducted simultaneously with normal training after 20 epochs of warmup. As for ImageNet, we use an SGD optimizer (weight decay 4e−5, batch size 256) to optimize the models for 120 epochs. The learning rate is 0.1 and decayed by 10 at epochs 50, 80, 110. The first 15 epochs remains plain training without pruning.

In the sparsity allocation process, 10% of the training data are used as the validation data for updating the keep ratios, while 90% are used for tuning the weights. For the optimization of $\mathcal{A}$, the penalty parameters ρ_1, ρ_2 is set to 0.01, and the scaling coefficient of $L_v(\Theta)$ is 1e+5. z is updated for 50 steps with learning rate 1e-3 in the inner optimization. In practice, to reach the budget faster, we project the gradients of Θ to be nonnegative. After each update of $\mathcal{A}$, the weights are tuned for 20 steps for adaption. After acquiring the budget, the whole training set is used for updating the weights.

In the differentiable pruning process, the L_1 norms of BN scales are chosen as the base importance scores $b_i = |\gamma_i|$. $\beta_1(\alpha)$ is found by solving Eq. 2 with the bisection method. $\beta_2(T)$ follows a increasing schedule: starts at 0.05 and gets multiplied by 1.1 every epoch. As $\beta_2 \rightarrow \infty$, the soft pruning process becomes a hard pruning process, and the inexactness $\mathcal{E}$ goes to 0.

5.2 Results on CIFAR-10 and ImageNet

On CIFAR-10, for SSL [20] and MorphNet [4], the regularization coefficients on the convolution weights or BN scaling parameters are adjusted to meet various budget constraints.

Table 2 and Fig. 4 show the results of pruning ResNet-20 and ResNet-56 on CIFAR-10. The pruned models obtained by DSA meet the budget constraints with smaller accuracy drops than the baseline methods. Compared with the regularization based methods (e.g., SSL and MorphNet), due to the explicit budget modeling, DSA guarantees that the resulting models meet different budget constraints, without hyperparameter trials. Compared with the iterative pruning methods (e.g., AMC), DSA allocates the sparsity in a gradient-based way and is more efficient (See Sect. 5.3). We also apply DSA to compress ResNet-18 and VGG-16, and the results are included in Appendix Table 1 and Fig. 2. It shows that DSA outperforms the recent work based on a discrete search [14]. For ResNet-18, DSA achieves 94.19% versus 93.91% of the baseline with roughly the same FLOPs ratio. For VGG-16, DSA achieves 90.16% with 20.4× FLOPs reduction, which is significantly better than the baseline [14] (88.78% with 14.0× FLOPs reduction).

Table 3 shows the results of applying DSA to prune ResNet-18 and ResNet-50 on ImageNet. As could be seen, DSA consistently outperforms other methods across different FLOPs ratios and network structure. For example, DSA could achieve a small accuracy drop of 1.11% while keeping 60% FLOPs of ResNet-18, which is significantly better than the baselines.

5.3 Analysis and Discussion

Computational Efficiency. Some recent studies [17,19] suggest that starting with a pre-trained model might not be necessary for pruning. Unlike current methods which rely on a pre-trained model, DSA could work in a "pruning from scratch" manner.

As shown in Fig. 1, traditional budgeted pruning consists of 3 stages: pretraining, sparsity allocation, and finetuning. The iterative pruning methods conduct hundreds of search-evaluation steps for sparsity allocation, e.g., AMC takes about 3 GPU hours for pruning ResNet-56 on CIFAR-10 [19]. After learning the structure, the finetuning stage takes 100–200 epochs for CIFAR-10 (60 epochs for ImageNet), which accounts for about 2–3 GPU hours for ResNet-56 on CIFAR-10 and 150 GPU hours for ResNet-18 on ImageNet. Moreover, these two

Table 3. Pruning results on ImageNet. "TG" stands for Topological Grouping

Network	TG	Method	FLOPs Ratio	Top-1 Acc Drop	Top-5 Acc Drop
ResNet18		Baseline	100%	**69.72%**	**89.07%**
		MiL [3]	65.4%	−3.65%	−2.30%
		SFP [5]	60.0%	−3.18%	−1.85%
		FPGM [6]	60.0%	−2.47%	−1.52%
	✓	Ours	60.0%	**−1.11%**	**−0.718%**
ResNet50		Baseline	100%	**76.02%**	**92.86%**
		APG [9]	69.0%	−1.94%	−1.95%
		GDP [12]	60.0%	−2.52%	−1.25%
		SFP [5]	60.0%	−1.54%	−0.81%
		FPGM [6]	60.0%	−1.12%	−0.47%
	✓	Ours	60.0%	**−0.92%**	**−0.41%**
		ThiNet [18]	50.0%	−4.13%	–
		CP [8]	50.0%	−3.25%	−1.40%
		FPGM [6]	50.0%	−2.02%	−0.93%
		PFS [19]	50.0%	−1.60%	–
		Hinge [11]	46.55%	−1.33%	–
	✓	Ours	50.0%	**−1.33%**	**−0.8%**

stages should be repeated for multiple rounds to achieve the maximum pruning rates [14,22], thus further increase the computational costs by several times. What's more, these methods need to be applied to the pre-trained models, and the pre-training stage takes about 300 and 120 epochs for models on CIFAR-10 and ImageNet. To summarize, the 3 stages can take up to 10 GPU hours for ResNet-56 on CIFAR-10, and 450 GPU hours for ResNet-18 on ImageNet.

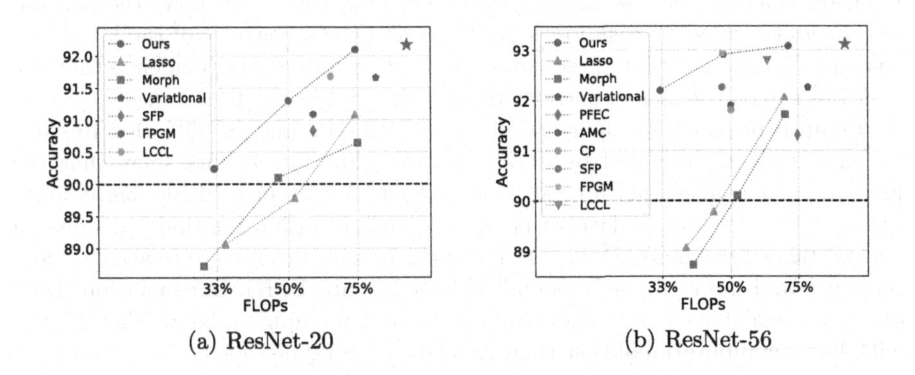

Fig. 4. Pruning results on CIFAR-10

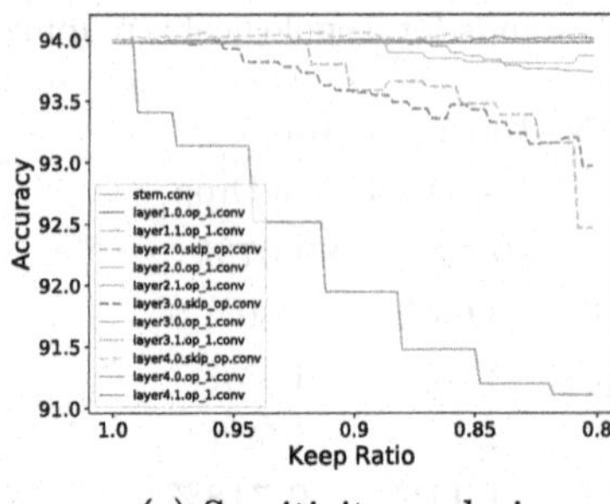

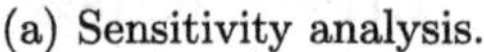

(a) Sensitivity analysis.

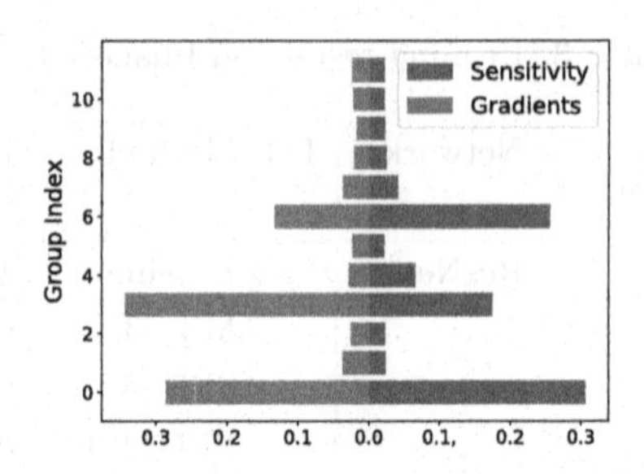

(b) The normalized magnitudes for layer-wise sensitivity and task loss's gradients in the first epoch.

Fig. 5. The alignment between sensitivity analysis and gradient magnitudes of ResNet-18 on CIFAR-10. The magnitudes are normalized by $\hat{v} = \text{softmax}(v/\text{std}(v))$

In contrast, the sparsity allocation in DSA is carried out in a more efficient gradient-based way, without the need of the pre-trained models. The extra cost of the sparsity allocation is small, since all the ADMM updates can be merged into the optimization of weights, and are conducted only once every tens of weight optimization steps. The whole DSA flow runs for 300 and 120 epochs on CIFAR-10 and ImageNet (5/300 GPU hours), thus speed up the overall pruning process by about 1.5×.

Rationality of the Differentiable Sparsity Allocation. In DSA, the task loss's gradient w.r.t. layer-wise pruning ratios directly guides the budget allocation. To see whether the gradient magnitudes align well with the local sensitivity, we conduct an empirical sensitivity analysis for ResNet-18 on CIFAR-10. We prune each layer (topological group) independently with different pruning ratios according to the L_1 norm, and show the test accuracy in Fig. 5a. Although this sensitivity analysis is heuristic and approximate, the accuracy drop could be interpreted as the local sensitivity for each group. Figure 5b shows the normalized magnitudes of the task loss's gradients and the sensitivity of the layer-wise sparsity. We can see that these two entities align well, giving evidence that the task loss's gradient indeed encodes the relative layer-wise importance.

Figure 6 presents the sparsity allocation (FLOPs budget 25%) for ResNet-18 on CIFAR-10 obtained by DSA and SSL [20]. The results show that the first layer for primary feature extraction should not be pruned too aggressively (group A), so do shortcut layers that are responsible for information transmission across stages (groups A, D, G, J). The strided convolutions are relatively more sensitive, and more channels should be kept (groups E, H). In conclusion, DSA obtains reasonable sparsity allocation that matches empirical knowledge [15,17], with lower computational cost than iterative pruning methods.

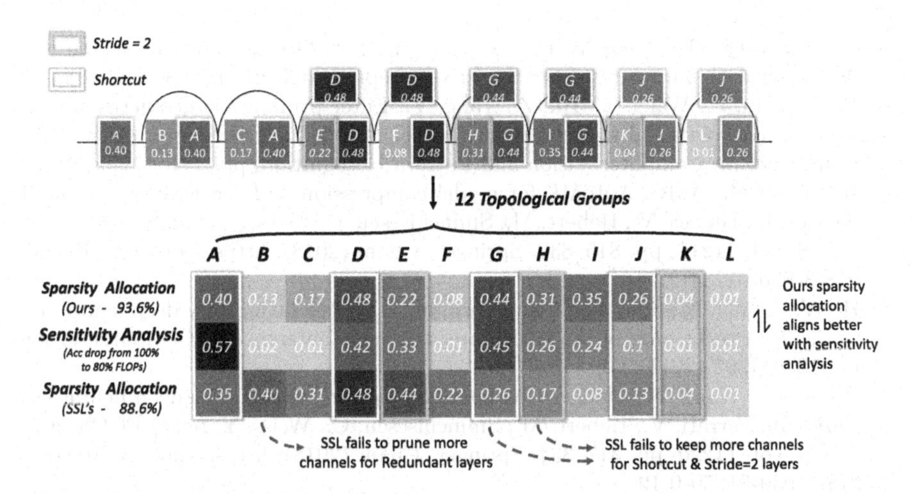

Fig. 6. The comparison between the normalized sensitivity, and the sparsity allocation of DSA and SSL [20] for ResNet-18 on CIFAR-10

6 Conclusion

In this paper, we propose Differentiable Sparsity Allocation (DSA), a more efficient method for budgeted pruning. Unlike traditional discrete search methods, DSA optimizes the sparsity allocation in a *gradient-based way*. To enable the gradient-based sparsity allocation, we propose a novel *differentiable pruning process*. Experimental results show that DSA could achieve superior performance than iterative pruning methods, with significantly lower training costs.

Acknowledgments. This work was supported by National Natural Science Foundation of China (No. 61832007, 61622403, 61621091, U19B2019), Beijing National Research Center for Information Science and Technology (BNRist). The authors thank Novauto for the support.

References

1. Boyd, S., Parikh, N., Chu, E., Peleato, B., Eckstein, J., et al.: Distributed optimization and statistical learning via the alternating direction method of multipliers. Found. Trends® Mach. Learn. **3**(1), 1–122 (2011)
2. Chin, T.W., Ding, R., Zhang, C., Marculescu, D.: Towards efficient model compression via learned global ranking. In: Proceedings of the IEEE Conference on Computer Vision and Pattern Recognition (2020)
3. Dong, X., Huang, J., Yang, Y., Yan, S.: More is less: a more complicated network with less inference complexity. In: Proceedings of the IEEE Conference on Computer Vision and Pattern Recognition, pp. 5840–5848 (2017)
4. Gordon, A., et al.: MorphNet: fast & simple resource-constrained structure learning of deep networks. In: Proceedings of the IEEE Conference on Computer Vision and Pattern Recognition, pp. 1586–1595 (2018)

5. He, Y., Kang, G., Dong, X., Fu, Y., Yang, Y.: Soft filter pruning for accelerating deep convolutional neural networks. arXiv preprint arXiv:1808.06866 (2018)
6. He, Y., Liu, P., Wang, Z., Hu, Z., Yang, Y.: Filter pruning via geometric median for deep convolutional neural networks acceleration. In: Proceedings of the IEEE Conference on Computer Vision and Pattern Recognition, pp. 4340–4349 (2019)
7. He, Y., et al.: AMC: AutoML for model compression and acceleration on mobile devices. In: Ferrari, V., Hebert, M., Sminchisescu, C., Weiss, Y. (eds.) ECCV 2018. LNCS, vol. 11211, pp. 815–832. Springer, Cham (2018). https://doi.org/10.1007/978-3-030-01234-2_48
8. He, Y., Zhang, X., Sun, J.: Channel pruning for accelerating very deep neural networks. In: Proceedings of the IEEE International Conference on Computer Vision, pp. 1389–1397 (2017)
9. Huang, Z., Wang, N.: Data-driven sparse structure selection for deep neural networks. In: Ferrari, V., Hebert, M., Sminchisescu, C., Weiss, Y. (eds.) ECCV 2018. LNCS, vol. 11220, pp. 317–334. Springer, Cham (2018). https://doi.org/10.1007/978-3-030-01270-0_19
10. Li, H., Kadav, A., Durdanovic, I., Samet, H., Graf, H.P.: Pruning filters for efficient convnets. arXiv abs/1608.08710 (2016)
11. Li, Y., Gu, S., Mayer, C., Gool, L.V., Timofte, R.: Group sparsity: the hinge between filter pruning and decomposition for network compression. In: Proceedings of the IEEE/CVF Conference on Computer Vision and Pattern Recognition (CVPR), June 2020
12. Lin, S., Ji, R., Li, Y., Wu, Y., Huang, F., Zhang, B.: Accelerating convolutional networks via global & dynamic filter pruning. In: IJCAI, pp. 2425–2432 (2018)
13. Liu, H., Simonyan, K., Yang, Y.: Darts: differentiable architecture search. arXiv preprint arXiv:1806.09055 (2018)
14. Liu, N., Ma, X., Xu, Z., Wang, Y., Tang, J., Ye, J.: AutoCompress: an automatic DNN structured pruning framework for ultra-high compression rates. In: AAAI, pp. 4876–4883 (2020)
15. Liu, Z., Mu, H., Zhang, X., Guo, Z., Yang, X., Cheng, K.T., Sun, J.: MetaPruning: meta learning for automatic neural network channel pruning. In: Proceedings of the IEEE International Conference on Computer Vision, pp. 3296–3305 (2019)
16. Liu, Z., Li, J., Shen, Z., Huang, G., Yan, S., Zhang, C.: Learning efficient convolutional networks through network slimming. In: Proceedings of the IEEE International Conference on Computer Vision, pp. 2736–2744 (2017)
17. Liu, Z., Sun, M., Zhou, T., Huang, G., Darrell, T.: Rethinking the value of network pruning. arXiv preprint arXiv:1810.05270 (2018)
18. Luo, J.H., Wu, J., Lin, W.: ThiNet: a filter level pruning method for deep neural network compression. In: Proceedings of the IEEE International Conference on Computer Vision, pp. 5058–5066 (2017)
19. Wang, Y., Zhang, X., Xie, L., Zhou, J., Su, H., Zhang, B., Hu, X.: Pruning from scratch. arXiv abs/1909.12579 (2019)
20. Wen, W., Wu, C., Wang, Y., Chen, Y., Li, H.: Learning structured sparsity in deep neural networks. In: Advances in Neural Information Processing Systems, pp. 2074–2082 (2016)
21. Yang, H., Zhu, Y., Liu, J.: ECC: platform-independent energy-constrained deep neural network compression via a bilinear regression model. In: Proceedings of the IEEE Conference on Computer Vision and Pattern Recognition, pp. 11206–11215 (2019)

22. Yang, T.-J., Howard, A., Chen, B., Zhang, X., Go, A., Sandler, M., Sze, V., Adam, H.: NetAdapt: platform-aware neural network adaptation for mobile applications. In: Ferrari, V., Hebert, M., Sminchisescu, C., Weiss, Y. (eds.) ECCV 2018. LNCS, vol. 11214, pp. 289–304. Springer, Cham (2018). https://doi.org/10.1007/978-3-030-01249-6_18
23. Zhao, C., Ni, B., Zhang, J., Zhao, Q., Zhang, W., Tian, Q.: Variational convolutional neural network pruning. In: Proceedings of the IEEE Conference on Computer Vision and Pattern Recognition, pp. 2780–2789 (2019)

Circumventing Outliers of AutoAugment with Knowledge Distillation

Longhui Wei[1,2(✉)], An Xiao[1], Lingxi Xie[1], Xiaopeng Zhang[1], Xin Chen[1,3], and Qi Tian[1]

[1] Huawei Inc., Shenzhen, China
weilh2568@gmail.com, {xiaoan1,tian.qi1}@huawei.com, 198808xc@gmail.com, zxphistory@gmail.com
[2] University of Science and Technology of China, Hefei, China
[3] Tongji University, Shanghai, China
1410452@tongji.edu.cn

Abstract. AutoAugment has been a powerful algorithm that improves the accuracy of many vision tasks, yet it is sensitive to the operator space as well as hyper-parameters, and an improper setting may degenerate network optimization. This paper delves deep into the working mechanism, and reveals that AutoAugment may remove part of discriminative information from the training image and so insisting on the ground-truth label is no longer the best option. To relieve the inaccuracy of supervision, we make use of knowledge distillation that refers to the output of a teacher model to guide network training. Experiments are performed in standard image classification benchmarks, and demonstrate the effectiveness of our approach in suppressing noise of data augmentation and stabilizing training. Upon the cooperation of knowledge distillation and AutoAugment, we claim the **new state-of-the-art** on ImageNet classification with a top-1 accuracy of **85.8%**.

Keywords: AutoML · AutoAugment · Knowledge distillation

1 Introduction

Automated machine learning (AutoML) has been attracting increasing attentions in recent years. In standard image classification tasks, there are mainly two categories of AutoML techniques, namely, neural architecture search (NAS) and hyper-parameter optimization (HPO), both of which focus on the possibility of using automatically learned strategies to replace human expertise. AutoAugment [4] belongs to the latter, which goes one step beyond conventional data augmentation techniques (*e.g.*, horizontal flipping, image rescaling & cropping, color jittering, *etc.*) and tries to combine them towards generating more training data without labeling new images. It has achieved consistent accuracy gain in

L. Wei and A. Xiao—Contributed equally to this work.

A. Vedaldi et al. (Eds.): ECCV 2020, LNCS 12348, pp. 608–625, 2020.
https://doi.org/10.1007/978-3-030-58580-8_36

image classification [4], object detection [11], *etc.*, and meanwhile efficient variants of AutoAugment have been proposed to reduce the computational burden in the search stage [13,17,25].

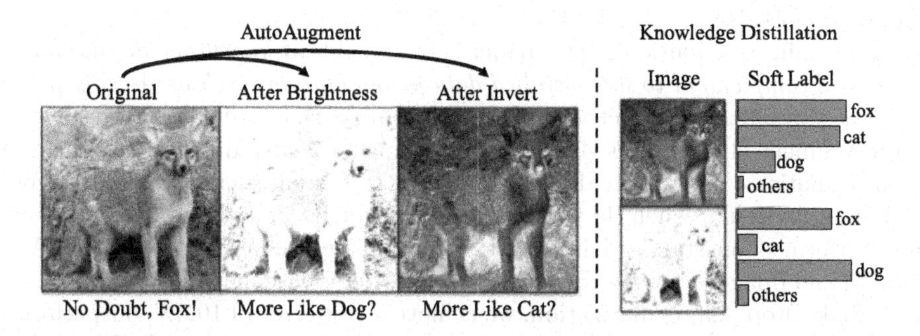

Fig. 1. Left: an image and its augmented copies generated by AutoAugment. The original image is clean and there is no doubt to use the ground-truth label, while the augmented counterparts look more like other classes which the annotation is not aware of. This phenomenon is called **augment ambiguity**. **Right**: We leverage the idea of knowledge distillation to provide softened signals to avoid ambiguity.

Despite their ability in improving recognition accuracy, we note that AutoAugment-based methods often require the search space to be well-designed. Without careful control (*e.g.*, in an expanded search space or with an increased distortion magnitude), these methods are not guaranteed to perform well – as we shall see in Sect. 3.2, an improper hyper-parameter may deteriorate the optimization process, resulting in even lower accuracy compared to the baseline. This puts forward a hard choice between more information (seeing a wider range of augmented images) and safer supervision (restricting the augmented image within a relatively small neighborhood around the clean image), which downgrades the upper-bound of AutoAugment-based methods.

In this paper, we investigate the reason of this contradictory. We find that when heavy data augmentation is added to the training image, it is probable that part of its semantic information is removed. An example of changing image brightness is shown in Fig. 1, and other transformation such as image translation and shearing can also incur information loss and make the image class unrecognizable (refer to Fig. 3). We name this phenomenon **augment ambiguity**. In such contaminated training images, insisting on the ground-truth label is no longer the best choice, as the inconsistency between input and supervision can confuse the network. Intuitively, complementary information that relates the augmented image to similar classes may serve as a better source of supervision.

Motivated by the above, we leverage the idea of knowledge distillation which uses a standalone model (often referred to as the *teacher*) to guide the target network (often referred to as the *student*). For each augmented image, the student receives supervision from both the ground-truth and the teacher signal, and in case that part of semantic information is removed, *e.g.*, an ambiguous (Fig. 1) or an out-of-distribution (Fig. 3) instance, the teacher can provide softened labels to avoid confusion. The extra loss between the teacher and student is measured by the KL-divergence between the score distributions of their top-ranked classes, and the number of involved classes is positively correlated to the magnitude of augmentation, since a larger magnitude often eliminates more semantics and causes smoother score distributions.

The main contribution of this paper is to reveal that *knowledge distillation is a natural complement to uncontrolled data augmentation, such as AutoAugment and its variants.* The effectiveness of our approach is verified in the space of AutoAugment [4] as well as that of RandAugment [5] with different strengths of transformations. Knowledge distillation brings consistent accuracy gain to recognition, in particular when the distortion magnitude becomes larger. Experiments are performed on standard image classification benchmarks, namely, CIFAR-10/100 and ImageNet. On CIFAR-100, with a strong baseline of PyramidNet [12] and ShakeDrop [49] regularization, we achieve a test error of 10.6%, outperforming all competitors with similar training costs. On ImageNet, in the RandAugment space, we boost the top-1 accuracy of EfficientNet-B7 [41] from 84.9% to 85.5%, with a significant improvement of 0.6%. Note that without knowledge distillation, RandAugment with a large distortion magnitude may suffer unstable training. Moreover, on top of EfficientNet-B8 [45], we set a **new record** on ImageNet classification (without extra training data) by claiming a top-1 accuracy of **85.8**%, surpassing the previous best by a non-trivial margin.

2 Related Work

Deep learning [24], in particular training deep neural networks, has been the standard methodology in computer vision. Modern neural networks, either manually-designed [14,19,22,34,38] or automatically searched [27,31,40,41,58,59], often contain a very large number of trainable parameters and thus raise the challenge of collecting more labeled data to avoid over-fitting. Data augmentation is a standard strategy to generate training data without additional labeling costs. Popular options of transformation include horizontal flipping, color/contrast jittering, image rotation/shearing, *etc.*, each of which slightly alters the geometry and/or pattern of an image but keeps its semantics (*e.g.*, the class label) unchanged. Data augmentation has been verified successful in a wide range of visual recognition tasks, including image classification [7,52,54], object detection [35], semantic segmentation [8], person re-identification [57], *etc.* Researchers have also discussed the connection between data augmentation and network regularization [10,36,49] methods.

With the rapid development of automated machine learning (AutoML) [43], researchers proposed to learn data augmentation strategies in a large search space to replace the conventional hand-designed augmentation policies. AutoAugment [4] is one of the early efforts that works on this direction. It first designed a search space with a number of transformations and then applied reinforcement learning to search for powerful combinations of the transformations to arrive at a high validation accuracy. To alleviate the heavy computational costs in the search stage, researchers designed a few efficient variants of AutoAugment.

Fast AutoAugment [25] moved the costly search stage from training to evaluation through bayesian optimization, and population-based augmentation [17] applied evolutionary algorithms to generate policy schedule by only a single run of 16 child models. Online hyper-parameter learning [26] combined the search stage and the network training process, and faster AutoAugment [13] formulated the search process into a differentiable function, following the recently emerging weight-sharing NAS approaches [2,28,30]. Meanwhile, some properties of AutoAugment have been investigated, such as whether aggressive transformations need to be considered [17] and how transformations of enriched knowledge are effectively chosen [55]. Recently, RandAugment [5] shared another opinion that the search space itself may have contributed most: based on a well-designed set of transformations, a random policy of augmentation works sufficiently well.

Knowledge distillation was first introduced as an approach to assist network optimization [16]. The goal is to improve the performance of a target network (often referred to as the student model) using two sources of supervision, one from the ground-truth label, and the other from the output signal of a pre-trained network (often referred to as the teacher model). Beyond its wide application on model compression (large teacher, small student [16,33]) and model initialization (small teacher, large student [3,34]), researchers later proposed to use it for standard network training, with the teacher and student models sharing the same network architecture [9,50,51], and sometimes under the setting of semi-supervised learning [42,56]. There have been discussions on the working mechanism of knowledge distillation, and researchers advocated for the so-called 'dark knowledge' [16] being some kind of auxiliary supervision, obtained from the pre-trained model [1,50] and thus different from the softened signals based on label smoothing [29,39].

In this paper, we build the connection between knowledge distillation and AutoAugment by showing that the former is a natural complement to the latter, which filters out noises introduced by overly aggressive transformations.

3 Our Approach

3.1 Preliminaries: Data Augmentation with AutoML

Let $\mathcal{D} = \{(\mathbf{x}_n, y_n)\}_{n=1}^{N}$ be a labeled image classification dataset with N samples, in which $\mathbf{x}_n$ denotes the raw pixels and y_n denotes the annotated label within the range of $\{1, 2, \ldots, C\}$, C is the number of classes. $\mathbf{y}_n$ is the vectorized version of y_n with the dimension corresponding to the true class assigned a value of 1 and all others being 0. The goal is to train a deep network, $\mathbb{M} : \mathbf{y}_n = \mathbf{f}(\mathbf{x}_n; \boldsymbol{\theta})$, where $\boldsymbol{\theta}$ denotes the trainable parameters. The dimensionality of $\boldsymbol{\theta}$ is often very large, *e.g.*, tens of millions, exceeding the size of dataset, N, in most of cases. Therefore, network training is often a ill-posed optimization problem and incurs over-fitting without well-designed training strategies.

The goal of data augmentation is to enlarge the set of training images without actually collecting and annotating more data. It starts with defining a transformation function, $\mathbf{x}_n^{\boldsymbol{\tau}} \doteq \mathbf{g}(\mathbf{x}_n, \boldsymbol{\tau})$, in which $\boldsymbol{\tau} \sim \mathcal{T}$ is a multi-dimensional vector parameterizing how the transformations are performed. Note that each dimension of $\boldsymbol{\tau}$ can take either discrete (*e.g.*, whether the image is horizontally flipped) or continuous (*e.g.*, the image is rotated for a specific angle), and different transformations can be applied to an image towards richer combinations. The idea of AutoAugment [4] is to optimize the distribution, $\mathcal{T}$, so that the model optimized on the augmented training set performs well on the validation set:

$$\mathcal{T}^{\star} = \arg\min_{\mathcal{T}} \mathcal{L}(\mathbf{g}(\mathbf{x}_n; \boldsymbol{\tau} \sim \mathcal{T}), \mathbf{y}_n; \boldsymbol{\theta}_{\mathcal{T}}^{\star} \mid (\mathbf{x}_n, \mathbf{y}_n) \sim \mathcal{D}_{\mathrm{val}}), \quad (1)$$

$$\text{in which} \quad \boldsymbol{\theta}_{\mathcal{T}}^{\star} = \arg\min_{\boldsymbol{\theta}} \mathcal{L}(\mathbf{g}(\mathbf{x}_n; \boldsymbol{\tau} \sim \mathcal{T}), \mathbf{y}_n; \boldsymbol{\theta} \mid (\mathbf{x}_n, \mathbf{y}_n) \sim \mathcal{D}_{\mathrm{train}}).$$

Here, $\mathcal{D}_{\mathrm{train}}$ and $\mathcal{D}_{\mathrm{val}}$ are two subsets of $\mathcal{D}$, used for training and validating the quality of $\mathcal{T}$, respectively. The loss function follows any conventions, *e.g.*, the cross-entropy form, $\mathcal{L}(\mathbf{x}_n^{\mathcal{T}}, \mathbf{y}_n; \boldsymbol{\theta}) = \mathbf{y}_n^{\top} \cdot \ln \mathbf{f}(\mathbf{x}_n^{\mathcal{T}}; \boldsymbol{\theta})$. Equation (1) is a two-stage optimization problem, for which existing approaches either applied reinforcement learning [4,17,25] or weight-sharing methods [13] which are often more efficient.

We follow the convention to assign each dimension in $\boldsymbol{\tau}$ to be an individual transformation, with the complete list shown below:

- `invert`
- `autoContrast`
- `equalize`
- `rotate`
- `solarize`
- `color`
- `posterize`
- `contrast`
- `brightness`
- `sharpness`
- `shear-x`
- `shear-y`
- `translate-x`
- `translate-y`

Therefore, $\boldsymbol{\tau}$ is a 14-dimensional vector and each dimension of $\boldsymbol{\tau}$ represents the magnitude of the corresponding transformation. For example, the fourth dimension of $\boldsymbol{\tau}$ represents the magnitude of `rotate` transformation, and a value of zero indicates the corresponding transformation being switched off. Each time a transformation is sampled from the distribution, $\boldsymbol{\tau} \sim \mathcal{T}$, at most two dimensions in it are set to be non-zero, and each selected transformation is assigned a probability that it is applied after each training image is sampled online.

3.2 AutoAugment Introduces Noisy Training Images

AutoAugment makes it possible to generate infinitely many images which do not exist in the original training set. On the upside, this reduces the risk of over-fitting during the training process; on the downside, it can introduce a considerable amount of outliers to the training process. Typical examples are shown in Fig. 2. When an image with its upper part occupied by main content (*e.g.*, *bee*) is sampled, the transformation of `translate-y` (shifting the image along the vertical direction) suffers risk of removing all discriminative contents within it outside the visible area, and thus the augmented image becomes meaningless in semantics. Nonetheless, the training process is not always aware of such noises and still uses the ground-truth signal, a one-hot vector, to supervise and thus confuse the deep network.

In Fig. 2, we also show how the training loss and validation accuracy curves change along with the magnitude of transformation. When the magnitude is 0 (*i.e.*, no augmentation is used), it is easy for the network to fit the training set

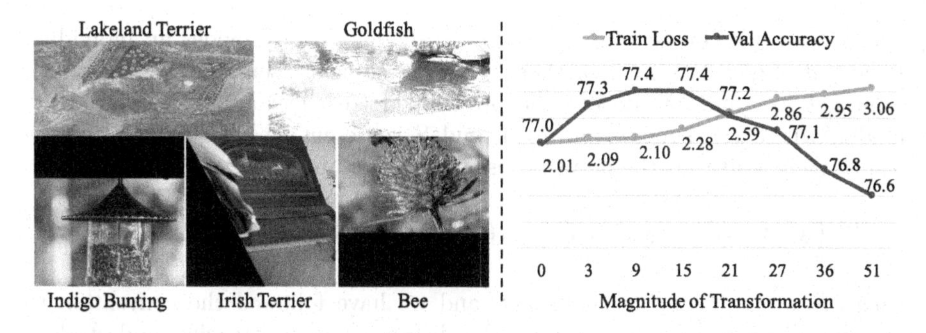

Fig. 2. Left: AutoAugment can generate meaningless training images but still assigns deterministic class labels to them. **Right**: The results of EfficientNet-B0 with different magnitudes of transformation on ImageNet. The training difficulty increases gradually with enlarging the magnitude of transformation, while the validation accuracy rises initially but drops at last. This phenomenon reveals the model starts from over-fitting to under-fitting.

and thus the training loss quickly drops, but the validation accuracy remains low which indicates over-fitting. With a relatively low magnitude of augmentation, the training loss increases gradually meanwhile the validation accuracy arrives at a higher plateau, *i.e.*, over-fitting is alleviated. However, if the magnitude of augmentation continues growing, it becomes more and more difficult to fit the training set, *i.e.*, the model suffers under-fitting. In particular, when the magnitude is set to be 36, the noisy data introduced to the training set is sufficiently high to bias the model training, *i.e.*, the results is lower than the baseline without AutoAugment.

From the above analysis, we realize that AutoAugment is indeed balancing between richer training data and heavier noises. Researchers provided comments from two aspects: some of them argued that the transformation strategies may have been overly aggressive and need to be controlled [17], while some others advocated for the benefit of exploring aggressive transformations so that richer information is integrated into the trained model [55]. We deal with this issue from a new perspective. We believe that aggressive transformations are useful to training, yet treating all augmented images just like they are clean (non-augmented) samples is not the optimal choice. Moreover, the same transformations operated on different images will cause different results, *i.e.*, some generated images can enrich the diversity of training set but the others are biased. Therefore, we treat every image differently for preserving richer information but filtering out noises.

3.3 Circumventing Outliers with Knowledge Distillation

Our idea is very simple. For a training image generated by AutoAugment, $\mathbf{g}(\mathbf{x}_n;\boldsymbol{\tau})$, we provide two-source supervision signals to guide network optimization. The first one remains the same as the original training process, with the standard cross-entropy loss computed based on the ground-truth class, $\mathbf{y}_n$.

The second one comes from a pre-trained model which provides an individual judgment of $\mathbf{g}(\mathbf{x}_n;\boldsymbol{\tau})$, *i.e.*, whether it contains sufficient semantics for classification. Let $\mathbb{M}^{\mathrm{T}}$ and $\mathbb{M}^{\mathrm{S}}$ denote the pre-trained (teacher) and target (student) model, where the superscripts of T and S represent 'teacher' and 'student', respectively, and thus the network loss function is upgraded to be:

$$\mathcal{L}^{\mathrm{KD}}\left(\mathbf{x}_n^{\boldsymbol{\tau}},\mathbf{y}_n;\boldsymbol{\theta}^{\mathrm{S}}\right)=\mathbf{y}_n^{\top}\cdot\ln\mathbf{f}^{\mathrm{S}}\left(\mathbf{x}_n^{\boldsymbol{\tau}};\boldsymbol{\theta}^{\mathrm{S}}\right)+\lambda\cdot\mathrm{KL}\left[\mathbf{f}^{\mathrm{S}}\left(\mathbf{x}_n^{\boldsymbol{\tau}};\boldsymbol{\theta}^{\mathrm{S}}\right)\|\mathbf{f}^{\mathrm{T}}\left(\mathbf{x}_n^{\boldsymbol{\tau}};\boldsymbol{\theta}^{\mathrm{T}}\right)\right], \tag{2}$$

where λ is the balancing coefficient, and we have followed the convention to use the KL-divergence to compute the distance between teacher and student outputs, two probabilistic distributions over all classes.

Intuitively, when the semantic information of an image is damaged by data augmentation, the teacher model that is 'unaware' of augmentation should produce less confident probabilistic outputs, *e.g.*, if an original image, $\mathbf{x}_n$, contains a specific kind of *bird* and some parts of the *bird* is missing or contaminated by augmentation, $\boldsymbol{\tau}$, then we expect the probabilistic scores of the augmented image, $\mathbf{x}_n^{\boldsymbol{\tau}}$, to be distributed over a few classes with close relationship to the true one. We introduce a hyper-parameter, K, and consider the K classes with the highest scores in $\mathbf{f}^{\mathrm{T}}\left(\mathbf{x}_n^{\boldsymbol{\tau}};\boldsymbol{\theta}^{\mathrm{T}}\right)$, forming a set denoted by $\mathcal{C}_K\left(\mathbf{x}_n^{\boldsymbol{\tau}};\boldsymbol{\theta}^{\mathrm{T}}\right)$. Due to the reason that the class probability decays rapidly with ranking, and low-ranked scores may contain much noise, it is often unsafe to force the student model to fit all teacher scores, so most often, we have $K \ll C$, and the choice of K will be discussed empirically in the experimental section. The KL-divergence between $\mathbf{f}^{\mathrm{T}}\left(\mathbf{x}_n^{\boldsymbol{\tau}};\boldsymbol{\theta}^{\mathrm{T}}\right)$ and $\mathbf{f}^{\mathrm{S}}\left(\mathbf{x}_n^{\boldsymbol{\tau}};\boldsymbol{\theta}^{\mathrm{S}}\right)$ is thus modified as:

$$\mathrm{KL}\left[\mathbf{f}^{\mathrm{S}}\left(\mathbf{x}_n^{\boldsymbol{\tau}};\boldsymbol{\theta}^{\mathrm{S}}\right)\|\mathbf{f}^{\mathrm{T}}\left(\mathbf{x}_n^{\boldsymbol{\tau}};\boldsymbol{\theta}^{\mathrm{T}}\right)\right]=\sum_{c\in\mathcal{C}_K\left(\mathbf{x}_n^{\boldsymbol{\tau}};\boldsymbol{\theta}^{\mathrm{T}}\right)} f_c^{\mathrm{T}}\left(\mathbf{x}_n^{\boldsymbol{\tau}};\boldsymbol{\theta}^{\mathrm{T}}\right)\cdot\ln\frac{f_c^{\mathrm{S}}\left(\mathbf{x}_n^{\boldsymbol{\tau}};\boldsymbol{\theta}^{\mathrm{S}}\right)}{f_c^{\mathrm{T}}\left(\mathbf{x}_n^{\boldsymbol{\tau}};\boldsymbol{\theta}^{\mathrm{T}}\right)}, \tag{3}$$

where f_c denotes the c-th dimension of $\mathbf{f}$.

3.4 Discussions and Relationship to Prior Work

A few prior work [1,50] studied how knowledge distillation works in the scenarios that teacher and student models have the same capacity. They argued that the teacher model should be strong enough so as not to provide low-quality supervision to the student model. However, this work provides a novel usage of the teacher signal: suppressing noises introduced by data augmentation. From this perspective, the teacher model can be considerably weaker than the student model but still contribute to recognition accuracy. Experimental results on CIFAR-100 (setting and details are provided in Sect. 4.1) show that a pre-trained Wide-ResNet-28-10 [53] with AutoAugment (test set error rate of 17.1%) can reduce the test set error rate of a Shake-Shake (26 2x96D) [10] trained with AutoAugment from 14.3% to 13.8%.

We noticed prior work [15] argued that data augmentation may introduce uncertainty to the network training process because the training data distribution is changed, and proposed to switch off data augmentation at the end of the training stage to alleviate the empirical risk of optimization. Our method provides an alternative perspective that the risk is likely to be caused by the noises of data augmentation and thus can be reduced by knowledge distillation. Moreover, the hyper-parameters in [15] (*e.g.*, when to switch off data augmentation) is difficult to tune. In training Wide-ResNet-28-10 [53] with AutoAugment on CIFAR-100, we follow the original paper to prevent data augmentation by adding 50 epochs to train the clean images only, but the baseline error rate (17.1%) is only reduced to 16.8%. In comparison, when knowledge distillation is added to these 50 epochs, the error rate is significantly reduced to 16.2%.

This work is also related to prior efforts that applied self-training to semi-supervised learning, *i.e.*, only a small portion of training data is labeled [23, 42,46]. These methods often started with training a model on the labeled part, then used this model to 'guess' a pseudo label for each of the unlabeled samples, and finally updated the model using all data with either ground-truth or pseudo labels. This paper verifies the effectiveness of knowledge distillation in the fully-supervised setting in which augmented data can be noisy. Therefore, we draw the connection between exploring unseen data (data augmentation) and exploiting unlabeled data (semi-supervised learning), and reveal the potential of integrating AutoAugment and/or other hyper-parameter optimization methods to assist and improve semi-supervised learning.

4 Experiments

4.1 Results on the CIFAR-10/100 Datasets

Dataset and Settings. CIFAR-10/100 [21] contain tiny images with a resolution of 32×32. Both of them have 50K training and 10K testing images, uniformly distributed over 10 or 100 classes. They are two commonly used datasets for validating the basic properties of learning algorithms. Following the convention [4,5], we train three types of networks, namely, wide ResNet (Wide-ResNet-28-10) [53], Shake-Shake (three variants with different feature dimensions) [10], and PyramidNet [12] with ShakeDrop regularization [49].

Knowledge Distillation Stabilizes AutoAugment. The core idea of our approach is to utilize knowledge distillation to restrain noises generated by severe transformations. This is expected to stabilize the training process of AutoAugment. To verify this, we start with training Wide-ResNet-28-10 on CIFAR-100. Note that the original augmentation space of AutoAugment involves two major kinds of transformations, namely, geometric or color-based transformations, on which AutoAugment as well as its variants limited the distortion magnitude of each transformation in a relatively small range so that the augmented images are mostly safe, *i.e.*, semantic information is largely preserved. In order to enhance the benefit brought by suppressing noises of aggressive augmentations, we design

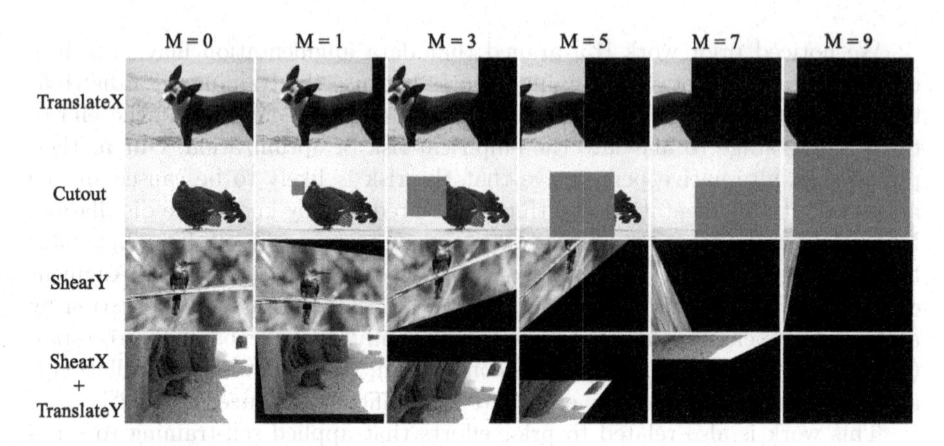

Fig. 3. Examples of transformations involved in our self-designed augment space. The distortion magnitude, M, is divided into 10 levels. The deformation introduced by transformations increases along with the magnitude. First three rows are examples of the deformation produced by each type of transformation with different magnitudes. The last row represents applying two consecutive transformations on a single image, which is the real case in our training scenario.

Table 1. Comparison between RandAugment with or without knowledge distillation in our self-designed augment space on CIFAR-100 based on Wide-ResNet-28-10. All numbers in the table are error rates (%). **M** indicates the distortion magnitude of each transformation. **RA** for RandAugment [5], and **KD** for knowledege distillation.

Model	Distortion magnitude, M							
	0	1	2	3	4	5	6	7
RA	18.4	19.5	20.7	22.4	25.7	31.6	40.3	55.1
RA+KD	18.0	17.6	18.5	19.9	21.9	27.0	34.9	48.0
Gain	+0.4	+1.9	+2.2	+2.5	+3.8	+4.6	+5.4	+7.1

a new augment space in which the restriction in distortion magnitude is much weaker. To guarantee that large magnitudes lead to complete damage of semantic information, we only preserve a subset of geometric transformations (`shear-x`, `shear-y`, `translate-x`, `translate-y`) as well as `cutout`, and set 10 levels of distortion, so that $M = 0$ implies no augment, and $M = 10$ of any transformation destroys the entire image. Note that the range of M here is specifically designed for the modified augment space, which is incomparable with the original definition of M in RandAugment (experimented in Sect. 4.2). Regarding knowledge distillation, we set $K = 3$ (computing KL-divergence between the distributions of top-3 classes, determined by the teacher model) for CIFAR-10 and $K = 5$ for CIFAR-100. The balancing coefficient, λ, and the softmax temperature, T, is set to be 1.0 and 2.0, respectively.

In this modified augment space, we experiment with the strategy of RandAugment [5] which controls the strength of augmentation by adjusting the distortion magnitude, M. For example, on the `translate-x` transformation, a magnitude of 3 allows the entire image to be shifted, to the left or right, by at most 30% of the visible field, and a magnitude of 10 enlarges the number into 100%, *i.e.*, the visible area totally disappears. More examples are shown in Fig. 3. Note that RandAugment performs two consecutive transformations on each image, therefore, a magnitude of 8 is often enough to destroy all semantic contents. Hence, M is constrained within the range of 0–7 in our experiments.

Table 2. Comparison between our approach and other data augmentation methods on CIFAR-10 and CIFAR-100. The teacher for all networks is Wide-ResNet-28-10, except for PyramidNet+ShakeDrop with itself as teacher on CIFAR-100 (due to the huge performance gap). All numbers in the table are error rates (%). **NA** indicates no augmentation is used, **AA** for AutoAugment [4], **FAA** for fast AutoAugment [25], **PBA** for population-based augmentation [17], and **RA** for RandAugment [5].

Dataset	Network	NA	AA	FAA	PBA	RA	Ours
CIFAR-10	Wide-ResNet-28-10	3.9	2.6	2.7	2.6	2.7	**2.4**
	Shake-Shake (26 2x32D)	3.6	2.5	2.5	2.5	–	**2.3**
	Shake-Shake (26 2x96D)	2.9	2.0	2.0	2.0	2.0	**1.8**
	Shake-Shake (26 2x112D)	2.8	**1.9**	**1.9**	2.0	–	**1.9**
	PyramidNet+ShakeDrop	2.7	**1.5**	1.7	**1.5**	**1.5**	**1.5**
CIFAR-100	Wide-ResNet-28-10	18.8	17.1	17.3	16.7	16.7	**16.2**
	Shake-Shake (26 2x96D)	17.1	14.3	14.6	15.3	–	**13.8**
	PyramidNet+ShakeDrop	14.0	10.7	11.7	10.9	–	**10.6**

Results of different distortion magnitudes are summarized in Table 1. With the increase of the magnitude, a larger portion of semantic information is expected to be removed from the training image. In this scenario, if we continue forcing the model to fit the ground-truth, one-hot supervision of each training sample, the deep network may get confused and 'under-fit' the training data. This causes consistent accuracy drop, especially in the modified augment space with only geometric transformations. Even when the full augment space is used (in which some transformations are not very sensitive to M), this factor persists and hinders the use of larger M values, and thus restricts the degree of freedom of AutoAugment.

Knowledge distillation offers an opportunity that each augmented image is checked beforehand, and a soft label is provided by a pre-trained teacher model to co-supervise the training process so that the deep network is not forced to fit the one-hot label. This is especially useful when the training image is contaminated by augmentation. As shown in Table 1, knowledge distillation provides consistent accuracy gain over RandAugment, as it slows down the accuracy drop with aggressive augmentation (the gain is larger as the distortion magnitude increases). More importantly, under a magnitude of $M = 1$, knowledge distillation produces an accuracy gain of 1.9%, assisting the RandAugment-only model with a deficit of 1.1% to surpass the baseline, claiming an advantage of 0.4%. This proves that the benefit mainly comes from the cooperation of RandAugment and knowledge distillation, not only from the auxiliary information provided by knowledge distillation itself [1,9,50].

Comparison with State-of-the-Arts. To make fair comparisons to the previous AutoAugment-based methods, we directly inherit the augmentation policies found on CIFAR by AutoAugment. In this full space, all transformations listed in Sect. 3.1, not only the geometric transformations, can appear. Results are summarized in Table 2.

On CIFAR-10, our method outperforms other augmentation methods consistently, in particular, on top of smaller networks (*e.g.*, the error rates of Wide-ResNet-28-10 and two Shake-Shake models are reduced by 0.2%). For larger models, in particular PyramidNet with ShakeDrop regularization, the room of improvement on CIFAR-10 is very small, yet we can observe improvement on very large models on the more challenging CIFAR-100 and ImageNet datasets (see the next part for details).

A side comment is that we have used the same teacher model (*i.e.*, Wide-ResNet-28-10, reporting a 2.6% error) which is relatively weak. We find this model can assist training much stronger students (*e.g.*, the Shake-Shake series, in which the error of the 2x96D model, 2.0%, is reduced to 1.8%). **In other words, weaker teachers can assist training strong students.** This delivers a complementary opinion to prior research which advocates for extracting 'dark knowledge' as some kind of auxiliary supervision [50] from stronger [16] or at least equally-powerful [9] teacher models, and further verifies the extra benefits brought by integrating knowledge distillation and AutoAugment together.

On CIFAR-100, we evaluate a similar set of network architectures, *i.e.*, Wide-ResNet-28-10, Shake-Shake (26 2x96D), and PyramidNet+ShakeDrop. As shown in Table 2, our results consistently outperform the previous state-of-the-arts. For example, on a relatively smaller Wide-ResNet-28-10, the error of AutoAugment decreases from 17.1% to 16.2% and significantly outperforms other methods, *e.g.*, PBA and RA. On Shake-Shake (26 2x96D), our approach also surpasses the previous best performance (14.3%) by a considerable margin of 0.5%. On pyramidNet with ShakeDrop, although the baseline accuracy is sufficiently high, knowledge distillation still brings a slight improvement (from 10.7% to 10.6%).

4.2 Results on the ImageNet Dataset

Dataset, Setting, and Implementation Details. ImageNet [6] is one of the largest visual recognition datasets which contains high-resolution images. We use the competition subset which has 1K classes, 1.3M training and 50K validation images. The number of images in each class is approximately the same for training data.

We build our baseline upon EfficientNet [41] and RandAugment [5]. EfficientNet contains a series of deep networks with different depths, widths and scales (*i.e.*, the spatial resolution at each layer). There are 9 variants of EfficientNet [45], named from B0 to B8. Equipped with RandAugment, EfficientNet-B7 reports a top-1 accuracy of 85.0% which is close to the state-of-the-art. We start with EfficientNet-B0 to investigate the impact of different knowledge distillation parameters on ImageNet, and finally compete with state-of-the-art results on EfficientNet-B4, EfficientNet-B7, and EfficientNet-B8.

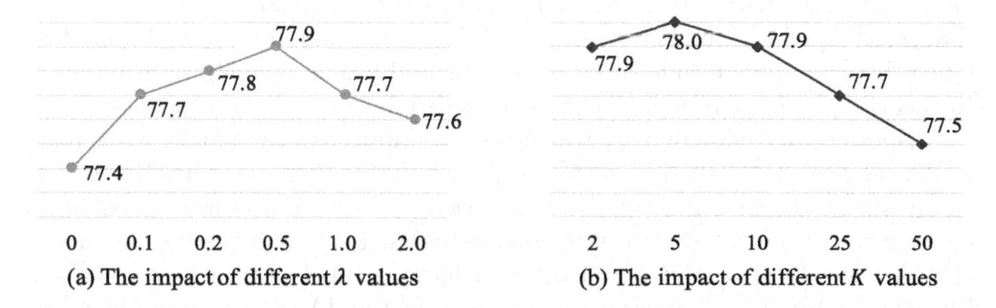

Fig. 4. Training EfficientNet-B0 with different KD parameters. All numbers reported are top-1 accuracy (%). **Left**: The testing accuracy of different λ values, while K is set as 10. **Right**: The testing accuracy of different K values, while λ is set to be 0.5.

We follow the implementation details provided by the authors[1], and reproduce the training process using PyTorch. For EfficientNet-B0, it is trained through 500 epochs with an initial learning rate to be 0.256 and decayed by a factor of 0.97 every 2.4 epochs. We use the RMSProp optimizer with a decay factor of 0.9 and a momentum of 0.9. The batch-normalization decay factor is set to be 0.99 and the weight decay 10^{-5}. We use 32 GPUs (NVIDIA Tesla-V100) to train EfficientNet-B0/B4, and 256 GPUs for EfficientNet-B7/B8, respectively.

The Impact of Different Knowledge Distillation Parameters. We start with investigating the impact of λ and K, two important hyper-parameters of knowledge distillation. Note that we fix the softmax temperature, T, to be 10.0 in all ImageNet experiments. We perform experiments on EfficientNet-B0 with a moderate distortion magnitude of $M = 9$, which, as we have shown in the right-hand side of Fig. 2, is a safe option on EfficientNet-B0. For λ, we set different values including 0.1, 0.2, 0.5, 1.0, and 2.0. For K, the optional values include 2, 5, 10, 25, and 50. To better evaluate the effect of each parameter, we fix one parameter value when changing the other.

Results are shown in Fig. 4. It is clear that a moderate λ performs best. While setting λ with a small value, *e.g.*, 0.1, knowledge distillation is only expected to affect a small part of training samples. Yet, it obtains a 0.3% accuracy gain, implying that these samples, though rarely seen, can make the training process

[1] https://github.com/tensorflow/tpu/tree/master/models/official/efficientnet.

unstable. On the other hand, when λ is overly large, *e.g.*, knowledge distillation can dominate the training process and force the student model to have a very similar behavior to the teacher model, which limits its ability and harms classification performance.

Regarding K, we note that $K = 5$ achieves the best performance, indicating that on average, each class is connected to 4 other classes. This was also suggested in [50]. Yet, we find that setting $K = 2$ or $K = 10$ reports similar accuracy, but the performance gradually drops as K increases. This implies including too many classes for KL-divergence computation is harmful, because each training image, after augmented with a relatively small distortion magnitude, is not likely to be connected to a large number of classes. However, to train more powerful models, larger distortion magnitudes need to be used and heavier ambiguity introduced. In this case, a larger K will be better, as we shall see in the next section.

Regardless of tuning hyper-parameters, we emphasize that all tested λ's, lying in the range of $[0.1, 2.0]$, and all tested K's, in $[2, 50]$, can bring positive effects on classification. This indicates that knowledge distillation is usually useful in training with augmented data. With the best setting, *i.e.*, a distortion magnitude of 9, a fixed K of 5, and $\lambda = 0.5$, we achieve a top-1 accuracy of 78.0% on EfficientNet-B0. This surpasses the accuracy of RandAugment (reproduced by us) and AdvProp [45] by margins of 0.6% and 0.4%, respectively.

Comparison to the State-of-the-Arts. To better evaluate the effectiveness of our approach, we further conduct experiments on more challenging large models, *i.e.*, EfficientNet-B4, EfficientNet-B7, and EfficientNet-B8. Given the fact that larger network is expected to over-fit more easily, for EfficienNet-B4 and EfficientNet-B7, we lift the magnitude of transformations on RandAugment from 9 in EfficientNet-B0 to 15 and 28, respectively. As discussed above, increasing the distortion magnitude brings more ambiguity to the training images so that each of them should be connected to more classes, and the knowledge distillation supervision should take a heavier weight. Hence, we increase K to 50 and λ to 2.0 in all experiments in this part.

Results are summarized in Table 3. By restraining the inevitable noises generated by RandAugment, our approach significantly boosts the baseline models. As shown in Table 3, the top-1 accuracy of EfficientNet-B4 is increased from 83.0% to 83.6%, and that of EfficientNet-B7 from 84.9% to 85.5%. The margin of 0.6% is considered significant in such powerful baselines. Both numbers surpass the current best, AdvProp [45], without using adversarial examples to assist training. Besides, when we simply double the training epochs of EfficientNet-B4, the top-1 accuracy is slightly improved from 83.0% to 83.2%, which is still much lower than 83.6% reported by applying another KD-guided training procedure.

Following AdvProp [45], we also move towards training EfficientNet-B8. The hyper-parameters remain the same as in training EfficientNet-B7. Due to GPU memory limit, we use the best trained EfficientNet-B7 (with a 85.5% accuracy) as the teacher model. We report a top-1 accuracy of 85.7%, which sets the **new state-of-the-art** on the ImageNet dataset (without extra training data). With the test image size increased from 672 to 800, the accuracy is slightly improved to 85.8%. We show the comparison with previous best models in Table 4.

Table 3. Comparison between our approach and other data augmentation methods on ImageNet. All numbers in the table are top-1 accuracy (%). **AA** indicates AutoAugment [4] is used, **RA** for RandAugment [5], and **AdvProp** for Adversarial Propagation method [45]. **RA**† denotes the results of RandAugment produced by ourselves in PyTorch. EfficientNet-B7* denotes the student model in the penultimate row, which achieves a top-1 accuracy of 85.5%.

Teacher network	Student network	AA	RA	RA†	AdvProp	Ours
EfficientNet-B0	EfficientNet-B0	77.3	–	77.4	77.6	**78.0**
EfficientNet-B4	EfficientNet-B4	83.0	–	83.0	83.3	**83.6**
EfficientNet-B7	EfficientNet-B7	84.5	85.0	84.9	85.2	**85.5**
EfficientNet-B7*	EfficientNet-B8	84.8	85.4	–	85.5	**85.7**

Table 4. Comparison to the state-of-the-arts on ImageNet. In the middle panel, we list three approaches with extra training data (a large number of weakly tagged or unlabeled images). Italic and bold texts highlight the best results to date without and with extra training data, respectively.

Method	Params	Extra training data	Top-1 (%)
ResNet-152 [14]	60M	–	77.8
Inception-v4 [37]	48M	–	80.0
ResNeXt-101 [47]	84M	–	80.9
SENet [18]	146M	–	82.7
AmoebaNet-C [32]	155M	–	83.5
GPipe [20]	557M	–	84.3
EfficientNet-B7 [5]	66M	–	85.0
EfficientNet-B8 [5]	88M	–	85.4
EfficientNet-L2 [41]	480M	–	85.5
AdvProp (EfficientNet-B8) [45]	88M	–	85.5
ResNeXt-101, Billion-scale [48]	193M	3.5B tagged images	84.8
FixRes ResNeXt-101, WSL [44]	829M	3.5B tagged images	86.4
Noisy Student (EfficientNet-L2) [46]	**480M**	**300M unlabeled images**	**88.4**
Ours (EfficientNet-B7 w/ KD)	66M	–	85.5
Ours (EfficientNet-B8 w/ KD)	*88M*	–	*85.8*

5 Conclusions

This paper integrates knowledge distillation into AutoAugment-based methods, and shows that the noises introduced by aggressive data augmentation policies can be largely alleviated by referring to a pre-trained teacher model. We adjust the computation of KL-divergence, so that the teacher and student models share similar probabilistic distributions over the top-ranked classes. Experiments show that our approach indeed suppresses noises introduced by data augmentation, and thus stabilizes the training process and enables more aggressive AutoAugment policies to be used. Our approach sets the new state-of-the-art, a 85.8% top-1 accuracy, on the ImageNet dataset (without extra training data).

Our research leaves several open problems. For example, it remains unclear whether useful information only exists in the top-ranked classes determined by the teacher model, and whether mimicking the class-level distribution is the optimal choice. Moreover, the balancing coefficient, λ, is a constant during training, which we believe there is room of improvement. We will continue investigating these topics in our future research.

References

1. Bagherinezhad, H., Horton, M., Rastegari, M., Farhadi, A.: Label refinery: improving ImageNet classification through label progression. arXiv preprint arXiv:1805.02641 (2018)
2. Brock, A., Lim, T., Ritchie, J.M., Weston, N.J.: SMASH: one-shot model architecture search through hypernetworks. In: International Conference on Learning Representations (2018)
3. Chen, T., Goodfellow, I., Shlens, J.: Net2net: accelerating learning via knowledge transfer. In: International Conference on Learning Representations (2016)
4. Cubuk, E.D., Zoph, B., Mane, D., Vasudevan, V., Le, Q.V.: AutoAugment: Learning augmentation strategies from data. In: Computer Vision and Pattern Recognition (2019)
5. Cubuk, E.D., Zoph, B., Shlens, J., Le, Q.V.: RandAugment: practical automated data augmentation with a reduced search space. In: Proceedings of the IEEE/CVF Conference on Computer Vision and Pattern Recognition Workshops, pp. 702–703 (2020)
6. Deng, J., Dong, W., Socher, R., Li, L.J., Li, K., Fei-Fei, L.: ImageNet: a large-scale hierarchical image database. In: Computer Vision and Pattern Recognition (2009)
7. DeVries, T., Taylor, G.W.: Improved regularization of convolutional neural networks with cutout. arXiv preprint arXiv:1708.04552 (2017)
8. Fang, H.S., Sun, J., Wang, R., Gou, M., Li, Y.L., Lu, C.: InstaBoost: boosting instance segmentation via probability map guided copy-pasting. In: International Conference on Computer Vision (2019)
9. Furlanello, T., Lipton, Z., Tschannen, M., Itti, L., Anandkumar, A.: Born again neural networks. In: International Conference on Machine Learning (2018)
10. Gastaldi, X.: Shake-shake regularization. arXiv preprint arXiv:1705.07485 (2017)
11. Ghiasi, G., Lin, T.Y., Le, Q.V.: NAS-FPN: learning scalable feature pyramid architecture for object detection. In: Computer Vision and Pattern Recognition (2019)
12. Han, D., Kim, J., Kim, J.: Deep pyramidal residual networks. In: Computer Vision and Pattern Recognition (2017)
13. Hataya, R., Zdenek, J., Yoshizoe, K., Nakayama, H.: Faster autoaugment: learning augmentation strategies using backpropagation. arXiv preprint arXiv:1911.06987 (2019)
14. He, K., Zhang, X., Ren, S., Sun, J.: Deep residual learning for image recognition. In: Computer Vision and Pattern Recognition (2016)
15. He, Z., Xie, L., Chen, X., Zhang, Y., Wang, Y., Tian, Q.: Data augmentation revisited: Rethinking the distribution gap between clean and augmented data. arXiv preprint arXiv:1909.09148 (2019)
16. Hinton, G., Vinyals, O., Dean, J.: Distilling the knowledge in a neural network. arXiv preprint arXiv:1503.02531 (2015)

17. Ho, D., Liang, E., Chen, X., Stoica, I., Abbeel, P.: Population based augmentation: efficient learning of augmentation policy schedules. In: International Conference on Machine Learning (2019)
18. Hu, J., Shen, L., Sun, G.: Squeeze-and-excitation networks. In: Computer Vision and Pattern Recognition (2018)
19. Huang, G., Liu, Z., Van Der Maaten, L., Weinberger, K.Q.: Densely connected convolutional networks. In: Computer Vision and Pattern Recognition (2017)
20. Huang, Y., Cheng, Y., Chen, D., Lee, H., Ngiam, J., Le, Q.V., Chen, Z.: GPipe: efficient training of giant neural networks using pipeline parallelism. In: Advances in Neural Information Processing Systems (2019)
21. Krizhevsky, A., Hinton, G., et al.: Learning multiple layers of features from tiny images (2009)
22. Krizhevsky, A., Sutskever, I., Hinton, G.E.: Imagenet classification with deep convolutional neural networks. In: Advances in Neural Information Processing Systems (2012)
23. Laine, S., Aila, T.: Temporal ensembling for semi-supervised learning. In: International Conference on Learning Representations (2017)
24. LeCun, Y., Bengio, Y., Hinton, G.: Deep learning. Nature **521**(7553), 436–444 (2015)
25. Lim, S., Kim, I., Kim, T., Kim, C., Kim, S.: Fast autoaugment. In: Advances in Neural Information Processing Systems (2019)
26. Lin, C., et al.: Online hyper-parameter learning for auto-augmentation strategy. In: International Conference on Computer Vision (2019)
27. Liu, C., et al.: Progressive neural architecture search. In: European Conference on Computer Vision (2018)
28. Liu, H., Simonyan, K., Yang, Y.: DARTS: differentiable architecture search. In: International Conference on Learning Representations (2019)
29. Pereyra, G., Tucker, G., Chorowski, J., Kaiser, Ł., Hinton, G.: Regularizing neural networks by penalizing confident output distributions. arXiv preprint arXiv:1701.06548 (2017)
30. Pham, H., Guan, M.Y., Zoph, B., Le, Q.V., Dean, J.: Efficient neural architecture search via parameter sharing. In: International Conference on Machine Learning (2018)
31. Real, E., et al.: Large-scale evolution of image classifiers. In: International Conference on Machine Learning (2017)
32. Real, E., Aggarwal, A., Huang, Y., Le, Q.V.: Regularized evolution for image classifier architecture search. In: AAAI Conference on Artificial Intelligence (2019)
33. Romero, A., Ballas, N., Kahou, S.E., Chassang, A., Gatta, C., Bengio, Y.: FitNets: hints for thin deep nets. arXiv preprint arXiv:1412.6550 (2014)
34. Simonyan, K., Zisserman, A.: Very deep convolutional networks for large-scale image recognition. In: International Conference on Learning Representations (2015)
35. Singh, K.K., Lee, Y.J.: Hide-and-seek: forcing a network to be meticulous for weakly-supervised object and action localization. In: International Conference on Computer Vision (2017)
36. Srivastava, N., Hinton, G., Krizhevsky, A., Sutskever, I., Salakhutdinov, R.: Dropout: a simple way to prevent neural networks from overfitting. J. Mach. Learn. Res. **15**(1), 1929–1958 (2014)
37. Szegedy, C., Ioffe, S., Vanhoucke, V., A Alemi, A.: Inception-v4, Inception-ResNet and the impact of residual connections on learning. In: AAAI Conference on Artificial Intelligence (2017)

38. Szegedy, C., et al.: Going deeper with convolutions. In: Computer Vision and Pattern Recognition (2015)
39. Szegedy, C., Vanhoucke, V., Ioffe, S., Shlens, J., Wojna, Z.: Rethinking the inception architecture for computer vision. In: Proceedings of the IEEE Conference on Computer Vision And Pattern Recognition, pp. 2818–2826 (2016)
40. Tan, M., et al.: MnasNet: platform-aware neural architecture search for mobile. In: Computer Vision and Pattern Recognition (2019)
41. Tan, M., Le, Q.V.: EfficientNet: rethinking model scaling for convolutional neural networks. In: International Conference on Machine Learning (2019)
42. Tarvainen, A., Valpola, H.: Mean teachers are better role models: weight-averaged consistency targets improve semi-supervised deep learning results. In: Advances in Neural Information Processing Systems (2017)
43. Thornton, C., Hutter, F., Hoos, H.H., Leyton-Brown, K.: Auto-WEKA: combined selection and hyperparameter optimization of classification algorithms. In: SIGKDD International Conference on Knowledge Discovery and Data Mining (2013)
44. Touvron, H., Vedaldi, A., Douze, M., Jégou, H.: Fixing the train-test resolution discrepancy. In: Advances in Neural Information Processing Systems, pp. 8252–8262 (2019)
45. Xie, C., Tan, M., Gong, B., Wang, J., Yuille, A.L., Le, Q.V.: Adversarial examples improve image recognition. In: Proceedings of the IEEE/CVF Conference on Computer Vision and Pattern Recognition, pp. 819–828 (2020)
46. Xie, Q., Luong, M.T., Hovy, E., Le, Q.V.: Self-training with noisy student improves ImageNet classification. In: Proceedings of the IEEE/CVF Conference on Computer Vision and Pattern Recognition, pp. 10687–10698 (2020)
47. Xie, S., Girshick, R., Dollar, P., Tu, Z., He, K.: Aggregated residual transformations for deep neural networks. In: Computer Vision and Pattern Recognition (2017)
48. Yalniz, I.Z., Jegou, H., Chen, K., Paluri, M., Mahajan, D.: Billion-scale semi-supervised learning for image classification. arXiv preprint arXiv:1905.00546 (2019)
49. Yamada, Y., Iwamura, M., Akiba, T., Kise, K.: ShakeDrop regularization for deep residual learning. IEEE Access **7**, 186126–186136 (2019)
50. Yang, C., Xie, L., Qiao, S., Yuille, A.L.: Training deep neural networks in generations: a more tolerant teacher educates better students. In: AAAI Conference on Artificial Intelligence (2019)
51. Yang, C., Xie, L., Su, C., Yuille, A.L.: Snapshot distillation: teacher-student optimization in one generation. In: Computer Vision and Pattern Recognition (2019)
52. Yun, S., Han, D., Oh, S.J., Chun, S., Choe, J., Yoo, Y.: CutMix: regularization strategy to train strong classifiers with localizable features. In: International Conference on Computer Vision (2019)
53. Zagoruyko, S., Komodakis, N.: Wide residual networks. In: British Machine Vision Conference (2016)
54. Zhang, H., Cisse, M., Dauphin, Y.N., Lopez-Paz, D.: mixup: beyond empirical risk minimization. In: International Conference on Learning Representations (2018)
55. Zhang, X., Wang, Q., Zhang, J., Zhong, Z.: Adversarial autoaugment. In: International Conference on Learning Representations (2020)
56. Zhang, Y., Xiang, T., Hospedales, T.M., Lu, H.: Deep mutual learning. In: Computer Vision and Pattern Recognition (2018)

57. Zhong, Z., Zheng, L., Kang, G., Li, S., Yang, Y.: Random erasing data augmentation. In: AAAI, pp. 13001–13008 (2020)
58. Zoph, B., Le, Q.V.: Neural architecture search with reinforcement learning. In: International Conference on Learning Representations (2017)
59. Zoph, B., Vasudevan, V., Shlens, J., Le, Q.V.: Learning transferable architectures for scalable image recognition. In: Computer Vision and Pattern Recognition (2018)

S2DNet: Learning Image Features for Accurate Sparse-to-Dense Matching

Hugo Germain[1(✉)], Guillaume Bourmaud[2], and Vincent Lepetit[1,2]

[1] LIGM, École des Ponts, Univ Gustave Eiffel, CNRS, Marne-la-vallée, France
{hugo.germain,vincent.lepetit}@enpc.fr
[2] Laboratoire IMS, Université de Bordeaux, Bordeaux, France
guillaume.bourmaud@u-bordeaux.fr

Abstract. Establishing robust and accurate correspondences is a fundamental backbone to many computer vision algorithms. While recent learning-based feature matching methods have shown promising results in providing robust correspondences under challenging conditions, they are often limited in terms of precision. In this paper, we introduce S2DNet, a novel feature matching pipeline, designed and trained to efficiently establish both robust and accurate correspondences. By leveraging a sparse-to-dense matching paradigm, we cast the correspondence learning problem as a supervised classification task to learn to output highly peaked correspondence maps. We show that S2DNet achieves state-of-the-art results on the HPatches benchmark, as well as on several long-term visual localization datasets.

Keywords: Feature matching · Classification · Visual localization

1 Introduction

Establishing both accurate and robust correspondences across images is an underpinning step to many computer vision algorithms, such as Structure-from-Motion (SfM) [21,48,49,55], visual tracking [23,64] and visual localization [45,53,54]. Yet, obtaining such correspondences in long-term scenarios where extreme visual changes can appear remains an unsolved problem, as shown by recent benchmarks [44,56]. In particular, illumination (e.g. daytime to night-time), cross-seasonal and structural changes are very challenging factors for feature matching.

The accuracy of the correspondences plays a major role in the performance of the aforementioned algorithms. Indeed, the noise perturbation experiment of Fig. 1 (left) shows the highly damaging impacts of errors of a few pixels on visual localization. A traditional and very commonly used paradigm for feature matching between two images consists in detecting a set of keypoints [7,13,

Electronic supplementary material The online version of this chapter (https://doi.org/10.1007/978-3-030-58580-8_37) contains supplementary material, which is available to authorized users.

A. Vedaldi et al. (Eds.): ECCV 2020, LNCS 12348, pp. 626–643, 2020.
https://doi.org/10.1007/978-3-030-58580-8_37

15,20,27,29,30,37,47,65], followed by a description stage [6,9,13–15,27,33,34, 37,50,51,65] in each image. Sparse sets of keypoints and their descriptors are then matched using for instance approximate nearest neighbours. This *sparse-to-sparse* matching approach has the main advantage of being both computationally and memory efficient. For long-term scenarios, this requires detecting repeatable keypoints and computing robust descriptors, which is very challenging.

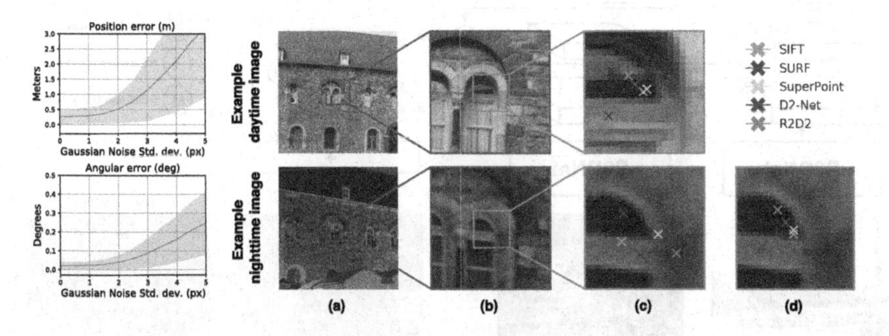

Fig. 1. Learning accurate correspondences. On the left, we report the impact of adding a gaussian noise of increasing variance on ground-truth 2D-3D correspondences for the task of visual localization, on Aachen Day-Night [44,46] images. This experiment highlights the importance of having very accurate correspondences, as offsets of a few pixels can lead to localization errors of several meters. Yet as shown on the right, sparse-to-sparse methods fail to make such accurate predictions. We show in (*a*) and (*b*) local regions of interest for a day-night image pair. In (*c*) [top], we display the keypoint detections being the nearest to the center of the patch in the daytime image for each detector; [bottom] we show the closest correspondent detected keypoints for each detector in the nighttime image. In (*d*), we show the correspondent image locations found by S2DNet in the nighttime image for daytime keypoint detections. S2DNet manages to find much more accurate correspondences than sparse-to-sparse methods.

With the advent of convolutional neural networks (CNNs), learning-based sparse-to-sparse matching methods have emerged, attempting to improve robustness of both detection and description stages in an end-to-end fashion. Several single-CNN pipelines [13,15,37] were trained with pixel-level supervision to jointly detect *and* describe interest points. These methods have yielded very competitive results especially in terms of number of correct matches, but fail to deliver highly accurate correspondences [15,37]. Indeed detecting the same keypoints repeatably across images is very challenging under strong visual changes, as illustrated in Fig. 1 (right). Thus, the accuracy of such methods becomes highly reliant on the feature detector's precision and repeatability.

A recently proposed alternative [18] to solve the repeatable keypoint detection issue is to shift the sparse-to-sparse paradigm into a *sparse-to-dense* approach: Instead of trying to detect consistent interest points across images, feature detection is performed asymmetrically and correspondences are searched exhaustively in the other image. That way, all the information in the challenging image is preserved, allowing each pixel to be a potential correspondent candidate.

Thanks to the popularization and development of GPUs, this exhaustive search can be done with a small computational overhead compared to other learning-based methods. This approach has showed to give competitive results [18] when trained in a weakly supervised fashion for the task of image retrieval.

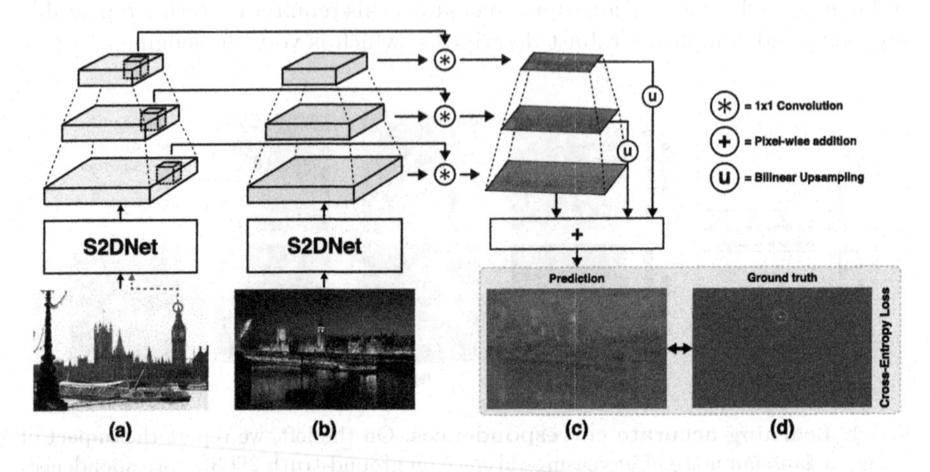

Fig. 2. S2DNet feature matching pipeline overview. Given an image and a set of detections coming from an off-the-shelf keypoint detector (*a*), we first extract a set of sparse multi-level descriptors with S2DNet. We then compute dense feature maps for a covisible image (*b*), and compute multi-level correspondence maps (*c*), which we aggregate using bilinear upsampling and addition. Correspondences can the be retrieved using a simple argmax operator. We explicitly train S2DNet to generate accurate and discriminative correspondence maps using a supervised classification approach (*d*).

In this paper, we reuse the sparse-to-dense idea while also addressing the accuracy issue. We introduce a novel feature matching pipeline which we name *S2DNet.*

Contributions: (i) We propose to cast the feature learning problem as a classification one in a sparse-to-dense paradigm. This is in contrast with State-of-the-art (SoA) algorithms (e.g. D2-Net, R2D2) which learn features using metric learning (with contrastive losses) and a sparse-to-sparse paradigm. We show that our approach delivers significant improvements on both image matching and visual localization popular benchmarks. (ii) Even when used in a classical sparse-to-sparse framework, the image features learned by S2DNet lead to much more accurate correspondences than SoA features. (iii) Our formalism also enables probabilistic interpretation, which allows us to perform outlier rejection through confidence thresholding. As demonstrated by our experiments, this significantly improves the performance. (iv) Rather than aiming to jointly learn a feature detector and a feature descriptor like SoA approaches, we show we can keep using classical keypoint detectors and rather focus on the feature learning problem. This leads to a single loss, while previous methods have to combine multiple loss terms. Moreover, we can also operate at single-scale. Our approach thus yields a fundamental simpler approach, in addition to improving performance.

2 Related Work

Establishing 2D to 2D correspondences between images is a key step for many applications in computer vision, whose performance often directly relies on the quantity and accuracy of such correspondences [17,62]. We can distinguish three categories for obtaining such correspondences, which either rely on a bilateral, no keypoint or asymmetrical detection stage.

Sparse-to-Sparse Feature Matching. The most popular and studied approach for feature matching is a two-stage pipeline that first detects interest point locations and assigns a patch-based descriptor to each of them. Detection is applied on both images to be matched, and we refer to these *detect-then-describe* approaches as *sparse-to-sparse* feature matching methods. To perform keypoint detection, a variety of both hand-crafted [7,20,27,29] and learning-based [34,65] detectors have been developed, each aiming to detect accurate keypoints in both a repeatable as well as illumination, scale and affine invariant fashion. For feature description, methods using histograms of local gradients [7,10,27,39] or learning-based patch description [4,11,34,58,59,65] have been widely used.

Yet, when working on long-term scenarios where very strong visual changes can appear, such methods fail to give reliable correspondences [44], motivating the need for data-driven methods leveraging information beyond patch-level. Among them, end-to-end learning-based pipelines such as LIFT [65] propose to jointly learn the detection and description stages. Methods like LF-Net [34] or SuperPoint [13] learn detection and description in a self-supervised way, using spatial augmentation of images through affine transformations. With D2-Net [15], Dusmanu *et al.* showed that a single-branch CNN could both perform detection and description, in a paradigm referred to as *detect-and-describe*. Their network is trained in a supervised way with a contrastive loss on the deep local features, using ground-truth pixel-level correspondences provided by Structure-from-Motion reconstructions [26]. R2D2 [37] builds on the same paradigm and formulates the learning of keypoint reliability and repeatability together with the detection and description, using this time a listwise ranking loss.

In order to preserve the accuracy of their correspondences, both D2-Net [15] and R2D2 [37] use dilated convolutions. Still when looking at feature-matching benchmarks like HPatches [5], the mean matching accuracy at error thresholds of one or two pixels is quite low. This indicates that their detection stage is often off by a couple of pixels. As shown in Fig. 1, these errors have direct repercussions on the subsequent localization or reconstruction algorithms.

Dense-to-Dense Feature Matching. Dense-to-dense matching approaches get rid of the detection stage altogether by finding mutual nearest neighbors in dense feature maps. This can be done using densely extracted features from a pre-trained CNN, combined with guided matching from late layers to earlier ones [56]. NCNet [38] trains a CNN to search in the 4D space of all possible correspondences, with the use of 4D convolutions. While they can be trained with weak supervision, dense-to-dense approaches carry high computational cost and memory consumption which make them hardly scalable for computer vision applications. Besides, the quadratic complexity of this approach limits the resolution of the images being used, resulting in correspondences with low accuracy.

Sparse-to-Dense Feature Matching. Very recently [18] proposed to perform the detection stage asymmetrically. In such setting, correspondences are searched exhaustively in the counterpart image, by running for instance a cross-correlation operation on dense feature maps with a sparse set of local hypercolumn descriptors. While this exhaustive search used to be a costly operation, it can now be efficiently computed on GPUs, using batched 1×1 convolutions. The main appeal for this approach is that under strong visual changes, the need for repeatability in keypoint detection is alleviated, allowing each pixel to be a detection. Instead, finding corresponding keypoint locations is left to the dense descriptor map. Preliminary work on sparse-to-dense matching [18] has shown that reusing intermediate CNN representations trained with weak supervision for the task of image retrieval can lead to significant gains in performance for visual localization. However, these features are not explicitly learned for feature matching, and thus fail to give pixel-accurate correspondences. We will refer to this method as S2DHM, for Sparse-to-Dense Hypercolumn Matching, in the rest of the paper.

In this paper, we propose to explicitly learn correspondence maps for the task of pixel-level matching.

3 Method

In this section we introduce and describe our novel sparse-to-dense feature-matching pipeline, which we call S2DNet.

3.1 The Sparse-to-Dense Paradigm

Given an image pair $(\mathtt{I}_\mathrm{A}, \mathtt{I}_\mathrm{B})$, our goal is to obtain a set of 2D correspondences which we write as $\{(\mathbf{p}_\mathrm{A}^n, \mathbf{p}_\mathrm{B}^n)\}_{n=1}^N$. Let us consider the case where a feature detector (*e.g.* the SuperPoint detector [13]) has been applied on image A, producing a set of N keypoints $\{\mathbf{p}_\mathrm{A}^n\}_{n=1}^N$. In this case, the feature matching problem reduces to a *sparse-to-dense* matching problem of finding a correspondent $\mathbf{p}_\mathrm{B}^n$ in image B for each detection $\mathbf{p}_\mathrm{A}^n$. We propose to cast this correspondence learning problem as a supervised classification task by restricting the set of admissible locations to the pixel coordinates of $\mathtt{I}_\mathrm{B}$. This leads to the following categorical distribution:

$$p\left(\mathbf{p}_\mathrm{B}^n | \mathbf{p}_\mathrm{A}^n, \mathtt{I}_\mathrm{A}, \mathtt{I}_\mathrm{B}, \Theta\right) = \frac{\exp\left(\mathtt{C}_n[\mathbf{p}_\mathrm{B}^n]\right)}{\sum_{\mathbf{q}\in\Omega} \exp\left(\mathtt{C}_n[\mathbf{q}]\right)}, \tag{1}$$

where $\mathtt{C}_n$ is a correspondence map of the size of $\mathtt{I}_\mathrm{B}$ produced by our novel S2DNet matching pipeline and Ω is the set of pixel locations of $\mathtt{I}_\mathrm{B}$. S2DNet takes as input $\mathbf{p}_\mathrm{A}^n$, $\mathtt{I}_\mathrm{A}$, $\mathtt{I}_\mathrm{B}$ and its parameters Θ. Equation 1 describes the likeliness of a pixel $\mathbf{p}_\mathrm{A}^n$ in $\mathtt{I}_\mathrm{A}$ to correspond to pixel $\mathbf{p}_\mathrm{B}^n$ in $\mathtt{I}_\mathrm{B}$.

3.2 S2DNet Matching Pipeline

We introduce S2DNet, a pipeline built specifically to perform sparse-to-dense matching which we illustrate in Fig. 2. Given a pair of images $(\mathtt{I}_{\mathrm{A}}, \mathtt{I}_{\mathrm{B}})$, we apply a convolutional backbone $\mathcal{F}$ on both images using shared network weights *i.e.* $\{\mathtt{H}_{\mathrm{A}}^{m}\}_{m=1}^{M} = \mathcal{F}(\mathtt{I}_{\mathrm{A}}; \Theta)$ and $\{\mathtt{H}_{\mathrm{B}}^{m}\}_{m=1}^{M} = \mathcal{F}(\mathtt{I}_{\mathrm{B}}; \Theta)$, where $\{\mathtt{H}_{\mathrm{A}}^{m}\}_{m=1}^{M}$ and $\{\mathtt{H}_{\mathrm{B}}^{m}\}_{m=1}^{M}$ correspond to intermediate feature maps extracted at multiple levels. Θ denotes the parameters of $\mathcal{F}$. Such representations are sometimes referred to as hyper-columns [19,42]. While the earlier layers encode little semantic meaning, they preserve high-frequency local details which is crucial for retrieving accurate key-points. Conversely in the presence of max-pooling, later layers loose in resolution but benefit from a wider receptive field and thus context.

For each detected keypoint $\mathbf{p}_{\mathrm{A}}^{n}$ in $\mathtt{I}_{\mathrm{A}}$, we extract a set of sparse descriptors in the dense intermediate feature maps $\mathtt{H}_{\mathrm{A}}^{m}$ and compute the dense correspondence map $\mathtt{C}_{n}$ against $\mathtt{H}_{\mathrm{B}}^{m}$, by processing each level independently, in the following way:

$$\mathtt{C}_{n} = \sum_{m=1}^{M} \mathcal{U}\left(\mathtt{H}_{\mathrm{A}}^{m}\left[\mathbf{p}_{\mathrm{A}}^{n,m}\right] * \mathtt{H}_{\mathrm{B}}^{m}\right), \tag{2}$$

where $\mathcal{U}$ refers to the bilinear upsampling operator to $\mathtt{I}_{\mathrm{B}}$ resolution, $\mathbf{p}_{\mathrm{A}}^{n,m}$ corresponds to downscaling the 2D coordinates $\mathbf{p}_{\mathrm{A}}^{n}$ to the resolution of $\mathtt{H}_{\mathrm{A}}^{m}$, and $*$ is the 1×1 convolution operator.

3.3 Training-Time

While state of the art approaches employ either a local contrastive or a listwise ranking loss [15,34,37] to train their network, we directly optimize for the task of sparse-to-dense correspondence retrieval by maximizing the log-likelihood in Eq. (1) which results in a single multi-class cross-entropy loss. From a practical point of view, for every training sample, this corresponds to computing the softmax of the correspondence map and evaluate the cross-entropy loss using the ground truth correspondence $\mathbf{p}_{\mathrm{B}}^{n}$. This strongly penalizes wrong predictions, regardless of their closeness to the ground-truth, forces the network to generate highly localized and peaked predictions and helps computing accurate correspondences.

3.4 Test-Time

At test-time, to retrieve the correspondences in $\mathtt{I}_{\mathrm{B}}$, we proceed as follows for each detected keypoint $\mathbf{p}_{\mathrm{A}}^{n}$:

$$\mathbf{p}_{\mathrm{B}}^{n*} = \underset{\mathbf{p}_{\mathrm{B}}^{n}}{\operatorname{argmax}}\; p\left(\mathbf{p}_{\mathrm{B}}^{n} | \mathbf{p}_{\mathrm{A}}^{n}, \mathtt{I}_{\mathrm{A}}, \mathtt{I}_{\mathrm{B}}, \Theta\right) = \underset{\mathbf{p}}{\operatorname{argmax}}\; \mathtt{C}_{n}\left[\mathbf{p}\right], \tag{3}$$

where $\mathtt{C}_{n} = \mathrm{S2DNet}(\mathbf{p}_{\mathrm{A}}^{n}, \mathtt{I}_{\mathrm{A}}, \mathtt{I}_{\mathrm{B}}; \Theta)$. By default, S2DNet does not apply any type of filtering and delivers one correspondence for each detected keypoint in the source image. Since we do not explicitly deal with co-visibility issues, we filter out some ambiguous matches if the following condition is not satisfied:

$$p\left(\mathbf{p}_{\mathrm{B}}^{n*} | \mathbf{p}_{\mathrm{A}}^{n}, \mathrm{I}_{\mathrm{A}}, \mathrm{I}_{\mathrm{B}}, \Theta\right) > \tau, \tag{4}$$

where τ is a threshold between 0 and 1.

3.5 S2DNet Architecture

As [15,18], we use a VGG-16 [52] architecture as our convolutional backbone. We place our intermediate extraction points at three levels, in `conv_1_2`, `conv_3_3` and `conv_5_3`, after the ReLU activations. Note that `conv_1_2` comes before any spatial pooling layer, and thus preserves the full image resolution. To both help with the convergence and reduce the final descriptors sizes, we feed these intermediate tensors to adaptation layers. They consist of two convolutional layers and a final batch-normalization [22] activation, with an output size of 128 channels. An illustration of our architecture can be seen in the supplementary material.

3.6 Differences with Sparse-to-Dense Hypercolumn Matching [18]

Sparse-to-Dense Hypercolumn Matching (S2DHM) [18] described a weakly supervised approach to learn hypercolumn descriptors and efficiently obtain correspondences using the sparse-to-dense paradigm. In this paper, we propose a supervised alternative which aims at directly learning accurate correspondence maps. As we will show in our experiments, this leads to significantly superior results. Moreover, in its pipeline S2DHM upsamples and concatenates intermediate feature maps before computing correspondence maps. In comparison, S2DNet computes correspondence maps at multiple levels before merging the results by addition. We will later show that the latter approach is much more memory and computationally efficient.

4 Experiments

In this section, we evaluate S2DNet on several challenging benchmarks. We first evaluate our approach on a commonly used image matching benchmark, which displays changes in both viewpoint and illumination. We then evaluate the performance of S2DNet on long-term visual localization tasks, which display even more severe visual changes.

4.1 Training Data

We use the same training data as D2-Net [15] to train S2DNet, which comes from the MegaDepth dataset [26]. This dataset consists of 196 outdoor scenes and 1, 070, 568 images, for which *SfM* was run with COLMAP [48,49] to generate a sparse 3D reconstruction. A depth-check is run using the provided depth maps to remove occluded pixels. As D2-Net, we remove scenes which overlap with the PhotoTourism [1,57] test set. Compared to D2-Net and to provide strong scale changes, we train S2DNet on image pairs with an arbitrary overlap. At each training iteration, we extract random crops of size 512×512, and randomly sample a maximum of 128 pixel correspondences. We train S2DNet for 30 epochs using Adam [24]. We use an initial learning rate of 10^{-3} and apply a multiplicative decaying factor of $e^{-0.1}$ at every epoch.

4.2 Image Matching

We first evaluate our method on the popular image matching benchmark HPatches [5]. We use the same 108 sequences of images as D2-Net [15], each sequence consisting of 6 images. These images either display changes in illumination (for 52 sequences) or changes in viewpoint (for 56 sequences). We consider the first frame of each sequence to be the reference image to be matched against every other, resulting in 540 pairs of images to match.

Evaluation Protocol. We apply the SuperPoint [13] keypoint detector on the first image of each sequence. For each subsequent pair of images, we perform sparse-to-dense matching using S2DNet (see Sect. 3.4). Additionally, we filter out correspondences which do not pass the cyclic check of matching back on their source pixel, which is equivalent to performing a mutual nearest-neighbor verification as it is done with D2-Net [15] and R2D2 [37].

We compute the number of matches which fall under multiple reprojection error thresholds using the ground-truth homographies provided by the dataset, and report the Mean Matching Accuracy (or MMA) in Fig. 3.

We compare S2DNet to multiple sparse-to-sparse matching baselines. We report the performance of RootSIFT [3,35] with a Hessian Affine detector [29] (Hes.det. + RootSIFT), HardNet++ [31] coupled with HesAffNet regions [32] (HAN + HN++), DELF [33], LF-Net [34], SuperPoint [13], D2-Net [15] and R2D2 [37]. We also include results from the sparse-to-dense method S2DHM [18].

Results. We find that the best results were achieved when combining SuperPoint [13] with a threshold of $\tau = 0.20$ (see Eq. 4), which are the results reported in Fig. 3. We experimentally found that above this threshold, some sequences obtain very few to no correspondence at all, which biases the results. We show that overall our method outperforms every baselines at any reprojection threshold. The gain in performance is particularly noticeable at thresholds of 1 and 2 pixels, indicating the correspondences we predict tend to be much more accurate. DELF [33] achieves competitive results under changes in illumination, which can be explained by the fact that keypoints are sampled on a fixed grid and that the images undergo no changes in viewpoint. On the other hand, it performs poorly under viewpoint changes.

Keypoint Detector Influence. We run an ablation study to evaluate the impact of different feature detectors, confidence thresholds as well as using a sparse-to-sparse approach, and report the results in Table 1 (left). We find that

S2DNet tends to work best when combined with SuperPoint [13]. We also experimentally find $\tau = 0.2$ to be a good compromise of correspondence rejection while also maintaining a high number of matches.

Sparse-to-Sparse vs. Sparse-to-Dense. We find that using S2DNet in a sparse-to-sparse setting (*i.e.* applying a detector on the image undergoing illumination or viewpoint changes) damages the results (see Table 1, left). This phenomenon translates the errors made by keypoint detectors, and motivates the sparse-to-dense setting. S2DNet efficiently leverages this paradigm and can find corresponding keypoints that would not have been detected otherwise. Conversely, we study the impact of using sparse-to-sparse learning-based methods D2-Net [15] and R2D2 [37] in a sparse-to-dense setting (see Table 1, right). In this setting, we define S2DNet sparse descriptors as the concatenated multi-scale feature vectors at the corresponding pixel locations. We find that using the sparse-to-dense paradigm systematically improves their performance under illumination changes, where images are aligned. This suggests that their descriptor maps are robust to illumination perturbations. On the other hand, performance is damaged for both methods under viewpoint changes, suggesting that their descriptor maps are not highly localized and discriminative. Concerning S2DHM, which was trained in a weakly supervised manner, running it in a sparse-to-sparse setting improves the accuracy. This highlights the importance of our main contribution, *i.e.* casting the sparse-to-dense matching problem as a supervised classification task.

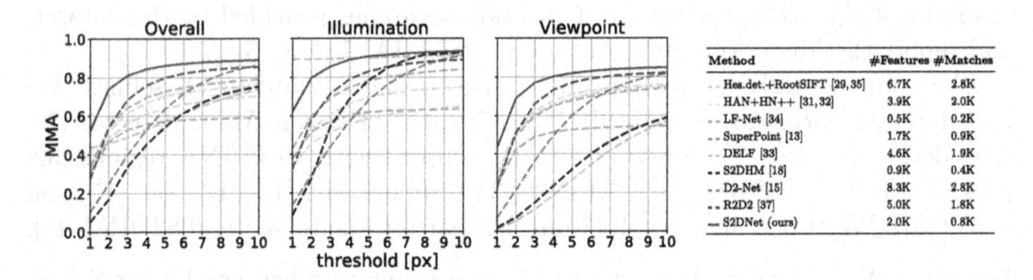

Method	#Features	#Matches
Hes.det.+RootSIFT [29,35]	6.7K	2.8K
HAN+HN++ [31,32]	3.9K	2.0K
LF-Net [34]	0.5K	0.2K
SuperPoint [13]	1.7K	0.9K
DELF [33]	4.6K	1.9K
S2DHM [18]	0.9K	0.4K
D2-Net [15]	8.3K	2.8K
R2D2 [37]	5.0K	1.8K
S2DNet (ours)	2.0K	0.8K

Fig. 3. HPatches Mean Matching Accuracy (MMA) comparison. We report in this table the best results for S2DNet, obtained when combined with SuperPoint detections. S2DNet outperforms all other baselines, especially at thresholds of one or two pixels. This study highlights the power of working in a sparse-to-dense setting, where every pixel in the target image becomes a candidate keypoint.

4.3 Long-Term Visual Localization

We showed that S2DNet provides correspondences which are overall more accurate than other baselines. We will now study its impact for the task of visual localization under challenging conditions. We report visual localization results under day-night changes and complex indoor scenes.

Table 1. Ablation study on HPatches. In **(a)**, we evaluate the performance of several detectors in both a sparse-to-dense (S2S) and sparse-to-sparse (S2S) setting using S2DNet descriptors. We find that S2DNet works best in the S2D setting, coupled with SuperPoint (SP) [13] detections, and a confidence threshold of $\tau = 0.20$. In **(b)**, we study the impact of using sparse-to-sparse learning-based methods in a sparse-to-dense setting. Results lead to the conclusion that D2-Net [15] and R2D2 [37] descriptor maps are robust to illumination changes but not highly discriminative locally.

Detector	Matching	τ	MMA@1	MMA@2	MMA@3	MMA@10
Harris [20]	S2D	0.20	**0.511**	**0.733**	**0.805**	**0.888**
	S2D	0.0	0.441	0.626	0.690	0.787
	S2S	-	0.278	0.464	0.565	0.763
SURF [7]	S2D	0.20	**0.511**	**0.742**	**0.823**	**0.902**
	S2D	0.0	0.436	0.639	0.718	0.828
	S2S	-	0.302	0.506	0.619	0.829
SIFT [27]	S2D	0.20	**0.487**	**0.700**	**0.771**	**0.851**
	S2D	0.0	0.441	0.626	0.690	0.787
	S2D	-	0.386	0.559	0.642	0.818
SuperPoint [13]	S2D	0.20	**0.563**	**0.747**	**0.815**	**0.895**
	S2D	0.0	0.469	0.623	0.686	0.788
	S2S	-	0.373	0.599	0.709	0.847
D2-Net [15]	S2D	0.20	**0.467**	**0.716**	**0.805**	**0.911**
	S2D	0.0	0.330	0.522	0.604	0.764
	S2S	-	0.118	0.285	0.425	0.777
R2D2 [37]	S2D	0.20	**0.478**	**0.715**	**0.799**	**0.901**
	S2D	0.0	0.341	0.522	0.598	0.746
	S2S	-	0.316	0.546	0.652	0.819

(a) S2S vs. S2D - S2DNet descriptors

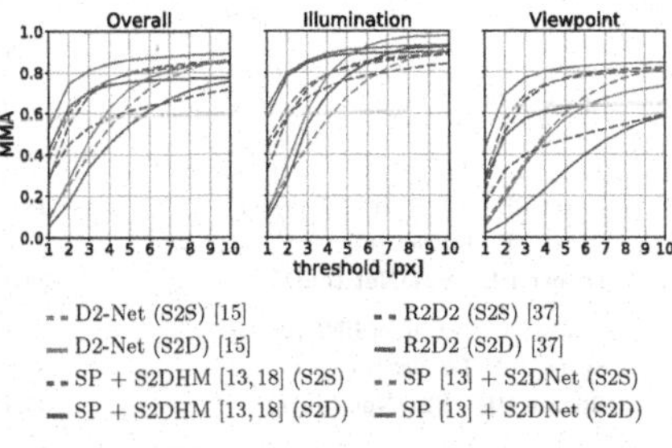

(b) S2S vs. S2D - Other descriptors

Datasets. We evaluate our approach on two challenging outdoor localization datasets which feature day-to-night changes, and one indoor dataset. The first dataset is Aachen Day-Night [44,46]. It features 4,328 daytime reference images taken with a handheld smartphone, for which ground truth camera poses are provided. The dataset also provides a 3D reconstruction of the scene [44], built using SIFT [27] features and *SfM*. The evaluation is done on 824 daytime and 98 nighttime images taken in the same environment. The second dataset is RobotCar Seasons [28]. It features 6,954 daytime reference images taken with a rear-facing camera mounted on a car driving through Oxford. Similarly, ground truth camera poses and a sparse 3D model of the world is provided [44] and we localize 3,978 images captured throughout a year. These images do not only exhibit nighttime conditions, but also cross-seasonal evolutions such as snow or rain. Lastly, we evaluate our pipeline on the challenging InLoc [56,63] dataset. This indoor dataset is difficult because of its large scale, illumination and long-term changes as well as the presence of repetitive patterns such as corridors (see Fig. 4). It contains 9,972 database and 356 high-resolution query images, as well as dense depth maps which can be used to perform dense pose verification. We report for each datasets the pose recall at three position and orientation thresholds for daytime and nighttime query images, as per [44].

Indoor Localization. The InLoc [56] localization benchmark comes with a predefined code base and several pipelines for localization. The first one is called Direct Pose Estimation (Direct PE) and performs hierarchical localization using the set of top-ranked database images obtained using image retrieval, followed by P3P-LO-RANSAC [16,25]. The second variant applies an intermediate spatial

Table 2. InLoc [56] (left) and Local Features Benchmark [44] (right) results. We report localization recalls in percent, for three translation and orientation thresholds. On InLoc, S2DNet outperforms both baselines at the finest threshold for the sparse categories. We also include Dense PE baseline results for reference. R2D2 authors did not provide results on this benchmark. On the local features benchmark (a predefined localization pipeline), S2DNet achieves state-of-the-art results at the medium precision threshold. Due to the relatively small number of query images however, recent methods like D2-Net and R2D2 are saturating around the same performance. *Note that R2D2 was trained on Aachen database images.

Method	*InLoc(fixed pipeline)* Threshold Accuracy		
	0.25m / 2°	0.5m / 5°	5m / 10°
Direct PE - Aff. RootSIFT [29,35]	18.5	26.4	30.4
Direct PE - D2-Net [15]	27.7	40.4	**48.6**
Direct PE - S2DNet (ours)	**29.3**	**40.9**	48.5
Sparse PE - Aff. RootSIFT [29,35]	21.3	32.2	44.1
Sparse PE - D2-Net [15]	35.0	48.6	62.6
Sparse PE - S2DNet (ours)	**35.9**	**49.0**	**63.1**
Sparse PE + Dense PV - Aff. RootSIFT [29,35]	29.5	42.6	54.5
Sparse PE + Dense PV - D2-Net [15]	38.0	**56.5**	65.4
Sparse PE + Dense PV - S2DNet (ours)	**39.4**	53.5	**67.2**
Dense PE + Dense PV - InLoc [56]	38.9	56.5	69.9

Method	*Aachen Day-Night (fixed pipeline)* Threshold Accuracy		
	0.25m 2°	0.5m 5°	5m 10°
RootSIFT [35]	3.7	52.0	65.3
HAN+HN [32]	37.8	54.1	75.5
SuperPoint [13]	42.8	57.1	75.5
DELF [33]	39.8	61.2	85.7
D2-Net [15]	44.9	66.3	**88.8**
R2D2 [37]*	**45.9**	66.3	**88.8**
S2DNet (ours)	**45.9**	**68.4**	**88.8**

verification step [36] to reject outliers, referred to as (Sparse PE). On top of this second variant, Dense Pose Verification (Dense PV) can be applied to re-rank pose candidates by using densely extracted RootSIFT [35] features. In each variant, we use S2DNet to generate 2D-2D correspondences between queries and database images, which are then converted to 2D-3D correspondences using the provided dense depth maps. We use a SuperPoint [13] detector and mutual nearest-neighbour filtering.

InLoc localization results are reported in Table 2. We compare our approach to the original InLoc baseline which uses affine covariant [29] detections and RootSIFT [35] descriptors, as well as results provided by D2-Net [15]. We find that S2DNet outperforms both sparse baselines at the finest threshold, and is on par with other methods at the medium and coarse thresholds. In the sparse setting, best results are achieved when combined with geometrical and dense pose verification (Sparse PE + Dense PV). In addition we include localization results that were computed by the benchmark authors using dense-to-dense feature matching (Dense PE). Due to the nature of our pipeline and the very high memory and computational consumption of this variant, we choose to limit our study to sparse correspondence methods. It is interesting to note however that S2DNet outperforms the original (Dense PE + Dense PV) InLoc baseline at the finest precision threshold, using a much lighter computation.

Day-Night Localization. We report day-night localization results with S2DNet using two localization protocols. Localization results reported in Table 2 show that S2DNet achieves state-of-the-art results, outperforming all other

Table 3. Localization results. We report localization recalls in percent, for three translation and orientation thresholds (*high*, *medium*, and *coarse*) as in [44]. We put in bold the **best** and underline the <u>second-best</u> performances for each threshold. S2DNet outperforms every baseline in nighttime conditions, except at the finest threshold of RobotCar Seasons. This can be explained by the extreme visual changes and blurriness that these images undergo. At daytime, S2DNet performance is on par with D2-Net [15]. *Note that S2DHM [18] was trained directly on RobotCar sequences, which explains the high nighttime performance. SMC [60] also uses additional semantic data and assumptions. R2D2 [37] authors did not provide localization results on these benchmarks.

		RobotCar Seasons						*Aachen Day-Night*					
		Day-All			*Night-All*			*Day*			*Night*		
	Method	Threshold Accuracy			Threshold Accuracy			Threshold Accuracy			Threshold Accuracy		
		0.25m 2°	0.5m 5°	5m 10°	0.25m 2°	0.5m 5°	5m 10°	0.25m 2°	0.5m 5°	5m 10°	0.25m 2°	0.5m 5°	5m 10°
Structure -based	CSL [53]	45.3	73.5	90.1	0.6	2.6	7.2	52.3	80.0	94.3	24.5	33.7	49.0
	AS [43]	35.6	67.9	90.4	0.9	2.1	4.3	57.3	83.7	<u>96.6</u>	19.4	30.6	43.9
	SMC [60] *	50.3	79.3	95.2	7.1	22.4	45.3	-	-	-	-	-	-
Retrieval -based	FAB-MAP [12]	2.7	11.8	37.3	0.0	0.0	0.0	0.0	0.0	4.6	0.0	0.0	0.0
	NetVLAD [2]	6.4	26.3	90.9	0.3	2.3	15.9	0.0	0.2	18.9	0.0	2.0	12.2
	DenseVLAD [61]	7.6	31.2	91.2	1.0	4.4	22.7	0.0	0.1	22.8	0.0	2.0	14.3
Hierar -chical	HF-Net [40]	53.0	79.3	95.0	5.9	17.1	29.4	79.9	88.0	93.4	40.8	56.1	74.5
	S2DHM [18] *	45.7	78.0	95.1	22.3	61.8	94.5	56.3	72.9	90.9	30.6	56.1	78.6
	D2-Net [15]	**54.5**	<u>80.0</u>	<u>95.3</u>	**20.4**	<u>40.1</u>	<u>55.0</u>	**84.8**	**92.6**	**97.5**	<u>43.9</u>	<u>66.3</u>	<u>85.7</u>
	S2DNet (ours)	<u>53.9</u>	**80.6**	**95.8**	<u>14.5</u>	**40.2**	**69.7**	<u>84.3</u>	<u>90.9</u>	95.9	**46.9**	**69.4**	**86.7**

methods at the medium precision threshold. It is important to note that R2D2 was finetuned on Aachen database images.

We then report in Table 3 localization results using a hierarchical approach, similar to [18,40,44]. Contrary to Table 2, these results do not allow to compare the keypoint matching approaches alone since localization pipelines are different. Even the comparison with D2Net is difficult to interpret since their full localization pipeline was not released. Still, S2DNet achieves state-of-the-art results in Aachen nighttime images, and outperforms all baselines that were not trained on RobotCar nighttime images at medium and coarse precision thresholds. At daytime, where detecting repeatable and accurate keypoints is easier, S2DNet is on par with other learning-based methods. At the finest nighttime RobotCar threshold it is likely that S2DNet features struggle to compute accurate correspondences, which can be explained by the extreme visual changes these images undergo (see Fig. 4). Overall, this study shows that S2DNet achieves better performance in particularly challenging conditions such as nighttime, compared to other sparse-to-sparse alternatives.

Table 4. Computational Time Study for visual localization. We compare the time performance of our method against other learning-based approaches in a visual localization scenario. For a given query image $\mathbf{I}_q$ and N reference images of size 1200×1600 with K detections each, we report the average measured time to perform image matching against each of them. In this standard setting, keypoint locations and descriptors have already been extracted offline from the reference images.

Method	Network backbone	Descriptor size	Forward pass on $\mathbf{I}_q$	Detection step on $\mathbf{I}_q$	Matching per keypoint	Total online computational time		
			t_A	t_B	t_C	$t_A + t_B + N \times K \times t_C$		
						$N = 1$ $K = 1000$	$N = 5$ $K = 1000$	$N = 15$ $K = 1000$
D2-Net [15]	VGG-16	512	17.8 ms	5.474 s	0.4 µs	5.492 s	5.494 s	5.498 s
R2D2 [37]	L2-Net	128	19.1 ms	479.6 ms	0.2 µs	0.499 s	**0.499 s**	**0.501 s**
S2DHM [18]	VGG-16	2048	326 ms	–	0.33 ms	0.656 s	1.976 s	5.276 s
S2DNet	VGG-16 + adap.	3 × 128	28.2 ms	–	0.31 ms	**0.338 s**	1.578 s	4.678 s

5 Discussion

Runtime Performance. To compare S2DNet against state-of-the-art approaches, we time its performance for the scenario of visual localization. We run our experiments on a machine equipped with an Intel(R) Xeon(R) E5-2630 CPU at 2.20 GHz, and an NVIDIA GeForce GTX 1080Ti GPU. We report the results in Table 4. In a localization setting, we consider the keypoint detection and description step to be pre-computed offline for reference images. Thus for an incoming query image, only sparse-to-sparse methods need to perform the keypoint detection and descriptor extraction step. We find this very step to be the bottleneck of learning-based methods like D2-Net [15] or R2D2 [37]. Indeed, these methods are slowed down by the non-maxima suppression operations, which are in addition run on images of multiple scales. For S2DHM [18] and S2DNet, no keypoint detection is performed on the incoming query image and most of the computation lies in the keypoint matching step. As expected however, the matching step is much more costly for these sparse-to-dense methods. Still, for 1000 detections and 1 retrieved image, S2DNet is the fastest method while for 15 retrieved image, it is on par with D2-Net.

Current Limitations of the Sparse-to-Dense Paradigm. One limitation of our current sparse-to-dense matching formulation appears for the task of multi-view 3D reconstruction. Indeed, the standard approach to obtain features tracks consists in 1) detecting and describing keypoints in each image, 2) matching pairs of images using the previously extracted keypoints descriptors and 3) creating tracks from these matches. In our S2D matching paradigm, every pixel becomes a detection candidate which is not compatible with the standard 3D reconstruction pipeline previously described. This limitation opens novel directions of research for rethinking the standard tracks creation pipeline and enabling the use of S2D matching in 3D reconstruction frameworks.

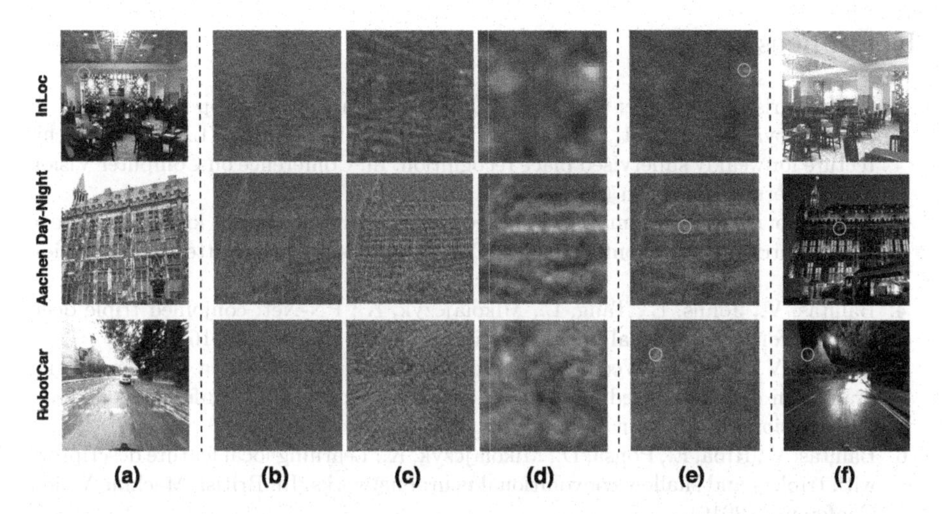

Fig. 4. Correspondence maps examples. From left to right: Reference image with a keypoint detection (a), intermediate correspondence maps predicted by S2DNet (b, c, d), aggregated pre-softmax correspondence map (e) and retrieved correspondent in the query image (f). From top to bottom: images from InLoc [56], Aachen Day-Night [46] and RobotCar Seasons [28].

Compatibility with Learning-Based Matchers. Learning-based matching methods like NG-RANSAC [8], OANet [66] or SuperGlue [41] process putative correspondences to return inlier confidence scores. These methods could easily be work as a post-processing step of S2DNet, to further improve matching results.

6 Conclusion

In this paper we presented S2DNet, a new sparse-to-dense learning-based keypoint matching architecture. In contrast to other sparse-to-sparse methods we showed that this novel pipeline achieves superior performance in terms of accuracy, which helps improve subsequent long-term visual localization tasks. Under visually challenging conditions, S2DNet reaches state-of-the-art performance for image matching and localization, and advocates for the development of sparse-to-dense methods.

Acknowledgement. This project has received funding from the Bosch Research Foundation (*Bosch Forschungsstiftung*).

References

1. Phototourism Challenge: CVPR 2019 Image Matching Workshop (2019)
2. Arandjelovic, R., Gronát, P., Torii, A., Pajdla, T., Sivic, J.: NetVLAD: CNN architecture for weakly supervised place recognition. In: Conference on Computer Vision and Pattern Recognition (2016)
3. Arandjelovic, R., Zisserman, A.: Three things everyone should know to improve object retrieval. In: Conference on Computer Vision and Pattern Recognition (2012)
4. Balntas, V., Johns, E., Tang, L., Mikolajczyk, K.: PN-Net: conjoined triple deep network for learning local image descriptors. arXiv Preprint (2016)
5. Balntas, V., Lenc, K., Vedaldi, A., Mikolajczyk, K.: HPatches: a benchmark and evaluation of handcrafted and learned local descriptors. In: Conference on Computer Vision and Pattern Recognition (2017)
6. Balntas, V., Riba, E., Ponsa, D., Mikolajczyk, K.: Learning local feature descriptors with triplets and shallow convolutional neural networks. In: British Machine Vision Conference (2016)
7. Bay, H., Tuytelaars, T., Gool, L.V.: SURF: speeded up robust features. In: European Conference on Computer Vision (2006)
8. Brachmann, E., Rother, C.: Neural-guided RANSAC: learning where to sample model hypotheses. In: ICCV (2019)
9. Brown, M.A., Hua, G., Winder, S.A.J.: Discriminative learning of local image descriptors. IEEE Trans. Pattern Anal. Mach. Intell. **33**, 43–57 (2011)
10. Calonder, M., Lepetit, V., Strecha, C., Fua, P.: BRIEF: binary robust independent elementary features. In: European Conference on Computer Vision (2010)
11. Choy, C.B., Gwak, J., Savarese, S., Chandraker, M.: Universal correspondence network. In: Advances in Neural Information Processing Systems, vol. 30 (2016)
12. Cummins, M.J., Newman, P.: FAB-MAP: probabilistic localization and mapping in the space of appearance. Int. J. Robot. Res. **27**, 647–665 (2008)
13. DeTone, D., Malisiewicz, T., Rabinovich, A.: SuperPoint: self-supervised interest point detection and description. In: CVPR Workshop (2018)
14. Dong, J., Soatto, S.: Domain-size pooling in local descriptors: DSP-SIFT. In: Conference on Computer Vision and Pattern Recognition (2014)
15. Dusmanu, M., et al.: D2-Net: a trainable CNN for joint description and detection of local features. In: Conference on Computer Vision and Pattern Recognition (2019)
16. Fischler, M.A., Bolles, R.C.: Random sample consensus: a paradigm for model fitting with applications to image analysis and automated cartography. Commun. ACM **24**, 381–395 (1981)
17. Gauglitz, S., Höllerer, T., Turk, M.: Evaluation of interest point detectors and feature descriptors for visual tracking. Int. J. Comput. Vision **94**, 335 (2011)
18. Germain, H., Bourmaud, G., Lepetit, V.: Sparse-to-dense hypercolumn matching for long-term visual localization. In: International Conference on 3D Vision (2019)
19. Hariharan, B., Arbeláez, P.A., Girshick, R.B., Malik, J.: Hypercolumns for object segmentation and fine-grained localization. In: Conference on Computer Vision and Pattern Recognition (2014)
20. Harris, C., Stephens, M.: A combined corner and edge detector. In: Proceedings of Fourth Alvey Vision Conference (1988)
21. Heinly, J., Schönberger, J.L., Dunn, E., Frahm, J.M.: Reconstructing the world* in six days *(as captured by the Yahoo 100 million image dataset). In: Conference on Computer Vision and Pattern Recognition (2015)

22. Ioffe, S., Szegedy, C.: Batch normalization: accelerating deep network training by reducing internal covariate shift. arXiv Preprint (2015)
23. Kim, H., Lee, D., Sim, J., Kim, C.: SOWP: spatially ordered and weighted patch descriptor for visual tracking. In: International Conference on Computer Vision (2015)
24. Kingma, D.P., Ba, J.: Adam: a method for stochastic optimization. CoRR abs/1412.6980 (2014)
25. Lebeda, K., Matas, J.E.S., Chum, O.: Fixing the locally optimized RANSAC. In: British Machine Vision Conference (2012)
26. Li, Z., Snavely, N.: MegaDepth: learning single-view depth prediction from internet photos. In: Conference on Computer Vision and Pattern Recognition (2018)
27. Lowe, D.G.: Distinctive image features from scale-invariant keypoints. Int. J. Comput. Vision **60**(2), 91–110 (2004)
28. Maddern, W., Pascoe, G., Linegar, C., Newman, P.: 1 year, 1000 km: the Oxford RobotCar dataset. Int. J. Robot. Res. **36**, 3–15 (2017)
29. Mikolajczyk, K., Schmid, C.: Scale & affine invariant interest point detectors. Int. J. Comput. Vision **60**, 63–86 (2004)
30. Mikolajczyk, K., et al.: A comparison of affine region detectors. Int. J. Comput. Vision **65**, 43–72 (2005)
31. Mishchuk, A., Mishkin, D., Radenović, F., Matas, J.: Working hard to know your neighbor's margins: local descriptor learning loss. In: Advances in Neural Information Processing Systems (2017)
32. Mishkin, D., Radenović, F., Matas, J.: Repeatability is not enough: learning affine regions via discriminability. In: Ferrari, V., Hebert, M., Sminchisescu, C., Weiss, Y. (eds.) ECCV 2018. LNCS, vol. 11213, pp. 287–304. Springer, Cham (2018). https://doi.org/10.1007/978-3-030-01240-3_18
33. Noh, H., Araujo, A., Sim, J., Weyand, T., Han, B.: Large-scale image retrieval with attentive deep local features. In: International Conference on Computer Vision (2016)
34. Ono, Y., Trulls, E., Fua, P., Yi, K.M.: LF-Net: learning local features from images. In: NeurIPS (2018)
35. Perdoch, M., Chum, O., Matas, J.: efficient representation of local geometry for large scale object retrieval. In: Conference on Computer Vision and Pattern Recognition (2009)
36. Philbin, J., Chum, O., Isard, M., Sivic, J., Zisserman, A.: Object retrieval with large vocabularies and fast spatial matching. In: Conference on Computer Vision and Pattern Recognition (2007)
37. Revaud, J., et al.: R2D2: repeatable and reliable detector and descriptor. In: Advances in Neural Information Processing Systems (2019)
38. Rocco, I., Cimpoi, M., Arandjelović, R., Torii, A., Pajdla, T., Sivic, J.: Neighbourhood consensus networks. In: Advances in Neural Information Processing Systems (2018)
39. Rublee, E., Rabaud, V., Konolige, K., Bradski, G.R.: ORB: an efficient alternative to SIFT or SURF. In: International Conference on Computer Vision (2011)
40. Sarlin, P.E., Cadena, C., Siegwart, R., Dymczyk, M.: From coarse to fine: robust hierarchical localization at large scale. In: Conference on Computer Vision and Pattern Recognition (2019)
41. Sarlin, P.E., DeTone, D., Malisiewicz, T., Rabinovich, A.: SuperGlue: learning feature matching with graph neural networks. In: CVPR (2020). https://arxiv.org/abs/1911.11763

42. Sattler, T., Havlena, M., Radenovic, F., Schindler, K., Pollefeys, M.: Hyperpoints and fine vocabularies for large-scale location recognition. In: International Conference on Computer Vision (2015)
43. Sattler, T., Leibe, B., Kobbelt, L.: Efficient & effective prioritized matching for large-scale image-based localization. IEEE Trans. Pattern Anal. Mach. Intell. **39**, 1744–1756 (2017)
44. Sattler, T., et al.: Benchmarking 6DOF outdoor visual localization in changing conditions. In: Conference on Computer Vision and Pattern Recognition (2018)
45. Sattler, T., et al.: Are large-scale 3D models really necessary for accurate visual localization? In: Conference on Computer Vision and Pattern Recognition (2017)
46. Sattler, T., Weyand, T., Leibe, B., Kobbelt, L.: Image retrieval for image-based localization revisited. In: British Machine Vision Conference (2012)
47. Savinov, N., Seki, A., Ladicky, L., Sattler, T., Pollefeys, M.: Quad-networks: unsupervised learning to rank for interest point detection. In: Conference on Computer Vision and Pattern Recognition (2016)
48. Schönberger, J.L., Frahm, J.M.: Structure-from-motion revisited. In: Conference on Computer Vision and Pattern Recognition (2016)
49. Schönberger, J.L., Zheng, E., Pollefeys, M., Frahm, J.M.: Pixelwise view selection for unstructured multi-view stereo. In: European Conference on Computer Vision (2016)
50. Simo-Serra, E., Trulls, E., Ferraz, L., Kokkinos, I., Fua, P., Moreno-Noguer, F.: Discriminative learning of deep convolutional feature point descriptors. In: International Conference on Computer Vision (2015)
51. Simonyan, K., Vedaldi, A., Zisserman, A.: Learning local feature descriptors using convex optimisation. IEEE Trans. Pattern Anal. Mach. Intell. **36**, 1573–1585 (2014)
52. Simonyan, K., Zisserman, A.: Very deep convolutional networks for large-scale image recognition. CoRR abs/1409.1556 (2014)
53. Svärm, L., Enqvist, O., Kahl, F., Oskarsson, M.: City-scale localization for cameras with known vertical direction. IEEE Trans. Pattern Anal. Mach. Intell. **39**(7), 1455–1461 (2017)
54. Svärm, L., Enqvist, O., Oskarsson, M., Kahl, F.: Accurate localization and pose estimation for large 3D models. In: Conference on Computer Vision and Pattern Recognition (2014)
55. Sweeney, C., Fragoso, V., Höllerer, T., Turk, M.: Large scale SfM with the distributed camera model. In: International Conference on 3D Vision (2016)
56. Taira, H., et al.: InLoc: indoor visual localization with dense matching and view synthesis. CoRR abs/1803.10368 (2018)
57. Thomee, B., et al.: YFCC100M: the new data in multimedia research. Commun. ACM **59**, 64–73 (2016)
58. Tian, Y., Fan, B., Wu, F.: L2-Net: deep learning of discriminative patch descriptor in euclidean space. In: Conference on Computer Vision and Pattern Recognition (2017)
59. Tian, Y., Yu, X., Fan, B., Wu, F., Heijnen, H., Balntas, V.: SOSNet: second order similarity regularization for local descriptor learning. In: Conference on Computer Vision and Pattern Recognition (2019)
60. Toft, C., et al.: Semantic match consistency for long-term visual localization. In: European Conference on Computer Vision (2018)
61. Torii, A., Arandjelovic, R., Sivic, J., Okutomi, M., Pajdla, T.: 24/7 place recognition by view synthesis. IEEE Trans. Pattern Anal. Mach. Intell. **40** (2015)
62. Tuytelaars, T., Mikolajczyk, K.: Local invariant feature detectors: a survey. Found. Trends Comput. Graph. Vision **3** (2007)

63. Wijmans, E., Furukawa, Y.: Exploiting 2D floorplan for building-scale panorama RGBD alignment. In: Conference on Computer Vision and Pattern Recognition (2016)
64. Yang, H., Shao, L., Zheng, F., Wang, L., Song, Z.: Recent advances and trends in visual tracking: a review. Neurocomputing **74**, 3823–3831 (2011)
65. Yi, K.M., Trulls, E., Lepetit, V., Fua, P.: LIFT: learned invariant feature transform. In: European Conference on Computer Vision (2016)
66. Yi, K.M., Trulls, E., Ono, Y., Lepetit, V., Salzmann, M., Fua, P.: Learning to find good correspondences. In: Proceedings of the IEEE Conference on Computer Vision and Pattern Recognition (2018)

RTM3D: Real-Time Monocular 3D Detection from Object Keypoints for Autonomous Driving

Peixuan Li[1,2,3,4,5], Huaici Zhao[1,2,4,5](✉), Pengfei Liu[1,2,3,4,5], and Feidao Cao[1,2,3,4,5]

[1] Shenyang Institute of Automation, Chinese Academy of Sciences, Shenyang, China
hczhao@sia.cn
[2] Institutes for Robotics and Intelligent Manufacturing, Chinese Academy of Sciences, Shenyang, China
[3] University of Chinese Academy of Sciences, Beijing, China
[4] Key Laboratory of Opto-Electronic Information Processing, Chinese Academy of Sciences, Shenyang, China
[5] Key Lab of Image Understanding and Computer Vision, Shenyang, Liaoning, China

Abstract. In this work, we propose an efficient and accurate monocular 3D detection framework in single shot. Most successful 3D detectors take the projection constraint from the 3D bounding box to the 2D box as an important component. Four edges of a 2D box provide only four constraints and the performance deteriorates dramatically with the small error of the 2D detector. Different from these approaches, our method predicts the nine perspective keypoints of a 3D bounding box in image space, and then utilize the geometric relationship of 3D and 2D perspectives to recover the dimension, location, and orientation in 3D space. In this method, the properties of the object can be predicted stably even when the estimation of keypoints is very noisy, which enables us to obtain fast detection speed with a small architecture. Training our method only uses the 3D properties of the object without any extra annotations, category-specific 3D shape priors, or depth maps. Our method is the first real-time system (FPS > 24) for monocular image 3D detection while achieves state-of-the-art performance on the KITTI benchmark.

Keywords: Real-time monocular 3D detection · Autonomous driving · Keypoint detection

1 Introduction

3D object detection is an essential component of scene perception and motion prediction in autonomous driving [2,9]. Currently, most powerful 3D detectors

---

Electronic supplementary material The online version of this chapter (https://doi.org/10.1007/978-3-030-58580-8_38) contains supplementary material, which is available to authorized users.

A. Vedaldi et al. (Eds.): ECCV 2020, LNCS 12348, pp. 644–660, 2020.
https://doi.org/10.1007/978-3-030-58580-8_38

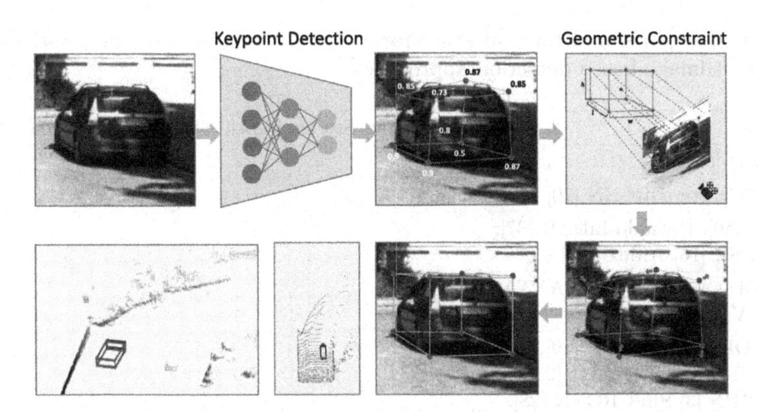

Fig. 1. Overview of proposed method: We first predict ordinal keypoints projected in the image space by eight vertexes and a central point of a 3D object. We then reformulate the estimation of the 3D bounding box as the problem of minimizing the energy function by using geometric constraints of perspective projection.

heavily rely on 3D LiDAR laser scanners for the reason that it can provide scene locations [8,29,41,46]. However, the LiDAR-based systems are expensive and not conducive to embedding into the current vehicle shape. In comparison, monocular camera devices are cheaper and convenient which makes it drawing an increasing attention in many application scenarios [6,26,40]. In this paper, the scope of our research lies in 3D object detection from only monocular RGB image.

Monocular 3D object detection methods can be roughly divided into two categories by the type of training data: one imposes complex features, such as instance segmentation, category-specific shape prior and even depth map to select best proposals in multi-stage fusion module [6,7,40]. These features require additional annotation work to train some stand-alone networks which will consume plenty of computing resources in the training and inferring stages. Another one only employs 2D bounding box and properties of a 3D object as the supervised data [3,20,33,42]. In this case, the most straightforward way build a deep regression network to directly predict the 3D information of the object, which caused the performance bottlenecks due to the large search space. To address this challenge, recent works have clearly pointed out that apply geometric constraints from 3D box vertexes to 2D box edges to refine or directly predict object parameters [3,20,23,26,28]. However, four edges of a 2D bounding box provide only four constraints on recovering a 3D bounding box while each vertex of a 3D bounding box might correspond to any edges in the 2D box, which will takes 4,096 of the same calculations to get one result [26]. Meanwhile, the strong reliance on the 2D box causes a sharp decline in 3D detection performance when predictions of 2D detectors even have a slight error. Therefore, most of these methods take advantage of two-stage detectors [10,11,32] to ensure the accuracy of 2D box prediction, which limit the upper-bound of the detection speed.

In this paper, we propose an efficient and accurate monocular 3D detection framework in the form of one-stage, which be tailored for 3D detection without

Table 1. Comparison of the real-time status and the requirements of additional data in different image-based detection approaches.

Method	Real time	Stereo	Depth	Shape/CAD	Segmentation
Mono3D [6]					✓
3DOP [7], stereoRCNN [20]		✓			
MF3D [40], Peseudo-LiDAR [37], MonoPSR [16] AM3D [25]			✓		
Mono3D++ [13], Deep-MANTA [5], 3DVP [38]				✓	
Deep3DBox [26], GS3D [19], MonoGRNet [31], FQNet [23], M3D-RPN [3] Shift-RCNN [28], MonoDIS [34]					
Ours (RTM3D)	✓				

relying on extra annotations, category-specific 3D shape priors, or depth maps. The framework can be divided into two main parts, as shown in Fig. 1. First, we perform a one-stage fully convolutional architecture to efficiently predict 9 of the 2D keypoints which are projected points from 8 vertexes and central point of 3D bounding box. This 9 keypoints provides 18 geometric constrains on the 3D bounding box. Inspired by CenterNet [45], we model the relationship between the eight vertexes and the central point to solve the keypoints grouping and the vertexes order problem. The SIFT, SUFT and other traditional keypoint detection methods [1,24]computed an image pyramid to solve the scale-invariant problem. A similar strategy was used by CenterNet as a post-processing step to further improve detection accuracy, which slows the inference speed. Note that the Feature Pyramid Network (FPN) [21] in 2D object detection is not applicable to the network of keypoint detection, because adjacent keypoints may overlap in the case of small-scale prediction. We propose a novel multi-scale pyramid of keypoint detection to generate a scale-space response. The final activate map of keypoints can be obtained by means of the soft-weighted pyramid. Given the 9 projected points, the next step is to minimize the reprojection error over the perspective of 3D points that parameterized by the location, dimension, and orientation of the object. We formulate the reprojection error as the form of multivariate equations in $\mathfrak{se}_3$ space, which can generate the detection results accurately and efficiently. We also discuss the effect of different prior information, such as dimension, orientation, and distance, predicted in parallel from our keypoint detection network. The prerequisite for obtaining this information is not to add too much computation so as not to affect the final detection speed. We model these priors and reprojection error term into an overall energy function in order to further improve 3D estimation.

To summarize, our main contributions are the following:

- We formulate the monocular 3D detection as the keypoint detection problem and combine the geometric constrains to generate properties of 3D objects more efficiently and accurately.

- We propose a novel one-stage and multi-scale network for 3D keypoint detection which provide the accurate project points for multi-scale object.
- We propose an overall energy function that can jointly optimize the prior and 3D object information.
- Evaluation on the KITTI benchmark, We are the first real-time 3D detection method using only images and achieves better accuracy under the same running time in comparing other competitors.

2 Related Work

Extra Data or Network for Image-based 3D Object Detection. In the last years, many studies develop the 3D detection in an image-based method for the reason that camera devices are more convenient and much cheaper. To complement the lacked depth information in image-based detection, most of the previous approaches heavily relied on the stand-alone network or additional labeling data, such as instance segmentation, stereo, wire-frame model, CAD prior, and depth, as shown in Table 1. Among them, monocular 3D detection is a more challenging task due to the difficulty of obtaining reliable 3D information from a single image. One of the first examples [6] enumerate a multitude of 3D proposals from pre-defined space where the objects may appear as the geometrical heuristics. Then it takes the other complex prior, such as shape, instance segmentation, contextual feature, to filter out dreadful proposals and scoring them by a classifier. To make up for the lack of depth, [40] embed a pre-trained stand-alone module to estimate the disparity. The disparity map concatenates the front view representation to help the 2D proposal network and the 3D detection can be boosted by fusing the extracted feature after RoI pooling and point cloud. As a followup, [25] combines the 2D detector and monocular depth estimation model to obtain the 2D box and corresponding point cloud. The final 3D box can be obtained by the regression of PointNet [30] after the aggregation of the image feature and 3D point information through attention mechanism, which achieves the best performance in the monocular image. Intuitively, these methods would certainly increase the accuracy of the detection, but the additional network and annotated data would lead to more computation and labor-intensive work.

Image-only in Monocular 3D Object Detection. Recent works have tried to fully explore the potency of RGB images for 3D detection. Most of them include geometric constraints and 2D detectors to explicitly describe the 3D information of the object. [26] uses CNN to estimate the dimension and orientation extracted feature from the 2D box, then it proposes to obtain the location of an object by using the geometric constraints of the perspective relationship between 3D points and 2D box edges. This contribution is followed by most image-based detection methods either in refinement step or as direct calculation on 3D objects [3,20]. All we know in this constraint is that certain 3D points are projected onto 2D edges, but the corresponding relationship and the exact location of the projection are not clear. Therefore, it needs to exhaustively enumerate $8^4 = 4096$ configurations to determine the final correspondence and can only provide four constraints, which is not sufficient for fully 3D representation in 9 parameters. It led to the need to estimate other prior information. Nevertheless, possible inaccuracies in the 2D bounding boxes may result in a grossly inaccurate solution with a small number of constraints. Therefore, most of these methods obtain more accurate 2D box through a two-stage detector, which is difficult to get real-time speed.

Keypoints in Monocular 3D Object Detection. It is believed that the detection accuracy of occluded and truncated objects can be improved by deducing complete shapes from vehicle keypoints [5,27,44]. They represent the regular-shape vehicles as a wire-frame template, which is obtained from a large number of CAD models. To train the keypoint detection network, they need to re-label the data set and even use depth maps to enhance the detection capability. [13] is most related to our work, which also considers the wire-frame model as prior information. Furthermore, It jointly optimizes the 2D box, 2D keypoints, 3D orientation, scale hypotheses, shape hypotheses, and depth with four different networks. This has limitations in run time. In contrast to prior work, we reformulate the 3D detection as the coarse keypoints detection task. Instead of predicting the 3D box based on an off-the-shelf 2D detectors or other data generators, we build a network to predict 9 of 2D keypoints projected by vertexes and center of 3D bounding box while minimize the reprojection error to find an optimal result.

3 Proposed Method

In this section. We first describe the overall architecture for keypoint detection and prior property prediction. Then we detail how to estimate the 3D bounding box of the object by maintaining 2D-3D consistency.

3.1 Keypoint Detection Network

Our keypoint detection network takes an only RGB image as the input and simultaneously generates 2D-related information, such as perspective points and 2D size, and 3D-related information, such as dimension, orientation and distance. As shown in Fig. 2, it consists of three components: backbone, keypoint feature pyramid, and detection head. The main architecture adopts a one-stage strategy that shares a similar layout with the anchor-free 2D object detector [15,18,36,45], which allows us to get a fast detection speed. Details of the network are given below.

Backbone. For the trade-off between speed and accuracy, we use two different structures as our backbones: ResNet-18 [12] and DLA-34 [43]. All models take a single RGB image $I \in \mathbb{R}^{W \times H \times 3}$ and downsample the input with factor $S = 4$. The ResNet-18 and DLA-34 build for image classification network, the maximal downsample factor is $\times 32$. We upsample the bottleneck thrice by three bilinear interpolations and 1×1 convolutional layer. Before the upsampling layers, we concatenate the corresponding feature maps of the low level while adding one

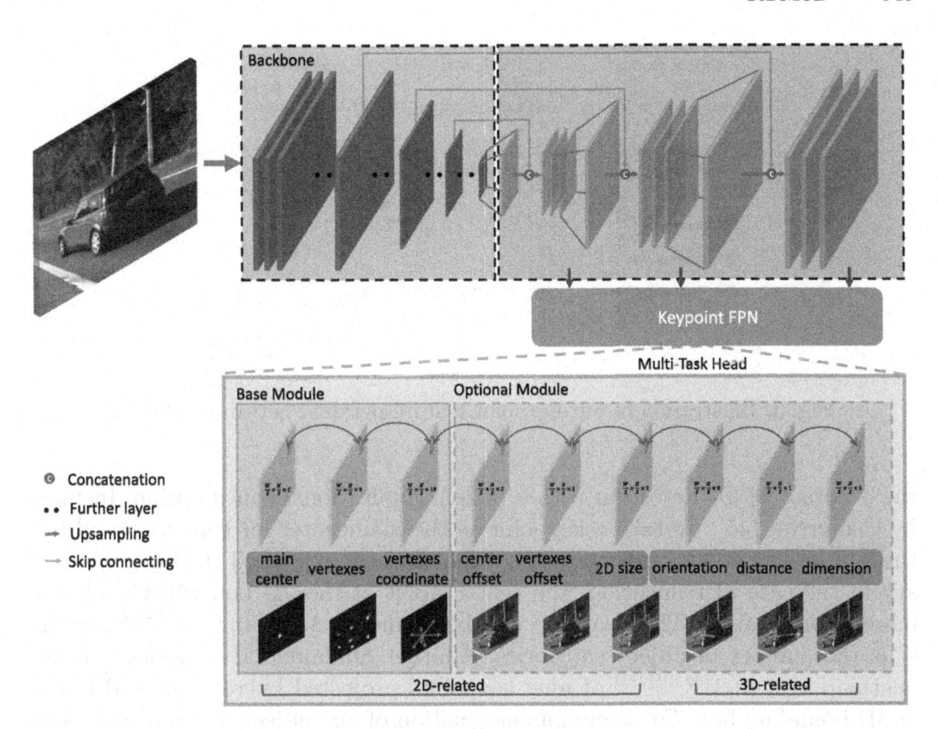

Fig. 2. An overview of proposed keypoint detection architecture: It takes only the RGB images as the input and outputs main center heatmap, vertexes heatmap, and vertexes coordinate as the base module to estimate 3D bounding box. It can also predict other alternative priors to further improve the performance of 3D detection.

1×1 convolutional layers for channel dimension reduction. After three upsampling layers, the channels are 256, 128, 64, respectively.

Keypoint Feature Pyramid. Keypoint in the image have no difference in size. Therefore, the keypoint detection is not suitable for using the Feature Pyramid Network (FPN) [21], which detect multi-scale 2D box in different pyramid layers. We propose a novel approach Keypoint Feature Pyramid Network (KFPN) to detect scale-invariant keypoints in the point-wise space, as shown in Fig. 3. Assuming we have F scale feature maps, we first resize each scale $f, 1 < f < F$ back to the size of maximal scale, which yields the feature maps $\hat{f}_{1<f<F}$. Then, we generate soft weight by a softmax operation to denote the importance of each scale. The finally scale-space score map S_{score} is obtained by linear weighing sum. In detail, it can be defined as:

$$S_{score} = \sum_{f} \hat{f} \odot softmax(\hat{f}) \tag{1}$$

where $\odot$ denote element-wise product.

Detection Head. The detection head is comprised of three fundamental components and six optional components which can be arbitrarily selected to boost

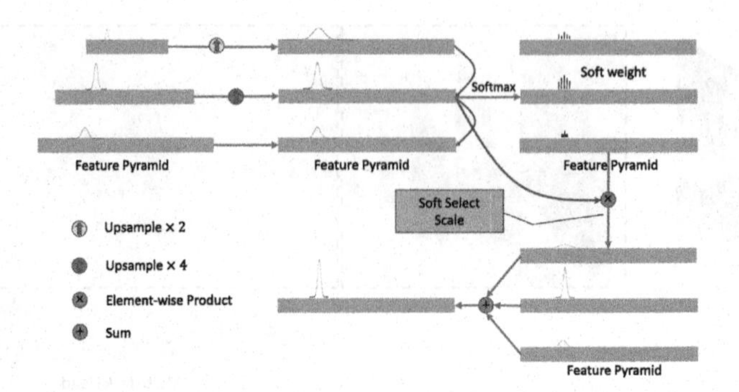

Fig. 3. Illustration of our keypoint feature pyramid network (KFPN).

the accuracy of 3D detection with a little computational consumption. Inspired by CenterNet [45], we take a keypoint as the maincenter for connecting all features. Since the 3D projection point of the object may exceed the image boundary in the case of truncation, the center point of the 2D box will be selected more appropriately. The heatmap can be define as $M \in [0,1]^{\frac{H}{S} \times \frac{W}{S} \times C}$, where C is the number of object categories. Another fundamental component is the heatmap $V \in [0,1]^{\frac{H}{S} \times \frac{W}{S} \times 9}$ of nine keypoints projected by vertexes and center of 3D bounding box. For keypoints association of one object, we also regress an local offset $V_c \in R^{\frac{H}{S} \times \frac{W}{S} \times 18}$ from the maincenter as an indication. Keypoints of V closest to the coordinates from V_c are taken as a group of one object.

Although the 18 constraints by the 9 keypoints have an ability to recover the 3D information of the object, more prior information can provide more constraints and further improve the detection performance. We offer a number of options to meet different needs for accuracy and speed. The center offset $M_{os} \in R^{\frac{H}{S} \times \frac{W}{S} \times 2}$ and vertexes offset $V_{os} \in R^{\frac{H}{S} \times \frac{W}{S} \times 2}$ are discretization error for each keypoint in heatmaps. The dimension $D \in \mathbb{R}^{\frac{H}{S} \times \frac{W}{S} \times 3}$ of 3D object have a smaller variance, which makes it easy to predict. The rotation $R(\theta)$ of an object only by parametrized by orientation θ (yaw) in the autonomous driving scene. We employ the *Multi-Bin* based method [26] to regress the local orientation. We classify the probability with cosin and sine offset of the local angle in one bin, which generates feature map of orientation $O \in \mathbb{R}^{\frac{H}{S} \times \frac{W}{S} \times 8}$ with two bins. We also regress the depth $Z \in \mathbb{R}^{\frac{H}{S} \times \frac{W}{S} \times 1}$ of 3D box center, which can be used as the initialization value to speed up the solution in Sect. 3.2.

Training. All the heatmaps of keypoint and maincenter training strategy follow the [18,45]. The loss solves the imbalance of positive and negative samples with focal loss [22]:

$$L_{kp}^{K} = -\frac{1}{N} \sum_{k=1}^{K} \sum_{x=1}^{H/S} \sum_{y=1}^{W/S} \begin{cases} (1-\hat{p}_{kxy})^{\alpha} log(\hat{p}_{kxy}) & if \quad p_{kxy} = 1 \\ (1-p_{kxy})^{\beta} \hat{p}_{kxy} log(1-\hat{p}_{kxy}) & otherwise \end{cases} \tag{2}$$

where K is the channels of different keypoints, $K = C$ in maincenter and $K = 9$ in keypoints. N is the number of maincenter or keypoints in an image, and

α and β are the hyper-parameters to reduce the loss weight of negative and easy positive samples. We set is $\alpha = 2$ and $\beta = 4$ in all experiments following [18,45]. p_{kxy} can be defined by Gaussian kernel $p_{xy} = exp\left(-\frac{x^2+y^2}{2\sigma}\right)$ centered by ground truth keypoint $\tilde{p}_{xy}$. For σ, we find the max area A_{max} and min area A_{min} of 2D box in training data and set two hyper-parameters σ_{max} and σ_{min}. We then define the $\sigma = A(\frac{\sigma_{max}-\sigma_{min}}{A_{max}-A_{min}})$ for a object with size A. For regression of dimension and distance, we define the residual term as:

$$\begin{aligned} L_D &= \frac{1}{3N}\sum_{x=1}^{H/S}\sum_{y=1}^{W/S} \mathbb{1}_{xy}^{obj}\left(D_{xy} - \Delta\widetilde{D}_{xy}\right)^2 \\ L_Z &= \frac{1}{N}\sum_{x=1}^{H/S}\sum_{y=1}^{W/S} \mathbb{1}_{xy}^{obj}\left(log(Z_{xy}) - log(\widetilde{Z}_{xy})\right)^2 \end{aligned} \tag{3}$$

We set $\Delta\widetilde{D}_{xy} = log\frac{\widetilde{D}_{xy}-\bar{D}}{D_\sigma}$, where $\bar{D}$ and D_σ are the mean and standard deviation dimensions of training data. $\mathbb{1}_{xy}^{obj}$ denotes if maincenter appears in position x, y. The offset of maincenter, vertexes are trained with an L1 loss following [45]:

$$\begin{aligned} L_{off}^m &= \frac{1}{2N}\sum_{x=1}^{H/S}\sum_{y=1}^{W/S} \mathbb{1}_{xy}^{obj}\left|M_{os}^{xy} - \left(\frac{p^m}{S} - \tilde{p}_m\right)\right| \\ L_{off}^v &= \frac{1}{2N}\sum_{x=1}^{H/S}\sum_{y=1}^{W/S} \mathbb{1}_{xy}^{ver}\left|V_{os}^{xy} - \left(\frac{p^v}{S} - \tilde{p}_v\right)\right| \end{aligned} \tag{4}$$

where p^m, p^v are the position of maincenter and vertexes in the original image. The regression coordinate of vertexes with an L1 loss as:

$$L_{ver} = \frac{1}{N}\sum_{k=1}^{8}\sum_{x=1}^{H/S}\sum_{y=1}^{W/S} \mathbb{1}_{xy}^{ver}\left|V_c^{(2k-1):(2k)xy} - \left|\frac{p^v - p^m}{S}\right|\right| \tag{5}$$

The finial multi-task loss for keypoint detection define as:

$$\begin{aligned} L = {} & \omega_{main}L_{kp}^C + \omega_{kpver}L_{kp}^8 + \omega_{ver}L_{ver} + \omega_{dim}L_D \\ & + \omega_{ori}L_{ori} + \omega_Z L_{dis} + \omega_{off}^m L_{off}^m + \omega_{off}^v L_{off}^v \end{aligned} \tag{6}$$

We empirical set $\omega_{main} = 1, \omega_{kpver} = 1, \omega_{ver} = 1, \omega_{dim} = 1, \omega_{ori} = 0.5, \omega_{dis} = 0.1, \omega_{off}^m = 0.5$ and $\omega_{off}^v = 0.5$ in our experimental.

3.2 3D Bounding Box Estimate

We estimate the 3D bounding box by enforcing the 2D-3D consistency between estimated 2D-related and 3D-related information, given by our keypoint detection network. Considering an image I, a set of $i = 1...N$ object are represented by 9 keypoints and other optional prior, keypoints as $\widehat{kp}_{ij}$ for $j \in 1...9$, dimension as $\widehat{D}_i$, orientation as $\hat{\theta}_i$, and distance as $\widehat{Z}_i$. The corresponding 3D bounding box B_i can be defined by its rotation $R_i(\theta)$, position $T_i = [T_i^x, T_i^y, T_i^z]^T$, and dimensions $D_i = [h_i, w_i, l_i]^T$. Our goal is to estimate the 3D bounding box B_i, whose projections of 3D center and vertexes on the image space best fit the corresponding 2D keypoints $\widehat{kp}_{ij}$. This can be solved by minimize the reprojection error of 3D keypoints and 2D keypoints. We formulate it and other prior errors as a nonlinear least squares optimization problem:

$$\begin{aligned} R^*, T^*, D^* = \underset{\{R,T,D\}}{\arg\min} \sum_{R_i, T_i, D_i} \left\| e_{cp}\left(R_i, T_i, D_i, \widehat{kp}_i\right) \right\|_{\Sigma_i}^2 \\ + \omega_d \left\| e_d\left(D_i, \widehat{D}_i\right) \right\|_2^2 + \omega_r \left\| e_r\left(R_i, \hat{\theta}_i\right) \right\|_2^2 \end{aligned} \tag{7}$$

where $e_{cp}(..), e_d(..), e_r(..)$ are measurement error of camera-point, dimension prior and orientation prior respectively. We set $\omega_d = 1$ and $\omega_r = 1$ in our experimental. Σ is the covariance matrix of keypoints projection error. It is the confidence extracted from the heatmap corresponding to the keypoints:

$$\Sigma_i = diag(softmax(V(\widehat{kp}_i)) \tag{8}$$

In the rest of the section, we will first define this error item, and then introduce the way to optimize the formulation.

Camera-Point. Following the [9], the homogeneous coordinate of eight vertexes and 3D center can be parametrized as:

$$\begin{aligned} P_{3D}^i = diag(D_i)Cor \\ Cor = \begin{bmatrix} 0 & 0 & 0 & 0 & -1 & -1 & -1 & -1 & -1/2 \\ 1/2 & -1/2 & -1/2 & 1/2 & 1/2 & -1/2 & -1/2 & 1/2 & 0 \\ 1/2 & 1/2 & -1/2 & -1/2 & 1/2 & 1/2 & -1/2 & -1/2 & 0 \\ 1 & 1 & 1 & 1 & 1 & 1 & 1 & 1 & 1 \end{bmatrix} \end{aligned} \tag{9}$$

Given the camera intrinsics matrix K, the projection of these 3D points into the image coordinate is:

$$kp_i = \frac{1}{s_i} K \begin{bmatrix} R & T \\ 0^T & 1 \end{bmatrix} diag(D_i)Cor = \frac{1}{s_i} K \exp(\xi^\wedge) diag(D_i)Cor \tag{10}$$

where $\xi \in \mathfrak{se}_3$ and exp maps the $\mathfrak{se}_3$ into SE_3 space. The projection coordinate should fit tightly into 2D keypoints detected by the detection network. Therefore, the camera-point error is then defined as:

$$e_{cp} = \widehat{kp}_i - kp_i \tag{11}$$

Minimizing the camera-point error needs the Jacobians in $\mathfrak{se}_3$ space. It is given by:

$$\begin{aligned} \frac{\partial e_{cp}}{\partial \delta \xi} = - \begin{bmatrix} \frac{f_x}{Z'} & 0 & -\frac{f_x X'}{Z'^2} \\ 0 & \frac{f_y}{Z'} 0 & -\frac{f_y Y'}{Z'^2} \end{bmatrix} \cdot [I, -P'^\wedge] \\ \frac{\partial e_{cp}}{\partial D_i} = -\frac{1}{9} \sum_{col=1}^{9} \begin{bmatrix} \frac{f_x}{Z'} & 0 & -\frac{f_x X'}{Z'^2} \\ 0 & \frac{f_y}{Z'} 0 & -\frac{f_y Y'}{Z'^2} \end{bmatrix} \cdot R \cdot Cor_{col} \end{aligned} \tag{12}$$

where $P^{'} = [X^{'}, Y^{'}, Z^{'}]^T = (\exp(\xi^{\wedge} P))_{1:3}$.

Dimension-Prior: The e_d is simply defined as:

$$e_d = \widehat{D}_i - D_i \tag{13}$$

Rotation-Prior: We define e_r in $SE3$ space and use log to map the error into its tangent vector space:

$$e_r = \log(R^{-1}R(\hat{\theta}))^{\vee}_{\mathfrak{se}_3} \tag{14}$$

These multivariate equations can be solved via the Gauss-newton or Levenberg-Marquardt algorithm in the g2o library [17]. A good initialisation is mandatory using this optimization strategy. We adopt the prior information generated by keypoint detection network as the initialization value, which is very important in improving the detection speed.

4 Experimental

4.1 Implementation Details

Our experiments were evaluated on the KITTI 3D detection benchmark [9], which has a total of 7481 training images and 7518 test images. We follow the [7] and [39] to split the training set as $train1, val_1$ and $train2, val_2$ respectively, and comprehensively compare our framework with other methods on this two validation as well as test set.

Our deep neural network implemented by using PyTorch with the machine i7-8086K CPU and 2 1080Ti GPUs. All the original image are padded to 1280×384 for training and testing. We project the 3D bounding box of Ground Truths in the left and right images to obtain Ground Truth keypoints and use the horizontal flipping as the data augmentation, which makes our dataset is quadruple with the origin training set. We run Adam [14] optimizer with a base learning rate of 0.0002 for 300 epochs and reduce $10\times$ at 150 and 180 epochs. For standard deviation of Gaussian kernel, we set $\sigma_{max} = 19$ and $\sigma_{min} = 3$. Based on the statistics of KITTI dataset, we set $\tilde{l} = 3.89, \tilde{w} = 1.62, \tilde{h} = 1.53$ and $\sigma_{\tilde{l}} = 0.41, \sigma_{\tilde{w}} = 0.1, \sigma_{\tilde{h}} = 0.13$. In the inference step, after 3×3 max pooling, we filter the maincenter and keypoints with threshold 0.4 and 0.1 respectively, and only keypoints that in the image size range are sent into the geometric constraint module. The backbone networks are initialized by a classification model pre-trained on the ImageNet data set. Finally, The ResNet-18 takes about three days with batch size 30 and DLA-34 for four days with batch size 16 in the training stage.

4.2 Comparison with Other Methods

To fully evaluate the performance of our keypoint-based method, for each task three official evaluation metrics be reported in KITTI: average precision

Table 2. Comparison of our framework with current image-based 3D detection methods for car category evaluated using metric AP_{3D} on val_1/val_2 of KITTI data set. "Extra" means the extra data used in training. Red denotes the highest result, blue for the second highest, and cyan for the third.

Method	Extra	Time	AP_{3D} (IoU=0.5)			AP_{3D} (IoU=0.7)		
			Easy	Moderate	Hard	Easy	Moderate	Hard
Mono3D [6]	Mask	4.2 s	25.19 / -	18.20 / -	15.52 / -	2.53 / -	2.31 / -	2.31 / -
3DOP [7]	Stereo	3 s	46.04 / -	34.63 / -	30.09 / -	6.55 / -	5.07 / -	4.10 / -
MF3D [40]	Depth	-	47.88 / 45.57	29.48 / 30.03	26.44 / 23.95	10.53 / 7.85	5.69 / 5.39	5.39 / 4.73
Mono3D++ [13]	Depth+Shape	>0.6s	42.00 / -	29.80 / -	24.20 / -	10.60 / -	7.90 / -	5.70 / -
GS3D [19]	None	2.3s	32.15 / 30.60	29.89 / 26.40	26.19 / 22.89	13.46 / 11.63	10.97 / 10.51	10.38 / 10.51
Deep3DBox [26]	None	-	27.04 / -	20.55 / -	15.88 / -	5.85 / -	4.10 / -	3.84 / -
MonoGRNet [31]	None	**0.06s**	**50.51**/ -	**36.97**/ -	30.82 / -	13.88 / -	10.19 / -	7.62 / -
FQNet[23]	None	3.33s	28.16 / 28.98	21.02 / 20.71	19.91 / 18.59	5.98 / 5.45	5.50 / 5.11	4.75 / 4.45
M3D-RPN [3]	None	0.16s	**48.96/49.89**	**39.57/36.14**	**33.01**/ 28.98	**20.27/20.40**	**17.06/16.48**	**15.21/13.34**
Ours (ResNet18)	None	**0.035s**	47.43 /**46.52**	33.86 /**32.61**	**31.04/30.95**	**18.13/18.38**	**14.14/14.66**	**13.33/12.35**
Ours (DLA34)	None	**0.055s**	**54.36/52.59**	**41.90/40.96**	**35.84/34.95**	**20.77/19.47**	**16.86/16.29**	**16.63/15.57**

Table 3. Comparison of our framework with current image-based 3D detection frameworks for car category, evaluated using metric AP_{BEV} on val_1/val_2 of KITTI data set.

Method	Extra	Time	AP_{BEV} (IoU=0.5)			AP_{BEV} (IoU=0.7)		
			Easy	Moderate	Hard	Easy	Moderate	Hard
Mono3D [6]	Mask	4.2 s	30.50 / -	22.39 / -	19.16 / -	5.22 / -	5.19 / -	4.13 / -
3DOP [7]	Stereo	3 s	55.04 / -	41.25 / -	34.55 / -	12.63 / -	9.49 / -	7.59 / -
MF3D [40]	Depth	-	55.02 / 54.18	36.73 / 38.06	31.27 / 31.46	22.03 / 19.20	13.63 / 12.17	11.60 / 10.89
Mono3D++ [13]	Depth+Shape	>0.6s	46.70 / -	34.30 / -	28.10 / -	16.70 / -	11.50 / -	10.10 / -
GS3D [19]	None	2.3s	- / -	- / -	- / -	- / -	- / -	- / -
Deep3DBox [26]	None	-	30.02 / -	23.77 / -	18.83 / -	9.99 / -	7.71 / -	5.30 / -
MonoGRNet [31]	None	**0.06s**	- / -	- / -	- / -	- / -	- / -	- / -
FQNet[23]	None	3.33s	32.57 / 33.37	24.60 / 26.29	21.25 / 21.57	9.50 / 10.45	8.02 / 8.59	7.71 / 7.43
M3D-RPN [3]	None	0.16s	**55.37/55.87**	**42.49/41.36**	**35.29/34.08**	**25.94/26.86**	**21.18/21.15**	**17.90/17.14**
Ours(ResNet18)	None	**0.035s**	**52.79/41.91**	**35.92/34.28**	**33.02/28.88**	**20.81/21.34**	**16.60/16.48**	**15.80/15.45**
Ours (DLA34)	None	**0.055s**	**57.47/56.90**	**44.16/44.69**	**42.31/41.75**	**25.56/24.74**	**22.12/22.03**	**20.91/18.05**

for 3D intersection-over-union (AP_{3D}), average precision for Birds Eye View (AP_{BEV}), and Average Orientation Similarity (AOS) if 2D bounding box available. We evaluate our method at three difficulty settings: easy, moderate, and hard, according to the object's occlusion, truncation, and height in the image space [9].

***AP_{3D}* and *AP_{BEV}*.** We compare our method with current image-based SOTA approaches and also provide a comparison about running time. However, it is not realistic to list the running times of all previous methods because most of them do not report their efficiency. The results AP_{3D}, AP_{BEV} and running time are shown in Table 2 and 3, respectively. ResNet-18 as the backbone achieves the best speed while our accuracy outperforms most of the image-only method. In particular, it is more than 100 times faster than Mono3D [6] while outperforms over 10% for both AP_{BEV} and AP_{3d} across all datasets. In addition, our ResNet-18 method is more than 75 times faster while having a comparable accuracy than 3DOP [7], which employs stereo images as the input. DLA-34 as the backbone achieves the best accuracy while having relatively good speed. It is faster about 3 times than the recently proposed M3D-RPN [3] while achieves the improvement in most of the metrics. Note that comparing our method with this all approaches is unfair because most of these approaches rely on extra stand-alone network or data in addition to monocular images. Nevertheless, we achieve the best speed with better performance.

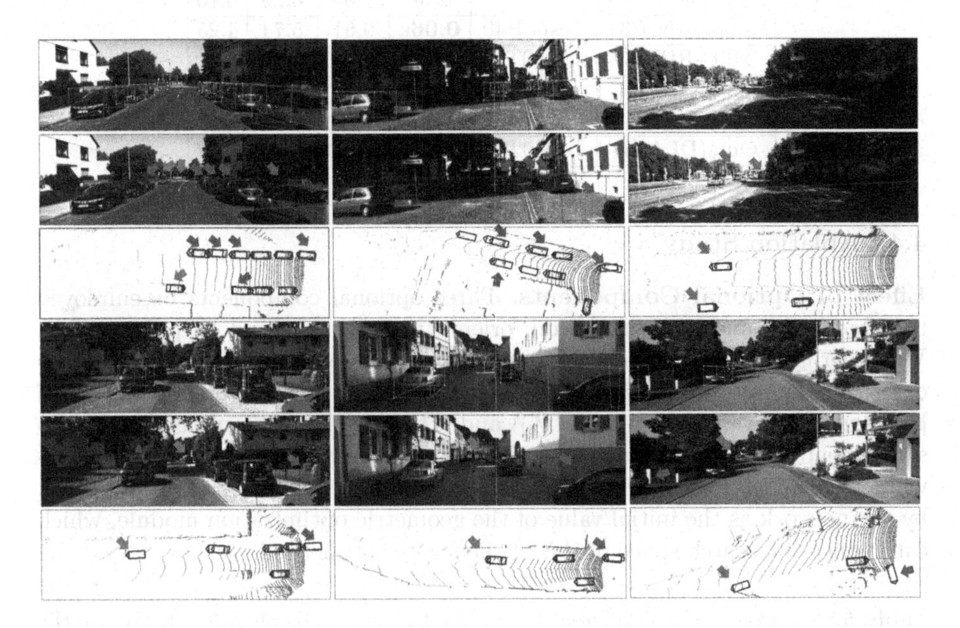

Fig. 4. Qualitative results of our 3D detection. From top to bottom are keypoints, projections of the 3D bounding box and bird's eye view image, ground truths in green and our results in blue. The crimson arrows, green arrows, and red arrows point to occluded, distant, and truncated objects, respectively. (Color figure online)

Results on the KITTI Testing Set. We also evaluate our results on the KITTI testing set, as shown in Table 4.

4.3 Qualitative Results

Figure 4 shows some qualitative results of our method. We visualize the keypoint detection network outputs, geometric constraint module outputs and BEV images. The results of the projected 3D box on image demonstrate than our method can handle crowded and truncated objects. The results of the BEV image show that our method has an accuracy localization in different scenes.

Table 4. Comparing 3D detection AP_{3D} on KITTI testing set. We use the DLA-34 as the backbone.

Method	Accelerater	Time	AP3D(IoU=0.7)		
			Easy	Mode	Hard
GS3D[19]	-	2.3s	7.69	6.29	6.16
MonoGRNet[31]	Tesla P40	**0.06s**	9.61	5.74	4.25
M3D-RPN[3]	1080Ti	0.16s	**14.76**	**9.71**	**7.42**
FQNet[23]	1080Ti	3.33s	2.77	1.51	1.01
MonoDIS[34]	V100	-	10.37	7.94	6.40
Ours(DLA34)	1080Ti	**0.055s**	**14.41**	**10.34**	**8.77**

4.4 Ablation Study

Effect of Optional Components. Three optional components be employed to enhance our method: dimension, orientation, distance and keypoints offset. We experiment with different combinations to demonstrate their effect on 3D detection. The results are shown in Table 5, we train our network with DLA-34 backbone and evaluate it using AP_{3D} and AP_{BEV}. The combinations of dimension, orientation, distance and keypoints offset achieve the best accuracy meanwhile have a faster running speed. This is because we take the output predicted by our network as the initial value of the geometric optimization module, which can reduce the search space of the gradient descent method.

Table 5. Ablation study of different optional selecting results on val_1 set. We use the DLA-34 as the backbone.

dim	ori	dist	off	Time (s)	AP3D (IoU = 0.5)			AP3D (IoU = 0.7)		
					Easy	Mode	Hard	Easy	Mode	Hard
√				0.058	51.21	40.73	35.00	18.23	17.05	15.94
	√			0.061	25.35	22.33	21.18	3.12	3.43	2.97
√	√			0.057	54.18	41.34	34.89	20.23	16.02	15.94
√	√	√		0.055	54.20	41.56	35.13	20.76	16.80	16.25
√	√	√	√	0.055	54.36	41.90	35.84	20.77	16.86	16.36

Effect of Keypoint FPN. We propose keypoint FPN as a strategy to improve the performance of multi-scale keypoint detection. To better understand its effect, we compare the AP_{3D} and AP_{BEV} with and without KFPN. The details are shown in Table 6, using KFPN achieves the improvement across all sets while no significant change in time consumption.

2D Detection and Orientation. Although our focus is on 3D detection, we also report the performance of our methods in 2D detection and orientation evaluation in order to better understand the comprehensive capabilities of our approach. We report the AOS and AP with a threshold IoU = 0.7 for comparison.

Table 6. Comparing 3D detection AP_{3D} of w/o KFPN and w/ KFPN for car category on val_1 set. We use the DLA-34 as the backbone.

KFPN	Time	AP_{3D} (IoU = 0.7)			AP_{3D} (IoU = 0.5)		
		Easy	Mode	Hard	Easy	Mode	Hard
w/o	0.054	50.14	40.73	34.94	17.47	15.99	15.36
w/	0.055	54.36	41.90	35.84	20.77	16.86	16.36

Table 7. Comparing of 2D detection AP_{2D} with IoU = 0.7 and orientation AOS results for car category evaluated on val_1/val_2 of KITTI data set. Only the results under the moderate criteria are shown. Ours (2D) represents the results from the keypoint detection network, and Ours (3D) is the 2D bounding box of the projected 3D box.

Method	AP_{2D}	AOS
Mono3D [6]	88.67/-	86.28/-
3DOP [7]	88.07/-	85.80/-
Deep3DBox [26]	-/**97.20**	-/**96.68**
DeepMANTA [5]	**90.89**/91.01	**90.66**/90.66
GS3D [19]	88.85/90.02	87.52/89.13
Ours (2D)	90.14/91.85	89.58/89.22
Ours (3D)	90.41/92.08	89.95/89.40

The results are shown in Table 7, the Deep3DBox train MS-CNN [4] in KITTI to produce 2D bounding box and adopt VGG16 [35] for orientation prediction, which gives him the highest accuracy. Deep3Dbox takes advantage of better 2D detectors, however, our AP_{3D} outperforms it by about 20% in moderate sets, which emphasize the importance of customizing the network specifically for 3D detection. Another interesting finding is that the 2D accuracy of back-projection 3D results is better than the direct prediction, thanks to our method that can infer the occlusive area of the object.

5 Conclusion

In this paper, we have proposed a faster and more accurate monocular 3D object detection method for autonomous driving scenarios. We reformulate 3D detection as the keypoint detection problem and show how to recover the 3D bounding box by using keypoints and geometric constraints. We specially customize the point detection network for 3D detection, which can simultaneously predict keypoints of the 3D box and other prior information of the object using only images. Our geometry module formulates this prior to easy-to-optimize loss functions. Our approach generates a stable and accurate 3D bounding box without containing stand-alone networks, additional annotation while achieving real-time running speed.

References

1. Bay, H., Tuytelaars, T., Van Gool, L.: SURF: speeded up robust features. In: Leonardis, A., Bischof, H., Pinz, A. (eds.) ECCV 2006. LNCS, vol. 3951, pp. 404–417. Springer, Heidelberg (2006). https://doi.org/10.1007/11744023_32
2. Behl, A., Hosseini Jafari, O., Karthik Mustikovela, S., Abu Alhaija, H., Rother, C., Geiger, A.: Bounding boxes, segmentations and object coordinates: how important is recognition for 3D scene flow estimation in autonomous driving scenarios? In: Proceedings of the IEEE International Conference on Computer Vision, pp. 2574–2583 (2017)
3. Brazil, G., Liu, X.: M3D-RPN: monocular 3D region proposal network for object detection. In: Proceedings of the IEEE International Conference on Computer Vision, Seoul, South Korea (2019)
4. Cai, Z., Fan, Q., Feris, R.S., Vasconcelos, N.: A unified multi-scale deep convolutional neural network for fast object detection. In: Leibe, B., Matas, J., Sebe, N., Welling, M. (eds.) ECCV 2016. LNCS, vol. 9908, pp. 354–370. Springer, Cham (2016). https://doi.org/10.1007/978-3-319-46493-0_22
5. Chabot, F., Chaouch, M., Rabarisoa, J., Teulière, C., Chateau, T.: Deep MANTA: a coarse-to-fine many-task network for joint 2D and 3D vehicle analysis from monocular image. In: Proceedings of the IEEE Conference on Computer Vision and Pattern Recognition, pp. 2040–2049 (2017)
6. Chen, X., Kundu, K., Zhang, Z., Ma, H., Fidler, S., Urtasun, R.: Monocular 3D object detection for autonomous driving. In: Proceedings of the IEEE Conference on Computer Vision and Pattern Recognition, pp. 2147–2156 (2016)
7. Chen, X., Kundu, K., Zhu, Y., Ma, H., Fidler, S., Urtasun, R.: 3D object proposals using stereo imagery for accurate object class detection. IEEE Trans. Pattern Anal. Mach. Intell. **40**(5), 1259–1272 (2017)
8. Chen, X., Ma, H., Wan, J., Li, B., Xia, T.: Multi-view 3D object detection network for autonomous driving. In: Proceedings of the IEEE Conference on Computer Vision and Pattern Recognition, pp. 1907–1915 (2017)
9. Geiger, A., Lenz, P., Urtasun, R.: Are we ready for autonomous driving? The KITTI vision benchmark suite. In: 2012 IEEE Conference on Computer Vision and Pattern Recognition, pp. 3354–3361. IEEE (2012)
10. Girshick, R.: Fast R-CNN. In: Proceedings of the IEEE International Conference on Computer Vision, pp. 1440–1448 (2015)
11. Girshick, R., Donahue, J., Darrell, T., Malik, J.: Rich feature hierarchies for accurate object detection and semantic segmentation. In: Proceedings of the IEEE Conference on Computer Vision and Pattern Recognition, pp. 580–587 (2014)
12. He, K., Zhang, X., Ren, S., Sun, J.: Deep residual learning for image recognition. In: Proceedings of the IEEE Conference on Computer Vision and Pattern Recognition, pp. 770–778 (2016)
13. He, T., Soatto, S.: Mono3D++: monocular 3D vehicle detection with two-scale 3D hypotheses and task priors. arXiv preprint arXiv:1901.03446 (2019)
14. Kingma, D.P., Ba, J.: Adam: a method for stochastic optimization. arXiv. Learning (2014)
15. Kong, T., Sun, F., Liu, H., Jiang, Y., Shi, J.: FoveaBox: beyond anchor-based object detector. arXiv preprint arXiv:1904.03797 (2019)
16. Ku, J., Pon, A.D., Waslander, S.L.: Monocular 3D object detection leveraging accurate proposals and shape reconstruction. In: Proceedings of the IEEE Conference on Computer Vision and Pattern Recognition, pp. 11867–11876 (2019)

17. Kummerle, R., Grisetti, G., Strasdat, H., Konolige, K., Burgard, W.: G 2 o: a general framework for graph optimization, pp. 3607–3613 (2011)
18. Law, H., Deng, J.: CornerNet: detecting objects as paired keypoints. In: Proceedings of the European Conference on Computer Vision (ECCV), pp. 734–750 (2018)
19. Li, B., Ouyang, W., Sheng, L., Zeng, X., Wang, X.: GS3D: an efficient 3D object detection framework for autonomous driving. In: The IEEE Conference on Computer Vision and Pattern Recognition (CVPR), June 2019
20. Li, P., Chen, X., Shen, S.: Stereo R-CNN based 3D object detection for autonomous driving. In: Proceedings of the IEEE Conference on Computer Vision and Pattern Recognition, pp. 7644–7652 (2019)
21. Lin, T.Y., Dollár, P., Girshick, R., He, K., Hariharan, B., Belongie, S.: Feature pyramid networks for object detection. In: Proceedings of the IEEE Conference on Computer Vision and Pattern Recognition, pp. 2117–2125 (2017)
22. Lin, T.Y., Goyal, P., Girshick, R., He, K., Dollár, P.: Focal loss for dense object detection. In: Proceedings of the IEEE International Conference on Computer Vision, pp. 2980–2988 (2017)
23. Liu, L., Lu, J., Xu, C., Tian, Q., Zhou, J.: Deep fitting degree scoring network for monocular 3D object detection. In: Proceedings of the IEEE Conference on Computer Vision and Pattern Recognition, pp. 1057–1066 (2019)
24. Lowe, D.G.: Distinctive image features from scale-invariant keypoints. Int. J. Comput. Vision **60**(2), 91–110 (2004)
25. Ma, X., Wang, Z., Li, H., Zhang, P., Ouyang, W., Fan, X.: Accurate monocular 3D object detection via color-embedded 3D reconstruction for autonomous driving. In: Proceedings of the IEEE International Conference on Computer Vision, pp. 6851–6860 (2019)
26. Mousavian, A., Anguelov, D., Flynn, J., Kosecka, J.: 3D bounding box estimation using deep learning and geometry. In: Proceedings of the IEEE Conference on Computer Vision and Pattern Recognition, pp. 7074–7082 (2017)
27. Murthy, J.K., Krishna, G.S., Chhaya, F., Krishna, K.M.: Reconstructing vehicles from a single image: shape priors for road scene understanding. In: 2017 IEEE International Conference on Robotics and Automation (ICRA), pp. 724–731. IEEE (2017)
28. Naiden, A., Paunescu, V., Kim, G., Jeon, B., Leordeanu, M.: Shift R-CNN: deep monocular 3D object detection with closed-form geometric constraints. In: 2019 IEEE International Conference on Image Processing (ICIP), pp. 61–65 (2019)
29. Qi, C.R., Liu, W., Wu, C., Su, H., Guibas, L.J.: Frustum PointNets for 3D object detection from RGB-D data. In: Proceedings of the IEEE Conference on Computer Vision and Pattern Recognition, pp. 918–927 (2018)
30. Qi, C.R., Su, H., Mo, K., Guibas, L.J.: PointNet: deep learning on point sets for 3D classification and segmentation. In: Proceedings of the IEEE Conference on Computer Vision and Pattern Recognition, pp. 652–660 (2017)
31. Qin, Z., Wang, J., Lu, Y.: MonoGRNet: a geometric reasoning network for monocular 3D object localization. In: Proceedings of the AAAI Conference on Artificial Intelligence, vol. 33, pp. 8851–8858 (2019)
32. Ren, S., He, K., Girshick, R., Sun, J.: Faster R-CNN: towards real-time object detection with region proposal networks. In: Advances in Neural Information Processing Systems, pp. 91–99 (2015)
33. Rubino, C., Crocco, M., Del Bue, A.: 3D object localisation from multi-view image detections. IEEE Trans. Pattern Anal. Mach. Intell. **40**(6), 1281–1294 (2017)

34. Simonelli, A., Bulo, S.R., Porzi, L., López-Antequera, M., Kontschieder, P.: Disentangling monocular 3D object detection. In: Proceedings of the IEEE International Conference on Computer Vision, pp. 1991–1999 (2019)
35. Simonyan, K., Zisserman, A.: Very deep convolutional networks for large-scale image recognition. Computer Science (2014)
36. Tian, Z., Shen, C., Chen, H., He, T.: FCOS: fully convolutional one-stage object detection. arXiv preprint arXiv:1904.01355 (2019)
37. Wang, Y., Chao, W.L., Garg, D., Hariharan, B., Campbell, M., Weinberger, K.Q.: Pseudo-lidar from visual depth estimation: bridging the gap in 3D object detection for autonomous driving. In: Proceedings of the IEEE Conference on Computer Vision and Pattern Recognition, pp. 8445–8453 (2019)
38. Xiang, Y., Choi, W., Lin, Y., Savarese, S.: Data-driven 3D voxel patterns for object category recognition. In: Proceedings of the IEEE Conference on Computer Vision and Pattern Recognition, pp. 1903–1911 (2015)
39. Xiang, Y., Choi, W., Lin, Y., Savarese, S.: Subcategory-aware convolutional neural networks for object proposals and detection, pp. 924–933 (2017)
40. Xu, B., Chen, Z.: Multi-level fusion based 3D object detection from monocular images. In: Proceedings of the IEEE Conference on Computer Vision and Pattern Recognition, pp. 2345–2353 (2018)
41. Yang, B., Luo, W., Urtasun, R.: PIXOR: real-time 3D object detection from point clouds. In: Proceedings of the IEEE Conference on Computer Vision and Pattern Recognition, pp. 7652–7660 (2018)
42. Yang, S., Scherer, S.: CubeSLAM: monocular 3-D object slam. IEEE Trans. Robot. **35**, 925–938 (2019)
43. Yu, F., Wang, D., Shelhamer, E., Darrell, T.: Deep layer aggregation. In: Proceedings of the IEEE Conference on Computer Vision and Pattern Recognition, pp. 2403–2412 (2018)
44. Zeeshan Zia, M., Stark, M., Schindler, K.: Are cars just 3D boxes?-Jointly estimating the 3D shape of multiple objects. In: Proceedings of the IEEE Conference on Computer Vision and Pattern Recognition, pp. 3678–3685 (2014)
45. Zhou, X., Wang, D., Krähenbühl, P.: Objects as points. arXiv preprint arXiv:1904.07850 (2019)
46. Zhou, Y., Tuzel, O.: VoxelNet: end-to-end learning for point cloud based 3D object detection. In: Proceedings of the IEEE Conference on Computer Vision and Pattern Recognition, pp. 4490–4499 (2018)

Video Object Segmentation with Episodic Graph Memory Networks

Xiankai Lu[1], Wenguan Wang[2](✉), Martin Danelljan[2], Tianfei Zhou[1], Jianbing Shen[1], and Luc Van Gool[2]

[1] Inception Institute of Artificial Intelligence, Abu Dhabi, UAE
carrierlxk@gmail.com
[2] ETH Zurich, Zürich, Switzerland
wenguanwang.ai@gmail.com
https://github.com/carrierlxk/GraphMemVOS

Abstract. How to make a segmentation model efficiently adapt to a specific video as well as online target appearance variations is a fundamental issue in the field of video object segmentation. In this work, a graph memory network is developed to address the novel idea of "learning to update the segmentation model". Specifically, we exploit an episodic memory network, organized as a fully connected graph, to store frames as nodes and capture cross-frame correlations by edges. Further, learnable controllers are embedded to ease memory reading and writing, as well as maintain a fixed memory scale. The structured, external memory design enables our model to comprehensively mine and quickly store new knowledge, even with limited visual information, and the differentiable memory controllers slowly learn an abstract method for storing useful representations in the memory and how to later use these representations for prediction, via gradient descent. In addition, the proposed graph memory network yields a neat yet principled framework, which can generalize well to both one-shot and zero-shot video object segmentation tasks. Extensive experiments on four challenging benchmark datasets verify that our graph memory network is able to facilitate the adaptation of the segmentation network for case-by-case video object segmentation.

Keywords: Video segmentation · Episodic graph memory · Learn to update

1 Introduction

Video object segmentation (VOS), as a core task in computer vision, aims to predict the target object in a video at the pixel level. Typically, according to

---

Electronic supplementary material The online version of this chapter (https://doi.org/10.1007/978-3-030-58580-8_39) contains supplementary material, which is available to authorized users.

A. Vedaldi et al. (Eds.): ECCV 2020, LNCS 12348, pp. 661–679, 2020.
https://doi.org/10.1007/978-3-030-58580-8_39

whether or not annotations are provided for the first frame during testing, VOS can be categorized into one-shot video object segmentation (O-VOS) [5,48] and zero-shot video object segmentation (Z-VOS) [52]. Provided with only first-frame annotations, O-VOS is to identify and segment the labeled object instances in the rest of the video [7,15,17,27,33,69]; whereas Z-VOS targets at automatically inferring the primary objects without any test-time indications [44,45].

O-VOS is a challenging task because there are no further assumptions regarding to the specific target object and the application scenes often contain similar distractor objects. To tackle these challenges, earlier methods typically perform network finetuning over each annotated object [5,50]. Though effective, this is quite time-consuming. Current popular solutions [11,31,49,51,70] are instead built upon an efficient *matching* based framework; they formulate the task as a differentiable matching procedure between the support set (*i.e.*, the first labeled frame or prior segmented frames) and query set (*i.e.*, current frame). Thus they can directly assign labels to the query frame, according to the pixel-wise similarity to the annotated first frame and/or previous processed frames.

Although omitting first-frame finetune and improving the performance to some extent, matching based O-VOS methods still suffer from several limitations. First, they typically learn a generic matching network and then apply it to test videos directly, failing to make full use of first-frame target-specific information. As a result, they cannot efficiently adapt to the input video. Second, as the segmentation targets may undergo appearance variation (*i.e.*, fast motion, occlusion), it is meaningful to perform online model updating. Third, matching based methods only modeling pair-relations between the query and each support frame, neglecting the rich context within the support set.

To address these issues, we take inspiration from the recent development of memory-augmented networks for few-shot learning [39,58] and develop a graph memory network to online adapt the segmentation model to a specific target in one single feed-forward pass. Specifically, by regarding O-VOS as episodic memory reasoning, our approach equips with the ability to slowly learn high-level knowledge for extracting useful representations from the offline training data, and the ability to rapidly fuse the unseen information from the first-frame annotation in the test video, via an external memory module. In this way, our model can internally modulate the output by learning to rapidly cache representation in the memory stores. During the segmentation, to maintain the variations of the object appearance, we perform memory updating by storing and recalling target information from the external memory. Therefore, we can implement the online model updating easily without extensive parameter optimization. In addition, the memory module, built upon the end-to-end memory network [43], is endowed with a graph structure to better mine the relations among memory cells.

The proposed graph memory network is neat and fast. For memory updating, instead of some prior matching based O-VOS models [31] inserting a new element into newly allocated position, our model performs message passing on the fixed-size graph memory without increasing memory consumption. Our model provides a principled framework; it generalizes Z-VOS task well, in which mainstream methods also lack the adaptation capability. As far as we know, this work represents the first effort that addresses both O-VOS and Z-VOS in a unified network design. Experiments on representative O-VOS datasets show the proposed method performs favorably against state-of-the-arts. Without bells and whistles, it also outperforms other competitors on Z-VOS datasets. These promising results demonstrate the efficacy and generalizability of our graph memory network.

2 Related Work

One-shot Video Object Segmentation (O-VOS) is to track the first-frame annotation to subsequent frames at pixel level. Traditional methods usually formulate this task as a *label propagation* process [1,35,40,53,57]. With the renaissance of the connectionism, deep learning based solutions become the domaniant [11,27,33,35] in the field of O-VOS. Among them are three representative strategies. One is the *segmentation-by-detection* scheme [5,6,10,26] that learns a video-specific representation about the first-frame annotated objects and then performs pixel-wise detection in the rest frames. Another one is the *propagation* based pipeline [10,45,60], which propagates segmented masks to fit objects in the upcoming frames. The third, *i.e.*, the more advanced, is the *matching* based strategy [7,11,15,17,25,31,49,51,69,70] which usually trains a prototypical Siamese matching network to find the most matching pixel (or embedding in the feature space) between the first frame (or a segmented frame) and the query frame, and then achieves label assignment accordingly. Some matching-based methods employ internal memory (*e.g.*, ConvLSTM [48,64]) or external memory (*e.g.*, [31,48]) to implicitly or explicitly store previously computed segmentation information for facilitating learning the evolution of objects over time. However, our utilization of memory differs from these methods substantially: **i)** we employ an external memory with learnable read-write controller to rapidly encode new video information for quick segmentation network updating; **ii)** compared to vanilla memory network [28,43], our graph memory network stores memory content in a structured manner that explicitly captures context in cells; **iii)** instead of writing new input to a newly located position [31], our memory is dynamically updated by iterative cell state renewing without increasing the memory size.

Zero-shot Video Object Segmentation (Z-VOS) aims to segment primary objects in unconstrained videos. This task has been widely studied over several decades which also called unsupervised video object segmentation [19,22,35,71]. Traditional methods usually leverage motion [18,29] or saliency cues [12,54] to obtain a heuristic representation for inferring the primary objects. Recent methods were built upon fully convolutional networks. Early methods explored two-stream architectures [8,16,41,72] or variants of recurrent neural networks [42,45,55]. Recent ones address comprehensive foreground reasoning from a global view by non-local structures [24,52]. In this work, rather than these methods learning a universal video foreground object representation and hoping it could generalize well to all unseen scenarios, our episodic memory design allows target-adaption on-the-fly by learning to update the segmentation network.

Learning to Update in VOS. To learn more video-specific information, a direct way is to perform iterative network finetune in the first frame [5,50]. Some recent efforts instead applied meta-learning for model updating [4,61,66], whose basic ideas are similar: learning segmentation model parameters on training videos. This paper, in contrast, uses a graph memory network with learnable, lightweight controllers to assimilate a new video. Thus the model can quickly adapt to unseen scenes and appearance variants through the memory node (cell) updating rather than sensitive and expensive network parameter generation [61,66].

3 Proposed Algorithm

3.1 Preliminary: Episodic Memory Networks

Memory networks augment neural networks with an external memory component [28,43,58], which allow the network to explicitly access the past experiences. They have been shown effective in few-shot learning [39,62,63] and object tracking [67]. Recently, episodic external memory networks have been explored to solve reasoning issues in visual question answering and visual dialog [21,28,43]. The basic idea is to retrieve the information required to answer the question from the memory with trainable read and write operators. Given a collection of input representations, the episodic memory module chooses which parts of the inputs to focus on through the neural attention. It then produces a "memory summarization" representation taking into account the query as well as the stored memory. Each iteration in the episode provides the memory module with newly relevant information about the input. As a result, the memory module has the ability to retrieve new information in each iteration and obtain a new representation about the input.

3.2 Learning to Update

In the context of O-VOS [5], the goal is to learn from the annotated objects in the first frame (*support set*) and predict them in the subsequent frames (*query set*). To this end, traditional methods usually finetune the network and perform online learning for each specific video. In contrast, we construct an episodic memory based learner on variety of tasks (videos), sampled from the distribution of training tasks [4], such that the learned model performs well on new unseen tasks (test videos). Thus we tackle O-VOS as a **"learning to update"** segmentation network procedure [38]: **i)** extracting a task representation from the one-shot support set, and **ii)** updating the segmentation network for the query given the task representation. As shown in Fig. 1, we enhance an episodic memory network with graph structure (*i.e.*, graph memory network) to: **i)** instantly adapt the segmentation network to a specific object, rather than performing lots of finetuning iterations; and **ii)** make full use of context within video sequences. As a result, our graph memory network has two abilities: learn to adjust the segmentation network from one-shot support set during the model initialization phase and learn to take advantage of segmented frames to update the segmentation during frame processing phase. Our model thus can make efficient case-by-case adaption and online updating within one feed-forward process.

3.3 Graph Memory Network

The graph memory network consists of an external graph memory and learnable controllers for memory operating. The memory provides short-term storage for new knowledge encoding and its graph structure allows to fully explore the context. The controllers interact with the graph memory through a number of read and write operations and they are capable of long-term storage via slow updates of the weights. Through the controllers, our model learns a general strategy for the types of representations it should place into the memory and how it should later use these representations for segmentation predictions.

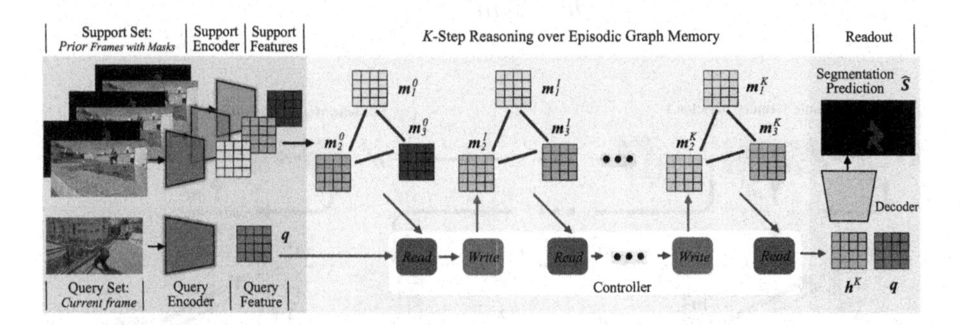

Fig. 1. Illustration of our graph memory based O-VOS method. Previous frames are fed together with the pre-defined or self-segmented masks to the support encoder to initialize graph memory nodes $\{\boldsymbol{m}_i^0\}_i$. Current frame is fed into the query encoder to output the query embedding $\boldsymbol{q}$. The graph memory interacts with $\boldsymbol{q}$ under several episodic reasoning (with learnable read and write controllers) to mine support context and generate video specific features. After K-step episodic reasoning, the decoder predicts segmentation mask $\hat{\boldsymbol{S}}$, based on the episodic feature $\boldsymbol{h}^K$ and query embedding $\boldsymbol{q}$.

The core idea of our graph memory network is to perform K steps of episodic reasoning to efficiently mine the structures in the memory and better capture target-specific information. Specifically, the memory is organized as a size-fixed, fully connected graph $\mathcal{G} = (\mathcal{M}, \mathcal{E})$, where node $m_i \in \mathcal{M}$ denotes i^{th} memory cell, and edge $e_{i,j} = (m_i, m_j) \in \mathcal{E}$ indicates the relation between cell m_i and m_j.

Given a query frame, the support set is considered as the combination of the first annotated frame and previously segmented frames. The graph memory is initialized from $N (= |\mathcal{M}|)$ frames, sampled from the support set. For each memory node m_i, its initial embedding $\boldsymbol{m}_i^0 \in mathbbR^{W \times H \times C}$ is generated by applying a fully convolutional memory encoder to the corresponding support frame to capture both the spatial visual feature as well as segmentation mask information.

Graph Memory Reading. A fully convolutional query encoder is also applied to the query frame to extract the visual feature $\boldsymbol{q} \in \mathbb{R}^{W\times H\times C}$. A learnable *read controller* first takes $\boldsymbol{q}$ as input and generates its initial state $\boldsymbol{h}^0$:

$$\boldsymbol{h}^0 = f_{\mathrm{P}}(\boldsymbol{q}) \in \mathbb{R}^{W\times H\times C}, \tag{1}$$

where $f_{\mathrm{P}}(\cdot)$ indicates a projection function.

At each episodic reasoning step $k \in \{1, ..., K\}$, the read controller interacts with the external graph memory by reading the content. Following the key-value retrieval mechanism in [21,28,43], we first compute the similarity between the query and each memory node m_i:

$$s_i^k = \frac{\boldsymbol{h}^{k-1} \cdot \boldsymbol{m}_i^{k-1}}{\|\boldsymbol{h}^{k-1}\|\|\boldsymbol{m}_i^{k-1}\|} \in [-1, 1]. \tag{2}$$

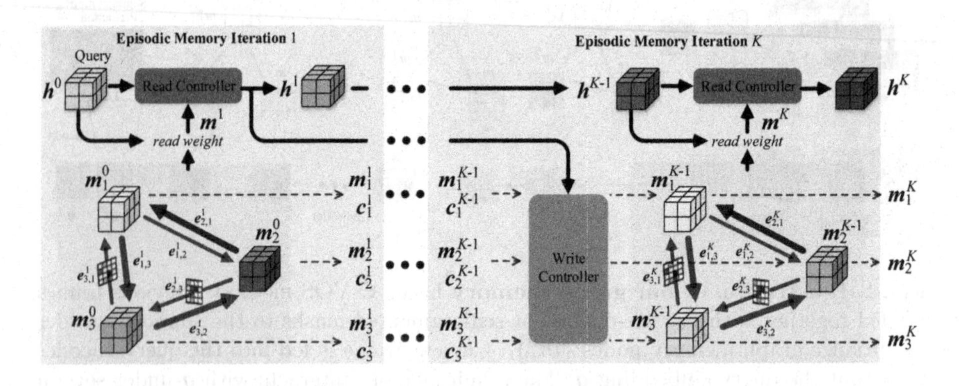

Fig. 2. Illustration of iterative reasoning over the episodic graph memory.

Next, we compute the read weight w_i^k by a softmax normalization function:

$$w_i^k = \exp(s_i^k) \Big/ \sum\nolimits_j \exp(s_j^k) \in [0, 1]. \tag{3}$$

Considering some nodes are noisy due to the underlying camera shift or out-of-view, w_i^k measures the confidence of memory cell m_i. Then the memory summarization $\boldsymbol{m}^k$ is retrieved using this weight to linearly combine the memory cell:

$$\boldsymbol{m}^k = \sum\nolimits_i w_i^k \boldsymbol{m}_i^{k-1} \in \mathbb{R}^{W\times H\times C}. \tag{4}$$

Through Eqs. (2–4), the memory module retrieves the memory cell most similar to $\boldsymbol{h}^k$ to obtain the memory summarization $\boldsymbol{m}^k$. Once reading the memory summarization, the read controller updates its state as follows:

$$
\begin{aligned}
\widetilde{\boldsymbol{h}}^k &= \boldsymbol{W}_r^h * \boldsymbol{h}^{k-1} + \boldsymbol{U}_r^h * \boldsymbol{m}^k &&\in \mathbb{R}^{W\times H\times C},\\
\boldsymbol{a}_r^k &= \sigma(\boldsymbol{W}_r^a * \boldsymbol{h}^{k-1} + \boldsymbol{U}_r^a * \boldsymbol{m}^k) &&\in [0,1]^{W\times H\times C},\\
\boldsymbol{h}^k &= \boldsymbol{a}_r^k \circ \widetilde{\boldsymbol{h}}^k + (\boldsymbol{1} - \boldsymbol{a}_r^k) \circ \boldsymbol{h}^{k-1} &&\in \mathbb{R}^{W\times H\times C},
\end{aligned} \tag{5}
$$

where $\boldsymbol{W}$s and $\boldsymbol{U}$s are convolution kernels and σ indicates the *sigmoid* activation function. '$*$' and '$\circ$' represent the convolution operation and Hadamard product, respectively. The update gate $\boldsymbol{a}^k$ controls how much previous hidden state $\boldsymbol{h}^{k-1}$ to be kept. In this way, the hidden state of the controller encodes both the graph memory and query representations, which are necessary to produce an output.

Episodic Graph Memory Updating. After each pass through the memory summarization, we need to update the episodic graph memory with the new query input. At each step k, a learnable memory *write controller* updates each memory cell (*i.e.*, graph node) m_i by considering its previous state $\boldsymbol{m}_i^{k-1}$, current content from the read controller $\boldsymbol{h}^k$, and the states from other cells $\{\boldsymbol{m}_j^{k-1}\}_{j\neq i}$. Specifically, following [52], we first formulate the relation $\boldsymbol{e}_{i,j}^k$ from m_j to m_i as the inner-product similarity over their feature matrices:

$$
\boldsymbol{e}_{i,j}^k = \boldsymbol{m}_i^{k-1}\, \boldsymbol{W}_e\, \boldsymbol{m}_j^{k-1\top} \in \mathbb{R}^{(WH)\times(WH)}, \tag{6}
$$

where $\boldsymbol{W}_e \in \mathbb{R}^{C\times C}$ indicates a learnable weight matrix, and $\boldsymbol{m}_i^{k-1} \in \mathbb{R}^{(WH)\times C}$ and $\boldsymbol{m}_j^{k-1} \in \mathbb{R}^{(WH)\times C}$ are flattened into matrix representations. $\boldsymbol{e}_{i,j}^k$ stores similarity scores corresponding to all pairs of positions in $\boldsymbol{m}_i$ and $\boldsymbol{m}_j$.

Then, for m_i, we compute the summarized information $\boldsymbol{c}_i^k$ from other cells, weighted by their normalized inner-product similarities:

$$
\boldsymbol{c}_i^k = \sum\nolimits_{j\neq i} \mathrm{softmax}(\boldsymbol{e}_{i,j}^k)\boldsymbol{m}_j^{k-1} \in \mathbb{R}^{W\times H\times C}, \tag{7}
$$

where softmax$(\cdot)$ normalizes each row of the input.

After aggregating the information from neighbors, the memory write controller updates the state of m_i as:

$$
\begin{aligned}
\widetilde{\boldsymbol{m}}_i^k &= \boldsymbol{W}_u^m * \boldsymbol{h}^k + \boldsymbol{U}_u^m * \boldsymbol{m}_i^{k-1} + \boldsymbol{V}_u^m * \boldsymbol{c}_i^k &&\in \mathbb{R}^{W\times H\times C},\\
\boldsymbol{a}_u^k &= \sigma(\boldsymbol{W}_u^a * \boldsymbol{h}^k + \boldsymbol{U}_u^a * \boldsymbol{m}_i^{k-1} + \boldsymbol{V}_u^a * \boldsymbol{c}_i^k) &&\in [0,1]^{W\times H\times C},\\
\boldsymbol{m}_i^k &= \boldsymbol{a}_u^k \circ \widetilde{\boldsymbol{m}}^k + (\boldsymbol{1} - \boldsymbol{a}_u^k) \circ \boldsymbol{m}^{k-1} &&\in \mathbb{R}^{W\times H\times C}.
\end{aligned} \tag{8}
$$

The graph memory updating allows each memory cell to embed the neighbor information into its representation so as to fully explore the context in the support set. Moreover, by iteratively reasoning over the graph structure, each memory cells encode the new query information and yield progressively improved representations. Compared with traditional memory network [58], the proposed graph memory network brings two advantages: **i)** the memory writing operation is fused into the memory updating procedure without increasing the memory size, and **ii)** avoiding designing complex memory writing strategies [21,39,58]. Figure 2 shows a detailed diagram of memory reading and updating.

Final Segmentation Readout. After K steps of the episodic memory updating, we leverage the final state $\boldsymbol{h}^K$ from the memory read controller to support the prediction on the query:

$$\hat{\boldsymbol{S}} = f_{\mathrm{R}}(\boldsymbol{h}^K, \boldsymbol{q}) \in [0,1]^{W \times H \times 2}, \quad (9)$$

where the readout function $f_{\mathrm{R}}(\cdot)$ gives the final segmentation probability maps.

3.4 Full Network Architecture

Network Configuration. Our whole model is end-to-end fully convolutional. Both the query encoder and memory encoder have the same network architecture, except for the inputs. The query encoder takes an RGB query frame as input, while, for the memory encoder, input is an RGB support frame concatenated with the one-channel softmax object mask and one-hot label map [31]. For the graph memory, the *read controller* (Eq. (5)) and *write controller* (Eq. (8)) are all implemented using ConvGRU [2], with 1×1 convolutional kernels. The project function f_{P} (Eq. (1)) is also realized with 1×1 convolutional layer. Similar to [59], the readout function f_{R} (Eq. (9)) is implemented with a decoder network, which consists of four blocks with skip connections to the corresponding ResNet50 [14] blocks. The kernel size of each convolution layer in the decoder is set as 3×3, excepting the last 1×1 convolution layer. The final 2-channel segmentation prediction is obtained by a softmax operation. The query and memory encoders are implemented as the four convolution blocks of ResNet50, initialized by the weights pretrained on ImageNet. The other layers are randomly initialized. Considering memory encoder takes binary mask and instance label maps as input, extra 1×1 convolutional layers are used for encoding these inputs. The resulting features are added with RGB features at the first blocks of ResNet50.

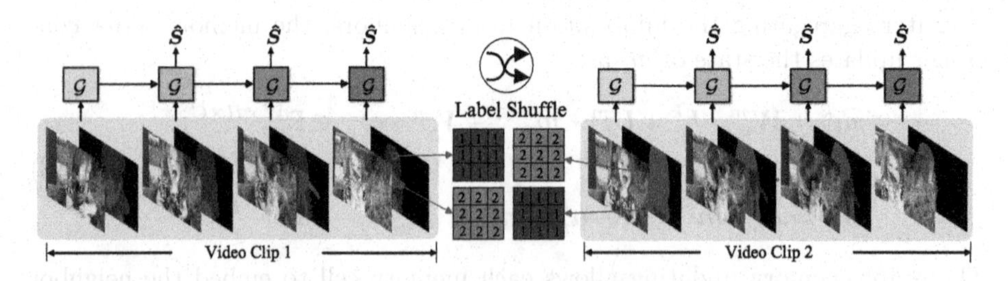

Fig. 3. From video clip to video clip, the instances with associated labels are shuffled. $\mathcal{G}$ denotes the graph memory network and $\hat{\boldsymbol{S}}$ is the output prediction.

Training. For O-VOS, we train our model following the "recurrence training" procedure [13,59]. Each training pass is formed by sampling a support set to build the graph memory and a relevant query set. The core heart of recurrence training is to mimic the inference procedure [4]. For each video, we sample $N + 1$

frames to build a support set (first N frames) and a query set (last frame). Thus, the N support frames can be represented by a N-node memory graph. We apply a cross entropy loss for supervising training.

To prevent the graph memory from only memorizing the relation between the instance and the one-hot vector label, we employ the label shuffling strategy [39]. As shown in Fig. 3, every time we run a segmentation episode, we shuffle the one-hot instance labels [49], meaning sometimes the label of specific instance becomes 2 instead of 1 and vice versa. This encourages the segmentation network to learn to distinguish the specific instance in current frame by considering the current training samples rather than memorizing specific relation between the target and the given label. See Sect. 4.3 for detailed experiments for the efficacy of our label shuffling strategy.

To further boost performance, we extend the training set with synthetic videos [31,49,59]. Specifically, for a static image, the video generation technique [33] is adopted to obtain simulated video clips, through different transformation operations (*e.g.*, rotation, scaling, translation and sheering). The static images come from existing image segmentation datasets. After pre-training on the synthetic videos, we use the real video data for finetuning.

For Z-VOS, we follow a similar training protocol as O-VOS, but the input modality only has RGB data. We do not use the label shuffling strategy, as we focus on an object-level Z-VOS setting (*i.e.*, do not discriminate different object instances). More training details for Z-VOS and O-VOS can be found in Sect. 3.5.

Inference. After training, we directly apply the learned network to unseen test videos without online finetuning. For O-VOS, we process each testing video in a sequential manner. For the first N frames, we compute the memory summarization (Eq. (4)) directly and write these frames into the memory. From $(N+1)^{th}$ frame, after segmentation, we would use this frame to update the graph memory. Considering the first frame and its annotation always provide the most reliable information, we re-initialize the node which stores the information about the first frame. Therefore, we use the first annotated frame, last segmented frame and $N-2$ frames sampled from previous segmented frames, as well as their pre-defined or segmented masks to build the memory. For multiple-instances cases, we run our model for each instance independently and obtain a soft-max probability mask for each one. Considering the underlying probability overlap between different instances, we combine these results together with a *soft-aggregation* strategy [59]. Our network achieves fast segmentation speed of 0.2 s per frame.

For Z-VOS, we randomly sample N frames from the same video to build the graph memory, then we process each frame based on the constructed memory. Considering the global information is more important than local information for handling underlying object occlusions and camera movements, we process each frame independently by re-initializing the graph memory with globally sampled frames. Following common practice [8,42,45], we employ CRF [20] binarization and the whole processing speed is about 0.3 s per frame.

3.5 Implementation Details

During pre-training, we randomly crop 384 × 384 patches from static image samples, and the video clip length is $N + 1 = 4$. During the main training, we crop 384 × 640 patches from real training videos. We randomly sample four temporal ordered frames from the same video with a maximum skip of 25 frames to build a training clip. Data augmentation techniques, like rotation, flip and saturation, are adopted to increase the data diversity. For O-VOS, we select a saliency dataset, MSRA10K [9], and semantic segmentation dataset, COCO [23], to synthesize videos. We use all the training videos in DAVIS_{17} [36] and Youtube-VOS [64] for the main training. For Z-VOS, we use two image saliency datasets, MSRA10K [9] and DUT [65], to generate simulated videos. These two datasets are selected following conventions [24,45] for fair comparison. After that, we take advantage of the training set of DAVIS_{16} [34] to finetune the network.

Our model is implemented on *PyTorch* and trained on four NVIDIA Tesla V100 GPUs with 32 GB memory per card. The batch size is set to 16. We optimize the loss function with Adam optimizer using "poly" learning schedule, with the base learning rate of $1e{-}5$ and power of 1.0. The pre-training stage takes about 24 h and the main training stage takes about 16 h for O-VOS.

4 Experiments

To verify the effectiveness and generic applicability of the proposed method, we perform experiments on different VOS settings. In concrete, we first evaluate

Table 1. Evaluation of O-VOS on DAVIS_{17} `val` set (Sect. 4.1), with region similarity $\mathcal{J}$, boundary accuracy $\mathcal{F}$ and average of $\mathcal{J}\&\mathcal{F}$. Speed is also reported.

Method		OSMN [66]	SIMMASK [51]	FAVOS [7]	RVOS [48]	OSVOS [5]	AGAME [17]	OnAVOS [50]	RGMP [59]
$\mathcal{J}\&\mathcal{F}$	Mean ↑	54.8	65.4	58.2	60.6	60.3	71.0	65.4	66.7
$\mathcal{J}$	Mean ↑	52.5	54.3	54.6	57.5	56.6	68.5	61.6	64.8
	Recall↑	60.9	62.8	61.1	65.2	63.8	78.4	67.4	74.1
	Decay↓	21.5	19.3	14.1	24.9	26.1	14.0	27.9	18.9
$\mathcal{F}$	Mean ↑	57.1	58.5	61.8	63.6	63.9	73.6	69.1	68.6
	Recall↑	66.1	67.5	72.3	73.2	73.8	83.4	75.4	77.7
	Decay↓	24.3	21.0	18.0	28.2	27.0	15.8	26.6	19.6
Times (s)		0.13	0.028	1.8	1.8	7.0	0.07	13	0.13
Method		OSVOS-S [27]	RANet [56]	FEELVOS [49]	CINM [3]	PReMVO [26]	DMMNet [70]	STM [31]	**Ours**
$\mathcal{J}\&\mathcal{F}$	Mean ↑	68.0	65.7	71.5	70.6	77.8	70.7	81.8	**82.8**
$\mathcal{J}$	Mean ↑	64.7	63.2	69.1	67.2	73.9	68.1	79.2	**80.2**
	Recall↑	74.2	73.7	79.1	74.5	83.1	77.3	88.7	**90.1**
	Decay↓	15.1	18.6	17.5	24.6	16.2	16.8	**8.0**	6.0
$\mathcal{F}$	Mean ↑	71.3	68.2	74.0	74.0	81.8	73.3	84.3	**85.2**
	Recall↑	80.7	78.8	83.8	81.6	88.9	82.7	91.8	**93.3**
	Decay↓	18.5	19.7	20.1	26.2	19.5	23.5	**10.5**	8.4
Times (s)		4.5	0.13	0.5	38	70	2.7	0.18	0.2

our model on two O-VOS datasets (Sect. 4.1) and then test it on two Z-VOS datasets (Sect. 4.2). Finally, in Sect. 4.3, agnostic experiments are conducted for in-depth analysis.

4.1 Performance for O-VOS

Datasets. Experiments are conducted on two well-known O-VOS benchmarks: DAVIS_{17} [36] and Youtube-VOS [64]. DAVIS_{17} comprises 60 videos for training and 30 videos for validation. Each video contains one or multiple annotated object instances. Youtube-VOS is a large-scale dataset which is split into a `train` set (3,471 videos) and a `val` set (474 videos). The validation set is further divided into `seen` subset which has the same categories (65) as the `train` set and `unseen` subset with 26 unseen categories.

Evaluation Criteria. Following the standard evaluation protocol of DAVIS_{17}, the mean region similarity $\mathcal{J}$ and contour accuracy $\mathcal{F}$ are reported. For Youtube-VOS, these two metrics are separately computed for the `seen` and `unseen` sets.

Quantitative Results. The performance of our network on DAVIS_{17} is shown in Table 1 with both online learning and offline approaches. Overall, our model outperforms all the contemporary methods and sets a new state-of-the-art in terms of mean $\mathcal{J}\&\mathcal{F}$ (82.8%), mean $\mathcal{J}$ (80.2%) and mean $\mathcal{F}$ (85.2%). Notably, our method obtains a much higher score for both region similarity and contour accuracy compared to several representative online learning methods: OSVOS [5], OnAVOS [50], AGAME [17] and DMMNet [70]. Furthermore, we report the segmentation speed and memory comparison by averaging the inference times for all instances. We observe that most segmentation-by-detection methods (*e.g.*, OSVOS [5]) consume small GPU memory but need a long time for first frame finetuning and online learning. Meanwhile, most matching based methods (*e.g.*, AGAME [17], FEELVOS [49], and RGMP [59]) achieve fast inference yet suffer from heavy memory cost. However, our method achieves better performance with fast speed and acceptable memory usage.

Moreover, we report the segmentation results on Youtube-VOS in Table 2. Our approach obtains a final score of 80.2%, significantly outperforming state-of-the-arts. Compared to memory-based method S2S [64], our model achieves much higher performance (*i.e.*, 80.2% *vs* 64.4%), which verifies the effectiveness of our external graph memory design. Moreover, our method performs favorably on both *seen* and *unseen* categories. Overall, our method achieves huge performance promotion over time-consuming online learning base methods without invoking online finetuning. This demonstrates the efficacy of our core idea of formulating O-VOS as a procedure of learning to update segmentation network.

Qualitative Results. In Fig. 4, we show qualitative results of our method at different time steps (uniformly sampled percentage w.r.t. the whole video length) on a few representative videos. Specifically, many instances in the first two DAVIS_{17} videos undergo fast motion and background clutter. However, through the graph

Table 2. Evaluation of O-VOS on Youtube-VOS `val` set (Sect. 4.1), with region similarity $\mathcal{J}$ and boundary accuracy $\mathcal{F}$. "Overall": averaged over the four metrics

Method		MSK [33]	OSMN [66]	RGMP [59]	OnAVOS [50]	RVOS [48]	OSVOS [5]	S2SS2S [64]	AGAME [17]	PreMVOS [26]	DMMNet [70]	STM [31]	Ours
Overall		53.1	51.2	53.8	55.2	56.8	58.8	64.4	66.1	66.9	58.0	79.4	**80.2**
Seen	Mean $\mathcal{J}$ ↑	59.9	60.0	59.5	60.1	63.6	59.8	71.0	67.8	71.4	60.3	79.7	**80.7**
	Mean $\mathcal{F}$ ↑	59.5	60.1	–	62.7	67.2	60.5	70.0	–	75.9	63.5	84.2	**85.1**
Unseen	Mean $\mathcal{J}$ ↑	45.0	40.6	45.2	46.6	45.5	54.2	55.5	60.8	56.5	50.6	72.8	**74.0**
	Mean $\mathcal{F}$ ↑	47.9	44.0	–	51.4	51.0	60.7	61.2	–	63.7	57.4	80.9	**80.9**

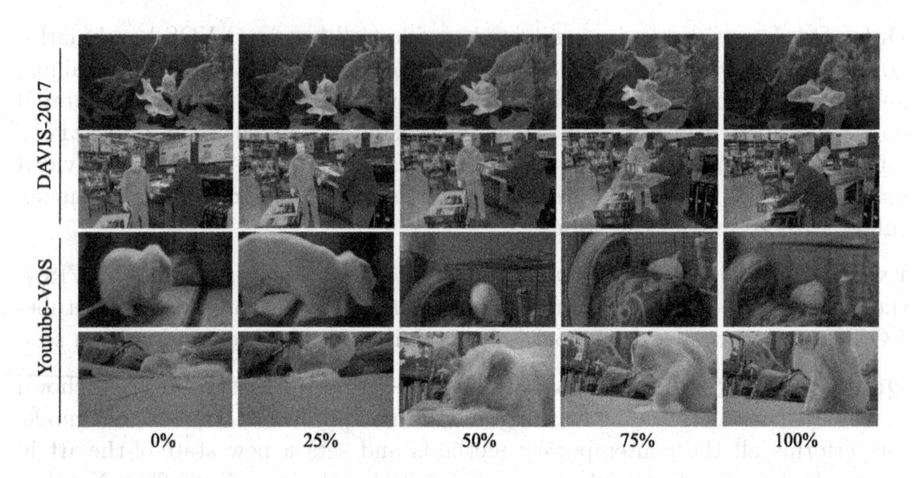

Fig. 4. Qualitative O-VOS results on DAVIS_{17} and Youtube-VOS (Sect. 4.1).

memory mechanism, our segmentation network can handle these challenging factors well. The last two Youtube-VOS videos present challenges that the instances suffer occlusion and out-of-view. Once the occlusion ends, our graph memory allows the segmentation network to re-detect the target and segment it successfully.

4.2 Performance for Z-VOS

Datasets. Experiments are conducted on two challenging datasets: DAVIS_{16} [34] and Youtube-Objects [37]. DAVIS_{16} contains 50 videos with 3,455 frames, covering a wide range of challenges, such as fast motion and occlusion. It is split into a `train` set (30 videos) and a `val` set (20 videos). Youtube-Objects has 126 videos belonging to 10 categories and 25,673 frames in total. The `val` set of DAVIS_{16} and whole Youtube-Objects are used for evaluation.

Evaluation Criteria. We follow the official evaluation protocols [32,34] and report the region similarity $\mathcal{J}$, boundary accuracy $\mathcal{F}$ and time stability $\mathcal{T}$ for DAVIS_{16}. Youtube-Objects is evaluated in terms of region similarity $\mathcal{J}$.

Quantitative Results. For DAVIS_{16} [34], we compare our method with 17 state-of-the-arts from DAVIS_{16} benchmark[1] in Table 3. Our method outperforms other competitors across most metrics. In particular, compared with recent

[1] https://davischallenge.org/davis2016/soa_compare.html.

Table 3. Evaluation of Z-VOS on DAVIS$_{16}$ val set [34] (Sect. 4.2), with region similarity $\mathcal{J}$, boundary accuracy $\mathcal{F}$ and time stability $\mathcal{T}$.

Method		MSG [29]	NLC [12]	CUT [18]	FST [32]	SFL [8]	LMP [44]	FSEG [16]	LVO [45]	UOVOS [73]
$\mathcal{J}$	Mean ↑	53.3	55.1	55.2	55.8	64.7	70.0	70.7	75.9	73.9
	Recall ↑	61.6	55.8	57.5	64.9	81.4	85.0	83.0	89.1	88.5
	Decay ↓	2.4	12.6	2.2	**0.0**	6.2	1.3	1.5	**0.0**	0.6
$\mathcal{F}$	Mean ↑	50.8	52.3	55.2	51.1	66.7	65.9	65.3	72.1	68.0
	Recall ↑	60.0	61.0	51.9	51.6	77.1	79.2	73.8	83.4	80.6
	Decay ↓	5.1	11.4	3.4	2.9	5.1	2.5	1.8	1.3	0.7
$\mathcal{T}$	Mean ↓	54.8	65.4	58.2	60.6	60.3	71.0	65.4	66.7	39.0
Method		ARP [19]	PDB [42]	MotAdapt [41]	LSMO [46]	AGS [55]	COSNet [24]	AGNN [52]	AnDiff [68]	**Ours**
$\mathcal{J}$	Mean ↑	76.2	77.2	77.2	78.2	79.7	80.5	80.7	81.7	**82.5**
	Recall ↑	89.1	91.1	93.1	87.8	91.1	93.1	94.0	90.9	**94.3**
	Decay ↓	7.0	0.9	5.0	4.1	1.9	4.4	0.03	2.2	4.2
$\mathcal{F}$	Mean ↑	65.3	72.1	70.6	74.5	77.4	79.4	79.1	80.5	**81.2**
	Recall ↑	83.4	83.5	84.4	84.7	85.8	89.5	**90.5**	85.1	90.3
	Decay ↓	7.9	-0.2	3.3	3.5	0.0	5.0	0.03	0.6	5.6
$\mathcal{T}$	Mean ↓	39.3	29.1	27.9	21.2	26.7	**18.4**	33.7	21.4	19.8

matching-based methods: COSNet [24], AGNN [52] and AnDiff [73], our method achieves an average $\mathcal{J}$ score of 82.5% which is 0.8% better than the second best method, AnDiff [73] despite the fact that it utilizes more training samples than ours. Compared with COSNet [24], our method achieves significant performance promotion of 2.0% and 1.8% in terms of mean $\mathcal{J}$ and mean $\mathcal{F}$, respectively. Notably, our method outperforms online learning based methods (*i.e.*, SFL [8], UVOS [73] and LSMO [46]) by a large margin.

We further report results on Youtube-Objects in Table 4 with detailed category-wise performance as well as the final average $\mathcal{J}$ score. As seen, our method surpasses other competitors significantly (reaching 71.4% mean $\mathcal{J}$), especially compared with recent matching based methods [24,52]. Overall, our model consistently yields promising results over different datasets, which clearly illustrates its superior performance and powerful generalizability.

Table 4. Evaluation of Z-VOS on Youtube-Objects [37] (Sect. 4.2). "Mean $\mathcal{J}$↑" denotes the results averaged over all the categories.

Method	LTV [30]	FST [32]	COSEG [47]	ARP [19]	LVO [45]	PDB [42]	FSEG [16]	SFL [8]	MotAdapt [41]	LSMO [46]	AGS [55]	COSNet [24]	AGNN [52]	**Ours**
Airplane (6)	13.7	70.9	69.3	73.6	86.2	78.0	81.7	65.6	77.2	60.5	87.7	81.1	81.1	86.1
Bird (6)	12.2	70.6	76.0	56.1	81.0	80.0	63.8	65.4	42.2	59.3	76.7	75.7	75.9	75.7
Boat (15)	10.8	42.5	53.5	57.8	68.5	58.9	72.3	59.9	49.3	62.1	72.2	71.3	70.7	68.6
Car (7)	23.7	65.2	70.4	33.9	69.3	76.5	74.9	64.0	68.6	72.3	78.6	77.6	78.1	82.4
Cat (16)	18.6	52.1	66.8	30.5	58.8	63.0	68.4	58.9	46.3	66.3	69.2	66.5	67.9	65.9
Cow (20)	16.3	44.5	49.0	41.8	68.5	64.1	68.0	51.1	64.2	67.9	64.6	69.8	69.7	70.5
Dog (27)	18.2	65.3	47.5	36.8	61.7	70.1	69.4	54.1	66.1	70.0	73.3	76.8	77.4	77.1
Horse (14)	11.5	53.5	55.7	44.3	53.9	67.6	60.4	64.8	64.8	65.4	64.4	67.4	67.3	72.2
Motorbike (10)	10.6	44.2	30.5	48.9	60.8	58.3	62.7	52.6	44.6	55.5	62.1	67.7	68.3	63.8
Train (5)	19.6	29.6	53.4	39.2	66.3	35.2	62.2	34.0	42.3	38.0	48.2	46.8	47.8	47.8
Mean $\mathcal{J}$↑	15.5	53.8	58.1	46.2	67.5	65.4	68.4	57.0	58.1	64.3	69.7	70.5	70.8	**71.4**

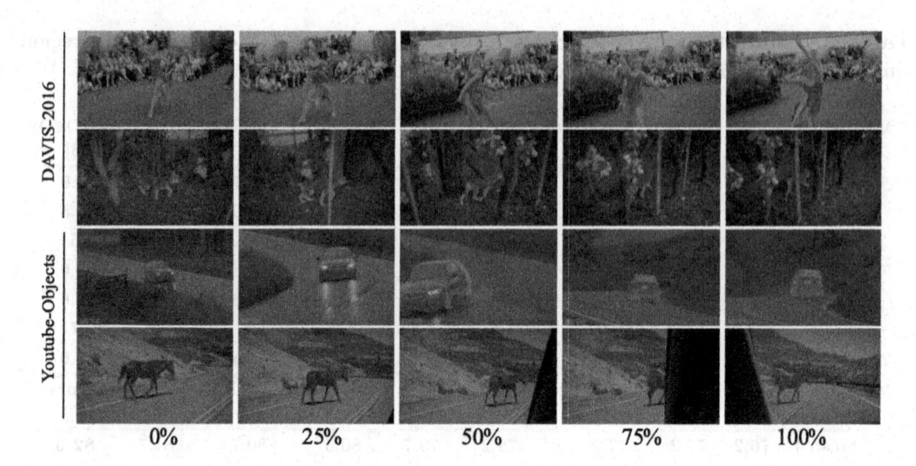

Fig. 5. Qualitative Z-VOS results on $DAVIS_{16}$ and Youtube-Objects (Sect. 4.2).

Quantitative Results. Figure 5 depicts some representative visual results on $DAVIS_{16}$ and Youtube-Objects. As can be observed, our method handles well these challenging scenes, typically with fast motion, partial occlusion and view changes, even without explicit foreground object indication.

4.3 Diagnostic Experiments

In this section, we analyze the contribution of different model components to the final performance. Specifically, we take O-VOS and Z-VOS as exemplar and evaluate all ablated versions on $DAVIS_{17}$ [36] and $DAVIS_{16}$ [34], respectively. The experimental results are evaluated by mean $\mathcal{J}$ and mean $\mathcal{F}$. For each ablated version, we retrain the model from scratch using the same protocol. From the whole results comparison in Table 5, we can draw several essential conclusions.

Table 5. Ablation study of our graph memory network (Sect. 4.3).

Aspect	Method	One-shot VOS $DAVIS_{17}$ [36]		Zero-shot VOS $DAVIS_{16}$ [34]	
		Mean $\mathcal{J}\uparrow$	Mean $\mathcal{F}\uparrow$	Mean $\mathcal{J}\uparrow$	Mean $\mathcal{F}\uparrow$
Full model	Graph memory (3 nodes, 3 episodes)	80.0	85.9	82.5	81.2
Backbone	Direct infer. *w/o* graph memory	73.5	78.4	73.2	72.5
Graph structure	2 nodes	76.0	81.4	78.5	76.8
	4 nodes	79.5	84.6	82.5	81.2
	5 nodes	80.0	85.9	82.5	81.2
State updating	$K = 0$	78.1	82.2	81.2	79.7
	$K = 1$	78.9	83.3	81.6	80.3
	$K = 2$	79.3	84.8	82.0	80.8
	$K = 4$	80.0	85.9	82.5	81.2
Training	*w/o* label shuffling	78.5	82.7	–	–

Graph Memory Network: First, with the proposed graph memory network, our method yields significant performance improvements (+6.5%, +7.5% and +9.3%, +8.7% in terms of mean $\mathcal{J}$ and mean $\mathcal{F}$) than the backbone over different VOS settings. This supports our view that the graph memory network with differential controllers can learn to update the segmentation network effectively.

Memory Size: We next investigate the influence of memory size on the final performance (1st and 3rd–5th rows). We find only a 3-node memory is enough for achieving good performance, further verifying the efficacy of our memory design.

Iterative Memory Reasoning: It is also of interest to assess the impact of our iterative memory updating strategy. When $K = 0$, it means no update for graph memory network, therefore the state of the network is fixed without online learning. In this case, the results deteriorate significantly. We further observe that more steps can boost the performance ($1 \rightarrow 3$) and when the step is increased to certain extent ($K = 4$), the performance remains almost unchanged.

Label Shuffling: Finally we study the effect of our label shuffling strategy. Comparing results on the first and last rows, we can easily observe that shuffling instance labels during network training indeed promotes O-VOS performance.

5 Conclusion

This paper integrates a novel graph memory mechanism to efficiently adapt the segmentation network to specific videos without catastrophic inference/finetune. Through episodic reasoning the memory graph, the proposed model is capable of generating video-specific memory summarization which benefits the final segmentation prediction significantly. Meanwhile, the online model updating can be implemented via learnable memory controllers. Our method is effective and principle, which can be easily extended to Z-VOS setting. Extensive experimental results on four challenging datasets demonstrate its promising performance.

Acknowledgements. This work was partially supported by Zhejiang Lab's Open Fund (No. 2019KD0AB04), Zhejiang Lab's International Talent Fund for Young Professionals, CCF-Tencent Open Fund, ETH Zürich Fund (OK), an Huawei Technologies Oy (Finland) project, an Amazon AWS grant, and Nvidia..

References

1. Badrinarayanan, V., Galasso, F., Cipolla, R.: Label propagation in video sequences. In: CVPR (2010)
2. Ballas, N., Yao, L., Pal, C., Courville, A.: Delving deeper into convolutional networks for learning video representations. In: ICLR (2016)
3. Bao, L., Wu, B., Liu, W.: CNN in MRF: video object segmentation via inference in a CNN-based higher-order spatio-temporal MRF. In: CVPR (2018)
4. Behl, H.S., Najafi, M., Arnab, A., Torr, P.H.: Meta learning deep visual words for fast video object segmentation. In: NeurIPS Workshop (2019)

5. Caelles, S., Maninis, K.K., Pont-Tuset, J., Leal-Taixé, L., Cremers, D., Van Gool, L.: One-shot video object segmentation. In: CVPR (2017)
6. Chen, Y., Pont-Tuset, J., Montes, A., Van Gool, L.: Blazingly fast video object segmentation with pixel-wise metric learning. In: CVPR (2018)
7. Cheng, J., Tsai, Y.H., Hung, W.C., Wang, S., Yang, M.H.: Fast and accurate online video object segmentation via tracking parts. In: CVPR (2018)
8. Cheng, J., Tsai, Y.H., Wang, S., Yang, M.H.: SegFlow: joint learning for video object segmentation and optical flow. In: ICCV (2017)
9. Cheng, M.M., Mitra, N.J., Huang, X., Torr, P.H., Hu, S.M.: Global contrast based salient region detection. IEEE TPAMI **37**(3), 569–582 (2015)
10. Ci, H., Wang, C., Wang, Y.: Video object segmentation by learning location-sensitive embeddings. In: ECCV (2018)
11. Duarte, K., Rawat, Y.S., Shah, M.: CapsuleVOS: semi-supervised video object segmentation using capsule routing. In: ICCV (2019)
12. Faktor, A., Irani, M.: Video segmentation by non-local consensus voting. In: BMVC (2014)
13. Finn, C., Abbeel, P., Levine, S.: Model-agnostic meta-learning for fast adaptation of deep networks. In: ICML (2017)
14. He, K., Zhang, X., Ren, S., Sun, J.: Deep residual learning for image recognition. In: CVPR (2016)
15. Hu, Y.-T., Huang, J.-B., Schwing, A.G.: VideoMatch: matching based video object segmentation. In: Ferrari, V., Hebert, M., Sminchisescu, C., Weiss, Y. (eds.) ECCV 2018. LNCS, vol. 11212, pp. 56–73. Springer, Cham (2018). https://doi.org/10.1007/978-3-030-01237-3_4
16. Jain, S.D., Xiong, B., Grauman, K.: FusionSeg: learning to combine motion and appearance for fully automatic segmentation of generic objects in videos. In: CVPR (2017)
17. Johnander, J., Danelljan, M., Brissman, E., Khan, F.S., Felsberg, M.: A generative appearance model for end-to-end video object segmentation. In: CVPR (2019)
18. Keuper, M., Andres, B., Brox, T.: Motion trajectory segmentation via minimum cost multicuts. In: ICCV (2015)
19. Koh, Y.J., Kim, C.: Primary object segmentation in videos based on region augmentation and reduction. In: CVPR (2017)
20. Krähenbühl, P., Koltun, V.: Efficient inference in fully connected CRFs with gaussian edge potentials. In: NIPS (2011)
21. Kumar, A., et al.: Ask me anything: dynamic memory networks for natural language processing. In: ICML (2016)
22. Lee, Y.J., Kim, J., Grauman, K.: Key-segments for video object segmentation. In: ICCV (2011)
23. Lin, T.-Y., et al.: Microsoft COCO: common objects in context. In: Fleet, D., Pajdla, T., Schiele, B., Tuytelaars, T. (eds.) ECCV 2014. LNCS, vol. 8693, pp. 740–755. Springer, Cham (2014). https://doi.org/10.1007/978-3-319-10602-1_48
24. Lu, X., Wang, W., Ma, C., Shen, J., Shao, L., Porikli, F.: See more, know more: unsupervised video object segmentation with co-attention Siamese networks. In: CVPR (2019)
25. Lu, X., Wang, W., Shen, J., Tai, Y.W., Crandall, D.J., Hoi, S.C.: Learning video object segmentation from unlabeled videos. In: CVPR (2020)
26. Luiten, J., Voigtlaender, P., Leibe, B.: PReMVOS: proposal-generation, refinement and merging for video object segmentation. In: ACCV (2018)
27. Maninis, K.K., et al.: Video object segmentation without temporal information. IEEE TPAMI **41**(6), 1515–1530 (2018)

28. Miller, A., Fisch, A., Dodge, J., Karimi, A.H., Bordes, A., Weston, J.: Key-value memory networks for directly reading documents. In: EMNLP (2016)
29. Ochs, P., Brox, T.: Object segmentation in video: a hierarchical variational approach for turning point trajectories into dense regions. In: ICCV (2011)
30. Ochs, P., Malik, J., Brox, T.: Segmentation of moving objects by long term video analysis. IEEE TPAMI **36**(6), 1187–1200 (2014)
31. Oh, S.W., Lee, J.Y., Xu, N., Kim, S.J.: Video object segmentation using space-time memory networks. In: ICCV (2019)
32. Papazoglou, A., Ferrari, V.: Fast object segmentation in unconstrained video. In: ICCV (2013)
33. Perazzi, F., Khoreva, A., Benenson, R., Schiele, B., Sorkine-Hornung, A.: Learning video object segmentation from static images. In: CVPR (2017)
34. Perazzi, F., Pont-Tuset, J., McWilliams, B., Van Gool, L., Gross, M., Sorkine-Hornung, A.: A benchmark dataset and evaluation methodology for video object segmentation. In: CVPR (2016)
35. Perazzi, F., Wang, O., Gross, M., Sorkine-Hornung, A.: Fully connected object proposals for video segmentation. In: CVPR (2015)
36. Pont-Tuset, J., Perazzi, F., Caelles, S., Arbeláez, P., Sorkine-Hornung, A., Van Gool, L.: The 2017 DAVIS challenge on video object segmentation. arXiv preprint arXiv:1704.00675 (2017)
37. Prest, A., Leistner, C., Civera, J., Schmid, C., Ferrari, V.: Learning object class detectors from weakly annotated video. In: CVPR (2012)
38. Rakelly, K., Shelhamer, E., Darrell, T., Efros, A.A., Levine, S.: Meta-learning to guide segmentation. In: ICLR (2019)
39. Santoro, A., Bartunov, S., Botvinick, M., Wierstra, D., Lillicrap, T.: Meta-learning with memory-augmented neural networks. In: ICML (2016)
40. Shankar Nagaraja, N., Schmidt, F.R., Brox, T.: Video segmentation with just a few strokes. In: ICCV (2015)
41. Siam, M., et al.: Video segmentation using teacher-student adaptation in a human robot interaction (HRI) setting. In: ICRA (2019)
42. Song, H., Wang, W., Zhao, S., Shen, J., Lam, K.-M.: Pyramid dilated deeper ConvLSTM for video salient object detection. In: Ferrari, V., Hebert, M., Sminchisescu, C., Weiss, Y. (eds.) ECCV 2018. LNCS, vol. 11215, pp. 744–760. Springer, Cham (2018). https://doi.org/10.1007/978-3-030-01252-6_44
43. Sukhbaatar, S., Szlam, A., Weston, J., Fergus, R.: End-to-end memory networks. In: NIPS (2015)
44. Tokmakov, P., Alahari, K., Schmid, C.: Learning motion patterns in videos. In: CVPR (2017)
45. Tokmakov, P., Alahari, K., Schmid, C.: Learning video object segmentation with visual memory. In: ICCV (2017)
46. Tokmakov, P., Schmid, C., Alahari, K.: Learning to segment moving objects. IJCV **127**(3), 282–301 (2019)
47. Tsai, Y.-H., Zhong, G., Yang, M.-H.: Semantic co-segmentation in videos. In: Leibe, B., Matas, J., Sebe, N., Welling, M. (eds.) ECCV 2016. LNCS, vol. 9908, pp. 760–775. Springer, Cham (2016). https://doi.org/10.1007/978-3-319-46493-0_46
48. Ventura, C., Bellver, M., Girbau, A., Salvador, A., Marques, F., Giro-i Nieto, X.: RVOS: end-to-end recurrent network for video object segmentation. In: CVPR (2019)
49. Voigtlaender, P., Chai, Y., Schroff, F., Adam, H., Leibe, B., Chen, L.C.: FEELVOS: fast end-to-end embedding learning for video object segmentation. In: CVPR (2019)

50. Voigtlaender, P., Leibe, B.: Online adaptation of convolutional neural networks for video object segmentation. In: BMVC (2017)
51. Wang, Q., Zhang, L., Bertinetto, L., Hu, W., Torr, P.H.: Fast online object tracking and segmentation: a unifying approach. In: CVPR (2019)
52. Wang, W., Lu, X., Shen, J., Crandall, D.J., Shao, L.: Zero-shot video object segmentation via attentive graph neural networks. In: ICCV (2019)
53. Wang, W., Shen, J., Porikli, F., Yang, R.: Semi-supervised video object segmentation with super-trajectories. IEEE TPAMI **41**(4), 985–998 (2018)
54. Wang, W., Shen, J., Yang, R., Porikli, F.: Saliency-aware video object segmentation. IEEE TPAMI **40**(1), 20–33 (2017)
55. Wang, W., et al.: Learning unsupervised video object segmentation through visual attention. In: CVPR (2019)
56. Wang, Z., Xu, J., Liu, L., Zhu, F., Shao, L.: Ranet: Ranking attention network for fast video object segmentation. In: ICCV (2019)
57. Wen, L., Du, D., Lei, Z., Li, S.Z., Yang, M.H.: JOTS: joint online tracking and segmentation. In: CVPR (2015)
58. Weston, J., Chopra, S., Bordes, A.: Memory networks. ICLR (2015)
59. Wug Oh, S., Lee, J.Y., Sunkavalli, K., Joo Kim, S.: Fast video object segmentation by reference-guided mask propagation. In: CVPR (2018)
60. Xiao, H., Feng, J., Lin, G., Liu, Y., Zhang, M.: MoNet: deep motion exploitation for video object segmentation. In: CVPR (2018)
61. Xiao, H., Kang, B., Liu, Y., Zhang, M., Feng, J.: Online meta adaptation for fast video object segmentation. IEEE Trans. Pattern Anal. Mach. Intell. **42**, 1205–1217 (2019)
62. Xie, G.S., et al.: Attentive region embedding network for zero-shot learning. In: CVPR (2019)
63. Xie, G.S., et al.: Region graph embedding network for zero-shot learning. In: ECCV (2020)
64. Xu, N., et al.: YouTube-VOS: sequence-to-sequence video object segmentation. In: Ferrari, V., Hebert, M., Sminchisescu, C., Weiss, Y. (eds.) ECCV 2018. LNCS, vol. 11209, pp. 603–619. Springer, Cham (2018). https://doi.org/10.1007/978-3-030-01228-1_36
65. Yang, C., Zhang, L., Lu, H., Ruan, X., Yang, M.: Saliency detection via graph-based manifold ranking. In: CVPR (2013)
66. Yang, L., Wang, Y., Xiong, X., Yang, J., Katsaggelos, A.K.: Efficient video object segmentation via network modulation. In: CVPR (2018)
67. Yang, T., Chan, A.B.: Learning dynamic memory networks for object tracking. In: Ferrari, V., Hebert, M., Sminchisescu, C., Weiss, Y. (eds.) ECCV 2018. LNCS, vol. 11213, pp. 153–169. Springer, Cham (2018). https://doi.org/10.1007/978-3-030-01240-3_10
68. Yang, Z., Wang, Q., Bertinetto, L., Bai, S., Hu, W., Torr, P.H.: Anchor diffusion for unsupervised video object segmentation. In: ICCV (2019)
69. Yoon, J.S., Rameau, F., Kim, J., Lee, S., Shin, S., Kweon, I.S.: Pixel-level matching for video object segmentation using convolutional neural networks. In: ICCV (2017)
70. Zeng, X., Liao, R., Gu, L., Xiong, Y., Fidler, S., Urtasun, R.: DMM-Net: differentiable mask-matching network for video object segmentation. In: ICCV (2019)
71. Zhang, D., Javed, O., Shah, M.: Video object segmentation through spatially accurate and temporally dense extraction of primary object regions. In: CVPR (2013)

72. Zhou, T., Wang, S., Zhou, Y., Yao, Y., Li, J., Shao, L.: Motion-attentive transition for zero-shot video object segmentation. In: AAAI (2020)
73. Zhuo, T., Cheng, Z., Zhang, P., Wong, Y., Kankanhalli, M.: Unsupervised online video object segmentation with motion property understanding. IEEE TIP **29**, 237–249 (2019)

Rethinking Bottleneck Structure for Efficient Mobile Network Design

Daquan Zhou[1,2,3(✉)], Qibin Hou[1(✉)], Yunpeng Chen[2], Jiashi Feng[1], and Shuicheng Yan[2]

[1] National University of Singapore, Singapore, Singapore
zhoudaquan21@gmail.com, andrewhoux@gmail.com, elefjia@nus.edu.sg
[2] Yitu Technology, Singapore, Singapore
{yunpeng.chen,shuicheng.yan}@yitu-inc.com
[3] Institute of Data Science, NUS, Singapore, Singapore

Abstract. The inverted residual block is dominating architecture design for mobile networks recently. It changes the classic residual bottleneck by introducing two design rules: learning inverted residuals and using linear bottlenecks. In this paper, we rethink the necessity of such design changes and find it may bring risks of information loss and gradient confusion. We thus propose to flip the structure and present a novel bottleneck design, called the sandglass block, that performs identity mapping and spatial transformation at higher dimensions and thus alleviates information loss and gradient confusion effectively. Extensive experiments demonstrate that, different from the common belief, such bottleneck structure is more beneficial than the inverted ones for mobile networks. In ImageNet classification, by simply replacing the inverted residual block with our sandglass block without increasing parameters and computation, the classification accuracy can be improved by more than 1.7% over MobileNetV2. On Pascal VOC 2007 test set, we observe that there is also 0.9% mAP improvement in object detection. We further verify the effectiveness of the sandglass block by adding it into the search space of neural architecture search method DARTS. With 25% parameter reduction, the classification accuracy is improved by 0.13% over previous DARTS models. Code can be found at: https://github.com/zhoudaquan/rethinking_bottleneck_design.

Keywords: Sandglass block · Residual block · Efficient architecture design · Image classification

D. Zhou and Q. Hou—Authors contributed equally.
D. Zhou—Work done during an internship at Yitu Tech.

Electronic supplementary material The online version of this chapter (https://doi.org/10.1007/978-3-030-58580-8_40) contains supplementary material, which is available to authorized users.

A. Vedaldi et al. (Eds.): ECCV 2020, LNCS 12348, pp. 680–697, 2020.
https://doi.org/10.1007/978-3-030-58580-8_40

1 Introduction

A common belief behind the design principles of most popular light-weight models (either manually designed or automatically searched) [27,31,34,35] is to adopt the inverted residual block [31]. Compared to the classic residual bottleneck block [12,14], this block shifts the identity mapping from high-dimensional representations to low-dimensional ones (*i.e.,* the bottlenecks). However, connecting identity mapping between thin bottlenecks would inevitably lead to information loss since the residual representations are compressed as shown in Fig. 2(b). Moreover, it would also weaken the propagation capability of gradients across layers, due to gradient confusion arising from the narrowed feature dimensions [32]. Therefore, despite the wide use of the inverted residual block, how to design residual blocks for mobile devices is worthy of studying.

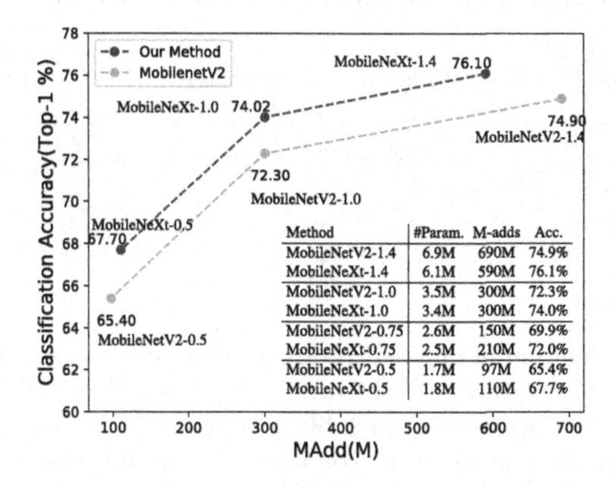

Method	#Param.	M-adds	Acc.
MobileNetV2-1.4	6.9M	690M	74.9%
MobileNeXt-1.4	6.1M	590M	76.1%
MobileNetV2-1.0	3.5M	300M	72.3%
MobileNeXt-1.0	3.4M	300M	74.0%
MobileNetV2-0.75	2.6M	150M	69.9%
MobileNeXt-0.75	2.5M	210M	72.0%
MobileNetV2-0.5	1.7M	97M	65.4%
MobileNeXt-0.5	1.8M	110M	67.7%

Fig. 1. Top-1 classification accuracy comparisons between the proposed MobileNeXt and MobileNetV2 [31]. We use different width multipliers to trade-off between model complexity and accuracy. Here, four widely-used multipliers are chosen, including 0.5, 0.75, 1.0, and 1.4. As can be seen, under each width multiplier, our MobileNeXt surpasses the MobileNetV2 baseline by a large margin, especially for the models with less learnable parameters.

In this paper, in view of the above concerns, we rethink the rationality of shifting from the classic bottleneck structure (Fig. 2(a)) to the popular inverted residual block (Fig. 2(b)) in developing mobile networks. In particular, we consider the following three fundamental questions. (i) What are the effects if we position the identity mapping (*i.e.,* shortcuts) at the high-dimensional representations as done in the classic bottleneck structure? (ii) While the linear activation can reduce information loss, should it only be applied to the bottlenecks? (iii) The previous questions remind us of the classic bottleneck structure which suffers high computational complexity. This cost can be reduced by replacing the dense spatial convolutions with depthwise ones, but, regarding the bottlenecks, should the depthwise convolution be still added in the low-dimensional bottleneck as conventional?

Motivated by the above questions, we present and evaluate a new bottleneck design, termed the sandglass block. Unlike the inverted residual block that builds shortcuts between linear bottlenecks, our sandglass block puts shortcut connections between linear high-dimensional representations, as shown in Fig. 2(c). Such structure preserves more information delivered between blocks compared to the inverted residual block and propagates more gradients backward to better optimize network training because of the high-dimensional residuals [32]. Furthermore, to learn more expressive spatial representation, instead of putting the spatial convolutions in the bottleneck with compressed channels, we propose to apply them in the expanded high dimensional feature space, which we find is an effective way of improving the model performance. In addition, we maintain the channel reduction and expansion process with pointwise convolutions to reduce computational cost. This makes our block quite different from the inverted residual block but more similar to the classic residual bottleneck.

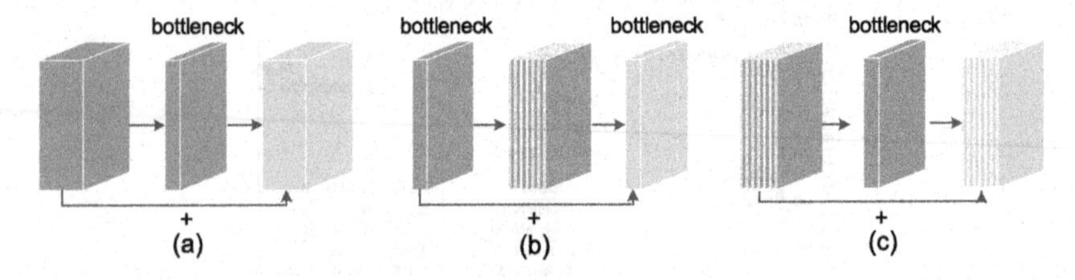

Fig. 2. Conceptual diagram of different residual bottleneck blocks. (a) Classic residual block with bottleneck structure [13]. (b) Inverted residual block [31]. (c) Our proposed sandglass block. We use thickness of each block to represent the corresponding relative number of channels. As can be seen, compared to the inverted residual block, the proposed residual block reverses the thought of building shortcuts between bottlenecks and adds depthwise convolutions (detached blocks) at both ends of the residual path, both of which are found crucial for performance improvement.

We stack the sandglass blocks in a modularized way to build the proposed MobileNeXt. Our network achieves more than 1.7% top-1 classification accuracy improvement over MobileNetV2 on ImageNet with slightly less computation and a comparable number of parameters as shown in Fig. 1. When applying the sandglass block on the EfficientNet topology to replace the inverted residual block, the resulting model surpasses the previous state-of-the-art by 0.5% with a comparable amount of computation but 20% parameter reduction. Particularly, in object detection, when taking SSDLite [25,31] as the object detector, using our MobileNeXt as backbone gains 0.9% in mAP on the Pascal VOC 2007 test set over MobileNetV2. More interestingly, we also experimentally find the proposed sandglass block can be used to enrich the search space of neural architecture search algorithms [24]. By adding the sandglass block into the search space as a 'super' operator, without changing the search algorithm, the resultant model can improve classification accuracy by 0.13% but with 25% less parameters compared to models searched from the vanilla space.

In summary, we make the following contributions in this paper:

- Our results advocate a rethinking of the bottleneck structure for mobile network design. It seems that the inverted residuals are not so advantageous over the bottleneck structure as commonly believed.
- Our study reveals that building shortcut connections along higher-dimensional feature space could promote model performance. Moreover, depthwise convolutions should be conducted in the high dimensional space for learning more expressive features and learning linear residuals is also crucial for bottleneck structure.
- Based on our study, we propose a novel sandglass block, which substantially extends the classic bottleneck structure. We experimentally demonstrate that this structure is more suitable for mobile applications in terms of both accuracy and efficiency and can be used as 'super' operators in architecture search algorithms for better architecture generation.

2 Related Work

Modern deep neural networks are mostly built by stacking building blocks, which are designed based on either the classic residual block with bottleneck structure [12] or the inverted residual block [31]. In this section, we categorize all related networks based on above two types of building blocks and briefly describe them below.

Classic Residual Bottleneck Blocks. The bottleneck structure was first introduced in ResNet [12]. A typical bottleneck structure consists of three convolutional layers: an 1×1 convolution for channel reduction, a 3×3 convolution for spatial feature extraction, and another 1×1 convolution for channel expansion. A residual network is often constructed by stacking a sequence of such residual blocks. The bottleneck structure was further developed in later works by widening the channels in each convolutional layer [41], applying group convolutions to the middle bottleneck convolution for aggregating richer feature representations [39], or introducing attention based modules to explicitly model inter-dependencies between channels [18,23]. There are also other works [3,37] combining residual blocks with dense connections to boost the performance. However, in spite of the success in heavy-weight network design, it is rarely used in light-weight networks due to the model complexity. Our work demonstrates that by reasonably adjusting the residual block, this kind of classic bottleneck structure is also suitable for light-weight networks and can yield state-of-the-art results.

Inverted Residual Blocks. The inverted residual block, which was first introduced in MobileNetV2 [31], reverses the idea of the classic bottleneck structure and connects shortcuts between linear bottlenecks. It largely improves performance and optimizes the model complexity compared to the classic MobileNet [17] which is composed of a sequence of 3×3 depthwise separable convolutions. Because of high efficiency, the inverted residual block has been widely adopted in the later mobile network architectures. ShuffleNetV2 [27] inserts a channel split module before the inverted residual block and adds another channel shuffle module after it. In HBONet [22], down-sampling operations are introduced into inverted residual blocks for modeling richer spatial information. MobileNetV3 [16] proposes to search for optimal activation functions and the expansion rate of inverted residual blocks at each stage. More recently, MixNet [36] proposes to search for optimal kernel sizes of the depthwise separable convolutions in the inverted residual block. EfficientNet [35] is also based on the inverted residual block but differently it uses a scaling method to control the network weight in terms of input resolution, network depth, and network width. Different from all the above approaches, our work advances the standard bottleneck structure and demonstrates the superiority of our building block over the inverted residual block in mobile settings.

Model Compression and Neural Architecture Search. Model compression algorithms are effective for removing redundant parameters for neural networks, such as network pruning [2,11,26,30], quantization [5,19], factorization [20,43], and knowledge distillation [15]. Despite efficient networks, the performance of the compressed networks is still closely related to the original networks' architectures. Thus, designing more efficient network architectures is essential for yielding efficient models. Neural architecture search achieves so by automatically searching efficient network architectures [1,9,34] or even parameters [42]. However, the search space (SS) requires human expertise and the performance of the searched networks is largely dependent upon SS as pointed out in [6,40]. In this paper, we show that our proposed building block is complementary to existing SS design principles and can further improve the performance of searched networks if added to existing search spaces.

3 Method

In this section, we first review preliminaries about the bottleneck structure used in previous residual networks and then describe our proposed block and network architecture.

3.1 Preliminaries

Residual Block with Bottleneck Structure. The classic residual block with bottleneck structure [12], as shown in Fig. 2(a), consists of two 1×1 convolution layers for channel reduction and expansion respectively and one 3×3 convolution layer between them for spatial information encoding. In spite of its success in heavy-weight network design [12], this conventional bottleneck structure is not suitable for building light-weight neural networks because of its large amount of parameters and computation cost in the standard 3×3 convolutional layer.

Depthwise Separable Convolutions. To reduce computational cost and make the network more efficient, depthwise separable convolutions [4,17] are developed to replace the standard one. As demonstrated in [4], a convolution with a $k \times k \times M \times$

N weight tensor, where $k \times k$ is the kernel size and M and N are the number of input and output channels respectively, can be factorized into two convolutions. The first is an M-channel $k \times k$ depthwise (*a.k.a* channel-wise) convolution to learn the spatial correlations among locations within each channel separately. The second is a pointwise convolution that learns to linearly combine channels to produce new features. As the combination of a pointwise convolution and a $k \times k$ depthwise convolution has significantly less parameters and computations, using depthwise separable convolutions in basic building blocks can remarkably reduce the parameters and computational cost. Our proposed architecture also adopts such separable convolutions.

Inverted Residual Block. The inverted residual block is specifically tailored for mobile devices, especially those with limited computational resource budget. More specifically, unlike the classic bottleneck structure as shown in Fig. 2(a), to save computations, it takes as input a low-dimensional compressed tensor and expands it to a higher dimensional one by a pointwise convolution. Then it applies depthwise convolution for spatial context encoding, followed by another pointwise convolution to generate a low-dimensional feature tensor as input to the next block. The inverted residual block presents two distinct architecture designs for gaining efficiency without suffering too much performance drop: the shortcut connection is put between the low-dimensional bottlenecks if necessary (as shown in Fig. 2(b)); and linear bottleneck is adopted.

Despite good performance [31], in inverted residual blocks, feature maps encoded by the intermediate expansion layer should be first projected to low-dimensional ones, which may not preserve enough useful information due to channel compression. Moreover, recent studies have unveiled that wider architecture is more favorable for alleviating gradient confusion [32] and hence can improve network performance. Putting shortcut connections between bottlenecks may prevent the gradients from top layers from being successfully propagated to bottom layers during model training because of the low-dimensionality of representations between adjacent inverted residual blocks.

3.2 Sandglass Block

In view of the aforementioned limitations of the inverted residual block, we rethink its design rules and present a sandglass block that can tackle the above issues by flipping the thought of inverted residuals.

Our design principle is mainly based on the following insights: (i) To preserve more information from the bottom layers when transiting to the top layers and to facilitate the gradients propagation across layers, the shortcuts should be positioned to connect high-dimensional representations. (ii) Depthwise convolutions with small kernel size (*e.g.*, 3×3) are light-weight, so we can appropriately apply a couple of depthwise convolutions onto the higher-dimensional features such that richer spatial information can be encoded to generate more expressive representations. We elaborate on these design considerations in the following.

Rethinking the Positions of Expansion and Reduction Layers. Originally, the inverted residual block performs expansion at first and then reduction. Based on the aforementioned design principle, to make sure the shortcuts connect high-dimensional representations, we propose to reverse the order of the two pointwise convolutions first. Let $\mathbf{F} \in \mathbb{R}^{D_f \times D_f \times M}$ be the input tensor and $\mathbf{G} \in \mathbb{R}^{D_f \times D_f \times M}$ the output tensor of a building block[1]. We do not consider the depthwise convolution and activation layers at this moment. The formulation of our building block can be written as follows:

$$\mathbf{G} = \phi_e(\phi_r(\mathbf{F})) + \mathbf{F}, \tag{1}$$

where ϕ_e and ϕ_r denote the two pointwise convolutions for channel expansion and reduction, respectively. In this way, we can keep the bottleneck in the middle of the residual path for saving parameters and computation cost. More importantly, this allows us to use the shortcut connection to connect representations with a large number of channels instead of the bottleneck ones.

High-Dimensional Shortcuts. Instead of putting shortcuts between bottlenecks, we put shortcuts between higher-dimensional representations as shown in Fig. 3(b). The 'wider' shortcut delivers more information from the input $\mathbf{F}$ to the output $\mathbf{G}$ compared to the inverted residual block and allows more gradients to propagate across multiple layers.

Learning Expressive Spatial Features. Pointwise convolutions can be used to encode the inter-channel information but fail to capture spatial information. In our building block, we follow previous mobile networks and adopt depthwise spatial convolutions. The inverted residual block adds depthwise convolutions between pointwise convolutions to learn expressive spatial context information. However, in our case, the position between two pointwise convolutions is the

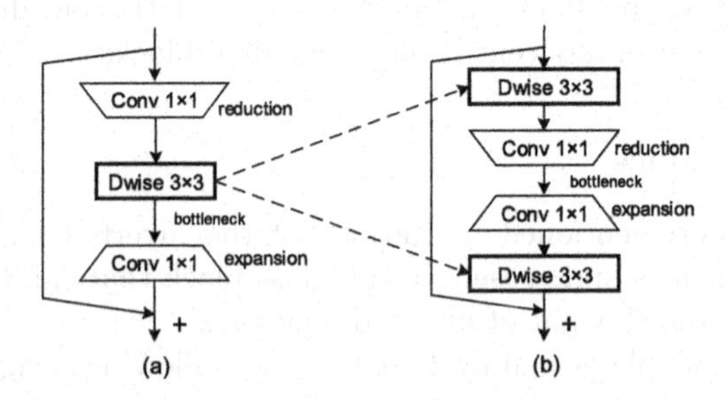

Fig. 3. Different types of residual blocks. (a) Classic bottleneck structure with depthwise spatial convolutions. (b) Our proposed sandglass block with bottleneck structure. To encode more expressive spatial information, instead of adding depthwise convolutions in the bottleneck, we propose to move them to the ends of the residual path with high-dimensional representations.

[1] For simplicity, we assume that the input and output of the building block share the same number of channels and resolution.

Table 1. Basic operator description of the proposed sandglass block. Here, 't' and 's' denote the channel reduction ratio and the stride, respectively.

Input dimension	Operator type	Output dimension
$D_f \times D_f \times M$	3 × 3 Dwise conv, ReLU6	$D_f \times D_f \times M$
$D_f \times D_f \times M$	1 × 1 conv, linear	$D_f \times D_f \times \frac{M}{t}$
$D_f \times D_f \times \frac{M}{t}$	1 × 1 conv, ReLU6	$D_f \times D_f \times N$
$D_f \times D_f \times N$	3 × 3 Dwise conv, linear, stride = s	$\frac{D_f}{s} \times \frac{D_f}{s} \times N$

bottleneck. Directly adding depthwise convolutions in the bottleneck as shown in Fig. 3(a) makes them have fewer filters and thus, less spatial information can be encoded. We experimentally found that this structure largely degrades the performance compared to MobileNetV2 by more than 1%.

Regarding the positions of the pointwise convolutions, instead of directly putting the depthwise convolution between the two pointwise convolutions, we propose to add depthwise convolutions at the ends of the residual path as shown in Fig. 3(b). Mathematically, our building block can be formulated as follows:

$$\hat{\mathbf{G}} = \phi_{1,p}\phi_{1,d}(\mathbf{F}) \tag{2}$$

$$\mathbf{G} = \phi_{2,d}\phi_{2,p}(\hat{\mathbf{G}}) + \mathbf{F} \tag{3}$$

where $\phi_{i,p}$ and $\phi_{i,d}$ are the i-th pointwise and depthwise convolutions, respectively. In this way, since both depthwise convolutions are conducted in high-dimensional spaces, richer feature representations can be extracted compared to the inverted residual block. We will give more explanations on the advantages of such design.

Activation Layers. It has been demonstrated in [31] that using linear bottlenecks helps in preventing the feature values from being zeroed. Following this suggestion, we do not add any activation layer after the reduction layer (the first pointwise convolutional layer). It should also be noted that though the output of our building block is high-dimensional, we empirically found adding an activation layer after the last convolution can negatively influence the classification performance. Therefore, activation layers are only added after the first depthwise convolutional layer and the last pointwise convolutional layer. We will give more explanations in our experiments on this.

Block Structure. Taking the above considerations, we design a novel residual bottleneck block. The structure details are given in Table 1, and the diagram can also be found in Fig. 3(b). Note that when the input and output have different channel numbers, we do not add the shortcut connection and for blocks with stride 2 we remove the first depthwise convolution. For depthwise convolutions, we always use kernel size 3 × 3 as done in other works [12,31]. We also utilize batch normalization and ReLU6 activation if necessary during training.

Relation to the Inverted and Classic Residual Blocks. Albeit both architectures exploit the bottlenecks, the design intuition and the internal structure are quite different. Our goal is to demonstrate that the idea of building shortcut connections between high-dimensional representations as in the classic bottleneck structure [12] is suitable for light-weight networks as well. To the best of our knowledge, this is the first work that attempts to investigate the advantages of the classic bottleneck structure over the inverted residual block for efficient network design. On the other hand, we also attempt to demonstrate that adding depthwise convolutions to the ends of the residual path in our structure can encourage the network to learn more expressive spatial information and hence yield better performance. In our experiment section, we will show more numerical results and provide detailed analysis.

3.3 MobileNeXt Architecture

Based on our sandglass block, we develop a modularized architecture, MobileNeXt. At the beginning of our network, there is a convolutional layer with 32 output channels. After that, our sandglass blocks are stacked together. Detailed information about the network architecture can be found in Table 2. Following [31], the expansion ratio used in our network is set to 6 by default. The output of the last building block is followed by a global average pooling layer to transform 2D feature maps to 1D feature vectors. A fully-connected layer is finally added to predict the final score for each category.

Table 2. Architecture details of the proposed MobileNeXt. Each row denotes a sequence of building blocks, which is repeated 'b' times. The reduction ratio used in each building block is denoted by 't'. The stride of the first building block in each stage is set to 2 and all the others are with stride 1. Each convolutional layer is followed by a batch normalization layer and the kernel size for all spatial convolutions is set to 3 × 3. We do not add identity mappings for those blocks have different input and output channels. We suppose there are totally k categories.

No.	t	Output dimension	s	b	Input dimension	Operator
1	–	112 × 112 × 32	2	1	224 × 224 × 3	conv2d 3 × 3
2	2	56 × 56 × 96	2	1	112 × 112 × 32	Sandglass block
3	6	56 × 56 × 144	1	1	56 × 56 × 96	Sandglass block
4	6	28 × 28 × 192	2	3	56 × 56 × 144	Sandglass block
5	6	14 × 14 × 288	2	3	28 × 28 × 192	Sandglass block
6	6	14 × 14 × 384	1	4	14 × 14 × 288	Sandglass block
7	6	7 × 7 × 576	2	4	14 × 14 × 384	Sandglass block
8	6	7 × 7 × 960	1	3	7 × 7 × 576	Sandglass block
9	6	7 × 7 × 1280	1	1	7 × 7 × 960	Sandglass block
10	–	1 × 1 × 1280	-	1	7 × 7 × 1280	avgpool 7 × 7
11	–	k	–	1	1 × 1 × 1280	conv2d 1 × 1

Identity Tensor Multiplier. The shortcut connections in residual blocks have been shown essential for improving the capability of propagating gradients across layers [12,31]. According to our experiments, we find that there is no need to keep the whole identity tensor to combine with the residual path. To make our network more friendly to mobile devices, we introduce a new hyper-parameter—identity tensor multiplier, denoted by $\alpha \in [0,1]$, which controls what portion of the channels in the identity tensor is preserved. For convenience, let ϕ be the transformation function of the residual path in our block. Originally, the formulation of our block can be written as $G = \phi(F) + F$. After applying the multiplier, our building block can be rewritten as

$$G_{1:\alpha M} = \phi(F)_{1:\alpha M} + F_{1:\alpha M}, \quad G_{\alpha M:M} = \phi(F)_{\alpha M:M}, \tag{4}$$

where the subscripts index the channel dimension.

The advantages of using α are mainly two-fold. First, after reducing the multiplier, the number of element-wise additions in each building block can be reduced. As pointed out in [27], the element-wise addition is time consuming. Users can choose a lower identity tensor multiplier to yield better latency with nearly no performance drop. Second, the number of memory access times can be reduced. One of the main factors that affect the model latency is the memory access cost (MAC). As the shortcut identity tensor is from the output of the last building block, its recurrent nature hints an opportunity to cache it on the chip in order to avoid the excessive off-chip memory access. Therefore, reducing the channel dimension of the identity tensor can effectively encourage the processors to store it in the cache or other faster memory near the processors and hence improve the latency. We will give more details on how this multiplier affects the performance and model latency in the experiment section.

4 Experiments

4.1 Experiment Setup

We adopt the PyTorch toolbox [29] to implement all our experiments. We use the standard SGD optimizer to train our models with both decay and momentum of 0.9 and the weight decay is 4×10^{-5}. We use the cosine learning schedule with an initial learning rate of 0.05. The batch size is set to 256 and four GPUs are used for training. Without special declaration, we train all the models for 200 epochs and report results on the ImageNet [21] for classification and Pascal VOC dataset [7] for object detection. We use distributed training with three epochs of warmup.

Table 3. Comparisons with MobileNetV2 using different width multipliers with input resolution 224 × 224. As can be seen, the smaller the multiplier is set to the better performance gain we achieve over MobileNetV2 with comparable latency (*e.g.,* 210 ms for both models with width multiplier 1.0) tested on Google Pixel 4XL under the PyTorch environment setting.

No.	Models	Param. (M)	MAdd(M)	Top-1 acc. (%)
1	MobileNetV2-1.40	6.9	690	74.9
2	MobileNetV2-1.00	3.5	300	72.3
3	MobileNetV2-0.75	2.6	150	69.9
4	MobileNetV2-0.50	2.0	97	65.4
5	MobileNetV2-0.35	1.7	59	60.3
6	MobileNeXt-1.40	6.1	590	76.1
7	MobileNeXt-1.00	3.4	300	74.0
8	MobileNeXt-0.75	2.5	210	72.0
9	MobileNeXt-0.50	2.1	110	67.7
10	MobileNeXt-0.35	1.8	80	64.7

4.2 Comparisons with MobileNetV2

In this subsection, we extensively study the advantages of our MobileNeXt over MobileNetV2 under various settings. Besides comparing performance of their full models (*i.e.,* , with weight multiplier of 1) for classification, we also compare their performance with other weight multipliers and quantization. This can help unveil the performance advantage of our model w.r.t. the full spectrum of model architecture configurations.

Comparison Under Different Width Multipliers. We use the width multiplier as a scaling factor to trade off the model complexity and accuracy of the model as used in [16,17,31]. Here, we adopt five different multipliers, including 1.4, 1.0, 0.75, 0.5, and 0.35, to show the superiority of our network over MobileNetV2. As shown in Table 3[2], our networks with different multipliers all outperform MobileNetV2 with comparable numbers of parameters and computations. The performance gain of our model over MobileNetV2 is especially high when the multiplier is small. This demonstrates that our model is more efficient since our model performance is much better at small sizes.

Comparison Under Post-training Quantization. Quantization algorithms are often used in real-world applications as a kind of effective compression tool with subtle performance loss. However, the performance of the quantized model is significantly affected by the original base model. We experimentally show that the MobileNeXt can achieve better performance than the MobileNetV2 when

[2] We also conduct latency measurements with TF-Lite on Pixel 4XL and the measured latency for MobileNeXt and MobileNetV2 are 66 ms and 68 ms respectively.

Table 4. Performance of our proposed MobileNeXt and MobileNetV2 after post-training quantization. In bites configurations, 'W' denotes the number of bits used to represent the weights of the model and 'A' denotes the number of bits used to represent the activations.

Model	Precision (W/A)	Method	Top-1 acc. (%)
MobileNetV2	INT8/INT8	Post Training Quant.	65.07
MobileNeXt	INT8/INT8	Post Training Quant.	$68.62_{+3.55}$
MobileNetV2	FP32/FP32	–	72.25
MobileNeXt	FP32/FP32	–	$74.02_{+1.77}$

Table 5. Performance of our proposed network and MobileNetV2 when adding the number of spatial convolutions (Dwise convs) in each building block. Obviously, our MobileNeXt performs much better than the improved MobileNetV2 with less learnable parameters and computational cost.

Method	#Dwise convs	Param. (M)	M-Adds (M)	Top-1 acc. (%)
MobileNetV2	2 (middle)	3.6	340	73.02
MobileNeXt	2 (top, bottom)	3.5	300	74.02

combined with the quantization algorithm. Here, we use a widely-used post-training linear quantization method introduced in [28]. We apply 8-bit quantization on both weights and activations as 8-bit is the most common scheme used on hardware platforms. The results are shown in Table 4. Without quantization, our network improves MobileNetV2 by more than 1.7% in terms of top-1 accuracy. When the parameters and activations are quantized to 8 bits, our network outperforms MobileNetV2 by 3.55% under the same quantization settings. The reasons for this large improvement are two-fold. First, compared to MobileNetV2, we move the shortcut in each building block from low-dimensional representations to high-dimensional ones. After quantization, more informative feature representations can be preserved. Second, using more depthwise spatial convolutions can help preserve more spatial information, which we believe is beneficial to the classification performance.

Comparison with MobileNetV2 on Structure. As shown in Fig. 3(b), our sandglass block contains two 3×3 depthwise convolutions for encoding rich spatial context information. To demonstrate the benefit of our model comes from our novel architecture rather than leveraging one more depthwise convolution or larger receptive field, in this experiment, we attempt to compare with an improved version of MobileNetV2 with one more depthwise convolution inserted in the middle of each inverted residual block. The results are shown in Table 5. Obviously, after adding one more depthwise convolution, the performance of MobileNetV2 increases to 73%, which is still far worse than ours (74%) with even more learnable parameters and complexity. This indicates that structurally our network does have an edge over MobileNetV2.

4.3 Comparison with State-of-the-Art Mobile Networks

To further verify the superiority of our proposed sandglass block over the inverted residual blocks, we add squeeze and excite modules into our MobileNeXt as done in [16,35]. We do not apply any searching algorithms on the architecture design and data augmentation policy. We directly take the EfficientNet-b0 architecture [35] and replace the inverted residual block with sandglass block with the basic augmentation policy. As shown in Table 6, with a comparable amount of computation and 20% parameter reduction, replacing the inverted residual block with sandglass block results in 0.4% top-1 classification accuracy improvement on ImageNet-1k dataset.

4.4 Ablation Studies

In Sect. 4.2, we have shown the importance of connecting high-dimensional representations with shortcuts. Here, we study how other model design choices contribute to the model performance and efficiency, including the effect of using wider transformation, the importance of learning linear residuals, and the role of identity tensor multiplier.

Importance of Using Wider Transformation. As described in Sect. 3, we apply spatial transformation and shortcut connections to high-dimensional representations. To demonstrate the importance of such operations, we follow the inverted residual block to use the shortcuts to connect the bottleneck representations. This operation leads to an accuracy decrease of 1%, which indicates the advantage of using wider transformation.

Importance of Linear Residuals. According to MobileNetV2 [31], its classification performance will be degraded when replacing the linear bottleneck with the non-linear one because of information loss. From our experiment, we obtain a more general conclusion. We find that though the shortcuts connect high-dimensional representations in our model, adding non-linear activations (ReLU6) to the last convolutional layer decreases the performance by nearly 1% compared to the setting using linear activations (no ReLU6). This indicates that learning linear residual (*i.e.,* adding no non-linear activation layer on the top of the residual path) is essential for light-weight networks with shortcuts connecting either expansion layers or reduction layers.

Effect of Identity Tensor Multiplier. Here, we investigate how the identity tensor multiplier (Sect. 3.3) would trades-off the model accuracy and latency. We use pytorch to generate the model and run it on Google Pixel 4XL. For each model, we measure the average inference time of 10 images as the final inference latency. As shown in Table 7, the reduction of the multiplier has subtle impacts on the classification accuracy. When half of the identity representations are removed, the performance has no drop but the latency is improved. When the multiplier is set to 1/6, the performance decreases by 0.34%, but with further improvement in terms of latency. This indicates that introducing such a hyper-parameter does matter for balancing the model performance and latency.

Table 6. Comparisons with other state-of-the-art models. MobileNeXt denotes the model based on our proposed sandglass block and MobileNeXt† denotes the models with sandglass block and the SE module [18] added for a fair comparison with other state-of-the-art models such as EfficientNet. We do not apply any searching algorithms.

Models	Param. (M)	MAdd (M)	Top-1 acc. (%)
MobilenetV1-1.0 [17]	4.2	575	70.6
MobilenetV2-1.0 [31]	3.5	300	72.3
MnasNet-A1 [34]	3.9	312	75.2
MobilenetV3-L-0.75 [16]	4.0	155	73.3
ProxylessNAS [1]	4.1	320	74.6
FBNet-B [38]	4.5	295	74.1
GhostNet-1.3 [10]	7.3	226	75.7
EfficientNet-b0 [35]	5.3	390	76.3
MobileNeXt-1.0	3.4	300	74.02
MobileNeXt-1.0†	3.94	330	76.05
MobileNeXt-1.1†	**4.28**	420	**76.7**

Table 7. Model performance and latency comparisons with different identity tensor multipliers. As can be seen, the latency can be improved by using lower identity tensor multipliers with only negligible sacrifice on the classification accuracy.

No.	Models	Tensor multiplier	Param. (M)	Top-1 acc. (%)	Latency (ms)
1	MobileNeXt	1.0	3.4	74.02	211
2	MobileNeXt	1/2	3.4	74.09	196
3	MobileNeXt	1/3	3.4	73.91	195
4	MobileNeXt	1/6	3.4	73.68	188

4.5 Application for Object Detection

To explore the transferable capability of the proposed approach against MobileNetV2, in this subsection, we apply our classification model to the object detection task as pretrained models. We use both the proposed network and MobileNetv2 as feature extractors and report results on the Pascal VOC 2007 test set [8] following [25] using SSDLite [25,31]. Similar to [31], the first and second layers of SSDLite are connected to the last pointwise convolution layer with output stride of 16 and 32, respectively. The rest of SSDLite layers are attached on top of the last convolutional layer with output stride of 32. During training, we use a batch size of 24 and all the models are trained for 240,000 iterations. For more detailed settings, readers can refer to [25,31].

In Table 8, we show the results when different backbone networks are used. Obviously, with the nearly the same number of parameters and computation, SSDLite with our backbone improves the one with MobileNetV2 by nearly 1%.

This demonstrates that the proposed network has better transferable capability compared to MobileNetV2.

4.6 Improving Architecture Search as Super-Operators

It has been verified in previous subsections that our proposed sandglass block is more effective than the inverted residual block in both the classification task and the object detection task. From a holistic perspective, we can also regard a residual block as a 'super' operator with more powerful transformation power than a regular convolutional operator. To further investigate the superiority of the proposed sandglass block over the inverted residual block, we separately add it into the search space of the differentiable searching algorithm (DARTS) [24] to see the network performance after architecture search and report the corresponding results on CIFAR-10 dataset. As shown in Table 9, by adding our sandglass block as an new operator into the DARTS search space without changing the cell structure, the resulting model achieves higher accuracy than the model with the original DARTS search space with about 25% parameter reduction. However, the searched model with the inverted residual block added in the search space decreases the original performance. This demonstrates that our proposed sandglass block can generate more expressive representations than the inverted residual block and can also be used in architecture search algorithms as a kind of 'super' operator. For more details on the searched cell structure, please refer to our supplementary materials.

Table 8. Detection results on the Pascal VOC 2007 test set. As can be seen, using the same SSDLite320 detector, replacing the MobileNetV2 backbone with our network achieves better results. Note that the multipliers of both MobileNetV2 and our network are set to 1.0.

No.	Method	Backbone	Param. (M)	M-Adds (B)	mAP (%)
1	SSD300	VGG [33]	36.1	35.2	77.2
2	SSDLite320	MobileNetV2 [31]	4.3	0.8	71.7
3	SSDLite320	MobileNeXt	4.3	0.8	72.6

Table 9. Results produced by different network architectures searched by DARTS [24]. For Lines 2 and 3, we separately add the inverted residual (IR) block and our sandglass block into the original search space of DARTS. We report results on CIFAR-10 dataset as in [24].

No.	Search space	Test error (%)	Param. (M)	Search method	#Operators
1	DARTS original	3.11	3.25	Gradient based	7
2	DARTS + IR Block	3.26	3.29	Gradient based	8
3	DARTS + sandglass block	2.98	2.45	Gradient based	8

5 Conclusions

In this paper, we deeply analyze the design rules and shortcomings of the previous inverted residual block. Based on the analysis, we propose to reverse the thought of adding shortcut connections between low-dimensional representations and present a novel building block, called the sandglass block, that connects high-dimensional representations instead. Experiments in both classification, object detection, and neural architecture search demonstrate the effectiveness of the proposed sandglass block and its potential to be used in more contexts.

Acknowledgement. Jiashi Feng was partially supported by MOE Tier 2 MOE2017-T2-2-151, NUS_ECRA_FY17_P08, AISG-100E-2019-035.

References

1. Cai, H., Zhu, L., Han, S.: ProxylessNAS: direct neural architecture search on target task and hardware. arXiv preprint arXiv:1812.00332 (2018)
2. Caron, M., Morcos, A., Bojanowski, P., Mairal, J., Joulin, A.: Pruning convolutional neural networks with self-supervision. arXiv preprint arXiv:2001.03554 (2020)
3. Chen, Y., Li, J., Xiao, H., Jin, X., Yan, S., Feng, J.: Dual path networks. In: Advances in Neural Information Processing Systems, pp. 4467–4475 (2017)
4. Chollet, F.: Xception: deep learning with depthwise separable convolutions. In: Proceedings of the IEEE Conference on Computer Vision and Pattern Recognition, pp. 1251–1258 (2017)
5. Choukroun, Y., Kravchik, E., Yang, F., Kisilev, P.: Low-bit quantization of neural networks for efficient inference. arXiv preprint arXiv:1902.06822 (2019)
6. Dong, X., Yang, Y.: NAS-Bench-102: extending the scope of reproducible neural architecture search. arXiv preprint arXiv:2001.00326 (2020)
7. Everingham, M., Van Gool, L., Williams, C.K.I., Winn, J., Zisserman, A.: The PASCAL Visual Object Classes Challenge 2012 (VOC2012) Results. http://www.pascal-network.org/challenges/VOC/voc2012/workshop/index.html
8. Everingham, M., Eslami, S.A., Van Gool, L., Williams, C.K., Winn, J., Zisserman, A.: The pascal visual object classes challenge: a retrospective. Int. J. Comput. Vision **111**(1), 98–136 (2015)
9. Guo, Z., et al.: Single path one-shot neural architecture search with uniform sampling. arXiv preprint arXiv:1904.00420 (2019)
10. Han, K., Wang, Y., Tian, Q., Guo, J., Xu, C., Xu, C.: GhostNet: more features from cheap operations. arXiv preprint arXiv:1911.11907 (2019)
11. Han, S., Mao, H., Dally, W.J.: Deep compression: compressing deep neural networks with pruning, trained quantization and Huffman coding. arXiv preprint arXiv:1510.00149 (2015)
12. He, K., Zhang, X., Ren, S., Sun, J.: Deep residual learning for image recognition. In: Proceedings of the IEEE Conference on Computer Vision and Pattern Recognition, pp. 770–778 (2016)
13. He, K., Zhang, X., Ren, S., Sun, J.: Deep residual learning for image recognition. In: CVPR (2016)

14. He, K., Zhang, X., Ren, S., Sun, J.: Identity mappings in deep residual networks. In: Leibe, B., Matas, J., Sebe, N., Welling, M. (eds.) ECCV 2016. LNCS, vol. 9908, pp. 630–645. Springer, Cham (2016). https://doi.org/10.1007/978-3-319-46493-0_38
15. Hinton, G., Vinyals, O., Dean, J.: Distilling the knowledge in a neural network. arXiv preprint arXiv:1503.02531 (2015)
16. Howard, A., et al.: Searching for mobilenetv3. In: Proceedings of the IEEE International Conference on Computer Vision, pp. 1314–1324 (2019)
17. Howard, A.G., et al.: MobileNets: efficient convolutional neural networks for mobile vision applications. arXiv preprint arXiv:1704.04861 (2017)
18. Hu, J., Shen, L., Sun, G.: Squeeze-and-excitation networks. In: Proceedings of the IEEE Conference on Computer Vision and Pattern Recognition, pp. 7132–7141 (2018)
19. Hubara, I., Courbariaux, M., Soudry, D., El-Yaniv, R., Bengio, Y.: Quantized neural networks: training neural networks with low precision weights and activations. J. Mach. Learn. Res. **18**(1), 6869–6898 (2017)
20. Jaderberg, M., Vedaldi, A., Zisserman, A.: Speeding up convolutional neural networks with low rank expansions. arXiv preprint arXiv:1405.3866 (2014)
21. Krizhevsky, A., Sutskever, I., Hinton, G.E.: ImageNet classification with deep convolutional neural networks. In: Advances in Neural Information Processing Systems, pp. 1097–1105 (2012)
22. Li, D., Zhou, A., Yao, A.: HBONet: harmonious bottleneck on two orthogonal dimensions. In: Proceedings of the IEEE International Conference on Computer Vision, pp. 3316–3325 (2019)
23. Li, X., Wang, W., Hu, X., Yang, J.: Selective kernel networks. In: Proceedings of the IEEE Conference on Computer Vision and Pattern Recognition, pp. 510–519 (2019)
24. Liu, H., Simonyan, K., Yang, Y.: DARTS: differentiable architecture search. arXiv preprint arXiv:1806.09055 (2018)
25. Liu, W., et al.: SSD: single shot MultiBox detector. In: Leibe, B., Matas, J., Sebe, N., Welling, M. (eds.) ECCV 2016. LNCS, vol. 9905, pp. 21–37. Springer, Cham (2016). https://doi.org/10.1007/978-3-319-46448-0_2
26. Liu, Z., Li, J., Shen, Z., Huang, G., Yan, S., Zhang, C.: Learning efficient convolutional networks through network slimming. In: Proceedings of the IEEE International Conference on Computer Vision, pp. 2736–2744 (2017)
27. Ma, N., Zhang, X., Zheng, H.T., Sun, J.: ShuffleNet V2: practical guidelines for efficient CNN architecture design. In: ECCV, pp. 116–131 (2018)
28. Migacz, S.: Nvidia 8-bit inference width TensorRT. In: GPU Technology Conference (2017)
29. Paszke, A., et al.: PyTorch: an imperative style, high-performance deep learning library. In: Advances in Neural Information Processing Systems, pp. 8024–8035 (2019)
30. Radu, V., et al.: Performance aware convolutional neural network channel pruning for embedded GPUs (2019)
31. Sandler, M., Howard, A., Zhu, M., Zhmoginov, A., Chen, L.C.: MobileNetV2: inverted residuals and linear bottlenecks. In: CVPR, pp. 4510–4520 (2018)
32. Sankararaman, K.A., De, S., Xu, Z., Huang, W.R., Goldstein, T.: The impact of neural network overparameterization on gradient confusion and stochastic gradient descent. arXiv preprint arXiv:1904.06963 (2019)
33. Simonyan, K., Zisserman, A.: Very deep convolutional networks for large-scale image recognition. arXiv preprint arXiv:1409.1556 (2014)

34. Tan, M., et al.: MnasNet: platform-aware neural architecture search for mobile. In: Proceedings of the IEEE Conference on Computer Vision and Pattern Recognition, pp. 2820–2828 (2019)
35. Tan, M., Le, Q.V.: EfficientNet: rethinking model scaling for convolutional neural networks. In: ICML (2019)
36. Tan, M., Le, Q.V.: MixConv: mixed depthwise convolutional kernels. CoRR, abs/1907.09595 (2019)
37. Touvron, H., Vedaldi, A., Douze, M., Jégou, H.: Fixing the train-test resolution discrepancy. In: Advances in Neural Information Processing Systems, pp. 8250–8260 (2019)
38. Wu, B., et al.: FBNet: hardware-aware efficient convnet design via differentiable neural architecture search. In: CVPR, pp. 10734–10742 (2019)
39. Xie, S., Girshick, R., Dollár, P., Tu, Z., He, K.: Aggregated residual transformations for deep neural networks. In: Proceedings of the IEEE Conference on Computer Vision and Pattern Recognition, pp. 1492–1500 (2017)
40. Ying, C., Klein, A., Real, E., Christiansen, E., Murphy, K., Hutter, F.: NAS-Bench-101: towards reproducible neural architecture search. arXiv preprint arXiv:1902.09635 (2019)
41. Zagoruyko, S., Komodakis, N.: Wide residual networks. arXiv preprint arXiv:1605.07146 (2016)
42. Zhou, D., Jin, X., Hou, Q., Wang, K., Yang, J., Feng, J.: Neural epitome search for architecture-agnostic network compression. In: International Conference on Learning Representations (2019)
43. Zhou, M., Liu, Y., Long, Z., Chen, L., Zhu, C.: Tensor rank learning in CP decomposition via convolutional neural network. Signal Process. Image Commun. **73**, 12–21 (2019)

Side-Tuning: A Baseline for Network Adaptation via Additive Side Networks

Jeffrey O. Zhang[1(✉)], Alexander Sax[1], Amir Zamir[3], Leonidas Guibas[2], and Jitendra Malik[1]

[1] UC Berkeley, Berkeley, USA
jozhang@berkeley.edu
[2] Stanford University, Stanford, USA
[3] Swiss Federal Institute of Technology (EPFL), Lausanne, Switzerland
http://sidetuning.berkeley.edu

Abstract. When training a neural network for a desired task, one may prefer to adapt a **pre-trained** network rather than starting from randomly initialized weights. Adaptation can be useful in cases when training data is scarce, when a single learner needs to perform multiple tasks, or when one wishes to encode priors in the network. The most commonly employed approaches for network adaptation are **fine-tuning** and using the pre-trained network as a fixed **feature extractor**, among others. In this paper, we propose a straightforward alternative: **side-tuning**. Side-tuning adapts a pre-trained network by training a lightweight "side" network that is fused with the (unchanged) pre-trained network via summation. This simple method works as well as or better than existing solutions and it resolves some of the basic issues with fine-tuning, fixed features, and other common approaches. In particular, side-tuning is less prone to overfitting, is asymptotically consistent, and does not suffer from catastrophic forgetting in incremental learning. We demonstrate the performance of side-tuning under a diverse set of scenarios, including incremental learning (iCIFAR, iTaskonomy), reinforcement learning, imitation learning (visual navigation in Habitat), NLP question-answering (SQuAD v2), and single-task transfer learning (Taskonomy), with consistently promising results.

Keywords: Sidetuning · Finetuning · Transfer learning · Representation learning · Lifelong learning · Incremental learning · Continual learning

1 Introduction

The goal of *side-tuning* (and generally network adaptation) is to capitalize on a pretrained model to better learn one or more novel tasks. The *side-tuning*

Electronic supplementary material The online version of this chapter (https://doi.org/10.1007/978-3-030-58580-8_41) contains supplementary material, which is available to authorized users.

A. Vedaldi et al. (Eds.): ECCV 2020, LNCS 12348, pp. 698–714, 2020.
https://doi.org/10.1007/978-3-030-58580-8_41

approach is straightforward: it assumes access to a given (base) model $B : \mathbb{X} \to \mathbb{Y}$ that maps the input x onto some representation y. *Side-tuning* then *learns* a side model $S : \mathbb{X} \to \mathbb{Y}$, so that the curated representations for the target task are

$$R(x) \triangleq B(x) \oplus S(x), \tag{1}$$

for some combining operation $\oplus$. For example, choosing $B(x) \oplus S(x) \triangleq \alpha B(x) + (1 - \alpha)S(x)$ (commonly called α-blending) reduces the side-tuning approach to: fine-tuning, feature extraction, and stage-wise training, depending on α (Fig. 2, right). Hence those can be viewed as special cases of the side-tuning approach (Fig. 1).

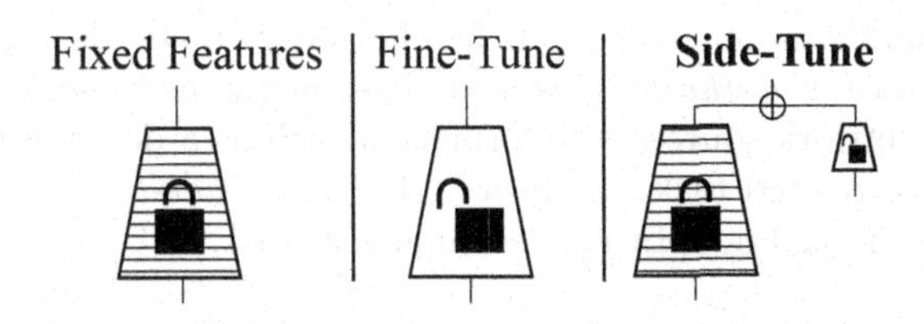

Fig. 1. The *side-tuning* framework vs. common alternatives *fine-tuning* and *fixed features*. Given a pre-trained network that should be adapted to a new task, *fine-tuning* re-trains the weights in the pretrained network and *fixed feature extraction* trains a readout function with no re-training of the pre-trained weights. In contrast, *Side-tuning* adapts the pre-trained network by training a lightweight conditioned "side" network that is fused with the (unchanged) pre-trained network using summation.

Side-tuning is an example of an *additive learning* approach, one that adds new parameters for each new task. Since *side-tuning* does not change the base model, it, by design, adapts to a target task without degrading performance on the base task. Unlike many other additive approaches, *side-tuning* places no constraints on the structure of the base model or side network, allowing for the architecture and sizes to vary independently. In particular, while other approaches require the side network to scale with the base network, *side-tuning* can use tiny networks when the base only requires minor updates. By adding fewer parameters per task, *side-tuning* can learn more tasks before the model grows large enough to require parameter consolidation.

Substitutive methods instead opt for a single large model that is updated on each task. These methods often require adding additional constraints per-task in order to prevent inter-task interference [12,30]. *Side-tuning* does not require such regularization since the base remains untouched.

Compared to existing state-of-the-art network adaptation and incremental learning[1] approaches, we find that the more complex methods perform no better—and often worse—than *side-tuning*.

This straightforward mechanism deals with the key challenges of incremental learning (Sect. 4.2). Namely, it does not suffer from either:

[1] Also referred to as lifelong or continual learning.

- **Catastrophic forgetting**: tendency of a network to lose previously learned knowledge upon learning new information.
- **Rigidity**: Increasing inability of a network to adapt to new problems as it accrues constraints from previous problems. Note: Incremental learning literature sometimes calls this *intransigence* [2]. We prefer *rigidity* as it is clear and widely used in psychology, dating back over 70 years [13,27].

We test *side-tuning* for incremental learning on iCIFAR and the more challenging iTaskonomy dataset, which we introduce, finding that incremental learning methods that work on iCIFAR often do not work as well in the more demanding setup. On these datasets, *side-tuning* uses side networks that are much smaller than the base. Consequently, even without consolidation, *side-tuning* uses fewer learnable parameters than the alternative methods.

Finally, because *side-tuning* treats the base model as a black-box, it can be used with non-network sources of information such as a decision trees or oracle information on a related task (see Sect. 4.4). Thus, *side-tuning* can be applied even when other model adaptation techniques cannot.

Table 1. Advantages of *side-tuning* vs. representative alternatives. Fixed features provide a fixed representation and if the pretrained model discarded important information then there is no way to recover the lost details. This often leads to modest performance. On the other hand, fine-tuning has a large number of learnable parameters which leads to overfitting. *Side-tuning* is a simple method that addresses these limitations with a small number of learnable parameters.

Method	1 Target Task Low Data	1 Target Task High Data	> 1 Target Tasks (incremental)
Fixed features	✓	✗ (Info Loss)	✗ (Info Loss)
Fine-tuning	✗ (Overfit)	✓	✗ (Forgetting)
Side-tuning	✓	✓	✓

2 Related Work

Broadly speaking, network adaptation methods either overwrite existing parameters (*substitutive* methods) or freeze them and add new parameters (*additive* learning).

Substitutive Methods modify an existing network to solve a new task by updating some or all of the network weights (simplest approach being fine-tuning). A large body of constraint-based methods focuses on how to regularize these updates in order to prevent forgetting earlier tasks. Methods such as [12,15,30] impose additional constraints for each new task, which slows down learning on later tasks (see Sect. 4.2 on rigidity, [3]). Other methods such as [4] relegate each task to approximately orthogonal subspaces but are then unable to transfer information across tasks. *Side-tuning* does not require such regularization since the base remains untouched.

Additive Methods circumvent forgetting by freezing the weights and adding a small number of new parameters per task. One economical approach is to use off-the-shelf-features with one or more readout layers [32]. However, off-the-shelf features cannot be updated for the new task, and so recent work has focused on how features can be modulated by applying per-task learned weight masks [17,28], by pruning [18], or by hard attention [31].

If information is missing from the original features, then recovering that information might require adding additional weights. Works such as [14,29] introduce a new network with independent access to the input and connect to various layers from the original network. Other works like [1,24,25] learn task-dependent parameters (e.g. separate batch norm, linear layers) that are inserted directly into the existing network. Tying the new weights directly into the original network architecture often requires making restrictive assumptions about the original network architecture (e.g. that it must be a ResNet [8]).

Unlike these previous works, *side-tuning* uses only late fusion and makes no assumptions about the base network. This means it can be applied on a larger class of models. While simpler, the results suggest that *side-tuning* offers similar or better performance to the more complex alternatives and calls into question whether that complexity buys much in practice.

Residual Learning exploits the fact that it is often easier to approximate a difference rather than the original function. This has been successfully used in ResNets [8] where a residual is learned on top of the identity function prior. Some network adaptation methods insert new residual-modeling parameters directly into the base architecture [14,24]. Residual learning has also been explored in robotics as residual RL [10,33], in which we train an agent for a single task by first taking a coarse policy (e.g. behavior cloning) and then training a residual network on top (using RL). For a single task, iteratively learning residuals is known as gradient boosting, but *side-tuning* adds a side network to adapt a base representation for a *new* task. We discuss the relationship in Sect. 4.4.

Meta-learning, unlike network adaptation approaches, seeks to create networks that are inherently adaptable. Typically this proceeds by training on tasks sampled from a known task distribution. *Side-tuning* is fundamentally compatible with this formulation and with existing approaches (e.g. [6]). Recent work suggests that these approaches work primarily by feature adaptation rather than rapid learning [22], and feature adaptation is also the motivation for our method.

3 Side-Tuning: The Simplest Additive Approach

Side-tuning learns a side model $S(x)$ and combines it with a pre-trained base model $B(x)$. The representation for the target task is $R(x) \triangleq B(x) \oplus S(x)$.

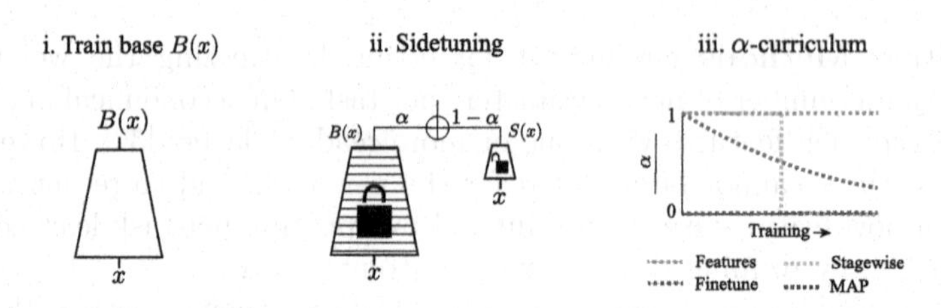

Fig. 2. Mechanics of *side-tuning*. (i) *Side-tuning* takes some core network and adapts it to a new task by training a side network. (ii) Connectivity structure when using *side-tuning* along with α-blending. (iii) Some of the existing common adaptation methods turn out to be special cases of an alpha blending with a side network. In particular: fine-tuning, feature extraction, and other approaches are *side-tuning* with a fixed curriculum on the blending parameter α, as shown in the plot.

3.1 Architectural Elements

Base Model. The base model $B(x)$ provides some core cognition or perception, and we put no restrictions on how $B(x)$ is computed. We never update $B(x)$, and in our approach it has zero learnable parameters. $B(x)$ could be a decision tree or an oracle for another task (experiments with this setup shown in Sect. 4.4). We consider several choices for $B(x)$ in Sect. 4.4, but the simplest choice is just a pretrained network.

Side Model. Unlike the base model, the side network, $S(x)$, is updated during training; learning a residual that we apply on top the base representation. One crucial component of the framework is that the complexity of the side network can scale to the difficulty of the problem at hand. When the base is relevant and requires only a minor update, a very simple side network can suffice.

Since the side network's role is to amend the base network to a new task, we initialize the side network as a copy of the base. When the forms of the base and side networks differ, we initialize the side network through knowledge distillation [9]. We investigate side network design decisions in Sect. 4.4. In general, we found *side-tuning* to perform well in a variety of settings and setups.

Combining Base and Side Representations
Side-tuning admits many options for the combination operator, $\oplus$, and we compare several in Sect. 4.5. We observe that alpha blending, $B(x) \oplus S(x) \triangleq \alpha B(x) + (1-\alpha)S(x)$, where α is treated as a learnable parameter works well and α correlates with task relevance (see Sect. 4.4).

While simple, alpha blending is expressive enough that it encompasses several common transfer learning approaches. As shown in Fig. 2(iii), *side-tuning* is equivalent to feature extraction when $\alpha = 1$. When $\alpha = 0$, *side-tuning* is instead equivalent to fine-tuning if the side network has the same architecture the base. If we allow α to vary during training, then switching α from 1 to 0 is equivalent to the common (stage-wise) training curriculum in RL where a policy is trained on top of some fixed features that are unlocked partway through training.

When minimizing estimation error there is often a tradeoff between the bias and variance contributions [7] (see Table 1). Feature extraction locks the weights and corresponds to a point-mass prior while fine-tuning is an uninformative prior yielding a low-bias high-variance estimator. *Side-tuning* aims to leverage the (useful) bias from those original features while making the representation *asymptotically consistent* through updates to the residual side-network[2].

Given the bias-variance interpretation, a notable curriculum for α during training is $\alpha(N) = \frac{k}{k+N}$ for $k > 0$ (hyperbolic decay) where N is the number of training epochs. This curriculum, placing less weight on the prior as more evidence accumulates, is suggestive of a *maximum a posteriori* estimate and, like the MAP estimate, it converges to the MLE (fine-tuning).

3.2 Side-Tuning for Incremental Learning

We often care about the performance not only on the current target task but also on the previously learned tasks. This is the case for *incremental learning*, where we want an agent that can learn a sequence of tasks $\mathcal{T}_1, ..., \mathcal{T}_m$ and is capable of reasonable performance across the entire set at the end of training. In this paradigm, catastrophic forgetting (diminished performance on $\{\mathcal{T}_1, ..., \mathcal{T}_{m-1}\}$ due to learning $\mathcal{T}_m$) becomes a major issue.

In our experiments, we dedicate one new side network to each task. We define a task $\mathcal{T} : x \mapsto P(Y)$ as a mapping from inputs, x, to a probability distribution over the output space, Y. For example, x is an RGB image mapped to probabilities over object classes, Y. Datasets for a task are a set of pairs $\{(x, y) \mid y \sim \mathcal{T}(x)\}$. For task $\mathcal{T}_t$, our loss function is

$$L(x_t, y_t) = \|D_t(\alpha_t B(x_t) + (1 - \alpha_t) S_t(x_t)) - y_t\| \quad (2)$$

where t is the task number and D_t is some decoder readout of the *side-tuning* representation. This simple approach leads to the training curve in Fig. 3 with no possible catastrophic forgetting. Furthermore, since *side-tuning* is independent of task order, training does not slow down as training progresses. We observe that this approach provides a strong baseline for incremental learning, outperforming existing approaches in the literature while using fewer parameters on more tasks (in Sect. 4.2).

[2] Sidetuning is one way of making features obey Cromwell's rule: "I beseech you, in the bowels of Christ, think it possible that you may be mistaken.".

Side-tuning naturally handles other continuous learning scenarios besides incremental learning. A related problem is that of continuous adaptation, where the agent needs to perform well (e.g. minimizing regret) on a stream of tasks with undefined boundaries and where there might very little data per task and no task repeats. As we show in Sect. 4.2, inflexibility becomes a serious problem for constraint-based methods and task-specific performance declines after learning more than a handful of tasks. Moreover, continuous adaptation requires an online method as task boundaries must be detected and data cannot be replayed (e.g. to generate constraints for EWC).

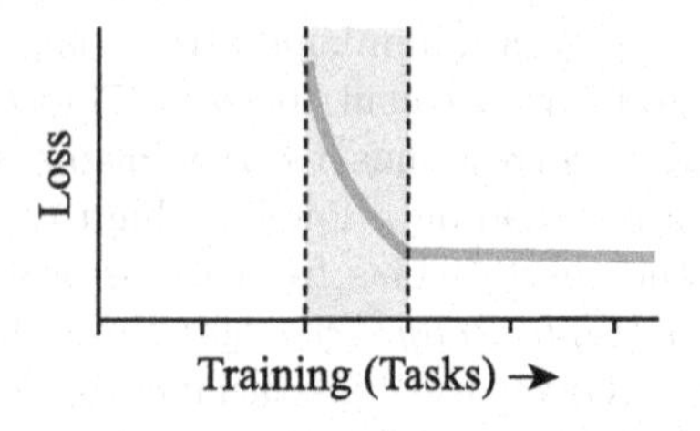

Fig. 3. Theoretical learning curve of *side-tuning*. The model learns during task-specific training and those weights are subsequently frozen, preserving performance.

Side-tuning could be applied to continuous adaptation by keeping a small working memory of cheap side networks that constantly adapt the base network to the input task. These side networks are small, easy to train, and when one

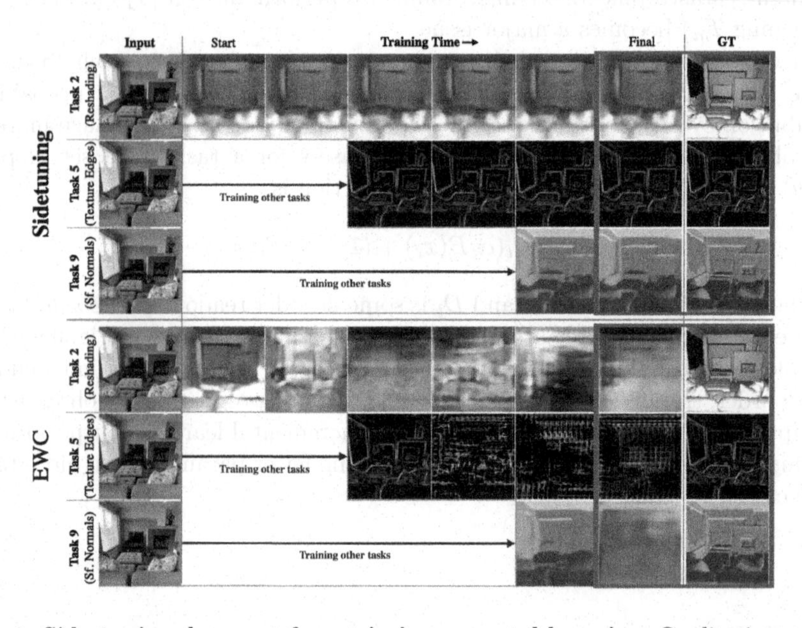

Fig. 4. *Side-tuning* does not forget in incremental learning. Qualitative results for iTaskonomy with additive learning (*side-tuning*, top 3 rows) and constraint-based substitutive learning (EWC, bottom 3 rows). Each row contains results for one task and columns show how predictions change over the course of training. Predictions from EWC quickly degrade over time, showing that EWC catastrophically forgets. Predictions from *side-tuning* do not degrade, and the initial quality is better in later tasks (e.g. compare the table in surface normals). We provide additional comparisons (including for PSP, PNN) in the supplementary.

of the networks begins performing poorly (e.g. signaling a distribution shift) that network can simply be discarded. This is an online approach, and online adaptation with small cheap networks has found recent success (e.g. in [20]).

4 Experiments

In the first section we show that when applied to the incremental learning setup, *side-tuning* compares favorably to existing approaches on both iCIFAR and the more challenging iTaskonomy dataset. We then extend this to multiple domains (computer vision, RL, imitation learning, NLP) in the simplified scenario for $m = 2$ tasks (transfer learning). Finally, we interpret *side-tuning* in a series of analysis experiments.

4.1 Baselines

We provide comparisons of *side-tuning* against the following methods:

Scratch: The network is initialized with appropriate random weights and trained using minibatch SGD with Adam [11].
Feature extraction (features): The pretrained base network is used as-is and is not updated during training.
Fine-tuning: An umbrella term that encompasses a variety of techniques, we consider a more narrow definition where pretrained weights are used as initialization and then training proceeds as in *scratch*.
Elastic Weight Consolidation (EWC). A constraint-based substitutive approach from [12]. We use the formulation from [30] which scales better.
Parameter Superposition (PSP): A parameter-masking substitutive approach from [4] that attempts to make tasks independent from one another by mapping the weights to approximately orthogonal spaces.
Progressive Neural Network (PNN): An additive approach from [29] which utilizes many lateral connections between the base and side networks. Requires the architecture of the base and side networks to be the same or similar.
Piggyback (PB): Learns task-dependent binary weight masks [17].
Residual Adapters (RA): An additive approach which learns task-dependent batch-norm and linear layers between layers in an existing network [24,25].
Independent: Each task uses a pretrained network trained independently for that task. This method uses far more learnable parameters than all the alternatives (e.g. saving a separate ResNet-50 for each task) and achieves strong performance.

4.2 Incremental Learning

On both the incremental Taskonomy [34] (iTaskonomy) and incremental CIFAR (iCIFAR [26]) datasets, *side-tuning* performs competitively against existing incremental learning approaches while using fewer parameters[3]. On the more challenging Taskonomy dataset, it outperforms other approaches.

- **iCIFAR.** Comprises 10 subsequent tasks by partitioning CIFAR-100 [26] into 10 disjoint sets of 10-classes each. Images are 32×32 RGB. First, we pretrain the base network (ResNet-44) on CIFAR-10. We then train on each subtask for 20k steps before moving to the next one. The SotA substitutive baselines (EWC and PSP) update the base network for each task (683K parameters), while *side-tuning* updates a four layer convolutional network per task (259K parameters after 10 tasks).
- **iTaskonomy.** Taskonomy [34] is significantly more challenging than CIFAR-100 and includes multiple computer vision tasks beyond object classification: including 2D (e.g. edge detection), 3D (e.g. surface normal estimation), and semantic (e.g. object classification) tasks. We note that approaches which work well on iCIFAR often do quite poorly in the more realistic setting. We created iTaskonomy by selecting all (12) tasks that make predictions from a single RGB image, and then created an incremental learning setup by selecting a random order in which to learn these tasks (starting with *curvature*). The images are 256×256 and we use a ResNet-50 for the base network and a 5-layer convolutional network for the *side-tuning* side network.

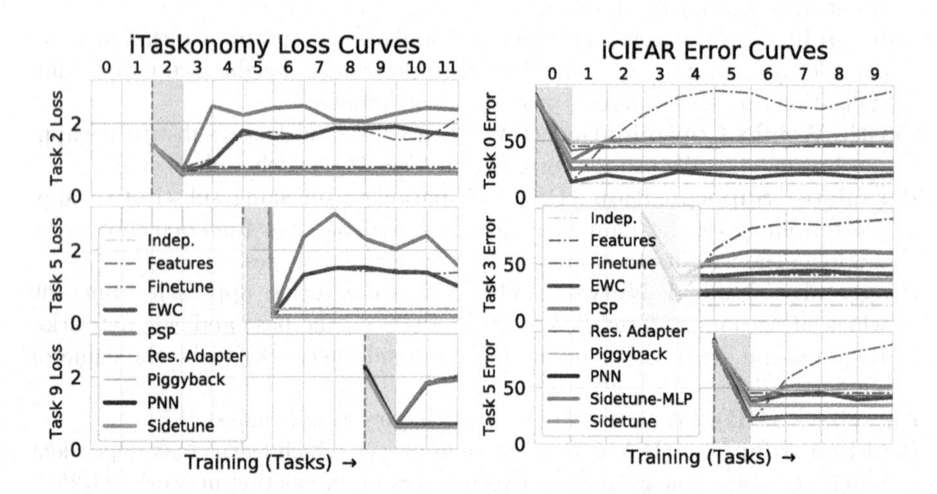

Fig. 5. Incremental Learning on iTaskonomy and iCIFAR. The above curves show loss and error for three tasks on iTaskonomy (left) and iCIFAR (right) datasets. The fact that *side-tuning* losses are flat after training (as we go right) shows that it does not forget previously learned tasks. That performance remains consistent even on later tasks (as we go down), showing that *side-tuning* does not become rigid. Substitutive methods show clear forgetting (e.g. PSP) and/or rigidity (e.g. EWC). In iTaskonomy, PNN and Independent are hidden under Sidetune. (Color figure online)

[3] Full experimental details (e.g. architecture) provided in the supplementary.

The number of learnable network parameters used across all tasks is 24.6M for EWC and PSP, and 11.0M for *side-tuning*[4].

Catastrophic Forgetting. As expected, there is no catastrophic forgetting in *side-tuning* and other additive methods. Figure 5 shows that the error for *side-tuning* does not increase after training (blue shaded region), while it increases sharply for the substitutive methods on both iTaskonomy and iCIFAR.

The difference is meaningful, and Fig. 4 shows sample predictions from *side-tuning* and EWC for a few tasks during and after training. As is evident from the bottom rows, EWC exhibits catastrophic forgetting on all tasks (worse image quality as we move right). In contrast, *side-tuning* (top) shows no forgetting and the final predictions are significantly closer to the ground-truth (boxed red).

Rigidity. *Side-tuning* learns later tasks as easily as the first, while constraint-based methods such as EWC stagnate. The predictions for later tasks are significantly better using *side-tuning* **even immediately after training and before any forgetting can occur** (e.g., *surface normals* in Fig. 4).

Figure 6 quantifies this slowdown. We measure rigidity as the log-ratio of the *actual loss* of the ith task over the loss when that task is *instead trained first* in the sequence. As expected, *side-tuning* experiences effectively zero slowdown on both datasets. For EWC, the added constraints make learning new tasks increasingly difficult and rigidity increases with the number of tasks (Fig. 6, left). PNN shows some positive transfer in iCIFAR (negative ratio value), but becomes rigid on the more challenging iTaskonomy, where tasks are more diverse.

Final Performance. Overall, *side-tuning* significantly outperforms the substitutive methods while using fewer than half the number of trainable parameters. It is comparable with additive methods while remaining remarkably simpler. On iCIFAR, vanilla *side-tuning* achieves a strong average rank (2.20 of 6, see

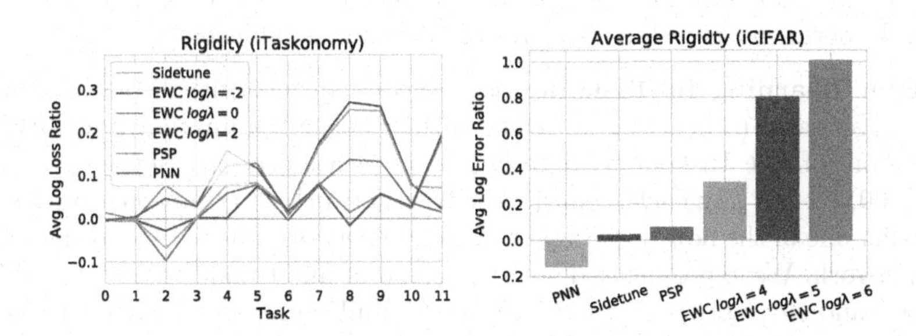

Fig. 6. Rigidity (Intransigence) on iTaskonomy and iCIFAR. *Side-tuning* always learns new tasks easily; EWC becomes increasingly unable to learn new tasks as training progresses. The same trend holds on iCIFAR (right).

[4] Numbers not counting readout parameters, which are common between all methods.

Table 2) and, when using the same combining operator (MLP) as PNN, is as good as the best-performing model without the additional lateral connections (see Fig. 5 right). On iTaskonomy, vanilla *side-tuning* achieves the best average rank (1.33 of 6, while the next best is 2.42 by PNN, see Table 2).

This is a direct result of the fact (shown above) that *side-tuning* does not suffer from catastrophic forgetting or rigidity (intransigence). It is not due to the fact that the sidetuning structure is specially designed for these types of image tasks; it is not (we show in Sect. 4.3 that it performs well on other domains). In fact, the much larger networks used in EWC and PSP *should* achieve better performance on any single task. For example, EWC produces sharper images early on in training, before it has had a chance to accumulate too many constraints (e.g. *reshading* in Fig. 4). But this factor was outweighed by *side-tuning's* immunity from the effects of catastrophic forgetting and compunding rigidity.

Table 2. Average rank on iTaskonomy and iCIFAR. Despite being simpler than alternatives, *side-tuning* generally achieved a better average rank than other approaches. The difference increases on the more challenging Taskonomy dataset, where *side-tuning* significantly outperformed all tested alternatives.

Method	Avg. rank (↓)	
	iTaskonomy	iCIFAR
EWC ($\lambda = 10^0, 10^5$)	5.25	2.70
PSP	5.25	5.60
Res. Adapter	3.58	4.40
Piggyback	3.17	5.00
PNN	2.42	**1.10**
Side-tune	**1.33**	2.20

4.3 Universality of the Experimental Trends

In order to address the possibility that *side-tuning* is somehow domain- or task-specific, we provide results showing that it is well-behaved in other settings. As the concern with *additive learning* is mainly that it is too inflexible to learn new tasks, we compare with fine-tuning (which outperforms other incremental learning tasks when forgetting is not an issue). For extremely limited amounts of data, feature extraction can outperform fine-tuning. We show that *side-tuning* generally performs as well as features or fine-tuning–whichever is better.

Transfer Learning in Taskonomy. We trained networks to perform one of three target tasks (object classification, surface normal estimation, and curvature estimation) on the Taskonomy dataset [34] and varied the size of the training set $N \in \{100, 4 \times 10^6\}$. In each scenario, the base network was trained (from scratch) to predict one of the non-target tasks. The side network was a copy of the original base network. We experimented with a version of fine-tuning that updated both the base and side networks; the results were similar to standard fine-tuning[5]. In all scenarios, *side-tuning* successfully matched the adaptiveness of fine-tuning, and significantly outperformed learning from scratch, as shown in Table 3a. The additional structure of the frozen base did not constrain performance with

[5] We defer remaining experimental details (learning rate, full architecture, etc.) to the supplementary materials. See provided code for full details.

Table 3. Side-tuning comparisons in other domains. Sidetuning matched the adaptability of fine-tuning on large datasets, while performing as well or better than the best competing method in each domain: **(a)** In Taskonomy, for *Normal Estimation* or *Object Classification* using a base trained for *Curvatures* and either 100 or 4M images for transfer. Results using *Obj. Cls.* base are similar and provided in the supplementary materials. **(b)** In SQuAD v2 question-answering, using BERT instead of a convolutional architecture. **(c)** In Habitat, learning to navigate by imitating expert navigation policies, using inputs based on either *Curvature* or *Denoising*. Finetuning does not perform as well in this domain. **(d)** Using RL (PPO) and direct interaction instead of supervised learning for navigation.

	Transfer Learning in Taskonomy		QA on SQuAD		Navigation (IL)		Navigation (RL)	
	From Curvature (100/4M ims.)		Match (↑)		Nav. Rew. (↑)		Nav. Rew. (↑)	
Method	Normals (MSE ↓)	Obj. Cls. (Acc. ↑)	Exact	F1	Curv.	Denoise	Curv.	Denoise
Fine-tune	**0.200 / 0.094**	**24.6 / 62.8**	**79.0**	**82.2**	10.5	9.2	10.7	**10.0**
Features	**0.204** / 0.117	24.4 / 45.4	49.4	49.5	**11.2**	8.2	**11.9**	8.3
Scratch	0.323 / **0.095**	19.1 / 62.3	0.98	4.65	9.4	**9.4**	7.5	7.5
Side-tune	**0.199 / 0.095**	**24.8 / 63.3**	**79.6**	**82.7**	**11.1**	**9.5**	**11.8**	**10.4**
	(a)		(b)		(c)		(d)	

large amounts of data (4M images), and *side-tuning* performed as well as (and sometimes slightly better than) fine-tuning.

Question-Answering in SQuAD v2. We also evaluated *side-tuning* on a question-answering task (SQuAD v2 [23]) using a non-convolutional architecture. We use a pretrained BERT [5] model for our base, and a second for the side network. Unlike in the previous experiments, BERT uses attention and no convolutions. Still, *side-tuning* adapts to the new task just as well as fine-tuning, outperforming features and scratch (Table 3b).

Imitation Learning for Navigation in Habitat. We trained an agent to navigate to a target coordinate in the Habitat environment. The agent is provided with both RGB input image and also an occupancy map of previous locations. The map does not contain any information about the environment—just previous locations. In this section we use Behavior Cloning to train an agent to imitate experts following the shortest path on 49k trajectories in 72 buildings. The agents are evaluated in 14 held-out validation buildings. Depending on what the base network was trained on, the source task might be useful (*Curvature*) or harmful (*Denoising*) for imitating the expert and this determines whether features or learning from scratch performs best. Table 3c shows that regardless of the which approach worked best, *side-tuning* consistently matched or beat it.

Reinforcement Learning for Navigation in Habitat. Using a different learning algorithm (PPO) and direct interaction instead of expert trajectories, we observe identical trends. We trained agents directly in Habitat (74 buildings). Table 3d shows performance in 14 held-out buildings after 10M frames of training. *Side-tuning* performs comparably to the max of competing approaches.

4.4 Learning Mechanics in Side-Tuning

Using Non-network Base Models. Since *side-tuning* treats the base model is a black box, it can be used even when the base model is not a neural network. On iTaskonomy, we show that side-tuning can effectively use ground truth curvature as a base for incremental learning whereas all the methods we compare against cannot use this information (with the exception of feature extraction). Specifically, we resize the curvature image from $256 \times 256{\times}2$ to $32 \times 32{\times}2$ and reshape it to $16 \times 16{\times}8$, the same size as the output of other base models. Side-tuning with ground truth curvature achieves a better rank (4.3) on iTaskonomy than *all 20* other methods (excluding Independent, 4.2)

Benefits for Intermediate Amounts of Data. We showed in the previous section that *side-tuning* performs like the best of {features, fine-tuning, scratch} in domains with abundant or scant data.

In order to test whether *side-tuning* could profitably synthesize the features with *intermediate* amounts of data, we evaluated each approach's ability to learn to navigate using 49, 490, 4900, or 49k expert trajectories and pretrained *denoising* features. *Side-tuning* was always the best-performing approach and, on intermediate amounts of data (e.g. 4.9k trajectories), outperformed the other techniques (*side-tune* 9.3, fine-tune: 7.5, features: 6.7, scratch: 6.6), Fig. 7).

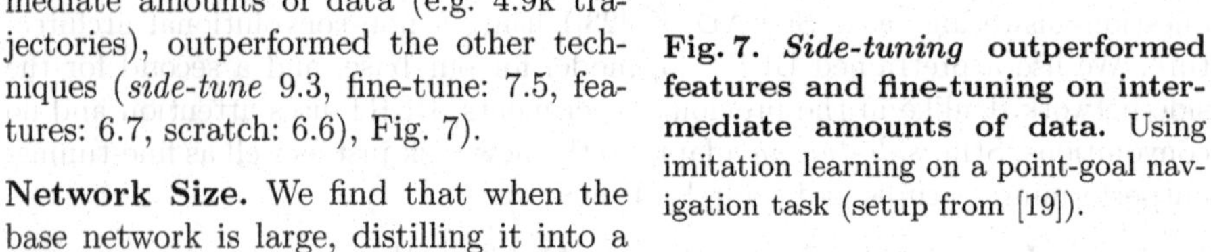

Fig. 7. ***Side-tuning* outperformed features and fine-tuning on intermediate amounts of data.** Using imitation learning on a point-goal navigation task (setup from [19]).

Network Size. We find that when the base network is large, distilling it into a smaller network and *sidetuning* will still retain most of the performance. In Fig. 8, we explore this in Habitat (RL using {curvature, denoise} $\rightarrow$ navigation), with other results in the supplementary.

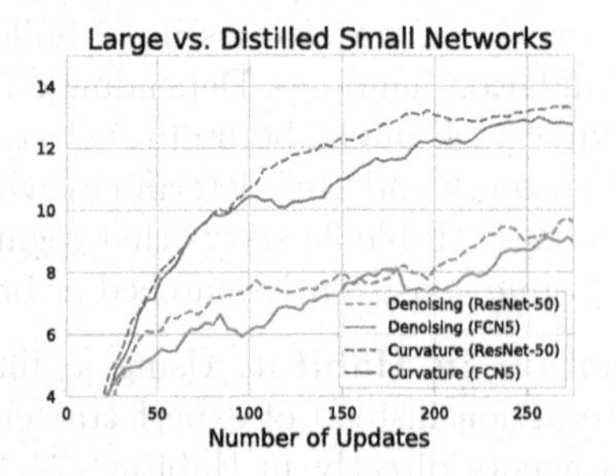

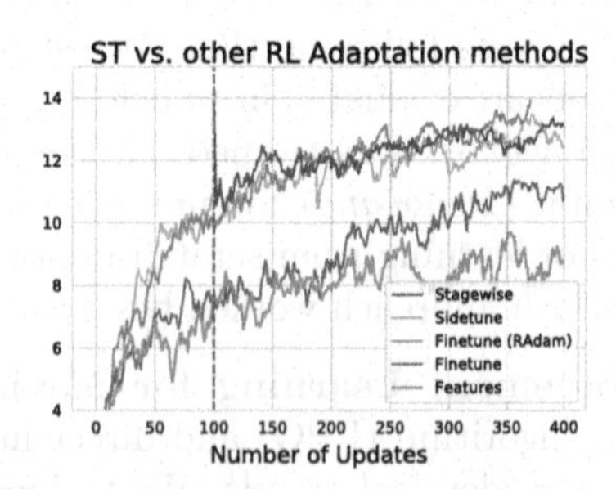

Fig. 8. Effect of network size in RL. (Left) The distilled (5-layer) networks perform almost as well as the original ResNet-50 base networks. The choice of which base to use (Curvature vs. Denoising) has a much larger impact. **(Right)** Fine-tuning, even with RAdam, performed significantly worse than the alternative approaches.

More than Just Stable Updates. In RL, fine-tuning often fails to improve performance. One common rationalization is that the early updates in RL are 'high variance'. The common stage-wise solution is to first train using fixed features and then unfreeze the weights. We found that this approach performs as well (but no better than) using fixed features–and *side-tuning* performed as well as both while not being domain-specific (Fig. 8). We tested the 'high-variance' theory by fine-tuning with both gradient clipping and an optimizer designed to prevent such high-variance updates (RAdam [16]). This provided no benefits over vanilla fine-tuning, suggesting that the benefits of side-tuning are not solely due to gradient stabilization early in training.

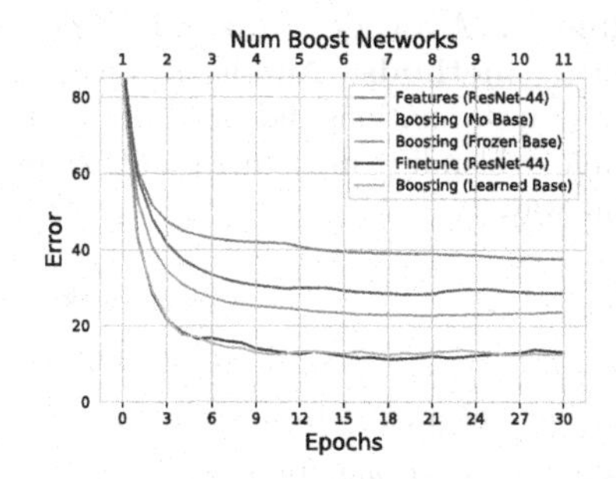

Fig. 9. Boosting. Deeper networks outperform many shallow learners.

Not Boosting. Since the side network learns a residual on top of the base network, could *side-tuning* be used for boosting? Although network boosting does improve performance on iCIFAR (Fig. 9), the parameters would've been better used in a deeper network rather than many shallow networks.

4.5 Analysis of Design Choices

We evaluate the effect of our architectural design decisions on task performance.

Base and Side Elements. *Side-tuning* uses two streams of information - one from the base model and one from the side model. Are both streams necessary? Table 5 shows that on the iTaskonomy experiment performance improves when using both models.

Merge Methods. Section 3.1 described different ways to merge the base and side networks. Table 4 evaluates a few of these approaches. Product and alpha-blending are two of the simplest approaches and have little overhead in terms of compute and parameter count. [29] (MLP) and [21] (FiLM) use multi-layer perceptrons to adapt the base network to the new task. Table 4 shows that alpha-blending, MLP, and FiLM are comparable, though the FiLM-based methods achieve marginally better average rank on iTaskonomy. We use alpha-blending as it adds fewer parameters and achieves similar performance.

Side Network Initialization. A good side network initialization can yield a minor boost in performance. We found that initializing from the base network slightly outperforms a low-energy initialization[6], which slightly outperforms Xavier initialization. However, we found that these differences were not statistically significant across tasks (H_0 : *pretrained* = *xavier*; $p = 0.07$, Wilcoxon

[6] Side network is trained to not impact the output. Full details in the supplementary.

Table 4. Average rank of various merge methods. Alternative feature-wise transformations did not outperform simple α-blending in a statistically significant way.

Method	Avg. rank (↓)
	iTaskonomy
Product (element-wise)	3.64
Summation (α-blending)	2.27
MLP [29]	2.18
FiLM [21]	1.91

Table 5. Both base and side networks contribute. Performance (average rank on iTaskonomy) improved when using *both* the base and side model in *side-tuning.*

Method	Avg. rank (↓)
	iTaskonomy
Base-only	2.55
Side-only	2.10
Side-tuning	1.36

signed-rank test). We suspect that initialization might be more important on harder problems. We test this by repeating the analysis without the simple texture-based tasks (2D keypoint + edge detection and autoencoding) and find the difference in initialization is now significant ($p = 0.01$).

5 Conclusions and Limitations

We have introduced the *side-tuning* framework, a simple yet effective approach for *additive learning.* Since it does not suffer from catastrophic forgetting or rigidity, it is naturally suited to incremental learning. The theoretical advantages are reflected in empirical results, and we found *side-tuning* to perform on par with *or better than* many current incremental learning approaches, while being significantly simpler. Experiments demonstrated this in challenging contexts and with various state-of-the-art neural networks across multiple domains.

More complex methods should need to demonstrate clear improvements over simply doing this naïve approach. We see several natural ways to improve it:

Better Forward Transfer: Our experiments used only a single base and single side network. Leveraging previously trained side networks could yield better performance on later tasks.

Learning When to Deploy Side Networks: Like most incremental learning setups, we assumed that the tasks are presented in a sequence and that task identities are known. Using several active side networks in tandem would provide a natural way to identify task change or distribution shift.

Using Side-Tuning to Measure Task Relevance: We found that α tracked task relevance in [34], but a more rigorous treatment of the interaction between the base, side, α and final performance could yield insight into how tasks relate.

Acknowledgements. This material is based upon work supported by ONR MURI (N00014-14-1-0671), Vannevar Bush Faculty Fellowship, an Amazon AWS Machine

Learning Award, NSF (IIS-1763268), a BDD grant and TRI. Toyota Research Institute ("TRI") provided funds to assist the authors with research but this article solely reflects the opinions and conclusions of its authors and not TRI or any other Toyota entity.

References

1. Bilen, H., Vedaldi, A.: Universal representations: the missing link between faces, text, planktons, and cat breeds. arXiv preprint arXiv:1701.07275 (2017)
2. Chaudhry, A., Dokania, P.K., Ajanthan, T., Torr, P.H.S.: Riemannian walk for incremental learning: understanding forgetting and intransigence. In: Ferrari, V., Hebert, M., Sminchisescu, C., Weiss, Y. (eds.) ECCV 2018. LNCS, vol. 11215, pp. 556–572. Springer, Cham (2018). https://doi.org/10.1007/978-3-030-01252-6_33
3. Chaudhry, A., Ranzato, M., Rohrbach, M., Elhoseiny, M.: Efficient lifelong learning with a-GEM. In: International Conference on Learning Representations (2019). https://openreview.net/forum?id=Hkf2_sC5FX
4. Cheung, B., Terekhov, A., Chen, Y., Agrawal, P., Olshausen, B.: Superposition of many models into one. In: Advances in Neural Information Processing Systems, pp. 10868–10877 (2019)
5. Devlin, J., Chang, M., Lee, K., Toutanova, K.: BERT: pre-training of deep bidirectional transformers for language understanding. In: Conference of the North American Chapter of the Association for Computational Linguistics (NAACL) (2019)
6. Finn, C., Abbeel, P., Levine, S.: Model-agnostic meta-learning for fast adaptation of deep networks. In: Proceedings of the International Conference on Machine Learning (ICML) (2017)
7. Geman, S., Bienenstock, E., Doursat, R.: Neural networks and the bias/variance dilemma. Neural Comput. **4**(1), 1–58 (1992)
8. He, K., Zhang, X., Ren, S., Sun, J.: Deep residual learning for image recognition. In: Proceedings of the IEEE Conference on Computer Vision and Pattern Recognition, pp. 770–778 (2016)
9. Hinton, G., Vinyals, O., Dean, J.: Distilling the knowledge in a neural network. arXiv preprint arXiv:1503.02531 (2015)
10. Johannink, T., et al.: Residual reinforcement learning for robot control. In: 2019 International Conference on Robotics and Automation (ICRA), pp. 6023–6029. IEEE (2019)
11. Kingma, D.P., Ba, J.: Adam: a method for stochastic optimization. In: International Conference on Learning Representations (2015)
12. Kirkpatrick, J., et al.: Overcoming catastrophic forgetting in neural networks. Proc. Nat. Acad. Sci. U.S.A. **114**(13), 3521–3526 (2017)
13. Leach, P.J.: A critical study of the literature concerning rigidity. Br. J. Soc. Clin. Psychol. **6**(1), 11–22 (1967)
14. Lee, J., Joo, D., Hong, H.G., Kim, J.: Residual continual learning. In: AAAI (2020)
15. Li, Z., Hoiem, D.: Learning without forgetting. IEEE Trans. Pattern Anal. Mach. Intell. **40**(12), 2935–2947 (2017)
16. Liu, L., et al.: On the variance of the adaptive learning rate and beyond. In: Proceedings of the Eighth International Conference on Learning Representations (ICLR 2020), April 2020
17. Mallya, A., Davis, D., Lazebnik, S.: Piggyback: adapting a single network to multiple tasks by learning to mask weights. In: Proceedings of the European Conference on Computer Vision (ECCV), pp. 67–82 (2018)

18. Mallya, A., Lazebnik, S.: PackNet: adding multiple tasks to a single network by iterative pruning. In: Proceedings of the IEEE Conference on Computer Vision and Pattern Recognition, pp. 7765–7773 (2018)
19. Savva, M., et al.: Habitat: a platform for embodied AI research. In: Proceedings of the IEEE/CVF International Conference on Computer Vision (ICCV) (2019)
20. Mullapudi, R.T., Chen, S., Zhang, K., Ramanan, D., Fatahalian, K.: Online model distillation for efficient video inference. In: Proceedings of the IEEE International Conference on Computer Vision, pp. 3573–3582 (2019)
21. Perez, E., Strub, F., De Vries, H., Dumoulin, V., Courville, A.: FILM: visual reasoning with a general conditioning layer. In: Thirty-Second AAAI Conference on Artificial Intelligence (2018)
22. Raghu, A., Raghu, M., Bengio, S., Vinyals, O.: Rapid learning or feature reuse? Towards understanding the effectiveness of MAML. In: International Conference on Learning Representations (ICLR) (2020)
23. Rajpurkar, P., Jia, R., Liang, P.: Know what you don't know: unanswerable questions for squad. In: Proceedings of the 56th Annual Meeting of the Association for Computational Linguistics (ACL) (2018)
24. Rebuffi, S.A., Bilen, H., Vedaldi, A.: Learning multiple visual domains with residual adapters. In: Advances in Neural Information Processing Systems, pp. 506–516 (2017)
25. Rebuffi, S.A., Bilen, H., Vedaldi, A.: Efficient parametrization of multi-domain deep neural networks. In: Proceedings of the IEEE Conference on Computer Vision and Pattern Recognition, pp. 8119–8127 (2018)
26. Rebuffi, S., Kolesnikov, A., Lampert, C.H.: iCaRL: incremental classifier and representation learning. In: IEEE Conference on Computer Vision and Pattern Recognition (CVPR). IEEE (2017)
27. Rigidity (psychology): Rigidity (psychology)a – Wikipedia, the free encyclopedia (2020). https://en.wikipedia.org/wiki/Rigidity_(psychology). Accessed 12 July 2020
28. Rosenfeld, A., Tsotsos, J.K.: Incremental learning through deep adaptation. IEEE Trans. Pattern Anal. Mach. Intell. (2018)
29. Rusu, A.A., et al.: Progressive neural networks. CoRR abs/1606.04671 (2016). http://arxiv.org/abs/1606.04671
30. Schwarz, J., et al.: Progress & compress: a scalable framework for continual learning. In: Proceedings of the International Conference on Machine Learning (ICML) (2018)
31. Serra, J., Suris, D., Miron, M., Karatzoglou, A.: Overcoming catastrophic forgetting with hard attention to the task. In: International Conference on Machine Learning (2018)
32. Sharif Razavian, A., Azizpour, H., Sullivan, J., Carlsson, S.: CNN features off-the-shelf: an astounding baseline for recognition. In: Proceedings of the IEEE Conference on Computer Vision and Pattern Recognition Workshops, pp. 806–813 (2014)
33. Silver, T., Allen, K.R., Tenenbaum, J., Kaelbling, L.P.: Residual policy learning. CoRR abs/1812.06298 (2018). http://arxiv.org/abs/1812.06298
34. Zamir, A.R., Sax, A., Shen, W.B., Guibas, L.J., Malik, J., Savarese, S.: Taskonomy: disentangling task transfer learning. In: IEEE Conference on Computer Vision and Pattern Recognition (CVPR). IEEE (2018)

Towards Part-Aware Monocular 3D Human Pose Estimation: An Architecture Search Approach

Zerui Chen[1,3](✉), Yan Huang[1], Hongyuan Yu[1,3], Bin Xue[3], Ke Han[1], Yiru Guo[5], and Liang Wang[1,2,4]

[1] Center for Research on Intelligent Perception and Computing, NLPR, CASIA, Beijing, China
{zerui.chen,hongyuan.yu,ke.han}@cripac.ia.ac.cn, {yhuang,wangliang}@nlpr.ia.ac.cn

[2] Center for Excellence in Brain Science and Intelligence Technology, CAS, Beijing, China

[3] School of Artificial Intelligence, University of Chinese Academy of Sciences, Beijing, China
xuebin2018@ia.ac.cn

[4] Chinese Academy of Sciences, Artificial Intelligence Research (CAS-AIR), Beijing, China

[5] School of Astronautics, Beihang University, Beijing, China
guoyiru@buaa.edu.cn

Abstract. Even though most existing monocular 3D pose estimation approaches achieve very competitive results, they ignore the heterogeneity among human body parts by estimating them with the same network architecture. To accurately estimate 3D poses of different body parts, we attempt to build a part-aware 3D pose estimator by searching a set of network architectures. Consequently, our model automatically learns to select a suitable architecture to estimate each body part. Compared to models built on the commonly used ResNet-50 backbone, it reduces 62% parameters and achieves better performance. With roughly the same computational complexity as previous models, our approach achieves state-of-the-art results on both the single-person and multi-person 3D pose estimation benchmarks.

Keywords: 3D pose estimation · Body parts · Neural architecture search

1 Introduction

3D human pose estimation plays a crucial role to unlock widespread applications in human-computer interaction, robotics, surveillance, and virtual reality. Compared with multi-view methods [19,41,43,52,61], monocular 3D human pose estimation is more flexible for deployment in outdoor environments. However, given its ill-posed nature, estimating 3D human poses from a single RGB

A. Vedaldi et al. (Eds.): ECCV 2020, LNCS 12348, pp. 715–732, 2020.
https://doi.org/10.1007/978-3-030-58580-8_42

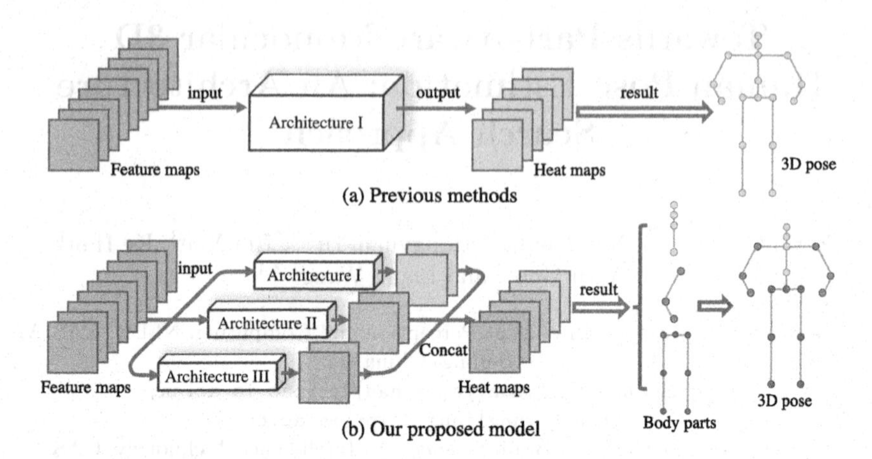

Fig. 1. Motivation. Most of the previous methods employ a single network architecture to deal with intrinsically heterogeneous human body parts (as shown in (a)). Instead, we are motivated to search for a suitable network architecture for a group of parts and estimate their 3D locations with a part-specific architecture (as shown in (b)).

image remains a challenging problem. Thanks to Convolutional Neural Networks (CNNs), many effective approaches are proposed and formulate the problem as joint coordinate regression [28,47] or heat maps learning [57,65]. Recently, many approaches [39,40,48,62] have followed a popular paradigm in predicting per voxel likelihood for each human joint and achieved competitive performance.

In most previous approaches shown in Fig. 1(a), CNNs share the same network architecture for predicting all human body parts with different degrees of freedom (DOFs), ranging from parts with higher DOFs like the wrists to parts with lower DOFs like the torso. However, a single network architecture might be sub-optimal to deal with various body parts. Because different parts might have various movement patterns and shapes, estimating their locations might require different network topologies (*e.g.*, different kernel sizes and distinct receptive fields). A recent effort [54] also demonstrates that it is effective to estimate different body parts by explicitly taking their DOFs into account.

As shown in Fig. 1(b), we approach the problem from a different angle and propose to estimate different body parts with part-specific network architectures. However, looking for optimal architectures for various body parts is an intractable and time-consuming job even for an expert. Therefore, instead of designing them manually, we consult the literature of neural architecture search (NAS) [4,14,17,23,31,49,56] and propose to search part-specific network architectures for different parts. In fact, the idea of searching network architectures for certain tasks is not new. Specifically, it has been applied in semantic segmentation [7,30,60] and object detection [8,13,42].

However, applying NAS into 3D human pose estimation is non-trivial, because current NAS approaches mainly focus on 2D visual tasks. Different from them, 3D human poses are commonly estimated in a higher-order volumetric space [11,40,48,52]. It consists of 2D spatial and depth axes and greatly increases the uncertainty during optimization. More importantly, how to use prior information about the human body structure to facilitate the architecture search and achieve a trade-off between accuracy and complexity is another issue.

To deal with these issues, we introduce the fusion cell in the context of NAS to increase the resolution of feature maps and generate desired volumetric heat maps efficiently. The fusion cell has multiple head networks that are various convolutional architectures, consisting of different kernels and operations. To improve the part-awareness of our model, we attempt to generate the volumetric heat map for each part with a specially optimized head network. Considering the symmetry prior of the human body structure, it is inefficient to search a different head network for each part. Our approach classifies all body parts into several groups and assigns each group with a part-specific architecture. In the search stage of our approach, all the architectures, including the fusion cell, are optimized by gradient descent. Then, we stack these optimized computational cells to construct our part-aware 3D pose estimator. In the evaluation stage, our part-aware 3D human pose estimator can select optimized head networks encoded in the fusion cell to estimate different groups of body parts.

Through extensive experiments, we show that our approach can achieve a good trade-off between complexity and performance. With 62% fewer parameters and 24% fewer FLOPs (multiply-adds), our approach outperforms the model using ResNet-50 backbone and achieves 53.6 mm in Mean Per Joint Position Error (MPJPE). By stacking more computational cells, it can further advance the state-of-the-art accuracy on Human3.6M by 2.3 mm with 41% fewer parameters.

Our contributions can be summarized as follows:

- Our work shows that it might be sub-optimal to estimate 3D poses of all body parts with a single network architecture. To the best of our knowledge, we make the first attempt to search part-specific architectures for different parts.
- We introduce the fusion cell to generate volumetric heat maps efficiently. In the fusion cell, we classify all body parts into several groups and estimate each group of parts with a distinct head network.
- Our part-aware 3D pose estimator is both compact and efficient. It achieves state-of-the-art accuracy on both the single-person and multi-person 3D human pose benchmarks using much fewer parameters and FLOPs.

2 Related Work

3D Human pose estimation has been studied widely in the past. In this section, we only focus on previous works that can be relevant to our work.

Estimate 3D poses from 2D Joints: Some approaches divide the task of 3D human pose estimation into first predicting 2D joint locations and then back-projecting them to estimate 3D human poses. The practice of inferring 3D human poses from their 2D projections can be traced back to the classic work [27]. Given the bone lengths, the problem boils down to a binary decision tree where each branch corresponds to two possible states of a joint concerning its parent. Jiang *et al.* [20] generate a set of hypothesis of 3D poses using Taylor's algorithm [50] and use them to query a large database of motion capture data to find the nearest neighbor. Similarly, the idea of exploiting nearest neighbor queries has been revisited by [15]. Chen *et al.* [6] also share the idea of using the detected 2D pose to query a large database of exemplary poses. Another common approach [3,63] is to learn an over-complete dictionary of basis 3D poses from a large database of motion capture data. Moreno-Noguer *et al.* [36] employ the pair-wise distance matrix of 2D joints to learn a distance matrix for 3D joints. Martinez *et al.* [32] design a fully-connected network to estimate 3D joint locations relative to the pelvis from 2D poses. Hossain *et al.* [16] exploit temporary information to calculate a sequence of 3D poses from a sequence of 2D joint locations. Ci *et al.* [10] combine the advantage of graph convolution network and fully-connected network and equip the model with strong generalization power. Cai *et al.* [5] introduce a graph-based local-to-global network to recover 3D poses from 2D pose sequences. These methods focus on estimating 3D poses from 2D poses, and we attempt to estimate 3D poses from monocular images.

Estimate 3D poses from Monocular Images: Recently, many approaches have been proposed to estimate 3D poses from monocular images in an end-to-end fashion. Li *et al.* [28] and Park *et al.* [38] exploit the 2D pose information to benefit 3D pose estimation. Rogez *et al.* [44] and Varol *et al.* [53] augment the training data with synthetic images and train CNNs to predict 3D poses from real images. Sun *et al.* [47] adopt a reparameterized pose representation using bones instead of joints. Pavlakos *et al.* [40] extend 2D heat maps to 3D volumetric heat maps and predict per voxel likelihood for each joint. Tome *et al.* [51] generalize Convolutional Pose Machine (CPM) [55] to the task of monocular 3D human pose estimation. Chen *et al.* [9] propose to decompose the volumetric representation into 2D depth-aware heat maps and joint depth estimation. Zhou *et al.* [65] propose a weakly-supervised transfer learning method that uses mixed 2D and 3D labels in a unified deep neural network. By introducing a simple integral operation, Sun *et al.* [48] unify heat maps learning and regression learning for pose estimation. Kocabas *et al.* [25] propose to train the 3D pose estimator with the multi-view triangulation in a self-supervised manner. Instead of estimating root-relative 3D poses, Moon *et al.* [35] propose to estimate 3D poses in the camera coordinate system directly. More recent works [1,21,22,26,37] tend to focus on reconstructing fine-grained 3D human shapes. Nevertheless, all works are limited in estimating all body parts with a single head network, and we attempt to search part-specific head networks for different body parts.

3 The Proposed Approach

In the literature of NAS, differential architecture search (DARTS) [30] is a representative method that can search effective network architectures using fewer computing resources. Therefore, we build our model on DARTS. First, we introduce some basic knowledge about DARTS. Then, we describe our approach to search part-specific head networks for intrinsically heterogeneous body parts.

3.1 Preliminaries: Differential Architecture Search (DARTS)

The framework of DARTS decomposes the searched network architecture into a number of (L) computational cells. There are two types of cells: the normal cell and the reduction cell. Both of them have typical convolution architectures to transform feature maps. Additionally, the reduction cell has another function to downsample the feature map. Each computational cell can be represented as a directed acyclic graph (DAG), consisting of an ordered sequence of N nodes ($\mathcal{N} = \{x^{(i)}|i = 1, ..., N\}$). In the DAG, each node $x^{(i)}$ ($i \in \{1, ..., N\}$) is a hidden representation (*i.e.*, feature map), and each edge $o^{(i,j)}(\cdot)$ denotes the transformation from $x^{(i)}$ to $x^{(j)}$ and is associated with an operation (*i.e.*, pooling and convolution). In each cell, there are two input nodes (*i.e.*, $x^{(1)}$ and $x^{(2)}$ receive outputs from the previous two cells) and one output node $x^{(N)}$ (*i.e.*, the concatenation of all intermediate nodes $(x^{(3)}, x^{(4)}, ..., x^{(N-1)})$). The output of an intermediate node $x^{(j)}$ is computed as:

$$x^{(j)} = \sum_{i<j} o^{(i,j)}(x^{(i)}) \tag{1}$$

Where the node $x^{(i)}$ is one predecessor of the node $x^{(j)}$. There is a pre-defined space of operations denoted by $\mathcal{O}$, each element of which is a fixed operation (*e.g.*, identity connection, convolution and max pooling). In the search stage, our goal is to automatically select one operation from $\mathcal{O}$ and assign the operation to $o^{(i,j)}(\cdot)$ for each pair of nodes.

The core idea of DARTS is to make the search space continuous, and formulate the choice of an operation as a softmax over all possible operations:

$$\bar{o}^{(i,j)}(x) = \sum_{o \in \mathcal{O}} \frac{exp(\alpha^{o}_{i,j})}{\sum_{o' \in \mathcal{O}} exp(\alpha^{o'}_{i,j})} o(x) \tag{2}$$

Where $\alpha^{o}_{i,j}$ denotes the learnable score of the operation $o(\cdot)$ on the edge from $x^{(i)}$ to $x^{(j)}$. $\alpha_{i,j} \in \mathbb{R}^{|\mathcal{O}|}$ represents the scores of all candidate operations over the edge. The architecture of a cell is denoted as $\alpha = \{\alpha_{i,j}\}$, consisting of $\alpha_{i,j}$ for all edges connecting pairs of nodes. Then, DARTS formulates architecture search as finding α to minimize the loss function on the validation set:

$$\min_{\alpha} \ L_{val}(w^*(\alpha), \alpha) \tag{3}$$

$$\text{s.t. } w^*(a) = \text{argmin}_\text{w} \ L_{train}(w, \alpha) \tag{4}$$

Where $w^*(\alpha)$ denotes the network weights associated with the architecture α, which is optimized on the training set. The architecture parameter α can be optimized via gradient descent by approximating Eq. 3 as:

$$\nabla_{\alpha} L_{val}(w^*(\alpha), \alpha) \approx \nabla_{\alpha} L_{val}(w - \xi \nabla_{w} L_{train}(w, \alpha), \alpha) \tag{5}$$

Where w denotes the current network weights, $\nabla_w L_{train}(w, \alpha)$ is the a gradient step of w and ξ is the step's learning rate. When we finish optimizing α in the search stage, we assign $o^{(i,j)}(\cdot)$ with the most likely operation candidate according to $\alpha^{(i,j)}$. For each intermediate node in a computational cell, DARTS retains its two strongest predecessors.

3.2 DARTS for Monocular 3D Human Pose Estimation

Since the framework of DARTS is originally designed for image classification, neither the normal cell nor the reduction cell can increase the resolution of feature maps. However, it is a common practice for 3D pose estimators to upsample feature maps from the size of 8×8 to the size of 64×64 consecutively and generate volumetric heat maps for all body parts. To this end, as shown in Fig. 2, we propose to introduce another type of cell, namely fusion cell, in the context of DARTS. It can upsample and transform feature maps propagated from previous cells. Just like the reduction cell performs downsampling at input nodes, the fusion cell also upsamples feature maps at input nodes as a preprocessing step. Then, we employ edges between two nodes (*i.e.*, convolution, pooling, *etc.*) to transform upsampled feature maps and produce volumetric heat maps for all parts at the output node. As shown in Fig. 2, it is interesting to note that the output node is the concatenation of all intermediate nodes and each intermediate node represents volumetric heat maps for a certain group of body parts. Through intermediate nodes in the fusion cell, we automatically divide all body parts into several groups, and the number of groups is equal to the number of intermediate nodes in the fusion cell. As shown in Fig. 2(a), there exist many candidate operations between nodes in the search stage, and we obtain the optimized architecture upon finishing the search process. In the optimized architecture shown in Fig. 2(b), we can observe that each intermediate node has been transformed by a different set of operations. In other words, we learn part-specific architectures in the search stage and employ them to estimate different groups of body parts in the evaluation stage.

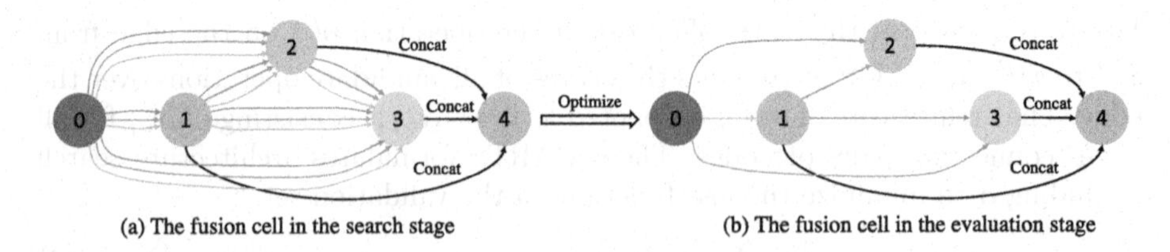

Fig. 2. An illustration of the fusion cell. Node 0 is the input node, and Node 1, 2, 3 are intermediate nodes. Node 4 is the output node and concatenates all intermediate nodes. Each edge represents one operation between two nodes. For simplicity, we only draw one input node here instead of two.

We follow a popular baseline [48] to build our part-aware 3D pose estimator. It predicts per voxel likelihood for each part and uses the soft-argmax operator to extract the 3D coordinate from the volumetric heat map. Instead of using ResNet-50 backbone and deconvolution layers, we search the whole network architecture. In the search stage, we stack the normal cell, the reduction cell, the fusion cell to construct our model with a total of N_c cells. We fix the number of reduction cells and fusion cells to N_r and N_f, respectively. Because the fusion cell is designed to generate volumetric heat maps at last, we first interweave $(N_c - N_r - N_f)$ normal cells and N_r reduction cells. Following the original DARTS, we organize the position of the reduction cell as:

$$P_r^i = \text{floor}(\frac{N_c - N_f}{N_r + 1}) \times i + 1 \tag{6}$$

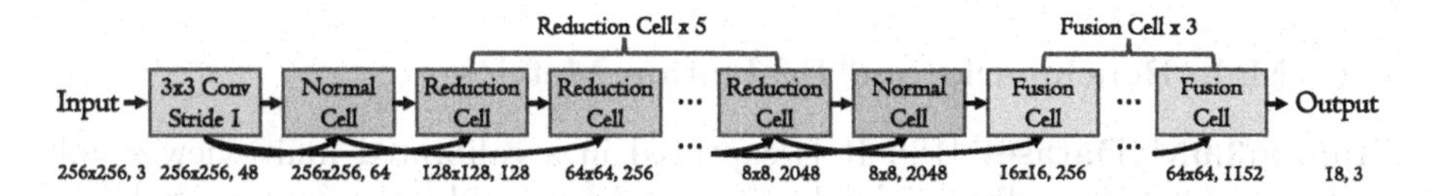

Fig. 3. An overview of our network architecture. We take the 256 × 256 input image as an example. It consists of ten computation cells: two normal cells, five reduction cells, and three fusion cells. The architecture of all types of cells are optimized in the search stage, and each cell receives inputs from the outputs of the previous two cells.

Where $i \in \{1, 2, ..., N_r\}$ denotes the ith reduction cell. P_r^i denotes the position of the ith reduction cell. floor(·) represents the function that discards the decimal point of a given number. After arranging normal cells and reduction cells, we append N_f fusion cells behind them. In the search stage, our model has a total of ten cells. We set N_r and N_f as 5 and 3, respectively. As illustrated in Fig. 3, out of the top seven cells, we interweave two normal cells and five reduction cells. Then, we append three fusion cells consecutively behind them to generate volumetric heat maps for all parts. We employ L1 loss to supervise estimated 3D poses and update network parameters w on the training set and architectures for all types of cells α on the validation set alternately.

When we finish the search process, we obtain the optimized normal cell, reduction cell, and fusion cell, as in Fig. 2(b). To evaluate the effectiveness of our searched architectures, we re-train our model constructed with these optimized cells. When our model is built with ten computational cells, the overview of its architecture is the same as what it was in the search stage. As shown in Fig. 3, given an input image, it first goes through a 3 × 3 convolution layer and a normal cell to generate the feature map. Then, we append five consecutive reduction cells to downsample the feature map and double its channel with a total stride of 2^5. After a series of reduction cells, the feature map is 8 × 8 × 2048 in size, and we use a normal cell to refine it further. To generate the volumetric heat map, we use the proposed fusion cell to upsample the feature map. Except for the last one, we set the output channel of remaining fusion cells to 256 as a common practice. Three consecutive fusion cells upsample the feature map with a total stride of 2^3 and generate the volumetric heat map of size 64 × 64 × 64 for all body parts. For each part, we extract its 3D coordinate from the corresponding volumetric heat map via the differential soft-argmax operation [48]. As we do in the search stage, we still employ L1 loss to train our model.

4 Experimental Evaluation

In this section, we present a detailed evaluation of our proposed approach. First, we introduce main benchmarks and present our experimental settings. Then, we conduct rigorous ablation analysis about our approach. Finally, we build our strongest part-aware estimator upon the knowledge obtained in ablation studies and compare it with state-of-the-art performance.

4.1 Main Benchmarks and Evaluation Metrics

Human3.6M Dataset [18]**:** It is captured in a calibrated multi-view studio and consists of 3.6 millions of video frames. Eleven subjects are recorded from four camera viewpoints, performing 15 activities. Previous works widely use two evaluation metrics. The first one is mean per joint position error (MPJPE), which first aligns the pelvis joint between estimated and ground-truth 3D poses and computes the average joint error among all human joints. The second metric uses Procrustes Analysis (PA) to align MPJPE further, and it is called PA MPJPE.

MuCo-3DHP and MuPoTS-3D Datasets [34]**:** These datasets are designed for multi-person 3D pose estimation. The training set is the MuCo-3DHP dataset, and it is generated by compositing the MPI-INF-3DHP dataset [33]. MuPoTS-3D dataset acts as the test set and contains 20 in-the-wild scenes. The evaluation metric is the 3D percentage of correct keypoints (3DPCK).

4.2 Experimental Settings and Implementation Details

Human3.6M Dataset: Two evaluation protocols are widely used. Protocol 1 uses six subjects (S1, S5, S6, S7, S8, S9) in training and reports the evaluation result on every 64th frame of Subject 11's videos using PA MPJPE. Protocol 2 uses six subjects (S1, S5, S6, S7, S8) in training and reports the evaluation result on every 64th frame of two subjects (S9, S11) using MPJPE. In the evaluation stage of our approach, we use additional MPII [2] 2D pose data during training.

In the search stage, we train the network only with Human3.6M data. We split three subjects (S1, S5, S6) as the training set to update the network parameter w and use two subjects (S7, S8) as the validation set to update the network architecture α. We include following eight operations in the pre-defined space $\mathcal{O}$: 3×3 and 5×5 separable convolutions, 3×3 and 5×5 dilated separable convolutions, 3×3 max pooling, 3×3 average pooling, identity and *zero*.

MuCo-3DHP and MuPoTS-3D Datasets: We create 400K composite frames of the MuCo-3DHP dataset, of which half are without appearance augmentation. We use additional COCO [29] 2D pose data during training.

Table 1. Quantitative evaluation of the number of intermediate nodes within each fusion cell on Human3.6M using Protocol 2. N_i denotes the number of intermediate nodes within each fusion cell. Lower is better, best in bold, second-best underlined.

Methods	Search space		Params	FLOPs	Direct.	Dicuss	Eating	Greet	Phone	Pose
	dil. conv.	sep. conv.								
Ours, $N_i = 1$	✓	✓	14.7M	22.9G	52.6	60.9	50.8	54.3	62.0	53.4
Ours, $N_i = 2$	✓	✓	13.0M	10.7G	**46.3**	**55.3**	**47.2**	**49.0**	**55.0**	**48.2**
Ours, $N_i = 3$	✓	✓	9.9M	**7.8G**	53.5	62.2	54.1	56.5	62.7	55.5
Ours, $N_i = 4$	✓	✓	**9.9M**	7.9G	50.8	60.0	53.2	53.3	60.7	50.8
Ours, $N_i = 2$	–	✓	15.9M	12.8G	55.8	61.3	52.5	55.3	63.0	54.6
Ours, $N_i = 2$	✓	–	10.4M	8.7G	52.0	60.0	51.4	53.9	61.5	52.6
Methods	Purch.	Sitting	SitD.	Smoke	Photo	Wait	Walk	WalkD.	WalkT.	Avg
Ours, $N_i = 1$	56.0	68.8	76.7	60.0	65.8	53.8	44.6	62.7	51.8	58.9
Ours, $N_i = 2$	**52.6**	**64.6**	**70.8**	**54.4**	**60.0**	**48.8**	**40.9**	**58.3**	**46.7**	**53.6**
Ours, $N_i = 3$	59.0	73.1	81.5	60.7	66.9	55.8	46.9	63.7	53.3	60.9
Ours, $N_i = 4$	55.9	69.8	74.3	58.7	64.8	53.0	43.4	61.2	49.5	58.0
Ours, $N_i = 2$	57.0	69.9	76.8	61.3	67.6	54.4	45.9	64.2	52.9	59.9
Ours, $N_i = 2$	56.6	68.8	76.7	59.9	66.3	53.5	44.8	62.6	51.6	58.7

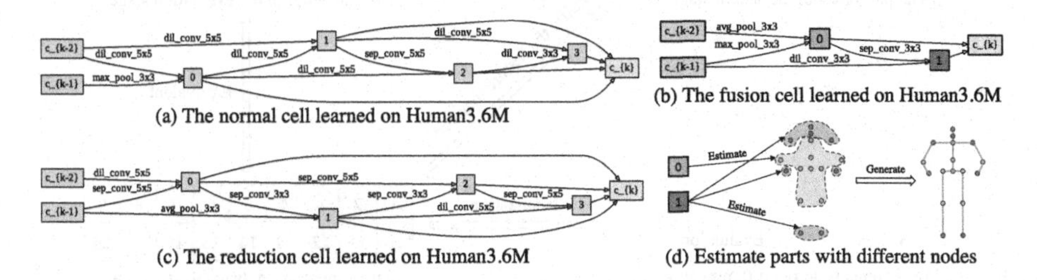

(a) The normal cell learned on Human3.6M
(b) The fusion cell learned on Human3.6M
(c) The reduction cell learned on Human3.6M
(d) Estimate parts with different nodes

Fig. 4. Cells found on Human3.6M dataset when we set N_i to 2. Our model uses two intermediate nodes encoded in the fusion cell to estimate different groups of body parts.

Implementation Details: In the search stage, to save GPU memory, we set the size of the input image and the volumetric heat map to 128×128 and $32 \times 32 \times 32$, respectively. The total training epoch is 25, and the parameter w is updated by the Adam optimizer [24] with a batch size of 40. The initial learning rate is 1×10^{-3} and reduced by a factor of 10 at the 15th and the 20th epoch. We start to optimize the network architecture α at the 8th epoch. Its learning rate and weight decay are 8×10^{-4} and 3×10^{-4}, respectively. The search process lasts two days on a single NVIDIA TITAN RTX GPU. In the evaluation stage, the size of the input image and the volumetric heat map are 256×256 and $64 \times 64 \times 64$, respectively. The total epoch is 20. We train our network with Adam with a batch size of 64. The initial learning rate is 1×10^{-3} and reduced by ten at the 12th and the 16th epoch. Training samples are augmented via rotation ($\pm 30°$), horizontal flip, color jittering, and synthetic occlusion [46]. The training process takes two days on four NVIDIA P100 GPUs. We run each experiment three times with different random seeds, and the confidence interval is about ± 0.3 mm.

4.3 Ablation Experiments

The Number of Intermediate Nodes in the Fusion Cell
As we explain in Sect. 3, the number of intermediate nodes in the fusion cell is equal to the number of groups that we divide all body parts into. In this set of experiments, by adjusting the number of intermediate nodes, we are motivated to explore how many groups all body parts are divided into is an optimal choice. In the search stage, we optimize the network architecture where the fusion cell can have $N_i \in \{1, 2, 3, 4\}$ intermediate nodes, and the model has a total of ten computational cells, as in Fig. 3. In Table 1, we can observe that the model with two intermediate nodes outperforms all the others on every action. Compared to dividing all parts into more or fewer groups, it achieves a better trade-off between performance and computational complexity. With only 13.0M parameters and 10.7G FLOPs, it encouragingly reduces MJPJE to 53.6 mm.

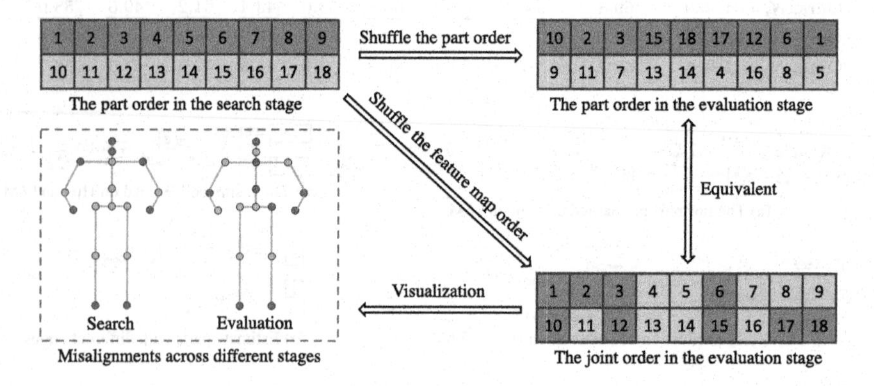

Fig. 5. Illustration of the equivalence between shuffling the part order and shuffling the heat map order. The number in the box denotes the part id. There are a total of eighteen parts. As shown in Fig. 4(d), within the last fusion cell, orange boxes indicate parts estimated by Node 0, and pink boxes indicate ones estimated by Node 1. (Color figure online)

To investigate what makes our architecture efficient when N_i is 2, we visualize searched architectures in Fig. 4. As a comparison, when N_i is 1, our model estimates all body parts with a single head network. It is computationally intensive, having 14.7M parameters and 22.9G FLOPs, but its performance is not satisfactory. Towards a better solution shown in Fig. 4(d), we employ two intermediate nodes encoded in the fusion cell to estimate the torso and limbs, respectively. Specifically, Node 0 is transformed from pooling layers and is robust to estimate parts with relatively low DOFs. On the other side, dilated convolutional layers empower Node 1 to capture long-range context information, which is helpful to estimate parts with higher DOFs, such as the wrist and ankle. The normal cell, shown in Fig. 4(a), consists of many dilated convolutional layers, which greatly increase the receptive field of our model, and are critical to performance improvement. As shown in Table 1, if we remove dilated convolution from our search space $\mathcal{O}$, our searched model has more parameters and FLOPs, and its performance drops from 53.6 mm to 59.9 mm. The reduction cell employs many depth-wise convolution layers to fuse multi-scale features efficiently. Similarly, we validate their importance by removing these operations from $\mathcal{O}$, and it leads to a 5.1 mm decline in performance.

The Part-Awareness of Our Model

We begin to validate the part-awareness of our approach from two perspectives. First, to investigate whether searched head networks are part-specific, we intend to shuffle the order of parts when we re-train our model in the evaluation stage. However, it is a little troublesome to do this since we would have to modify the data augmentation policy according to the shuffled order. Alternatively, as shown in Fig. 5, we propose to shuffle the order of heat maps produced in the last fusion cell. The implementation of the shuffle operation is the same as ShuffleNet [59], which is efficient and GPU-friendly. If our model trained with the shuffled order behaves obviously worse than the original one, we can validate that our optimized head networks are part-aware. We run experiments three times and train our model with different shuffled orders. As shown in Table 2, we observe that all models trained with shuffled orders suffer from a significant drop in performance, more than 3 mm in MPJPE. As we take a closer look, the decline in performance also reflects on every individual part, especially parts with higher DOFs (*e.g.*, ankle, knee), and their estimation accuracy might drop by more than 5 mm. By comparing models trained with shuffled orders, we validate that our approach learns part-specific head networks for specific body parts in the search stage.

In our model, the fusion cell plays a pivotal role in learning part-specific head networks. To evaluate the importance of the fusion cell, we replace them with deconvolution layers and only search the backbone network. The backbone network only consists of normal cells and reduction cells. For a fair comparison, all constructed networks have two normal cells and five reduction cells, and their

Table 2. Quantitative evaluation of the shuffled part order on Human3.6M using Protocol 2. We set N_c and N_i to 10 and 2 respectively. We compute part-wise MPJPE to report performance. Bold values indicate parts estimated by Node 0 and italic values denote ones estimated by Node 1.

Methods	Pelvis	R Hip	R Kn.	R An.	L Hip	L Kn.	L An.	Torso	Neck
Ours, original	**0.0**	**23.2**	**53.4**	**74.4**	**22.6**	**47.3**	**75.5**	**37.2**	**44.7**
Ours, shuffled 1	**0.0**	*24.3*	**56.2**	*78.3*	**23.7**	*50.3*	**78.6**	*38.9*	**48.3**
Ours, shuffled 2	**0.0**	**25.1**	*56.1*	**83.8**	**24.9**	*52.1*	**82.8**	**39.4**	*46.5*
Ours, shuffled 3	**0.0**	**24.3**	**58.2**	*81.7*	*24.2*	*53.6*	**82.7**	**40.4**	**45.4**
Methods	Nose	Head	L Sh.	L El.	L Wr.	R Sh.	R El.	R Wr.	Avg
Ours, original	*46.9*	*50.8*	*51.7*	*72.1*	*92.2*	*50.6*	*76.0*	*93.9*	53.6
Ours, shuffled 1	*53.7*	**57.8**	*57.0*	**74.7**	*95.3*	**55.4**	*80.1*	**98.2**	57.2
Ours, shuffled 2	**50.7**	*55.7*	*54.2*	**74.1**	*93.3*	*53.5*	**79.9**	*96.0*	56.9
Ours, shuffled 3	*49.3*	*53.1*	*55.0*	**75.8**	**95.3**	**54.2**	*80.8*	*97.7*	57.1

Table 3. Quantitative evaluation of the importance of the fusion cell on Human3.6M using Protocol 2. BS and WS denote the backbone search and the whole architecture search, respectively. We compute action-wise MPJPE to report the network performance. Lower is better, best in bold, second-best underlined.

Methods	Backbone	Pretrain	Params	FLOPs	Direct.	Dicuss	Eating	Greet	Phone	Pose
Ours, ResNet	ResNet50	✓	34.3M	14.1G	50.8	**52.3**	54.8	57.9	**52.8**	**47.0**
Ours, BS	Searched	–	20.5M	12.5G	49.0	59.9	49.8	53.5	58.0	51.0
Ours, WS	–	–	**13.0M**	**10.7G**	**46.3**	55.3	**47.2**	**49.0**	55.0	48.2
Methods	Purch.	Sitting	SitD.	Smoke	Photo	Wait	Walk	WalkD.	WalkT.	Avg
Ours, ResNet	**52.1**	**62.0**	73.7	**52.6**	**58.3**	50.4	40.9	**54.1**	**45.1**	53.9
Ours, BS	56.0	65.8	77.5	56.3	63.8	52.9	44.4	62.7	50.0	57.1
Ours, WS	52.6	64.6	**70.8**	54.4	60.0	**48.8**	**40.9**	58.3	46.7	**53.6**

only difference is whether they have fusion cells. In Table 3, compared to the backbone search, searching the whole network architecture improves performance by 3.5 mm and reduces 37% parameters and 14% FLOPs. In comparison with the model built on the commonly used ResNet-50 backbone, we advance estimation accuracy by 0.3 mm with 62% fewer parameters and 24% fewer FLOPs. Through our experiments, we show that fusion cells significantly contribute to the compactness and efficiency of our approach and exhibit more competitive performance over models using the ResNet-50 backbone.

The Number of Computational Cells

Instead of stacking only ten computation cells, we attempt to construct a deeper part-aware 3D pose estimator, according to Eq. 6. As shown in Table 4, as we increase the number of computational cells, our model becomes better in performance but has more parameters and FLOPs. When N_c is 20, our model achieves the best performance, 47.3 mm in MPJPE. As we increase N_c from 10 to 20, the gain in network parameters (from 13.0M to 20.4M) and FLOPs (from 10.7G to 14.1G) also leads to an improvement in performance (from 53.6 mm to 47.3 mm). This phenomenon also demonstrates that the network architecture optimized during the search process is computationally efficient.

Table 4. Quantitative evaluation of the number of cells on Human3.6M using Protocol 2. N_c denotes the number of computational cells. We compute action-wise MPJPE to report the network performance. Lower is better, best in bold, second-best underlined.

Methods	Params	FLOPs	Direct.	Discuss	Eating	Greet	Phone	Pose	Purch.
Ours, $N_c = 10$	**13.0M**	**10.7G**	46.3	55.3	47.2	49.0	55.0	48.2	52.6
Ours, $N_c = 15$	14.7M	12.7G	45.8	53.7	43.4	49.4	52.0	46.4	51.4
Ours, $N_c = 20$	20.4M	14.1G	**41.4**	**48.6**	**42.0**	**45.3**	**47.1**	**42.3**	**46.0**
Methods	Sitting	SitD.	Smoke	Photo	Wait	Walk	WalkD.	WalkT.	Avg
Ours, $N_c = 10$	64.6	70.8	54.4	60.0	48.8	40.9	58.3	46.7	53.6
Ours, $N_c = 15$	60.8	63.4	50.9	55.6	45.7	40.8	55.4	44.5	50.9
Ours, $N_c = 20$	**57.9**	**62.1**	**47.8**	**51.2**	**43.6**	**36.1**	**51.1**	**41.5**	**47.3**

Table 5. Comparison with state-of-the-art methods on Human3.6M using Protocol 1. S denotes our small part-aware model with ten cells, and L denotes our large model with twenty cells. Lower is better, best in bold, second-best underlined.

Methods	Dire.	Dis.	Eat	Gre.	Phe.	Pose	Pur.	Sit	SitD.	Smo.	Phot..	Wait	Walk	WD.	WT.	Ave.
Yasin [58]	88.4	72.5	108.5	110.2	97.1	81.6	107.2	119.0	170.8	108.2	142.5	86.9	92.1	165.7	102.0	108.3
Chen [6]	71.6	66.6	74.7	79.1	70.1	67.6	89.3	90.7	195.6	83.5	93.3	71.2	55.7	85.9	62.5	82.7
Moreno [36]	67.4	63.8	87.2	73.9	71.5	69.9	65.1	71.7	98.6	81.3	93.3	74.6	76.5	77.7	74.6	76.5
Zhou [64]	47.9	48.8	52.7	55.0	56.8	49.0	45.5	60.8	81.1	53.7	65.5	51.6	50.4	54.8	55.9	55.3
Sun [47]	42.1	44.3	45.0	45.4	51.5	43.2	41.3	59.3	73.3	51.0	53.0	44.0	38.3	48.0	44.8	48.3
Fang [12]	38.2	41.7	43.7	44.9	48.5	40.2	38.2	54.5	64.4	47.2	55.3	44.3	36.7	47.3	41.7	45.7
Sun [48]	36.9	36.2	40.6	40.4	41.9	34.9	35.7	50.1	59.4	40.4	44.9	39.0	30.8	39.8	36.7	40.6
Moon [35]	31.9	**30.6**	39.9	<u>35.5</u>	<u>34.8</u>	**30.2**	<u>32.1</u>	<u>35.0</u>	<u>43.8</u>	**35.7**	**37.6**	<u>30.1</u>	<u>24.6</u>	**35.7**	<u>29.3</u>	<u>34.0</u>
Ours, S	<u>31.8</u>	33.4	<u>38.9</u>	37.9	36.4	36.6	32.6	36.2	47.8	38.9	43.0	32.6	26.5	39.8	30.8	36.4
Ours, L	**27.5**	<u>30.9</u>	**34.0**	**35.5**	**32.4**	<u>30.8</u>	**31.9**	**32.7**	**41.9**	<u>36.3</u>	<u>39.1</u>	**28.4**	**23.3**	<u>37.1</u>	**27.0**	**32.7**

4.4 Comparison with the State-of-the-Art

To demonstrate the effectiveness and the generalization ability of our approach, we conduct our experiments on both single-person and multi-person 3D pose estimation benchmarks. Previous works have different experimental settings, and we summarize comparison results in Tables 5, 6 and 7, respectively. In Fig. 6, we show qualitative results produced by our model with ten cells. It can generalize well for in-the-wild images, even on challenging poses and crowded scenes.

Single-Person 3D Human Pose Estimation: We compare our approach on Human3.6M with state-of-the-art methods in Tables 5 and 6. By reducing about 40% parameters, our large part-aware model advances the-state-of-the-art accuracy by 1.3 mm and 2.3 mm in protocol 1 and protocol 2, respectively. If we add supervision on intermediate feature maps, the performance of our small

Table 6. Comparison with state-of-the-art methods on Human3.6M using Protocol 2. S denotes our small part-aware model with ten cells, and L denotes our large model with twenty cells. Lower is better, best in bold, second-best underlined.

Methods	Dire.	Dis.	Eat	Gre.	Phe.	Pose	Pur.	Sit	SitD.	Smo.	Phot.	Wait	Walk	WD.	WT.	Ave.
Chen [6]	89.9	97.6	90.0	107.9	107.3	93.6	136.1	133.1	240.1	106.7	139.2	106.2	87.0	114.1	90.6	114.2
Tome [51]	65.0	73.5	76.8	86.4	86.3	68.9	74.8	110.2	173.9	85.0	110.7	85.8	71.4	86.3	73.1	88.4
Moreno [36]	69.5	80.2	78.2	87.0	100.8	76.0	69.7	104.7	113.9	89.7	102.7	98.5	79.2	82.4	77.2	87.3
Zhou [64]	68.7	74.8	67.8	76.4	76.3	84.0	70.2	88.0	113.8	78.0	98.4	90.1	62.6	75.1	73.6	79.9
Mehta [33]	57.5	68.6	59.6	67.3	78.1	56.9	69.1	98.0	117.5	69.5	82.4	68.0	55.3	76.5	61.4	72.9
Fang [12]	50.1	54.3	57.0	57.1	66.6	53.4	55.7	72.8	88.6	60.3	73.3	57.7	47.5	62.7	50.6	60.4
Omran [37]	–	–	–	–	–	–	–	–	–	–	–	–	–	–	–	59.9
Sun [47]	52.8	54.8	54.2	54.3	61.8	53.1	53.6	71.7	86.7	61.5	67.2	53.4	47.1	61.6	63.4	59.1
Kanazawa [47]	–	–	–	–	–	–	–	–	–	–	–	–	–	–	–	56.8
Moon [35]	50.5	55.7	50.1	51.7	53.9	46.8	50.0	61.9	68.0	52.5	55.9	49.9	41.8	56.1	46.9	53.3
Sun [48]	47.5	**47.7**	49.5	50.2	<u>51.4</u>	<u>43.8</u>	<u>46.4</u>	<u>58.9</u>	<u>65.7</u>	<u>49.4</u>	<u>55.8</u>	<u>47.8</u>	<u>38.9</u>	**49.0**	<u>43.8</u>	<u>49.6</u>
Ours, S	<u>46.3</u>	55.3	<u>47.2</u>	<u>49.0</u>	55.0	48.2	52.6	64.6	70.8	54.4	60.0	48.8	40.9	58.3	46.7	53.6
Ours, L	**41.4**	<u>48.6</u>	**42.0**	**45.3**	**47.1**	**42.3**	**46.0**	**57.9**	**62.1**	**47.8**	**51.2**	**43.6**	**36.1**	<u>51.1</u>	**41.4**	**47.3**

Table 7. Comparison with state-of-the-art methods on MuPoTS-3D using all ground truths. S denotes our small part-aware model with ten cells, and L denotes our large model with twenty cells. Higher is better, best in bold, second-best underlined.

Methods	S1	S2	S3	S4	S5	S6	S7	S8	S9	S10
Rogez [45]	67.7	49.8	53.4	59.1	67.5	22.8	43.7	49.9	31.1	78.1
Mehta [34]	81.0	60.9	64.4	63.0	69.1	30.3	65.0	59.6	64.1	83.9
Moon [35]	<u>94.4</u>	77.5	79.0	<u>81.9</u>	<u>85.3</u>	<u>72.8</u>	<u>81.9</u>	<u>75.7</u>	<u>90.2</u>	<u>90.4</u>
Ours, S	93.1	<u>76.7</u>	<u>79.9</u>	78.2	83.6	64.6	79.0	72.5	87.6	88.3
Ours, L	**95.8**	**80.2**	**81.3**	**84.6**	**87.1**	**74.5**	**82.7**	**79.4**	**91.2**	**93.3**
Methods	S11	S12	S13	S14	S15	S16	S17	S18	S19	S20
Rogez [45]	50.2	51.0	51.6	49.3	56.2	66.5	65.2	62.9	66.1	59.1
Mehta [34]	68.0	68.6	62.3	59.2	70.1	80.0	79.6	67.3	66.6	67.2
Moon [35]	<u>79.2</u>	<u>79.9</u>	<u>75.1</u>	<u>72.7</u>	<u>81.1</u>	<u>89.9</u>	<u>89.6</u>	<u>81.8</u>	<u>81.7</u>	<u>76.2</u>
Ours, S	76.1	79.4	71.1	70.6	77.7	86.6	87.1	80.3	79.5	72.0
Ours, L	**83.4**	**82.0**	**78.6**	**76.5**	**84.3**	**92.1**	**91.1**	**85.3**	**82.4**	**77.8**

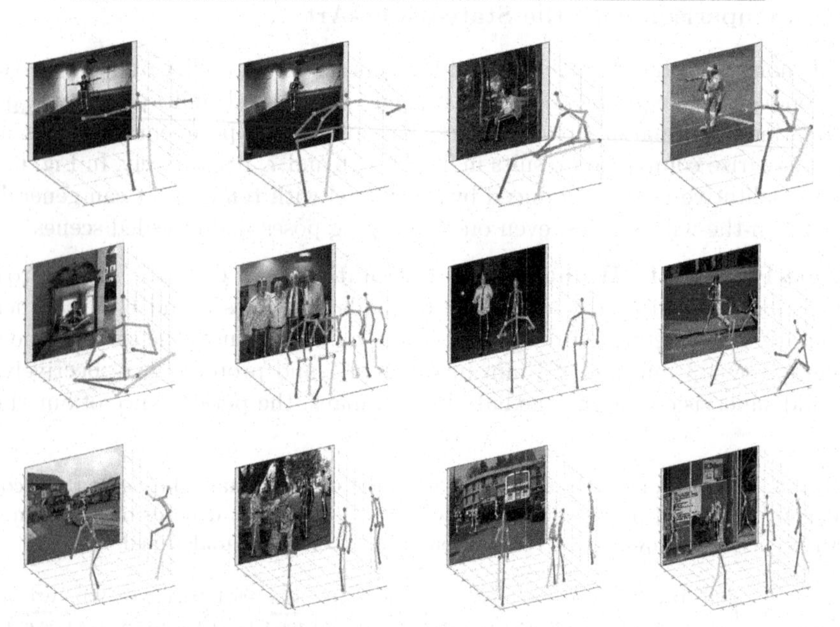

Fig. 6. Qualitative results on different datasets. Our small model produces convincing results even on challenging poses and crowded scenes.

model can be significantly improved, achieving 50.4 mm in Protocol 2. Moreover, our method is also compatible with some efficient learning frameworks [19,25,62].

Multi-person 3D Human Pose Estimation: For multi-person 3D pose estimation, we use RootNet [35] to estimate absolute depth for the root joint of each person. As shown in Table 7, we compare our model with previous state-of-the-art multi-person pose estimation methods on MuPoTS-3D, and our large part-aware 3D pose estimator achieves more superior performance on every sequence.

5 Conclusion and Future Works

In this work, we propose to estimate 3D poses of different parts with part-specific neural architectures. In the search stage, we optimize the architectures of different types of cells via gradient descent. Then, we interweave optimized computational cells to construct our part-aware 3D pose estimator, which is compact and efficient. Our model advances the state-of-the-art accuracy on both the single-person and multi-person 3D human pose estimation benchmarks. In the future, we attempt to explore other NAS methods to search 3D pose estimators in a larger space, which may open up the possibility for a global optimization.

Acknowledgements. This work is jointly supported by National Key Research and Development Program of China (2016YFB1001000), Key Research Program of Frontier Sciences, CAS (ZDBS-LY-JSC032), National Natural Science Foundation of China (61525306, 61633021, 61721004, 61806194, U1803261, 61976132), Shandong Provincial Key Research and Development Program (2019JZZY010119), HW2019SOW01, and CAS-AIR.

References

1. Alldieck, T., Pons-Moll, G., Theobalt, C., Magnor, M.: Tex2shape: detailed full human body geometry from a single image. In: ICCV (2019)
2. Andriluka, M., Pishchulin, L., Gehler, P., Schiele, B.: 2D human pose estimation: new benchmark and state of the art analysis. In: CVPR (2014)
3. Bogo, F., Kanazawa, A., Lassner, C., Gehler, P., Romero, J., Black, M.J.: Keep it SMPL: automatic estimation of 3D human pose and shape from a single image. In: Leibe, B., Matas, J., Sebe, N., Welling, M. (eds.) ECCV 2016. LNCS, vol. 9909, pp. 561–578. Springer, Cham (2016). https://doi.org/10.1007/978-3-319-46454-1_34
4. Cai, H., Zhu, L., Han, S.: ProxylessNAS: direct neural architecture search on target task and hardware. In: ICLR (2019)
5. Cai, Y., et al.: Exploiting spatial-temporal relationships for 3D pose estimation via graph convolutional networks. In: ICCV (2019)
6. Chen, C.H., Ramanan, D.: 3D human pose estimation = 2D pose estimation + matching. In: CVPR (2017)
7. Chen, L.C., et al.: Searching for efficient multi-scale architectures for dense image prediction. In: NeurIPS (2018)
8. Chen, Y., Yang, T., Zhang, X., Meng, G., Xiao, X., Sun, J.: DetNAS: backbone search for object detection. In: NeurIPS (2019)
9. Chen, Z., Guo, Y., Huang, Y., Liang, W.: Learning depth-aware heatmaps for 3D human pose estimation in the wild. In: BMVC (2019)
10. Ci, H., Wang, C., Ma, X., Wang, Y.: Optimizing network structure for 3D human pose estimation. In: ICCV (2019)
11. Fabbri, M., Lanzi, F., Calderara, S., Alletto, S., Cucchiara, R.: Compressed volumetric heatmaps for multi-person 3D pose estimation. In: CVPR (2020)
12. Fang, H., Xu, Y., Wang, W., Liu, X., Zhu, S.C.: Learning knowledge-guided pose grammar machine for 3D human pose estimation. In: AAAI (2018)
13. Ghiasi, G., Lin, T.Y., Le, Q.V.: NAS-FPN: learning scalable feature pyramid architecture for object detection. In: CVPR (2019)

14. Guo, Z., et al.: Single path one-shot neural architecture search with uniform sampling. In: NeurIPS (2019)
15. Gupta, A., Martinez, J., Little, J.J., Woodham, R.J.: 3D pose from motion for cross-view action recognition via non-linear circulant temporal encoding. In: CVPR (2014)
16. Hossain, M.R.I., Little, J.J.: Exploiting temporal information for 3D human pose estimation. In: Ferrari, V., Hebert, M., Sminchisescu, C., Weiss, Y. (eds.) ECCV 2018. LNCS, vol. 11214, pp. 69–86. Springer, Cham (2018). https://doi.org/10.1007/978-3-030-01249-6_5
17. Howard, A., et al.: Searching for MobileNetV3. In: ICCV (2019)
18. Ionescu, C., Papava, D., Olaru, V., Sminchisescu, C.: Human3.6M: large scale datasets and predictive methods for 3D human sensing in natural environments. In: TPAMI (2014)
19. Iskakov, K., Burkov, E., Lempitsky, V., Malkov, Y.: Learnable triangulation of human pose. In: ICCV (2019)
20. Jiang, H.: 3D human pose reconstruction using millions of exemplars. In: ICPR (2010)
21. Jiang, W., Kolotouros, N., Pavlakos, G., Zhou, X., Daniilidis, K.: Coherent reconstruction of multiple humans from a single image. In: CVPR (2020)
22. Kanazawa, A., Black, M.J., Jacobs, D.W., Malik, J.: End-to-end recovery of human shape and pose. In: CVPR (2018)
23. Kasim, M., et al.: Up to two billion times acceleration of scientific simulations with deep neural architecture search. arXiv preprint arXiv:2001.08055 (2020)
24. Kingma, D.P., Ba, J.: Adam: a method for stochastic optimization. In: ICLR (2014)
25. Kocabas, M., Karagoz, S., Akbas, E.: Self-supervised learning of 3D human pose using multi-view geometry. In: CVPR (2019)
26. Kolotouros, N., Pavlakos, G., Daniilidis, K.: Convolutional mesh regression for single-image human shape reconstruction. In: CVPR (2019)
27. Lee, H.J., Chen, Z.: Determination of 3D human body postures from a single view. In: CVGIP (1985)
28. Li, S., Chan, A.B.: 3D human pose estimation from monocular images with deep convolutional neural network. In: ACCV (2014)
29. Lin, T.-Y., et al.: Microsoft COCO: common objects in context. In: Fleet, D., Pajdla, T., Schiele, B., Tuytelaars, T. (eds.) ECCV 2014. LNCS, vol. 8693, pp. 740–755. Springer, Cham (2014). https://doi.org/10.1007/978-3-319-10602-1_48
30. Liu, C., et al.: Auto-DeepLab: hierarchical neural architecture search for semantic image segmentation. In: CVPR (2019)
31. Liu, H., Simonyan, K., Yang, Y.: DARTS: differentiable architecture search. In: ICLR (2019)
32. Martinez, J., Hossain, R., Romero, J., Little, J.J.: A simple yet effective baseline for 3D human pose estimation. In: ICCV (2017)
33. Mehta, D., et al.: Monocular 3D human pose estimation in the wild using improved CNN supervision. In: 3DV (2017)
34. Mehta, D., et al.: Single-shot multi-person 3D pose estimation from monocular RGB. In: 3DV (2018)
35. Moon, G., Chang, J.Y., Lee, K.M.: Camera distance-aware top-down approach for 3D multi-person pose estimation from a single RGB image. In: ICCV (2019)
36. Moreno-Noguer, F.: 3D human pose estimation from a single image via distance matrix regression. In: CVPR (2017)

37. Omran, M., Lassner, C., Pons-Moll, G., Gehler, P., Schiele, B.: Neural body fitting: unifying deep learning and model based human pose and shape estimation. In: 3DV (2018)
38. Park, S., Hwang, J., Kwak, N.: 3D human pose estimation using convolutional neural networks with 2D pose information. In: Hua, G., Jégou, H. (eds.) ECCV 2016. LNCS, vol. 9915, pp. 156–169. Springer, Cham (2016). https://doi.org/10.1007/978-3-319-49409-8_15
39. Pavlakos, G., Zhou, X., Daniilidis, K.: Ordinal depth supervision for 3D human pose estimation. In: CVPR (2018)
40. Pavlakos, G., Zhou, X., Derpanis, K.G., Daniilidis, K.: Coarse-to-fine volumetric prediction for single-image 3D human pose. In: CVPR (2017)
41. Pavlakos, G., Zhou, X., Derpanis, K.G., Daniilidis, K.: Harvesting multiple views for marker-less 3D human pose annotations. In: CVPR (2017)
42. Peng, J., Sun, M., ZHANG, Z.X., Tan, T., Yan, J.: Efficient neural architecture transformation search in channel-level for object detection. In: NeurIPS (2019)
43. Qiu, H., Wang, C., Wang, J., Wang, N., Zeng, W.: Cross view fusion for 3D human pose estimation. In: ICCV (2019)
44. Rogez, G., Schmid, C.: Mocap-guided data augmentation for 3D pose estimation in the wild. In: NeurIPS (2016)
45. Rogez, G., Weinzaepfel, P., Schmid, C.: LCR-Net: localization-classification-regression for human pose. In: CVPR (2017)
46. Sárándi, I., Linder, T., Arras, K.O., Leibe, B.: Synthetic occlusion augmentation with volumetric heatmaps for the 2018 ECCV PoseTrack challenge on 3D human pose estimation. In: ECCVW (2018)
47. Sun, X., Shang, J., Liang, S., Wei, Y.: Compositional human pose regression. In: ICCV (2017)
48. Sun, X., Xiao, B., Wei, F., Liang, S., Wei, Y.: Integral human pose regression. In: Ferrari, V., Hebert, M., Sminchisescu, C., Weiss, Y. (eds.) ECCV 2018. LNCS, vol. 11210, pp. 536–553. Springer, Cham (2018). https://doi.org/10.1007/978-3-030-01231-1_33
49. Tan, M., et al.: MnasNet: platform-aware neural architecture search for mobile. In: CVPR (2019)
50. Taylor, C.J.: Reconstruction of articulated objects from point correspondences in a single uncalibrated image. In: CVIU (2000)
51. Tome, D., Russell, C., Agapito, L.: Lifting from the deep: convolutional 3D pose estimation from a single image. In: CVPR (2017)
52. Tu, H., Wang, C., Zeng, W.: VoxelPose: towards multi-camera 3D human pose estimation in wild environment. In: Vedaldi A., Bischof H., Brox T., Frahm JM. (eds) ECCV, pp. 197–212 . Springer, Cham (2020). https://doi.org/10.1007/978-3-030-58452-8_12
53. Varol, G., et al.: Learning from synthetic humans. In: CVPR (2017)
54. Wang, J., Huang, S., Wang, X., Tao, D.: Not all parts are created equal: 3D pose estimation by modeling bi-directional dependencies of body parts. In: ICCV (2019)
55. Wei, S.E., Ramakrishna, V., Kanade, T., Sheikh, Y.: Convolutional pose machines. In: CVPR (2016)
56. Xu, Y., et al.: PC-DARTS: partial channel connections for memory-efficient architecture search. In: ICLR (2020)
57. Yang, W., Ouyang, W., Wang, X., Ren, J., Li, H., Wang, X.: 3D human pose estimation in the wild by adversarial learning. In: CVPR (2018)
58. Yasin, H., Iqbal, U., Kruger, B., Weber, A., Gall, J.: A dual-source approach for 3D pose estimation from a single image. In: CVPR (2016)

59. Zhang, X., Zhou, X., Lin, M., Sun, J.: ShuffleNet: an extremely efficient convolutional neural network for mobile devices. In: CVPR (2018)
60. Zhang, Y., Qiu, Z., Liu, J., Yao, T., Liu, D., Mei, T.: Customizable architecture search for semantic segmentation. In: CVPR (2019)
61. Zhang, Z., Wang, C., Qin, W., Zeng, W.: Fusing wearable IMUs with multi-view images for human pose estimation: a geometric approach. In: CVPR (2020)
62. Zhou, K., Han, X., Jiang, N., Jia, K., Lu, J.: HEMlets pose: learning part-centric heatmap triplets for accurate 3D human pose estimation. In: ICCV (2019)
63. Zhou, X., Zhu, M., Leonardos, S., Derpanis, K.G., Daniilidis, K.: Sparseness meets deepness: 3D human pose estimation from monocular video. In: CVPR (2016)
64. Zhou, X., Zhu, M., Pavlakos, G., Leonardos, S., Derpanis, K.G., Daniilidis, K.: MonoCap: monocular human motion capture using a CNN coupled with a geometric prior. In: TPAMI (2018)
65. Zhou, X., Huang, Q., Sun, X., Xue, X., Wei, Y.: Towards 3D human pose estimation in the wild: a weakly-supervised approach. In: ICCV (2017)

REVISE: A Tool for Measuring and Mitigating Bias in Visual Datasets

Angelina Wang(✉), Arvind Narayanan, and Olga Russakovsky

Princeton University, Princeton, USA
angelina.wang@princeton.edu

Abstract. Machine learning models are known to perpetuate and even amplify the biases present in the data. However, these data biases frequently do not become apparent until after the models are deployed. To tackle this issue and to enable the preemptive analysis of large-scale dataset, we present our tool. REVISE (REvealing VIsual biaSEs) is a tool that assists in the investigation of a visual dataset, surfacing potential biases currently along three dimensions: (1) object-based, (2) gender-based, and (3) geography-based. Object-based biases relate to size, context, or diversity of object representation. Gender-based metrics aim to reveal the stereotypical portrayal of people of different genders. Geography-based analyses consider the representation of different geographic locations. REVISE sheds light on the dataset al.ong these dimensions; the responsibility then lies with the user to consider the cultural and historical context, and to determine which of the revealed biases may be problematic. The tool then further assists the user by suggesting actionable steps that may be taken to mitigate the revealed biases. Overall, the key aim of our work is to tackle the machine learning bias problem early in the pipeline. REVISE is available at https://github.com/princetonvisualai/revise-tool.

Keywords: Dataset bias · Dataset analysis · Computer vision fairness

1 Introduction

Computer vision dataset bias is a well-known and much-studied problem. In 2011, Torralba and Efros [60] highlighted the fact that every dataset is a unique slice through the visual world, representing a particular distribution of visual data. Since then, researchers have noted the under-representation of object classes [9,44,47,49,54,65], object contexts [13,16,52], object sub-categories [71], scenes [70], gender [10,35], gender contexts [11,69], skin tones [10,63], geographic locations [16,55] and cultures [16]. The downstream effects of these under-representations range from the more innocuous, like limited generalization of car classifiers [60], to the much more serious, like having deep societal

Electronic supplementary material The online version of this chapter (https://doi.org/10.1007/978-3-030-58580-8_43) contains supplementary material, which is available to authorized users.

A. Vedaldi et al. (Eds.): ECCV 2020, LNCS 12348, pp. 733–751, 2020.
https://doi.org/10.1007/978-3-030-58580-8_43

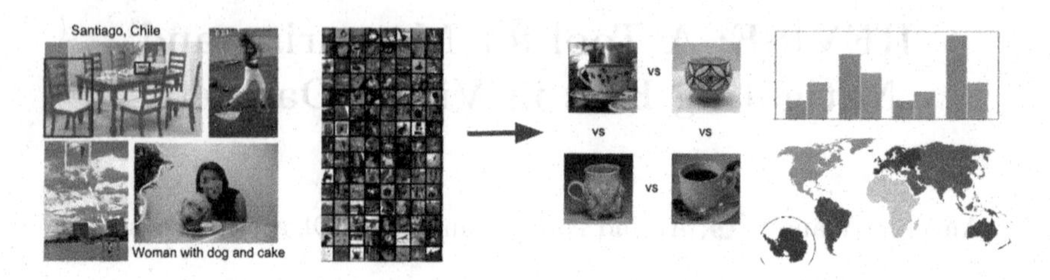

Fig. 1. Our tool takes in as input a visual dataset and its annotations, and outputs metrics, seeking to produce insights and possible actions.

implications in automated facial analysis [10,26]. Efforts such as Datasheets for Datasets [23] have played an important role in encouraging dataset transparency through articulating the intent of the dataset creators, summarizing the data collection processes, and warning downstream dataset users of potential biases in the data (Fig. 1).

However, this is only the beginning. It is often impossible to foresee the biases hiding in the data, and manual review is certainly not a feasible strategy given the scale of modern datasets.

Bias Detection Tool. To mitigate this issue, we provide an automated tool for REvealing VIsual biaSEs (REVISE) in datasets. REVISE is a broad-purpose tool for surfacing the under- and different- representations hiding within visual datasets. For the current exploration we limit ourselves to three sets of metrics: (1) object-based, (2) gender-based and (3) geography-based. Object-based analysis is most familiar to the computer vision community [60], considering statistics about object frequency, scale, context, or diversity of representation. Gender-based analysis considers the representation of people of different genders within the dataset [23,69]; such issues are gaining attention within the computer vision community. Future iterations of REVISE will include analysis of additional axes of identity. Finally, geography-based analysis considers the portrayal of different geographic regions within the dataset; this is a new but very important conversation within the community [55].

We imagine two primary use cases: (1) dataset builders can use the actionable insights produced by our tool during the process of dataset compilation to guide the direction of further data collection, and (2) dataset users who train models can use the tool to understand what kinds of biases their models may inherit as a result of training on a particular dataset.

Example Findings. REVISE automatically surfaces a variety of metrics that highlight unrepresentative or anomalous patterns in the dataset. To validate the usefulness of the tool, we have used it to analyze several datasets commonly used in computer vision: COCO [43], OpenImages [39], YFCC100m [58]. Some examples of the kinds of automatic insights our tool has found include:

- In the object detection dataset COCO [43], some objects, e.g., `airplane`, `bed` and `pizza`, are frequently large in the image. This is because fewer images of

airplanes appear in the sky (far away; small) than on the ground (close-up; large). One way for the dataset creator to mitigate the problem is to query for images of airplane appearing in scenes of mountains, desert, sky.
- The OpenImages dataset [39] depicts a large number of people who are too small in the image for human annotators to determine their gender; nevertheless, we found that annotators infer that they are male 69% of the time, and especially in scenes of outdoor sports fields, parks. Computer vision researchers might want to exercise caution with these gender annotations so they don't propagate into the model.
- In the computer vision and multimedia dataset YFCC100m (Yahoo Flickr Creative Commons 100 million) [58] images come from 196 different countries. However, we estimate that for around 47% of those countries—especially in developing regions of the world—the images are predominantly photos taken by visitors to the country rather than by locals, potentially resulting in a stereotypical portrayal.

A benefit of our tool is that a user doesn't need to have specific biases in mind, as these can be hard to enumerate. Rather, the tool automatically surfaces unusual patterns. REVISE cannot automatically say which of these patterns, or lack of patterns, are problematic, and leaves that analysis up to the user's judgment and expertise. It is important to note that "bias" is a contested term, and while our tool seeks to surface a variety of findings that are interesting to dataset creators and users, not all may be considered forms of bias by everyone.

2 Related Work

Data Collection. Visual datasets are constructed in various ways, with the most common being through keyword queries to search engines, whether singular (e.g., ImageNet [53]) or pairwise (e.g., COCO [43]), or by scraping websites like Flickr (e.g., YFCC100m [58], OpenImages [39]). There is extensive preprocessing and cleaning done on the datasets. Human annotators, sometimes in conjunction with automated tools [70], then assign various labels and annotations. Dataset collectors put in significant effort to deal with problems like long-tails to ensure a more balanced distribution, and intra-class diversity by doing things like explicitly seeking out non-iconic images beyond just the object itself in focus.

Dataset Bias. Rather than pick a single definition, we adopt an inclusive notion of bias and seek to highlight ways in which the dataset builder can monitor and control the distribution of their data. Proposed ways to deal with dataset bias include cross-dataset analysis [60] and having the machine learning community learn from data collection approaches of other disciplines [8,32]. Recent work [51] has looked at dataset issues related to consent and justice, and motivate enforcing Institutional Review Boards (IRB) approval for large scale datasets. Constructive solutions will need to combine automated analysis with human judgement as automated methods cannot yet understand things like the historical context of a statistical imbalance in the dataset. Our work takes this approach by automatically supplying a host of new metrics for analyzing a dataset along with actions that can be taken to mitigate these findings. However, the responsibility lies with the user to select next steps. The tool is open-source, lowering the resource and effort barrier to creating ethical datasets [32].

Computer Vision Tools. Hoiem et al.[27] built a tool to diagnose the weaknesses of object detector models in order to help improve them. More recently, tools in the video domain [4] are looking into forms of dataset bias in activity recognition [56]. We similarly in spirit hope to build a tool that will, as one goal, help dataset curators be aware of the patterns and biases present in their datasets so they can iteratively make adjustments.

Algorithmic Fairness. In addition to looking at how models trained on one dataset generalize poorly to others [59,60], many more forms of dataset bias are being increasingly noticed in the fairness domain [12,45,66]. There has been significant work looking at how to deal with this from the algorithm side [17,18, 36,62] with varying definitions of fairness [21,24,37,50,68] that are often deemed to be mathematically incompatible with each other [14,38], but in this work, we look at the problem earlier in the pipeline from the dataset side.

Automated Bias Detectors. IBM's AI Fairness 360 [6] is an open-source toolkit that discovers biases in datasets and machine learning models. However, its look into dataset biases is limited in that it first trains a model on that dataset, then interrogates this trained model with specific questions. On the other hand, REVISE looks directly at the dataset and its annotations to discover model-agnostic patterns. The Dataset Nutrition Label [28] is a recent project that assesses machine learning datasets. Differently, our approach works on visual rather than tabular data which allows us to use additional computer vision techniques, and goes deeper in terms of presenting a variety of graphs and statistical results. Swinger et al. [57] looks at automatic detection of biases in word embeddings, but we look at patterns in visual images and their annotations.

3 Tool Overview

REVISE is intended to be general enough to yield insights at varying levels of granularity, depending on the annotations available. We do use external tools and pretrained models [1,30,33,34,70] to derive some of our metrics, and acknowledge these models themselves may contain biases.

REVISE takes the form of a Jupyter notebook interface that allows exploration and customization of metrics. The analyses that can be performed depend on the annotations available:

Object-based insights require instance labels and, if available, their corresponding bounding boxes and object category. Datasets are frequently collected together with manual annotations, but we are also beginning to use automated computer vision techniques to infer some semantic labels, like scenes.

Gender-based insights require gender labels of the people in the images. The tool is general enough that given labels of any groupings of people, such as racial groups, the corresponding analyses can be performed. In this paper we've limited our analyses to a grouping based on perceived binary gender because these labels already exist in the datasets we look at, even though it is not at all inclusive of all gender categories. We use the terms male and female to refer to binarized socially-perceived gender expression, and not gender identity nor sex assigned at birth, neither of which can be inferred from an image.

Table 1. Object-based summary: for image content and object annotations of COCO

Metric	Example insight	Example action
Object counts	Within the supercategory `appliance`, `oven` and `refrigerator` are overrepresented and `toaster` is underrepresented	Query for more `toaster` images
Duplicate annotations	The same object is frequently labeled as both `doughnut` and `bagel`	Manually reconcile the duplicate annotations
Object scale	`Airplane` is overrepresented as very large in images, as there are few images of airplanes smaller and flying in the sky	Query more images of `airplane` with `kite`, since they're more likely to have a small `airplane`
Object co-occurrences	`Person` appears more with unhealthy `food` like `cake` or `hot dog` than `broccoli` or `orange`	Query for more images of people with a healthier `food`
Scene diversity	`Baseball glove` doesn't occur much outside of sports fields	Query images of `baseball glove` in different scenes like a sidewalk
Appearance diversity	The appearance of `furniture` objects become more varied when they come from scenes like `water, ice, snow` and `outdoor sports fields, parks` rather than predominantly from `home or hotel`	Query more images of `furniture` in `outdoor sports fields, parks`, since this scene is more common than `water, ice, snow`, and still contributes appearance diversity

Geography-based insights require country- or subregion-level annotations on where each image is taken. Information about the user who took each image would also be helpful—for example to determine if they were a local or a tourist. We do not have this user information in the datasets we analyze and instead infer it from the language and content of the tag captions.

In the rest of the paper we will describe some insights automatically generated by our tool on various datasets, and potential actions that can be taken.

4 Object-Based Analysis

We begin with an object-based approach to gain a basic understanding of a dataset. Much visual recognition research has centered on recognizing objects as the core building block [19], and a number of object recognition datasets have been collected e.g., Caltech101 [20], PASCAL VOC [19], ImageNet [15,53]. In Sect. 4.1 we introduce 6 such metrics reported by REVISE; in Sect. 4.2 we dive into the actionable insights we surface as a result, all summarized in Table 1.

4.1 Object-Based Metrics

Object Counts. Object counts in the real world tend to naturally follow a long-tail distribution [49,54,65]. But for datasets, there are two main views: reflecting the natural long-tail distribution (e.g., in SUN [64]) or approximately equal balancing (e.g., in ImageNet [53]). Either way, the first-order statistic when analyzing a dataset is to compute the per-category counts and verify that they are consistent with the target distribution. Objects can also be grouped into hierarchical *supercategories*: e.g., an `appliance` supercategory encompasses the more granular instances of `oven`, `refrigerator`, and `microwave` in COCO [43]. By computing how frequently an object is represented both within its supercategory, as well as all objects, this allows for a fined-grained look at frequency statistics: for example, while the `oven` and `refrigerator` objects fall below the median number of instances for an object class in COCO, it is nevertheless notable that both of these objects are around double as represented as the average object from the `appliance` class (Fig. 2).

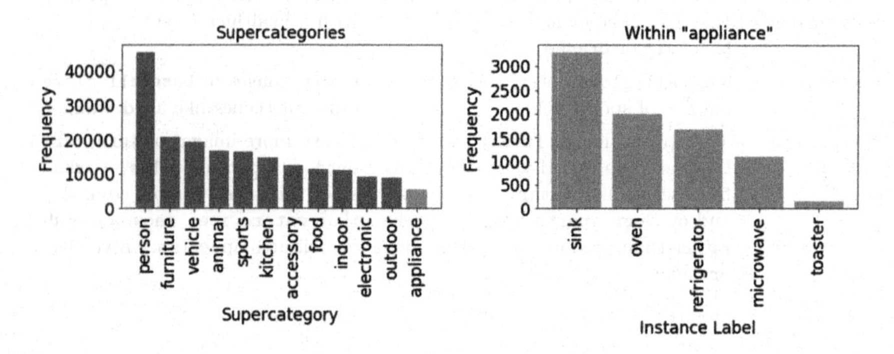

Fig. 2. `Oven` and `refrigerator` counts fall below the median of object classes in COCO; however, they are actually over-represented within the appliance category.

Duplicate Annotations. A common issue with object dataset annotation is the labeling of the same object instance with two names (e.g., `cup` and `mug`), which is especially problematic in free-form annotation datasets such as Visual Genome [40]. In datasets with closed-world vocabulary, image annotation is commonly done for a single object class at a time causing confusion when the same object is labeled as both `trumpet` and `trombone` [53]. While these occurrences are manually filtered in some datasets, automatic identification of such pairs is useful for both dataset curators (to remove errors) and to dataset users (to avoid over-counting). REVISE automatically identifies such object instances, and in the

OpenImages dataset [39] some examples of automatically detected pairs include bagel and doughnut, jaguar and leopard, and orange and grapefruit.

Object Scale. It is well-known that object size plays a key role in object recognition accuracy [27,53], as well as semantic importance in an image [7]. While many quantizations of object scale have been proposed [27,43], we aim for a metric that is both comparable across object classes and invariant to image resolution to be suitable for different datasets. Thus, for every object instance we compute the fraction of image area occupied by this instance, and quantize into 5 equal-sized bins across the entire dataset. This binning reveals, for example, that rather than an equal 20% for each size, 77% of airplanes and 73% of pizzas in COCO are extra large (>9.3% of the image area).

Object Co-occurrence. Object co-occurrence is a known contextual visual cue exploited by object detection models [22,48], and thus can serve as an important measure of the diversity of object context. We compute all pairwise object class co-occurrence statistics within the dataset, and use them both to identify surprising co-occurrences as well as to generate potential search queries to diversify the dataset, as described in Sect. 4.2. For example, we find that in COCO, person appears in 43% of images containing the food category; however, person appears in a smaller percentage of images containing broccoli (15%), carrot (21%), and orange (29%), and conversely a greater percentage of images containing cake (55%), donut (55%), and hot dog (56%).

Scene Diversity. Building on quantifying the common context of an object, we additionally strive to measure the scene diversity directly. To do so, for every object class we consider the entropy of scene categories in which the object appears. We use a ResNet-18 [25] trained on Places [70] to classify every image into one of 16 scene groups,[1] and identify objects like person that appear in a higher diversity of scenes versus objects like baseball glove that appear in fewer kinds of scenes (almost all baseball fields). This insight may guide dataset creators to further augment the dataset, as well as guide dataset users to want to test if their models can support out-of-context recognition on the objects that appear in fewer kinds of scenes, for example baseball gloves in a street.

Appearance Diversity. Finally, we consider the appearance diversity (i.e., intra-class variation) of each object class, which is a primary challenge in object detection [67]. We use a ResNet-110 network [30] trained on CIFAR-10 [41] to extract a 64-dimensional feature representation of every instance bounding box, resized to 32 $\times$ 32 pixels. We first validate that distances in this feature space correspond to semantically meaningful measures of diversity. To do so, on the COCO dataset we compute the average distance with $n = 500,000$ between two object instances of the same class (5.91 ± 1.44), and verify that it is smaller than

[1] Because top-1 accuracy for even the best model on all 365 scenes is 55.19%, but top-5 accuracy is 85.07%, we use the less granular scene categorization at the second tier of the defined scene hierarchy here. For example, aquarium, church indoor, and music studio fall into the scene group of indoor cultural.

the average distance between two object instances belonging to different classes but the same supercategory (6.24 ± 1.42) and further smaller than the average distance between two unrelated objects (6.48 ± 1.44). This metric allows us to identify individual object instances that contribute the most to the diversity of an object class, and informs our interventions in the next section.

4.2 Object-Based Actionable Insights

The metrics of Sect. 4.1 help surface biases or other issues, but it may not always be clear how to address them. We strive to mitigate this concern by providing examples of meaningful, actionable, and useful steps to guide the user.

For duplicate annotations, the remedy is straight-forward: perform manual cleanup of the data, e.g., as in Appendix E of [53]. For the others the path is less straight-forward. For datasets that come from web queries, following the literature [19,43,53] REVISE defines search queries of the form "`XX` and `YY`," where `XX` corresponds to the target object class, and `YY` corresponds to a contextual term (another object class, scene category, etc.). REVISE ranks all possible queries to identify the ones that are most likely to lead to the target outcome, and we investigate this approach more thoroughly in Appendix C.

For example, within COCO, `airplanes` have low diversity of scale and are predominantly large in the images. Our tool identifies that smaller airplanes co-occurred with objects like `surfboard` and scenes like `mountains, desert, sky` (which are more likely to be photographed from afar). In other words, size matters by itself, but a skewed size distribution could also be a proxy for other types of biases. Dataset creators aiming to diversify their dataset towards a more uniform distribution of object scale can use these queries as a guide. These pairwise queries can similarly be used to diversify appearance diversity. `Furniture` objects appear predominantly in indoor scenes like `home or hotel`, so querying for furniture in scenes like `water, ice, snow` would diversify the dataset. However, this combination is quite rare, so we want to navigate the tradeoff between a pair's commonness and its contribution to diversity. Thus, we are more likely

Fig. 3. The left shows the tradeoff for `furniture` in COCO between how much scenes increase appearance diversity (our goal) and how common they are (ease of collecting this data). To maximize both, `outdoor sports fields, parks` would be the most efficient way of augmenting this category. `Water, ice, snow` provides the most diversity but is hard to find, and `home or hotel` is the easiest to find but provides little diversity. On the right are sample images of `furniture` from these scenes.

to be successful if we query for images in the more common `outdoor sports fields, parks` scenes, which also brings a significant amount of appearance diversity. The tool provides a visualization of this tradeoff (Fig. 3), allowing the user to make the final decision.

5 Gender-Based Analysis

We next look into potential discrepancies in various aspects of how each gender is represented, summarized in Table 2. The two datasets we have gender labels for are COCO and OpenImages. The gender labels in COCO are from [69], and their methodology in determining the gender for an image is that if at least one caption contains the word "man" and there is no mention of "woman", then it is a male image, and vice versa for female images. We use the same methodology along with other gendered labels like "boy" and "girl" on OpenImages using pre-existing annotations of individuals. In Sect. 5.1 we explain some of the metrics that we collect, and in Sect. 5.2 we discuss possible actions.

Table 2. Gender-based summary: investigating representation of different genders

Metric	Example insight	Example action
Contextual representation	Males occur in more outdoors scenes and with `sports` objects. Females occur in more indoors scenes and with `kitchen` objects	Collect more images of females in outdoors scenes with `sports` objects, and vice versa for males
Interactions	In images with musical instrument `organ`, males are more likely to be actually playing the `organ`	Collect more images of females playing `organs`
Appearance differences	Males in `sports uniforms` tend to be playing outdoor sports, while females in `sports uniforms` are often indoors or in swimsuits	Collect more images of each gender with `sports uniform` in their underrepresented scenes
Gender label inference	When gender is unlikely to be identifiable, people in images are by default labeled as male	Prune these gender labels from the dataset so as not to reinforce societal stereotypes

5.1 Gender-Based Metrics

Contextual Representation. We look into the contexts different genders tend to be featured in through object groups and scenes, with results in Appendix A.

Instance Counts and Distances. Analyzing the object instances themselves allows a more granualar understanding of gender biases in the dataset. In OpenImages we find that objects like `cosmetics`, `doll`, and `washing machine` are overrepresented with females, and objects like `rugby ball`, `beer`, `bicycle` are overrepresented with males. However, beyond just looking at the number of times objects appear, we also look at the distance an object is from a person. We use a scaled distance measure as a proxy for understanding if a particular person, p, and object, o, are actually interacting with each other in order to derive more meaningful insight than just quantifying a mutual appearance in the same image. The distance measure we define is $dist = \frac{\text{distance between p and o centers}}{\sqrt{\text{area}_p * \text{area}_o}}$ to estimate distance in the 3D world. In Appendix B we validate this notion that our distance measure can be used as a proxy interaction. We consider these distances in order to disambiguate between situations where a person is merely in an image with an object in the background, rather than directly interacting with the object, revealing biases that were not clear from just looking at the frequency differences. For example, `organ` (the musical instrument) did not have a statistically significant difference in frequency between the genders, but does in distance, or under our interpretation, relation. In Fig. 4 we investigate what accounts for this difference and see that when a male person is pictured with an organ, he is likely to be playing it, whereas a female person may just be near it but not necessarily directly interacting with it. Through this analysis we discover something more subtle about how an object is represented.

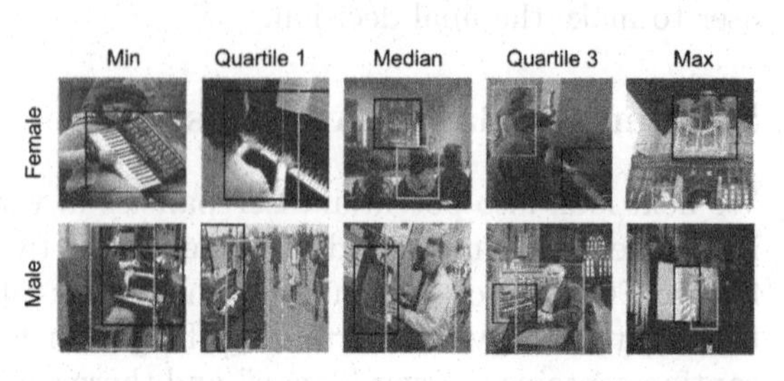

Fig. 4. 5 images from OpenImages for a person (red bounding box) of each gender pictured with an organ (blue bounding box) along the gradient of inferred 3D distances. Males tend to be featured as actually playing the instrument, whereas females are oftentimes merely in the same space as the instrument. (Color figure online)

Fig. 5. Qualitative interpretation of what the visual model has learned for the `sports uniform` and `flower` objects between the two genders in OpenImages.

Appearance Differences. We also look into the appearance differences in images of each gender with a particular object. This is to further disambiguate situations where numbers, or even distances, aren't telling the whole story. This analysis is done by (1) extracting FC7 features from AlexNet [42] pretrained on Places [70] on a randomly sampled subset of the images to get scene-level features, (2) projecting them into $\sqrt{\text{number of samples}}$ dimensions (as is recommended in [29,31]) to prevent over-fitting, and then (3) fitting a Linear Support Vector Machine to see if it is able to learn a difference between images of the same object with different genders. To make sure the female and male images are actually linearly separable and the classifier isn't over-fitting, we look at both the accuracy as well as the ratio in accuracy between the SVM trained on the correctly labeled data and randomly shuffled data. In Fig. 5 we can see what the Linear SVM has learned on OpenImages for the `sports uniform` and `flower` categories. For `sports uniform`, males tend to be represented as playing outdoor sports like baseball, while females tend to be portrayed as playing an indoor sport like basketball or in a swimsuit. For `flower`, we see another drastic difference in how males and females are portrayed, where males pictured with a `flower` are in formal, official settings, whereas females are in staged settings or paintings.

Gender Label Inference. Finally, we examine the concerning practice of assigning gender to a person in the case where the person is far too small to be identifiable, or no face is even detected in the image. This is not to say that if these cases are not met it is acceptable to assign gender, but merely that assigning gender when one of these two cases is applicable is a particularly egregious practice. For example, it's been shown that in images where a person is fully clad with snowboarding equipment and a helmet, they are still labeled as male [11] due to preconceived stereotypes. We investigate the contextual cues annotators rely on to assign gender, and consider the gender of a person unlikely to be identifiable if the person is too small (below 1000 pixels, which is the number of dimensions that humans require to perform certain recognition tasks in color images [61]) or if automated face detection (using Amazon Rekognition [1]) fails. For COCO, we find that among images with a human whose gender is unlikely to be identifiable, 77% are labeled male. In OpenImages,[2] this fraction is 69%. Thus, annotators seem to default to labeling a person as male when they cannot identify the gender; the use of male-as-norm is a problematic practice [46]. Further, we find that annotators are most likely to default to male as a gender label in `outdoor sports fields, parks` scenes, which is 2.9x the rate of female. Similarly, the rate for `indoor transportation` scenes is 4.2x and `outdoor transportation` is 4.5x, with the closest ratio being in `shopping and dining`, where male is 1.2x as likely as female. This suggests that in the absence of gender cues from the person themselves, annotators make inferences based on image context.

[2] Random subset of size 100,000.

5.2 Gender-Based Actionable Insights

Compared to object-based metrics, the actionable insights for gender-based metrics are less concrete and more nuanced. There is a tradeoff between attempting to represent the visual world as it is versus as we think it should be. In contemporary societies, gender representation in various occupations, activities, etc. is unequal, so it is not obvious that aiming for gender parity across all object categories is the right approach. Gender biases that are systemic and historical are more problematic than others [5], and this analysis cannot be automated. Further, the downstream impact of unequal representation depends on the specific models and tasks. Nevertheless, we provide some recommendations.

A trend that appeared in the metrics is that images frequently fell in line with common gender stereotypes. Each gender was under- or over-represented in a particular way, and dataset collectors may want to adjust their datasets to account for these by augmenting in the direction of the underrepresentations. Dataset users may want to audit their models, and look into to what extent their models have learned the dataset's biases before they are deployed.

For the metric of gender label inference, this brings up a larger question of which situations, if any, gender labels should ever be assigned. However, that is outside the scope of this work, where we simply recommend that dataset creators should give clearer guidance to annotators, and remove the gender labels on images where gender can definitely not be determined (Table 3).

Table 3. Geography-based summary: looking into the geo-representation of a dataset, and how that differs between countries and subregions

Metric	Example insight	Example action
Country distribution	Most images are from the USA, with very few from the countries of Africa	Collect more images from the countries of Africa
Local language analysis	Countries in Africa and Asia that are already underrepresented are frequently represented by non-locals rather than locals	Collect more images taken by locals rather than visitors in underrepresented countries
Tag counts, appearances	`Wildlife` is overrepresented in Kiribati, and `mosque` in Iran	Collect other kinds of images representing these countries

6 Geography-Based Analysis

Finally, we look into the geography of the images, and the cultural biases that arise. We use the YFCC100m dataset [58] because of the geo-location data it contains. However, we use a different subset of the dataset for metrics that require more annotations, and explain each below.

6.1 Geography-Based Metrics

Country Distribution[3]**.** The first thing we look at is the geographical distribution of where images come from. Researchers have looked at OpenImages and ImageNet and found these datasets to be amerocentric and eurocentric [55], with models dropping in performance when being run on images from underrepresented locales. In the left side of Fig. 6 it immediately stands out that the USA is drastically overrepresented compared to other countries, with the continent of Africa being very sparsely represented.

Local Language Analysis[4]**.** However, the locale of an image can be misleading, since if all the images taken in a particular country are only by tourists, this would not necessarily encompass the geo-representation one would hope for. The right side of Fig. 6 shows the percentage of images taken in a country and captioned in something other than the national language(s), as detected by the fastText library [33,34]. We use the lower bound of the binomial proportion confidence interval in the figure so that countries with only a few images total which happen to be mostly taken by tourists are not shown to be disproportionately imaged as so. Even with this lower bound, we see that many countries that are represented poorly in number are also under-represented by locals. To determine the implications in representation based on who is portraying a country, we categorize an image as taken by a local, tourist, or unknown, using a combination of language detected and tag content as an imperfect proxy. We then investigate if there are appearance differences in how locals and tourists portray a country by automatically running visual models. Although our tool does not find any such notable difference, this kind of analysis can be useful on other datasets where a local's perspective is dramatically different than that of a tourist's.

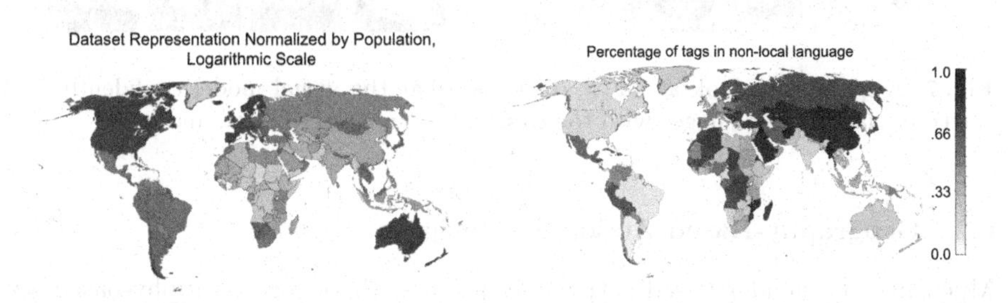

Fig. 6. Geographic distribution normalized by population (left) and percentage of tags in a non-local language (right) in YFCC100m. Even when underrepresented countries are imaged, it is not necessarily by someone local to that area.

[3] Data subset: images with geo-location metadata.
[4] Data subset: images with geo-location metadata and Flickr tags.

Tag Counts, Appearances[5]**.** To gain insight into the actual content of what is being portrayed in images from country to country, we look at the tags assigned to each image. This allows us to discern if certain tags are over/under-represented between countries. We consider the frequency with which each tag appears in the set of a country's tags, compared to the frequency that same tag makes up in the rest of the countries. Some examples of over- and under- representations include Kiribati with `wildlife` at 86x, North Korea with `men` at 76x, Iran with `mosque` at 30x, Egypt with `politics` at 20x, and United States with `safari` at .92x. But because, as we've seen in previous sections, numbers don't always tell the full story, we also look into the appearances of how different subregions, as defined by the United Nations geoscheme [3], represent certain tags. DeVries et al. [16] showed that object-recognition systems perform worse on images from countries that are not as well-represented in the dataset due to appearance differences within an object class, so we look into such appearance differences within a Flickr tag. We perform the same analysis as in Sect. 5.1 where we run a Linear SVM on the featurized images, this time performing 17-way classification between the different subregions. In Fig. 7 we show an example of the `dish` tag, and what images from the most accurately classified subregion, Eastern Asia, look like compared to images from the other subregions. Images with the `dish` tag tend to refer to food items in Eastern Asia, rather than a satellite dish or plate, which is a more common practice in other regions. While this is a more innocent discrepancy, one could imagine how it may be important to know if other tags are represented differently across subregions so that models do not overfit to one particular subregion's representation of an object.

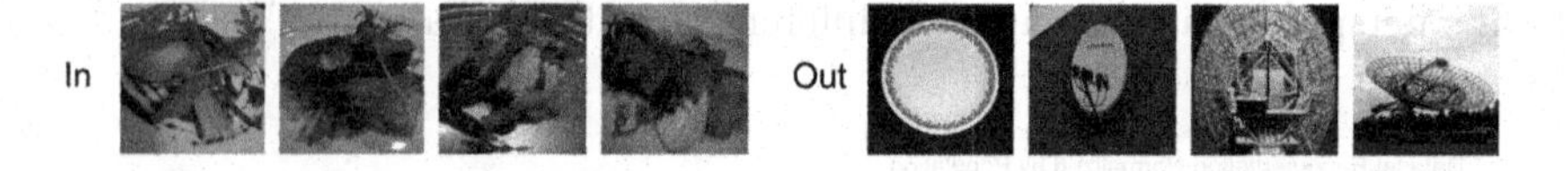

Fig. 7. A qualitative look at YFCC100m for what the visual model confidently and correctly classifies for images with the `dish` tag as in Eastern Asia, and out.

6.2 Geography-Based Actionable Insights

Much like the gender-based actionable insights, those for geography-based are also more general and dependent on what the model trained on the data will be used for. Under- and over- representations can be approached in ways similar to before by augmenting the dataset, an important step in making sure we do not have a one-sided perspective of a country. Dataset users should validate that their models are not overfitting to a particular country's representation by

[5] Data subset: images with geo-location metadata and cleaned English tags from a list of 1540 from the Tag Prediction competition [2]. Because we are using this prexisting dataset of tags in order to meaningfully relate different images, we are excluding a large variety of images that have captions in a non-English language.

testing on more geographically diverse data. It is clear that as we deploy more and more models into the world, there should be some form of either equal or equitable geo-representation. This emphasizes the need for data collection to explicitly seek out more diversity in locale, and specifically from the people that live there. Technology has been known to leave groups behind as it makes rapid advancements, and it is crucial that dataset representation does not follow this trend and base representation on digital availability. It requires more effort to seek out images from underrepresented areas, but as Jo et al. [32] discuss, there are actions that can and should be taken, such as explicitly collecting data from underrepresented geographic regions, to ensure a more diverse representation.

7 Conclusions

In conclusion, we present the REVISE tool, which automates the discovery of potential biases in visual datasets and their annotations. We perform this investigation along three axes: object-based, gender-based, and geography-based, and note that there are many more axes along which biases live. What cannot be automated is determining which of these biases are problematic and which are not, so we hope that by surfacing anomalous patterns as well as actionable next steps to the user, we can at least bring these biases to light.

Acknowledgments. This work is partially supported by the National Science Foundation under Grant No. 1763642 and No. 1704444. We would also like to thank Felix Yu, Vikram Ramaswamy, Zhiwei Deng, and Sadhika Malladi for their helpful comments, and Zeyu Wang, Deniz Oktay, and Nobline Yoo for testing out the tool and providing feedback.

References

1. Amazon rekognition. https://aws.amazon.com/rekognition/
2. A Yahoo Flickr grand challenge on tag and caption prediction (2016). https://multimediacommons.wordpress.com/tag-caption-prediction-challenge/
3. United Nations statistics division - methodology (2019). https://unstats.un.org/unsd/methodology/m49/
4. Alwassel, H., Heilbron, F.C., Escorcia, V., Ghanem, B.: Diagnosing error in temporal action detectors. In: European Conference on Computer Vision (ECCV) (2018)
5. Bearman, S., Korobov, N., Thorne, A.: The fabric of internalized sexism. J. Integr. Soc. Sci. **1**(1), 10–47 (2009)
6. Bellamy, R.K.E., et al.: AI fairness 360: an extensible toolkit for detecting, understanding, and mitigating unwanted algorithmic bias. arXiv:1810.01943 (2018)
7. Berg, A.C., et al.: Understanding and predicting importance in images. In: Conference on Computer Vision and Pattern Recognition (CVPR) (2012)
8. Brown, C.: Archives and Recordkeeping: Theory Into Practices. Facet Publishing, London (2014)
9. Buda, M., Maki, A., Mazurowski, M.A.: A systematic study of the class imbalance problem in convolutional neural networks. arXiv:1710.05381 (2017)

10. Buolamwini, J., Gebru, T.: Gender shades: intersectional accuracy disparities in commercial gender classification. In: ACM Conference on Fairness, Accountability, Transparency (FAccT) (2018)
11. Burns, K., Hendricks, L.A., Saenko, K., Darrell, T., Rohrbach, A.: Women also snowboard: Overcoming bias in captioning models. In: European Conference on Computer Vision (ECCV) (2018)
12. Caliskan, A., Bryson, J.J., Narayanan, A.: Semantics derived automatically from language corpora contain humanlike biases. Science **356**(6334), 183–186 (2017)
13. Choi, M.J., Torralba, A., Willsky, A.S.: Context models and out-of-context objects. Pattern Recogn. Lett. **33**, 853–862 (2012)
14. Chouldechova, A.: Fair prediction with disparate impact: a study of bias in recidivism prediction instruments. Big Data (2017)
15. Deng, J., et al.: ImageNet: a large-Scale Hierarchical Image Database. In: CVPR (2009)
16. DeVries, T., Misra, I., Wang, C., van der Maaten, L.: Does object recognition work for everyone? In: Conference on Computer Vision and Pattern Recognition Workshops (CVPRW) (2019)
17. Dwork, C., Hardt, M., Pitassi, T., Reingold, O., Zemel, R.: Fairness through awareness. In: Proceedings of the 3rd Innovations in Theoretical Computer Science Conference (2012)
18. Dwork, C., Immorlica, N., Kalai, A.T., Leiserson, M.: Decoupled classifiers for fair and efficient machine learning. arXiv:1707.06613 (2017)
19. Everingham, M., Gool, L.V., Williams, C.K.I., Winn, J., Zisserman, A.: The pascal visual object classes (VOC) challenge. Int. J. Comput. Vis. (IJCV) **88**, 303–338 (2010)
20. Fei-Fei, L., Fergus, R., Perona, P.: Learning generative visual models from few training examples: an incremental Bayesian approach tested on 101 object categories. In: IEEE CVPR Workshop of Generative Model Based Vision (2004)
21. Gajane, P., Pechenizkiy, M.: On formalizing fairness in prediction with machine learning. arXiv:1710.03184 (2017)
22. Galleguillos, C., Rabinovich, A., Belongie, S.: Object categorization using co-occurrence, location and appearance. In: Conference on Computer Vision and Pattern Recognition (CVPR) (2008)
23. Gebru, T., et al.: Datasheets for datasets. In: ACM Conference on Fairness, Accountability, Transparency (FAccT) (2018)
24. Hardt, M., Price, E., Srebro, N.: Equality of opportunity in supervised learning. In: Advances in Neural Information Processing Systems (NeurIPS) (2016)
25. He, K., Zhang, X., Ren, S., Sun, J.: Deep residual learning for image recognition. In: European Conference on Computer Vision (ECCV) (2016)
26. Hill, K.: Wrongfully accused by an algorithm. The New York Times (2020). https://www.nytimes.com/2020/06/24/technology/facial-recognition-arrest.html
27. Hoiem, D., Chodpathumwan, Y., Dai, Q.: Diagnosing error in object detectors. In: European Conference on Computer Vision (ECCV) (2012)
28. Holland, S., Hosny, A., Newman, S., Joseph, J., Chmielinski, K.: The dataset nutrition label: a framework to drive higher data quality standards. arXiv:1805.03677 (2018)
29. Hua, J., Xiong, Z., Lowey, J., Suh, E., Dougherty, E.R.: Optimal number of features as a function of sample size for various classification rules. Bioinformatics **21**, 1509–1515 (2005)
30. Idelbayev, Y.: (2019). https://github.com/akamaster/pytorch_resnet_cifar10

31. Jain, A.K., Waller, W.: On the optimal number of features in the classification of multivariate gaussian data. Pattern Recogn. **10**, 365–374 (1978)
32. Jo, E.S., Gebru, T.: Lessons from archives: strategies for collecting sociocultural data in machine learning. In: ACM Conference on Fairness, Accountability, Transparency (FAccT) (2020)
33. Joulin, A., Grave, E., Bojanowski, P., Douze, M., Jégou, H., Mikolov, T.: Fasttext.zip: compressing text classification models. arXiv preprint arXiv:1612.03651 (2016)
34. Joulin, A., Grave, E., Bojanowski, P., Mikolov, T.: Bag of tricks for efficient text classification. arXiv preprint arXiv:1607.01759 (2016)
35. Kay, M., Matuszek, C., Munson, S.A.: Unequal representation and gender stereotypes in image search results for occupations. In: Human Factors in Computing Systems, pp. 3819–3828 (2015)
36. Khosla, A., Zhou, T., Malisiewicz, T., Efros, A.A., Torralba, A.: Undoing the damage of dataset bias. In: European Conference on Computer Vision (ECCV) (2012)
37. Kilbertus, N., Rojas-Carulla, M., Parascandolo, G., Hardt, M., Janzing, D., Schölkopf, B.: Avoiding discrimination through causal reasoning. In: Advances in Neural Information Processing Systems (NeurIPS) (2017)
38. Kleinberg, J., Mullainathan, S., Raghavan, M.: Inherent trade-offs in the fair determination of risk scores. In: Proceedings of Innovations in Theoretical Computer Science (ITCS) (2017)
39. Krasin, I., et al.: Openimages: a public dataset for large-scale multi-label and multi-class image classification. Dataset (2017). https://github.com/openimages
40. Krishna, R., et al.: Visual genome: connecting language and vision using crowdsourced dense image annotations (2016). https://arxiv.org/abs/1602.07332
41. Krizhevsky, A.: Learning multiple layers of features from tiny images. Technical report (2009)
42. Krizhevsky, A., Sutskever, I., Hinton, G.E.: ImageNet classification with deep convolutional neural networks. In: Advances in Neural Information Processing Systems (NeurIPS), pp. 1097–1105 (2012)
43. Lin, T.-Y., et al.: Microsoft COCO: common objects in context. In: Fleet, D., Pajdla, T., Schiele, B., Tuytelaars, T. (eds.) ECCV 2014. LNCS, vol. 8693, pp. 740–755. Springer, Cham (2014). https://doi.org/10.1007/978-3-319-10602-1_48
44. Liu, X.Y., Wu, J., Zhou, Z.H.: Exploratory undersampling for class-imbalance learning. IEEE Trans. Syst. Man Cybern. **39**, 539–550 (2009)
45. Mehrabi, N., Morstatter, F., Saxena, N., Lerman, K., Galstyan, A.: A survey on bias and fairness in machine learning. arXiv:1908.09635 (2019)
46. Moulton, J.: The myth of the neutral 'man'. In: Sexist Language: A Modern Philosophical Analysis, pp. 100–116 (1981)
47. Oksuz, K., Cam, B.C., Kalkan, S., Akbas, E.: Imbalance problems in object detection: a review. arXiv e-prints arXiv:1909.00169, August 2019
48. Oliva, A., Torralba, A.: The role of context in object recognition. Trends Cogn. Sci. **11**, 520–527 (2007)
49. Ouyang, W., Wang, X., Zhang, C., Yang, X.: Factors in finetuning deep model for object detection with long-tail distribution. In: Conference on Computer Vision and Pattern Recognition (CVPR) (2016)
50. Pleiss, G., Raghavan, M., Wu, F., Kleinberg, J., Weinberger, K.Q.: On fairness and calibration. In: Advances in Neural Information Processing Systems (NeurIPS) (2017)

51. Prabhu, V.U., Birhane, A.: Large image datasets: a pyrrhic win for computer vision? arXiv:2006.16923 (2020)
52. Rosenfeld, A., Zemel, R., Tsotsos, J.K.: The elephant in the room. arXiv:1808.03305 (2018)
53. Russakovsky, O., et al.: ImageNet large scale visual recognition challenge. Int. J. Comput. Vis. (IJCV) **115**(3), 211–252 (2015). https://doi.org/10.1007/s11263-015-0816-y
54. Salakhutdinov, R., Torralba, A., Tenenbaum, J.: Learning to share visual appearance for multiclass object detection. In: Conference on Computer Vision and Pattern Recognition (CVPR) (2011)
55. Shankar, S., Halpern, Y., Breck, E., Atwood, J., Wilson, J., Sculley, D.: No classification without representation: assessing geodiversity issues in open datasets for the developing world. In: NeurIPS workshop: Machine Learning for the Developing World (2017)
56. Sigurdsson, G.A., Russakovsky, O., Gupta, A.: What actions are needed for understanding human actions in videos? (2017)
57. Swinger, N., De-Arteaga, M., Heffernan IV, N.T., Leiserson, M., Kalai, A.: What are the biases in my word embedding? In: Proceedings of the AAAI/ACM Conference on Artificial Intelligence, Ethics, and Society (AIES) (2019)
58. Thomee, B., et al.: YFCC100M: the new data in multimedia research. Commun. ACM **59**, 64–73 (2016)
59. Tommasi, T., Patricia, N., Caputo, B., Tuytelaars, T.: A deeper look at dataset bias. In: German Conference on Pattern Recognition (2015)
60. Torralba, A., Efros, A.A.: Unbiased look at dataset bias. In: Conference on Computer Vision and Pattern Recognition (CVPR) (2011)
61. Torralba, A., Fergus, R., Freeman, W.T.: 80 million tiny images: a large dataset for non-parametric object and scene recognition. IEEE Trans. Pattern Anal. Mach. Intell. **30**(11), 1958–1970 (2008)
62. Wang, Z., et al.: Towards fairness in visual recognition: effective strategies for bias mitigation. In: Conference on Computer Vision and Pattern Recognition (CVPR) (2020)
63. Wilson, B., Hoffman, J., Morgenstern, J.: Predictive inequity in object detection. arXiv:1902.11097 (2019)
64. Xiao, J., Hays, J., Ehinger, K.A., Oliva, A., Torralba, A.: Sun database: large-scale scene recognition from abbey to zoo. In: Conference on Computer Vision and Pattern Recognition (CVPR) (2010)
65. Yang, J., Price, B., Cohen, S., Yang, M.H.: Context driven scene parsing with attention to rare classes. In: Conference on Computer Vision and Pattern Recognition (CVPR) (2014)
66. Yang, K., Qinami, K., Fei-Fei, L., Deng, J., Russakovsky, O.: Towards fairer datasets: filtering and balancing the distribution of the people subtree in the ImageNet hierarchy. In: ACM Conference on Fairness, Accountability, Transparency (FAccT) (2020)
67. Yao, Y., Zhang, J., Shen, F., Hua, X., Xu, J., Tang, Z.: Exploiting web images for dataset construction: a domain robust approach. IEEE Trans. Multimedia **19**, 1771–1784 (2017)
68. Zhang, B.H., Lemoine, B., Mitchell, M.: Mitigating unwanted biases with adversarial learning. In: Proceedings of the 2018 AAAI/ACM Conference on AI, Ethics, and Society (2018)

69. Zhao, J., Wang, T., Yatskar, M., Ordonez, V., Chang, K.W.: Men also like shopping: reducing gender bias amplification using corpus-level constraints. In: Proceedings of the Conference on Empirical Methods in Natural Language Processing (EMNLP) (2017)
70. Zhou, B., Lapedriza, A., Khosla, A., Oliva, A., Torralba, A.: Places: a 10 million image database for scene recognition. IEEE Trans. Pattern Anal. Mach. Intell. **40**, 1452–1464 (2017)
71. Zhu, X., Anguelov, D., Ramanan, D.: Capturing long-tail distributions of object subcategories. In: Conference on Computer Vision and Pattern Recognition (CVPR) (2014)

Contrastive Learning for Weakly Supervised Phrase Grounding

Tanmay Gupta[1(✉)], Arash Vahdat[3], Gal Chechik[2,3], Xiaodong Yang[3], Jan Kautz[3], and Derek Hoiem[1]

[1] University of Illinois Urbana-Champaign, Champaign, USA
tgupta6@illinois.edu
[2] Bar Ilan University, Ramat Gan, Israel
[3] NVIDIA, Santa Clara, USA

Abstract. Phrase grounding, the problem of associating image regions to caption words, is a crucial component of vision-language tasks. We show that phrase grounding can be learned by optimizing word-region attention to maximize a lower bound on mutual information between images and caption words. Given pairs of images and captions, we maximize compatibility of the attention-weighted regions and the words in the corresponding caption, compared to non-corresponding pairs of images and captions. A key idea is to construct effective negative captions for learning through language model guided word substitutions. Training with our negatives yields a $\sim$ 10% absolute gain in accuracy over randomly-sampled negatives from the training data. Our weakly supervised phrase grounding model trained on COCO-Captions shows a healthy gain of 5.7% to achieve 76.7% accuracy on Flickr30K Entities benchmark. Our code and project material will be available at http://tanmaygupta.info/info-ground.

Keywords: Mutual information · InfoNCE · Grounding · Attention

1 Introduction

Humans can learn from captioned images because of their ability to associate words to image regions. For instance, humans perform such word-region associations while acquiring facts from news photos, making a diagnosis from MRI scans and radiologist reports, or enjoying a movie with subtitles. This word-region association problem is called word or phrase *grounding* and is a crucial capability needed for downstream applications like visual question answering, image captioning, and text-image retrieval.

Existing object detectors can detect and represent object regions in an image, and language models can provide contextualized representations for noun phrases

Electronic supplementary material The online version of this chapter (https://doi.org/10.1007/978-3-030-58580-8_44) contains supplementary material, which is available to authorized users.

A. Vedaldi et al. (Eds.): ECCV 2020, LNCS 12348, pp. 752–768, 2020.
https://doi.org/10.1007/978-3-030-58580-8_44

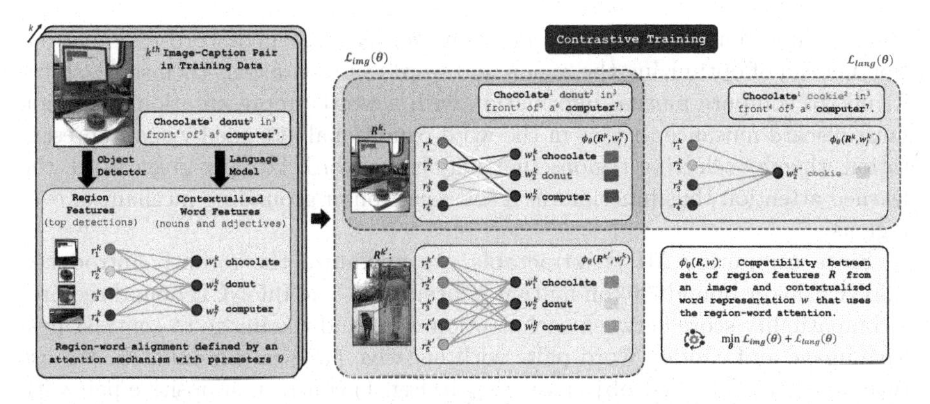

Fig. 1. Overview of our contrastive learning framework. We begin by extracting region and word features using an object detector and a language model respectively. Contrastive learning trains a word-region attention mechanism as part of a compatibility function ϕ_θ between the set of region features from an image and individual contextualized word representations. The compatibility function is trained to maximize a lower bound on mutual information with two losses. For a given caption word, $\mathcal{L}_{\texttt{img}}$ learns to produce a higher compatibility for the true image than a negative image in the mini-batch. $\mathcal{L}_{\texttt{lang}}$ learns to produce a higher compatibility of an image with a true caption-word than with a word in a negative caption. We construct negative captions by substituting a noun word like "donut" in the true caption with contextually plausible but untrue words like "cookie" using a language model.

in the caption. However, learning a mapping between these continuous, independently trained visual and textual representations is challenging in the absence of explicit region-word annotations. We focus on learning this mapping from weak supervision in the form of paired image-caption data without requiring laborious grounding annotations.

Current state-of-the-art approaches [1,11,33] formulate weakly supervised phrase grounding as a multiple instance learning (MIL) problem [18,25]. The image can be viewed as a bag of regions. For a given phrase, all images with captions containing the phrase are treated as positive bags while remaining images are treated as negatives. Models aggregate per region features or phrase scores to construct image-level predictions that can be supervised with image-level labels in the form of phrases or captions. Common aggregation approaches include max or mean pooling, noisy-OR [13], and attention [11,18]. Popular training objectives include binary classification loss [13] (whether the image contain the phrase) or caption reconstruction loss [33] (generalization of binary classification to caption prediction) or ranking objectives [1,11] (do true image-caption or image-phrase pairs score higher than negative pairs).

Figure 1 provides an overview of our proposed contrastive training. We propose a novel formulation of the weakly supervised phrase grounding problem as that of maximizing a lower bound on mutual information between set of region features extracted from an image and contextualized word representations. We use pretrained region and word representations from an object detector and a language model and perform optimization over parameters of word-region attention instead of optimizing the region and word representations themselves. Intuitively, to compute mutual information with a word's representation, attention must discard nuisance regions in the word-conditional attended visual representation, thereby selecting regions that match the word. For any given word, the learned attention thus functions as a soft selection or grounding mechanism over regions.

Since computing MI is intractable, we maximize the recently introduced InfoNCE lower bound [30] on mutual information. The InfoNCE bound requires a compatibility score between each caption word and the image to contrast positive image and caption word pairs with negative pairs in a minibatch. We use two objectives. The first objective ($\mathcal{L}_{\texttt{img}}$ in Fig. 1) contrasts a positive pair with negative pairs with the same caption word but different image regions. The second objective ($\mathcal{L}_{\texttt{lang}}$ in Fig. 1) contrasts a positive pair with negative pairs with the same image but different captions. We show empirically that sampling negative captions randomly from the training data to optimize $\mathcal{L}_{\texttt{lang}}$ does not yield any gains over optimizing $\mathcal{L}_{\texttt{img}}$ only. Instead of random sampling, we propose to use a language model to construct context-preserving negative captions by substituting a single noun word in the caption.

We design the compatibility function using a `query-key-value` attention mechanism. The `queries` and `keys`, computed from words and regions respectively, are used to compute a word-specific attention over each region which acts as a soft alignment or grounding between words and regions. The compatibility score between regions and word is computed by comparing attended visual representation and the word representation.

Our key contributions are: (i) a novel MI based contrastive training framework for weakly supervised phrase grounding; (ii) an InfoNCE compatibility function between a set of regions and a caption word designed for phrase grounding; and (iii) a procedure for constructing context-preserving negative captions that provides $\approx 10\%$ absolute gain in grounding performance.

1.1 Related Work

Our work is closely related to three active areas of research. We now provide an overview of prior arts in each.

Weakly Supervised Phrase Grounding. Weakly supervised phrase localization is typically posed as a multiple instance learning (MIL) problem [18,25] where each image is considered as a bag of region proposals. Images whose captions mention a word or a phrase are treated as positive bags while rest of the images are treated as negatives for that word or phrase. Features or scores for a phrase or the entire caption are aggregated across all regions to make a prediction for the image. Common methods of aggregation are max or average pooling, noisy-OR [13], or attention [18,33]. With the ability to produce image-level scores for pairs of images and phrases or captions, the problem becomes an image-level fully-supervised phrase classification problem [13] or an image-caption retrieval problem [1,11]. An alternatives to the MIL formulations is the approach of Ye *et al.* [44] which uses statistical hypothesis testing approach to link concepts detected in an image and words mentioned in the sentence. While all the above approaches assume paired image-caption data, Wang *et al.* [42] recently address the problem of phrase grounding without access to image-caption pairs. Instead they assume access to a set of scene and color classifiers, and object detectors to detect concepts in the scene and use word2vec [27] similarity between concept labels and caption words to achieve grounding.

MI-Based Representation Learning. Recently MI-based approaches have shown promising results on a variety representation learning problems. Computing the MI between two representations is challenging as we often have access to samples but not the underlying joint distribution that generated the samples. Thus, recent efforts rely on variational estimation of MI [3,6,20,30]. An overview of such estimators is discussed in [31,40] while the statistical limitations are reviewed in [26,34].

In practice, MI-based representation learning models are often trained by maximizing an estimation of MI across different *transformations* of data. For example, deep InfoMax [17] maximizes MI between local and global representation using MINE [6]. Contrastive predictive coding [16,30] inspired by noise contrastive estimation [14,29] assumes an order in the features extracted from an image and uses summary features to predict future features. Contrastive multiview coding [39] maximizes MI between different color channels or data modalities while augmented multiscale Deep InfoMax [5] and SimCLR [8] extract views using different augmentations of data points. Since the infoNCE loss is limited by the batch size, several previous work rely on memory banks [15,28,43] to increase the set of negative instances.

Joint Image-Text Representation Learning. With the advances in both visual analysis and natural language understanding, there has been a recent shift towards learning representation jointly from both visual and textual domains [2,9,22–24,35–38,45]. ViLBERT [24] and LXMERT [38] learn representation from both modalities using two-stream transformers, applied to image and text independently. In contrast, UNITER [9], VisualBERT [23], Unicoder-VL [22], VL-BERT [35] and B2T2 [2] propose a unified single architecture that learns representation jointly from both domains. Our method is similar to the first group, but differs in its fundamental goal. Instead of focusing on learning a task-agnostic representation for a range of downstream tasks, we are interested in the quality of region-phrase grounding emerged by maximizing mutual information. Moreover, we rely on the language modality as a weak training signal for grounding, and we perform phrase-grounding without any further finetuning.

2 Method

Consider the set of region features and contextualized word representation as two multivariate random variables. Intuitively, estimating MI between them requires extracting the information content shared by these two variables. We model this MI estimation as maximizing a lower bound on MI with respect to parameters of a word-region attention model. This maximization forces the attention model to downweight regions from the image that do not match the word, and to attend to the image regions that contain the most shared information with the word representation.

Section 2.1 describes MI and the InfoNCE lower bound. Section 2.2 introduces notation and InfoNCE based objective for learning phrase grounding from paired image caption data. Section 2.3 presents the design of a word-region attention based compatibility function which is part of the InfoNCE objective.

2.1 InfoNCE Lower Bound on Mutual Information

Let $x \in \mathcal{X}$ and $y \in \mathcal{Y}$ be random variables drawn from a joint distribution with density $p(x, y)$. The MI between x and y measures the amount of information that these two variables share:

$$\mathrm{MI}(x, y) = \mathbb{E}_{(x,y)\sim p(x,y)} \left[\log \frac{p(x, y)}{p(x)p(y)} \right], \tag{1}$$

which is also the Kullback–Leibler Divergence from $p(x, y)$ to $p(x)p(y)$.

However, computing MI is intractable in general because it requires a complete knowledge of the joint and marginal distributions. Among the existing MI estimators, the InfoNCE [30] lower bound provides a low-variance estimation of MI for high dimensional data, albeit being biased [31]. The appealing variance properties of this estimator may explain its recent success in representation learning [8,16,30,36]. InfoNCE defines a lower bound on MI by:

$$\mathrm{MI}(x, y) \geq \log(k) - \mathcal{L}_k(\theta). \tag{2}$$

Here, $\mathcal{L}_k$ is the InfoNCE objective defined in terms of a compatibility function ϕ parametrized by θ: $\phi_\theta : \mathcal{X} \times \mathcal{Y} \to \mathbb{R}$. The lower bound is computed over a mini-batch $\mathcal{B}$ of size k, consisting of one positive pair $(x, y) \sim p(x, y)$ and $k-1$ negative pairs $\{(x'_i, y)\}_{i=1}^{k-1}$ where $x' \sim p(x)$:

$$\mathcal{L}_k(\theta) = \mathbb{E}_{\mathcal{B}} \left[-\log \left(\frac{e^{\phi_\theta(x,y)}}{e^{\phi_\theta(x,y)} + \sum_{i=1}^{k-1} e^{\phi_\theta(x'_i,y)}} \right) \right]. \tag{3}$$

Oord *et al.* [30] showed that maximizing the lower bound on MI by minimizing $\mathcal{L}_k$ with respect to θ leads to a compatibility function ϕ_{θ^*} that obeys

$$e^{\phi_{\theta^*}(x,y)} \propto \frac{p(x|y)}{p(x)} = \frac{p(x, y)}{p(x)p(y)}, \tag{4}$$

where θ^* is the optimal θ obtained by minimizing $\mathcal{L}_k$.

2.2 InfoNCE for Phrase Grounding

Recent work [11] has shown that pre-trained object detectors such as Faster-RCNN [32] and language models such as BERT [12] provide rich representations in the visual and textual domains for the phrase grounding problem. Inspired by this, we aim to maximize mutual information between region features generated by an object detector and contextualized word representation extracted by a language model (Fig. 2).

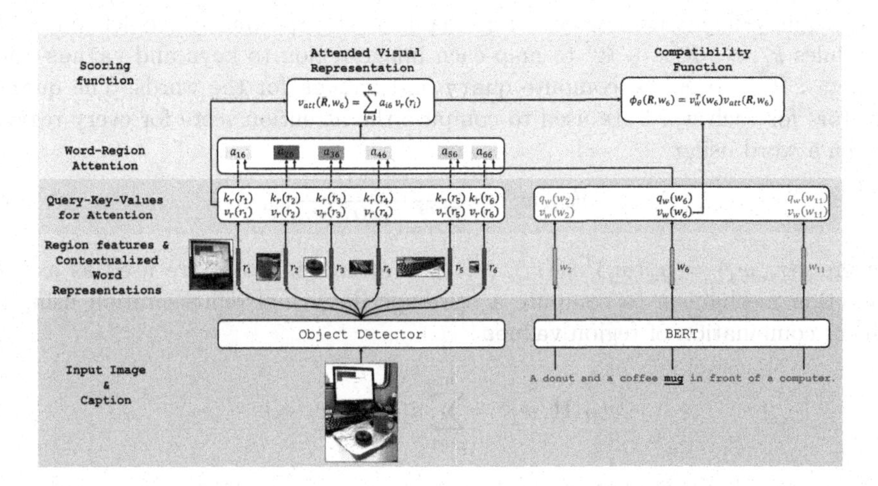

Fig. 2. Compatibility function ϕ_θ with word-region attention. The figure shows compatibility computation between the set of image regions and the word "mug" in the caption. The compatibility function consists of learnable `query-key-value` functions k_r, v_r, q_w, v_w. The `query` constructed from contextualized representation of the word "mug" is compared to `keys` created from region features to compute attention scores. The attention scores are used as weights to linearly combine `values` created from region features to construct an attended visual representation for "mug". The compatibility is defined by the dot product of the attended visual representation and `value` representation for "mug".

Let us denote image region features for an image by $\mathbf{R} = \{r_i\}_{i=1}^m$ where m is the number of regions in the image with each $r_i \in \mathbb{R}^{d_r}$. Similarly, caption word representations are denoted as $\mathbf{W} = \{w_j\}_{j=1}^n$ where n is the number of words in the caption with each word represented as $w_j \in \mathbb{R}^{d_w}$.

We maximize the InfoNCE lower bound on MI between image regions and each individual word representation denoted by $\text{MI}(\mathbf{R}, w_j)$. Thus using Eq. 2 we maximize the following lower bound:

$$\sum_{j=1}^{n} \text{MI}(\mathbf{R}, w_j) \geq n \log(k) - \sum_{j=1}^{n} \mathcal{L}_{kj}(\theta). \tag{5}$$

We empirically show that maximizing the lower bound in Eq. 5 with an appropriate choice of compatibility function ϕ_θ results in learning phrase grounding without strong grounding supervision. The following section details the design of the compatibility function.

2.3 Compatibility Function with Attention

The InfoNCE loss in our phrase grounding formulation requires a compatibility function between the *set* of region feature vectors $\mathbf{R}$ and the contextualized word representation w_j. To define the compatibility function, we propose to use a `query-key-value` attention mechanism [41]. Specifically, we define neural modules $k_r, v_r : \mathbb{R}^{d_r} \rightarrow \mathbb{R}^d$ to map each image region to `keys` and `values` and $q_w, v_w : \mathbb{R}^{d_w} \rightarrow \mathbb{R}^d$ to compute `query` and `values` for the words. The `query` vectors for each word are used to compute the attention score for every region given a word using

$$a(r_i, w_j) = \frac{e^{s(r_i, w_j)}}{\sum_{i'=1}^{m} e^{s(r_{i'}, w_j)}}, \tag{6}$$

where $s(r_i, w_j) = q_w(w_j)^T k_r(r_i)/\sqrt{d}$. The attention scores are used as a soft selection mechanism to compute a word-specific visual representation using a linear combination of region `values`

$$v_{att}(\mathbf{R}, w_j) = \sum_{i=1}^{m} a(r_i, w_j) v_r(r_i). \tag{7}$$

Finally, the compatibility function is defined as $\phi_\theta(\mathbf{R}, w_j) = v_w^T(w_j)\ v_{att}(\mathbf{R}, w_j)$, where θ refers to the parameters of neural modules k_r, v_r, q_w, and v_w, implemented using simple feed-forward MLPs. Following Eqs. 3 & 5, the InfoNCE loss for phrase grounding is defined as

$$\mathcal{L}_{\texttt{img}}(\theta) = \mathbb{E}_{\mathcal{B}}\left[-\sum_{j=1}^{n} \log\left(\frac{e^{\phi_\theta(\mathbf{R}, w_j)}}{e^{\phi_\theta(\mathbf{R}, w_j)} + \sum_{i=1}^{k-1} e^{\phi_\theta(\mathbf{R}'_i, w_j)}}\right)\right]. \tag{8}$$

which is marked using subscript `img` as negative pairs are created by replacing image regions from a positive pair with regions extracted from negative instance in the mini-batch.

Remark: We enforce compatibility between each word and all image regions using $\text{MI}(\mathbf{R}, w_j)$ in Eq. 5, but not between a region and all caption words $(\text{MI}(r_i, \mathbf{W}))$. This is because the words only describe part of the image, so there will be regions with no corresponding word in the caption.

2.4 Context-Preserving Negative Captions

The objective in Eq. 8 trains the compatibility function by contrasting positive regions-word pairs against pairs with replaced image regions. We now propose a complementary objective function that contrasts the positive pairs against negative pairs whose captions are replaced with plausible negative captions. However, extracting negative captions that are related to a captions is challenging as it requires semantic understanding of words in a caption. Here, we leverage BERT as a pretrained bidirectional language model to extract such negative captions.

Caption	Negatives Selected After Reranking	Candidates Rejected After Reranking
A man is seated at a counter with all types of delicious looking foods, yet is completely unaffected, casually reading his newspaper.	menu, books, phone, scripts, email, messages, bible, tablet	newspaper, paper, journal, article, magazine
A BMX bike rider in red clothing and a helmet is riding his bike next to a wooden fence.	bench, pole, statue, door, table, chair, sign, platform, piano	fence, gate, wall, railing, screen
A man in a blue jumpsuit stands next to a red van pulling a trailer.	bike, sedan, horse, jeep, cart, car, tractor, bull, engine, motorcycle	van, trailer, vehicle, light, truck
A man and a boy are playing with a dog in the evening.	girl, lady, mother, woman, teenager, child, teacher, mom	man, boy, guy, couple, youth
A woman in a brown sweater sits at a table covered with food.	boy, guy, gentleman, kid, nurse, soldier, waiter, priest, child	woman, female, person, face, lady
A man with shorts and a hat is holding onto a little boy and a dog.	gloves, glasses, coat, trousers, bags, apron, moustache, beard	shorts, ties, pants, stripes, jeans

Fig. 3. Context-preserving negative captions. We construct negative captions which share the same context as the true caption but substitute a noun word. We choose the substitute using a language model such that it is plausible in the context but we reject potential synonyms or hypernyms of the original word by a re-ranking procedure.

For a caption with a noun word s and context c, we define a context-preserving negative caption as one which has the same context c but a different noun s' with the following properties: (i) s' should be plausible in the context; and (ii) the new caption defined by the pair (s', c) should be untrue for the image. For example, consider the caption `''A man is walking on a beach''` where s is chosen as `''man''` and c is defined by `''A [MASK] is walking on a beach''` where `[MASK]` is the token that denotes a missing word. A potential candidate for a context-preserving negative caption might be `''A woman is walking on a beach''` where s' is `woman`. However, `''A car is walking on a beach''` and `''A person is walking on a beach''` are not negative captions because `car` is not plausible given the context, and the statement with `person` is still true given that the original caption is true for the image.

Constructing Context-Preserving Negative Captions. We propose to use a pre-trained BERT language model to construct context-preserving negative captions for a given true caption. Our approach for extracting such words consists of two steps: First, we feed the context c into the language model to extract 30 most likely candidates $\{s'_l\}_{l=1}^{30}$ for the masked word using probabilities $p(s'|c)$ predicted by BERT. Intuitively, these words correspond to those that fill in the masked word in caption according to BERT. However, the original masked word or its synonyms may be present in the set as well. Thus, in the second step, we pass the original caption into BERT to compute $q(s'_l|s, c)$ which we use as a proxy for how true (s'_l, c) is given that (s, c) is true. We re-rank the candidates using the score $\frac{p(s'|c)}{q(s'|s,c)}$ and we keep the top 25 captions $\{(s'_l, c)\}_{l=1}^{25}$ as negatives for the original caption (s, c).

We empirically find that the proposed approach is effective in extracting context-preserving negative captions. Figure 3 shows a context-preserving negatives for a set of captions along with candidates that were rejected after re-ranking. Note that the selected candidates match the context and the rejected candidates are often synonyms or hypernyms of the true noun.

Training with Context-Preserving Negative Captions. Given the context-preserving negative captions, we can train our compatibility function by contrasting the positive pairs against negative pairs with plausible negative captions. We use a loss function similar to InfoNCE to encourage higher compatibility score of an image with the true caption than any negative caption. Let w and $\{w'_l\}_{l=1}^{25}$ denote the contextualized representation of the positive word s and the corresponding negative noun words $\{s'_l\}_{l=1}^{25}$. The language loss is defined as

$$\mathcal{L}_{\mathbf{lang}}(\theta) = \mathbb{E}_{\mathcal{B}}\left[-\log\left(\frac{e^{\phi_\theta(\mathbf{R},w)}}{e^{\phi_\theta(\mathbf{R},w)} + \sum_{l=1}^{25} e^{\phi_\theta(\mathbf{R},w'_l)}}\right)\right]. \tag{9}$$

For captions with multiple noun words, we randomly select s from the noun words for simplicity.

2.5 Implementation Details

Regions and Visual Features. We use the Faster-RCNN object detector provided by Anderson *et al.* [4] and used for extracting visual features in the current state-of-the-art phrase grounding approach Align2Ground [11]. The detector is trained jointly on Visual Genome object and attribute annotations and we use a maximum of 30 top scoring bounding boxes per image with 2048 dimensional ROI-pooled region features.

Contextualized Word Representations. We use a pretrained BERT language model to extract 768 dimensional contextualized word representations for each caption word. Note that BERT is trained on a text corpora using masked language model training where words are randomly replaced by a `[MASK]` token in the input and the likelihood of the masked word is maximized in the distribution over vocabulary words predicted at the output. Thus, BERT is trained to model distribution over words given context and hence suitable for modeling $p(s|c)$ defined in Sect. 2.4 for constructing context-preserving negative captions.

Query-Key-Value Networks. We use an MLP with 1 hidden layer for each of k_r, v_r, q_w, v_w for all experiments except the ablation in Fig. 4. We use BatchNorm [19] and ReLU activations after the first linear layer. The hidden layer has the same number of neurons as the input dimensions of these networks which are 2048 for (k_r, v_r), and 768 for (q_w, v_w). The output layer is 384 (= 768/2) for all networks.

Losses. Since we only care about grounding noun phrases, we compute $\mathcal{L}_{\texttt{img}}$ only for noun and adjective words in the captions as identified by a POS tagger instead of all caption words for computation efficiency.

Optimization. We optimize $\mathcal{L}_{img} + \mathcal{L}_{lang}$ computed over batches of 50 image-caption pairs using the ADAM optimizer [21] with a learning rate of 10^{-5}. We compute $\mathcal{L}_{\texttt{img}}$ for each image using other images in the batch as negatives.

Attention to Phrase Grounding. We use the BERT tokenizer to convert captions into individual word or sub-word tokens. Attention is computed per token. For evaluation, the phrase-level attention score for each region is computed as the maximum attention score assigned to the region by any of the tokens in the phrase. The regions are then ranked according to this phrase level score.

3 Experiments

Our experiments compare our approach to state-of-the-art on weakly supervised phrase localization (Sect. 3.2), ablate gains due to pretrained language representations and context-preserving negative sampling using a language model (Sect. 3.3), and analyse the relation between phrase grounding performance and the InfoNCE bound that we optimize as a proxy for phrase grounding (Sect. 3.4).

3.1 Datasets and Metrics

We train our models on image-caption pairs from COCO training set which consists of $\sim$ 83K training images. We use the validation set with $\sim$ 41K images for part of our analysis. Each image is accompanied with 5 captions. For evaluation, we use the Flickr30K Entities validation set for model selection (early stopping) and test set for reporting final performance. Both sets consist of 1K images with 5 captions each. We report two metrics:

Recall@k which is the fraction of phrases for which the ground truth bounding box has an IOU ≥ 0.5 with any of the top-k predicted boxes.

Pointing accuracy which requires the model to predict a single point location per phrase and the prediction is counted as correct if it falls within the ground truth bounding box for the phrase. Unlike recall@k, pointing accuracy does not require identifying the extent of the object. Since our model selects one of the detected regions in the image, we use center of the selected bounding box as the prediction for each phrase for computing pointing accuracy.

3.2 Performance on Flickr30K Entities

Table 1 compares performance of our method to existing weakly supervised phrase grounding approaches on the Flickr30K Entities test set. A few existing approaches train on Flickr30K Entities train set and report recall@1 while recent methods use COCO train set and report pointing accuracy. Further, all approaches use different visual features making direct comparison difficult. For a fair comparison to state-of-the-art, we use Faster-RCNN trained on Visual Genome object and attribute annotations used in Align2Ground [11] and report performance for models trained on either datasets on both recall and pointing accuracy metrics.

Table 1. Grounding performance on Flickr30K Entities test set. We make our approach directly comparable to the current state-of-the-art, Align2Ground [11]. The performance of older methods are reported for completeness but the use of different visual features makes direct comparison difficult.

Method	Training data	Visual features	R@1	R@5	R@10	Accuracy
GroundeR (2015) [33]	Flickr30K Entities	VGG-det (VOC)	28.94	–	–	–
Yeh *et al.* (2018) [44]	Flickr30K Entities	VGG-cls (IN)	22.31	–	–	–
Yeh *et al.* (2018) [44]	Flickr30K Entities	VGG-det (VOC)	35.90	–	–	–
Yeh *et al.* (2018) [44]	Flickr30K Entities	YOLO (COCO)	36.93	–	–	–
KAC Net+Soft KBP (2018) [7]	Flickr30K Entities	VGG-det (VOC)	38.71	–	–	–
Fang *et al.* (2015) [13]	COCO	VGG-cls (IN)	–	–	–	29.00
Akbari *et al.* (2019) [1]	COCO	VGG-cls (IN)	–	–	–	61.66
Akbari *et al.* (2019) [1]	COCO	PNAS Net (IN)	–	–	–	69.19
Align2Ground (2019) [11]	COCO	Faster-RCNN (VG)	–	–	–	71.00
Ours	Flickr30K Entities	Faster-RCNN (VG)	**47.88**	**76.63**	**82.91**	**74.94**
Ours	COCO	Faster-RCNN (VG)	**51.67**	**77.69**	**83.25**	**76.74**

Table 2. Benefits of language modeling. The first two rows show the gains due to pretrained language representations. The next three rows show gains from each step in our proposed context-preserving negative caption construction.

Negative captions	Language model	R@1	R@5	R@10	Accuracy
None	BERT (Random)	25.66	59.57	75.16	57.37
None	BERT (Pretrained)	35.74	72.91	82.07	66.89
Random	BERT (Pretrained)	36.32	72.42	81.81	66.92
Contextually plausible	BERT (Pretrained)	**48.05**	**76.78**	**82.97**	**74.91**
Excluding near-synonyms & hypernyms	BERT (Pretrained)	**51.67**	**77.69**	**83.25**	**76.74**

Using the same training data and visual feature architecture, our model shows a 5.7% absolute gain in pointing accuracy over Align2Ground. Learning using our contrastive formulation is also quite sample efficient as can be seen by only a 2 to 3 points drop in performance when the model is trained on the much smaller Flickr30K Entities train set which has approximately one-third as many image-caption pairs as COCO.

3.3 Benefits of Language Modeling

Our approach benefits from language modeling in two ways: (i) using the pre-trained language model to extract contextualized word representations, and (ii) using the language model to sample context-preserving negative captions. Table 2 evaluates along both of these dimensions.

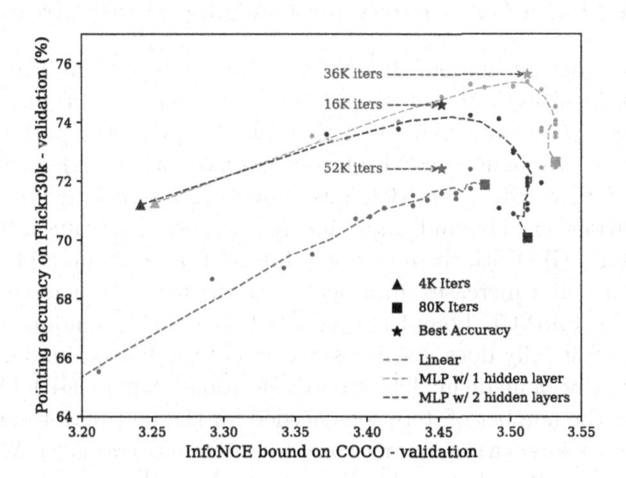

Fig. 4. Relation between `InfoNCE` lower bound and phrase grounding performance with training iterations for 3 different choices of `key-value` modules in the compatibility function ϕ_θ. Each epoch is $\sim$ 8K iterations. The scattered points visualize the measured quantities during training. The dashed lines are created by applying moving average to highlight the trend.

Gains from Pretrained Word Representations. In Table 2, `BERT (Random)` refers to the BERT architecture initialized with random weights and finetuned on COCO image-caption data along with parameters of the attention mechanism. `BERT (Pretrained)` refers to the off-the-shelf pretrained BERT model which is used as a contextualized word feature extractor during contrastive learning without finetuning. We observe a ~10% absolute gain in both recall@1 and pointing accuracy by using pretrained word representations from BERT.

Gains from Context-Preserving Negative Caption Sampling. Our context-preserving negative sampling has two steps. The first step is drawing negative noun candidates given the context provided by the true caption. The second step is re-ranking the candidates to filter out likely synonyms or hypernyms that are also true for the image.

First, note that randomly sampling negative captions from training data for computing $\mathcal{L}_{\text{lang}}$ performs similarly to only training using $\mathcal{L}_{\text{img}}$. Model trained with contextually plausible negatives significantly outperforms random sampling by $\geq$8% gain in recall@1 and pointing accuracy. Excluding near-synonyms and hypernyms yields another $\sim$3 points gain in recall@1 and accuracy.

3.4 Is InfoNCE a Good Proxy for Learning Phrase Grounding?

The fact that optimizing our InfoNCE objective results in learning phrase grounding is intuitive but not trivial. Figure 4 shows that maximizing the InfoNCE lower bound correlates well with phrase grounding performance on a heldout dataset. We make several interesting observations: **(i)** As training progresses (from left to right), InfoNCE lower bound (Eq. 5) mostly keeps increasing on the validation set. This indicates that there is no overfitting in terms of the InfoNCE bound. **(ii)** With the increase in InfoNCE lower bound, phrase grounding performance first increases until peak performance and then decreases. This shows that the InfoNCE bound is correlated with the grounding performance but maximizing it fully does not necessarily yield the best grounding. A similar observation has been made in [39] for representation learning. **(iii)** The peak performance and the number of iterations needed for the best performance depends on the choice of `key-value-query` modules. One and two layer MLPs hit the peak faster and perform better than linear functions. We refer the reader to the supplementary material for a discussion of limitations of our approach.

3.5 Qualitative Results

Figure 5 visualizes the word-region attention learned by our model. The qualitative results demonstrate the following abilities: **(i)** localizing different objects mentioned in the same caption with varying degrees of semantic relatedness, e.g., `man` and `canine` in row 1 vs. `man` and `woman` in row 3; **(ii)** disambiguation between two instances of the same object category using caption context. For example, `boy` and `another` in row 4 and bride and groom from other men and women in row 3; **(iii)** localizing object parts such as toddler's `shirt` in row 2 and instrument's `mouthpiece` in row 5; **(iv)** handling occlusion, e.g., `table` covered with toys in row 6; **(v)** handling uncommon words or categories like `ponytail` and `mouthpiece` in row 5 and `hose` in row 7.

(1)
A man and a canine both stand on a snowy plane looking out into the distance.
#boxes:27
man:0.26,0.17,0.09
canine:0.33,0.25,0.06
(2)
A toddler in a yellow shirt standing in front of a living complex next to a baby carriage.
#boxes:34
toddler:0.59,0.20,0.06
shirt:0.63,0.17,0.08
(3)
A man in a tuxedo and a woman in a bride's gown are leaving a church.
#boxes:50
man:0.17,0.09,0.09
woman:0.19,0.15,0.08
(4)
One boy follows another at the park.
#boxes:37
boy:0.25,0.18,0.18
another:0.12,0.10,0.08
(5)
A man with a ponytail wearing a blue collared shirt is playing an instrument's mouthpiece.
#boxes:12
ponytail:0.29,0.19,0.10
mouthpiece:0.32,0.16,0.09
(6)
Two kids sitting at a table full of toys.
#boxes:34
kids:0.22,0.17,0.16
table:0.33,0.15,0.10
(7)
A curly-haired little girl watering plants with a hose.
#boxes:24
haired:0.27,0.17,0.16
hose:0.29,0.16,0.16

Fig. 5. Visualization of attention. We show all detected regions and top-3 attended regions with attention scores for two words highlighted in each caption. More qualitative results can be found on our project page http://tanmaygupta.info/info-ground/

4 Conclusion

In this work, we offer a novel perspective on weakly supervised phrase grounding from paired image-caption data which has traditionally been cast as a multiple instance learning problem. We formulate the problem as that of estimating mutual information between image regions and caption words. We demonstrate that maximizing a lower bound on mutual information with respect to parameters of a region-word attention mechanism results in learning to ground words in images. We also show that language models can be used to generate context-preserving negative captions which greatly improve learning in comparison to randomly sampling negatives from training data.

Acknowledgement. This work was done partly at NVIDIA and is partly supported by ONR MURI Award N00014-16-1-2007

References

1. Akbari, H., Karaman, S., Bhargava, S., Chen, B., Vondrick, C., Chang, S.F.: Multi-level multimodal common semantic space for image-phrase grounding. In: CVPR (2018)
2. Alberti, C., Ling, J., Collins, M., Reitter, D.: Fusion of detected objects in text for visual question answering. In: EMNLP (2019)
3. Alemi, A.A., Fischer, I., Dillon, J.V., Murphy, K.: Deep variational information bottleneck. In: ICLR (2017)
4. Anderson, P., et al.: Bottom-up and top-down attention for image captioning and visual question answering. In: CVPR (2017)
5. Bachman, P., Hjelm, R.D., Buchwalter, W.: Learning representations by maximizing mutual information across views. arXiv preprint arXiv:1906.00910 (2019)
6. Belghazi, M.I., et al.: Mutual information neural estimation. In: ICML (2018)
7. Chen, K., Gao, J., Nevatia, R.: Knowledge aided consistency for weakly supervised phrase grounding. In: CVPR (2018)
8. Chen, T., Kornblith, S., Norouzi, M., Hinton, G.: A simple framework for contrastive learning of visual representations. arXiv preprint arXiv:2002.05709 (2020)
9. Chen, Y.C., et al.: UNITER: Learning universal image-text representations. arXiv preprint arXiv:1909.11740 (2019)
10. Choe, J., Oh, S.J., Lee, S., Chun, S., Akata, Z., Shim, H.: Evaluating weakly supervised object localization methods right. ArXiv (2020)
11. Datta, S., Sikka, K., Roy, A., Ahuja, K., Parikh, D., Divakaran, A.: Align2Ground: weakly supervised phrase grounding guided by image-caption alignment. In: ICCV (2019)
12. Devlin, J., Chang, M.W., Lee, K., Toutanova, K.: BERT: pre-training of deep bidirectional transformers for language understanding. In: NAACL-HLT (2018)
13. Fang, H., et al.: From captions to visual concepts and back. In: CVPR (2014)
14. Gutmann, M., Hyvärinen, A.: Noise-contrastive estimation: A new estimation principle for unnormalized statistical models. In: AISTATS (2010)
15. He, K., Fan, H., Wu, Y., Xie, S., Girshick, R.: Momentum contrast for unsupervised visual representation learning. arXiv preprint arXiv:1911.05722 (2019)
16. Hénaff, O.J., Razavi, A., Doersch, C., Eslami, S., Oord, A.V.D.: Data-efficient image recognition with contrastive predictive coding. arXiv preprint arXiv:1905.09272 (2019)
17. Hjelm, R.D., et al.: Learning deep representations by mutual information estimation and maximization. In: ICLR (2019)
18. Ilse, M., Tomczak, J.M., Welling, M.: Attention-based deep multiple instance learning. In: ICML (2018)

19. Ioffe, S., Szegedy, C.: Batch normalization: accelerating deep network training by reducing internal covariate shift. In: ICML (2015)
20. Kim, H., Mnih, A.: Disentangling by factorising. In: ICML (2018)
21. Kingma, D.P., Ba, J.: Adam: a method for stochastic optimization. In: ICLR (2015)
22. Li, G., Duan, N., Fang, Y., Jiang, D., Zhou, M.: Unicoder-VL: a universal encoder for vision and language by cross-modal pre-training. arXiv preprint arXiv:1908.06066 (2019)
23. Li, L.H., Yatskar, M., Yin, D., Hsieh, C.J., Chang, K.W.: VisualBERT: a simple and performant baseline for vision and language. arXiv preprint arXiv:1908.03557 (2019)
24. Lu, J., Batra, D., Parikh, D., Lee, S.: VilBERT: pretraining task-agnostic visiolinguistic representations for vision-and-language tasks. In: NeurIPS (2019)
25. Maron, O., Lozano-Pérez, T.: A framework for multiple-instance learning. In: NeurIPS (1998)
26. McAllester, D., Stratos, K.: Formal limitations on the measurement of mutual information. arXiv preprint arXiv:1811.04251 (2018)
27. Mikolov, T., Sutskever, I., Chen, K., Corrado, G.S., Dean, J.: Distributed representations of words and phrases and their compositionality. In: NIPS (2013)
28. Misra, I., van der Maaten, L.: Self-supervised learning of pretext-invariant representations. arXiv preprint arXiv:1912.01991 (2019)
29. Mnih, A., Kavukcuoglu, K.: Learning word embeddings efficiently with noise-contrastive estimation. In: NeurIPS (2013)
30. Oord, A.V.D., Li, Y., Vinyals, O.: Representation learning with contrastive predictive coding. arXiv (2018)
31. Poole, B., Ozair, S., Van Den Oord, A., Alemi, A., Tucker, G.: On variational bounds of mutual information. In: ICML (2019)
32. Ren, S., He, K., Girshick, R., Sun, J.: Faster R-CNN: towards real-time object detection with region proposal networks. In: NeurIPS (2015)
33. Rohrbach, A., Rohrbach, M., Hu, R., Darrell, T., Schiele, B.: Grounding of textual phrases in images by reconstruction. In: Leibe, B., Matas, J., Sebe, N., Welling, M. (eds.) ECCV 2016. LNCS, vol. 9905, pp. 817–834. Springer, Cham (2016). https://doi.org/10.1007/978-3-319-46448-0_49
34. Song, J., Ermon, S.: Understanding the limitations of variational mutual information estimators. In: ICLR (2020)
35. Su, W., et al.: VL-BERT: pre-training of generic visual-linguistic representations (2020)
36. Sun, C., Baradel, F., Murphy, K., Schmid, C.: Contrastive bidirectional transformer for temporal representation learning. arXiv preprint arXiv:1906.05743 (2019)
37. Sun, C., Myers, A., Vondrick, C., Murphy, K., Schmid, C.: VideoBERT: a joint model for video and language representation learning. In: ICCV (2019)
38. Tan, H., Bansal, M.: LXMERT: learning cross-modality encoder representations from transformers. In: EMNLP (2019)
39. Tian, Y., Krishnan, D., Isola, P.: Contrastive multiview coding. arXiv preprint arXiv:1906.05849 (2019)
40. Tschannen, M., Djolonga, J., Rubenstein, P.K., Gelly, S., Lucic, M.: On mutual information maximization for representation learning. In: ICLR (2020)
41. Vaswani, A., et al.: Attention is all you need. In: NeurIPS (2017)
42. Wang, J., Specia, L.: Phrase localization without paired training examples. In: ICCV (2019)

43. Wu, Z., Xiong, Y., Yu, S.X., Lin, D.: Unsupervised feature learning via non-parametric instance discrimination. In: CVPR (2018)
44. Yeh, R.A., Do, M.N., Schwing, A.G.: Unsupervised textual grounding: linking words to image concepts. In: CVPR (2018)
45. Zhou, L., Palangi, H., Zhang, L., Hu, H., Corso, J.J., Gao, J.: Unified vision-language pre-training for image captioning and VQA. In: AAAI (2020)

Collaborative Learning of Gesture Recognition and 3D Hand Pose Estimation with Multi-order Feature Analysis

Siyuan Yang[1,2], Jun Liu[3](✉), Shijian Lu[4], Meng Hwa Er[2], and Alex C. Kot[2]

[1] Rapid-Rich Object Search (ROSE) Lab, Interdisciplinary Graduate Programme, Nanyang Technological University, Singapore, Singapore
siyuan005@e.ntu.edu.sg

[2] School of Electrical and Electronic Engineering, Nanyang Technological University, Singapore, Singapore
{emher,eackot}@ntu.edu.sg

[3] Information Systems Technology and Design Pillar, Singapore University of Technology and Design, Singapore, Singapore
jun_liu@sutd.edu.sg

[4] School of Computer Science and Engineering, Nanyang Technological University, Singapore, Singapore
shijian.Lu@ntu.edu.sg

Abstract. Gesture recognition and 3D hand pose estimation are two highly correlated tasks, yet they are often handled separately. In this paper, we present a novel collaborative learning network for joint gesture recognition and 3D hand pose estimation. The proposed network exploits joint-aware features that are crucial for both tasks, with which gesture recognition and 3D hand pose estimation boost each other to learn highly discriminative features. In addition, a novel multi-order multi-stream feature analysis method is introduced which learns posture and multi-order motion information from the intermediate feature maps of videos effectively and efficiently. Due to the exploitation of joint-aware features in common, the proposed technique is capable of learning gesture recognition and 3D hand pose estimation even when only gesture or pose labels are available, and this enables weakly supervised network learning with much reduced data labeling efforts. Extensive experiments show that our proposed method achieves superior gesture recognition and 3D hand pose estimation performance as compared with the state-of-the-art.

Keywords: Gesture recognition · 3D hand pose estimation · Multi-order multi-stream feature analysis · Slow-fast feature analysis · Multi-scale relation

---

Electronic supplementary material The online version of this chapter (https://doi.org/10.1007/978-3-030-58580-8_45) contains supplementary material, which is available to authorized users.

A. Vedaldi et al. (Eds.): ECCV 2020, LNCS 12348, pp. 769–786, 2020.
https://doi.org/10.1007/978-3-030-58580-8_45

1 Introduction

Gesture recognition and 3D hand pose estimation are both challenging and fast-growing research topics which have received contiguous attention recently due to their wide range of applications in human-computer interaction, robotics, virtual reality, augmented reality, etc. The two tasks are closely correlated as they both leverage heavily on joint-aware features, i.e. features related to the hand joints [19,40]. On the other hand, the two tasks are often tackled separately by dedicated systems [1,4,9,23,24]. Though some recent efforts [13,22,29] attempt to handle the two tasks at one go, it does not consider to iteratively gain benefits from mutual learning of them.

In this paper, we propose to perform gesture recognition and 3D hand pose estimation mutually. We design a novel collaborative learning strategy to exploit joint-aware features that are crucial for both tasks, with which gesture recognition and 3D hand pose estimation can learn to boost each other progressively, as illustrated in Fig. 1.

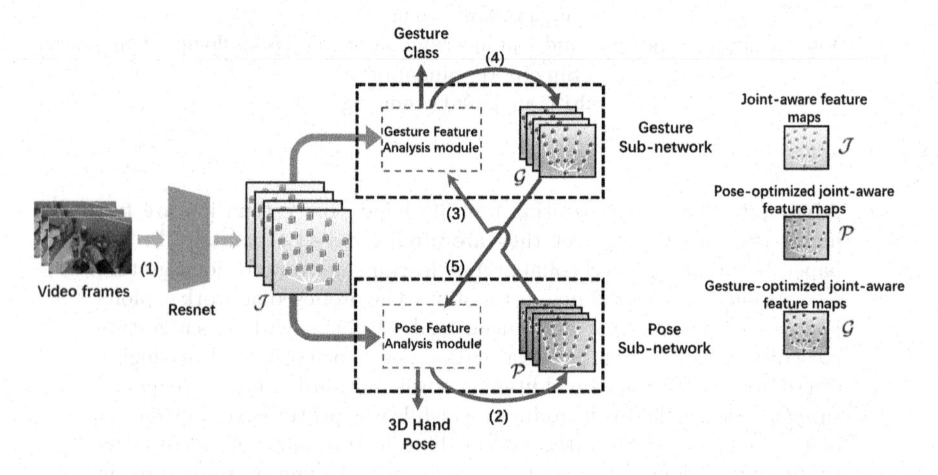

Fig. 1. Overview of our proposed network architecture for gesture recognition and 3D hand pose estimation from videos. The input is video frames, and the output is predicted gesture class of the video and 3D hand joint locations of each video frame. The process flow of our network can be divided into 5 stages: (1) Generating $\mathcal{J}$. (2) Generating $\mathcal{P}$ and Predicting 3D Hand Pose. (3) Aggregating input to Gesture Sub-Network. (4) Generating $\mathcal{G}$ and Recognizing Gesture Class. (5) Aggregating input to Pose Sub-Network. (As shown by the (1)–(5) in this Figure). Stage (2) to (5) are operated in an iterative way (details are introduced in Sect. 3.1).

Inspired by the successes [28,34] that use motion information for human activity recognition in videos, we exploit motion information for better gesture recognition by focusing more on joint-aware features. Specifically, we distinguish slowly and fast-moving hand joints and exploit such motion information in the intermediate network layers to learn enhanced and enriched joint-aware features. Beyond that, we propose a multi-order multi-stream feature analysis module that exploits more discriminative and representative joint motion information according to the intermediate joint-aware features.

Additionally, annotating 3D hand poses is often very laborious and time-consuming. To address this issue, we propose a weakly supervised 3D pose estimation technique that can learn accurate 3D pose estimation models from the gesture labels which are widely available in many video data. We observe that the weakly supervised learning improve the 3D pose estimation significantly when only a few samples with 3D pose annotations are included, largely because the exploited joint-aware features that are useful for both gesture recognition and 3D hand pose estimation tasks. At the other end, the weakly supervised learning can also learn accurate gesture estimation models from hand image sequences with 3D pose annotations with similar reasons.

The contributions of this work can be summarized from four aspects. *First*, we propose a novel collaborative learning network that leverage joint-aware features for both gesture recognition and 3D hand pose estimation simultaneously. To the best of our knowledge, this is the first network that exploits and optimizes the joint-aware features for both gesture recognition and 3D hand pose estimation. *Second*, it designs a multi-order feature analysis module that employs a novel slow-fast feature analysis scheme to learn joint-aware motion features which improves the gesture recognition greatly. *Third*, it designs a multi-scale relation module to learn hierarchical hand structure relations at multiple scales which enhances the performance of gesture recognition clearly. *Fourth*, we propose a weakly supervised learning scheme that is capable of leveraging hand pose (or gesture) annotations to learn powerful gesture recognition (or pose estimation) model. The weakly supervised learning greatly relieves the data annotation burden especially considering the very limited annotated 3D pose data and the wide availability of annotated hand gesture data.

2 Related Work

Gesture and Action Recognition. In the early stage, many gesture and action recognition methods were developed based on handcrafted features [14,15,32,33]. With the advance of deep learning, Convolutional neural networks (CNNs) [7,28,30,31,34,36,38,39] have been applied to gesture recognition and action recognition. Simonyan and Zisserman [28] proposed a two-stream architecture, where one stream operates on RGB frames, and the other on optical flow. Many works follow and extend their framework [7,30,36]. They all use the optical flow as the motion information. Wang *et al.* [34] built a new motion representation: RGB difference, which stacks the differences between consecutive frames, to save the time of optical flow extraction. The calculation process of optimal flow [28,34] and RGB difference [34] are all pre-processed which is outside of the learning process.

Inspired by the above-mentioned works, in our work, we propose a new multi-order multi-stream feature analysis module, which is conducted at the intermediate features that capture more discriminative and representative motion information as compared to the original video data. Specifically, a slow-fast feature analysis module is added to consolidate the features of both the slowly and fast-moving joints at multiple orders which significantly enhances the gesture-aware features for more reliable gesture recognition.

3D Hand Pose Estimation. 3D hand pose estimation from RGB images has received much attention recently [2,5,6,23,26,40]. However, only a few works [22,29] focused on performing gesture recognition and 3D hand pose estimation from the RGB videos jointly. Tekin *et al.* [29] predicted hand pose and action categories first, and then use the predicted information to do the gesture recognition.

We propose to leverage the joint-aware features for mutual 3D pose estimation and gesture recognition. A novel collaborative learning method is proposed which iteratively boosts the performance of the two tasks by optimizing the joint-aware features which are crucial for both tasks. It also enables the weakly-supervised learning for 3D hand pose estimation.

Joint Gesture/Action Recognition and 3D Pose Estimation. Gesture (or action) recognition and 3D pose estimation are highly related, thus many works performed gesture (or action) recognition based on the results of pose estimation. In the Skeleton-based gesture (or action) recognition [9,16,17,19,20,24], joints' location (pose) information is used for recognizing the gesture (or action) categories. In the RGB-based action recognition, Liu *et al.* [21] also proposed to recognize human actions based on the pose estimation maps. Nie *et al.* [35] and Luvizon *et al.* [22] performed pose estimation and action recognition in a single network, yet they did not consider these two tasks mutually to optimize the performance of each other, i.e., they performed the two tasks either in a parallel way or in a sequential way.

Different from the aforementioned methods, we design a new collaborative learning method that boosts the learning of gesture recognition and 3D hand pose estimation in an *iterative* manner as shown in Fig. 1. To the best of our knowledge, our method is the first that learns gesture-aware and hand pose-aware information for boosting the two tasks progressively.

Weakly-Supervised Learning on 3D Hand Pose Estimation. In the past few years, several works focus on weakly-supervised learning in 3D pose estimation and 3D hand pose estimation areas, since it is hard to obtain the 3D pose annotations. Cai *et al.* [3,4] proposed a weakly-supervised adaptation method by bridging the gap between fully annotated images and weakly-labelled images. Zhou *et al.* [37] transformed knowledge from 2D pose to 3D pose estimation network using re-projection constraint to 2D results. Chen *et al.* [8] used the multi-view 2D annotation as the weak supervision to learn a geometry-aware 3D representations.

All aforementioned methods still used 2D joint information as the weak supervision to generate 3D hand poses. Differently, we propose that the gesture label can also be used as the weak supervision for 3D hand pose estimation. Our experiments show that this weak-supervised learning method is efficient.

3 Methodology

We predict gesture categories and 3D hand joint locations directly from RGB image sequences as illustrated in Fig. 1. Specifically, the input is a sequence of RGB images centered on hand which is fed to a pre-trained ResNet [11] to learn joint-aware feature maps $\mathcal{J}$ (as shown in Fig. 1). The learned $\mathcal{J}$ are then fed to pose sub-network and gesture sub-network which learn collaboratively for more discriminative features. The whole network is trained in an end-to-end manner, more details to be presented in the following subsections.

3.1 Collaborative Learning for Gesture Recognition and 3D Hand Pose Estimation

Gesture recognition and 3D hand pose estimation are both related to the joint-level features. Joints' locations have been used for skeleton-based action recognition and gesture recognition, while gesture classes also contain potential hand posture information that is useful for hand pose estimation.

We propose a collaborative learning method that simultaneously learns the gesture features and 3D hand pose features mutually in an iterative way, as illustrated in Fig. 1. As described above, the pre-trained ResNet [11] is used to learn the joint-aware feature maps $\mathcal{J}$. Specifically, we equally divide the joint-aware feature maps $\mathcal{J}$ to N groups, where N is the number of hand joints, *i.e.* $\mathcal{J} = \{\mathcal{J}_i | i = 1, ..., N\}$, and $\mathcal{J}_i$ is the subset of feature maps representing the joint i ($i \in [1, N]$).

Pose Sub-network: Following the previous works [18,37,40], we first use a Pose Feature Analysis module to estimate the 2D heatmaps based on the intermediate features for generating the 3D hand pose. The Pose Feature Analysis module is composed by two parts: 2D hand pose estimation part and depth regression part, which is similar to [18,37,40]. For the **2D hand pose estimation part**, its input are the joint-aware feature maps $\mathcal{J}$ and its output are N heatmaps (denoted by $\mathcal{H}$). Each map $\mathcal{H}_i$ is a $H \times W$ matrix, representing a 2D probability distribution of each joint in the image.

Follow the deep regression module in [18,37]. We aggregate the joint-aware feature maps $\mathcal{J}$ and the generated 2D heatmaps $\mathcal{H}$ with 1×1 convolution by a summation operation, the summed feature maps are input of the **deep regression module**. Here the 1×1 convolution is used to map the generated 2D heatmaps $\mathcal{H}$ and the joint-aware feature maps $\mathcal{J}$ to the same size. The deep regression module contains a sequence of convolutional layers with pooling and a fully connected layer in order to regress the depth values $D = \{D_i | i = 1, ..., N\}$, where D_i denotes the depth value of the i_{th} joint.

Since the output of pose sub-network is the input of the gesture sub-network, and pose sub-network and gesture sub-network operate iteratively (as shown in Fig. 1), we set the input and output of pose sub-network the same size. To keep the size constant, we first duplicate the depth values to the same size of the heatmaps, and concatenate them with 2D heatmaps. For each joint, its depth value is a scalar, while heatmap's size is $H \times W$. Thus, we duplicate depth value HW to match heatmaps' size to facilitate feature concatenation. Secondly, the 1×1 convolution is used to map the concatenated feature maps and the joint-aware feature maps $\mathcal{J}$ to the same size to generate the output of pose sub-network, named pose-optimized joint-aware feature maps $\mathcal{P}$ (see Fig. 1).

Gesture Sub-network: The input of Gesture Sub-Network is obtained by aggregating the joint-aware feature maps $\mathcal{J}$ and pose-optimized joint-aware feature maps $\mathcal{P}$ with 1×1 convolution followed by a summation. The resultant feature maps are fed to the Gesture Feature Analysis module to generate the gesture-optimized joint-aware feature maps $\mathcal{G}$ and gesture category y (see Fig. 1). Where the Gesture Feature Analysis module contains a sequence of convolutional layers as well as temporal convolution (TCN) layers to get the temporal relation, TCN layers are used here to predict the gesture class y.

Collaborative Learning Method: As shown in Fig. 1, we design a collaborative learning strategy to perform gesture recognition and 3D hand pose estimation in an iterative way. Our proposed framework's learning processes can be described in the following stages:

(1) **Generating** $\mathcal{J}$: The pre-trained ResNet [11] is used to learn the joint-aware feature maps $\mathcal{J}$.
(2) **Generating** $\mathcal{P}$ **and Predicting 3D Hand Pose**: The learned feature maps $\mathcal{J}$ are fed to Pose Feature Analysis module (shown in Fig. 1) to generate 3D hand poses (2D Heatmaps $\mathcal{H}$ and depth values D), and also the pose-optimized joint-aware feature maps $\mathcal{P}$.
(3) **Aggregating input to Gesture Sub-Network**: The 1×1 convolution is used to generate intermediate feature maps by aggregating the joint-aware feature maps $\mathcal{J}$ and the pose-optimized joint-aware feature maps $\mathcal{P}$.
(4) **Generating** $\mathcal{G}$ **and Recognizing Gesture Class**: The intermediate feature maps are fed to Gesture Feature Analysis module as input to generate the gesture-optimized joint-aware feature maps $\mathcal{G}$ and to recognize gesture category y.
(5) **Aggregating input to Pose Sub-Network**: We aggregate the gesture-optimized joint-aware feature maps $\mathcal{G}$ and the joint-aware feature maps $\mathcal{J}$ with 1×1 convolution followed by a summation. The aggregated feature maps are fed to next iteration's Pose Sub-Network as input for further feature learning.
(6) Stage **2** to **5** repeat in an iterative way to perform gesture recognition and hand pose estimation collaboratively for further improving the performance.

3.2 Multi-order Multi-stream Feature Analysis

As discussed in Sect. 2, prior studies have shown that motion information such as optical flow [28,34] is crucial in video-based recognition. As we aim to learn joint-aware features, we propose a multi-order multi-stream feature analysis module as shown in Fig. 2 to learn the motion information based on the joint-aware features. The proposed multi-order multi-stream module participates in the Gesture Feature Analysis module (see Fig. 1).

Since the pre-trained ResNet [11] and our pose sub-network operate at the image level, the corresponding feature maps belonging to hand joints in an image. We name the image-level features as **Zero-Order Features** (denote by Zo, which stand for pose information and static information), as shown in the top line of Fig. 2, the cubes in it are feature maps of the corresponding hand joints. Zero-Order features form $N \times C \times H \times W$ tensors, where N is the total number

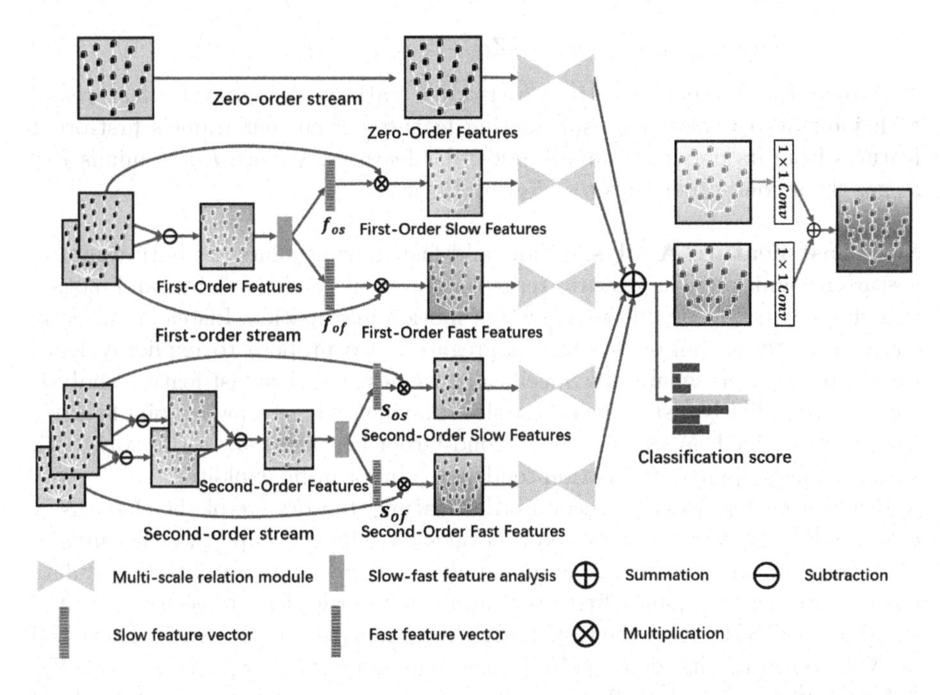

Fig. 2. Illustration of the multi-order multi-stream feature analysis module: With the zero-order features Zo as input, the multi-order multi-stream analysis generates motion information on the intermediate features including first-order slow & fast features and second-order slow & fast features. These four motion features, together with the zero-order features, are fed to five multi-scale relation modules (more details in Fig. 3), respectively, to generate gesture-optimized joint-aware feature maps $\mathcal{G}$ and gesture category y. The generated $\mathcal{G}$ are aggregated with joint-aware feature maps $\mathcal{J}$ and fed to the pose sub-network for pose feature learning. Our multi-order multi-stream feature analysis module participates in the Gesture Feature Analysis module, as shown in Fig. 1. (More description of Fig. 2 are illustrated in supplementary material.)

of hand joints, C is the number of channels for each hand joint, H and W are the height and width of feature maps, respectively.

First-Order Features can be seen as velocity features. A temporal neighborhood pair of feature maps is constructed from the entire Zero-Order Features as follows:

$$\mathcal{U}_1 = \{\langle Zo_{t-1}, Zo_t \rangle : t \in T\} \tag{1}$$

$$Fo_t = Zo_t - Zo_{t-1} \tag{2}$$

where T is the length of input image sequences. First-order features of each joint are calculated by subtracting features of one frame from the previous frame. We use Zo_t minus Zo_{t-1} to get the first-order features (denote by Fo) as in Eq. 2.

Second-Order Features can be seen as the acceleration features. We construct a triplet subset for each frame's features:

$$\mathcal{U}_2 = \{\langle Zo_{t-1}, Zo_t, Zo_{t+1} \rangle : t \in T\} \tag{3}$$

$$So_t = (Zo_{t+1} - Zo_t) - (Zo_t - Zo_{t-1}) = (Fo_{t+1} - Fo_t) \tag{4}$$

Similar to the manner of getting first-order features, second-order features of each joint are calculated by subtracting features of current frame's first-order features from its previous frame's first-order features. We use Fo_{t+1} minus Fo_t to get the second-order features (So) by Eq. 4.

Slow-Fast Feature Analysis: Slow and fast moving joints are both useful in gesture recognition. The features representing static tendency joints and motion tendency joints encode different levels of motion information. Instead of directly considering these motion features aggregately, we propose to explicitly learn these motion levels separately. Specifically, we design a slow-fast feature analysis method to explicitly distinguish these slow-moving and fast-moving joint features from First-Order Features Fo and Second-Order Features So. In this way, both static tendency joints and motion tendency joints can be exploited.

First-order features and second-order features tensors are of the shape $N \times C \times H \times W$ (the same as the zero-order ones). We first reshape these features to $N \times CHW$ matrices (where N is the number of hand joints), and then calculate the L_2 norm on each joint's first-order and second-order feature vector (with the shape of $1 \times CHW$) from the reshaped features matrices, respectively. There will be N L_2 norm results, denoted by Feature Difference ($FD = \{FD_i | i = 1, ...N\}$, a $N \times 1$ vector). Each FD_i is a value representing the motion magnitude of each corresponding joint. We adopt Gaussian distributions to obtain the feature maps of slow-moving and fast-moving joints. For slow motion analysis, we aim to enhance features from the *more static* joints, i.e., assign larger weights to joints that move more slowly. We use a Gaussian function (with FD_{min} as mean and $(FD_{max} - FD_{min})/3$ as standard deviation) to map FD values to weights (FD_{min}/FD_{max} denotes the min/max FD values). With this mapping, the weight of the joint with the min/max motion magnitude (FD_{min}/FD_{max}) will be close to 1/0. As there are N hand joints, we will obtain a $N \times 1$ *slow vector* that contains weights for the features of N joints. Similarly, we aim to enhance features from the *more dynamic* joints using the fast motion analysis module. We thus set FD_{max} and $(FD_{min} - FD_{max})/3$ as the mean and standard deviation. In this way, the joint that has min/max motion magnitude will have a weight around 0/1.

When the slow and fast motion analysis modules apply on the first-order and second-order features Fo and So, we obtain four $N \times 1$ vectors that contain weights of features of N joints as shown in Fig. 2: 1) First-order slow vector (f_{os}); 2) First-order fast vector (f_{of}); 3) Second-order slow vector (s_{os}); and 4) Second-order fast vector (s_{of}). All these four vectors are used to refine the zero-order features Zo which are first reshaped to an $N \times CHW$ matrix and then multiplied with these four vector separately. The embedding features are then reshaped back to $N \times C \times H \times W$ tensors, namely, first-order-slow features, first-order-fast features, second-order-slow features and second-order-fast features as shown in Fig. 2. These four features together with the zero-order features are fed to the multi-scale relation module (details to be discussed in the Sect. 3.3), respectively. Finally, the results of each stream are averaged to obtain the gesture-optimized joint-aware feature maps $\mathcal{G}$ and the gesture category y.

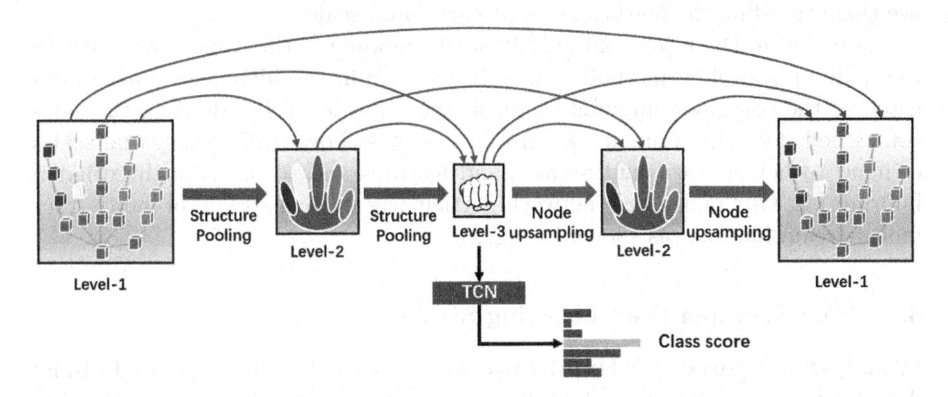

Fig. 3. Illustration of the multi-scale relation module: The multiple scale analysis process the feature maps from the slow-fast feature analysis at three different levels to generate relations at each level. It interact with the Gesture Sub-network by applying temporal convolution (TCN) on Level 3 (containing global information) to generate the classification scores. Node up-sampling is applied to keep the input and output of the same shape. (Color figure online)

3.3 Multi-scale Relation Module

Considering different levels of semantic information contained in the hierarchical structure of hand, human hand can be defined with different scales. As shown in Fig. 3, we show three levels, where the level-1 is the local level consisting of the hand joints, and the level-2 is the middle level representing five fingers and palm. For the level-3, we see the hand globally as complete holistic information. Following the connection between contiguous scale, we use the structure pooling to perform feature aggregation across these three scales, and recognize gesture class y at the Level-3 using TCN, since it contains the global information.

Structure pooling means we use average pooling over the hand joints by following hierarchical physical structure of hand to perform step-wise feature aggregation. We first average features of the joints that belong to each finger or palm, in order to get features for the five fingers and palm (see Fig. 3), then average features of five fingers and palm to obtain the final global features representing full hand.

Additionally, we calculate a relation matrix for each level to better learn the features at each scale. Take the first level as the example; the whole feature maps size is $N \times C \times H \times W$. We first activate it through two embedding function ($1 \times 1 \times 1$ convolution). The two embedding features are rearranged and reshape to a $N \times CHW$ matrix and $CHW \times N$ matrix. They are then multiplied to obtain a $N \times N$ relation matrix. The values of the matrix mean the degree of relation between each pair of joints. The softmax function is used here to do the normalization. In this way, we can calculate relation matrices for each level and use them to refine the feature maps at each hand scale.

To maintain the input and output of this module in the same shape, we use the node up-sampling method: joints' features from the higher level are duplicated to the corresponding child joint in the lower level. In addition, the skip-connections (see thin blue arrows in Fig. 3) are used over different spatial scales of hand to better learn multi-scale hand features and to preserve the original information. Our multi-scale network participates in each stream of multi-order multi-stream module (as shown in Fig. 2).

3.4 Weakly-Supervised Learning Strategy

Weakly-Supervised 3D Hand Pose Estimation Using Gesture Labels: Annotating 3D poses is often laborious, and it's difficult to have a large amount of video samples with 3D pose annotations for training. In the supervised learning, the pose-optimized joint-aware feature maps $\mathcal{P}$ and the gesture-optimized joint-aware feature maps $\mathcal{G}$ are learned based on the joint-aware feature maps $\mathcal{J}$. We therefore propose a weakly-supervised learning method that use gesture labels as weak supervision for 3D hand pose estimation. We provide different ratios of training data with 3D pose annotations in training process.

Weakly-Supervised Gesture Recognition Using Pose Labels: When only a few videos have gesture labels, we can similarly use 3D hand pose annotations as weak supervision for gesture recognition. We provide different ratios of training data with gesture labels in training to make our method more applicable.

3.5 Training

We use the following losses in training. **2D Heatmaps loss.** $L_{2d} = \sum_{n=1}^{N} \|\mathcal{H}_n - \hat{\mathcal{H}}_n\|_2^2$, This loss measures the $L2$ distance between the predicted heatmaps $\mathcal{H}_n$ and the ground-truth heatmaps $\hat{\mathcal{H}}_n$. **Depth Regression loss.** $L_{3d} = \sum_{n=1}^{N} \|D_n - \hat{D}_n\|_2^2$, where D_n and $\hat{D}_n$ are the estimated and the ground truth depth values, respectively. L_{3d} is also based on the $L2$ distance. **Classification loss.** We use the standard categorical cross-entropy loss to supervise the gesture classification process, which is $L_c = CrossEntropy(y, \tilde{y})$, where y is the class predicted score and $\tilde{y}$ is the ground truth category.

Fully-Supervised Training Strategy. In our implementation, we first fine-tune the ResNet-50 to make it sensitive to human joint information. We then train the entire network in an end-to-end manner with the objective function:

$$L = \lambda_{2d} L_{2d} + \lambda_{3d} L_{3d} + \lambda_c L_c \tag{5}$$

Weakly-Supervised Training Strategy. Based on the Eq. 5, we set $\lambda_{2d} = 0$ and $\lambda_{3d} = 0$ when the samples do not have 3D pose annotations and we use gesture categories as weak supervision for 3D hand pose estimation. Similarly, we set $\lambda_c = 0$ for video sequences without gesture labels, where we use 3D pose annotations as weak supervision for gesture recognition.

4 Experiment

Implementation Details: We implement our method with the PyTorch framework, and optimize the objective function with the Adam optimizer with mini-batches of size 4. The learning rate starts from 10^{-4}, with a 10 times reduction when the loss is saturated. Following the same setting in [18,37], the input image is resized to 256×256, and the heatmap resolution is set at 64×64. In the experiment, the parameters in the objective function are set as follows: $\lambda_{2d} = 1$, $\lambda_{3d} = 0.001$ and $\lambda_c = 0.001$. For the **weakly-supervised learning**, we choose the 15% to 40% samples as the weakly supervision samples and set $\lambda_{2d} = 0$ and $\lambda_{3d} = 0$ when the samples do not have 3D pose annotations (gesture categories are used as weak supervision for 3D hand pose estimation). Additionally, we set $\lambda_c = 0$ for video sequences without gesture labels, where 3D pose annotations are used as weak supervision for gesture recognition as described in Sect. 3.4.

Following [34], each input video is divided into K segments and a short clip is randomly selected from each segment in training. On testing, each video is similarly divided into K segments and one frame is selected from each segment to make sure that temporal space between adjacent frames is equal to T/K. The final classification scores are computed by the average over all clips from each video, and the pose estimation is presented on image level.

Datasets: We perform extensive experiments on the large-scale and challenging dataset: First-Person Hand Action (FPHA) [10] for simultaneous gesture recognition and 3D hand pose estimation. To the best of our knowledge, this is the only publicly available dataset that provides labels of accurate 2D & 3D hand poses and gesture labels. The dataset consists of 1175 gesture videos with 45 gesture classes. The videos are performed by 6 actors under 3 different scenarios. A total of 105, 459 video frames are annotated with accurate hand pose and action classes. Both 2D and 3D annotations of the total 21 hand keypoints are provided for each frame. We follow the protocol in [10,29] and use 600 video sequences for training and the remaining 575 video sequences for testing.

Evaluation Metrics: We adopt the widely used metrics for evaluation of gesture recognition and 3D hand pose estimation. For gesture recognition, we directly evaluate the accuracy of video classification. For 3D pose estimation, we use the percentage of correct keypoints (PCK) score that evaluates the pose estimation accuracy with different error thresholds.

4.1 Experimental Results

Gesture Recognition: Table 1 shows the comparison with state-of-the-art gesture recognition methods. It can be seen that our method outperforms the state-of-the-art by up to 3%, showing its effectiveness gesture recognition. Additionally, augmenting each of our proposed module (multi-scale relation, multi-order multi-stream and collaborative learning strategy) yield improved gesture recognition performance.

3D Hand Pose Estimation: We compare our method with prior works on FPHA as shown in the first graph in Fig. 4. Table 2 shows three 3D PCK results at three specific error threshold. It can be seen that our method outperforms the state-of-the-art with a large range between 0 mm and 30 mm. Even though we use color images, our results are better than [10] that uses depth images which demonstrates the advantage of our proposed method.

Qualitative Results on 3D Hand Pose Estimation: Figure 5 illustrates 3D pose estimations by our method. We compare the ground truth 3D poses (in blue-color structures) and the predicted 3D pose (in red-color structures) in the same 3D coordinate system. We also provide the predicted 2D poses in the original RGB image. As Fig. 5 shows, our method is capable of accurately predicting 3D poses of different orientations with different backgrounds.

Table 1. Comparisons to state-of-the-art gesture recognition methods: "Baseline" means 1-iteration network with no multi-order feature analysis and multi-scale relation.

Model	Input modality	Accuracy
Joule-depth [12]	Depth	60.17%
Novel view [27]	Depth	69.21%
HON4D [25]	Depth	70.61%
FPHA + LSTM [10]	Depth	72.06%
Two-stream-color [28]	Color	61.56%
Joule-color [12]	Color	66.78%
Two-stream-flow [28]	Color	69.91%
Two-stream-all [28]	Color	75.30%
[29] - HP	Color	62.54%
[29] - HP + AC	Color	74.20%
[29] - HP + AC + OC	Color	82.43%
Baseline	Color	72.17%
Baseline + multi-scale	Color	78.26%
Baseline + multi-scale + multi-order	Color	83.83%
Baseline + multi-scale + multi-order + 2-iterations	Color	85.22%

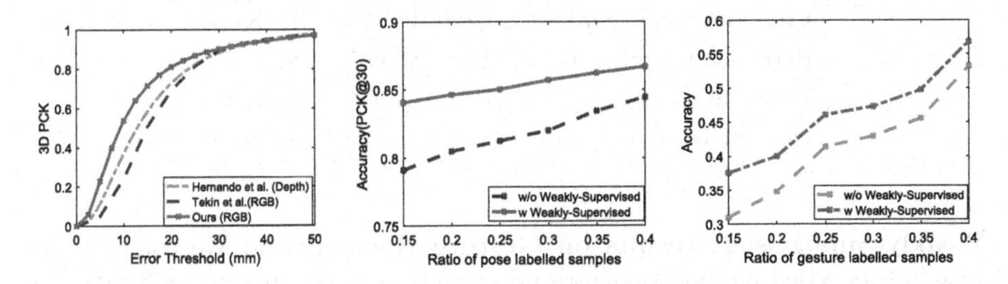

Fig. 4. Left: Comparing our method with [10] and [29] for 3D hand pose estimation with 3D PCK metric. **Middle:** Comparing our weakly supervised method with the baseline (with 3D PCK@30) when different amounts of pose labels are used. **Right:** Comparing our weakly supervised method with the baseline (with classification accuracy) when different amount of gesture labels are used.

4.2 Weakly-Supervised Learning

Weakly-Supervised Results on 3D Hand Pose Estimation: We present multiple experiments on our weakly-supervised method by providing different ratios (15% to 40%) of samples with pose labels (gesture labels are provided for all training samples) and compare with the baseline that does not use gesture labels. Figure 4 (middle) shows 3D PCK@30 (percentage of correct keypoint when error threshold smaller than 30 mm) results of the baseline and our weakly-supervised method. It can be seen that the 3D hand pose estimation is improved significantly for all labeled ratios when weak supervision is included. This validates that joint-aware features in the gesture can benefit 3D hand pose estimation.

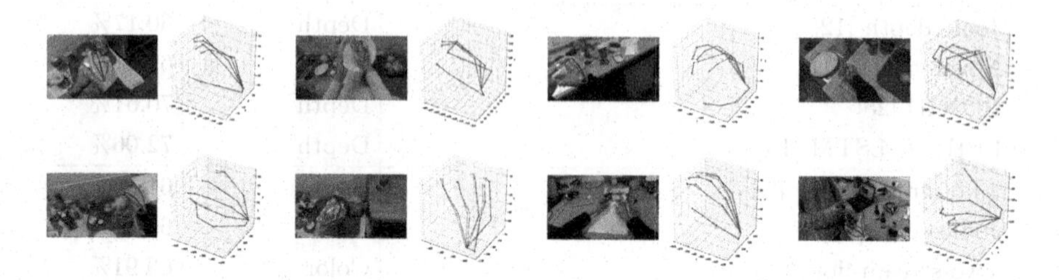

Fig. 5. Qualitative illustration of our proposed method: It shows the predicted 2D poses shown on the original image. It also compares the predicted 3D poses (the blue-color structures) and the Ground Truth 3D poses (the red-color structures). (Color figure online)

Table 2. Comparisons on 3D pose estimation: Numbers are percentage of correct keypoint (PCK) over respective error threshold, more results available in Fig. 4 (left). Our results are based on the proposed 2-iterations multi-order structure.

Error threshold (mm)	PCK@20	PCK@25	PCK@30
Hernando (Depth) [10]	72.13%	82.08%	87.87%
Tekin (RGB) [29]	69.17%	81.25%	89.17%
Ours (RGB)	81.03%	86.61%	90.11%

Weakly-Supervised Results on Gesture Recognition: We compare our weakly supervised method that uses pose labels as weak supervision for gesture recognition with the baseline which does not use pose labels. We conduct experiments by providing different ratios of training samples with gesture labels, while the pose labels of all samples are given. As Fig. 4 (right) shows, our weakly-supervised learning improves the gesture recognition significantly for all labeled ratios. This validates that joint-aware features in hand poses can improve the gesture recognition performance greatly.

4.3 Ablation Studies

Impact of Number of Network Iterations: Table 3 shows the 3D PCK results and classification results of our method under different iterations of collaborative learning. It can be seen that our method improves with increasing iterations. This can be expected since hand pose estimation and gesture recognition learn in a collaborative manner and boost each other. Note that the improvement of 3D PCK and gesture recognition slows down with the increase of iterations.

We use the two-iteration network in the experiment for the balance between accuracy and computational complexity. Note all these comparisons are based on the zero-order framework. We cannot evaluate multi-order network for the 3-itr, 4-itr, 5-itr due to our GPU's memory limitation.

Effect of the Multi-order Module: We analyze the advantage of our proposed multi-order module by implementing four variants as shown in Table 4 (part 1, 2, and 4). It can be seen that adding first-order and second-order slow-fast features leads to an accuracy improvement by 1.7% and 2.9%, respectively. Our multi-order module (Zero-order + First and Second order slow-fast) achieves the best accuracy at 85.22%, demonstrating its effectiveness.

Table 3. Evaluation of our proposed network on gesture recognition and pose estimation with respect to different iteration numbers.

Iteration (itr) number	1-itr	2-itr	3-itr	4-itr	5-itr
Pose estimation (PCK@30)	87.2%	89.3%	89.8%	89.9%	**89.9%**
Gesture recognition accuracy	78.3%	80.9%	81.7%	81.9%	**82.0%**

Table 4. Evaluation of our proposed gesture recognition network with different combinations of motion features of different orders and slow-fast patterns. (All experiments below are based on the 2-iteration network.

	Network setting	Accuracy	Δ
1	Zero-order	80.87%	
2	Zero-order + First-order slow-fast	82.61%	1.74%
	Zero-order + Second-order slow-fast	83.80%	2.93%
3	Zero-order + First and Second order slow	82.96%	2.09%
	Zero-order + First and Second order fast	82.09%	1.22%
4	Zero-order + First and Second order slow-fast	85.22%	4.35%

Effect of the Slow Feature and Fast Feature: We also evaluate the impact of the slow-fast features and Table 4 (part 3) shows the results. It can be seen that the slow features and the fast features can improve the accuracy by 2.1% and 1.2%, respectively, and the best accuracy is obtain when both are included.

Effect of the Multi-scale Relation: We also assess the effectiveness of the our multi-scale relation module and Table 1 shows experimental results. As Table 1 shows, removing the multi-scale relation module leads to around 6% accuracy drop as compared with the "Baseline" and "Baseline + multi-scale", showing the benefit of the proposed multi-scale relation.

5 Conclusion

In this paper, we have presented a collaborative learning method for joint gesture recognition and 3D hand pose estimation. Our model learns in a collaborative way to recurrently exploit the joint-aware feature to progressively boost the performance of each task. We have developed a multi-order multi-stream model to learn motion information in the intermediate feature maps and designed a multi-scale relation module to extract semantic information at hierarchical hand structure. To learn our model in scenarios that lack labeled data, we leverage one fully-labeled task's annotations as weak supervision for the other very few labeled task. The proposed collaborative learning network achieves state-of-the-art performance for both gesture recognition and 3D hand pose estimation tasks.

Acknowledgement. The research was carried out at the Rapid-Rich Object Search (ROSE) Lab, Nanyang Technological University, Singapore. This research work was partially supported by SUTD projects PIE-SGP-Al-2020-02 and SRG-ISTD-2020-153.

References

1. Abavisani, M., Joze, H.R.V., Patel, V.M.: Improving the performance of unimodal dynamic hand-gesture recognition with multimodal training. In: The IEEE Conference on Computer Vision and Pattern Recognition (CVPR), June 2019
2. Boukhayma, A., Bem, R.D., Torr, P.H.: 3D hand shape and pose from images in the wild. In: Proceedings of the IEEE Conference on Computer Vision and Pattern Recognition, pp. 10843–10852 (2019)
3. Cai, Y., Ge, L., Cai, J., Magnenat-Thalmann, N., Yuan, J.: 3D hand pose estimation using synthetic data and weakly labeled RGB images. IEEE Trans. Pattern Anal. Mach. Intell. **PP**, 1 (2020)
4. Cai, Y., Ge, L., Cai, J., Yuan, J.: Weakly-supervised 3D hand pose estimation from monocular RGB images. In: Proceedings of the European Conference on Computer Vision (ECCV), pp. 666–682 (2018)
5. Cai, Y., et al.: Exploiting spatial-temporal relationships for 3D pose estimation via graph convolutional networks. In: Proceedings of the IEEE International Conference on Computer Vision, pp. 2272–2281 (2019)
6. Cai, Y., Huang, L., et al.: Learning progressive joint propagation for human motion prediction. In: Proceedings of the European Conference on Computer Vision (ECCV) (2020)
7. Carreira, J., Zisserman, A.: Quo vadis, action recognition? A new model and the kinetics dataset. In: Proceedings of the IEEE Conference on Computer Vision and Pattern Recognition, pp. 6299–6308 (2017)
8. Chen, X., Lin, K.Y., Liu, W., Qian, C., Lin, L.: Weakly-supervised discovery of geometry-aware representation for 3D human pose estimation. In: Proceedings of the IEEE Conference on Computer Vision and Pattern Recognition, pp. 10895–10904 (2019)
9. De Smedt, Q., Wannous, H., Vandeborre, J.P.: Skeleton-based dynamic hand gesture recognition. In: Proceedings of the IEEE Conference on Computer Vision and Pattern Recognition Workshops, pp. 1–9 (2016)

10. Garcia-Hernando, G., Yuan, S., Baek, S., Kim, T.K.: First-person hand action benchmark with RGB-D videos and 3D hand pose annotations. In: The IEEE Conference on Computer Vision and Pattern Recognition (CVPR), June 2018
11. He, K., Zhang, X., Ren, S., Sun, J.: Deep residual learning for image recognition. In: Proceedings of the IEEE Conference on Computer Vision and Pattern Recognition, pp. 770–778 (2016)
12. Hu, J.F., Zheng, W.S., Lai, J., Zhang, J.: Jointly learning heterogeneous features for RGB-D activity recognition. In: The IEEE Conference on Computer Vision and Pattern Recognition (CVPR), June 2015
13. Iqbal, U., Garbade, M., Gall, J.: Pose for action-action for pose. In: 2017 12th IEEE International Conference on Automatic Face & Gesture Recognition (FG 2017), pp. 438–445. IEEE (2017)
14. Klaser, A., Marszałek, M., Schmid, C.: A spatio-temporal descriptor based on 3D-gradients (2008)
15. Laptev, I.: On space-time interest points. Int. J. Comput. Vis. **64**(2–3), 107–123 (2005). https://doi.org/10.1007/s11263-005-1838-7
16. Liu, J., Shahroudy, A., Xu, D., Kot, A.C., Wang, G.: Skeleton-based action recognition using spatio-temporal LSTM network with trust gates. IEEE Trans. Pattern Anal. Mach. Intell. **40**(12), 3007–3021 (2018)
17. Liu, J., Wang, G., Duan, L., Abdiyeva, K., Kot, A.C.: Skeleton-based human action recognition with global context-aware attention LSTM networks. IEEE Trans. Image Process. **27**(4), 1586–1599 (2018)
18. Liu, J., et al.: Feature boosting network for 3D pose estimation. IEEE Trans. Pattern Anal. Mach. Intell. **42**, 494–501 (2019)
19. Liu, J., Shahroudy, A., Xu, D., Wang, G.: Spatio-temporal LSTM with trust gates for 3D human action recognition. In: Leibe, B., Matas, J., Sebe, N., Welling, M. (eds.) ECCV 2016. LNCS, vol. 9907, pp. 816–833. Springer, Cham (2016). https://doi.org/10.1007/978-3-319-46487-9_50
20. Liu, J., Wang, G., Hu, P., Duan, L.Y., Kot, A.C.: Global context-aware attention LSTM networks for 3d action recognition. In: Proceedings of the IEEE Conference on Computer Vision and Pattern Recognition, pp. 1647–1656 (2017)
21. Liu, M., Yuan, J.: Recognizing human actions as the evolution of pose estimation maps. In: The IEEE Conference on Computer Vision and Pattern Recognition (CVPR), June 2018
22. Luvizon, D.C., Picard, D., Tabia, H.: 2D/3D pose estimation and action recognition using multitask deep learning. In: The IEEE Conference on Computer Vision and Pattern Recognition (CVPR), June 2018
23. Mueller, F., et al.: Ganerated hands for real-time 3D hand tracking from monocular RGB. In: Proceedings of the IEEE Conference on Computer Vision and Pattern Recognition, pp. 49–59 (2018)
24. Nguyen, X.S., Brun, L., Lezoray, O., Bougleux, S.: A neural network based on SPD manifold learning for skeleton-based hand gesture recognition. In: The IEEE Conference on Computer Vision and Pattern Recognition (CVPR), June 2019
25. Oreifej, O., Liu, Z.: HON4D: histogram of oriented 4D normals for activity recognition from depth sequences. In: The IEEE Conference on Computer Vision and Pattern Recognition (CVPR), June 2013
26. Rad, M., Oberweger, M., Lepetit, V.: Domain transfer for 3D pose estimation from color images without manual annotations. In: Jawahar, C.V., Li, H., Mori, G., Schindler, K. (eds.) ACCV 2018. LNCS, vol. 11365, pp. 69–84. Springer, Cham (2019). https://doi.org/10.1007/978-3-030-20873-8_5

27. Rahmani, H., Mian, A.: 3D action recognition from novel viewpoints. In: The IEEE Conference on Computer Vision and Pattern Recognition (CVPR), June 2016
28. Simonyan, K., Zisserman, A.: Two-stream convolutional networks for action recognition in videos. In: Advances in Neural Information Processing Systems, pp. 568–576 (2014)
29. Tekin, B., Bogo, F., Pollefeys, M.: H+O: unified egocentric recognition of 3D hand-object poses and interactions. In: The IEEE Conference on Computer Vision and Pattern Recognition (CVPR), June 2019
30. Tran, D., Wang, H., Torresani, L., Ray, J., LeCun, Y., Paluri, M.: A closer look at spatiotemporal convolutions for action recognition. In: Proceedings of the IEEE Conference on Computer Vision and Pattern Recognition, pp. 6450–6459 (2018)
31. Tu, Z., et al.: Multi-stream CNN: learning representations based on human-related regions for action recognition. Pattern Recogn. **79**, 32–43 (2018)
32. Wang, H., Kläser, A., Schmid, C., Liu, C.L.: Dense trajectories and motion boundary descriptors for action recognition. Int. J. Comput. Vis. **103**(1), 60–79 (2013). https://doi.org/10.1007/s11263-012-0594-8
33. Wang, H., Schmid, C.: Action recognition with improved trajectories. In: Proceedings of the IEEE International Conference on Computer Vision, pp. 3551–3558 (2013)
34. Wang, L., et al.: Temporal segment networks: towards good practices for deep action recognition. In: Leibe, B., Matas, J., Sebe, N., Welling, M. (eds.) ECCV 2016. LNCS, vol. 9912, pp. 20–36. Springer, Cham (2016). https://doi.org/10.1007/978-3-319-46484-8_2
35. Xiaohan Nie, B., Xiong, C., Zhu, S.C.: Joint action recognition and pose estimation from video. In: Proceedings of the IEEE Conference on Computer Vision and Pattern Recognition, pp. 1293–1301 (2015)
36. Xie, S., Sun, C., Huang, J., Tu, Z., Murphy, K.: Rethinking spatiotemporal feature learning: speed-accuracy trade-offs in video classification. In: Proceedings of the European Conference on Computer Vision (ECCV), pp. 305–321 (2018)
37. Zhou, X., Huang, Q., Sun, X., Xue, X., Wei, Y.: Towards 3D human pose estimation in the wild: a weakly-supervised approach. In: The IEEE International Conference on Computer Vision (ICCV), October 2017
38. Zhu, H., Vial, R., Lu, S.: TORNADO: a spatio-temporal convolutional regression network for video action proposal. In: Proceedings of the IEEE International Conference on Computer Vision, pp. 5813–5821 (2017)
39. Zhu, H., et al.: YoTube: searching action proposal via recurrent and static regression networks. IEEE Trans. Image Process. **27**(6), 2609–2622 (2018)
40. Zimmermann, C., Brox, T.: Learning to estimate 3D hand pose from single RGB images. In: Proceedings of the IEEE International Conference on Computer Vision, pp. 4903–4911 (2017)

Author Index

Printed in the United States
By Bookmasters